# Approaches to Water Sensitive Urban Design

# Approaches to Water Sensitive Urban Design

## Potential, Design, Ecological Health, Urban Greening, Economics, Policies, and Community Perceptions

*Edited by*

**Ashok K. Sharma**

**Ted Gardner**

**Don Begbie**

Elsevier
Radarweg 29, PO Box 211, 1000 AE Amsterdam, Netherlands
The Boulevard, Langford Lane, Kidlington, Oxford OX5 1GB, United Kingdom
50 Hampshire Street, 5th Floor, Cambridge, MA 02139, United States



**Library of Congress Cataloging-in-Publication Data**
A catalog record for this book is available from the Library of Congress

**British Library Cataloguing-in-Publication Data**
A catalogue record for this book is available from the British Library

ISBN: 978-0-12-812843-5



*Publisher:* Candice Janco
*Acquisition Editor:* Louisa Hutchins
*Editorial Project Manager:* Emily Thomson
*Production Project Manager:* Kamesh Ramajogi
*Cover Designer:* Mark Rogers

*Front Cover Artwork:* Simon Beecham (Green wall), Frank Hanson (Swale), Deborah O'Bannon (Bioretention), Stephen Cook (Urban lake), Ted Gardner (Raintank)

Typeset by TNQ Technologies

# Contents

# List of Contributors

**Simon Beecham**, Natural and Built Environments Research Centre, University of South Australia, Adelaide, SA, Australia

**Don Begbie**, Urban Water Security Research Alliance, South East Queensland, Brisbane, Australia

**Beau B. Beza**, School of Architecture and Built Environment, Deakin University, Geelong, VIC, Australia

**Rosmina Bustami**, Natural and Built Environments Research Centre, University of South Australia, Adelaide, SA, Australia

**Josh Byrne**, School of Design and Built Environment, Curtin University, Bentley, WA, Australia

**Rob Catchlove**, Wave Consulting, Southbank, VIC, Australia

**Elizelle Juanee Cilliers**, Urban and Regional Planning, Research Unit for Environmental Sciences and Management, North-West University, Potchefstroom, South Africa

**Stephen Cook**, CSIRO Land and Water, Clayton, VIC, Australia

**Stewart Dallas**, School of Engineering and Information Technology, Murdoch University, Murdoch, WA, Australia

**Graeme C. Dandy**, School of Civil, Environmental and Mining Engineering, University of Adelaide, Adelaide, SA, Australia

**Peter Dillon**, Honorary Fellow, CSIRO Land and Water, Glen Osmond SA, Australia; Adjunct Chair, National Centre for Groundwater Research and Training, Flinders University, SA, Australia

**Michael Di Matteo**, School of Civil, Environmental and Mining Engineering, University of Adelaide, Adelaide, SA, Australia; Water Technology Pty. Ltd., Adelaide, SA, Australia

**Meredith Dobbie**, School of Geography and Environmental Science, Monash University, Melbourne, VIC, Australia

**Jago Dodson**, Centre for Urban Research, RMIT University, Melbourne, VIC, Australia

**Hildegard Edith Rohr**, Urban and Regional Planning, Research Unit for Environmental Sciences and Management, North-West University, Potchefstroom, South Africa

**Lisa Ehrenfried**, Yarra Valley Water, Mitcham, VIC, Australia

**Casey Furlong**, Centre for Urban Research, RMIT University, Melbourne, VIC, Australia

**Ted Gardner**, Institute for Innovation and Sustainability, College of Engineering and Science, Victoria University, Melbourne, Vic, Australia

**Simon Gehrmann**, Technical University of Darmstadt, Dept. for Urban design and Development, Faculty of Architecture, Darmstadt, Germany

**Ashantha Goonetilleke**, Queensland University of Technology, Brisbane, QLD, Australia

**Melissa Green**, Josh Byrne & Associates, Fremantle, WA, Australia

**Wade Hadwen**, Australian Rivers Institute, Griffith University, Nathan, QLD, Australia

**David Hamlyn-Harris**, Bligh Tanner Pty Ltd., Fortitude Valley, QLD, Australia

**Frank Hanson**, Victorian Planning Authority, Melbourne, VIC, Australia

**Robert J. Hawley**, Sustainable Streams LLC, Louisville, KY, United States

**Alan Hoban**, Bligh Tanner P/L, Brisbane, QLD, Australia

**Sayed Iftekhar**, Centre for Environmental Economics & Policy (CEEP), University of Western Australia

**Elmira Jamei**, Course Chair of Building Design, College of Engineering and Science, Victoria University, Melbourne, VIC, Australia

**Anthony Ladson**, Victorian University, College of Engineering and Science, Melbourne, VIC, Australia; Moroka Pty Ltd., Clifton Hill, VIC, Australia

**James LaGro, Jr.**, University of Wisconsin-Madison, Madison, WI, United States

**Jane-Louise Lampard**, University of the Sunshine Coast, Sippy Downs, QLD, Australia

**Catherine Leigh**, Australian Rivers Institute, Griffith University, Nathan, QLD, Australia

**Rosemary Leonard**, School of Social Sciences and Psychology, University of Western Sydney, Penrith, NSW, Australia

**Terry Lucke**, Stormwater Research Group, University of the Sunshine Coast, Sippy Downs, QLD, Australia

**Yanan Ma**, AECOM-Water, Kansas City, MO, United States

**Holger R. Maier**, School of Civil, Environmental and Mining Engineering, University of Adelaide, Adelaide, SA, Australia

**Tony McAlister**, Water Technology Pty Ltd, West End, QLD, Australia

**Peter Morison**, School of Ecosystem and Forest Sciences, University of Melbourne, Melbourne, VIC, Australia

**Jack Mullaly**, Ideanthro, PO Box 386, Sherwood, QLD, Australia

**Baden R. Myers**, School of Natural and Built Environments, University of South Australia, Adelaide, SA, Australia

**Wendy Neilan**, Australian Rivers Institute, Griffith University, Nathan, QLD, Australia

**Michael Newham**, Australian Rivers Institute, Griffith University, Nathan, QLD, Australia

**Deborah J. O'Bannon**, University of Missouri-Kansas City, Kansas City, MO, United States

**Micah Pendergast**, City of Port Phillip, Melbourne, VIC, Australia

**David Pezzaniti**, School of Natural and Built Environments, University of South Australia, Adelaide, SA, Australia

**Sam Phillips**, Department of Environment, Water and Natural Resources, for the Adelaide and Mt Lofty Ranges NRM Board, SA, Australia

**Robert Pitt**, Emeritus Cudworth Professor of Urban Water Systems, University of Alabama, Tuscaloosa, AL, United States

**Carolyn Polson**, Australian Rivers Institute, Griffith University, Nathan, QLD, Australia

**John C. Radcliffe**, Commonwealth Scientific and Industrial Research Organisation (CSIRO), Glen Osmond, SA, Australia

**Samira Rashetnia**, Institute for Innovation and Sustainability, College of Engineering and Science, Victoria University, Melbourne, Vic, Australia

**Mostafa Razzaghmanesh**, ORISE Fellow at US Environmental Protection Agency, Edison, NJ, United States

**Leon Rowlands**, Switchback Consulting, Caloundra, QLD, Australia

**Annette Rudolph-Cleff**, Technical University of Darmstadt, Dept. for Urban design and Development, Faculty of Architecture, Darmstadt, Germany

**Engelbert Schramm**, ISOE – Institute for Social-Ecological Research, Frankfurt/Main, Germany

**Ashok K. Sharma**, Institute for Innovation and Sustainability, College of Engineering and Science, Victoria University, Melbourne, Vic, Australia

**Fran Sheldon**, Australian Rivers Institute, Griffith University, Nathan, QLD, Australia

**Leila Talebi**, Senior Water Resources Engineer, Paradigm Environmental, San Diego, CA, United States

**Nigel Tapper**, Urban Climate Research Group Leader, School of Geography and Environmental Science, Monash University, Melbourne, VIC, Australia

**Grace Tjandraatmadja**, Institute of Sustainability and Innovation, Victoria University, Melbourne, VIC, Australia

**Susan van de Meene**, School of Social Sciences, Monash University, Clayton, VIC, Australia

**Marjorie van Roon**, University of Auckland, Auckland, New Zealand

**Geoff J. Vietz**, Streamology Pty Ltd., The University of Melbourne, VIC, Australia

**Christopher Walker**, Stormwater Research Group, University of the Sunshine Coast, Sippy Downs, QLD, Australia; Covey Associates Pty Ltd, Sunshine Coast, QLD, Australia

**Andrea Walton**, Josh Byrne & Associates, Fremantle, WA, Australia; CSIRO Land and Water, Brisbane, QLD, Australia

**James Ward**, Natural and Built Environments Research Centre, University of South Australia, Adelaide, SA, Australia

**Kym Whiteoak**, RMCG, Melbourne, Australia

**Martina Winker**, ISOE – Institute for Social-Ecological Research, Frankfurt/Main, Germany

**Qian Yu**, China Institute of Water Resources and Hydropower Research, Beijing, China

**Joshua Zeunert**, UNSW, Sydney, NSW, Australia

**Martin Zimmermann**, ISOE – Institute for Social-Ecological Research, Frankfurt/Main, Germany

# Editors

**Ashok K. Sharma**

Dr. Ashok K. Sharma is an Associate Professor at the Institute of Sustainability and Innovation, College of Engineering and Science, Victoria University, Melbourne, Australia. He has 30 years of research, teaching, and industrial experience on planning and design of centralized and decentralized water, wastewater, and stormwater systems; integrated urban water management; and water sensitive urban design. As Principal Research Engineer, CSIRO, Australia, he led research on alternative water, wastewater, and stormwater systems to address knowledge gaps in their mainstream uptake. He also worked as a Planning Engineer at the Department of Natural Resources and Mines, Queensland, Australia, an Engineer at Uttar Pradesh State Water Corporation, and an Assistant Professor at Delhi College of Engineering in India. He has coauthored 3 books, 11 book chapters, 70 journal and 69 conference publications, and 45 technical reports. He completed his B. Tech (Civil Eng.) at G B P Agriculture and Technology University, Pantnagar, India and ME (Environmental Eng.) and PhD (Civil Eng.) at the Indian Institute of Technology, Roorkee, India. He is a Fellow of the Institution of Engineers (Australia) and CP Eng. (Australia) (ashok.sharma@vu.edu.au; asharma2006@gmail.com).

**Ted Gardner**

Adjunct Professor Ted Gardner holds adjunct appointments at Victoria University, Melbourne and at a number of Australian regional universities. He chairs the technical advisory committee of the Australian Water Association's e-Water journal. Prior to his retirement in 2010, Ted was a Principal Research Scientist, Integrated Urban Water Systems, CSIRO, where he led research projects into decentralized water technologies and stormwater harvesting and reuse in South East Queensland. He was also the Principal Scientist with the Queensland Department of Environment and Resource Management, leading the Urban Water Cycle group, which focused on urban water sustainability. In 2005, Ted was awarded the Australia Day Award Public Service Medal for his work on water recycling and urban water supply. In 2014, he was awarded the biennial McLean-Idema award from Irrigation Australia for his career work on irrigation using recycled water. He has an extensive publication record including over 200 peer-reviewed journal and conference papers, 4 book chapters, coeditor of a scientific monograph on purified recycled water, and an IWA book on rainwater systems, and he has made numerous presentations to technical and community groups. Ted completed his Bachelor of Agricultural Science and Master of Agricultural Science at the University of Queensland (tedandkayegardner@bigpond.com).

**Don Begbie**

Don Begbie was Executive Officer, Australian Water Recycling Centre of Excellence, and Program Manager, Research and Development, until the Center's closure in December 2016. Prior to that, Don was the Director of the Urban Water Security Research Alliance in South East Queensland, Australia, where he managed and coordinated the delivery of research for urban water security with a focus on integrated water management and alternative water sources such as rainwater tanks and stormwater harvesting. He was previously the Director of Water Science, Queensland Department of Natural Resources and Water, where he managed the delivery of research into urban water systems, groundwater and surface water modeling, and freshwater quality and aquatic ecosystem health. Don completed both his Bachelor of Agricultural Science and Master of Agricultural Studies at The University of Queensland (donaldbegbie@bigpond.com).

# Foreword

Increasingly, we are living in big cities and managing stormwater is becoming a major challenge, thanks to the increase in hard, impermeable surfaces of roofs, roads, footpaths, and car parks.

Systems that can quickly drain runoff to minimize flooding may inadvertently create new problems: these systems can also very effectively transfer contaminants from the urban environment to receiving creeks and rivers. Pollutants such as nitrogen, phosphorus, heavy metals, and sediments substantially reducing their ecological health may indeed increase flooding through increased sedimentation.

Designing better systems to manage stormwater through water-sensitive urban design therefore makes a lot of sense—with its emphasis on stormwater reuse, rainwater tanks, and installing more vegetation, our precious environment and water resources are better protected.

There is much written about water-sensitive urban design from across the globe—this book helps to distill this knowledge for both the student and the urban water practitioner.

The challenges of achieving the right mix of hydrology, water quality, and design esthetics to protect or restore natural ecosystem functioning of complex urban water systems should not be underestimated in helping create a sense of place for new urban communities.

This book covers a wide range of topics and I congratulate the editors in getting such a skilled mix of international authors together to share their knowledge in a holistic and integrated manner, while keeping a fine balance between theory and practice.

I recommend this book to you, whether you are a student, a designer, a planner, or a local authority.

The principles and insights in this book will no doubt open your eyes to new concepts. Water is a limited and valuable resource, and we must cherish by adopting smart management techniques.

Happy reading!

**Dr. Christine Williams**
**A/Queensland Chief Scientist**

# Preface

Conventional stormwater systems in cities were designed to quickly drain the stormwater runoff from urban areas to minimize flooding. However, this hydrologically efficient system of gutters and big pipes was also very efficient in transferring contaminants and sediment to receiving creeks and waterways. This invariably caused a substantial reduction in their ecological health, and a destruction of their stream morphology by erosion and/or sediment smothering. Stormwater is essentially a diffuse pollution source and, as such, it is much more challenging to manage than point sources such as the discharge from sewage treatment plants and factories.

Over the last few decades Australia has invested many hundreds of millions of dollars into sewage treatment to reduce the contaminant loads into the bays and estuaries that surrounded most of its major cities. The attention of society is now turning to urban creeks and rivers that provide such important ecosystem services to their communities. Many of these waterways have been straightened and lined with concrete to make them more efficient conduits to transport the extra rainfall runoff from rapid urbanization.

Urban society has also developed the aspiration to be more locally self-sufficient and to protect the remaining natural urban ecosystem, involving effluent reuse, stormwater capture and reuse, rainwater tanks, combined with more energy-efficient technologies. Hence the concept of water sensitive urban design (WSUD) started to take off in Australia in the 1990s, with new ways of designing and building suburbs, which do not rely on the direct drainage of runoff from impervious surfaces to waterways. Moreover, there was an emphasis on alternative urban water supplies, renaturalization of water courses and associated riparian areas, and installing vegetative technologies that not only looked attractive in the urban street, but also delivered a much-improved stormwater quality.

Given the connectedness of the global community, it's not surprising that this WSUD concept emerged in other countries of the world, although each had their own nomenclature and drivers. Hence the terms: best management practices (BMPs), green infrastructure (GI), integrated urban water management (IUWM), low impact development (LID), low impact urban design and development (LIUDD), source control (SC), stormwater control measures (SCMs), sustainable urban drainage systems (SUDS), Sponge City, and experimental sewer systems (ESS). The specific drivers for this innovation also varied between countries, with North America initially focusing on water quality improvement, while much of Europe was driven by the need to reduce local flooding and overflows from their "combined sewers," which carry both stormwater and sewage. Australia focused on water quality protection, waterway ecosystem protection, and littoral zone conservation, while other countries, such as China, are facing urban water shortages that somewhat perversely are accompanied by regular flooding, and impaired stormwater quality.

Even though these approaches are comparatively new, we find ourselves today with a wide range of WSUD technologies, design models, descriptive terms, driving objectives, guidelines, regulations, effectiveness metrics, and economic values as part of societies' journey to urban sustainability.

WSUD approaches are implemented in existing and new developments to address impacts from climate change, urbanization, and population growth. Incorporating WSUD as a mainstream practice in urban developments can play a significant role in the transition from the current water, wastewater, and stormwater systems to a more sustainable paradigm including mitigating impacts from climate change and urbanization. WSUD systems can deliver multiple benefits including water supply, stormwater quality improvements, flood control, landscape amenity, healthy living environment, and ecosystem health improvement of urban waterways.

So, if we know so much about WSUD, why do we need to write another book on it? The answer we think is in the vast store of data, information, and social drivers that can make distilling "the knowledge" a very difficult task for the student, the water practitioner, and the urban planner.

It is also important to understand what WSUD cannot do, especially for protection from low-frequency flooding events and the high erosion losses and stream degradation that can occur during civil construction before WSUD measures are implemented. The challenge is getting the right mix of hydrology with water quality and design aesthetics to protect, or restore, the natural functioning of a complex urban water ecosystem, which helps create a sense of place for new urban communities.

In this book, we aim to provide a holistic overview of WSUD technologies, their applications, and successes using Australian and international studies (mainly North America and Northern Europe). The book has 27 chapters, each written by different authors, and has been divided into several themes. These chapters are described in brief to provide overview of the associated themes.

## 1. HISTORY OF WSUD AND WSUD APPROACHES

Chapter 1 sets the scene for water sensitive urban development, both historically and geographically. It considers the evolution of ecologically sustainable stormwater management in Europe, North America, United Kingdom, Asia, Australia/NZ and introduces terms used in those countries, such as LID, SuDS, Sponge City, and GI. A key underlying principle of WSUD/LID is to emulate the natural hydrology of a site by using decentralized management measures. However, the drivers for adopting WSUD can be quite different between countries, and includes: sewer overflow protection, flood management, access to green space, water quality protection, waterway ecosystem protection and littoral zone conservation, and stormwater harvesting and reuse.

Chapter 2 focuses on the WSUD technologies used in Australia. Avoidance measures (such as permeable pavement) avoid the generation of contaminated stormwater runoff from allotments. Mitigation measures (including gross pollutant traps, swales, bioretention basins, wetlands, and smart street trees) are typically implemented to detain and treat stormwater runoff. The selection of technologies is heavily influenced by the preferences of local authorities and site-specific considerations such as soil type and slope and existing assets. Despite over 20 years of WSUD practice in Australia, there is still much to be learned about the performance of many of the treatment technologies, as installed in the field. Nonetheless, simple visual assessment of healthy plant growth is a very useful criterion of the operational effectiveness of vegetated, treatment devices.

## 2. STORMWATER QUALITY

Chapter 3 discusses the chemical and microbiological characteristics of stormwater and the types and efficacies of typical stormwater quality mitigation measures. Catchment characteristics including stormwater and wastewater infrastructure, land-use activities, traffic characteristics, and climate are key influencers of water quality. The authors give a detailed description of pollutants' build up and wash off processes, and how these may be modeled. Detections of pharmaceuticals, human pathogens, and human-specific biomarkers in stormwater from catchments with **separate** sewers highlight the need for further research on pollutant transport processes.

## 3. DESIGN GUIDELINES AND REGULATIONS

Chapter 4 discusses the international, national, regional, and local planning and design guidelines that have been developed by various agencies for the sustainable implementation of WSUD/LID systems. These guidelines help water professionals to plan, design, and implement these approaches based on urban development requirements, water quality and hydrology criteria, catchment characteristics, local climatic conditions, local regulations, and environmental and community considerations.

Chapter 5 reviews WSUD policy and regulation in Australia and internationally. Case studies from Australia, Europe, the United States, and Singapore show how the mix of policies, incentives, regulation, capacity building, and institutional perceptions at various levels impact the institutional culture and context in each jurisdiction. Municipal government has typically been the key agent for WSUD implementation. However, collaboration is required across discipline areas and stakeholders to support and empower local government and community in the implementation of WSUD.

## 4. POTENTIAL FOR WSUD

Chapter 6 reminds us that most WSUD features are designed to be multifunctional elements that provide benefits to runoff volume, peak flow rate, water quality, and stream ecology. The systems are also intended to reduce flooding and peak flows from small and frequent storms. Critical parameters for successful flood mitigation performance are detention storage size, the portion of catchment impervious area connected to the storage, and the rate at which the storage is emptied. Their efficacy to reduce flooding and peak flows from larger, less frequent storms at the broader catchment scale has yet to be confirmed.

Chapter 7 discusses the use of GI stormwater controls such as rain gardens, swales, and porous pavement to alleviate the magnitude and frequency of combined sewer overflows (CSO). Although many modeling studies have demonstrated the potential of large-scale use of these controls for CSO reduction, there have been few monitoring efforts. This chapter reviews two such large-scale projects in the United States, which monitored the performance of retrofitted GI in combined sewer catchments with areas of 8–40 ha.

Chapter 8 examines the impacts and magnitude of sediment loads generated during the construction phase of subdivision and compares this with the loads generated during the operational phase of development—traditionally the major focus of WSUD in Australia. Without application of erosion and sediment control measures, sediment export from the construction phase is orders of magnitude greater than the sediment export from unmitigated operational-phase runoff. Even with application of conventional best practice measures, the construction-phase loads are still equivalent to nearly a decade of operational-phase sediment exports. The author recommends that much greater emphasis should be placed on the construction phase in regulation and research.

Chapter 9 introduces the role of stormwater and roof water harvesting for beneficial use as part of an integrated WSUD approach to urban development. An effective scheme must combine sufficient rainfall, a suitable catchment, opportunities for diversion and storage, adequate demands, and water treatment suitable for the proposed end uses. Other issues discussed include stormwater contamination, validation and verification, and governance issues. However, the main impediments for operators to develop harvesting schemes with regulatory and financial confidence are the uncertainties with the long-term operation, governance, and compliance requirements.

## 5. ECOLOGICAL HEALTH COVERING IMPACTS AND BENEFITS FROM WSUD

Urban development changes the hydrology of catchments (including runoff volume, frequency, and peak flow) and the transport of sediment, nutrients, and pollutants. Consequently, it has a degrading impact on urban stream morphology and in-stream biota. A key question is whether, and to what extent, WSUD can prevent these changes. Chapter 10 suggests that WSUD can restore hydrology at small scales; however, restoration at the catchment scale is much more challenging, and there is limited evidence that existing techniques are effective. Chapter 11 finds that even when WSUD measures are implemented to help restore more natural flow patterns, degraded water quality can have an overriding influence on stream ecosystem health.

Chapter 12 discusses the changes to stream morphology and the opportunities for WSUD to ameliorate the impact. WSUD has been successful in reducing pollutant loads and providing some reductions in flow volume. However, current practice has commonly failed to arrest the geomorphic degradation of streams, due in part to the fact that WSUD has rarely been applied at a catchment scale, sufficient to mitigate the increased magnitude and frequency of runoff from connected impervious areas.

In Chapter 13 we learn that engineered urban lakes primarily increase amenity and property values and provide a flood-mitigation purpose. The failure of many urban lakes to remain in a healthy ecosystem usually stems from poor design and a lack of runoff pretreatment. Once a lake changes to a degraded state, it is very difficult to recover the initial healthy state.

## 6. WSUD ECONOMICS AND OPTIMIZATION

Economic assessment of WSUD investments is challenging. Data shortages and the broad range of nonfinancial benefits provided by WSUD make it difficult to rigorously quantify economic benefit. Chapter 14 provides a framework to overcome these difficulties. Cost–benefit analysis (CBA) transparently provides a decision-maker with a decision metric for proceeding with an investment, or otherwise. Total Economic Value (TEV) identifies and categorizes all benefits accruing from an investment, including environmental and social benefits that may be difficult to quantify. A remaining challenge to rigorous economic assessment is the data availability of environmental and social benefits produced by WSUD investments.

Chapter 15 discusses how optimization methods can be used to plan and design WSUD schemes to achieve the best outcomes and identify system trade-offs between a range of economic, social, and environmental indicators. Two case studies consider the selection, sizing, and layout of WSUD components for water quality improvement and stormwater harvesting. Future developments in optimization are also discussed.

## 7. WSUD IN INTEGRATED URBAN WATER MANAGEMENT AND URBAN PLANNING

Chapter 16 outlines a case study from Melbourne, Australia, where water industry experts discuss the practical infrastructure and urban planning processes to achieve the vision of IUWM and WSUD. Effective coordination of policy development, strategy, planning, and implementation of WSUD approaches is required to overcome the primary barriers to achieving these visions.

Chapter 17 focuses on the lessons from South Africa in WSUD and GI, planning, application, and implementation. It describes the need for context-driven design guidelines and for emerging middle class South Africa to become familiar with WSUD approaches, the importance of social benefits, and the integration of WSUD into mainstream spatial planning.

WSUD is currently applied at the local municipal level, with much of the current planning and design-related WSUD material focused on stormwater harvesting, management, and maintenance-related issues such as greening roads and street verges, open space areas, and a cities' landscape features. Chapter 18 highlights the opportunity for WSUD to contribute to and enhance urban sustainability through the relatively new concepts of healthy and liveable cities, which can be used to promote sustainability and provide economic and social benefits to communities.

## 8. URBAN HEAT ISLAND AND GREENING THE CITY

Urbanization can lead to the development of the urban heat island effect, whereby public health and thermal comfort are adversely affected. Chapter 19 provides examples from various climates to illustrate how the application of GI (including parks, street trees, green roofs, and green walls) and WSUD approaches can be effective in mitigating increased urban air temperature.

Chapter 20 reviews the key elements of resilient green roof and living wall systems. Green roofs and living walls are becoming an important component of WSUD systems and provide many environmental, economic, and social benefits such as: reduced temperatures both inside and outside of buildings, reduce building energy usage, improved air quality, and reduced pollution levels. This chapter will assist urban planners and designers in developing resilient GI for cities, particularly those located in dry climates.

Chapter 21 provides a European perspective on the use of novel urban water systems in greening and cooling the urban environment. Although WSUD design principles usually focus on stormwater management, this chapter provides examples of the integration of urban wastewater into WSUD.

## 9. CAPACITY BUILDING AND COMMUNITY PERCEPTION FOR WSUD

As the stormwater components of WSUD have gained traction, large numbers of SCMs have been constructed as new assets. However, failure to appropriately maintain and operate these WSUD assets runs the risk of reducing public support for the implementation and adoption of WSUD approaches. Chapter 22 describes the challenges, operation, and maintenance requirements, and an eight-step process is described to develop WSUD asset management plans for the ongoing operation and maintenance of SCMs as a mainstream activity in local authorities.

Capacity-building programs are a critical component for the successful delivery and operation of WSUD systems. As WSUD systems are comparatively new, different skills for their planning, design, operation, and maintenance are required, and the associated capacity building programs are still evolving. Chapter 23 uses a case study from South Australia to follow the process of developing a business case and implementing a capacity building program. Successful capacity building results in the efficient delivery of assets, an improved return on the investment, and reduces the risk of asset failure.

The community can easily recognize the improved aesthetics, greenspace, and recreational amenity features of above ground WSUD systems. However, there is a need to educate the community about the benefits of other less visual outcomes such as water quality improvement and flood mitigation. Chapter 24 explores five dimensions of people's attitudes to, and engagement with, WSUD systems: visibility; recreation and other amenity; economic benefits for residents; place attachment; and social capital and community engagement. Interventions that increase awareness of WSUD benefits strengthen social capital within a community and helps support WSUD over the long term.

## 10. WSUD POST IMPLEMENTATION ASSESSMENT AND CASE STUDIES

Postimplementation assessment of developments designed with WSUD features is essential to learn from on-ground implementation of such systems to better inform future developments. Chapter 25 describes a case study from Kansas City, Missouri, USA, where a linear regression model was developed and verified with field data using a limited palette of SCM installations. The model was demonstrated to reliably estimate stormwater removal/capture by SCMs in the catchment. The performance of SCMs over a range of rainfall events during a 3-year monitoring period was shown to be effective in preventing CSO and supported the efficacy of green solutions in reducing urban runoff.

Chapter 26 provides a precinct-scale case study of an infill development near Perth, Western Australia. The development implemented a range of sustainable water, energy, and urban greening initiatives in a medium density site of mixed building typologies. Understanding the delivery process and learnings from the on-ground implementation experience are an important factor for the success of future such developments.

Chapter 27 provides Australian and international case studies of some leading edge WSUD approaches and discusses the challenges and benefits from implementing WSUD. The findings reinforce the importance of WSUD being integrated across different urban functions, stakeholders, and levels of government. The benefits of WSUD often extend beyond the primary objective of improved urban stormwater management, reflecting the multifunctional nature of many WSUD approaches. Case study findings can be used to refine standards and guidelines, build confidence in the WSUD approaches, and help build public understanding and engagement in the benefits of WSUD. The studies also identified the importance of using economic instruments that reflect the true cost of different stormwater management approaches, thereby helping create financial incentives for the adoption of WSUD.

**Ashok K. Sharma**
**Ted Gardner**
**Don Begbie**

Chapter 1

# History of Water Sensitive Urban Design/Low Impact Development Adoption in Australia and Internationally

John C. Radcliffe

*Commonwealth Scientific and Industrial Research Organisation (CSIRO), Glen Osmond, SA, Australia*

## Chapter Outline



**ABSTRACT**

Since the 1980s, urban planning and development has increasingly taken account of natural water and nutrient cycles, with water flows managed as would have occurred on the original greenfield site, emulating the original ecosystem. However, other drivers deriving from the impact of urban development and increased human activity include

- managing stormwater quality,
- mitigating a risk of increased flooding, especially where there had been some previous history of them in the greenfield environment,
- harvesting of rainwater and stormwater for potable and nonpotable use,
- greening the urban environment to reduce the heat island effect generated by intensive urban development and increased pavements, and
- improving the aesthetics of the urban environment to encourage a feeling of well-being in the community.

These philosophies were developed in North America as *Low Impact Development* and later *Green Infrastructure* (which sometimes also assumes a consideration of energy management) in response to the 1972 passage of the *Clean Water Act* (US). Planning may incorporate green roofs, rain gardens, swales, permeable pavements, wetlands, green spaces, and urban natural vegetation corridors. Britain adopted *Sustainable Urban Drainage Systems (SuDS)* techniques. In Europe, planners and local government respond to the European Union *Water Management* and *Flooding Directives* on a river basin basis. In Australia, similar policy developments took place under the

**Approaches to Water Sensitive Urban Design. https://doi.org/10.1016/B978-0-12-812843-5.00001-0**

philosophy of *Water Sensitive Urban Design* (*WSUD*). The components and drivers vary between communities, each choosing emphases appropriate to their catchments, infrastructure, seasonal climate, local water cycle, and social expectations.
Cities in East Asia have been undergoing rapid urbanization over the past 40 years, often accompanied by increased flooding, especially in China. In 2013, China introduced new urban policies which included the concept of *Sponge Cities* where "stormwater can be naturally conserved, infiltrated, and purified" for potential reuse, thereby reducing flood risks and increasing water availability. Construction guidelines were issued. Thirty major cities are participating as pilot cities. Each is eligible for central government subsidies. In Africa, the term *Water-Sensitive Settlements* has been suggested as the WSUD approach as originally envisaged does not take cognizance of the "developmental" or "equity" issues, which are particularly important in developing countries that may have legacies from their colonial or apartheid past.
Many countries are now beginning to consider the necessity for WSUD in their cities and settlements. Addressing technical integration problems, legislative constraints, social equity, and community acceptance will be necessary for them to develop *Water Sensitive Cities*. This chapter summarizes the history of WSUD adoption and examples of the approach that has been taken across the globe.

**Keywords:** Decentralized urban design; Green infrastructure; Low impact development; Low impact urban design and development; Sistemas Urbanos de Drenaje Sostenible; Sponge cities; Sustainable urban drainage systems; Techniques alternatives; Water sensitive settlements; Water sensitive urban design.

## 1.1 INTRODUCTION

As humans moved into settlements, they inevitably located near to rivers and smaller watercourses for ease of access to water, the fount of life. After heavy rain, floods sometimes occurred, but that was part of the normal water cycle. As buildings developed, there was an increase in the impermeability of surfaces, thereby increasing runoff. Watercourses began to erode from this additional water flow, or else channels were specifically constructed to take away the water. By the end of the 18th century, London, then the world's largest city, housed 10% of the population of Britain. Open channels carried not only stormwater but also wastewater to the River Thames (White, 2009). One London water company secured its supply below this point reinforcing the epidemiological observations about cholera by John Snow (Paneth, 2004). As cities further developed, piped systems of sewage collection were installed. In many older cities, these were linked with the stormwater channels that had been developed, producing what have come to be known as "combined sewers." Frequently these discharged to the nearest river or other receiving waters, adding to the pollution already occurring from enhanced overland flows following closer settlement.

Many older cities in Britain, Europe, Japan, Asia, and eastern North America still depend on combined sewer/stormwater systems, even though in many cities, new such installations have not been permitted since the mid-twentieth century. A typical American example is Portland, Maine, incorporated in 1876. This city and its surrounding greater metropolitan area has a population of half a million and is served by a sewerage system which began at the time of incorporation. It now consists of 151 km of sewers, 227 km of stormwater pipe, and 192 km of combined sewers. During wet weather, raw sewage can overflow from the combined sewer system into Casco Bay before reaching a sewage treatment works, making swimmers sick and contaminating seafood. There is currently no treatment system for polluted stormwater runoff from the land (Gallinaro, 2015).

By contrast, combined sewers are not used in any major Australian city with the exception of central Launceston (Jessup, 2015). Consequently sewer overflows and contamination of urban watercourses are much less of a problem in Australia. Nevertheless, stormwater ingress into the sewerage system during storm events can lead to sewage overflows into creeks, rivers, and bays. Sydney, which initiated a separate stormwater system from 1890, can be vulnerable to this because of its sandstone geology (Aird, 1961).

A consequence of increased population density in cities has been an inevitable increase in impermeable surfaces leading to an increase in stormwater runoff. Because of urban development, the natural water cycle has been disrupted, causing such problems as urban flooding, droughts, poor water quality, and reduced groundwater-sourced base flow in creeks. The historical response has been based on a philosophy to build facilities that carry away stormwaters as fast as possible, often using concrete conveyance structures that are themselves impermeable, leading to increased pollution, sedimentation, and measurable suspended solids that can exceed the environmental thresholds of the receiving waters.

A response to this has been one of the **detentions**, which involves the construction of large capacity holding basins or wetlands to hold runoff for relatively short periods to reduce peak flow rates until released into natural or artificial watercourses to continue the hydrological cycle as channel flow, evaporation, groundwater recharge, and input to lakes and marine water bodies. The volume of surface runoff involved in the temporary ponding process is relatively unchanged (Argue, 2005), but attenuation of the flow rate allows the water to leave the detention basin over a longer time interval with reduced risk of damaging flood levels and erosion. However, such approaches are well removed from the original pre-settlement hydrological cycle because of the increased volumes involved, the increased frequency of runoff events, flood risks, reduced groundwater infiltration, and evapotranspiration associated with the increasing fraction of impervious surfaces (Konrad, 2003).

Since the 1980s, management of stormwater has undergone progressive change—in parts driven by changes in the nature of cities, the increased proportion of a country's population living in them, and the frequency and severity of flooding. However, the widespread creation of Environment Protection Authorities also had an important impact. Although these gave particular attention to the composition of waters from wastewater treatment plants discharged to receiving waters, attention also was directed to the composition of stormwaters and the nonpoint pollution consequences of them.

Progressively, understanding broadened to encompass the whole hydrological cycle bringing in the concept of **retention** which refers to procedures and schemes whereby stormwater is held for relatively long periods and reused in the urban water cycle via the natural processes of infiltration, percolation, evaporation, evapotranspiration, domestic use (in house and outdoors), and industrial uses. The aim is to minimize direct discharge to natural or artificial watercourses (Argue, 2005). This seeks as far as possible to retain precipitation on the area where it falls, embodying a framework of *Integrated Catchment Management* or *Total Catchment Management*. The concept has led to changing the entire development philosophy and the evolution of cities. These developments resulted in a variety of technology descriptors such *as Low Impact Development (LID), Green Infrastructure (GI), Sustainable Urban Drainage Systems (SuDS), Water Sensitive Urban Design (WSUD), Low Impact Urban Design and Development (LIUDD), and Water Sensitive Cities (WSC).* Although the concepts are similar, different titles have been adopted in different countries (Fletcher et al., 2015).

To review the effectiveness of WSUD/LID technologies, Ahiablame et al. (2012) examined the global literature on the topic and also commented on some examples of computational models used for developing options. They observed that the literature has focused on evaluating examples of bioretention basins rather than other technologies, but evaluation should be strengthened for other techniques such as green roofs and swale systems. They also noted the need for greater attention to be given to microbial removal and the assessment of new and emerging contaminants such as trace organics and pharmaceuticals.

A total of 14 LID structures can be identified and are classified into three types: point, linear, and area. These can be incorporated into the *System for Urban Stormwater Treatment and Analysis Integration* (SUSTAIN) stormwater model (Shoemaker et al., 2009), a tool for evaluating, selecting, and placing best management practices (BMPs) in an urban watershed on the basis of user-defined cost and effectiveness criteria. SUSTAIN is capable of evaluating the optimal location, type, and cost of stormwater practices needed to meet water quality goals and has improved effectiveness in considering the separate sediment fractions of sand, silt, and clay. Bioretention cells, cisterns, constructed wetlands, dry ponds, infiltration basins, rain barrels, sand filters (surface), and wet ponds are classified as point installations. Grassed swales, infiltration trenches, and sand filters (nonsurface) are linear structures; green roofs and porous pavements are defined as area structures. Bioretention and bioinfiltration technologies have become the principal control mechanisms at source (Davis et al., 2009), rather than relying on end-of-pipe solutions such as large sedimentation dams. These systems are described in detail in Chapter 2.

However, these technologies cannot just be used in some sort of universal urban design as there is no one definable problem and no one big solution. There are many small, mutually dependent problems and solutions (Wong, 2016). The appropriate solutions will require consideration of the location, hydrology, infrastructural and planning options, and sociological circumstances. There will usually be a need to protect the ecosystem health of urban watercourses. This will require inter alia management of total suspended solids (TSS), total nitrogen (TN), total phosphorus (TP), and other pollutants. Not least of the considerations will be whether the city has a combined sewer system that also conveys stormwater.

## 1.2 NORTH AMERICA

### 1.2.1 United States of America—*Low Impact Development/Green Infrastructure*

Within the United States, the term LID has been adopted. Its first use appears to have been by Barlow et al. (1977), covering land-use planning in Vermont following the introduction of the US Federal *Clean Water Act* in 1972 (USEPA, 2017). This was encouraged by a seminal LID manual prepared for use in St George's County, Maryland, later reprinted for wider use (Coffman, 1999), seeking to emulate the original "natural" hydrology that had been present before urban development had taken place. The vaguely defined expression, *BMP* then came into use to encompass pollution issues, but following a review of stormwater practice (National Research Council, 2008), was replaced by the term *Stormwater Control Measures (SCM).*

**FIGURE 1.1** Bioretention infiltration system.

However, the regulatory environment for implementing LID is managed by the individual states with great variability. Over 770 US cities have combined sewer systems that lead to the discharge of uncontrolled overflows (including untreated sewage) into receiving waters during storm events (USEPA, 2016a). Earles et al. (2009) listed major barriers to the incorporation of LID into urban planning, viz., (1) LID is typically not integrated early in the planning process; (2) LID is recommended but not mandated; and (3) there is no consensus on LID protocols between the different governmental departments. To help overcome these impediments, a National Municipal Stormwater Alliance has been formed in the United States to coordinate the activities of state and regional municipal stormwater organizations, with a vision to provide clean water for the nation (NMSA, 2016). This is based on meeting stormwater quality runoff standards for stormwater separately transported through ***Municipal Separate Storm Sewer Systems***, (not combined), so-called "MS4s," which often discharge untreated stormwater into local water bodies. The United States uses a concept of *Total Daily Maximum Load*, required under section 303(d) of the *Clean Water Act* (*US*), for managing waterways and the discharges of natural and human-induced pollution sources to them. It is a calculation of the maximum amount of pollutants that a waterbody can receive from point pollution sources, nonpoint pollution sources, and allowing a margin of error to permit the water resource to still meet water quality standards required for its beneficial uses, be it for a drinking water reservoir, industrial use, or irrigation (USEPA, 2016b). To assist regional planning practitioners, the US Environment Protection Authority maintains a LID Urban Design Tools Website (USEPA, 2007).

LID with its focus on stormwater quality/peak discharge has been complemented with the concept of GI which links landscape architecture and urban ecosystem services with water cycle management and on occasions linking to achieving energy efficiency. The incorporation of green roofs, rain gardens, swales, permeable pavements, improved infiltration (Fig. 1.1), wetlands, green spaces, and urban natural vegetation corridors aims to improve urban amenity and reduce flood and pollution risks. The initiative is supported by the Environment and Water Resources Institute of the American Society of Civil Engineers, which seeks to assist the integration of public policy and technical expertise into the planning, design, construction, operation, management, and regulation of environmentally sound, sustainable infrastructure (ASCE-EWRI, 2016).

New York, the most densely developed city in the United States, has responded through its Department of Environment Protection to the changing approach toward stormwater management. Under its Protection Strategic Plan 2011−14, it aims to maximize GI, control other water sources, and reduce runoff from existing and newly developed areas (Bloomberg, 2011). The aim is to progressively develop over the next 20 years, installations to collect the first 25 mm of rainfall on 10% of the impermeable areas to reduce combined sewer overflows. A GI Plan will increase public and private investment in swales, green roofs, and other source controls, as well as in managing stormwaters. Other benefits will flow from cooler urban temperatures, better air quality, more green space, lower energy bills, expected higher property values, and a reduction in combined sewer overflows (OECD, 2014).

## 1.2.2 Canada

Canada began to develop stormwater planning in the late 1970s, including detention basins built on ditch and pipe systems. Like elsewhere in North America, planning was often driven by a cycle in which local governments typically proceeded from flooding to panic to planning and then to procrastination and the next flood (BC MWLAP, 2002). In 1992, the province of British Columbia generated a Stormwater Planning Guidebook that recognized detention ponds can mitigate flooding but usually do not prevent the ongoing channel erosion that creates adverse impacts on property and fisheries. Detention solutions also often do not allow the sustained *stream base flow* that is ecologically critical in dry months. The Guidebook recognized five principles, viz., agree that stormwater is a resource; design for the complete spectrum of rainfall events; act on a priority basis in at-risk drainage catchments; plan at four scales—regional, watershed, neighborhood, and site; and that best solutions and reduced costs are achieved by adaptive management (BC MWLAP, 2002).

Ontario issued its Stormwater Management Planning and Design Manual in 2003 (Ontario Ministry of Environment, 2003) and focused on the opportunities for infiltration at the site scale and recognized that stormwater management solutions need to consider specific site conditions. However, it continued to be primarily oriented to a combination of allotment level, conveyance, and end-of-pipe stormwater management practices. The multiple objectives of stormwater management were maintaining the hydrologic cycle, protection of water quality, and preventing increased erosion and flooding. End-of-pipe stormwater management practices were still required to control urbanization's adverse effects, which remained after preventative techniques and allotment level and conveyance measures had been applied.

By 2010, Ontario observed that the practice of managing stormwater was continuing to evolve as the science of watershed management and understanding of watersheds grew. The *Low Impact Development Stormwater Management Planning and Design Guide* (T&RCA and CVCA, 2010) was released to augment the 2003 Ontario Design Manual. The ultimate goal of LID was defined as to maintain natural or predevelopment hydrologic conditions, including minimizing the volume of runoff produced at the site (i.e., neighborhood, subdivision, or individual lot). Runoff reduction is defined as the total runoff volume reduced through urban tree canopy interception, evaporation, rainwater harvesting, and engineered infiltration and evapotranspiration stormwater BMPs. LID comprises a set of site design strategies that minimize runoff with **distributed**, small-scale structural practices that mimic natural or predevelopment hydrology through the processes of infiltration, evapotranspiration, harvesting, filtration, and detention of stormwater. These practices can effectively remove nutrients, pathogens, and metals from runoff. They reduce the volume and peak discharge of stormwater flows. Opportunities were recognized at the community scale, the neighborhood scale with infill, and redevelopment opportunities and opportunities at the site scale. Thirteen LID structures were suggested for consideration. These include

- Reduced lot grading
- Roof discharge to surface ponding areas
- Roof discharge to soakaway pits
- Pervious pipes
- Pervious catch basins
- Sand filters
- Infiltration trenches
- Enhanced grass swales
- Vegetated filter strips
- Wet ponds
- Dry ponds
- Wetlands, and
- Infiltration basins.

Several of these also provided space to store snow. Structures were adapted to withstand cold weather conditions in the region, withstand freeze–thaw conditions, and where possible treat the quality of snowmelt runoff.

There have been concerns that LIDs may be less effective in cold climates where the ground can be frozen, with poor substrate permeability and low biological growth rates coinciding with high flows from subsequent snow melts. However, Roseen et al. (2009) demonstrated that while impacts due to cold climate had been observed, they did not substantially change hydraulic efficiency.

## 1.3 EUROPE

Within the EU as a whole, a revised Water Directive in 2000 (transposed into United Kingdom legislation in 2003), sought to extend water protection to all surface and groundwaters. The aims were to achieve "good status" for all waters by a set deadline; develop water management based on river basins; establish a "combined approach" of emission limit values and quality standards; get the prices right; ensure citizens were closely involved; and to streamline legislation. Objectives are set for each river basin encompassing ecological status, quantitative status, chemical status, and protected area objectives (EU, 2016a). Flood risk management plans were required to be developed by 2015, but no specific technologies were advised.

However, the Director-General Environment of the European Commission identified preferred options for the flood risk management by working with nature, rather than against it. Building up GI—which requires investing in ecosystems—offers triple-win measures through (1) contribution to the protection and restoration of floodplain and coastal ecosystems; (2) mitigation of climate change impacts by conserving or enhancing carbon stocks or by reducing emissions caused by wetland and river ecosystem degradation and loss; and (3) by provision of cost-effective protection against some of the threats that may result from climate change such as increased floods. The key ecosystem services from floodplains are water retention, clearance of water, and prevention of soil erosion. These services can contribute significantly to flood prevention and mitigation if the delivering ecosystems are in good health (D-G Environment, 2011).

A *Multiuse water Services* (MUS) approach for the more effective planning and management of the urban environment has been developing in Europe. The objective has been to optimize solutions for greenfield (new) and retrofitted urban planning developments. Components encompass the reinstatement of natural flows, increased infiltration, better use of rainwater, reuse of wastewater, and improved urban planning that incorporates local characteristics into the plan. The expectation is that local water management is more efficient and effective than if managed centrally. This potentially global MUS approach is being led through Germany, France, the Netherlands, and the United Kingdom (Maksimovic et al., 2015).

The EU also sponsored a WSUD manual through its "Software Workbench for Interactive, Time Critical, and Highly self-adaptive Cloud applications" (SWITCH) program. This joint venture research project is focused on "Innovative Water Management for the City of the Future," to achieve a more integrated approach in urban water management (Hoyer et al., 2011). To achieve this goal, SWITCH improved the scientific basis and shared knowledge to ensure that future water systems are robust, flexible, and adaptable to a range of global change pressures. The SWITCH project (2006—11) defined and reported on specific water issues in 12 cities (Fig. 1.2), namely Accra (Ghana); Alexandria (Egypt); Beijing (China); Belo Horizonte (Brazil); Birmingham (UK); Bogotá (Colombia); Cali (Colombia); Hamburg (Germany); Lima (Peru); Lódź (Poland); Tel Aviv (Israel); and Zaragoza (Spain) (SWITCH, 2012).

Because of the variability of European flooding events, much flexibility in objectives and measures is being left to the European states in line with "subsidiarity," a principle that directs attention to those levels of government where policy objectives can best be formulated and implemented (EU, 2016b). Thirty-four EU guideline documents are available. Water resource managers are expected to review plans and maps every 6 years. The number of inhabitants and economic activity potentially at risk and the environmental damage potential must be indicated (EU, 2016b).

## 1.3.1 Britain—*Sustainable Urban Drainage Systems (SuDS)*

The concept of *Sustainable Urban Drainage Systems (SuDS)* was developed in Britain and was published in 2000 as definitive guidance documents for Scotland and Northern Ireland (Martin et al., 2000a) and separately for England and Wales (Martin et al., 2000b). Details have since been progressively revised as *The SuDS Manual* (Woods Ballard et al., 2016). SuDS is a key part of WSUD, integrating the management of surface water runoff into the urban form. However, WSUD considers more broadly the whole water cycle, including wastewater and water supply, and the wider integration of watercourses and flood pathways within urban planning and design. The four primary benefits of SuDS are focused on water quantity, water quality, amenity, and biodiversity. The philosophy is to try to mimic the natural hydrology that is usually adversely impacted by urban development. Priority is primarily on water quantity rather than quality by mitigating peak flow rates and runoff volumes. Reducing potential pollution impacts and stream bank erosion risks is also important.

Many strategies are described in detail in *The SuDS Manual*, including rainwater collection as a resource, green roofs, pervious pavements, bioretention systems such as rain gardens (Fig. 1.3) that allow temporary ponding and soil infiltration, increasing evapotranspiration from revegetation such as tree planting, and the use of swales, detention basins, ponds, and wetlands to achieve water quality remediation and to slow runoff rates. Most strategies encourage infiltration and recharge to groundwater. Runoff events are controlled to ensure that peak flows do not affect the morphology or ecology of receiving waters. For previously established drainage systems, structures will have been designed to accommodate a 1:100 year flood event. There should be no runoff from small, frequent (5 mm) rainfall events—these should be intercepted on site, especially as they may contain the "first flush" of nonpoint pollution, thereby minimizing impact on downstream ecosystem health. *The SuDS Manual* has methods to estimate runoff characteristics by referring to the earlier Flood Estimation Handbook and the regional variations that can be used in Britain and Ireland to estimate peak flow frequency curves. Design criteria requirements include that there is no flooding on site for up to a 1:30 year event **unless** there is community acceptance of temporary overflow facilities, such as car parks that also serve other community functions.

The World Bank has adopted the SuDS philosophy as a component of its guide to integrated urban flood risk management (Jha et al., 2012).

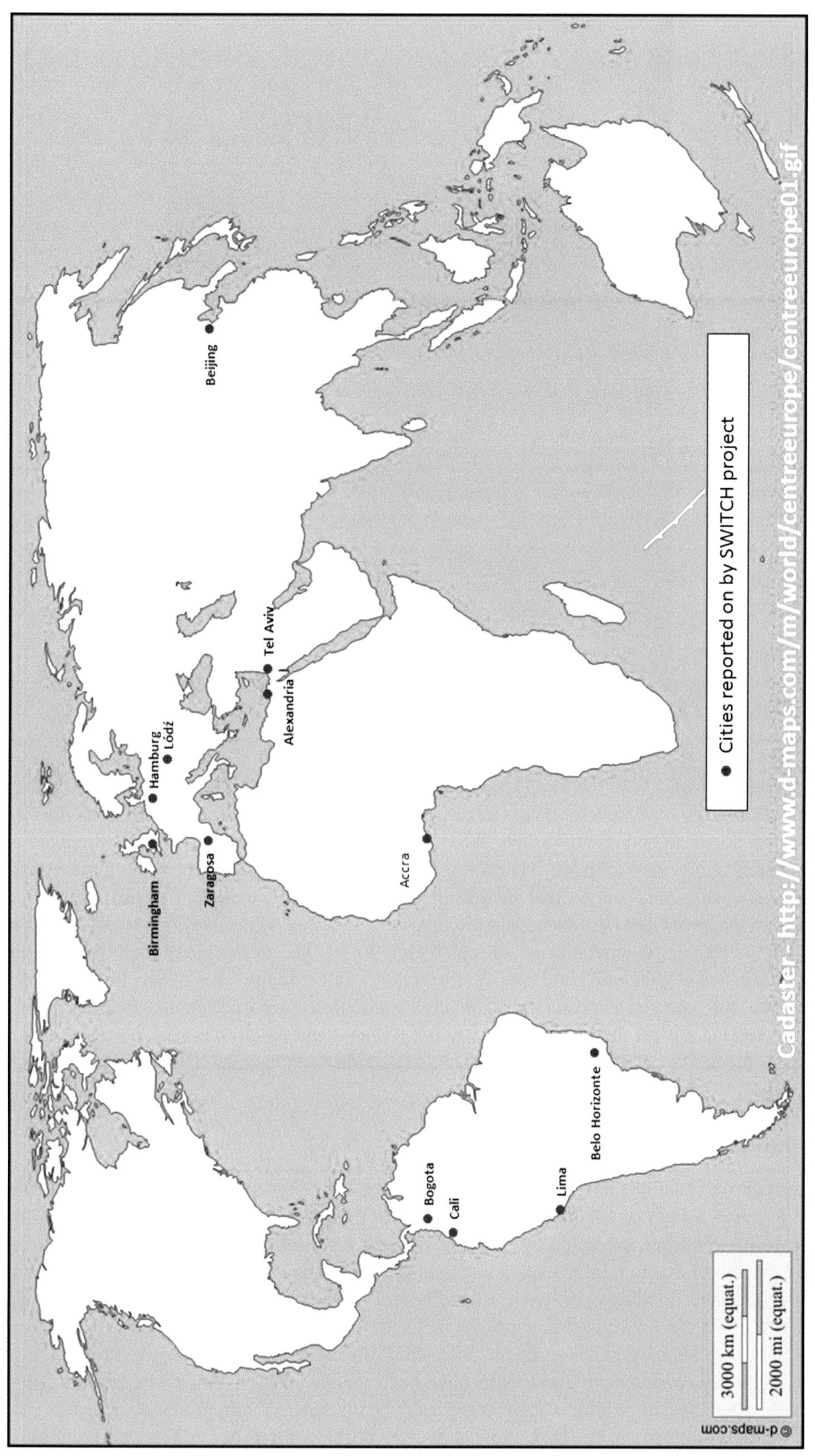


FIGURE 1.2 Cities whose water system issues were reviewed in the EU SWITCH project.

FIGURE 1.3 Median strip rain gardens suitable for small rainfall events.

### 1.3.2 Germany—*Decentralized Urban Design*

Somewhat similar philosophies have developed in Germany under the title of *Decentralized Urban Design* (Imbe, 2013), but Germany has also begun adopting SuDS and WSUD. Decentralized sustainable urban drainage systems for the treatment of stormwater runoff are becoming increasingly prevalent. Traditionally, most stormwater has been gathered together irrespective of quality and source and treated in an "end-of-pipe" treatment system. This is changing. New German federal regulations have been introduced and progressively adopted by states which stipulate that heavily polluted stormwater runoff should be treated separately at source (WHG, 2009). The most common stormwater pollutant, TSS, is classified as an "authoritative evaluation" parameter (*Leitparameter* in German). This means that TSS concentrations can be used as an indicator for potential concentrations of other particulate-based pollutants. Regulations being developed suggest that any SuDS devices must treat all stormwater, and unlike some other countries, no bypassing of high flows is permitted. In addition, the SuDS device must demonstrate a reduction in the annual TSS load by at least 92% during the testing procedure (Dierkes et al., 2015).

### 1.3.3 The Netherlands

The Netherlands provides an example where the specific objectives of adopting LID or WSUD are defined by the local geographical circumstances. Much of the Netherlands comprises polders which have been reclaimed from the sea and whose surfaces are below sea level, protected by banks (dykes). It has become necessary to develop measures for more sustainable water management due to rainfall events threatening the expanding urban areas. In September 1998, rainfall of 130 mm in 24 h caused severe flooding over much of the Dutch lowlands because the design drainage capacity of Dutch polders was sized for removing 14 mm rainfall per 24 h. Consequently, the decentralized retention and collection of rainwater (for example, retention by green roofs or collected from building and traffic areas) and the local storage of collected rainwater in ditches, ponds, lakes, and tanks have been mandated (Schuetze and Chelleri, 2013). Only surplus water, which can be neither retained, infiltrated, nor stored, may be discharged from polders to rivers or canals, and thence to the sea. New urban developments in the Dutch lowlands require the compulsory construction and integration of open surface water bodies, representing 5%–10% of the land area of the new development (Schuetze and Chelleri, 2013).

### 1.3.4 France—*Techniques alternatives*

The flexibility left to EU member states can be exemplified by the necessity for major cities to respond in terms of the catchments and hydrogeology on which they are located. For example, in Paris, the objective has changed from sending water outside the city through a pipe to managing it on site or at a river basin level by *Techniques alternatives*, involving retention and infiltration, including with green roofs, gardens, and lagoons, to encourage stormwater filtration and reuse (OECD, 2014). Maigne (2006) outlined initial approaches by several other French cities. In 2014, the International Office of Water, in Limoges, established a demonstration portfolio of "alternative techniques"—stormwater structures based on retention and infiltration (OIEau, 2014).

### 1.3.5 Spain—*Sistemas Urbanos de Drenaje Sostenible (SUDS)*

Spain is a country with alternating periods of long droughts and torrential rains, especially in the Mediterranean regions. This situation had led to the urgent need to manage stormwater properly to avoid lack of water resources in dry areas in the southern part of the country and to manage high volumes of rainfall in northern regions. In 1926, Spain was divided into Hydrographic Confederations which control their own watersheds. However, although the Spanish state establishes minimum requirements, different entities of regional governments are largely responsible for water management. The first reference to SUDS urban application occurred in 1997, when CLABSA (Clavegueram de Barcelona S.A) included recommendations for the implementation of Sustainable Urban Drainage Techniques (Técnicas de Drenaje Urbano Sostenible—TEDUS). Although SuDS technology has existed elsewhere since the 1960s, its application in Spain as SUDS was delayed until after 2000. It has since been promoted for implementation in Madrid, Barcelona, San Sebastian, Oleiros, Gijón, Zaragoza, Santander, and Valencia. Much initial research was undertaken by the University of Cantabria, Santander, primarily in the area of permeable pavements. This evolution of SUDS in Spain has been described in detail by Castro-Fresno et al. (2013).

The EU LIFE program is the EU's funding instrument for the environment and climate action. Within it, AQUAVAL ("The efficient management of rainwater in urban environments") was a 2010–13 program in Valencia that aimed to boost a more sustainable management of rainwater in Spanish municipalities, ensuring that rainwater and stormwater are included in water resources and land-use planning policies (Perales-Momparler et al., 2013). The AQUAVAL's main target was to find, implement, and promote innovative solutions to decrease the impacts of developments on quantity and quality of urban runoff (e.g., flooding, combined sewer overflow spills, pollution, drought, etc.). The municipal scale was chosen for the project implementation, with case studies in two different municipalities within the province of Valencia: Xàtiva and Benaguasil. Its subject focus was to demonstrate that innovative stormwater management approaches like Sustainable Drainage Systems (SuDS) could be effectively used in Mediterranean cities (where they were highly unknown), adding social and environmental values (Perales-Momparler et al., 2014). The hydrological and water quality results for swales and the basin demonstrated significant attenuation of flows, volumes, and concentrations. The project showed that SuDS can reduce runoff volumes and peak flow drainage systems under Mediterranean climatic conditions. A Transition Manual covering urban water, energy, urban planning, etc., has since been produced for Benaguasil. It is intended for decision makers at the local level, water utilities, and practitioners (Perales-Momparler et al., 2015).

## 1.4 AUSTRALIA—*WATER SENSITIVE URBAN DESIGN*

Following an initial use of the term WSUD by Mouritz (1992), the first formal guidance to it appears to be by Whelans and Halpern Glick Maunsell (1994). WSUD became summarized as formulating development plans that incorporate multiple stormwater management objectives and involve a proactive process, which recognizes the opportunities for urban design, landscape architecture, and stormwater management infrastructure to be intrinsically linked (Wong, 2000).

From 1994, various Ministerial Councils initiated the National Water Quality Management Strategy (ARMCANZ and ANZECC, 1994), which now encompasses 24 guidelines that include water quality, groundwater, diffuse and point pollution, sewerage systems, effluent management, and water recycling. In consequence of the Commonwealth/States/Territories commitment to the Intergovernmental Agreement on the National Water Initiative (NWI, 2004), guidelines were issued for WSUD (Joint Steering Committee for Water Sensitive Cities, 2009). These guidelines integrate the design of components of the urban water cycle, incorporating water supply, wastewater, stormwater and groundwater management, urban design, and environmental protection. The guidelines suggest that WSUD treatments should be aiming for 80% reduction in TSS, 60% reduction in TP, 45% reduction in TN, and a 90% reduction in gross pollutants when compared with untreated stormwater runoff. By using a variety of WSUD techniques, the postdevelopment, peak 1-year average recurrence interval (ARI) event discharging to the receiving waterway should not exceed the predevelopment condition.

Model for Urban Stormwater Improvement Conceptualization is a software program based on WSUD principles originally developed by the Cooperative Research Center for Catchment Hydrology. It has since been upgraded to model a wide range of treatment devices to identify the best way to capture and innovatively manage and/or reuse stormwater, reduce its contaminants load, and reduce the frequency of runoff. It evaluates alternative treatment devices until the best combination of cost, hydrology, and water quality improvement is achieved (eWater, 2016).

Australian thinking has also developed the concept of WSC in which the cumulative sociopolitical drivers for a set of evolutionary steps in the development of a city have been identified (Brown et al., 2009). Six progressive achievement levels of the urban water transition framework are recognized as a city develops. The first three stages describe the evolution of the water system to provide essential services such as secure access to potable water ("Water Supply City"), public health protection ("Sewered City"), and flood protection ("Drained City"). These are followed by the "Waterways City," "Water Cycle City," and ultimately a "Water Sensitive City." The latter provides a palette of services, including social amenity (for example—green spaces and ameliorating city heat island effects) and environmental protection, reliable water services under constrained resources (including accessing nonclimate dependant sources), intergenerational equity, and resilience to climate change (Fig. 1.4). The Cooperative Research Center for WSC, established in 2012, has since been developing an index for benchmarking WSC (Chesterfield et al., 2016).

The potential for stormwater control measures (SCMs) dispersed throughout catchments for improving both water quality and flow regimes and their scope to restore more natural hydrology, water quality, and consequently, ecological condition in the receiving stream must be investigated before detailed design and installation programs commence. Such issues are being pursued in an experimental study by Walsh et al. (2015). The project has involved monitoring during a period of continuing urban development of the 450 ha Little Stringybark Catchment, near Melbourne, while also considering retrofitting of existing installations. The experiment was designed as an extension of a Beyond Before-After-Control-Impact design. In addition to comparing ecological patterns in the Little Stringybark Catchment stream with those in other equally degraded urban streams as controls, a comparison of reference streams in forested catchments that are not affected by urban stormwater runoff has been included. The design allows control for differential responses to factors, such as climate between control and reference streams. The work has involved the integrated participation of catchment community stakeholders (residents and property owners), local government, and evolving planning regulations. Monitoring commenced in 2001, with joint action beginning in 2008, effectively becoming a long-term social–ecological study. The objectives were to reduce the frequency and magnitude of polluted high flows from impervious surfaces by ensuring that control measures had sufficient retention capacity to avoid overflow of untreated stormwater in 95% of rain events.

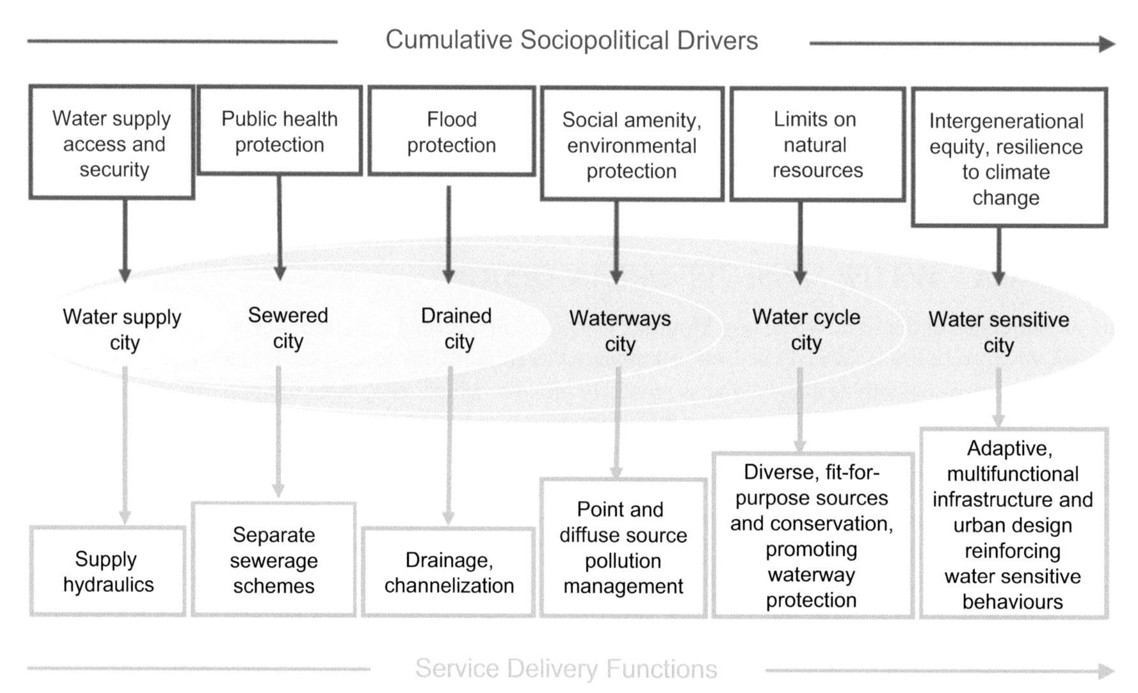


**FIGURE 1.4** Urban water management transitions framework. *Reproduced from Brown, R.R., Keath, N., Wong, T.H.F., 2009. Urban water management in cities: historical, current and future regimes. Water Science and Technology 59 (5), 847–855, with permission from the copyright holders, IWA Publishing.*

The measures were to restore the quantity and quality of base flows lost by the construction of impervious surfaces and ensure substantial harvesting of stormwater for internal uses (e.g., toilet flushing, hot water, laundry) or irrigation of open space to allow evapotranspiration losses. Access to public funding allowed residents to install 237 SCMs (5 installed independently of the project) and 58 installed by Yarra Ranges Council. Most implementation of control measures was achieved in July 2012—October 2013, and improvements to water quality were observed in October 2013—April 2014. The rate and trajectory of ecological recovery in streams following removal or mitigation of stressors remains uncertain. Responses probably will span multiple years. A website gives progress to the community and links to science outcomes from the project at https://urbanstreams.net/lsc/.

Most Australian States and Territories have implemented policies based on the principles of WSUD. There is great variability in implementation across Australia's capital cities which are the seats of the state governments. The state governments have constitutional responsibility for environment, water, land use, and public open space planning policies (Commonwealth of Australia Constitution Act, 1900). All except New South Wales have statutory State Planning Policies but are yet to harmonize and incentivize WSUD practices (Choi, 2016). Application in terms of state legislation is usually delegated to the third tier of government, local government. As a result, the extent of WSUD implementation can be discretionary, often impacted by other competing priorities.

## 1.4.1 Queensland

Queensland has a comprehensive approach within its State Planning Policy regulatory instrument (Queensland, 2016) with especial emphasis on environmental values, stormwater management, and acid sulfate soils. Components in its stormwater planning include controlling the rate and volume of runoff to approximate predevelopment values, including the one in one year average recurrence interval (ARI) flood frequency, impervious area first flush systems, infiltration techniques, and stream rehabilitation (Queensland, 2010). Within South East Queensland, following a detailed nutrient study of Moreton Bay (Dennison and Abal, 1999), a program was implemented to conserve the ecological quality of Moreton Bay, focusing initially on reducing the nitrogen discharge from wastewater treatment plants. This has been followed by a unique community-based program to protect and improve waterways with a collaborative model, involving industry, government, and research partnerships (Healthy Land and Water 2018). The focus has been on diffuse pollutants (especially suspended solids) from urban areas and up-catchment rural areas.

## 1.4.2 New South Wales

In New South Wales, a WSUD Program called "Splash", facilitated by NSW Local Land Services seeks to identify and address capacity needs within the Sydney Metropolitan region and country cities, assisting their transition to become Water Sensitive Cities (Sydney Water 2018). There is no mandatory requirement for WSUD to be adopted through any enacted state legislation or policies, but the state has a strong thrust of increasing urban water use efficiency and managing stormwater. The *Environmental Planning and Assessment Act 1979* (*NSW*) and the *Local Government Act 1993* (*NSW*) established a planning framework using State Environment Protection Policies and Regional Environmental Plans to set objectives, policies, and requirements for developments of defined state or regional significance. The main WSUD-related SEPP is the Building and Sustainability Index (BASIX) scheme, driven through the Environmental Planning and Assessment Regulation of 2000 and the 2004 State Environmental Planning Policy. BASIX, among other attributes, requires a 40% reduction in potable mains water use for all new residential developments and redevelopments compared with an average NSW annual potable water consumption from the residential sector, measured on a per capita basis (90,340 L of water per person per year). Stormwater management includes strong orientation to erosion and sediment control, especially during the construction phases of infrastructure and urban development (New South Wales, 2015). This broad legislative base has encouraged local governments to integrate their policy responsibilities covering planning and land management, stormwater, water conservation, water quality and supply, and wastewater management services. The *Water Industry Competition Act 2006* (*NSW*) aims to encourage competition in water supply and sewage services in NSW and to facilitate the development of infrastructure supporting production and reticulation of water, including recycled water, for which local governments have not previously been involved. An example of this is the Decentralized Water Master Plan 2012—30 for the City of Sydney, which manages the Central Business District of the city. This Master Plan is founded on the principles of integrated water cycle management and WSUD, and it differs from traditional management of water in three key ways. Sewage and stormwater are treated as a water resource rather than waste to be discharged. Recycled water can be used many times locally—a less wasteful way of using water. Stormwater is retained, slowed down, and treated with vegetated systems, rather than being disposed of swiftly through concrete pipes and channels into the receiving waters of Sydney Harbor (City of Sydney, 2012).

### 1.4.3 Victoria

The Victorian *State Environment Protection Policy for Waters of Victoria* identifies beneficial uses of Victoria's waterways, including natural aquatic ecosystems and associated wildlife; water-based recreation; agricultural water supply; potable water supply; production of molluscs for human consumption; commercial and recreational use of edible fish and crustaceans; and industrial water use. There had been concern that stormwater and industrial pollution issues could be affecting the ecology of Port Phillip Bay, which receives runoff from one of Australia's most densely populated catchments of nearly 10,000 $km^2$. The Port Phillip Bay Environmental Study of 1996, which was a seminal contribution to WSUD policy development, established that the Bay was remarkably resilient principally because the entering toxicants (heavy metals, pesticides, and petrochemicals) were being largely "locked up" in the sediments. Nitrogen had a rapid turnover in the populations of aquatic microorganisms and seafloor organisms (Harris et al., 1996), leading to substantial denitrification rather than eutrophication. However, the capacity is finite and has led to a state policy to reduce nitrogen load to the Bay either from diffuse (i.e., stormwater) or point sources (sewage treatment plants).

WSUD is now recognized as an alternative to the traditional conveyance approach to stormwater management (Melbourne Water, 2016a). The city seeks to minimize the extent of impervious surfaces and mitigates changes to the natural water balance through on-site reuse of the water and through temporary storage (City of Melbourne, 2008). Specific Victorian guidelines for stormwater management cover 80% retention of the typical urban annual load for TSS, 45% retention for TP, 45% retention for TN, and 70% retention of the typical urban annual load for gross pollutants (i.e., litter). The guidelines also prescribe that discharges for 1.5 year ARI be maintained at predevelopment levels (Victoria Stormwater Committee, 1999). Melbourne Water operates a stormwater offset service, which involved in 2016, a financial contribution by residential developers of AUD $6645 per kilogram of annual TN load for stormwater management works to be undertaken in another location. These alternative WSUD works, constructed elsewhere, "offset" stormwater impacts not treated within the development (Melbourne Water, 2016b).

### 1.4.4 South Australia

South Australia has a Technical Manual (South Australia, 2010), prepared with the objective of minimizing demand on the reticulated water supply system; protecting and restoring aquatic and riparian ecosystems and habitats; protecting the scenic, landscape, and recreational values of streams; minimizing treated wastewater discharges to the natural environment; and integrating water into the landscape to enhance visual, social, cultural, biodiversity, and ecological values. It seeks to reduce greenhouse gas emissions by reducing water consumption, increasing rainwater harvesting, and "natural" treatment alternatives. The Manual recognizes all water sources in the total water cycle as valuable resources, including rainwater (collected from the roof); runoff (including stormwater collected from all impervious surfaces); potable mains water (drinking water); groundwater; graywater (from bathroom taps, showers, and laundries); and blackwater (from kitchen sinks and toilets). The South Australian Manual reflects the dry Mediterranean climate, limited catchment water harvesting opportunities, and the importance of groundwater for high-value agriculture. It includes an emphasis on wastewater and the associated need to meet discharge standards to receiving waters, especially to St Vincent Gulf adjacent to which Adelaide is built. During the period 1949–1995, some 4000 ha of seagrass were lost between Aldinga and Largs Bay. Seagrass along the Adelaide coastline continued to decrease with 720 hectares lost between 1995 and 2002 (South Australia, 2003). Mangroves also decreased, even though pollutant loads from treated wastewater discharged into St Vincent Gulf were also reduced.

An Adelaide Coastal Waters Study was initiated in 2002. This showed that there was no evidence that toxicants or nutrients other than nitrogen played a key role in the ecosystem degradation. However, sediment movement inshore of the seagrass beds was sufficient to prevent regrowth of seagrasses (Fox et al., 2007). Results were similar to other studies such the Port Phillip Bay and Moreton Bay studies. In consequence, an Adelaide Coastal Water Quality Improvement Plan (South Australia, 2013) has been developed. The interagency complexities of preparing such a plan are reflected in the participation of 15 departments, authorities, nongovernment organizations, and a public company in the steering committee to oversee the process, which also involved substantial community consultation. The Plan provides a long-term strategy that is consistent with community expectations to achieve and sustain water quality improvement for Adelaide's coastal waters and creates conditions to facilitate ecosystem restoration and the return of seagrass along the Adelaide coastline. This is to be achieved through reducing nutrient, sediment, and colored dissolved organic matter discharges. Promoting integrated management and use of wastewater and stormwater across Adelaide is crucial. This integrated management has been led by the City of Salisbury and the South Australian Water Corporation with use of Managed Aquifer Recharge. The program began with a wetland constructed with the subdivision that created the suburb of Para Hills. This wetland (Fig. 1.5) provides flood protection, sedimentation, a pleasant urban amenity with walking and bicycle tracks, and a small

**FIGURE 1.5** This City of Salisbury, South Australia, "Paddocks" wetland, was developed with the suburb of Para Hills.

wildlife refuge and serves as a source for managed aquifer recharge (Fig. 1.6) of an aquifer from which water is withdrawn for irrigation of nearby playing fields. The council now has 40 wetlands, eight of which include managed aquifer recharge. They are linked through a ring main to service 500 customers, including 31 schools, with recycled water (Radcliffe et al., 2017). The same techniques are being adopted by other local government areas. All of these initiatives involve encouraging implementation of WSUD, though it is not mandatory at this stage (South Australia, 2013).

**FIGURE 1.6** Managed aquifer recharge well adjacent to the Salisbury "Paddocks" wetland shown in Fig. 1.5.

### 1.4.5 Western Australia

Much of Perth, the capital city of Western Australia, is built on a sand plain so that stormwater infiltration to groundwater is the default stormwater management. In such locations, discharge of stormwater off residential blocks to the street is prohibited. Western Australia therefore accepts WSUD as it strives to have Perth become a "waterwise city." Policies integrate management of catchments to maintain or improve water resources; manage risks to human life and property, including adequate flood clearance from 100-year ARI flooding, and surface or groundwater inundation/waterlogging; ensure the efficient use of water resources; and recognize and maintain economic, social, and cultural values. An extensive set of policies and guidelines links water and state planning processes (New WAter Ways, 2016a,b).

Southwest Western Australia has a demonstrated drying climate. Perth is dependent on groundwater for about 45% of its scheme water (potable supply). It prohibits access to domestic groundwater bores in winter to ensure infiltration is maximized. Upper level mounded aquifers located above aquitards in the Yarragadee Basin are important for maintaining urban wetland ecosystems. Perth has initiated desalination and is adding groundwater replenishment from recycled water. Perth's urban water management strives to maintain appropriate aquifer levels, recharge and surface water characteristics in accordance with assigned beneficial uses by managing groundwater recharge sustainably, minimizing seawater intrusion, minimizing the export of pollutants such as phosphorus and nitrogen to surface, groundwater or the marine environment, and preventing groundwater acidification processes (Western Australia, 2008).

It can be concluded that while WSUD is widely recognized across Australia, the emphasis varies in response to local circumstances, as it should. Melbourne and Brisbane are particularly oriented to protecting urban creek ecosystem health. All capitals are conserving their riparian environments. Erosion and sediment control are important in Brisbane, Sydney, and Melbourne. Nitrogen management is highlighted in Brisbane, Melbourne, and Adelaide, particularly from wastewater treatment plants to ensure protection of inshore riparian ecosystems. Perth is particularly keen to minimize use of its potable water resources and achieve effective management of groundwater resources on which its potable resources depend. Most capitals have invested in nonclimate dependent supplemental drinking water sources.

## 1.5 NEW ZEALAND—*LOW IMPACT URBAN DESIGN AND DEVELOPMENT (LIUDD)*

The concept of sustainability was embodied in the New Zealand *Resource Management Act 1991*. The Auckland Council has developed an updated Low Impact Design Manual for Auckland (Lewis et al. (2015)). New Zealand progressed the concept of LIUDD with a nationwide LIUDD research and implementation program. The uniquely New Zealand approach has adapted WSUD to encompass the cultures of the indigenous Māori community. The six-year program from 2003 was funded by the New Zealand Foundation for Research, Science, and Technology and led by Landcare New Zealand and the University of Auckland (Puddephatt and Heslop, 2008). Van Roon and van Roon (2009) described LIUDD as operating within a hierarchy of principles. These involve working on a catchment basis to maintain the integrity of mauri (the unique personality of all things animate and inanimate) in ecosystems, whereas ensuring selection of sites that minimize impact and adverse effects. Ecosystem services and infrastructure are used to efficiently maximize local resource use and minimize waste. Plans are to promote and support alternative development forms that create space for nature; restore, enhance, and protect biodiversity; reduce and contain contaminants; recognize natural soil, water, and nutrient cycles; and ensure energy efficiency. LIUDD also recognizes and provides for aspirations by Māori groups for biodiversity protection and enhancement.

Criteria for design and construction of land development and subdivision infrastructure for use by local government and developers include LID and LIUDD principles. They are covered in New Zealand Standard NZ4404 (2010). The Standard encourages sustainable development and is applicable to greenfield sites, infill development, and brownfield redevelopment projects. NZS4404 has been in part or more generally embedded in many of the Infrastructure Development Standards or Codes of Practice for local authorities in New Zealand, albeit they are not legally binding. More recently, the Auckland Council has commissioned a review of where it stands in relation to the WSC transitions framework (Ferguson et al., 2014).

## 1.6 EAST ASIA

The cities of Eastern Asia are evolving and urbanizing at a far greater rate than those of the developed world. Green Growth philosophies, which include consideration of urban hydrology, have been supported through Southeast Asia by the Organization for Economic Cooperation and Development and United Nations Environment Program (Spies and Dandy, 2012). Green growth recognizes the interdependency between economic and environmental systems and the risks posed by increased water scarcity, resource bottlenecks, air and water pollution, soil degradation, climate change, and biodiversity loss.

### 1.6.1 South Korea

South Korea has been a leader in the philosophy of "Green Growth," responding to environmental degradation brought about by rapid economic growth. It was 36% urbanized in the 1960s, but by 2005 was 86% urbanized. The percentage of impervious surfaces in the capital, Seoul, had increased from 8% in 1962 to 48% in 2010. Surface runoff increased from 11% to 52%, whereas evapotranspiration decreased from 43% to 18% (Kim et al., 2015a). With more than 90% of the downtown area urbanized, major flooding occurred in 2011 from the heaviest rain in 100 years (Yoon, 2014). The floods increased interest in the application of LID (Lee et al., 2013). The increase, both in the number of heavy rainfall days and impervious surfaces from new development, means that the existing drainage systems can no longer cope with the increase in runoff. Improving GI in urban areas is critical to reducing the reliance on "end-of-pipe" treatment systems (OECD, 2014). Seoul has attempted a number of initiatives, such as building 77 more drainage facilities with the capacity to retain over 550,000 $m^3$ of water, but there is room to improve urban planning (Lee, 2012).

One approach has been the continued development by Kim Y.J. et al. (2015b) of the Rainfall—Storage—Drain modeling system for use with buildings in high density, impermeable areas. Rainfall is collected from the roofs of the buildings into large capacity underground tanks, effectively holding basins, which can be slowly emptied to storm sewers after the storm event and whose capacity then becomes available for the next storm. The model showed that in Seoul City, a tank of 29 L/$m^2$ roof area can control the runoff of a 30-year frequency storm with the drainage pipes of 10-year design period (Kim and Han, 2008). In the case where a storage tank of 10 $m^3$ per 100 $m^2$ roof area is installed for the roof surface of all buildings in a small area, the results show that the peak flow and total outflow are reduced by up to 20% and 18%, respectively.

Increasingly, new developments are modeling the hydrology that will result from change and how it will be managed. Lee et al. (2013) examined the predevelopment condition of a proposed new town of 175 ha in Cheonan city, South Korea, and the impact on a 15 section plan whose imperviousness increased from 9% to 83%. Sixty urban constructed wetlands were incorporated into road reserves, with provision for 463 lateral ditch infiltration devices, 845 infiltration trenches, and 80 grass swales. Using the SWMM5 model (USEPA, 2016c), it was estimated that LID would reduce runoff by 55%—66% (peak discharge) and 25%—121% (flow volume) compared with a development without LID features. The impact of LID runoff reduction under peak flows for 50-, 80-, and 100-year return period conditions was estimated to be 6%—16% (peak runoff) and 33%—37% (runoff volume).

To address the issue further, also encompassing South Korean pollution issues, Son et al. (2017) have created a Low Impact Development district planning model (LID-DP) that incorporates LID into urban district planning, rather than just development site planning. The resulting model provides (1) a set of district unit planning processes that consider LID standards; and (2) a set of evaluation methods that measure the benefits of the LID-DP model over standard urban development practices. The model was applied to a test site in Cheongju city. The test simulation showed that LID-DP created a maximum 290% increase of infiltration and 40% decrease of surface runoff compared with the conventional urban development method. The volume of nonpoint source pollutant generated with LID-DP showed a maximum reduction of TN, TP, and Biological Oxygen Demand (BOD) of 37%, 56%, and 72%, respectively. Consequently, South Korea has begun considering LID features over traditional engineering structural techniques for stormwater management in urban planning, taking account of catchment ecology.

### 1.6.2 Japan

Japan is a country that is generally mountainous but with high-density cities built on low-lying areas increasingly given to flooding. Nearly 30% of Japanese cities have received localized torrential rain (>100 mm/h) in the past. Japan introduced, in 1973, the use of permeable sidewalks as an approach to what has become LID. In 1993, this was extended for possible use on low-traffic roads and has since been introduced elsewhere (Fig. 1.7). A *Road Surface Stormwater Treatment Manual* was released in 2005 (Jeong et al., 2015). The Japanese Association for Rainwater Storage and Infiltration Technology is an industry association with 22 full members and 48 supporting members. The association seeks to promote rainwater storage and infiltration facilities and rainwater utilization at a river basin scale, achieving a balance between flood control, water use, and maintaining the aquatic environment. Techniques encouraged include wet vegetation ditches under permeable sidewalks, gravel void storage and infiltration under permeable roads, permeable street boxes from drainage culverts, and rainwater storage and infiltration adjacent to roads (ARSIT, 2016). Any system must be maintained and kept clear, with "first flush" pollution needing particular attention. Maintenance highlights the need for cooperation between local government agencies, building owners, and roads authorities (Imbe, 2013).

FIGURE 1.7 Example of an Australian permeable pavement suitable for a low-traffic road.

### 1.6.3 China—*Sponge Cities*

China presents an even more dramatic situation. The urban population increased from 29% in 1995 to 46% in 2010 and is anticipated to reach 70% by 2020 (Shi et al., 2016). Shortage of water resources and pollution of the water environment have become constraining factors in the sustainable development of China. About 80% of water resources are in the southern area of the Yangtze River where there is 36% of the cultivated land, whereas 19% of water resources are in the northern area of the Huaihe River, which has 64% of the cultivated land (Fig. 1.8). The rainfall in the northern area is concentrated in the three months of summer with a major risk of floods (Qian, 2007). The demand for water is increasing fast with population growth and industrial demand. However, water use efficiency is low, with the leakage rate of urban piping systems higher than 20%. Half of the 600 largest cities are suffering from inadequate water supply. Forty cities suffer from acute water shortage in that only 60%−70% of peak demand supply can be met (Qian, 2007, 2016). Ensuring water availability to cities is of primary importance, whereas managing runoff quality and peak discharge is of secondary, but still major importance. Note that this order of priority is usually reversed for Australian cities.

China's water resource management system involves policies and laws being set centrally but administered through national, provincial, prefectural, and county levels of administration. Urban water management represents the most challenging component of water resource management. Chinese public policy has a continuing history of establishing independent management systems with limited coordination of their related functions and often with limited communication between agencies responsible for complementary management functions (Cosier and Shen, 2009).

In 2013, Chinese President Xi Jinping spoke at the Central Working Conference of Urbanization where he highlighted the significance of "building *Sponge Cities* where stormwater can be naturally conserved, infiltrated, and purified" for potential reuse (Tu and Tian, 2015). These authors also describe how, since this announcement, "Sponge City" planning and construction has been pursued throughout the country. In 2014, a "Sponge City" pilot program was initiated. To implement the plan, the Ministry of Housing and Urban Rural Development, and Ministries of Finance and Water released the "Guideline to promote building Sponge Cities" (State Council, 2015). Under the guideline, cities in China are to collect and utilize 70% of the stormwater, with 20% of urban areas meeting the target by 2020, and the proportion will increase to 80% by 2030. As a consequence, 16 cities (Baicheng, Qian'an, Jinan, Hebi, Changde, Wuhan, Pingxiang, Guian New Area, Zhengjiang, Jiaxing, Chizhou, Xiamen, Nanning, Chongqing, Suining, and Xixian New Area) were designated as pilot cities in 2015. A further 14 cities (Beijing, Dalian, Guyuan, Qingyuan, Qingdao, Xining, Yuxi, Yuncheng, Ningbo, Shanghai, Fuzhou, Sanya, Shenzhen, and Zhuhai) were added in 2016 (NL, 2016). These cities are shown in Fig. 1.8. The government will oversee the construction of *Sponge Cities* and let the market play a decisive role in allocating resources. Various fund-raising methods, including public−private partnership and franchising, are to be promoted according to the guideline. The Chinese central government will allocate each *Sponge City* between 400 and 600 million RMB (approximately AUD85 million to AUD128 million) toward developing ponds, filtration basins, and wetlands; and to build permeable roads and public spaces that enable stormwater to be infiltrated and reused. The plan is to manage 60% of stormwater in these pilot cities (Austrade, 2016).

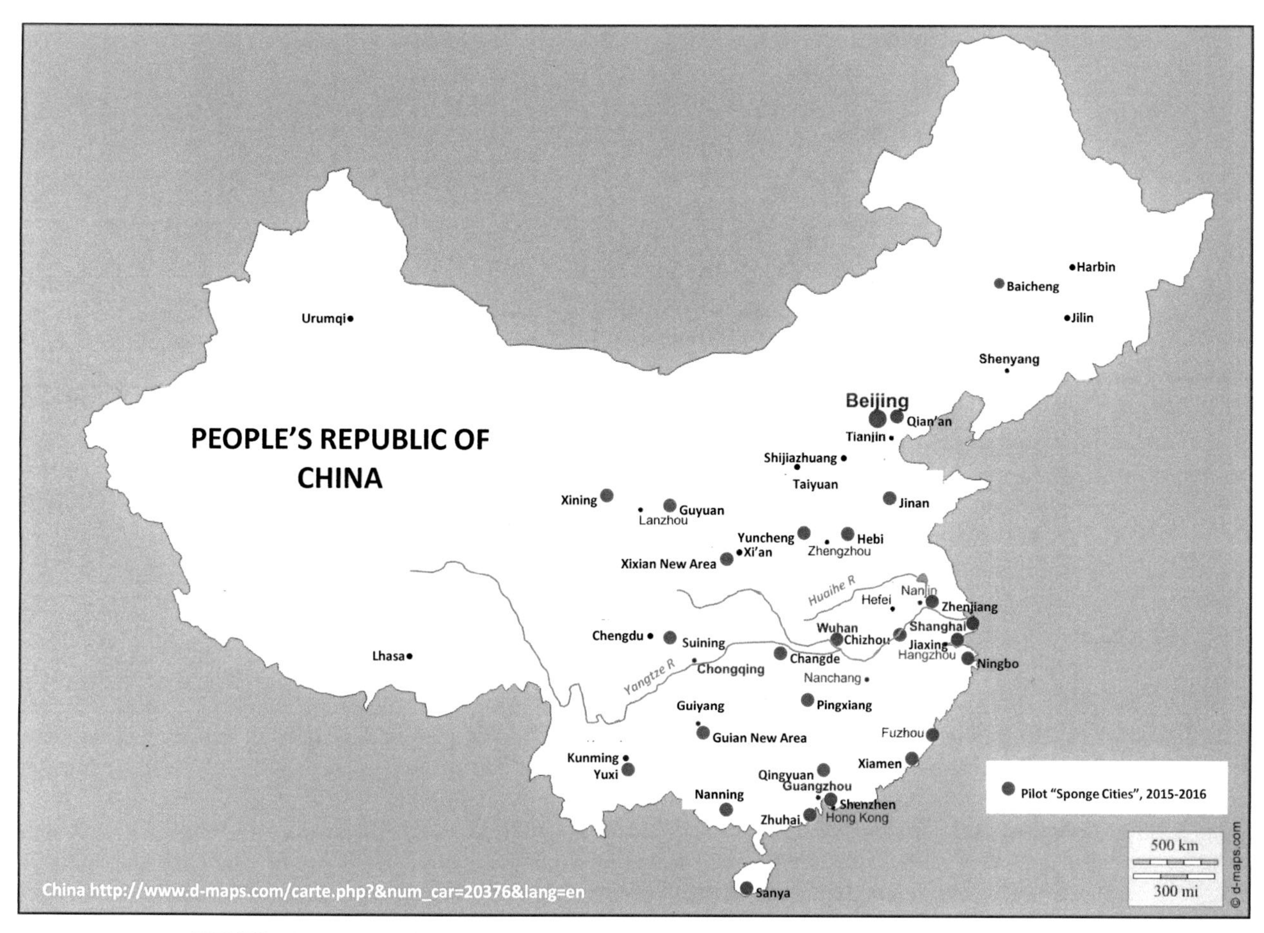


FIGURE 1.8 Location of the 30 Chinese pilot "Sponge Cities" chosen in 2015 and 2016 for investment.

Water shortage is restricting economic development and along with pollution, is harmful to human health, whereas flooding and waterlogging are disasters for human settlement. An LID strategy is needed for

- prioritizing and controlling water demand and encouraging water saving;
- controlling pollution at source;
- increasing investment in wastewater treatment;
- developing nontraditional water resources, including rainwater harvesting;
- reclaiming wastewater and desalinating seawater;
- preventing or reducing the impacts of major floods and waterlogging; and
- reframing rainwater and wastewater as resources, including as a source of energy (Qian, 2016).

But implementation is not without difficulties. Li et al. (2016) noted that the diversity of climate and rainfall in China, particularly the storm characteristics (depth/intensity/frequency), is the main factors that influence the performances of individual and combined LID devices. The 2- and 50-year design rainfalls (200-min duration) for Hong Kong are 122 and 260 mm, respectively, whereas for Seattle (USA), the similar values are only 20 and 37 mm, respectively. In China, these large intense storms cause urban floods. A 6-h duration rainfall intensity of 30 mm/hour is sufficient to cause the flash floods of 20-year design storm in steep terrain cities such as Jinan, Chongqing, and Qingdao (Fig. 1.8). However, a similar intensity storm would need a duration of over 10 h in Shanghai to cause floods because the river density is high, and the terrain is flat.

Although Chinese water resource planners and urban planners are still coming to grips with what will be required to implement the "Sponge City" philosophy, it is evident that modeling skills will be important. In addition to using well-known models such as soil and water assessment tool and soil and water integrated model, there has been considerable interest in developing a distributed hydrological model, named hydroinformatic modeling system, now widely tested in China, and subsequently extending it to encompass ecosystems and nutrients (Liu et al., 2009).

FIGURE 1.9 Visitors explore the wetlands in Yanqi Park, Beijing.

China also faces the problem of retrofitting the 'Sponge City" philosophy to areas developed in the boom construction years of the 1990s onward, where *hutongs* (alleys formed by lines of traditional single-storey residences) were replaced by high-rise apartments. Even recently constructed facilities can be improved. Jia et al. (2012) reworked the design of the 36 ha Beijing Olympic Village comprising 42 high-rise residential buildings. The village was originally built with some LID characteristics, including porous pavements, roof gardens, infiltration trenches, green spaces, and rainwater tanks. By using the Storm Water Management Model (SWMM) (USEPA, 2016c), the authors showed that with further LID improvements, including rerouting roof runoff through green spaces, increasing detention times in storage facilities, and properly designed bioretention cells, total runoff volume could be reduced by 27% and the peak flow rate reduced by 21% compared with that which occurred in the original construction.

As part of encouraging the development of LID facilities, the Chinese government has established to Yanqi Lake demonstration area in Beijing, encompassing swales and a major wetland (Fig. 1.9).

Although China has started the LID/WSUD journey a little later than western countries, it is likely to soon overcome its constraints of inadequate regulations and control. Discrepancies exist between different laws and regulations. Administration procedures and tools will be needed to control compliance with existing regulations and to safeguard that each on-site LID design fits into the overall Sponge City concept. Policies will be needed to promote integration between government agencies. Some legislative barriers may need to be overcome (Geiger, 2015). However, China has the potential to adopt a "leapfrogging" pathway for its developing cities that are not yet "locked-in" to traditional monofunctional, single-purpose infrastructure (Wong, 2016). It has the potential to become the most authoritative proponent of what are otherwise evolving as LID, Sustainable Urban Drainage Systems, GI, and WSUD, as it embarks on the path to deliver "Sponge Cities" as WSC.

## 1.6.4 Taiwan

Taiwan is also demonstrating the importance of a quite specific landscape-driven approach to WSUD. The island has a hilly to mountainous east coast topography with short, fast-flowing monsoon-filled rivers discharging to the west coast, which contains the lowlands and deltas on which most Taiwanese cities are built. An example is Hsinchu, a fast-growing science-based city that suffers from both droughts and floods. Located 100 km south of the capital, Taipei, the basic concept for the sustainable water management for Hsinchu requires the integration of a network of both below ground and open surface water bodies for the retention and storage of rainwater in new urban development areas, as well as in the existing floodplains and agricultural land. Areas that have acted as floodplains during extreme precipitation events which occurred in the past, perhaps only once in 100 years, but which will almost certainly become more common due to the effects of climate change, are reserved for agricultural purposes. They will not be used for urban development.

These extended floodplains are affected by flooding after 450 mm rainfall in 24 h, serve as stormwater retention and infiltration basins in the wet season, and may only be used for agricultural purposes in the dry season (Schuetze and Chelleri, 2013).

### 1.6.5 Singapore

This city-state in Southeast Asia is probably the epitome of integrated water management. As an island nation of very small area, it brings together four water sources—imported water from Malaysia, desalination, water recycling, and catchment water. Its success can be attributed to having one authority in charge, the Public Utilities Board of Singapore. It also highlights that different cities will have different objectives for why they choose to adopt WSUD/LID principles—water security being the primary motivator for Singapore. LID facilities include rain gardens (bioremediation cells), which include rock and wood chips to facilitate nitrogen removal before water discharges to a drainage canal (Ong et al., 2012). Monitoring of storms in 2009–10 demonstrated their effectiveness, with measured reductions in export of 46% in TN, 21% in TP, and 57% in TSS (Ong et al., 2012). Stormwater planters have been established in major city streets, porous pavements that surround housing blocks, and are also used in parking lots. Many wetlands are managed using islands of hydroponic plants growing in a floating substrate but carrying out the same functions as wetlands species conventionally rooted into the floor of the wetland. Furthermore, such floating islands easily accept changing water levels in the lakes and reservoirs. Numerous constructed wetlands have been built throughout Singapore, removing significant levels of nitrogen, phosphorus, TSS, zinc, and lead (Irvine et al., 2014). The use of green roofs and wall greenery has been encouraged by a 50% subsidy for their installation. Apart from their impact on water conservation and quality and the mitigation of "heat island" effects, all these initiatives also mitigate flooding in the lower levels of Singapore (Irvine et al., 2014).

## 1.7 DEVELOPING COUNTRIES AND THEIR CITIES

Developing cities in the emerging nations have much less well-developed water management systems but are beginning to consider the possibilities of LID or WSUD.

**Phnom Penh**, the capital of Cambodia, has a combined sewer system that discharges into open channels. 10% of the effluent flows straight into the Mekong River, whereas the remaining 90% flows (or is pumped) into naturally occurring wetlands, which are used for fishing and crop cultivation. There is no sewage treatment facility, but the wetlands have proven to be remarkably efficient, reducing *E*scherichia *coli* levels by 2 to 4 $Log_{10}$ and TN and TP by up to 71%. However, some of the wetlands are being filled in for construction purposes, potentially leading to a "pave-all" end use, thereby substantially increasing flooding risks. An LID approach has been suggested to offset these problems (Irvine et al., 2015).

**Bangkok** is located on the lower Chao Phraya river delta, which has a contributing basin of ~160 000 $km^2$, representing 30% of the entire area of the country. The southward flow of water toward Bangkok is managed through a dozen dams, which serve for both water conservation and flood control. Management of such facilities with conflicting objectives can prove difficult. The city has been suffering from an increased frequency of flooding with its storage dams being occasionally overwhelmed. Contributing factors are land subsidence due to overpumping groundwater and construction activities, increased runoff from urbanization, lack of systemic operation of the dams, infilling of privately owned klongs (drainage canals) due to urbanization, increased sedimentation, and encroaching building above the canals, causing flow restrictions (Davivongs et al., 2012). Traditional engineering solutions to get flood waters away quicker have been pursued. However, proposals have been explored to introduce LIDs to manage local flooding by trying to emulate the predevelopment hydrology. It has been argued that most LID research has occurred in temperate climates and is of doubtful relevance to monsoonal climates. For example, a two-year recurrent storm event in Bangkok was similar to a 20–100-year frequency storm event in upstate New York. Yet, Singapore, described earlier, has shown that there is a place for LID in high-rainfall monsoonal environments. Modeling showed that provision in a Thai periurban village of rain barrels on every house, and bioretention basins in each subcatchment could reduce combined sewage volume by up to 40% and pollutant load by 34%–40%, but with questionable economic attractiveness (Chaosakul et al., 2013).

**In African countries**, WSUD is beginning to be adopted. The South African Water Research Commission has published guidelines that have been adapted to the South African social environment (Armitage et al., 2014). These guidelines note that the WSUD approach as originally envisaged does not take cognizance of the "developmental" or "equity" issues, which are particularly important in developing countries, including the Republic of South Africa as a result of the country's apartheid legacy. It would be difficult for urban development and redevelopment to "address the sustainability of water" where a substantial proportion of the population still does not have access to basic water supply or sanitation. These considerations have led to their suggestion to move from WSUD to a new South African term *Water*

*Sensitive Settlements* (WSS). Thus, while "leapfrogging" through developmental states is an attractive ideal (Wong, 2016), there are fundamental social (equity) and practical issues that need to be considered.

## 1.8 CONCLUSION

The commonality of principles encompassed within LID, SUDS, GI, WSUD, LIUDD and now Sponge City, and WSS are increasingly being recognized. The emphases appropriate to each location will be dependent on the hydrological, infrastructural, planning, and sociological circumstances in each case.

Sewer overflow protection and flood management, together with access to green space, are important in Europe. Water quality protection and green space aesthetics, as well as flood mitigation, are significant in the United States. In Australia, waterway ecosystem protection and littoral zone conservation have been drivers in coastal Brisbane, Melbourne, and Adelaide, whereas groundwater protection is a dominant consideration in Perth. Flood mitigation is a major consideration in much of Asia, whereas China is also adopting the importance of stormwater harvesting and environmental protection. Social equity is a driver in Africa. Most of these communities will also be seeking to assess possible changes due to the effects of climate change. All will seek an improved understanding for accommodating the water cycle within their urban planning and development.

As has been identified by Jia et al. (2016), there are challenges to be faced in introducing the WSUD/LID techniques, especially in a developing environment. These include overcoming inertia derived from familiarity with traditional approaches and measures of success, the need to ensure interagency coordination at local levels, adequate capital and economic return, availability of site-specific technical guidance, and proof of concept or product before adoption. To these may be added the important need to provide adequate resources for postconstruction maintenance.

The underlying WSUD/LID principles seek to emulate the natural hydrology of a site by using decentralized management measures. They highlight the need to integrate stormwater management at an early stage of planning for new developments, recognize its value as a natural resource, manage it as close as possible to its origins, support as far as possible conservation of the existing natural environment, and maintain the hydrological functions of the landscape with an orientation toward prevention rather than mitigation and repair. Costs may well be lower than less effective traditional "end-of-pipe" installations. The community should be a co-owner of the philosophy being pursued.

## REFERENCES


Ahiablame, L.M., Engel, B.A., Chaubey, I., 2012. Effectiveness of low impact development practices: literature review and suggestions for future research. Water Air and Soil Pollution 223, 4253–4273. https://doi.org/10.1007/s11270-012-1189-2.

Aird, W.V., 1961. The Water Supply, Sewerage and Drainage of Sydney. Sydney Metropolitan Water, Sewerage and Drainage Board, Sydney, 347 pp. https://www.google.com.au/search?q=Sydney+Watwer+-+sewer+overflows+-+sandstone+geolofgy&ie=utf-8&oe=utf-8&client=firefox-b&gfe_rd=cr&ei=BOsHWa7iBMLu8wezsZfgBA#.

AQUAVAL, 2010. Sustainable Urban Water Management Plans, Promoting Sustainable Urban Drainage Systems (SUDS) and Considering Climate Change, in the Province of Valencia. http://www.aquavalproject.eu/index.asp?accADesplegar=001.

Argue, J. (Ed.), 2005. Water Sensitive Urban Design: Basic Procedures for 'Source Control' of Stormwater. Urban Water Resources Centre, University of South Australia, Adelaide, 273 pp.

ARMCANZ, ANZECC, 1994. National Water Quality Management Strategy: Water Quality Management - An Outline of the Policies. https://www.environment.gov.au/water/quality/publications/nwqms-water-quality-management-outline-policies.

Armitage, N., Fisher-Jeffes, L., Carden, K., Winter, K., Naidoo, V., Spiegel, A., Mauck, B., Coulson, D., 2014. Water Sensitive Urban Design (WSUD) for South Africa: Framework and Guidelines. Water Research Commission, Pretoria, 198 pp. http://www.wrc.org.za/Pages/DisplayItem.aspx?ItemID=10843&FromURL=%2fPages%2fKH_DocumentsList.aspx%3fdt%3d1%26ms%3d4%3b5%3b%26d%3dWater+Sensitive+Urban+Design+(WSUD)+for+South+Africa%3a+Framework+and+guidelines%26start%3d61.

ARSIT, 2016. Association for Rainwater Storage and Infiltration Technology, Japan, Membership. http://arsit.or.jp/membership_list.

ASCE-EWRI, 2016. http://www.asce.org/environmental-and-water-resources-engineering/environmental-and-water-resources-institute/.

Austrade, February 17, 2016. China's Sponge City Program - Making China's Cities Watertight Presents Opportunities for Australian Water Technologies. http://www.austrade.gov.au/ArticleDocuments/6585/China Sponge CityProgram.pdf.aspx.

Barlow, D., Burrill, G., Nolfi, J., 1977. Research Report on Developing a Community Level Natural Resource Inventory System. Center for Studies in Food Self-Sufficiency, Vermont Institute of Community Involvement, Burlington, VT, 49 p.

BC MWLAP, 2002. Stormwater Planning: A Guidebook for British Columbia. British Columbia Ministry of Water, Land and Air Protection. http://bc.waterbalance.ca/resources/guidance-documents/stormwater.

Bloomberg, M., 2011. NYC Environment Protection – Strategy 2011-2014. www.nyc.gov/html/dep/pdf/strategic_plan/dep_strategy_2011.pdf.

Brown, R.R., Keath, N., Wong, T., 2009. Urban water management in cities: historical, current and future regimes. Water Science and Technology 59 (5), 847–855.

Castro-Fresno, D., Andrés-Valeri, V.C., Sañudo-Fontaneda, L.A., Rodriguez-Hernandez, J., 2013. Sustainable drainage practices in Spain, specially focused on pervious pavements. Water 5, 67–93. https://doi.org/10.3390/w5010067.

Chaosakul, T., Koottatep, T., Irvine, K., 2013. Low impact development modeling to assess localized flood reduction in Thailand. Journal of Water Management Modeling. https://doi.org/10.14796/JWMM.R246-18, 2466-18.

Chesterfield, C., Urich, C., Beck, L., Burge, K., Charette-Castonguay, A., Brown, R.R., Dunn, G., de Haan, F., Lloyd, S., Rogers, B.C., Wong, T., 2016. A water sensitive cities index – benchmarking cities in developed and developing countries. In: LID 2016 Conference, Beijing, Paper 387. https://www.researchgate.net/publication/308312783_A_Water_Sensitive_Cities_Index_-Benchmarking_cities_in_developed_and_developing_countries.

Choi, L., 2016. Towards a Water Sensitive Policy Framework for Australia's Cities. LID 2016 Conference, Beijing, Paper 381.

City of Melbourne, 2008. City of Melbourne WSUD Guidelines - Applying the Model WSUD Guidelines. Government of Victoria, Melbourne Water, City of Melbourne, 165 p. https://www.melbourne.vic.gov.au/SiteCollectionDocuments/wsud-full-guidelines.pdf.

City of Sydney, 2012. Decentralised Water Master Plan 2012–2030, 80 p. http://www.cityofsydney.nsw.gov.au/__data/assets/pdf_file/0005/122873/Final-Decentralised-Water-Master-Plan.pdf.

Coffman, L., 1999. Low-Impact Development Design Strategies - An Integrated Design Approach Prince George's County. Department of Environmental Resources, Maryland, USA. www.princegeorgescountymd.gov/DocumentCenter/Home/View/86.

Commonwealth of Australia Constitution Act, 1900. (Imp) Constituting the Commonwealth of Australia. http://www.austlii.edu.au/au/legis/cth/consol_act/coaca430/.

Cosier, M., Shen, D., 2009. Urban water management in China. International Journal of Water Resources Development 25 (2), 249–268. https://doi.org/10.1080/07900620902868679.

D-G Environment, March 8, 2011. Towards Better Environmental Options for Flood Risk Management. European Commission, DG ENV D.1 (2011) 236452, Brussels. http://ec.europa.eu/environment/water/flood_risk/better_options.htm.

Davis, A.P., Hunt, W.F., Traver, R.G., Clar, M., 2009. Bioretention technology: overview of current practice and future needs. Journal of Environmental Engineering 135, 109–117.

Davivongs, V., Yokohari, M., Hara, Y., 2012. Neglected canals: deterioration of indigenous irrigation system by urbanization in the west peri-urban area of Bangkok metropolitan region. Water 4, 12–27. https://doi.org/10.3390/w4010012.

Dennison, W.C., Abal, E.G., 1999. Moreton Bay Study: A Scientific Basis for the Healthy Waterways Campaign. South East Queensland Regional Water Quality Management Strategy, Brisbane, 246 p. www.ian.umces.edu/pdfs/ian_book_421.pdf.

Dierkes, c., Lucke, T., Helmreich, B., 2015. General technical approvals for decentralised sustainable urban drainage systems (SUDS)—the current situation in Germany. Sustainability 7, 3031–3051. https://doi.org/10.3390/su7033031.

Earles, A., Rapp, D., Clary, J., Lopitz, J., 2009. Breaking down the barriers to low impact development in Colorado. In: Proceedings of the World Environmental and Water Resources Congress 2009, Kansas City, MO, USA, 17–21 May 2009.

EU, 2016a. The EU Water Directive. http://ec.europa.eu/environment/water/water-framework/info/intro_en.htm.

EU, 2016b. The EU Floods Directive. http://ec.europa.eu/environment/water/flood_risk/implem.htm.

eWater, 2016. MUSIC by eWater – Market Leading Stormwater Management Software. http://ewater.org.au/products/music/.

Ferguson, B.C., Brown, R.R., Werbeloff, L., 2014. Benchmarking Auckland's stormwater management practice against the water sensitive cities framework. In: Prepared by the Cooperative Research Centre for Water Sensitive Cities for Auckland Council. Auckland Council Technical Report, TR2014/007. http://www.aucklandcouncil.govt.nz/SiteCollectionDocuments/aboutcouncil/planspoliciespublications/technicalpublications/tr2014007benchmarkingaucklandsstormwatermanagementpractice.pdf.

Fletcher, T.D., Shuster, W., Hunt, W.F., Ashley, R., Butler, D., Arthur, S., Trowsdale, S., Barraud, S., Semadeni-Davies, A., Bertrand-Krajewski, J.-L., Mikkelsen, P.S., Rivard, G., Uhl, M., Dagenais, D., Viklander, M., 2015. SUDS, LID, BMPs, WSUD and more – the evolution and application of terminology surrounding urban drainage. Urban Water Journal 12, 525–542. https://doi.org/10.1080/1573062X.2014.916314.

Fox, D.R., Batley, G.E., Blackburn, D., Bone, Y., Bryars, S., Cheshire, A., Collings, G., Ellis, D., Fairweather, P., Fallowfield, H., Harris, G., Henderson, B., Kämpf, J., Nayar, S., Pattiaratchi, C., Petrusevics, P., Townsend, M., Westphalen, G., Wilkinson, J., 2007. Adelaide coastal water study, final report. In: Summary of Study Findings, November 2007, vol. 1. CSIRO, Glen Osmond, SA, 56 p. www.epa.sa.gov.au/files/477350_acws_report.pdf.

Gallinaro, N., 2015. Paying for stormwater. In: Maine Stormwater Conference, Portland ME, 16–18 November 2015. https://maineswc.files.wordpress.com/2015/12/2015_swc_municipal_2-0_gallinaro.pdf.

Geiger, W.F., 2015. Sponge city and LID technology — vision and tradition. Landscape Architecture Frontiers 3 (2), 10–20. V3/I2/10. www.journal.hep.com.cn/laf/EN/Y2015/V3/I2/10.

Harris, G., Batley, G., Fox, D., Hall, D., Jernakoff, P., Molloy, R., Murray, A., Newell, B., Parslow, J., Skyring, G., Walker, S., 1996. Port Phillip Bay Environmental Study: Final Report. CSIRO, Dickson, ACT, 248 p.

Healthy Land and Water, 2018. Water Sensitive Urban Design. http://healthywaterways.org/initiatives/waterbydesign/wsud.

Hoyer, J., Dickhaut, W., Kronawitter, L., Weber, B., 2011. Water Sensitive Urban Design - Principles and Inspiration for Sustainable Stormwater Management in the City of the Future – Manual. Jovis Verlag GmbH, Berlin, 115 p. http://switchurbanwater.lboro.ac.uk/outputs/pdfs/W5-1_GEN_MAN_D5.1.5_Manual_on_WSUD.pdf.

Imbe, M., 2013. Stormwater treatment from the road in Japan. In: Regional Workshop on Eco-Efficient Water Infrastructure towards Sustainable Urban Development and Green Economy in Asia and the Pacific, 12–13 December 2013, Bangkok, Thailand. http://www.unescap.org/sites/default/files/3.2%20Japan_Imbe.pdf.

Irvine, K.N., Chua, L.H.C., Eikass, H.S., 2014. The four national taps of Singapore: a holistic approach to water resource management from drainage to drinking water. Journal of Water Resource Modelling C375. https://doi.org/10.14796/JWMM.C375.

Irvine, K.N., Sovann, C., Suthipong, S., Kok, S., Chea, E., 2015. Application of PCSWMM to access wastewater treatment and urban flooding scenarios in Phnom Penh, Cambodia. A tool to support eco-city planning. Journal of Water Resource Modelling C389. https://doi.org/10.14796/JWMM.C389.

Jeong, H., Kim, H., Teodosio, B., Ramirez, R., Ahn, J., 2015. A review of test beds and performance criteria for permeable pavements. In: Chang, et al. (Eds.), Advances in Civil Engineering and Building Materials IV. Taylor and Francis, London.

Jessup, C., 2015. "Launceston's Combined Sewerage System – Investigation and Strategy Development" - A Dissertation Submitted towards the Degree of Bachelor of Engineering Honours (Civil). University of Southern Queensland Faculty of Health, Engineering & Sciences. https://eprints.usq.edu.au/29255/1/Jessup_C_Aravinthan.pdf.

Jha, A.K., Bloch, R., Lamond, J., 2012. Cities and Flooding - A Guide to Integrated Urban Flood Risk Management for the 21st Century. The World Bank, Washington, USA, 635 p.

Jia, H.F., Lu, Y.W., Yu, S.L., Chen, Y.R., 2012. Planning of LID-BMPs for urban runoff control: the case of Beijing Olympic Village. Separation and Purification Technology 84, 112–119. https://doi.org/10.1016/j.seppur.2011.04.026.

Jia, H.F., Wang, Z., Yu, S.L., 2016. Opportunity and challenge, China's sponge city plan. Hydrolink 2016 (4), 100–102.

Joint Steering Committee for Water Sensitive Cities, 2009. Evaluating Options for Water Sensitive Urban Design – A National Guide. http://webarchive.nla.gov.au/gov/20130904201040/. http://www.environment.gov.au/water/publications/urban/water-sensitive-design-national-guide.html.

Kim, Y.J., Han, M.Y., 2008. Rainfall-Storage-Drain (RSD) Model for Runoff Control Rainwater Tank System Design in Building Rooftop.

Kim, R., Lee, S.W., Lee, J.-H., Lee, D., Shafique, M., 2015a. Challenges in Seoul Metropolitan for restoring urban water cycle. In: Maine Stormwater Conference, Portland ME, 16–18 November 2015. https://maineswc.files.wordpress.com/2015/12/2015_swc_s-korea_5-1_kim-lee-lee-shafique.pdf.

Kim, Y.J., Kim, T., Park, H., Han, M.Y., 2015b. Design method for determining rainwater tank retention volumes to control runoff from building rooftops. KSCE Journal of Civil Engineering 19, 1585–2590. https://doi.org/10.1007/s12205-013-0269-1. https://link.springer.com/article/10.1007/s12205-013-0269-1.

Konrad, C.P., November 2003. Effects of Urban Development on Floods. Fact Sheet FS-076–03. US Geological Service. https://pubs.usgs.gov/fs/fs07603/.

Lee, D., 2012. How to install stormwater management facilities for sustainable urban developments. In: Policy and Issues Environment Information, Korea, October 2012 [quoted in OECD (2014), Compact City Policies: Korea: Towards Sustainable and Inclusive Growth, OECD Green Growth Studies]. OECD Publishing. https://doi.org/10.1787/9789264225503-en.

Lee, J.M., Hyun, K.H., Choi, J.S., 2013. Analysis of the impact of low impact development on run-off from a new district in Korea. Water Science and Technology 68 (6), 1215–1221. https://doi.org/10.2166/wst.2013.346.

Lewis, M., James, J., Shaver, E., Blackbourn, S., Leahy, A., Seyb, R., Simcock, R., Wihongi, P., Sides, E., Coste, C., 2015. Water sensitive design for stormwater. Auckland Council Guideline Document GD2015/004.

Li, N., Yu, Q., Wang, J., Du, X., 2016. The effects of low impact development practices on urban stormwater management. In: LID 2016 Conference, Beijing, Paper 354.

Liu, C.M., Yang, S.T., Wen, Z.Q., Wang, X.L., Wang, Y.J., Li, Q., Sheng, H.R., 2009. Development of ecohydrological assessment tool and its application. Science China Technological Sciences 52 (7), 1947–1957.

Maigne, J., 2006. Sustainable Management of 'Alternative Techniques' in Stormwater Purification. École Nationale du Génie Rural des Eaux et Forêts (ENGREF - National School of Rural Water and Forestry Engineering). Centre de Montpellier, France. https://www.agroparistech.fr/IMG/doc/Version_finale_anglais_Maigne.doc.

Maksimovic, C., Kurian, M., Ardakanian, R., 2015. Rethinking Infrastructure Design for Multi-Use Water Services. Springer.

Martin, P., Turner, B., Waddington, K., Pratt, C., Campbell, N., Payne, J., Reed, B., 2000a. Sustainable Urban Drainage Systems: Design Manual for Scotland and Northern Ireland. Construction Industry Research and Information Association (CIRIA). Publication 521.

Martin, P., Turner, B., Dell, J., Pratt, C., Campbell, N., Payne, J., Reed, B., 2000b. Sustainable Urban Drainage Systems – Design Manual for England and Wales. Construction Industry Research and Information Association (CIRIA). Publication C522.

Melbourne Water, 2016a. Stormwater Management (WSUD). http://www.melbournewater.com.au/planning-and-building/stormwater-management/pages/stormwater-management.aspx.

Melbourne Water, 2016b. Stormwater Offsets Explained. http://www.melbournewater.com.au/Planning-and-building/schemes/offset/Pages/What-are-stormwater-quality-offsets.aspx.

Mouritz, M., 1992. Sustainable Urban Water Systems: Policy & Professional Praxis. Murdoch University, Perth, Australia (quoted by Fletcher et al., 2015).

National Research Council, 2008. Urban Stormwater in the United States. National Academies Press, Washington, DC. https://www.nap.edu/catalog/12465/urban-stormwater-management-in-the-united-states.

Sydney Water, 2018. Splash. https://www.sydneywatertalk.com.au/splash-network?page=1 (accessed 15 May 2018).

New South Wales, 2015. Stormwater Publications. http://www.environment.nsw.gov.au/stormwater/publications.htm.

New WAter Ways, 2016a. What Is Water Sensitive Urban Design? http://www.newwaterways.org.au/About-Us/What-is-water-sensitive-urban-design.

New WAter Ways, 2016b. Policy and Guidelines. http://www.newwaterways.org.au/Resources/Policy-and-guidelines.

NL, 2016. Factsheet Sponge City Construction in China, Embassy of the Kingdom of the Netherlands. I&M Department, 4 p.

NMSA, 2016. National Municipal Stormwater Alliance. http://nationalstormwateralliance.org/.

NWI, 2004. Intergovernmental Agreement on the National Water Initiative. www.agriculture.gov.au/water/policy/nwi.

NZ4404, 2010. Land Development and Subdivision Infrastructure. Standards New Zealand, Ministry of Business, Innovation and Employment, Wellington, New Zealand.

OECD, 2014. Compact City Policies: Korea: Towards Sustainable and Inclusive Growth, OECD Green Growth Studies. OECD Publishing. https://doi.org/10.1787/9789264225503-en.

OIEau, 2014. Centre National de Formation aux Métiers de l'Eau (CNFME) – De nouvelles plates-formes pédagogiques dédiées à la gestion intégrée des eaux pluviales (in French). http://www.oieau.fr/oieau/notre-actualite-et-avancement-de/article/cnfme-de-nouvelles-plates-formes?lang=fr.

Ong, G.S., Kalyanaraman, G., Wong, K., Wong, T.H.F., 2012. Monitoring Singapore's first bioretention system: raingarden at Balam estate. In: WSUD 2012: Water Sensitive Urban Design: Building the Water Sensitive Community. 7th International Conference on WSUD, February 21-23-2012, Melbourne. Engineers Australia, Canberra.

Ontario Ministry of Environment, 2003. Stormwater Management Planning and Design Manual. Queen's Printer for Ontario, 379 p. https://www.ontario.ca/document/stormwater-management-planning-and-design-manual.

Paneth, N., 2004. Assessing the contributions of John snow to epidemiology, 150 years after removal of the broad street pump handle. Epidemiology 15 (5), 514–516.

Perales-Momparler, S., Jefferies, C., Perigüell-Ortega, E., Peris-Garcia, P.P., Muñoz-Bonet, J.L., 2013. Inner-city SUDS retrofitted sites to promote sustainable stormwater management in the Mediterranean region of Valencia: AQUAVAL (Life+ EU Programme) Novatech Conference 2013. http://documents.irevues.inist.fr/bitstream/handle/2042/51238/2A44-306PER.pdf;sequence=1 (accessed 20 June 2018).

Perales-Momparler, S., Hernández-Crespo, C., Vallés-Morán, F., Martín, M., Andrés-Doménech, I., Andreu-Álvarez, J., Jefferies, C., 2014. SuDS efficiency during the start-up period under Mediterranean climatic conditions. CLEAN Soil Air Water 42 (2), 178–186. https://doi.org/10.1002/clen.201300164.

Perales-Momparler, S., Duffy, A., Morales-Torres, A., García, P.P.P., 2015. E2STORMED Transition Manual, Municipality of Benaguasil, 41 p. www.e2stormed.eu/wp-content/uploads/2013/02/3.-Transition-Manual-Malta.pdf.

Puddephatt, J., Heslop, V., 2008. Low Impact Urban Design and Development – Concepts, Policy, Practice – Guidance on an Integrated Process. University of Auckland/Landcare Research, 18 p. www.landcareresearch.co.nz/publications/.../LIUDD_Maintenance_Puddephat_2008.

Qian, Y., 2007. Sustainable management of urban water resources in China. Water Science and Technology: Water Supply 7 (2), 23–30. https://doi.org/10.2166/ws.2007.037.

Qian, Y., 2016. Water and sustainable development. In: Keynote Presentation. LID 2016 Conference, Beijing.

Queensland, 2010. Urban Stormwater Quality Planning Guidelines 2010. Department of Environment and Resource Management, 242 p.

Queensland, April 2016. State Planning Policy—State Interest Guideline – Water Quality. Department of Infrastructure, Local Government and Planning, 28 p.

Radcliffe, J.C., Page, D., Naumann, B., Dillon, P., 2017. Fifty years of water sensitive urban design, Salisbury, south Australia. Frontiers of Environmental Science and Engineering 11 (4), 7. https://doi.org/10.1007/s11783-017-0937-3.

Roseen, R.M., Ballestero, T.P., Houle, J.J., Avellaneda, P., Briggs, J., Fowler, G., Wildey, R., 2009. Seasonal performance variations for storm-water management systems in cold climate conditions. Journal of Environmental Engineering ASCE 135 (3), 128–137.

Schuetze, T., Chelleri, L., 2013. Integrating decentralized rainwater management in urban planning and design: flood resilient and sustainable water management using the example of coastal cities in The Netherlands and Taiwan. Water 5, 593–616. https://doi.org/10.3390/w5020593.

Shi, Y., Zhou, J.P., Liao, L., Zhao, J., 2016. Incorporating green infrastructures for internationally funded development of small towns in China. In: LID 2016 Conference, Beijing, June 2016, Paper 533.

Shoemaker, L., Riverson, J., Alvi, K., Zhen, J.X., Paul, S., Rafi, T., 2009. SUSTAIN – A Framework for Placement of Best Management Practices in Urban Watersheds to Protect Water Quality. U.S. Environmental Protection Agency, Washington, DC. EPA/600/R-09/095.

Son, C.H., Hyun, K.H., Kim, D., Baek, J.I., Ban, Y.U., 2017. Development and application of a low impact development (LID)-based district unit planning model. Sustainability 9, 145. https://doi.org/10.3390/su9010145.

South Australia, 2003. State of Environment Report. Environment Protection Authority, Adelaide, 108 p. www.epa.sa.gov.au/files/4771457_soe_index.pdf.

South Australia, 2010. Water Sensitive Urban Design Technical Manual. https://www.sa.gov.au/topics/property-and-land/land-and-property-development/planning-professionals/water-sensitive-urban-design.

South Australia, 2013. Adelaide Coastal Water Quality Improvement Plan (ACWQIP). Environment Protection Authority, Adelaide, 163 p. http://www.epa.sa.gov.au/files/477449_acwqip_final.pdf.

Spies, B., Dandy, G., 2012. Sustainable Water Management: Securing Australia's Future in a Green Economy. Australian Academy of Technological Sciences and Engineering (ATSE), Melbourne, 148 p.

State Council, 2015. Guidance for Promoting the State Council Sponge Urban Construction - Provinces, Autonomous Regions and Municipalities, Members, Directly under the State Council: State Council Journal Office No. 75, October 16 2015 (国务院办公厅关于推进 海绵城市建设的指导意见 国办发 (2015) 75号 – in Chinese). http://www.gov.cn/zhengce/content/2015-10/16/content_10228.htm. http://english.gov.cn/policies/latest_releases/2015/10/16/content_281475212984264.htm.

SWITCH, 2012. Managing Water for the Future – Cities. http://www.switchurbanwater.eu/cities/3.php.

T&RCA, CVCA, 2010. Low Impact Development Stormwater Management Planning and Design Guide. Version 1.0. Toronto and Region Conservation Authority and Credit Valley Conservation Authority. http://www.creditvalleyca.ca/low-impact-development/low-impact-development-support/stormwater-management-lid-guidance-documents/low-impact-development-stormwater-management-planning-and-design-guide/.

Tu, X., Tian, T., 2015. Six questions towards a sponge city — report on power of public policy: sponge city and the trend of landscape architecture. Landscape Architecture Frontiers 3 (2), 22–31.

USEPA, 2007. Low Impact Development (LID) Urban Design Tools Website. http://www.lid-stormwater.net/.

USEPA, 2016a. What are combined sewer overflows (CSOs)? https://www3.epa.gov/region1/eco/uep/cso.html.

USEPA, 2016b. Implementing Clean Water Act Section 303(d): Impaired Waters and Total Maximum Daily Loads (TMDLs). https://www.epa.gov/tmdl.

USEPA, 2016c. Storm Water Management Model (SWMM) Version 5.1.011 with Low Impact Development (LID) Controls. https://www.epa.gov/water-research/storm-water-management-model-swmm.

USEPA, 2017. History of the Clean Water Act. https://www.epa.gov/laws-regulations/history-clean-water-act.

Van Roon, M., van Roon, H., 2009. Low impact urban design and development - the big picture - an introduction to LIUDD principles and methods framework. In: Landcare Research Science Series No. 37. Manaaki Whenua Press, Lincoln, NZ.

Victoria Stormwater Committee, 1999. Urban Stormwater - Best Practice Environmental Management Guidelines. CSIRO, Melbourne, 268 p. http://www.publish.csiro.au/ebook/3822.

Walsh, C.J., Fletcher, T.D., Bos, D.G., Imberger, S.J., 2015. Restoring a stream through retention of urban stormwater runoff: a catchment-scale experiment in a social–ecological system. Freshwater Science 34 (3), 1161–1168.

Western Australia, October 2008. Better Urban Water Management Western Australian Planning Commission, 45 p.

Whelans, Maunsell, H.G., 1994. Planning and Management Guidelines for Water Sensitive Urban (Residential) Design, Report Prepared for the Department of Planning and Urban Development of Western Australia, ISBN 0 64615 468 0.

WHG, 2009. Gesetz zur Ordnung des Wasserhaushalts (Wasserhaushaltsgesetz WHG). http://www.gesetze-im-internet.de/bundesrecht/whg_2009/gesamt.pdf (In German).

White, M., 2009. The Rise of Cities in the 18th Century. https://www.bl.uk/georgian-britain/articles/the-rise-of-cities-in-the-18th-century.

Wong, T.H.F., 2000. Improving urban stormwater quality – from theory to implementation. Water – Journal of the Australian Water Association 27 (6), 28–31.

Wong, T., 2016. Human Settlements - A Framing Paper for the High-Level Panel on Water. Australian Water Partnership, Canberra.

Woods Ballard, B., Wilson, S., Udale-Clarke, H., Illman, S., Scott, T., Ashley, R., Kellagher, R., 2016. The SuDS Manual, v. 6 – CIRIA Report 753. CIRIA, London. http://www.ciria.org/Resources/Free_publications/SuDS_manual_C753.aspx.

Yoon, D.K., 2014. Disaster management and emergency management policies in Korea. In: Kapucu, N., Liou, K.T. (Eds.), Disaster and Development – Examining Global Issues and Cases. Springer.

Chapter 2

# Water Sensitive Urban Design Approaches and Their Description

Alan Hoban
*Bligh Tanner P/L, Brisbane, QLD, Australia*

## Chapter Outline



**ABSTRACT**

Water sensitive urban design (WSUD) is a broad area of practice, with a wide range of objectives from managing hydrology and water quality through to improving urban amenity and mitigating urban heat island impacts. Unsurprisingly, there is a wide set of management practices that can be applied. This chapter outlines some of the key approaches used in WSUD and provides a summary of current knowledge about practical application and performance.

Key approaches covered in this chapter are as follows:

Measures to reduce the generation of polluted stormwater runoff: erosion and sediment control, rainwater tanks, downpipe diverters, litter management, street sweeping, green walls and roofs, and permeable pavement.

Measures to mitigate (treat) polluted stormwater runoff: gully baskets, gross pollutant traps, swales, bioretention, rain gardens, wetlands, floating wetlands, and street trees.

The practices covered are by no means exhaustive, and the body of knowledge in relation to these grows year by year; however, it is hoped that this chapter will provide the reader with a useful introduction to the range of WSUD approaches to assist with specifying and designing projects.

**Keywords:** Bioretention; Downpipe diverters; Erosion and sediment control; Floating wetlands; Green roofs; Green walls; Gross pollutant traps; Gully baskets; Litter; Permeable pavement; Porous pavement; Rain gardens; Rainwater tanks; Stormwater; Stormwater treatment; Street sweeping; Street trees; Swales; Urban design; Water sensitive urban design; Wetlands; WSUD.

**Approaches to Water Sensitive Urban Design. https://doi.org/10.1016/B978-0-12-812843-5.00002-2**

## 2.1 INTRODUCTION

Water sensitive urban design (WSUD) is a broad area of practice with a wide range of objectives, including to

- minimize impacts on existing natural features and ecological processes
- minimize impacts on natural hydrologic behavior of catchments
- protect water quality of surface and groundwaters
- minimize demand on the reticulated water supply system
- improve the quality of and minimize polluted water discharges to the natural environment
- incorporate collection treatment and/or reuse of runoff, including roofwater and other stormwater
- reduce runoff and peak flows from urban development
- reuse treated effluent and minimize wastewater generation
- increase social amenity in urban areas through multipurpose green space, landscaping, and integrating water into the landscape to enhance visual, social, cultural, and ecological values
- add value while minimizing development costs (e.g., drainage infrastructure costs)
- account for the nexus between water use and wider social and resource issues
- harmonize water cycle practices across and within the institutions responsible for waterway health, flood management, pollution prevention, and protection of social amenity (National Water Commission, 2010).

With such a broad set of principles, it's unsurprising then that there is a wide range of practices and technologies available to help contribute toward achieving them. This chapter introduces a selection of those practices and technologies.

## 2.2 OVERVIEW OF WATER SENSITIVE URBAN DESIGN APPROACHES

For all the practices and technologies addressed in this chapter, there are many different variants, configurations, and situations in which they can be applied. The guidance in this chapter is general in nature and should not be considered as limiting the possible use of various technologies. WSUD is a relatively new field of practice, and ongoing innovation is important to increase knowledge and improve practice.

For the purposes of this chapter, WSUD approaches have been grouped into two categories:

1. Avoidance measures are those which avoid the generation of contaminated stormwater runoff from allotments. These include measures such as erosion and sediment control (ESC), rainwater tanks, downpipe diverters, and permeable pavements.
2. Mitigation measures are those which are typically implemented to detain and treat stormwater runoff.

The distinction is subjective, and some measures span both categories. However, it is a useful concept for grouping stormwater treatment measures as it is aligned with the principles of the waste management hierarchy (see NSW EPA, 2018). Many stormwater practitioners are trained to think about mitigation options first and foremost, which means avoidance measures are rarely given their due consideration.

For most technologies, there is a relative paucity of reliable field data considering the extent of implementation, but there are some exceptions (see Hunt et al., 2008; Parker, 2010; Hatt et al., 2009a,b; Mangangka et al., 2015a,b; Lucke and Nichols, 2015; Nichols and Lucke, 2016; Drapper and Hornbuckle, 2015). Reasons for this include

- For many treatment systems, it is difficult to set up equipment to sample inflows and outflows;
- Robust monitoring of stormwater quality is expensive as it often requires event-based sampling of inflow and outflow throughout multiple storm events and then ensuring appropriate sample handling, chain of custody, and laboratory analyses;
- There are different criteria for evaluating the efficacy of stormwater practices, including several differing statistical methods to determine percentage removal, for example, summation of loads, concentration removal efficiency, efficiency ratio, etc. (see Kelly and Bardak, 2015; and USEPA, 2002); and
- Results from one study may not always be relevant in other situations where there are differences in rainfall, stormwater quality, catchment characteristics, or treatment measure configuration or condition. However, Egodawatta et al. (2014) provide a methodology for assessing compatibility of stormwater treatment performance data between different geographic regions using the same treatment system. Another taxonomy method to classify rainfall events in terms of their influence on stormwater quality is described in Liu et al. (2012).

**TABLE 2.1 Alignment Between Various WSUD Approaches and Different Management Objectives**

| | Peak Flow Reduction | Runoff Volume Reduction | Gross Pollutants | TSS/ TP/TN | Hydrocarbons | Amenity |
|---|---|---|---|---|---|---|
| Preserve and maintain waterways and riparian areas | | | | | | |
| Urban design/housing design | High | High | | | | High |
| Erosion and sediment control | | | | High | | |
| Permeable paving | High | High | | Med | Med | |
| Rainwater tanks | High | High | | | | |
| Downpipe diverters | Med | Med | | | | |
| Green roofs | | Med | | | | High |
| Street sweeping | | | Med | Low | | |
| Litter control | | | High | | | High |
| Gully baskets | | | High | Med | | High |
| Vegetated swales | Med | | | Med | Med | Med |
| Gross pollutant traps | | | High | * | Med | |
| Wetlands | Med | Low | | * | Low | High |
| Floating wetlands | | | | * | | Med |
| Bioretention (rain gardens) | Med | High | Med | Med | Med | High |
| Watersmart street trees | Low | Med | | Med | Med | High |
| Proprietary filtration devices | | | Med | Med* | Med* | |

*TN*, total nitrogen; *TP*, total phosphorus; *TSS*, total suspended solids; *WSUD*, water sensitive urban design; *, highly variable.

## 2.3 DRIVERS FOR IMPLEMENTING WATER SENSITIVE URBAN DESIGN APPROACHES

A major emphasis of WSUD has been to reduce stormwater pollutant loads, including all forms of sediment. However, there is a growing school of thought that coarse sediment is relatively inert and may in fact be beneficial for urban streams to help balance the increased erosivity of urban runoff (e.g., Russell et al., 2017; Hawley and Vietz, 2015). Ensuring adequate coarse sediment supply could become a major theme in WSUD in coming years.

Table 2.1 lists a range of WSUD approaches and a semiquantitative ranking of the various management objectives that they contribute toward.

## 2.4 AVOIDING STORMWATER POLLUTION

The following measures can be used to reduce the generation of stormwater and stormwater pollutants in urban areas.

### 2.4.1 Preserve and maintain waterways and riparian areas

Intact waterways and riparian corridors provide excellent water treatment. Once waterways are degraded, it's very difficult to restore them, and it can take many decades before riparian water quality processes recover after disturbance (Dosskey et al., 2010).

New developments should avoid direct impacts on waterways and riparian areas. Buffers, either side of the channel centerline, should be preserved to protect environmental values, allow for natural geomorphic process of erosion and accretion, and provide long-term flood resilience to surrounding areas. In the absence of more site-specific data, as a general guide, gullies should have at least a 10 m buffer either side, increasing by 10 m for each stream order (i.e., 20 m for a first-order stream, 30 for a second-order stream, etc.).

## 2.4.2 Erosion and sediment control

ESC is a broad term for the suite of practices used to limit sediment laden runoff leaving construction sites.

Various opinions exist as to whether ESC is part of, or separate to, WSUD. Given the significant environmental impact caused by poor ESC and the deleterious impact it can have on downslope stormwater treatment measures (such as smothering bioretention systems), it is considered important to include it within this chapter. Chapter 8, by Leon Rowlands, provides more detailed discussion of sediment and erosion control.

Russell et al. (2017) undertook a global literature review and found that newly urbanizing catchments with high rates of active construction generate large amounts of sediment, within the range 1−270 t/ha year, with a median value of 7.4 t/ha year. They found yields tend to be 3−420 times higher than background rates, with a median increase of around 60 times forest yields and 10 times agricultural yields. At the site scale, fine-grained sediment loads can be 21−12,000 times higher than background rates.

To give a sense of perspective to these ratios, the Revised Universal Sediment Loss Equation, based on typical South East Queensland rainfall and slopes averaging 5%, estimates soil loss is in the order of 100−160 t/ha year increasing to around 400 t/ha year on slopes of about 10% (Bligh Tanner, 2017). This may be compared with a median loss of 7.4 t/ha year during urban construction (Russell et al., 2017) or/and 285 tonnes/ha over an 18-month construction window from a small (8 ha) subdivision in South East Queensland (Gardner, 2015).

McIntosh et al. (2013) posed the question "Does most of the negative ecological impact from urbanisation occur during or shortly after the process of construction? Assuming that the primary degrading force is from on-going runoff, then water sensitive urban design features, and careful planning of urban layout and imperviousness seems the appropriate management focus. If however, the primary impacts occur during or shortly after construction, then management by means of improved sediment erosion control, changes to the way land is cleared, built on, and then re-vegetated, could well be the more effective intervention strategy."

Erosion control opportunities should always be considered before sediment control, as it is often simpler and easier to avoid sediment being entrained in runoff, than it is to get sediment to deposit out of runoff.

Key practices include avoiding earthworks activities during expected high-rainfall periods, careful construction scheduling, good site drainage, including diverting "clean" upslope runoff around areas of disturbance, maintaining soil cover, hydromulch, polymer sprays, stockpile management, sediment fences, and sediment basins (Witheridge, 2012).

High efficiency sediment basins (Fig. 2.1) can significantly improve water quality outcomes (for example, see Butler et al., 2004; Dudgeon, 2015; Auckland Regional Council, 2012), aided by flocculants such as alum, or polyelectrolytes, such as poly aluminum chloride, although care must be taken to ensure there is no adverse change to the pH or toxicity of the receiving water when using these.

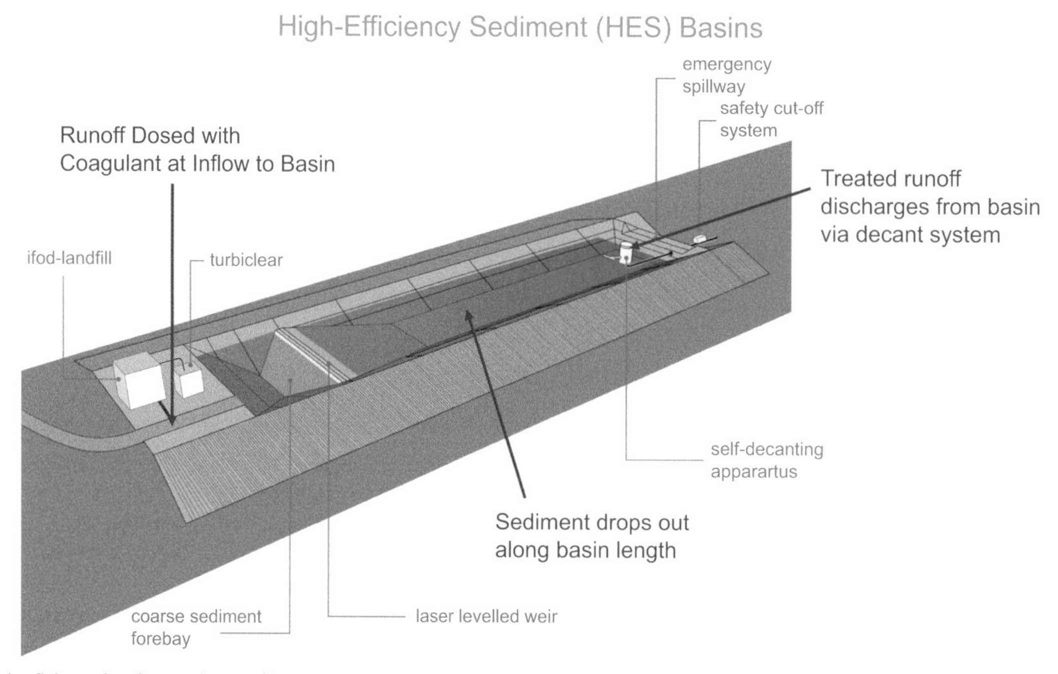


**FIGURE 2.1** Schematic of a continuous-flow high efficiency sediment basin layout with automated flocculant dosing. *Source: TURBID Stormwater Solutions, http://www.turbid.com.au/.*

## 2.4.3 Street sweeping and litter control

Street sweeping can help reduce general litter loads (anthropogenic litter, organic litter, and coarse sediment) but is less effective at addressing fine sediment and associated pollutants. For example, Terstriep (1982), in a study of street sweeping activities in Illinois, United States, found mechanical street sweeping at frequencies as great as twice weekly is neither effective in reducing the mean concentrations nor total loads of pollutants in urban storm runoff.

Street litter can also be managed through local regulation, for example, prohibiting supply of lightweight, nonbiodegradable, plastic shopping bags (introduced in Tasmania, Australia, in 2013), and through drink container deposit schemes, such as that operating in South Australia since 1977.

Another relevant source control measure is now the widespread requirements for dog owners to pick up and dispose of droppings.

## 2.4.4 Urban design/housing design

Urban design influences road and lot layouts and building styles and has a direct impact on fraction of impervious surfaces and hence the amount of stormwater that is generated. Australia has seen a trend larger homes being built on smaller lots, with lower occupancy rates and smaller backyards over the past several decades (Kelly and Donegan, 2015) (Fig. 2.2). In contrast, housing models being adopted in the United Kingdom are achieving both higher urban densities and larger (communal) backyards (Hall, 2010).

Changing urban design and housing designs is a considerable task, and urban planning advocates have been trying to affect change for years. Nonetheless, it remains one of the more powerful ways of influencing stormwater outcomes. For example, building two-storey homes instead of single-storey homes of similar living area can halve the footprint of a house. The smaller roof area and increased garden result in less stormwater runoff, as well as a range of household benefits associated with a larger yard, provided the allotment size remains the same.

## 2.4.5 Rainwater tanks

Rainwater tanks can be a simple lot-scale solution for reducing stormwater runoff and supplementing mains water supply to households (Fig. 2.3). A common size in urban Australia is 5 kL.

A typical rainwater tank system involves capturing, screening, and storing roof runoff in tanks for subsequent (non-potable) reuse such as clothes washing, toilet flushing, and garden irrigation. Overflows from the tanks can be directed to landscaped areas, other stormwater treatment measures (e.g., swales or bioretention systems), or the street stormwater drainage system.

Rainwater tanks can provide an alternative source of water for end uses such as hot water systems, toilets, laundry, and gardens, thereby reducing the demand on centralized mains water supplies and minimizing the overall volume of stormwater runoff. Other benefits are discussed further in Sharma et al. (2015, 2016).

**FIGURE 2.2** Typical greenfield housing development in Australia dominated by impervious surfaces.

FIGURE 2.3 Rainwater tanks can convert roof runoff into a valuable resource.

Rainwater tanks in warm climate areas have been linked to the breeding of the dengue fever vector, the *Aedes aegypti* mosquito, in rainwater tanks with missing or faulty insect screens (enHealth, 2004). Some roof material types may be unsuitable for roofwater harvesting if there is potential for human ingestion of the collected roofwater. For example, roof junctions sealed with lead flashing, or roofs coated in bitumen, or treated timber roofs are not suitable for roofwater harvesting. Similarly, roof areas subject to discharges from wood burner flues or air-conditioning units should also be avoided (Water by Design, 2009). More details on managing the chemical and microbiological risks from using rainwater can be found in the International Water Association (IWA) rainwater systems book by Sharma et al. (2015).

Rainwater tanks can also be configured to provide varying levels of flood detention by having a dedicated "airspace" or void in the top of the tank. When appropriately configured, and with the necessary regulatory controls, this can reduce the need for flood detention systems.

A recent innovation is the *Talking Tanks* concept. Developed by Iota P/L, the commercial arm of South East Water (a water utility in Melbourne, Victoria), *Talking Tanks* monitors tank water levels and automatically releases water at a controlled rate, if required. The system allows the release of water from set points that are chosen by the user, according to rain or storm predictions that are received via a web link from the Bureau of Meteorology. The system developers state "The system automatically releases water, creates storage capacity and prevents overflows of stormwater. With unique self-learning, these intelligent systems are paving the way forward for efficient management of rainwater tanks" (for further information see: http://www.iota.net.au/).

The idea of centrally controlled rainwater tank levels has also been successfully deployed in Seoul, Korea, (Han and Mun, 2011) and in Washington, DC, USA, (Quigley and Brown, 2014) to control peak flow discharges from stormwater. Performance of rainwater harvesting systems is highly contextual and influenced by the local climate, the connected roof area, tank size, and the actual demands on the water. This can easily be modeled using long-term computer simulation using local rainfall data at either short (subhourly) or daily time steps. See, for example, the IWA Rainwater book edited by Sharma et al. (2015) (chapters by Vieritz et al., 2005).

As with any distributed technology, end-user uptake and maintenance diligence will be variable, and this should be factored into any planning analyses. For example, a study of 223 detached dwellings in South East Queensland, Australia, with newly installed rainwater tanks, found only 24% of households complied with the requirement to have at least 50% of roof area connected to that rainwater tank (Biermann and Butler, 2015).

Regular maintenance of roofwater harvesting systems is important to manage water quality (i.e., avoid excessive ingress of organic matter into storage systems from roofs and gutter systems) and mitigate mosquito risk (Moglia et al., 2013). Maintenance should therefore encompass the roof, guttering, leaf screens, downpipes, tanks, mosquito screens, top up switching valves (to mains water), and pumping systems.

The effective service life of a roofwater harvesting system depends on the type of storage system used (i.e., above ground or below ground and materials such as plastic vs. concrete vs. zincalume or galvanized steel). Typically, a well-maintained roofwater harvesting system should have an effective tank service life of 20–30 years, with electric pumps typically requiring replacement every 5–10 years, depending on their construction quality and the intensity of their use (Water by Design, 2009).

## 2.4.6 Downpipe diverters

Downpipe diverters are a simple way of adapting existing downpipes so that rainfall is diverted to irrigate gardens and lawns (Fig. 2.4). This uses water that would otherwise create stormwater. The devices are low cost, have wide applicability, and can be easily retrofitted.

Downpipe diverters use a manually controlled flap valve to direct roofwater from minor rainfall events to gardens or grassed areas, whereas large storm flows are automatically diverted into the conventional drainage system via the inbuilt passive bypass plumbing system. The device should be installed with an angled mesh leaf screen located above the diversion device. They are an alternative to rainwater tanks, particularly in areas where there are concerns with mosquitoes or limited area is available for tanks. They are best suited to use on downpipes where the runoff can infiltrate into soils and on slopes that fall away from the building.

A robust design has recently been developed by Melbourne Water and Master Plumbers Association of Victoria, Australia (https://www.melbournewater.com.au/sites/default/files/DOWNPIPE%20DIVERSION.pdf), and system packages have been provided to residents via local governments. The device costs AU$135 (in 2016), with the installation cost depending on whether it was installed during initial house construction or as a retrofit. Over 500 of these devices have been installed in Melbourne by 2016.

Periodic cleaning of the diversion device is needed to remove any accumulated debris.

## 2.4.7 Permeable paving

There is a broad range of paving technologies that allow water to permeate through a trafficable surface. Four main categories of permeable paving are listed below (Eisenberg et al., 2015; Mullaney and Lucke, 2014) and shown in Fig. 2.5.

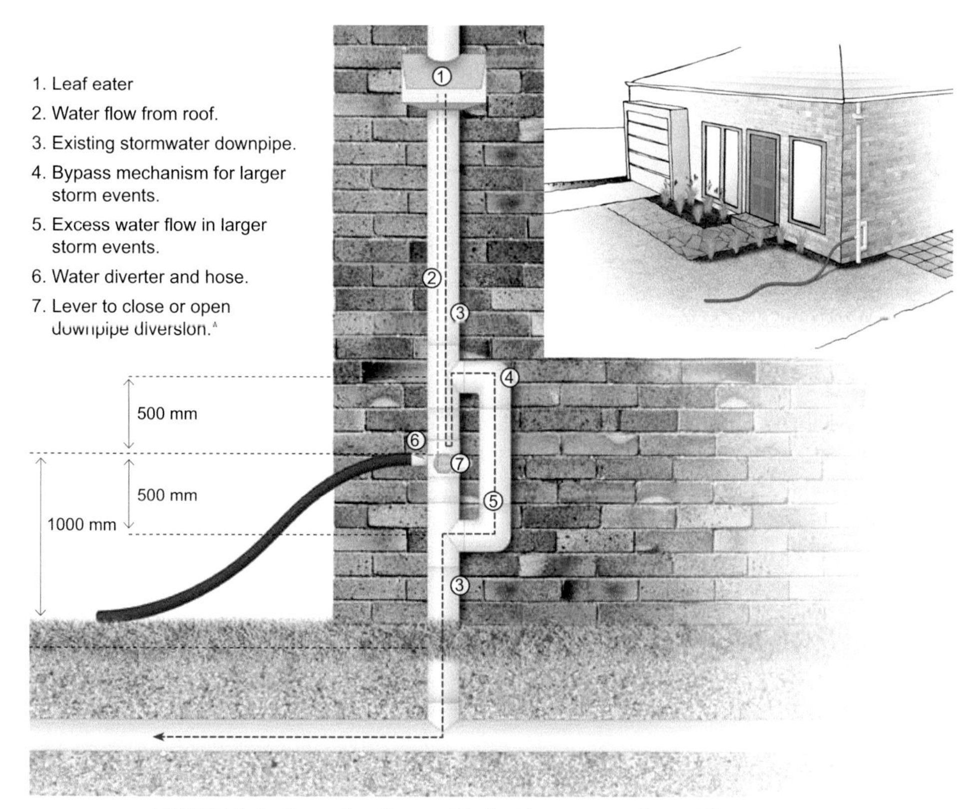


**FIGURE 2.4** Downpipe diverter. *Used with permission from Melbourne Water.*

FIGURE 2.5 (a–d) Permeable pavement systems come in many styles, including permeable interlocking concrete pavers and grid pavement systems. *(c) Used with permission from Amalie Wright, Landscapology.*

**Porous asphalt (PA)**: PA is similar to conventional asphalt, except the fines are removed to create greater void space. Additives and higher-grade binders are typically used to provide greater durability and prevent breakdown.

**Pervious concrete (PC)**: PC is produced by reducing the fines in the mix to maintain interconnected void space and has a coarser appearance than standard concrete.

**Permeable interlocking concrete pavement (PICP)**: PICP is made of interlocking concrete pavers that maintain drainage through aggregate-filled gaps between the pavers. The pavers themselves are not permeable.

**Grid pavement systems (plastic or concrete)**: Grid pavement systems are modular grids filled with turf and/or gravel. Open-celled concrete or plastic structural units are typically filled with small uniformly graded gravel that allows infiltration through the surface.

Permeable pavements can be designed with underdrainage systems that collect water for reuse or discharge, but more commonly, allow water to infiltrate into the subsoil (Fig. 2.6). They can be designed for a range of traffic loadings, varying from pedestrian foot traffic to trucks. Like any pavement, poor engineering design that fails to provide adequate structural support for heavy vehicles can lead to uneven subsidence. Useful guidance on structural engineering design of permeable pavements is provided in Eisenberg et al. (2015).

Rainfall falling on the surface infiltrates into the voids between the pavement elements, with primary treatment by filtration. Hence, the stormwater is treated at source. They can obviate the need for additional drainage or flood detention systems in urban areas; they hydrate soils in urban environments that may lead to healthier urban tree growth, and they recharge local aquifers.

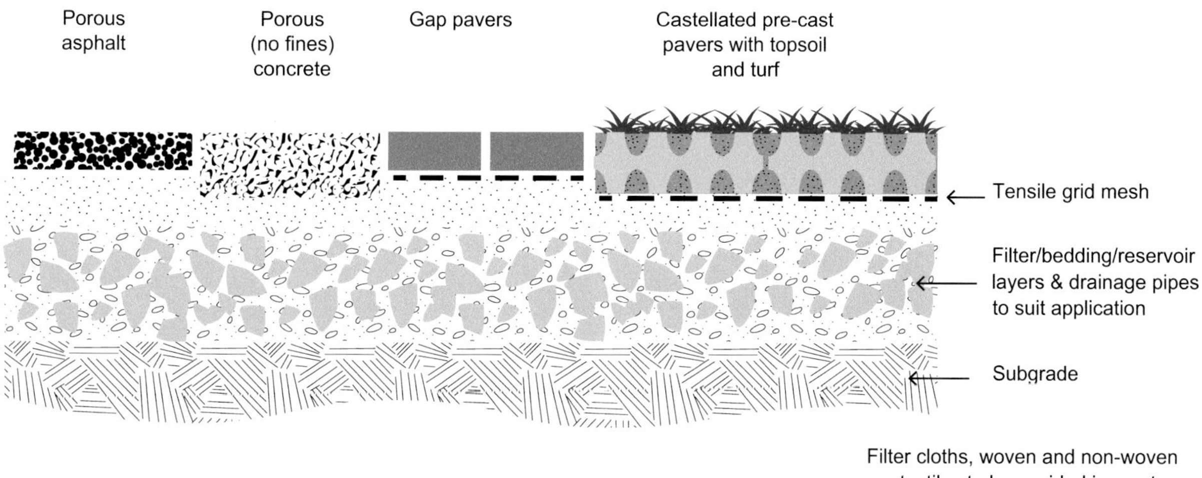


**FIGURE 2.6** Permeable pavements can have various profiles, engineered to suit the specific design application.

Permeable pavements are best suited for low traffic loads, which are subject to direct rainfall only, rather than receiving runoff from high sediment areas. As such, car parks, driveways, and pedestrian areas are well suited for this technology.

A key risk with permeable pavement is clogging from the sediment they retain, thereby substantially reducing their own permeability. PC produces higher permeability and better clogging resistance than PA, and there are significant gains in permeability and clogging resistance when the porosity is raised beyond 20% (Fwa et al., 2015).

Permeable pavements can reduce the amount of stormwater runoff and pollutants being generated by urban areas, although most studies tend to focus on infiltration responses. For example, average runoff reduction from porous pavements varies between 50% and 93% (Ahiablame et al., 2012).

Boogaard et al. (2014) examined 55 sites located in the Netherlands and Australia, which ranged in age from 1 to 12 years old. They tested the Australian systems for a 4 exceedance per year (4EY) storm and found 90% of the pavements performed at this standard. Another study on an 8-year-old permeable interlocking concrete paving system found it to be very effective at filtering and removing sediment from stormwater, with its overall infiltration performance still satisfactory (Lucke and Beecham, 2011).

Bean et al. (2007) undertook a field survey of the infiltration rates of concrete grid pavers, permeable interlocking concrete pavers, and porous concrete. They found that higher surface infiltration rates in concrete grid pavers could be maintained by regular street sweeping. Removing the top layer of residual material (13–19 mm) from the larger voids within the concrete grid and replacing it with sand increased infiltration from 49 to 86 mm/h.

There are limited data on water quality performance—although pollutant load reductions would be expected to be commensurate with volumetric load reductions in runoff. In one study, turbidity was reduced (42%–95%), and effluent was measured at <10 NTU for the first 3 months after maintenance (Sansalone et al., 2012).

Maintenance is essential to achieve long-term permeability in areas with moderate to fine sediment loads. Maintenance by vacuuming or sonication (agitation with sound waves) has been found to restore at least 96% of the initial hydraulic conductivity of clogged permeable concrete pavements (Sansalone et al., 2012). Philadelphia Water (USA) has found vacuuming to be effective in maintaining permeability of porous pavements (Stephen White, Philadelphia Water, pers comm).

Fine clays and silts in stormwater runoff pose a serious clogging risk. Permeable pavers have been shown to be effective in reducing very fine particles (Sansalone et al., 2012), and over time, some of this material accumulates deeper within the subgrade of the pavement and is difficult to remove. Overall, the durability of permeable pavements is moderate—but also is highly context dependent.

## 2.4.8 Green roofs and walls

Green roofs and walls cover a broad range of approaches to add greenery to the rooftops and walls of buildings in the urban environment. Systems include shallow ("extensive") and deep ("intensive") substrate roof plantings, trellis systems growing vines on facades, and vertically supported growing media (Fig. 2.7). The role of green roofs and walls is examined in more detail in Chapters 20 and 21.

**FIGURE 2.7** Green roof and green wall systems come in a variety of styles, matching plant species, substrate, and irrigation.

Green roof systems can help mitigate urban heat island effects, reduce runoff, improve urban amenity, and provide insulation to buildings. They are suited to a range of urban environments, especially where the amenity benefits can be realized (i.e., the systems are visible and accessible). Installations must be easily accessible for maintenance, and roofs and/or walls need to have sufficient load-bearing capacity (this may limit the ability to retrofit them on some buildings). Solar aspect and shading need to be considered in site selection and system design.

Weed invasion is a key risk, especially from wind-blown weed seeds. Other issues include plant dieback due to heat stress, lack of moisture, or poor plant selection. If the growing matrix is poorly designed, there are also risks of structural damage due to inadequate waterproofing and membrane penetration by roots.

Ahiablame et al. (2012) cite 13 studies with runoff reduction ranging from 23% to 100%. Water quality outcomes are likely to be highly dependent on the fertilizer regime of the system. Chapter 20 provides a good summary of the water quality constituents measured in green roof outflows across the world. Berndtsson et al. (2010) concluded that the factors potentially influencing green roof runoff quality can be summarized as the growing media, applied fertilizers, applied water for summer irrigation, and air pollutants.

Maintenance requirements will vary depending on the configuration of the system and may include horticultural care and weed management, irrigation system maintenance, and growing media renewal.

There are limited long-term data on the durability performance of green roofs. For green walls, trellis systems are likely to be the most durable, as the soil and irrigation systems are more compact.

## 2.5 MITIGATION OPTIONS

The following measures can be used to mitigate the level of pollutants in urban stormwater.

## 2.5.1 Gully baskets

Gully baskets are simple mesh baskets that can be retrofitted into existing stormwater gully (side entry) pits (Fig. 2.8). When fitted with a 200-micron mesh (0.2 mm), they can be an effective way of capturing and storing gross pollutants (litter) and particulate-bound pollutants.

Gully baskets work by filtering stormwater through a mesh basket located within stormwater gully inlets. The sizes are customizable to the size of the gully pit, and there are different grades of mesh inserts available, ranging from 100 to 3000 μm (i.e., 0.1–3 mm).

Gully baskets provide a low-cost way of targeting pollutants in key hot spot areas and can be easily retrofitted into existing urban areas. They retain captured pollutants in a dry state, thereby avoiding the pollutant leaching problem associated with wet-sump pollutant traps.

Supply and installation costs are typically about AUD$500 (in 2017) and can be removed and relocated to other locations if needed. They are best suited for litter "hot spot" areas such as town centers, commercial and industrial areas, school precincts, and recreational facilities.

Areas with street trees that generate high leaf litter loads (especially deciduous trees) may be unsuitable for gully baskets unless they are coupled with regular street sweeping or an enhanced maintenance regime.

Gully baskets need to have an effective bypass flow mechanism so that if the basket is full of litter, the normal drainage functions of the gully inlet, linked to the kerb and channel, are not adversely affected.

Litter capture is likely to be highly effective, provided they are cleaned out regularly. Buoyant litter has the potential to be reliberated under bypass conditions. Pollutant removal rates of 50% total suspended solids (TSS), 30% total phosphorus (TP), and 20% total nitrogen (TN) are likely (Butler et al., 2004; Bligh Tanner, 2014b; DesignFlow, 2015), although actual performance will be sensitive to the particle size composition of local stormwater.

Periodic clean out by vacuum truck is required. There are currently no documented studies on the required frequency, although clean out at least twice per year would be a minimum, with a higher frequency in higher litter generating areas.

## 2.5.2 Gross pollutant traps

Gross pollutants include coarse sediments (particles > 5 mm), anthropogenic litter, and leaf litter. Leaves and other forms of organic material make up a large proportion of the gross pollutants in urban stormwater and form a substantial "bycatch" in devices installed to prevent anthropogenic litter entering waterways.

Gross pollutant traps (GPTs) are usually propriety devices and come in a wide range of styles and configurations, including generic sumps, trash racks, and floating litter traps (Fig. 2.9). GPTs are sometimes used to reduce coarse sediment and litter entering downstream treatment measures, such as wetlands, as a way of optimizing maintenance requirements.

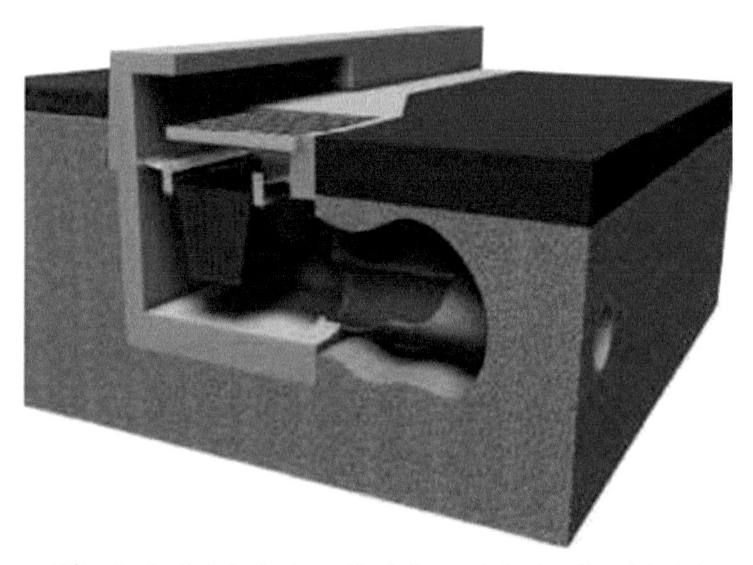

**FIGURE 2.8** Gully basket insert. *Used with permission from Ecosol Pty Ltd.*

**FIGURE 2.9** (a–d) Various styles of gross pollutant traps. *(c) Used with permission from LafargeHolcim, (d) Used with permission from Ecosol Pty Ltd.*

Some devices work on basic sedimentation or mechanical screening, whereas others work on hydrodynamic separation using the centrifugal forces created from the energy of the stormwater flow. GPTs can provide valuable pretreatment for stormwater harvesting schemes and have an advantage over bioretention systems in this context because there is negligible loss of runoff volume through the device that can help maximize yields.

They can be used to help educate the community about litter, as they capture it in a visible manner. For example, Melbourne Water uses floating litter traps in the Yarra River to both collect buoyant litter and increase public awareness of litter issues.

There are a range of trash racks and litter traps that can hold collected material in a dry state, which is preferable in high-litter locations. In wet-sump style installations (the most common type of GPT), organic matter can decompose and create anaerobic conditions, leading to the leaching and desorption of nutrients and metals if the sumps are not cleaned out within a few days of the storm event. For example, Walker et al. (1999) found increases in TP and TSS downstream of a wet-sump gross pollutant trap in dry weather conditions. Abood and Riley (1997) also found that gross pollutants had a deleterious effect on water quality, with the decomposition of the gross pollutants increasing with time whenever they were kept in a wet, anaerobic environment. Significant increases were observed in phosphorus, nitrogen, chemical oxygen demand (COD), and suspended solids in downstream pools during the first 100 days after collection. Cigarette butts in particular were found to increase phosphorus, suspended sediment, and COD concentrations within about 10 days of the samples being immersed. For this reason, where nutrient management is important for the health of a receiving environment, it is advisable to couple wet-sump GPTs with another treatment measure that is capable of reliable nutrient reduction.

Because of the wide range of treatment options that fall within the GPT category, it is not possible to provide generalized treatment performance. Hunter (2001) recommends that wet vaults should be cleaned out as soon as possible after a rainfall event and/or at intervals not exceeding 30 days. Inspection of the device should occur after 10 mm of rainfall, or every 2 weeks, whichever comes first. Such maintenance regimes are unrealistic, given that many GPTs in Australia are subject to an annual clean out at best.

## 2.5.3 Vegetated swales

Swales are simple vegetated drains that convey runoff, while providing an opportunity for sedimentation, seepage, and reduction in flow velocity. Swales can come in a wide variety of styles, ranging from formal turf-lined systems through to densely vegetated channels.

Swales work by infiltrating runoff into underlying soils, enhancing sedimentation from slowing the flow of water and by filtration through the vegetation. Swales can be integrated with an underlying biofiltration trench for enhanced water quality performance.

Because swales provide stormwater conveyance, they can reduce or remove the need for underground pipe drainage. This can result in capital cost savings of up to $5000 per allotment (Bligh Tanner, 2014a). The vegetation in swales is passively irrigated by runoff, minimizing the need for irrigation in urban areas. As a green infrastructure measure, they contribute to mitigating urban heat island effects. If installed in open space or parkland, maintenance costs are minimized.

Swales can be used in a wide range of urban settings, including center median strips, road verges, within allotment landscaping, and in parklands. They should be considered wherever stormwater conveyance is needed and where it is appropriate to have vegetation (Fig. 2.10).

**FIGURE 2.10** Examples of swales in Australia.

Swales are best suited for slopes between 2% and 5%. On flatter grades, systems may require underdrainage and should generally be mass planted, not turfed, to avoid the need for mowing in potentially boggy areas. On steeper slopes (>5%), check dams and/or rock linings may be needed to minimize bed scour. Care needs to be taken when designing swales in locations with multiple driveway crossovers, as this can increase both installation costs and the risk of blockages. Wide verges that allow driveways to cross swales at-grade (rather than using a culvert) are preferable.

Key risks associated with swales are bogginess and ponding, especially if they are installed on a relatively flat grade, on heavy soils, or in areas subjected to regular vehicle traffic and hence wheel rutting. In contrast, scour and erosion can occur if swales are located on steep grades or on sandy soil. Localized flooding can occur if the flow capacity of swales is reduced because of debris, accumulated sediment, or the dumping of waste.

Reductions in stormwater concentrations can vary from 30% to 98% for TSS, from 30% to 99% for TP, and from 14% −60% for TN (Barrett et al., 1998; Lloyd et al., 2002; Yu et al., 2001; Bäckström, 2003; Ahiablame et al., 2012). Performance is highly linked to the swale configuration and the composition of stormwater, with greater removal of particulate-bound pollutants compared with soluble pollutants.

Maintenance requirements for swales are simple and involve mowing, weeding or supplementary planting (depending on swale type), litter and debris removal, and periodic reprofiling depending on sediment loads. Swales are highly resilient, and many existing swales in Australian urban and rural areas are more than 20 years old.

## 2.5.4 Bioretention (rain gardens)

Bioretention systems, also known as rain gardens, are vegetated soil filters that treat stormwater by vertical percolation through a soil filter media. The use of saturated zones underneath bioretention systems is likely to assist in helping plants survive extended dry periods and provides anaerobic conditions for denitrification. Fig. 2.11 provides a cross-sectional diagram of a typical bioretention system.

Water quality is managed through a combination of filtration, sorption, transformation, denitrification, plant uptake, and exfiltration. One of the major benefits of bioretention systems is in the slowing and reducing of runoff. A summary of recent research on the hydrologic benefits of bioretention reported significant attenuation of peak flows and runoff volumes, with an average volumetric loss of 59% across 116 monitored events (Hoban, 2017).

Bioretention systems are very versatile in terms of shape and size and can be integrated into a wide range of urban settings (Fig. 2.12). Ideally, bioretention systems are located on sites where there is about 1 m of vertical fall between the inlet and outlet, to enable adequate filter media depth and underdrainage. However, bioretention systems can also be designed without underdrainage, allowing them to be created in a wide range of topographic circumstances.

To manage the risk of failure and promote better urban design integration, individual bioretention systems should ideally be limited in size to less than 500 $m^2$ and service catchments no larger than 3 ha.

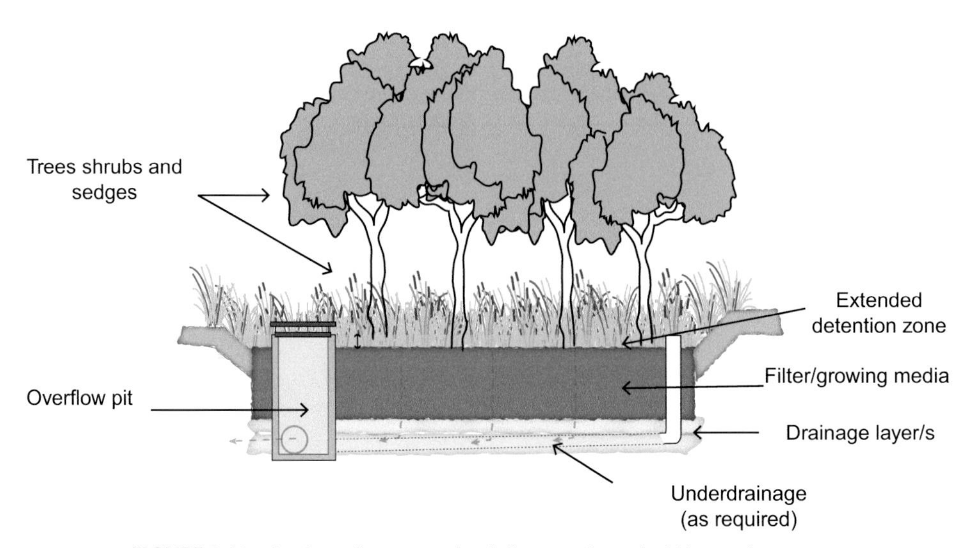


**FIGURE 2.11** A schematic cross-sectional diagram of a typical bioretention system.

FIGURE 2.12 Bioretention systems (rain gardens) can be integrated into many urban settings.

Key risks for bioretention systems include

- Smothering with sediment from erosion during upslope construction;
- Blocking of the filter media with retained sediment;
- Weed infestations;
- Excessive wetness that can result in slimes clogging the filter media;
- Plant death during long dry spells or regular plant water stress because of using a filter media with a low water holding capacity. This can be mitigated by adding organic media to the filter media and including a saturated zone underneath to provide irrigation via capillary rise;
- Vulnerable to poor construction practices given the number of elements that need to be constructed correctly; and
- Herbicides and other toxicants in urban runoff.

Short-term studies in both controlled and field conditions indicate good pollutant reduction potential. However, there are few long-term field trials. The treatment performance of bioretention systems is variable, and while controlled laboratory studies suggest reliable reductions in nutrient concentrations (Fletcher et al., 2007), most field studies show highly variable performance, often with negligible reductions in nutrient concentrations (for example, Hatt et al., 2009a,b; Lucke et al., 2018; Parker, 2010).

Of course, the pollutant loads must go somewhere, and these sinks are either the matrix of the biofilter (for P, heavy metals and sediment), plant uptake, or nitrification/denitrification for the N-load. The field studies suggest the primary treatment mechanism is the reduction in the volume of runoff, thereby reducing pollutant loads. A review of field studies by Hoban (2017) found average reductions in stormwater runoff volumes of 59% (Fig. 2.13).

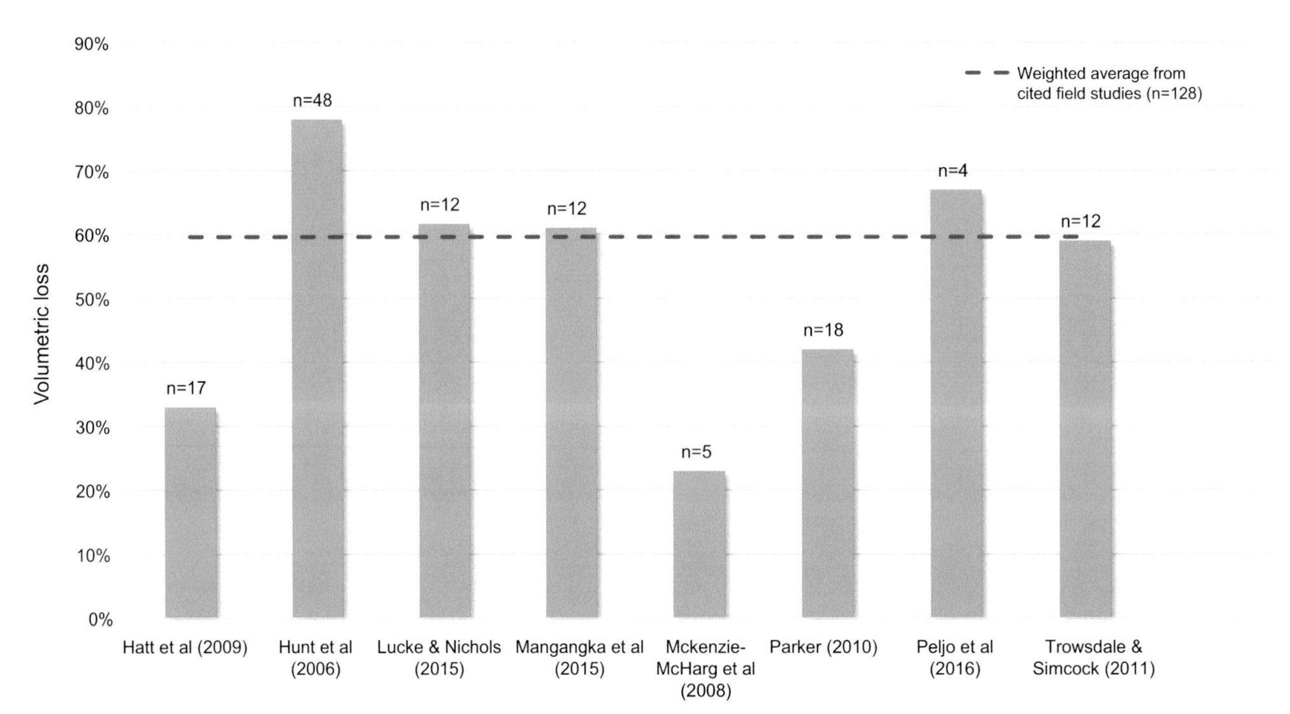


**FIGURE 2.13** Reduction in runoff volume in bioretention systems reported by a range of authors (Hoban, 2017).

Typical maintenance requirement for bioretention systems include

- Weed management and supplementary planting;
- Litter removal;
- Coarse sediment removal; and
- Periodic flushing of underdrainage.

Maintenance efforts can be minimized by

- Ensuring there are trees and shrubs in the system to provide shade and help minimize weeds;
- Focusing on the effective establishment of a locally appropriate plant community;
- Ensuring the filter media has adequate organic carbon for denitrification reactions;
- Organic material in the media to increase water holding capacity;
- Providing a coarse sediment forebay where high sediment loads are expected; and
- Ensuring the system is accessible for maintenance.

There is still insufficient scientific and anecdotal evidence to be confident about the long-term viability of bioretention systems in a range of climatic contexts, especially in arid and dry tropical environments. Success has been variable, with a 15-year-old system in South East Queensland, Australia, having good vegetation growth and low-maintenance requirements. However, there are many more examples of very poor design and implementation. Some systems that appeared to perform well for several years have then suffered vegetation dieback. There have been some prominent failures with very large bioretention systems ($>1000\ m^2$). Smaller systems, tailored into the urban landscape design, appear to have the best prospects of long-term success as they are more likely to receive ongoing maintenance and plant replacement. There are encouraging signs that systems with saturated zones beneath them have better plant performance and presumably better pollutant reduction.

## 2.5.5 Wetlands

Stormwater treatment wetlands are shallow vegetated waterbodies that detain stormwater runoff and slowly release it after rainfall events. Stormwater wetlands differ from wastewater treatment wetlands in that they tend to have more variable water levels (a result of the variable nature of rainfall) and also treat influent water with much lower and more variable nutrient concentrations.

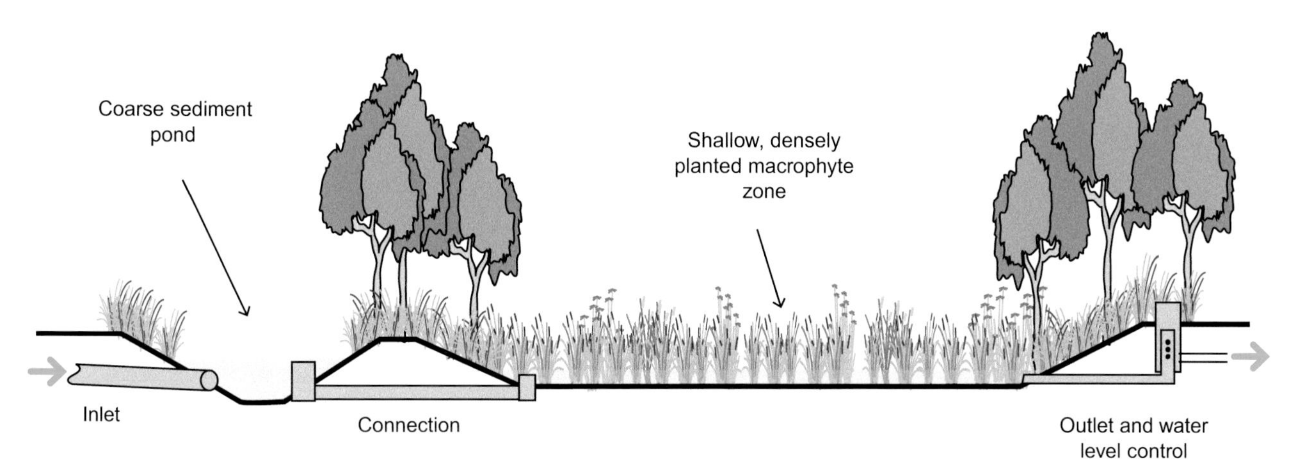


**FIGURE 2.14** A schematic diagram of a typical configuration of a constructed wetland.

Stormwater wetlands use forebay sedimentation, fine filtration, adhesion, biological uptake, and transformation processes to remove pollutants from stormwater. A typical configuration for a constructed wetland is shown in Fig. 2.14 and includes

- A trash rack to capture anthropogenic litter;
- A sediment pond that facilitates the deposition of coarser sediment and is designed to allow easy access for sediments removal;
- A shallow depth macrophyte zone, planted with dense stands of erect reeds and sedges;
- A high-flow bypass to protect the wetland from high-velocity flows during major storm events; and
- An outlet pool that regulates maximum and minimum water levels in the wetland and controls the rate of water release after storm events. Best practice outlet designs can easily be adapted to fine tune the wetland's hydrologic regime, typically by using a removable steel plate or riser pipe. They also have a submerged connection to the macrophyte zone to minimize blockage from buoyant debris.

Wetlands can have multiple benefits including

- Improving water quality (contingent on good design and widespread plant cover).
- Attenuation (smoothing) of peak flows from urban storms. Wetlands tend to flatten and prolong runoff hydrographs but are unlikely to reduce overall runoff volumes. Whether this is beneficial to the local receiving waterway and its ecology needs to be evaluated on a case-by-case basis.
- Wetlands can be beautiful and engaging places, providing opportunities for adventure, exploration, and delight by the community. They can also provide opportunities to help reconnect communities with nature and the water cycle and educate people about stormwater issues (Fig. 2.15).

Wetlands are best located on low-lying flat land and are preferably sized at about 3%–10% of the contributing catchment area. When wetland proportions are much smaller than this, they become fully charged with water after only small amounts of rainfall and may not have adequate time to drain between successive rainfall events. This can result in excessive inundation stress on the plants and potentially plant failure (Hoban et al., 2006). Wetland sizing is often undertaken using Model for Urban Stormwater Improvement Conceptualization (eWater, 2017). However, designers should be mindful that software tools do not provide guidance on plant viability, and it is possible to model excellent pollutant removal for a wetland configuration that, in practice, will not sustain healthy plants.

Wetlands are also susceptible to aquatic weed infestations, roosting, and grazing from large nesting birds and high evaporation rates during extended dry seasons. Care also needs to be taken when designing wetlands in proximity to airfields, due to the risk of aircraft bird strikes. Suspended agricultural bird netting is one solution to this problem.

Many constructed wetlands do not establish and/or sustain adequate plant cover. Some potential causes of this include

- Large areas having water levels too deep for healthy plant growth;
- Damage to plants from waterfowl;

FIGURE 2.15 Constructed stormwater wetlands can come in many forms and provide key opportunities for engagement and interaction; however, they are susceptible to major plant dieback.

- Prolonged inundation during rainfall events (due to undersizing or deficiencies in outlet pool hydraulics and blockages, Hoban et al., 2006); and
- Herbicides and other chemical stressors in urban runoff.

Other key risks associated with wetlands include aquatic weed infestations, mosquitos, and elevated nutrients and pathogens associated with waterfowl, which is often exacerbated by duck-feeding by local residents.

Water quality outcomes from stormwater wetlands are rarely monitored due to the complexity of instrumentation, cost of chemical analyses, and challenges in dealing with the large inter- and intrastorm variability. Some of the key nitrogen processing, for example, is believed to occur in the residual water held in the wetland between storm events (Carleton et al., 2000), which may not be apparent in a typical event monitoring program. Most studies have shown that both natural and constructed wetlands retain nutrients and sediments, whereas some other studies have shown they have little effect or can even increase nutrient and sediment loads to receiving waterbodies (Fisher and Acreman, 2004; Coleman, 2007; Lucas et al., 2015). For example, Parker et al. (2009) and Mangangka et al. (2015a,b) measured reasonable pollutant reductions at a 10-year-old constructed wetland at Coomera Waters, Queensland. Conversely, Moreton Bay Regional Council in South East Queensland reported higher cyanobacterial counts at the wetland outlet compared with the inlet, and this was coupled with very high increases in nutrient concentrations (Karen Waite, MBRC, pers comm).

Key maintenance activities for wetlands are weed removal and supplementary planting, ensuring inlets and outlets are free from debris and functioning correctly, removal of litter, and periodic desilting the inlet pond.

Of all the common wetland performance criteria, appearance is probably the most variable due to the difficulty in establishing and maintaining good plant cover. The majority of constructed stormwater wetlands visited by the author (in Australia) showed moderate to major plant dieback.

Further research is needed into how to design resilient constructed wetlands that deliver reliable water quality improvement. A reasonable approach is to design future wetlands by mimicking natural wetlands in the local area, maintaining shallow standing water depths (<100 mm), and limiting transient ponding depths to <300 mm.

## 2.5.6 Floating wetlands

Floating rafts of vegetation occur naturally in many waterbodies. They often form where buoyant vegetation traps sediment and other organic material and forms a substrate on which emergent macrophytes grow. Where such rafts are able to be found, portions can be translocated into other waterbodies to propagate new floating wetlands.

Manufactured floating wetlands are a relatively new technology, involving plants growing in manufactured floating rafts, with roots extending into the water column (Tanner and Headley, 2011). They can potentially be retrofitted into existing urban waterbodies. In theory, floating wetlands work by reducing flow velocities, which promotes the settlement of suspended solids. The dense root mass also filters suspended solids and supports a complex biofilm that reduces pollutant loads via processes of adhesion, filtration, nutrient uptake, and sequestration (Stewart and Downing, 2008).

A major risk is that waterfowl may colonize the wetland, affecting plants and lake water quality because of increased defecation. Redlands City Council and the City of Gold Coast (in South East Queensland) have removed floating wetlands, as they became a haven for ibis and waterfowl, both nuisance birds in an urban setting. Other potential issues are short circuiting of flow paths around dense root zones unless flow is constrained by floating, vertical training walls, and anoxic conditions in the benthos beneath them. In shallow waterbodies, roots may anchor to the base of the waterbody, which could lead to plant death by inundation as the water level rises with stormwater inflow.

There are only a limited number of credible peer-reviewed papers on floating wetlands. Borne et al. (2013) undertook one of the few paired field trials on floating wetlands and reported enhanced TSS reduction (41%) compared with the paired conventional pond. The interim report (Borne, pers comm) also examined nitrogen removal and noted the reductions were not statistically significant.

Tanner and Headley (2011) reported that floating wetlands reduced **wastewater** TSS by up to 80%, TN by 35%, and TP by 20%. Winston et al. (2013) monitored the water quality of two ponds in North Carolina (USA) before and after installation of floating wetlands. The floating rafts occupied 10%–20% of the pond areas. The benefit to pond performance was modest. They reported no significant reduction for TSS, N, P (and its forms) in one pond but observed significant reduction for TSS and TP in the other smaller pond. They also noted an increase in aquatic weeds following installation.

Suki et al. (2010) undertook large-scale laboratory tests of floating wetlands with eutrophied water (not stormwater) and reported significant nutrient reductions under low loading rates but noted that harvesting of plant material would be needed to sustain such removal rates.

## 2.5.7 Water smart street trees

Simple designs allow runoff from kerb and channel to provide water into the root zone of street trees (Fig. 2.16). Passively irrigated street trees are increasingly being adopted by local governments such as Melbourne City Council (which has been trialing various designs for over 10 years), City of Sydney, Brisbane City Council, New York City, and in the United Kingdom.

Research by Denman et al. (2011) shows that trees irrigated with stormwater grow about twice as fast as those irrigated with tap water. This was found to be the case across a range of tree species and soil types with saturated hydraulic conductivities from 4 to 170 mm/h. Given that street trees often receive no irrigation at all, this approach provides not only a source of water but also allows nutrients in stormwater to be sequestered by the trees, rather than being discharged to waterways.

The stormwater quality benefits of each individual tree are modest; however, there is potential for widespread application, with a large number of trees having a cumulative impact, given that a typical street tree density in the inner city is about 100 trees/ha of street, corresponding to a 25% canopy cover.

FIGURE 2.16 Examples of water smart street trees.

## 2.6 CONCLUSION

WSUD is a broad area of practice, with a wide range of objectives, and an even wider set of management practices. The matching of management practices to design objectives and site context is very complex and requires an interdisciplinary method of practice. This chapter has outlined some of the key approaches—grouped into avoidance and mitigation options—to highlight the range of options available, how they function, what is currently known about their performance, and some of the key risks and unresolved issues.

In practice, the selection of technologies is heavily influenced by the preferences of local authorities and the jurisdiction of the planning and development system, which can often limit the use of small and on-lot management measures in favor of larger end-of-pipe systems. Site-specific considerations such as soils, slope, and existing assets also play a key role in influencing the preferred outcome. Practitioners are advised to consult with the relevant local authority to determine which treatment measures are acceptable. It is also very important to control sediment export during civil construction as losses during this period can exceed losses during the long-term, postconstruction WSUD period, by orders of magnitude.

Despite over 20 years of WSUD practice in Australia, there is still much to be learned about the stormwater performance of many of the treatment technologies. This view is supported by recent findings on the hydrologic performance of bioretention systems, showing they function more as transpiring sponges rather than biogeochemical filters. This highlights that we are still early on in the journey of WSUD practice, and the spirit of innovation and experimentation needs to continue for many years to come. Nonetheless, simple visual assessment of healthy plant growth is a very useful criterion of the operational effectiveness of vegetated, treatment devices.

## REFERENCES


Abood, M., Riley, S.J., 1997. Impact on Water Quality of Gross Pollutants: Research Report No. 121. Urban Water Research Association of Australia.

Ahiablame, L.M., Engel, B.A., Chaubey, I., 2012. Effectiveness of low impact development practices: literature review and suggestions for future research. Water, Air, and Soil Pollution 223 (7), 4253–4273.

Auckland Regional Council, 2012. Proprietary Devices Evaluation Protocol (PDEP) for Stormwater Quality Treatment Devices, Version 03, December 2012.

Bäckström, M., 2003. Grassed swales for stormwater pollution control during rain and snowmelt. Water Science and Technology 48 (9), 123–132.

Barrett, M.E., Irish Jr., L.B., Malina Jr., J.F., Charbeneau, R.J., 1998. Characterization of highway runoff in Austin, Texas, area. Journal of Environmental Engineering 124 (2), 131–137.

Bean, E.Z., Hunt, W.F., Bidelspach, D.A., 2007. Field survey of permeable pavement surface infiltration rates. Journal of Irrigation and Drainage Engineering 133 (3), 249–255.

Berndtsson, J.C., Bengtsson, L., Jinno, K., 2010. Runoff water quality from intensive and extensive vegetated roofs. Ecological Engineering 35, 369–380.

Biermann, S., Butler, R., 2015. Physical verification of household rainwater tank systems. In: Sharma, A.K., Begbie, D., Gardner, T. (Eds.), Rainwater Tank Systems for Urban Water Supply – Design, Yield, Energy, Health Risks, Economics and Community Perceptions. IWA Publishing. ISBN13: 9781780405353.

Bligh Tanner, 2014a. Review of State Infrastructure Standards. A Report for the Queensland Government Dept. State Development and Infrastructure Planning.

Bligh Tanner, 2014b. Peer Review of the SPELFilter and SPELStormSack for Use in Mackay Regional Council.

Bligh Tanner, 2017. Fraser Coast Regional Water Quality Strategy – A Report for Fraser Coast Council June 2017.

Boogaard, F., Lucke, T., Beecham, S., 2014. Effect of age of permeable pavements on their infiltration function. CLEAN – Soil, Air, Water 42 (2), 146–152.

Borne, K.E., Fassman, E.A., Tanner, C.C., 2013. Floating treatment wetland retrofit to improve stormwater pond performance for suspended solids, copper and zinc. Ecological Engineering 54, 173–182.

Butler, K., Ockleston, G., Foster, M., 2004. Auckland city's field and laboratory testing of stormwater catchpit filters. In: New Zealand Water and Waste Association Stormwater Conference, Rotorua, May 2004.

Carleton, J.N., Grizzard, T.J., Godrej, A.N., Post, H.E., Lampe, L., Kenel, P.P., 2000. Performance of a constructed wetlands in treating urban stormwater runoff. Water Environment Research 72 (3), 295–304.

Coleman, P., 2007. Do Stormwater Wetlands Affect Urban Stream Health? (Honours dissertation) Monash University School of Chemistry.

Denman, E.C., May, P.B., Moore, G.M., 2011. The use of trees in urban stormwater management. In: 12th National Street Tree Symposium.

DesignFlow, 2015. Independent Peer Review of Proprietary Device: A Report for City of Gold Coast. Designflow.

Dosskey, M., Michael, G.D., Philippe, V., Noel, P.G., Craig, J.A., 2010. The role of riparian vegetation in protecting and improving chemical water quality in streams. Journal of the American Water Resources Association 46 (2), 261.

Drapper, D., Hornbuckle, A., February 2015. Field monitoring of a stormwater treatment train with pit baskets and filter media cartridges: a review of protocol criteria relating to a treatment train at a townhouse development in South-east Queensland. Water (AWA) 1–6.

Dudgeon, S., 2015. High Efficiency Sediment Basins Could Reduce Stormwater Quality Treatment Infrastructure Costs by $5.4 Billion in South East Queensland over the Next 30 Years.

Egodawatta, P., Mcgree, J., Wijesiri, B., Goonetilleke, A., 2014. Compatibility of stormwater treatment performance data between different geographical areas. Water (AWA) 41, 53–57.

Eisenberg, B., Lindow, K.C., Smith, D.R.E.D.S., 2015. Permeable Pavements. American Society of Civil Engineers.

enHealth, 2004. Guidance on Use of Rainwater Tanks. Commonwealth of Australia.

eWater, 2017. www.ewater.org.au (for latest product documentation on MUSIC).

Fisher, J., Acreman, M.C., 2004. Water quality functions of wetlands. Hydrology and Earth System Sciences 8 (4), 673–685.

Fletcher, T., Zinger, Y., Deletic, A., Bratières, K., 2007. Treatment efficiency of biofilters; results of a large-scale column study. In: Rainwater and Urban Design 2007. Engineers Australia, pp. 266–273.

Fwa, T.F., Lim, E., Tan, K.H., 2015. Comparison of permeability and clogging characteristics of porous asphalt and pervious concrete pavement materials. Transportation Research Record: Journal of the Transportation Research Board 2511, 72–80.

Gardner, T., 2015. WSUD: has the doing exceeded the knowing? Water (AWA) 42 (4), 4–6.

Hall, T., 2010. The Life and Death of the Australian Backyard. CSIRO Publishing.

Han, M.Y., Mun, J.S., 2011. Operational data of the Star city rainwater harvesting system and its role in climate change adaption and a social influence. Water Science and Technology 63, 2796–2801.

Hatt, B.E., Fletcher, T.D., Deletic, A., 2009a. Pollutant removal performance of field-scale stormwater bioretention systems. Water Science and Technology 59 (8), 1567–1576. https://doi.org/10.2166/wst.2009.173.

Hatt, B.E., Fletcher, T.D., Deletic, A., 2009b. Hydrologic and pollutant removal performance of stormwater biofiltration systems at the field scale. Journal of Hydrology 365, 310–321.

Hawley, R.J., Vietz, G.J., 2015. Addressing the urban stream disturbance regime. Freshwater Science 35 (1), 278–292.

Hoban, A.T., 2017. Facing the MUSIC: a review of bioretention performance. In: 2017 Joint IECA National Conference and Stormwater Queensland Conference, 11–12 Oct 2017 Brisbane.

Hoban, A.T., Breen, P.F., Wong, T.H.F., 2006. Relating water level variation to vegetation design in constructed wetlands. In: Fletcher, T., Deletic, A. (Eds.), 4th International Conference on Water Sensitive Urban Design, 2006 Melbourne.

Hunt, W.F., Smith, J.T., Jadlocki, S.J., Hathaway, J.M., Eubanks, P.R., 2008. Pollutant removal and peak flow mitigation by a bioretention cell in urban Charlotte, NC. Journal of Environmental Engineering 134 (5), 403–408.

Hunter, G., 2001. Stormwater Quality Improvement Devices Issues for Consideration. IPWEA Conference.

Kelly, C., Bardak, A., 2015. A statistical approach to assess stormwater treatment device performance data. Water (AWA) 42, 61–67.

Kelly, J.F., Donegan, P., 2015. City Limits: Why Australia's Cities are Broken and How We Can Fix Them. Univ. Publishing, Melbourne.

Liu, A., Goonetilleke, A., Egodawatta, P., 2012. Taxonomy for rainfall events based on pollutant wash-off potential in urban areas. Ecological Engineering 47, 110–114.

Lloyd, S.D., Wong, T.H.F., Porter, B., 2002. The planning and construction of an urban stormwater management scheme. Water Science and Technology 45 (7), 1–10.

Lucas, R., Earl, E.R., Babatunde, A.O., Bockelmann-Evans, B.N., 2015. Constructed wetlands for stormwater management in the UK: a concise review. Civil Engineering and Environmental Systems 32 (3), 251–268.

Lucke, T., Beecham, S., 2011. Field investigation of clogging in a permeable pavement system. Building Research and Information 39, 603–615.

Lucke, T., Nichols, P., 2015. The pollution removal and stormwater reduction performance of street-side bioretention basins after ten years in operation. Science of the Total Environment 536, 784–792.

Lucke, T., Drapper, D., Hornbuckle, A., 2018. Urban stormwater characterisation and nitrogen composition from lot-scale catchments — new management implications. Science of the Total Environment 619–620 (2018), 65–71.

Mangangka, I., Liu, A., Egodawatta, P., Goonetilleke, A., 2015a. Performance characterisation of a stormwater treatment bioretention basin. Journal of Environmental Management 150, 173–178.

Mangangka, I.R., Liu, A., Egodawatta, P., Goonetilleke, A., 2015b. Sectional analysis of stormwater treatment performance of a constructed wetland. Ecological Engineering 77, 172–179.

McIntosh, B.S., Aryal, S., Ashbolt, S., Sheldon, F., Maheepala, S., Gardner, T., Chowdhury, R., Gardiner, R., Hartcher, M., Pagendam, D., Hodgson, G., Hodgen, M., Pelzer, L., 2013. Urbanisation and Stormwater Management in South East Queensland - Synthesis and Recommendations. Urban Water Security Research Alliance Technical Report No. 106.

Moglia, M.G., Tjandraatmadja, G., Sharma, A.K., 2013. Exploring the need for rainwater tank maintenance: survey, review and simulations. Water Science and Technology: Water Supply 13, 191–201.

Mullaney, J., Lucke, T., 2014. Practical review of pervious pavement designs. CLEAN – Soil, Air, Water 42, 111–124.

National Water Commission, 2010. What Is Water Sensitive Urban Design Webpage. http://webarchive.nla.gov.au/gov/20110225204208/http://www.nwc.gov.au/www/html/216-water-sensitive-urban-design.asp?intSiteID=1. www.nwc.gov.au.

Nichols, P., Lucke, T., 2016. Evaluation of the long-term pollution removal performance of established bioretention cells. GEOMATE 11, 22363–22369.

NSW EPA, 2018. http://www.epa.nsw.gov.au/your-environment/recycling-and-reuse/warr-strategy/the-waste-hierarchy.

Parker, N., 2010. Assessing the Effectiveness of Water Sensitive Urban Design in Southeast Queensland (M.Eng thesis). Faculty of Built Environment and Engineering, Queensland University of Technology.

Parker, N., Gardner, T., Goonetilleke, A., Egodawatta, P., Giglio, D., 2009. Effectiveness of WSUD in the 'real world'. In: The 6th International Water Sensitive Urban Design Conference and the 3rd Hydropolis: Towards Water Sensitive Cities and Citizens, Perth, Western Australia. Available at: http://eprints.qut.edu.au/26520/.

Quigley, M., Brown, B., 2014. Transforming Our Cities: High-Performance Green Infrastructure. Water Environment Research Foundation, Alexandria, VA. ISBN: 978-1-78040-673-2/1-78040-673-8, 96 p.

Russell, K.L., Vietz, G.J., Fletcher, T.D., 2017. Global sediment yields from urban and urbanizing watersheds. Earth-Science Reviews 168, 73–80.

Sansalone, J., Kuang, X., Ying, G., Ranieri, V., 2012. Filtration and clogging of permeable pavement loaded by urban drainage. Water Research 46 (20), 6763–6774.

Sharma, A.K., Begbie, D., Gardner, T., 2015. Rainwater Tank Systems for Urban Water Supply – Design, Yield, Energy, Health Risks, Economics and Community Perceptions. IWA Publishing. http://www.iwapublishing.com/books/9781780405353/rainwater-tank-systems-urban-water-supply.

Sharma, A.K., Cook, S., Gardner, T., Tjandraatmadja, G., 2016. Rainwater tanks in modern cities: a review of current practices and research. Journal of Water and Climate Change 7 (3), 445–466. https://doi.org/10.2166/wcc.2016.039.

Stewart, T.W., Downing, J.A., 2008. Macroinvertebrate communities and environmental conditions in recently constructed wetlands. Wetlands 28 (1), 141–150.

Suki, J.P.S., Yates, C.R., Tanner, C.C., 2010. Assessment of Floating Treatment Wetlands for Remediation of Eutrophic Lake Waters - MAero Stream.

Tanner, C.C., Headley, T.R., 2011. Components of floating emergent macrophyte treatment wetlands influencing removal of stormwater pollutants. Ecological Engineering 37 (3), 474–486.

Terstriep, M.L., 1982. Hydrologic Design of Impounding Reservoirs in Illinois. Bulletin (Illinois State Water Survey) No. 67.

Trowsdale, Simcock, 2011. Urban stormwater treatment using bioretention. Journal of Hydrology 397 (2011), 167–174.

USEPA, 2002. Urban Stormwater BMP Performance Monitoring – A Guidance Manual for Meeting the National Stormwater BMP Database Requirements, April 2002.

Vieritz, A.M., Neumann, L.E., Cook, S., 2015. Rainwater tank modelling. In: Sharma, A.K., Begbie, D., Gardner, T. (Eds.), Rainwater Tank Systems for Urban Water Supply - Design, Yield, Energy, Health risks, Economics and Community perceptions. IWA Publishing. ISBN13: 9781780405353.

Walker, T.A., Allison, R.A., Wong, T.H.F., Wootton, R.M., 1999. Removal of Suspended Solids and Associated Pollutants by a CDS Gross Pollutant Trap. Cooperative Research Centre for Catchment Hydrology.

Water By Design, 2009. Concept Design Guidelines for Water Sensitive Urban Design. South East Queensland Healthy Waterways Partnership, ISBN 978-0-9806278-1-7.

Winston, R.J., Hunt, W.F., Kennedy, S.G., Merriman, L.S., Chandler, J., Brown, D., 2013. Evaluation of floating treatment wetlands as retrofits to existing stormwater retention ponds. Ecological Engineering 54, 254–265.

Witheridge, G., 2012. Principles of Construction Site Erosion and Sediment Control. © Catchments & Creeks Pty. Ltd., Brisbane, Queensland. www.catchmentsandcreeks.com.au.

Yu, S.L., Kuo, J.T., Fassman, E.A., Pan, H., 2001. Field test of grassed-swale performance in removing runoff pollution. Journal of Water Resources Planning and Management 127 (3), 168–171.

## FURTHER READING

Al-Hamdan, A.Z., Nnadi, F.N., Romah, M.S., 2007. Performance reconnaissance of stormwater proprietary best management practices. Journal of Environmental Science and Health, Part A 42 (4), 427–437.

Auckland Regional Council, 1999. Technical Publication 90 – Erosion & Sediment Control Guidelines for Land Disturbing Activities in the Auckland Region. http://www.aucklandcouncil.govt.nz/EN/planspoliciesprojects/reports/technicalpublications/Pages/technicalpublications51-100.aspx.

Bartley, R., Henderson, A., Wilkinson, S., Whitten, S., Rutherford, I., 2015. Stream Bank Management in the Great Barrier Reef Catchments: A Handbook. CSIRO, p. 80.

Biermann, S., Sharma, A., Chong, M.N., Umapathi, S., Cook, S., 2012. Assessment of the Physical Characteristics of Individual Household Rainwater Tank Systems in South East Queensland. Urban Water Security Research Alliance Technical Report No. 66. http://www.urbanwateralliance.org.au/publications/technicalreports/index.html.

Burns, M.J., Fletcher, T.D., Duncan, H.P., Hatt, B.E., Ladson, A.R., Walsh, C.J., 2015. The performance of rainwater tanks for stormwater retention and water supply at the household scale: an empirical study. Hydrological Processes 29 (1), 152–160.

Burton, A., Pitt, R., 2015. Stormwater Effects Handbook: A Toolbox for Watershed Managers, Scientists and Engineers. Available Online: http://unix.eng.ua.edu/~rpitt/Publications/BooksandReports/Stormwater%20Effects%20Handbook%20by%20%20Burton%20and%20Pitt%20book/hirezhandbook.pdf.

Dierkes, C., Lucke, T., Helmreich, B., 2015. General technical approvals for decentralised sustainable urban drainage systems (SUDS) - the current situation in Germany. Sustainability 2015 (7), 3031–3051. https://doi.org/10.3390/su7033031.

Duncan, H.P., 1999. Urban Stormwater Quality: A Statistical Overview. Cooperative Research Centre for Catchment Hydrology, Melbourne, Australia (Report 99/3).

Geosyntec Consultants, Wright Water Engineers, Inc., 2009. Urban Stormwater BMP Performance Monitoring, October 2009.

Hatt, B.E., Fletcher, T.D., Deletic, A., 2008. Hydraulic and pollutant removal performance of fine media stormwater filtration systems. Environmental Science and Technology 42, 2535–2541.

Hoban, A.T., 2015. Then and Now. Stormwater Australia Bulletin, 218 ed. Stormwater Australia.

Hunt, W., Jarrett, A.R., Smith, J.T., Sharkey, L.J., 2006. Evaluating bioretention hydrology and nutrient removal at three field sites in North Carolina. Journal of Irrigation and Drainage Engineering 132, 600–608.

Li, L., Davis, A., 2014. Urban stormwater runoff nitrogen composition and fate in bioretention systems. Environmental Science and Technology 48, 3403–3410.

Mckenzie-McHarg, A., Smith, N., Hatt, B., 2008. Stormwater Gardens to Improve Urban Stormwater Quality in Brisbane.

Nichols, P., Lucke, T., Drapper, D., 2015. Field and evaluation methods used to test the performance of a stormceptor class 1 stormwater treatment device in Australia. Sustainability 7, 16311–16323. https://doi.org/10.3390/su71215817.

Peljo, L., Dubowski, P., Dalrymple, B., 2016. The Performance of Streetscape Bioretention Systems in South East Queensland. Stormwater 2016.

Stormwater Australia, 2014. Evaluation Protocol for Stormwater Quality Treatment Devices (SQIDEP). Consultation Draft Release Nov 2014.

Thompson, R., Parkinson, S., 2011. Assessing the local effects of riparian restoration on urban streams. New Zealand Journal of Marine and Freshwater Research 45, 625–636.

Vaze, J., Chiew, F., 2004. Nutrient loads associated with different sediment sizes in urban stormwater and surface pollutants. Journal of Environmental Engineering 130, 391–396.

Washington Department of Ecology, 2011. Technical guidance manual for evaluating emerging stormwater treatment technologies. In: Technology Assessment Protocol – Ecology (TAPE), Publication No. 02-10-037, August 2011.

Water by Design, 2010. MUSIC Modelling Guidelines. SEQ Healthy Waterways Partnership, Brisbane, Queensland, ISBN 978-0-9806278-4-8.

Wong, T.H.F., Engineers Australia, National Committee on Water, 2006. Australian Runoff Quality: A Guide to Water Sensitive Urban Design. Engineers Media, Crows Nest, NSW.

Wong, G., Ansen, J., Fassman, E., 2012. Proprietary Devices Evaluation Protocol for Stormwater Treatment. Auckland Council Guideline Document GD003.

Zinger, Y., Blecken, G.T., Fletcher, T.D., Viklander, M., Deletic, A., 2013. Optimising nitrogen removal in existing stormwater biofilters: benefits and tradeoffs of a retrofitted saturated zone. Ecological Engineering 2013 (51), 75–82.

Chapter 3

# Stormwater Quality, Pollutant Sources, Processes, and Treatment Options

Ashantha Goonetilleke[1] and Jane-Louise Lampard[2]
[1]Queensland University of Technology, Brisbane, QLD, Australia; [2]University of the Sunshine Coast, Sippy Downs, QLD, Australia

## Chapter Outline



**ABSTRACT**

Increasing urbanization is a common phenomenon around the world, which results in the conversion of previously vegetated areas into impervious surfaces. Natural sources and anthropogenic activities common to urban areas deposit a range of physical, chemical, and microbial pollutants on these impervious surfaces. These include, gross pollutants, sediments, nutrients, oxygen-demanding waste, metals, hydrocarbons, and microbial pollutants. These pollutants undergo continuous physical and chemical transformations after deposition, which will result in changes to their toxicity, mobility, and bioavailability. These transformations are due to factors such as abrasion by vehicle tyres, resuspension by wind turbulence, and exposure to atmospheric moisture and photolysis. Pollutant deposition or build-up and consequent pollutant wash-off with stormwater runoff results in the discharge of pollutants to receiving waters, often exceeding their assimilation capacity. Additionally, the presence of impervious surfaces and hydraulically efficient stormwater conveyance systems result in the discharge of increased stormwater runoff volumes. The quality and quantity impacts on stormwater due to urbanization are influenced by a range of factors, including, urban form, land-use activities, traffic characteristics, and climate characteristics. These impacts pose a threat to human and ecosystem health. Mitigation of these adverse impacts is essential for improving receiving water quality and thereby human well-being. In the context of safeguarding urban water ecosystems, Water Sensitive Urban Design (WSUD) offers an effective solution where a range of devices are available for targeting specific pollutants. However, the design of WSUD devices is constrained by the lack of understanding of how natural and anthropogenic factors influence the underlying mechanisms of pollutant export, transformation, and removal. Furthermore, stormwater pollutant processes and treatment

Approaches to Water Sensitive Urban Design. https://doi.org/10.1016/B978-0-12-812843-5.00003-4

performance of WSUD devices are expected to be significantly influenced by the predicted impacts of climate change. Consequently, it is important to quantitatively assess the changes to stormwater quantity and quality, and the resulting performance of WSUD devices to ensure adaptability to a changing climate.

**Keywords:** Stormwater pollutant processes; Stormwater pollutants; Stormwater quality; Stormwater treatment; Urbanization; Water Sensitive Urban Design.

## 3.1 INTRODUCTION

The increasing conversion of rural land use into urban land use is a common phenomenon in most parts of the world because of perceived benefits of urban living as opposed to rural living. Urbanization involves the outward expansion of population centers beyond their original limits to accommodate a growing population. The United Nations have projected that the global population living in urban areas will reach 66% by 2050 (UNDESA, 2014). Urbanization results in irrevocable changes to the landscape, a shift in demographic patterns, and economic, social, and environmental impacts on a region. Among these transformations, the spread of the built environment and increase in anthropogenic activities common to urban areas can result in significant pollutant inputs to urban receiving waters, thereby degrading water quality (Jacobson, 2011; Miller et al., 2014). This in turn can pose risks to human and ecosystem health (Bocca et al., 2004; Hamers et al., 2002). As such, mitigating stormwater pollution is a major requirement in urban water management for enhancing urban liveability.

Stormwater pollution is inherently complex, posing significant technical challenges in the design of effective mitigation measures. An in-depth understanding of the factors and processes that influence the pollutant characteristics of stormwater is essential for reducing the uncertainty in choosing appropriate stormwater pollution mitigation measures (WWAP, 2012; Wijesiri et al., 2016). This chapter discusses the types and sources of stormwater pollutants, pollutant processes, the chemical and microbial characteristics of stormwater, the impacts of these pollutants, potential impacts of climate change, and commonly adopted stormwater pollution mitigation approaches.

## 3.2 IMPACTS OF URBANIZATION ON AQUATIC ECOSYSTEMS

As a consequence of transforming the natural environment into the built environment, vegetated lands are replaced by impervious surfaces such as roads, parking lots, and rooftops. The increase in impervious areas reduces the volume of rainfall infiltration during storms, resulting in increased volume of stormwater runoff compared to previously vegetated lands. Greater uniformity in the slope of impervious surfaces, and reduced hydraulic roughness, results in increased runoff velocity (Jacobson, 2011; Marsalek et al., 2007). These changes are reflected in the shape of the runoff hydrograph with increased peak flows, reduced lag time, and reduced baseflow (Wen et al., 2014).

As the fraction of impervious surfaces within a catchment increases, a relatively larger portion of streamflow is delivered by stormwater runoff rather than baseflow. Additionally, groundwater recharge is reduced as a result of reduced stormwater infiltration. This leads to the decrease in baseflow levels in streams during longer dry weather periods (Bell et al., 2016). During dry weather periods, significant pollutant loads can accumulate on urban impervious surfaces, and the accumulated pollutants are subsequently mobilized and entrained in runoff during storm events, and transported to receiving waters. The transport of pollutants is enhanced because of the increase in stormwater runoff volume and flow velocity, and the improved drainage system, thereby degrading the quality of urban receiving waters (Meyer et al., 2005).

Consequently, the physical, chemical, and microbial quality of urban receiving waters changes because of the discharge of stormwater runoff carrying a range of pollutants. These pollutants undergo continuous physical and chemical transformations. These transformations are due to factors such as abrasion by vehicle tyres, resuspension by wind turbulence, and exposure to atmospheric moisture and photolysis (Gunawardana et al., 2012a,b; Jayarathne et al., 2017, 2018).

Urban stormwater runoff contains a substantial amount of particulate solids primarily contributed by roadside soil. Additionally, automobile-use activities and abrasion products generated from different impervious surfaces, such as asphalt and concrete, also produce particulate solids (Gunawardana et al., 2012a,b; Hvitved-Jacobsen et al., 2010; Mummullage et al., 2016a,b). In addition to increasing the turbidity and sedimentation in receiving water bodies, biologically active suspended solids can result in low dissolved oxygen levels and reduced photosynthesis, which directly affect aquatic fauna (Erickson et al., 2013).

A range of research studies have noted the important role that solids play in the transport of other pollutants which are adsorbed to particulate surfaces. These can be toxic and adversely impact flora and fauna diversity in urban water ecosystems (Beach, 2005; Duong and Lee, 2011; Gunawardana et al., 2012a). These toxicants are primarily heavy metals and hydrocarbons such as polycyclic aromatic hydrocarbons (PAHs) (Brown and Peake, 2006; Gobel et al., 2007; Herngren et al., 2005).

Organic matter deposited on urban surfaces can be found in two forms: as organic macropollutants such as vegetation debris and as organic micropollutants emitted as dust particles from combustion systems (Gobel et al., 2007). Once discharged and deposited in urban receiving waters, organic matter is subject to decomposition through microbial action. This contributes to the depletion of dissolved oxygen in the waterbody, thereby increasing risks to species diversity.

Similar to organic matter, the enrichment of nutrients such as nitrogen and phosphorus in urban waters can also contribute to the reduction in dissolved oxygen because of the occurrence of algal blooms and nuisance weeds, which leads to the decrease in the aquatic species population (Heisler et al., 2008; Oliver and Boon, 1992). The main sources of nutrients are fertilizer applied to lawns, industrial discharges, detergents, animal waste, and septic tank and sewerage system leakages (Chiew et al., 1997). Human and animal waste contributes to microbial contamination of receiving waters because of the presence of pathogenic organisms (Jochimsen et al., 1998; NHMRC and NRMMC, 2011).

## 3.3 STORMWATER POLLUTANTS AND SOURCES

Pollutants in stormwater originate from natural sources and anthropogenic activities. Vegetation debris and roadside soils, as discussed in Section 3.2, are the most frequent sources of naturally occurring stormwater pollutants, which include suspended solids/sediments, metals, nutrients, oxygen-demanding waste (decomposing organic matter), and microorganisms. In addition to containing minerals and metals inherently associated with the specific geology of a catchment, sediments originating from anthropogenic activities bind with pollutants and act as a mobile substrate in transporting these pollutants to receiving waters.

Fecal matter from humans and animals contribute to nutrient and microbial loads present in urban stormwater. Pathogens in stormwater originate from human and animal feces, litter, and resuspended sediment (Schiff and Kinney, 2001; Arnone and Walling, 2007). Human sewage is considered to be the pollutant source of most concern to human health (Gaffield et al., 2003; Sauer et al., 2011; Sidhu et al., 2013). However, domestic and wild animals can also be reservoirs of microorganisms that are pathogenic to humans. For example, dogs, cats, squirrels, and possums are reservoirs of *Cryptosporidium parvum* (Atwill et al., 2001; Ahmed et al., 2012; Ryan et al., 2014), dogs and cats are also reservoirs of *Giardia duodenalis* (Monis et al., 2009). The range of animal feces present in stormwater is influenced by the geographic location, the ratio of green space to the built environment, housing density and style, and local laws relating to domestic pet ownership, and waste recovery and disposal.

Microbial indicators, human biomarkers, and human pathogenic bacteria, protozoa, and viruses have been detected in stormwater from urban areas. *Enterococci*, *Escherichia coli*, and thermotolerant coliforms found in the feces of warm blooded animals, including humans, are frequently used indicators of fecal contamination in stormwater because of the low cost and speed of detection methods. The presence of fecal indicator bacteria in stormwater can provide a trigger to investigate potential sewage ingress into stormwater. However, it does not necessarily infer the presence of human pathogens. Although animals can be reservoirs of zoonotic pathogens, many microorganisms associated with animals are not capable of initiating disease in humans (Monis et al., 2009; Ryan et al., 2014).

The range of pollutants entering stormwater from anthropogenic sources varies across and within catchments. Factors influencing the range of pollutants present include the design of drainage infrastructure; the mixture of land-use practices (e.g., industrial, commercial, residential or parkland); the intensity of each land-use practice; road and building materials used; the range of transport systems and traffic density; the percentage of impervious surfaces; human behavior; and waste disposal practices (Goonetilleke et al., 2005; Brown and Peake, 2006; Gobel et al., 2007). Common stormwater pollutants originating from anthropogenic practices include hydrocarbons and surfactants, metals, insecticides, herbicides, pharmaceuticals and personal care products, oxygen demanding materials, nutrients, sewage, and litter (Makepeace et al., 1995; Sidhu et al., 2012; Ma et al., 2016; Gernjak et al., 2017).

The most significant factor influencing the presence of sewage in stormwater is the design and age of the sewer infrastructure within an urban area. Sewer design in urban precincts varies internationally and is influenced by topography, climate, and the technology available at the time the area was urbanized. For example, in Australia, stormwater and sewage are conveyed separately. However, in many parts of Europe and the United States, stormwater and sanitary waste are

**TABLE 3.1 Typical Stormwater Pollutants and Their Primary Sources**

| Pollutant | Primary Sources |
|---|---|
| Litter | Roadside vegetation, waste disposal practices |
| Suspended solids/sediments | Roadside soil, traffic activities, construction, and demolition activities |
| Oxygen-demanding waste | Roadside vegetation, industrial activities |
| Nutrients | Roadside vegetation, fertilizer application |
| Hydrocarbons | Traffic and industrial activities |
| Metals | Traffic and industrial activities, roadside soil |
| Microbial matter | Human and animal waste |

combined. Sewage contains chemical compounds and microorganisms that are potentially hazardous to human health. Pharmaceuticals, recreational drugs, and other chemicals such as caffeine and artificial sweeteners enter sewer systems via human excretion or intentional flushing of the compounds down toilets as a disposal method. Ingress of human sewage into stormwater in catchments with separate sewers occurs because of cross-connections, damage to sewerage infrastructure which allows stormwater ingress, sewer overflows during wet weather, or illegal dumping of human waste. Enteric pathogens, pharmaceuticals, and caffeine have been detected in stormwater not only from catchments where stormwater and sewage are combined (as expected) but also in catchments where they are separated (Sidhu et al., 2012; de Man et al., 2014; Gernjak et al., 2017).

In terms of other stormwater pollutants, the primary sources include vehicular traffic, roadside soil and vegetation, and industrial activities (Gunawardana et al., 2012a; Adachi and Tainosho, 2004; Xie et al., 2000; Jullien and François, 2006). Emissions and abrasion wear products are the most important traffic-related pollutants. These can be categorized as follows:

- exhaust emissions;
- fuel and lubrication leakages;
- vehicle component wear, including tires, brakes, and chassis; and
- pavement wear.

Table 3.1 provides a summary of typical stormwater pollutants and their primary sources.

## 3.4 INFLUENCE OF CATCHMENT AND RAINFALL CHARACTERISTICS ON STORMWATER QUALITY AND QUANTITY

Stormwater quality and quantity characteristics vary across catchments, influenced by their inherent natural and built characteristics. Natural elements of a catchment influencing stormwater quality and quantity include rainfall patterns (frequency, intensity, duration); landscape and land-use characteristics, and soil characteristics. The volume and flow rate of runoff entering a stormwater drainage system is also influenced by the range of surfaces and the percentage of impervious surfaces such as sealed roads, roofs, and driveways in a catchment (Eriksson et al., 2007). Land-use practices are an acknowledged influencer of stormwater quality. However, to date, the degree of influence resulting from different land uses remains poorly understood. A review of stormwater studies before 1999 found lead (Pb), biochemical oxygen demand, chemical oxygen demand, total coliforms, fecal coliforms, and fecal streptococci counts to be greater in high-density urban catchments compared with low-density urban catchments (Duncan, 1999). Higher concentrations of metals have been observed in stormwater from industrial catchments and catchments with mining (sand and clay) activities, than in predominantly commercial or residential catchments (Tiefenthaler et al., 2008; Page et al., 2013). Statistically significant positive relationships have been observed between population density and bacterial counts (Duncan, 1999; Selvakumar and Borst, 2006).

Goonetilleke et al. (2005) noted that urban form, rather than land use, was the primary factor influencing urban stormwater quality. For example, detached single dwellings with gardens were found to contribute higher pollutant loads to stormwater than multi-family dwellings, because of relatively larger landscaped areas and high percentages of road surface area, even though both land uses are classified as residential. Liu (2011), in an in-depth investigation into the influence of rainfall and catchment characteristics on urban stormwater quality found that in addition to conventional catchment characteristics (i.e., land use and impervious surface fraction), other factors such as impervious area layout, urban form, and site specific characteristics are also important in pollutant processes such as build-up and wash-off.

Liu et al. (2016) further noted that in conventional stormwater treatment design, stormwater quality is considered as a stochastic variable and not related to the characteristics of the rainfall events. This approach can reduce the effectiveness of a treatment system as stormwater quality can also be influenced by the rainfall characteristics (Liu et al., 2012a). Based on the suspended solids loads generated, Liu (2011) classified rainfall events into high average intensity-short duration (Type 1), high average intensity-long duration (Type 2), and low average intensity-long duration (Type 3). It was found that cumulatively Type 1 rainfall events generate a greater fraction of the annual suspended solids load compared with the other rainfall types. Based on a modeling study of three urban catchments in the Gold Coast, Australia, these outcomes were further confirmed by Liu et al. (2016). Type 1 events were found to produce 58% of the total solids load, but generated only 29% of the total runoff volume. Furthermore, rainfall events smaller than 6-month average recurrence interval (ARI) were found to generate 68% of the total annual runoff volume and 69% of total solids load exported compared with rainfall events larger than 6-month ARI. The study further found that sizing treatment systems for a 6-month ARI will treat 70% of the annual runoff and total suspended solids load, which can be considered as a good outcome for not over-capitalizing on treatment system size. In the case of the catchments investigated in the study noted above, the threshold levels for high average intensity-short duration were 31 mm/h and 0.4 h. However, it is important to note that the threshold levels for the three rainfall categories can be location-specific.

## 3.5 POLLUTANT PROCESSES

Stormwater pollutants undergo two fundamental processes: build-up and wash-off. Furthermore, the pollutants are also subject to re-distribution as a result of resuspension, aggregation, and re-deposition during build-up and wash-off (Fig. 3.1). These processes are influenced by a range of natural and anthropogenic factors. In the context of stormwater management, it is necessary to understand the underlying mechanisms and factors influencing pollutant processes in order to design effective pollution mitigation strategies.

### 3.5.1 Pollutant build-up

Pollutant accumulation on urban surfaces is driven by the rate of deposition, length of the antecedent dry period, and redistributional processes such as re-suspension, aggregation, and re-deposition. The build-up process is influenced by a

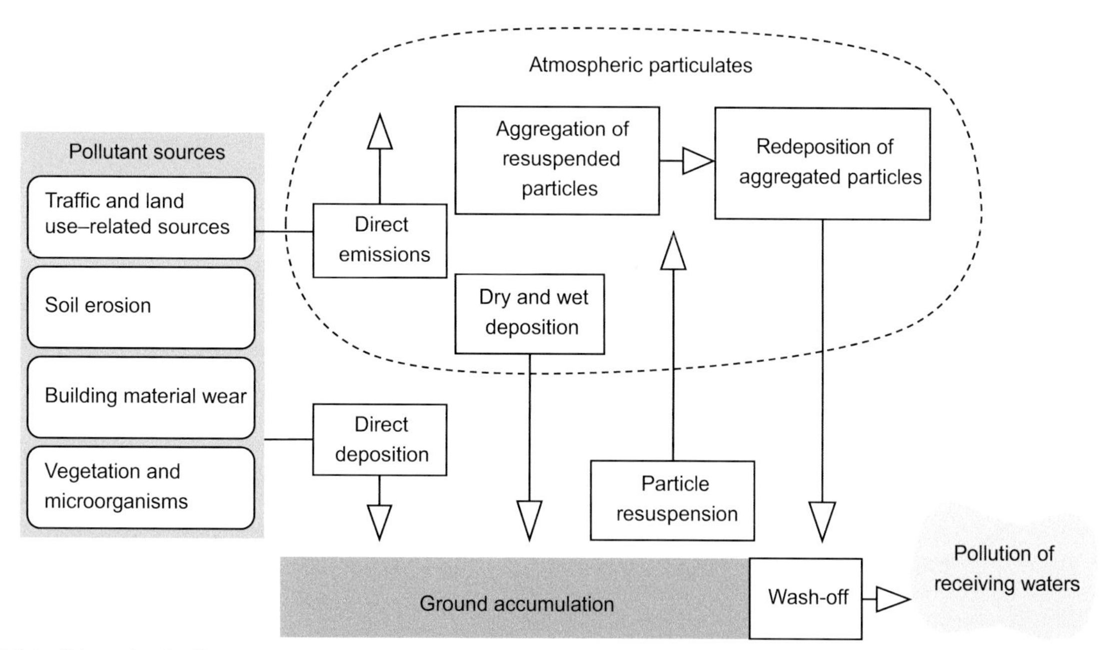


**FIGURE 3.1** Schematic of pollutant processes. *Adapted from Mummullage, S.W.N., 2015. Source Characterisation of Urban Road Surface Pollutants for Enhanced Water Quality Predictions (Ph.D.). Queensland University of Technology.*

range of natural (e.g., climatic) and anthropogenic factors (e.g., traffic volume and congestion, speed, type of automobile use, and land use). It is also important to note that some of these factors are interrelated. For example, automobile-use activities can be specific to land use in a given area. For example, a relatively high fraction of diesel-operated heavy-duty vehicles is present in industrial and commercial areas, whereas petrol-operated light-duty vehicles is present most frequently in residential areas.

Although build-up of particulate solids is influenced by land use, vehicular traffic influences the buildup of particle-bound pollutants such as toxic heavy metals and hydrocarbons (Mummullage et al., 2016a,b). Moreover, the physical and chemical characteristics of pollutants can change as pollutants from different sources interact through physical processes such as aggregation, and chemical reactions such as chemisorption and ion exchange. Consequently, the affinity of pollutants such as heavy metals and hydrocarbons to particulate solids changes with time, resulting in variations in the load and composition of pollutants during build-up (Gunawardana et al., 2015; Ziyath et al., 2016; Jayarathne et al., 2017, 2018).

Potential changes to chemical characteristics of pollutants can occur from photolysis, oxidation–reduction, and hydrolysis reactions, with the significance depending on the length of the antecedent dry period. As the pollutants are exposed to light (e.g., infrared, visible, ultraviolet) over a period of time, the molecules adsorb the energy from the light and transform into different pollutants. Fig. 3.2 illustrates the potential transformations of a PAH molecule. Moreover, the excited molecules (molecules with elevated energy) of a particular pollutant species can react with the molecular forms of other pollutants and transform into different pollutants (Miller and Olejnik, 2001; Pirjola et al., 2012).

In past research studies, mathematical modeling has been employed to better understand the underlying mechanisms of pollutant buildup, and how natural and anthropogenic factors influence process mechanisms, with a view to predicting stormwater quality. The process models developed in these studies are commonly based on the temporal variations in pollutant build-up. Among several phenomenological models (e.g., linear, power, exponential, reciprocal, and hyperbolic), the power function (Eq. 3.1) is widely used in stormwater quality modeling. The empirical coefficients of this equation are found to be influenced by a range of factors such as pollutant type, particle size, and characteristics of impervious surfaces (Ball et al., 1998; Hvitved-Jacobsen et al., 2010). It is important to note that these basic phenomenological models do not

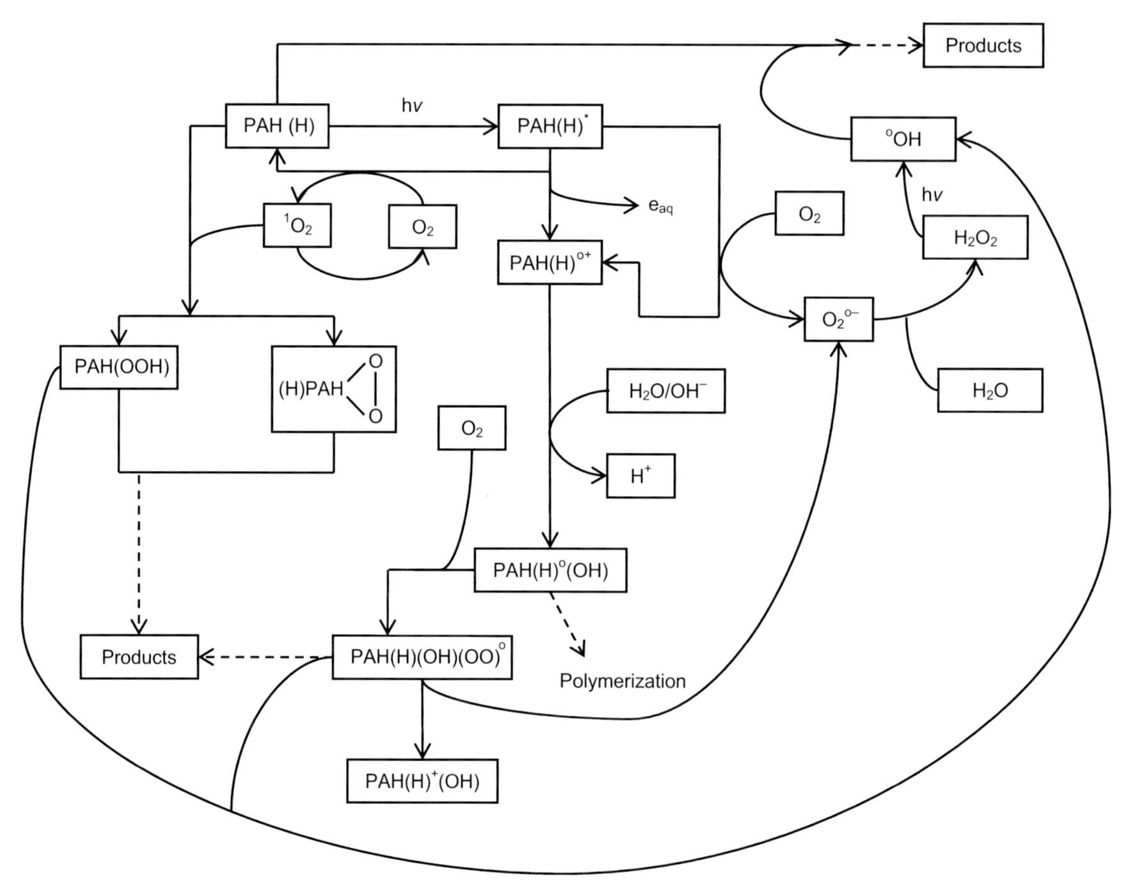


FIGURE 3.2 Pathways of the photodegradation of polycyclic aromatic hydrocarbons (PAHs) in the $O_2/H_2O$ system. *Adapted from Miller, J.S., Olejnik, D., 2001. Photolysis of polycyclic aromatic hydrocarbons in water. Water Research 35 (1), 233–243.*

accurately replicate the characteristics of the build-up process. Build-up process variability is one such characteristic that must be accounted for in modeling to accurately predict stormwater quality for the design of effective stormwater pollution mitigation strategies (Wijesiri et al., 2015a).

$$B = \alpha t^{\beta} \tag{3.1}$$

where: B—buildup load (g/m$^2$), t—antecedent dry days, $\alpha$ and $\beta$—empirical buildup coefficients.

### 3.5.2 Pollutant wash-off

The process of mobilizing and transporting particle-bound pollutants accumulated on urban surfaces is referred to as pollutant wash-off. The mobilization of pollutants adhering to surfaces is driven by the kinetic energy of raindrops and the shear stress created by stormwater runoff (Egodawatta et al., 2007). Although rainfall characteristics (intensity, duration, runoff volume, and velocity) play a significant role in influencing the mechanisms of pollutant wash-off, the amount of pollutants available on the surface before the surface runoff event primarily influences the wash-off load (Duncan, 1999; Wijesiri et al., 2015b).

Fig. 3.3 illustrates the two primary concepts in relation to the wash-off process based on the initially available pollutants build-up load: the source-limiting concept and the transport-limiting concept (Vaze and Chiew, 2002). According to the source-limiting concept, a rainfall event will wash-off all pollutants available on a surface, and the consequent build-up event will start from a zero amount of pollutants on the surface. According to the transport-limiting concept, a rainfall event has the capacity to wash-off only a fraction of the initially available pollutant load. Past research studies, including Vaze and Chiew (2002), have confirmed the more frequent occurrence of the transport-limiting phenomenon.

Moreover, Zhao et al. (2016) noted that fine and coarse particles will primarily undergo either the source-limiting or transport-limiting process. During the initial part of a rainfall event, the source-limiting process is found to govern the wash-off of fine particles (i.e., first flush), where only a very limited amount of coarse particles is removed. During the latter part of a rainfall event, where only relatively a limited amount of fine particles remain, the transport-limiting process governs the wash-off of coarse particles. The "first flush" phenomenon relates to the occurrence of a relatively high-pollutant concentration at the early stage of a stormwater runoff event and a concentration peak before the stormwater runoff peak (Alias et al., 2014; Lee et al., 2002). The high concentration can result in "shock" load of pollutants to receiving waters, exceeding the assimilative capacity of a waterbody. Alias et al. (2014) investigated the occurrence of first flush based on different sectors of a runoff "pollutograph." It was found that rainfall intensity and rainfall depth play key roles in determining the pollutant wash-off, with a higher magnitude first flush often occurring during those rainfall events that had higher rainfall intensity during the initial period. Therefore, the design of stormwater treatment devices needs to target specific parts of the runoff hydrograph if they are to address extreme cases, such as first flush.

The empirical exponential wash-off function defined by Eq. (3.2), which does not accurately replicate process variability (similar to the case of buildup), is commonly used in stormwater quality modeling tools (MikeUrban, 2014; Rossman, 2009). Modifications have been introduced to this basic model by Egodawatta et al. (2007) as described by Eq. (3.3), to account for the capacity of a storm event to mobilize pollutants. This required an additional parameter, the

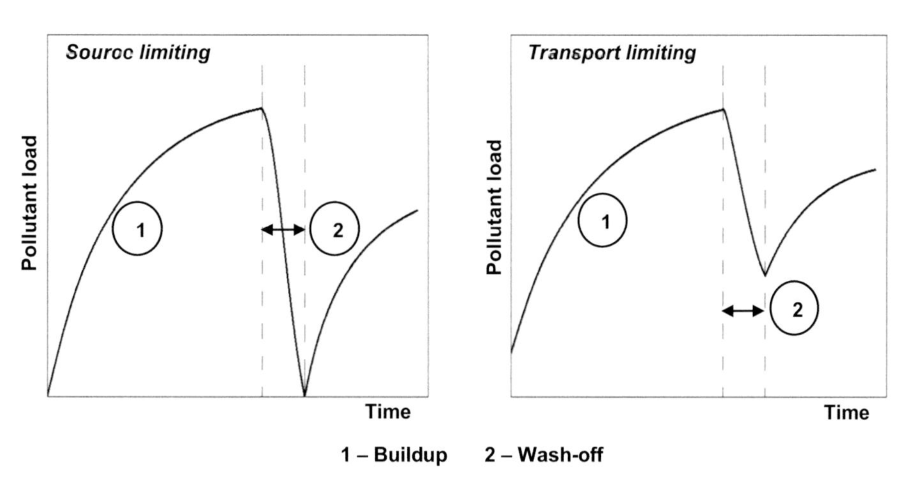


FIGURE 3.3 Conceptual depiction of the source-limiting and transport-limiting concepts in pollutant wash-off (Wash-off occurs in the period between the two *dashed lines*).

"capacity factor" ($C_F$), which is influenced by rainfall intensity, particle size, and the characteristics of the impervious surface. However, further investigations are needed to incorporate new knowledge into stormwater quality modeling tools and to improve the replication accuracy.

$$W = W_0(1 - e^{-kIt}) \tag{3.2}$$

$$F_w = \frac{W}{W_0} = C_F(1 - e^{-kIt}) \tag{3.3}$$

where: W—amount of particles washed off, $W_0$—initially available particulate buildup, $F_w$—fraction wash-off, $C_F$—capacity factor, t—rainfall duration (min), I—rainfall intensity (mm/h), k—wash-off coefficient.

### 3.5.3 Impact of climate change on pollutant processes

The impact of climate change on urban stormwater quality can be viewed in the context of three commonly predicted phenomena: (1) increase in the antecedent dry period between rainfall events; (2) increase in intensity of typical rainfall events in a given area; and (3) decrease in rainfall duration (AGO, 2003; Delpla et al., 2009). Over longer dry periods, pollutants can accumulate on urban surfaces in greater amounts while being subject to greater re-distribution (re-suspension, aggregation, re-deposition) from traffic-generated turbulence and wind, and other climatic factors. Consequently, the resulting changes to particle characteristics can influence the adsorption of other pollutants to particulate solids.

Another impact of an increased antecedent dry period is the chemical transformation of pollutants. Longer exposure to natural and anthropogenic activities will result in changes to pollutant toxicity, mobility, and bioavailability during build-up (Jayarathne et al., 2017, 2018). Consequently, pollutant load and composition will vary over the antecedent dry period, significantly influencing stormwater quality characteristics (Wijesiri et al., 2015a,b).

In relation to pollutant wash-off, the kinetic energy of raindrops plays a major role in pollutant mobilization. Therefore, the predicted increase in high-intensity rainfall events will exert a significant influence on the pollutant wash-off load. Over short-duration rainfall events, finer particles would contribute most to the total wash-off load, thereby increasing the first flush effect (discussed in Section 3.5.2).

## 3.6 CHEMICAL AND MICROBIAL CHARACTERISTICS OF STORMWATER

International stormwater quality data sets include samples collected from a broad range of locations. Research into the quality characteristics of stormwater before the 1980s focused primarily on physicochemical characteristics, heavy metals, and fecal indicator bacteria. Samples were generally collected from where people swim: in fresh or marine waters that have stormwater inflows or the stormwater outlets nearest to the swimming location. In the late 1970s and early 1980s, the focus shifted to understanding the impact of a range of priority pollutants on the ecological health of aquatic environments. Consequently, marine waters, urban creeks, and rivers receiving stormwater were sampled. Increased harvesting and reuse of stormwater runoff since the 1990s to augment drinking water supplies, or for nonpotable end uses such as irrigation and toilet flushing, resulted in an increased focus on understanding the chemical compounds and microorganisms in stormwater which have the potential to adversely impact human health (Fletcher et al., 2008; Hering et al., 2013; Lim et al., 2011; Segal, 2004; Wong et al., 2013). An emerging body of research is investigating stormwater that has not mixed with other waters and hence, has not been subjected to any treatment, microbial transformations, or dilution.

Driven by water scarcity concerns, Singapore and Australia are two of the earliest adopters of stormwater harvesting and reuse schemes for indirect potable and nonpotable reuse. As such, a significant proportion of the stormwater quality data focusing on human health hazards originates from these countries. Microbial studies from 2000 onward have assumed ingestion to be the exposure route of greatest concern during stormwater reuse and have focused heavily on pathogens found in human sewage (AWQC, 2008; Sidhu et al., 2012; Page et al., 2013). However, with the exception of Singapore, the majority of stormwater reuse globally involves nonpotable end uses such as irrigation, road washing, and household supply for gardening or toilet flushing. A small number of studies in Australia and The Netherlands have sought to identify pathogens of concern via inhalation, such as *Legionella pneumophila*, which individuals may be exposed to via aerosols generated during irrigation or water-based recreational activities, involving stormwater (Lampard et al., 2012; de Man et al., 2014).

Limited discernible patterns for stormwater quality have been derived to-date within events at single catchments or between events across different catchments (Page et al., 2013; Howitt et al., 2014; Reeve et al., 2015; Gernjak et al., 2016). Furthermore, the degree of influence exerted by catchment characteristics on pollutant concentrations, and overall stormwater quality, remains poorly understood (Liu et al., 2012b, 2013). An evaluation of Australian data sets for stormwater and waters that receive stormwater undertaken by one of the authors to this publication found that pollutant concentrations in stormwater are difficult to standardize for a range of reasons. Firstly, there are relatively only a small number of data sets for stormwater compared with other water matrixes (e.g., wastewater or groundwater). Secondly, there is notable variability across data sets regarding:

- sample collection methods (grab or composite samples);
- timing of sample collection within an event (first flush, timed intervals, or undescribed collection points across an event);
- range and number of pollutants investigated;
- small number of samples collected per site per data set;
- laboratory analytical methods (e.g., culture-based or polymerase chain reaction, presence/absence or quantification); and
- data reporting format (qualitative or quantitative).

Table 3.2 presents lognormal summary statistics for stormwater quality in Australian urban catchments as published by NRMMC et al. (2009a). It is important to note that for two reasons the data provided in Table 3.2 should be used with caution and primarily for purposes of guidance rather than for drawing definitive conclusions. Firstly, all data relate to catchments with separate sewer and stormwater water infrastructure. Secondly, an appreciable portion of data sets is based on samples from water bodies that receive stormwater rather than from stormwater directly.

The studies by Sidhu et al. (2012), Page et al. (2013), Reeve et al. (2015), and Gernjak et al. (2016) represent extensive investigations of stormwater pollutants in Australian urban catchments. These studies investigated physicochemical parameters and the presence of a broad range of chemicals and microorganisms across catchments of varying sizes, with

**TABLE 3.2 Lognormal Summary Statistics for Untreated Stormwater Quality in Australian Urban Catchments**

| | | | | | | Percentile | | |
|---|---|---|---|---|---|---|---|---|
| **Contaminant** | **Unit** | **Mean** | **SD** | **5th** | **25th** | **50th** | **75th** | **95th** |
| **Pathogens** | | | | | | | | |
| *Campylobacter* (bacteria) | #/100 mL | 3.31 | 1.97 | 1.00 | 1.93 | 2.89 | 4.21 | 7.02 |
| *Cryptosporidium* (protozoa) | #/10 L | 176 | 211 | 12 | 52 | 112 | 222 | 546 |
| *Giardia* (protozoa) | #/10 L | 1.81 | 2.08 | 0.12 | 0.55 | 1.17 | 2.29 | 5.55 |
| **Bacteria—Indicators** | | | | | | | | |
| Coliforms | #/100 mL | 97,665 | 170,197 | 3,369 | 17,668 | 44,884 | 106,860 | 355,988 |
| *Clostridium perfringens* | #/100 mL | 925 | 1,016 | 103 | 315 | 614 | 1,153 | 2,748 |
| *Escherichia coli* | #/100 mL | 59,339 | 71,939 | 3,835 | 17,203 | 37,511 | 74,564 | 184,382 |
| *Enterococci* | #/100 mL | 13,792 | 10,928 | 1,621 | 6,043 | 11,229 | 18,586 | 34,465 |
| Fecal coliforms | #/100 mL | 69,429 | 82,740 | 4,694 | 20,440 | 44,168 | 87,235 | 215,568 |
| Fecal streptococci | #/100 mL | 29,771 | 21,717 | 3,829 | 13,991 | 25,212 | 40,317 | 70,894 |
| Somatic coliphages | #/100 mL | 17,530 | 20,917 | 1,154 | 5,088 | 11,115 | 22,083 | 54,704 |
| **Metals** | | | | | | | | |
| Aluminum | mg/L | 1.19 | 0.60 | 0.49 | 0.78 | 1.07 | 1.47 | 2.29 |
| Arsenic | mg/L | 0.009 | 0.001 | 0.006 | 0.008 | 0.009 | 0.009 | 0.011 |
| Barium | mg/L | 0.028 | 0.005 | 0.021 | 0.025 | 0.028 | 0.031 | 0.038 |
| Cadmium | mg/L | 0.0198 | 0.0242 | 0.0015 | 0.0061 | 0.0127 | 0.0248 | 0.0606 |
| Chromium | mg/L | 0.009 | 0.005 | 0.002 | 0.005 | 0.008 | 0.011 | 0.017 |
| Copper | mg/L | 0.055 | 0.047 | 0.012 | 0.025 | 0.041 | 0.068 | 0.141 |
| Iron | mg/L | 2.842 | 1.246 | 1.126 | 1.956 | 2.674 | 3.540 | 5.100 |
| Lead | mg/L | 0.073 | 0.048 | 0.017 | 0.040 | 0.063 | 0.095 | 0.162 |
| Manganese | mg/L | 0.111 | 0.046 | 0.054 | 0.079 | 0.103 | 0.134 | 0.197 |

*Continued*

**TABLE 3.2 Lognormal Summary Statistics for Untreated Stormwater Quality in Australian Urban Catchments—cont'd**

| | | | | Percentile | | | | |
|---|---|---|---|---|---|---|---|---|
| **Contaminant** | **Unit** | **Mean** | **SD** | **5th** | **25th** | **50th** | **75th** | **95th** |
| Mercury | g/L | 0.218 | 0.105 | 0.080 | 0.143 | 0.201 | 0.273 | 0.411 |
| Nickel | mg/L | 0.009 | 0.004 | 0.004 | 0.007 | 0.009 | 0.011 | 0.017 |
| Zinc | mg/L | 0.293 | 0.153 | 0.080 | 0.183 | 0.272 | 0.379 | 0.570 |
| **Nutrients** | | | | | | | | |
| Oxidized nitrogen | mg/L | 0.680 | 0.446 | 0.132 | 0.361 | 0.592 | 0.900 | 1.523 |
| Total dissolved nitrogen | mg/L | 3.28 | 2.61 | 0.68 | 1.55 | 2.59 | 4.19 | 8.22 |
| Total kjeldahl nitrogen | mg/L | 2.84 | 4.14 | 0.60 | 0.95 | 1.59 | 3.04 | 8.82 |
| Total organic nitrogen | mg/L | 0.623 | 0.828 | 0.160 | 0.233 | 0.367 | 0.669 | 1.874 |
| Total nitrogen | mg/L | 3.09 | 2.33 | 0.62 | 1.52 | 2.51 | 4.00 | 7.46 |
| Filtered reactive phosphorus | mg/L | 0.664 | 0.762 | 0.050 | 0.204 | 0.430 | 0.839 | 2.037 |
| Total phosphorus | mg/L | 0.480 | 0.413 | 0.075 | 0.207 | 0.367 | 0.620 | 1.261 |
| **Organics** | | | | | | | | |
| Polycyclic aromatic hydrocarbons | μg/L | 0.262 | 0.306 | 0.017 | 0.078 | 0.168 | 0.331 | 0.811 |
| **Physicochemical Indicators** | | | | | | | | |
| Ammonia | mg/L | 1.135 | 1.187 | 0.102 | 0.394 | 0.793 | 1.464 | 3.281 |
| Bicarbonate alkalinity as $CaCO_3$ | mg/L | 35.21 | 3.36 | 29.99 | 32.87 | 35.04 | 37.37 | 40.97 |
| Biochemical oxygen demand | mg/L | 54.28 | 45.58 | 6.56 | 22.87 | 42.53 | 72.03 | 140.77 |
| Chemical oxygen demand | mg/L | 57.67 | 17.22 | 32.90 | 45.41 | 55.75 | 67.85 | 88.72 |
| Chloride | mg/L | 11.40 | 1.05 | 9.75 | 10.67 | 11.35 | 12.08 | 13.20 |
| Oil and grease | mg/L | 13.13 | 8.11 | 3.43 | 7.45 | 11.47 | 16.93 | 28.25 |
| pH | | 6.35 | 0.54 | 5.50 | 5.98 | 6.33 | 6.70 | 7.27 |
| Sodium | mg/L | 10.63 | 2.82 | 6.58 | 8.62 | 10.31 | 12.29 | 15.72 |
| Suspended solids | mg/L | 99.73 | 83.60 | 19.01 | 45.41 | 77.24 | 127.19 | 254.47 |
| Total dissolved solids | mg/L | 139.6 | 17.30 | 112.89 | 127.44 | 138.54 | 150.58 | 169.60 |
| Total organic carbon | mg/L | 16.90 | 3.33 | 11.99 | 14.54 | 16.60 | 18.92 | 22.80 |
| Turbidity | NTU | 50.93 | 40.46 | 7.98 | 23.21 | 40.74 | 66.78 | 127.79 |

*NTU*, nephelometric turbidity unit; *SD*, standard deviation. Component nitrogen and phosphorus concentrations may be greater than total nitrogen and phosphorus due to a statistical aberration.
Adapted from NRMMC, EPHC, NHMRC, 2009a. Australian Guidelines for Water Recycling: Managing Health and Environmental Risks (Phase 2) - Stormwater Harvesting and Reuse, National Resource Management Ministerial Council (NRMMC), Environmental Protection and Heritage Council (EPHC) & National Health and Medical Research Council (NHMRC). National Water Quality Management Strategy, Canberra, ACT, Australia.

differing land uses and climates. Although the metals and pharmaceuticals investigated in each of the studies were reasonably consistent, the pesticides[1] investigated varied, influenced by the type of weeds and/or insects likely to be present in the catchment, and relevant ecosystem protection guidelines or stormwater storage location. For example, harvested stormwater intended for storage in aquifers has comparatively more stringent guidelines than other storage locations (ANZECC and ARMCANZ, 2000; NRMMC et al., 2009a,b).

1. Herbicides and insecticides have been grouped together under the heading of pesticides for consistency with Australian and international water quality.

The presence of more than 200 chemical compounds was investigated in these studies, yet only a relatively small number of compounds were detected at, or above, concentrations of concern to human health or the environment. The majority of these compounds were metals or pesticides. Simazine, a herbicide used mainly for grass weeds was the only pesticide detected in all catchments where it was investigated ($n = 15$). Caffeine and mestranol, an active ingredient in contraceptive pills, were the only pharmaceuticals detected at concentrations exceeding drinking water guidelines (Gernjak et al., 2017). Gernjak et al. (2017) proposed that a catchment-specific land-use link exists for the industrial chemicals, 4-t-octylphenol, 4-nonylphenol, and bisphenol A, noting that while a strong variation in concentrations existed, all detections were well below drinking water guideline values. Five endocrine disrupting chemicals likely to enter stormwater via human sewage (17-α-estradiol, 17-α-ethinylestradiol, 17β-estradiol, estriol, or estrone) were not detected above the limit of quantification in the same study (Gernjak et al., 2017).

As discussed above, investigations into the quality of undiluted and untreated stormwater is an emerging area of research, and to-date, there are limited published data available. Table 3.3 provides a summary of the chemical compounds detected at concentrations, at or above drinking water or aquatic ecosystem guideline values, in untreated stormwater from 19 catchments (Page et al., 2013; Howitt et al., 2014; Reeve et al., 2015; Gernjak et al., 2016). The majority of exceedances are in relation to aquatic ecosystem protection guidelines, rather than drinking water guidelines. It is also important to note that some of the drinking water guideline values that were exceeded are based on aesthetic criteria only (e.g., iron) and not related to adverse human health outcomes.

Comparison of results for microorganisms across research studies is hindered by the characteristics of existing stormwater data sets described earlier (e.g., analytical methods and data reporting styles), recovery ratios, and the presence of inhibiting substances such as humics. Before 2000, studies utilizing both, culture and molecular methods usually reported either the presence/absence or identification at genus level only (e.g., *Campylobacter* spp. or *Cryptosporidium*). Only a small number of studies provide quantifications of detected pathogens and there is notable variation in reporting styles across the data sets (e.g., range, median, 95th percentile). Microorganisms detected in raw Australian stormwater are listed in Table 3.4.

Advancements in molecular detection methods related to DNA and RNA sequencing has enabled microbial source tracking to determine the species of origin of microorganisms in water (Ahmed et al., 2007). These techniques have been applied in Australia and the United States to explore sources of fecal contamination in stormwater and waters that receive stormwater (Sidhu et al., 2013; Reeve et al., 2015; Ahmed et al., 2018). Using human-specific biomarkers and viruses, Sidhu et al. (2013) concluded that sewage ingress into stormwater was occurring frequently in some catchments with separate storm sewer infrastructure. While Reeve et al. (2015) concluded that *Cryptosporidium* detected in 82% of samples from a mixed industrial/residential catchment were of environmental rather than human origin, and therefore not of human health concern. Such advancements in analytical capabilities enable more informed assessment of human health risks associated with the reuse of harvested stormwater and facilitate determination of appropriate treatment processes required to protect public health across different reuse scenarios.

## 3.7 OVERVIEW OF WATER SENSITIVE URBAN DESIGN AND CURRENT TREATMENT APPROACHES

### 3.7.1 The philosophy of Water Sensitive Urban Design

Utilizing limited resources to fulfill the needs of an increasing human population in urban areas is a challenge. The adverse impacts of urbanization on urban water ecosystems is a major issue faced by regulatory agencies. The fundamental concepts of Water Sensitive Urban Design (WSUD) play an important role here, as it integrates the main aspects of urban water management (urban water cycle, water supply, wastewater, stormwater, groundwater and potable water management, and urban design) and treat all sources of water generated in a catchment as a resource (BMY_WBM, 2009). Although the term "WSUD" is commonly used in Australia, the approach is known by different terms in other countries such as low impact development (the United States and China), best management practices (the United States), and sustainable urban drainage system (the United Kingdom) (see Chapter 1 for further details and discussion).

WSUD is based on eight key principles (Donofrio et al., 2009; Waltert and Schläpfer, 2010):

1. Protection of water ecosystems such as creeks, rivers, and wetlands, while transforming vegetated lands into built environment.
2. Improving the quality of water discharged into urban receiving waters.

**TABLE 3.3 Chemical Compounds Detected, at or Above, a Human Health or Aquatic Ecosystem[a] Guideline, in Untreated Stormwater From Australian Catchments[b]**

| Chemical Compounds | Guidelines Exceeded in One or More Samples |
|---|---|
| **Heavy Metals** | |
| Aluminum | Drinking water (aesthetic), aquatic ecosystems |
| Arsenic | Drinking water, aquatic ecosystems (As III) |
| Cadmium | Drinking water, aquatic ecosystems |
| Chromium | Drinking water (Cr VI), aquatic ecosystems (Cr VI) |
| Copper | Aquatic ecosystems |
| Iron | Drinking water (aesthetic)[c] |
| Lead | Drinking water, aquatic ecosystems |
| Mercury | Drinking water, aquatic ecosystems |
| Molybdenum | Drinking water[c] |
| Nickel | Drinking water, aquatic ecosystems |
| Silver | Aquatic ecosystems |
| Vanadium | Drinking water[c] |
| Zinc | Aquatic ecosystems |
| **Pesticides[d]** | |
| Azinphos-methyl | Aquatic ecosystems |
| Chlorpyrifos | Aquatic ecosystems |
| Diazinon | Aquatic ecosystems |
| Fenitrothion | Aquatic ecosystems |
| Malathion (Maldison) | Aquatic ecosystems |
| MCPA | Drinking water[c] |
| Parathion | Aquatic ecosystems |
| Simazine | Drinking water, aquatic ecosystems |
| **Pharmaceuticals** | |
| Caffeine | Drinking water[c] |
| Mestranol | Drinking water |

[a]*Aquatic ecosystem guideline of 90% species protection freshwater, slight to moderately disturbed systems (ANZECC and ARMCANZ, 2000).*
[b]*Stormwater drainage is separate to sewerage infrastructure in all catchments.*
[c]*No aquatic ecosystem guideline.*
[d]*Herbicides and insecticides have been grouped together under the heading of pesticides for consistency with Australian and International water quality.*
Adapted from Page, D., Gonzalez, D., Dillon, P., Vanderzalm, J., Vadakattu, G., Toze, S., Sindhu, J., Miotlinski, K., Torkzaban, S., Barry, K., 2013. Managed Aquifer Recharge Stormwater Use Options: Public Health and Environmental Risk Assessment. Goyder Institute for Water Research Technical Report, Adelaide, SA, Australia; Howitt, J.A., Mondon, J., Mitchell, B.D., Kidd, T., Eshelman, B., 2014. Urban stormwater inputs to an adapted coastal wetland: role in water treatment and impacts on wetland biota. Science of the Total Environment 485–486, 534–544; Reeve, P., Monis, P., Lau, M., Reid, K., van den Akker, B., Humpage, A., King, B., Leusch, F., Keegan, A., 2015. Quantifying water quality characteristics of stormwater. In: Water Research Australia Final Report, Project 3015–11; Gernjak, W., Lampard, J., Tang, J., 2016. CRCWSC Stormwater Database. Cooperative Research Centre for Water Sensitive Cities, Melbourne, Australia.

3. Enhancing the reuse of stormwater, wastewater, and recycled water to restore the urban water balance.
4. Conserving water resources by reducing potable water use (use of efficient household water systems and water reuse).
5. Landscaping incorporated with stormwater treatment devices, while utilizing land space as wildlife habitats and for recreational purposes.

**TABLE 3.4 Microorganisms Detected in Untreated/Unmixed Stormwater From Australian Catchments**

| | |
|---|---|
| Microbial indicators and human biomarkers | *Escherichia coli, Enterococci,* Fecal coliforms (including thermotolerant), somatic bacteriophages, somatic coliphages, F-RNA bacteriophages, *Bacteroides* HF183, and *nifH* gene |
| Bacteria | *Campylobacter* spp., *Campylobacter jejuni, Clostridium perfringens, Legionella* spp., *Legionella pneumophila, Pseudomonas aeruginosa, Salmonella* spp., *Salmonella enterica, Staphylococcus aureus* |
| Protozoa | *Cryptosporidium* spp., *Giardia* spp., *Giardia lamblia* |
| Viruses | Adenovirus (gene copies), polyomavirus (gene copies) |

Adapted from Sidhu, J., Gernjak, W., Toze, S., 2012. Health Risk Assessment of Urban Stromwater. Urban Water Security Reserach Alliance Technical Report No. 102, Brisbane, Australia; Lampard, J., Chapman, H., Stratton, H., et al., 2012. Pathogenic bacteria in urban stormwater drains from inner-city precincts. In: Proceedings of WSUD 2012: Water Sensitive Urban Design: Building the Water Sensitve Community: 7th International Conference on Water Sensitive Urban Design. Melbourne, Australia; Page, D., Gonzalez, D., Dillon, P., Vanderzalm, J., Vadakattu, G., Toze, S., Sindhu, J., Miotlinski, K., Torkzaban, S., Barry, K., 2013. Managed Aquifer Recharge Stormwater Use Options: Public Health and Environmental Risk Assessment. Goyder Institute for Water Research Technical Report, Adelaide, SA, Australia; Reeve, P., Monis, P., Lau, M., Reid, K., van den Akker, B., Humpage, A., King, B., Leusch, F., Keegan, A., 2015. Quantifying water quality characteristics of stormwater. In: Water Research Australia Final Report, Project 3015–11; Lampard, J., Sidhu, J., Agulló-Barceló, M., Larsen, E., Gernjak, W., 2017. Microbial Quality of Untreated Stormwater in Australian Catchments: Human Health Perspectives. Cooperative Research Centre for Water Sensitive Cities, Melbourne, Australia.

6. Mitigating extreme changes to the runoff hydrograph (increase of peak flow and runoff volume) by improving infiltration and groundwater recharge.
7. Integrating water amenities into urban landscape design, and thereby enhancing visual, social, cultural, and ecological values.
8. Designing economically feasible WSUD devices that can be easily implemented in the field.

In this chapter, these principles are discussed in relation to the application of WSUD approaches for stormwater pollution mitigation.

## 3.7.2 Common Water Sensitive Urban Design approaches in stormwater management

WSUD is being incorporated into urban developments in many parts in the world as most of the practices are capable of mitigating the adverse impacts of urbanization on both, stormwater quantity and quality. However, WSUD should be implemented by combining different devices to take advantage of the strengths in performance of a specific device (Mangangka et al., 2016). Such combinations of WSUD devices are referred to as "treatment trains." A typical stormwater treatment train is shown in Fig. 3.4. The following discussion encompasses common structural treatment measures. Chapter 2 provides a more detailed description and application of the major types of WSUD features.

### *3.7.2.1 Rain gardens*

Rain gardens are shallow vegetated beds typically used for stormwater management in residential areas (Fig. 3.5). These devices enhance stormwater infiltration and groundwater recharge, while reducing peak flow. Rain gardens are also known as bioretention systems as they provide biological treatment of stormwater using plants and microorganisms. The diversity of vegetation plays an important role in both, hydrologic and treatment performance of rain gardens. Stormwater treatment is also attributed to several other processes such as adsorption, ion exchange, and plant uptake (Jennings, 2016; Stovin et al., 2017; Yang et al., 2013).

### *3.7.2.2 Pervious pavements*

The spread of impervious surfaces leads to the increase in stormwater runoff volume and velocity. Therefore, pervious pavements (Fig. 3.6) are an effective alternative to impermeable surfaces such as roads, driveways, and parking lots to enhance stormwater infiltration through a coarse subbase into the soil. Pervious pavements are constructed in both, monolithic (single continuous porous material of concrete or asphalt) and modular forms (concrete grid or individual concrete, ceramic, or plastic paving blocks). However, it is important to install measures (e.g., underlying drainage system) to direct the captured stormwater runoff into other treatment devices, such as rain gardens and swales, to remove traffic-related pollutants (Ferguson, 2005; Kumar et al., 2016).

**FIGURE 3.4** A typical Water Sensitive Urban Design treatment train.

**FIGURE 3.5** A typical rain garden. *Photographs supplied by Samira Rashetnia, Victoria University, Australia.*

### 3.7.2.3 Tree pits

Stormwater tree pits are an alternative approach to the traditional practice of planting street trees, enabling stormwater treatment and infiltration. Similar to rain gardens, tree pits are also referred to as bioretention systems as stormwater management involves filtration, detention, and biological uptake. Therefore, the selection of trees is critical for the performance of tree pits and should take into account soil type and other site-specific conditions. This approach can be

**FIGURE 3.6** Pervious pavements. *Photographs supplied by Oriana Sanicola, Stormwater Research Group (SWRG), School of Science and Engineering. University of the Sunshine Coast, Australia.*

**FIGURE 3.7** Typical tree pits.

incorporated in new constructions, re-developments, and also for retrofitting (Fig. 3.7). The treatment performance can be improved by connecting with other WSUD devices such as pervious pavements. Frequent maintenance of these devices is necessary for litter and sediment removal (USEPA, 2013).

### *3.7.2.4 Biofilters/bioretention basins*

Bioretention basins are WSUD devices that primarily treat stormwater pollutants such as nutrients through the combined effect of a surface layer of vegetation and filter media (Chen et al., 2013; Mangangka et al., 2015) (Fig. 3.8). The main purpose of vegetation is to maintain the porosity and high permeability of the treatment matrix and provide a carbon source for denitrification reactions. The surface layer is used as a detention zone, where stormwater is detained, allowing infiltration into and percolation through the filter media. The layers underlying the vegetation typically consist of filter media (coarse sand/fine gravel), a transition layer, and a drainage layer surrounding a perforated underdrain pipe. In the case where fine gravel is used as filter media, a transition layer is introduced between filter media and drainage layer, preventing the migration of fine material into the bottom layers.

FIGURE 3.8 A typical bioretention basin.

Bioretention basins also have the potential to control stormwater runoff quantity. This depends on the porosity, the soil moisture characteristics (available water capacity), and the pre-storm moisture content of the filter media. This latter value is influenced by the length of the antecedent dry period and transpiration potential of the vegetation. Longer dry periods and plant species suitable for dry weather conditions reduce the moisture content, increase the soil water deficit, and consequently, the filtration and detention/retention time (Hunt et al., 2008; Parker et al., 2009).

Once installed, the filter media needs to be kept permeable to maintain satisfactory performance. Therefore, timely replacement of the filter media is necessary. This will prevent clogging due to sediment and flushing and leaching of native materials (filter media) and previously accumulated pollutants (Hatt et al., 2008; WBD, 2010).

### 3.7.2.5 Swales

Swales are constructed as a pretreatment device for downstream structures of a treatment train, primarily focusing on improving stormwater quality with limited flow control through detention. The vegetated top layer of a swale initially removes the coarser fraction of particulate solids in stormwater runoff. As the runoff enters the bioretention component, which is a layered filter media of soil, sand, and gravel layers, a number of treatment mechanisms such as filtration, infiltration, adsorption, and biological uptake are involved (Davis et al., 2012; Kazemi et al., 2011). The treatment capacity of vegetated/bioretention swales depends on the longitudinal slope (controls the flow velocity and avoids erosion of swale surface); species type and height of the vegetation; filter media; and cross-sectional area of the swale (controls the detention time) (Leroy et al., 2016). Fig. 3.9 shows a typical roadside vegetated/bioretention swale.

### 3.7.2.6 Gross pollutant traps

Gross pollutants, which are typically larger than 5 mm, include litter and vegetation debris. These relatively large debris are washed-off during storm events, resulting in offensive odours and poor aesthetics of the aquatic ecosystem. Gross pollutant traps (GPTs) are commonly used as pretreatment structures to prevent downstream treatment devices from clogging (Madhani and Brown, 2015; Madhani et al., 2009). Fig. 3.10 shows a typical GPT device.

### 3.7.2.7 Sediment traps/sedimentation ponds

Sedimentation ponds are used in areas where there is a high sediment load (e.g., construction sites) to prevent deposition of sediments in downstream treatment devices such as constructed wetlands and rain gardens. These pretreatment structures commonly consist of fully/partially separated basins (Fig. 3.11), allowing adequate retention time for the sedimentation of fine and coarse particles (DPLG, 2010; Karlsson et al., 2010). Sedimentation ponds also control peak flows as they provide temporary storage for stormwater runoff (Hvitved-Jacobsen et al., 2010).

### 3.7.2.8 Constructed wetlands

Constructed wetlands are stormwater treatment devices that remove pollutants by combining a number of mechanisms such as sedimentation, vegetation uptake, adsorption, filtration, and microbial decomposition within a shallow waterbody. Both,

**FIGURE 3.9** A typical roadside swale.

**FIGURE 3.10** A typical gross pollutant trap device (trash rack).

submerged and emergent plant species in the wetland play a significant role in improving stormwater quality as they facilitate most of the pollutant removal mechanisms (Headley and Tanner, 2012; Moshiri, 1993). Fig. 3.12 shows a typical constructed wetland.

Before stormwater is discharged into the wetland, the flow is controlled using a high-flow bypass system and only the design flow is treated. The inlet of the wetland consists of a relatively deep basin, which is fed with stormwater through a flow spreader. This inlet basin discharges the pretreated stormwater (after sedimentation of coarser particulate solids) into a main wetland area through a porous wall. The relatively shallow (<300 mm) main treatment area is vegetated with submerged and

FIGURE 3.11 A typical sedimentation pond.

FIGURE 3.12 A typical constructed wetland system.

extensive emergent macrophyte species. Before being discharged, the stormwater treated in the main area subjected to natural ultraviolet radiation that acts as a partial disinfection measure in an open water zone (Mangangka et al., 2016).

Constructed wetlands not only improve stormwater quality, but also reduce the runoff volume and peak flow through infiltration and evaporation as the water is retained in the wetland. However, the reduction in runoff volume gradually decreases as the wetland reaches saturated conditions, with time (Persson et al., 1999).

### 3.7.3 Performance of Water Sensitive Urban Design treatment devices

Conventional stormwater treatment devices such as ponds and detention basins are incorporated into management strategies to meet stormwater quantity and quality objectives for discharge into receiving waters. However, these treatment devices may not be able to achieve stormwater quality suitable for reuse. The objectives for stormwater treatment in Australia are commonly based on a minimum reduction in the annual load of pollutants such as suspended solids and nutrients destined for discharge to receiving waters. WSUD devices, if appropriately designed, can provide an additional level of treatment to enable stormwater reuse. Typical WSUD devices and the pollutants they are designed to remove or treat are listed in Table 3.5.

**TABLE 3.5 Typical Water Sensitive Urban Design (WSUD) Devices and Target Pollutants**

| WSUD System | Target Pollutants |
|---|---|
| Gross pollutant traps | • Litter and coarse sediments larger than 5 mm |
| Pervious pavements | • Coarse sediments |
| Sediment traps/sedimentation ponds | • Coarse and medium sediment |
| Swales | • Coarse sediments<br>• Nutrients |
| Biofilters/bioretention basins, rain gardens and tree pits | • Litter and sediments<br>• Toxicants<br>• Nutrients<br>• Microorganisms |
| Constructed wetlands | • Fine particles<br>• Particle-bound pollutants<br>• Nutrients<br>• Microorganisms |

Adapted from BMY_WBM., 2009. Evaluating Options for Water Sensitive Urban Design - A National Guide. Joint Steering Committee for Water Sensitive Cities (JSCWSC), Australia.

To obtain optimum performance, stormwater treatment devices that provide different levels of treatment are commonly combined, forming a treatment train. Standard treatment trains include primary, secondary, and tertiary treatment levels, and a particular treatment structure may overlap two or more treatment levels depending on specific design features.

According to NRMMC et al. (2009a), constructed wetlands and biofilters have the highest potential for addressing pollutants relevant to risk management (i.e., pathogens, nutrients, and suspended solids). Although the pollutant removal efficiency can be highly variable, constructed wetlands effectively remove nutrients, suspended solids, and metals (Malaviya and Singh, 2012). Overall, constructed wetlands function well as pretreatment structures for applications such as stormwater harvesting and aquifer storage and recovery (Page et al., 2013).

Biofiltration is commonly used in urban areas as the amount of land involved is relatively smaller than for constructed wetlands. Laboratory studies have shown that biofilters have the potential to remove a large proportion of total suspended solids and heavy metal loads from stormwater (Blecken et al., 2010a,b; Hatt et al., 2008). These results have also been confirmed by several field studies (e.g., Hatt et al., 2009; Brown and Hunt, 2011). Although the removal of nutrients can be beneficial, the above studies have also found that biofilters can also be producers of nutrients. A summary of the performance of stormwater biofilters based on field studies is presented in Table 3.6.

Research into removal of microorganisms by biofilters has grown, as the recognition of the potential of stormwater as an alternate water supply for nonpotable and indirect potable applications has increased. Early laboratory research and pilot studies have sought to identify the effectiveness of differing vegetation, submerged zones, dry periods, and filter media type and configuration for removal of fecal indicator organisms (e.g., *E. coli*) using synthetic stormwater (Deletić et al., 2014). The focus is now expanding to investigate the removal of reference pathogens outlined in stormwater and drinking water guidelines (e.g., *Campylobacter jejuni*, *Cryptosporidium oocysts*, and Adenovirus). The removal efficacy of biofilters is influenced by the design and operational conditions, with the selection of vegetation species, filter media configuration, and the length of dry periods between events being important. Laboratory testing of different biofilter media configurations has shown $\log_{10}$ removal in excess of 3 log (99.9% removal), with a mean of 2 log (99% removal) for different pathogens (Li et al., 2012). The inclusion of copper-zeolite into biofilter media has been shown to increase the log removal rate of *E. coli* by 53% (Li et al., 2016). Biofilters containing submerged zones have been shown to reduce human reference pathogens by log reductions ranging from 0.7 to 1.7. However, removal effectiveness was found to reduce following extended dry periods and large runoff volumes (Chandrasena et al., 2017).

## 3.7.4 Potential impacts of climate change on the performance of Water Sensitive Urban Design devices

The potential impacts of climate change on the performance of WSUD devices can be hypothesized in terms of the changes to their treatment efficiency (Burge et al., 2012). As discussed in Section 3.5.3, the predicted changes to stormwater

**TABLE 3.6** Performance of Stormwater Biofilters in Australia and the United States

| | Total Suspended Solids | | | Total Phosphorus | | | Total Nitrogen | |
|---|---|---|---|---|---|---|---|---|
| References | Reduction in Median Conc. (%) | Mean Mass Removal (Load Reduction) (%) | Reduction in Mean Conc. (%) | Reduction in Median Conc. (%) | Mean Mass Removal (Load Reduction) (%) | Reduction in Mean Conc. (%) | Reduction in Median Conc. (%) | Mean Mass Removal (Load Reduction) (%) |
| Hatt et al. (2009) | 87 | 76 | | −214 | −398 | | −18 | −7 |
| | 89 | 93 | | 83 | 86 | | 19 | 37 |
| | | | 80 | | | 50 | | |
| Passeport et al. (2009) | | | | 63 | 53 | | 54 | 56 |
| | | | | 58 | 68 | | 55 | 47 |
| Parker et al. (2009) | | 72 | | | 49 | | | 42 |
| Brown and Hunt (2011) | 59 | | | −13 | | | 58 | |
| Brown and Hunt (2012),[a] | | 71 | | | 5 | | | 12 |
| | | 79 | | | 12 | | | 35 |

Negative values indicate export of pollutants.
[a]*Systems tested pre- and postrestorative maintenance, prerestorative presented first.*

quantity and quality due to climate change can also reduce the performance of WSUD devices. Consequently, in the evaluation of WSUD performance, it is important to ensure their adaptability under future climate change scenarios (Brudler et al., 2016; Howe et al., 2005; Hunter, 2012).

Similar to stormwater quality, the impacts of climate change on pollutant processes can be significant because of the predicted increase in the antecedent dry period between rainfall events, increase in rainfall intensity, and decrease in rainfall duration. Burge et al. (2012) modeled the performance of a number of WSUD devices under different climate scenarios based on the potential changes in rainfall and evaporation. They reported that appropriately designed and installed WSUD devices, including bioretention systems, swales, and constructed wetlands, are capable of coping with the changing climate. The study (Burge et al., 2012) outcomes revealed that constructed wetlands tend to slightly improve both, hydrologic and pollutant removal performances under severe climate scenarios (i.e., extreme changes in characteristics and patterns of rainfall and evaporation). However, it is important to note that these modeling studies were not based on highly accurate data sets, such as high-resolution rainfall time series, due to technical constraints (e.g., computational difficulties associated with dynamic downscaling). Furthermore, the influence of future urban developments on the hydrologic regime and pollutant generation was not adequately accounted. Therefore, these modeling outcomes need to be interpreted with some caution.

## 3.8 CONCLUSIONS

Urbanization is increasing around the world, resulting in the conversion of previously vegetated areas to impervious surfaces, with consequential detrimental impacts on stormwater runoff quality and quantity. Stormwater pollution impacts ecosystem health through the degradation of physical and chemical quality and microbial contamination of water. Increase in suspended solids, which carry toxic pollutants such as heavy metals and hydrocarbons, and nutrient enrichment, results in a decrease in aquatic biodiversity. Mitigation of these adverse impacts is essential for improving receiving water quality and thereby, human well-being. Quantification of pollutant loads and concentrations is the basis for designing effective measures to improve stormwater quality before being discharged to receiving water or before storage and reuse.

The quality characteristics of stormwater vary across catchments. Although the degree of influence of specific catchment characteristics is yet to be clearly understood, the design of urban form, including stormwater and wastewater infrastructure, land-use activities, traffic characteristics, and climate characteristics, is considered as key influencers. In some instances, heavy metals, petroleum hydrocarbons, pesticides, pharmaceuticals, and microbial pollutants have been detected in stormwater above ecosystem protection and drinking water guideline values. Advances in fecal source tracking methods are enabling new insights into the microbial profile of stormwater, enabling a clearer understanding of potential human health risks. Varying detections across catchments, of pharmaceuticals, human pathogens, and human-specific biomarkers in stormwater, even in catchments with separate storm and sewer infrastructure, highlight the need for further research at a catchment level to understand how pollution is occurring.

On urban surfaces, pollutants such as particulate solids and associated toxicants, including heavy metals and hydrocarbons, accumulate during dry weather periods. The accumulated pollutants are subsequently washed off by rainfall events and transported to receiving water bodies through drainage networks. Pollutant mobilization and transport are significantly enhanced by the increase in stormwater runoff volume and velocity because of the presence of impervious surfaces.

In the context of safeguarding urban water ecosystems, WSUD is recognized as an effective way to minimizing the impacts of stormwater pollution. WSUD devices need to be implemented in combination, referred to as a treatment train, to take advantage of the specific strengths and to mitigate the inherent weaknesses of individual devices. Common structural measures used for stormwater management include GPTs, sedimentation ponds, pervious pavements, rain gardens, tree pits, swales, bioretention basins, and constructed wetlands. Effectively managed WSUD devices can reduce the impact of pollutants on the receiving water environment and provide pretreatment to stormwater intended for reuse for nonpotable and potable purposes. However, the design of WSUD devices is constrained by the lack of understanding of how natural and anthropogenic factors influence the underlying mechanisms of pollutant export, transformation, and removal.

Stormwater pollutant characteristics and the performance of WSUD devices are expected to be significantly influenced by the likely impacts of climate change. In relation to stormwater quality, pollutant processes, and consequently, pollutant loads and concentrations are expected to undergo significant changes. In relation to WSUD devices, the impacts of climate change can be identified in terms of the changes in treatment performance. Consequently, it is important to quantitatively assess the changes to stormwater quantity and quality, and the resulting performance of WSUD devices to ensure adaptability to a changing climate.

## REFERENCES


Adachi, K., Tainosho, Y., 2004. Characterization of heavy metal particles embedded in tire dust. Environment International 30 (8), 1009–1017.

AGO, 2003. In: Pittock, B. (Ed.), Climate Change: An Australian Guide to the Science and Potential Impacts. Australia Australian Greenhouse Office, Canberra.

Ahmed, W., Stewart, J., Gardner, T., Powell, D., Brooks, P., Sullivan, D., Tindale, N., 2007. Sourcing faecal pollution: a combination of library-dependent and library-independent methods to identify human faecal pollution in non-sewered catchments. Water Research 41, 3771–3779.

Ahmed, W., Hodgers, L., Sidhu, J.P.S., Toze, S., 2012. Fecal indicators and zoonotic pathogens in household drinking water taps fed from rainwater tanks in Southeast Queensland, Australia. Applied and Environmental Microbiology 78, 219–226.

Ahmed, W., Lobos, A., Senkbeil, J., Peraud, J., Gallard, J., Harwood, V., 2018. Evaluation of the novel crAssphage marker for sewage pollution tracking in storm drain outfalls in Tampa, Florida. Water Research 131, 142–150.

Alias, N., Liu, A., Goonetilleke, A., Egodawatta, P., 2014. Time as the critical factor in the investigation of the relationship between pollutant wash-off and rainfall characteristics. Ecological Engineering 64 (0), 301–305.

ANZECC and ARMCANZ, 2000. Australian and New Zealand guidelines for fresh and marine water quality. In: National Water Quality Management Strategy Paper No 4. Australian and New Zealand Environment and Conservation Council & Agriculture and Resource Management Council of Australia and New Zealand, Canberra.

Arnone, R.D., Walling, J.P., 2007. Waterborne pathogens in urban watersheds. Journal of Water and Health 5, 149–162.

Atwill, E.R., Camargo, S.M., Phillips, R., Alsono, L.H., Tate, K.W., Jensen, W.A., Bennet, J., Little, S., Salmon, T.P., 2001. Quantitative shedding of two genotypes of *Cryptosporidium parvum* in California ground squirrels (*Spermophilus beecheyi*). Applied and Environmental Microbiology 67, 2840–2843.

AWQC, 2008. Pathogens in Stormwater. Australian Water Quality Centre Report for the NSW Department of Environment and Climate Change and the Sydney Metropolition Catchment Management Authority. Adelaide, Australia.

Ball, J.E., Jenks, R., Aubourg, D., 1998. An assessment of the availability of pollutant constituents on road surfaces. Science of the Total Environment 209 (2–3), 243–254.

Beach, J., 2005. Safe drinking water. In: Hrudey, S.E., Hrudey, E.J. (Eds.), Lessons from Recent Outbreaks in Affluent Nations. International Water Association Publishing, ISBN 1843390426, 486 pp.

Bell, C.D., McMillan, S.K., Clinton, S.M., Jefferson, A.J., 2016. Hydrologic response to stormwater control measures in urban watersheds. Journal of Hydrology 541 (Part B), 1488–1500.

Blecken, G.-T., Zinger, Y., Deletić, A., Fletcher, T.D., Hedström, A., Viklander, M., 2010a. Laboratory study on stormwater biofiltration: nutrient and sediment removal in cold temperatures. Journal of Hydrology 394 (3), 507–514.

Blecken, G.-T., Zinger, Y., Deletić, A., Fletcher, T.D., Viklander, M., 2010b. Laboratory studies on metal treatment efficiency of stormwater biofilters. In: International Short Course: Advances in Knowledge of Urban Drainage from the Catchment to the Recieving Water. University of Calabria, Italy, 15/06/2010-15/06/2010.

BMY_WBM, 2009. Evaluating Options for Water Sensitive Urban Design - A National Guide. Joint Steering Committee for Water Sensitive Cities (JSCWSC), Australia.

Bocca, B., Alimonti, A., Petrucci, F., Violante, N., Sancesario, G., Forte, G., Senofonte, O., 2004. Quantification of trace elements by sector field inductively coupled plasma mass spectrometry in urine, serum, blood and cerebrospinal fluid of patients with Parkinson's disease. Spectrochimica Acta Part B: Atomic Spectroscopy 59 (4), 559–566.

Brown, R., Hunt, W., 2011. Underdrain configuration to enhance bioretention exfiltration to reduce pollutant loads. Journal of Environmental Engineering 137 (11), 1082–1091.

Brown, R.A., Hunt, W.F., 2012. Improving bioretention/biofiltration performance with restorative maintenance. Water Science and Technology 65 (2), 361–367.

Brown, J.N., Peake, B.M., 2006. Sources of heavy metals and polycyclic aromatic hydrocarbons in urban stormwater runoff. Science of the Total Environment 359, 145–155.

Brudler, S., Arnbjerg-Nielsen, K., Hauschild, M.Z., Rygaard, M., 2016. Life cycle assessment of stormwater management in the context of climate change adaptation. Water Research 106, 394–404.

Burge, K., Browne, D., Breen, P., Wingad, J., 2012. Water sensitive urban design in a changing climate: estimating the performance of WSUD treatment measures under various climate change scenarios. In: Paper Presented at the WSUD 2012: Water Sensitive Urban Design; Building the Water Sensiitve Community; 7th International Conference on Water Sensitive Urban Design, Melbourne.

Chandrasena, G., Deletić, A., Lintern, A., Henry, R., McCarthy, D., 2017. Stormwater biofilters as barriers against *Campylobacter jejuni, Cryptosporidium oocysts* and adenoviruses; results from a laboratory trial. Water 9, 949.

Chen, X., Peltier, E., Sturm, B.S., Young, C.B., 2013. Nitrogen removal and nitrifying and denitrifying bacteria quantification in a stormwater bioretention system. Water Research 47 (4), 1691–1700.

Chiew, F.H.S., Mudgway, L.B., Duncan, H.P., McMahon, T.A., 1997. Urban Stormwater Pollution. Coorporative Research Centre for Catchment Hydrology, Australia.

Davis, A.P., Stagge, J.H., Jamil, E., Kim, H., 2012. Hydraulic performance of grass swales for managing highway runoff. Water Research 46 (20), 6775–6786.

Deletić, A., McCarthy, D., Chandrasena, G., Li, Y., Hatt, B., Payne, E., Zhang, K., Henry, R., Kolotelo, P., Rangjelovic, A., Meng, Z., Glaister, B., Pham, T., Ellerton, J., 2014. Biofilters and Wetlands for Stormwater Treatment and Harvesting. Cooperative Research Centre for Water Sensitive Cities, Monash University, Melbourne.

Delpla, I., Jung, A.V., Baures, E., Clement, M., Thomas, O., 2009. Impacts of climate change on surface water quality in relation to drinking water production. Environment International 35 (8), 1225–1233.

de Man, H., van den Berg, H.H.J.L., Leenen, E.J.T.M., Schijven, J.F., Schets, F.M., van der Vliet, J.C., van Knapen, F., de Roda Husman, A.M., 2014. Quantitative assessment of infection risk from exposure to waterborne pathogens in urban floodwater. Water Research 48, 90–99.

Donofrio, J., Kuhn, Y., McWalter, K., Winsor, M., 2009. Water-sensitive urban design: an emerging model in sustainable design and comprehensive water-cycle management. Environmental Practice 11 (3), 179–189.

DPLG, 2010. Water Sensitive Urban Design Technical Manual for the Greater Adelaide Region. Adelaide: Department of Planning and Local Government, SA, Australia.

Duncan, H.P., 1999. Urban Stormwater Quality: A Statistical Overview. Cooperative Research Centre for Catchment Hydrology, Melbourne, Australia.

Duong, T.T.T., Lee, B.-K., 2011. Determining contamination level of heavy metals in road dust from busy traffic areas with different characteristics. Journal of Environmental Management 92 (3), 554–562.

Egodawatta, P., Thomas, E., Goonetilleke, A., 2007. Mathematical interpretation of pollutant wash-off from urban road surfaces using simulated rainfall. Water Research 41 (13), 3025–3031.

Eriksson, E., Baun, A., Scholes, L., Ledin, A., Ahlman, S., Revitt, M., et al., 2007. Selected stormwater priority pollutants - a European perspective. Science of the Total Environment 382, 41–51.

Erickson, A.J., Weiss, P.T., Gulliver, J.S., 2013. Impacts and composition of urban stormwater. In: Erickson, A.J., Weiss, P.T., Gulliver, J.S. (Eds.), Optimizing Stormwater Treatment Practices: A Handbook of Assessment and Maintenance. Springer New York, New York, NY, pp. 11–22.

Ferguson, B.K., 2005. Porous Pavements. CRC Press.

Fletcher, T.D., Deletić, A., Mitchell, V.G., Hatt, B.E., 2008. Reuse of urban runoff in Australia: a review of recent advances and remaining challenges. Journal of Environmental Quality 37 (5_Suppl.). S-116-S-127.

Gaffield, S.J., Goo, R.L., Richards, L.A., Jackson, R.J., 2003. Public health effects of inadequately managed stormwater runoff. American Journal of Public Health 93 (9), 1527–1533.

Gernjak, W., Lampard, J., Tang, J., 2016. CRCWSC Stormwater Database. Cooperative Research Centre for Water Sensitive Cities, Melbourne Australia.

Gernjak, W., Lampard, J., Tang, J.Y.M., 2017. Characterisation of Chemical Hazards in Stormwater. Cooperative Research Centre for Water Sensitive Cities, Melbourne, Australia.

Gobel, P., Dierkes, C., Coldewey, W.G., 2007. Stormwater runoff concentration matrix for urban areas. Journal of Contaminant Hydrology 91, 26–42.

Goonetilleke, A., Thomas, E., Ginn, S., Gilbert, D., 2005. Understanding the role of land use in urban stormwater quality management. Journal of Environmental Management 74, 31–42.

Gunawardana, C., Goonetilleke, A., Egodawatta, P., Dawes, L., Kokot, S., 2012a. Role of solids in heavy metals build-up on urban road surfaces. Journal of Environmental Engineering 138 (4), 490–498.

Gunawardana, C., Goonetilleke, A., Egodawatta, P., Dawes, L., Kokot, S., 2012b. Source characterisation of road dust based on chemical and mineralogical composition. Chemosphere 87 (2), 163–170.

Gunawardana, C., Egodawatta, P., Goonetilleke, A., 2015. Adsorption and mobility of metals in build-up on road surfaces. Chemosphere 119 (0), 1391–1398.

Hamers, T., Smit, L.A.M., Bosveld, A.T.C., van den Berg, J.H.J., Koeman, J.H., van Schooten, F.J., Murk, A.J., 2002. Lack of a distinct gradient in biomarker responses in small mammals collected at different distances from a highway. Archives of Environmental Contamination and Toxicology 43 (3), 0345–0355. https://doi.org/10.1007/s00244-002-1230-3.

Hatt, B.E., Fletcher, T.D., Deletić, A., 2008. Hydraulic and pollutant removal performance of fine media stormwater filtration systems. Environmental Science and Technology 42 (7), 2535–2541.

Hatt, B.E., Fletcher, T.D., Deletić, A., 2009. Hydrologic and pollutant removal performance of stormwater biofiltration systems at the field scale. Journal of Hydrology 365 (3), 310–321.

Headley, T., Tanner, C., 2012. Constructed wetlands with floating emergent macrophytes: an innovative stormwater treatment technology. Critical Reviews in Environmental Science and Technology 42 (21), 2261–2310.

Heisler, J., Glibert, P.M., Burkholder, J.M., Anderson, D.M., Cochlan, W., Dennison, W.C., Suddleson, M., 2008. Eutrophication and harmful algal blooms: a scientific consensus. Harmful Algae 8 (1), 3–13.

Hering, J.G., Waite, T.D., Luthy, R.G., Drewes, J.E., Sedlak, D.L., 2013. A changing framework for urban water systems. Environmental Science and Technology 47 (19), 10721–10726.

Herngren, L., Goonetilleke, A., Ayoko, G.A., 2005. Understanding heavy metal and suspended solids relationships in urban stormwater using simulated rainfall. Journal of Environmental Management 76, 149–158.

Howe, C., Jones, R.N., Maheepala, S., Rhodes, B., 2005. Implications of Potential Climate Change for Melbourne's Water Resources. CSIRO Urban Water and CSIRO Atmospheric Research, Australia.

Howitt, J.A., Mondon, J., Mitchell, B.D., Kidd, T., Eshelman, B., 2014. Urban stormwater inputs to an adapted coastal wetland: role in water treatment and impacts on wetland biota. Science of the Total Environment 485–486, 534–544.

Hunt, W., Smith, J., Jadlocki, S., Hathaway, J., Eubanks, P., 2008. Pollutant removal and peak flow mitigation by a bioretention cell in urban Charlotte, NC. Journal of Environmental Engineering 134 (5), 403–408.

Hunter, G., 2012. Impacts of Climate Change on Urban Stormwater Infrastructure in Metropolitan Sydney. The Sydney Metropolitan Catchment Management Authority, Parramatta, NSW.

Hvitved-Jacobsen, T., Vollertsen, J., Nielsen, A.H., 2010. Urban and Highway Stormwater Pollution: Concepts and Engineering. CRC Press, Taylor and Francis Group.

Jacobson, C.R., 2011. Identification and quantification of the hydrological impacts of imperviousness in urban catchments: a review. Journal of Environmental Management 92, 1438–1448.

Jayarathne, A., Egodawatta, P., Ayoka, G.A., Goonetilleke, A., 2017. Geochemical phase and particle size relationships of metals in urban road dust. Environmental Pollution 230, 218–226.

Jayarathne, A., Egodawatta, P., Ayoko, G.A., Goonetilleke, A., 2018. Intrinsic and extrinsic factors which influence metal adsorption to road dust. Science of the Total Environment 618 (Suppl. C), 236–242.

Jennings, A.A., 2016. Residential rain garden performance in the climate zones of the contiguous United States. Journal of Environmental Engineering 142 (12), 04016066.

Jochimsen, E.M., Carmichael, W.W., An, J., Cardo, D.M., Cookson, S.T., Holmes, C.E., Barreto, V.S.T., 1998. Liver failure and death after exposure to microcystins at a hemodialysis center in Brazil. New England Journal of Medicine 338 (13), 873–878.

Jullien, A., François, D., 2006. Soil indicators used in road environmental impact assessments. Resources, Conservation and Recycling 48, 101–124.

Karlsson, K., Viklander, M., Scholes, L., Revitt, M., 2010. Heavy metal concentrations and toxicity in water and sediment from stormwater ponds and sedimentation tanks. Journal of Hazardous Materials 178 (1–3), 612–618.

Kazemi, F., Beecham, S., Gibbs, J., 2011. Streetscape biodiversity and the role of bioretention swales in an Australian urban environment. Landscape and Urban Planning 101 (2), 139–148.

Kumar, K., Kozak, J., Hundal, L., Cox, A., Zhang, H., Granato, T., 2016. In-situ infiltration performance of different permeable pavements in a employee used parking lot—A four-year study. Journal of Environmental Management 167, 8–14.

Lampard, J., Chapman, H., Stratton, H., et al., 2012. Pathogenic bacteria in urban stormwater drains from inner-city precincts. In: Proceedings of WSUD 2012: Water Sensitive Urban Design: Building the Water Sensitve Community: 7th International Conference on Water Sensitive Urban Design. Melbourne, Australia.

Lampard, J., Sidhu, J., Agulló-Barceló, M., Larsen, E., Gernjak, W., 2017. Microbial Quality of Untreated Stormwater in Australian Catchments: Human Health Perspectives. Cooperative Research Centre for Water Sensitive Cities, Melbourne, Australia.

Lee, J.H., Bang, K.W., Ketchum Jr., L.H., Choe, J.S., Yu, M.J., 2002. First flush analysis of urban storm runoff. Science of the Total Environment 293 (1–3), 163–175.

Leroy, M-c., Portet-Koltalo, F., Legras, M., Lederf, F., Moncond'huy, V., Polaert, I., Marcotte, S., 2016. Performance of vegetated swales for improving road runoff quality in a moderate traffic urban area. Science of the Total Environment 566, 113–121.

Li, Y.L., Deletić, A., Alcazar, L., Bratieres, K., Fletcher, T.D., McCarthy, D.T., 2012. Removal of *Clostridium perfinens, Escherichia coli* and F-RNA coliphages by stormwater biofilters. Ecological Engineering 49, 137–145.

Li, Y.L., McCarthy, D.T., Deletić, A., 2016. *Escherichia coli* removal in copper-zeolite-integrated stormwater biofilters: effects of vegetation, operational time, intermittent drying weather. Ecological Engineering 90, 234–243.

Lim, M.H., Leong, Y.H., Tiew, K.N., et al., 2011. Urban stormwater harvesting: a valuable water resource of Singapore. Water Practice and Technology 6 (4).

Liu, 2011. Influence of Rainfall and Catchment Charactersitics on Urban Stormwater Quality (Ph.D. thesis). Queensland University of Technology, Brisbane, Australia.

Liu, A., Goonetilleke, A., Egodawatta, P., 2012a. Taxonomy for rainfall events based on pollutant wash-off potential in urban areas. Ecological Engineering 47, 110–114.

Liu, A., Goonetilleke, A., Egodawatta, P., 2012b. Inadequacy of land use and impervious area fraction for determining urban stormwater quality. Water Resources Management 26, 2259–2265.

Liu, A., Egodawatta, P., Guan, Y., Goonetilleke, A., 2013. Influence of rainfall and catchment characteristics on urban stormwater quality. Science of the Total Environment 444, 255–262.

Liu, A., Guan, Y., Egodawatta, P., Goonetilleke, A., 2016. Selecting rainfall events for effective water sensitive urban design: a case study in Gold Coast City, Australia. Ecological Engineering 92, 67–72.

Ma, Y., Egodawatta, P., McGree, J., Liu, A., Goonetilleke, A., 2016. Human health risk assessment of heavy metals in urban stormwater. Science of the Total Environment 557–558, 764–772.

Madhani, J.T., Brown, R.J., 2015. The capture and retention evaluation of a stormwater gross pollutant trap design. Ecological Engineering 74, 56–59.

Madhani, J.T., Dawes, L.A., Brown, R.J., 2009. A perspective on littering attitudes in Australia. Journal of the Society for Sustainability and Environmental Engineering 9/10, 13–20.

Makepeace, D.K., Smith, D.W., Stanley, S.J., 1995. Urban stormwater quality: summary of contaminant data. Critical Reviews in Environmental Science and Technology 25 (2), 93–139.

Malaviya, P., Singh, A., 2012. Constructed wetlands for management of urban stormwater runoff. Critical Reviews in Environmental Science and Technology 42 (20), 2153–2214.

Mangangka, I.R., Liu, A., Egodawatta, P., Goonetilleke, A., 2015. Performance characterisation of a stormwater treatment bioretention basin. Journal of Environmental Management 150, 173–178.

Mangangka, I.R., Liu, A., Goonetilleke, A., Egodawatta, P., 2016. Enhancing the Storm Water Treatment Performance of Constructed Wetlands and Bioretention Basins. Springer Briefs in Water Science and Technology.

Marsalek, J., Cisneros, B.J., Karamouz, M., Malmquist, P.-A., Goldenfum, J.A., Chocat, B., 2007. Urban Water Cycle Processes and Interactions: Urban Water Series-UNESCO-IHP, vol. 2. Taylor & Francis.

Meyer, J.L., Paul, M.J., Taulbee, W.K., 2005. Stream ecosystem function in urbanizing landscapes. Journal of the North American Benthological Society 24 (3), 602–612.

MikeUrban, 2014. Mike Urban Collection System - User Guide User Guide. Danish Hydraulic Institute.

Miller, J.D., Kim, H., Kjeldsen, T.R., Packman, J., Grebby, S., Dearden, R., 2014. Assessing the impact of urbanization on storm runoff in a peri-urban catchment using historical change in impervious cover. Journal of Hydrology 515, 59–70.

Miller, J.S., Olejnik, D., 2001. Photolysis of polycyclic aromatic hydrocarbons in water. Water Research 35 (1), 233–243.

Monis, P.T., Caccio, S.M., Thompson, R.C.A., 2009. Variation in *giardia*: towards a taxonomic revision of the genus. Trends in Parasitology 25, 93–100.

Moshiri, G.A., 1993. Constructed Wetlands for Water Quality Improvement. CRC Press.

Mummullage, S., Egodawatta, P., Ayoko, G.A., Goonetilleke, A., 2016a. Sources of hydrocarbons in urban road dust: identification, quantification and prediction. Environmental Pollution 216, 80–85.

Mummullage, S., Egodawatta, P., Ayoko, G.A., Goonetilleke, A., 2016b. Use of physicochemical signatures to assess the sources of metals in urban road dust. Science of the Total Environment 541, 1303–1309.

Mummullage, S.W.N., 2015. Source Characterisation of Urban Road Surface Pollutants for Enhanced Water Quality Predictions. Ph.D. Queensland University of Technology.

NHMRC, NRMMC, 2011. National Water Quality Management Strategy Australian Drinking Water Guidelines 6, vol. 3.3. National Health and Medical Research Council and Natural Resource Management Ministerial Council, Australia.

NRMMC, EPHC, NHMRC, 2009a. Australian Guidelines for Water Recycling: Managing Health and Environmental Risks (Phase 2) - Stormwater Harvesting and Reuse, National Resource Management Ministerial Council (NRMMC), Environmental Protection and Heritage Council (EPHC) & National Health and Medical Research Council (NHMRC). National Water Quality Management Strategy, Canberra, ACT, Australia.

NRMMC, EPHC, NHMRC, 2009b. Australian Guidelines for Water Recycling: Managing Health and Environmental Risks (Phase 2) Managed Aquifer Recharge. Natural Resource Management Ministerial Council, the Environment Protection and Health Council and the National Health and Medical Research Council.

Oliver, R.L., Boon, P.I., 1992. Bioavailability of nutrients in turbid waters. Land and Water Resources Research and Development Corporation. Murray Darling Freshwater Research Centre, Albury, NSW, Australia.

Page, D., Gonzalez, D., Dillon, P., Vanderzalm, J., Vadakattu, G., Toze, S., Sindhu, J., Miotlinski, K., Torkzaban, S., Barry, K., 2013. Managed Aquifer Recharge Stormwater Use Options: Public Health and Environmental Risk Assessment. Goyder Institute for Water Research Technical Report, Adelaide, SA, Australia.

Parker, N., Gardner, T., Goonetilleke, A., Egodawatta, P., Giglio, D., 2009. Effectiveness of WSUD in the "real world". Paper Presented at the the 6th International Water Sensitive Urban Design Conference and the 3rd Hydropolis: Towards Water Sensitive Cities and Citizens, Perth, Australia.

Passeport, E., Hunt, W.F., Line, D.E., Smith, R.A., Brown, R.A., 2009. Field study of the ability of two grassed bioretention cells to reduce storm-water runoff pollution. Journal of Irrigation and Drainage Engineering 135, 505.

Persson, J., Somes, N., Wong, T., 1999. Hydraulics efficiency of constructed wetlands and ponds. Water Science and Technology 40 (3), 291–300.

Pirjola, L., Lähde, T., Niemi, J.V., Kousa, A., Rönkkö, T., Karjalainen, P., Hillamo, R., 2012. Spatial and temporal characterization of traffic emissions in urban microenvironments with a mobile laboratory. Atmospheric Environment 63, 156–167.

Reeve, P., Monis, P., Lau, M., Reid, K., van den Akker, B., Humpage, A., King, B., Leusch, F., Keegan, A., 2015. Quantifying water quality characteristics of stormwater, 2015. Water Research Australia Project 3015-11. Adelaide, SA.

Rossman, L.A., 2009. Stormwater Management Model User's Manual Version 5.0. U.S. Environmental Protection Agency, Washington, DC.

Ryan, U.N.A., Fayer, R., Xiao, L., 2014. *Cryptosporidium* species in humans and animals: current understanding and research needs. Parasitology 141, 1667–1685.

Sauer, E.P., VandeWalle, J.L., Bootsma, M.J., McLellan, S.L., 2011. Detection of the human specific Bacteroides genetic marker provides evidence of widespread sewage contamination of stormwater in the urban environment. Water Research 45 (14), 4081–4091.

Schiff, K., Kinney, P., 2001. Tracking sources of bacterial contamination is stormwater discharges to Misson Bay, California. Water Environment Research 73 (5), 534–542.

Segal, D., 2004. Singapore's Water Trade with Malaysia and Alternatives. John F. Kennedy School of Government, Harvard University, Boston.

Selvakumar, A., Borst, M., 2006. Variation of microorganism concentrations in urban stormwater runoff with land use and seasons. Journal of Water and Health 4, 109–124.

Sidhu, J., Gernjak, W., Toze, S., 2012. Health Risk Assessment of Urban Stormwater. Urban Water Security Reserach Alliance Technical Report No. 102, Brisbane, Australia.

Sidhu, J.P.S., Ahmed, W., Gernjak, W., Aryal, R., McCarthy, D., Palmer, A., Kolotelo, P., Toze, S., 2013. Sewage pollution in urban stormwater runoff as evident from the widespread presence of multiple microbial and chemical source tracking markers. Science of the Total Environment 463–464 (Suppl. C), 488–496.

Stovin, V., Dunnett, N., Yuan, J., 2017. The influence of vegetation on rain garden hydrological performance. Urban Water Journal 14 (10), 1083–1089.

Tiefenthaler, L.L., Stein, E.D., Schiff, K.C., 2008. Watershed and land use-based sources of trace metals in urban storm water. Environmental Toxicology and Chemistry 27, 227–287.

UNDESA, 2014. World Urbanization Prospects: The 2014 Revisions, Highlights. United Nations Department of Economic and Social Affairs, Population Division, New York.

USEPA, 2013. Stormwater to Street Tress. Office of Wetlands, Oceans and Watersheds, U.S. Environmental Protection Agency, Washington, DC.

Vaze, J., Chiew, F.H.S., 2002. Experimental study of pollutant accumulation on an urban road surface. Urban Water 4 (4), 379–389.

Waltert, F., Schläpfer, F., 2010. Landscape amenities and local development: a review of migration, regional economic and hedonic pricing studies. Ecological Economics 70 (2), 141–152.

WBD, 2010. Construction and establishment guidelines: swales, bioretention systems and wetlands. In: Water by Design, vol. 1.1. South East Queensland Healthy Waterways Partnership, Brisbane, Australia.

Wen, J-c., Lee, Y-j., Cheng, S-j., Lee, J-h., 2014. Changes of rural to urban areas in hydrograph characteristics on watershed divisions. Natural Hazards 74 (2), 887–909. https://doi.org/10.1007/s11069-014-1220-6.

Wijesiri, B., Egodawatta, P., McGree, J., Goonetilleke, A., 2015b. Influence of pollutant build-up on variability in wash-off from urban road surfaces. Science of the Total Environment 527–528 (0), 344–350.

Wijesiri, B., Egodawatta, P., McGree, J., Goonetilleke, A., 2015a. Process variability of pollutant build-up on urban road surfaces. Science of the Total Environment 518–519 (0), 434–440.

Wijesiri, B., Egodawatta, P., McGree, J., Goonetilleke, A., 2016. Understanding the uncertainty associated with particle-bound pollutant build-up and wash-off: a critical review. Water Research 101, 582–596.

Wong, T.H.F., Allen, R., Brown, R.R., Deletić, A., Gangadharan, L., Gernjak, W., Jakob, C., Johnstone, P., Reeder, M., Tapper, N., Vietz, G., Walsh, C.J., 2013. Blueprint 2013 – Stormwater Management in a Water Sensitive City. Cooperative Research Centre for Water Sensitive Cities, Melbourned Australia, 978-1-921912-02-03.

WWAP, 2012. The United Nations World Water Development Report 4: Managing Water under Uncertainty and Risk, vol. 1. UNESCO, Paris.

Xie, S., Dearing, J.A., Bloemendal, J., 2000. The organic matter content of street dust in Liverpool, UK, and its association with dust magnetic properties. Atmospheric Environment 34 (2), 269–275.

Yang, H., Dick, W.A., McCoy, E.L., Phelan, P.L., Grewal, P.S., 2013. Field evaluation of a new biphasic rain garden for stormwater flow management and pollutant removal. Ecological Engineering 54, 22–31.

Zhao, H., Chen, X., Hao, S., Jiang, Y., Zhao, J., Zou, C., Xie, W., 2016. Is the wash-off process of road-deposited sediment source limited or transport limited? Science of the Total Environment 563–564, 62–70.

Ziyath, A.M., Egodawatta, P., Goonetilleke, A., 2016. Build-up of toxic metals on the impervious surfaces of a commercial seaport. Ecotoxicology and Environmental Safety 127, 193–198.

Chapter 4

# WSUD Design Guidelines and Data Needs

Ashok K. Sharma[1], Samira Rashetnia[1], Ted Gardner[1] and Don Begbie[2]

[1]*Institute for Innovation and Sustainability, College of Engineering and Science, Victoria University, Melbourne, Vic, Australia;* [2]*Urban Water Security Research Alliance, South East Queensland, Brisbane, Australia*

## Chapter Outline



**ABSTRACT**

National, state, and local council guidelines are provided by various agencies for planning, technical design, implementation, operation, and maintenance of water sensitive urban design (WSUD). Also referred to as best management practices, low impact development, green infrastructure, sustainable urban drainage systems, and low impact urban design and development. Despite the different nomenclature in different parts of the world, their concepts and functionalities are similar. Chapters 1 and 2 have described these WSUD systems/approaches in detail. This chapter lists some of the publicly available national, state, regional, and local design guidelines. The WSUD systems covered in these guidelines come primarily from regions where these systems are frequently implemented. The guidelines cover WSUD system planning and design methods.

**Keywords:** BMPs; Data needs; Design guidelines; Flood management; Harvesting potential; LID; Stormwater quality and quantity; WSUD.

## 4.1 INTRODUCTION

Water Sensitive Urban Design (WSUD) was introduced as a new approach to sustainable water cycle management and integrated urban water management (IUWM) by a multidisciplinary group of professionals and academics in Perth, Western Australia, in the early 1990s (Argue, 2004). Since then, similar initiatives have appeared in other parts of the world, but with different terminologies (Fletcher et al., 2015), including best management practices (BMPs), green infrastructure (GI), IUWM, low impact development (LID), low impact urban design and development (LIUDD), source control, stormwater control measures, sustainable urban drainage systems (SUDS), and experimental sewer systems (ESS). These terms and initiatives have been described in Chapter 1.

National, regional, and local planning and design guidelines have been developed by various national, state, and local agencies for the sustainable implementation of WSUD systems. These guidelines help water professionals, designers,

**Approaches to Water Sensitive Urban Design. https://doi.org/10.1016/B978-0-12-812843-5.00004-6**

planners, and managers to plan, design, and implement these approaches based on urban development requirements, water quality and hydrology criteria, catchment characteristics, local climatic conditions, local regulations, and environmental and community considerations.

The aim of this chapter is to provide information on some of the easily accessible guidelines describing the various WSUD approaches and their design. Because of the wide knowledge and ready availability of information on WSUD systems design in the public domain, this chapter does not cover the actual design methods, but rather provides a list of such guidelines. While being comprehensive, the WSUD guidelines listed in this chapter are not claimed to be all inclusive. Readers may find additional guidelines, aimed at water professionals that are specific to their geographical location and context.

### 4.1.1 Type of water sensitive urban design systems

WSUD systems can be classified into two main categories based on their implementation location or scale:

- Household and streetscape scale WSUD approaches; and
- Urban development scale WSUD approaches.

A list of WSUD systems, their implementation scale, and functions are provided in Table 4.1. A similar list of SUDS systems and their functionalities is also provided by CIRIA (2007). Some of the system can be applied at both household and street scale, and generally have multifunctionalities. Not all the systems listed in Table 4.1 may be adopted in all the geographical locations around the world, and their functionalities can depend on the way they are implemented. Sometimes they can have both a primary and a secondary function. For example, the primary function of rainwater tanks is water harvesting for potable and nonpotable uses, however, they can also reduce urban runoff and flooding.

### 4.1.2 Functionalities of water sensitive urban design approaches and treatment train

WSUD systems have a single and multiple functions depending on the nature of the structure configuration. These functionalities include stormwater quality and quantity management (water quality improvement through treatment trains), flood management (peak flow and volume reduction), stormwater harvesting (alternative water supply source), and enhancing amenity and biodiversity (Table 4.1).

The stormwater quantity and peak flow management are managed through processes such as infiltration, retention, detention, attenuation, conveyance, and water harvesting. Stormwater detention refers to holding runoff for a short period to reduce peak flow rates, and then its slow release to receiving water bodies. Retention refers to systems where stormwater is captured, stored, and reused in the urban water cycle. End uses include domestic or industrial uses, irrigation of open space and GI, and recharge of groundwater systems, often by natural infiltration.

The stormwater quality improvement includes processes such as sedimentation and precipitation, filtration, adsorption, biodegradation, nutrient, uptake by plants, and gaseous losses such as denitrification (CIRIA, 2007). From a water quality treatment perspective, WSUD systems can be divided into three categories based on their applicability and effectiveness (adopted from Argue, 2004):

- Primary: Coase physical screening (>5 mm – Lewis et al., 2015) and sedimentation devices (>0.1 mm)—for example, side entry pits, trash racks, litter booms, and sedimentation basins.
- Secondary: Systems that remove fine particulate matter (<0.1 mm—Melbourne Water, 2013) by sedimentation and filtration—for example, swales, porous and permeable pavements, infiltration basins, and high efficiency sedimentation basins.
- Tertiary: Systems that provide filtration, adsorption, and biological uptake of very fine particles (<0.01 mm) and micro-pollutants such as heavy metals and nutrients—for example, reed beds, bioretention systems, and wetlands.

Single WSUD systems generally do not have the capability to achieve the desired design objectives by themselves. Rather, a group of WSUD approaches, operating in sequence, is usually selected based on their different functionalities, to act as a system to achieve the stormwater quality and quantity objectives. These systems are arranged in series or parallel to achieve the desired outcome and are referred to as a ***treatment train***.

The WSUD techniques listed in Table 4.1 are described in various guideline documents. An extensive list of these documents/guidelines is provided in Table 4.2.

**TABLE 4.1** **Commonly Adopted WSUD Tools/Systems and Their Functions**

| | Function | | | | |
|---|---|---|---|---|---|
| **WSUD Systems/Technique** | **Flow Rate Reduction** | **Water Quality Management** | **Flood Management** | **Harvesting Potential** | **Enhancing Amenity/Biodiversity** |
| **House and Streetscape WSUD Approaches** | | | | | |
| Rain tanks | X | | X | X | |
| Rain garden | X | | | | X |
| Soak pits/soak ways | X | | X | | |
| Porous pavements | X | X | | | |
| Vegetated filter strips | X | X | | | |
| Swales and bioretention swales/ bioswales | X | X | X | | X |
| Infiltration basins/trenches | X | | X | | X |
| Sand filters | | X | | | |
| Bioretention systems | X | X | X | | X |
| Buffer strips | X | X | | | |
| Gardens, landscape | X | X | | | X |
| Infiltration basins/trenches | X | | | X | |
| Water butts (small tanks) | X | | X | X | |
| Perforated chambers and pipes | X | | | X | |
| Roof systems (green/blue roofs) | X | X | X | | X |
| Leaky wells and infiltration trenches | X | | | X | |
| Underground storage/vaults | X | | X | X | |
| Siphonic roof water systems | | | X | X | |
| Geocellular/modular systems | X | X | | | |
| Green/living walls | X | | | | X |
| **Development Scale WSUD Approaches** | | | | | |
| Trash racks | | X | | | |
| Gross pollutant traps | | X | | | |
| Catch basins | | X | | | |
| Detention basins | | | X | | |
| Ponds and urban lakes | | X | X | X | X |
| Sedimentation basins | X | X | | | |
| Constructed wetlands | X | X | | X | X |
| Subsurface wetland | | X | | | X |
| Large stormwater storage units | X | | X | X | |
| Atlantis artificial aquifers | | | | X | |
| Natural aquifers—Managed Aquifer Recharge | | | | X | |

*WSUD*, water sensitive urban design.

**TABLE 4.2 Readily Available Guideline Documentation for WSUD Tools and Systems**

| WSUD System | References and Guidelines/Documents Where WSUD Systems are Described |
|---|---|
| Green and blue roofs | Department of Planning and Local Government SA (2010), Argue (2004), Centre for Watershed Protection (2009), Centre for Watershed Protection (2010), Dorman et al. (2013), TR&CCA (2010), Washington State University Pierce County Extension and Puget Sound Action Team (2005), CIRIA (2007), Armitage et al. (2013) and Australian National Guide (2009). |
| Siphonic roof water systems | Department of Planning and Local Government SA (2010). |
| Rainwater tanks | Sharma et al. (2015), Department of Planning and Local Government SA (2010), Department of Environment WA (2004), City of Melbourne (2006), CSIRO (2005), Tasmanian Government Living (2012), Australian National Guide (2009), Armitage et al. (2013), CIRIA (2007), Centre for Watershed Protection (2009), Dorman et al. (2013), Minnesota Stormwater Design Team (2013), TR&CCA (2010), Washington State University Pierce County Extension and Puget Sound Action Team (2005), Coombes et al. (2001) and Upper Parramatta River Catchment Trust (2004). |
| Simple downspout disconnection | Centre for Watershed Protection (2009). |
| Trash racks, baskets, and booms | Argue (2004) and Department of Environment WA (2004). |
| Catch basins | Argue (2004) and CIRIA (2007). |
| Leaky wells and infiltration trenches | Department of Planning and Local Government SA (2010), Argue (2004) and CSIRO (2005). |
| Soak pits, soakaways, wells | Department of Planning and Local Government SA (2010), Argue (2004), Armitage et al. (2013), Department of Environment WA (2004), CIRIA (2007), TR&CCA (2010) and Roldin et al. (2012). |
| Geocellular/modular systems | CIRIA (2007). |
| Dry wells | Centre for Watershed Protection (2009). |
| Permeable pavements | Argue (2004), Department of Planning and Local Government SA (2010), Department of Environment WA (2004), City of Melbourne (2006), Australian National Guide (2009), CSIRO (2005), CIRIA (2007), Transport Roads & Maritime (2017), Jones Edmunds & Associates (2013), Centre for Watershed Protection (2009), Dorman et al. (2013), TR&CCA (2010), Beecham et al. (2012), Ullate et al. (2011) and Upper Parramatta River Catchment Trust (2004). |
| Gross pollutant traps | Department of Planning and Local Government SA (2010), Department of Environment WA (2004), Melbourne Water (2010) and City of Melbourne (2006). |
| Bioretention basins | Argue (2004), Department of Planning and Local Government SA (2010), Department of Environment WA (2004), CSIRO (2005), Brisbane City Council (2006), Healthy Waterways (2006), City of Melbourne (2006), Tasmanian Government Living (2012), Australian National Guide (2009), Armitage et al. (2013), CIRIA (2007), Transport Roads & Maritime (2017), Jones Edmunds & Associates (2013), Centre for Watershed Protection (2009), Dorman et al. (2013), TR&CCA (2010), Washington State University Pierce County Extension and Puget Sound Action Team (2005), Davis et al. (2009), Trowsdale and Simcock (2011), Upper Parramatta River Catchment Trust (2004). |
| Rain gardens | Department of Planning and Local Government SA (2010), City of Melbourne (2006), Washington State University Pierce County Extension and Puget Sound Action Team (2005) and Centre for Watershed Protection (2009). |
| Constructed wetlands | Argue (2004), CSIRO (2005), Department of Planning and Local Government SA (2010), Department of Environment WA (2004), Brisbane City Council (2006), City of Melbourne (2006), Tasmanian Government Living (2012), Australian National Guide (2009), Armitage et al. (2013), CIRIA (2007); Transport Roads & Maritime (2017), Dorman et al. (2013) and Centre for Watershed Protection (2010). |
| Gardens, landscape | Argue (2004), Department of Planning and Local Government SA (2010), Australian National Guide (2009), Armitage et al. (2013) and CIRIA (2007). |

*Continued*

**TABLE 4.2 Readily Available Guideline Documentation for WSUD Tools and Systems—cont'd**

| WSUD System | References and Guidelines/Documents Where WSUD Systems are Described |
|---|---|
| Buffer strips, filter strips, and vegetated filter strips | Argue (2004), CSIRO (2005), Armitage et al. (2013), CIRIA (2007), Department of Planning and Local Government SA (2010), Department of Environment WA (2004), Brisbane City Council (2006), Tasmanian Government Living (2012), Australian National Guide (2009), Centre for Watershed Protection (2009), Dorman et al. (2013) and TR&CCA (2010). |
| Sedimentation basins | CSIRO (2005), Department of Planning and Local Government SA (2010), Brisbane City Council (2006), Tasmanian Government Living (2012), Australian National Guide (2009), Transport Roads & Maritime (2017) and Green Infrastructure (2012). |
| Swales—dry swales and bioretention swales | Argue (2004), CSIRO (2005), Department of Planning and Local Government SA (2010), Department of Environment WA (2004), Jones Edmunds & Associates (2013), Brisbane City Council (2006), City of Melbourne (2006), Tasmanian Government Living (2012), Australian National Guide (2009), Armitage et al. (2013), CIRIA (2007), Transport Roads & Maritime (2017), Dorman et al. (2013); TR&CCA (2010), Centre for Watershed Protection (2009), Stagge et al. (2012) and Upper Parramatta River Catchment Trust (2004). |
| Infiltration basins and trenches | Argue (2004), CSIRO (2005), Department of Planning and Local Government SA (2010), Department of Environment WA (2004), Brisbane City Council (2006), City of Melbourne (2006), Tasmanian Government Living (2012), Armitage et al. (2013), CIRIA (2007), Australian National Guide (2009), Dorman et al. (2013), TR&CCA (2010), Centre for Watershed Protection (2009), Centre for Watershed Protection (2010), Minnesota Stormwater Design Team (2013), CIRIA (2007) and Upper Parramatta River Catchment Trust (2004). |
| Sand filters | CSIRO (2005), Brisbane City Council (2006), Tasmanian Government Living (2012), Australian National Guide (2009), Armitage et al. (2013), CIRIA (2007), Dorman et al. (2013) and Upper Parramatta River Catchment Trust (2004). |
| Water butts | CIRIA (2007). |
| Detention basins | Argue (2004) and Centre for Watershed Protection (2010). |
| Ponds and lakes | CSIRO (2005), Tasmanian Government Living (2012), Australian National Guide (2009), Armitage et al. (2013), CIRIA (2007), Transport Roads & Maritime (NSW) (2017) and Centre for Watershed Protection (2010). |
| Subsurface wetland | Tasmanian Government Living (2012). |

*WSUD*, water sensitive urban design.

## 4.1.3 Design criteria

A wide range of performance criteria form the foundation of WSUD systems design. These concepts include WSUD performance goals and objectives; hydraulic design concepts; flood and peak discharge control strategies; water quality management strategies; erosion and sediment control; channel protection; natural heritage systems protection; groundwater recharge and base flow maintenance; ecosystem protection; water supply and wastewater; and stormwater harvesting (EPA, 2004; Argue, 2004; NRC, 2009; Australian National Guide, 2009; TR&CCA, 2010; Fewkes, 2012; Melbourne Water, 2013; Sharma et al., 2016).

Some examples of design criteria/objectives are listed in the following section:

- Reducing export of Total Suspended Solids (TSS), Total Phosphorus (TP) and Total Nitrogen (TN) by certain fraction…say 60%, 40%, 20%
- Maintaining the 1-year ARI runoff characteristics
- Retaining the first 10 mm runoff from impervious areas
- Maintaining the flow rate duration characteristics of the preurbanized catchment
- Reducing stormwater flow rate into combined sewers
- Ensuring directly connected impervious area is maintained at <5%
- Preventing "high" sediment loss during civil construction

- Preventing increased erosive flow velocities in urban creeks
- Maintaining or improving the ecosystem health of urban creeks
- Reducing flooding in high-density areas
- Pretreatment of stormwater before aquifer recharge
- Creating urban amenity along with WSUD functions

### 4.1.4 Selection of water sensitive urban design systems and treatment trains

There are number of factors and considerations that can help in selecting suitable WSUD (BMP) approaches, or treatment trains, to achieve design objectives for a given location, development, or a catchment. These factors can be arranged into groups: impact area and design objectives; onsite versus regional control; catchment/watershed factors; terrain factor; stormwater treatment suitability; physical feasibility factors; environmental and habitat factors; community acceptance (e.g., amenity and recreation values); and regulatory limits for stormwater discharge (EPA, 2004; TR&CCA, 2010).

The selection of stormwater management treatment measures also depends on the stormwater pollutant to be managed, as different WSUD approaches have a different treatment capacity. Other factors to consider in treatment system selection include site constraints (for example, catchment slope and land availability), sediment input, stormwater quality, and installation in tidal areas (Argue, 2004). All these factors impact on the design of WSUD measures.

## 4.2 DESIGN GUIDELINES

The design guidelines generally include local regulations, design criteria, hydraulic/hydrological design, water quality considerations, structural design, and construction, operation, and management requirements. These guidelines provide methods and approaches to estimate storage volumes, and inlet and outlet structure sizes for given stormwater quantity and quality management outcomes, based on the Annual Recurrence Interval (rainfall frequency, intensity, duration) and catchment area. They also consider local characteristics (soil type, topography, water table, etc.) and environmental conditions. The hydraulic/hydrological design aspects cover runoff rate and volume estimation, conveyance system design, storage design, and drainage system sizing (CIRIA, 2007). The key aspect is to avoid major or nuisance flooding. The availability of prestorm storage is of major importance in reducing flow rate (Chapter 6 by Myers and Pezzaniti).

Some of these guidelines also describe the generic methodology/decision process for designing a WSUD strategy to achieve desired outcomes (Department of Local Government and Planning SA, 2010; Melbourne Water, 2013). The methodology typically includes (Department of Local Government and Planning SA, 2010):

- understand the site;
- identify objectives;
- identify suitable WSUD measures;
- discuss with local regulators/council;
- conceptualize site design;
- select a base case and model for comparison;
- locate WSUD measures and identify and model options;
- check if the objectives are met;
- finalize measures;
- obtain required approvals from relevant authorities; and
- undertake detailed design.

Some design guidelines/documents also provide suggestions on the suitability/type of plant species for bioretention swales, bioretention basins, sediment basins, wetlands, and ponds. These guidelines are generally for local regions (CSIRO, 2005). Maintaining functioning vegetation over the long term is an essential component of WSUD treatment systems, providing surface area to trap suspended solids, removing pollutants, reducing flow velocities, maintaining infiltration rates, stabilizing banks, and enhancing landscape, including biodiversity.

National, regional, and local guidelines are provided by various agencies to plan, design, and implement WSUD systems to manage stormwater quality and quantity in a consistent way. A list of readily available guidelines from different parts of the world is provided in Table 4.3.

**TABLE 4.3** National, Local, and Regional Design Guidelines and Relevant Contents From Around the World

| Guideline | Country<br>Year | Design Relevant Contents |
|---|---|---|
| Stormwater best management practice design guide: Vol 1 General consideration | US<br>EPA (2004) | Regulations that impact stormwater BMP design, BMP design concepts and guidance, BMP types (source control and treatment BMPs—ponds, wetlands, infiltration, and vegetative biofilters), and BMP selection (impact area and design objectives, onsite vs. regional controls, watershed factors, terrain factors, stormwater treatment suitability, physical feasibility factors, community and environmental factors, locational and permitting factors). |
| San Antonio River Basin Low Impact Development Technical Design Guidance Manual | US<br>Dorman et al. (2013) | Regional considerations, LID selection—structural BMPs (infiltration BMPs, including bio-retention, bioswales, permeable pavement), filtration BMPs (including planter boxes, green roofs, sand filter), volume-storage and reuse BMPs (including stormwater wetlands and rainwater harvesting), and conveyance and pretreatment BMPs (including vegetated swales and vegetated filter strips). |
| Duval County Low Impact Development Design Manual | US<br>Jones Edmunds & Associates. Inc Florida (2013) | Evaluating your site and planning for LID, LID practices in Duval county (grassed conveyance swales, shallow bioretention, previous pavements) |
| State of Minnesota Stormwater Manual | US<br>Minnesota Stormwater Design Team (2013) | Stormwater concepts and stormwater management, stormwater issues, stormwater control practices (BMPs), by type (construction practices, pretreatment practices, postconstruction practices, nonstructural practices, structural practices, stormwater and rainwater harvest, and use/reuse), by treatment mechanism (pretreatment, filtration practices, infiltration practices, sedimentation practices, chemical practices), regulatory, permitting, models, calculations, methodologies, pollutant removal, credits. |
| Green Infrastructure for Southwestern Neighborhoods Version 1.2 (Revised) | US<br>Green Infrastructure (2012) | General GI practices (vegetation and mulch), StreetSide GI practices (curb cuts, curb cut and basin, rock-lined edges, curb cut and basin, shallow slope, sediment traps), in-street GI practices (Chicanes, Medians, Traffic circles, Street width reduction), parking lot GI practices, maintenance. |
| New York State Stormwater Management Design Manual | US<br>New York State, Department of Environmental Conservation (2015) | Stormwater management planning, GI for stormwater management—quality: pond, wetland, infiltration, filtering practices, and open channels; quantity: above ground systems (dry detention, blue roofs), underground systems (underground storage vaults, chambers, pipes, infiltration systems); unified stormwater sizing criteria; GI practices; performance criteria; SMP selection; and stormwater management design examples. |
| Coastal Stormwater Supplement to the Georgia Stormwater Management Manual | US<br>Commission for Watershed Protection (2009) | Stormwater management and site planning and design criteria, calculating the stormwater runoff volumes associated with the stormwater management criteria, satisfying the stormwater management and site planning and design criteria, GI practices (green roofs, permeable pavements, undisturbed pervious areas (i.e., aquatic buffer), vegetated filter strips, grass channels, simple downspout disconnection, rain gardens, stormwater planters, dry wells, rainwater harvesting, bioretention areas infiltration practices, dry swales). |
| Low Impact Development Technical Guidance Manual for Puget Sound | US<br>Washington State University and Pouget Sound action team (2005) | Site planning and layout, vegetation protection, reforestation and maintenance, clearing and grading, integrated management practices, LID design and flow modeling guidance (permeable pavements, dispersion, vegetated roofs, rainwater harvesting, reverse slope sidewalks, minimal excavation foundation, bioretention area, rain gardens), hydrology analysis. |

*Continued*

**TABLE 4.3** National, Local, and Regional Design Guidelines and Relevant Contents From Around the World—cont'd

| Guideline | Country<br>Year | Design Relevant Contents |
|---|---|---|
| South African Guidelines for Sustainable Drainage Systems | South Africa<br>Water Research Commission<br>University of Cape Town<br>Armitage et al. (2013) | Introduction to SUDS, design criteria, and methods, source controls (green roofs, rainwater harvesting, soakaways, permeable pavements), local controls (filter strips, swales, infiltration trenches, bioretention areas, sand filters), regional controls (detention ponds, retention ponds, constructed wetlands). |
| SUDS Manual<br>CIRIA United Kingdom | UK<br>London<br>CIRIA (2007) | Hydraulic design methods, SUDS selection, source control (green roofs, soakaways, water butts, rainwater storage), pretreatment (infiltration systems, catch basins, sedimentation manhole), filter strips, trenches, swales, bioretention, pervious pavements, geocellular/modular systems, sand filters, infiltration basins, detention basins, ponds, stormwater wetlands, inlets and outlets, landscape, construction, and operation and maintenance. |
| Low Impact Development Stormwater Management Planning and Design Guide Ver. 1 | Canada<br>Toronto and Region Conservation and Credit Valley Conservation<br>Authorities<br>TR&CCA (2010) | Integrating stormwater management into the planning process, LID practices, design of structural LID practices for stormwater management (rainwater harvesting, green roofs, roof downspout disconnection, soakaways, infiltration trenches and chambers, bioretention, vegetated filter strips, permeable pavement, enhanced grass swales, dry swales, perforated pipe systems). |
| Evaluating options for Water Sensitive Urban Design—A National Guide, Australia | Australian National Guide (2009) | WSUD objective setting, options for achieving WSUD objectives (potable water demand reduction techniques, stormwater management techniques—sediment basins, swales and buffer stripes, bioretention swales and basins, sand filters, constructed wetland, ponds and lakes, infiltration systems, aquifer storage and recovery, porous pavements, retarding basins, green roofs, stream, and riparian vegetation and rehabilitation), evaluation of WSUD options, WSUD risks and issues, WSUD monitoring considerations. |
| WSUD Engineering Procedures: Stormwater | Australia<br>CSIRO (2005) | Hydraulic design regions; design of WSUD systems covered: sediment basins, bioretention swales, bioretention basins, sand filters, swales/buffer systems, constructed wetlands, ponds, infiltration measures, rainwater tank, and aquifer storage and recovery; plant species for WSUD treatment elements. |
| Water Sensitive Urban Design Technical Manual Greater Adelaide Region (South Australia) | Australia<br>Department of Planning and Local Government SA (2010) | WSUD measures for different types and scale of development, designing a WSUD strategy for your development, demand reduction, rainwater tanks, rain gardens, green roofs, and infiltration systems (soakaways; infiltration trenches; infiltration basins; and leaky wells), pervious pavements, urban water harvesting and reuse, gross pollutant traps, bioretention systems for streetscapes, swales and buffer strips, sedimentation basins, constructed wetlands, wastewater management, modeling process and tools, and siphonic roof water systems. |
| Stormwater Management Manual for Western Australia | Australia<br>Department of Environment, WA (2004) | Understanding the context (water quality, water quantity, groundwater management, flood management, healthy ecological communities, and total water cycle management), best planning practice for stormwater management, integrating stormwater management approaches, stormwater management plans (storage and use: rainwater storage and aquifer recharge, including infiltration basins and trenches, soak wells, and pervious pavement; Conveyance systems: swales and buffer strips, bioretention systems, living streams), retrofitting, nonstructural controls, performance monitoring, and evaluation. |

| | | |
|---|---|---|
| Stormwater Strategy (A Melbourne Water strategy for managing rural and urban runoff) | Australia<br>City of Melbourne (2006), Melbourne Water (2013), and Clear Water (2017) | The history of stormwater management and achievements, delivering multiple community outcomes, Better on-ground outcomes, planning, policy and regulation, delivering the strategy, WSUD policy commitment, getting WSUD on the ground. |
| WSUD Technical design guidelines for South East Queensland ver. 1 | Australia<br>Healthy Waterways (2006) | Introducing WSUD, swales (incorporating buffer strips), bioretention swales, sediment basins, bioretention basins, constructed stormwater wetlands, infiltration measures, sand filters, aquifer storage and recovery, plant selection for WSUD systems. |
| WSUD—Engineering procedures for stormwater management in Tasmania | Australia<br>Tasmanian Government Living (2012) | Hydrologic design regions, sediment basins, bioretention systems, bioretention basins, sand filters, swales/buffer systems, constructed wetlands, ponds, infiltration measures, rainwater tanks, aquifer storage, and other measures (subsurface wetlands, porous pavements, natural areas, including reforestation and revegetation). |
| Concept design guidelines for Water Sensitive Urban Design | Australia<br>Water by Design (2009) | Conceptual design process, best planning practices, BMPs—rainwater harvesting, stormwater harvesting, wastewater treatment for reuse, gross pollutant capture devices, sand filters, bioretention systems, constructed wetlands, porous pavements, and infiltration measures. |
| Managing Urban Stormwater Harvesting and Reuse | Australia<br>Department of Environment and Conservation, NSW (2006) | Introduction, overview of stormwater harvesting, statutory requirements, risk management, planning considerations, design considerations, operational considerations. |
| Water sensitive urban design guideline | Australia<br>Transport Road & Maritime Services, (2017) | Integrating WSUD into projects, designing of WSUD systems (vegetated swales [table drains], bioretention swales, sediment basins, bioretention basins, ponds, porous and permeable pavements, water storage, irrigation, or other reuse opportunities, exfiltration/aquifer recharge). |
| WSUD: Basic procedures "source control" of stormwater—A handbook for Australian practice | University of South Australia, Australian Water Association, and Stormwater Industry Association<br>Argue (2004) | Devices and systems used in stormwater management (roof systems, leaky wells and infiltration trenches, filter stripes, permeable pavements, swales and bioretention swales, trash racks, fine sediment removal, retention device, catch basins and soak pits, baskets, booms, infiltration basins, extended detention basins, wet detention basins, constructed wetland), theory, data, and implementation issues, getting started with "source control" in WSUD, runoff quality control, aspects of stormwater quality, storages for pollution control, storages for stormwater harvesting. |
| WSUD Guidelines (Design, construction, and maintenance) | Australia<br>South East and growth area councils—Melbourne, Australia<br>Melbourne Water (2013) | Introduction (What are WSUD and Why use them), planning and design, construction and maintenance, councils' steps on WSUD design. |
| WSUD Reference Guidelines | Australia<br>Hornsby Shire Council, NSW (2015) | Hornsby Shire Council DCP 1C.1.2 Stormwater Management Requirements, General information on WSUD strategy preparation, MUSIC modeling parameters for Hornsby, bioretention systems as WSUD treatment, applicant lodgement checklist for WSUD strategy, case study—industrial and high-density residential. |
| Water Sensitive Design for Stormwater | New Zealand<br>Auckland Council<br>Lewis et al. (2015) | WSD principles for stormwater, objectives of WSUD implementation, site assessment, concept design phase planning, stormwater treatment train, source control, filtering and conveyance, bioretention, detention, and attenuation |

*BMPs*, best management practices; *GI*, green infrastructure; *LID*, low impact development; *SUDS*, sustainable urban drainage systems; *WSUD*, water sensitive urban design.

## 4.3 DATA REQUIREMENT FOR PLANNING AND DESIGN OF WATER SENSITIVE URBAN DESIGN SYSTEMS

Detailed local data are required for the planning and design of WSUD techniques, in particular for

- Assessment of the development and water receiving environment;
- Capacity of existing infrastructure, if any,
- Selection of suitable WSUD approaches; and
- Agreed design criteria.

The nature and type of data requirements, including their details, can vary with the nature of development, environmental conditions, and adopted design criteria. The data requirements have been listed in various WSUD guidelines (EPA, 2004; CIRIA, 2007). Table 4.4 describes the type of data generally required for WSUD system planning and design.

## 4.4 CONCLUSIONS

A wide range of guidelines is readily available and accessible in the public domain to support stormwater systems planners, designers, and water professionals to plan, design, and implement WSUD systems. Different agencies, based on their jurisdictional responsibilities for stormwater management, provide national, regional, and local guidelines aimed at managing stormwater quality and quantity and providing amenity and biodiversity values. In general, the guidelines support planners and designers by

- Describing local stormwater quality and quantity management regulations/guidelines.
- Describing WSUD system selection criteria and selection methodology.
- Providing options for retention/detention-based stormwater source control.
- Methods for the estimation of stormwater runoff based on a selected ARI and rainfall intensity. (i.e., a design storm)
- Providing overall conceptual design processes to incorporate WSUD in a development and to identify which professionals are required for input (where appropriate).
- Outlining methods for sizing of WSUD systems.
- Suggesting local plant species for WSUD treatment elements.

**TABLE 4.4 Data Requirements for WSUD System Planning and Design**

| Data Type | Data Description |
|---|---|
| Nature of the development | Existing or new (greenfield or infill), residential, industrial and/or commercial, rural or semirural. |
| Development area/catchment topography | Nature/shape, direction, and magnitude of ground slope and suitable elevation difference for gravity flow requirements. |
| Development characterizes | Low/high development density, pubic open space, roads, and other amenities. |
| Soil characteristics | Type of soil, geology, permeability and stability, erosivity. |
| Groundwater | Water table, seasonal variation in water table, local use of groundwater for potable or nonpotable end uses, groundwater recharge facilities. |
| Local climate data | Rainfall, evaporation, temperature, etc. |
| Reserved areas for WSUD systems | Land area reserved for the provision/construction of WSUD systems in a catchment/development. |
| Existing drainage system | Type of drainage system and nature/location of water receiving environment. |
| Local environmental issues | Environmental sensitivity of the area, declared groundwater or surface water catchment. |
| Receiving waters | Ecosystem health of local creeks, estuaries, etc. |

*WSUD*, water sensitive urban design.

## REFERENCES

Argue, R.J., 2004. WSUD: Basic Procedures for 'Source Control' of Stormwater – A Handbook for Australian Practice. The University of South Australia, Stormwater Industry Association and Australian Water Association. Sixth printing 2011. http://search.ror.unisa.edu.au/record/UNISA_ALMA11143570070001831.

Armitage, N., Vice, M., Fisher-Jeffes, L., Winter, K., Spiegel, A., Dunstan, J., 2013. Alternative Technology for Stormwater Management the South African Guidelines for Sustainable Drainage Systems. http://www.wrc.org.za/Knowledge%20Hub%20Documents/Research%20Reports/TT%20558-13.pdf.

Australian National Guide, 2009. Evaluating Options for Water Sensitive Urban Design - A National Guide. Joint Steering Committee for Water Sensitive Cities, Department of the Environment, Water, Heritage and the Arts, Canberra, Australia. http://155.187.2.69/water/publications/urban/water-sensitive-design-national-guide.html.

Beecham, S., Pezzaniti, D., Kandasamy, J., 2012. Stormwater treatments using permeable pavements. Water Management 165 (3), 161–170. https://search.proquest.com/openview/34cf4e70b36bdf90cdbd583df8791f4d/1?pq-origsite=gscholar&cbl=135357.

Brisbane City Council, 2006. Water Sensitive Urban Design Engineering Guidelines (Superseded) and Fact Sheets. https://www.brisbane.qld.gov.au/planning-building/planning-guidelines-and-tools/superseded-brisbane-city-plan-2000/water-sensitive-urban-design/engineering.

Centre for Watershed Protection, 2009. Coastal Stormwater Supplement to the Georgia Stormwater to the Georgia Stormwater Management Manual, first ed. Ellicott City, MD. http://www.dnr.sc.gov/marine/NERR/present/LIDmanual/Georgia-CSS-Final-Apr-09.pdf.

Centre for Watershed Protection, 2010. Stormwater Management & Design Manual for New York State. Department of Environmental Conservation, Ellicott City, MD. http://www.dec.ny.gov/docs/water_pdf/swdm2010entire.pdf.

CIRIA, 2007. The SUDS Manual. CIRIA, Classic House, London. WWW.ciria.org.

City of Melbourne, 2006. Applying the Model WSUD Guidelines. An Initiative of the Inner Melbourne Action Plan. WSUD Guidelines, pp. 1–165. https://www.melbourne.vic.gov.au/SiteCollectionDocuments/wsud-full-guidelines.pdf.

Clear Water, Melbourne Water, 2017. https://www.clearwater.asn.au/resource-library/.

Coombes, P.J., Frost, A., Kuczera, G., 2001. Impact of Rainwater Tank and On-Site Detention Options on Stormwater Management in the Upper Parramatta River Catchment. Upper Parramatta River Catchment Trust, NSW, Australia. https://www.scribd.com/document/74335540/Australia-Impact-of-Rainwater-Tank-and-On-Site-Detention-Options-on-Stormwater-Management-in-the-Upper-Parramatta-River-Catchment.

CSIRO, 2005. WSUD Engineering Procedures: Stormwater, Melbourne Water. CSIRO Publishing, Australia. http://www.publish.csiro.au/book/4974/.

Davis, A.P., Hunt, W.F., Traver, R.G., Clar, M., 2009. Bioretention technology: overview of current practice and future needs. Journal of Environmental Engineering 135 (3), 109–117. https://ascelibrary.org/doi/pdf/10.1061/%28ASCE%290733-9372%282009%29135%3A3%28109%29.

Department of Environment, 2004. Introduction, Stormwater Management Manual for Western Australia. Department of Environment, Perth, Western Australia (WA). http://www.water.wa.gov.au/urban-water/urban-development/stormwater/stormwater-management-manual.

Department of Environment and Conservation, 2006. Managing Urban Storm Water Harvesting and Reuse. NSW, Australia. http://www.environment.nsw.gov.au/resources/stormwater/managestormwatera06137.pdf.

Department of Environmental Conservation, 2015. New York State Stormwater Management Design Manual. http://www.dec.ny.gov/chemical/29072.html.

Department of Planning and Local Government, 2010. Water Sensitive Urban Design Technical Manual for the Greater Adelaide Region. Government of South Australia (SA), Adelaide. https://www.sa.gov.au/topics/planning-and-property/land-and-property-development/planning-professionals/water-sensitive-urban-design.

Dorman, T., Frey, M., Wright, J., Wardynski, B., Smith, J., Tucker, B., Riverson, J., Teague, A., Bishop, K., 2013. San Antonio River Basin Low Impact Development Technical Design Guidance Manual, v1. San Antonio River Authority, San Antonio, TX. https://www.sara-tx.org/wp-content/uploads/2015/05/Full-LID-Manual.pdf.

EPA, 2004. Stormwater Best Management Practice Design Guide: Vol. 1 General Consideration. US Environmental Protection Agency. https://cfpub.epa.gov/si/si_public_record_report.cfm?dirEntryId=99739.

Fewkes, A., 2012. A review of rainwater harvesting in the UK. Structural Survey 30 (2), 174–194. http://www.emeraldinsight.com/doi/pdfplus/10.1108/02630801211228761.

Fletcher, T.D., Shuster, W., Hunt, W.F., Ashley, R., Butler, D., Arthur, S., Trowsdale, S., Barraud, S., Semadeni-Davies, A., Bertrand-Krajewski, J., Mikkelsen, P.S., Rivard, G., Uhl, M., Dagenais, D., Viklander, M., 2015. SUDS, LID, BMPs, WSUD and more – the evolution and application of terminology surrounding urban drainage. Urban Water Journal 12 (7), 525–542.

Green Infrastructure for South-Western Neighbourhoods Version 1.2 (Revised), 2012. US Watershed Management Group, Tucson, Arizona. https://wrrc.arizona.edu/sites/wrrc.arizona.edu/files/WMG_Green%20Infrastructure%20for%20Southwestern%20Neighborhoods.pdf.

Healthy Waterways, 2006. Water Sensitive Urban Design Technical Design Guidelines for South East Queensland - Version 1. http://hlw.org.au/u/lib/mob/20151210164506_9581d6262ed405324/2006_wsudtechdesignguidelines-4mb.pdf.

Hornsby Shire Council, 2015. WSUD Reference Guidelines. NSW, Australia. http://www.hornsby.nsw.gov.au/__data/assets/pdf_file/0018/72351/WSUD-Reference-Guideline.pdf.

Jones Edmunds, Associates, 2013. Duval Country Low-Impact Development Design Manual. Gainesville, Florida 32641. http://www.coj.net/departments/planning-and-development/docs/development-services-division/reviewgroup/lidmanualfinal2013-09-06.aspx.

Lewis, M., James, J., Shaver, E., Blackbourn, S., Leahy, A., Seyb, R., Simcock, R., Wihongi, P., Sides, E., Coste, C., 2015. Water Sensitive Design for Stormwater. Auckland Council Guideline Document GD2015/004. Prepared by Boffa Miskell for Auckland Council. http://content.aucklanddesignmanual.co.nz/project-type/infrastructure/technical-guidance/Documents/GD04%20WSD%20Guide.pdf.

Melbourne Water, 2010. Design, Construction & Maintenance of WSUD (Growth Area). https://www.clearwater.asn.au/user-data/resource-files/WSUD_Growth-Area-Council_Guidelines_DEC_2010−Final.pdf.

Melbourne Water, 2013. Water Sensitive Urban Design Guidelines South Eastern Councils. https://www.melbournewater.com.au/sites/default/files/South-Eastern-councils-WSUD-guidelines.pdf.

Minnesota Stormwater Design Team, 2013. State of Minnesota Stormwater Manual. https://stormwater.pca.state.mn.us/index.php?title=Stormwater_and_rainwater_harvest_and_use/reuse.

NRC, 2009. Urban Stormwater Management in the Unites States, National Research Council. The National Academies Press, Washington, DC, USA. https://www.nap.edu/catalog/12465/urban-stormwater-management-in-the-united-states.

Roldin, M., Mark, O., Kuczera, G., Mikkelsen, P.S., Binning, P.J., 2012. Representing soakaways in a physically distributed urban drainage model − upscaling individual allotments to an aggregated scale. Journal of Hydrology 414−415, 530−538. https://www.sciencedirect.com/science/article/pii/S0022169411008158.

Sharma, A.K., Begbie, D., Gardner, T., 2015. Rainwater Tank Systems for Urban Water Supply − Design, Yield, Energy, Health Risks, Economics and Community Perceptions. IWA Publishing. ISBN13:9781780405353 http://www.iwapublishing.com/books/9781780405353/rainwater-tank-systems-urban-water-supply.

Sharma, A.K., Pezzaniti, D., Myers, B., Cook, S., Tjandraatmadja, G., Chacko, P., Chavoshi, S., Kemp, D., Leonard, R., Koth, B., Walton, A., 2016. Water sensitive urban design: an investigation of current systems, implementation drivers, community perceptions and potential to supplement urban water services. Water 8 (7), 272. https://doi.org/10.3390/w8070272.

Stagge, J.H., Davis, A.P., Jamil, E., Kim, H., 2012. Performance of grass swales for improving water quality from highway runoff. Water Research 20 (15), 6731−6742. https://www.ncbi.nlm.nih.gov/pubmed/22463860.

Tasmanian Government Living, 2012. Water Sensitive Urban Design. Engineering Procedures for Stormwater Management in Tasmania. http://epa.tas.gov.au/documents/wsud_manual_2012.pdf.

TR&CCA, 2010. Low Impact Development Stormwater Management Planning and Design Guide ver. 1. Toronto and Region Conservation and Credit Valley Conservation Authorities, Toronto, Canada. https://cvc.ca/wp-content/uploads/2014/04/LID-SWM-Guide-v1.0_2010_1_no-appendices.pdf.

Transport Roads, Maritime, 2017. Water Sensitive Urban Design Guideline Applying Water Sensitive Urban Design Principles to New South Wales (NSW) Transport Projects. http://www.rms.nsw.gov.au/documents/projects/planning-principles/urban-design/water-sensitive-urban-design-guideline.pdf.

Trowsdale, S.A., Simcock, R., 2011. Urban stormwater treatment using bioretention. Journal of Hydrology 397 (3−4), 167−174. https://ac.els-cdn.com/S0022169410007195/1-s2.0-S0022169410007195-main.pdf?_tid=12232aac-0bb8-11e8-803f-00000aab0f6c&acdnat=1517974764_5edce25d47c0ee6b9ebab22e140d5b8c.

Ullate, E.G., Lopez, E.C., Fresno, D.C., Bayon, J.R., 2011. Analysis and Contrast of Different Pervious Pavements for Management of Storm-water in Parking Area in Northern Spain. https://link.springer.com/content/pdf/10.1007%2Fs11269-010-9758-x.pdf.

Upper Parramatta River Catchment Trust, 2004. Water Sensitive Urban Design, Technical Guidelines for Western Sydney. Prepared by URS Australia Pty Ltd.. http://www.richmondvalley.nsw.gov.au/icms_docs/138067_Development_Control_Plan_No_9_-_Water_Sensitive_Urban_Design.pdf

Washington State University Pierce County Extension and Puget Sound Action Team, 2005. Low Impact Development: Technical Guidance Manual for Puget Sound. WA (Olympia and Tacoma). http://www.psp.wa.gov/downloads/LID/LID_manual2005.pdf.

Water by Design, 2009. Concept Design Guidelines for Water Sensitive Urban Design Version 1. South East Queensland Healthy Waterways Partnership, Brisbane, Australia. http://hlw.org.au/u/lib/mob/20141014090729_ab798e4cb34883b80/2009_wsudconceptguide_82mb.pdf.

## FURTHER READING

Melbourne Water, 2009. Water Sensitive Urban Design Guidelines, Melbourne, Australia. https://www.clearwater.asn.au/user-data/resource-files/WSUD_Guidelines_Jan2009.pdf.

Chapter 5

# The Role of Policy and Regulation in WSUD Implementation

**Grace Tjandraatmadja**
*Institute of Sustainability and Innovation, Victoria University, Melbourne, VIC, Australia*

## Chapter Outline



**ABSTRACT**
This chapter uses case studies to examine the impact of policy and regulation on the implementation of water sensitive urban design in Australia and the international scene. Policy, regulation, and institutional setup are key influences on the implementation of WSUD/low impact development/green infrastructure. Case studies from Australia and around the world (Europe, the United States, Singapore) show how the mix of policies, incentives, regulation, capacity building, and institutional perceptions at various levels impacts, and is influenced by, the institutional culture and context in each jurisdiction. The municipal government has typically been the key agent for WSUD implementation through local planning laws. However, the implementation process requires collaboration across multiple government levels, water agencies, and other societal stakeholders from multiple disciplines (planning, water and wastewater, roads, landscaping, and parks) to support and empower local government and community in the implementation of WSUD.

**Keywords:** Green infrastructure (GI); Institutional setup; Low impact development (LID); Low impact urban design and development (LIUDD); Policy; Regulation; Sustainable urban drainage systems (SUDS); Water sensitive urban design (WSUD).

**Approaches to Water Sensitive Urban Design. https://doi.org/10.1016/B978-0-12-812843-5.00005-8**

## 5.1 INTRODUCTION

Water sensitive urban design, low impact development, sustainable urban development, and green infrastructure (GI) are important tools for water resource management and adaptation to changing climate and increasing urbanization (Fletcher et al., 2015; Sharma et al., 2012).

The USEPA (2017) defines water sensitive urban design (WSUD) or low impact development (LID) as:

*an approach to planning and designing urban areas to make use of stormwater and reduce the harm it causes to surface waters. It typically adopts systems and practices that use or mimic natural processes that result in the infiltration, evapotranspiration or use of stormwater in order to protect water quality and associated aquatic habitat. The term GI refers to the management of wet weather flows using these processes, and refers to the patchwork of natural areas that provide habitat, flood protection, cleaner air and cleaner water. At both the site and regional scale, WSUD/LID/GI practices aim to preserve, restore and create green space using soils, vegetation, and rainwater management techniques.*

WSUD focuses on solutions that integrate the water cycle with the local context. This enables the development of tailored solutions for the management of water cycle elements and can generate multiple benefits: water quality protection, waterway ecosystem health, flood risk reduction, alternative water supplies, enhanced amenity, liveability, resilience, increase in biodiversity improvement, habitat for wildlife, etc.

WSUD implementation requires an understanding of the water cycle in the local physical and environmental context (climate, soil, geomorphology, etc.), the integration of multiple disciplines (engineering, urban planning, landscaping, hydrology, science, etc.), and coordination across agencies and sectors (government, community, and private) to enable the delivery of services. WSUD is better suited for a governance and service delivery model, characterized by participation and coordination of societal actors, including both formal (government and private sector) and informal agencies (community groups). Consequently, WSUD is often perceived to be more complex to adopt than gray infrastructure (i.e., traditional engineered infrastructure) installed for the rapid supply, transport, treatment and removal of water, wastewater, and drainage (Sharma et al., 2012; Dolowitz, 2015).

Internationally, there has been a shift toward sustainable urban practices and water sensitive urban design driven by urbanization, environmental degradation, climate change, and financial pressures (Carter et al., 2015). The transition of WSUD from a niche to a mainstream paradigm has been encouraged by scientists and government policy initiatives (Lim and Lu, 2016; Brudler et al., 2016). There is a wide range of implementation frameworks adopted around the world, including the USEPA framework (1972), the EU Framework Directive (2000), and the Australian ANZECC (2000).

Although our understanding of the philosophical concepts, technologies, and life-cycle costs and benefits from WSUD case studies has increased over time, its adoption is not yet a mainstream practice (Fletcher et al., 2015; Morrison and Brown, 2011).

This chapter examines WSUD implementation at an international scale, with emphasis on the role of legislative, regulatory, and institutional policies for WSUD uptake, using case study experiences from the United States, Europe, Australia, and Asia.

## 5.2 THE EVOLUTION OF INSTITUTIONAL ARRANGEMENTS TO PROMOTE TECHNOLOGICAL INNOVATION

Regulatory and institutional frameworks are aligned to conventionally accepted technologies and can act as barriers to the uptake of disruptive technologies. The existing water, wastewater, and drainage service models were based on the linear "big pipes in and big pipes out" transfer model, which aimed to protect public health and avoid nuisance impacts using large-scale technological solutions for narrowly defined service problems.

Moreover, water services governance is commonly centralized, rigidly regulated, and managed by government actors or licensed agencies (Pahl-Wostl, 2007). Often, institutional roles and rules were designed around specific elements of the water cycle, e.g., water supply, wastewater, or drainage services, considered in isolation.

On the other hand, WSUD promotes the "integration of urban planning with the management, protection, and conservation of the urban water cycle to ensure urban water management is sensitive to natural hydrological and ecological processes" (National Water Commission, 2004). WSUD offers a wide range of technologies at various scales (property, precinct, catchment), which can be managed using distributed, semicentralized, or centralized models, in contrast to the centralized governance model for traditional water infrastructure.

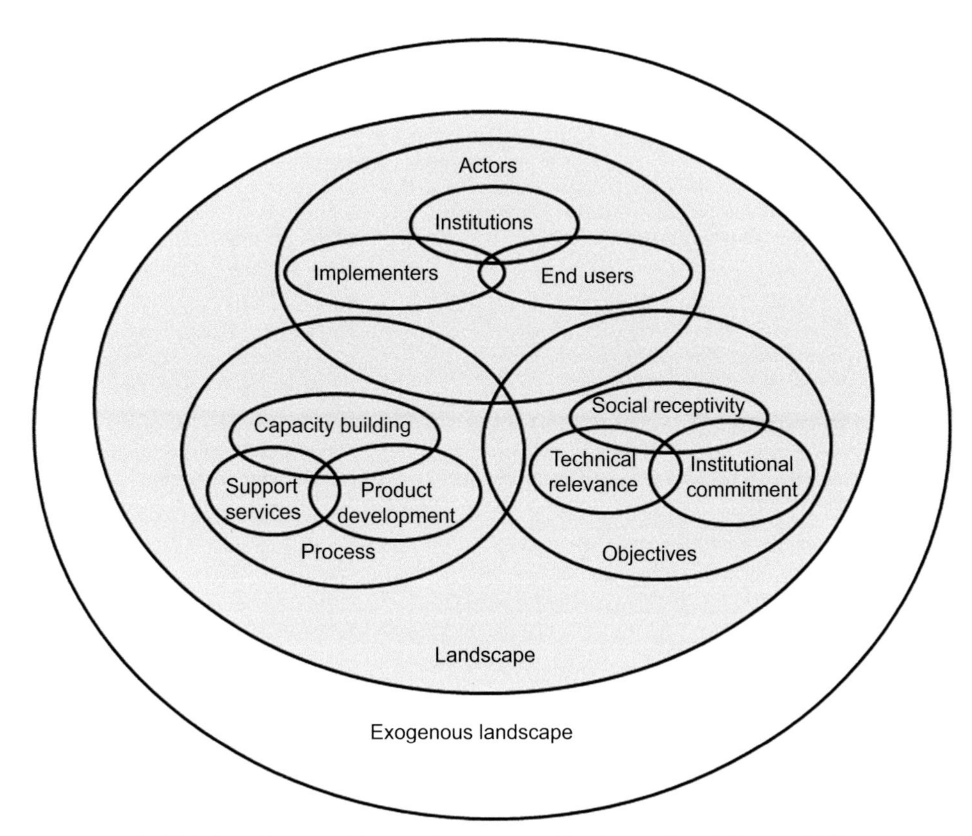


**FIGURE 5.1** Framework for enabling transition (Tjandraatmadja and Sharma, 2015).

WSUD implementation therefore requires a sociotechnical shift from the traditional governance model. Change policies need to address a complex landscape that combines exogenous factors (e.g., climate, economy, population) and endogenous societal traits (social, institutional, and technical) which are interlinked as shown in Fig. 5.1. A sustainable transition requires buy-in by key societal stakeholders, clear objectives that address stakeholder needs, and a process to attain such objectives (Tjandraatmadja et al., 2014; Tjandraatmadja and Sharma, 2015). Any transition process has to address expectations of market acceptance, service reliability, market imbalances, institutional and cultural barriers, institutional rules and inertia, and obligations and strategic goals of the organizations (Ferguson et al., 2013).

A diverse mix of tools and mechanisms can promote change and address in the different barriers (financial, technical, legislative, etc.) as illustrated in Table 5.1. Because WSUD applies to a wide range of spatial scales—regional, precinct, development, single plot—policy and legislation instruments need to consider the influence of the stakeholders and institutional actors at each scale. In addition, the design of the tools and their role in the implementation process will also evolve over time, requiring frequent reassessment and redesign, as needs change.

## 5.3 WSUD IMPLEMENTATION AROUND THE WORLD

Drawing on the experience of case studies in countries that have developed frameworks for sustainable stormwater management, we examine the regulatory and policy strategies used to implement WSUD in the United States, Europe (in particular, Germany), Australia and, more recently, Singapore.

## 5.4 AUSTRALIA

Australia is one of the developed nations most vulnerable to climate change. A highly urbanized country, more than 70% of the Australian population, lives in the capital cities shown in Fig. 5.2 (ABS, 2017).

Water infrastructure in Australia is resilient to the high seasonal variability in climate; however, climate change will increase the frequency of climate extremes that will exceed the capacity that Australia's infrastructure is designed to withstand (Preston and Jones, 2006; Garnaut, 2008; Saunders and Pearson, 2013).

**TABLE 5.1 Tools Used to Promote Technological Change**

| Tool | Examples | How Barriers are Addressed |
|---|---|---|
| Policy & legislation | National, state and municipal policies, legislation and by-laws, modification of planning, building and plumbing codes, application requirements, pollution abatement. | Policies enable technology uptake and integration into the institutional and societal landscape to ensure the fulfilment of societal needs. Legislation enables clarification of roles, responsibilities and mechanisms for government and non-government stakeholders. |
| Financial incentives and/or penalties | Rebates, tax exemptions, grant programs, low interest loans, discounts to products, fees. | Reduces the financial advantage derived from market dominance (scale) that incumbent technologies have over new technologies. |
| Capacity building | Training programs, investment in R&D, pilots and case studies, development of guidelines, professional accreditation programs, information campaigns. | Transfers technical knowledge, builds capacity and confidence of stakeholders in the adoption of new technologies or systems. |
| Technical standards and quality control | Standards, inspection and certification, certification and monitoring programs, product accreditation and service compliance programs. | Develops technical knowledge and ensures confidence of stakeholders in regards to the design, performance, implementation and operation of new technologies over its life. |

**FIGURE 5.2** Australian jurisdictions and capital cities. *Adapted from GEOCAT: 61756. 2005 Commonwealth of Australia (Geoscience Australia). https://d28rz98at9flks.cloudfront.net/61756/61756.pdf.*

Australian cities have separate stormwater drainage and sewerage systems. Water and wastewater services are typically provided by water utilities, and stormwater services are most commonly managed by local government or equivalent agencies (i.e., municipalities).

WSUD is a recognized measure for improving the resilience and the adaptability of Australian cities to increasing climate extremes and urban growth pressures.

## 5.4.1 Policy context

In Australia, the national effort toward WSUD commenced in 2004, with the National Water Initiative, which was a policy agreement between federal, state, and territory governments on urban and agricultural water issues (Australian Government, 2017). It facilitated, *inter alia*, the creation of water sensitivity cities, which lead onto the development of national guidelines for WSUD implementation (Choi et al., 2015).

States and territories progressed the next stage of reform at their own pace, driven by their priorities and particular context. As a result, each jurisdiction has developed its own WSUD policy and implementation strategy. This section will provide an overview of the Australian WSUD context, whereas Section 5.4.3 will provide details of the implementation framework in each jurisdiction.

The majority of Australian jurisdictions have state policies and tools that support WSUD as shown in Table 5.2. The implementation process, models, and tools adopted across jurisdictions vary, but a common feature is that the **implementation of WSUD occurs at the discretion of local government agencies**.

The main policy, regulatory, and legislative instruments used across Australia to support WSUD consist of

- State Planning Policy (SPP) frameworks;
- Water protection quality targets;
- Supporting tools for implementation at various levels; and
- Funding tools and mechanisms.

SPP sets a strategic framework for land use and development. In most Australian jurisdictions, the SPP is supportive of WSUD in the whole state; however, in New South Wales (NSW) and the Northern Territory (NT), WSUD policies apply to specific geographic locations (Piper, 2011; Tjandraatmadja et al., 2014). SPPs alone have limited enforceability and require support in the form of legal enforcement to enable their implementation (Choi et al., 2015).

Water quality protection targets and objectives enable environmental protection, and their importance is recognized in policy in all jurisdictions (Choi et al., 2015; Fortune and Drewry, 2011; McCallum and Boulot, 2015). However, each state adopts a different model for their setup and enforcement.

Queensland, Victoria, Tasmania, Western Australia, and the Australian Capital Territory have WSUD water quality targets that are incorporated in planning and development processes at both the state and local government levels. In addition, those jurisdictions also align water management with land-use planning, which is a key enabler for the implementation of WSUD through the development and approval process (Choi et al., 2015; Tjandraatmadja et al., 2014). Victoria, Queensland, and the Australian Capital Territory have mandatory water quality targets. In the other jurisdictions, water quality targets are incorporated under nonstatutory guidelines. However, with the exception of Victoria and the Australian Capital Territory, the mechanisms for implementation of stormwater quality and flow targets are discretionary, subject to the resources of local government and their local planning schemes.

The state of Victoria has a clear legal framework that requires the compliance with water quality targets and the use of best practice environmental management in stormwater management standards as the basis for planning decisions for all new developments. This greatly improved WSUD implementation (Cook et al., 2015). In addition, the policy and programs for stormwater management and capacity building are coordinated by a single entity, Melbourne Water (MW), responsible for waterway health and protection. MW works with local government to build capacity and support WSUD. This has enabled the integration of WSUD-centered stormwater quantity and quality management into the planning process and capacity building based on local government needs. This is particularly important for the successful operation and maintenance of WSUD features over the medium to long term (Cook et al., 2015).

The state of Queensland has mandatory environmental values and water quality targets under the SPP for certain scales of development. However, compliance with those targets is not mandated. Instead, the requirement is that the local governments' planning scheme "coordinates and integrates" with state and regional planning matters to the Minister's satisfaction (Cook et al., 2015). Some suggest these arrangements can have negative impacts on planning outcomes, as technical engineering solutions rather than the WSUD philosophy can often be adopted (Choi et al., 2015).

**TABLE 5.2** WSUD Policy and Support Instruments Across Australia (Choi et al., 2015; Tjandraatmadja et al., 2014; Cook et al., 2015)

| Tools | ACT | NSW | NT | SA | Qld | Tas | Vic | WA |
|---|---|---|---|---|---|---|---|---|
| Drivers | Waterway health | Water conservation. Waterway protection in selected areas* | Waterway protection in selected areas* | Water security with a focus on stormwater harvesting | Flood management, waterway and environmental health | Waterway health | Waterway health | Protection of groundwater and aquifers |
| WSUD targets | √ (M) | √ | √ | √ | √ | √ | √ (M) | √ |
| Integrated planning | √ | | | | √ | | √ | √ |
| Local WSUD guidelines | √ | √ (selected areas) | √ | √ | √ | √ | √ | √ |
| Legislated WSUD requirements for development | √ | √ (selected areas) | √ LGA | | √ | √ | √ | √ |
| State or territory Planning Policy | √ | Selected areas | Selected areas | √ | √ | √ | √ | √ |
| Capacity building program | No | WSUD.org and Splash Network (collaboration of Sydney Catchment Management authority and Greater Sydney Land Services) + Network of LGAs | No | Water sensitive SA (funded by Adelaide and Mount Lofty Ranges NRM Board, six LGAS, Australian Gov. Landcare Program, SA Water, LGA and Stormwater south Australia | Water by Design | No. | Clearwater program and Living Rivers | New Waterways program, hosted by WA Department of Water. Multi-agency partnership Dept. Planning, LGA, Swan River Trust and Urban Development Institute. |

Australian Capital Territory (ACT), New South Wales (NSW), Northern Territory (NT), South Australia (SA), Queensland(Qld), Tasmania (Tas), Victoria (Vic), Western Australia (WA), Mandatory (M), Local government authority (LGA).

Western Australia's policy framework supports the development and use of urban water management plans (UWMPs) and water management strategies as planning tools to set and manage water quality targets to prevent contamination of groundwater. The state relies on groundwater as a primary drinking water supply resource. Developments of various scales are required to be accompanied by a UWMP that incorporates WSUD principles. Preparation of a UWMP is generally a condition of subdivision approval by the local government with input from the Department of Water. However, the planning policy is less clear in stipulating how planning authorities are to give "due regard" to the state planning framework (Choi et al., 2015).

South Australia (SA) implemented mandatory reticulated water demand management through a building code. SA supports WSUD in policy and has quantitative water quality targets. However, these are not incorporated into the planning strategy and rely on local government to adopt WSUD requirements through local planning policy (Myers et al., 2014).

In NSW, the policy focus at state level is on mandatory targets for reducing water and energy consumption (e.g., up to 40% reduction in drinking water consumption) applicable to all new developments (BASIX, 2014). Proof of non-detrimental impact on receiving waters is required for specific growth areas under the State Environment Protection Policy (SEPP), and mandatory state stormwater quality objectives apply to selected areas, e.g., Sydney Growth Centres, the Sydney drinking water catchment, etc. (Tjandraatmadja et al., 2014). Although there is in-principle support for WSUD at state government level, each local government area is required to develop its own WSUD policy and targets.

All jurisdictions lack WSUD policy targeted at urban infill and lot scale development. Western Australia is the only jurisdiction that provides statutory planning control for all residential developments in relation to stormwater management via their SPP 3.1 Residential Design Code (WAPC, 2013). The Code requires all stormwater runoff to be retained on-site where possible (Choi et al., 2015). In other jurisdictions, individual lot scale developments are subject to various sustainability targets under building and plumbing regulations for each state. Despite the gaps in state policy, a number of local government areas across all jurisdictions have extended WSUD requirements to developments of various scales, including infill areas (Tjandraatmadja et al., 2014).

The NT has an array of policies and strategies that encourage WSUD in priority environmental management areas, with implementation driven by local government, such as the city of Darwin. However, it lacks a regulatory framework to support stormwater management for environmental protection (Cook et al., 2015; NTEPA, 2014; Tjandraatmadja et al., 2014).

### 5.4.1.1 *Institutional roles and responsibilities*

Local (or municipal) government is responsible for local planning and the approval of development applications. It is typically the main actor responsible for WSUD implementation. State government policy provides the overarching planning policy framework to legitimize WSUD, with high-profile and/or large-scale development planning decisions often made at the state government level, but implemented at local government level (Cook et al., 2015).

Local government areas have different approaches for WSUD implementation and promote WSUD at a range of scales (individual plot, street, precinct, and public open space), depending on specific local drivers (e.g., enhanced liveability, amenity, waterways protection) and constraints.

Location and scale of development impact WSUD implementation, with a wider pallet of WSUD technological options (property, street, precinct, subdivision scale), available for greenfield areas compared with infill areas. In particular, the number of feasible WSUD measures for small-scale infill developments and their impact on the stormwater objectives in a catchment can be limited. In such cases, the local government authority (LGA) often needs to evaluate a broader range of off-site WSUD measures that may be more biophysically effective and cost-efficient than local, small-scale solutions. Some LGAs also require higher levels of passive public open space[1] to be delivered through the planning process, compared to infill public open space requirements, thus providing land for delivery of WSUD at different scales (Choi et al., 2015).

However, infrastructure planning by water agencies and local government is often not aligned to facilitate the provision of WSUD in the public open spaces, although this is gradually changing, with examples of collaborative planning emerging across Australia.

1. Passive open spaces are vegetated areas that enable passive recreation, play and unstructured physical activity, and have the potential to provide benefits to stormwater quality and quantity. Examples include parks, gardens, linear corridors, conservation bushlands, nature reserves, public squares, and community gardens (Victorian and Tasmanian Parks and Leisure Australia, 2013).

### 5.4.1.2 Technical guidance and capacity building

Multistakeholder research agencies such as the Cooperative Research Centre for Water Sensitive Cities (https://watersensitivecities.org.au/) have been enablers in the promotion of WSUD, generating new ideas and tools in the technological, social, and policy spheres. However, the key challenge is the transition from research into practice.

Asset handover leads to maintenance responsibilities, and maintenance costs is a key concern to local government, as they are usually responsible for the long-term maintenance of WSUD assets in Australia. (as described in Chapter 22).

There are no national standards for WSUD infrastructure design, construction, and maintenance, although there is a wide range of technical guidance, which varies from state to state (Cook et al., 2015; Tjandraatmadja et al., 2014). A number of state agencies and local government agencies have developed implementation guidelines to encourage authorities, councils, and developers to achieve WSUD objectives (Cook et al., 2015). Choi et al. (2015) observed that the guidelines in most states are nonstatutory, poorly integrated, generally difficult to apply in an integrated and efficient manner, and that some have not been regularly updated. Thus, a key challenge is ensuring that the range of topics covered in WSUD guidelines and local standards (technical design, asset maintenance, and cost-benefit analysis) are relevant, tailored to the local physical, climatic, and geological conditions, and that the guidelines are well integrated into the planning framework for each region. In their absence, a number of local government areas have resorted to developing their own WSUD implementation guidelines. However, this can be challenging for local government areas with limited resources and is also inefficient.

Thus, capacity building programs are an important enabler for the sharing and dissemination of technical information, and best management practices (BMPs) in Australia. Most jurisdictions actively work on capacity building programs to support WSUD implementation as shown in Table 5.2. The majority of the capacity building programs are multiagency collaborations. In addition, Victoria also has a complimentary program, Living Rivers, which provides funding, guidance, and technical expertise to local councils interested in implementing pilot WSUD initiatives and building skills for waterway protection (Melbourne Water, 2012).

## 5.4.2 Funding

Financial constraints in local government can be a major challenge for WSUD implementation, particularly due to the need for institutional capacity building, the different maintenance regimens, and the governance required for WSUD treatment systems.

All jurisdictions, except SA, allow local government to levy developer contributions for stormwater under their Planning Acts. However, the approaches and the amount levied vary markedly within and across jurisdictions as shown in Table 5.3.

Queensland and NSW place a cap on the contribution levy, and any amount above the cap is regulated by the independent pricing regulator (Choi et al., 2015). Water quality offset programs are adopted in ACT and Victoria, which enable developers to pay money in lieu of providing physical works on their land. For example, the stormwater quality offset programs in Melbourne, Victoria, allows developers to pay money in lieu of providing physical works to achieve best standards for development if local conditions are not suitable. The funds are used to construct downstream stormwater treatment (Melbourne Water 2006, 2015, 2017a,b). A number of local government areas in Queensland and SA have developed their own offset schemes based on such a model (Cook et al., 2015).

## 5.4.3 Australian case studies

### 5.4.3.1 Australian capital territory

The Australian Capital Territory planning strategy (ACT Government, 2014) sets broad direction, targets, and objectives for water resources and water quality management. The *Waterways Water Sensitive Urban Design General Code* (the Code) adopts total urban water cycle management principles for the management of water resources (demand reduction, stormwater, and wastewater treatment and reuse) (ACT Government, 2009). The Code aims to promote predevelopment levels of stormwater export. It sets mandatory targets for potable water reduction and stormwater quality and quantity for all new residential, commercial, institutional, and industrial greenfield developments and redevelopments. The Code also proposes measures that can be adopted to achieve targets, tools to demonstrate that targets are being met, designs specific to Canberra conditions, and guidelines to assist developers in preparing submissions.

Under the Code, the onus for meeting water quality targets for a development lies with the developer, whereas regional and catchment-wide targets are the responsibility of government (Goyder Institute for Water Research, 2011). In case a

**TABLE 5.3 Fees and Charges for WSUD Development in Australia (Choi et al., 2015; Cook et al., 2015)**

| State | Development Contributions Under the Relevant Planning Act | Levies Under LG Act | Open Public Space Requirements | Off-set Schemes |
|---|---|---|---|---|
| ACT | | | | Developers can seek approval to contribute to offsite treatment if it is not feasible onsite. |
| NSW | Fee is capped at A$20K*/dwelling in established areas; A$30K/dwelling for greenfield areas. | LG can decide if onsite WSUD stormwater management can result in reduction or avoidance of the stormwater levy | Yes, but no value prescribed. | |
| Qld | Fee is Capped at A$20K* for 1-2 bedrooms; A$30K*/dwelling for 3 or more dwellings. | Yes | Infrastructure charges under an LGIP/PIP. No amount prescribed. | LGA can set offsets for water quality treatment in other parts of the catchment |
| SA | Development Act allows councils to provide basic subdivision infrastructure and dedicated open space. Some LGAs developed their own. | Yes | Max. 12-12.5% of the total area of the site depending on the size and number of allotments. | None. Some LGAs have implemented their own offset scheme. |
| Vic | Planning scheme may include one or more DCP for the purpose of levying community infrastructure or development infrastructure | Yes | 5% of site value under the sub-division Act, unless a different rate is specified in planning scheme or through PSP and DCP process −approximately 10% with 6% as active open space under the PSP Guidelines. | Offset program administered by Melbourne Water in service area. |
| WA | SPP 3.6 which forms part of all planning schemes | Yes | 10% under SPP 3.6 | |
| Tas | | | | None |

developer is unable to meet the water quality requirements, the planning authority has the discretion to allow the payment of a contribution by the developer for off-site control measures (ACT Government, 2009). A review of the WSUD code was initiated in 2014 to examine the strategic context for stormwater management and restructure the WSUD code such that it is more flexible and cost-effective, and simplifies the performance assessment (ACT Government, 2014).

### *5.4.3.2 New South Wales*

NSW has mandatory targets for water and energy consumption in all developments, including a 40% reduction in potable water consumption (BASIX) (https://www.basix.nsw.gov.au/planning-tools/basix), which has been a major driver for the installation of rainwater tanks in residential properties (NSW Government, 2004, NSW Government, 2014).

Stormwater regulations captured by the SEPP 2011 apply only to selected drinking water catchments and environmentally sensitive areas. In these areas, new developments must demonstrate *neutral or beneficial effect* on water quality and adopt recommended practices to attain this objective. Assessments are performed by local councils using standardized evaluation tools that are based on the risk posed to receiving waters. The Sydney Catchment Authority, the entity responsible for management of the main drinking water catchment, provides guidance, supporting tools and materials to interested local councils (SMCMA, 2012).

In all other areas of NSW, WSUD implementation is at the discretion of the local government, resulting in a fragmented WSUD policy across the State. Outside of the SEPP prescribed areas, a number of local government agencies have developed their own policies and objectives for WSUD implementation (Tjandraatmadja et al., 2014).

The Local Government Amendment (Stormwater) Act 2005 provides local governments with the option to charge a levy for managing stormwater services (OEH, 2011a). The upper limit for the annual levy on residential properties is AUD$25 per 350 $m^2$ of allotment area (DLG, 2006). Local governments can use their discretion in deciding if rebates for the levy apply to properties with on-site stormwater management (DLG, 2006). A review of the stormwater management service charge found that 77 out of 152 local councils implemented the charge in 2008/09 (OEH, 2011b).

#### 5.4.3.3 South Australia

In SA, the policy focus has been on water security and, in particular, stormwater recovery for reuse. There are also mandatory requirements for an alternate water source in new developments, which is commonly addressed by rainwater harvesting.

Stormwater management and infrastructure governance are shared across multiple agencies, with roles and responsibilities based on traditional disciplines. This fragments the components of the water cycle and poses challenges for WSUD implementation (Tjandraatmadja et al., 2014).

The state is highly supportive of WSUD through its strategic planning policy (DEWNR, 2013). The Adelaide Coastal Water Quality Improvement plan established the need to manage stormwater runoff quality (nitrogen, suspended solids, and colored matter) and quantity discharging into the Gulf St Vincent (McDowell and Pfennig, 2013). SA has quantitative water quality targets, but they are not connected to the planning strategy and are not mandatory (Cook et al., 2015).

SPP and legislation provide in-principle, nonbinding support for WSUD. Local government is responsible to "promote the provisions of the planning strategy under the Development Act" and has broad discretion to adopt WSUD policy (Choi et al., 2015).

The Development Act 1993 requires that "development complies with any requirements relating to sustainability of a building from an environmental perspective" and a range of nonstatutory policy and strategy documents have clauses supportive of WSUD (e.g., regional plans and structure plans for new growth areas) (Government of South Australia, 2011; Cook et al., 2015). The SA Planning Policy Library includes "water-sensitive design" principles of development control (Government of South Australia, 2011), but it has no reference to the supporting information. In the end, local government has to interpret the principles and define their application in local development plans (Cook et al., 2015).

Development plans and policies in each local government area are defined by the needs, resources, capacity, and strategies of each council. For common development, compliance with the Residential Code (Government of South Australia, 2008), which has no references to WSUD, is required. To implement any mandatory stormwater treatment objectives or WSUD requirements in the development plan, the local government needs to lodge a Development Plan Amendment to a statutory agency, except for prescribed areas, which have water quality target requirements set by the SA's EPA (Myers et al., 2013).

As a result, a number of local government areas have developed their own WSUD objectives and targets (e.g., City of Onkaparinga, City of Marion); the uptake of stormwater objectives and WSUD is highly fragmented across the state (Tjandraatmadja et al., 2014; Cook et al., 2015; Sharma et al., 2016). Local governments can also struggle to obtain an operation and maintenance budget for WSUD features. Water sector practitioners in SA noted that major impediments to WSUD uptake included a lack of knowledge around WSUD maintenance requirements and a paucity of WSUD post-implementation monitoring and performance assessment to verify if design objectives were achieved (Sharma et al., 2016).

#### 5.4.3.4 Queensland

The SPP—Water Quality (DSDIP, 2013) sets the framework and policy instruments in support of WSUD implementation, including quantitative water quality targets. The SPP covers the protection of environmental values of receiving waters, flow and quality controls of stormwater runoff for new developments ($>2500\ m^2$) to minimize nutrient release to waterways, and adaptation of stormwater management design to specific climatic zones in Queensland.

The SPP requires the local planning scheme to coordinate and integrate any state and regional dimensions of the matters dealt with by the planning scheme, including the management of stormwater and wastewater for protection of environmental values (Sustainable Planning Act, 2009 s88(1)(d); Choi et al., 2015).

The SPP—Water Quality guidelines state that WSUD principles can be incorporated into the planning and development process and refer to a number of best practice guidelines developed for specific Queensland context (Cook et al., 2015).

The SPP code sets water quality targets but is mandatory only for developments on land area greater or equal than 2500 $m^2$ and it is up to LGAs whether to adopt WSUD requirements for developments of other scales (Choi et al., 2015).

A number of local government areas (e.g., Mackay, Gold Coast, Townsville, and Brisbane) have developed technical guidelines for WSUD implementation at the local level.

The Healthy Waterways program in South East Queensland provided a sophisticated framework for the evaluation of the environmental health of waterways (hlw.org.au/report-card/report-card-methods) and was instrumental in the promotion of WSUD in the state. However, there are no binding targets for stormwater quality. Recent reviews have recommended more coordination and collaboration between state and local government in promoting compliance in the development industry and addressing uncertainty regarding WSUD management responsibilities (Jones et al., 2012; Cook et al., 2015).

### 5.4.3.5 Victoria

Victoria is considered the most successful state in mainstreaming WSUD within Australia (Choi et al., 2015; Cook et al., 2015). A seminal environmental report on Port Phillip Bay, adjacent to Melbourne, established the specific conditions for environmental protection of the Bay and focused on nitrogen and suspended sediment loads (Harris et al., 1996).

Nonetheless, there is no overarching WSUD policy in Victoria, but rather a number of complementary instruments to support WSUD implementation, including (1) planning rules (clause 56) in Victoria planning provisions requiring WSUD in new subdivisions; (2) best practice environmental guidelines and water targets linked to State Environmental Planning Policy; (3) clear responsibilities for WSUD implementation shared between MW and local government councils; (4) large number of demonstrations projects to prove concept and cost-effectiveness; (5) capacity building and seed funding programs; (6) champions and leading municipal councils; and (7) organizational leadership such as Stormwater Victoria, MW, Municipal Association of Victoria, specific urban developers and consulting firms (Morrison and Chesterfield, 2012). Many of the policy and programs addressing stormwater management are managed by MW, which is both the bulk water supplier and caretaker agency for waterways' health in catchments >60 ha.

Clause 56.07 under the Victorian Planning Provisions is a distinguishing and a key enabler for WSUD implementation in the state of Victoria. Clause 56.07 requires all new residential developments and subdivisions to comply with integrated water management provisions and urban runoff management objectives in accordance with the *Urban stormwater best practice environmental management guidelines* (CSIRO, 1999). These apply to all local government planning decisions. Consequently, the legislation effectively gives local government a mandate to enforce WSUD.

Exemptions can be granted in existing urban areas under certain conditions, but these only apply to small residential subdivisions (e.g., one lot into two lots) or "infill" development in an area of less than 1 ha. Under such exemptions, offset payments (to MW's Stormwater Quality Offset program) usually apply after subdivision approval by the local government. Developments >5 ha do not have the option of paying the offset and must meet best practice water quality targets within the development.

Commercial, industrial, and residential developments in **developed areas** are not required to adopt WSUD measures; however, a number of urban inner-city councils have developed their own requirements for infill as local planning requirements (Morrison and Chesterfield, 2012). In these council areas, best practice targets for WSUD applies to all infill applications (new buildings, extensions to buildings, and subdivisions) greater than 50 $m^2$ floor area in business zones.

The Stormwater Quality Offset Program managed by Melbourne Water (2006) is a financial contribution from developers for regional water quality works, to offset pollution not treated by WSUD in a development (Melbourne Water, 2015). The offset rate is currently (in 2017) $6645 per kilogram of nitrogen plus an 8.9% administration fee (Melbourne Water, 2014). The offset contribution is based on the area developed and the development type (available from https://www.melbournewater.com.au/planning-and-building/developer-guides-and-resources/drainage-schemes-and-contribution-rates-1). Higher density developments are charged a higher rate per $m^2$ due to the higher volume of stormwater runoff and higher nitrogen load discharged to receiving waters. The offset rate is also adjusted for rainfall: developments in higher rainfall zones pay a higher rate due to increased runoff volumes. Melbourne has a strong spatial rainfall gradient, with annual rainfall increasing from 540 to 1170 mm on a west to east trajectory (BOM, 2017).

An example stormwater quality contribution calculation is presented below (Melbourne Water, 2006):

Assumptions:

Standard contribution rate (i.e., for lots from 450 $m^2$ but less than 1000 $m^2$) = A$6645 per hectare.

Development density factor = standard residential = 1.0.

Development size = 1 ha.

Percentage reduction achieved on-site = 36% (treatment performance estimated using specialist water quality modeling)

- 36% reduction in total nitrogen (typical annual load t/year) achieves 80% of the best practice objective
- Offsets required for remaining 20%
- Offset contribution—$6645/ha × 1 ha × development density factor (1) × 20% (shortfall of best practice)
- Amount payable = $1329 (once off).

In addition, capacity building and grants programs, respectively, the Clearwater and the Living Rivers Programs, were created to support local government and industry professionals to plan, manage, and implement WSUD. Clearwater focuses on building the skills, knowledge, and networks to promote integrated water management practices and the understanding of the life-cycle requirements of WSUD features (design, planning, engagement, governance, and capacity building) (https://www.clearwatervic.com.au) (D'Aspromonte et al., 2012). The Living Rivers program provides funding, jointly with expertise and guidance to help municipalities build capacity, develop, and implement WSUD projects (https://www.melbournewater.com.au/community-and-education/apply-funding/living-rivers-funding).

Despite being considered the most successful state in WSUD implementation efforts (Choi et al., 2015), WSUD uptake is not uniform across the Melbourne Region. A survey of 38 local municipalities revealed strong municipal commitment to WSUD in coastal areas, and in areas with high vegetation cover. These coincided with communities of higher wealth and population (Morrison and Brown, 2011). More generally, WSUD was seen as a passive practice, mostly driven by a bottom-up process (i.e., not by senior council management), with a few exceptions where councils are strongly committed to WSUD practices (Eggleton, 2012). At the time only 44% of the 38 municipalities considered stormwater quality when conducting other infrastructure projects, and only 34% had budget for stormwater quality works (Morrison and Brown, 2011). A major challenge was the ability of local government staff to deal with the increased complexity of integrated water management, which can hinder smaller local councils from adopting more complex treatment features (Eggleton, 2012).

Notwithstanding the supportive policy and legislative framework, there is a strong need to continue to invest in building capacity in councils, particularly for adequate resourcing and long-term maintenance of WSUD features.

#### 5.4.3.6 Western Australia

Western Australia relies on groundwater as a major source of water supply in the state. Land use and water planning are integrated in state and local planning policy, which has contributed to significant investment in the development of local government capacity and resources for WSUD implementation (Government of Western Australia, 2011, 2013).

SPP-Water Resources (SPP2.9) (WAPC, 2006) supports a range of high level WSUD policy measures. However, the Planning and Development Act 2005 does not establish, with clarity, where planning authorities are required to give due regard to the State Planning framework.

Western Australia's policy framework supports the development and use of Urban Water Management Strategies as planning tools to set and manage water quality targets to prevent groundwater contamination (Choi et al., 2015).

The Liveable Neighbourhoods guidelines (DPI and WAPC, 2015) and the Better Urban Water Management framework (WAPC and DEPI, 2008) set the key WSUD policies for district[2] structure planning, urban subdivision in greenfield areas, and urban renewal developments (residential, commercial, industrial). Hence, consideration of urban water management requires the involvement of all planning agencies, from state to local government.

Large (>25 lots) or small sub-subdivisions within a priority catchment must demonstrate how the development complies with the state policy. New developments are required to maintain surface water concentrations at predevelopment levels and, if possible, improve on these conditions. If the stormwater discharges exceed the ambient conditions, the proponent must undertake water quality improvements in the development area or achieve an equivalent water quality improvement offset inside the catchment. The attainment of water quality and quantity objectives must be justified using modeling tools or other assessment methods acceptable to the Department of Water.

A developer must prepare a UWMP, which incorporates WSUD principles as a condition of subdivision approval by the Western Australian Planning Commission (WAPC), followed by approval from the local government, with input from the Department of Water. The WAPC approves local planning schemes and structure plans, subdivision, and UWMP, whereas local government approves engineering/construction drawings and specifications and monitors construction activities.

---

2. An area generally greater than 300 ha (but may not be in inner metropolitan areas) may be greater than one local government area (Government of Western Australia, 2008).

In practice, the enforcement and implementation of WSUD at local level is challenging because the size of the subdivisions and preapproval process by WAPC can restrict the flexibility of local government in setting WSUD strategies at local scales.

### *5.4.3.7 Tasmania*

The state government of Tasmania established water quality management targets and objectives for all estuarine and coastal surface waters in Tasmania, which was instrumental in enabling WSUD implementation (State Policy on Water Quality Management, State of Tasmania 1997). The State Policy on Water Quality Management provides a strategic framework for water quality management, including stormwater, and is implemented through local planning schemes. The state stormwater strategy (DPIPWE, 2010) supports the need to manage stormwater and recommends a range of best management WSUD practices and stormwater quality and quantity targets for new developments based on Integrated Water Cycle Management and WSUD principles. When the targets were introduced (in 2010), federal funds were provided to industry and local governments to implement various WSUD projects, which integrated stormwater quality and quantity management objectives.

The State Policy on Water Quality Management provides guidance on managing stormwater during the construction and operation stages of a development, and in established urban areas. Stormwater targets are consistent with those in other Australian states and apply to new developments >500 m$^2$ of impervious area. However, implementation at local level is at the discretion of local government.

The policy was further supported by WSUD design guidelines for various local conditions (Department of Primary Industries Parks Water and Environment, 2012), and capacity building programs for the local government, industry, community, and school sectors (Chrispijn and Weise, 2012). Furthermore, the state government created a stormwater task force to coordinate stormwater monitoring programs, and the development of technical guidance, e.g., Tasmania's water sensitive design manual and a model for the stormwater management plan (Government of Tasmania, 2012); training workshops, forums, and the creation of stormwater support officer positions.

A number of local governments and regions have developed WSUD development guidelines and practice notes to assist developers in implementing WSUD approaches. For example, Hobart City Council provides guidance for possible WSUD configurations for different development types (Hobart City Council, 2006).

Overall, 40 WSUD stormwater projects were implemented in the last 10 years by local government and industry. The funding for such projects came from a range of sources, but mostly from federal government grants given the absence of state funding.

The Tasmanian Local Government Act (1993) enables charges to be levied for stormwater management systems for pollution control. These levies can be used by local government to fund retrofitted systems. The trigger to implement these financing tools is the apparent deterioration of waterways due to contaminated runoff in an established area, and where the general works program is unable to cover the capital cost of the management systems.

### *5.4.3.8 Northern Territory*

Urban development in the NT is predominately in the wet tropics region in Australia (>1700 mm annual rainfall), characterized by a high-intensity seasonal rainfall and an extended dry season (May to October) with less than 10 mm rainfall per month (Saunders and Pearson, 2013)—a climate that poses specific challenges to WSUD.

WSUD approaches in new developments are not mandatory in the NT. There are no quantitative targets for developments to minimize the downstream impacts of runoff (Aurecon, 2011). However, a range of policies and strategies encourage the adoption of WSUD in developments located in priority environmental management areas, such as around Darwin Harbour, which is the receiving environment from the two most populated urban areas in the Territory, Darwin and Palmerston.

The NT Water Act 2004 regulates the use of water (including surface and groundwater) and defines applicable environmental values and associated water quality objectives (Northern Territory Government, 2016). Under the *Water Act*, a declared Water Control District requires enhanced management for the preservation of groundwater reserves, river flows, and wetlands.

The NT Planning Scheme (Clause 11.4.1) (https://nt.gov.au/property/building-and-development/northern-territory-planning-scheme-part-5.pdf) specifies that development applications that propose to subdivide rural or unzoned land must include a land suitability assessment and a stormwater management plan (Cook et al., 2015). The land suitability categories include drainage assessment, erosion risk, and flooding (Northern Territory Government, 2014). There is no explicit mention of managing nutrients in stormwater runoff. In addition, developments in priority Environmental Management areas may be required to prove that they will not a cause detrimental impact on the environment and may need to comply with conditions specified by a development consent authority's WSUD plan (Macmillan and McManus, 2009).

A WSUD planning guide was developed for the NT's Department of Planning and Infrastructure (McAuley and McManus, 2009). This guide proposed objectives for WSUD in new developments and recommendations on how WSUD strategies can be incorporated in the development assessment process.

The Darwin Harbour Strategy 2009–15 (Darwin Harbour Advisory Committee, 2010) was the basis for strategic development planning in the Darwin Harbour region. It established a vision, goals, and supporting mechanisms for WSUD implementation in new large, greenfield developments in the Darwin Harbour area (McAuley, 2009). The WSUD Urban Design Strategy led to the development of policy, tools, and resources to support planning and implementation of WSUD systems in the region.

A draft policy promoting the adoption of WSUD objectives and enforceable targets was developed in 2010, but the legislation was not finalized. However, the 2014 Stormwater Strategy for the Darwin Harbour region recognized the need for a regulatory framework to enable WSUD implementation (NTEPA, 2014).

Interim guidelines for design and assessment of WSUD features were initially developed by Territory agencies, but the major local councils in the region, City of Darwin and the City of Palmerston, developed their own guidelines to proceed with local WSUD implementation (McAuley and McManus, 2009).

WSUD features in a small number of developments in Darwin (~30 developments in total), which has allowed local government to follow-up the progress of WSUD implementation in each of the developments, which in turn enabled the building of in-house capacity on WSUD assessment (Tjandraatmadja et al., 2014).

The key challenges for the NT are the limited understanding of the performance and life-cycle costs of best management practice technologies, given the unique climate conditions of the region, as well as the usual candidates of awareness, technical capacity, and resources (Saunders and Pearson, 2013).

### 5.4.4 Lessons from Australia

WSUD implementation has progressed at different rates across Australian states, and there have been pockets of significant success regarding implementation. Although the WSUD implementation frameworks differ across states, practitioners in all jurisdictions agree that WSUD uptake in local government areas varies depending on the LGA's technical capacity and resources. Even in Victoria, there is recognition that further work is required before WSUD becomes a mainstream planning activity. Creating an enabling environment to support LGAs has been a critical step in implementation.

All the states and territories have made efforts to incorporate the principles of WSUD in the planning and development process at both state and local government levels. State policies and guidelines provided the overarching framework for consideration of WSUD, but its implementation occurs at the discretion of local government, through its local planning schemes. The exceptions are Victoria and the ACT, which developed mandatory WSUD targets for new developments. In other states, such as NSW and Western Australia, state governments have introduced amendments to planning schemes that formalize the consideration of WSUD in the development process but still require local government to develop their own policies and instruments.

In many instances, local government guidelines acknowledge WSUD principles and best practice approaches but specify neither performance targets of reduced pollutant export loads nor the maintenance of predevelopment runoff flows. Often the focus of WSUD in local government guidelines is on water quality. The quantity of water entering minor drainage systems and natural waterways is dealt with by Council's engineering technical specifications for drainage. Hence stormwater quality and quantity outcomes are not yet fully integrated.

The Victorian framework and model is considered an exemplar because of the clarity in the roles of agencies, alignment between policy, environmental objectives, capacity building, and the provision of ongoing funding to enable local government to implement WSUD. Brown et al. (2013) argue that the stormwater transition process in Melbourne is a sterling example of long-term systemic change, enabled by relationship developments at different organizational levels and spheres, and not the result of a top-down, master blue print policy process.

Once an enabling framework was established, a common challenge is the design and technical performance of WSUD. A common challenge is the lack of scientific verification for achieving water quality and quantity objectives in Australian jurisdictions (Adams and Jayasuriya, 2014; Cook et al., 2015; Fletcher et al., 2008; Hamel and Fletcher, 2014). In particular, because a large number of WSUD systems were first developed in the Eastern states of Australia (NSW and Victoria), the effectiveness of these BMPs in other climatic zones and soil conditions, such as the NT and Western Australia, requires further evaluation. Overall WSUD performance post-implementation requires verification to enable improvements and to ensure that investment to achieve water quality objectives in particular is appropriate (Gardner, 2015).

## 5.5 UNITED STATES OF AMERICA

In the United States, protection of water quality in streams and lakes from overland pollution has been the main driver for stormwater management since the 1970s, and a strong legislative approach for environmental protection is used.

### 5.5.1 Policy and legislative context

The 1972 Clean Water Act (CWA) requires all commercial, municipal, and industrial point source dischargers of wastewater or stormwater across the United States to focus on water quality by obtaining a permit, and complying with the effluent limits specified in the National Pollutant Discharge Elimination System (NPDES) (Boatwright et al., 2014; Hansen, 2013; USEPA, 1972).

The NPDES system covers a range of issues beyond WSUD, but in many states, WSUD has been the main focus for dealing with non–point source water quality problems. The NPDES applies to all municipalities operating separate stormwater and sewer systems. The NPDES includes provisions to encourage municipalities to use WSUD techniques in their stormwater management plans to stay within their permitted Total Maximum Daily pollution Load limit. Since its introduction, this has led to an exponential increase in the numbers and type of WSUD features across municipalities (Dolowitz et al., 2012; Dolowitz, 2015).

Under the NPDES, all regulated areas must submit a notice of intent on how they propose to comply with the USEPA (or equivalent state agencies) Minimum Control Measures. Under the CWA, a municipality responsible for compliance with wastewater and or stormwater permits can choose how to comply with its legal obligations to manage combined sewer overflows (CSOs). They can adopt WSUD infrastructure if it can demonstrate that the WSUD infrastructure will perform as effectively as the traditional gray infrastructure, such as pipes, culverts, and other engineered solutions, which usually involve concrete and steel (Hansen, 2013). To demonstrate CWA compliance, communities need to characterize baseline conditions in a catchment, implement WSUD options, and measure the results (Hansen, 2013).

The legislation requires municipalities to use WSUD approaches in stormwater management plans and stresses the importance of using relevant local authorities in the process. Federal technical services and financial aid is provided to the states and municipalities to help prevent, reduce, and eliminate pollution.

Under the CWA, the USEPA is required to create a set of technologies based on effluent limits to set by the NPDES permits system (USEPA, 2017) and is authorized to provide federal assistance to support demonstration projects, and control activities, that use WSUD technologies and techniques.

Furthermore, to support states and local organizations, a number of federal agencies collate, disseminate, and promote the adoption of WSUD stormwater BMPs from the United States and across the globe. The USEPA regularly publishes data on low impact stormwater BMPs (https://www.epa.gov/npdes/national-menu-best-management-practices-bmps-stormwater#edu) and implementation rates across the United States (https://www.epa.gov/water-research/geoplatform-stormwater-bmp-performance-database-0). The explicit purpose is to encourage municipalities and states to borrow the best practices and increase the pressure on local political leaders to be better than just average in the ranking tables. This has contributed to more than 30,000 demonstration sites that offer municipal governments the ability to learn from each other and share knowledge (Dolowitz, 2015; Zathmakesh et al., 2015).

### 5.5.2 Institutional roles and responsibilities

The USEPA is responsible for setting criteria, standards, and conditions for the implementation of programs for stormwater management at the national level. The responsibility for implementation is delegated to the states, provided the states' programs meet national requirements. If the states do not elect to participate or do not secure federal approval of their programs, implementation proceeds based on the national authority exercised by national officials. For states that do participate, implementation employs national and state authorities concurrently. In the end, implementation still requires state implementers, with additional leverage from federal government. In practice, the final responsibility for stormwater and CSO management resides with local government and municipalities (Breckenridge, 2014).

USEPA also conducts workshops and publicizes the best BMPs from around the United States and the world.

Many municipalities are now implementing GI programs to reduce stormwater runoff and non–point source pollution and to avoid the rising costs of replacing aging gray infrastructure due to changes in climate and population (Madden, 2010). To encourage WSUD uptake, the USEPA has entered into partnerships with selected municipalities, providing funds and technical expertise to support the setup of demonstration projects and research on WSUD practices in cities such as Portland, Oregon (Hansen, 2013). Cities such as Portland, Seattle (Washington), Chicago (Illinois), Austin (Texas),

Fairfax (Virginia), and Madison (Wisconsin) are early adopters of WSUD for stormwater management and are perceived as exemplars of what could be achieved (Dolowitz et al., 2012; Dolowitz, 2015).

### 5.5.3 Challenges to WSUD uptake

Despite the institutional and support framework developed, WSUD uptake has been slow and fragmented in the United States (Bowman and Thompson, 2009; Madden, 2010; Dolowitz, 2015). Notwithstanding demonstration cities, few American cities have invested in WSUD on a significant scale, preferring gray infrastructure (Madden, 2010). The communities that implemented WSUD practices to meet stormwater management goals often created their own guidelines, manuals, capacity building, and awareness programs (Benedict and McMahon, 2006; Maass, 2016; Chaffin et al., 2016).

The implementation of WSUD at municipal level depends on each municipalities' ability to access resources and familiarity with WSUD (White and Boswell, 2006; Maass, 2016; Dolowitz, 2015). This is seen in the WSUD programs from the municipalities of Oro Valley, Kirkland, Philadelphia, and Los Angeles as shown in Table 5.4. The municipalities differ in population size, resources, and capacity to implement WSUD. Oro Valley, the smallest, struggled to implement an effective WSUD program. Large cities with more resources, such as Philadelphia, are often able to negotiate better federal support and develop more comprehensive WSUD programs (Maass, 2016).

The barriers to WSUD uptake at the municipal and practitioner level include a diverse range of institutional, political, regulatory, and technical design challenges. In particular, budget priorities, institutional norms, path dependence, risk aversion, and reliability are strong influences on how municipal government interpret and implement laws and make decisions (Madden, 2010; Dolowitz, 2012; Jiang et al., 2015).

The perceived challenges to WSUD implementation include

1. The need to pass local planning laws and adjust plumbing and building codes;
2. Capacity for integration with existing infrastructure (e.g., sewers system) and their minimum load requirements to prevent blockage;
3. Health regulations related to waterborne pathogens;
4. The legislative approval process;
5. Uncertainty of legal challenges by those adversely affected by the new technology (e.g., road builders);
6. Water rights, allocation of financial and legal liability, contracting, and insurance;
7. Lack of community support (Chaffin et al., 2016);
8. Gap between government policymakers and technocrats in legislating/implementing WSUD;
9. Tax codes and incentives; and
10. Technical design limitations.

For instance, the southwestern/western United States faces multiple water challenges and rapid population growth, accompanied by a lack of regulatory and/or economic drivers, and a shortage of information on the performance of the BMPs in their arid/semiarid climates. These issues severely hinder the uptake of WSUD (Jiang et al., 2015). Many of the BMPs and techniques have been developed and tested in the humid eastern United States, with few studies focusing on the arid and semiarid regions of the western/southwestern United States. This information "bias" was contributing to the misperception that WSUD technologies were useless in a region with low annual precipitation (Jiang et al., 2015). Lack of support for WSUD in the Midwest is also influenced by relatively few installations, perceptions of higher costs, lack of willingness to pay by home owners, legal considerations, and financial risk (Bowman and Thompson, 2009; Hansen, 2013).

States and municipalities across the United States have different governing systems, legal codes, local governing laws, and regimes for services. Legal, zoning, and building codes are set locally and can become barriers to the transfer of WSUD BMPs across municipal and state boundaries (Dolowitz, 2015). Thus, municipalities often prefer solutions already adopted within their own state, in the belief that such programs or technologies require less effort to adapt to the local economic, social, legal, and political systems (Dolowitz, 2015).

However, findings suggest that the community expectations and resources in high-performing municipalities further influence the WSUD implementation. An analysis of the local government preparation for stormwater BMPs in California and Kansas demonstrated that coastal municipalities produced higher quality plans than their noncoastal counterparts (White and Boswell, 2006; Morrison and Brown, 2011). One explanation was that "stormwater quality problems were more salient to a coastal community due to their economic and quality of life interests in coastal recreation," Alternatively, coastal communities could also have higher expectations of stormwater quality improvement.

**TABLE 5.4 Examples of WSUD Policy and Implementation in the USA**

| Municipality | Implementation |
|---|---|
| Oro Valley, Arizona[a] (40,000 residents) | The stormwater utility is responsible for safety, water quality protection and environment before and after storm events. The Stormwater Utility Commission's Stormwater Management Plan Annual Report was created in compliance with The Arizona Department of Environmental Quality. The Plan provides information on the towns, stormwater program, performance measures and monitoring process. However, no monitoring occurs as the town's washes do not fit the criteria stipulated in the ADEQ permit.<br>The City conducts public education via the provision of: (a) brochures and fact sheets with utility bills, at public events, website and response to public enquiries to explain stormwater management, (b) household stormwater friendly activities and source control activities with information on BMPs, and (c) a partnership program where community groups or individuals supported and informed by council can choose to maintain a certain wash.<br>Funding of the program is via a flat stormwater fee: US$2.90 for residential properties and US$2.90 per 5,000 sq.ft of impervious surface for commercial, business and non-profit properties. Stormwater fee covers 50% of SWU's funding and the other half is funded by local. State and Federal agencies, volunteer groups and the Pima County Regional Flood Control District. |
| Kirkland, Washington[a] (85,000 residents) | Kirkland City has a Surface Water Master Plan and a Stormwater Management Plan. The city uses the 2009 King County Surface Water Design Manual which requires developments to evaluate the feasibility of WSUD facilities and to install at least one element to mitigate runoff onsite, based on lot size and the amount of impervious area. If BMPs are not feasible on a site a "Stormwater adjustment form" must be submitted to gain an exemption. Smaller developments are encouraged to undertake the assessment, but it is not mandatory.<br>Single-family properties pay a flat annual rate $217.62 ($16.87 per month plus 7.5% utility tax). Commercial and multi-family properties pay $16.87 per equivalent service unit (a measure of the impervious surface of the property) plus 7.5% utility tax. Commercial properties can reduce their annual stormwater annual fee by 10% if they implement a 'permissive rainwater harvesting system' to collect and use 95% of the annual runoff volume. |
| Philadelphia, Pennsylvania[b] (1.56million residents) | Philadelphia Water Department's plan, Green City, Clean Waters, is a 25year plan to develop a citywide network of green stormwater infrastructure to manage run-off, restore streams and upgrade wet-weather treatment plants. It is based on a watershed and GI approach and includes large scale WSUD on public land, requirements and incentives for WSUD on private land and large-scale tree program. PWD has also developed technical resources and tools, offers a WSUD grant program for non-residential property owners. Implementation is a collaboration between the PUD, the Office of Watersheds, Streets Department, Park commissions and various city departments. Stormwater fees for residential customers are based on the average surface area impervious cover, whilst commercial customers fees are based on the specific impervious area and total area of the property, fees are reduced runoff is reduced. The city entered into a 20-yr agreement with the USEPA to support and adapt the program thus enabling it not to face penalties in case of lack of compliance during the implementation period. |
| Los Angeles, California[a] (4 million residents) | City of Los Angeles (CoLA) created a 'Standard Urban Stormwater Mitigation Plan' (SUSMP) in 2002 to manage stormwater runoff from developments and collaborates with the greater Los Angeles region to develop specific watershed management plans, which integrate WSUD. CoLA has a Low Impact Development Ordinance, which requires the SUSMP to include WSUD BMPs in all development and redevelopments which need to capture, infiltrate or use 'rainwater from a three quarter inch rain-storm' using the techniques outlined in the City's LID Handbook (2011). The LID handbook provides WSUD requirements, processes, design criteria for BMPs. The LID ordinance requires a LID Plan for the approval of building and grading permits.<br>Technical requirements are based on the scale of the development: small-scale residential developments can use rain barrels, permeable pavers, raingardens to comply with the ordinance, larger developments choose from a more complex list of BMP options. It provides details on construction, O&M, size requirements and infiltration, soil and vegetation. If the development is unable to achieve ordinance requirements onsite, an offsite mitigation project that provides at least the same level of protection to maintain the original site can be negotiated.<br>Extensive information is provided to community via CoLA's website, Facebook, twitter, YouTube, Blog and a library of documents produced by the city and reports and links to other sources. CoLA also runs disposal of pollutants, volunteer clean-ups, classroom and clean-up event promotion. |

[a]*Maass (2016)*
[b]*Madden (2010), Hansen (2013)*

In addition, municipal stormwater performance and the relative economic conditions and education of the community were correlated. Municipalities that were early adopters of best management practice plans had higher levels of wealth, education, and larger size populations in their respective communities (White and Boswell, 2007).

In summary, the US experience demonstrates that despite the support of federal agencies, changes in regulations, funding for exemplar case studies, and the dissemination of information on WSUD, knowledge transfer and adoption at a practical level are strongly influenced by local and regional factors (parochialism). Barriers differ across regions and local municipalities, but there are some common challenges in implementation. Current barriers include a lack of unified policies and legislation for integrated water resource management across municipal boundaries or watersheds (Maass, 2016), lack of acceptance by residents, inadequate support from the multiple local stakeholders involved in decision-making in any municipal area and limited understanding of the potential role of WSUD (particularly in semiarid/areas) (Dolowitz, 2012, 2015). These emphasize the importance of building capacity and assisting local government to plan and build collaboration across local agencies and departments for the planning, implementation, funding, and operation of WSUD infrastructure (Dolowitz, 2015).

## 5.6 SINGAPORE

### 5.6.1 Policy and institutional context

Singapore is a small, high-population density island nation of 716 $km^2$ at risk of water scarcity despite rainfall in excess of 2400 mm/year (Lim and Lu, 2016). Until recently, it was highly dependent on imported water from neighboring countries. To reduce its water security risk, the Singapore Government decided to expand its water supply catchment to 90% of the country as part of its national water strategy, in addition to recycled, desalinated, and imported water sources (PUB, 2017). The Singapore Government drove the transition toward an integrated water island by embarking on a process of regulatory, administrative, and infrastructure reform based on WSUD principles and lessons from the Australian framework for integrated water management (Lim and Lu, 2016).

Singapore introduced the Active Beautiful Clean (ABC) Waters Program in 2006 via the Public Utilities Board (PUB). The ABC goal was to achieve urban stormwater and flood control using WSUD techniques. The Program has resulted in the installation of over 60 best management practice projects across Singapore to date, with another 100 projects scheduled by 2030 (Lim and Lu, 2016).

Singapore renovated its dense network of concrete drains and canals to transform two-thirds of its land into water catchments that supply 17 reservoirs and transport excess stormwater into the sea (https://www.pub.gov.sg/watersupply/fournationaltaps/localcatchmentwater). The country adopted BMPs to treat stormwater at source and to restore the natural hydrological process. Flood barriers were built in areas prone to flooding, and stormwater was integrated into the landscape, where possible, to improve water quality and create aesthetically pleasing community spaces around waterbodies. The features promoted included bioretention systems, infiltration systems, green roofs and walls, restoration of built canals to natural form, constructed wetlands, etc.

### 5.6.2 Institutional reform

The ABC Master Plan was launched in 2007 using a 3P approach (People, Public, Private) and brought together public agencies, private companies, and the community for the development of catchment planning policies, stormwater management plans at various scales, performance targets, and compliance tools.

Technical know-how was enhanced through investment in R&D by government agencies, delivered by tertiary institutions using field experiments and demonstration sites to test the performance of BMPs. The program developed new knowledge on the design, performance, and life-cycle costs of systems tailored to the local conditions.

Sharing of learnings was also a critical part of the process. For example, the PUB released technical guidelines to assist designers and practitioners in the technical implementation of BMPs and included

- ABC Waters Design Guidelines (2009, 2011, 2014);
- Handbook of Managing Urban Runoff (2013); and
- Engineering Procedures for ABC Waters Design Features (2009, 2011).

The National Parks Board also produced guidelines for selection of local plants for rooftop greenery and waterbodies (Tan and Sia, 2008; Yong et al., 2010; Tan and Chiang, 2011 in Lim and Lu, 2016).

The PUB invested in building the technical capacity of the sector, with training and certification programs (e.g., ABC certification, ABC professional programs, and ABC professionals registry), whereas the government required ABC principles and designs to be incorporated into the curriculum of local universities (Lim and Lu, 2016). To promote public awareness and develop environmental stewardship, the PUB invested heavily in community education and awareness campaigns, creating publications, educational activities, information signs at ABC project locations, and the promotion of citizenship participation in monitoring activities at ABC sites (Lim and Lu, 2016). In addition, the PUB actively promoted community acceptance of recycled water for drinking in their New Water initiative.

### 5.6.3 Transition challenges

The ABC Program has largely achieved its Active and Beautiful goals in only 10 years as evidenced by successful institutional reform and community awareness and engagement. The hydrological and environmental performance is still under implementation and is being actively monitored for verification of BMPs' performance in the tropical climate. Nonetheless, the program has generated a number of innovative technologies and has been recognized by a number of design and sustainability awards (Lim and Lu, 2016).

The ABC Program progress was driven by strong political will, generous resourcing, and the holistic framework adopted for its implementation. Success factors include

- The role of PUB as the major coordinator of the technical and institutional reform was clearly defined and enabled effective coordination of information, roles, and collaboration across agencies and sectors.
- The rapid on-ground delivery of ABC technologies is largely due to clear design guidelines that were enforced by the PUB. This is in contrast to, say, the United Kingdom where WSUD concepts have been practiced for a much longer time, but which lacked unified design codes and performance databases (Lucas et al., 2015).
- The development of institutional design guidelines maintained consistency in design and WSUD performance monitoring for tropical cities.
- Local community involvement played an important role in the success of WSUD, including citizenship science groups assisting with the monitoring, patronage, and active maintenance of WSUD sites.
- WSUD practitioners in Singapore sought advice from and collaborated closely with practitioners from countries and regions with similar climatic conditions and urban growth patterns, such as from Queensland, Australia (Biermann et al., 2012; Lim and Lu, 2016).

## 5.7 EUROPE

In Europe, especially in Nordic countries and Germany, there has been strong promotion of source control and stormwater management to achieve multiple environmental outcomes such as environmental protection, stormwater treatment and drainage, spatial and water planning for flood risk management, and alternative water sources.

### 5.7.1 Policy context

The European Water Framework Directive (WFD) regulates the protection, management, and use of water resources in Europe, including all surface waters and groundwater (EC, 2000). European Union members are responsible for the application and translation of the WFD to their own context, as shown in the examples in Table 5.5.

Water and sewage companies have been an integral stakeholder and agent in the promotion of source control and stormwater management through environmentally friendly techniques since 2007. In Sweden, Denmark, and the Netherlands, water and sewage providers are publicly owned, not-for-profit agencies. In Germany and France, they are public and/or private agencies, whereas in England, they are privatized service providers (Chouli et al., 2007).

Most cities in Europe have both combined and separate sewers. Large-scale stormwater management and treatment functions are distributed across stakeholders, based on the institutional setup particular to each country or region. The experience of European countries that "end-of-pipe" techniques were very expensive to manage for combined sewers contributed to a shift toward source control techniques for stormwater integrated in urban projects with multifunctional facilities (Chouli et al., 2007).

Table 5.5 provides a summary of the current institutional context for stormwater management in Sweden, Denmark, the Netherlands, Germany, and France, which have transitioned toward WSUD. Challenges in the transition toward integrated source control were experienced by all countries. The transition required the reorganization of the services, significant efforts

**TABLE 5.5 Examples of WSUD Policy and Regulatory Context Across Europe (Chouli et al., 2017)**

| Country | Institutional, Policy and Regulatory Context |
|---|---|
| Denmark | Each city has one public company responsible for energy, transport, water and sewerage, etc.<br>Majority of cities have constructed wetlands or multiple lagoons upstream to improve the ecological quality of urban streams and lakes.<br>Drainage fees are split 60:40 for wastewater and stormwater management. Drainage Departments offer incentives to property owners (lower drainage fees, 40% refund of drainage connection fees and technical assistance), public information campaigns and examples of source control techniques in high environmental value and public areas. |
| Sweden | Each city has one public company responsible for energy, transport, water and sewerage, etc.<br>Stockholm property owners and the Street Department pay a stormwater fee proportional to the amount of impervious surface area. The fee is reduced if the property owner adopts source control techniques.<br>The Sewage Department provides public information and technical guidance to property owners. New projects of construction and rehabilitation include pilot source control applications. |
| The Netherlands | The National policy target aimed to reduce by 50% the combined sewer overflow in terms of phosphorus and nitrates from 1995 to 2005; to disconnect 20% of urban areas from sewers; retain stormwater in rural areas and to use source control techniques in all new urban projects.<br>Municipalities and industries pay fees to the Water Boards (river managers) based on the pollution emitted. This incentivised many municipalities to build plants that treat both wastewater and stormwater. Some municipalities have disconnected the stormwater from the sewerage network.<br>Municipalities typically build aesthetic source control projects, offer technical assistance and organize public information campaigns and financial help to home owners who want to disconnect from the sewerage. |
| Germany | Policies vary across regions. For instance, in Dresden Water companies collect taxes based on imperviousness from property owners and municipalities (based on road surface area). Dresden uses rainwater for municipal applications and organizes public campaigns for the promotion of source control techniques.<br>The Water company offers technical guidance for the implementation of source control in private properties. |
| France | The 1992 Water Law requires municipalities to define future zones of run-off in their territory and requires treatment of the run-off discharge. Stormwater managers (municipalities, countries and inter-municipal services) prefer small local WSUD projects for treatment instead of retention basins.<br>All important projects concerning rainwater discharge, artificial infiltration and the creation of impervious areas of more than 5ha require a special permit.<br>Some demand managers set a mandatory requirement that land owners have to apply source control for approval of future construction. The criteria differs across stormwater managers (e.g. area of future projects, project importance, maximum outflow thresholds (L/s), critical rainfall) and depend on their capacity to control stormwater and the socio-political situation rather than national policy. For example, the Seine Saint Denis county, after 25years of experience, prefers open air multi-functional installations (e.g. flood facilities and open green spaces that can be flooded in case of rain) for a more efficient use of urban space, lower cost and ease to identify mal-functions. Whist in Limoges the lowest cost solution was the adoption of ponds and wetlands upstream of the city in rural areas, with installations are co-funded by the city and rural municipalities.<br>Water agencies, Counties and Regions act as big scale coordinators and sponsor source control techniques in existing urban areas and pilot projects. A new French water law is expected to introduce stormwater fees for future stormwater management. |
| United Kingdom | Most urban projects are fully managed by private development companies. The national policy focuses in funding only flood reduction vulnerability and not flood protection. All new urban projects are required to guarantee long-term stormwater management. The Environmental Agency develops a map of flood zones and sets local effluent restrictions (5-10L/ha), issues permits for all important stormwater discharges and collects yearly fees from polluters.<br>Fees are reduced if the developer adopts the Agency's technical guidance for stormwater management (e.g. infiltration and wetlands facilities). Municipalities also have the freedom to apply stricter obligations to developers. The final criteria are agreed through negotiation between the developer, the municipality, the sewer company and the environmental agency. Many municipalities demand source control techniques, rehabilitation of channelled streams and stormwater reuse. |

to establish collaboration between water agencies with other stakeholders (road services, parks services, urban planners, architects, and citizens), and the redefinition of priorities and roles regarding stormwater planning across agencies (especially if floods were not of high concern) (Cettner et al., 2013). It also required many years of adaption, the gradual upskilling of new specialists (in planning, environmental engineers, public relations specialists), and the building of new collaborations.

**TABLE 5.6 Examples of WSUD Policy, Regulatory and Financial Instruments in Use (Chouli et al., 2007)**

| | Pilot Projects | Regulation Restrictions | Discharge Control | Discharge Fees/ Penalties | Stormwater Fees | Tax Breaks/ Fees Discounts | Subsidies | Information Campaigns |
|---|---|---|---|---|---|---|---|---|
| Sweden | √ | √ | √ | | √ | √ | | √ |
| Denmark | √ | √ | √ | | √ | √ | | √ |
| The Netherlands | √ | √ | √ | √ | √ | | √ | √ |
| Germany | √ | √ | √ | √ | √ | √ | √ | √ |
| England | √ | √ | √ | √ | √ | √ | | √ |
| France | √ | √ | √ | | √ | | √ | √ |
| Australia | √ | √ | √ | | √ | | | √ |
| USA | √ | √ | √ | √ | √ | | | √ |

National and regional collaboration was enabled through new legislation, introduction of control and sanction systems on urban stormwater, and funding of research/pilot projects. Table 5.6 provides an indication of the mix of policy, regulatory, and financial instruments adopted in the promotion of WSUD features in selected European countries, Australia, and the United States. In particular, the introduction of long-term source control techniques on private property adopted a mix of legislation, information campaigns, technical assistance, and financial incentives. Many of the control techniques were still in the experimental phase and their whole life-cycle costs were not fully understood when first implemented (Chouli et al., 2007).

Financial levers have an important role in the sustainable implementation of WSUD. Funding models for stormwater management differed among countries and even from one city to another. Financial instruments adopted include public funding, local taxes, sewage levies based on water consumption, sewage rates based on property value, and sewage levies based on impervious surfaces, as shown in Table 5.6. A particular feature in many European countries was the adoption of visible stormwater facilities, such as wetlands, which reduced the risk of illegal connections and created aesthetic features for municipalities that were highly visible to rate payers (Chouli et al., 2007).

Germany in particular had a very comprehensive implementation program with a wide mix of instruments that will be further explored in Section 5.7.2.

## 5.7.2 Germany

Germany has over 40 years' experience with stormwater management using WSUD. Technologies, such as green roofs, swales, and constructed wetlands, are a common practice and are governed by German standards and norms (Nickel et al., 2013). Reduction in population densities, new climate challenges (such as the increasing frequency and intensity of storm events), increased risk of flooding, CSOs, and stormwater discharges from separate systems were strong incentives for the adoption of WSUD at the municipal level.

As in many other countries, urban stormwater and wastewater management are the responsibility of local government. The sewerage infrastructure is mostly combined sewers for collection, treatment, and disposal of sewage and stormwater at centralized treatment plants.

### *5.7.2.1 Policy and institutional context*

Germany adopts a complex tiered system of spatial planning instruments that uses GI to achieve multiple benefits, including stormwater management, habitat protection, and climate control at local and regional level. The instruments align ecological values with economic values and urban development (Chouli et al., 2007).

At a national level, the German Federal Water Act (Wasserhaushaltsgesetz) (WHG, 2009) and the State Water Laws set the legal framework for stormwater and wastewater management. The Water Act implements the European WFD and requires due care to maintain the functions of the natural hydrologic cycle and to avoid increases in runoff rates and flow volumes. Since 2008, the Water Act has prioritized stormwater management, with preference given to infiltration near the source (par 55, WHG, 2009 in Nickel et al., 2014).

The Federal Nature Conservation Act (Bundesnaturschutzgesetz) uses landscape planning as the key instrument for the conservation of natural resources (BNatSchG, 2009). This enables municipalities to use landscape planning to develop a full-coverage strategy tailored to the municipalities' interests. Municipalities develop a local landscape plan (Landschaftsplan), which is based on specifications articulated in regional and state landscape programs. At the local level, the local landscape plan supports the search for mitigation sites and the determination of mitigation and environmental compensation measures required under nature conservation law for impacts or intrusions on the natural environment, e.g., compensatory green spaces.

Urban land-use plans support the local plans according to the German Federal Building Code (BauGB, 1960). The **preparatory plan** shows the intended urban development for a municipality, whereas the **binding land** use plan defines the use of individual land parcels. If detrimental impacts on the environment are expected, mitigation or compensation measures are required to comply with the Federal Nature Conservation Act (Keeley, 2004; Buehler et al., 2011 in Nickel et al., 2014). These measures include the creation of green spaces and implementation of WSUD technologies. The areas and/or measures for compensation of impacts are included in the binding plans. Such a model allows greater integration of GI and integrated stormwater management at a range of scales in urban planning.

Water authorities collect fees for discharges to water bodies and the Verbände (river managers). Federal law allows each region to develop its own policy. In North Rhine-Westphalia, all new development projects are required to adopt stormwater infiltration. Funding is offered for research, the development of stormwater master plans, and the implementation of specific source control techniques. Funds are collected from the managing agencies (municipalities, water companies, Verbände, industries) responsible for wastewater treatment and CSOs according to the pollution discharge. The Verbände also offer technical guidance to municipalities regarding source control techniques and coordinate stormwater management (Chouli et al., 2007).

A municipality, in collaboration with its water company, is the key factor responsible for implementing source control techniques in public areas and for their promotion to the private sector. Infiltration and stormwater reuse for gardening and cleaning are quite popular in Germany (Chouli et al., 2007).

Permits are generally not required for infiltration of lightly polluted stormwater into groundwater. However, stormwater collected from impervious surfaces with heavy traffic requires treatment before its percolation or discharge to surface waters. Infiltration of stormwater in designated water protection areas also requires a permit. These regulations apply to new developments and to pipeline rehabilitation projects in developed areas.

The use of policy instruments to promote WSUD is widespread at municipal level. In a survey of 400 German municipalities conducted in 2003, 104 adopted stormwater policy instruments, 18 provided direct financial subsidies, over 50 had indirect subsidies, and 36 mandated the use of WSUD under selected conditions (FBB, 2003 in Nickel et al., 2014).

#### *5.7.2.2 Financing*

Germany adopts equity and user-pay principles for financing (Nickel et al., 2014). The annual average combined sewerage charge is €115 per year per capita (AUD$173). Stormwater fees are charged by over 60% of all German municipalities.

German law sets a stormwater fee estimated from the actual contribution to the total stormwater burden (Nickel et al., 2014). Stormwater fees are determined by assessing the surface area of individual land parcels and their stormwater contribution to the central drainage system. The average annual stormwater charge is €0.89/m$^2$ (AUD$1.33/m$^2$) impermeable surface.

In many German municipalities, the adoption of on-site WSUD measures results in reduced stormwater fees for the property owner (e.g., a green roof can result in a 50% discount on the stormwater fee) (Ansel et al., 2011), thereby incentivizing implementation of WSUD devices/practices. Stormwater fees and discounts are seen as an efficient and proven incentive for the implementation of WSUD (FBB, 2003 in Nickel et al., 2014) as the following example of Berlin city will illustrate.

#### *5.7.2.3 Berlin city*

In Berlin, a range of instruments, including financial subsidies, nonfinancial incentives, fees, and regulations, have been used to promote green spaces since the 1980s. During the initial period, there was significant investment in promoting green spaces in private properties via subsidies to increase green cover, reduce impervious areas, and promote green roofs and walls to enhance evapotranspiration. This resulted in a significant spread of green roofs, courtyards, and facades, mostly on private property. The program ended when the funds were exhausted, but a number of inner-city districts continue to run similar programs today. The impact on CSOs was not estimated as that was not the main motivation.

In the late 1980s, the city also supported ecological buildings and urban development projects, placing a strong focus on WSUD for stormwater management and optimization of new technologies in a typical local setting. This was supported by scientific investigation, which created a publicly accessible knowledge base, guidance in planning, implementing and maintaining GI installations, and regulatory standards. Mandatory rainwater management and ecological construction guidelines applied to all public construction projects (schools, administration offices, etc.), with the government sector leading the way in transferring research knowledge into practice.

Berlin introduced planning instruments to support GI at all scales, such as a city-wide strategic and binding planning instrument. Ecological objectives, the preservation of habitat, and open greenspace were incorporated in urban development to create a landscape program that includes nature conservation. This became the basis for future urban development and the creation of binding land use and landscape plans for other urban sections.

In 1997, Berlin introduced the Green Area Ratio (GAR), which incentivized GI on a smaller scale (Keeley, 2011). The GAR set binding values (0.3−0.6) for that portion of a specific land plot that serves an ecosystem function. It is estimated based on the allotment's ability to store, evaporate, or infiltrate rainwater and to provide environmental services. Weighting factors are defined based on the surface characteristics. The GAR can incorporate various WSUD techniques, allowing architects and property owners' greater flexibility in meeting requirements. Architects and property owners expressed positive feedback regarding the GAR ease of use and the immediate visual improvements that can provide benefits, such as energy savings. The GAR calculation logic was similar to other planning indexes and ratios, which helped planning staff to understand GAR. Advice on how to meet the GAR requirements and reduce stormwater fees was available and highly valued by planners. Currently, 12 of Berlin's 28 binding Landscape Plans have GAR goals, with a further 14 plans awaiting GAR approval. In summary, the GAR was politically positive, but the environmental impact has been difficult to measure (Chouli et al., 2007).

In contrast, the city of Stuttgart set mandatory WSUD requirements for stormwater management to achieve the environmental goals in the LaPro for new urban developments. However, urban planners and architects have shown resistance to such binding provisions compared to the GAR, which offers greater flexibility (Chouli et al., 2007).

The implementation of the EU WFD and whole of basin planning has strengthened the use of GI for stormwater management in Berlin and reduced pollution to waterbodies through better stormwater management. Furthermore, Berlin city and the state of Brandenburg, which surrounds the city, have agreed on strategic nutrient reduction goals at a subcatchment level, with a target of 50% reduction by 2021 (Senguv, 2011 in Chouli et al., 2007). Achieving this target will require decentralized stormwater management concepts for a number of subcatchments due to the insufficient/ineffective centralized treatment and storage capacities. The whole process will require communication across all relevant stakeholders, providing an evidence-based method based on cost-effectiveness for prioritizing measures among all relevant stakeholders (Nickel et al., 2014), which helps avoid decisions that may reduce future options for action (Rehfeld-Klein, 2011; Lem, 2011 in Nickel et al., 2014).

Given the split fees structure, the stormwater fees were adjusted to provide discounts or exemptions for surfaces based on their contributions to runoff. This incentivized property owners to comply with the technical rules and to adopt stormwater management measures that reduced impervious areas. The fee amount is high to encourage WSUD/GI, particularly for the public sector (schools, sports field, hospitals, administrative offices, etc.), industry, and businesses that operate large sites with considerable amounts of impervious cover. This led to a number of projects and strategic initiatives by municipalities to reduce stormwater runoff.

The key lesson was that fees worked well when combined with other instruments and incentives, particularly with respect to existing development.

Berlin also introduced rainwater management at source under the Berlin Water Act in 2000 (http://www.berlin-klimaschutz.de/en/climate-protection-berlin/laws-and-regulations-federal-state-berlin) to improve groundwater recharge and to reduce costs for infrastructure and treatment of polluted stormwater runoff. The provisions specify fees for infiltration of rain from different surfaces, according to the pollutant load. Permit-free infiltration applies only for lightly polluted roof runoff.

In conjunction with limits on stormwater discharges to predevelopment levels, innovative concepts that combine stormwater management with rainwater harvesting have been developed, such as the iconic 7 ha, Daimler-Chrysler site in Postdamer Platz (http://landarchs.com/potsdamer-platz-in-berlin-becomes-a-sustainable-ecofriendly-urban-square/). The multiple objectives of the Daimler-Chrysler site included flood management, savings in mains water and operating costs, and improvement of microclimate. However, a common barrier to the more widespread use of these innovations has been the late consultation with the water utility in the urban planning process. Their exclusion has inhibited the integration of the optimal stormwater management options into the whole urban water cycle (Rehfeld-Klein, 2011 in Nickel et al., 2014).

#### *5.7.2.4 Key lessons*

In Germany's case, national level policies, which promoted river basin management, integrated spatial planning, and ecological compensation measures, were merged with creative regional solutions for the financing and management of stormwater at source. This enabled the successful greening and regeneration of many areas, such as the Berlin and the Emscher (mid-western Germany) regions. A key feature of the German experience was the strong focus on the more holistic benefits associated with WSUD and GI, including amenity and liveability. Other key lessons include

- Successful implementation was achieved using an integrated environmental planning approach to help balance environmental and urban development, based on long-term goals using green infrastructure However, policy instruments that are biased toward a specific technology should be avoided.
- Leadership is needed from public authorities, while at the same time enabling stakeholders to participate in the transformative process. The early participation of water utilities in the urban planning process is essential.
- Emscher and Berlin highlighted the value of an integrated planning approach that identified the multiple benefits provided by GI.
- Political will and perseverance to pursue medium- to long-term goals is essential.
- Well defined, quantifiable, long-term goals enabled long-term and transboundary transformations.
- The ability to evaluate the cost-effectiveness and appropriateness of various management options is important. In Emscher, benchmarking was possible given the existence of clear goals, whereas Berlin did not have an equivalent measurable benchmarking system.
- Nutrient goals had a positive impact on stormwater management and protection of water resources.
- The need for a suite of instruments and different approaches, as most of the instrument options have a specific target group, a limited time frame, and are adopted at different stages of the urban cycle. However, in combination, their effectiveness was enhanced.
- Fees only work well if combined with other instruments and incentives, particularly with respect to existing development. On their own they are not effective.
- Working across sectors with industry champions and leaders was a key success attribute.
- The GAR was a quantifiable metric. Other cities such as Washington DC and Seattle in the United States have since implemented GAR based on the Berlin model (Keeley, 2011).

### 5.7.3 Lessons from Europe

Key lessons from the European WSUD experience were

- Stormwater master plans conceived nationwide, and in collaboration with all existing stakeholders, tend to be more effective and cost-efficient. Such plans require the evaluation of multiple inputs, including field data; funding possibilities; vulnerability of flooded areas; political objectives; acceptable pollution risks; and targets for planning. It also requires coordination for the management of WSUD features.
- The integration of source control is necessary to prevent the saturation of any hydraulic works and needs to be considered in future planning.
- Restrictions on imperviousness and mandatory source control for new construction were unable to achieve the results desired, due to difficulties with the control of behaviors on private properties. However, information campaigns and technical guidance from public services motivated many property owners to contribute to stormwater harvesting.
- Management of the larger schemes was better suited to the larger urban agencies.

## 5.8 LESSONS FROM THE CASE STUDIES

The case studies highlight the complex challenges associated with the transition toward WSUD: multiple barriers need to be addressed, a wide range of stakeholders and their roles identified, and processes created to empower actors and stakeholders to transition toward WSUD through addressing social, technical, and institutional factors in the societal landscape. The case studies illustrate how the initial societal landscape in each case study area can impact the design and change process outcomes. The four transition case studies highlighted are as follows:

1. In Australian, in a country with a federation governance model with a relatively small population of 24-million inhabitants concentrated in coastal, urban centers, transition was driven by climate, water security, and environmental protection challenges. Water reform at national level and WSUD transition is implemented using different models and tools in each state.
2. The United States has a large population spread over a large geographic area, and a strong federal governance model for public services, with a diverse range of climatic, institutional arrangements, and interests across different states and municipalities. The United States adopted a national, top-down policy and legislative approach to support WSUD transition, with technical support and capacity building developed at a national level. It faces substantial challenges in the transfer from state (policy) to municipal government (implementation).
3. The concerted and efficient transition process in Singapore, a small island nation with a cohesive and strong government, driven by water security challenges. Singapore implemented a sophisticated, well-resourced, and planned change framework involving
   a. whole of society and sophisticated technical support;
   b. capacity building for institutions and community to develop environmental stewardship and the promotion of citizenship participation;
   c. collaboration between academia, government, and private sectors in the transition;
   d. transfer and embedding of WSUD knowledge through standards and codes and in university curricula;
   e. clear and defined roles for different agencies, with the water utility as the central coordinator.
4. The successful transition in Germany, where WSUD practices and GI are now mainstreamed, was driven by environmental challenges. The transition process was characterized by strong alignment between environmental protection, landscape and land-use planning policies and legislation, technical support, and promotion of collaboration across agencies. It was supported with financial instruments and capacity development, including polluter-pay funding models over decades. The process was also based on a mix of mandatory and discretionary powers, shared between different government levels and binding tools. In particular, the process promoted the secondary and holistic benefits of GI, instead of focusing on stormwater alone.

Implementing a transition process is more complex for countries with large areas, multiple jurisdiction government (local, state, national), multiple climate zones, land morphology, and diverse regulatory and institutional setups. The main actor for WSUD implementation in all case studies, except for Singapore, was the local government, which had either the discretion to develop its own implementation framework or had a mandate with significant discretion, and support from state or national agencies. Common lessons can be derived from the frameworks and implementation processes examined in the case studies:

1. **Clarity in drivers and objectives**. Clear and defined drivers and objectives, such as the environmental protection objectives in Germany and Australia, and water security for Singapore, assisted the development of clear policies. In the case of Berlin, Germany, WSUD targeted multiple objectives and benefits, including GI, which helped articulate the value proposition to a wide range of stakeholders in the community. Targets and objectives required support in policy and legislation.
2. **Support in policy and enabling legislation on national, state, and local government levels**. The development and alignment of policy and legislation supporting the overall objectives was a key enabler and provided local government agencies with legitimacy to implement WSUD. In particular, the alignment and harmonization of policies and legislation across interdependent disciplines, such as environmental protection, planning, drainage, and land use (e.g., Victoria and Germany) enabled the development of consistent processes and mechanisms and provided clarity on roles and responsibilities in stormwater management across different government agencies and departments.
3. **Collaboration across agencies**. Alignment of roles and responsibilities in planning and maintenance of WSUD infrastructure required multidisciplinary teams (transport and road infrastructure, water, planning, landscape, wastewater, parks and gardens, etc.) and led to more effective outcomes. Many of the case studies created new roles and responsibilities across agencies to implement and maintain WSUD features.
   The benefits of building cooperation among WSUD stakeholders at all stages of planning and implementation were a key enabler for sustainable WSUD implementation. In a number of case studies, this enabled the identification of gaps and solutions to address the challenges associated with WSUD, and it could be driven by a top-down approach (e.g., Singapore) or a bottom-up approach with local champions (e.g., Philadelphia and Berlin) with support from the national or state level.
4. **Empowering local government as the implementer**. All case studies placed the final responsibility and discretion of WSUD implementation with local government via its local development planning rules. This provided local

government with the flexibility to tailor implementation to local needs, and is appropriate given the various scales to which WSUD applies (property, precinct, catchment). However, the local government's limited financial resources, technical capacity, and competing local priorities could often hinder WSUD implementation. Local government requires capacity building and resourcing for effective WSUD implementation.

Different mechanisms and levels of support were provided to local government through legislation, technical and financial resourcing, and capacity building. For example, the Victorian (Australia) framework is based on a model that required all new developments to manage stormwater to prescribed targets using Best Practice Environment Management as the default position. But the Victorian state government also provided local government with the flexibility for exemptions and changes within their jurisdictions through offsets and local planning laws. This allowed local authorities to adopt off-site treatment or select more effective treatment options if deemed appropriate. It also supported capacity building, knowledge dissemination, and funding for pilots via a dedicated agency. In contrast, each LGA in NSW was responsible for the development of its own stormwater objectives, local policies, and implementation mechanisms.

In a number of case studies, WSUD implementation required analysis and planning at a scale that was larger than a single local government area. Implementing actors were either local government or coalitions of local government in partnership with other agencies, such as a water utility (e.g., PUB, MW, water utilities in Germany), sharing expertise, and increasing scale. This required mechanisms for collaborative planning with multiple agencies and local government areas for the development of stormwater management solutions.

5. **Technical support**. Provision of support to the implementer is essential, via technical instruments (standards and best practices) tailored to local climatic, geological, and institutional conditions. This includes alignment of building and planning codes, clarity of roles and responsibilities, funding for pilots, and resources to enable information sharing. For information sharing and transfer, there are advantages in having a dedicated agency or body with a clear mandate that can streamline information capture, knowledge transfer, and assist with appropriate tools (standards, guidelines, etc.) for capacity building and support to LGAs.
6. **Funding models, adoption of a mix of financial instruments**. For a number of case studies, stormwater charges funded stormwater infrastructure based on polluter-pay principles, which linked the amount and quality of stormwater runoff, or the impermeable area, to the charges and fees. This provided a transparent link between cause and effect, with the pricing signal indicative of the costs of mitigation. This allowed polluters to decide on a trade-off between charges, and taking responsibility for stormwater management measures on their properties. It also provided local government with funding for investment in WSUD services. However, fees and charges must be set appropriately to foster action.
7. **Benefits of community education and support**. Germany and Singapore invested heavily over decades in raising community awareness about stormwater management projects, environmental impacts and issues, media, signaling, and community activities, which raised the interest and acceptance of WSUD projects. Likewise, increasing the visibility, and promoting the aesthetics of WSUD projects, helped to raise tax-payer awareness and acceptance by the community in Europe, Singapore, and parts of Australia. Dean et al. (2016) demonstrated a relationship between knowledge and support for policies related to water conservation and protection, water sensitive urban design, and alternative water sources. Hence, raising community awareness and communication about how WSUD creates tangible benefits that they can enjoy is important for creating societal expectations and the long-term sustainability.

## 5.9 CONCLUSIONS

Lessons have shown the importance of consistent messages and support at national, state, and local level, albeit different roles are bestowed on each government level. Most importantly, it is the enabling and empowering at the local level that is instrumental to the uptake of WSUD both for local government and community.

Most of WSUD is implemented at local government level. The drivers and conditions in each jurisdiction, catchment, and even specific local government, differ and thus need to be examined for each individual area and at the various relevant scales. However, this also requires policy and legislative instruments at the national and/or state level to support local implementers. Furthermore, policy and regulation need to be complemented with practical instruments that target the multiple barriers that can hinder change, which include capacity building, financing mechanisms, community support, technical assurance, and regulatory and institutional support.

WSUD implementation is an evolutionary change process. The uptake and maintenance of WSUD differs from traditional gray infrastructure, and consequently the funding and maintenance requirements differ. It requires multidisciplinary skills, and the roles and responsibilities may need to be reworked before WSUD is mainstreamed. Financing mechanisms are a key factor for the sustainability of WSUD.

Different frameworks and methodologies have been used in WSUD implementation, and while lessons can be shared, the direct transfer from one locality to another is not always feasible. The verification and improvement of WSUD technical features and performance based on the targeted outcomes is also an important factor to be considered.

Overall, the case studies highlighted the importance of flexibility and adaptability in the implementation of WSUD because needs and barriers for different segments and stakeholders will evolve over time. This will require the monitoring of outcomes, and the review of instruments (policy, regulatory, financial, technical) adopted over time. In particular, the institutional setup and capacity development for WSUD needs to be shared across various levels of government and sectors in society to develop enduring and sustainable change.

## ACKNOWLEDGMENTS

We would like to thank Dr. Judy Blackbeard and colleagues in Melbourne Water, for their advice and valuable comments in the review of the manuscript.

## REFERENCES

ABS, 2017. 2071.0-Census of Population and Housing: Reflecting Australia - Stories from the Census, Snapshot of Australia. http://www.abs.gov.au/ausstats/abs@.nsf/Lookup/2071.0main+features22016.

ACT Government, 2009. Waterways Water Sensitive Urban Design General Code. ACT Planning and Land Authority (ACTPLA), Australian Capital Territory Government. http://www.legislation.act.gov.au/ni/2008-27/copy/64663/pdf/2008-27.pdf.

ACT Government, 2014. ACT Water Strategy 2014-44: Striking the Balance. Environment and Sustainable Development Directorate, ACT Government.

Adams, R., Jayasuriya, N., 2014. Assessing the effectiveness of three water sensitive urban design (WSUD) measures in SE Melbourne'. In: 35th Hydrology and Water Resources Symposium. Engineers Australia.

Ansel, W., et al., 2011. Leitfaden Dachbergrünung fur Kommunen. Deutscher Dachgärtenerverband e.V, Nürtigen.

ANZECC, 2000. Australian and New Zealand Guidelines for Fresh and Marine Water Quality. Australian and New Zealand Environment and Conservation Council—ANZECC. http://www.mfe.govt.nz/fresh-water/tools-and-guidelines/anzecc-2000-guidelines.

Aurecon, December 20, 2011. Draft Stormwater Drainage Strategy, Part Lot 9765. Report 203655—MO01. Prepared for CIC Australia, Town of Palmerston.

Australian Government, 2017. National Water Initiative. http://agriculture.gov.au/water/policy/nwi.

BauGB, 1960. Baugesetzbuch in der Fassung der Bekanntmachung vom 23. September 2004 (BGBl. I S. 2414), das zuletzt durch Artikel 1 des Gesetzes vom 22. Juli 2011 (BGBl. I S. 1509) geandert worden ist. http://www.gesetze-im-internet.de/bbaug/index.html.

Benedict, M.E., McMahon, E.T., 2006. Green Infrastructure: Linking Landscapes and Communities. Island Press, Washington, DC.

Biermann, S., Sharma, A., Chong, M.N., Umapathi, S., Cook, S., 2012. Assessment of the Physical Characteristics of Individual Household Rainwater Tank Systems in South East Queensland. Urban Water Security Research Alliance. Technical Report No. 66. http://www.urbanwateralliance.org.au/publications/UWSRA-tr66.pdf.

BNatSchG, 2009. Gesetz uber Naturschutz und Landschaftspfledge (Bundesnaturschutzgesetz), 29 Juli 2009 (BGBI.I.S.2542). http://www.gesetze-im-internet.de/bnatschg_2009/BJNR254210009.html.

Boatwright, J., Stephenson, K., Boyle, K., Nienow, S., 2014. Subdivision infrastructure affecting storm water run off and residential property values. Journal of Water Resources Planning and Management 140 (4), 524—532.

Bowman, T., Thompson, J., 2009. Barriers to implementation of low-impact and conservation subdivision design: developer perceptions and resident demand. Landscape and Urban Planning 92 (2), 96—105.

Breckenridge, L.P., 2014. Green infrastructure in cities: expanding mandates under federal law. Trends 45 (6), 21—25.

Brown, R.R., Farrelly, M.A., Loorbach, D.A., 2013. Actors working the institutions in sustainability transitions: the case of Melbourne's stormwater management. Global Environmental Challenge 23, 701—718.

Brudler, S., Arnbjer-Nielsen, K., Hauschild, M.Z., Rygaard, M., 2016. Life cycle assessment of stormwater management in the context of climate change adaptation. Water Research 106, 394—404.

Buehler, R.M., et al., 2011. How Germany became Europe's green leader: a look at four decades of sustainable policymaking. Solutions 2 (5), 51—63.

Bureau of Meteorology (BOM), 2017. Climate Statistics for Australian Locations. Summary Statistics Melbourne Regional Office.

Carter, J.G., Cavan, G., Connelly, A., Guy, S., Handley, J., Kazmierczak, A., 2015. Climate change and the city: building capacity for urban adaptation. Progress in Planning 95, 1—66.

Cettner, A., Ashley, R., Viklander, M., Nilsson, K., 2013. Stormwater management and urban planning: lessons from 40years of innovation. Journal of Environmental Planning and Management 56 (6), 786—801. https://doi.org/10.1080/09640568.2012.706216.

Chaffin, B.C., Shuster, W.D., Garmestani, A.S., Furio, B., Albro, S.L., Gardiner, M., Spring, M., Green, O.O., 2016. A tale of two rain gardens: barriers and bridges to adaptive management of urban stormwater in Cleveland, Ohio. Journal of Environmental Management 183, 431—441.

Choi, L., Mcilrath, B., Williams, D., 2015. A review of the policy framework for WSUD in Five Australian cities. In: Project B5.1, Second Water Sensitive Cities Conference, September 2015. CRC for Water Sensitive Cities, Brisbane, Queensland, Australia. http://www.watersensitivesa.com/wp-content/uploads/04-Linda_Choi-Policy_framework_for_WSUD_in_five_Australian_cities.pdf.

Chouli, E., Aftias, E., Deutsch, J.C., 2007. Applying stormwater management in Greek cities: learning from the European experience. Desalination 210, 61–68.

Chrispijn, J., Weise, R., 2012. Mainstreaming WSUD in Tasmania – a model for achieving best practice in regional centers. In: Stormwater 2012–2nd National Conference on Urban Water Management, Melbourne 16-18 October 2012.

Cook, S., Myers, B., Newland, P., Pezzaniti, D., Kemp, D., 2015. Pathways for Implementation of Water Sensitive Urban Design Policy in South Australia. Goyder Institute of Water Research Technical Report No. 15/51, Adelaide, South Australia, pp. 1839–2725.

CSIRO, 1999. Urban Stormwater Best Practice Environmental Management Guidelines. CSIRO Publishing.

Dean, A.J., Fielding, K.S., Newton, F.J., 2016. Community knowledge about water: who has better knowledge and is it associated with water-related behaviours and support for water-related policies? PLoS One 11 (7), e0159063. https://doi.org/10.1371/journal.pone.0159063. on-line.

Department of Local Government (DLG), July 2006. Stormwater Management Services Charge Guidelines. Government of New South Wales. https://www.olg.nsw.gov.au/sites/default/files/Stormwater-Guidelines.pdf.

Department of Planning and Western Australia Planning Commission, 2015. Liveable Neighbourhoods – Draft. http://www.planning.wa.gov.au/dop_pub_pdf/LiveableNeighbourhoods_2015.pdf.

Department of Primary Industries Parks Water and Environment (DIPWE), December 2010. State Stormwater Strategy. Government of Tasmania, p. 46. http://epa.tas.gov.au/documents/state_stormwater_strategy_december_2010.pdf.

Department of Primary Industries Parks Water and Environment (DIPWE), 2012. Water Sensitive Urban Design – Engineering Procedures for Stormwater Management in Tasmania, State of Tasmania. http://epa.tas.gov.au/documents/wsud_manual_2012.pdf.

Department of State Development, December 2013. Infrastructure and Planning (DSDIP), 2013. State Planning Policy Guideline - State interest -water quality. Department of State Development, Infrastructure and Planning, State of Queensland, Brisbane, Queensland 30–31.

Dolowitz, D.P., 2015. Stormwater management the American way: why no policy transfer. AIMS Environmental Science 2 (3), 868–883.

Dolowitz, D., Keeley, M., Medearis, D., 2012. Stormwater management: can we learn from others? Policy Studies 33 (6), 501–521.

D'Aspromonte, D., Slater, T., Godfrey, M., 2012. The evolution of the WSUD guidelines for Melbourne councils. In: Stormwater 2012, Melbourne, Australia.

Eggleton, S., 2012. Assessing and responding to local governments capacity to deliver sustainable stormwater management. In: 7th International Conference on Water Sensitive Urban Design. Melbourne.

European Commission, 2000. The EU Water Framework Directive. http://ec.europa.eu/environment/pubs/pdf/factsheets/wfd/en.pdf.

European Union, October 2000. EU water framework directive 2000/60/EEC. Establishing a framework for community action in the field of water policy. Official Journal of the European Communities L327, 1–72. http://eur-lex.europa.eu/resource.html?uri=cellar:5c835afb-2ec6-4577-bdf8-756d3d694eeb.0004.02/DOC_1&format=pdf.

FBB, 2003. SchlagLicht 3. Fachvereinigung Bauwerksbegrünung e.V. https://www.gebaeudegruen.info/service/downloads/t//fbb-schlachtlicht-3.

Ferguson, B.C., Brown, R.R., Frantzeskaki, N., De Haan, F.J., Deletic, A., 2013. The enabling institutional context for integrated water management: lessons from Melbourne. Water Research 47, 7300–7314.

Fletcher, T.D., Deletic, A., Mitchell, V.G., Hatt, B., 2008. Reuse of urban runoff in Australia: a review of recent advances and remaining challenges. Journal of Environmental Quality 37 (Suppl. 5), S-116–S127. https://doi.org/10.2134/jeq2007.0411.

Fletcher, T.D., Shuster, W., Hunt, W.F., Ashley, R., Butler, D., Arthur, S., Trowsdale, S., Barraud, S., Semadeni-Davies, A., Bertrand-Krajewski, J.L., Mikkelsen, P.S., Rivard, G., Uhl, M., Dagenais, D., Viklander, M., 2015. SUDS, LID, BMPs, WSUD and more – the evolution and application of terminology surrounding urban drainage. Urban Water 12 (5), 525–542. https://doi.org/10.1080/1573062X.2014.916314.

Fortune, J., Drewry, J. (Eds.), 2011. Darwin Harbour Region Research and Monitoring 2011. Department of Natural Resources, Environment, The Arts and Sport, Palmerston, NT, Australia, pp. 20–21. Report Number 18/2011D.

Gardner, T., 2015. My point of view: WSUD: has the doing exceeded the knowing? Water 42 (4), 6–8. http://digitaledition.awa.asn.au/default.aspx?iid=122664&startpage=page0000003#folio=1.

Garnaut, R., 2008. Climate Change Review. http://www.garnautreview.org.au/update-2011/garnaut-review-2011/garnaut-review-2011.pdf.

Government of South Australia, 2008. Residential Development Code Updates. http://www.sa.gov.au/topics/planning-and-property/land-and-property-development/building-and-property-development-applications/streamlined-residential-development/residential-development-code-updates.

Government of South Australia, 2011. South Australian Planning Policy Library, Version 6, September 2011. Department of Planning and Local Government. http://www.sa.gov.au/upload/franchise/Housing%20property%20and%20land/PLG/SA_Planning_Policy_Library_version_6.pdf.

Government of Tasmania, 2012. WSUD Engineering Procedures for Stormwater Management in Tasmania. http://epa.tas.gov.au/documents/wsud_manual_2012.pdf.

Government of Western Australia, October 2008. Better Urban Water Management. Western Australian Planning Commission and Department for Planning and Infrastructure, p. ix. https://www.planning.wa.gov.au/dop_pub_pdf/Better_Urban_Water_Management.pdf.

Government of Western Australia, July 2011. Water Sensitive Urban Design Brochures. Department of Water. http://www.water.wa.gov.au/__data/assets/pdf_file/0018/1809/99294.pdf.

Government of Western Australia, 2013. Guidance Note 5-The Role of Local Government. Department of Water. http://www.water.wa.gov.au/__data/assets/pdf_file/0017/1781/104374.pdf.

Goyder Institute For Water Research, 2011. Interim Water Sensitive Urban Design Targets for Greater Adelaide. Goyder Institute for Water Research. Technical Report Series No. 11/7.

Hamel, P., Fletcher, T.D., 2014. The impact of stormwater source-control strategies on the (low) flow regime of urban catchments. Water Science and Technology 69 (4), 739–745. https://doi.org/10.2166/wst.2013.772.

Hansen, K., 2013. Green infrastructure and the law, planning and environmental law: issues and decisions that impact the built and natural environments. American Planning Association 65 (8), 4–7. https://doi.org/10.1080/15480755.2013.824791.

Harris, G., Batley, G., Fox, D., Hall, D., Jernakoff, P., Molloy, R., Murray, A., Newell, B., Parslow, J., Skyring, G., Walker, S., 1996. Port Phillip Bay Environmental Study Final Report. CSIRO Publishing.

Hobart City Council, 2006. Water Sensitive Urban Design Site Development Guidelines and Practice Notes. https://www.hobartcity.com.au/Development/Planning/Water-sensitive-urban-design.

Jiang, Y., Yuan, Y., Piza, H., 2015. A review of applicability and effectiveness of low impact development/green infrastructure practices in arid/semi-arid United States. Environments 2, 221–249. https://doi.org/10.3390/environments2020221.

Jones, S.E., Hoban, A.T., O'Neill, A.H., 2012. The policy and planning framework for water management in South-East Queensland: insights for Australia, paper 335. In: 7th International Conference on Water Sensitive Urban Design. Melbourne.

Keeley, M., 2004. Green roofs incentives: tried and true techniques from Europe. In: Proceedings from 2nd annual green roofs for healthy cities conference. http://www.greenroofs.org/index.php/component/content/article/10-miscarchive/107.

Keeley, M., 2011. The green area ratio: an urban site sustainability metric. Journal of Environmental Planning and Management 54 (7), 937–958. https://doi.org/10.1080/09640568.2010.547681.

Lem, K., Interview on 31 October 2011. Berlin: Berliner Wasserbetriebe. in by Nickel et al (2014).

Lim, H.S., Lu, X.X., 2016. Sustainable urban stormwater management in the tropics: an evaluation of Singapore's ABC waters program. Journal of Hydrology 538, 842–862.

Lucas, R., Earl, E.R., Babatunde, A.O., Bockelmann-Evans, B.N., 2015. Constructed wetlands for stormwater management in the UK: a concise review. Journal of Civil Engineering and Environmental Systems 32 (3), 251–268.

Maass, A., April 2016. Analysis of Best Management Practices for Addressing Urban Stormwater Runoff. University of Arizona (Master thesis). http://hdl.handle.net/10150/608332.

Macmillan, N., McManus, R., May 2009. Recommendation for Implementation of WSUD Strategy within Existing Legislation and Policy Framework – Discussion Paper. NT Department of Planning and Infrastructure.

Madden, S., June 2010. Choosing Green over Grey: Philadelphia's Innovative Stormwater Infrastructure Plan (Master thesis). Massachusetts Institute of Technology, pp. 33–39.

McAuley, A., May 2009. WSUD Technical Design Guidelines, p. 92. http://www.equatica.com.au/Darwin/reports-pdfs/Final%20Docs/8005_Darwin%20WSUD%20Technical%20Design%20Guideline%20FINAL%20_May09_.pdf.

McAuley, A., McManus, R., May 2009. WSUD Planning Guide, p. 24. http://www.equatica.com.au/Darwin/reports-pdfs/Final%20Docs/8005_Darwin%20WSUD%20 Planning%20Guide%20FINAL%20_May09_.pdf.

McCallum, T., Boulot, E., 2015. Becoming a Water Sensitive City: A Comparative Review of Regulation in Australia. Cooperative Research Centre for Water Sensitive Cities, Melbourne, Australia, ISBN 978-1-921912-29-0.

McDowell, L.M., Pfennig, P., July 2013. Adelaide Coastal Water Quality Improvement Plan (ACWQIP). Environment Protection Authority. http://www.epa.sa.gov.au/xstd_files/Water/Other/acwqip_final.pdf.

Melbourne Water, 2006. Stormwater Quality Offsets - A Guide for Developers, first ed. Melbourne Water https://www.melbournewater.com.au/Planning-and-building/Applications/Documents/Stormwater-quality-offset-scheme.pdf.

Melbourne Water, 2012. Living Rivers Program. Melbourne Water. https://www.melbournewater.com.au/getinvolved/applyforfunding/livingriversfunding/Pages/Living-Rivers-funding-and-support.aspx.

Melbourne Water, 2014. Stormwater Offset Rate Review. Melbourne Water, Melbourne. https://www.melbournewater.com.au/Planning-and-building/schemes/offset/Pages/Stormwater-offset-rate-review.aspx.

Melbourne Water, 2015. Schemes, Contributions and Offset Rates Explained, Melbourne. https://www.melbournewater.com.au/Planning-and-building/schemes/about/Pages/schemes-contributions-offset-rates explained aspx

Melbourne Water, 2017a. Waterways and Drainage Charge. https://www.melbournewater.com.au/aboutus/customersandprices/Pages/Waterways-and-drainage-charge.aspx.

Melbourne Water, 2017b. Calculate Your Stormwater Offset Contribution. Melbourne Water. https://www.melbournewater.com.au/planning-and-building/developer-guides-and-resources/drainage-schemes-and-contribution-rates-1.

Morrison, P., Chesterfield, C., 2012. Enhancing the management of urban stormwater in a new paradigm, paper 316. In: 7th International Conference on Water Sensitive Urban Design. Melbourne.

Morrison, P.J., Brown, R.R., 2011. Understanding the nature of publics and local policy commitment to water sensitive urban design. Landscape and Urban Planning 99, 83–92.

Myers, B., Chacko, P., Tjandraatmadja, G., Cook, S., Umapathi, S., Pezzaniti, D., Sharma, A.K., 2013. The Status of Water Sensitive Urban Design Schemes in South Australia. Goyder Institute for Water Research Technical Report Series No. 13/11. ISSN: 1839-2725. South Australia, Adelaide. http://www.goyderinstitute.org/_r100/media/system/attrib/file/91/WSUD_Task%201_Report_WSUD%20Inventory_final%20for%20web.pdf.

Myers, B., Chacko, P., Tjandraatmadja, G., Cook, S., Umapathi, S., Pezzaniti, D., Sharma, A.K., 2014. The Status of Water Sensitive Urban Design Schemes in South Australia. Goyder Institute for Water Research Technical Report Series No. 13/11, Adelaide, South Australia, pp. 1839–2725.

National Water Commission, 2004. Intergovernmental Agreement on a national water initiative -Between the Commonwealth of Australia and the Governments of New South Wales. The Australian Capital Territory and the Northern Territory, Victoria, Queensland, South Australia, p. 30. http://content.webarchive.nla.gov.au/gov/wayback/20160615061050/http://www.nwc.gov.au/__data/assets/pdf_file/0008/24749/Intergovernmental-Agreement-on-a-national-water-initiative.pdf.

New South Wales Government, 2004. BASIX. https://www.basix.nsw.gov.au/iframe/about-basix/legislation.html.
New South Wales Government, 2014. BASIX (Building Sustainability Index) -Legislation, NSW Government. Department of Planning and Infrastructure. http://www.basix.nsw.gov.au/basixcms/aboutbasix/legislation.html.
Nickel, D., Schoenfelder, W., Medearis, D., Dolowitz, D.P., Keeley, K., Shuster, W., 2014. German experience in managing stormwater with green infrastructure. Journal of Environmental Planning and Management 57 (3), 403–423.
Northern Territory Government, March 2014. Northern Territory Planning Scheme. Department of Lands, Planning and the Environment.
Northern Territory Government, July 2016. Water Act. 1. https://legislation.nt.gov.au/Legislation/WATER-ACT.
NTEPA, August 2014. A Stormwater Strategy for the Darwin Harbour Region. Northern Territory Environment Protection Authority. https://ntepa.nt.gov.au/__data/assets/pdf_file/0004/284872/stormwater_strategy_darwin_harbour.pdf.
Office of Environment and Heritage, 2011a. Stormwater Management Service Charge. http://www.environment.nsw.gov.au/stormwater/smsc.htm.
Office of Environment and Heritage, 2011b. Stormwater Management Service Charge Implementation Monitoring (Covering Financial Years 2006-07 to 2008-09). Government of New South Wales. http://www.environment.nsw.gov.au/stormwater/20110268stormwatercharge.htm.
Pahl-Wostl, C., 2007. Transitions towards adaptive management of water facing climate and global change. Water Resources Management 21 (1), 49–62.
Piper, P., 2011. Water sensitive urban design in the Northern Territory. In: Fortune, J., Drewry, J. (Eds.), Darwin Harbour Region Research and Monitoring 2011. Department of Natural Resources, Environment, the Arts and Sport, Palmerston, NT, Australia, pp. 20–21. Report number 18/2011D.
Preston, B.L., Jones, R.N., 2006. Climate Change Impacts on Australia and the Benefits of Early Action to Reduce Global Greenhouse Gas Emissions. CSIRO, Melbourne, p. 41.
PUB Singapore, 2017. Our Water Future. https://www.pub.gov.sg/Documents/PUBOurWaterOurFuture.pdf.
Rehfeld-Klein, M., Interview on 21 October 2011 by Nickel et al (2014).
Saunders, N.J., Pearson, W.L., 2013. Climate change adaptation of urban water management systems in the wet/dry tropics. Australian Journal of Water Resources 17 (2), 180–192. https://doi.org/10.7158/W13-016.2013.17.2.
SenGuv, 2011. Reduzierung der Nährst*off*belastungen von Dahme, Spree und Havel in Berlin sowie der Unteren Havel in Brandenburg. Senatsverwaltung für Gesundheit, Umwelt und Verbraucherschutz Berlin und Ministerium für Umwelt, Gesundheit und Verbraucherschutz Brandenburg, Berlin, Potsdam.
Sharma, A.K., Cook, S., Tjandraatmadja, G., Gregory, A., 2012. Impediments and constraints in the uptake of water sensitive urban design measures in greenfield and infill developments. Water, Science and Technology 65 (2), 340–352.
Sharma, A.K., Pezzaniti, D., Myers, B., Cook, S., Tjandraatmadja, G., Chacko, P., Chavoshi, S., Kemp, D., Leonard, R., Koth, B., Walton, A., 2016. Water sensitive urban design: an investigation of current systems, implementation drivers, community perceptions and potential to supplement urban water services. Water 8 (7), 272. https://doi.org/10.3390/w8070272.
SMCMA (Sydney Metropolitan Catchment Management Authority), October 2012. Sydney Metropolitan CMA 2012 Catchment Action Plan – A Plan for Sydney's Liveability. Sydney Metropolitan Catchment Management Authority, State of NSW.
South Australia Department of Environment Water and Natural Resources (DEWNR), 2013. Water Sensitive Urban Design - Creating More Liveable and Water Sensitive Cities for South Australia. Government of South Australia, Adelaide, SA, Australia. http://www.google.com.au/url?sa=t&rct=j&q=&esrc=s&source=web&cd=1&ved=0ahUKEwj52vCO_aXWAhUJa7wKHZmwDWIQFggpMAA&url=http%3A%2F%2Fwww.environment.sa.gov.au%2Ffiles%2F516f3ac2-16ff-43fd-b078-a26900b99a81%2Fwater-sensitive-urban-design-policy-gen.pdf&usg=AFQjCNF6Lptx2_lEJkKlDR0kx8rmFw7Fyw.
State of Tasmania, 1997. State Policy on water quality management 1997, p. 32. http://epa.tas.gov.au/documents/state_policy_on_water_quality_management_1997.
Tan, A., Chiang, K., 2011. Vertical greenery in the tropics. Centre for Urban Greenery and Ecology (CUGE), Singapore, p. 100.
Tan, P.Y., Sia, A., 2008. A selection of plants for green roofs in Singapore. Centre for Urban Greenery and Ecology (CUGE), Singapore, p. 118.
Tjandraatmadja, G., Sharma, A.K., 2015. Incentive policies and programmes for rainwater management in urban areas. In: Santos, D.B., Medeiros, S.S., Brito, L.T., Gnadlinder, J., Cohim, E.V., Paz, V.P.S., Gheyi, H.R. (Eds.), Captacao, manejo e uso de agua de chuva, PB. Insa, Brazil, pp. 441, p. 215.
Tjandraatmadja, G., Cook, S., Sharma, A., Chacko, P., Myers, B., Pezzaniti, D., 2014. Water Sensitive Urban Design Impediments and Potential: Contributions to the SA Urban Water Blueprint: Post-implementation Assessment and Impediments to WSUD. Goyder Institute for Water Research Technical Report Series No. 14/16, Adelaide, South Australia, pp. 1839–2725.
United States Environmental Protection Agency (USEPA), 1972. Clean Water Act, USA.
USEPA, 2017. Urban Runoff: Low Impact Development. https://www.epa.gov/nps/urban-runoff-low-impact-development.
Victorian/Tasmanian Division of Parks and Leisure Australia and Department of Planning and Community Development, August 2013. Open Space Planning and Design Guide, p. 123.
Western Australia Planning Commission and Western Australia Department of Planning and Infrastructure (WAPC and WADPI), October 2008. Better Urban Water Management. Western Australia Planning Commission, Western Australia Department of Planning and Infrastructure, State of Western Australia. http://www.water.wa.gov.au/__data/assets/pdf_file/0003/1668/82305.pdf.
Western Australia Planning Commission, 2006. State Planning Policy 2.9 – Water Resources. Government Gazette, WA. https://www.planning.wa.gov.au/dop_pub_pdf/SPP_2_9.pdf.
WHG, 2009. Wasserhaushaltsgesetz - WHG, 31 Juli 2009. http://www.gesetze-im-internet.de/whg_2009/WHG.pdf.
White, S.S., Boswell, M.R., 2006. Planning for water quality: implementation of the NPDES phase II stormwater program in California and Kansas. Journal of Environmental Planning and Management 49 (1), 141–160. https://doi.org/10.1080/09640560500373386.

White, S.S., Boswell, M.R., 2007. Stormwater quality and local government innovation. Journal of the American Planning Association 73 (2), 185–193. https://doi.org/10.1080/01944360708976152.

Yong, J.W.H., Tan, P.Y., Hassan, N.F., Tan, S.N., 2010. A selection of plants for waterways and waterbodies in the tropics. Centre for Urban Greenery and Ecology (CUGE), Singapore, p. 480.

Zathmakesh, Z., Burian, S.J., Karamouz, M., Tavakol-Davani, H., Goharian, E., 2015. Low impact development practices to mitigate climate change effects on urban stormwater run-off: case study of New York City: many communities in the USA implement LID to meet stormwater management goals. Journal of Irrigation and Drainage Engineering 141 (1), 1–13.

## FURTHER READING

ANZECC and ARMCANZ, 2000. Australian Guidelines for Urban Stormwater Management. Commonwealth of Australia, Canberra, Australia, p. 72.

Schweitzer, N., 2013. Greening the Streets: A Comparison of Sustainable Stormwater Management in Portland. Pomona Senior Theses, Oregon and Los Angeles, California. Paper 85.

Ward, S., Barr, S., Butler, D., Memon, F.A., 2012. Rainwater harvesting in the UK: socio-technical theory and practice. Technological Forecasting and Social Change 79, 1354–1361.

Wright, H., 2011. Understanding green infrastructure: the development of a contested concept in England. Local Environment 16 (10), 1003–1019.

Chapter 6

# Flood and Peak Flow Management Using WSUD Systems

**Baden R. Myers and David Pezzaniti**
*School of Natural and Built Environments, University of South Australia, Adelaide, SA, Australia*

## Chapter Outline



**ABSTRACT**

This chapter investigates the extent to which water sensitive urban design (WSUD) can be used for the management of flooding and peak flows in urban catchments. It provides background to the reader regarding the intent of WSUD for flood and flow management, followed by a discussion of key drainage principles, including how WSUD can be incorporated into them. Finally, the chapter presents peer-reviewed case study literature, which has examined the effectiveness of WSUD measures to manage flooding and peak flows.

In most statements of WSUD principles, it is generally understood that WSUD is intended to contribute to peak flow management and flood control for small and frequent storms. Literature broadly supports the effectiveness of WSUD measures for frequent storms (i.e., up to 2-year average recurrence interval events). "Natural flow management" and "Sponge Cities" are two design philosophies very much related to WSUD, and are intended to reduce flooding and peak flows from larger, less frequent storms at the broader catchment scale. However, the success of these concepts is yet to be confirmed.

Broadly speaking, the mitigation of flooding and peak flow rates by WSUD systems is through the incorporation of detention or retention processes. Detention stores water temporarily for controlled release downstream through an orifice outlet or weir, whereas retention reduces runoff volume by storing water for on-site disposal through processes such as reuse, infiltration, and evapotranspiration. The critical parameters for all WSUD storage mechanisms to mitigate peak flow and flooding are storage size, the portion of catchment impervious area connected to the storage, and the rate at which the storage is emptied. Notwithstanding the focus in the literature on WSUD system performance for volume reductions and water quality, the effectiveness of WSUD for flood and peak

**Approaches to Water Sensitive Urban Design. https://doi.org/10.1016/B978-0-12-812843-5.00006-X**

flow management has not been extensively measured in the field. Rather, case studies reporting the performance of WSUD to mitigate flooding and peak flows are largely based on computer simulation. Simulation is a valid technique, but it has generally been limited to design (synthetic) storm events. It is also lacking in sensitivity analysis of the key parameters described earlier—storage volume, connected impervious area, and emptying rate. In particular, research tends to assume that storages are empty prior to design storm simulation, which can overestimate actual performance. Nonetheless, the review indicates that WSUD can contribute to flood and flow management for frequent storms. It is recommended that research continue into this aspect of WSUD, particularly focusing on *continuous simulation processes* and the study of observed runoff and flooding data prior to, and following WSUD implementation.

**Keywords:** Detention; Flood management; Peak flow management; Retention.

## 6.1 INTRODUCTION

In this chapter, we investigate the extent to which water sensitive urban design (WSUD) can be or has been used for the management of flood and peak flows in urban catchments. The approach used is a review of literature: first to provide background to the reader and then to focus on peer-reviewed case studies on the use of WSUD measures to manage flooding and peak flows.

In Section 6.1, we explore the role of WSUD for flood and frequent flow management: why flooding and peak flows are important for stormwater management, and the extent to which WSUD is expected to perform in this regard. In Section 6.2, we explore critical knowledge that underpins our understanding of how WSUD can be developed, or assessed, to reduce flooding and peak flows. In Section 6.3, we present a review of literature which examines the use of WSUD for flood and peak flow management for major and minor storms. The outcomes of this are discussed in Section 6.4.

WSUD studies that focus on runoff volume management (or performance) without consideration of flood and/or peak flow is not relevant to the aim of this chapter. Moreover, the term "average recurrence interval" (ARI) has been employed to describe the frequency and magnitude of storms. ARI is related to the "return period" and represents the average periods between which a given rainfall or runoff total may occur over a particular duration (Ball et al., 2016). It is a term used in the majority of the literature we reviewed.

### 6.1.1 Is WSUD intended to consider flood and peak flow management?

It is apparent that WSUD is strongly related to flood and peak flow management. From its earliest use as terminology in the 1990s (Fletcher et al., 2014), WSUD practices have focused on sustainably managing flooding and peak flows for urban runoff. For example, in an early WSUD guideline produced by Whelans et al. (1994), four WSUD objectives were provided, the first of which was to:

*manage the water balance (including groundwater and streamflows, along with flood damage and waterway erosion).*

Subsequent WSUD guidelines have continued to include flow management principles among design objectives. For example, the Victorian Stormwater Committee (1999) provided five WSUD objectives, the fourth of which was:

*reduction of runoff and peak flows from urban developments by employing local detention measures and minimising impervious areas.*

The introduction to *Australian Runoff Quality* (Wong, 2005) also included flow management as a WSUD objective:

*Preserving the natural hydrological regime of catchments.*

The need to manage flooding and peak flows is also noted in international literature related to WSUD. For example, reducing urban flooding is a specific target of the "Sponge Cities" initiative in China. In the United Kingdom, Sustainable Urban Drainage Systems or Sustainable Drainage Systems (SuDS) is a term used in a similar way to WSUD in Australia and New Zealand. The most authoritative guide in the United Kingdom is *The SuDS manual* (Woods Ballard et al., 2015), which includes water quantity as one of the four major principles of SuDS, specifically noting the aim to:

*Control the quantity of runoff to support the management of flood risk [and] maintain and protect the natural water cycle.*

Likewise, in the United States, the terms low impact development (LID) and green infrastructure (GI) have grown in use and also consider similar principles to WSUD. Each is also intended to consider storm runoff flooding and peak flow management. According to the US Environmental Protection Agency[1]:

> *At the city or county scale, green infrastructure is a patchwork of natural areas that provides habitat, flood protection, cleaner air, and cleaner water.*

Based on these varied sources of WSUD objectives and the objectives of sustainable urban design approaches applied internationally, the management of flooding and peak flows is very much considered to be a major role of WSUD.

## 6.1.2 The importance of flooding in WSUD implementation and stormwater management

The focus on the protection of life and assets has historically involved the rapid conveyance of flooding flows downstream—what has come to be known as a "drained city" in the lexicon of modern stormwater management concepts (Fig. 6.1). This has historically led to the reliance on rapid conveyance of runoff from urban environments, with little regard for the other, later, phases of urban stormwater management.

More recently, stormwater management guidelines emphasize the need to adopt a more holistic approach to stormwater management, which includes management responses such as stormwater quality improvement, stormwater reuse, and integration with landscape (ARMCANZ and ANZECC, 2000). Flooding, however, still remains the most important consideration in the management of urban stormwater runoff—a "drained city" is a prerequisite component of a water-sensitive city (as per Fig. 6.1). For this reason, an assessment of the capacity of WSUD and related techniques (such as natural flood management and sponge cities, which are detailed in Section 6.3) to manage urban floods is important for stormwater managers to ensure that new techniques are "fit for purpose."

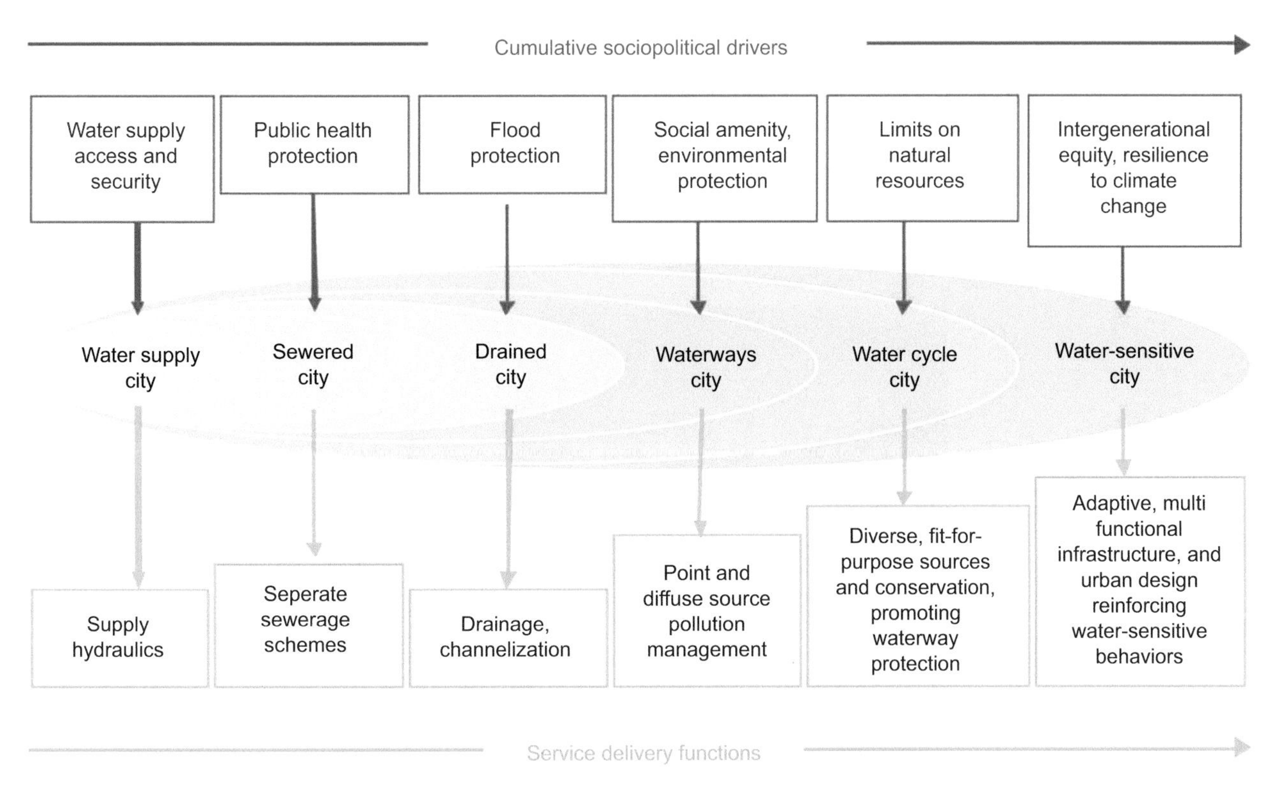


**FIGURE 6.1** The City State Continuum for urban water management. *From Brown, R.R., Keath, N., Wong, T.H.F., 2009. Urban water management in cities: historical, current and future regimes. Water Science and Technology 59, 847–855.*

1. https://www.epa.gov/green-infrastructure/what-green-infrastructure.

### 6.1.3 The extent to which WSUD is expected to contribute to flood and flow management

In many cases, the flow and flood management design goals of WSUD measures are limited to more frequent, low intensity storms. This understanding was evident in early considerations of sustainable stormwater management. For example, ASCE & WEF (1993) provided the following design criteria for the design of urban stormwater management systems:

> *(a) drainage systems are designed for large infrequent runoff events [10, 25, 50 or 100 year ARI]; (b) design events for runoff quality control are small frequent events [smaller than the 1 year ARI runoff event].*

This is because the scale of treatment systems required to intercept and treat the runoff volume from large, infrequent storms has generally been considered an uneconomic method to produce a reduction in annual pollutant load. As explained by ARMCANZ and ANZECC (2000):

> *a significant proportion of the long term (e.g., annual) runoff volume from urban catchments occurs from frequent, smaller flood events and from a proportion of larger events. Commonly in urban areas, approximately 90–95% of the mean annual runoff volume occurs at flows less than the 3 month average recurrence interval event (approximately 25–50% of the 1 year flow in temperate areas). These frequent events are the target of most stormwater quality controls.*

This approach is also adopted internationally. In France, the terms "alternative techniques" or "compensatory techniques" are used to refer to stormwater and flood management, such as retention and infiltration systems. According to Fletcher et al. (2014) (citing Petrucci, 2012), design rules for alternative techniques are still very much focused on hydraulics, particularly "the reduction of flooding with a high return period." Likewise, in the United States, GI stormwater control measures are typically implemented with the intent of reducing runoff flow rate and volume for smaller, more frequent storms (Lewellyn et al., 2016).

Design guidelines for individual WSUD elements also tend to recommend a focus on smaller more frequent storms. For example, in an effort to avoid consuming commercially valuable land area in urban developments, the design of rain gardens and bioretention for small to medium catchments is recommended to match the ≦1 in 1-year ARI runoff flow rate (Payne et al., 2015). Although designed to treat storm events larger in size and volume than the 1 in 1-year ARI, stormwater wetlands are "typically not feasible" to intercept stormwater runoff with a 1 in 100-year frequency (Kadlec and Wallace, 2009). Indeed, many design guidelines recommend sizing stormwater wetlands for flows up to the 1 in 1-year ARI (Melbourne Water, 2005; Wong et al., 1999), Above this flow rate, overflow mechanisms should be provided. It is important to be aware that without a bypass mechanism, WSUD systems can overflow in an uncontrolled manner. An example of such overflow is shown in Fig. 6.2, showing overflow from the berm of a wetland treatment system originally

FIGURE 6.2 A wetland at maximum capacity with an overflowing berm. The Paddocks Wetlands, Para Hills West, South Australia.

**FIGURE 6.3** Inlet basin and bypass mechanism for a modern wetland in Parafield, South Australia. The wetland inlet is center left of the image, and the bypass channel is on the upper right hand side.

constructed in the 1970s, and which still functions as a treatment and harvesting scheme. A bypass arrangement for a more recent wetland treatment system for stormwater harvesting is shown in Fig. 6.3.

There are some WSUD stormwater management solutions specifically for the control of flooding and peak flows. Detention systems, discussed in more detail in Section 6.2.2, have been used for centuries to temporarily detain runoff from a catchment or development site (Echols and Pennypacker, 2015). Some practitioners may argue that detention is not a form of WSUD, but we posit that traditional detention techniques do attempt to achieve the flow-related WSUD objectives quoted in Section 6.1.1. Detention does, however, have a number of disadvantages (adapted from Commonwealth of Australia, 2002):

- Regulations and design approaches can be over simplified
- Detention applied to sites in the lower parts of a catchment may in fact increase flow rates downstream due to lag in the system
- Maintenance of systems can present a financial burden
- There is usually little scope for water quality improvement using detention measures.

While WSUD measures other than detention predominantly target frequent flow events up to a 1-year ARI, they have also been used to manage less frequent, high-intensity storm flows, including major flood events. For example, WSUD guidelines (Argue, 2004), which focus on "source control" of stormwater runoff, provide design guidance and examples of stormwater runoff management up to the 100-year ARI design storm. A notable application of these guidelines includes Parfitt Square in South Australia, as described in Section 6.3.1.3.

## 6.2 FLOOD MANAGEMENT CONCEPTS

In this section, a brief overview of flood management concepts provides the context for the following review. It provides information about the following:

- the major/minor flood management approach used in many jurisdictions;
- the concept of detention and retention systems;
- the definition and importance of using event-based and continuous simulation to design and assess the performance of WSUD;
- the use of structural and nonstructural WSUD measures; and
- the application of distributed and end-of-catchment WSUD solutions.

### 6.2.1 The major/minor approach to urban drainage

In urban environments representative of the drained city in Fig. 6.1, urban stormwater flood and flow management is achieved through the implementation of major and minor drainage systems. Frequent storms are managed by the *minor* drainage system. This system collects storm runoff from roads and buildings and routes it into curbs and channels, which discharges into stormwater inlets, where it is transported via underground pipe systems toward outfalls joining receiving waters. In residential catchments, the minor system is recommended to have a design capacity of between a 2-year and 10-year ARI storms (State of Queensland, 2013). Some guidelines indicate that up to a 20-year ARI may be appropriate (Ball et al., 2016). The system is designed to prevent nuisance flooding at the catchment surface, and allows safe use of infrastructure such as roads and footpaths (Argue, 1986; Ball et al., 2016).

The *major* drainage system is designed to convey runoff that would otherwise exceed the capacity of the minor system. It has a design capacity of the order of a 50-year ARI or greater depending on the infrastructure it is protecting. The system typically includes roadways, walkways, and certain public open spaces, which are designed to convey water away from built infrastructure, whilst ensuring that velocity/depth conditions are below prescribed limits to allow for safe traffic conditions. Major flows are then conveyed to receiving waters. The minor system still contributes to flood and flow management during a major storm, but it conveys only a small proportion of the peak discharge.

### 6.2.2 Detention and retention systems for management of flooding and peak flow

In addition to conveyance, storage measures can be used in the management of flooding and peak flows. Storage represents the mechanism by which WSUD can contribute to flooding and peak flow management. Storage can be represented by detention storages, retention storages, or storages that apply elements of both. **Retention**-based measures are defined as those that intercept and permanently hold the water or divert it for another purpose, effectively preventing water from entering the downstream drainage system. Examples of retention storage include rainwater tanks (Fig. 6.4) and infiltration systems (Fig. 6.5).

**Detention systems** are those that temporarily hold runoff in a storage and release stored water into the downstream drainage system, usually at a restricted flow rate through an orifice or similar structure. At the single dwelling allotment scale, "on-site detention" is typically implemented in the form of a detention tank. For larger development sites, such as larger commercial sites and residential subdivisions, detention is generally applied in the form of detention basins or detention ponds (Fig. 6.6).

The design goals of detention systems, as well as some retention systems, are to ensure that postdevelopment runoff rates from an urban development remain equal to, or less than, the predevelopment peak flow rates for one or more target ARIs. The target ARI(s) are typically set by local or state government and range from 1 to 10 years in residential areas. Alternately, minimum detention tank sizes may be specified. For example, current policy for a local government in the Upper Parramatta River region in NSW, Australia, requires 470 kL of detention storage per hectare or approximately 23 kL for a 500 $m^2$ development site (e.g., Cumberland Council, 2015).

The impact of applying either detention or retention systems on the catchment hydrograph differs for a given storm. They are demonstrated by the hypothetical case study presented by Argue (2004). It describes a redevelopment scenario in Parramatta, Australia, where a 6 ha catchment of 60 home allotments (45% impervious area) is redeveloped to include 120

FIGURE 6.4 A typical rainwater tank that collects rainwater from the roof of a home in Rostrevor, South Australia. The overflow pipe discharges into the stormwater system.

**FIGURE 6.5** A 300 m$^2$ infiltration system collecting runoff from 13 medium-density residential allotments in Henley, South Australia. The basin is one of the two storages that collectively mitigate flows up the 10-year average recurrence interval peak discharge in an area where conveyance was difficult, as the site was situated at a level below the nearest stormwater trunk main.

**FIGURE 6.6** An "end of pipe" detention basin for a residential development site in Flagstaff Hill, South Australia (a) and the outflow control structure of that system (b).

allotments (80% impervious area). Using a 100-year, 1-h design storm, the analysis is used to determine the required storage size to maintain the peak flow rate from the catchment following densification. By assuming storages are empty at the commencement of the storm, the study found that the required on site **retention** (OSR) volume for infiltration systems on each new home (18.3 kL) is smaller than that required for conventional on site **detention** (OSD) systems (28 kL). This finding is somewhat compromised by the use of design storm event simulation and the assumption of empty storages (see Section 6.2.3). But of most interest here is the comparative hydrographs for the retention and detention cases, as shown in Fig. 6.7. Note that detention systems reduce peak flows by collecting runoff and releasing it slowly, thus extending the overall flow duration. Retention systems on the other hand collect a portion of flow (from the beginning of the event) until they are at capacity, after which bypass or overflow occurs.

Many constructed WSUD systems incorporate both storage and detention, including stormwater biofilters and sedimentation basins. A simple example of such a measure is a combined detention and rainwater tank such as that shown in Fig. 6.8. In this example, the detention storage zone is that volume above the 10 mm orifice outlet, which empties by gravity into the street gutter in between events. Below this outlet, the retention storage holds water until it can be used by the homeowner for irrigation or toilet flushing. There are also systems available that can not only store water permanently

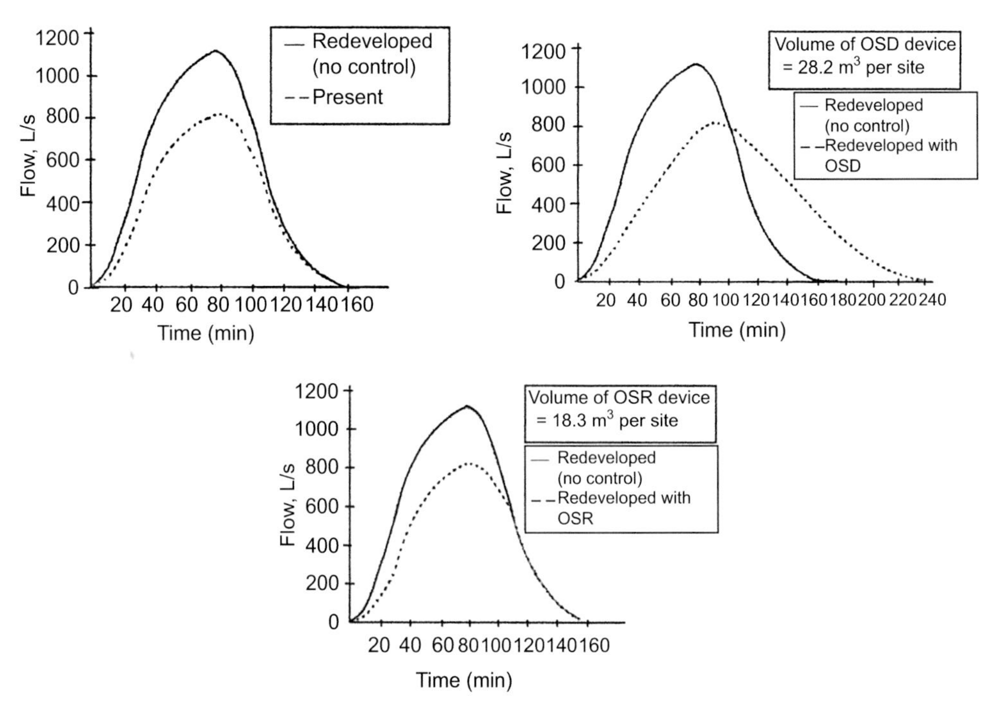


**FIGURE 6.7** The comparison of (a) a predevelopment and postdevelopment hydrograph for the 60 allotment catchments, (b) the same case with detention fitted, and (c) the same case with retention fitted (Argue, 2004).

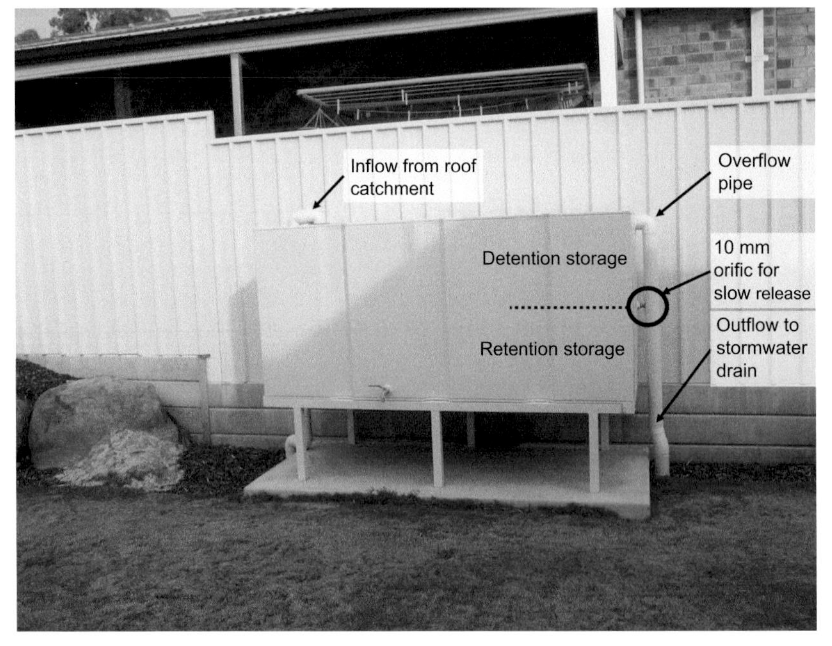


**FIGURE 6.8** A combined 1.5 kL retention and detention tank in Tranmere, South Australia.

but also empty relatively rapidly part or all of the storage zone in preparation for predicted storms. For example, IOTA services, a division of South East Water in Victoria, Australia,[2] provide off-the-shelf technology to meet local government or developer requirements. Their systems have been installed on a trial site in Lyndhurst, south east of Melbourne, Australia, but no performance data is available in the public domain.

2. http://www.iota.net.au/.

Stormwater biofilters/rain gardens may be designed to either detain or retain stormwater. As noted in Chapter 2, intercepted water is detained at the surface prior to infiltration through the filter media. Water in excess of the media water-holding capacity passes through the filter media and is usually collected using underlying pipes and conveyed downstream, thus acting like a detention system. However, water held by soil media evapotranspires between events, thereby providing an element of retention.

### 6.2.3 Event-based and continuous simulation

Much of the literature, which explores the effectiveness of WSUD for flood and/or peak flow management, to be reviewed in Section 6.3, is based on computer simulations. These simulation studies use two general approaches that are important to understand when the reader is interpreting study results. The first is **event-based simulation** and is undertaken using a single storm per model run. The storm is usually synthetically derived to represent a particular storm frequency (ARI) and duration based on observed local rainfall records. While this approach is useful for design purposes, it has drawbacks, especially for WSUD system simulation. This is because the main benefits of WSUD for flood and flow management are derived from the availability of storage volume **before** the storm event occurs. The size of this (variable) available storage volume for any given design event is a function of rainfall prior to and the storage drawdown characteristics in the lead up to real storm events. It is therefore difficult to specify in a design scenario. This problem has been long acknowledged by researchers in the field of urban stormwater management (e.g., James and Robinson, 1982; Coombes et al., 2003).

This problem is illustrated by Pezzaniti (2003) who investigated the impact of rainwater tanks in a small urban catchment in Adelaide, Australia. The volume of storage available in domestic rainwater tanks was determined for events with a duration and intensity equivalent to that of a 5-year ARI design storm. For some events, the rainwater tanks were full prior to the occurrence of the design rainfall event because preceding rainfall had filled the tank. Fig. 6.9 illustrates this condition, showing that the storage volume of a 4.5 kL rainwater tank reached capacity **before** the occurrence of a rain burst equivalent to a 5-year ARI with a 45-min duration.

The shortcomings of event-based simulation are overcome by adopting continuous simulation. **Continuous simulation** uses a longer-term time series of observed (or generated) rainfall data to produce a runoff hydrograph. This can then be analyzed using frequency analysis techniques (e.g., see Sharma et al., 2016). The technique allows for consideration of storage prior to events of interest. However, there are some disadvantages. Continuous simulation is more time-consuming in terms of finding appropriate records, checking data quality, and data processing time. Furthermore, not all models are capable of continuous simulation, and run time can be prohibitive, especially for complex models such as 2D flood studies. There is also little guidance available for design engineers on how to appropriately analyze simulation outputs from continuous modeling.

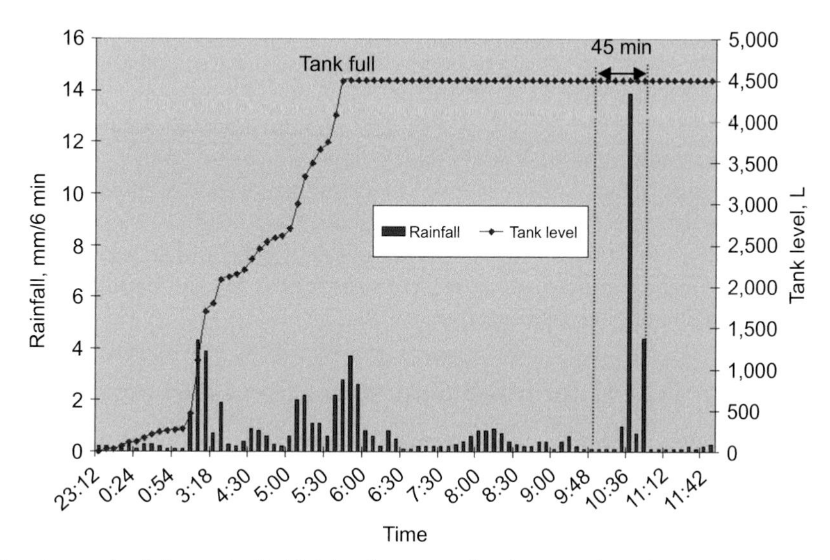


**FIGURE 6.9** Excerpt of continuous simulation results, highlighting the storage of a rainwater tank, a rain burst with a 5-year average recurrence interval occur on December 27, 1929. *Adapted from Pezzaniti, D., 2003. Drainage System Benefits of Catchment-Wide Use of Rainwater Tanks. Urban Water Resources Centre, University of South Australia, Adelaide, SA, Australia.*

### 6.2.4 Structural and nonstructural WSUD for flooding and peak flow control

As with the case for water quality improvement, pathways to manage flooding and peak flow rates using WSUD may be considered nonstructural or structural measures. Nonstructural WSUD manages flooding and peak flows without constructing a physical WSUD element. For example, in greenfield developments, local authorities may adopt a policy that encourages infiltration, restricts runoff by limiting allotment impervious area, encourages vegetation in upstream areas, and/or ensures that natural streams are protected in the development planning process. A public education campaign about WSUD concepts is also an important nonstructural WSUD measure to justify or support public investment. Structural WSUD measures on the other hand involve the construction of infrastructure for flood and flow management and can include detention basins, ponds, rainwater tanks, permeable paving, bioretention systems, and wetlands, to name a few examples.

### 6.2.5 Distributed and end-of-catchment WSUD solutions

There are two potential approaches to managing floods and peak flow rates using WSUD: *distributed* measures throughout a catchment, or a larger and concentrated measure at the *end of the catchment*. The distributed approach involves the use of multiple systems throughout a catchment to produce a cumulative impact on flow within or at the end of the catchment. It includes the use of on-site detention or retention systems, such as rainwater tanks, allotment rain gardens, or street-scale biofilters. The distributed approach is intended to influence the transmission of peak runoff flows through a catchment by dampening the flow rate accumulating from within the catchment boundary to catchment outlet. This is especially important in existing urbanized catchments where infill development (allotment subdivision into two or more lots) occurs, and the catchment's impervious area gradually increases.

End of catchment solutions are those that are intended to manage flood flows from a much larger catchment and therefore tend to be situated near the end of a development/catchment. These include large detention basins, retention/infiltration basins, and wetlands. They are typically installed in greenfield developments or by local/state authorities where public land is available to meet the needs of a developed catchment.

The effectiveness of distributed and end of catchment WSUD depends on the location(s) where flood and flow control is required, and whether the design is part of a new drainage system or retrofitted to an existing system. For example, an end of catchment solution will only mitigate flow conditions at the point of application, not upstream; indeed, they should be carefully designed to prevent backwater effects that can exacerbate conditions upstream. Likewise, a distributed solution, if improperly implemented, may reduce peak flows for subcatchment areas, but cumulative flow downstream may be worse than expected, as occurs for some detention policies described in Section 6.3.2.4.

## 6.3 A REVIEW OF WSUD-BASED FLOOD CONTROL CASE STUDIES

There is a broad body of literature available on the ability of WSUD to provide sustainable management of stormwater. This includes a strong focus on the ability of WSUD to improve urban stormwater quality (see Chapter 3), reduce runoff volume (e.g., Ahiablame et al., 2013; Lewellyn et al., 2016), and allow the collection and harvesting of stormwater as an alternative water supply to offset mains water use (Chapter 9). This section, however, reviews case studies where WSUD has been implemented to reduce flooding and peak flows during large, infrequent storms.

We note there is little evidence available in peer-reviewed literature that describes the impact of WSUD for managing major flood events. This may be because of the general understanding that WSUD is intended to target smaller, more frequent storm events (Section 6.1.3). Alternatively it may also be because of the difficulty of monitoring infrequent storms in the field. There is no guarantee that a high-intensity infrequent storm of interest will occur within a study timeframe, and hence such studies are limited and tend to be retrospective.

### 6.3.1 The effectiveness of WSUD for infrequent storm flood and peak flow management

#### 6.3.1.1 Flood control in Augustenborg, Malmö, Sweden

Sörensen (2016) explored the reduction in flooding due to WSUD and natural flood management in Augustenborg, a suburb of Malmö, Sweden. The subcatchment was retrofitted with WSUD and natural flow management techniques but surrounded by conventionally drained subcatchments. The treated subcatchment was retrofitted with detention ponds, temporary flooding areas, and infiltration surfaces, including green roofs, lawns, and permeable parking spaces. Surface flow rates were amended using surface swales and ditches. To investigate the impact on major flooding, Sörensen (2016) examined the number of insurance claims (per ha) in the retrofitted and surrounding catchments following a severe storm in August 2014, which had an estimated return period of between 50 and 200 years. The upgraded Augustenborg catchment had approximately 10% of the flood damage claims compared with surrounding areas. Although the study had a number of methodological limitations—for example, it did not present data for previous major storms prior to retrofitting the catchment or elaborate on the total insurance claim value versus the annualized construction costs for the retrofit—the investigation represents the first to report *observed* flood damage reductions. Other studies of infrequent, large storms have been limited to model predictions.

### *6.3.1.2 Flood control in Melbourne, Victoria, Australia*

Larger event magnitudes were also considered in the simulation-based study presented by Burns et al. (2015). The study used a coupled 1D (SWMM, US EPA, Cincinnati, Ohio, USA) and 2D (BreZo, University of California, Irvine, California, USA) modeling technique on a 2.8 ha urban catchment in the eastern suburbs of Melbourne, Australia, with a 23% impervious fraction. The authors aimed to explore the impact of installing rainwater tanks and infiltration trenches on simulated flooding in the catchment. The design storms had a 1-, 5-, 20-, and 100-year recurrence intervals of varying duration. The study assumed that a 5 kL tank collected runoff from 200 $m^2$ of roof per domestic dwelling. It assumed full connectivity of impervious ground area, including roads and other impervious areas, to the infiltration measures. The study also assumed that all storages were empty prior to the design storm burst, which may, or may not, be the case prior to an actual storm. The modeling results found that the WSUD measures reduced the extent of flooding for events up to the 20-year ARI. They were most effective for short duration events, with rainfall magnitudes similar to that of the available WSUD storage capacity—in this case, 18-min storms with rainfall depths up to 20 mm. The authors recommended that rainwater tanks that automatically empty based on predicted rainfall would allow the predictions to become reality.

This case study is illustrative of many WSUD studies focusing on flood and flow management, which are reviewed later in this chapter, but focusing on frequent storms. The study confirms that WSUD can have benefits to flood and flow management in a catchment, which is important for catchment managers striving for sustainable drainage solutions. The study findings are, however, limited to computational modeling, and the results, although positive, were produced using liberal estimates of WSUD design inputs. For example, based on physical verification of 223 households in Queensland, Australia, it was found that, on average, only 118 $m^2$ of roof area was connected to rainwater tanks, despite the 200 $m^2$ required under the regulations at that time (Biermann and Butler, 2015).

### *6.3.1.3 Flood control in Birmingham, UK*

Viavattene and Ellis (2013) investigated the use of a coupled 1D model, STORM (Ahlman et al., 2007), and a 2D Flood Area model in ArcGIS (ESRI, Redlands, California, USA) to explore the extent of flooding before and after retrofitting a fully urbanized 12 ha subcatchment in Birmingham, UK, with green roofs and infiltration systems. The event-based model applied a design storm with at least a 200-year ARI of 8-h duration and with two peak rainfall intensities. The study showed that a 30% reduction in surface flood volume was possible by strategically implementing permeable paving and green roofs at the upstream end of the catchment by taking advantage of roofs and parking areas for infiltration. Other opportunities to retrofit WSUD measures were limited in the densely urbanized catchment. The authors observed that rainfall from the first peak reduced the performance of green roofs for the remainder of the storm, as their storage capacity became saturated. Clearly antecedent storage capacity is an important factor to consider in such simulations.

### *6.3.1.4 Design case studies for major flow control*

Two additional examples of using WSUD to mitigate flooding and peak flow from major storms are Parfitt Square, Bowden, South Australia, and Lincoln Square, Melbourne, Australia.

Parfitt Square is a WSUD-based drainage solution for a 1 ha medium density development area with 27 residential allotments that drain to 0.6 ha of parkland (Fig. 6.10). The system incorporates runoff collection, treatment, storage, and reuse (Barton and Argue, 2007; Argue and Pezzaniti, 1999). A key feature of the system is a 100-year ARI surface storage capacity built into the parkland which drains via a gravel-filled trench. This directs water to a groundwater bore, where water passively recharges the underlying aquifer. The aquifer has been subsequently used for irrigation of the parkland. Unfortunately, system performance in terms of flood control and peak flow management has not been reported postconstruction.

Gray literature is also available describing a storage and reuse tank installed beneath Lincoln Square, part of Melbourne's CBD. It has been designed primarily for flood control of the downgradient Elizabeth Street precinct (Brown, 2016).

FIGURE 6.10 Upstream portion of the 0.6 ha Parfitt Square water sensitive urban design system in Adelaide, Australia. Image depicts the above ground storage area where water infiltrates into a subsurface drain.

The 2 ML underground tank was designed to intercept a 20-year ARI storm event from the contributing catchment (pers. comm. Urban Sustainability Branch, City of Melbourne). While intercepted runoff can be used for irrigation, the tank was fitted with a volume release mechanism whereby the tank can be emptied in the lead up to significant storm predictions. There are, however, no data reporting the effectiveness of the tank for flow or flood management at the time of writing due to the recent completion date of the project (2018).

### 6.3.1.5 Natural flood management

Section 6.1.3 indicated that WSUD literature does not typically place a strong focus on the management of infrequent, high-intensity storms. However, there are design approaches for managing large-scale flooding, which have similar objectives to WSUD. Natural flood management is a design philosophy, which is popular in the United Kingdom and parts of Europe (Bubeck et al., 2015). It broadly aims to reduce flooding while achieving multiple benefits throughout a catchment by "protecting, restoring, and emulating the natural regulating function of catchments, rivers, floodplains, and coasts." The means to achieve natural flood management includes measures such as promotion of interception, infiltration, and groundwater storage; reconnecting rivers to floodplains; restoring floodplain woodlands; restoring channels to increase length; revegetation to increase roughness of hillslopes, channels, and floodplains; and restoration of drained peatland. In the United Kingdom, natural flood management grew in popularity following a review of flooding in 2007, which found that natural processes should be considered alongside more traditional "flood defence" measures, such as control dams and barriers (Environment Agency, 2010).

There is a growing body of evidence that examines the effectiveness of natural flood management for improved flood management, but more research is required to achieve a better understanding of benefits and costs (Dadson et al., 2017). For example, a review of 25 natural flood management schemes in the United Kingdom indicated that while woodland coverage is beneficial for reducing peak flow and increasing biodiversity, it also incurs costs by reducing the land available for farming and food production.

### 6.3.1.6 Sponge Cities

The concept of Sponge Cities is another natural flood management technique which is being implemented in China (Radcliffe, 2017). Following recent urban flooding problems, which caused significant loss of life and severe economic impacts, the Chinese government has implemented the Sponge Cities program to improve flooding using LID processes. It aims to make the city act like a sponge to absorb rainfall, rather than cause runoff. The program is based in 30 cities, including Beijing and Shanghai, and is described in more detail in Chapter 1. Each city receives USD$60–90M per year over 3 years to implement sponge city initiatives, all of which align well with both natural flood management and WSUD. The initiative includes natural ecosystem conversion, degraded ecosystem restoration and remediation, and LID practices, as opposed to conventional gray infrastructure (Jiang et al., 2017). At this early stage, there is little available evidence of its effectiveness, and there is much to be done to finalize implementation of the strategy (Liu et al., 2017). However, interested practitioners are advised that the Sponge Cities initiative may well provide significant research opportunities in the realm of flood and peak flow management in the coming years.

## 6.3.2 The effectiveness of WSUD for frequent storm flood and peak flow management

In this section, the effectiveness of WSUD in managing flooding and peak flows from smaller and more frequent storms is examined. The section is broken up into studies that report on the performance of infiltration measures, rainwater tanks and reuse, stormwater bioretention/biofilters, and stormwater detention systems.

### *6.3.2.1 Infiltration measures—trenches, permeable paving, and green roofs*

The most comprehensive study on the performance of infiltration measures for managing frequent storm events was the one undertaken by Roldin et al. (2012) into the impact of constructing "soakaway" infiltration systems across a 300 ha catchment in Copenhagen, Denmark, to reduce combined sewer overflow (CSO) frequency. CSOs are a problem in catchments which convey stormwater and wastewater together, rather than in separate pipelines and channels (see Chapter 7). For context Fig. 6.11 shows a typical soakaway design similar to that assumed to have been used in the study.

Roldin et al. (2012) noted that CSOs occurred when combined wastewater and stormwater flow "overspilled" from drainage pits in the lower parts of the catchment area and into the Harrestrup Stream. One of the overflow points contributed 95% of the total CSO volume and was thus the focus of the study. Continuous modeling of the entire catchment was undertaken using MIKE URBAN CS/MOUSE (DHI Software, Denmark) using a 10-year rainfall record. The simulation included groundwater interactions. The calibrated model estimated the frequency and flow volume of CSOs under three proposed scenarios—the catchment with no soakaways; an "optimistic" distribution of soakaways (connecting 65% of impervious area to soakaways), and a "realistic" distribution of soakaways (connecting 8% of the catchment impervious area due to installation limitations presented by shallow water tables). The study found that the volume of CSOs was reduced by 24% and the mean frequency was reduced from 5.2 to 4.4 events per annum for the "realistic" soakaway scenario. The study commented on the importance of a continuous modeling approach to fully assess the impact of soakaway discharge to the main drainage system. It also recommended that future assessments should consider the impact of groundwater levels on both the soakaways and the drainage network. More detailed discussion on the role of WSUD approaches in sewer system overflow management is provided in Chapter 7.

Liao et al. (2013) reported on the impact of applying five different WSUD scenarios in a 374 ha catchment in Shanghai, China. Three base scenarios were simulated in EPA SWMM (US EPA, Cincinnati, Ohio, USA) including the predevelopment site, the postdevelopment site, and the postdevelopment site with WSUD. WSUD devices included porous pavement, bioretention systems, infiltration trenches, rain barrels, and swales, although little detail was provided on the assumed sizes and other simulation parameters. Simulations were run with 1-, 2-, and 5-year ARI design storms. The results showed that WSUD practices were able to significantly reduce the volume of runoff, peak flow rates, and flooding in the catchment. Rain barrels, permeable paving, and infiltration trenches generally performed best compared with

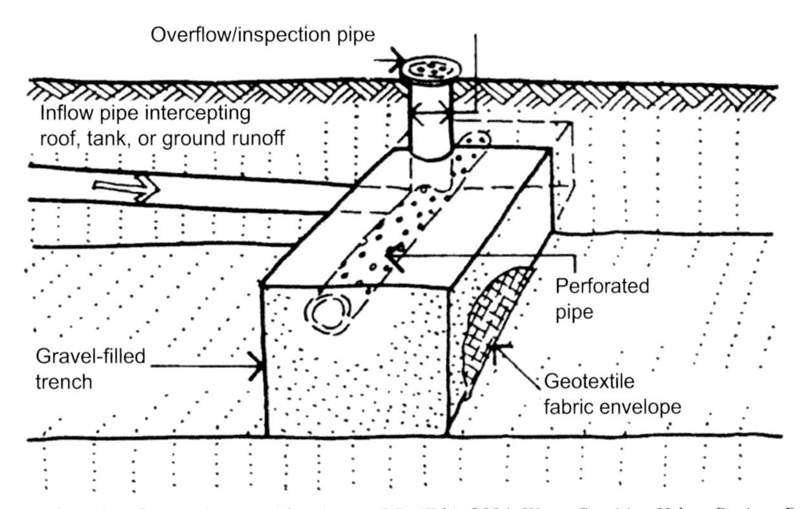


**FIGURE 6.11** Typical layout of a subsurface soakaway. *After Argue, J.R. (Ed.), 2004. Water Sensitive Urban Design: Basic Procedures for 'Source Control' of Stormwater - A Handbook for Australian Practice. Urban Water Resources Centre, University of South Australia, Adelaide, SA.*

bioretention systems and grass swales. However, we suggest the findings should be treated with caution—the performance of these WSUD measures is related to the assumed storage volume, connected catchment area, and rate of discharge for each system—none of which were defined. The ability of systems to reduce flooding and peak flows was found to decline with storm events of increasing duration, but the study did not provide any information on the assumed storage capacities of systems at the beginning of design storms.

Jato-Espino et al. (2016) presented information on the reduced runoff volume and combined sewer surcharge, using the EPA SWMM (US EPA, Cincinnati, Ohio, USA) model on a 31 ha catchment in Donostia, Spain. The study compared the impact of a design storm with a 10-year ARI on the catchment in its current state and for scenarios with green roofs or permeable pavement. The simulations showed that green roofs and permeable paving reduced runoff and sewer surcharge volumes by 38% and 68%, respectively. Permeable paving was most effective in this study for two reasons—as noted by the authors, it had a greater surface area, but it also had a greater total storage volume to collect runoff according to the EPA SWMM data input. It is unclear whether there was a contributing impervious area to the permeable paving in this study, which may also improve performance. As usual, no data were provided on the available volume of each WSUD storage prior to the simulated storm.

Based on this literature and the two infiltration-based studies for major storm events outlined in Section 6.3.1.2 and 6.3.1.3, it is evident that infiltration can provide an effective means of reducing peak flow rates and flood magnitude in urban environments. Results are, however, less positive for infrequent (major) events, and generous assumptions have been made regarding storage conditions prior to storms in most cases. More research is required to reliably assess infiltration system performance under a range of conditions, preferably using continuous simulation with some investigation of infiltration/storage size parameter sensitivity. Or by reporting the performance based on observed flood and/or flow data pre- and postimplementation.

#### 6.3.2.2 Rainwater tanks

Studies on the effectiveness of rainwater tanks for peak flow management are limited, and these limitations are similar to those of the infiltration system literature in Section 6.3.2.1. Petrucci et al. (2012) reported perhaps the most detailed investigation into the impact of rainwater tanks on peak flow and flood control. Their study examined retrofitting tanks to existing homes in a residential catchment in France. Participants were offered one or two tanks of either 0.6 or 0.8 kL each, resulting in the installation of 173 kL of tank volume on approximately 160 of the 450 homes in the 23 ha catchment. The authors used the United States Environment Protection Agency Storm Water Management model (EPA SWMM) (US EPA, Cincinnati, Ohio, USA) to assess flood response. The model was calibrated on observed data before and after tank installation. The rainwater tanks were generally ineffective for flood control because they were too small to influence runoff from events larger than a 3-year ARI. However, the tanks reduced peak flow rates from smaller and more frequent runoff events of lower magnitude. Although this study presented a less than optimistic view of the effectiveness of rainwater tanks, it should be noted that the study was based on retrofitting relatively small tanks to a limited number of homes. It did not explore alternate cases of reuse demand, where tanks may be emptied more frequently. We observe (yet again) that antecedent storage capacity is an important factor to consider in these types of experiments/simulations.

More positive results were reported by Gee and Hunt (2016) in a rare study where observed flow data have been used to explore the performance of a WSUD solution for flow management—albeit with the untreated "control catchment" assessed by simulation. The stored water volume and overflow from two rainwater tank systems were monitored in Craven County, North Carolina, USA. The designers were clearly cognizant of the need for available prestorm storage to maximize effectiveness—one tank (167 $m^2$ roof connection, 8 kL volume) was fitted with a passive release component where 2.2 kL of the tank volume was drained via a slow release orifice with a 3-day emptying time. The other tank (290 $m^2$ roof catchment, 12 kL volume) was fitted with an active release mechanism, which released up to 11 kL of stored water in the lead up to predicted rainfall. Two years of observed rainfall and tank spill flow rate data were compared with roof runoff data calculated in the absence of the tanks. Results showed that the peak flow rate was reduced by an average of 90% due to passive release and 93% due to active release storage. The largest event monitored was a 1-year ARI storm with a 24-h duration, which contained a 2-year, 5-min storm burst. The tank systems reduced the peak runoff from this storm by 78% (passive release) to 100% (active release).

Campisano et al. (2014) examined the ability of rainwater tanks to reduce roof runoff peak flow rate from a single dwelling at a case study site in the south of Italy using a self-developed modeling tool. The study assumed that all the roof area of the home was connected to rainwater tanks, with usage represented by toilet water demand. They found that rainwater tanks could reduce peak flows by intercepting flow before it became runoff. For a typical tank (e.g., 2 kL volume, connected to a 200 $m^2$ roof in a region with annual rainfall of 620 mm), the peak runoff rate was reduced by 30%–65% for about half the events in the simulated year. However, this statistics ignores events of interest for stormwater design and flood management, i.e., design events greater than 1-year recurrence interval, as noted in Section 6.2.1. Furthermore, the assumed connected roof area was very optimistic given the roof connectivity studies noted in Section 6.3.1.2. The study findings are also limited to roof runoff—although this is valuable information, the stated reduction does not give a complete picture of the contribution of the rainwater tank to total allotment runoff, which is of key interest for urban development policy and planning needs.

In recognition of the shortcomings of detention systems, the co-contribution of rainwater tank storage was explored by Coombes et al. (2003). Using a continuous simulation model on a case study single dwelling site in Parramatta, NSW, Australia, they found installing a 10 kL rainwater tank could mitigate peak flows and reduce other detention tank requirements for events up to the 2-year ARI. In a townhouse development scenario, where connected roof area represents a larger portion of the total catchment, the response carried over to even larger ARI events, including the 5- and 100-year ARI. The findings of this study further reinforced the key factors noted in this chapter that determine WSUD system performance for flow management—total storage volume, prestorm storage volume, and the proportion of connected catchment impervious area.

Overall, studies on the peak flow and flooding benefits of adopting rainwater tanks have presented mixed results. The results are all strongly influenced by the availability of prestorm storage. Where tanks were assumed empty prior to storms, the peak flow rate of frequent storm events (<2-year ARI) was usually effectively mitigated. However, small tank storages and/or a limited capacity to discharge (e.g., through dependence on reuse demand) inhibit their effectiveness.

#### *6.3.2.3 Stormwater bioretention and biofiltration*

The attenuation of peak flows by stormwater biofilters has been examined using case study field systems with positive results. Hatt et al. (2009) reported that there was a mean peak flow attenuation of 80% using a biofilter in Melbourne, Australia. However, 11 of the 28 events monitored over a 5-month period caused bypass flow to occur, which was not measured. Hence, although the reported 80% reduction presents positive performance for small, frequent events, they are clearly not indicative of storms with 1-year ARI or greater, as indicated by the frequency of bypass. Of course this may have been a function of the biofilter size, which was 1% of the contributing catchment area—between 1.5% and 3% is more commonly recommended in current design standards (Water by Design, 2012; Payne et al., 2015).

Strong performance was also reported by Yang et al. (2013) for a novel rain garden design, which could reduce peak flows rates by almost 90% for "heavy" storms, defined as >12 mm rainfall over a 24-h period. Although this performance was impressive, the filter size represented approximately 10% of the total catchment area much higher than the 1.5%–3% typically required in guidelines noted above. Furthermore, the ARI of rain events was not reported. Subsequent investigation by the authors revealed that the "heavy" event criterion fell well short of the 1-year ARI for all Australian capital cities.

Stormwater modeling has also been used to examine the efficacy of rain gardens for flow control. James and Dymond (2012) investigated the impact of distributed bioretention in a 154 ha urban catchment in Blacksburg, Virginia, USA, to ascertain whether bioretention could restore predevelopment peak flow rates. The study used the Bentley Sewergems (Bentley Systems Inc., PA, USA) software, and simulations were based on design storms. The study found that bioretention could restore peak flows, but not fully restore predevelopment runoff volumes when they were assumed to infiltrate stored water. However, the desired flow rates were achieved when the total bioretention area was 7% of the monitored catchment area again much larger than the 1.5%–3% in current guidelines for water quality control (Water by Design, 2012; Payne et al., 2015). Furthermore, the study did not report the available water storage capacity prior to the design event.

The effectiveness of bioretention for reducing peak flows was also investigated by Juan et al. (2016). These authors also used design storms as the basis for simulation: a small, frequent storm, and storms with a 2-year and a 10-year ARI, each with a 24-h duration. The study applied the VFlo simulation tool incorporated into a geographical information system (GIS) software package (Vieux, 2004) to investigate peak flow rate and total runoff volume prior to, and following, an extensive catchment retrofit with bioretention and green roofs. The study catchments (28,700 and 8800 ha) were located near Houston, Texas, USA, and results showed that bioretention was more effective than green roofs because of the larger contributing impervious surface area. Reductions in peak flow ranged from 11% to 25% for the frequent storm and 1%–9% for the 10-year storm. The reduction in peak flows decreased as the rainfall event magnitude increased.

Overall, we conclude that bioretention systems have an ability to mitigate runoff behavior from frequent storms, which is supported by observed and simulated data. While restoration of predevelopment peak flows is possible, research so far suggests that systems must be much larger than that typically adopted for water quality treatment (approximately

1.5%–3% of the contributing catchment area). The performance of these proportionally smaller biofilters remains unclear when assessed using both computational and monitoring techniques. Further research is recommended to explore this and the sensitivity of filter media infiltration, especially as infiltration rate degrades over time. There is also no literature that specifically explores flood control using biofilters or bioretention systems.

#### 6.3.2.4 Stormwater detention

As previously discussed, stormwater detention is not traditionally considered to be an example of WSUD. However, detention is implemented with similar design intents to those described for WSUD and related concepts in Section 6.1.1, i.e., to manage peak flow rates and preserve flow downstream. We argue that detention, such as that shown in Fig. 6.6, can be constructed in a way which contributes to urban GI, especially if a permanent pool volume is included. The performance of detention systems is therefore an important consideration in any discussion around WSUD for flow management.

Fennessey et al. (2001) explored the performance of detention ponds on the flow rates from a hypothetical 7.7 ha catchment in the United States, under pre- and posturban development conditions. The study used 33 years of subdaily rainfall data in a hydrology model to simulate the time series of **daily** flow data under conditions of predevelopment, postdevelopment, and postdevelopment with a range of stormwater pond design standards. The data were then used to calculate the 1-, 2-, 5-, 10-, 25-, 50-, and 100-year ARI of flows resulting from seven scenarios. The study found that while the design standards produced an outcome that was sufficient to detain flows from large events (>10-year ARI), none were effective at preserving the 1- or 2-year ARI predevelopment flow conditions. This is because the discharge rate from all the detention basin designs produced a catchment outflow larger than the 1–2-year ARI flow rate, but smaller than the 10-year ARI. Similar results were found in the simulation studies undertaken in Virginia, United States (Hixon and Dymond, 2015), and Washington, USA (Booth and Jackson, 1997).

These studies highlight a long recognized problem of blanket detention policies to control catchment flows—the incremental flow from multiple detention systems (designed to manage runoff from developing land parcels in a catchment) can cumulatively produce unintended exaggerated peak flow rates downstream (Yeh and Labadie, 1997). The problems are not limited to detention ponds. In Sydney, Australia, where on-site detention has been practiced for allotment subdivision in existing suburbs since 1991, up to 90% of the 3500 installations were compromised by construction faults (O'Loughlin et al., 1995). Subsequent computational modeling indicated that on-site detention systems, such as detention ponds, were not having the desired effect on catchment-wide peak flow rates, as flows were aggravated in some storm events (Beecham et al., 2005). Despite the long-term implementation of on-site detention, as with other forms of WSUD, there remains a need for monitoring studies to verify predicted performance.

## 6.4 DISCUSSION AND RESEARCH NEEDS

There are several key themes emerging from the literature on the use of WSUD for the management of flooding and peak flows. Broadly speaking, Roy et al. (2008) concluded that peer-reviewed literature on the management of peak flows and urban flooding is limited, particularly on the application of WSUD at the catchment scale. It would appear that despite numerous publications since then, there is still little research showing benefits at site or catchment scales. Moreover, the reported studies that do exist are lacking in field data to support findings, or to support calibrated and verified modeling. In respect of computer modeling data, there is no generally accepted, rigorous method for the assessment of WSUD modifications on flood and flow management. Most studies adopt event-based simulation, with only a few acknowledging the importance of the prestorm storage volume for design storms (Burns et al., 2015; Viavattene and Ellis, 2013).

Other assumptions regarding WSUD performance also limit the validity of the conclusions. Examples include studies that adopted liberal assumptions such as unusually large storages, unrealisticly large connected impervious areas, or high infiltration rates. The impact of varying these key parameters on predicted WSUD performance has not been well reported. For example, although it is evident that infiltration can provide an effective means of reducing peak flow rates and flood magnitude in urban environments, results are limited to small, frequent storm events and generous assumptions regarding prestorm storage conditions. More research is required to reliably assess infiltration system performance under a range of conditions (preferably using continuous simulation) with investigation of outcome sensitivity to infiltration/storage size parameter values. In addition, observed flood and/or flow data pre- and postimplementation should be reported to assess performance expectations.

Prediction of flow response to infiltration-based measures is also influenced by the assumed infiltration rate of the filter and/or native soil media or, in the case of permeable paving, the surface infiltration rate. However, while varying levels of infiltration are assumed with reasonable justification, there are no peer-reviewed studies that explore the sensitivity of performance to degrading infiltration values. Maintenance (or lack thereof), environmental conditions, and contributing impervious surface area all influence the design life of infiltration-based systems (Argue, 2004). A further limitation of infiltration-based systems research is that they were often applied as a "blanket" scenario across the whole study area. With the exception of Roldin et al. (2012), this assumption ignores limitations because of a shallow groundwater table.

There were some general trends evident in the literature regarding WSUD performance. For example, reduction in flooding and peak flows declined with storm events of increasing magnitude (e.g., Juan et al., 2016; Burns et al., 2015) and duration (Liao et al., 2015). It was also clear that measures that capture runoff from catchment areas greater than that of their own foot print (e.g., permeable paving and biofilters) were more effective for flood and peak flow management than measures that just intercept rainfall directly, such as green roofs. When examined using simulation techniques, a combination of linked WSUD systems tended to be more effective in reducing peak flows than a single WSUD measure.

The difficulty of implementing blanket policy to manage flow rates was evident, even for addressing a single development site (Fennessey et al., 2001) let alone larger flows on broader catchments (Fennessey et al., 2001; Iacob et al., 2014). This emphasizes the need to treat each catchment on an individual basis and develop policy that specifies requirements that maximize outcomes. In reality, this is difficult to achieve because of societal demands for equity and consistency in the implementation of development policy, including WSUD requirements (Sharma et al., 2016).

There are several research needs and recommendations for researchers with an interest in the management of flooding and peak flows using WSUD elements. First, there is a lack of case studies that examine the impact of WSUD implementation on flood frequency, flood volume, and peak flow using field observation and/or gauged flow data. The studies outlined above represent almost all of the available literature on this topic. And it is clear that there are no studies which present paired catchment data (e.g., Stephenson, 1994; Codner et al., 1988) or observed flow data for a catchment prior to and following the implementation of WSUD. Furthermore, many computational modeling studies were hypothetical in nature and often on ungauged catchments, limiting the ability to conduct model calibration.

Another research need is recognition of the influence of antecedent rainfall and/or storage conditions on the performance of WSUD systems. Although there are several studies that report simulated flow responses of a catchment with multiple scenarios (e.g., predevelopment, postdevelopment, and postdevelopment with WSUD measures), many of these are based on a design storm event simulation, which do not consider the prestorm storage capacity. A corollary to this is the lack of an accepted continuous modeling tool for catchment managers to design and assess the performance of WSUD systems to meet a given water quality/quantity/flow rate outcome.

On a positive note, the opportunities for increasing our understanding of how WSUD can contribute to urban flood and peak flow management are increasing. In particular, the ongoing implementation of the Sponge Cities Program in China should provide a variety of case studies for researchers to study their effectiveness.

## 6.5 CONCLUSIONS

The WSUD features such as those described in Chapter 2 are designed to be multifunctional elements that provide benefits to runoff volume, peak flow rate, water quality, and stream ecology. It is generally understood that WSUD systems are intended, *inter alia*, to reduce flooding and peak flows from small and frequent storms (i.e., <1-year ARI events). Research to date has tended to support this view, but there has been less beneficial impact demonstrated for larger, infrequent storms. Natural flow management and "Sponge Cities" are two design philosophies very much related to WSUD, which intend to reduce flooding and peak flows from larger, less frequent storms at the broader catchment scale. While they remain popular in the United Kingdom, Europe, and China, there is still a need for research that can definitively link natural flow management measures to flood management effectiveness. The paucity of experimental evidence may be due to the required scale of the research, and the fact that "Sponge Cities" is still in its implementation phase in China.

The performance of WSUD for reducing flooding and peak flows is known to hinge on storage- either in the retention or detention forms. The three key parameters that influence the effectiveness of these storages are the total storage volume, the connectivity of the catchment to the WSUD storage volume, and the prestorm storage volume (determined by the rate of empting). Current WSUD research tends to be largely based on computer simulation, with only few studies calibrated to observed conditions. There have also been few studies that evaluate the impact of WSUD on the **observed** flooding and peak flow, resulting from a catchment retrofitted with WSUD features. However, none reported a paired catchment case study. Although simulation is a valid technique, it has generally been limited to design storm events (single storm bursts) and has not explored the effect of available prestorm storage volume of a WSUD system. This is despite several authors having identified this to be an important consideration. Nonetheless, the key outcomes from simulation-based research can help practitioners optimize the performance of WSUD for managing stormwater. WSUD methods for reducing flooding and peak flow can be optimized using the following key design techniques:

- The total storage volume and the rate at which the storage volume(s) empties (by infiltration, reuse, discharge, or evaporation) are key factors in determining system outflow characteristics.
- The connectivity of a WSUD feature to impervious area has an important influence on peak flows. For example, devices that collect runoff from an impervious areas in addition to their own surface area (such as detention basins and bioretention systems) tend to be more effective for reducing peak flow rates than those based on interception (such as green roofs).
- WSUD storages for retention (including rainwater tanks, other reuse systems of infiltration-based measures) can be very effective at reducing peak flow for storm events, provided they are emptied before major storm occurs. Technology is available to achieve this type of distributed management. However, the cost of implementation and system reliability are still to be resolved. The effectiveness of rainwater tanks in reducing peak flow and flooding is limited by the often small size of the storages and their limited capacity to empty (e.g., by withdrawal for reuse). The size of the storm which can be retained, is a function of the storage, contributing catchment area, and the rainfall depth, for a given storm frequency.
- Large detention basins are a traditional method to reduce downstream peak discharge in urban areas for large, infrequent events. If multi tasked as green open areas, they could be considered to be part of the palette of WSUD techniques. Despite much experience, however, research has indicated that in some cases, undesirable flow conditions at the catchment outlet can result where multiple detention systems, designed in accordance with subcatchment scale flow release targets, combine to produce undesirable flow conditions at the catchment scale.

## REFERENCES

Ahiablame, L.M., Engel, B.A., Chaubey, I., 2013. Effectiveness of low impact development practices in two urbanized watersheds: retrofitting with rain barrel/cistern and porous pavement. Journal of Environmental Management 119, 151–161.

Ahlman, S., De Roo, C., Svensson, G., Sieker, H., 2007. Source and flux modelling in urban stormwater management (STORM/SEWSYS): application examples in Germany and Sweden. In: Thévenot, D.R. (Ed.), DayWater: An Adaptive Decision Support System for Urban Stormwater Management. IWA Publishing, London, UK.

American Society of Civil Engineers & Water Enviornment Federation, 1993. Design and construction of urban stormwater management systems. In: ASCE Manuals and Reports of Engineering Practice No 77. ASCE Manuals and Reports of Engineering Practice No 77. American Society of Civil Engineers, New York, NY, USA.

Argue, J., 1986. Storm Drainage Design in Small Urban Catchments: A Handbook for Australian Practice. Special Report SR 34. Australian Road Research Board, Vermont South, Victoria, Australia.

Argue, J.R. (Ed.), 2004. Water Sensitive Urban Design: Basic Procedures for 'Source Control' of Stormwater - A Handbook for Australian Practice. Urban Water Resources Centre, University of South Australia, Adelaide, SA.

Argue, J.R., Pezzaniti, D., 1999. Catchment "greening" using stormwater in Adelaide, South Australia. Water Science and Technology 39, 177–183.

ARMCANZ & ANZECC, 2000. Australian Guidelines for Urban Stormwater Management. Agriculture and Resource Management Council of Australia and New Zealand, Australian and New Zealand Environment and Conservation Council and the National Water Quality Management Strategy, Canberra, ACT, Australia.

Ball, J., Babister, M., Nathan, R., Weeks, W., Weinmann, E., Retallick, M., Testoni, I. (Eds.), 2016. Australian Rainfall and Runoff: A Guide to Flood Estimation. Commonwealth of Australia (Geoscience Australia), Canberra, ACT, Australia.

Barton, A., Argue, J., 2007. A review of the application of water sensitive urban design (WSUD) to residential development in Australia. Australian Journal of Water Resources 11, 31–39.

Beecham, S., Kandasamy, J., Khiadani, M., Trinh, D., 2005. Modelling on-site detention on a catchment-wide basis. Urban Water Journal 2, 23–32.

Biermann, S., Butler, R., 2015. Physical verification of household raintank water systems. In: Sharma, A.K., Begbie, D., Gardner, T. (Eds.), Rainwater Tank Systems for Urban Water Supply – Design, Yield, Energy, Health Risks, Economics and Community Perceptions. IWA Publishing, London, UK.

Booth, D.B., Jackson, C.R., 1997. Urbanization of aquatic ecosystems: degradation thresholds, stormwater detection, and the limits of mitigation. Journal of the American Water Resources Association 33, 1077–1090.

Brown, R.R., Keath, N., Wong, T.H.F., 2009. Urban water management in cities: historical, current and future regimes. Water Science and Technology 59, 847–855.

Brown, S.L., 2016. Massive Tank under Melbourne Park to Protect Flinders Street Station against Flooding. ABC News (Online). http://www.abc.net.au/news/2016-08-04/massive-tank-under-melbourne-park-to-protect-against-cbd-floods/7689276.

Bubeck, P., Kreibich, H., Penning-Rowsell, E.C., Botzen, W.J.W., De Moel, H., Klijn, F., 2015. Explaining differences in flood management approaches in Europe and in the USA – a comparative analysis. Journal of Flood Risk Management 10, 436–445.

Burns, M.J., Schubert, J.E., Fletcher, T.D., Sanders, B.F., 2015. Testing the impact of at-source stormwater management on urban flooding through a coupling of network and overland flow models. Wiley Interdisciplinary Reviews: Water 2, 291–300.

Campisano, A., Di Loberto, D., Modica, C., Reitano, S., 2014. Potential for peak flow reduction by rainwater harvesting tanks. Procedia Engineering 89, 1507–1514.

Codner, G.P., Laurenson, E.M., Mein, R.G., 1988. Hydrologic effects of urbanisation: a case study. In: Hydrology and Water Resources Symposium 1988. The Institution of Engineers, Australia, Canberra, ACT, Australia.

Commonwealth of Australia, 2002. Introduction to Urban Stormwater Management in Australia. Department of Environment and Heritage, Canberra, ACT, Australia.

Coombes, P.J., Kuczera, G., Frost, A., O'Loughlin, G., Lees, S., 2003. The impact of rainwater tanks in the upper Parramatta river catchment. Australasian Journal of Water Resources 7, 121–129.

Cumberland Council, 2015. On-site Stormwater Detenion Policy. Cumberland Council, Merrylands, NSW, Australia.

Dadson, S.J., Hall, J.W., Murgatroyd, A., Acreman, M., Bates, P., Beven, K., Heathwaite, L., Holden, J., Holman, I.P., Lane, S.N., O'connell, E., Penning-Rowsell, E., Reynard, N., Sear, D., Thorne, C., Wilby, R., 2017. A restatement of the natural science evidence concerning catchment-based 'natural' flood management in the UK. Proceedings of the Royal Society A: Mathematical, Physical and Engineering Science 473.

Echols, S., Pennypacker, E., 2015. The History of Stormwater Management and Background for Artful Rainwater Design. Artful Rainwater Design: Creative Ways to Manage Stormwater. Island Press/Center for Resource Economics, Washington, DC.

Environment Agency, 2010. Working with Natural Processes to Manage Flood and Coastal Erosion Risk - A Guidance Document. Environment Agency (UK), Bristol, UK.

Fennessey, L., Hamlett, J., Aron, G., Lasota, D., 2001. Changes in runoff due to stormwater management pond regulations. Journal of Hydrologic Engineering 6, 317–327.

Fletcher, T.D., Shuster, W., Hunt, W.F., Ashley, R., Butler, D., Arthur, S., Trowsdale, S., Barraud, S., Semadeni-Davies, A., Bertrand-Krajewski, J.-L., Mikkelsen, P.S., Rivard, G., Uhl, M., Dagenais, D., Viklander, M., 2014. SUDS, LID, BMPs, WSUD and more – the evolution and application of terminology surrounding urban drainage. Urban Water Journal 1–18.

Gee, K.D., Hunt, W.F., 2016. Enhancing stormwater management benefits of rainwater harvesting via innovative technologies. Journal of Environmental Engineering 142, 04016039.

Hatt, B.E., Fletcher, T.D., Deletic, A., 2009. Hydrologic and pollutant removal performance of stormwater biofiltration systems at the field scale. Journal of Hydrology 365, 310–321.

Hixon, L.F., Dymond, R.L., 2015. Comparison of stormwater management strategies with an urban watershed model. Journal of Hydrologic Engineering 20, 04014091.

Iacob, O., Rowan, J.S., Brown, I., Ellis, C., 2014. Evaluating wider benefits of natural flood management strategies: an ecosystem-based adaptation perspective. Hydrology Research 45, 774–787.

James, M., Dymond, R., 2012. Bioretention hydrologic performance in an urban stormwater network. Journal of Hydrologic Engineering 17, 431–436.

James, W., Robinson, M., 1982. In: Degroot, W. (Ed.), Stormwater Detention Facilities. New England College. ASCE, pp. 163–175.

Jato-Espino, D., Charlesworth, S., Bayon, J., Warwick, F., 2016. Rainfall–runoff simulations to assess the potential of SuDS for mitigating flooding in highly urbanized catchments. International Journal of Environmental Research and Public Health 13, 149.

Jiang, Y., Zevenbergen, C., Fu, D., 2017. Understanding the challenges for the governance of China's "sponge cities" initiative to sustainably manage urban stormwater and flooding. Natural Hazards 1, 521–529.

Juan, A., Hughes, C., Fang, Z., Bedient, P., 2016. Hydrologic performance of watershed-scale low-impact development in a high-intensity rainfall region. Journal of Irrigation and Drainage Engineering 143.

Kadlec, R.H., Wallace, S.D., 2009. Treatment Wetlands [Electronic Resource]. CRC Press, Boca Raton, FL.

Lewellyn, C., Lyons, C., Traver, R., Wadzuk, B., 2016. Evaluation of seasonal and large storm runoff volume capture of an infiltration green infrastructure system. Journal of Hydrologic Engineering 21, 04015047.

Liao, Z.L., He, Y., Huang, F., Wang, S., Li, H.Z., 2013. Analysis on LID for highly urbanized areas' waterlogging control: demonstrated on the example of Caohejing in Shanghai. Water Science and Technology 68, 2559–2567.

Liao, Z.L., Zhang, G.Q., Wu, Z.H., He, Y., Chen, H., 2015. Combined sewer overflow control with LID based on SWMM: an example in Shanghai, China. Water Science and Technology 71, 1136–1142.

Liu, H., Jia, Y., Niu, C., 2017. "Sponge city" concept helps solve China's urban water problems. Environmental Earth Sciences 76, 473.

Melbourne Water, 2005. WSUD Engineering Procedures: Stormwater. CSIRO Publishing, Melbourne, Victoria, Australia.

O'Loughlin, G., Beecham, S., Lees, S., Rose, L., Nicholas, D., 1995. On-site stormwater detention systems in Sydney. Water Science and Technology 32, 169–175.

Payne, E.G.I., Hatt, B.E., Deletic, A., Dobbie, M.F., Mccarthy, D.T., Chandrasena, G.I., 2015. Adoption Guidelines for Stormwater Biofiltration System. Cooperative Research Centre for Water Sensitive Cities, Melbourne, Australia.

Petrucci, G., 2012. The Diffusion of Source Control for Urban Stormwater Management: A Comparison between the Current Practices and the Hydrological Rationality (La diffusion du contrôle à la source des eaux pluviales urbaines: confrontation des pratiques à la rationalité hydrologique) (Unpublished doctoral dissertation). Universite Paris Est (Ecole des Ponts ParisTech), Paris, France.

Petrucci, G., Deroubaix, J.-F., De Gouvello, B., Deutsch, J.-C., Bompard, P., Tassin, B., 2012. Rainwater harvesting to control stormwater runoff in suburban areas. An experimental case-study. Urban Water Journal 9, 45–55.

Pezzaniti, D., 2003. Drainage System Benefits of Catchment-Wide Use of Rainwater Tanks. Urban Water Resources Centre, University of South Australia, Adelaide, SA, Australia.

Radcliffe, J.C., 2017. The evolution of low impact development - water sensitive urban design and its extention to China as "sponge cities". Water e-Journal 2.

Roldin, M., Fryd, O., Jeppesen, J., Mark, O., Binning, P.J., Mikkelsen, P.S., Jensen, M.B., 2012. Modelling the impact of soakaway retrofits on combined sewage overflows in a 3 $km^2$ urban catchment in Copenhagen, Denmark. Journal of Hydrology 452–453, 64–75.

Roy, A.H., Wenger, S.J., Fletcher, T.D., Walsh, C.J., Ladson, A.R., Shuster, W.D., Thurston, H.W., Brown, R.R., 2008. Impediments and solutions to sustainable, watershed-scale urban stormwater management: lessons from Australia and the United States. Environmental Management 42, 344–359.

Sharma, A., Pezzaniti, D., Myers, B., Cook, S., Tjandraatmadja, G., Chacko, P., Chavoshi, S., Kemp, D., Leonard, R., Koth, B., Walton, A., 2016. Water sensitive urban design: an investigation of current systems, implementation drivers, community perceptions and potential to supplement urban water services. Water 8, 272.

Sörensen, J., 2016. Open LID stormwater system tested during severe flood event. In: International Low Impact Development Conference 2016. Beijing, China.

State of Queensland, 2013. Queensland Urban Drianage Manual - Third Edition 2013, Provisional. The State of Queensland (Department of Energy and Water Supply); Brisbane City Council; Institute of Public Works Engineering Australia, Queensland Division Ltd., Brisbane, QLD, Australia.

Stephenson, D., 1994. Comparison of the water balance for an undeveloped and a suburban catchment. Hydrological Sciences Journal 39, 295–307.

Viavattene, C., Ellis, J.B., 2013. The management of urban surface water flood risks: SUDS performance in flood reduction from extreme events. Water Science and Technology 67, 99–108.

Victorian Stormwater Committee, 1999. Urban Stormwater: Best Practice Environmental Management Guidelines. CSIRO Publishing, Melbourne, VIC, Australia.

Vieux, B.E., 2004. Distributed Hydrologic Modeling Using GIS. Kluwer Academic, Norwell, MA, USA.

Water by Design, 2012. Bioretention Technical Design Guidelines (Version 1). Healthy Waterways Ltd., Brisbane, QLD, Australia.

Whelans, C., Maunsell, H.G., Thompson, P., 1994. Planning and Management Guidelines for Water Sensitive Urban (Residential) Design. Department of Planning and Urban Development of Western Australia, Perth, WA, Australia.

Wong, T.H.F., 2005. Introduction. In: Wong, T.H.F. (Ed.), Australian Runoff Quality. Engineers Australia, Canberra.

Wong, T.H.F., Breen, P.F., Somes, N.L.G., Lloyd, S.D., 1999. Managing Urban Stormwater Using Constructed Wetlands. Cooperative Research Centre for Catchment Hydrology, Monash, VIC.

Woods Ballard, B., Wilson, S., Udale-Clarke, H., Illman, S., Scott, T., Ashley, R., Kellagher, R., 2015. The SuDS Manual. CIRIA, London, UK.

Yang, H., Dick, W.A., Mccoy, E.L., Phelan, P.L., Grewal, P.S., 2013. Field evaluation of a new biphasic rain garden for stormwater flow management and pollutant removal. Ecological Engineering 54, 22–31.

Yeh, C.-H., Labadie, J.W., 1997. Multiobjective watershed-level planning of storm water detention systems. Journal of Water Resources Planning and Management 123, 336–343.

Chapter 7

# Water Sensitive Urban Design Approaches in Sewer System Overflow Management

Leila Talebi[1] and Robert Pitt[2]

[1]*Senior Water Resources Engineer, Paradigm Environmental, San Diego, CA, United States;* [2]*Emeritus Cudworth Professor of Urban Water Systems, University of Alabama, Tuscaloosa, AL, United States*

## Chapter Outline



## ABSTRACT

Early management of combined sewer overflows (CSOs) incorporated sewer separation, large-scale storage, and rapid, but partial, treatment. There is increasing interest focusing on green infrastructure (GI) stormwater controls to alleviate CSO magnitudes and frequencies. GI encompasses many elements of urban infrastructure and usually stresses water and wastewater systems. Mostly, GI components for CSO management incorporate infiltration of source stormwater flows, along with possible retention and beneficial use components.

There have been many monitoring projects examining the performance of small individual GI controls (such as biofilters, green roofs, and porous pavement for the treatment of stormwater flows). Many modeling efforts have demonstrated the potential of large-scale use of these controls for CSO control, but there have been few monitoring efforts examining complex implementations of these controls in large drainage systems. This chapter reviews two such large-scale projects, located in Cincinnati, Ohio, and Kansas City, Missouri, USA, which monitored GI performance in combined sewer systems, indicating expected performance and potential shortcomings. Recommendations are also provided to reduce some of the monitoring problems observed during these projects.

GI infiltration controls were retrofitted in seven different test locations, ranging in area from 8 to 40 ha and included porous pavement sidewalks and parking lots, curb cut biofilters, grass swales, rain gardens, bioretention systems, and stormwater harvesting systems. These projects not only demonstrated significant flow reductions but also highlighted the need for widespread implementation of stormwater runoff controls affecting most of the site runoff to achieve the large flow reductions necessary to meet typical CSO reduction objectives. The use of GI controls in new developments would be much more efficient as their use and placement can be better incorporated in the site design.

Approaches to Water Sensitive Urban Design. https://doi.org/10.1016/B978-0-12-812843-5.00007-1

**Keywords:** Beneficial uses; Bioretention performance; Conservation design; CSO; Green infrastructure; Stormwater; Urban watershed monitoring.

## 7.1 INTRODUCTION

Piped sewer systems are only one type of improved sanitation facility, serving only a small portion of the world's population, and exist mostly in developed countries. Many of the piped sewer systems that do exist in the world are combined sewers that carry both sanitary sewage and wet weather stormwater, with overflows occurring when excessive stormwater overloads the capacity of the sewer. The United States, in contrast, is served mostly by separate sanitary and stormwater systems but does have a number of combined systems.

Combined sewers are used in 772 US cities, mostly in the midwest and northeast, with a few on the west coast (Fig. 7.1). The National Research Council (NRC) states that improved stormwater capture and beneficial use provides a means to reduce combined sewer overflows (CSOs) and associated pollution discharges (NRC, 2015). Cities are increasingly turning to intensive stormwater management strategies, including stormwater capture and beneficial use, to decrease stormwater discharges to combined sewer systems.

The United Kingdom also has many combined sewers. The Chartered Institution of Water and Environmental Management (CIWEM, 2004) issued a policy statement on the environmental impact of CSOs. They state that in 1989, about 20,000 CSOs occurred in England and Wales. They found that where they could not be eliminated, the volumes and pollutant loads need to be controlled. Alternative drainage management techniques, which control flows at or as near as possible to their source, have become identified by the generic title of sustainable urban drainage systems (SUDS). These include flow reduction practices such as permeable pavements, grass filter swales, and filter strips, as well as peak flow reduction practices such as ponds, basins, and artificial wetlands. The CIWEM states that environmental regulators throughout the United Kingdom are actively encouraging the use of SUDS, where conditions allow, as a means of reducing stormwater flows entering combined sewers, thereby reducing the frequency and magnitude of CSO discharges.

In the United States, the control of CSOs has been a goal of regulators for many years. Early practices focused on separation of the combined sewers into separate stormwater and sanitary sewage systems. In areas where this was not feasible, peak flow reduction was used to moderate the high flows causing overflows. The US EPA's 1989 CSO Strategy included nine minimum controls (NMCs) for CSOs (EPA, 1995). These are the minimum technology-based controls to

FIGURE 7.1 Combined sewers are used in 772 cities in the United States, predominantly in the northeast and midwest, as shown by the dots on this map (EPA, 2001).

address CSO problems without the need for extensive engineering studies or significant construction costs. They are meant to be implemented before the construction of long-term control measures. The NMCs are listed as follows:

1. Proper operation and regular maintenance programs for the sewer system and CSO outfalls;
2. Maximum use of storage in the collection system;
3. Pretreatment to ensure that CSO impacts are minimized;
4. Maximization of flow to the wastewater treatment facility for treatment;
5. Elimination of CSOs during dry weather;
6. Control of solid and floatable materials in CSOs;
7. Pollution prevention programs to reduce contaminants in CSOs;
8. Public communication to ensure that the public receives adequate notification of CSO occurrences and CSO impacts; and
9. Monitoring to characterize the environmental impacts of CSOs and the efficacy of CSO controls.

The goal is to reduce the volume (and/or the timing) of stormwater flows before entering the combined sewer system. Green infrastructure (GI) encompasses many elements of urban infrastructure and usually stresses water and wastewater systems. Mostly, GI components for CSO management incorporate infiltration of source stormwater flows, along with possible retention and beneficial use components. GI options are even becoming recognized as options for CSO control strategies as part of court-mandated consent decrees in the United States. GI options are also encouraged by the US EPA.

In 2011, the US EPA released a memo titled "Achieving Water Quality through Integrated Municipal Stormwater and Wastewater Plans" (EPA, 2011). The memo states that "integrated planning also can lead to the identification of sustainable and comprehensive solutions, such as green infrastructure, that improve water quality as well as support other quality of life attributes that enhance the vitality of communities." The memo stresses the EPA's commitment to GI through community partnerships and guidance documents.

## 7.2 STORMWATER CONTROL MEASURES TO REDUCE FLOWS TO COMBINED SEWERS

There are several terms used internationally to describe the stormwater controls that have been applied to reduce wet weather surface flows to combined sewers (which usually incorporate other water infrastructure improvement goals), such as water sensitive urban design (WSUD), GI, and low impact development (LID), along with other terms such as conservation design and greenfield development. These terms are further covered in Chapter 1. Generally, these approaches focus on reducing stormwater flow volumes (in combined sewers) as their primary objective and may also consider delay in peak flows to allow the treatment facility to recover before additional flows are added to the combined system.

The NRC (2009) reviewed many stormwater control measures (SCMs) that can be applied to an urban area. Many of these are components of WSUD, GI, LID, etc., strategies that can be repurposed for CSO control. These 20 control measures are listed in Table 7.1. This table also identifies the *when, where, and who* of implementing these control measures.

The NRC (2009) also included reviews of the applicability of different types of SCMs, especially their suitability for different development types. Each type of development has a different footprint, impervious fractional cover, open space, land cost, and existing stormwater infrastructure. Consequently, stormwater controls that are ideally suited for one type of development may be impractical or uneconomic for another. The NRC report concludes that there are more options during initial land development than during redevelopment. Retrofitting solutions into existing development is often particularly difficult because of existing infrastructure and limited available land area. These two aspects are especially important when planning to apply these 20 CSO management controls to existing areas.

### 7.2.1 Green infrastructure integrated into CSO control strategy

Municipalities in the United States are implementing GI components to reduce CSOs. Some examples include infiltration facilities such as biofilters, rain gardens, porous pavement, and grass swales, along with practices emphasizing evapotranspiration such as green roofs, and local beneficial uses of stormwater for public open space irrigation. According to the EPA (2011), the motivation for including GI in CSO programs is based on the multiple objectives of meeting CSO regulatory mandates while improving water quality in a manner that will provide broader community benefits and potentially reduce costs. Table 7.2 summarizes characteristics of some of the cities in the United States (profiled by the EPA, 2011) that have implemented GI as part of their CSO control strategies, including their target levels of wet weather

**TABLE 7.1 Summary of SCMs—When, Where, and Who (NRC, 2009)**

| SCM[a] | When | Where | Who |
|---|---|---|---|
| *Product substitution* | Continuous | National, state, regional | Regulatory agencies |
| ***Watershed and land use planning*** | Planning stage | Watershed | Local planning agencies |
| ***Conservation of natural areas*** | Site and watershed planning stage | Site<br>Watershed | Developer, local planning agency |
| ***Impervious cover minimization*** | Site planning stage | Site | Developer, local review authority |
| *Earthwork minimization* | Grading plan | Site | Developer, local review authority |
| Erosion and sediment control | Construction | Site | Developer, local review authority |
| ***Reforestation and soil conservation*** | Site planning and construction | Site | Developer, local review authority |
| *Pollution prevention SCMs for stormwater hotspots* | Postconstruction or retrofit | Site | Operators and local and state permitting agencies |
| **Runoff volume reduction—reuse** | Postconstruction or retrofit | Rooftop | Developer<br>Local agencies |
| **Runoff volume reduction—vegetated** | Postconstruction or retrofit | Site | Developer<br>Local agencies |
| **Runoff volume reduction—subsurface** | Postconstruction or retrofit | Site | Developer<br>Local agencies |
| **Peak reduction and runoff treatment** | Postconstruction or retrofit | Site | Developer<br>Local agencies |
| Runoff treatment | Postconstruction or retrofit | Site | Developer<br>Local agencies |
| *Aquatic buffers and managed floodplains* | Planning, construction, and postconstruction | Stream corridor | Developer<br>Local agencies<br>Landowners |
| Stream rehabilitation | Postdevelopment | Stream corridor | Local agencies |
| *Municipal housekeeping* | Postdevelopment | Streets and stormwater infrastructure | MS4 permittee[b] |
| *Illicit discharge detection and elimination* | Postdevelopment | Stormwater infrastructure | MS4 permittee |
| *Stormwater education* | Postdevelopment | Stormwater infrastructure | MS4 permittee |
| *Residential stewardship* | Postdevelopment | Stormwater infrastructure | MS4 permittee |

*SCM*, stormwater control measure.
[a]*Nonstructural SCMs are in italics. Those in bold indicate a significant runoff volume and/or peak flow rate reduction, which are of most interest for combined sewer overflow strategies.*
[b]*MS4 is the acronym for Municipal Separate Storm Sewer Systems.*

flow control. The combined sewer service areas could benefit from GI controls, but to date the use of GI implementation is relatively small, although increasing with time. In most cases, GI components are allowed to replace traditional hard engineering gray infrastructure components (for example, concrete and steel materials used in traditional practices such as large storage tanks, storage tunnels, pumps, enlarged sewers, etc.) in the CSO control consent decrees, if they are shown to be equivalent or better. It is expected that combinations of green and gray infrastructure will be the most cost-effective CSO control solutions in most cases. Consequently, many communities are implementing pilot projects to demonstrate the benefits of the GI components.

**TABLE 7.2 Case Study of Combined Sewer System Initiatives in a Range of USA Cities (EPA, 2011)**

| City | Service Area Population (Millions) | Service Area (Square Kilometers) | Combined Sewer System Area (Square Kilometers) | Targeted Level of Wet Weather Flow |
|---|---|---|---|---|
| Cincinnati, Ohio | 0.86 | 750+ | 160 | 85% Capture of wet weather volume |
| Cleveland, Ohio | >1 | 900 | 210 | 2–4 Events annually; 98% volume reduction |
| Detroit, Michigan | 3 (2008) | 2700 | 560 | 100% of wet weather flow captured for treatment at the WWTP[a] Up to 13 events annually of treated CSO discharge. Estimated 29,500,000 $m^3$ treated discharge (first flush capture + screening and disinfection) |
| Kansas City, Missouri | 0.65 | 1100 | 150 | 88% Volume reduction |
| Louisville, Kentucky | 0.69 | 1000 | 95 | Overall 96% capture of wet weather flow; overflow frequencies of 0, 4, or 8 discharges per year, depending on location |
| Milwaukee, Wisconsin | 1.1 | 1100 | 62 | 100% Volume reduction Zero events annually by 2035 |
| New York City, New York | 8.4 | 870 | 390 | Reduction of 45,800,000 $m^3$ |
| Philadelphia, Pennsylvania | 1.5 | 350 | 165 | Elimination of the mass of pollutants that would be removed by the capture of 85% volumetric capture of wet weather flow; CSO reduction of approximately 29,900,000 $m^3$/year |
| Portland, Oregon | 0.53 | 350 | 110 | Average discharges: Four winter discharge events/year, one summer discharge event every 3 years 99% Volume reduction for Columbia slough 94% Volume reduction for Willamette river |
| Seattle, Washington | 0.63 | 220 | 140 | One event annually 99% Volume reduction |
| King County, WA | 1.5 | 1100 | 140 | One event annually |
| St. Louis, Missouri | 1.4 | 1400 | 190 | 0–4 Events annually, except for Mississippi river |
| Washington, DC | >0.5 million in the district of Columbia plus 1.6 million in the suburbs | 1900 | 50 | Overall 99% capture of wet weather volume |

*CSO*, combined sewer overflow.
[a]WWTP, *wastewater treatment plant.*

Besides these US examples, GI stormwater approaches are also popular in many countries. There have been many international conferences highlighting GI, which provide examples and implementation strategies. Many of these meetings include other aspects of urban infrastructure, but most emphasize water systems. According to *The Guardian* (October 1, 2015: China's sponge cities: soaking up water to reduce flood risks) "China has now chosen 16 urban districts across the country, including Wuhan, Chongqing and Xiamen, to become pilot sponge cities. Over the next 3 years, each will receive up to 600 m yuan (£62 m) to develop ponds, filtration pools and wetlands, as well as to build permeable roads and public spaces that enable stormwater to soak into the ground. Ultimately, the plan is to manage 60% of rainwater falling in the cities." More details on China's pilot Sponge City program can be found in Chapter 1.

This chapter is not intended to be a review of international GI efforts, but rather provide a focused summary of the monitoring experiences of large-scale GI application in two cities in the United States.

## 7.3 FULL-SCALE MONITORING OF WSUD/GI BENEFITS FOR CSO CONTROL

Few demonstration projects have been established to quantify the benefits associated with WSUD/GI controls at large scales, although there are many studies that have applied modeling approaches to quantify GI benefits (e.g., Lucas and Sample, 2015; Lukes and Kloss, 2008; Nasrin et al., 2016, 2017). The following discussion summarizes the large-scale monitoring results of GI benefits at two US demonstration projects. One in Kansas City, Missouri, and the other in Cincinnati, Ohio. These projects demonstrate and document the benefits and application limitations of GI for CSO.

Although many studies are available describing the monitoring performance of individual stormwater controls (documented in the International BMP Database, available at: http://bmpdatabase.org/), it is much rarer to have data showing how multiple numbers of these controls function together in larger areas, as is required for effective CSO reductions. These two projects, supported by the US EPA's Office of Research and Development and the Metropolitan Sewer District (MSD) of Greater Cincinnati, illustrate how extensive applications of GI controls can significantly reduce the flows in combined sewers during wet weather. The full data sets for these case studies may be found in Talebi (2014), Talebi and Pitt (2014), and Pitt and Talebi (2014).

GI infiltration controls were retrofitted in seven different test locations, ranging in area from 8 to 40 ha. They included porous pavement sidewalks and parking lots, curb cut biofilters, grass swales, rain gardens, bioretention systems, and stormwater harvesting systems for on-site irrigation. Although these projects demonstrated significant flow reductions, they also highlighted the need for the widespread implementation of stormwater runoff controls to achieve the large flow reductions necessary to meet typical CSO reduction objectives.

### 7.3.1 Flow and rainfall monitoring used in demonstration projects

The following describes the flow and rainfall monitoring at the Kansas City demonstration project sites. Similar equipment and procedures were also used by Cincinnati in their demonstration projects. Each station included an ISCO 2150 area velocity sensor that measured water flow rate through channels and pipes (http://www.teledyneisco.com/en-us/water-andwastewater/Pages/2150-Area-Velocity-Module.aspx). Multiple rain gages were located at several locations in and near the test and control watersheds that were used as part of the project analyses and quality assurance/quality control (QA/QC) activities. Automatic ISCO water samplers (http://www.teledyneisco.com/en-us/water-and-wastewater/portable-samplers) were also installed at several locations near individual controls (mainly biofilters) to measure water quality benefits of the biofilters, even though the subsurface discharges were drained to the combined sewers for treatment at the municipal treatment facility. The Cincinnati locations had similar flow sensors and rain gages but did not include the water samplers.

The flow data were subjected to extensive QA/QC analyses, as was the flow monitoring in the combined sewers. Most of the data occurred during periods of low flows, which are especially difficult to accurately monitor. After the basic flow QA/QC analyses, the combined sewer flow data were further analyzed to enable separation of direct stormflows from the combined flows. The first process was to prepare time series plots for all dry days of the week for each month to identify the base sanitary sewer flows. The time series analyses for all dry weekdays (Monday through Friday) within each month had similar trends, as did all dry weekends (Saturday and Sunday). An example is shown on Fig. 7.2 from a Run Chart analysis in Minitab. Therefore, all flow data patterns for the dry weekdays were combined and their average was used as the base flow pattern for weekdays for each month. Similarly, the average flow patterns for all dry weekends were used as the base flow for weekends for each month. This resulted in two base flow patterns for each month for each watershed—one for weekdays and another for weekends. These dry weather flow patterns were subtracted from the combined flows during wet weather (as illustrated in Fig. 7.3) to result in the direct runoff associated with each rain event as illustrated in Fig. 7.4. The direct runoff information was then used, along with the rain data for each event, to calculate the total stormwater runoff volumes at each monitoring location as described in Pitt and Talebi (2014).

In this project, box plots, supported by other statistical and graphical analyses, were used to compare runoff volumes for different study periods (before and after construction). These plots were used to graphically depict the distribution of the data sets and to indicate possible groupings of the data. Other basic data plots, including scatterplots, time series, and box-and-whisker plots, were used for demonstrating overall data trends, along with QA/QC analyses (such as finding data

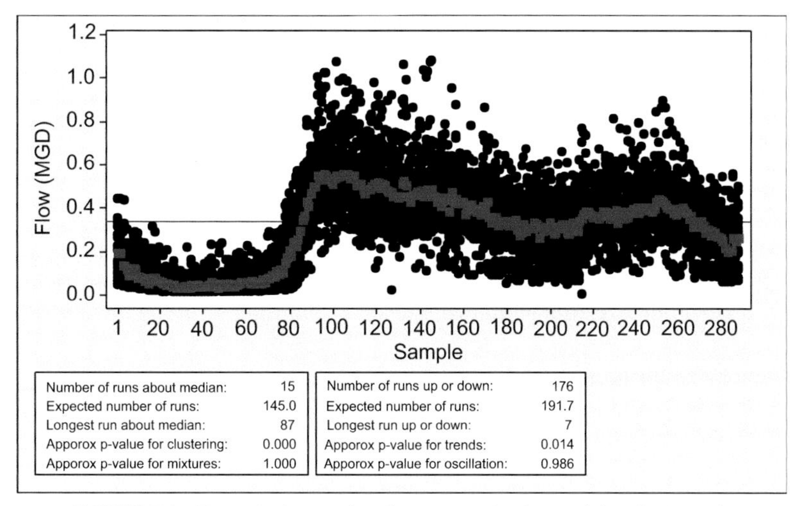


**FIGURE 7.2** Example dry weather flow pattern for dry weekdays in 1 month.

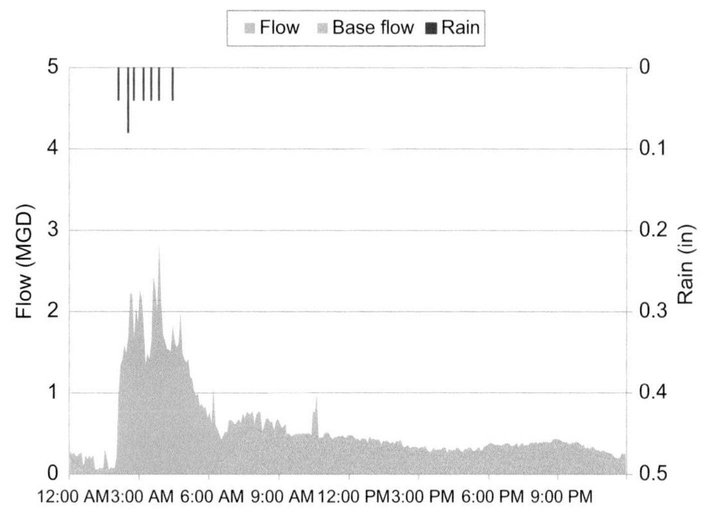


**FIGURE 7.3** Observed combined wet weather flow in the combined sewer at Kanas City.

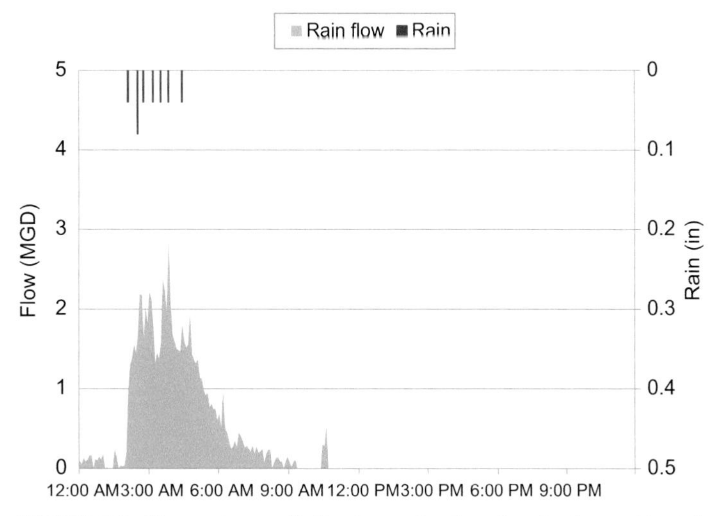


**FIGURE 7.4** Stormwater runoff after dry weather base flow has been subtracted.

gaps and meter errors). Time series plots were also used to show flows in the sewer lines over time for monthly and daily periods. These time series helped define trends, patterns, and possible clustering of data.

## 7.3.2 Cincinnati, Ohio, GI monitoring

GI has not been fully accepted by the regulatory reviewers of the Cincinnati's CSO control program (EPA, 2011). However, the Metropolitan Sewer District (MSD) budgets included $8M USD in 2009 and $15 M USD in 2010 toward GI planning and implementation activities. The approved Wastewater Implementation Plan allows MSD to propose revisions by adding or substituting GI for gray infrastructure where it is justified. About 20 demonstration projects are located in Cincinnati; the following text summarizes the monitoring results from three of these projects (Talebi, 2014; Talebi and Pitt, 2014).

The Cincinnati GI demonstration project evaluated the effectiveness of GI stormwater controls in areas served by combined sewer systems at three institutional locations, including Cincinnati State College, the Cincinnati Zoo, and the Clark Montessori High School sites. Multiple high-resolution (5-min) flow monitors were installed at the three sites. High-resolution flow measurements from in-system flow sensors were evaluated to measure the runoff volume reductions after GI facility construction at each study site. Flow data were available before, during, and after the construction of the GI stormwater controls at most locations as shown in Table 7.3. The two locations identified as separate sewer sampling locations (Cincinnati State College area and the Zoo main entrance) were located immediately before these flows entered the combined sewers. Hence, any observed flow reductions would reflect concurrent flow reductions in the combined sewers.

### 7.3.2.1 Cincinnati State Technical and Community College

This site is located on a hill and the main drainage patterns are to the north and south. Runoff from the southern portion of the campus flows south into a combined sewer system, whereas runoff from the northern part of the campus flows north into another separate combined sewer system. Fig. 7.5 shows the two areas studied at this location, and Table 7.4 indicates the associated land cover. Table 7.5 summarizes the stormwater controls used at the college. The northern combined sewer also drains a 120+ ha upland residential area. This upland drainage area had combined sewer flows that were much larger than the observed flows from the test area, making any observations of differences in the test area difficult to monitor and detect. Most of our analyses at the college site therefore focused on the southern drainage area where the site runoff could be isolated before entering the combined sewer.

### 7.3.2.2 Cincinnati Zoo

Two monitoring locations are located at the Cincinnati Zoo; one at the African Savannah exhibit area and the other at the main zoo entry area (Fig. 7.6). Both areas are served by combined sewers. The predevelopment conditions at the African

**TABLE 7.3 Availability of Sewer Flow Data During Different Green Infrastructure (GI) Construction Phases at Each of the Three Monitoring Projects in Cincinnati, Ohio**

| Location | Jan-10 | Feb-10 | Mar-10 | Apr-10 | May-10 | Jun-10 | Jul-10 | Aug-10 | Sep-10 | Oct-10 | Nov-10 | Dec-10 | Jan-11 | Feb-11 | Mar-11 | Apr-11 | May-11 | Jun-11 | Jul-11 | Aug-11 | Sep-11 | Oct-11 | Nov-11 | Dec-11 | Jan-12 | Feb-12 | Mar-12 | Apr-12 | May-12 | Jun-12 | Jul-12 | Aug-12 | Sep-12 | Oct-12 | Nov-12 | Dec-12 |
|---|---|---|---|---|---|---|---|---|---|---|---|---|---|---|---|---|---|---|---|---|---|---|---|---|---|---|---|---|---|---|---|---|---|---|---|---|
| Cincinnati State College Combined Sewer (above & below site monitoring) | | | | | | | | | | | | | | | | | | | | | | | | | | | | | | | | | | | | |
| Cincinnati State College Separate Sewer (single monitoring location) | | | | | | | | | | | | | | | | | | | | | | | | | | | | | | | | | | | | |
| Cincinnati Zoo - Main Entrance (separate sewer) | | | | | | | | | | | | | | | | | | | | | | | | | | | | | | | | | | | | |
| Cincinnati Zoo - African Savannah (combined sewer) | | | | | | | | | | | | | | | | | | | | | | | | | | | | | | | | | | | | |
| Clark Montessori High School (combined sewer) | | | | | | | | | | | | | | | | | | | | | | | | | | | | | | | | | | | | |

Before GI construction (pink), during GI construction (yellow), and after GI construction (green). *GI*, green infrastructure.

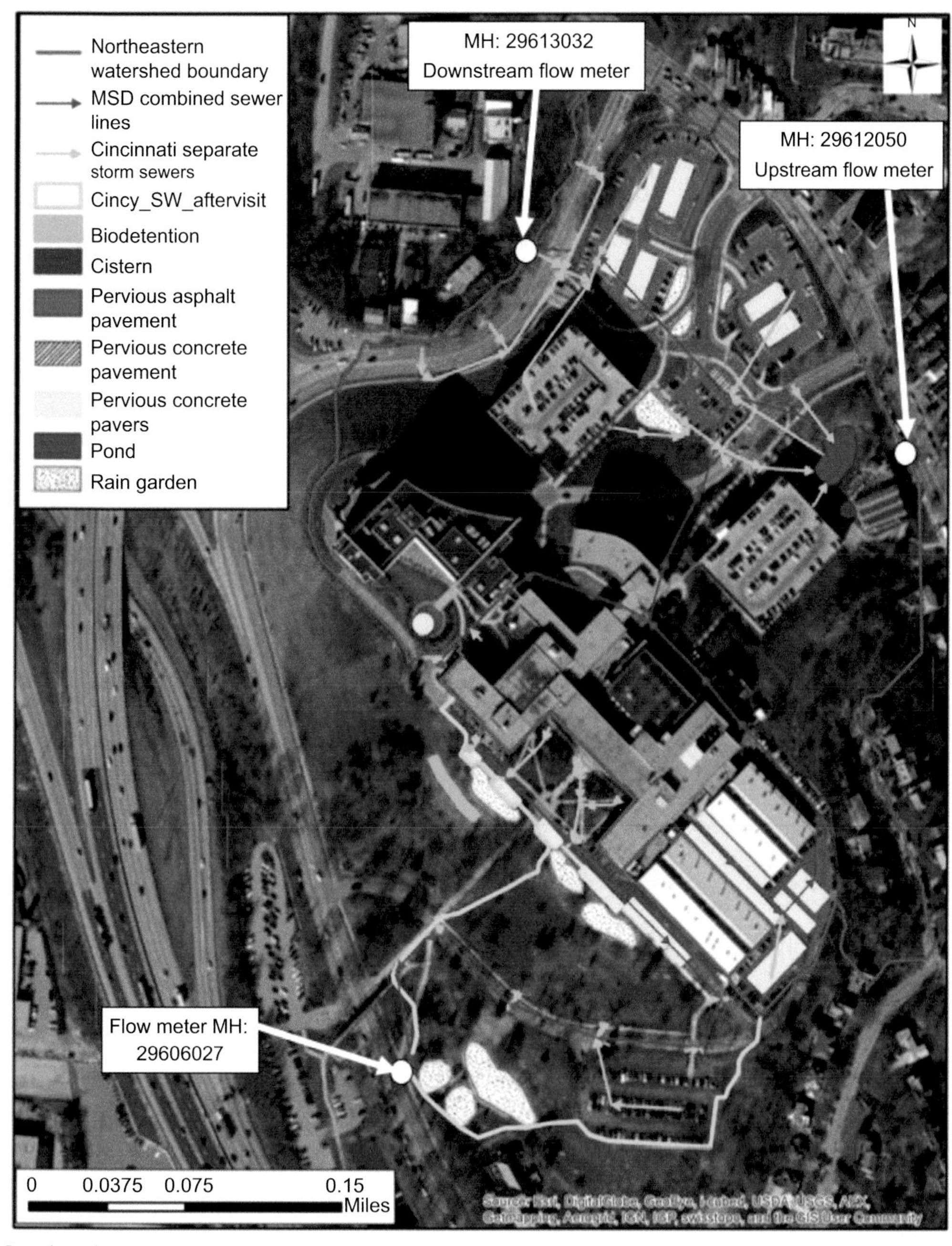


FIGURE 7.5 Location of green infrastructure stormwater controls at Cincinnati State College, OH, USA (Talebi and Pitt, 2014). *MSD*, Metropolitan Sewer District.

Savannah area were a large paved parking lot and small landscaped areas. The zoo entrance area now has a large area of porous pavement as well as a cistern. The new African Savannah area has a 1.6 ML cistern, together with compost-amended soil that was decompacted after the original pavement was removed. A flow monitoring station measured the flows in a 1 m diameter combined sewer pipe coming from the African Savannah area. The main entrance entry monitoring location measured separate stormwater flows in a 0.6 m diameter pipe, draining an area where large amounts of paver blocks replaced impervious pavement. The stormwater then enters a combined sewer. Therefore, the flow reductions observed at these locations directly affect the combined sewer flows for these areas. The zoos' two drainage areas that we monitored are shown in Fig. 7.6 with land cover characteristics described in Tables 7.6 and 7.7.

In the African Savannah area, the compost modified soil comprises about 40% of the total drainage area, whereas at the main entrance, the porous pavers comprise about 45% of the total drainage area.

**TABLE 7.4 Summary of Land Cover Characteristics for the Cincinnati State College Study Area (Talebi and Pitt, 2014)**

| | Northern Part of Campus | | Southern Part of Campus | |
|---|---|---|---|---|
| **Land Cover Type** | **Area ($m^2$)** | **Area (%)** | **Area ($m^2$)** | **Area (%)** |
| Landscaped area | 45,230 | 39.7 | 21,130 | 59.9 |
| Parking lot | 25,100 | 22.1 | 4,510 | 12.8 |
| Paved area | 250 | 0.2 | 0 | 0 |
| Roof | 22,450 | 19.7 | 3,300 | 9.3 |
| Street | 14,600 | 12.8 | 4,000 | 11.3 |
| Walkway | 6,370 | 5.6 | 2,330 | 6.7 |
| Total | 113,990 | 100.0 | 35,270 | 100 |

**TABLE 7.5 Summary of Green Infrastructure (GI) Installed at Cincinnati State College (Talebi and Pitt, 2014)**

| GI Feature | Size[1] | Comments |
|---|---|---|
| Pervious asphalt pavement | 186 $m^2$ | All have underdrains and located mostly at the parking lot in northeastern part of the campus |
| Pervious concrete pavement | 153 $m^2$ | |
| Pervious concrete pavers | 3720 $m^2$ | |
| Rain gardens | 5,223 $m^2$ | Ten rain gardens installed, mostly in the southwestern part of the campus |
| Cistern for rainwater harvesting | 96,000 L | Two 40,000 L in-ground storage tanks connected to irrigation systems plus one 16,000 L above ground cistern for greenhouse use |
| Infiltration trench | 143 $m^2$ | Located in southwestern part of the campus |
| Bioretention (level spreader) | 39 $m^2$ | Located in southwestern part of the campus |
| Pond | 641 $m^2$ | Located in northeastern part of the campus, close to the greenhouse |
| Retaining wall | 43 m | Planted with sedum |

[1]*The total GI components (not including the cistern and retaining wall) cover 10,105 $m^2$, treating a total site area of 149,260 $m^2$ (about 7% of total area).*

### *7.3.2.3 Clark Montessori High School*

This project area is surrounded by a residential area and is located near Hyde Park in Cincinnati, Ohio. There is one monitoring location that measures the combined sewer flows in a 0.5 m diameter pipe from this newly constructed area. Fig. 7.7 and Table 7.8 describe the monitored area and the location and type of stormwater controls. This site also had a residential area up-gradient of the school that contributed uncontrolled stormwater to the monitored location.

### *7.3.2.4 Observed flow reductions at the three Cincinnati demonstration project locations*

Fig. 7.8 illustrates the statistically significant flow reductions that occurred after GI construction at Cincinnati State College at the southern monitoring location. The volumetric runoff coefficient (Rv) is the fraction of the rainfall that occurs as direct runoff (ratio of runoff depth to rain depth). It is used in these analyses to illustrate reductions in runoff associated with the GI installations.

The before and during construction monitoring periods were combined, as statistical analyses did not indicate any significant differences during these project periods. At this study area, the statistical tests indicate that the "after" construction flows are much less than the "before and during" construction flows. This supports the hypothesis that areas with GI controls installed have reduced stormwater flows.

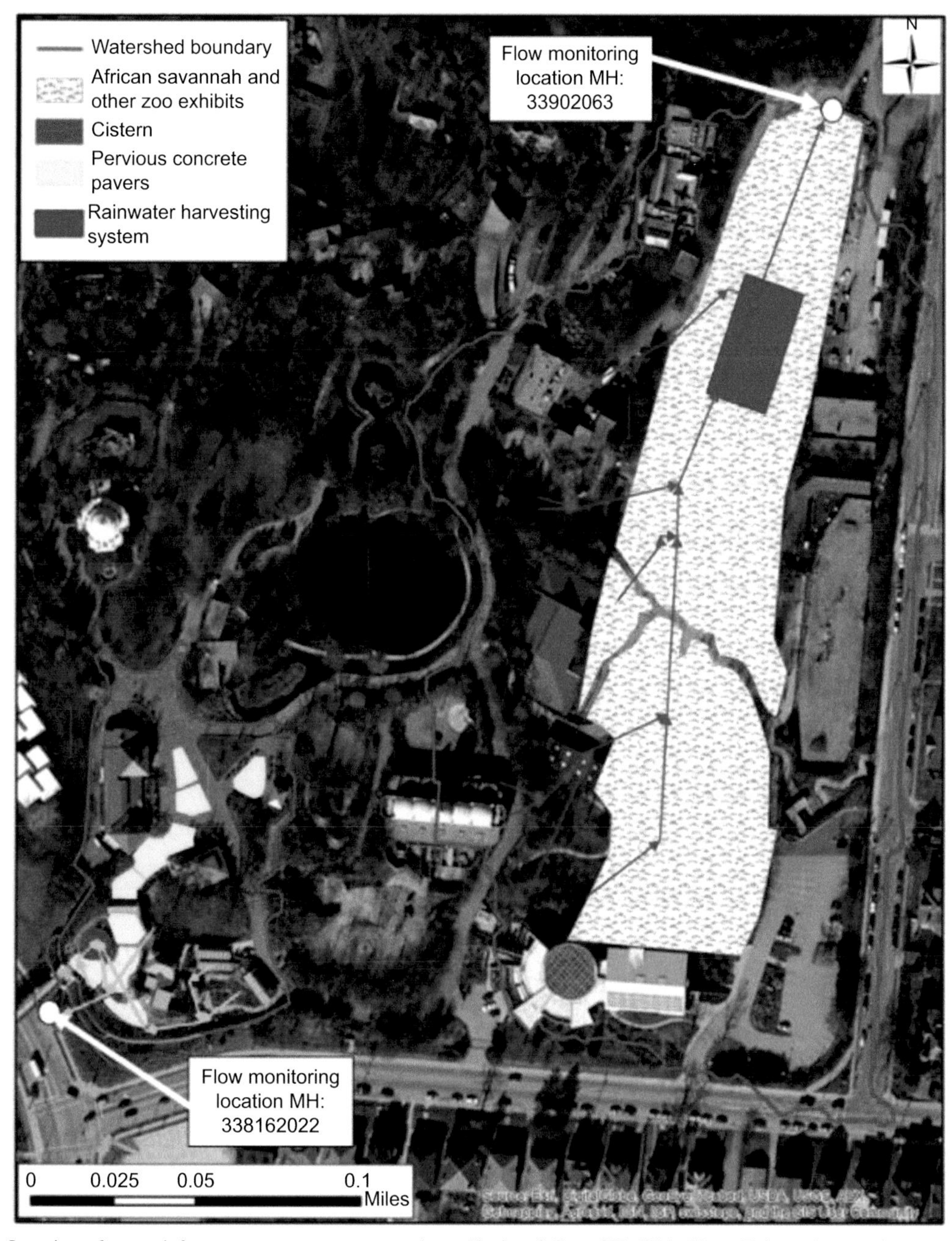


FIGURE 7.6 Location of green infrastructure stormwater controls at Cincinnati Zoo, OH, USA (Note: Enhanced vegetation area was still under construction when this aerial photograph was taken) (Talebi and Pitt, 2014).

The main entrance of the Cincinnati Zoo is about 1 ha and is served by a small separate sewer system before it connects to the main combined sewer. More than 60% of the paved area have been replaced by porous paver blocks. There is also a cistern with a capacity of about 40 $m^3$ collecting runoff from rooftops and reused for landscape irrigation. However, only postconstruction flow monitoring was available at this location. Fig. 7.9 is a plot of the postconstruction-monitored runoff depth compared with the rain depth. The $R_V$ is about 0.1 for this area (based on the slope term from the regression analysis), substantially less than the ratio of about 0.8, expected for preconstruction runoff conditions.

Fig. 7.10 illustrates the statistically significant reduction in runoff flows after construction compared with the preconstruction monitoring period at the Cincinnati Zoo African Savannah combined sewage monitoring location.

At the African Savannah area at the zoo, the Kruskal–Wallis one-way analysis of variance was used to indicate if any significant differences in runoff between these categories occurred. The results show that the "after construction" period

**TABLE 7.6 Summary of Land Cover Characteristics for African Savannah Area at Cincinnati Zoo (Talebi and Pitt, 2014)**

| Land Cover Type | Area ($m^2$) | Percentage of Total Area (%) |
|---|---|---|
| Landscaped area (newly modified soil, decompacted and with compost amendments, replacing prior paved parking area) | 21,240 | 39.2 |
| Active construction | 14,210 | 26.2 |
| Parking lot | 2,840 | 5.2 |
| Paved area | 930 | 1.7 |
| Roof | 7,120 | 13.1 |
| Street | 2,310 | 4.3 |
| Walkway | 5,520 | 10.2 |
| Total | 54,180 | 100 |

**TABLE 7.7 Summary of Land Cover Characteristics for Main Entrance of Cincinnati Zoo Area (Talebi and Pitt, 2014)**

| Land Cover Type | Area ($m^2$) | Percentage of Total Area (%) |
|---|---|---|
| Landscaped area | 4000 | 40.2 |
| Porous block pavement | 4460 | 44.8 |
| Roof | 1000 | 15.1 |
| Total | 9960 | 100 |

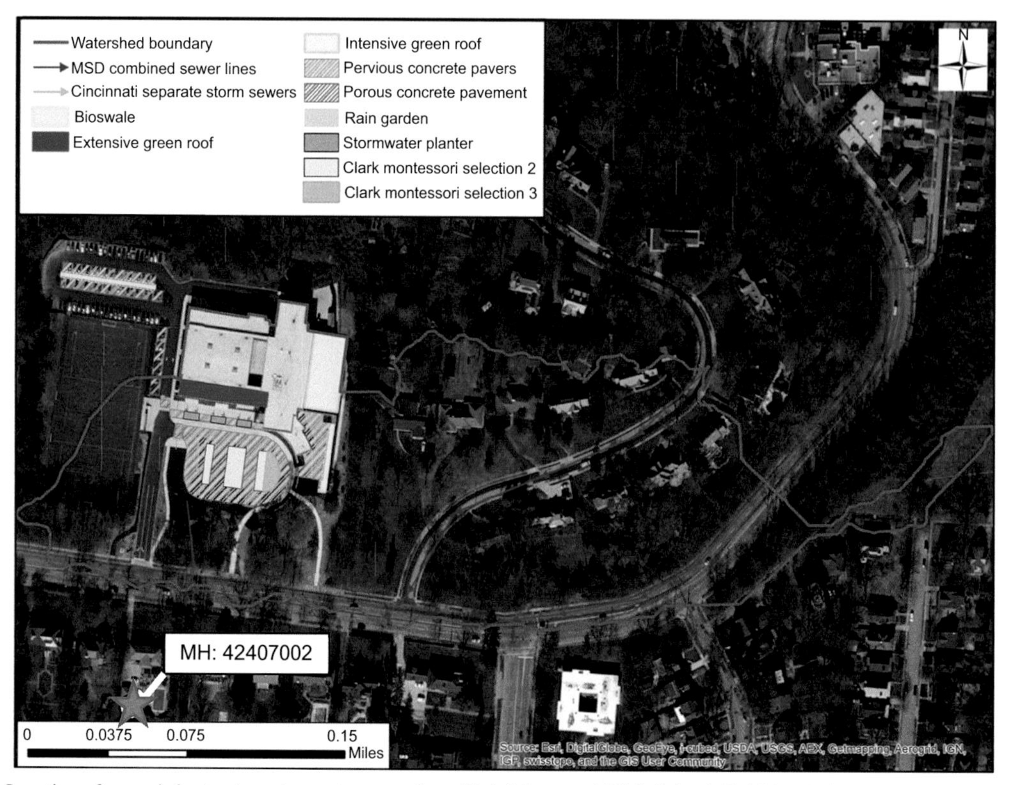


FIGURE 7.7 Location of green infrastructure stormwater controls at Clark Montessori High School (Talebi and Pitt, 2014). *Orange star* indicates the monitoring location at the designated manhole for flows from the entire watershed. *MSD*, Metropolitan Sewer District.

**TABLE 7.8 Summary of Land Cover Characteristics for Clark Montessori High School (Talebi and Pitt, 2014)**

| Land Cover Type | Area ($m^2$) | Fraction of Total Area (%) |
|---|---|---|
| Driveway | 2,120 | 3.6 |
| Landscaped area | 34,320 | 57.5 |
| Parking lot | 2,050 | 3.4 |
| Paved area | 1,390 | 2.3 |
| Roof | 8,050 | 13.5 |
| Soccer Field | 2,400 | 4.0 |
| Street | 8,000 | 13.4 |
| Walkway | 1,390 | 2.3 |
| Total | 59,740 | 100.0 |

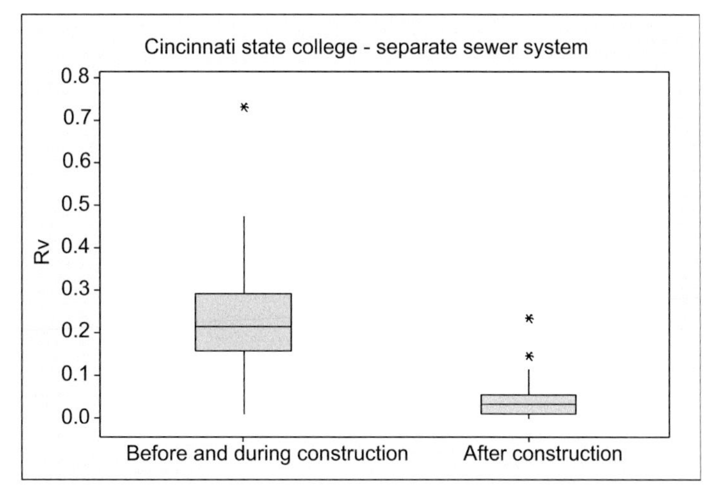


**FIGURE 7.8** $R_v$ values for before and during construction periods compared with the post green infrastructure construction period at Cincinnati State College separate sewer system (Talebi and Pitt, 2014).

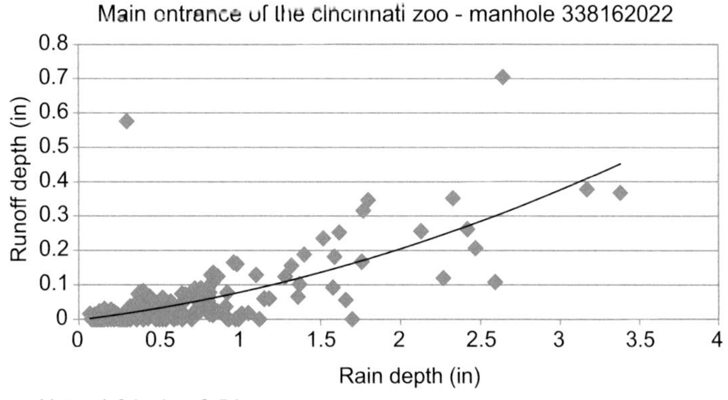


**FIGURE 7.9** Runoff depth versus the rain depth for 176 events monitored after the construction of the porous pavers at the main entrance of the Cincinnati Zoo (no monitoring data are available for the period before installation of the porous paver system) (Talebi and Pitt, 2014). (Note: 1.0 inch = 25.4 mm).

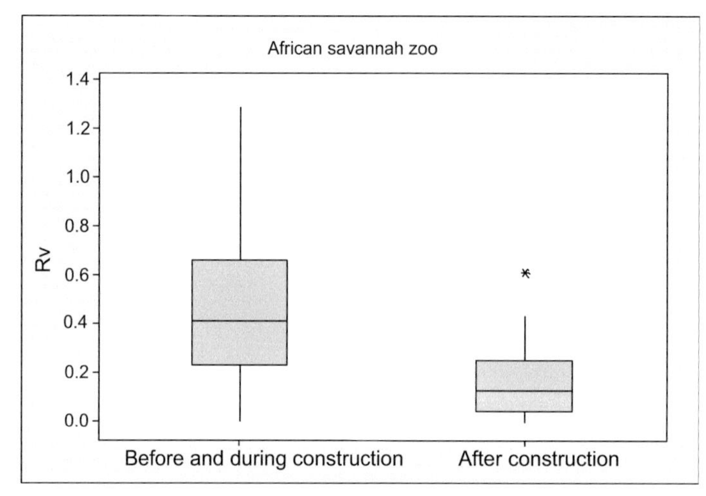


**FIGURE 7.10** $R_v$ values for the events monitored "before" green infrastructure (GI) construction compared with monitored events "after" the construction of the GI facilities at the African Savannah (Talebi and Pitt, 2014).

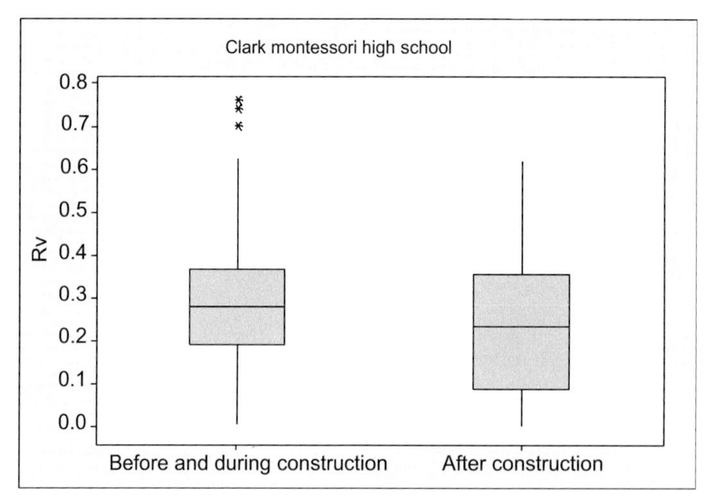


**FIGURE 7.11** $R_v$ values for before and during construction periods (combined) compared with the after construction period at Clark Montessori High School (Talebi and Pitt, 2014).

was statistically different from the "before construction" and "during construction" periods, while the "before" and "during" monitoring periods had similar runoff responses (hence the data for these two flow phases were combined). This supports the hypothesis that the flows being discharged from areas with GI stormwater controls are about half of that which would have occurred if the GI stormwater controls were absent.

Fig. 7.11 is a box-and-whisker plot that shows the $R_V$ values for before and during construction (combined) and after construction periods at the Clark Montessori High School. The difference is relatively small because of the small portion of the watershed being controlled by the GI facilities (due to the large, untreated, up-gradient residential area). Nonetheless, because of the large amount of data available, the small difference in runoff was statistically significant.

Table 7.9 is a summary of the monitored flow reductions for the three Cincinnati monitoring locations: Cincinnati State College, the Cincinnati Zoo, and the Clark Montessori High School. The stormwater reductions are related to the fraction of the catchment area treated with GI. Long-term runoff reductions of about 70%–80% were measured at the Cincinnati Zoo site, where more than 40% of the catchment was treated, and for the Cincinnati State College site, where bioinfiltration and rain gardens comprised about 7% of the total drainage area.

The runoff reductions shown in Table 7.9 are a function of the amount of runoff from the whole drainage areas and the amount infiltrated by the devices. At Cincinnati State College, southern area runoff reductions are large because most of the runoff in that part of the site is directed toward the biofilters and other infiltration devices. In addition, these devices were able to infiltrate most of the runoff. In contrast, the Clark Montessori High School runoff reductions were only about 20% because of the relatively small fraction of the total drainage area being directed to infiltration devices.

**TABLE 7.9 Summary of Runoff Volume Reductions for Site Runoff at Cincinnati Study Areas Using Green Infrastructure Controls (Talebi, 2014)**

| Location | Runoff Volume Reduction (%) Compared With Preconstruction Data (Statistical Significance of Difference Shown in Brackets) |
|---|---|
| Cincinnati State College—southern area (bioinfiltration and rain gardens comprising about 7% of the total drainage area) | 85% ($P < .001$) |
| Cincinnati Zoo—main entrance (extensive pervious concrete pavers, comprising about 45% of the total drainage area) | >80%; based on postconstruction $R_v$ values of about 0.1, compared to about 0.8 for conventional pavement) |
| Cincinnati Zoo—African Savannah (rainwater harvesting system and pavement removal, comprising about 40% of the total drainage area) | 70% ($P < .001$) |
| Clark Montessori High School (green roofs and parking lot biofilters, only treating about 20% of the total drainage area) | 20% ($P = .027$) |

## 7.3.3 Kansas City, Missouri, EPA demonstration project for green infrastructure

The Kansas City Water Services Department provides wastewater collection and treatment services for approximately 0.65 million people located within the City and in 27 tributary or "satellite" communities. This demonstration project was developed to demonstrate the application of GI for CSO control in the Middle Blue River. Kansas City, Missouri, embraced the concept of GI as part of its CSO control strategy at both the political and policy levels (EPA, 2011). The goals of the Overflow Control Plan are eliminating or treating 88% of the annual wet weather combined sewer system volume; reducing the CSO discharge volume from 24 million $m^3$/year to 5.3 million $m^3$/yr; and reducing, or otherwise controlling, infiltration and inflow that enters the separate stormwater system.

The Kansas City project is one of the largest projects in the United States using extensive GI controls. It is located in a fully monitored 100 acre (40 ha) neighborhood that encompasses about 200 GI-based stormwater controls (listed in Table 7.10) along with sewer rehabilitation. The case study therefore provides an opportunity to evaluate the benefits of GI stormwater controls at both small scale (using infiltration data from individual GI stormwater monitoring) and large scale (impacts of all GI controls on the 40 ha combined sewer system compared with an adjacent 35 ha catchment with no stormwater controls). The paired catchment monitoring and the "before" and "after" monitoring of flows in the test catchment were used to quantify the GI benefits.

Fig. 7.12 shows the test and control watershed boundaries and the locations of the main flow monitoring stations located in the combined sewer system. Monitoring stations UMKC02a, C02b, and C03 measured flows from the control watershed, while station UMKC01 measured the flows from the treated watershed alone. Other combined sewer flow locations (KCM001, KCM002, and KCM003) are also shown on Fig. 7.12, which were used as complimentary flow monitoring stations. The initial experimental design intended to compare the runoff from the same events from both the test and control watersheds, along with comparing the runoff from the test watershed during, before, and after GI construction rains for complimentary verification of the project results. Unfortunately, the flow sensors for the control watershed failed frequently, resulting in few complete paired data sets from both areas. There were also questions relating to the as-built combined sewers in the control watershed that could not be resolved during this project. Therefore, most of the performance evaluations relied on the preconstruction versus postconstruction data from the test watershed. The limited data from the control watershed were used for additional model verification (not presented in this chapter, which focusses on the direct monitoring results).

Fig. 7.13 indicates the placement of the GI stormwater controls that were required to be placed in the public rights-of-way to allow the city to conduct any required maintenance (according to the consent decree).

Table 7.10 summarizes the number and types of GI infiltrating stormwater controls that were used to minimize the discharge of stormwater to the combined sewer system. These controls were installed and maintained by the Kansas City Municipal Sewer District as part of the city's CSO consent decree. This table shows the different categories of infiltration devices, the total number of each type in the test watershed, total footprint areas compared with the total drainage areas for each type of control, the drainage area for each control unit, and the total area treated by each type of control unit. Overall, about 22 ha were serviced by the control devices. This is about 55% of the treated catchment.

**TABLE 7.10 Summary of Stormwater Control Components in the 40 ha Pilot Study Area in Kansas City, Missouri (Pitt and Talebi, 2014)**

| Design Plan Component | Structural Description | Number of Stormwater Control Units in Test Area | Drainage Area for Each Control Unit (ha) | Total Area Treated by the Control Units (ha) | Device Footprint Area as a % of the Total Drainage Catchment Area |
|---|---|---|---|---|---|
| Bioretention | Bioretention without curb extension | 24 | 0.16 | 3.9 | 1.6% |
| | Curb extensions with bioretention | 28 | 0.16 | 4.5 | 1.5% |
| | Shallow bioretention | 5 | 0.16 | 0.8 | 1.6% |
| Bioswale | Vegetated swale infiltrates to background soil | 1 | 0.20 | 0.2 | 8.9% |
| Cascade | Terraced bioretention cells in series | 5 | 0.16 | 0.8 | 1.9% |
| Porous sidewalk or pavement | With underdrain | 18 | 0.006 | 0.12 | 100.0% |
| | With underground storage cubes | 5 | 0.006 | 0.04 | 99.9% |
| Rain garden | Rain garden without curb extension | 64 | 0.16 | 10.4 | 2.8% |
| | Curb extensions with rain gardens | 8 | 0.16 | 1.3 | 1.5% |
| | Total number of control units (not including the porous pavement area) | 135 | | | |
| | Total area treated (ha) | | | 22.0 | |
| | Area per unit (ha) | | | 0.26 | |

Because the installation of the stormwater controls was restricted to the rights-of-way along the roads accessible to the City, not all the runoff could be treated by the GI controls. Fig. 7.14 shows that approximately 50% of the drainage area in the treated catchment was not affected by the controls. These excluded areas were mostly connected to yard drains which drained directly to the combined sewers. These drains were installed in the past to solve minor flooding and ponding in the backyards. Other reasons why the areas were not treated included exclusion of proposed structures, such as driveways and important trees.

### *7.3.3.1 Kansas City green infrastructure monitoring results*

Similar monitoring QA/QC and statistical analyses were conducted for the Kansas City project as described previously for Cincinnati.

One of the most critical site characteristics affecting the design and performance of stormwater infiltrating devices is the infiltration of the disturbed urban soils where they are to be located. Urban soils have little resemblance to available general soil maps because of grading during sub division construction and compaction. Therefore, in addition to the site surveys, the University of Missouri Kansas City (UMKC) students conducted site soils surveys for the area and used small-scale infiltrometers to measure infiltration rates in the disturbed urban soils of the test watershed area. These tests were supplemented using actual recording stage recorders in some of the rain gardens and curb cut biofilters during actual rains. The actual infiltration measurements in the constructed biofilters indicated that long-term system infiltration rates are generally between 25 and 100 mm/h (average of 64 mm/h), which is much greater than assumed during the initial designs that relied on traditional mapped soil information.

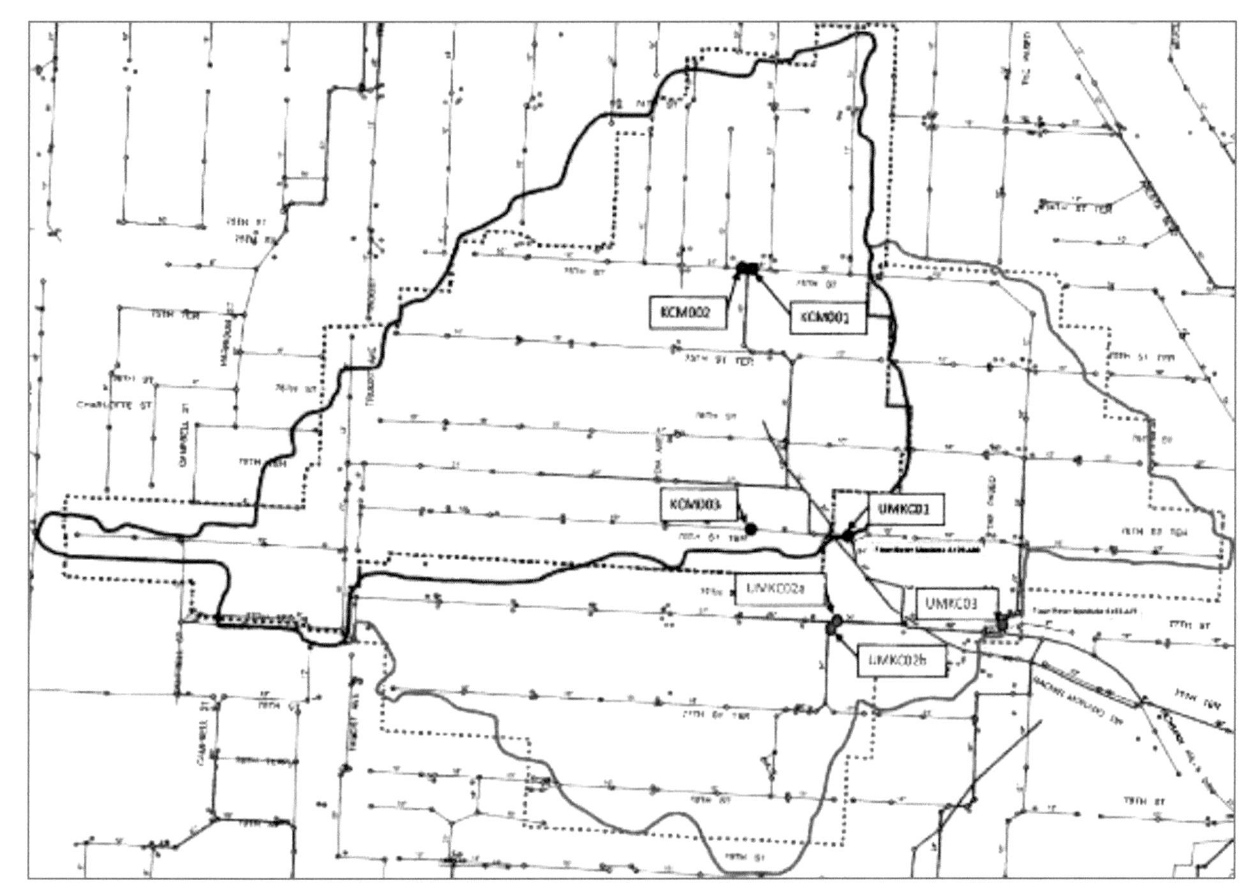


**FIGURE 7.12** Treated (40 ha) and control (34 ha) urban catchment in Kansas City, Missouri, showing the major combined sewer flow monitoring locations. The *red outlined area* is the green infrastructure treated watershed (with the *red boxes* showing the combined sewer flow monitoring locations). The *blue outlined area* (and *blue boxes*) indicates untreated (control) watershed.

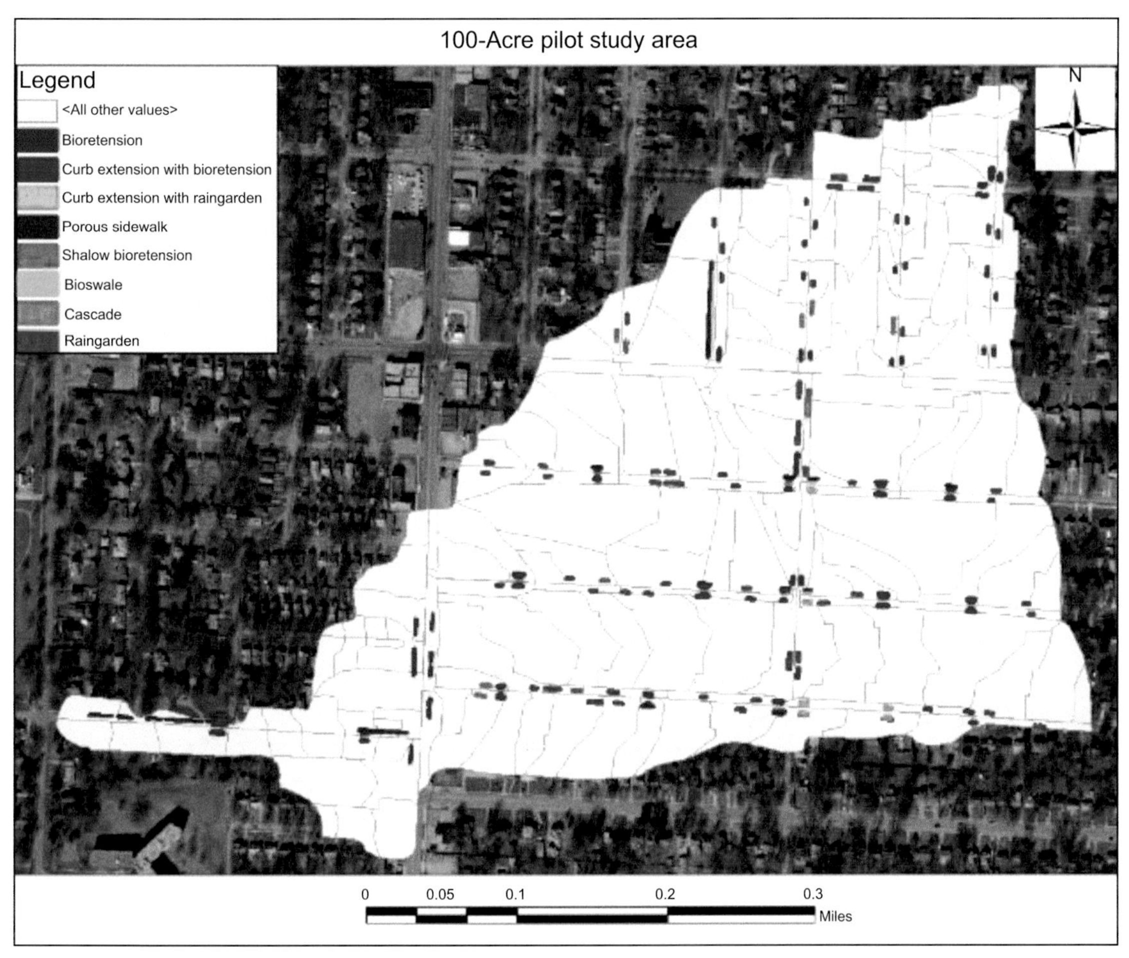


**FIGURE 7.13** Location of stormwater green infrastructure controls in the 40 ha pilot study area in Kanas City, Missouri (Pitt and Talebi, 2014).

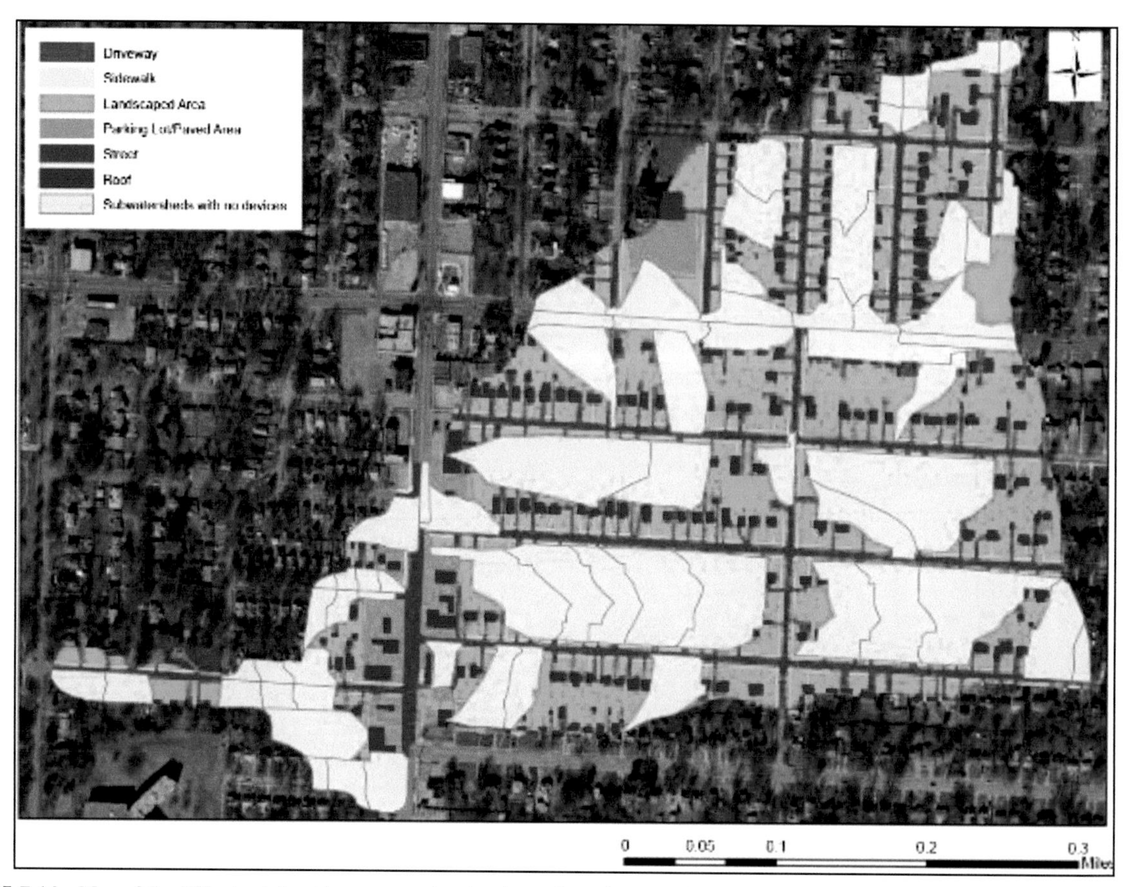


FIGURE 7.14 Map of the 40 ha treated catchment area showing the surface characteristics of those areas receiving green infrastructure stormwater treatment and subwatersheds with no treatment devices (Pitt and Talebi, 2014).

Runoff monitoring was conducted in the combined sewer system at several locations in the pilot area and the control urban catchments. This sampling arrangement enabled flows to be separated for the treated and the control watersheds. Fig. 7.15 illustrates how the volumetric runoff coefficients ($R_V$) significantly decreased after installation of the infiltration devices. Rain and runoff were monitored in 2009 and 2010 before the construction of the GI infrastructure. Additional events were monitored after the sewer in the pilot catchment was rehabilitated, and these data were compared with the flows observed before sewer relining. Construction of the stormwater controls started after sewer relining. A total of 75 events were available for the preconstruction baseline conditions and 37 events were available for the postconstruction

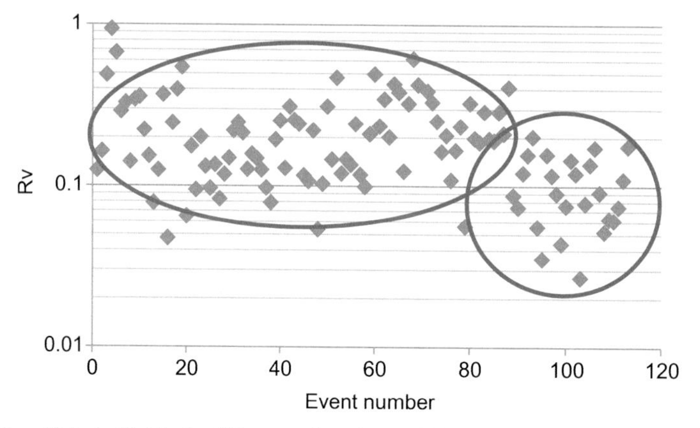


FIGURE 7.15 Volumetric runoff coefficients (Rv) in the 40 ha treated catchment (UNKC01) before (circled events to the left) and after green infrastructure construction (circled events to the right) (Pitt and Talebi, 2014).

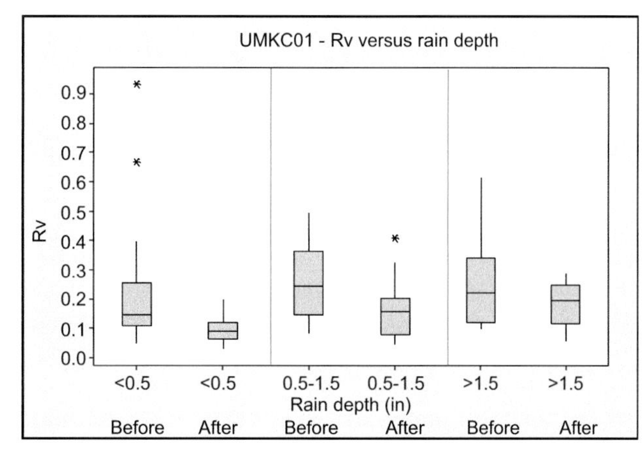


**FIGURE 7.16** Comparisons of the volumetric runoff coefficient values ($R_v$) at UMKC01 for rainfall events before and after green infrastructure construction. The comparisons have been stratified into <0.5, 0.5 to 1.5, and >1.5 inch rainfall events (Pitt and Talebi, 2014). (Note: 1.00 inch = 25.4 mm).

conditions. The flow-weighted $R_v$ decreased from about 0.26 before GI installations to about 0.18 after construction, reflecting a 32% decrease in runoff volume. This decrease, although significant, is relatively low because only about 55% of the drainage area was directed to the infiltration devices. Another reason for the reduced level of performance is that performance varies by runoff amounts.

Fig. 7.16 plots the $R_v$ values for before and after construction of the stormwater controls for different rain categories. The small (<13 mm) and intermediate (13–38 mm) rainfall events showed statistically significant decreases in $R_v$ after GI construction ($P = .002$ and .023, respectively). The large (>38 mm) rainfall events did not support this same conclusion due, in part, to their fewer numbers ($P = .57$). However, there is an apparent decrease in the **variability** in $R_v$ after GI construction for these large events. Obviously, large runoff results may rapidly consume the available storage in the biofilters, resulting in surface bypassing and reduced infiltration. Small events are more completely infiltrated as they may be completely stored in the treatment device, with no overflow, allowing increased fractions of their flow to be infiltrated.

Water quality improvements associated with biofilter operation were also examined by using ISCO automatic water samplers at inlets and at underdrains at eight biofilters. Few events resulted in subsurface underdrain discharges to the combined sewers and fewer still resulted in bypass overflows. The overall influent suspended sediment concentrations (SS) for the 49 influent samples available averaged 173 mg/L (26–711 mg/L range). Paired statistical tests indicated that the differences in the influent and underdrain SS concentrations for the six events that had underdrain samples (259 vs. 77 mg/L average) were significant ($P = .048$). However the differences in the influent and overflow SS concentrations for the four events that had overflows (126 vs. 89 mg/L) were not statistically significant ($P = .33$) for the few paired samples available. Therefore, complete pollutant removal from the surface flows occurred for more than 80% of the events (49 out of 59), with the biofilters trapping most of the particulates.

Table 7.11 summarizes the flow reductions observed at the Kansas City monitoring location. The flow reductions because of the GI facilities were highly significant ($P < .001$). The magnitude of the flow reduction volumes was only

**TABLE 7.11 Summary and Statistical Comparisons of the Runoff Percentages ($R_v$) Measured in the Combined Sewer Before and After GI Facilities Were Installed at Kansas City, Missouri (Talebi, 2014)**

| Monitoring Period | Monitoring Dates | Number of Monitored Storm Events | Flow-Weighted $R_v$ Values | Percentage of Change Compared With Baseline Monitoring |
|---|---|---|---|---|
| Initial baseline + postsewer relining | (March 23, 2009–June 16, 2010) and (February 24, 2011–March 19, 2011) | 75 | 0.26 | n/a |
| After construction of GI infrastructure | April 2013–October 2013 | 37 | 0.18 | 32% decrease ($P < .001$***) |

***P-value from Mann–Whitney Rank-Sum test. *GI*, green infrastructure.

about 32%, due to the limited fraction (50%) of the total catchment treated with GI devices, along with the relatively large uncertainty of the flow monitoring at that location.

The Kansas City GI demonstration project site is unique because a major portion of the test area receives direct treatment from many individual stormwater control devices, and a large area (40 ha) was monitored to demonstrate the actual flow reductions in the combined sewer. In addition, a suitable number of monitored events were available (75 events preconstruction and 37 events postconstruction) to provide sufficient confidence and statistical power for the flow reduction comparison.

As in most retrofit installations, stormwater controls cannot be placed to treat all the flows from the entire watershed area because of access exclusions caused by existing infrastructure, large trees, and surface drainage paths. Conversely some areas were treated by multiple control units, with overflows from up-gradient devices flowing into down-gradient devices. The total impervious portion of the treated area in the test catchment was about 45%, whereas the total impervious portion for the untreated catchment area is about 37%, indicating that greater unit area runoff is likely from the treated catchment (due to a larger percentage of impervious areas near the GI treatment areas).

## 7.4 BENEFICIAL USE OF STORMWATER IN CSO MANAGEMENT PLANS

A promising component of CSO management plans involves the diversion of stormwater to beneficial uses. The effectiveness of this practice is dependent on local rainfall, development characteristics, and especially opportunities for beneficial end use.

A report prepared by an NRC committee (NRC, 2015) examined how stormwater (along with graywater) can be used for beneficial uses. Although reducing reliance on domestic water supplies was the primary objective of this report, it was also noted that reducing CSO frequencies and magnitudes would also be associated with this practice. The report listed an array of on-site uses for stormwater for urban applications, including toilet and urinal flushing, public open space irrigation, vehicle wash water, air conditioner chiller water, firefighting, commercial laundries, vehicle washing, street cleaning, decorative fountains, and other water features. Table 7.12 is a summary of some of the common current beneficial uses for stormwater in the United States classified into domestic, neighborhood, and regional scale applications. The Water Environment Research Foundation (WERF) also has reported on the beneficial uses of stormwater, including many international case studies (Pitt et al., 2011).

The WERF nonpotable water beneficial use report (Pitt et al., 2011) calculated the hydrological benefits of storage and subsequent irrigation use of runoff collected from roofs and other impervious surfaces for many US locations. Unless the storage tanks are quite large (>40 kL for the Kansas City area, for example), they would have minimal benefits in reducing direct runoff to combined sewers. The Cincinnati Zoo example has a large storage tank (40 + kL) and the demand to drive sizeable runoff reductions.

One method to increase the effectiveness of storage tanks, the beneficial use of stormwater, and hence reduced CSO events, is real-time control to manage storage volumes compared with predicted runoff amounts. Quigley and Brown (2014) describe the use of distributed real-time control (DRTC) technologies for GI. These systems include advanced rainwater harvesting systems, dynamically controlled green roofs, wet detention basins, and bioretention systems. The intent of the WERF project was to demonstrate the potential application of the technologies, as well as their practical deployment and use. The use of the OptiRTC DRTC system (Quigley and Brown, 2014) is a suite of cloud-based web applications. Basically, this system tracks the amount of roof runoff stored in a tank and compares the available volume with the predicted runoff for the next rain event. If the available storage volume in the tank is insufficient to capture the expected runoff from a critical rain, the tank is automatically drained to the combined sewer before the rain event (when sewer and treatment capacity are available). The tank can then capture the runoff during the subsequent rain event, thereby diverting the flow that would otherwise be discharged into the combined sewer.

Quigley and Brown (2014) present several built case studies demonstrating the benefits of this approach compared with traditional runoff storage systems. One case study was located in Washington, DC, at two fire stations. The system consisted of two interconnected 7 kL concrete cisterns collecting runoff from 300 $m^2$ of roof area at Engine House 3. At Engine House 25, two interconnected 7 kL concrete cisterns collected runoff from 490 $m^2$ of roof area. Water from both stations was used for cleaning vehicles and parking bays and to supplement water storage on the firefighting trucks. A calibrated hydrology model demonstrated that real-time control of the tank levels prevented more than 260 kL of stormwater being discharged per year. The operating cost of the system was about $0.05/L.

**TABLE 7.12 Common Beneficial End Uses for Stormwater in the United States and Their Scale of Application (NRC, 2015)**

| | | Stormwater Application | | | |
|---|---|---|---|---|---|
| **Category of Use** | **Specific Types of Use** | **Household/ Small Building** | **Neighborhood/ Institutions** | **Regional** | **Limitations** |
| Urban applications | Toilet and urinal flushing | X | X | | Dual distribution system costs; Dual plumbing in buildings may be required; Greater burden on cross-connection control. |
| | Heating, ventilation, air conditioning (HVAC) makeup water or evaporative cooling | X | X | | |
| | Firefighting | | X | | |
| | Laundries | | X | | |
| | Vehicle washing | X | X | | |
| | Street cleaning | | X | | |
| | Decorative fountains/ other ornamental water features | X | X | | |
| Landscape irrigation | Lawns, flowerbeds | X | X | | Dual distribution system costs; Uneven seasonal demand; High salinity and sodium adsorption ratio of snowmelt can adversely affect plant health. |
| | Parks, golf courses | | X | | |
| | Playgrounds/schools | | X | | |
| | Agriculture | X | X | | |
| | Greenways | | X | | |
| Wildlife habitats and recreational uses | Wetlands | | X | X | Dual distribution system costs; Nutrient removal required to prevent algal growth; Potential ecological impacts depending on gray water quality and sensitivity of species. |
| | Ornamental or recreational water body | | X | X | |
| Large-scale water supply augmentation | Surface impoundments | | X | X | Treatment for contaminant removal may be needed; Recharge requires specific hydrogeological conditions; Potential for water quality degradation. |
| | Groundwater recharge | | X | X | |

## 7.5 CONCLUSIONS

During moderate to large rainfall events, uncontrolled combined systems overflow often results in discharges of raw or poorly treated sewage, together with untreated stormwater, into the urban environment. In developed countries, management of combined sewers focus on reducing the frequency and magnitude of CSOs.

One promising solution is the use of distributed infiltration (and storage) practices to reduce stormwater flows to the combined sewers. This is generally described WSUD or GI programs. Because of the high construction costs and experimental difficulties in monitoring these options, much of the "validation" has been undertaken using computer models. Indeed, few experimental studies have quantified the behavior of these systems at urban catchments scales. We have reported on two such experimental studies in this chapter.

The Cincinnati study monitored three institutions: a technical college, a zoo, and a high school. Up to 200 runoff events were monitored at these institutions, including before and after installation of the infiltration controls. Statistically significant flow reductions were observed after GI installations. Depending on the fraction of the flow treated by the GI devices, 20%–80% of the runoff was diverted from the combined sewer systems and locally infiltrated.

In the Kansas City study, a 40 ha urban catchment was monitored for over 100 events distributed over the pre- and post-GI construction periods. The reduction in rainfall runoff percentage was about 32% and statistically significant. Only about 55% of the total drainage area was captured by these infiltration devices, which reduced the effectiveness of the overall program. Nonetheless, the flows entering the infiltration devices were reduced by about 58%. As expected, flow reductions were greater for small- and moderate-sized rains than for large rainfall events. Retrofitting spatially dispersed GI controls throughout residential areas is very difficult when trying to capture most of the area flows because of the large number of property owners and the complex drainage systems. Institutional areas, in contrast, can be more effectively controlled with fewer owners, making the placement of fewer, but larger, infiltration controls more likely. Redeveloping areas also offer good opportunities for retrofitting GI controls, provided there is a good coordination between site planners and engineers for the whole area.

We would advise that experiments at this scale are not for the faint hearted because of the complexity of flow measurement in sewers, the difficulty of retrofitting GI over the whole of the designated catchment, and the confounding effects of runoff from untreated, up-gradient sewer catchments.

We conclude our chapter with a brief overview of the stormwater runoff reduction benefits from capturing and reusing roof runoff (i.e., rainwater). However, to be effective at catchment scales, the rainwater cisterns need to be large (>40 kL for the Kansas City area, for example), their prestormwater levels need sophisticated management, and the end use demands need to be substantial, such as public open space irrigation.

The following recommendations warrant consideration when trying to quantify the benefits of GI controls in combined sewers. They suggest solution paths for some of these confounding issues:

- Excellent as-built sewerage system data and mapping are needed. Over the years, remedial construction occurred in some of the Kansas City test and control areas to address local drainage problems. Typically, these included the use of yard drains constructed in the backyards in low-laying areas that collected runoff for extended periods. These yard drains were extended to the combined sewers. Other changes may have been made to provide parallel drainage lines in the area. It was not clear whether these were connected to the monitored drainage network and, if so, whether they were still in operation. Detailed sewer surveys are needed to clarify any uncertainties in the drainage areas. These questions are more difficult to resolve in large treated and control catchments.
- Monitor adjacent treated and control areas before and after construction of stormwater GI. This will account for typical year-to-year rainfall variations and also detect sensor problems early.
- Test areas should have the majority of their flows treated by the control practices to maximize measurable reductions. Any untreated, up-gradient areas should be small in comparison with the test areas as problems arise when subtracting flow from a large upstream area from the combined downstream area flows to estimate the runoff from the treated area. In other words, subtracting two large numbers from each other generates a high uncertainty error in the difference value.
- Low flows are challenging to monitor, especially if water depths are very small in relation to the pipe diameter. Acoustical depth sensors mounted on the pipe crowns could be used in conjunction with the area velocity sensors. Other devices include installing flumes in the pipes for more accurate discharge measurements and temporary plywood v-notch weirs in manholes without restricting peak flows.
- Some of the flow monitoring instruments experienced premature and frequent failures that went undetected for extended periods. The solutions include sensor redundancy and frequent inspection of the sensor data.
- The numbers of events to be monitored needs to be sufficient to detect the expected flow reductions, considering the variability in the data, and the data quality objectives (see Burton and Pitt, 2002, for experimental design guidance). The flow data collected during our projects had greater variability than expected, but the multiple/overlapping monitoring locations and scales, along with an extended postconstruction monitoring periods, allowed effective statistical analyses to achieve the project objectives.
- The excellent cooperation and participation of the cities of Cincinnati and Kansas was critical for the successful completion of this project. A high level of involvement and cooperation is essential for these types of complex projects to succeed.

## ACKNOWLEDGMENTS

Many personnel of the Metropolitan Sewer District of Greater Cincinnati assisted with monitoring and suppling the project data and their participation and support is gratefully acknowledged. The Kansas City National Demonstration of Advanced Drainage Concepts Using Green Solutions for CSO Control Project was funded and supported by the US EPA Office of Research and Development, Urban Watershed Management Branch. The direction and support of the EPA Project Officers, Michelle Simon and Richard Field, are gratefully acknowledged.


## REFERENCES


Burton Jr., G.A., Pitt, R., 2002. Stormwater Effects Handbook: A Tool Box for Watershed Managers, Scientists, and Engineers. CRC Press, Inc., Boca Raton, FL, ISBN 0-87371-924-7, 911 p.

CIWEM (Charter Institution of Water and Environmental Management), 2004. Policy Position Statement: Environmental Impacts of Combined Sewer Overflows (CSOs). London. http://www.ciwem.org/wp-content/uploads/2016/04/Environmental-impacts-of-combined-sewer-overflows.pdf.

EPA, 2001. Achieving Water Quality through Integrated Municipal Stormwater and Wastewater Plans. Memo from N. Stoner and C. Giles to EPA Regional Administrators and OW and OECA Directors. Washington, DC, October 27, 2011. 3 p.

EPA, May 1995. Combined Sewer Overflows; Guidance for Nine Minimum Controls. US Environmental Protection Agency, Office of Water, Washington, DC. EPA 832-B-95-003. 69 p.

EPA, July 21, 2011. Prioritizing and Selecting Green Infrastructure in Combined Sewer System Service Areas (draft report). 86 pgs.

Lucas, W.C., Sample, D.J., 2015. Reducing combined sewer overflows by using outlet controls for green stormwater infrastructure: case study in Richmond, Virginia. Journal of Hydrology 520, 473–488.

Lukes, R., Kloss, C., 2008. Managing Wet Weather with Green Infrastructure. Portland Oregon.

Nasrin, T., Sharma, A.K., Muttil, N., 2017. Impact of short duration intense rainfall events on sanitary sewer network performance. Water 9 (3), 225. https://doi.org/10.3390/w9030225. www.mdpi.com/journal/water.

Nasrin, T., Muttil, N., Sharma, A.K., 2016. WSUD strategies to minimise the impacts of climate change and urbanisation on urban sewerage systems: quantifying the effectiveness of rainwater tanks in reducing sanitary sewerage overflows in a case study in Melbourne, Victoria. AWA Water eJournal 1 (3). https://doi.org/10.21139/wej.2016.025.

NRC (National Research Council), Committee on Reducing Stormwater Discharge Contributions to Water Pollution, National Academy of Science, 2009. Urban Stormwater Management in the United States. ISBN: 13: 978–0-309-12539-0. National Academies Press, Washington, DC, 598 p.

NRC (National Research Council), Committee on Beneficial uses of Graywater and Stormwater, National Academy of Science, December 2015. Beneficial Uses of Graywater and Stormwater: An Assessment of Risks, Costs, and Benefits. National Academies Press, Washington, DC.

Pitt, R., Talebi, L., Bean, R., Clark, S., November 2011. Stormwater Non-Potable Beneficial Uses and Effects on Urban Infrastructure. Water Environment Research Foundation, Report No. INFR3SG09, Alexandria, VA, 224 p.

Pitt, R., Talebi, L., February 11, 2014. Modeling of Green Infrastructure Components and Large Scale Test and Control Watersheds. Prepared for USEPA Office of Research and Development, Urban Watershed Management Branch, 426 p. http://unix.eng.ua.edu/~rpitt/Publications/5_Stormwater_Treatment/GI_controls/Green_Infrastructure_and_WinSLAMM_Modeling_at_KC.pdf.

Quigley, M., Brown, B., 2014. Transforming Our Cities: High-Performance Green Infrastructure. ISBN: 978-1-78040-673-2/1-78040-673-8. Water Environment Research Foundation, Alexandria, VA, 96 p.

Talebi, L., Pitt, R., July 25, 2014. Retrofitted Green Infrastructure Stormwater Controls at Cincinnati State College, the Cincinnati Zoo, and the Clark Montessori High School: An Evaluation Using Monitoring Data and WinSLAMM. Prepared for Metropolitan Sewer District of Greater Cincinnati, 103 p. http://unix.eng.ua.edu/~rpitt/Publications/5_Stormwater_Treatment/GI_controls/Cincy_modeling_report_Final.pdf.

Talebi, L., September 2014. Assessment of Integrated Green Infrastructure-based Stormwater Controls in Small to Large Scale Developed Urban Watersheds (Ph.D. dissertation). Department of Civil, Construction, and Environmental Engineering, The University of Alabama, Tuscaloosa, AL, 595 p. http://unix.eng.ua.edu/~rpitt/Publications/11_Theses_and_Dissertations/Leila_Dissertation.pdf.


## FURTHER READING


World Health Organization and UNICEF, 2012. Progress on drinking water and sanitation: 2012 update. In: WHO/UNICEF Joint Monitoring Programme for Water Supply and Sanitation. New York, NY. https://www.unicef.org/media/files/JMPreport2012.pdf.

Chapter 8

# Erosion and Sediment Control—WSUD During the Construction Phase of Land Development

**Leon Rowlands**

*Switchback Consulting, Caloundra, QLD, Australia*

## Chapter Outline



**ABSTRACT**

The construction phase is often overlooked when considering the impacts of land development on water quality. This aspect of water sensitive urban design (WSUD) is known as "erosion and sediment control" (ESC) and its name reflects the key processes that need to be managed. This chapter examines the impacts and magnitude of sediment loads generated during the construction phase and compares this with the loads generated during the operational phase of development, which has traditionally been the major focus of WSUD in Australia.

The methods used to manage these impacts is explored, firstly through an examination of the policy approaches used in different jurisdictions and secondly at a technical level through a review of management measures and technical innovations. The success of these approaches is then considered by benchmarking compliance levels and examining different approaches to regulation.

Finally, the chapter discusses overlaps with the operational phase of WSUD infrastructure, which occurs during the conversion or decommissioning of ESC measures at the end of the construction phase.

**Keywords:** Construction; Drainage; Erosion; High-efficiency sediment basins; Land development; Regulation; Sediment; Urban development.

## 8.1 INTRODUCTION

The term "erosion and sediment control" (ESC) is used in Australia to describe the suite of management interventions applied to mitigate sediment pollution during the construction phase of land development. In this chapter, the term "construction phase" refers to the relatively short period in which the conversion of "undeveloped" land to an urban form occurs. Typically, this involves exposing the earth to undertake earthworks and construct essential services and infrastructure such as roads, stormwater drainage, water supply, sewer, and electricity and communications.

The term "operational phase" refers to the long-term and ongoing period following construction when the developed urban form has been completed. Traditionally in Australia, water quality management practices have been referred to as ESC during the construction phase and water sensitive urban design (WSUD) during the operational phase. Although ESC has often been viewed in the past as a separate process to WSUD, there are obvious areas of overlap. For example, both are concerned with protection of waterway health and often temporary ESC devices are converted to permanent WSUD infrastructure at the end of construction (i.e., conversion of a sediment basin to a bioretention device). For these reasons, ESC should be viewed as part of the overall WSUD philosophy of land development.

There are many publications that aim to guide practitioners on the technical implementation of ESCs. This chapter does not aim to be another "how to" manual. Rather the aim of this chapter is to provide the reader with a broader understanding of how ESC fits within the overall aims and philosophy of WSUD and the issues preventing successful adoption.

## 8.2 EROSION PROCESSES AND MANAGEMENT APPROACHES

### 8.2.1 Erosion processes

Although erosion is a naturally occurring process, accelerated erosion often occurs during land development due to the removal of surface vegetation cover and disturbance of topsoil and subsoils combined with changes in the patterns of surface flows (Fig. 8.1). Leersnyder et al. (2016) identifies the following as the different types of water-borne erosion that can occur.

### 8.2.2 Management hierarchy and potential impacts

Within areas of active earthworks, it is generally considered that splash and sheet erosion are unavoidable because of the exposure of bare soil to the impact of rainfall mobilizing fine-grained sediments (in particular, clay and silt fractions). However, drainage controls and soil management should be implemented during *all* stages of construction to ensure that rill, gully, tunnel, and channel erosion are prevented or minimized. Sediment controls are implemented to reduce the remaining sediment export to streams once appropriate erosion and drainage controls have been implemented on site.

In this way, the combined application of erosion, drainage, and sediment control measures should be viewed as a **treatment train**. Within this treatment train, sediment controls are devices for the removal of sediment (rather than the prevention of erosion) and will always have limited efficacy. For this reason, the priority should always be on establishing a stable landform through surface cover (erosion control) and drainage controls so that erosion is prevented to the maximum extent possible, rather than placing an overreliance on sediment control.

If left unmitigated, sediment pollution from land development sites can result in a range of adverse off-site impacts to infrastructure, the economy, and social values. Such impacts include blockage to drainage system (and associated flooding), reduced traffic safety (from debris deposited on roadways), and direct damage to private property (such as swimming pools filling with dirty water).

In terms of environmental impacts, sediment pollution is well established as a major degrading factor of aquatic environments within the scientific literature. The range of impacts is summarized in Russell et al. (2017) as the following:

- Smothering of the channel bed—resulting in loss of primary production and impacts on fish and macroinvertibrates that rely on benthic habitat;
- Increased suspended sediment in the water column—impacting fish and filter feeders; and
- Increased delivery of sediment-associated contaminants such as nutrients and heavy metals—resulting in eutrophication of receiving waters and toxicity to aquatic organisms.

Such environmental impacts are equally applicable to both freshwater and poorly flushed marine environments such as bays and estuaries. Within Queensland, local examples of impacts include the smothering of seagrasses through human-induced sediment loads (Walker and McComb, 1992). Similarly, recent Queensland research on freshwater urban streams has established sedimentation of pool habitat as a major degrading process of their ecological health and suggests reducing loads through improved ESC as a key management intervention (McIntosh et al., 2013).

The link between human activities and declines in waterway health has been confirmed through catchment and receiving water modeling, such as the modeling undertaken as part of the Moreton Bay Regional Council Total Water Cycle Management Strategy (BMT-WBM, 2010). This study identified that continuing with "business-as-usual" land development practices will yield further declines in waterway health for the Pumicestone Passage (a poorly flushed estuary in South East Queensland, Australia).

FIGURE 8.1 Types of erosion. (a) **Splash or raindrop erosion**—It is caused by the direct impact of rainfall on the exposed soil surface, which causes the soil particles to be dislodged. Splash erosion is addressed by covering the soil surface **with erosion controls**. Note the stone in the adjacent image, which has protected the underlying soil from raindrop erosion and resulted in a "pillar" of soil as the surrounding unprotected soils have been eroded. There are many examples of erosion controls such as mulch, chemical soil binders, and vegetative cover. (b) **Sheet erosion**—It is the movement of soil caused by the uniform sheet flow of runoff over a surface and is addressed by covering the soil surface **with erosion controls**. (c) **Rill erosion**—It occurs when sheet flows concentrate into small channels, or rills, and the resulting velocity causes concentrated areas of soil scour. Rill erosion is addressed through appropriate **drainage controls**. Such controls aim to intercept sheet flows before they concentrate into rills and manage the flow in a controlled manner. For example, contour banks or catch drains are often implemented parallel to the contour to direct flow to a diversion drain flowing perpendicular to the contour. (d) **Gully erosion**—It is a form of rill erosion, where the rills exceed 300 mm in depth (Leersnyder et al., 2016), and is addressed through appropriate **drainage controls** as noted above. (e) **Tunnel erosion**—It typically occurs as a consequence of dispersive subsoils where infiltrating water causes dispersion of the finer particles (clay + silt) in the subsoil, and the low strength slurry allows tunneling to occur, as interflow moves downslope. Eventually, structural collapse of the whole soil profile occurs. Tunnel erosion can require a range of interventions such as soil amelioration (gypsum), ripping, compaction, and drainage management. (f) **Channel erosion**—The accelerated channel erosion within a watercourse occurs as a result of increased runoff rates following urbanization of the catchment. It causes bank collapse and steam bed scouring. It is addressed through **permanent water sensitive urban design infrastructure or stream bank armoring**. (g) **Mass movement**—It occurs as a result of slope instability that may be caused by geotechnical issues such as increased groundwater pressure, a low soil strength layer, removal of the toe of a batter, or vegetation clearing.

## 8.3 CONSTRUCTION AND OPERATIONAL-PHASE SEDIMENT LOADS

Although the impacts of sediment on receiving waters is generally well acknowledged, the relative contributions and significance of sediment export at the construction and operational phases of land development are often debated. The following sections examine the loads of sediment occurring from unmitigated land development during the construction and operational phases and then examine the residual loads that occur when best practice management approaches are applied.

### 8.3.1 Unmitigated loads

There are a range of models available for estimation of soil loss rates from construction sites, with perhaps the most widely applied being the Revised Universal Soil Loss Equation (RUSLE). The RUSLE is described in IECA (2008) as having the following form:

$$A = R \cdot K \cdot LS \cdot C \cdot P \tag{8.1}$$

where A is the annual soil loss due to erosion [t/ha year]; R the rainfall erosivity factor; K the soil erodibility factor; LS the topographic factor derived from slope length and slope gradient; C the cover and management factor; and P the erosion control practice factor.

The limitations of RUSLE are that it only accounts for soil loss through sheet and rill erosion and ignores the effects of gully erosion and dispersive soils. There is also no accounting for deposition of sediment before reaching the waterway. Due to these limitations, RUSLE is at best a coarse predictor of total suspended solids (TSS) loads delivered to waterways.

To compare operational and construction-phase sediment loads, a case study location in Nambour in South East Queensland was selected. Based on IECA (2008) guidance for the selection of parameters, the RUSLE predicts soil loss rates for the construction phase for this location in the order of 290 t/ha year. This soil loss rate has been supported by field measurements by Gardner (2010) for the same locality, who estimated sediment deposition within the waterway, downstream of the 8 ha land development, at 285 t/ha during the construction phase.

For operational-phase sediment loads, estimation is most commonly undertaken in Australia using the Model for Urban Stormwater Conceptualisation or MUSIC (eWater, 2014). Using accepted model parameters (hydrologic and pollutant export) for this same locality (Healthy Waterways Ltd., 2010), a TSS load of 1640 kg/ha year is predicted during the operational phase of land development based on an urban residential land use. The construction-phase loads estimated from RUSLE are therefore more than 170 times the loads predicted for the operational phase, on a per time basis. Or expressed as another way, more than 170 years of erosion prevention during resident occupation at the operational phase is required to offset the sediment load that occurs during 12 months of civil construction at the construction phase.

Although the predictions of construction-phase and operational-phase sediment loads are not directly comparable for the reasons listed above (i.e., RUSLE model limitations), the estimates do provide a clear indication that untreated construction-phase sediment loads are orders of magnitude greater than operational-phase loads on a unit time basis.

While sediment yield for a particular location will differ because of variations in topography and climate, the general trends described above are applicable to all locations. Russell et al. (2017) compared sediment yields for urban and urbanizing catchments based on published data from 48 international studies and also found that total sediment yield is likely to greatly increase during urbanization (the construction phase) and then decline during the operational phase but remain elevated above predevelopment conditions.

### 8.3.2 Residual loads from conventional management and control measures

It is clear from the preceding section that unmitigated construction-phase sediment loads equate to many decades of postconstruction loads and have the potential to overshadow any beneficial actions implemented to reduce environmental impacts during the postconstruction phase. However, the question remains of how significant the loads of each phase are once typical best management practices are applied.

For the purposes of this analysis, best management practices[1] (IECA, 2008) are simplified to typical end of line controls, being listed as follows:

- Sediment basins during construction
- Bioretention basins postconstruction

1. More detail on the range of construction-phase erosion and sediment control measures can be found in IECA (2008), while water sensitive urban design measures applicable to the operational phase can be found at http://hlw.org.au/initiatives/waterbydesign/water-sensitive-urban-design-wsud.

The residual load (i.e., the load remaining once best practice is applied) for the postconstruction phase is obtained by noting that current WSUD policy requires TSS loads to be reduced by 80% compared with an unmitigated development (DILGP, 2017). The residual load of TSS postconstruction is therefore 20% of the unmitigated load or about 330 kg/ha year in the example used above.

The residual load of TSS discharged from a sediment basin during construction has been derived by Rowlands and Leinster (2014) using the following equation:

$$L_{residual} = 0.01 \times R \times C_v \times [C_1 \times u + C_2 \times (1 - u)] \tag{8.2}$$

where L is the residual load of TSS discharged from basin (kg TSS/ha year); R the average annual rainfall (mm/year); $C_v$ the volumetric runoff coefficient; $C_1$ the event mean concentration of TSS for flows dewatered from basin (mg/L); $C_2$ the event mean concentration of TSS for bypass flows over basin spillway (mg/L); and u the basin hydrologic effectiveness.

The hydrologic effectiveness of a sediment basin is the proportion of average annual runoff that is able to be treated to the design discharge target ($C_1$). There have been a number of studies that have established that conventional batch sediment basins have low hydrologic effectiveness, with a value of about 30% applicable to Nambour (Rowlands and Leinster, 2015).

The design discharge concentration ($C_1$) is typically set at 50 mg/L by the regulatory agencies and is achieved through either gravity settlement or enhanced gravity settlement through the addition of chemical flocculants (IECA, 2008; Landcom, 2004). The parameter, therefore, with by far the greatest uncertainty is the concentration of TSS ($C_2$) in the flow of water over the basin spillway. This water is only partially treated, as the spillway flow occurs when the design rainfall event for the basin is exceeded.

Values for $C_2$ are difficult to obtain as they are either rarely monitored or the parties holding such data (i.e., the urban developers or their consultants) are reluctant to disclose them because of regulatory issues. However, with recent advances of automated flocculant dosing systems and automated monitoring, such datasets are now becoming available. Based on review of these datasets, typical values of the order of 300 mg/L can be expected in spillway flow (Butch Uechtritz, pers. comm.), though such values can be highly site- and event specific. Based on construction duration of 12 months, this results in a residual load (L) from Eq. (8.2) of about 2400 kg/ha year.

The temporal pattern of estimated TSS loads from urban development, after best practice management measures are applied, is shown in Fig. 8.2. This figure demonstrates that even with best practice controls in place, the TSS loads during construction is equivalent to about 7 years of postconstruction phase loads.

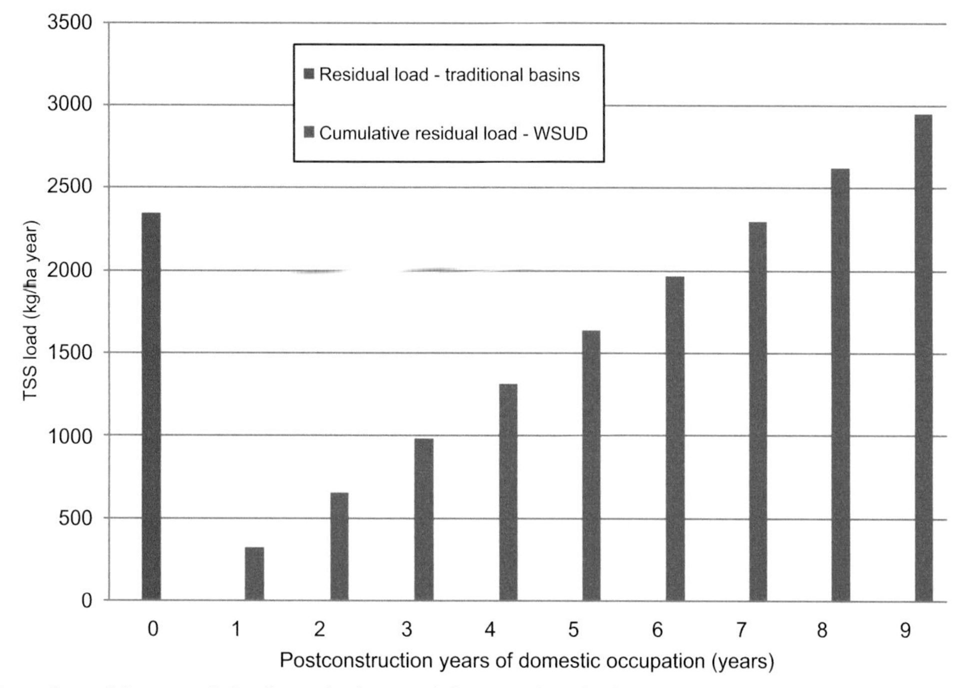


**FIGURE 8.2** Comparison of the suspended sediment load exported from an urban development area with best practice erosion prevention with the *cumulative* yearly load from an occupied subdivision after implementation of best practice water sensitive urban design (WSUD) technologies. The example applies to a typical subdivision in South East Queensland (Australia).

The key message from this analysis is that without application of control measures, sediment export from the construction phase is orders of magnitude greater than the sediment export from unmitigated operational-phase runoff. However, even with application of conventional best practice measures, the construction-phase loads are still equivalent to nearly a decade of operational-phase sediment exports. It is clear that advances in the efficacy of best practice measures for construction-phase controls are required before their sediment loads approach a similar magnitude to that of operational-phase annual loads. Suggested advances are discussed in Section 8.5.

## 8.4 COMPLIANCE LEVELS AND INDUSTRY PERFORMANCE

There are few published studies in Australia documenting the level of compliance of construction sites implementing best practice ESCs, although a number of such studies have been conducted by various local governments. Healthy Waterways Ltd. (2014) provides one of the few published surveys, though it is noted that this study was also removed from public access after being available on the organization's website for a short period. This apparent reluctance of organizations to be transparent on levels of ESC performance may in itself be one of the main barriers to improved performance.

The Healthy Waterways Ltd. (2014) study undertook field-based audits of 57 construction sites in 6 local government areas of South East Queensland. The study concluded that the overall standard of ESC implementation was very poor, with only 5% of the sites found to be substantially compliant with best practice.

There has been significant work by the scientific expert community on examining the reasons for poor ESC performance throughout South East Queensland. The Integrated Urban Water Scientific Expert Panel (IUWSEP) of Healthy Waterways produced a paper on this specific issue (IUWSEP, 2013).

The IUWSEP found that there was a significant lack of enforcement being applied to noncompliant development sites. Hence, there was a strong financial incentive for developers/contractors to not fully implement ESC. In fact, the IUWSEP found that the costs of fully implementing ESC outweighed the likely costs of enforcement (due to low detection levels and low enforcement penalties) by a factor of 14. In a competitive urban real estate market, this situation significantly disadvantages the developers who want to implement ESC by making them less cost-effective than those who choose to ignore, or only partially implement, best practice sediment control as required by law.

Unless local governments undertake consistent proactive enforcement on this issue and use enforcement tools that are appropriate to the scale of the offenses, it is inevitable that market forces will continue to deliver the observed low levels of ESC implementation. A review of political donations registered for local government elections in high-growth regional areas in Queensland indicates why this situation is unlikely to change. Donors for successful mayoral candidates in such localities are often dominated by the development and construction industry. Even without the influence that inevitably comes from political donations, the development industry is often a major component of the economy of these localities, and any direct action to regulate the industry is perceived as harmful to the local economy, and initiatives to improve performance are therefore either short-lived or focused on "education" or other nonpunitive actions.

## 8.5 DESIGN STANDARDS AND INNOVATION

In the context of chronically low levels of implementation of best practice, it is particularly important that design objectives and policy requirements are easily understood and facilitate the continuous improvement in water quality outcomes. Design objectives can take two possible forms:

- **Prescriptive standards**—where a set of "rules" or required actions are specified, which, when implemented, are assumed to achieve an acceptable environmental outcome; or
- **Performance-based standards**—where a specific environmental "outcome" or quantitative limit is set by the regulator based on a desired environmental outcome, and flexibility is allowed in terms of how that outcome is achieved.

Within Australia and New Zealand, ESC standards and design objectives have mainly fallen into the category of being prescriptive standards (DILGP, 2017; DEHP, 2010; Leersnyder et al., 2016). The key problem with prescriptive standards is that they tend to enshrine the status quo and inhibit industry innovation. This has been more of a problem in Australia than in New Zealand where Auckland Regional Council has heavily invested in ESC research[2] and has been able to advance both prescriptive standards and the technology underpinning those standards (Leersnyder et al., 2016).

2. Auckland Regional Council have published many reports relating to ESC, which have advanced the understanding and management of ESC. The full catalog of technical reports can be found at the following link: http://www.aucklandcouncil.govt.nz/EN/planspoliciesprojects/reports/technicalpublications/Pages/home.aspx#technical.

An example of a design objective, which has inhibited innovation, is the sediment basin design standards used in Australia before 2017 for fine and dispersive sediments (type-F and type-D basins, respectively), as documented in IECA (2008). These basins operate as a "batch" process, whereby runoff fills the basin; the basin is then treated with a chemical flocculant/coagulant (if necessary) before allowing time for the entire volume in the basin to settle. Once acceptable water quality is achieved in the basin and the basin is then emptied in readiness for the next event.

Current guidelines (IECA, 2008) mandate that sediment basin sizing in Queensland is designed to capture the runoff volume from the 5-day 80th percentile rainfall depth. Discussions with the author of the original standard, which began "life" in NSW in the late 1990s as the 5-day 75th percentile rainfall depth, confirmed that this value was chosen by selecting a percentile which gave a basin size that was thought to be economically tolerable to the development industry at the time (Ian Joliffee, pers. comm.). The expectation was that this standard would become more rigorous over time. This principle of a set of constantly evolving "economic best practice" objectives is consistent with the postconstruction phase objectives commonly employed in WSUD.

The sediment basin design objectives are frequently misunderstood by practitioners. Specifically, the 80th percentile criteria gives the impression that a very high proportion of annual runoff is able to be treated to the design standard of 50 mg/L; however, analysis by Rowlands and Leinster (2014), which is shown in Fig. 8.3, has demonstrated that this is not the case. In the example shown in Fig. 8.3 for Caloundra in South East Queensland, it can be seen that a sediment basin sized to the current design standard is capable of treating only 34% of the annual runoff volume to the release criteria of 50 mg/L TSS (i.e., a hydrologic effectiveness of 34%). This analysis has been repeated in other locations in Australia and it has been shown that basin effectiveness decreases (i.e., lower hydrologic effectiveness) at higher latitudes. This is due to the more pronounced seasonality of rainfall in higher latitude coastal areas of Australia, where rainfall often occurs in consecutive events that do not allow sufficient time for the above batch treatment and dewatering process to occur.

The overall effect of this current prescriptive sediment basin design standard has been to lock in a specific management measure as accepted best practice (batch sediment basins) that has a low treatment efficacy. This has inhibited uptake of new technologies such as continuous flow or high-efficiency sediment (HES) basins.

These HES basins operate as continuous flow systems rather than as batch systems, with an ongoing addition of flocculant and release of treated water throughout the runoff event (see Fig. 8.4). The use of a continuous flow process means these systems can treat a much greater volume of runoff than a similar sized batch system (i.e., a traditional type-F/D basin). Manual handling costs associated with flocculating and dewatering traditional basins are also eliminated by such systems.

Despite the apparent benefits of these systems and the presence of examples in New Zealand for decades (Auckland Regional Council, 1999) as well as installations in South East Queensland in public sector works for almost 10 years,

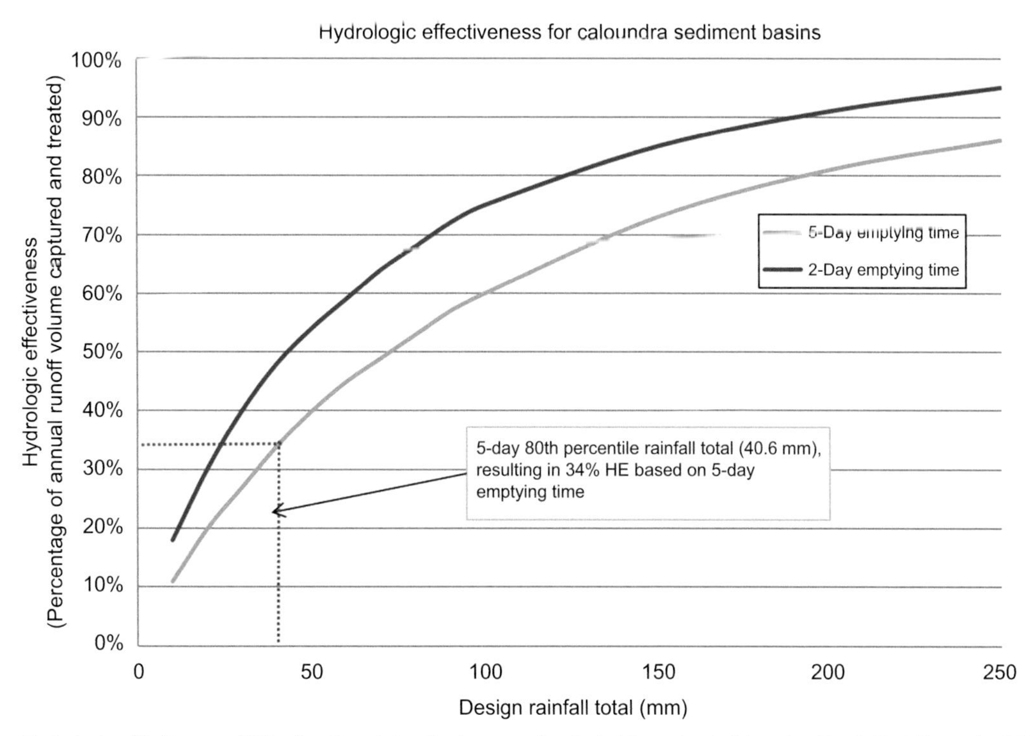


**FIGURE 8.3** Hydrologic effectiveness (HE) of sedimentation basins operating in batch mode at Caloundra, South East Queensland (Rowlands and Leinster, 2014).

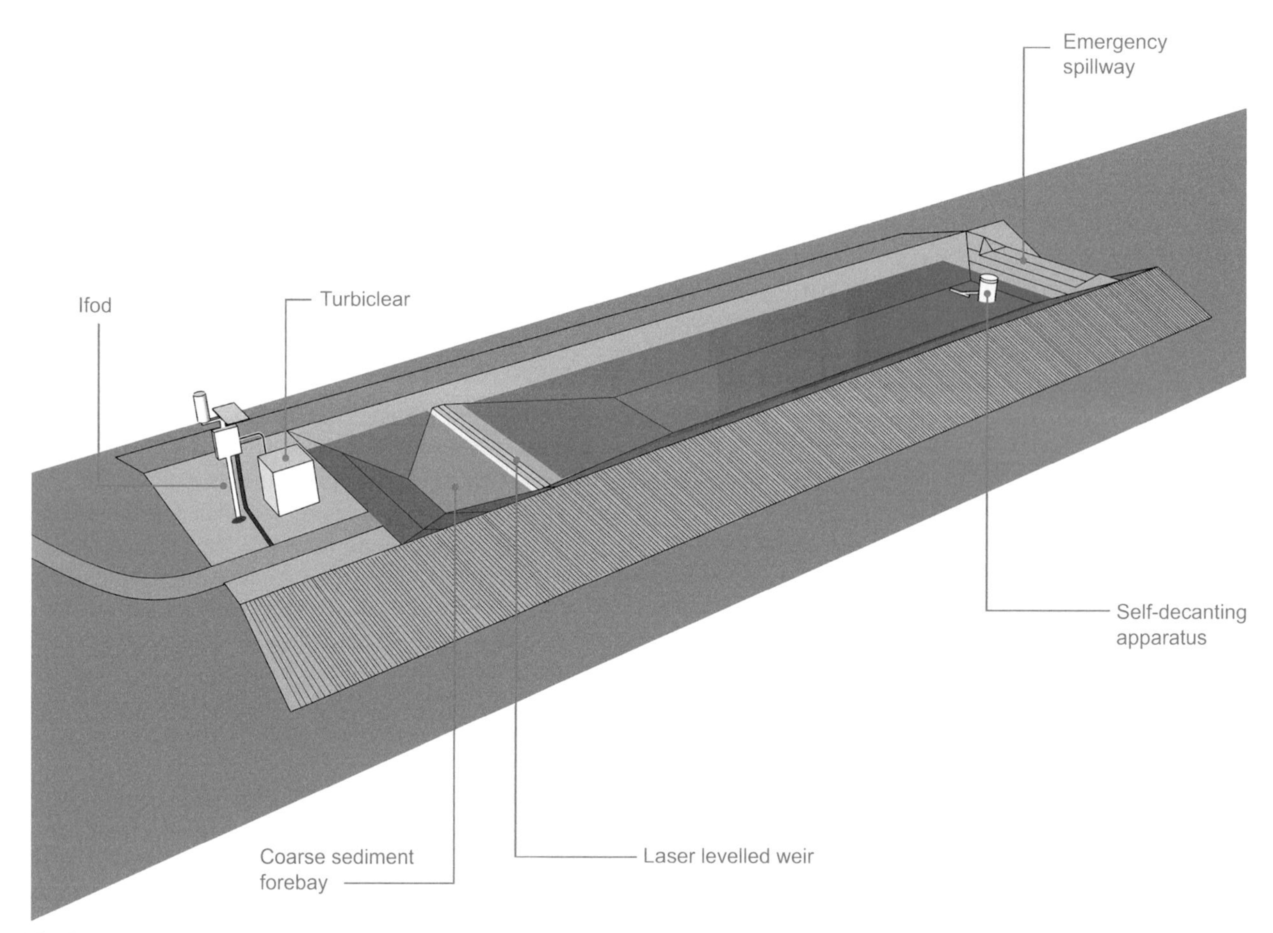


**FIGURE 8.4** Schematic of a continuous flow or high-efficiency sediment basin layout with automated flocculant dosing. *Source: TURBID Stormwater Solutions, http://www.turbid.com.au/.*

take-up of this technology by the private sector in Australia has been very poor. We argue this is largely due to an absence of an appropriate head of power mandating their use.

The deficiencies of current objective suites have been recognized in Queensland and have led to development of performance-based sediment control standards (DILGP, 2017). The aims of such standards are to

- Clearly articulate the required level of environmental protection and significantly advance protection beyond the current standards; and
- Provide a standard that facilitates innovation by being independent of any single specific technology.

An example of this new generation of design objectives is the sediment control standard contained in the Queensland State Planning Policy (DILGP, 2017), which states the following:

> *Sediment controls are designed, implemented, and maintained for all exposed site areas to a standard that would achieve at least 80% of the average annual runoff volume of the contributing catchment treated to 50 mg/L TSS or less (i.e., 80% hydrologic effectiveness).*

Such an objective would see a major shift in environmental performance due to the large increase in hydrologic effectiveness of sediment controls compared with current sediment basin design (i.e., an increase from about 34% to 80% hydrologic effectiveness). This higher standard can be achieved by a range of approaches depending on the location and climate, including the following:

- Enlarged batch sediment basins and/or reduced treatment and emptying times;
- Perimeter bunding and total capture of runoff on flat sites; or
- Adoption of HES basins.

## 8.6 REGULATORY APPROACHES

Chapter 5 examines in detail the impact of policy and regulation on the implementation of WSUD through case studies from Australia and the international scene. This section provides a brief overview of the application of process-based and outcome-based regulatory approaches adopted in different jurisdictions.

Regulation of ESC may be undertaken by different levels of government (federal, state, or local) depending on the location and type of work and the applicable legislation. In some situations (such as Queensland, Australia) it may be more than one level of government with legislative responsibilities. There are a range of regulatory approaches that have been adopted in different jurisdictions and that can be broadly categorized as being either of the following:

- Prescriptive or process-based regulation or
- Outcome-based regulation

  Many jurisdictions adopt a blend of these broad categories, though the defining features of each system are as follows:
- Process-based regulation:
  - Requires preparation and endorsement of a detailed ESC plan;
  - References a prescriptive standard or guideline;
  - Enforced through conditions of approval; and
  - Compliance and auditing against approved/endorsed detailed ESC plan.
- Outcome-based regulation:
  - Does not approve detailed ESC plans;
  - Requires specific outcomes (through legislation or performance-based conditions or standards); and
  - Compliance and auditing against performance standard.

The system of regulation adopted for a particular jurisdiction will be informed by the legislation, available resources, and willingness to undertaken enforcement. The key differences between the above alternate regulatory approaches are that process-based regulation is much more resource intensive and also focuses resources at the planning stage of development. Outcome-based regulation requires fewer resources and focuses those resources on auditing compliance during construction. While both systems can be effective, if there is an unwillingness to undertake enforcement or if penalties are relatively trivial, then outcome-based regulation risks being completely ineffectual.

Within the Auckland region in New Zealand, there has been a strong commitment to ESC and accordingly the regulator (Auckland Regional Council) has been able to commit resources to a process-based system of regulation. This system is supported by a range of publications including clear "how to" guidelines and a well-organized system of construction auditing.

By contrast, local governments in Queensland, Australia, have often adopted a mishmash of process-based and outcome-based systems. This may be a consequence of ESC having two legislative heads of power in Queensland:

- The ***Sustainable Planning Act 2009***—which essentially prescribes the process of development applications and planning decisions and allows conditions to be placed on approvals; and
- The ***Environmental Protection Act 1994***—which specifies offenses relating to water contamination and environmental harm.

A pure outcome-based system of regulation in Queensland would be one where the development approvals are *silent* on ESC and regulation is undertaken entirely under the water contamination and environmental harm provisions of the *Environmental Protection Act 1994*. This system would result in no ESC plans being reviewed or approved and all regulation taking place in the field through compliance auditing and enforcement. This system is essentially the same as that employed by the Queensland Department of Environment and Heritage Protection when regulating development that falls outside the jurisdiction of local government (DEHP, 2010).

To date, no local government in Queensland has adopted the pure outcome-based system of regulation described above. Conversely, a general lack of resources combined with a development approval process, which is not conducive for detailed plan review, has also meant that few have adopted a pure process-based system. The resulting confusion of approaches and general reluctance to undertake meaningful enforcement has contributed to the low levels of compliance noted in Section 8.4.

## 8.7 ROLES, RESPONSIBILITIES, AND CONTRACTS

Preceding sections have examined issues surrounding the way local and state governments set policy and design objectives, as well as undertaking regulation. The private sector has also often failed to incorporate processes and system that would improve ESC outcomes.

There is often confusion in civil construction projects around roles and responsibilities for design and implementation of ESCs, including which party carries the legal liability should controls not be adequately implemented. The issue of liability depends on the laws of the particular jurisdiction in which the works occur. In Queensland, Australia, the issue is complicated by the existence of the two legislative heads of power for ESC discussed in Section 8.6, namely the *Sustainable Planning Act 2016* and the *Environmental Protection Act 1994*.

The principal contractor can be held accountable under the *Environmental Protection Act 1994* as the party who bears the earth, while the developer has responsibility under the *Sustainable Planning Act 2009* to ensure the development complies with the conditions of approval. Both Acts contain Executive Officer liability provisions that can make the managers who control the organization guilty of an offense if the company commits an offense. The effect of this legislation is that both the developer and principal contractor have exposure to enforcement and it is therefore in the interests of both parties for ESC to be properly designed and implemented.

Given the above, it is surprising that a cooperative approach to ESC is not more common on construction sites and that many construction contracts continue to include ESC as part of the lump-sum tender and then put compliance with all statutory provisions on the principal contractor. Such an approach does little to manage the enforcement risk to the developer and creates a situation where contractors are forced to inadequately cost ESC to maintain a competitive tender. Inevitably this process leads in many cases of ESC, which are reactive, lacks any overall strategy and has not undergone formal design, and is likely to result in water contamination when rainfall occurs (refer Fig. 8.5).

A better system would be one that appropriately manages the legal risk for all the parties involved, as well as the risk to the environment. The following processes are suggested:

1. Developer provides ESC plans that identify the major controls. This then forms part of the tender package for pricing;
2. Tender response identifies the ESC component of the lump-sum price;
3. Developer excludes tenderers who provide inadequate allocation for ESC;
4. Contract requires compliance with ESC standard (e.g., DEHP, 2010), which sets performance outcomes;
5. Contract requires the contractor to prepare and adapt "for construction" ESC plans to suit the particular construction methodology; and
6. Developer funds independent audits throughout construction.

The process ensures adequate budgeting for ESC and accountability in design and implementation. This ensures that the contractor has a workable ESC strategy to follow and that controls are properly designed and proactively implemented. Ultimately this leads to less conflict between the developer, contractor, and regulator, as well as better water quality outcomes for the environment.

**FIGURE 8.5** An example of inadequate erosion and sediment control outcomes in South East Queensland resulting from deficient planning and contract processes.

## 8.8 CONVERSION OF CONSTRUCTION-PHASE MEASURES TO PERMANENT WSUD INFRASTRUCTURE

It is common practice that the footprint allocated for permanent end of line stormwater treatment devices are used during construction for siting sediment basins for ESC. The key issue with this approach is determining an appropriate time for the conversion of the sediment basin into the permanent stormwater treatment device such as a bioretention basin.

Premature conversion, while construction- or building-phase sediment loads remain high, risks damaging the permanent stormwater treatment device or discharging high-sediment loads to the environment. Conversely, excessively delaying conversion means that the benefits of the vegetated permanent devices such as nutrient reduction are not delivered in a timely manner.

For smaller developments or where an end of line device services a single development stage, it has generally been accepted that there are four options regarding conversion that are imperfect but acceptable in terms of balancing the competing needs for waterway protection and asset delivery. These options vary by the type of permanent device, but for bioretention devices, they include (Healthy Waterways Ltd., 2009) the following:

- **Option 1—Surface protection**: The bioretention basin is constructed and the filter media surface covered with geofabric and turf during the building phase. The device acts as a shallow sediment basin during this phase.
- **Option 2—Bypass flows and early plant establishment**: Flows bypass the bioretention basin during the house building phase so that the filter media can be planted. The bioretention basin provides no sediment removal under this scenario and hence provides no environmental benefit until the end of the building phase.
- **Option 3—Sediment basin and bioretention function**: This approach combines Option 1 with a sediment basin installed upstream, thereby providing greater sediment removal.
- **Option 4—Leave as a sediment basin**: This approach delays construction of the bioretention basin until the end of the house building phase.

For larger developments, it is common that an end of line bioretention basin or wetland may service a number of development stages. Options 1 and 2 therefore become untenable for larger developments as they do not provide adequate sediment control during construction of the subsequent development stages. Option 4 is the approach most often taken for larger developments (i.e., leave as a sediment basin until the final stage). However, this can mean that there are many years where the completed stages are not appropriately treated, and runoff from these completed stages mixes with construction runoff and consequently receives only sediment removal via the sediment basin (i.e., limited or no nutrient removal).

Option 3 would therefore appear to be the preferred approach for larger developments but has rarely been used due to the following:

- The low hydrologic effectiveness of conventional sediment basins (refer Section 8.5) means that the bioretention basin will still receive high sediment loads, so the filter media will still need to have surface protection;
- Nutrient removal of the bioretention device will be low because of the need for surface protection and so offers little benefit;
- The space required for both a temporary sediment basin and a permanent bioretention basin is rarely available.

It can be seen from the above that there is a serious and urgent need for a better approach for managing the transition from construction to operational-phase stormwater management on larger developments involving multiple stages. The availability of HES basins provides new options to address this issue.

Fig. 8.6 provides a conceptual layout of an integrated stormwater treatment train that is capable of providing enhanced treatment for multistage developments while also allowing for early vegetation establishment of the permanent WSUD elements. Each element of the treatment train is described as follows:

1. **Temporary construction water dam**—This component would ultimately be decommissioned at the end of construction. While construction is occurring, it provides a range of functions. Initially, it allows for capture of construction water. By reusing water for construction purposes at this point, it minimizes chemical flocculant use in the next process. The dam also provides detention storage upstream of the HES basin, which allows the flowrate to this process to be controlled and hence the size of the HES device to be minimized. Lastly, the dam allows for high flows to bypass the downstream treatment train and hence avoid excessive sediment loads to any vegetated components.
2. **HES basin**—This is a temporary measure while construction is occurring in the catchment and will ultimately be converted to an inlet pond to the constructed wetland once construction is complete. Provision of the detention storage in the dam upstream of the HES basin allows the basin size to remain at the size required ultimately as a wetland inlet

FIGURE 8.6 Conceptual layout of resilient stormwater treatment train, which achieves both construction- and operational-phase discharge water quality objectives for multistage developments. *HES*, high-efficiency sediment basin.

pond. The HES basin is provided with flow-based dosing of chemical flocculant and is the primary means of removing sediment loads while construction is occurring.

3. **Constructed Wetland or bioretention basins**—This device will receive treated flows from the HES basin, with all other flows bypassing this device. Due to the provision of the HES basin upstream, the wetland vegetation can be established from the outset with the confidence that the wetland will be protected from high construction-phase sediment loads. This means that the wetland vegetation will be fully established by the time construction is completed in the catchment. The inclusion of the constructed wetland ensures that the operational-phase design objectives (load reductions) are achieved for the early stages of completed development in the catchment.

The benefits of adopting this integrated treatment approach include the following:

- Achieves water quality design objectives for construction- (ESC) and operational phase (WSUD) for all phases of site development without compromising one for the sake of the other;
- Manages environmental risk through a multiple barrier approach;
- Treatment elements are resilient to variable water quality inputs;
- Achieves high standard of landscape amenity within a small overall treatment area "footprint";
- Allows for early vegetation of permanent WSUD features, resulting in significantly shorter overall establishment and delivery timeframes;
- Saves on imported water costs for construction water demands by facilitating supply of both untreated "raw" stormwater and primary treated "clean" stormwater to meet construction water requirements; and
- Saves on construction-phase stormwater treatment costs by minimizing the volume of stormwater requiring flocculant treatment to achieve TSS discharge criteria (50 mg/L).

## 8.9 CONCLUSION

The construction phase of land development presents a significant risk to aquatic ecosystems based on the human-induced erosion and sediment transport processes. The following conclusions are provided in relation to the management of these risks:

- Sediment yields delivered to aquatic ecosystems during construction are both qualitatively and quantitatively different from that which occurs during the postconstruction or operational phase of land development (when house occupation occurs). It is this postconstruction phase that is currently the focus of most WSUD technology, research, and regulations in Australia.
- Modeling suggest that without application of control measures, sediment export from the construction phase is orders of magnitude greater than the sediment export from unmitigated operational-phase runoff. However, even with application of conventional best practice measures, the construction-phase loads are still equivalent to nearly a decade of

operational-phase sediment exports. Based on this relative magnitude of impacts, we believe greater emphasis should be placed on the construction phase in regulation and research.

- Compliance statistics for ESC performance are rarely recorded by regulators. The statistics available for South East Queensland, Australia, confirm anecdotal reports of very low levels of implementation of best practice ESC. This is believed to be directly linked to the low levels of detection and enforcement of ESC-related offenses by local government. Greater responsibility needs to be taken by local government for the active regulation of the development industry if meaningful improvements are to be achieved.
- Design standards for ESC have in the past stifled innovation and resulted in sediment basins capable of capturing and treating only around 30% of the annual average runoff volume to the design discharge standard (50 mg/L TSS). New design standards, which are outcome-focused and not linked to any particular technology, are now appearing and are encouraging adoption of newer technology such as HES basins. Such technologies are able to achieve leaps in performance to around 80% of average annual runoff being treated to the discharge standard.
- Regulators often adopt a mix of process and outcome-based approaches to regulate ESC that creates considerable confusion among stakeholders. A clear approach should be adopted by regulators based on consideration of the resources available and particular legislative framework.
- Responsibilities between the contractors, consultants, and developers for ESC are often poorly understood and poorly defined and this confusion contributes to poor ESC outcomes. Given that elements of legal liability generally attach to all parties, it is suggested that a more cooperative and equitable approach to ESC should be adopted, and simple amendments to the contract and tendering process are suggested to achieve this.
- The conversion of construction-phase ESC measures to permanent WSUD devices has in the past created a tension between the competing priorities of early asset delivery, protecting vegetated WSUD devices from sediment loads and also protecting downstream water quality. New technologies such as HES basins allow for new ways to managing both the construction and operational phases of development, which result in fewer compromises and better overall outcomes.

## REFERENCES


Auckland Regional Council, 1999. Technical Publication 90 – Erosion & Sediment Control Guidelines for Land Disturbing Activities in the Auckland Region. http://www.aucklandcouncil.govt.nz/EN/planspoliciesprojects/reports/technicalpublications/Pages/technicalpublications51-100.aspx.

BMT-WBM, 2010. Total Watercycle Management Strategy for Moreton Bay Regional Council. https://www.moretonbay.qld.gov.au/uploadedFiles/moretonbay/development/planning/TWCM-Strategy.pdf.

Department of Infrastructure Local Government and Planning, 2017. State Planning Policy. https://planning.dilgp.qld.gov.au/planning/better-planning/state-planning.

Department of Environment and Heritage Protection, 2010. EHP Procedural Guideline: Standard Work Method for the Assessment of the Lawfulness of Releases to Waters from Construction Sites in South East Queensland EM1135. http://www.ehp.qld.gov.au/licences-permits/businessindustry/pdf/lawful-water-release-construction-sitesem1135.pdf.

eWater, 2014. MUSIC User Manual. https://wiki.ewater.org.au/display/MD6/MUSIC+Version+6+Documentation+and+Help+Home.

Gardner, T., June 30, 2010. Urban stormwater quality management (USQM) strategic actions. In: Presentation to the Erosion and Sediment Control Steering and Technical Advisory Group Meeting. North Quay, Brisbane.

Healthy Waterways Ltd., 2009. Construction and Establishment Guideline – Swales, Bioretention Systems and Wetlands. http://hlw.org.au/u/lib/mob/20141110115715_e0d7137a06637493e/ce_guidelines_v11_frontend.pdf.

Healthy Waterways Ltd., 2010. MUSIC Modelling Guidelines Version 1.0. http://hlw.org.au/u/lib/mob/20141110114128_5aed87c313f50ca3d/2010_musicmodellingguidelines_v10-025mb.pdf.

Healthy Waterways Ltd., 2014. Benchmarking Erosion and Sediment Control Performance in SEQ – 2013. http://hlw.org.au/u/lib/mob/20141110105925_befa5d35e0fb94da3/wbd_esc-benchmarking_online.pdf.

International Erosion Control Association (IECA), 2008. Best Practice Erosion and Sediment Control.

Landcom, 2004. Managing Urban Stormwater: Soils and Construction.

Leersnyder, H., Bunting, K., Parsonson, M., Stewart, C., 2016. Erosion and Sediment Control Guide for Land Disturbing Activities in the Auckland Region. Auckland Council Guideline Document GD2016/005. Prepared by Beca Ltd. and Southern Skies Environmental for Auckland Council. http://content.aucklanddesignmanual.co.nz/project-type/infrastructure/technical-guidance/Documents/GD05%20Erosion%20and%20Sediment%20Control.pdf.

McIntosh, B.S., Aryal, S., Ashbolt, S., Sheldon, F., Maheepala, S., Gardner, T., Chowdhury, R., Gardiner, R., Hartcher, M., Pagendam, D., Hodgson, G., Hodgen, M., Pelzer, L., 2013. Urbanisation and Stormwater Management in South East Queensland – Synthesis and Recommendations. Urban Water Security Research Alliance. Technical Report No. 106. http://www.urbanwateralliance.org.au/publications/UWSRA-tr106.pdf.

Rowlands, Leinster, 2014. Sediment basin performance – is it as good as you think and does it really matter?. In: Stormwater Queensland State Conference. Toowomba. 2014.

Rowlands, L., Leinster, S., 2015. Sediment basin performance – is it as good as you think and does it really matter? Stormwater Queensland Conference 2015, Toowoomba.

Russell, K., Vietz, G., Fletcher, T., 2017. Global sediment yields for urban and urbanizing watersheds. Earth-Science Reviews 168, 73–80. ISSN 0012-8252. https://doi.org/10.1016/j.earscirev.2017.04.001.

Scientific Expert Panel for Integrated Urban Water Management, 2013. Working Paper - Environmental Regulation of SPA Assessable Residential Development in South East Queensland in 2013: Implications for Improving Erosion and Sediment Control on Construction Sites.

Walker, D.I., McComb, A.J., 1992. Seagrass degradation in Australian coastal waters. Marine Pollution Bulletin 25 (5), 191–195. ISSN 0025-326X. https://doi.org/10.1016/0025-326X(92)90224-T. http://www.sciencedirect.com/science/article/pii/0025326X9290224T.

Chapter 9

# Water Harvesting Potential of WSUD Approaches

**David Hamlyn-Harris[1], Tony McAlister[2] and Peter Dillon[3,4]**

*[1]Bligh Tanner Pty Ltd., Fortitude Valley, QLD, Australia; [2]Water Technology Pty Ltd, West End, QLD, Australia; [3]Honorary Fellow, CSIRO Land and Water, Glen Osmond SA, Australia; [4]Adjunct Chair, National Centre for Groundwater Research and Training, Flinders University, SA, Australia*

## Chapter Outline



**ABSTRACT**

This chapter provides an introduction to the role of stormwater and roofwater harvesting for beneficial use as part of an integrated water sensitive urban design approach to urban development. Stormwater harvesting provides a useful alternative water source as well as significant stormwater management benefits by reducing runoff volumes and the associated pollutants. The different types of harvesting and potential uses for the water are discussed, as are the important design considerations to be addressed in developing a successful scheme. Analytical tools for assessing potential stormwater yield are identified. Other related issues that may affect scheme development are discussed, including water treatment, validation and verification, and jurisdictional and governance issues.

**Approaches to Water Sensitive Urban Design. https://doi.org/10.1016/B978-0-12-812843-5.00009-5**

**Keywords:** Guidelines; Managed aquifer recharge; Regulation; Roofwater harvesting; Storage; Stormwater harvesting; Treatment; Water balance; Water recycling.

# 9.1 INTRODUCTION

## 9.1.1 What is stormwater harvesting?

Stormwater is excess rainfall (after initial losses and ongoing infiltration) converted to runoff from land surfaces such as buildings, paved areas, roads, and open space. Roofwater is the component of stormwater that is generated from building roofs before it reaches the ground and mixes with other sources of stormwater.

Stormwater harvesting is the process of capturing and storing rainfall runoff for beneficial reuse within a household or community. Integrated into a broader WSUD strategy, stormwater harvesting provides far greater benefits than simply being a source of water supply. It reduces runoff rates and volumes and the associated export of pollutants from a catchment.

Stormwater harvesting is not new. In fact, all surface water supplies are essentially stormwater harvesting on a grand scale, and many rural communities and households around the world have relied on rainwater capture from roofs as their sole source of drinking water for centuries. In this discussion, stormwater harvesting is considered in an urban context, i.e., capturing runoff from urbanized catchments to provide a source of nonpotable, and in some cases potable, water supply, most commonly as a supplementary source to the main urban reticulated supply. Roofwater harvesting, as a subset of stormwater harvesting, is the capture of runoff from the roofs of houses and other buildings for potable and nonpotable purposes, usually undertaken privately, and typically using reticulated water supply (where available) as a backup in times of low rainfall.

As a potential water supply source, stormwater harvesting is obviously climate dependent, that is, it must rain for stormwater to be generated. This can introduce concerns about long-term reliability and highlights the importance of a reticulated supply as a backup in periods of low rainfall. As a counter to this concern, urban catchments are far more hydrologically efficient than forested or rural catchments (i.e., they generate more runoff from rainfall, especially for those small to medium events that are often more prevalent in periods of low rainfall). Hence, even under prolonged periods of low rainfall, urban areas can still generate significant quantities of runoff, which is often not the case for nonurbanized catchments. A study byv Clark et al. (2015) for an urban catchment in Salisbury, South Australia, showed that water supply security was diminished by less than the decline in mean rainfall, and that this decline was more than offset by the projected increase in impervious area. This contrasts with nearby rural catchments where the percentage decline in annual runoff was several times the percentage decline in annual rainfall. Thus, as climate change eventuates, urban areas are likely to become increasingly attractive as water supply catchments, provided that relevant water quality issues can be economically and practically addressed.

In Australia, there are numerous examples of stormwater harvesting schemes, some of which have been in existence for many years. A large number of these schemes were instituted because of pressures on more conventional potable water supplies during the recent (2001–2009) Millennium Drought in Australia. Typical examples of Australian schemes include:

- Golf course and public park harvesting from dams on local watercourses, e.g., Sydney's Centennial Park and nearby golf courses (http://www.centennialparklands.com.au/about/planning/environmental_management/water_conservation);
- Urban parkland irrigation schemes, such as Brisbane's South Bank Rain Bank (http://southbankcorporation.com.au/our-places-and-projects/completed-projects/rain-bank/);
- Supplementary supplies to improve town water security, e.g., the Blackman's Swamp Creek and Ploughman's Creek stormwater harvesting schemes, Orange, New South Wales (Vanderzalm et al., 2014b);
- Domestic and industrial water supply schemes such as Fig Tree Place, Newcastle, New South Wales (Coombes et al., 2014);
- Local alternative water sources, such as the Fitzgibbon Stormwater Harvesting (FiSH) nonpotable water supply and the Fitzgibbon Potable Roofwater Harvesting (PotaRoo) demonstration projects in Brisbane, Queensland (Queensland Department of Environment, 2014b); and
- Use of natural aquifers for stormwater storage and recovery, including confined aquifers in Adelaide (18 GL/year installed capability), Canberra and Melbourne and the unconfined superficial aquifer underlying Perth (200 GL/year) and to a much smaller extent, the Botany sand aquifer in Sydney; all predominantly used for nonpotable uses such as irrigation.

A significant international example is the Singapore stormwater reuse system, which supplies harvested stormwater for potable use (refer Section 9.2.2).

In the context of stormwater harvesting, there are several important distinctions to be made between stormwater and roofwater. These relate to catchment efficiency, water quality, and ownership. A roofwater catchment is an efficient impermeable catchment yielding around 95% of rainfall as runoff, whereas broader urban catchments may only yield 25%–40% of rainfall as runoff, depending on the amount of permeable open space. Although not contaminant free, roofwater is typically a far better-quality source than stormwater, as it is not subject to many of the pollutants originating from urban land uses, e.g., vehicle emissions, domestic animal feces, sewer leaks, etc. (see for example Chapter 3 and Duncan, 1999). As a roof catchment is invariably privately owned, roofwater harvesting can be considered as a private activity, with privately owned infrastructure (storage, treatment, and distribution) and private ownership of the water captured. Stormwater harvesting, or communal roofwater harvesting such as the PotaRoo project mentioned above, is more likely to be undertaken on a community scale, with communal provision and ownership of infrastructure. Such communal infrastructure is often easier to regulate, control, and maintain than private infrastructure. Development of a community-based scheme may also raise the question of who owns the water, i.e., the individual or the State, and who has the rights to use the harvested water and to sell it, though this is more commonly related to the interception of overland or creek flow rather than rainwater.

There are several ideal, though not mutually exclusive, prerequisites for a viable stormwater harvesting system:

- Reliable and regular rainfall;
- Sufficient catchment area with land uses that do not adversely affect water quality;
- A suitable location, such as an impoundment or weir, from which to divert or pump stormwater to a storage facility;
- A suitably sized retention storage, which could include tanks, dams, and/or aquifers;
- Appropriate water treatment to ensure that the water supplied is fit for purpose;
- Sufficient demand for the water to ensure that the captured water is reused beneficially;
- A distribution system to the point of use; and
- Customers who are prepared to regularly use the additional supply, for either nonpotable or potable uses, depending on the degree of treatment.

The requirements for roofwater harvesting are essentially the same as the above, although clearly at a different scale, with potable qualities being easier to achieve because of the better and more consistent quality of roofwater runoff. It should be noted that many water authorities in Australia recommend treatment (disinfection) of roofwater prior to potable consumption (Magyar and Ladson, 2015; Ahmed et al., 2011). There are, however, many people in Australia who use roofwater for this purpose without such treatment with no apparent adverse health impacts.

A centrally operated stormwater harvesting scheme supplying water to a diverse range of customers will also be required to operate within an appropriate governance structure that clearly defines:

- Who owns the infrastructure and accepts long-term responsibility for its continued safe operation;
- Who regulates the activity and the laws/regulations and guidelines in place to formalize that; and
- What monitoring, testing, and reporting is required to comply with water quality and public health regulatory requirements.

Finally, there is the question of jurisdictional costs and benefits, e.g., if a stormwater harvesting scheme is developed by one jurisdiction (say a water utility), but also provides benefits to a second jurisdiction (say a Catchment Management Authority or local government authority), there are opportunities for these "external" benefits to be capitalized to the advantage of the scheme developer. This introduces the concepts of "avoided costs," i.e., expenditure not required because of the alternative supply, and "value capture," i.e., gaining financial benefit from third parties indirectly benefitted by a scheme, often called an externality.

The range of stormwater harvesting systems is illustrated in Figs. 9.1–9.5.

## 9.1.2 Why harvest stormwater in the urban environment?

There are multiple reasons and multiple benefits for harvesting stormwater:

- Traditionally, roofwater may have been the only source of drinking water, or tanks may have been used to enhance otherwise unreliable or poor quality reticulated supplies. In modern industrialized societies with reliable potable supplies, the motivation for harvesting is more likely to be related to reticulated water substitution, i.e., providing an alternative water source to reduce demand on the reticulated water system. In Australia, notably in South East Queensland, this has been

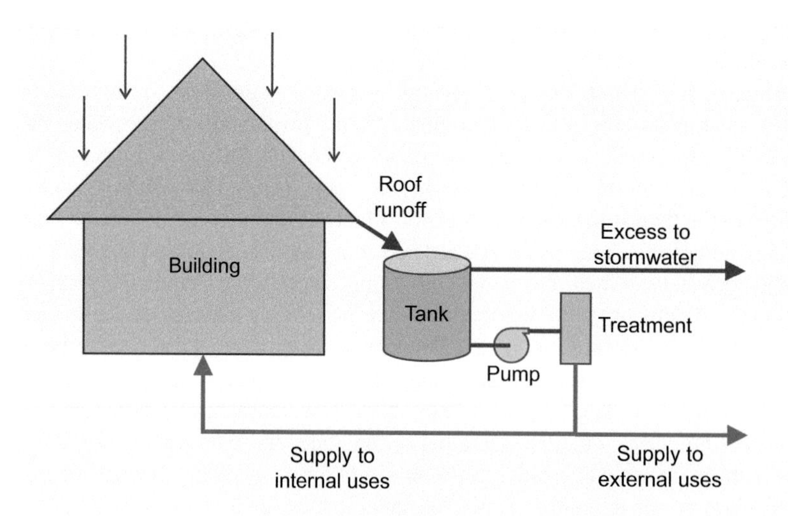


FIGURE 9.1 Household roofwater harvesting.

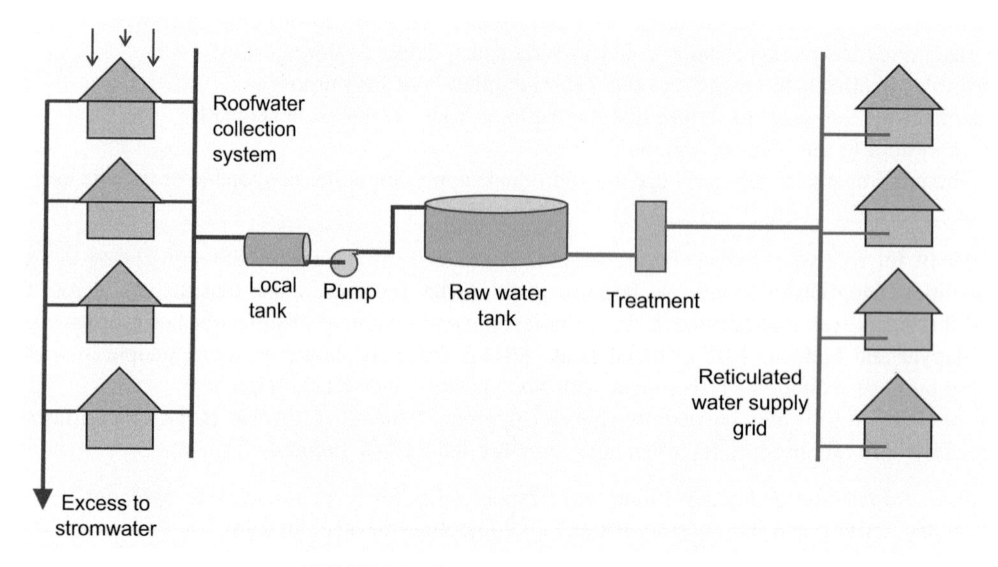


FIGURE 9.2 Communal roofwater harvesting.

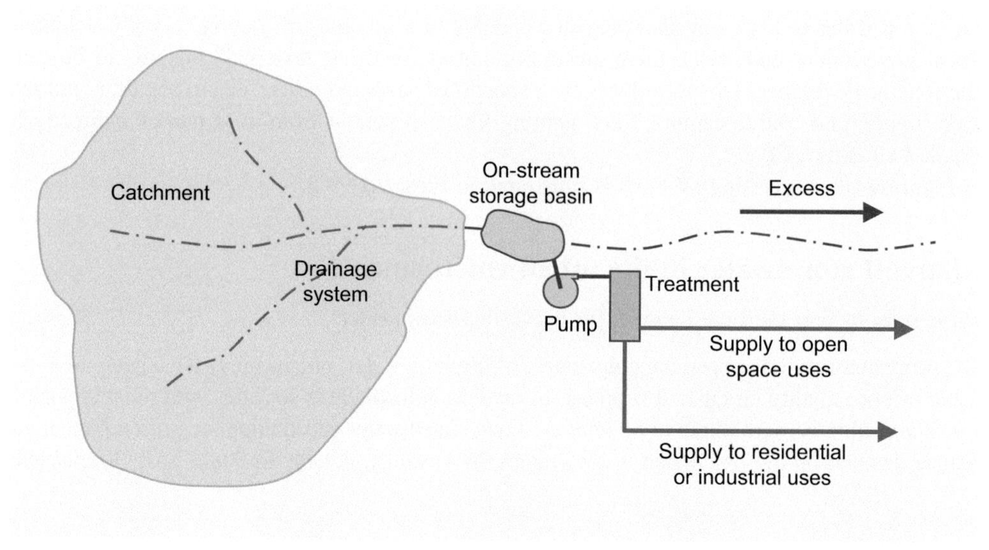


FIGURE 9.3 Stormwater harvesting on-stream storage.

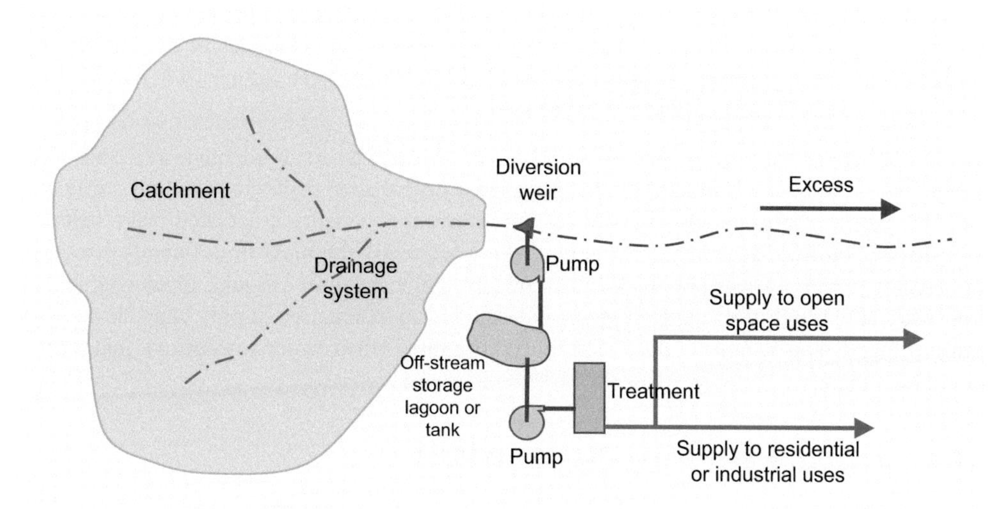


**FIGURE 9.4** Stormwater harvesting off-stream storage.

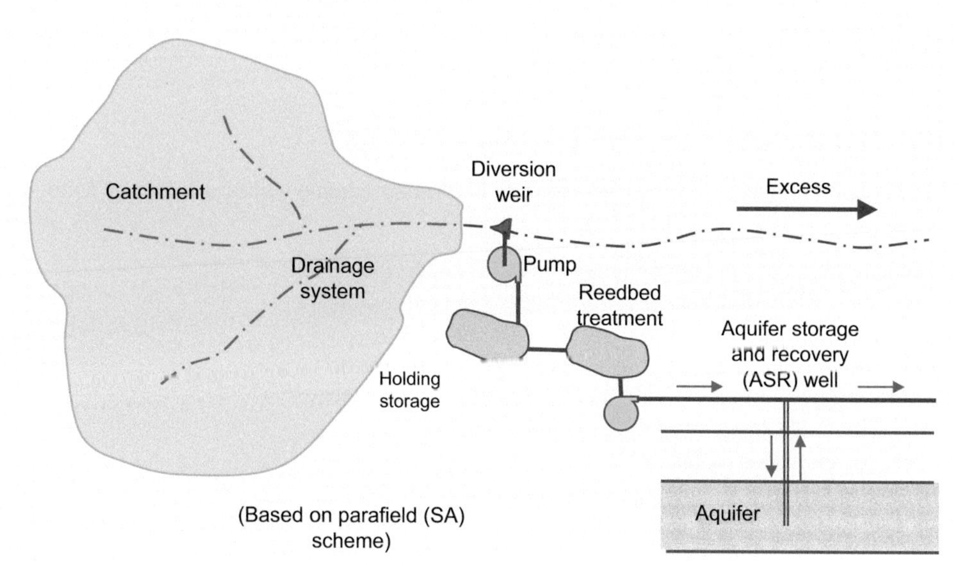


**FIGURE 9.5** Stormwater harvesting managed aquifer recharge.

used to good effect in combination with other reuse and conservation measures to significantly reduce overall water demands (refer Head, 2014; Van Dijk et al., 2013; Beal et al., 2011; Beal et al., 2012; Beal et al., 2014; Chong et al., 2011). Australian Bureau of Statistics (ABS) data show that in 2013, 34% of Australian households living in a dwelling suitable for a rainwater tank had a rainwater tank. Rainwater tanks were a more prevalent feature for households residing outside capital cities (44%) compared with those living in capital cities (28%). Around 86% of South Australian households living outside of Adelaide had a rainwater tank installed at their dwelling, followed by 56% of Victorian households living outside of Melbourne (http://www.abs.gov.au/ausstats/abs@.nsf/Lookup/4602.0.55.003main+features4Mar%202013);

- Urbanization increases the proportion of impermeable surface in a catchment, and hence the rate and volume of runoff. Stormwater harvesting takes advantage of the additional "local" water resource created by urbanization and helps to mitigate the increase in runoff, with downstream waterway stability and flood mitigation benefits;
- Another reason to harvest stormwater is to increase water self-sufficiency and community resilience. This is particularly relevant for smaller and peri-urban communities where the utilization of water from a range of sources is essential for their viability; and

- Stormwater harvesting is often driven by the need to keep public spaces green in the hot dry summers typical of southern Australia, and to conserve potable reticulated water for potable uses. Examples of this include Brisbane's South Bank Rain Bank and South Australia's Water Proofing Adelaide projects (http://www.naturalresources.sa.gov.au/adelaidemtloftyranges/water/managing-water/stormwater).

The benefits of stormwater harvesting encompass all of the above, i.e., to supplement available water supplies, for potable water substitution, to take advantage of a local resource, to enhance resilience and self-sufficiency, and to keep the urban landscape green. There are also significant environmental benefits associated with reducing the rate and volume of runoff discharged from an urban catchment, and consequently the mass of pollutants discharged to receiving waterways. Any local alternative water supply will also benefit the community through greater reliability of supply for enhanced green space and liveability; depending on the source, an alternative supply may be less subject to water restrictions during drought, and therefore allow continued irrigation when water restrictions make the use of potable water for this purpose illegal.

### 9.1.3 Relevance of stormwater harvesting to WSUD

Water sensitive urban design and stormwater harvesting have a potential symbiotic relationship. WSUD can create opportunities for harvesting in terms of improved water quality, and also opportunities for diversion and storage. Stormwater harvesting also provides benefits to WSUD through reducing the volumes of water requiring treatment. WSUD measures may also complement stormwater harvesting, for example through improved water quality from swales, bioretention filters, and wetlands, or the possible use of wetlands or detention basins as harvesting locations, or as water storages for a harvesting system. Roofwater capture and reuse in particular can significantly benefit WSUD measures through the reduction in stormwater flows at the source.

## 9.2 STORMWATER RUNOFF IN THE URBAN CONTEXT

The technical viability and volumetric yield of a stormwater harvesting scheme will be affected by a range of factors. These include:

- Catchment area and the nature of the catchment, i.e., the impervious fraction;
- Rainfall and its temporal distribution;
- Drainage systems characteristics;
- The rate at which runoff can be diverted to a storage during an event;
- Requirement to maintain minimum environmental flows in the waterway;
- The volume and nature of storage available, including aquifers;
- Potential demand for the water produced;
- The water quality required for the intended end uses; and
- Characteristics and land uses of the catchment—as these are expected to strongly influence runoff water quality.

### 9.2.1 Potential uses of harvested stormwater

In principle, harvested stormwater or roofwater could be used for any end use, subject to appropriate treatment. With modern treatment technologies, there are essentially no technical constraints to producing water of any quality. In practice however, because of factors related to cost, risk, regulatory provisions, and public perception, stormwater harvesting is often restricted to nonpotable or indirect potable uses. This said, roofwater is used extensively as a household potable water supply in Australia, especially in rural areas, and increasingly so in urban areas.

Typical nonpotable uses include:

- Public open space, playing field, and landscape irrigation;
- Garden watering;
- Toilet flushing;
- Laundry supply;
- Hot water supply;
- Firefighting;
- Environmental applications (e.g., groundwater recharge).

Examples of nonpotable use schemes include:

- The South Bank Rain Bank in Brisbane, Queensland, which captures harvested stormwater from a 30 ha urban catchment for the irrigation of public parklands and to supply water to water features (South Bank Corporation, 2012; http://southbankcorporation.com.au/our-places-and-projects/completed-projects/rain-bank/);
- Seven separate Brisbane City Council stormwater harvesting schemes that capture water via on-stream and off-stream storages, (followed in some cases by treatment in wetlands) for irrigation of sporting fields (https://www.brisbane.qld.gov.au/environment-waste/natural-environment/brisbanes-creeks-rivers/protect-our-waterways/stormwater-harvesting-reuse-projects);
- More than 30 aquifer storage and recovery systems using wells targeting confined limestone or fractured rock aquifers in the Adelaide plains and Willunga basin, South Australia; and
- Perth, Western Australia, where the majority of rooftop rainwater and stormwater infiltrates via sumps and infiltration basins into the superficial (shallow unconfined) aquifer, which is a major source of domestic and municipal irrigation supplies.

Typical indirect potable uses include the placement of harvested and potentially treated stormwater into a water supply storage (e.g., a dam or aquifer) for subsequent potable reuse.

Examples of indirect potable use schemes include:

- The City of Mount Gambier, South Australia, relies on stormwater injection wells to replenish and freshen the limestone aquifer that supplies Blue Lake, the source of the city's potable water supply (Vanderzalm et al., 2014a). Stormwater infiltration wells have recharged local aquifers since the 1880s to augment the city's current 3 GL/year drinking water supply. Recent studies and revision of water quality management show compliance with National Water Quality Management Strategy Guidelines for Water Recycling, including for potable use (Vanderzalm et al., 2014a,b);
- In Orange, New South Wales, urban stormwater harvesting was developed during the Millennium Drought (van Dijk et al., 2013) to supplement its potable water supply. Stormwater was diverted to off-stream storages, pretreated in wetlands and transferred in batch lots to Suma Park Dam (the main water supply source for Orange) to be mixed with dam water, treated to potable standard and only then reticulated to consumers. The harvesting system and water quality aspects are documented in Vanderzalm et al. (2014b). More recently, stormwater harvesting for potable use has been discontinued and the water is instead used to supplement the town's dual reticulation recycled water supply. See also http://www.orange.nsw.gov.au/site/index.cfm?display=492834, http://www.orange.nsw.gov.au/site/index.cfm?display=158554, and http://www.orange.nsw.gov.au/site/index.cfm?display=158542;
- The Warrnambool, Victoria, roofwater harvesting scheme is an indirect potable reuse system supplying roofwater into an existing raw water storage, where it is treated in the town's water supply plant and distributed to customers. The initial scheme, which had been developed as a demonstration site, harvests water from 254 homes though a 4.4 km network of gravity pipelines, saving 37 ML of water per year. The initial scheme included a trunk main that has been sized to serve a future catchment of 3300 house lots. The system is expected to have very low operating costs as it uses gravity drainage and does not include any additional pumping or extra treatment (http://www.wannonwater.com.au/2015/june/roof-water-harvesting-project-expanded-in-warrnambool.aspx); and
- Aura, on south east Queensland's Sunshine Coast, is a 50,000 person master-planned community, which commenced construction early in 2015 and which will be developed over the next 20–30 years. The 2360 ha site is located on Bells Creek and immediately upstream of Pumicestone Passage, a marine waterway of national environmental significance. The site is close to Ewen Maddock Dam, part of the South East Queensland's water supply system, thereby providing an opportunity to augment the dam supply with stormwater harvested from Aura. Constraints to development of stormwater harvesting included: environmental guidelines requiring protection of water quality in Pumicestone Passage; many environmental issues that require special management, including habitats for acid frogs, shallow groundwater tables, and freshwater wetlands; and a high level of public scrutiny and State and Federal appraisal. The techniques that are being adopted at Aura represent the benchmark for "best practice" in environmental protection for greenfield developments, and will deliver significantly enhanced water related sustainability credentials compared with conventional development (McAlister et al., 2017; Water Technology, 2016).

Examples of direct potable use schemes include:

- The Fitzgibbon Potable Roofwater Harvesting (PotaRoo) scheme (Fig. 9.6) in Brisbane, Queensland, was developed to demonstrate the concept of harvesting roofwater from private homes, reticulated to a central treatment plant, treated to potable standard, and injected into the traditional water supply reticulation grid (Queensland Department of Environment, 2014b);

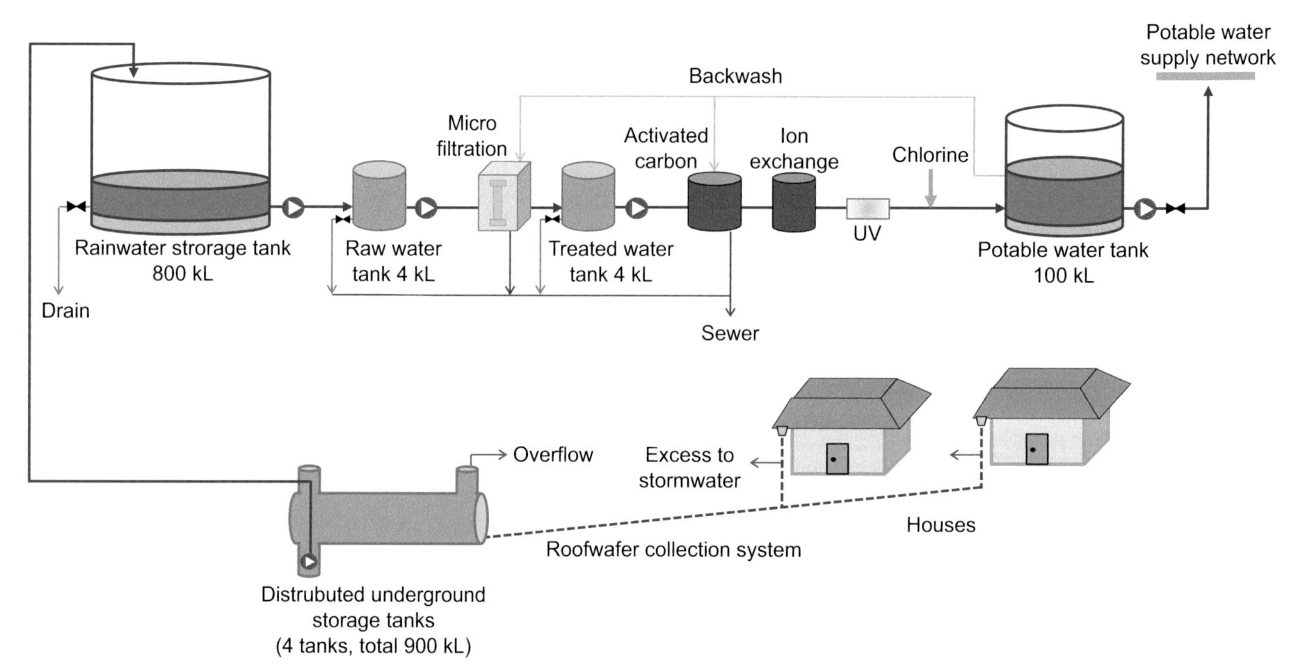


**FIGURE 9.6** Schematic of the Fitzgibbon Potable Roofwater Harvesting (PotaRoo) scheme, Brisbane, Queensland.

- Capo di Monte, a 110 dwelling retirement village at Mt Tamborine, Queensland, captures roofwater from household roofs, transfers it to a central storage, where it is treated and distributed back to the dwellings for household potable use (http://www.urbanwateralliance.org.au/publications/UWSRA-tr68.pdf and http://www.urbanwateralliance.org.au/publications/forum-2nd-2010/1-2-Steven-Cook.pdf);
- The Kalkallo water harvesting scheme, approximately 28 km north of the Melbourne CBD, Victoria, has been designed to harvest stormwater runoff from an industrial estate, with the long-term aspiration of treating it to potable water standard. The project involves harvesting stormwater from a 160 ha catchment (approximately 1 ML/day), which will be **batch treated** to a potable water standard using an advanced treatment train, tested, and proven safe before release to consumers. The treated stormwater will be used to supplement the existing potable water supply. The treatment train uses a wetland, powdered activated carbon, dissolved air flotation and filtration, granular activated carbon filters, microfiltration, advanced oxidation, and chlorination. Space has been allowed for future installation of reverse osmosis if required to achieve potable compliance. Fluoridation will also be required at that stage. Construction and commissioning of the treatment plant was completed in late 2013. However, as of 2017, the scheme had not become operational because of slower than anticipated development, and poor runoff water quality (especially turbidity) during the land development phase. Further information on the Kalkallo project can be found in McCallum (2015) and Wilson et al. (2010); and
- In Singapore, stormwater is harvested from 620 $km^2$ of catchments in the Sungei Seletar and Bedok stormwater harvesting scheme, treated with advanced processes including ozonation, and blended into Singapore's drinking water supply as described in Vanderzalm et al. (2014b). Singapore commissioned its first potable stormwater reuse schemes in 1986, having actively captured and reused urban runoff for irrigation of public green spaces since the 1970s (Lim and Lu, 2016; Fong and Nazarudeen, 1996). Two thirds of Singapore's (mostly urban) land area is now active drinking water supply catchment (Vincent et al., 2014; PUB, 2017). Urban runoff from predominantly high-density residential areas is transferred to 17 reservoirs for storage prior to treatment to potable standard. Collection and transfer methods vary across the catchments, being influenced by topography, and the stage of development in each catchment at the time the scheme was commissioned (Fong and Nazarudeen, 1996). In a recent development, the storage reservoir was situated in a central location to provide aesthetic value and a focal point for recreation. Stormwater is either gravity-fed directly into drinking water reservoirs or captured in concrete detention ponds and pumped to a reservoir. The design of diversion systems to avoid dry weather flows and first flush runoff varies across catchments, dependent on topography and the degree of tidal influence. Land use planning restrictions and community education programs have been put in place to limit pollutants entering the reservoirs (Luan, 2010). Importantly, stormwater is handled separately to sewage in Singapore (PUB, 2004). Data from the first 6 years of operation show that the quality of the raw stormwater from the Singapore catchments was similar to that of runoff from a forested catchment upstream that had

limited development (Fong and Nazarudeen, 1996). Analysis of 20 years of data showed little variability, with the quality of untreated stormwater in reservoirs generally similar to that of a protected upland reservoir in Singapore, with lower pollutant concentrations than those reported in a comprehensive review of international chemical and microbial stormwater data from 1967 to 1992 (Lim and Lu, 2016; Fletcher et al., 2014). This is expected to be an outcome somewhat unique to Singapore, which has very strict land use controls and a high degree of community compliance in terms of limiting litter and other sources of pollution.

## 9.2.2 Quantifying stormwater runoff in the urban water balance

Urbanization has a dramatic effect on catchment runoff, increasing the frequency, rate and volume of runoff. The quantum of these changes is related to factors such as development density, existing land use, soil and topographical characteristics, climate, etc. and no single definition of change can be provided (see Chapter 9 of Australian Rainfall and Runoff—http://arr.ga.gov.au/arr-guideline and also Australian Runoff Quality—https://www.eabooks.com.au/Australian-Runoff-Quality-Guide-to-Water-Sensitive-Urban-Design).

In Australia, stormwater harvesting is increasingly being seen as a valuable addition to the development and provision of a resilient set of urban water supply sources, which is considerably under utilized (The Senate - Environment and Communications References Committee, 2015).

To demonstrate each of the above key points, the Cooperative Research Centre for Water Sensitive Cities developed an urban water balance for South East Queensland (Fig. 9.7), which demonstrates that urban stormwater runoff exceeds the total volume of water used by the region (Philp et al., 2008), yet stormwater supplies less than 1% of total water use (Farooqui et al., 2016).

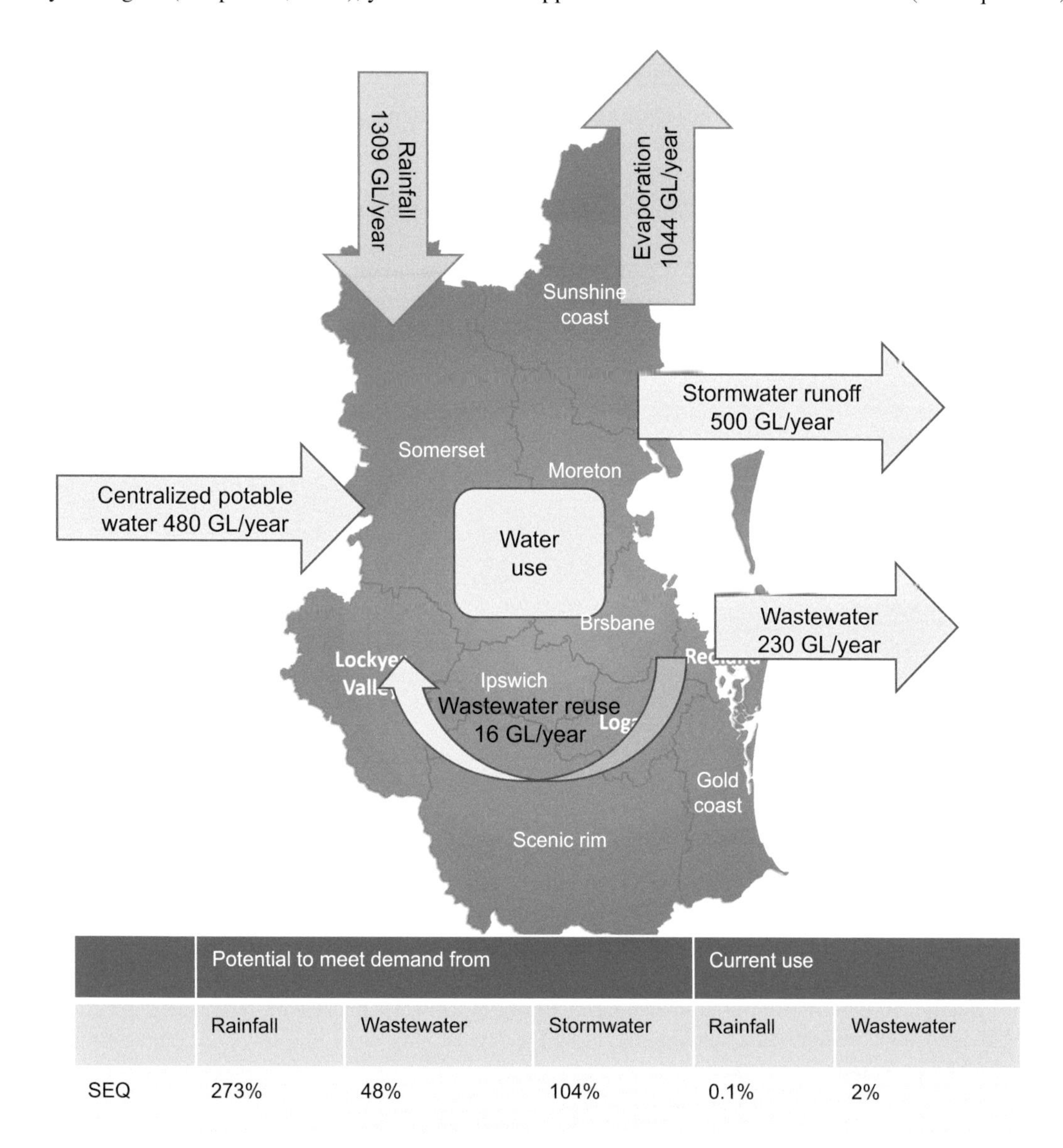


| | Potential to meet demand from | | | Current use | |
|---|---|---|---|---|---|
| | Rainfall | Wastewater | Stormwater | Rainfall | Wastewater |
| SEQ | 273% | 48% | 104% | 0.1% | 2% |

**FIGURE 9.7** Schematic of the water mass balance of urban South East Queensland. *Based on Farooqui, T.A., Renouf, M.A., Kenway, S.J., December 1, 2016. A metabolism perspective on alternative urban water servicing options using water mass balance. Water Research 106, 415–428.*

### 9.2.3 Other alternative water sources and their uses

Stormwater is not the only local alternative water source. Other sources could include recycled wastewater, recycled greywater, untreated surface water, or groundwater. Stormwater harvesting needs to be considered in the context of these other sources of water that may be available, and which in fact, may compete for the available demand. For example, there may be little value in providing stormwater harvesting if a reliable recycled water scheme supplying nonpotable water needs is already available. Similarly, a stormwater harvesting scheme servicing a community where each house already has a reasonably sized roofwater tank (say 5 kL) for appropriate domestic uses may not be viable. There may also be advantages in integrating stormwater with treated wastewater to increase harvestable volume and improve water quality or water security of separate systems, especially where large buffering storages such as aquifers or reservoirs are available.

### 9.2.4 Water quality considerations

With respect to water quality, an important consideration is that stormwater runoff is unlikely to be immediately fit for purpose, and harvested stormwater will need to be of a quality that is suitable for the intended use. Further, the desired end use water quality, and the extent of treatment required will increase as the risk of human contact or ingestion increases. Water quality and treatment is discussed further in Section 9.6, and in Chapter 2.

### 9.2.5 The significance of catchment land uses past and present

Runoff contaminants are sourced from the catchment and are a function of land uses, past and present. Examples include:

- Urbanization—sediment, organic material, hydrocarbons, organic chemicals, microbiological contamination (animal faeces or sewerage system leakage or overflows);
- Rural activities—sediment, nutrients, pesticides;
- Unsewered catchments—nutrient and microbiological contamination from poorly treated sewage effluent such as overflows from septic absorption trenches;
- Current or former industrial uses—metals, organic chemicals, hydrocarbons;
- Former use for landfill—nutrients, metals, organic chemicals, hydrocarbons;
- Presence of major roads or transport corridors—sediment, hydrocarbons, organic chemicals; and
- Fire risk—sediment or ash from fires or fire-fighting chemicals.

Catchment land uses and associated water quality impacts may prevent harvesting for certain uses. If possible, source water quality monitoring should be undertaken to identify potential contaminants of concern, which will then guide the suitability of the water for harvesting, or at least confirm treatment requirements for acceptable uses (e.g., Page et al., 2013a). In lieu of, or as a supplement to locally available data, information sources such as Chemical Hazard Assessment of Stormwater Micropollutants (Leusch et al., 2016) can be used.

It is important to consider potential catchment management opportunities to enhance water quality. Examples include:

- Implementation of WSUD systems throughout the catchment to enhance water quality prior to harvesting;
- Identification and ideally the removal of all actual or potential sewage overflows; and
- Coordination with emergency services to identify and mitigate water quality risk from industrial fires (such as chemical manufacturers/packagers) or transport spills.

## 9.3 YIELD ASSESSMENT

### 9.3.1 Scheme yield and the water balance

This section addresses the importance of scheme yield and the water balance, and the various factors that affect yield. It also introduces the range of analytical approaches available to designers. The ability to develop reliable yield estimates is an essential input to scheme design, and there is a range of techniques and tools available to evaluate potential yield.

Yield is the effective volumetric supply of usable water from a stormwater harvesting scheme for beneficial uses. Yield is important as it defines the benefits of a scheme, the basis for determining costs and revenues, and ultimately whether a scheme is worthwhile pursuing.

## 9.3.2 The water balance

The water balance is illustrated in Fig. 9.8. The mass balance considers all the inputs to a proposed scheme and balances these against the system losses and uses to determine the amount of water available to meet demand and, most importantly, the infrastructure required to optimize the water availability.

| Inputs | Outputs |
|---|---|
| • Precipitation<br>• Other supply sources, e.g., a backup water supply | • Runoff not diverted to storage (including environmental flows)<br>• Evaporation and leakage from storage<br>• Storage overflow<br>• Plant uptake and evaporation (evapotranspiration)<br>• Irrigation losses to evaporation, deep drainage, or runoff<br>• Use and discharge to wastewater |

A water balance is not a static calculation, and must consider the variability of rainfall and demand over time. Historically, water balances have been simplistic, using for example, average monthlyrainfall data to provide a coarse annual model. Such analyses do not account for the variability of rainfall within each month, and from year to year. They tend to underestimate potential demand. For example, a month with an average rainfall of 100 mm may apparently have a low irrigation water requirement. However, if the rain falls only over the first few days or last few days of the month, then a substantial irrigation demand may well occur over most of the month. Similarly, irrigation demand in a dry year is almost certainly higher than the demand in a wet year.

In Australia today, we have ready access to at least daily historical climate data for any location for periods of up to 100 years (e.g., SILO data base https://www.longpaddock.qld.gov.au/silo). In most cities, more detailed 6-min rainfall data are available (http://www.bom.gov.au/climate/data/).

Water balance modeling should be undertaken on at least a daily timescale. This can be done with relatively simple models or spreadsheets. The use of a finer timescale requires more granular data and more sophisticated analysis techniques, which are available via many contemporary modeling packages. Such analyses may be warranted when there are

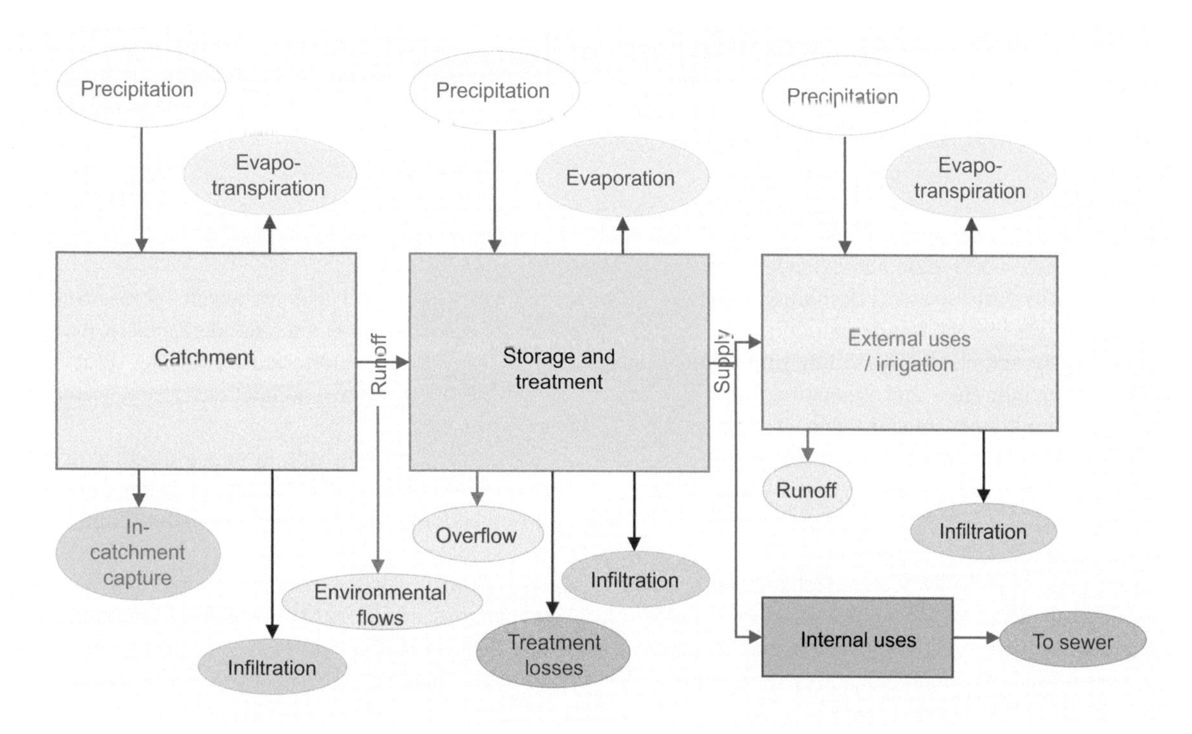


**FIGURE 9.8** Water balance schematic of a catchment considering source, treatment/storage, and reuse.

water cycle element (e.g., stormwater–water supply–wastewater) feedback loops to be considered, or dependencies in how a stormwater harvesting system may behave. For example, how does rainwater harvesting affect water supply peaking factors given that, for small tanks, intraday rainfall and demand patterns may affect yield calculations if daily averages are used? Careful attention to such issues and the use of contemporary and appropriate analysis techniques is recommended. Refer to Lucas et al. (2010) and Vieritz et al. (2015) for further discussion.

### 9.3.3 Analytical tools

Modeling is an essential activity in understanding and designing a stormwater harvesting scheme. Like all real-world situations, the complexity of the modeling tool and the techniques that are applied need to match the scale and complexity of the system being evaluated. For example, simple lot-scale evaluations may be able to be undertaken using spreadsheet water balance models, informed by appropriate water cycle configurations and data sets. By comparison, the evaluation of a "whole of suburb" system including combinations of land uses, water uses, and long-term climate cycle evaluations will require a far more complex and thorough system simulation.

We are fortunate in Australia to have a range of modeling tools that can be used to support and inform stormwater investigations, and a summary of these tools is provided below:

- **MUSIC** (http://www.ewater.org.au/products/music/)—Model for Urban Stormwater Improvement Conceptualization—MUSIC is designed to help urban stormwater professionals assess possible strategies to manage urban stormwater hydrology and diffuse sourced pollution impacts. It operates on a subdaily time step and allows sizing of WSUD devices for contaminant removal and peak flow reduction.
- **Source** (http://www.ewater.org.au/products/ewater-source/)—Source allows for scalable solutions to be developed and incorporates the impact that uncertainties in models and inputs have on the outcomes. Source is a suite of models and tools that have been incorporated into a single flexible adaptable modeling environment that recognizes the practical, political, and technical issues in developing water policy, and the need for transparency and sustainability. It can readily be customized by users to address specific local problems, or can be preconfigured for typical integrated water usage situations. Rainfall–runoff models, pollutant export models, sediment/nutrient retention models, link routing models, link decay models, and nodal process models are all selectable from a wide range of alternatives.
- **Urban Developer** (https://toolkit.ewater.org.au/Tools/Urban-Developer)—Urban Developer is a decision-support tool, based on the best available science, which allows consistent cost–benefit assessment of urban water management activities. Urban Developer allows users to improve understanding and management of the various impacts of urban water management options from the lot to the suburb scale. It allows performance assessment of integrated urban water management options across the entire urban water cycle.
- **InfoWorks** (http://www.innovyze.com/products/infoworks_ws/)—InfoWorks provides a model management solution that enables the analysis and management of catchments and water distribution network models. The combination of a relational data base, hydraulic engine, and spatial analysis tools provides a single, flexible network modeling application for both steady state and extended period dynamic simulations.
- **PURRS** (https://urbanwatercyclesolutions.com/)—Probabilistic Urban Rainwater and wastewater Reuse Simulator—PURRS utilizes climate inputs (rainfall, temperature, and potential evaporation) to simulate demands of mains water, recycled water and electricity, soil moisture, financial, health risks, stormwater runoff, and wastewater discharges. The model can include trees and vegetation, rainwater harvesting, on-site wastewater treatment and reuse, water efficient appliances, infiltration trenches, rain gardens, and on-site detention.
- **GOLDSIM** (http://www.goldsim.com/Home/)—GoldSim is a dynamic, probabilistic simulation software package. This general-purpose simulator is a hybrid of several simulation approaches, combining an extension of system dynamics with some aspects of discrete event simulation, and embedding the dynamic simulation engine within a Monte Carlo simulation framework.
- **WaterCress** (http://www.waterselect.com.au/watercress/watercress.html) –WaterCress is a simple and flexible daily hydrological model with graphics interface to allow ready comparisons of alternative harvesting configurations, climatic data sequences, and water demands to produce statistical data on stocks and flows of user-specified interest. It has been used to evaluate reliability of stormwater harvesting water supplies under climate change and urbanization (Clark et al., 2015).
- **Aquacycle** (https://toolkit.ewater.org.au/Tools/Aquacycle)—Aquacycle is a daily urban water balance model, which was developed to simulate the total urban water cycle as an integrated whole and to provide a tool for investigating

the use of locally generated stormwater and wastewater as a substitute for imported water alongside water use efficiency. The model is intended as a gaming tool rather than a design tool, giving an overall impression on the feasibility for using stormwater and wastewater at a particular site.

In addition to these existing models, the CRC for Water Sensitive Cities is in the process of developing analytical tools that should further assist the water industry in better simulating and understanding the nature and value of stormwater harvesting schemes. These modeling tools, which will be rolled out to the marketplace in coming years, are summarized below:

- **UrbanBEATS** (https://watersensitivecities.org.au/content/urbanbeats-strategic-planning-tool-for-exploring-water-sensitive-futures/)—Urban Biophysical Environments and Technologies Simulator (UrbanBEATS) is a spatial model for testing scenarios of planning, design, and placement of WSUD and stormwater harvesting measures in urban catchments under regulatory/policy constraints. The model reconstructs urban form with planning rules and assesses many possible WSUD interventions (including stormwater harvesting) to meet various targets (e.g., pollution, recycling). It provides recommended locations for suitable WSUD measures and stormwater harvesting facilities, and a possible "blueprint" on how to implement suitable strategies over time.
- **DAnCE4Water** (https://watersensitivecities.org.au/content/project-b4-1/)—Dynamic Adaptation for enabling City Evolution for Water (DAnCE4Water) aims to support planners and decision makers to use the complexity of the urban water system to their advantage, and to better understand which combination of strategies (which may include stormwater harvesting) are robust under uncertain future scenarios. DAnCE4Water is a collaborative decision-support tool that is embedded in a strategic planning process. It is anticipated that it will allow users to test the effectiveness of different water management strategies in achieving specified performance outcomes under a wide range of future climate scenarios.

Regardless of which model is used, it is imperative that due consideration be given to the data sources used in informing the model (see more below) and importantly, on the application of robust and relevant model algorithms and model parameters. Wherever possible, models should be informed by relevant model guideline documents (e.g., http://hlw.org.au/resources/documents/music-modelling-guidelines-doc-10119) or preferably, be calibrated to data from nearby urban developments such that certainty can be gauged as to the accuracy of model predictions.

### 9.3.4 Data sources

To understand and model a stormwater harvesting system, several key data sets are required. These data sets, and the likely sources of information from which they can be drawn, are discussed below:

- **Meteorological data**—Rainfall, evaporation, and temperature data are required to inform most contemporary stormwater harvesting assessments. In Australia, the majority of these data are readily available from the Bureau of Meteorology (http://www.bom.gov.au/climate/data/ and https://www.longpaddock.qld.gov.au/silo/);
- **Catchment details**—Data describing the extent, land uses, and topography of a catchment are generally readily available for most contemporary projects from State and Local Governments;
- **Land Use Characteristics**—For an existing development, these data will be able to be derived from relevant strategic planning documents. For a proposed or greenfield development, these data will need to be obtained from the project proponent;
- **Development Demographics**—For an existing development, these data will be able to be obtained from the ABS (http://www.abs.gov.au/). For a proposed or greenfield development, these data will need to be obtained from the project proponent; and
- **Water Use and Wastewater Generation Data**—For an existing development, these data may be able to be obtained from the local water/wastewater utility or authority. For a greenfield development, modeling will be required to estimate these parameters. Some end use studies have been completed that provide further information about water usage, e.g., Beal and Stewart (2013) and Thyer et al. (2008).

## 9.4 RUNOFF CAPTURE AND DIVERSION

### 9.4.1 The importance of diversion systems

A primary objective of stormwater harvesting is to capture as much water as possible, as quickly as practicable when the opportunity arises, so that it is available for use when needed. The location, type, and capacity of diversion systems are therefore a key factor in determining the success of a harvesting scheme.

Stormwater harvesting is essentially an opportunistic activity. The configuration of a scheme will very much depend on the topographic opportunities present in the catchment and adjacent landscape in question.

After capture, stormwater is transferred to a storage. If this storage is online, i.e., a retention basin within a waterway or drainage system, then no diversion is required, and full capture, up to the storage capacity, is theoretically possible. If the storage is offline, water from the waterway must be transferred into the storage at a suitable flow rate to get the best benefit from the runoff event.

The optimum rate at which water is diverted will depend on the water balance and the scheme configuration. For example, if the catchment is relatively small and reacts quickly to a rainfall event, and a high proportion of catchment runoff is to be captured, then the diversion system must have a relatively high capacity, ideally similar to the flow rate (L/s) in the waterway. Conversely, a larger catchment with a reliable baseflow allows diversion at a lower, more constant, rate.

## 9.4.2 Opportunities to look for

In identifying diversion opportunities, issues to be considered include:

- Suitable upstream catchment area;
- Accessible channel or stormwater pipe;
- An existing pit or structure suitable for use as a diversion structure, or the opportunity to construct one;
- Reliable access for operation and maintenance of the diversion structure and pumps; and
- Access to electricity supply.

Things to be avoided include sites that are subject to excessive flooding and diversion structures that create dangerous confined spaces for workers.

## 9.4.3 Design considerations

In designing a diversion, the following aspects need to be addressed:

- Design of diversion structures to avoid increasing the risk of upstream flooding;
- Provision of a pump well of sufficient storage to match pump requirements, thereby avoiding an excessive number of pump starts and stops;
- Most appropriate pump configuration, i.e., number of pumps, pumping rate;
- Need for screening/pretreatment prior to pumping, to exclude gross solids and limit pumping of sediments; and
- Requirement for water quality controls, such as an in-line oil and sediment trap to improve water quality entering the storage.

A pumped diversion system will include most, if not all, of the components illustrated in Fig. 9.9.

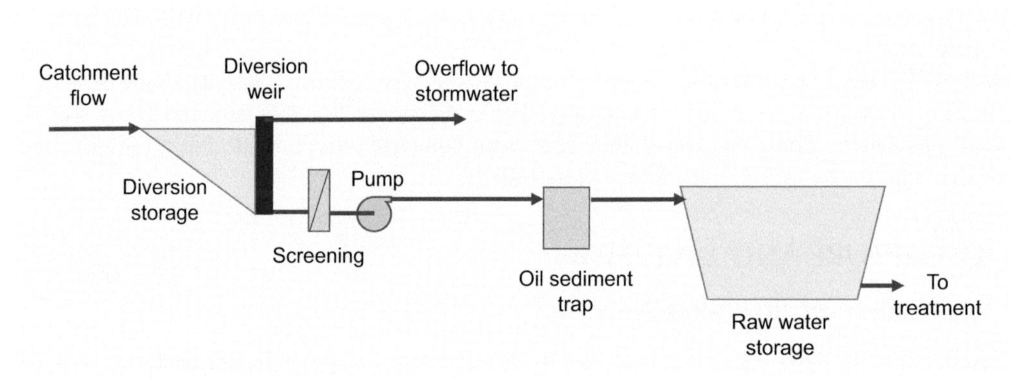


FIGURE 9.9 Schematic of a stormwater diversion, pumping, and storage system.

## 9.5 STORING HARVESTED STORMWATER

### 9.5.1 Why storage is important?

The primary objective of stormwater harvesting is to capture as much water as possible during a runoff event, so that it is available for use when needed. The location, type, and capacity of a storage system are of fundamental importance in determining the success of a harvesting scheme. Providing substantial storage in an urban environment is usually a major and costly challenge to urban water harvesting and can easily limit the capacity or financial viability of a scheme. In many cases, harvesting has only been possible where natural or man-made storage opportunities already exist, for example if a lake or a disused quarry is located in or near the catchment, or suitable aquifers exist in close proximity. In other cases, specific opportunities exist that have allowed construction of a storage to be integrated into the site landscaping or development, for example a buried storage beneath a playing field or recreation park (refer to Section 9.5.2.5).

Storages require space and, in an urban environment, the availability and value of land will be a major consideration. For example, a building owner will be reluctant to sacrifice car parking spaces for a tank, and a developer will be reluctant to dedicate saleable land for a storage lagoon. The best location will always be where it does not impact on the commercial success of a development, i.e., within land not suitable for other purposes, such as a power line easement.

### 9.5.2 Infrastructure requirements/types of storage

Experience has shown that the types and configuration of storages available are limited only by the proponent's imagination, and many different systems have been developed in recent years. Storage types can be grouped as follows:

- Natural surface storages;
- Disused quarries;
- Man-made lagoons;
- Tanks;
- Cellular systems; and
- Aquifers.

#### *9.5.2.1 Natural surface storages*

The use of an existing lake or wetland, or one created primarily for another purpose such as for aesthetic purposes, or for runoff treatment, clearly has cost advantages as a water storage, provided that this does not detract from the main purpose of the water body.

A wetland used for water storage will have limited capacity as the allowable water level variation will be small to protect plant health. An aesthetic water feature may also limit acceptable drawdown to avoid exposing muddy banks. This could be solved however by edge treatments such as vertical walls or extended edge planting (Fig. 9.10).

**FIGURE 9.10** Wetland Zone, Halpine lake, Mango Hill, Queensland. *Image: Alan Hoban.*

### 9.5.2.2 Disused quarries

Disused quarries can provide substantial storage volumes in a restricted footprint because of the available impoundment depth and the absence of plant health and aesthetic concerns that apply to other systems. Care needs to be taken to not underestimate the costs of their adaptation. For example, contaminated site remediation, stabilization of rock faces, dental concrete to seal fissures, and unforeseen environmental considerations. A good example of these complications occurred with a disused brick pit at Sydney's Olympic Park at Homebush Bay for use as a water storage (http://www.sopa.nsw.gov.au/our_park/environment/water). The pit had become the home of the endangered green and golden bell frog, thereby limiting its use. The storage also required extensive rock face stabilization to make it suitable, and safe for use (Fig. 9.11).

### 9.5.2.3 Man-made lagoons

The most cost-effective way to construct a substantial storage volume will be an earth lagoon. A typical rural earth storage, i.e., a farm dam or "turkey's nest," will be unlined and constructed with limited engineering controls. However, for urban stormwater harvesting it will be necessary to construct it to more stringent engineering standards to ensure stability and safety. The following considerations need to be taken into account:

- The need for a membrane liner to control seepage water losses and to protect water quality;
- The need for a floating membrane cover to limit evaporative losses, control algal growth, and exclude birdlife;
- Fencing to prevent unauthorized access, and exclude wildlife;
- If unfenced, low gradient batter slopes and/or edge treatments to protect public safety;
- Flow inlet arrangements;
- Emergency overflow provisions;
- Rainwater removal from the surface of floating covers;
- Water level monitoring instrumentation; and
- Water draw-off arrangements, e.g., a floating intake to minimize the risk of pumping sediment from the storage.

For equal working volumes, a lagoon will have a larger footprint than a tank because of the relatively small depth to volume ratio, and the space required for embankments and access (Figs. 9.12–9.14).

### 9.5.2.4 Tanks

Tank storages include a broad range of systems and include:

- Manufactured tanks commonly constructed of molded polyethylene or membrane-lined corrugated steel: 1–50 kL;
- Membrane-lined, bolted steel "panel" tanks: 50–1000 kL;
- Cast in-situ reinforced concrete tanks: 50–400 kL;

**FIGURE 9.11** Disused brick pit at Sydney Olympic Park, Sydney, Australia repurposed as a stormwater storage resevoir. *Image: Wikimedia Commons.*

**FIGURE 9.12** Water storage basin, Norfolk Lakes, Narangba, Queensland. *Image: Alan Hoban.*

**FIGURE 9.13** Man-made urban lake, Lake Eden, Northlakes, Queensland. *Image: Alan Hoban.*

**FIGURE 9.14** 5 ML lined and covered lagoon to store harvested stormwater at Fitzgibbon, Brisbane. *Image: David Hamlyn-Harris.*

FIGURE 9.15 3 kL household polyethylene water tank Melbourne, Victoria. *Image: David Hamlyn-Harris.*

FIGURE 9.16 800 kL membrane-lined bolted steel tank to store harvested rainwater at Fitzgibbon, Brisbane. *Image: David Hamlyn-Harris.*

- Purpose designed and constructed reinforced concrete tanks: 1–10 ML; and
- Large diameter reinforced concrete pipe sub surface storage systems, such as Humes Rainvault: 50–300 kL.

Tanks are appropriate for small-scale systems requiring a small stored volume, for medium-scale schemes where space is available for large aboveground storage, and for large schemes requiring large underground storages when surface space is not available (Figs. 9.15–9.17).

#### 9.5.2.5 Cellular systems

Cellular or modular systems comprise proprietary plastic cells, similar to large milk crates, which are placed in a membrane-lined and covered excavation, and then buried. They are most suitable for use under playing fields and landscaping that is not subject to high traffic loads.

**FIGURE 9.17** 240 kL Humes Rainvault tanks under construction, Fitzgibbon, Brisbane. *Image: David Hamlyn-Harris.*

**FIGURE 9.18** Atlantis Flo-Tank cellular storage installation. *Image: Atlantis Corporation.*

Examples of cellular systems include the Atlantis + Flo-Tank (http://atlantiscorporation.com.au/product-2/) or the Novaplus Drainwell (http://www.novaplas.com.au/products/drainwell/drainwell-stormwater-detention-underground-tanks-soakwells/) and SDS GEOlight (http://www.sdslimited.com/products/geolight/).

The benefits of cellular systems are that they allow for dual use of open space and they provide flexibility in terms of shape and configuration. On the other hand, they may not be suitable for areas with very high surface loads. The cell configuration also restricts access for maintenance and desilting (Fig. 9.18).

### 9.5.2.6 *Managed aquifer recharge*

Aquifer storage is usually substituted for tanks and off-stream storages where aquifers are suitable. Aquifer storage and recovery are usually only a small fraction of the cost of tanks, especially for large storages.

Managed aquifer recharge (MAR) is the intentional recharge of groundwater for subsequent recovery or for environmental benefit (refer to Fig. 9.5). An introduction to MAR including the variety of objectives, methods, applications, economics, and regulatory considerations is documented in Dillon et al. (2009). A range of water sources are used, and in Australia urban stormwater has been prominent with about 200 GL/year in Perth and more than 20 GL/year in Adelaide being recharged for subsequent recovery for irrigation.

MAR depends on the presence of a suitable aquifer, that is, geological strata capable of storing and transmitting water. When present, these provide large natural storages beneath cities, with minimal surface footprint.

To store stormwater in an aquifer from winter to summer, and across years, there is a need to construct access to the aquifer, which is normally as simple as sumps, infiltration basins, or wells. The rates of recharge to most aquifers through these systems are generally slower than the rate of stormwater runoff, so surface detention storage is also needed to give time for recharge to occur. The design of the system will depend on runoff rates and volumes, land available for detention systems such as wetlands, recharge method, soil, and aquifer hydraulic and geochemical properties, and water use requirements.

Suitable aquifers can play a key role in providing buffering capacity to withstand seasons and years of low rainfall, and the required characteristics of aquifers for interyear storage are more strict than for seasonal storage (Dillon, 2016).

To determine the presence of a suitable aquifer, the first step is to examine hydrogeological reports and maps, and existing well records (bore logs) that are in close proximity. Most Australian states have this information available online. If the prospects look favorable, the next step is to drill a well to the target aquifer to confirm the lithology of the profile, and the target aquifer(s) depth, thickness, composition, and confinement. A well also allows pumping tests to reveal aquifer hydraulic characteristics and enables sampling for groundwater quality. Having a qualified hydrogeologist supervise the driller and provide a report will give the specifications for the recharge well, or the dimensions of the infiltration basin, and should include the information necessary for estimating pretreatment requirements. This information will also help to predict the proportion of recharged water recoverable with acceptable quality when recharging brackish aquifers (see Ward et al., 2009).

### 9.5.3 Comparative costs of storage types

An indication of the relative costs of the various storage types is provided in Table 9.1.

## 9.6 THE NEED FOR WATER TREATMENT

### 9.6.1 Why is treatment required?

Stormwater runoff can contain a range of contaminants. As a generalization, harvested stormwater cannot be reused for urban purposes without some level of treatment, with the extent of treatment dependent on the proposed end use. Harvested roof runoff has traditionally been used for potable purposes in rural areas with little or no treatment. However, increasing awareness of the potential health risks associated with untreated roofwater has led to greater use of in-line treatment systems, including sediment filters, activated carbon filters, and ultraviolet disinfection. The use of untreated surface runoff is not acceptable except for the lowest risk uses, such as irrigation of nonfood crops, or restricted access landscaping.

**TABLE 9.1 Indicative Costs (AUD$ in 2017) and Land Area Requirements of Alternative Stormwater Storages**

| Type of Storage | Storage Size Range Costed (ML) | Unit Capital Cost of Storage[a] ($'000/ML) | Land Surface Area Required ($m^2$/ML) |
|---|---|---|---|
| Rainwater tank—polyethylene | 0.002–0.010 | 240 | 500 |
| Concrete tank—trafficable | 1–4 | 1200 | 200 |
| Precast concrete panel tank | 4–8 | 300 | 250 |
| Lined earthen dam impoundment | 4–8 | 15 | 600 |
| Large dam—gravity or concrete | 350–200,000 | 5–12 | 100–200 |
| Pond infiltration/soil aquifer treatment | 200–600 | 1.2–2.4 | 20–60[b] |
| Aquifer storage and recovery | 75–2000 | 5–12 | 1[c] |

[a]*Excluding land cost.*
[b]*For hydraulic loading rates of 17–50 m/year.*
[c]*1 $m^2$/ML for ASR system, but if detention storage is required to capture stormwater, size may be 20–100 $m^2$/ML depending on runoff from catchment and capture efficiency.*
Adapted from Dillon, P., Pavelic, P., Page, D., Beringen H., Ward, J., February 2009. Managed Aquifer Recharge: An Introduction, Waterlines Report No 13, 65 p. https://recharge.iah.org/files/2016/11/MAR_Intro-Waterlines-2009.pdf.

### 9.6.2 Risk assessment

A detailed risk assessment should be undertaken to identify the key risks associated with a proposed scheme, the control systems required to manage the risks, and the key points in the system where the risks should be managed. Risk management frameworks have been adopted as the basis for developing water supply systems in Australia, including in the *Australian Drinking Water Guidelines* (2011) and the *Australian Guidelines for Water Recycling* (2006). Other risk management frameworks that could be used for stormwater harvesting, including AS/NZS 4360:2004 *Risk Management*, ISO 31000:2009 *Risk management — principles and guidelines*, and Hazard and Critical Control Point analysis.

The Guidelines for MAR (NRMMC, EPHC, and NHMRC, 2009b) give guidance for assessing risk for any source of water, including stormwater, harvested and stored in aquifers for any type of future use. Supporting these guidelines is a compendium of case studies demonstrating risk assessments for actual projects (Page et al., 2010).

### 9.6.3 Stormwater quality

Stormwater quality is highly variable and indicative data is provided in Table 9.2 and discussed in more detail in Chapter 6. Stormwater quality is highly catchment specific, and varies during and between runoff events. In principle, the best data will be that obtained by sampling runoff from the proposed harvesting location, though event based stormwater sampling is very challenging. Nevertheless, some local data will be useful to compare with data from other similar catchments.

### 9.6.4 Treatment requirements

Treated water quality objectives must be based on factors such as the availability of guideline target values, the sensitivity of proposed uses to specific contaminants, the reliability of available information on health and environmental impacts, and the ease, accuracy, and cost of monitoring. Treated water quality requirements will be defined by the intended use for the water and the extent to which users will come into contact with the water. Table 9.3 provides indicative quality requirements for a range of different contact levels.

For nonpotable uses, the key water treatment requirements (see Fig. 9.19) could include:

- Prescreening at the diversion structure to exclude gross solids;
- Oil and sediment separation prestorage to minimize hydrocarbons and coarser sediment fractions entering the storage;
- Solids removal post storage, e.g., chemical coagulation and sedimentation to reduce suspended solids load (including algae) on the filters;
- Media filtration (sand or glass media) to reduce water turbidity to levels suitable for effective disinfection (NTU $< 1$);
- Activated carbon filtration to remove residual amounts of organic chemicals and hydrocarbons;
- Primary disinfection using chlorine (liquid sodium hypochlorite) (organic carbon sensitive); and
- Second barrier disinfection using ultraviolet (UV) irradiation (turbidity sensitive).

For potable uses, the key water quality and treatment requirements will include all of the above processes for non-potable use with all or some of the following additional requirements:

- Use of membrane filtration (micro- or ultrafiltration) to improve solids/turbidity removal;
- Use of advanced oxidation processes (i.e., hydrogen peroxide + UV) to remove organic chemicals; and
- Reverse osmosis treatment to reduce dissolved solids (salinity) and remove high molecular weight compounds.

### 9.6.5 Treatment systems

In Australia, there are specialist water treatment manufacturers and suppliers who will be capable of delivering a stormwater treatment system to required specifications. The important consideration when procuring a plant will be to clearly define the minimum functional and performance requirements for the system, and the responsibilities of the parties if the specified requirements are not met. Requirements could include total plant throughput and/or total treated water produced (ML/day), minimum water quality requirements, and plant redundancy provisions.

### 9.6.6 Validation and verification

Validation is the upfront process of confirming that a treatment process is capable of meeting specified and claimed treatment requirements. For example, if a plant requires 3-log reduction of *Cryptosporidium* from the filtration system, the

**TABLE 9.2 Typical Untreated Urban Stormwater Quality Compared to the Australian Guidelines for Water Recycling (Phase 2) Values for Stormwater Harvesting and Reuse**

| | | Reported Concentrations | |
|---|---|---|---|
| **Constituent in Raw Stormwater** | **Units** | **Various Sources[a]** | **Australian Guidelines[b]** |
| **Pathogens (Bacteria)** | | | |
| Coliforms | cfu/100 mL | – | $3400–3.6 \times 10^5$ |
| *Escherichia coli* | cfu/100 mL | $100–10^6$ | $3800–1.8 \times 10^5$ |
| Thermotolerant (faecal) coliforms | cfu/100 mL | $0–10^5$ | $4700–2.1 \times 10^5$ |
| Faecal streptococci | cfu/100 mL | $100–10^5$ | $3800–7.1 \times 10^4$ |
| Enterococci | cfu/100 mL | $100–10^5$ | $1600–3.4 \times 10^4$ |
| Salmonella | cfu/100 mL | 0–10 | – |
| *Clostridium perfringens* | cfu/100 mL | $100–10^4$ | $1.0 \times 10^2–2.7 \times 10^3$ |
| *Campylobacter* | cfu/100 mL | 0–10 | 1.0–7.0 |
| Pathogens (Viruses) | per 100 mL | 10–1000 | – |
| Enteroviruses | cfu/100 mL | 10–100 | – |
| Adenoviruses | cfu/100 mL | 10–100 | – |
| Rotavirus | pfu/100 mL | – | 12 |
| Somatic Coliphage | pfu/100 mL | $10–10^5$ | $1200–5.5 \times 10^4$ |
| F-RNA bacteriophage | pfu/100 mL | 0–100 | – |
| **Pathogens (Protozoa)** | | | |
| *Cryptosporidium* | cfu/10 L | – | 12–550 |
| *Giardia* | cfu/10 L | – | 0.12–5.6 |
| **General Parameters** | | | |
| Total organic carbon (TOC) | mg/L | 13–40 | 12–23 |
| Electrical conductivity | SmS/cm | 74–2900 | – |
| Total dissolved solids | mg/L | 44–210 | 110–170 |
| Suspended solids | mg/L | 5.0–180 | 19–250 |
| Turbidity | NTU | 12–370 | 8–130 |
| pH | | 6.3–8.5 | 5.5–7.3 |
| **Hydrocarbons** | | | |
| Oil and grease | mg/L | 0.11–100 | 3.4–28 |
| PAHs | mg/L | 0.2–4.4 | 0.017–0.81 |
| **Nutrients** | | | |
| Total nitrogen | mg N/L | 0.4–33 | 0.62–7.5 |
| Total phosphorus | mg P/L | 0.034–8.8 | 0.075–1.3 |
| **Metals/Metalloids/Halides** | | | |
| Total aluminum | mg/L | 0.1–4.9 | 0.49–2.3 |
| Total chromium | mg/L | 0.06–0.42 | 0.002–0.017 |
| Total iron | mg/L | 0.08–9 | 1.1–5.1 |
| Total lead | mg/L | 0–0.53 | 0.17–0.16 |
| Total zinc | mg/L | 0.09–5.8 | 0.08–0.57 |

*Continued*

**TABLE 9.2** Typical Untreated Urban Stormwater Quality Compared to the Australian Guidelines for Water Recycling (Phase 2) Values for Stormwater Harvesting and Reuse—cont'd

| Constituent in Raw Stormwater | Units | Reported Concentrations | |
|---|---|---|---|
| | | Various Sources[a] | Australian Guidelines[b] |
| **Pesticides[c]** | | | |
| BHC | mg/L | $2.0 \times 10^{-5}$ | – |
| PCB-1254 (industrial runoff) | mg/L | $6.3 \times 10^{-4}$ | – |
| PCB-1260 (industrial runoff) | mg/L | $4.4 \times 10^{-4}$ | – |
| Aldrin (OC) | mg/L | – | – |
| Chlordane | mg/L | $1.5 \times 10^{-5}$-0.0017 | – |
| Dieldrin (OC) | mg/L | $2.0 \times 10^{-6}$–$6.0 \times 10^{-6}$ | – |
| Endrin (OC) | mg/L | $4.5 \times 10^{-5}$ | – |
| Methoxychlor (OC) | mg/L | $2.0 \times 10^{-5}$ | – |
| *N*-Nitrosodimethylamine (NDMA/DMN) | mg/L | 0.003 | – |

[a]*Data ranges sourced from: Monash University (2006a,b), CRC for Catchment Hydrology (1999), DEC NSW (2006) and Burton and Pitt (2002).*
[b]*Data sourced from NRMMC, EPHC, and NHMRC, 2009a.*
[c]*OC, organochlorine pesticides; OP, organophosphorus pesticides;* HGCMS, *herbicides by GCMS;* HLCMS, *herbicides by LCMS;* OTHER, *other pesticides;* SP, *synthetic pyrethroids;* PH, *phenoxyacid herbicides;* SURR, *surrogate.*

**TABLE 9.3** Treated Water Quality Requirements as a Function of End Uses

| Use | Log Reductions | Water Quality Criteria | Preventive Measures |
|---|---|---|---|
| **Potable[a]** | | | |
| Potable consumption; swimming pools | Not specified | Risk based approach adopted to determine project specific health-based targets, includes microbiological and chemical/pharmaceutical health risks | Multiple barrier approach |
| **Nonpotable[b]** | | | |
| Dual reticulation with indoor use; commercial food crops | Virus—2.4<br>Protozoa—1.9<br>Bacteria—2.4 | Turbidity <25 NTU (median) & <10 NTU (95th %ile); <2 NTU (target)<br>*Escherichia coli* <1/100 mL | Strengthened cross-connection controls with potable water supply |
| Municipal use—unrestricted access | Virus—1.3<br>Protozoa—0.8<br>Bacteria—1.3 | Turbidity <25 NTU (median) & <100 NTU (95th %ile)<br>*E. coli* <10/100 mL | |
| Municipal use—restricted access | Virus—1.3<br>Protozoa—0.8<br>Bacteria—1.3 | | Public access restricted during irrigation; buffer distances |
| Municipal use—drip irrigation | Virus—1.3<br>Protozoa—0.8<br>Bacteria—1.3 | | Drip irrigation |

[a]*Australian Guidelines for Water Recycling: Augmentation of Drinking Water Supplies (EPHC, NHMRC, and NRMMC, 2008).*
[b]*Australian Guidelines for Water Recycling (Phase 2): Stormwater Harvesting and Reuse—(NRMMC, EPHC, and NHMRC, 2009a).*

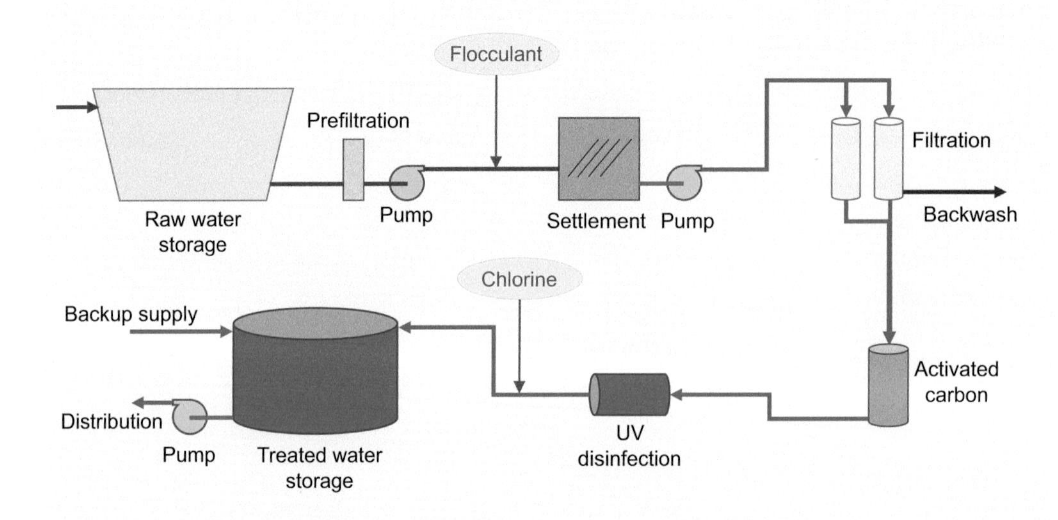


FIGURE 9.19 Indicative treatment system requirements for nonpotable uses.

proposed equipment must be independently challenge tested to demonstrate the log reduction prior to the water being distributed to users. Validation is required before equipment is accepted for incorporation into the plant design.

Verification is the ongoing process of monitoring treated water quality to confirm continuing compliance with specified requirements. This would normally be undertaken using a combination of online instrumentation and laboratory analysis of samples, depending on the parameters of interest.

For more detailed information refer to New South Wales Dept. Primary Industries, Office of Water (2015) and Victorian Department of Health (2013).

## 9.7 DISTRIBUTION AND PLUMBING

A significant and often underestimated challenge of any local alternative water scheme is distribution, i.e., delivery of the water from the water treatment plant to the customer or point of use. At its simplest expression, a golf course or park using captured water for landscape irrigation might supply water directly into their irrigation system. At its most complex, distribution of nonpotable water to an urban community would necessitate a dual reticulation network (i.e., 3rd pipe), as well as separate internal household plumbing to service toilet flushing, etc.

The design requirements for a dual reticulation system will be similar to the conventional potable water system in terms of infrastructure alignment, valving and supply pressures, and meeting similar levels of service regarding reliability and continuity of supply. A large system will also require service reservoirs to maintain system pressure. A small system on the other hand is likely to adopt a pumped pressurized system using centrally located, variable speed pumps.

The cost of a dual reticulation network may be "lost" in the overall costs of the urban development. Nevertheless, it will be significant and may be higher than many of the other water/sewerage components. An example of this is the FiSH scheme, at Fitzgibbon, Brisbane, Australia, ( see Section 9.1.1 and Queensland Department of Environment, 2014b) which cost $AUD3.2M for the diversion, storage, and treatment, plus an additional $AUD2.9M for the reticulation system to 1500 homes, i.e., around $4100 per home (Queensland Department of Environment, 2014a). This does not include the marginal cost of providing dual plumbing to all the houses serviced by the nonpotable supply. This is similar to the cost of installing a 3−5 kL rainwater tank, including the tank, pressure pump, power supply, and plumbing, though at Fitzgibbon this would not have been an option because of the very small lot sizes.

The cost of retrofitting an existing suburb will be even higher because of the cost of working within operating road corridors, avoiding other services, and site restoration. These costs are likely to preclude its adoption as a supply to individual houses. In addition, there would be significant costs to the individual to modify existing household plumbing to accommodate the supply. On the other hand, retrofitting to service parks, gardens, or sporting fields where there are far fewer individual connections and less constrained utility alignments may be viable. A successful example of this is the Adelaide GAP project (Greening Adelaide Project), which supplies 3.8 GL/year of treated recycled water to the City of

Adelaide, South Australia, and its surrounds. The project, which required construction of a new recycled water treatment plant adjacent to the existing wastewater treatment plant at Glenelg, transfers 35 ML/day of treated wastewater to irrigate the parklands and maintain gardens of Adelaide (http://www.cityofadelaide.com.au/planning-development/sustainable-adelaide/water/).

The Managed Aquifer Recharge and Stormwater Use Options (MARSUO) study in Salisbury, SA, Dandy et al. (2013) found that public open space irrigation was the cheapest option for stormwater harvesting, closely followed by additional treatment and water quality management for augmenting drinking water supplies through existing water mains. Third pipe systems for household water supplies were considerably more expensive even in new subdivisions. It was the most expensive option for brownfield development, i.e., retrofitting to existing suburban areas.

Arguments against dual reticulation often focus on the cost and the potential for the system to become a "stranded asset." In this context, the assets could become stranded if either an alternative delivery system were found or, for some reason, the scheme ceased to operate. A simple consideration of costs suggests that the cheapest approach to distribution would be to treat the stormwater to a higher standard and inject directly into the potable water supply network making the dual reticulation redundant, i.e., direct potable reuse as described in Khan (2013). The argument is that the marginal cost of treatment is likely to be small compared with the cost of the dual reticulation. This would still not be without its own considerable costs (and risks) associated with water treatment validation, and ongoing verification and compliance. However, it could be seen as the ultimate outcome for the use of local alternative water sources within an integrated water management system, i.e., a large number of small systems feeding into the water grid, providing increased capacity and drought resilience.

The risk of a small-scale scheme failing for some reason, and ceasing operation, is an argument often used by water utilities against adoption of local, alternative water supplies. Their view is that, regardless of the question of ownership and legal responsibilities, the utility may become the "service provider of last resort" and end up carrying the supply responsibility and associated costs.

## 9.8 OPPORTUNITIES TO INTEGRATE WATER HARVESTING INTO WSUD SYSTEMS

### 9.8.1 What are the opportunities?

WSUD recommends the collection, storage, treatment, and reuse of urban stormwater. Runoff is collected and treated in devices such as rain gardens, bioretention systems, lakes, and wetlands, the "building blocks" of the stormwater treatment network, which is the fundamental "core" of a WSUD system. Refer to Chapters 2 and 4 for further detail. There are numerous opportunities for stormwater harvesting to interface with these building blocks of a WSUD system due to the multiplicity of storages provided by them. The water quality improvements associated with these actions are also important in delivering stormwater, which may be fit for purpose for various potential reuse opportunities.

### 9.8.2 Benefits of water harvesting in the WSUD context

Key issues that WSUD systems typically aim to address relate to the rate and volume of stormwater runoff generation and the contaminants that are present within this stormwater. Regarding these issues, the key benefits of water harvesting in a WSUD context can be summarized as:

- At-source stormwater harvesting (rainwater tanks) reduces the volume and rates of stormwater generation, making downstream water treatment both more efficient and effective, as there is less stormwater to deal with, and the runoff rates are lower. This key opportunity has been realized at the Aura project on the Sunshine Coast, south east Queensland, (McAlister et al., 2017) where, for this reason, rainwater tanks (5 kL for detached dwelling and 3 kL for attached dwellings) have been mandated for all 25,000 dwellings at build out;
- Implicit in capturing the downstream benefits delivered by rainwater tanks is the need to regularly draw down the stored water, such that rainwater tanks have storage capacity for the next storm event. A key tenet of WSUD is to reduce the pressures on centralized supplies, and hence lot-scale stormwater harvesting adds value in this regard. Again, in the context of the Aura project, to achieve this key attribute, it has been mandated that all dwellings have their roofwater tanks plumbed into the house to supply toilet flushing, laundry, hot water, and external uses;
- Where there are unconfined aquifers containing fresh groundwater (such as in Perth), recharge takes place directly from the WSUD device, and the aquifer provides the storage and distribution system for subsequent use for suburban irrigation supplies;

- Where there are confined aquifers, even brackish ones (e.g., in Adelaide), these can be used for storage and recovery for local use to support summer irrigation for sporting and amenity fields, and to maintain green infrastructure during dry periods;
- Roofwater often contains moderate levels of nitrogen due to atmospheric deposition. Water harvesting and reuse can capture this material and prevent this nitrogen from affecting sensitive downstream environments, though this will vary from location to location due to rainfall, and also depend on factors such as how much of the roof runoff is captured and used; and
- Ensuring that environmental flow and downstream habitats are protected. For example in the Aura project, stormwater harvesting was required to ensure downstream RAMSAR wetlands (linked to Pumicestone Passage) were not affected by major changes in salinity range or environmental flows (McAlister et al., 2017).

### 9.8.3 Hydrological impacts of WSUD

The impacts of water sensitive urban design on groundwater are frequently overlooked where recharge is not a primary objective, but may consequently result in unexpected harm. For example, in making urban catchments more permeable via permeable pavements and water detention that increases infiltration, groundwater levels may rise and impact on underground infrastructure, or cause waterlogging or down slope salinization of soil. It is recommended that groundwater studies are embedded in water sensitive urban design to ensure that WSUD methods used are appropriate for the hydrogeological conditions. There will be different palettes of solutions where water tables are shallow or deep, fresh or saline. Disposing of water to aquifers is not MAR. Rather, the risks should be assessed, and the MAR Guidelines (NRMMC, EPHC, and NHMRC, 2009b) provide a framework for such assessment. They help guide the possibilities for the recovery and safe use, or environmental benefit, of intentionally infiltrated water.

## 9.9 JURISDICTIONAL AND GOVERNANCE CONSIDERATIONS

### 9.9.1 Why is this important?

A significant challenge for any alternative water source relates to governance questions, i.e., who owns the system, who is responsible for its management, who regulates the scheme, and who is the responsible entity for compliance and long-term liability obligations? There is also the question of “who pays” for the scheme. With today’s technologies, there is always a technical solution to water harvesting challenges, and, subject to cost, there is nearly always a way to make any water suitable for any use. So, it is technically straightforward to create a scheme. However, in the risk-averse world of large-scale water systems, serious challenges arise in finding a willing long-term owner, as well as defining the operational rules that are both commercially viable, and acceptable to the regulator.

A similar type of question applies to the different jurisdictions that may be involved in the development of stormwater harvesting, or those that may benefit from it. Challenges arise when an investment within one jurisdiction leads to savings in another, and the ability of the investing agency to capture tangible benefits from that saving.

### 9.9.2 Ownership and operation

Ideally, any local alternative water scheme could be owned and operated by a water utility as part of its integrated water management strategy. However, experience in Australia shows that the interest in this type of approach is greatest in periods of drought when there is uncertainty about the security of traditional water supplies. Alternative water sources are then actively sought to increase overall system capacity. Historically, this interest wanes quickly when drought is replaced by floods, as is typically experienced in Australia (Head, 2010). When reservoir storage levels are high, local systems are more likely to be privately developed, and operated at an individual development scale. Alternatively, they are developed for a specific purpose such as irrigation of playing fields.

### 9.9.3 Regulation and guidelines

In Australia, the National Water Quality Management Strategy has generated the Australian Guidelines for Water Recycling, including guidelines for augmentation of drinking water supplies, stormwater harvesting, and MAR—refer to Table 9.4. Most states adopt these guidelines, but also have their own specific local requirements.

Some states also regulate requirements for recycling, for example the Queensland Public Health Regulation (2005), (www.legislation.qld.gov.au/view/pdf/2017-06-30/sl-2005-0281) which regulates the minimum water quality requirements for wastewater recycling, including potable and nonpotable uses.

**TABLE 9.4 Guidelines Relevant to Stormwater Harvesting and Reuse Applications**

| | Application | | | | | |
|---|---|---|---|---|---|---|
| **Guideline** | **Potable Consumption** | **Toilet Flushing** | **General External Wash-Down** | **Swimming Pool Filter Backwash and Top Up** | **Public Space Irrigation** | **Refilling Water Features** |
| Australian Guidelines for Water Recycling (Phase 1) (NRMMC, EPHC, and AHMC, 2006) | √ | √ | √ | √ | √ | √ |
| Australian Guidelines for Water Recycling: Augmentation of Drinking Water Supplies (Phase 2) (EPHC, NHMRC, and NRMMC, 2008) | √ | √ | √ | √ | × | × |
| Australian Guidelines for Water Recycling: Managing Health and Environmental Risks (Phase 2): Stormwater Harvesting and Reuse (NRMMC, EPHC, and NHMRC, 2009a) | × | √ | √ | x | √ | √ |
| Australian Guidelines for Water Recycling: Managing Health and Environmental Risks (Phase 2): Managed Aquifer Recharge (NRMMC, EPHC, and NHMRC, 2009b) | √ | √ | √ | √ | √ | √ |
| Managing Urban Stormwater: Harvesting and Reuse (DEC NSW, 2006) | √ | √ | √ | × | √ | √ |
| Australian and New Zealand Guidelines for Fresh and Marine Water Quality (ANZECC, ARM-CANZ, 2000) | × | × | × | √ | √ | √ |
| Australian Drinking Water Guidelines 2011 | √ | √ | √ | × | × | × |

The *Australian Guidelines for Water Recycling: Stormwater Harvesting and Reuse—Phase 2* (NRMMC, EPHC, and NHMRC, 2009a) is the most relevant reference guideline for specific water quality targets for stormwater intended for nonpotable uses.

Note that although recycled water is regulated, this generally does not include the actual stormwater harvesting. This raises the issue of risk for service providers, and they often prefer to use an alternative water source, such as recycled wastewater, where ownership and regulatory requirements are clearly defined. If stormwater is used where there are no regulated requirements, more risk and legal responsibility is placed on the provider to meet reasonable standards. As a generalization, investors in large, capital-intensive urban developments go to considerable lengths to minimize their risks, especially from those factors beyond their immediate control.

## 9.9.4 Regulation of managed aquifer recharge

Water quality considerations need to address protection of groundwater quality, providing a supply that meets the requirements for the intended end use (e.g., drinking or irrigation) and ensuring the sustainability of the recharge operation, that is, avoiding clogging.

Australian Guidelines for MAR (NRMMC, EPHC, and NHMRC, 2009b) (refer Table 9.4) provide the instructions that, if followed, will satisfy the human and environmental health considerations of the MAR project. In some states

(Victoria, Western Australia, and South Australia), these are enshrined in state policies and regulations (Victoria Environment Protection Authority, 2009; WA Department of Water, 2011; South Australia Department of Environment Water and Natural Resources, 2014; South Australia Environment Protection Authority, 2015). In other states adherence to the national guidelines is regarded as "deemed to comply" with state requirements, as all state and territory governments, as well as the Commonwealth, signed off on these Guidelines under the National Water Quality Management Strategy (http://www.agriculture.gov.au/water/quality/nwqms). The Guidelines lead the proponent through a staged risk assessment to form a risk management plan for their operation that includes preventive measures such as treatment requirements.

A series of example risk assessments for MAR case studies, including for stormwater harvesting, are documented by Page et al. (2010).

A framework for water entitlements for MAR operations was developed by Ward and Dillon (2011) based on Australia's National Water Initiative (http://agriculture.gov.au/water/policy/nwi). This utilizes the robust separation of processes for setting equitable shares in a resource, and the periodic allocation of water in response to changes in the available resource. This is applied to entitlements to take water for MAR, to recharge aquifers, to recover water from aquifers, and for end uses of recovered water.

In all states of Australia, there are existing policies for taking natural waters from streams that are based on catchment water allocation plans. It is only in the Australian Capital Territory that entitlements to urban stormwater are determined in the context of environmental flows (ACT Government, 2006, 2014). In other states, stormwater is regarded as a local government issue, and until recently, there has been little competition to harvest it. There also seems to be little effort to account for urban stormwater in sustaining environmental flows downstream.

Until recently, competition between multiple aquifer storage and recovery sites has not occurred, and states have been slow to adopt principles that address the issue of cumulative impacts of recharge operations. However, the right to recover water produced through MAR operations is the key uncertainty issue affecting investment by MAR proponents, and certainly needs to be addressed. In some states, such as South Australia, transferability of these recovery rights is established on a case-by-case basis. However, the principles laid out by Ward and Dillon (2011) that address entitlements to harvesting of urban stormwater are not yet enshrined in water resources management policy, such as cap and trade, and are not yet on the horizon for less developed water resources.

The MAR Guidelines address constraints on pressures, flow rates, and volumes to avoid adverse impacts on aquifers and existing groundwater users. There is of course some dependency between the MAR Guidelines and entitlement issues, and the Australian MAR Guidelines adopt the approach that the possibility for entitlements to be affected must be addressed before submitting risk assessments relating to human health and environment protection.

### 9.9.5 Compliance and liability

Notwithstanding that stormwater harvesting is largely unregulated in Australia, the challenge of regulation is to ensure that (1) a scheme meets minimum requirements in terms of water quality, the protection of public health and the environment; and (2) the cost of complying with regulations is not so large that it makes all but the largest schemes unviable.

The burden of compliance can be time-consuming and expensive. This includes responsibility for meeting minimum performance requirements; validation of treatment processes to ensure suitability to meet water quality requirements; ongoing verification testing to confirm compliance with water quality requirements; system auditing to confirm no cross-connections with the potable water supply; impact monitoring of soils and groundwater; complaints management; environmental reporting; and external auditing. An example is the Pimpama Coomera water recycling scheme established under the Pimpama Coomera Waterfutures project (http://www.goldcoast.qld.gov.au/pimpama-coomera-recycled-water-master-plan-public-reports-8063.html), which has been mothballed because of high operating and compliance costs.

As stormwater harvesting is not regulated in Australia (aside from via MAR), there is potential to minimize these costs, notwithstanding the proponent's legal duty of care to protect community health.

Examples of stormwater management plans for existing stormwater harvesting systems involving aquifers for non-potable and potable use are given by Page et al. (2013b) and Vanderzalm et al. (2014a). An example of an audit report of the stormwater harvesting system against its plan (for nonpotable use) is given by Stevens (2014).

### 9.9.6 Who pays?

Establishing new small water schemes, including stormwater harvesting, can be expensive and full cost recovery from the users may be difficult to achieve.

The costs for stormwater harvesting using MAR are explored to an extent in Dandy et al. (2013) for a local council and a state-owned water utility that have different drivers and objectives. For large-scale irrigation or industrial use the levelized cost to local governments of stormwater harvesting with aquifer storage and recovery is less than one-third of the price of mains water. A study in Adelaide has also shown that the cost of improving stormwater quality for drinking water supplies following aquifer storage is about half the cost of seawater desalination (Dillon et al., 2014).

Subsidy grant funding has been available in the past to help offset capital costs. For example, the AUD$200M Australian Federal Government's Water for the Future—National Urban Water and Desalination Plan: Stormwater Harvesting and Reuse Projects Fund (http://www.agriculture.gov.au/water/urban/completed-programmes/national-urban-water-and-desalination-plan), established during the Millennium Drought of 2001–09, supported 52 stormwater reuse demonstration schemes. Scheme establishment costs are best considered as part of a mix of integrated water management measures providing additional capacity and supply security, thereby allowing deferral of other major capital-intensive infrastructure such as dams and seawater desalination plants.

Allocating the benefits accruing to different parties or jurisdictions from stormwater harvesting can be difficult, i.e., how can the developer of a scheme that provides benefits to a third party receive some financial benefit to help offset his costs?

An example of this could be a stormwater harvesting scheme being developed by a water utility in a particular catchment to provide an alternative water source. In so doing, some water and pollutants are removed from the catchment, thereby providing an environmental benefit to a developer who is subdividing urban land within the same catchment. The developer is required by development approval conditions to manage the flow rate and quality of the stormwater discharged from his land. How can the developer take advantage of the utility's stormwater harvesting in determining the design requirements, and how can the utility get some financial benefit from the consequential environmental benefits of its scheme? If the developer installs flow detention and treatment systems to meet the full environmental requirements, the environmental benefits of both schemes will exceed that required to protect the environment.

Establishing a market for trading of environmental credits may be one solution to this question. Some jurisdictions already allow developers to purchase environmental credits in place of undertaking specific works. For example in Victoria, Melbourne Water charges AUD$15/kg of N if a developer chooses not to install their own WSUD devices; the levy is used to build central stormwater treatment facilities.

## 9.10 CONCLUSION

Stormwater runoff in urban areas represents a very substantial potential water resource, equal for example, to the total annual potable water demand for South East Queensland. Urban stormwater runoff is also a more reliable source than that from a rural catchment, due in part to its higher runoff coefficients from impervious areas. Moreover, in the coastal cities of Australia, rainfall tends to be higher than for the inland catchments supplying the major water storages. On the other hand, stormwater capture will be geographically dispersed, unlike a major supply dam.

An effective stormwater harvesting scheme must combine sufficient rainfall, a suitable catchment, opportunities for diversion and storage, adequate demands, and water treatment suitable for the proposed end uses. Technologies are available to produce water suitable for any end use, including potable consumption. A major challenge in an urban environment is identifying cost-effective storage. Pre existing storages, such as existing lakes, suitable aquifers, or disused quarries, provide the opportunity to create storages at minimal cost. Suitable aquifers can play a key role in providing buffering capacity to withstand seasons and years of low rainfall.

Stormwater reuse in Australia is currently dominated by nonpotable, public open space irrigation of parks and playing fields. Domestic use requires dual reticulation and dual household plumbing, both of which significantly increase the cost. Annual compliance costs for potable substitution schemes will also be significant, and considerably higher than for nonpotable uses, though the significant costs of a separate distribution system are avoided.

Stormwater runoff will be contaminated with sediment, organic materials, pathogens, trace organic chemicals (pesticides and herbicides), and hydrocarbons from motor vehicles. Microbiological contaminants will derive from animal faeces as well as leakage from sewerage systems, particularly in wet weather if constructed overflows are present.

As stormwater runoff is rainfall dependent and is unlikely to coincide with the periods of greatest water demand, it is very important to evaluate the water balance to ensure a successful design, i.e., balancing seasonal runoff and storages, with seasonally variable water demand. A range of modeling tools are available to evaluate the potential yield from a stormwater harvesting scheme and to help prove up the design.

Stormwater reuse has multiple benefits other than water substitution. In particular, stormwater harvesting has the potential to play a major role as part of the overall WSUD stormwater management strategy for a development.

By removing water from the catchment, along with its associated contaminants, harvesting reduces runoff volumes and contaminant loads, and thereby reduces the treatment requirements from WSUD facilities within the catchment. WSUD devices can be seen as primary building blocks for stormwater harvesting, providing opportunities to improve harvested water quality, as well as for flow diversion and storage. Multiple harvesting and treatment locations in an urban catchment are likely to provide the best ecological benefits for creeks, but at the expense of multiple installations and dispersed management obligations.

In Australia, there are extensive national and state guidelines available to inform the development of stormwater harvesting schemes. However, stormwater harvesting and reuse tends to be less regulated than recycled wastewater, causing uncertainty in the long-term governance and compliance obligations for a potential operator. The real challenges for stormwater harvesting do not relate to how to do it, nor the technologies available to be used. Rather the uncertainty about the long-term operation, governance, and compliance requirements is a major impediment for operators to develop schemes with regulatory and financial confidence.

Rainwater capture is another form of stormwater harvesting and reuse, most often applied at an individual domestic scale, but also at a communal scale. Domestic rainwater harvesting has a range of long-term maintenance issues associated with leaf accumulation in gutters, integrity of tank insect screens, and pump and control systems operation. Centralized schemes, where management is more informed and technically skilled, will be in a better position to effectively manage these challenges. Rainwater is used as the main source of potable supply in rural communities, and for potable substitution in urban communities. It also provides catchment runoff benefits (reduced runoff volumes and pollutant loads) allowing better performance of WSUD devices.

## REFERENCES


ACT Government, 2006. Think water, act water: strategy for sustainable water resource management in the ACT. In: 2004-2005 Progress Report. ACT Government, Canberra. www.thinkwater.act.gov.au/documents/TWAW04-05progress_report_web.pdf.

ACT Government, August 2014. ACT Water Strategy 2014–44. Striking the Balance. ACT Government, Environment and Planning Directorate. http://www.environment.act.gov.au/__data/assets/pdf_file/0019/621424/ACT-Water-Strategy-ACCESS.pdf.

Ahmed, W., Gardner, T., Toze, S., 2011. Microbiological quality of roof harvested rainwater and health risks: a review. J. Env. Qual. 40 (1), 13–21.

Australian and New Zealand Environment and Conservation Council (ANZECC), Agriculture and Resource Management Council of Australia and New Zealand, 2000. Australian and New Zealand Guidelines for Fresh and Marine Water Quality. Australian Water Association, Canberra.

Beal, C.D., Stewart, R.A., 2013. Identifying residential water end-uses underpinning peak day and peak hour demand. Journal of Water Resources Planning and Management. https://doi.org/10.1061/(ASCE)WR.1943-5452.000035.

Beal, C., Stewart, R., Huang, T., Rey, E., 2011. SEQ residential end use study. Water (AWA) 38, 92–96.

Beal, C., Sharma, A., Gardner, T., Chong, M., 2012. A desktop analysis of potable water savings from internally plumbed rainwater tanks in south east Queensland. Water Resources Management 26, 1577–1590.

Beal, C., Makki, A., Stewart, R., 2014. What does rebounding water use look like? An examination of post-drought and post flood water end use demand in Queensland, Australia. Water Science and Technology: Water Supply 14, 561–568.

Burton, G., Pitt, R., 2002. Stormwater Effects Handbook: A Toolbox for Watershed Managers, Scientists, and Engineers. CRC Press, USA (from U.S. Environmental Protection Agency).

Chong, M., Umapathi, S., Mandak, A., Sharma, A., Gardner, T., 2011. A Benchmark Analysis of Water Savings from Mandated Rainwater Tank Users in South East Queensland ( Phase2). Urban Water Security Research Alliance. Technical Report No 49. http://www.urbanwateralliance.org.au/publications/technicalreports/.

Clark, R., Gonzalez, D., Dillon, P., Charles, S., Cresswell, D., Naumann, B., 2015. Reliability of water supply from stormwater harvesting and managed aquifer recharge with a brackish aquifer in an urbanising catchment and changing climate. Environmental Modelling and Software 72, 117–125.

Coombes, P., Argue, J., Kuczera, G., 2014. Figtree Place: A Case Study in Water Sensitive Urban Development (WSUD).

CRC for Catchment Hydrology, 1999. Publications: Urban Stormwater Quality. https://ewater.org.au/archive/crcch/archive/pubs/prog4.html.

Dandy, G., Ganji, A., Kandulu, J., Hatton Macdonald, D., Marchi, A., Maier, H., Mankad, A., Schmidt, C.E., 2013. Managed Aquifer Recharge and Stormwater Use Options: Net Benefits Report. Goyder Institute for Water Research, Adelaide, South Australia, pp. 1839–2725. Technical Report Series No. 14/1. http://www.goyderinstitute.org/publications/technical-reports/.

Department of Environment and Conservation NSW, 2006. Managing Urban Stormwater: Harvesting and Reuse. DEC NSW, South Sydney.

Dillon, P., 2016. Managing aquifer recharge in integrated solutions to groundwater challenges. Ch 2 p3–16. In: Vogwill, R. (Ed.), Solving the Groundwater Challenges of the 21st Century. CRC Press/Balkema, Taylor & Francis Group, London, UK. Intl. Assoc. Hydrogeologists Selected Papers No 22.

Dillon, P., Pavelic, P., Page, D., Beringen, H., Ward, J., February 2009. Managed Aquifer Recharge: An Introduction, Waterlines Report No 13, 65 p. https://recharge.iah.org/files/2016/11/MAR_Intro-Waterlines-2009.pdf.

Dillon, P., Page, D., Dandy, G., Leonard, R., Tjandraatmadja, G., Vanderzalm, J., Rouse, K., Barry, K., Gonzalez, D., Myers, B., 2014. Managed Aquifer Recharge Stormwater Use Options: Summary of Research Findings. Goyder Institute for Water Research. Technical Report 14/13. http://www.goyderinstitute.org/publications/technical-reports/.

Duncan, H.P., February 1999. Urban Stormwater Quality: A Statistical Overview. Report 99/3. Cooperative Research Centre for Catchment Hydrology.

Environmental Protection and Heritage Council (EPHC), National Health and Medical Research Council (NHMRC), Natural Resource Management Ministerial Council (NRMMC), 2008. Australian Guidelines for Water Recycling: Augmentation of Drinking Water Supplies (Phase 2), Canberra. http://www.agriculture.gov.au/water/quality/nwqms.

Farooqui, T.A., Renouf, M.A., Kenway, S.J., December 1, 2016. A metabolism perspective on alternative urban water servicing options using water mass balance. Water Research 106, 415–428.

Fletcher, T.D., Vietz, G., Walsh, C.J., 2014. Protection of stream ecosystems from urban stormwater runoff: the multiple benefits of an ecohydrological approach. Progress in Physical Geography 38, 543–555.

Fong, L.M., Nazarudeen, H., 1996. Collection of Urban Stormwater for Potable Supply in Singapore. Water Quality International, pp. 36–40. May/June.

Head, B., 2010. Wicked Problems in Water Governance: Paradigm Changes to Promote Water Sustainability and Address Planning Uncertainty. Urban Water Security Research Alliance. Technical Report No. 38.

Head, B., 2014. Managing urban water crises: adaptive policy responses to drought and flood in Southeast Queensland, Australia. Ecology and Society 19, 33.1–33.14.

Khan, S., October 2013. Drinking water through recycling: the benefits and costs of supplying direct to the distribution system. Australian Academy of Technological Sciences and Engineering (ATSE). ISBN 978-1-921388-25-5.

Leusch, F.M., Van Der Merwe, J., Lampard, J.-L., Khan, S., Hawker, D., Humpage, A., 2016. Chemical Hazard Assessment of Stormwater Micropollutants (CHASM) Guidance Manual. Water RA project 3023. www.waterra.com.au/publications/document-search/?download=1335.

Lim, H.S., Lu, X.X., 2016. Sustainable urban stormwater management in the tropics: an evaluation of Singapore's ABC waters program. Journal of Hydrology 538, 842–862.

Luan, I.O.B., 2010. Singapore water management policies and practices. International Journal of Water Resources Development 26, 65–80.

Lucas, S.A., Coombes, P.J., Sharma, A.K., 2010. The impact of diurnal water reuse patterns, demand management and rainwater tanks on the supply network design. Journal of Water Supply and Technology-Water Supply-WSTWS 10 (1), 69–80.

Magyar, M.I., Ladson, A.R., 2015. Chemical quality of rainwater in rain tanks. In: Sharma, A. (Ed.), Donald Begbie and Ted Gardner. Rainwater Tank Systems for Urban Water Supply. IWA Publishing, London, ISBN 9781780405353.

McAlister, T., Stephens, M., Allen, A., 2017. Aura, the city of colour – Australia's shining example of widescale integrated water cycle management. Water Practice and Technology 12 (3).

McCallum, T., January 2015. Kalkallo: A Case Study in Technological Innovation amidst Complex Regulation. Cooperative Research Centre for Water Sensitive Cities, Melbourne, Australia, ISBN 978-1-921912-24-5. https://watersensitivecities.org.au/wp-content/uploads/2016/05/TMR_A3-2_KalkalloCaseStudy.pdf.

Monash University, 2006a. Integrated Stormwater Treatment and Harvesting: Technical Guidance Report. http://iswr.eng.monash.edu.au/research/projects/stormwater.

Monash University, 2006b. Quantifying Stormwater Recycling Risks and Benefits: Heavy Metals Review. http://iswr.eng.monash.edu.au/research/projects/stormwater.

New South Wales Dept. Primary Industries, Office of Water, 2015. Validation and verification-what's the difference? In: Recycled Water Information Sheet No 7 www.water.nsw.gov.au/__data/assets/pdf_file/0009/560475/IS7_Validation-and-verification_final.pdf.

National Health and Medical Research Council (NHMRC), Natural Resource Management Ministerial Council (NRMMC), 2011. Australian Drinking Water Guidelines. NHMRC and NRMMC, Canberra.

Natural Resource Management and Ministerial Council (NRMMC), Environmental Protection and Heritage Council (EPHC), Australian Health Ministers' Conference (AHMC), 2006. Australian Guidelines for Water Recycling: Managing Health and Environmental Risks (Phase 1). NHMRC and NRMMC, Canberra. http://www.agriculture.gov.au/water/quality/nwqms.

Natural Resource Management Ministerial Council (NRMMC), Environmental Protection and Heritage Council (EPHC), National Health and Medical Research Council (NHMRC), 2009a. Australian Guidelines for Water Recycling (Phase 2). Stormwater Harvesting and Reuse, Canberra. NWQMS Guideline 23. http://www.agriculture.gov.au/water/quality/nwqms.

Natural Resource Management Ministerial Council (NRMMC), Environmental Protection and Heritage Council (EPHC), National Health and Medical Research Council (NHMRC), 2009b. Australian Guidelines for Water Recycling (Phase 2). Managed Aquifer Recharge, Canberra. NWQMS Guideline 24. http://www.agriculture.gov.au/water/quality/nwqms.

Page, D., Dillon, P., Vanderzalm, J., Bekele, E., Barry, K., Miotlinski, K., Levett, K., December 2010. Managed aquifer recharge case study risk assessments. In: CSIRO Water for a Healthy Country Flagship Report, 144 p. https://publications.csiro.au.

Page, D., Gonzalez, D., Dillon, P., Vanderzalm, J., Vadakattu, G., Toze, S., Sidhu, J., Miotlinski, K., Torkzaban, S., Barry, K., 2013a. Managed Aquifer Recharge and Stormwater Use Options: Public Health and Environmental Risk Assessment Final Report. Goyder Institute for Water Research. Technical Report 13/17. http://www.goyderinstitute.org/publications/technical-reports/.

Page, D., Gonzalez, D., Naumann, B., Dillon, P., Vanderzalm, J., Barry, K., 2013b. Stormwater Managed Aquifer Recharge Risk-Based Management Plan, Parafield Stormwater Harvesting System, Stormwater Supply to the Mawson Lakes Recycled Water Scheme, Industrial Uses and Public Open Space Irrigation. Goyder Institute for Water Research. Technical Report 13/18. http://www.goyderinstitute.org/publications/technical-reports/.

Philp, M., Mcmahon, J., Heyenga, S., Marinoni, O., Jenkins, G., Maheepala, S., Greenway, M., 2008. Review of Stormwater Harvesting Practices. Urban Water Security Research Alliance. Technical Report No. 9.

PUB, 2004. Code of practice on sewerage and sanitary works, Addendum No. 2. In: Water Reclamation (Network) Departments. Public Utilities Board, Singapore, pp. 1–56.

PUB, 2017. Stormwater Management. Available at: https://www.pub.gov.sg/drainage/stormwatermanagement.

Queensland Department of Environment, 2014a. Fitzgibbon Stormwater Harvesting (FiSH) Scheme - Final Report. Economic Development Queensland, Brisbane. http://www.environment.gov.au/system/files/pages/cf7aacbd-c546-4f41-81f4-f8ecc5bcf30e/files/fitzgibbon-stormwater-report.docx.

Queensland Department of Environment, 2014b. Fitzgibbon Potable Roofwater Harvesting (PotaRoo) Scheme - Final Report. Economic Development Queensland, Brisbane. http://www.environment.gov.au/system/files/pages/0fa50583-b33a-4ed2-a150-a19d991a021b/files/fitzgibbon-potaroo-final-report.doc.

South Australia, Environment Protection Authority, 2015. Environment Protection (Water Quality) Policy 2015. www.epa.sa.gov.au/files/11255_wqepp_policy2015.pdf.

South Australia, Department of Environment, Water and Natural Resources, 2014. MAR Regulations. http://www.environment.sa.gov.au/managing-natural-resources/water-use/water-resources/stormwater/managed-aquifer-recharge/mar-regulations.

South Bank Corporation, May 2012. The South Bank Rain Bank (Storm Water Harvesting Project) Final Report. https://www.environment.gov.au/system/files/pages/3c42fcfd-75df-40af-844d-a8e2b9446f95/files/south-bank-final-report.pdf.

Stevens, D., 2014. Audit of the Parafield Stormwater Harvesting and Managed Aquifer Recharge System for Non - Potable Use against the Stormwater Risk - Based Management Plan. Goyder Institute for Water Research. Occasional Paper Series No. 14/1 ISSN: 2204–0528. http://www.goyderinstitute.org/publications/occasional-papers/.

The Senate - Environment and Communications References Committee, 2015. Stormwater Management in Australia. Commonwealth of Australia, Canberra.

Thyer, M., Hardy, M., Coombes, P., Patterson, C., 2008. The impact of end-use dynamics on urban water system design criteria. Australian Journal of Water Resources 12 (2), 161–170.

Van Dijk, A., Beck, H., Crosbie, R., De Jeu, R., Liu, Y., Podger, G., Timbal, B., Viney, N., 2013. The Millennium Drought in southeast Australia (2001-2009):Natural and human causes and implications for water resources, ecosystems, economy and society. Water Resources Research 49, 1–18.

Vanderzalm, J., Page, D., Dillon, P., Lawson, J., Grey, N., Sexton, D., Williamson, D., 2014a. A Risk-Based Management Plan for Mount Gambier Stormwater Recharge System: Stormwater Recharge to the Gambier Limestone Aquifer. Goyder Institute for Water Research. Technical Report 14/7. http://www.goyderinstitute.org/publications/technical-reports/.

Vanderzalm, J., Page, D., Gonzalez, D., Barry, K., Toze, S., Bartak, R., Shisong, Q., Weiping, W., Dillon, P., Lim, M.H., 2014b. Managed Aquifer Recharge and Stormwater Use Options: Satellite Sites Stormwater Quality Monitoring and Treatment Requirements Report. Goyder Institute for Water Research. Technical Report 14/10. http://www.goyderinstitute.org/publications/technical-reports/.

Victorian Department of Health, 2013. Guidelines for Validating Treatment Processes for Pathogen Reduction: Supporting Class a Water Recycling Schemes in Victoria. https://www2.health.vic.gov.au/.

Victorian Environment Protection Authority, 2009. Guidelines for Managed Aquifer Recharge (MAR) – Health and Environmental Risk Management. Publication 1290. http://www.epa.vic.gov.au/~/media/Publications/1290.pdf.

Vieritz, A.M., Neumann, L.E., Cook, S., 2015. Rainwater tank modelling…in Rainwater Tank Systems for Urban Water Supply. Edited by Sharma, A. Donald Begbie & Ted Gardner, IWA London, ISBN 9781780405353.

Vincent, L., Michel, L., Catherine, C., et al., 2014. The energy cost of water independence: the case of Singapore. Water Science and Technology 70, 787–794.

Ward, J., Dillon, P., January 2011. Robust policy design for managed aquifer recharge. In: Waterlines Report Series No 38, 28 p. http://webarchive.nla.gov.au/gov/20160615084848/http://archive.nwc.gov.au/library/waterlines/38.

Ward, J.D., Simmons, C.T., Dillon, P.J., Pavelic, P., 2009. Integrated assessment of lateral flow, density effects and dispersion in aquifer storage and recovery. J of Hydrology 370, 83–99.

Water Technology, 2016. Aura Stormwater Harvesting Scheme Strategic Implementation Report, Report Prepared for Stockland.

Western Australia, Department of Water, 2011. Department of Water 2010, Operational Policy 1.01 – Managed Aquifer Recharge in Western Australia, Perth. http://www.water.wa.gov.au/__data/assets/pdf_file/0016/1564/96686.pdf.

Wilson, G., Pamminger, F., Narangala, R., Knight, K., Tucker, S., Mcgrath, J., 2010. Stormwater for Potable Reuse Can Be Part of a Greenfield Urban Water Solution – Kalkallo Case Study. Ozwater 2010.

## FURTHER READING

Coombes, P., Barry, M., 2007. Climate change, efficiency of water supply catchments and integrated water cycle management in Australia. In: 13th International Rainwater Catchment Systems and 5th International Water Sensitive Urban Design Conference. Sydney Australia.

PMSEIC, 2007. Water for Our Cities: Building Resilience in a Climate of Uncertainty. Prime Minister's Science, Engineering and Innovation Council (PMSEIC). http://catalogue.nla.gov.au/Record/4196896.

Sharma, A.K., Begbie, D., Gardner, T., 2015. Rainwater Tank Systems for Urban Water Supply - Design, Yield, Energy, Health risks, Economics and Community perceptions. IWA Publishing.

Chapter 10

# Using WSUD to Restore Predevelopment Hydrology

Anthony Ladson[1,2]

[1]*Victorian University, College of Engineering and Science, Melbourne, VIC, Australia;* [2]*Moroka Pty Ltd., Clifton Hill, VIC, Australia*

## Chapter Outline



**ABSTRACT**

Urban development changes hydrology. A key question is whether, and to what extent, water sensitive urban design (WSUD) can prevent these changes. The impact of urbanization includes changes to the catchment water balance and alteration of flow regimes, including increased volume, runoff frequency, and high flows. Seasonality is altered and low flows are changed, either increased or decreased, depending on local features. There is evidence from monitoring and modeling studies that WSUD can restore hydrology at small scales; however, restoration at the catchment scale is much more challenging, and there is limited evidence that existing techniques are effective. New ways of designing and building suburbs are required, which do not rely on the direct drainage of flow from impervious surfaces to waterways.

**Keywords:** Flow regime restoration; Low impact design; Stormwater control measures; Urbanization; Water sensitive urban design.

## 10.1 INTRODUCTION

This chapter considers the use of water sensitive urban design (WSUD) to restore predevelopment hydrology and protect the ecosystem health of urban waterways. Urban development changes hydrology. The key question explored here is whether, and to what extent, WSUD measures can prevent or minimize these changes.

The effect of urbanization on catchment water balance is considered first, in Section 10.2, as this underpins and provides the explanation for observed changes in streamflows. It also seems likely that hydrologic restoration will require

**Approaches to Water Sensitive Urban Design. https://doi.org/10.1016/B978-0-12-812843-5.00010-1**

changing the water balance of an urban catchment to become closer to what it would have been before development. Section 10.3 summarizes the effect of urbanization on flow regimes. In Section 10.4, the physical drivers of these changes are discussed. In Section 10.5, the ability of WSUD to address the myriad effects of urbanization is explored at the site scale, whereas Section 10.6 addresses restoration of flow regimes at the catchment scale.

## 10.2 IMPACT OF URBANIZATION ON A CATCHMENT WATER BALANCE

A water balance in an urban catchment equates the change in storage with the difference between inputs (precipitation and mains water) and outputs (evaporation, stormwater runoff, and wastewater discharge). The implications of urbanization for a catchment water balance are listed as follows:

- Inputs are changed. Mains water is supplied to urban catchments and some of this additional water can end up in streams because of irrigation or leakage (Woolmington and Burgess, 1983; Burgess et al., 1984; Puust et al., 2010).
- The water stored in the catchment changes. The increase in impervious area means less infiltration of rainwater. Imported water may contribute to groundwater storage through leakage from water supply and sewage pipes; or water may leak into pipes or enter the gravel filled trenches surrounding pipes, depleting groundwater (Bonneau et al., 2017).
- There are changes in the way that water leaves a catchment. Runoff volumes are often substantially increased and are disposed of through hydraulically efficient drainage networks. There may be less opportunity for water to evaporate if it is quickly drained from a catchment. Conversely, evapotranspiration may increase if water is supplied for irrigation of parks, gardens, and playing fields.

The change in the rate and volume of inputs, outputs, and storage is the basis for the hydrologic behavior we see in urban areas: the rapid response to rainfall and increased flood magnitude and frequency, which co-occur with development.

Following Mitchell et al., (2003), the water balance for an urban catchment can be expressed as

$$\Delta S = (P + I) - (E_a + R_s + R_w) \tag{10.1}$$

where $\Delta S$ is the change in catchment storage, $P$ is precipitation, $I$ is imported water, $E_a$ is actual evapotranspiration, $R_s$ is stormwater runoff, $R_w$ is wastewater discharge.

Water balance studies carried out for urban catchments in Australia are summarized in Table 10.1.[1] These studies show that

- Imported water substantially increases catchment inflows as it is 30%–40% of precipitation; and
- Not all imported water is exported as wastewater via the sewerage system (wastewater is 60%–85% of imported water), which means imported water contributes to evaporation, stormwater, and/or groundwater.

These conclusions are consistent with a range of published water balances for urban catchments (Grimmond and Oke, 1986; Barron et al., 2013).

The reported average catchment water balances (Table 10.1) are useful, but further insights can be gained from considering the variation between years. Mitchell et al. (2003) provide detailed water balance information for Curtin, a suburb of Canberra in the Australian Capital Territory (ACT) (Fig. 10.1), for the 16 years between 1979 and 1995. Water balances for average, driest, and wettest years are shown in Table 10.2. Wettest and driest years are also graphed in Fig. 10.1.

Climate has a substantial influence on the water balance. Annual precipitation in Canberra was highly variable over the 16-year study period, ranging from 247 to 914 mm/year. On average, rainfall was three times greater than mains inputs, but in the driest year, mains water imports exceeded rainfall. That is, urbanization more than doubled catchment inflows in the driest year. In the wettest year, imported mains water made up only 13% of water input.

In the driest year, evapotranspiration was greater than precipitation. That is, imported mains water provided a substantial contribution to evapotranspiration. Urban landscapes in summer are often hot and dry, compared with rural landscapes (Coutts et al., 2007), but in the driest years, this is being mitigated by imported water. The excess of evapotranspiration over precipitation in some years has also been reported by Grimmond and Oke (1986) in their review of urban water balances.

1. The National Water Accounts reported by the Bureau of Meteorology (Bureau of Meteorology, 2015) contain information on water use in regions that include the urban areas of Adelaide, Canberra, Melbourne, Perth, South East Queensland, and Sydney. However, these accounts also include substantial rural water use in surrounding areas so are less useful for isolating urban influences (www.bom.gov.au/water/nwa).

**TABLE 10.1** Annual Water Balance Data From Suburbs of Australian Cities (Units are mm), Where Studies Include Multiple Years, Average Values are Listed

| | | Input | | | Output | | | | |
|---|---|---|---|---|---|---|---|---|---|
| Location | Study Area | Precipitation | Imported Water | Imported Water as a Percentage of Precipitation (%) | Actual Evapotranspiration | Stormwater Runoff | Wastewater Export | Change in Store (Miss-Close)[d] | Wastewater/Imported Water (%) |
| Canberra (Curtin, ACT) (Mitchell et al., 2003) (1979–96) | 27 $km^2$ | 630 | 200 | 32 | 508 | 203 | 118 | 1 | 59 |
| Sydney, NSW (Bell, 1972) (1962–71) | 1035 $km^2$ | 1150 | 349[a] | 30 | 736 | 501 | 262 | 0 | 75 |
| Sydney, NSW (Kenway et al., 2011) (2004–05) | 1420 $km^2$ | 952 | 370 | 39 | 766 | 281 | 319 | −44 | 86 |
| Perth (Subiaco-Shenton Park, WA) (McFarlane, 1984) | NA | 788 | 285 + 96[b] | 36 | 766 | 104 | 154 | 117[c] | 54 |
| Melbourne, Victoria (Kenway et al., 2011) (2004–05) | 1818 $km^2$ | 763 | 237 | 31 | 688 | 165 | 190 | −43 | 80 |
| Brisbane and surrounding areas (South East Queensland) (Kenway et al., 2011) (2004–05) | 1281 $km^2$ | 1021 | 374 | 37 | 814 | 390 | 179 | 12 | 49 |

[a] *Includes imported water and use of groundwater.*
[b] *Inflow of stormwater from upstream area.*
[c] *Adjusted for change in groundwater storage.*
[d] *Calculated by difference between inputs and outputs; see original studies for details.*

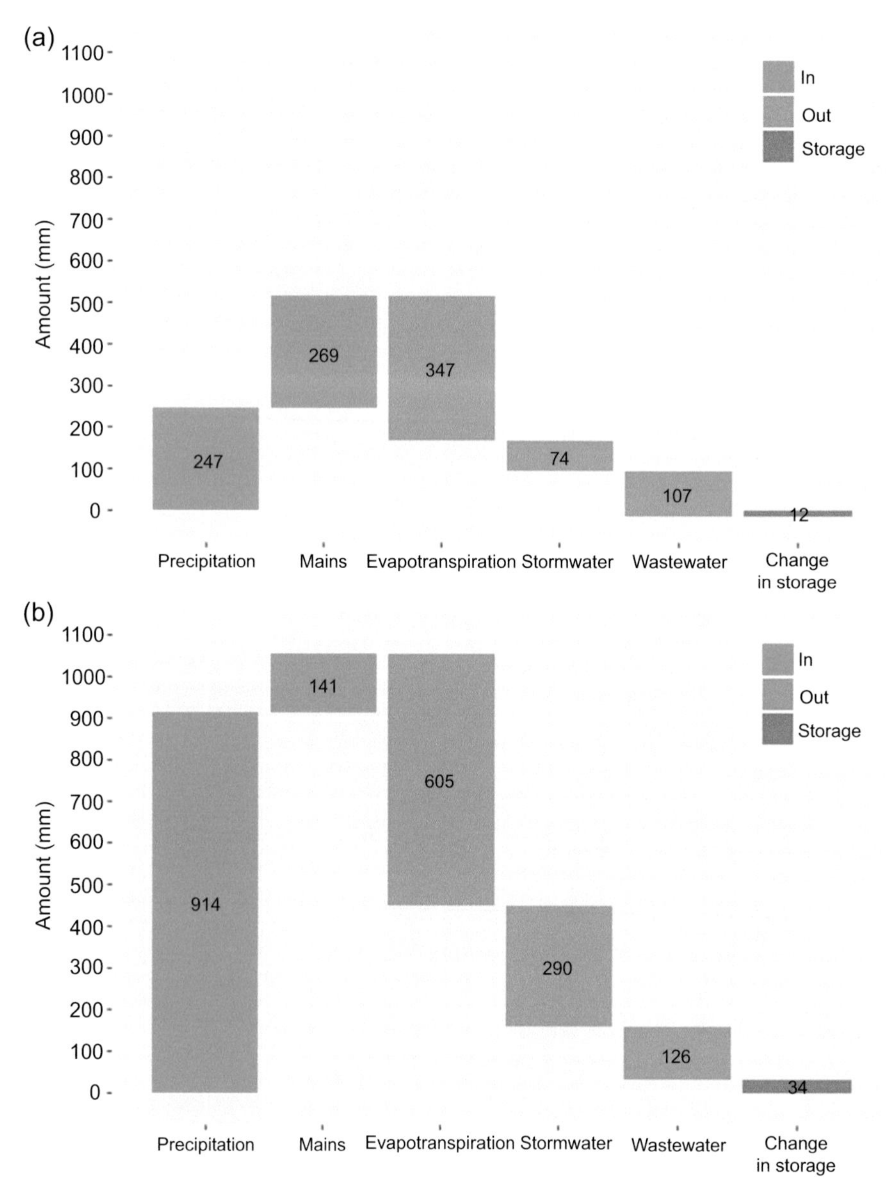


**FIGURE 10.1** Water balance for Curtin (Canberra ACT) for (a) the driest and (b) the wettest year as estimated by Mitchell et al. (2003).

The largest output was always evapotranspiration, ranging from 57% to 67% of total input. Wastewater outputs were reasonably constant between these years (107–126 mm) and do not seem strongly influenced by climate. In contrast, stormwater runoff, as expected, was highly influenced by climate, changing by a factor of about 4 from 74 mm in the driest year to 290 mm in the wettest. Stormwater is augmented by mains water through excess garden watering and leakage of reticulation pipes (Woolmington and Burgess, 1983; Burgess et al., 1984).

These water balances clearly demonstrate the impact of urbanization. Water imports through the drinking water system exceed wastewater discharge, so there must be a net increase in the outputs of stormwater and/or evaporation. The data from Canberra (Curtin, ACT) show that in dry years, mains water more than doubles water inputs, so the effect on stormwater and evapotranspiration is substantial.

## 10.2.1 Comparison of rural and urban water balances

There are a few studies that contrast water balances for urban and neighboring natural catchments (Grimmond and Oke, 1986; Stephenson, 1994; Claessens et al., 2006; Bhaskar and Welty, 2012). As expected, urbanization usually causes an increase in runoff, which we explore in the next section.

**TABLE 10.2** Urban Water Balance in Wet, Dry, and Average Years Curtin (Canberra, ACT) Over a 16-Year Study Period (Mitchell et al., 2003)

| Quantity | Driest Year | Average Year | Wettest Year |
|---|---|---|---|
| Precipitation (mm) | 247 | 630 | 914 |
| Mains Water (mm) | 269 | 200 | 141 |
| **Total input (mm)** | **516** | **830** | **1055** |
| Ratio precipitation to mains water | 0.9 | 3.2 | 6.5 |
| Increase in catchment water inputs from mains water | 209% | 132% | 115% |
| Actual evapotranspiration (ET) (mm) | 347 | 508 | 605 |
| Ratio ET to precipitation | 1.4 | 0.8 | 0.7 |
| Ratio ET to total inputs | 67% | 61% | 57% |
| Stormwater runoff (mm) | 74 | 203 | 290 |
| Wastewater export (mm) | 107 | 118 | 126 |
| Excess of mains water over wastewater (mm) | 162 | 82 | 15 |
| Change in storage (mm) | −12 | 1 | 34 |

The effect on the evapotranspiration term of the water balance is complex and variable. Energy balance studies usually show that evapotranspiration is higher in urban compared with rural areas, although this appears to vary by location, as it depends on the availability of water in the landscape (Coutts et al., 2007). In turn, water availability is linked to the fraction of vegetated area and the extent and quantity of irrigation of sports fields, parkland, and domestic gardens (Grimmond and Oke, 1986). Where stormwater is rapidly drained and where watering restrictions are in place, there is likely to be less water available to be evaporated.

The amount of evaporation has implications for urban microclimates. Where there are low levels of evaporation, solar radiation is converted to sensible rather than latent heat, which can cause urban areas to be hot and dry (Coutts et al., 2009). There are moves to increase the availability of water for evaporation to provide cooling and increase thermal comfort (see Chapters 19−21). Increased watering using mains water is one way to achieve this, but the result would be to move the water balance further away from predevelopment conditions and exacerbate water scarcity. There are opportunities to use WSUD structures for stormwater harvesting and reuse, rather than discharging it to creeks and waterways (Coutts et al., 2012).

## 10.2.2 Restoring the water balance

The path to restoring flow regimes begins with restoring the water balance. Mains inputs would be reduced if stormwater could be used as an alternative supply (Mitchell et al., 2002; Coombes and Mitchell, 2006). This would save mains water, decrease stormwater runoff, and move the water balance closer to predevelopment conditions. It is also important to consider the availability of water in the urban landscape to encourage a favorable microclimate for people during hot, dry weather. Reuse of stormwater and wastewater could contribute to enhanced evapotranspiration and therefore cooling (Coutts et al., 2009).

# 10.3 EFFECT OF URBANIZATION ON FLOW REGIMES

The previous section considered the overall water balance of an urban area. Here, the focus is on the impact of urbanization on flow regimes. We can broadly divide impacted streams into two types: (1) those that have substantial flow reduction because of water harvesting for urban supply and (2) those that accept urban drainage flows. Some streams have both types of impacts. For example, flow in the Yarra River in Melbourne, Australia, not only is subject to a 38% reduction because of up-catchment water harvesting but also receives stormwater inflows from highly urbanized tributaries in lower reaches (SKM, 2005).

The effect of urbanization on flow regimes where waterways accept urban drainage has been widely reported and was well recognized by the 1960s (Leopold, 1968). Knowledge of urban influences has been confirmed in paired catchment studies (e.g., Codner et al., 1988; Miller et al., 2014; Prosdocimi et al., 2015) from comparisons of flow regimes before and after development (e.g., Beighley and Moglen, 2002; Dougherty et al., 2007) and the response of catchments where there is a gradient of urbanization (e.g., Rose and Peters, 2001; Heejun, 2007; Smith et al., 2013; Schwartz and Smith, 2014; Bell et al., 2016).

The effect of urbanization on flow regimes can be summarized as follows with details provided in the following sections:

- Increase in the amount of rainfall that is converted to streamflow, resulting in an increased number of runoff events per year, increased number of peaks, increased peak discharge for a given annual exceedance probability (AEP) (e.g., 1 in 1 year AEP), and increased flow volume;
- More rapid rises and falls in streamflow, increased "flashiness";
- Changes in low flows including reduced time when the stream is not flowing—often, urban streams flow all the time—but decreased baseflow volume;
- Increased flood magnitudes; and
- Increased volume and rate of flow per unit catchment area.

In brief, runoff frequency, high flows, and the total volume of runoff increase. Baseflow is likely to decrease, but cease-to-flow periods are reduced. Examples of these changes are given below, with the physical causes of these changes reviewed in Section 10.4.

## 10.3.1 Increased runoff volume

More rainfall is converted to runoff in urban catchments both because of the increased impervious areas and increased runoff from **pervious** areas, which are usually wetter from irrigation using imported water and possibly more compacted and hence less permeable than predevelopment soils (Harris and Rantz, 1964; Cordery, 1976; Ferguson and Suckling, 1990; Trudeau and Richardson, 2015).

As an example in a paired catchments study in Canberra with similar rainfall totals, runoff from the urban *Giralang* catchment was approximately six times greater than that from the rural *Gungahlin* catchment (Fig. 10.2). The annual volumetric runoff coefficient for the urban catchment was 41% compared with 13% for the rural catchment (Table 10.3). In general, volumetric runoff coefficients are higher in urban catchments for both annual and event-based time scales (Fleming, 1994; Smith et al., 2005). The runoff coefficient for small areas of impervious surfaces is often over 80% (Walsh et al., 2012).

Similar results were produced in a study in South East Queensland. T. Webber (pers. Comm.) used the Source model framework (Welsh et al., 2012) to develop hydrologic models for two subcatchments of Bundamba Creek. One subcatchment consisted of 1102 ha of native vegetation and the other was an 1147 ha urban area. A simulation was undertaken from 1980 to 2013 (average rainfall 918 mm/year; average Potential Evapotranspiration [PET] 1574 mm/year). The differences in runoff are shown in Fig. 10.3. The urban area produced approximately three times the runoff volume mainly because of an increase in quick flow (direct response to rainfall), but slow flow (baseflow) also increased. The runoff coefficients were 11.5% for the catchment with native vegetation and 35.8% for the urban catchment. Runoff was produced on 49 days per year from the urban catchment and 5 days per year from the catchment with native vegetation.

## 10.3.2 Faster response to rainfall

Urban streams respond more rapidly to rainfall compared with rural catchments. An early literature review by Cordery (1976) showed that urbanization decreased catchment response times by a factor between 2 and 12 depending on the extent of impervious surfaces (Fig. 10.4 see also Kibler, 1982).

Recent studies and reviews confirm the magnitude of these increases (e.g., Miller et al., 2014; Smith et al., 2005; Smith and Smith, 2015). A study of the urbanizing Giralang catchment in Canberra showed that the lag time (time from centroid of rainfall excess to centroid of surface runoff) decreased from 11hr under rural conditions to 0.4 hr when the catchment was urbanized, a factor of 27 (Codner et al., 1988).

## 10.3.3 Increased runoff frequency

Runoff occurs more frequently because of urbanization. Small rainfall events of 1−2 mm will cause runoff from impervious surfaces (ASCE, 1975), but much more rainfall is usually required to produce runoff from grassland or

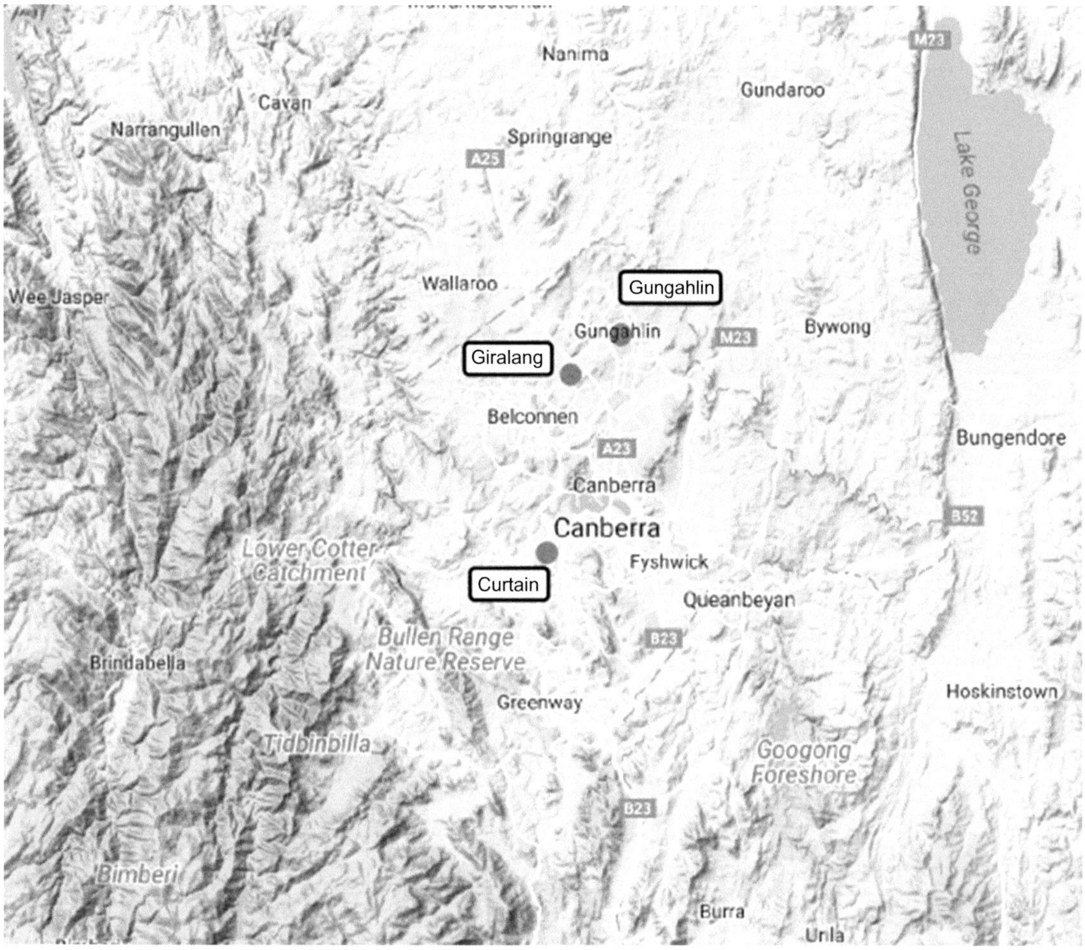

FIGURE 10.2 Location of Canberra suburbs Curtin, Giralang, and Gungahlin in the Australian Capital Territory. *Map data © 2017 Google.*

**TABLE 10.3 Increased Runoff Volume With Urbanization. Paired Catchments, Canberra**

| Quantity | Urban[b] (Giralang) | Rural[c] (Gungahlin) |
|---|---|---|
| Average annual rainfall (mm)[a] | 565 | 535 |
| Average annual runoff (mm) | 231 | 41 |
| Proportion of rainfall converted to runoff (%) | 41 | 13 |

[a]*Based on rainfall and runoff August 1, 1976 to July 31, 1984.*
[b]*Catchment area 94 ha.*
[c]*Catchment area 112 ha.*
From Codner, G.P., Laurenson, E.M., Mein, R.G., 1988. Hydrologic effects of urbanisation: a case study. In: Hydrology and Water Resources Symposium, ANU Canberra, 1–3 Feb 1988, Institution of Engineers Australia. pp. 201–205; Laurenson, E.M., Codner, G.P., Mein, R.G., 1985. Giralang/gungahlin Paired Catchment Study: A Review. Department of Housing and Construction, Canberra.

forest (Pilgrim and Cordery, 1993). Larger rainfall events that produce runoff in natural catchments occur much less often than the small events that produce runoff from impervious surfaces. For example, in forested areas to the east of Melbourne, Australia, daily rainfalls of 10–30 mm are required to produce runoff (Walsh et al., 2005a). Rainfall events of this magnitude only occur about 5–15 days a year at this location. In comparison, runoff from impervious surfaces such as roofs, roads, and driveways is produced from only about 1 mm of rain, which occurs about 10 times as often per year (see Section 10.3.5 below) (Walsh et al., 2005a; Ladson et al., 2006).

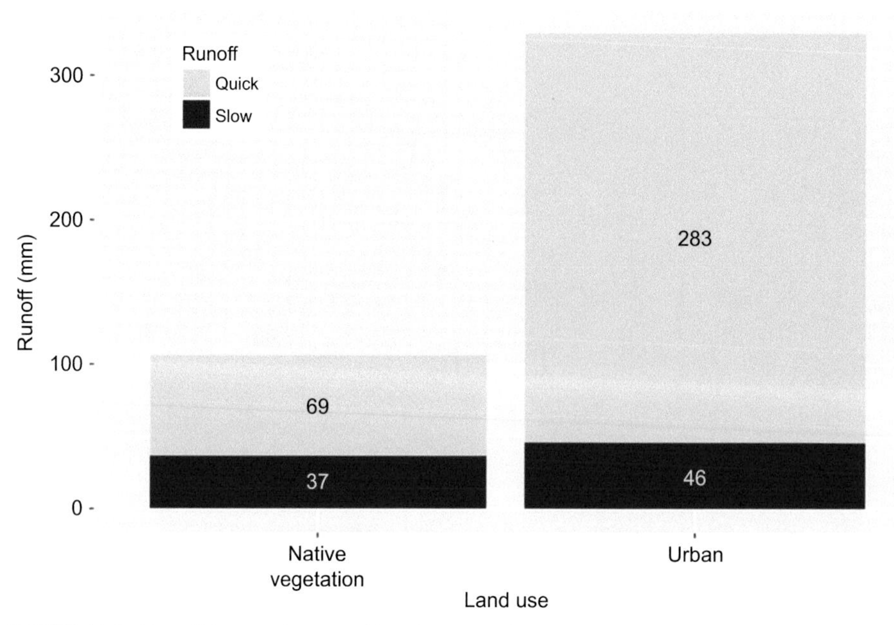


FIGURE 10.3 Runoff from catchments with native vegetation and urban land uses (T. Webber pers. comm.).

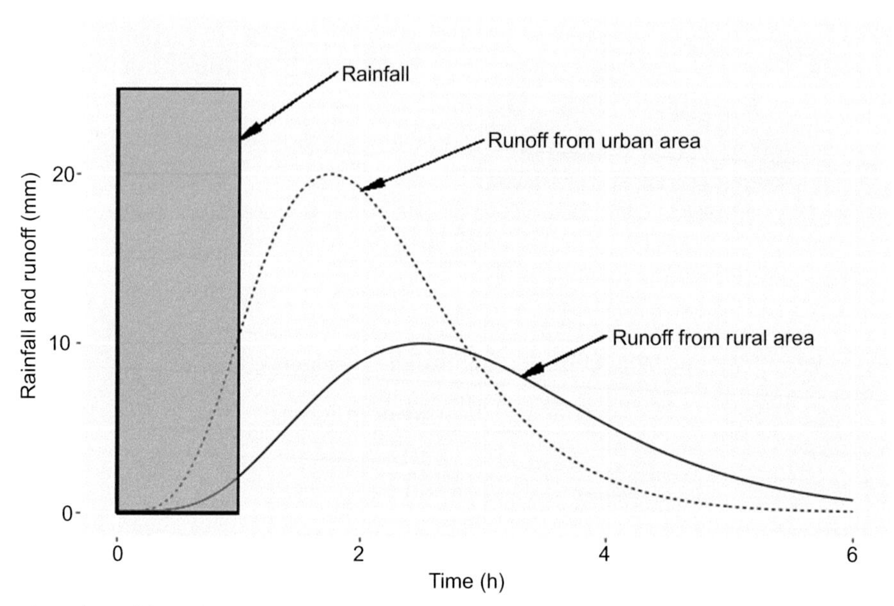


FIGURE 10.4 Comparison of runoff from similar urban and rural catchments. *Adapted from Cordery, I., 1976. Some effects of urbanisation on streams. Civil Engineering Transactions, The institution of Engineers, Australia CE18 (1), 7–11.*

## 10.3.4 Increased number and magnitude of flood peaks

The faster response to rainfall and greater runoff volume contribute to an increased number and size of flood peaks following urbanization. This behavior is considered to increase the flashiness of the stream hydrograph.

As an example, consider the flow in Brushy and Olinda Creeks, neighboring streams in eastern Melbourne. Brushy Creek has a much higher proportion of connected impervious area than Olinda Creek (Table 10.4). An example of the different responses to similar rainfall events is shown in Fig. 10.5. The greater number and size of runoff peaks in Brushy Creek is evident.

**TABLE 10.4** Catchment Details for Brushy Creek and Olinda Creek

| | Brushy Creek | Olinda Creek |
|---|---|---|
| Gauge number | 229,249 | 229,690 |
| Latitude of gauge | 37°46′ 55.83″S | 37°47′ 51.24″S |
| Longitude of gauge | 145°18′ 23.73″E | 145°22′ 28.67″E |
| Catchment area (km$^2$) | 14.16 | 22.82 |
| Forest cover[a] | 23% | 84% |
| Grass/pasture cover | 48% | 8% |
| Horticulture | 2% | 3% |
| Impervious surfaces | 28% | 5% |
| Attenuated imperviousness[b] | 13.2% | 0.3% |
| Connected impervious area | 28% | 5% |
| Annual average catchment rainfall (mm) | 963 | 1408 |

[a]*Land use data source from Melbourne Water.*
[b]*Calculated using the optimal model of Walsh and Kunapo (2009). This is an estimate of connected imperviousness that inversely weights impervious areas by their distance from the nearest stormwater drain or stream.*

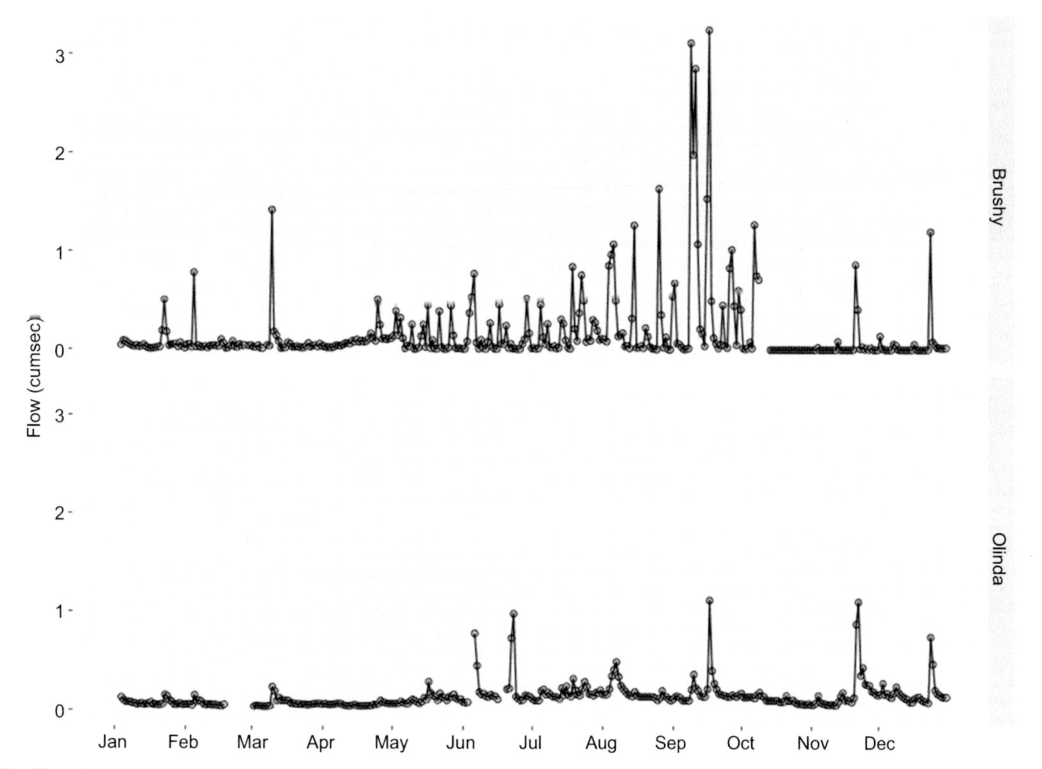


**FIGURE 10.5** Flow in Brushy Creek (top) and Olinda Creek (bottom) near Melbourne, Australia, in response to similar rainfall events over a 12-month period (1988).

A measure of flashiness was proposed by Baker et al. (2004), which was termed the Richards-Baker Flashiness Index (RBFI). It is the sum of absolute values of change in mean daily flows divided by the sum of the mean daily flows. Flashy catchments have RBFI ratios of around 1, whereas constant flow catchments (a hypothetical case) have an RBFI of 0. For the flow shown in Fig. 10.5, the urbanized Brushy Creek has an RBFI of 1.0, whereas the less urbanized Olinda Creek has an RBFI of 0.31.

Another commonly used measure of flashiness is the "$TQ_{mean}$"—the proportion of time that flows in each year is greater than mean flow for that year. This decreases with urbanization and has been shown to be linked to ecological condition of a stream (Booth et al., 2004). As an example, for Brushy Creek, which has greater urban development, the $TQ_{mean}$ is 0.21, which can be compared with the less urbanized Olinda Creek, where the value is 0.37 (for the period 1988–2016).

Often, the value of $TQ_{mean}$ is calculated and then averaged for several years. However, the time of the year when flows are greater than the mean is also altered by urbanization. Using $TQ_{mean}$ as the metric, urbanization results in high flows occurring more often, but for shorter periods, and is dispersed throughout the year. Comparing the plots of Olinda Creek and Brushy Creek in Fig. 10.6, higher flows (flows above the mean) are clustered in the winter (June–August) for Olinda Creek, as its catchment wets up because of the higher rainfall and reduced evaporation, which occur seasonally in this area. For Brushy Creek, with the same climate, short bursts of high flow occur throughout the whole year. A key issue revealed by this analysis is the changed seasonality of high flow, which is a result of urbanization. This is just one of the many changes in flow regime caused by urbanization that leads to poor stream condition (Burns et al., 2014).

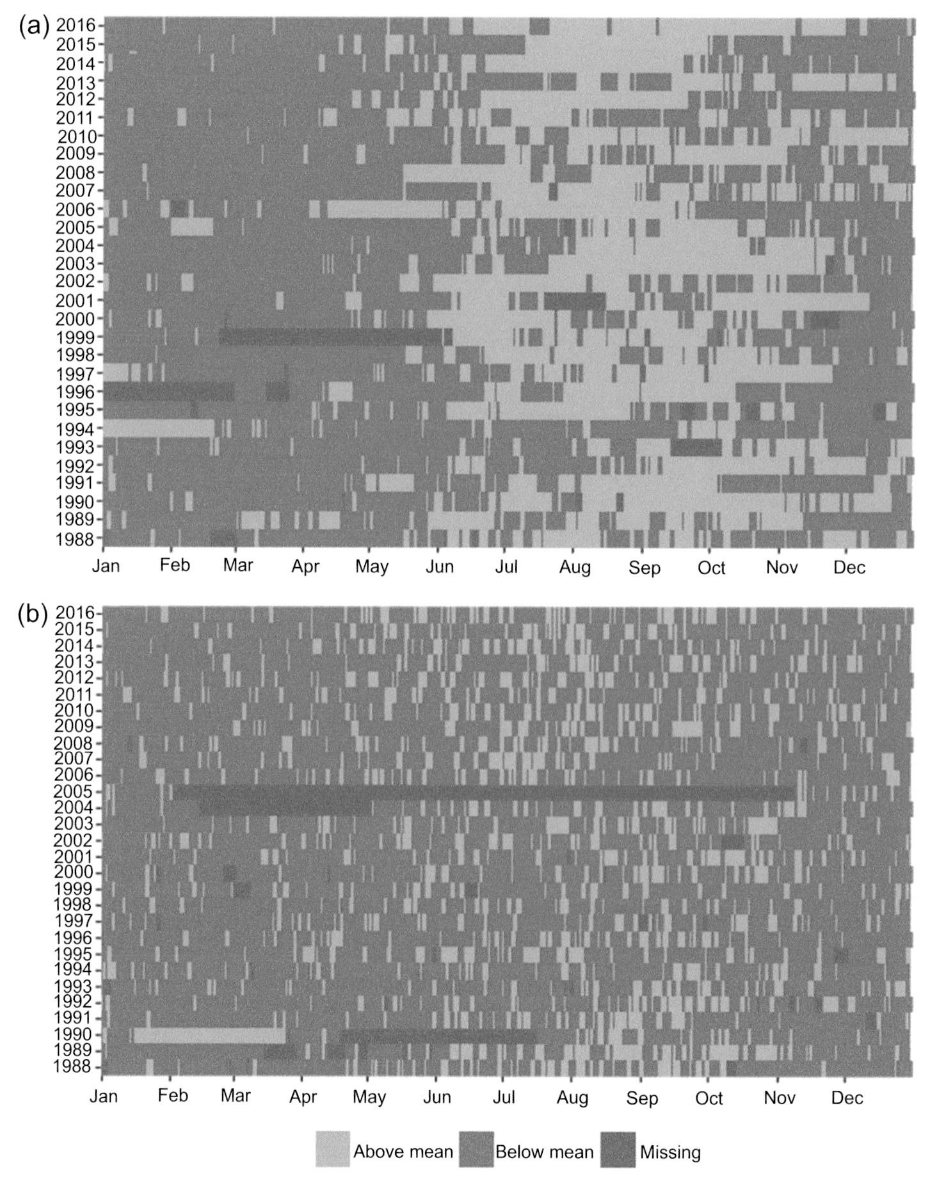


FIGURE 10.6 Time of year and periods when flows are above the mean flow value in each year for (a) Olinda Creek—less urbanized and (b) Brushy Creek—more urbanized. Both creeks are located east of Melbourne, Australia.

### 10.3.5 Decreased initial loss

In flood hydrology, the initial loss is the amount of rainfall required to wet up a catchment before runoff commences. Initial loss can have a major influence on the size of the flood peak produced by a storm. Although it has long been known that initial loss is lower in urban compared with rural areas, a compilation of loss data has recently become available for Australia, which illustrates the differences (Ball et al., 2016). For this large dataset, the mean initial loss for urban catchments is 1.1 mm with a standard deviation of 1.17 mm. For the rural catchments, initial loss is highly variable, with a mean initial loss of 32 mm and a standard deviation of 16.8 mm. These values are consistent with other Australian studies (El-Kafagee and Rahman, 2011; Tularam and Ilahee, 2007). A probability density plot of initial loss in Fig. 10.7 shows the contrast in magnitude and variability for urban and rural catchments. These data support the observation that urban catchments will usually produce runoff from rainfall totals as small 1−2 mm, which are usually sufficient to satisfy the initial loss, with further event rainfall being converted to runoff.

### 10.3.6 Increased flood frequency and magnitude

Decreased initial loss and subsequent increased number and size of flood peaks from urbanization are expressed as increased flood frequency. Urban development causes up to a tenfold increase in peak flows for frequent floods (which occur once a year or more often), with diminishing impacts on larger and less frequent floods (Leopold, 1968; ASCE, 1975; Hollis, 1975; Cordery, 1976; Mein and Goyen, 1988; Shuster et al., 2005; Brath et al., 2006; Jacobson, 2011).

### 10.3.7 Impact on low flows

The impact of urbanization on low flow is complex, with some features of urbanization leading to increases in low flows, whereas others contributing to reduced low flows. Factors that may increase low flows include increased deep drainage from irrigation of gardens, parks, and playing fields; leakage from pipes carrying mains water, wastewater, or stormwater; and detention and retention of stormwater in pervious storages. Features contributing to reduced low flow include rapid and efficient drainage of water reducing infiltration; increased impervious areas; and extraction of shallow groundwater (Price, 2011). Increases, decreases, and variable trends in low flow measurements are widely reported (Price, 2011).

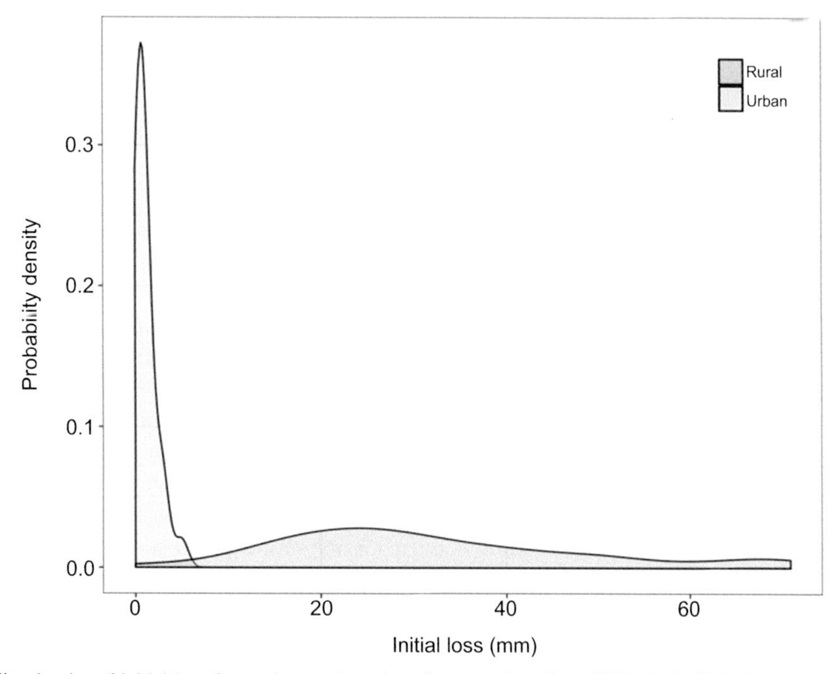


**FIGURE 10.7** Probability density of initial loss from urban and rural catchments. *Data from 2016. In: Ball, J., Babister, M., Nathan, R., Weeks, W., Weinmann, E., Retallick, M., Testoni, I. (Eds.), Australian Rainfall and Runoff: A Guide to Flood Estimation, Commonwealth of Australia (Geoscience Australia).*

Reduced baseflow in urban streams is associated with rapid recession, but the stream may not dry out entirely. Instead, very low flows may be maintained by leakage of mains water, which can be traced by testing for fluoride where this is added to water supplies (Kaushal and Belt, 2012).

In the Giralang/Gungahlin paired catchment study in Canberra (Fig. 10.2), the creek in the rural catchment was ephemeral, only flowing on 14% of days (Codner et al., 1988). This compares to continuous flow from the urbanized catchment. This effect was attributed to overwatering of parks and sports fields, leading to surface and subsurface flow entering the urban drainage system and subsequently into the waterways. This effect was further explored by Woolmington and Burgess (1983) who showed that the timing of flow increases in urban streams in Canberra was linked to domestic garden watering that occurred outside of office hours. Small rainfall events decreased low flows because people avoided watering. Garden watering and leakage of water from the mains system resulted in permanent flow in the urban streams they studied (Burgess et al., 1984). Continuous flow in urban creeks has also been noted in other studies (for example, Bell and Vorst, 1976), where it was attributed to leakage of household and septic tank waste. The recent focus on reducing nonrevenue water and water saving rules has probably reduced this contribution to low flows in waterways.

## 10.4 PHYSICAL CAUSES OF FLOW REGIME CHANGE

This section outlines the physical changes that result from urbanization which lead to the flow changes noted above. These include the following:

- Increased rainfall intensity
- Catchment changes
- Drainage system changes
- Importation of water.

### 10.4.1 Increased rainfall intensity

In summer, cities tend to be warmer than surrounding nonurban land, which can lead to more intense afternoon rainfall. This can increase flooding and shift the pattern of flooding so that summer floods become more common (Burian and Shepherd, 2005; Zhang et al., 2014).

The impact of urbanization on annual rainfall totals is more variable and depends on specific conditions such as wind direction, the extent of the urban heat island effect (see Chapter 19), and the amount and type of fine particles or liquid droplets from pollution (Shepherd, 2005).

### 10.4.2 Catchment changes

Catchment changes resulting from urbanization include removal of trees and vegetation, increased impervious area, and extension of the drainage network.

The link between forest cover and flow yield is well known, with decreasing forest cover resulting in increased total volume of flow at the catchment scale, probably because of reduced evapotranspiration and decreased interception of rainfall (Brown et al., 2013). The extent of tree cover has also been shown to be an important factor in the determining the runoff volumes from urban catchments at the annual time scale (Bell et al., 2016).

A second key catchment change is the increase in impervious surfaces leading to increases in the proportion of rainfall that runs off, the rapid response to rainfall, and decreased baseflow (Shuster et al., 2005).

The increase in drainage density is the third major influence of urbanization at the catchment scale (Walsh et al., 2016a). In an undeveloped catchment, many sources of runoff are remote from the stream network, allowing opportunities for flow to re-infiltrate or evaporate before reaching a stream. In highly urbanized areas, every impervious surface is connected to a drain, no matter how distant from the main stream network. That means almost all runoff will contribute to streamflow. Walsh et al. (2016a) argue that while most studies of urban impacts on steam ecosystems have focused on impervious cover as the key cause of degradation, it could be argued that it is the extension of the drainage network that is the more important factor. That is, directly connected impervious areas are far more hydrologically responsive than isolated impervious areas.

### 10.4.3 Drainage system changes

As part of urbanization, drainage systems are modified to be more efficient, and as a result there is less attenuation of storm hydrographs, increased flow velocities, and reduced detention storage of water for a given flow. There is also reduced floodplain storage, at least up to the level where flows are no longer contained within pipes. Modeling suggests that when considering the increase in peak flows that follow urbanization, improved hydraulic efficiency of watercourses is more important than the increase in impervious surfaces (Wong et al., 2000). In catchments in Singapore and Canberra, storage of flowing water, for a given discharge, was reduced by factors of 8 and 10 following urbanization (Selvalingham et al., 1987). Reduced storage contributes to the flashy nature of urban streams (see Figs. 10.4 and 10.5).

### 10.4.4 Importing water

The discussion of the urban water balance in Section 10.2 highlighted the significance of imported water, especially in low-rainfall years. Imported water contributes to low flows (Woolmington and Burgess, 1983; Burgess et al., 1984), evapotranspiration from urban areas (Grimmond and Oke, 1986), and to groundwater (Bhaskar et al., 2016). It is likely that ongoing efforts by water authorities to detect leaks, reduce nonrevenue water, and encourage water saving have reduced the contribution of imported water to streamflow.

## 10.5 TOWARD RESTORATION OF PREDEVELOPMENT FLOW REGIMES

The previous sections have considered the impact of urbanization on the water balance of a suburb, the changes in flow regimes caused by urbanization, and the physical aspects of development that lead to these changes. Flow regime changes are extensive and varied with multiple overlapping causes (Walsh et al., 2005b).

WSUD or low impact development is often intended to return stream hydrology toward its predevelopment state. This includes efforts to reduce runoff volume, increase infiltration, reduce peak flows, extend lag time, and increase baseflow (Eckart et al., 2017). There are a range of techniques available to achieve these outcomes.

### 10.5.1 Overview of WSUD approaches

WSUD systems are commonly divided into infiltration-based approaches and retention techniques (Fletcher et al., 2013; Eckart et al., 2017).

Measures to increase infiltration are intended to reduce the volume of runoff during storm events, recharge groundwater, and restore baseflow. Approaches include swales, infiltration trenches, rain gardens (bioretention systems), sand filters, and permeable pavements.

Retention techniques aim to retain water, for short or long periods, to decrease storm flow peaks and, perhaps, volumes. Methods include retarding basins, wetlands, and ponds. Some systems, for example, wetlands and rain gardens, can contribute to retention and infiltration depending on how they are designed and constructed. Systems may be combined with stormwater harvesting to further reduce flows and provide an alternative source of water for other uses. Examples include centralized stormwater collection systems and harvesting from wetlands or rainwater tanks at the household scale, which are plumbed to some regular demand, e.g., toilet flushing. Leaky tanks have also been proposed. These capture and store rainwater, supply some demand, and provide a low flow to "leak" water to the stream, thus increasing baseflows (Duncan et al., 2014).

Design guidance for WSUD techniques is widely available with guides produced for cities in various climates. For example, guides for southern Australia include Melbourne Water (2005) and Argue and Allen (2004); for the United States WEF and ASCE (2012) and EPA (2004); and for Canada (CVC, 2010). Further details of WSUD design guidelines are provided in Chapter 4.

Streamflow restoration requires that individual WSUD techniques are combined and integrated (Burns et al., 2012). Treatment trains are commonly used to improve water quality (Wong, 2006) but similar approaches can be used to restore the flow regime. For example, a rain tank could be used to retain flow with overflows routed to an infiltration system. Rainwater harvesting and reuse of stormwater could also be included in WSUD systems. Chapter 2 provides a more detailed description of WSUD techniques and their application to achieve sound integrated water management outcomes.

### 10.5.2 Impact on hydrology

The focus of this section is on the effectiveness of WSUD measures at the scale of individual treatment devices. At this small scale, there is consistent evidence that WSUD can mitigate at least some of the effects of urbanization. The situation is much less clear at the catchment scale, which is discussed in the next section.

Recent reviews have considered the performance of WSUD in detail (e.g., Ahiablame et al., 2012; Askarizadeh et al., 2015; Jarden et al., 2016; Eckart et al., 2017; Li et al., 2017). These reviews provide separate discussion of modeling studies and field assessments. Modeling studies are much more numerous, although less convincing.

Summarizing the results from modeling studies first, these suggest that the following WSUD techniques are effective in reducing *runoff volume*: porous pavements, bioretention, green roofs, rainwater tanks, infiltration basins, swales, and tree planting. Stormwater harvesting is a key strategy to reduce annual runoff volumes, particularly when applied to regular demands or demands that have a similar seasonal pattern to rainfall (Fletcher et al., 2013).

Modeling has identified that the following WSUD techniques are effective in reducing *peak flow rates*: rainwater tanks for retention and reuse, permeable pavements, bioretention systems, and flow routing from pervious to impervious surfaces. Any approach that increases initial and/or continuing losses during a storm event is likely to be effective in reducing flood peaks. However, there is an interaction between WSUD performance in reducing peak flow and rainfall patterns during storm events. For example, if high rainfall intensity occurs early in a storm, then peaks are more likely to be mitigated by those WSUD measures that increase initial loss. These measures are less effective if rainfall intensity occurs later in a storm (Qin et al., 2013; Loveridge and Babister, 2016). Therefore, it is important to use a range of rainfall temporal patterns to design WSUD measures and assess their effectiveness (Testoni et al., 2016).

Field studies have confirmed the effectiveness of WSUD at small scales. For example, a treatment train of porous pavement and bioretention reduced runoff volumes from a car park by 70% and runoff frequency by 60% (Brown et al., 2012). A study of biofilters undertaken in Victoria, Australia, showed they attenuated peak runoff flow rates by at least 80% and reduced runoff volumes by 33% on average (Hatt et al., 2009). Hamel et al. (2013) reviewed stormwater control measures (SCMs) that increase infiltration and so mitigated the impacts of urbanization on baseflow.

## 10.6 RESTORATION OF FLOW REGIMES AT THE CATCHMENT SCALE

Field experiments and modeling studies have shown that WSUD has the potential to mitigate most of the effects of urbanization at small scales. However, demonstrating WSUD effectiveness at the catchment scale, and over the whole flow regime, is much more challenging. The review by Li et al. (2017) focuses on the question of whether SCMs can restore altered flow regimes for whole watersheds. They considered the results of 45 studies, dividing these into research based on computer simulation (31 studies) and research that included field measurements of hydrologic effectiveness (14 studies). These studies are discussed in the following sections:

### 10.6.1 Modeling studies

The modeling studies not only, showed that SCMs were generally effective in improving aspects of flow regimes but also highlighted limitations. Integrated approaches were shown to be more effective than individual measures, but complete restoration to predevelopment hydrology is experimentally challenging and was examined in only a few cases (7 of 31 studies). The effectiveness of WSUD in improving low flows was not commonly modeled. Promising approaches to flow regime restoration include storage and harvesting of rainwater combined with distributed infiltration measures. This combination has the potential to improve both high and low flows.

A key result is that restoration of the flow regime of a stream affected by urbanization likely requires a water balance that is close to what it was before development (Duncan et al., 2014). This is particularly true of ephemeral streams. Modeling suggests that water balance restoration is challenging as it requires harvesting and use of large volumes of water (Aryal et al., 2016). Restoring predevelopment runoff volumes requires intercepting or preventing 80%—95% of runoff—the actual amount depending on rainfall and catchment physiography (Burns et al., 2013). The critical element is having sufficient and regular demand for water along with cost-effective storage.

One modeling study showed that large volume reductions could be achieved using leaky tanks, bioretention, and stormwater harvesting. Bioretention systems must be set up to promote infiltration. Stormwater harvesting, storage, and reuse were required at both household and precinct scales (Duncan et al., 2014).

Designing an integrated system to achieve large volume reductions is very challenging on both engineering and economic grounds. There are numerous ways to achieve moderate volume reductions (say 30%), which provide the designer with lots of options to select WSUD measures. When large volume reductions are required, there are only a few measures that are appropriate (e.g., large rainwater tanks leaking and spilling to large bioretention systems along with reuse from all streams). The ability to tailor a system to a budget or optimize cost effectiveness is limited (Duncan et al., 2014).

## 10.6.2 Monitoring studies

Li et al. (2017) found only three comprehensive catchment-scale experiments that were reported in detail: Shepherd Creek, Cincinnati, Ohio, USA (Shuster and Rhea, 2013); Tributary 104, Clarksburg, Maryland, USA (Loperfido et al., 2014; Rhea et al., 2015; Bhaskar et al., 2016); and Little Stringybark Creek, Melbourne, Australia (Walsh et al., 2015).

The Shepherd Creek study involved the implementation of household-scale rainwater tanks and rain gardens that focused on capturing roof runoff in small catchments (25–69 ha). Comparisons before and after development and with neighboring controls showed a small decrease in runoff volume. This small effect was attributed to the focus of treatment on roofs, which were a small proportion of the catchment impervious area. Runoff from other impervious surfaces remained untreated.

In the "Tributary 104" experiment, both distributed and centralized SCMs were installed to treat the additional runoff that resulted from urbanization. Results showed increased baseflow and decreases in peak flow, but interventions could not replicate the flow regimes from the reference catchment. Treatment effects were small compared with the influence of rapid urbanization and were probably not of practical significance (Rhea et al., 2015). Bhaskar et al. (2016) report on the effect of 73 infiltration-focused measures in the 110 ha catchment. Results showed that baseflow was larger than for the forested reference catchment. Bhaskar et al. (2016) described this as an *unintended consequence* of the intervention that was aimed at reducing runoff volumes and peak flows.

The Little Stringybark Creek experiment on the outskirts of Melbourne, Australia, implemented 300 SCMs in a catchment of 450 ha (Walsh et al., 2015). Only preliminary results are available at this stage but these show small reductions in runoff coefficients for individual storm events (Burns et al., 2016).

Li et al. (2017) conclude that these catchment-scale monitoring studies do provide some evidence of the effect of WSUD measures in restoring flow regimes, but that the outcomes are small and uncertain, although additional analysis is yet to be published for Little Stringybark Creek.

## 10.6.3 Designing WSUD interventions to restore flow regimes

A major challenge is the design of WSUD systems to restore flow regimes. The designer needs to have a target for restoration and a clear understanding of the capability of a range of WSUD measures.

Comparing predevelopment and postdevelopment flow regimes requires consideration of hydrologic metrics that provide information on all the aspects of hydrology, which are impacted by urbanization (see Section 10.2). This is a challenging area. There are many metrics considered in the literature; 171 are reviewed by Olden and Poff (2003). Gordon et al. (2004) comment that there is no limit to the number of hydrological parameters that can be developed. The task is to determine those that have strong associations with stream hydraulics, ecology, and geomorphology. Important reviews include Gao et al., 2009; DeGasperi et al., 2009; Kennen et al., 2010; and Steuer et al., 2010.

A useful framework is to consider changes in flow in five categories: magnitude, frequency, duration, rate of change, and timing (Duncan et al., 2014). Flow *magnitude* influences velocity, sediment transport, and the wetted area of the stream channel. The *frequency* of flow events is important (Stewardson and Gippel, 2003). For example, if there is a change in hydrology such that the time between events that scour vegetation is shorter than the time it takes for vegetation to recover, then vegetation will not survive. Event *duration* is significant in many cases. For example, long periods of zero flow may result in the loss of pools, which provide refuge during drought. The *rate of change* influences the time available for invertebrates to adjust or seek shelter. This is important for communities in terms of flood warning and affects the ability of fish and macroinvertebrates to respond to increased flow stress during high flow events and move back to pools when flows decrease. The seasonality of high flows, low flows, and other significant flow events is also ecologically important (Duncan et al., 2014).

Some example metrics that have been used to assess urban flow regimes are listed in Table 10.5. Two of these, $TQ_{mean}$ and the RBFI, were explored in Section 10.3.4. Duncan et al. (2014) attempted to use similar metrics to devise interventions to maintain the flow regime in an ephemeral catchment earmarked for development in Victoria, Australia. They found that the maintenance of low flows and runoff volumes to be particularly challenging because of the "superabundance" of flow caused by urbanization.

Along with metrics based on flow measurements, catchment-based indicators have been proposed, which can be related to the frequency of small runoff events and which have been shown to be an important factor in stream condition (Walsh et al., 2005a). For example, Walsh et al. (2009) developed a retention capacity index that quantified the degree of drainage connection of impervious surfaces to waterways.

**TABLE 10.5 Example of Flow Metrics to Guide Design of Water Sensitive Urban Design Systems to Restore a Flow Regime Altered by Urbanization**

| Name | Description | References |
|---|---|---|
| Low-flow magnitude | Seven-day minimum flow divided by the mean daily flow | Olden and Poff (2003) |
| High-flow frequency | Number of events per year greater than three times the median flow | Clausen and Biggs (1997) |
| Low-flow frequency | Number of low flow events per year <25th percentile flow | Kennen et al. (2010) |
| High-flow duration | Proportion of days in a year that the daily mean flow is greater than the annual mean flow ($TQ_{mean}$—time above flow mean) | Booth et al. (2004) |
| Rate of change | Sum of absolute values of change in mean daily flows divided by the sum of the mean daily flows (Richards-Baker Flashiness Index) | Baker et al. (2004) |
| Timing | The time of year when the 7-day minimum flow occurs | Cassin et al. (2005) |

Complete restoration of a flow regime affected by urbanization will require consideration of a broad range of flow metrics. This contrasts with earlier guidelines that focused on narrower objectives such as the maintenance of the 1.5-year average recurrence interval (ARI) flow at predevelopment magnitude (Stormwater committee, 1999). Managing flows to meet specific objectives by considering a small number of indicators will still have merit, but it is not likely to result in treatment of the entire "urban stream syndrome" because of the multifaceted effects of development on the flow regime (Walsh et al., 2005b; Burns et al., 2014).

## 10.6.4 Maintenance and performance over time

There are concerns that the effectiveness of WSUD measures may decrease with time. For example, a 15-year study of infiltration trenches showed that infiltration rates decreased because of clogging by fine particles (Bergman et al., 2010). Future projections from measured declines suggest that as infiltration decreases, runoff will increase tenfold over a 100-year design life.

Maintenance of rainwater tanks and household systems for water harvesting is known to be a major issue (ABS, 2013; Moglia et al., 2013; Mankad and Greenhill, 2014; Sharma et al., 2015). This suggests that these systems will become less effective at reducing volumes and peak flows without regular maintenance. These declines in performance are generally not included in modeling studies, and few monitoring studies are based on long-term data.

## 10.6.5 Design of monitoring studies

An additional challenge for the assessment of the effect of WSUD is the design of monitoring studies. Quality studies use some version of the before-after-control-impact approach, which includes assessment of change before and after intervention in both catchments where intervention takes place and in neighboring control catchments (Stewart-Oaten et al., 1986; Underwood, 1991). For the Little Stringybark Creek study, assessment of reference streams was also included. These streams were in forested catchments that were not affected by urban stormwater runoff. Three reference and three control streams were monitored, along with the stream where interventions took place, meaning seven streams were assessed in total. This before-after-control-reference-impact approach is perhaps the current "gold standard," as it allows for differential responses to factors, such as climate, between intervention, control, and reference streams (Walsh et al., 2015).

## 10.7 CONCLUSION

Confirmation of the ability of WSUD to restore hydrologic regimes to predevelopment conditions will need to await further research, but there are many challenges. Suburban development creates an enormous number of small, impervious subcatchments that all drain to streams, and, unfortunately, even small amounts of connected impervious area can cause significant hydro-ecological changes (Walsh et al., 2005b). It is standard practice for every roof, every driveway, and all roads to be connected to the drainage system. The task of retrofitting each of these connections to incorporate detention, harvesting, and infiltration seems overwhelming. In addition, new connected impervious surfaces are being created, even in established suburbs, through infill development; civil works that are undertaken to improve drainage in response to complaints; and the ongoing formalization of parkland, creation of bike paths, and the general works undertaken by local government. These all tend to improve drainage efficiency and hence modify runoff characteristics.

This suggests that an evolutionary approach to restoring flow regimes will be too slow and too expensive and will always be playing catchup. Real improvement in flow regimes seems unlikely, or at best transitory, as the effectiveness of WSUD measures declines with time. Instead, a revolutionary approach is required in the way water-related aspects of suburbs are designed and constructed so that no impervious surfaces are **directly drained** to waterways and that all the excess runoff these surfaces create is managed through harvest, reuse, and infiltration. This is consistent with the principles for urban stormwater management (Walsh et al., 2016b). These principles include that (1) the postdevelopment balance of evapotranspiration, streamflow, and infiltration should mimic the predevelopment balance; (2) SCMs should deliver flow regimes that mimic the predevelopment regime in quality and quantity; (3) SCMs should have capacity to store rain events for all storms that would not have produced widespread surface runoff in the predevelopment state, thereby avoiding increased frequency of disturbance to biota; and (4) SCMs should be applied to all impervious surfaces in the catchment of concern.

Key tasks are to develop the WSUD technologies to achieve these principles along with the institutional arrangements to support their adoption.

## REFERENCES


ABS, 2013. 4602.0.55.003 Environmental issues: water use and conservation. Australian Bureau of Statistics, Commonwealth of Australia.

Ahiablame, L.M., Engel, B.A., Chaubey, I., 2012. Effectiveness of low impact development practices: literature review and suggestions for future research. Water Air Soil Pollution 223, 4253–4273.

Argue, J.R., Allen, M.D., 2004. Water Sensitive Urban Design: Basic Procedures for 'source Control' of Stormwater: A Handbook for Australian Practice. University of South Australia.

Aryal, S., Ashbolt, S., McIntosh, B.S., Petrone, K.P., Maheepala, S., Chowdhurry, R.K., Gardner, T., Gardiner, R., 2016. Assessing and mitigating the hydrologic impacts of urbanisation in Semi-urban catchments using the stormwater management model. Water Resources Management 30 (4), 5437–5454.

ASCE (American Society of Civil Engineers), 1975. Aspects of hydrological effects of urbanization. ASCE Task Committee on the effects of urbanization on low flow, total runoff, infiltration, and groundwater recharge of the Committee on surface water hydrology of the Hydraulics Division. Journal of the Hydraulics Division HY5 449–468.

Askarizadeh, A., Rippy, M.A., Fletcher, T.D., Feldman, D.L., Peng, J., Bowler, P., Mehring, A.S., Winfrey, B.K., Vrugt, J.A., AghaKouchak, A., Jiang, S.C., Sanders, B.F., Levin, L.A., Taylor, S., Grant, S.B., 2015. From rain tanks to catchments: use of low-impact development to address hydrologic symptoms of the urban stream syndrome. Environmental Science and Technology 49 (19), 11264–11280.

Baker, D., Richards, R.P., Loftus, T.T., Kramer, J.W., 2004. A new flashiness index: characteristics and applications to Midwestern rivers and streams. Journal of the American Water Resources Association 40 (2), 503–522.

Ball, J., Babister, M., Nathan, R., Weeks, W., Weinmann, E., Retallick, M., Testoni, I. (Eds.), 2016. Australian Rainfall and Runoff: A Guide to Flood Estimation, Commonwealth of Australia (Geoscience Australia).

Barron, O.V., Barr, A.D., Donn, M.J., 2013. Effect of urbanization on the water balance of a catchment with shallow groundwater. Journal of Hydrology 485, 162–176.

Beighley, R.E., Moglen, G.E., 2002. Trend assessment in rainfall-runoff behaviour in urbanising watersheds. Journal of Hydrologic Engineering 7 (1), 27–34.

Bell, C.D., McMillan, S., Clinton, S., Jefferson, A.J., 2016. Hydrologic response to stormwater control measures in urban watersheds. Journal of Hydrology 541, 1488–1500.

Bell, F.C., 1972. The acquisition, consumption and elimination of water by the Sydney Urban System. Proceedings of the Ecology Society of Australia 7, 161–176.

Bell, F.C., Vorst, P.C., 1976. Catchment Hydrology and Urban Development: A Case Study. University of New South Wales.

Bergman, M., Hedegaard, M.R., Peterson, M.F., Binning, P., Ole, M., Mikkelsen, P.S., 2010. Evaluation of Two Stormwater Infiltration Trenches in Central Copenhagen after 15 Years of Operation. Novatech 2010. IWA Publishing, Lyon, France.

Bhaskar, A., Welty, C., 2012. Water balances along an urban-to-rural gradient of metropolitan Baltimore, 2001–2009. Environmental and Engineering Geoscience 28 (1), 37–50.

Bhaskar, A.S., Hogan, D.M., Archfield, S.A., 2016. Urban base flow with low impact development. Hydrological Processes 30, 3156–3171.

Bonneau, J., Fletcher, T.D., Costelloe, J.F., Burns, M.J., 2017. Stormwater infiltration and the 'urban karst' - a review. Journal of Hydrology 552, 115–141.

Booth, D.B., Karr, J.R., Schauman, S., Konrad, C.P., Morley, S.A., Larson, M.G., Burges, S.J., 2004. Reviving urban streams: land use, hydrology, biology and human behaviour. Journal of the American Water Resources Association 40 (5), 1351–1364.

Brath, A., Montanari, A., Moretti, G., 2006. Assessing the effect on flood frequency of land use change via hydrological simulation (with uncertainty). Journal of Hydrology 324 (1–4), 141–153.

Brown, A.E., Western, A.W., McMahon, T.A., Zhang, L., 2013. Impact of forest cover changes on annual streamflow and flow duration curves. Journal of Hydrology 483 (0), 39–50.
Brown, R.A., Line, D.E., Hunt, W.F., 2012. LID treatment train: pervious concrete with subsurface storage in series with bioretention and care with seasonal high water tables. Journal of Environmental Engineering 138 (6), 689–697.
Bureau of Meteorology, 2015. National Water Accounts. http://www.bom.gov.au/water/nwa/.
Burgess, J.S., Woolmington, E., Henderson, A., 1984. Leakages in urban water systems: Village Creek, Canberra, Australia. Journal of Environmental Management 19 (2), 133–146.
Burian, S.J., Shepherd, J.M., 2005. Effect of urbanization on the diurnal rainfall pattern in Houston. Hydrologic Processes 19 (5), 1089–1103.
Burns, M.J., Fletcher, T.D., Walsh, C.J., Ladson, A.R., Hatt, B.E., 2013. Setting Objectives for Hydrologic Restoration: From Site-scale to Catchment Scale. NOVATECH, Lyon, France.
Burns, M.J., Walsh, C.J., Fletcher, T.D., Ladson, A.R., Hatt, B.E., 2014. A landscape measure of urban stormwater runoff effects is a better predictor of stream condition than a suite of hydrologic factors. Ecohydrology. https://doi.org/10.1002/eco.1497.
Burns, M., Fletcher, T.D., Walsh, C., Bos, D., Imberger, S., Duncan, H., Li, C., 2016. Hydrologic and water quality responses to catchment-wide implementation of stormwater control measures. In: Novatech 9th International Conference, Lyon, France.
Burns, M., Fletcher, T.D., Walsh, C.J., Ladson, A., Hatt, B.E., 2012. Hydrological shortcomings of conventional urban stormwater and opportunities for reform. Landscape and Urban Planning 105 (3), 230–240.
Cassin, J.R., Fuerstenberg, R.L., Tear, K., Whiting, D., St John, B., Muray, J., Burkey, J., 2005. Development of Hydrological and Biological Indicators of Flow Alternation in Puget Sound Lowland Streams. Department of Natural Resources and Parks, Seattle, Washington, King County.
Claessens, L., Hopkinson, C., Rastetter, E., Vallino, J., 2006. Effect of historical changes in land use and climate on the water budget of an urbanizing watershed. Water Resources Research 42 (3), W03426.
Clausen, B., Biggs, B.J.F., 1997. Relationships between benthic biota and hydrological indices in New Zealand streams. Freshwater Biology 38, 327–342.
Codner, G.P., Laurenson, E.M., Mein, R.G., 1988. Hydrologic effects of urbanisation: a case study. In: Hydrology and Water Resources Symposium, ANU Canberra, 1–3 Feb 1988, Institution of Engineers Australia, pp. 201–205.
Coombes, P.J., Mitchell, V.G., 2006. Urban water harvesting and reuse. In: Wong, T.H.F. (Ed.), Australian Runoff Quality: A Guide to Water Sensitive Urban Design, Engineers Australia, Canberra, Australian Capital Territory, 6.1 – 6.15(2006).
Cordery, I., 1976. Some effects of urbanisation on streams. Civil Engineering Transactions, The institution of Engineers, Australia CE18 (1), 7–11.
Coutts, A.M., Beringer, J., Tapper, N., 2007. Impact of increasing urban density on local climate: spatial and temporal variations in the surface energy balance in Melbourne, Australia. Journal of Applied Meteorology and Climatology 46, 477–493.
Coutts, A.M., Beringer, J., Jimi, S., Tapper, N.J., 2009. In: The Urban Heat Island in Melbourne: Drivers, Spatial and Temporal Variability, and the Vital Role of Stormwater. Stormwater 2009. Stormwater Industry Association.
Coutts, A.M., Tapper, N.J., Beringer, J., Loughnan, M., Demuzere, M., 2012. The capacity for Water Sensitive Urban Design to support urban cooling and improve human thermal comfort in the Australian context. Progress in Physical Geography 37 (1), 2–28.
CVC, 2010. Credit Valley Conservation. Low Impact Development Stormwater Management Planning and Design Guide.
DeGasperi, C.L.H., Berge, B.K., Whiting, R.J., Burkey, J.J., Cassin, L., Fuerstenberg, R., 2009. Linking hydrologic alteration to biological impairment in urbanizing streams of the Puget Lowland, Washington, USA. Journal of the American Water Resources Association 45 (2), 512–533.
Dougherty, M., Dymond, R.L., Grizzard, T.J., Godrej, A.N., Zipper, C.E., Randolf, J., 2007. Quantifying long-term hydrologic response in an urbanizing basin. Journal of Hydrologic Engineering 12 (1), 33–41.
Duncan, H.P., Fletcher, T.D., Vietz, G., Urrutiaguer, M., 2014. The feasibility of maintaining ecologically and geomorphically important elements of the natural flow regime in the context of a superabundance of flow. In: September 2014. Melbourne Waterway Research-Practice Partnership Technical Report. 14.5.
Eckart, K., McPhee, Z., Bolisetti, T., 2017. Performance and implementation of low impact development - a review. The Science of the Total Environment 607–608, 413–432.
El-Kafagee, M., Rahman, A., 2011. A study on initial and continuing losses for design flood estimation in New South Wales. In: 19th International Congress on Modelling and Simulation, Perth, Australia, 12–16 December 2011.
EPA, 2004. Stormwater Best Management Practice Design Guide. US Environmental Protection Agency.
Ferguson, B.K., Suckling, P.W., 1990. Changing rainfall-runoff relationships in the urbanizing Peachtree Creek Watershed, Atlanta, Georgia. Water Resources Bulletin 26 (2), 313–322.
Fleming, N.S., 1994. An Investigation into Rainfall-runoff Relationships. Research Report No. R 119. Department of Civil and Environmental Engineering. The University of Adelaide.
Fletcher, T.D., Andrieu, H., Hamel, P., 2013. Understanding, management and modelling of urban hydrology and its consequences for receiving waters: a state of the art. Advances in Water Resources 51, 261–279.
Gao, Y., Vogel, R.M., Kroll, C.N., Poff, N.L., Olden, J.D., 2009. Development of representative indicators of hydrologic alteration. Journal of Hydrology 374, 136–147.
Gordon, N.D., McMahon, T.A., Finlayson, B.L., Gippel, C.J., Nathan, R.J., 2004. In: Stream Hydrology: An Introduction for Ecologists. Wiley.
Grimmond, C.S.B., Oke, T.R., 1986. Urban water balance: 2. Results from a suburb of Vancouver, British Columbia. Water Resources Research 22 (10), 1404–1412.

Hamel, P., Daly, E., Fletcher, T.D., 2013. Source-control stormwater management for mitigating the impacts of urbanisation on baseflow: a review. Journal of Hydrology 485, 201–211.

Harris, E.E., Rantz, S.E., 1964. Effect of urban growth on streamflow regimen of Permanente Creek Santa Clara County, California: hydrological effects of urban growth. In: Geological Survey Water-supply Paper 1591-B. US Department of the Interior.

Hatt, B.E., Fletcher, T.D., Deletic, A., 2009. Hydrologic and pollutant removal performance of stormwater biofiltration systems at the field scale. Journal of Hydrology 365 (3–4), 310–321.

Heejun, C., 2007. Comparative streamflow characteristics in urbanizing basins in the Portland Metropolitan Area, Oregon, USA. Hydrological Processes 21 (2), 211–222.

Hollis, G.E., 1975. The effect of urbanization on floods of different recurrence interval. Water Resources Research 11, 431–435.

Jacobson, C.R., 2011. Identification and quantification of the hydrological impacts of imperviousness in urban catchments: a review. Journal of Environmental Management 92, 1438–1448.

Jarden, K.M., Jefferson, A.J., Grieser, J.M., 2016. Assessing the effects of catchment-scale urban green infrastructure retrofits on hydrograph characteristics. Hydrological Processes 30 (10), 1536–1550.

Kaushal, S.S., Belt, K.T., 2012. The urban watershed continuum: evolving spatial and temporal dimensions. Urban Ecosystems 15, 409–435.

Kennen, J.G., Riva-Murray, K., Beaulieu, K.M., 2010. Determining hydrologic factors that influence stream macroinvertebrate assemblages in the northeastern US. Ecohydrology 3, 88–106.

Kenway, S., Gregory, A., McMahon, J., 2011. Urban water mass balance analysis. Journal of Industrial Ecology 15 (5), 693–706.

Kibler, D.F., 1982. Urban Stormwater Hydrology. American Geophysical Union.

Ladson, A.R., Walsh, C.J., Fletcher, T.D., 2006. Improving stream health in urban areas by reducing runoff frequency from impervious surfaces. Australian Journal of Water Resources 10 (1), 23–33.

Laurenson, E.M., Codner, G.P., Mein, R.G., 1985. Giralang/Gungahlin Paired Catchment Study: A Review. Department of Housing and Construction, Canberra.

Leopold, L.B., 1968. Hydrology for urban land planning - A guidebook on the hydrologic effect of urban land use. US Geological Survey Circular 554 (18). http://pubs.usgs.gov/circ/1968/0554/report.pdf.

Li, C., Fletcher, T.D., Duncan, H.P., Burns, M.J., 2017. Can stormwater control measures restore altered urban flow regimes at the catchment scale? Journal of Hydrology 549, 631–653.

Loperfido, J.V., Noe, G., Jarnagin, B.S.T., Hogan, D.M., 2014. Effects of distributed and centralized stormwater best management practices and land cover on urban stream hydrology at the catchment scale. Journal of Hydrology 519, 2584–2595.

Loveridge, M., Babister, M., 2016. Interdependence of design losses and temporal patterns in design flood estimation. In: 37th Hydrology and Water Resources Symposium. Queenstown, NZ, 28 Nov – 2 Dec. Engineers Australia, pp. 276–286.

Mankad, A., Greenhill, M., 2014. Motivational indictors predicting the engagement, frequency and adequacy of rainwater tank maintenance. Water Resources Research 50 (1), 29–38.

McFarlane, D.J., 1984. The Effect of Urbanisation on Ground-water Quality and Quantity in Perth, Western Australia (Ph.D. thesis). The University of Western Australia.

Mein, R.G., Goyen, A.G., 1988. Urban runoff. Civil Engineering Transactions CE30 (4), 225–238.

Melbourne Water, 2005. WSUD: Engineering Procedures – Stormwater. CSIRO Publishing. http://www.publish.csiro.au/book/4974/.

Miller, J.D., Kim, H., Kjeldsen, T.R., Packman, J., Grebby, S., Dearden, R., 2014. Assessing the impact of urbanization on storm runoff in a peri-urban catchment using historical change in impervious cover. Journal of Hydrology 515, 59–70.

Mitchell, V.G., Mein, R.G., McMahon, T.A., 2002. Utilising stormwater and wastewater resources in urban areas. Australasian Journal of Water Resources 6 (1), 31–43.

Mitchell, V.G., McMahon, T.A., Mein, R.G., 2003. Components of the total water balance of an urban catchment. Environmental Management 32 (6), 735–746.

Moglia, M.G., Tjandraatmadja, G., Sharma, A.K., 2013. Exploring the need for rainwater tank maintenance: survey, review and simulations. Water Science and Technology: Water Supply 13, 191–201.

Olden, J.D., Poff, N.L., 2003. Redundancy and the choice of hydrologic indices for characterizing streamflow regimes. River Research and Applications 19 (2), 101–121.

Pilgrim, D., Cordery, I., 1993. Flood runoff. In: Maidment, D. (Ed.), Handbook of Hydrology. McGraw-Hill.

Price, K., 2011. Effects of watershed topography, soils, land use, and climate on baseflow hydrology in humid regions: a review. Progress in Physical Geography 35 (4), 465–492.

Prosdocimi, I., Kjeldsen, T.R., Miller, J.D., 2015. Detection and attribution or urbanization effect on flood extremes using nonstationary flood-frequency models. Water Resources Research 51 (6), 4244–4262.

Puust, R., Kapelan, Z., Savic, D.A., Koppel, T., 2010. A review of methods for leakage management in pipe networks. Urban Water Journal 7 (1), 25–45.

Qin, H., Li, Z., Fu, G., 2013. The effects of low impact development on urban flooding under different rainfall characteristics. Journal of Environmental Management 129, 577–585.

Rhea, L., Jarnagin, T., Hogan, T., Loperfido, J.V., Shuster, W.D., 2015. Effects of urbanization and stormwater control measures on streamflows in the vicinity of Clarksburg, Maryland, USA. Hydrological Processes 29 (20), 4413–4426.

Rose, S.C., Peters, N.E., 2001. Effects of urbanization on streamflow in the Atlanta area (Georgia, USA): a comparative hydrological approach. Hydrological Processes 15 (8), 1441–1457.

Schwartz, S.S., Smith, B., 2014. Slowflow fingerprints of urban hydrology. Journal of Hydrology 515, 116–128.
Selvalingham, S., Liong, S.Y., Manoharan, P.C., Laurenson, E.M., 1987. Tests of RORB parameter adjustments with urbanization. Civil Engineering Transactions 29 (2), 63–70.
Sharma, A.K., Begbie, D., Gardner, T., 2015. Rainwater Tank Systems for Water Supply: Design, Yield, Energy, Health Risks, Economics and Community Perceptions. IWA Publishing.
Shepherd, J.M., 2005. A review of current investigations of urban-induced rainfall and recommendations for the future. Earth Interactions 9 (12), 1–27.
Shuster, W.D., Rhea, L., 2013. Catchment-scale hydrologic implications of parcel-level stormwater management (Ohio USA). Journal of Hydrology 485, 177–187.
Shuster, W.D., Bonta, J., Thurston, H., Warnemuende, E., Smith, D.R., 2005. Impacts of impervious surface on watershed hydrology: a review. Urban Water Journal 2 (4), 263–275.
SKM, 2005. Determination of the minimum environmental water requirement for the Yarra River. In: Report by Sinclair Knight Merz for Melbourne Water. http://www.vewh.vic.gov.au/__data/assets/pdf_file/0008/368738/Yarra-EWR-flow-recommendations-report_final-211105.pdf.
Smith, B.K., Smith, J.A., 2015. The flashiest watersheds in the contiguous United States. American Meteorological Society 16, 2365–2381.
Smith, B.K., Smith, J.A., Baeck, M.L., Villarini, G., Wright, D.B., 2013. Spectrum of storm event hydrologic response in urban watersheds. Water Resources Research 49, 2649–2663.
Smith, J.A., Miller, A.J., Baeck, M.L., Nelson, P.A., Fisher, G.T., Meierdiercks, K.L., 2005. Extraordinary flood response of a small urban watershed to short-duration convective rainfall. Journal of Hydrometeorology 6 (5), 599–617.
Stephenson, D., 1994. Comparison of the water balance for an undeveloped and a surburban catchment. Hydrological Sciences Journal 39 (4), 295–307.
Steuer, J.J., Stensvold, K.A., Gregory, M.B., 2010. Determination of biologically significant hydrologic condition metrics in urbanizing watersheds: an empirical analysis over a range of environmental settings. Hydrobiologia 654, 27–55.
Stewardson, M.J., Gippel, C.J., 2003. Incorporating flow variability into environmental flow regimes using the flow events method. River Research and Applications 19, 459–472.
Stewart-Oaten, A., Murdoch, W.W., Parker, K.R., 1986. Environmental impact assessment: "psuedoreplication" in time? Ecology 67, 929–940.
Stormwater Committee, 1999. Urban stormwater: best practice environmental management guidelines. CSIRO.
Testoni, I., Babister, M., Retallick, M., Loveridge, M., 2016. Regional temporal patterns for Australia. In: 37th Hydrology and Water Resources Symposium 2016. Queenstown, NZ, 28 Nov – 2 Dec. Engineers Australia, pp. 541–551.
Trudeau, M.P., Richardson, M., 2015. Change in event-scale hydrologic response in two urbanizing watersheds of the Great Lakes St Lawrence Basin 1969–2010. Journal of Hydrology 523, 650–662.
Tularam, G.A., Ilahee, M., 2007. Initial loss estimates for tropical catchments of Australia. Environmental Impact Assessment Review 27 (6), 493–504.
Underwood, A.J., 1991. Beyond BACI: experimental designs for detecting human environmental impacts on temporal variations in natural populations. Australian Journal of Marine and Freshwater Research 42, 569–587.
Walsh, C.J., Kunapo, J., 2009. The importance of upland flow paths in determining urban impacts on stream ecosystems. Journal of the North American Benthological Society 28 (4), 977–990.
Walsh, C.J., Fletcher, T.D., Ladson, A.R., 2005a. Stream restoration in urban catchments through redesigning stormwater systems: looking to the catchment to save the stream. Journal of the North American Benthological Society 24 (3), 690–705.
Walsh, C.J., Roy, A., Feminella, H.J.W., Cottingham, P., Groffman, P.M., Morgan, R., 2005b. The urban stream syndrome: current knowledge and the search for a cure. Journal of the North American Benthological Society 24 (3), 706–723.
Walsh, C., Fletcher, T., Ladson, A., 2009. Retention capacity: a metric to link stream ecology and storm-water management. Journal of Hydrologic Engineering 14 (4), 399–406.
Walsh, C.J., Fletcher, T.D., Burns, M.J., 2012. Urban stormwater runoff: a new class of environmental flow problem. PLoS One. https://doi.org/10.1371/journal.pone.0045814.
Walsh, C.J., Fletcher, T.D., Bos, D., Imberger, S., 2015. Restoring a stream through retention of urban stormwater runoff: a catchment-scale experiment in a social-ecological system. Freshwater Science 34 (3), 1161–1168.
Walsh, C.J., Fletcher, T.D., Vietz, G.J., 2016a. Variability in stream ecosystem response to urbanization: unraveling the influences of physiography and urban land and water management. Progress in Physical Geography 40 (5), 714–731.
Walsh, C.J., Booth, D.B., Burns, M.J., Fletcher, T.D., Hale, R.L., Hoang, L.N., Livingston, G., Rippy, M.A., Roy, A.H., Scoggins, M., Wallace, A., 2016b. Principles for urban stormwater management to protect stream ecosystems. Freshwater Science 35 (1), 398–411.
WEF and ASCE, 2012. Design of Urban Stormwater Controls. Water Environment Federation Press.
Welsh, W.D., Vaze, J., Dutta, D., Rassam, D., Rahman, J.M., Jolly, I.D., Wallbrink, P., Podger, G.M., Bethune, M., Hardy, M., Teng, J., Lerat, J., 2012. An integrated modelling framework for regulated river systems. In: Environmental Modelling and Software. https://doi.org/10.1016/j.envsoft.2012.02.022.
Wong, T.H.F. (Ed.), 2006. Australian Runoff Quality: A Guide to Water Sensitive Urban Design. Engineers Media.
Wong, T., Breen, P.F., Lloyd, S., 2000. Water Sensitive Road Design - Design Options for Improving Stormwater Quality of Road Surfaces. Cooperative Research Centre for Catchment Hydrology, Melbourne.
Woolmington, E., Burgess, J.S., 1983. Hedonistic water use and low-flow runoff in Australia's national capital. Urban Ecology 7 (3), 215–227.
Zhang, Y., Smith, J.A., Luo, L., Wang, Z., Baeck, M.L., 2014. Urbanization and rainfall variability in the Beijing metropolitan region. Journal of Hydrometeorology 15 (6), 2219–2245.

Chapter 11

# Urbanization: Hydrology, Water Quality, and Influences on Ecosystem Health

**Fran Sheldon, Catherine Leigh, Wendy Neilan, Michael Newham, Carolyn Polson and Wade Hadwen**
*Australian Rivers Institute, Griffith University, Nathan, QLD, Australia*

## Chapter Outline



**ABSTRACT**
Urbanization significantly alters the hydrology of catchments, the transport of sediment, nutrients, and pollutants and consequently has a degrading impact on urban stream biota. In response to these impacts, water sensitive urban design (WSUD) is often seen as an approach to reduce the hydrological and pollutant risks for receiving waterbodies. We explored the impacts of urbanization across three stream types: an urban system, one subject to upstream WSUD development, and a forested stream. Our results suggested that there were strong differences in the hydrology, water quality, and macroinvertebrate assemblage composition between the FOREST site and the two sites impacted by urbanization; the WSUD and the URBAN sites. Despite the upstream WSUD development, both the WSUD and URBAN sites had hydrology characterized by higher rates of rise and fall of flood peaks. However, very little variation in macroinvertebrate assemblage composition across the three streams could be directly attributed to the changed hydrology. Rather it was differences in water quality that showed the strongest influence on the biota, suggesting that even when measures are implemented to help restore more natural flow patterns, degraded water quality can have an overriding influence on stream ecosystem health.

**Keywords:** Hydrology; Macroinvertebrates; Urbanization; Water quality.

## 11.1 INTRODUCTION

Urbanization has significantly altered the hydrology of catchments and consequently the transport of sediment, nutrients, and pollutants. These changes begin with the initial impacts of deforesting catchments into spaces for agriculture or urban development, which significantly increases runoff and the transport of sediment, nutrient, and organic matter into receiving waterways (Heaney and Huber, 1984; Paul and Meyer, 2001; Hawley and Vietz, 2016). Thereafter, conventionally developed urban catchments tend to be characterized by having very high proportions of impervious area (Fig. 11.1; Walsh et al., 2005). This affects the hydrology and ecology of urban streams in a number of ways. First, the impervious surfaces inhibit the infiltration of rainfall into soils and, therefore, can limit the recharge of groundwater resources (Brown et al., 2009).

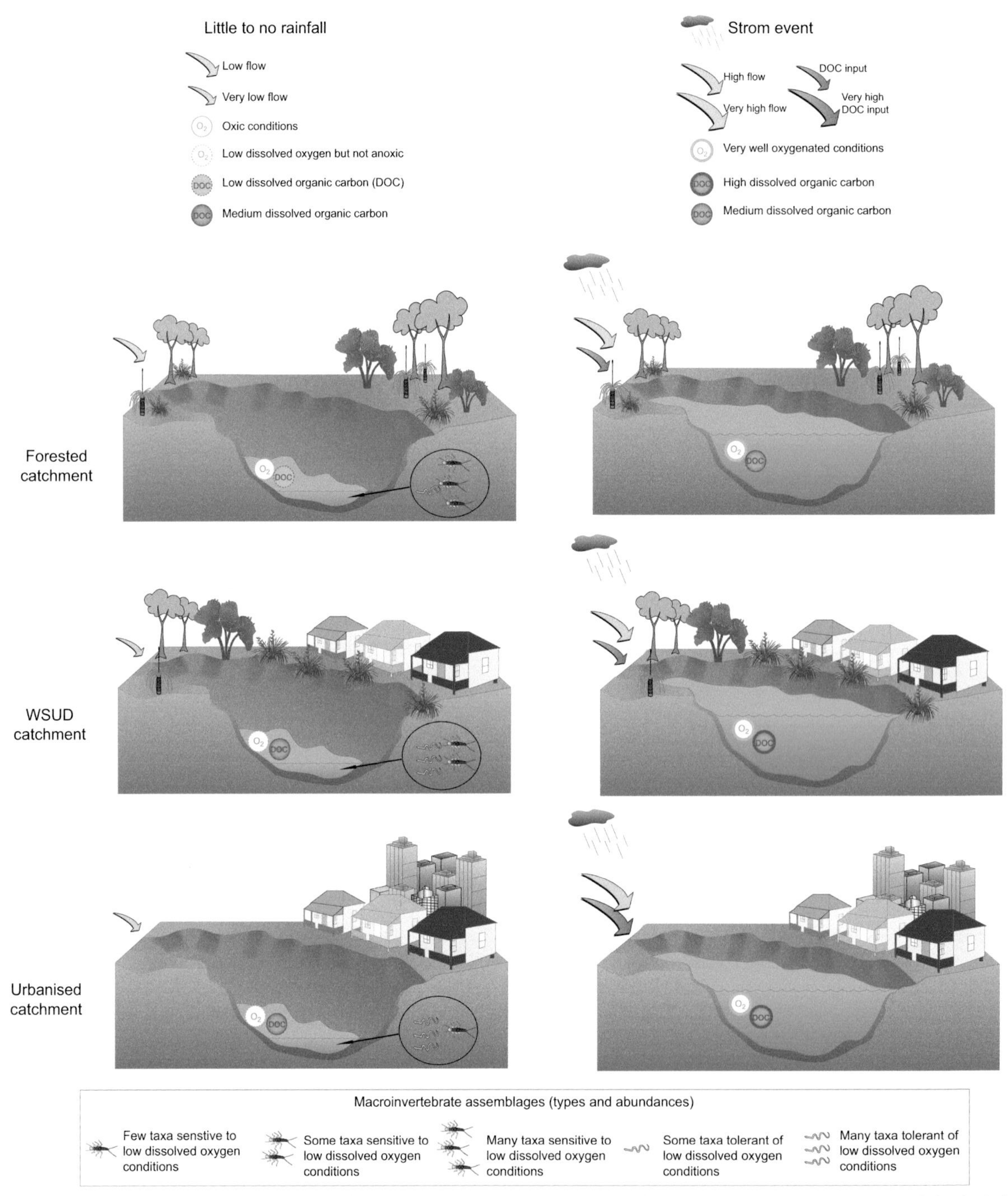


FIGURE 11.1 Conceptual model of the impacts of urbanization in stream ecosystems. Within the model, the two extremes of the hydrological continuum are represented (No Flow and Storm events) for each of the three upstream catchment conditions FOREST, WSUD and URBAN). WSUD developments vary in their extent and are mostly designed to reduce the hydrological impacts of urbanization but removing the flashiness of flows. *From Sheldon, F., Pagendam, D., Newham, M., McIntosh, B., Hartcher, M., Hodgson, G., Leigh, C., Neilan, W., 2012b. Critical Thresholds of Ecological Function and Recovery Associated with Flow Events in Urban Streams. Urban Water Security Research Alliance Technical Report No. 99.*

Second, impervious surfaces capture and direct rainfall into networks of drains and pipes, which can quickly discharge into urban creeks. The consequences of this networking are classically observed in the hydrographs of urban streams, where rainfall events are much flashier, with higher flow peaks and shorter durations than equivalent events in forested catchments (Walsh et al., 2005; Burns et al., 2012; see also Chapter 10).

The consequences of urban hydrology on the ecology of urban streams have received considerable attention in the literature. Not only does the stormwater delivered through piped networks represent an incredibly powerful erosive force (Burns et al., 2012) but also carries a wide range of contaminants and pollutants that stress receiving waterbodies and their biota (Herlihy et al., 1998; Grimm et al., 2005). The responses of biota in urban streams have been studied at multiple scales, with researchers highlighting how changes in hydrology, sediment, and nutrient loads can have a transformational effect on the way that stream ecosystems function (Grimm et al., 2005; Kaye et al., 2006; Walsh et al., 2007; Tsoi et al., 2011; Burns et al., 2012).

Given the evidence of the impacts of urbanization on stream ecosystems and the critical role of impervious surfaces in changing the routing and speed of water from rainfall events entering waterways (Fletcher et al., 2013), there has been a huge investment in urban stormwater management since the 1990s (Burns et al., 2012). Most of these efforts are commonly referred to as water sensitive urban design (WSUD), which is an approach to manage stormwater in urban developments to reduce the hydrological and pollutant risks for receiving waterbodies (Roy et al., 2008). WSUD designs typically are engineered systems that seek to retain stormwater and treat it before discharge into urban streams (Roy et al., 2008). The pollutant reduction targets for WSUD assets can vary, but typically the stated objectives focus on reducing total suspended sediments entering waterways by around 80% and reducing total nitrogen (N) and total phosphorus (P) loads by 45% (Parker, 2010).

There are a wide range of WSUD features that can be implemented to address stormwater management in urban contexts. These include, but are not restricted to, infiltration trenches, porous paving, rain gardens, rainwater tanks, swales, sediment basins, gross pollutant traps, and constructed wetlands (Parker, 2010). Despite the proliferation of designs and their implementation over the past 20 years, there are surprisingly few field-based assessments of WSUD system performance (but see Parker, 2010 and Adyel et al., 2016). This lack of field assessments creates two issues. First, new designs and approaches are not being improved using real-world experience and evidence. This is a problem, particularly given the rapid evolution of WSUD, because all designs are predicated on laboratory or pilot-scale performance studies that have not always been adequately scaled up, or repeated in different contexts, to provide support for implementation (Parker, 2010). Second, it remains unclear as to whether the condition of receiving waterbodies has responded, in either a negative or a positive way, to WSUD implementation. This is less a question of engineering, design, and monitoring of WSUD assets and more a challenge around environmental monitoring and ecosystem health assessment. There are many technical challenges in this space, including the mismatch in scale of WSUD implementation and the scale of threats to the condition of urban streams (Peterson et al., 2011) and the fact that long lag times are common, as we seek to understand how ecosystems respond to management interventions (Tsoi et al., 2011).

In this chapter, we seek to examine urban stream ecology in the context of urban development by contrasting the hydrology, water quality, and ecosystem health, as measured by macroinvertebrate assemblages, across three stream reaches with different upstream conditions—forested, WSUD, and urban. We used indices based on macroinvertebrate family presence and their abundance at each site as our biological measure of ecosystem health. Macroinvertebrates as a broad group show a diverse range of tolerances to anthropogenic disturbance, and changes in the presence and abundance of specific groups, particularly the sensitive EPT taxa, provide good indications of disturbance (Clarke et al., 2003). The EPT taxa, insects of the orders Ephemeroptera (mayflies), Plecoptera (stoneflies), and Trichoptera (caddisflies), are a particularly sensitive group often used in ecosystem health assessment (Lenat and Crawford, 1994; Sponseller et al., 2001). The aim here is to add to our knowledge base around how streams might respond to WSUD implementation, at a large scale, and to inform our future approaches to WSUD performance assessment.

Fig. 11.1 presents a conceptual model of the impacts of urbanization in stream ecosystems. Within the model, two extremes are represented (FOREST and URBAN). WSUD developments vary in their extent and are mostly designed to reduce the hydrological impacts of urbanization by removing the flashiness of flows. The capacity for any WSUD development to restore, or maintain, ecosystem health should depend on the extent to which it was able to recreate the natural flow regime of the stream ecosystem. Hence, we would expect water quality and ecosystem health parameters to vary along this continuum, reflecting the degree to which WSUD is implemented in any given catchment.

## 11.2 METHODS

### 11.2.1 Study sites and study design

Three broad reaches of three streams in South East Queensland, Australia, with different upstream catchment conditions (FOREST, WSUD, and URBAN) were chosen to explore temporal trends in hydrology, water quality, and macroinvertebrate assemblage composition over a 12-month period, from winter (dry) through to summer (wet) (June 2011—May 2012),

**TABLE 11.1 Description and Location of the Three Gaged Catchments**

| | FOREST | WSUD | URBAN |
|---|---|---|---|
| Stream and suburb | Tingalpa Creek, Sheldon | Blunder Creek, Daintree Crescent, Forest Lake | Stable Swamp Creek, Sunnybank |
| Latitude (degree) | −27.57 | −27.62 | −27.58 |
| Longitude (degree) | 153.18 | 152.97 | 153.05 |
| Upstream catchment area (ha) | 2785 | 360 | 442 |
| Total impervious area upstream (%) | 1 | 42 | 38 |
| Catchment slope (%) | 1 | 1.1 | 1.5 |

From Chowdhury, R., Gardner, T., Gardiner, R., Hartcher, M., Aryal, S., Ashbolt, S., Petrone, K., Tonks, M., Ferguson, B., Maheepala S., McIntosh, B.S., 2013. SEQ Catchment Modelling for Stormwater Harvesting Research: Instrumentation and Hydrological Model Calibration and Validation. Urban Water Security Research Alliance Technical Report No. 83.

reflecting the annual change from dry to wet conditions. As we were aiming to reduce site-based differences and maximize upstream catchment differences, sites were selected so they had similar site-based riparian cover (>60%). All sites were instrumented with hydrological monitoring stations that collected both hourly flow and water quality data. The three stream reaches were Tingalpa Creek (FOREST treatment), Blunder Creek tributary (WSUD treatment), and Stable Swamp Creek (URBAN treatment) (Table 11.1; Fig. 11.2). This design may be considered pseudoreplicated as each treatment was isolated to a specific stream. To overcome this problem, we chose three stream reaches, separated by approximately 50 m of stream, which were considered to be "independent" sites, and assessed our findings against the predictions from the conceptual model (Fig. 11.1). We were limited to three streams through the availability of continual data loggers.

Tingalpa Creek (FOREST) is an upland creek in the Redlands region of South East Queensland; the upstream catchment was completely forested. At the sampling sites along the creek reach, the channel comprised bedrock riffles, runs and pools with complex microhabitats of snags (fallen timber), tree roots, and macrophytes (Fig. 11.2). Stable Swamp Creek (URBAN) and the tributary to Blunder Creek (WSUD) were both in the Oxley−Blunder Creek subcatchment of the lower Brisbane River. The upstream reaches of Stable Swamp Creek (URBAN) are highly urbanized, the channel is degraded and deeply incised, and pools have deep silt with poor habitat quality, mostly dominated by introduced macrophytes. Upstream of the sampling reach, there is a large off-stream wetland that most likely plays a role in baseline hydrology, particularly after large flow events. The WSUD site, a tributary of Blunder Creek, was downstream of Forest Lake, a large stormwater retention basin constructed during the 1990s as part of a masterplan residential development; the channel is deeply incised and degraded with mostly poor habitat quality. The presence of good riparian vegetation cover on both banks, and associated in-channel woody debris and relatively stable bank structure, has allowed some riffle development along the channel. Given the nature of the WSUD development at Forest Lake, we would expect the major impacts on the downstream tributary to Blunder Creek to be associated with changes in hydrology, with most of the impervious surfaces associated with urbanization upstream of the Forest Lake impoundment. Based on the conceptual model (Fig. 11.1), we would predict the hydrological patterns evident at the WSUD site to resemble those of the FOREST site more than those of the URBAN site. However, despite the expected positive impact on hydrology of Forest Lake on the WSUD site and the presence of good riparian vegetation in the reach, the immediate surrounding area was highly urbanized and likely to have a strong influence on both water quality and macroinvertebrate assemblage diversity (Walsh et al., 2007).

## 11.2.2 Hydrology

Hydrological data for the three catchments was obtained from a larger dataset of 12 catchments, which had been instrumented (for full details, see Chowdhury et al., 2013). A tipping bucket rain gage (0.2 mm) and pressure transducer with data logger measured continuous 6-min rainfall and water height data, respectively, at each site from 2009 until the end of 2012. Gaps in the hydrology data were filled using linear interpolation, which fills the gaps in the data by drawing a straight line between the ends of the data where the gap occurs (Marsh, 2004). A series of hydrological metrics (see Table 11.2) were calculated using daily flow data and summarized by calendar month for the period January 2009 until December 2011. These metrics were chosen based on their relevance to the conceptual understanding of hydrological impacts on ecological health in urban streams (Walsh et al., 2005; Walsh and Kunapo, 2009). To explore patterns in the

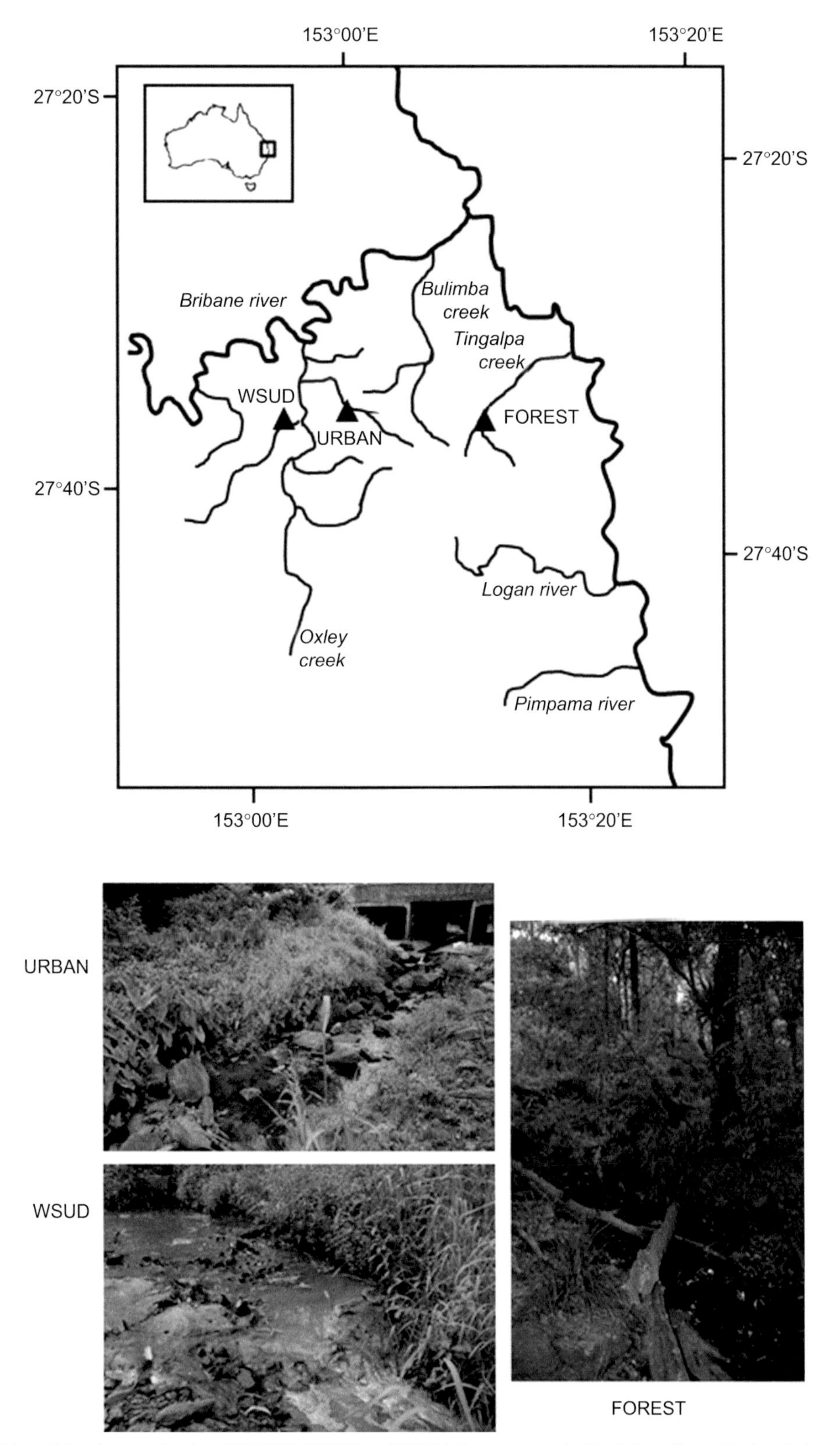


**FIGURE 11.2** Position of the three study sites, FOREST, WSUD and URBAN, on streams in South East Queensland and photos taken at each site.

**TABLE 11.2 Description of the Variables Calculated From the Daily Flow Data (Marsh, 2004)**

| Variable Name | Variable Acronym | Variable Description |
|---|---|---|
| Minimum | Min | Minimum is the smallest value for flow recorded for the time period. |
| Maximum | Max | Maximum is the largest value for flow recorded for the time period. |
| Percentile 10 | P 10 | The 10th percentile is the value that is exceeded by 10% of the records. |
| Percentile 90 | P 90 | The 90th percentile is the value that is exceeded by 90% of the records. |
| Mean daily flow | MDF | The mean daily flow is a measure of central tendency and is calculated as the average of the records (sum of values/number of days in the time period). |
| Median daily flow | Med | The median is the "middle" value for the entire record: it is the value exceeded 50% of the time. For flow data, the median is usually much lower than the mean daily flow because the distribution of discharge data is negatively skewed with a lower limit of zero and no upper limit. |
| Coefficient of variation | CV | The CV of daily flow is the mean of all daily flow values divided by the standard deviation for the daily flow values. |
| Standard deviation | STD | The standard deviation is a measure of how widely the values are dispersed from the mean value. The standard deviation has the same units as the input data. |
| Skewness | Skw | Skewness is a measure of how different the mean and median are. Skew = mean/median. |
| Number of zero flow days | Zer | The number of zero days counted. Note that the number of zero flow days does not include days with a missing record unless they are filled with zero values. |
| Number of high spell (5) | HS(5)Num | Number of times the flow exceeded five times the mean flow value for the time period. |
| Number of high spell (10) | HS(10)Num | Number of times the flow exceeded 10 times the mean flow value for the time period. |
| Number of low spell (0.5) | LS(0.5)Num | Number of times the flow fell below half the mean flow value for the time period. |
| Number of low spell (0.1) | LS(0.1)Num | Number of times the flow fell below one 10th of the mean flow value for the time period. |
| Longest low spell (0.5) | LS(0.5)Long | Duration (in days) of longest low flow event below half the mean flow value for the time period. |
| Mean of low spell (0.5) troughs | LS(0.5)Peak | Mean of low spell troughs. Low spell threshold was set at half the mean flow value for the time period. |
| Mean duration of low spell (0.5) | LS(0.5)MeanDur | Mean duration (in days) of low spell events. Low spell threshold was set at half the mean flow value for the time period. |
| LS(0.5)TotDur | LS(0.5)TotDur | Total duration (in days) of low spell events for the time period. Low spell threshold was set at half the mean flow value for the time period. |
| Number of rises | NumRise | Number of continuous periods of rise. |
| Mean magnitude of rises | MMagRise | Mean difference in the flow values between the start and end of the rise. |
| Mean duration of rises | MDurRise | Mean duration of periods of rise. |
| Total duration of rises | TotDurRise | Total duration of periods of rise. |
| Mean rate of rise | MRateRise | Mean rate of rise. |
| Greatest rate of rise | GreatRatRise | The fastest rate of rise. |
| Number of falls | NumFall | Number of continuous periods of fall. |
| Mean magnitude of falls | MMagFall | Mean difference in the flow values between the start and end of the fall. |

*Continued*

**TABLE 11.2** Description of the Variables Calculated From the Daily Flow Data (Marsh, 2004)—cont'd

| Variable Name | Variable Acronym | Variable Description |
|---|---|---|
| Mean duration of falls | MDurFall | Mean duration of periods of fall. |
| Total duration of falls | TotDurFall | Total duration of periods of fall. |
| Mean rate of fall | MRateFall | Mean rate of fall. |
| Greatest rate of fall | GreatRateFall | The fastest rate of fall. |
| Baseflow index | BFI | Ratio of baseflow to total flow in a period. |
| Flood flow index | FFI | Ratio of nonbaseflow to total flow in a period. |
| Mean daily baseflow | MDBF | Mean of baseflow in a period. |

hydrological data, we used a multivariate analysis technique, principal component analysis (PCA), to help visualize hydrological similarities, or differences, between the study streams (Quinn and Keough, 2002). PCA can summarize complex multidimensional data into fewer dimensions allowing the strong drivers of difference (in our case, specific calculated hydrological metrics) to be determined. Differences in hydrological metrics were explored between treatment streams.

## 11.2.3 Water quality

A water quality sonde was installed in each catchment at the same site as the gage (for full details, see Chowdhury et al., 2013) to measure pH, dissolved oxygen (DO), turbidity, and electrical conductivity at hourly intervals. Water quality data for each metric were converted into daily maximum, minimum, mean, median, and range values. To understand the influence of water quality on macroinvertebrate assemblage composition, the median monthly data for the sample months were calculated from the daily data because macroinvertebrates are more likely to respond to the background water quality regime, rather than specific water quality measurements on any one sample day.

## 11.2.4 Macroinvertebrate assemblages

Urbanization, through changes in hydrology, can adversely impact habitat diversity and complexity within streams (Walsh et al., 2005). Recognizing this, we mapped each site into distinct microhabitats, including "riffle," "pool," "macrophyte (aquatic plants)," and "snag (woody debris)." Replicate samples from each microhabitat were taken by sweeping a 250 μm mesh pond net over an area approximating 5 $m^2$ for 20 s. Samples were returned to the laboratory where they were sorted and macroinvertebrates identified to family using a stereo microscope and the abundance of each distinct taxa recorded. For analyses, both the complete community data and the subset of EPT: Ephemeroptera (mayflies), Plecoptera (stoneflies), and Trichoptera (caddisflies) abundance data were used.

To explore differences in assemblage composition between the three sites, a suite of multivariate analysis techniques was used (see Quinn and Keough, 2002). To reduce the influence of extremely abundant or rare taxa on patterns, data (abundance of each macroinvertebrate family in each sample) were square root transformed before analysis. To explore changes in assemblage composition between sites and through time, the Bray—Curtis similarity coefficient (Quinn and Keough, 2002) was applied to the transformed data. This calculates a value of similarity between each sample based on the presence and abundance of each macroinvertebrate family. All taxa occurring in one or more samples were retained in the subsequent analyses. Using the Bray—Curtis similarity matrix, analysis of similarities (ANOSIM) was used to statistically explore differences between each stream, with these differences displayed visually using nonmetric multidimensional scaling (MDS) ordination. To explore which families might be contributing to the differences between sites, the similarity percentages procedure (SIMPER) was used (Clarke and Gorley, 2001). Finally, to explore the influence of background water quality and hydrology on assemblage composition, monthly median values for recorded water quality parameters and calculated hydrology metrics were generated. The BIOENV routine was then used to explore which variables from each of the two environmental datasets best explained the observed patterns in the macroinvertebrate assemblage dataset. All multivariate analyses were conducted in the Primer-E Software package (PRIMER-E: www.primer-e.com).

## 11.3 RESULTS AND DISCUSSION

### 11.3.1 Hydrology

In keeping with the conceptual model on the influence of urbanization on stream hydrology (Walsh et al., 2005), both the URBAN and WSUD sites had higher total discharge and more frequent discharge events than the FOREST site (Fig. 11.3). When the hydrological metrics (Table 11.2) were calculated from the monthly data (2009–11) from the three sites and analyzed using a PCA analysis, there was clear overlap in the hydrological signature of all three sites. The distribution of the sample months along PC1 (x-axis) reflects the flashiness of the monthly hydrology as measured by mean daily flow and the magnitude of the rate of rise and fall of flood flows. Only some months from the URBAN and WSUD sites showed clear differences through their distribution along PC1 (x-axis) (Fig. 11.4), associated with variation in the number of high flow events, the magnitude and rate of the flood rising limb, and the magnitude of the flood falling limb (Fig. 11.4). These hydrological metrics describe how flashy any individual event is. The position of the three streams on the PCA plot suggests that the URBAN and WSUD sites had, overall, more variable flows between months, with some months having high flows that were characterized by rapid rates of rise and fall. The spread of monthly hydrology points along PC2 (y-axis) suggested similar variability between all three sites in monthly baseflow conditions, as measured by the baseflow index (BFI), and flood conditions, as measured by the flood flow index (see Table 11.2).

To better understand the influence of upstream catchment on hydrology, the sequence of very high flow months (December 2010 and January 2011) in the URBAN and WSUD sites was removed. In the resulting reduced PCA, separation of monthly hydrology was more apparent (Fig. 11.5). The distribution of the sample months along PC1 again reflects the flashiness of the monthly hydrology, as measured by mean daily flow, and the magnitude of the rate of rise and fall of flood flows, whereas the distribution along PC2 reflects monthly minimum and baseflow conditions. The distribution of hydrological sample months from the FOREST stream is less variable across both PC1 and PC2, suggesting lower temporal variability in flood magnitude and rate of rise and fall, as well as less temporal variability in low flows and BFI. The URBAN site tended to have the greatest temporal variability, as indicated by the more dispersed hydrological sample months across both axes. Overall, using the suite of hydrological metrics in a PCA, the hydrological changes imposed by the upstream WSUD intervention on the Blunder Creek WSUD site appeared to contribute to it being much more similar to the FOREST site than to the URBAN site.

### 11.3.2 Water quality

#### *11.3.2.1 Dissolved oxygen*

DO varied throughout the year with mean daily DO generally much lower in the summer months (December to February) compared with the winter months, across all sites (Fig. 11.6a). Interestingly, the URBAN site had much higher mean daily

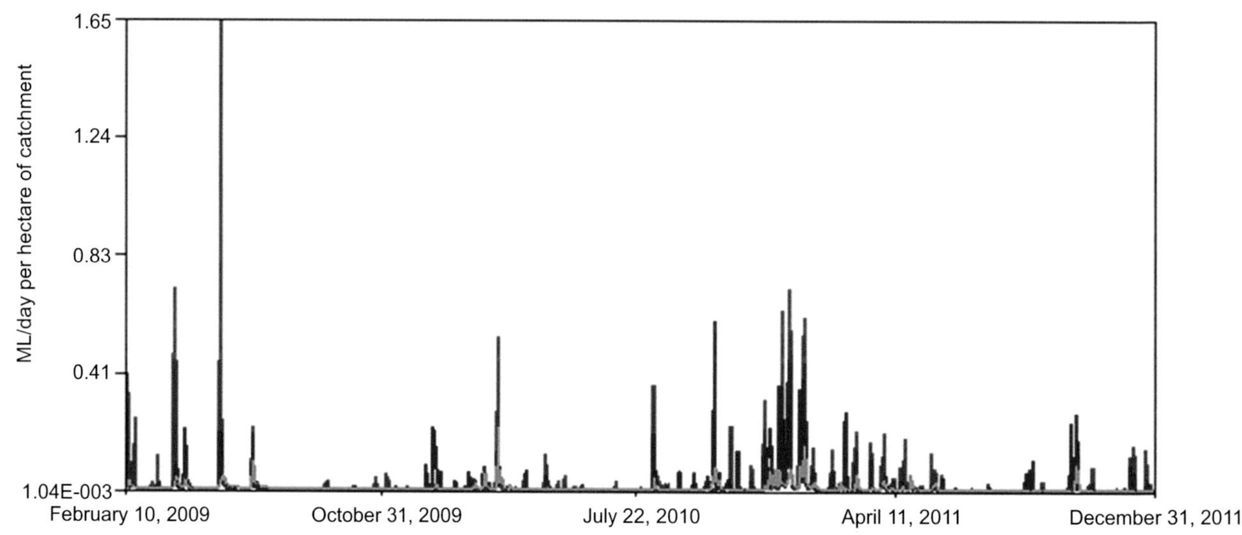


**FIGURE 11.3** Hydrograph of daily flow (ML/day per hectare of catchment) for FOREST (Green), WSUD (Blue), and URBAN (Red) sites between 2009 and 2011. *From Sheldon, F., Pagendam, D., Newham, M., McIntosh, B., Hartcher, M., Hodgson, G., Leigh, C., Neilan, W., 2012b. Critical Thresholds of Ecological Function and Recovery Associated with Flow Events in Urban Streams. Urban Water Security Research Alliance Technical Report No. 99.*

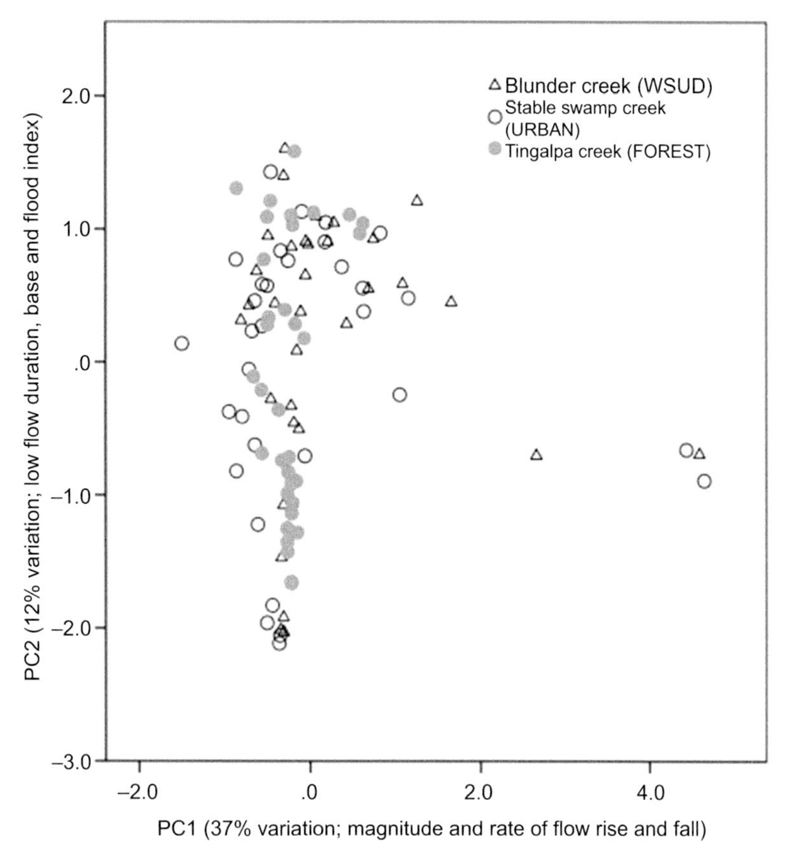


**FIGURE 11.4** Principal component analysis (varimax rotation) of hydrological metrics calculated from monthly flow data from 2009 until 2011 for three streams in South East Queensland; an URBAN site (Stable Swamp Creek), a WSUD site (upstream Blunder Creek), and a FOREST site (Tingalpa Creek) (see Table 11.2).

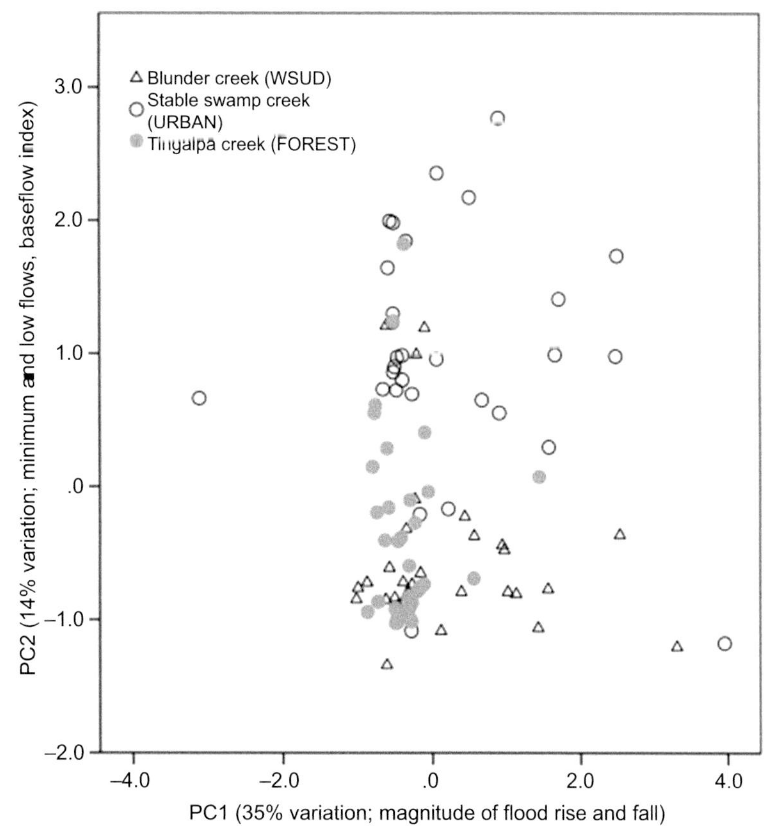


**FIGURE 11.5** Principal component analysis (varimax rotation) of hydrological metrics calculated from monthly flow data from 2009 until 2011 for three streams in South East Queensland with the extremely high flow months from Fig. 11.3 removed; an URBAN site (Stable Swamp Creek), a WSUD site (upstream Blunder Creek), and a FOREST site (Tingalpa Creek) (See Table 11.2).

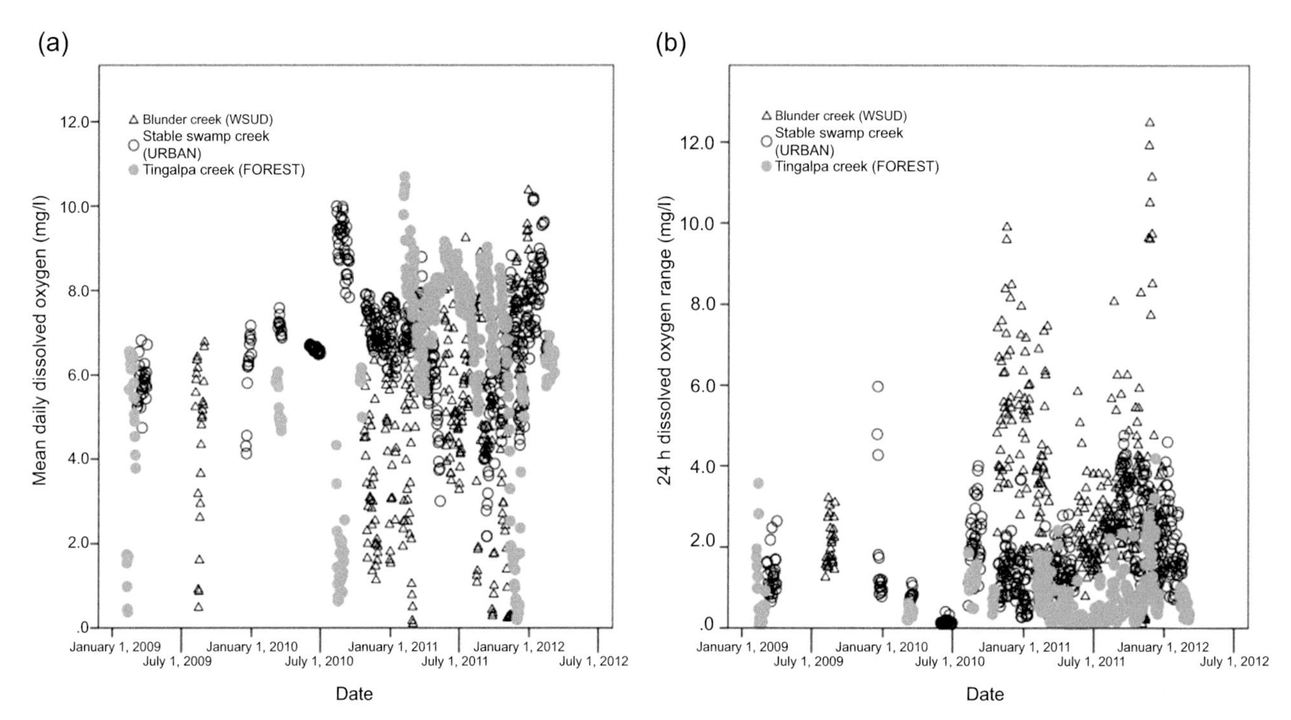


**FIGURE 11.6** (a) Mean daily dissolved oxygen (DO) (mg/L) levels across all three sites for the period of data logging (January 2009–June 2012) and (b) daily DO range (mg/L) across all three sites for the period of data logging (January 2009–June 2012). Note that continual logging of all sites commenced in January 2011.

DO levels across most of the year compared with either the WSUD or FOREST sites, possibly reflecting the higher baseflow at this site. The daily range in DO is often used as an indicator of stream health (Bunn et al., 2010), with lower daily ranges seen to be more typical of healthy streams. Across our sites, both the WSUD and URBAN sites had larger daily ranges across all seasons compared with the FOREST site (Fig. 11.6b). These DO results are consistent with how we understand water quality in urban creeks. Generally DO is much lower across the seasons because of increased respiration rates associated with higher organic loads from impervious surfaces in urban catchments (Blunder Creek) unless water flow is maintained (Stable Swamp Creek). Intuitively, upstream WSUD should reduce nutrient and organic inputs to urban streams and therefore reduce the likelihood of large DO fluctuations. However, our WSUD site is downstream of Forest Lake, a large stormwater retention basin, and while this system appears to positively influence some aspects of hydrology (as outlined previously), it is eutrophic and likely increases nutrient levels in the downstream tributaries. Interestingly, daily mean DO levels in the FOREST site were often extremely low, particularly in the summer (Fig. 11.6a) possibly reflecting the position of the sonde in a pool within the creek, the high organic load from riparian leaf litter, and the often low-flow conditions at the FOREST site between storm events. This suggests that at times during the year, conditions in the forested catchment will be just as severe as in the urbanized catchments. However, in the urbanized catchments, there are likely to be other drivers of poor health, including increased loads of labile organic carbon from road runoff, high heavy metals, and other pollutants that will exacerbate the ecosystem health impacts of low levels of DO.

### *11.3.2.2 Water temperature*

Water temperature at the three sites varied across the year as expected, with maximum daily temperatures around 10–15°C during the winter months (June, July, August) and summer temperatures around 25–30°C (Fig. 11.7a). Maximum daily temperature and daily temperature range (over 24 h) at a site are considered to be good indicators of stream health (Bunn et al., 2010), with higher maximums and temperature ranges indicative of poorer stream health. However, despite similar riparian conditions across the three sites, which should have mediated water temperature through shading, higher maximum daily temperatures were found at both WSUD and URBAN sites when compared with the FOREST site (Fig. 11.7a). Although the three sites in this study were chosen because of their good riparian cover, they differed in the extent of upstream riparian cover. Only the forested Tingalpa Creek (FOREST) has continuous riparian cover upstream of the sample site, and the influence of this extensive shading can be seen in its mean daily temperature

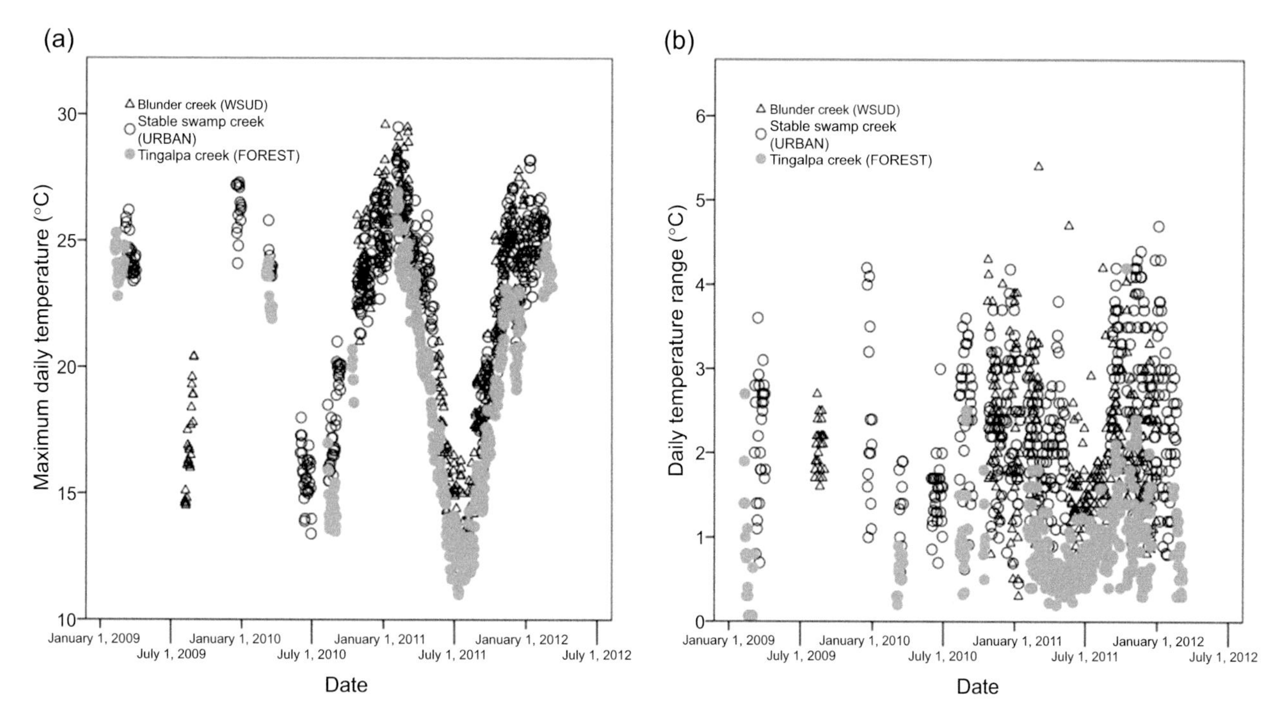


**FIGURE 11.7** (a) Maximum daily temperature (°C) across all three sites for the period of data logging (January 2009–June 2012) and (b) daily temperature range (°C) across all three sites for the period of data logging (January 2009–June 2012). Note that continual logging of all sites commenced in January 2011.

range that is nearly 1.5°C lower than either the URBAN or WSUD sites. The much higher daily temperature range in the WSUD and URBAN sites most likely reflects upstream discontinuous riparian cover. This suggests that for good stream health outcomes, even when WSUD is implemented, the condition of the upstream catchment can have an overriding influence (Sheldon et al., 2012a).

### 11.3.2.3 Electrical conductivity

High levels of water electrical conductivity can also pose risks for stream health. Increased conductivity has been associated with land use change, usually the result of converting forested catchments into agricultural or urban landscapes (Suárez et al., 2017). We would therefore expect higher conductivities in the more urbanized streams. However, across the three sites, the role of urbanization in increased stream conductivity was not clear. Highest conductivity values were observed in the WSUD site (Fig. 11.8) with little apparent difference between the URBAN and FOREST sites. The lower overall conductivity in the URBAN site may reflect the higher baseflow recorded in that system compared with the FOREST site, where low flows and isolation of instream pools were common during the drier months. The exact cause of the higher baseflow in the URBAN site is unclear; it may have been associated with deep drainage through garden watering or leaking local water mains. However, there was also an isolated off-stream wetland upstream from the URBAN site that may have contributed to baseflow through alluvial flow paths. The much higher conductivity observed in the tributary to Blunder Creek (WSUD) is likely the result of considerable dissolved iron entering the stream at that point. The exact cause of the iron floc in the stream was unclear, but probably reflected biogeochemical processes occurring within the soils surrounding the stream.

### 11.3.2.4 pH and impacts of iron precipitate at the Blunder Creek WSUD site

pH was not expected to vary greatly between the three streams, as pH has not been observed to vary greatly in streams across South East Queensland (Bunn et al., 2010; Sheldon et al., 2012a) and stream acidification is not a major issue. However, when the minimum daily pH from the three sites was compared, the Blunder Creek WSUD site had a significantly lower minimum pH (analysis of variance: $F_{2,1139} = 60.286$, $P < .001$) with average minimum values of <6.5 (Fig. 11.9). Interestingly, the FOREST site also had quite variable pH values during the observation period with a number of extremely low values (Fig. 11.9); the sonde at the Tingalpa site was positioned in a pool, and it is likely that under

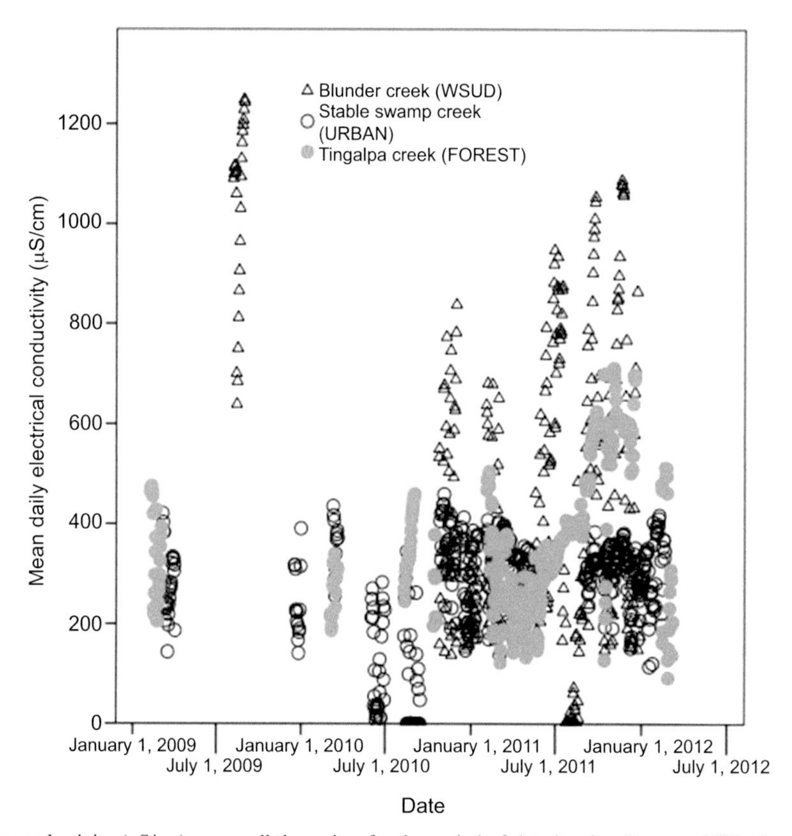


**FIGURE 11.8** Mean daily conductivity (μS/cm) across all three sites for the period of data logging (January 2009–June 2012). Note that continual logging of all sites commenced in January 2011.

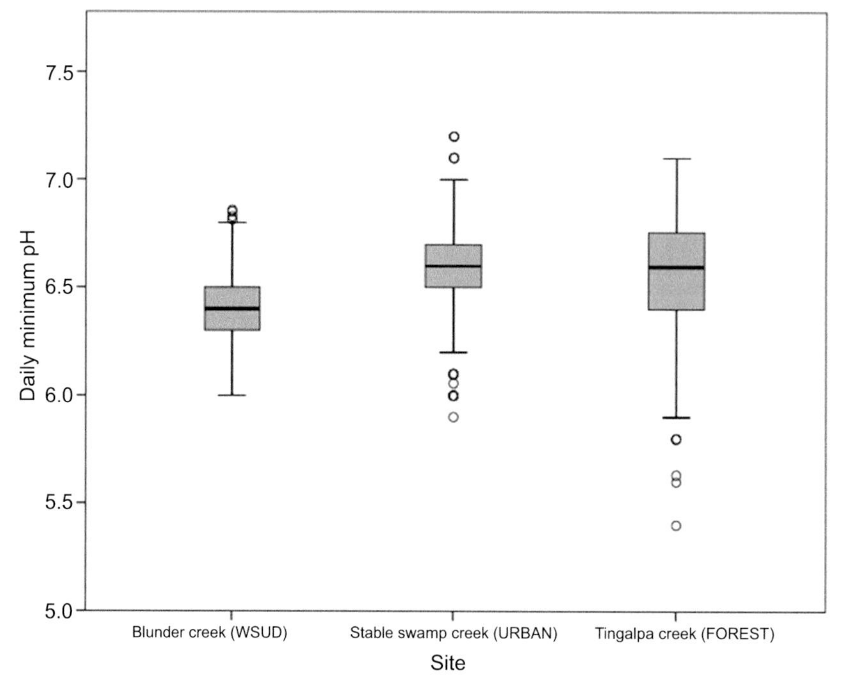


**FIGURE 11.9** Box plots (minimum, first quartile, median, third quartile, and maximum) of recorded daily minimum pH across all three sites for the period of data logging (January 2009–June 2012). Note that continual logging of all sites commenced in January 2011.

certain low-flow conditions, respiration increased within the pool may have caused these occasional drops in pH. The extremely low average minimum pH at the WSUD Blunder Creek site is likely to have been caused by an increase in iron floc and associated dissolved iron at the site.

Iron precipitates are a natural and common occurrence even in pristine streams (Abesser et al., 2006; Duckworth et al., 2009). They occur where dissolved ferrous iron from groundwater enters streams and comes in contact with an environment where oxygen is abundant (Fig. 11.10; Schwertmann, 1991; Rhoton et al., 2002). Ferrous iron oxidation in

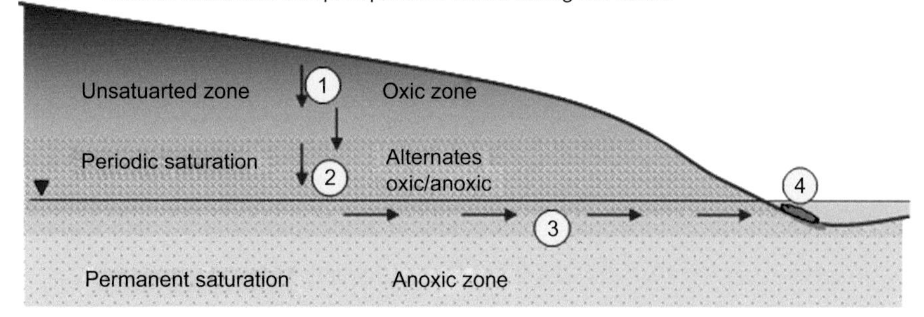


**FIGURE 11.10** Conceptual model of iron precipitation in a natural stream and exacerbated conditions in an incised urban stream. *From Sheldon, F., Pagendam, D., Newham, M., McIntosh, B., Hartcher, M., Hodgson, G., Leigh, C., Neilan, W., 2012b. Critical Thresholds of Ecological Function and Recovery Associated with Flow Events in Urban Streams. Urban Water Security Research Alliance Technical Report No. 99.*

streams can have negative impacts at sites where it occurs. The oxidation and hydrolysis reactions of ferrous iron produce $H^+$ ions that can cause both acidification and salinization of stream waters (Singer and Stumm, 1970):

$$Fe^{2+} + 1/4O_2 + H^+ = Fe^{3+} + 1/2H_2O$$

$$Fe^{3+} + 3H_2O = Fe(OH)_3(s) + 3H^+$$

The release/oxidation of ferrous iron into streams in excess can contribute to water quality and habitat degradation, and it has been known for more than 20 years that high levels of iron in streams reduce macroinvertebrate abundance and density (Rasmussen and Lindegaard, 1988). Circumstances that may enhance the release of ferrous iron into streams include forest clearing and reduced evapotranspiration, which increases deep drainage, runoff, and dissolved organic matter loads, leading to a greater load of available iron (Albert et al., 2005). Industrial and urban effluents and runoff may also contribute to excess iron loads in streams, causing degradation and in extreme cases smothering the benthic habitat (Nedeau et al., 2003).

From observations made during fieldwork at the Blunder Creek WSUD site, we suggest that symptoms of urbanization on streams may also lead to increased loads of iron and precipitates. In particular, we noticed that incision and downcutting of streams, caused by increased runoff from impervious surfaces, expose previously buried subsoils, rich in ferrous iron minerals. Because subsurface flow is released from these deeper soil layers into the stream channel, it may carry with it a greater load of iron than water discharged from shallower soil layers. Fig. 11.10 presents a conceptual model of iron precipitation in a natural stream. The model describes the changes that occur as water moves through the soil, changing its percent oxygen saturation and thereby mobilizing iron that occurs within the soil due to bacterial anaerobic respiration.

## 11.4 MACROINVERTEBRATE ASSEMBLAGE COMPOSITION

### 11.4.1 General patterns

Almost 75,000 individuals were collected from 106 taxa across the 12-month sampling period. The mean species richness per sample (number of species per sample) differed between sites (analysis of variance $F_{2,101} = 5.679$, $P < .01$); the FOREST site had a mean richness per sample of 18 ($\pm$0.8), the WSUD site had a mean richness of 21 ($\pm$1.5), while the URBAN site had a mean richness of 22 ($\pm$1.0) (Table 11.3; Fig. 11.11). Similarities in species richness do not represent similarity in species composition; rather an increase in species richness is often associated with disturbance because factors that cause disturbance in streams, such as higher nutrient loads, often create conditions for large numbers of invasive "pest" species. To better understand the influence of urbanization on stream assemblage composition, the composition of the overall assemblage was considered, along with the presence and abundance of sensitive taxa such as the EPT taxa (Sponseller et al., 2001).

### 11.4.2 Assemblage composition

Macroinvertebrate assemblage composition differed significantly between the three sites (ANOSIM Global $R = 0.678$, $P < .001$). MDS allows a two-dimensional visualization of the similarity in assemblage composition for any two samples. Samples (depicted by points) that are closer together in two-dimensional space are more similar in terms of assemblage composition. For an explanation of MDS, see Quinn and Keough (2002). Differences can be seen in the visual representation of the similarity between samples from each site (Fig. 11.12). The assemblage of macroinvertebrates from the FOREST site was distinctive compared with that from either the URBAN or WSUD site, with only minor overlap in similarity between samples from the WSUD and URBAN sites (Fig. 11.12).

Samples from the URBAN site, regardless of habitat, had a within-site similarity of 43.5% and were dominated by Dipterans from the family Chironominae, or midges, with the Chironominae contributing 33%, while the Oligochaeta (worms) contributed a further 12%. Microcrustaceans contributed a further 6.5% and the hydrobiid gastropods (snails) a further 6%. The within-site similarity of all samples from the WSUD site was 44.5% with the major contribution to within-site similarity from the microcrustaceans (14%) and gastropods (snails) of the families Planorbidae and Physidae (24%) (Fig. 11.13). In contrast, the within-site similarity of all samples from the FOREST site was 40%, with the Leptophlebiid mayflies contributing 30%, the chironomids (midges) a further 22%, and the Aytid shrimps a further 8.5% (Fig. 11.13).

Of the long-term logged water quality variables, median daily DO (mg/L), mean daily pH, and mean daily temperature range (°C) explained 33% of the variation in assemblage composition. The long-term hydrology metrics, however, explained little variation in the assemblage composition, with only 15% explained by a combination of the rate of

**TABLE 11.3 Average Number of Individuals Per Sample Collected From the Three Sites Over the 12-Month Sampling Period. Sensitive EPT Taxa are Highlighted in Gray**

| Scientific Name | | | | | | |
|---|---|---|---|---|---|---|
| Phyla | Class | | Common Name | Blunder Ck (WSUD) | Stable Swamp Ck (URBAN) | Tingalpa Ck (FOREST) |
| Cnidaria | Hydrozoa | | Hydras | 5.4 | 1.0 | 1.2 |
| Platyhelminthes | | | Flatworms | 32.8 | 4.2 | 0.6 |
| Nemertea | | | Proboscis worms | 1.2 | 0.2 | 0.0 |
| Annelida | Oligochaeta | | Worms | 99.4 | 148.9 | 8.1 |
| | Hirudinea | | Leeches | 0.4 | 1.3 | 0.0 |
| Mollusca | Gastropoda | | Snails | 577.5 | 75.5 | 1.5 |
| Nematoda | | | Roundworms | 56.3 | 2.0 | 0.1 |
| Arthropoda | Arachnida | Acari | Mites | 101.0 | 3.3 | 1.3 |
| | Crustacea | Microcrustaceans | Water fleas | 377.1 | 60.5 | 3.8 |
| | | Decapoda | Shrimps | 1.1 | 0.6 | 26.3 |
| | | Isopoda | Slaters | 0.0 | 0.0 | 0.3 |
| | Insecta | Coleoptera | Beetles | 0.5 | 5.6 | 13.6 |
| | | Hemiptera | Bugs | 1.3 | 24.5 | 14.1 |
| | | Diptera | Flies | 115.6 | 301.1 | 90.5 |
| | | Ephemeroptera | Mayflies | 1.1 | 85.7 | 151.7 |
| | | Epiproctophora | Dragonflies | 1.5 | 2.3 | 0.6 |
| | | Lepidoptera | Moths | 0.6 | 1.0 | 0.5 |
| | | Megaloptera | Alderflies | 0.0 | 0.0 | 1.4 |
| | | Neuroptera | Lacewings | 0.4 | 0.0 | 0.2 |
| | | Plecoptera | Stoneflies | 0.0 | 0.0 | 4.1 |
| | | Trichoptera | Caddisflies | 5.7 | 113.9 | 46.9 |
| | | Zygoptera | Damselflies | 4.6 | 6.3 | 0.6 |

stormflow rise, rate of stormflow fall, and the Base Flow Index (Table 11.2). While this is essentially at odds with that observed in more temperate systems (Walsh et al., 2005; Fletcher et al., 2013), it may not be a surprising result for a subtropical system. Streams in South East Queensland are characterized by a period of relatively dry weather (June–October) where catchments can become "hardened" and in many respects resemble impervious surfaces, which coincides with the onset of intense storm events at the commencement of summer (November–December) (Leigh et al., 2013). During this time, flows in all stream types, even streams with extensive upstream forested areas, may resemble the "flashy" pattern of urban streams. This suggests that the endemic fauna of subtropical streams may have both resistance and resilience traits to "flashy" hydrology.

## 11.4.3 Diversity of EPT taxa

The proportion of the insect orders Ephemeroptera (mayflies), Plecoptera (stoneflies), and Trichoptera (caddisflies)—the EPT taxa—is a commonly used metric for assessing the health of streams. A higher proportion of EPT taxa suggest "better" stream health (Sponseller et al., 2001). EPT taxa are relatively more sensitive to a range of pressures associated with poor stream health, including habitat degradation, reduced water quality, increased pollutants, and changed hydrological regime, compared with other aquatic macroinvertebrate orders (e.g., Diptera [flies], Odonata [dragonflies],

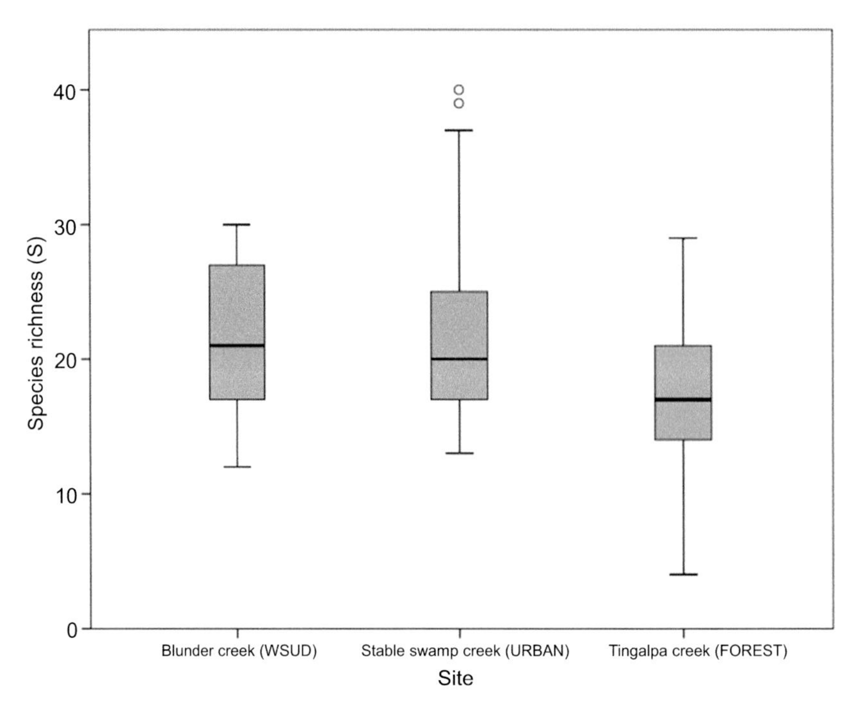


**FIGURE 11.11** Box plots (minimum, first quartile, median, third quartile, and maximum) of species richness from all samples collected from each of the treatment streams between June 2011 and May 2012.

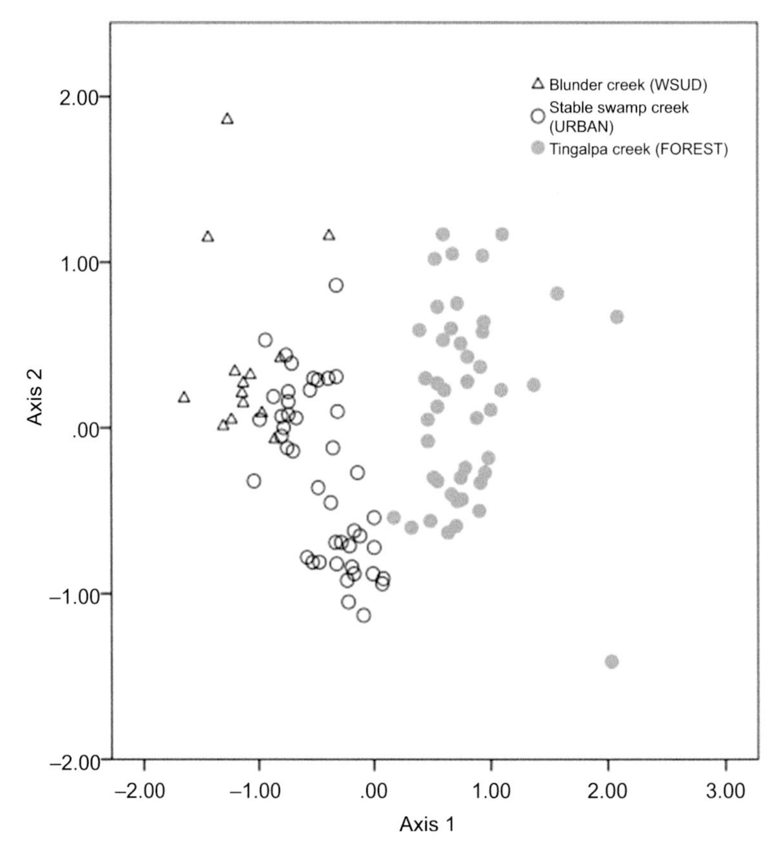


**FIGURE 11.12** Two-dimensional multidimensional scaling ordination plot based on the Bray−Curtis similarity measure showing the distribution of all samples for the three different streams. Stress = 0.14.

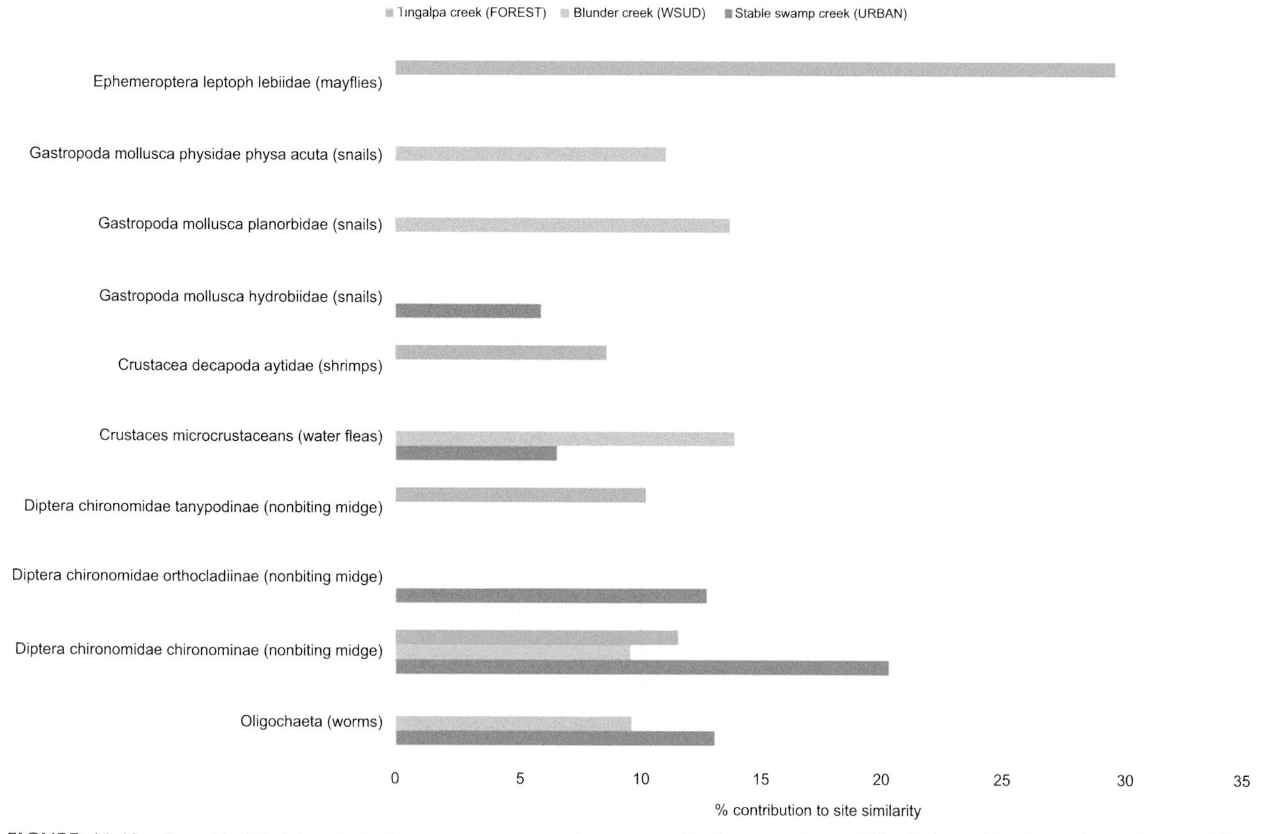


**FIGURE 11.13** Taxa identified by similarity percentages procedure as contributing to at least 50% of the within-site variation in assemblage composition.

Coleoptera [beetles], Heteroptera [true bugs]). Hence, the diversity of EPT taxa throughout the sampling period for the three sites was compared. We predicted that the FOREST site would have a higher diversity of EPT taxa than either the WSUD or URBAN streams.

Fig. 11.14 shows there was a significant difference between the three sites in the abundance per sample of mayflies (Ephemeroptera: ANOVA $F = 6.142$; $P < .01$), stoneflies (Plecoptera: ANOVA $F = 8.603$; $P < .01$), and caddisflies (Trichoptera: ANOVA $F - 5.150$, $P < .001$). There were significantly more Ephemeroptera and Plecoptera in the FOREST site compared with the URBAN or WSUD site (Fig. 11.14). However, there were significantly more Trichoptera in the URBAN site compared with either the FOREST or WSUD site. The patterns of EPT abundance across the three sites were in accordance with the conceptual model predictions, with a greater number of EPT found in the FOREST site, while the WSUD site had the lowest number, possibly reflecting the poor water quality (low pH and high conductivity) at this site.

## 11.5 CONCLUSIONS

The differences in hydrology found between the sites we monitored can be related to urbanization. The two more urbanized sites, Stable Swamp Creek (URBAN) and the Blunder Creek tributary (WSUD), had a larger number of flow rises across nearly all months, higher average daily flows, and greater rates of fall during flow recession. The one parameter that was markedly different to that expected was the minimum daily flow and the BFI. The URBAN site had continual baseflow throughout the year, which may assist in diluting some of the extreme water quality changes associated with urbanization—such as increased nutrients and organic matter loads (Walsh et al., 2007). The cause of the continual baseflows at the Stable Swamp Creek site is unclear. However, there is a wetland not far upstream from the gaging station, which may act like a "sponge" holding water during wet times and continually releasing it downstream during periods of low flow within the stream (see Leigh et al., 2010). Overall, using a suite of hydrological metrics, the hydrological changes imposed by the upstream WSUD intervention on the Blunder Creek WSUD site appeared to contribute to it being much more similar to the FOREST site than the URBAN site.

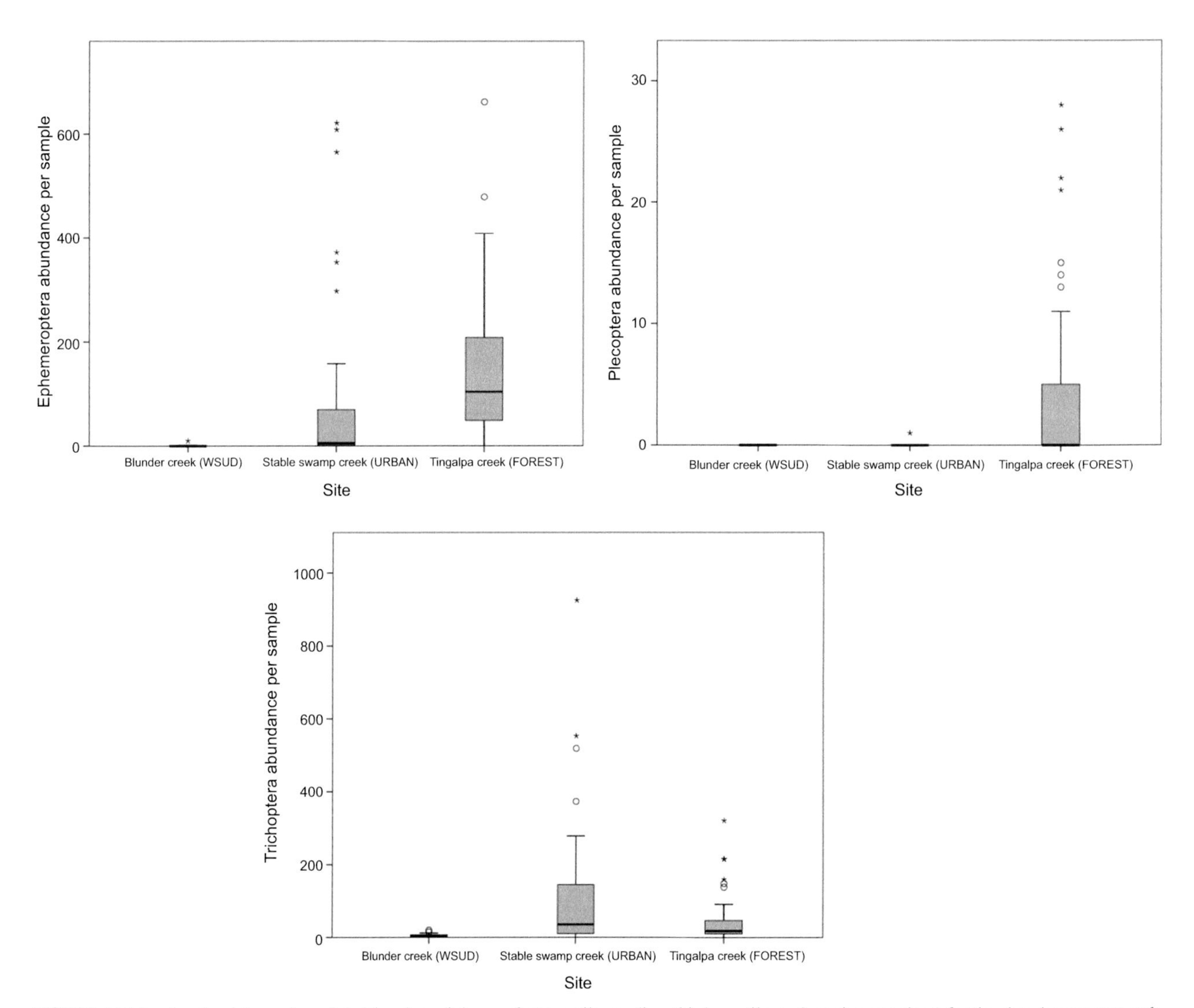


**FIGURE 11.14** Species richness box plots (showing minimum, first quartile, median, third quartile, and maximum values) for the abundance per sample of (a) Ephemeroptera, (b) Plecoptera, and (c) Trichoptera for each of the stream types.

As with the hydrological metrics, there were clear water quality differences between sites that could be related to urbanization. Stable Swamp Creek (URBAN) and the Blunder Creek tributary (WSUD) had larger daily ranges in DO and temperature. The higher conductivity and low pH observed at the WSUD site were likely associated with groundwater inputs containing dissolved Fe, which can cause both acidification and salinization of stream waters (Singer and Stumm, 1970). The multiple water quality impacts at the WSUD site, including high daily ranges in DO and temperature, high conductivity, and acidic conditions, make it a harsh environment for macroinvertebrates. Water quality parameters in the forested Tingalpa Creek, although occasionally harsh, are more often typical of forested catchments, with low daily ranges in DO and temperature and low conductivity.

Species richness was similar across all three sites, but assemblage composition was significantly different, suggesting that each stream type supported a distinct assemblage characterized by species presence and abundance. Stable Swamp Creek (URBAN) and the tributary to Blunder Creek (WSUD) were dominated by taxa common in degraded streams, including midges (Chironominae) and worms (Oligochaeta) (Suren and McMurtrie, 2005). In comparison, the Tingalpa Creek (FOREST) assemblage was dominated by mayflies (Ephemeroptera: Leptophlebiidae) and other insect orders—more typical of headwater streams with good riparian and catchment cover (Sponseller et al., 2001; Sheldon et al., 2012a). When the assemblage data were compared with the water quality and hydrology data, it was stream electrical conductivity, daily temperature range, and aspects of hydrograph shape (rates of rise and fall) that explained the most variation in assemblage differences between the three sites. These physical parameters are known to be heavily influenced by urbanization (Walsh et al., 2005; Hawley and Vietz, 2016).

Many of the biological and water quality differences observed between the FOREST site and the WSUD and URBAN sites could be attributed to the hydrological impacts of urbanization. In nearly every parameter measured in this study, the WSUD site was more closely aligned to the URBAN site than the FOREST site. Given the known negative impacts on stream systems that even small areas of upstream impervious area (>10%) can have (Walsh et al., 2005), the manner in which WSUD is implemented will have a major impact on the ecosystem health outcomes for stream ecosystems. If the implementation of WSUD measures focusses completely on hydrology without also considering water quality impacts, the ecosystem health outcomes for the stream may not be realized.

In summary, our results suggested that there were strong differences in the hydrology, water quality, and macroinvertebrate assemblage between the FOREST site and the two sites impacted by urbanization. Despite the upstream WSUD development of Forest Lake, both the WSUD and URBAN sites had hydrology characterized by high rates of rise and fall of flood peaks. However, very little variation in macroinvertebrate composition across the three streams could be directly attributed to the monthly hydrological metrics calculated from the daily data. Rather, DO, pH, and temperature range explained more variation, suggesting that even when measures are implemented to help restore natural flow patterns, poor water quality can have an overriding influence on stream ecosystem health.

The WSUD site measured in this study was implemented in the 1990s suggesting that the stream ecosystem should have had plenty of time to adjust to the new conditions, notwithstanding reports that the "ghost of land use past" can have a significant impact on contemporary stream conditions (Maloney et al., 2008). The Forest Lake area was a cleared agricultural catchment before urbanization, and the initial clearing most likely contributed to the significant stream incision, which is adversely impacting stream conditions today. To fully understand the role of WSUD developments in restoring streams, ecosystem health studies should cover the full range of WSUD types and control for factors such as prior land use, upstream riparian cover, and unique changes in the groundwater hydrogeochemistry.

## REFERENCES


Abesser, C., Robinson, R., Soulsby, C., 2006. Iron and manganese cycling in the storm runoff of a Scottish upland catchment. Journal of Hydrology 326, 59–78.

Adyel, T.M., Oldham, C.E., Hipsey, M.R., 2016. Stormwater nutrient attenuation in a constructed wetland with alternating surface and subsurface flow pathways: event to annual dynamics. Water Research 107, 66–82.

Albert, S., O'Niel, J.M., Udy, J.W., Ahern, K.S., O'Sullivan, C.M., Dennison, W.C., 2005. Blooms of the cyanobacterium *Lyngbya majuscula* in coastal Queensland, Australia: disparate sites, common factors. Marine Pollution Bulletin 51, 428–437.

Brown, R.R., Keath, N., Wong, T.H.F., 2009. Urban water management in cities: historical, current and future regimes. Water Science and Technology 59, 847–855.

Bunn, S.E., Abal, E.G., Smith, M.J., Choy, S.C., Fellows, C.S., Harch, B.D., Kennard, M.J., Sheldon, F., 2010. Integration of science and monitoring of river ecosystem health to guide investments in catchment protection and rehabilitation. Freshwater Biology 55 (Suppl. 1), 223–240.

Burns, M.J., Fletcher, T.D., Walsh, C.J., Ladson, A.B., Hatta, B.E., 2012. Hydrologic shortcomings of conventional urban stormwater management and opportunities for reform. Landscape and Urban Planning 105, 230–240.

Chowdhury, R., Gardner, T., Gardiner, R., Hartcher, M., Aryal, S., Ashbolt, S., Petrone, K., Tonks, M., Ferguson, B., Maheepala, S., McIntosh, B.S., 2013. SEQ Catchment Modelling for Stormwater Harvesting Research: Instrumentation and Hydrological Model Calibration and Validation. Urban Water Security Research Alliance. Technical Report No. 83.

Clarke, R.T., Wright, J.F., Furse, M.T., 2003. RIVPACS models for predicting the expected macroinvertebrate fauna and assessing the ecological quality of rivers. Ecological Modelling 160, 219–233.

Clarke, K.R., Gorley, R.N., 2001. Primer V5: User Manual/Tutorial. Plymouth Marine Laboratory, Plymouth.

Duckworth, O.W., Holmström, S.J.M., Peña, J., Sposito, G., 2009. Biogeochemistry of iron oxidation in a circumneutral freshwater habitat. Chemical Geology 260, 149–158.

Fletcher, T.D., Andrieu, H., Hamel, P., 2013. Understanding, management and modelling of urban hydrology and its consequences for receiving waters: a state of the art. Advances in Water Resources 51, 261–279.

Grimm, N.B., Sheibley, R.W., Crenshaw, C.L., Dahm, C.N., Roach, W.J., Zeglin, L.H., 2005. N retention and transformation in urban streams. Journal of the North American Benthological Society 24, 626–642.

Hawley, R.J., Vietz, G.J., 2016. Addressing the urban stream disturbance regime. Freshwater Science 35, 278–292.

Heaney, J.P., Huber, W.C., 1984. Nationwide assessment of urban runoff impact on receiving water-quality. Water Resources Bulletin 20, 35–42.

Herlihy, A.T., Stoddard, J.L., Johnson, C.B., 1998. The relationship between stream chemistry and watershed land cover data in the mid-atlantic region, U.S. Water, Air and Soil Pollution 105, 377–386.

Kaye, J.P., Groffman, P.M., Grimm, N.B., Baker, L.A., Pouyat, R.V., 2006. A distinct urban biogeochemistry? Trends in Ecology and Evolution 21, 192–199.

Leigh, C., Sheldon, F., Kingsford, R.T., Arthington, A.H., 2010. Sequential floods drive 'booms' and wetland persistence in dryland rivers: a synthesis. Marine and Freshwater Research 61, 896–908.

Leigh, C., Burford, M., Connolly, R.M., Olley, J.M., Saek, E., Sheldon, F., Smart, J.C.R., Bunn, S.E., 2013. Science to support management of receiving waters in an event-driven ecosystem: from land to river to sea. Water 5, 780–797.

Lenat, D.R., Crawford, J.K., 1994. Effects of land use on water quality and aquatic biota of three North Carolina piedmont streams. Hydrobiologia 294, 185–199.

Maloney, K.O., Feminella, J.W., Mitchell, R.M., Miller, S.A., Mulholland, P.J., Houser, J.N., 2008. Landuse legacies and small streams: identifying relationships between historical land use and contemporary stream conditions. Journal of the North American Benthological Society 27, 280–294.

Marsh, N., 2004. Time Series Analysis Module: River Analysis Package, Cooperative Research Centre for Catchment Hydrology. Monash University, Melbourne, Australia.

Nedeau, E.J., Merritt, R.W., Kaufman, M.G., 2003. The effect of industrial effluent on an urban stream benthic community: water quality vs. habitat quality. Environmental Pollution 123, 1–13.

Parker, N.R., 2010. Assessing the Effectiveness of Water Sensitive Urban Design in Southeast Queensland (Masters by Research thesis). Queensland University of Technology. https://eprints.qut.edu.au/34119/.

Paul, M.J., Meyer, J.K., 2001. Streams in the urban landscape. Annual Review of Ecology and Systematics 32, 333–365.

Peterson, E.E., Sheldon, F., Darnell, R., Bunn, S.E., Harch, B.D., 2011. A comparison of spatially explicit landscape representation methods and their relationship to stream condition. Freshwater Biology 56, 590–610.

Quinn, G., Keough, M., 2002. Experimental Design and Data Analysis for Biologists. Cambridge University Press, Cambridge.

Rasmussen, K., Lindegaard, C., 1988. Effects of iron compounds on macroinvertebrate communities in a Danish lowland river system. Water Research 22, 1101–1108.

Rhoton, F.E., Bigham, J.M., Lindbo, D.L., 2002. Properties of iron oxides in streams draining the Loess Uplands of Mississippi. Applied Geochemistry 17, 409–419.

Roy, A.H., Wenger, S.J., Fletcher, T.D., 2008. Impediments and solutions to sustainable, watershed-scale urban stormwater management: lessons from Australia and the United States. Environmental Management 42, 344–359.

Schwertmann, U., 1991. Solubility and dissolution of iron oxides. Plant and Soil 130, 1–25.

Sheldon, F., Peterson, E.E., Boone, E.L., Sippel, S., Bunn, S.E., Harch, B.D., 2012a. Identifying the spatial scale of land use that most strongly influences overall river ecosystem health score. Ecological Applications 22, 2188–2203.

Sheldon, F., Pagendam, D., Newham, M., McIntosh, B., Hartcher, M., Hodgson, G., Leigh, C., Neilan, W., 2012b. Critical Thresholds of Ecological Function and Recovery Associated with Flow Events in Urban Streams. Urban Water Security Research Alliance. Technical Report No. 99.

Singer, P.C., Stumm, W., 1970. The rate determining step in the production of acidic mine wastes. Science 167, 1121–1123.

Sponseller, R.A., Benfield, E.F., Valett, H.M., 2001. Relationships between land use, spatial scale and stream macroinvertebrate communities. Freshwater Biology 46, 1409–1424.

Suárez, M.L., Sánchez-Montoya, M.M., Gómez, R., Arce, M.I., del Campo, R., Vidal-Abarca, M.R., 2017. Functional response of aquatic invertebrate communities along two natural stress gradients (water salinity and flow intermittence) in Mediterranean streams. Aquatic Sciences 79, 1–12.

Suren, A.M., McMurtrie, S., 2005. Assessing the effectiveness of enhancement activities in urban streams: II. Responses of invertebrate communities. River Research and Applications 21, 439–453.

Tsoi, W.Y., Hadwen, W.L., Fellows, C.S., 2011. Spatial and temporal variation in the ecological stoichiometry of aquatic organisms in an urban catchment. Journal of the North American Benthological Society 30, 533–545.

Walsh, C.J., Roy, A.H., Feminella, J.W., Cottingham, P.D., Groffman, P.M., Morgan, R.P., 2005. The urban stream syndrome: current knowledge and the search for a cure. Journal of the North American Benthological Society 24, 706–723.

Walsh, C.J., Kunapo, J., 2009. The importance of upland flow paths in determining urban effects on stream ecosystems. Journal of the North American Benthological Society 28, 977–990.

Walsh, C.J., Walker, K.A., Gehling, J., MacNally, R., 2007. Riverine invertebrate assemblages are degraded more by catchment urbanisation than by riparian deforestation. Freshwater Biology 52, 574–587.

Chapter 12

# Protecting and Managing Stream Morphology in Urban Catchments Using WSUD

Geoff J. Vietz[1] and Robert J. Hawley[2]

[1]*Streamology Pty Ltd., The University of Melbourne, VIC, Australia;* [2]*Sustainable Streams LLC, Louisville, KY, United States*

## Chapter Outline



**ABSTRACT**

Streams in the urban landscape provide significant social, economic, and ecological benefits, yet, they are often physically degraded. The main culprit has been identified as excess stormwater runoff from conventional drainage systems, in addition to alterations to the sediment regime to streams. This commonly leads to the necessity for direct intervention to protect societal infrastructure. Water sensitive urban design (WSUD) provides an alternative by addressing the causes of urban-induced physical degradation of streams rather than patching the symptoms. In this chapter, we discuss the issues associated with conventional urban stormwater drainage, the changes to stream channel form and functioning, the opportunities for stormwater control measures to ameliorate the impact, and an approach to inform stormwater design. Applying WSUD principles to address the cause of physical degradation of streams in urban catchments can offer a more sustainable and economically viable option for engineers, managers and planners, and lead to streams capable of better supporting myriad benefits.

**Keywords:** Channel; Disturbance; Geomorphology; Sediment; Stormwater.

Approaches to Water Sensitive Urban Design. https://doi.org/10.1016/B978-0-12-812843-5.00012-5

## 12.1 INTRODUCTION

Considering the role rivers and streams play in human civilization, it is no coincidence that most of the great cities of the world lie on the banks of rivers (Mumford, 1961). As cities develop, they are commonly built along the tributary streams that pass through cities. This process of urbanization has the unintended consequence of severely degrading the physical and ecological condition of rivers and streams. In the worst case, small streams are obliterated by piping. For the rivers and streams that remain on the surface, runoff from impervious surfaces, commonly via stormwater pipes, translates rainfall to a streamflow regime that physically degrades streams.

It is the changes in channel form (morphology) and function (geomorphic processes) under urbanization of a catchment, which are the foci of this chapter. We also discuss opportunities to address stormwater runoff to reduce the impact of urbanization and protect streams. By better understanding the influence of urban stormwater runoff on stream geomorphology, we can more effectively and efficiently manage streams in our cities and suburbs for improved economic, social, and ecological values.

Alluvial stream channels are inherently dynamic in their physical form and are continually adjusting to inputs to attain the optimum configuration for the water and sediment delivered to them from the landscapes they drain. Similar to a human body in calorific balance, the size of streams under such dynamic equilibrium conditions remains relatively stable. That delicate balance is often abruptly ended when urbanization modifies inputs to the stream.

If stormwater runoff can be considered as the primary driver of stream channel change, then a secondary impact is the response from common engineering management intervention and channel reconfiguration. These engineering solutions often attempt to increase the stability of streams with hard materials such as rock. Channel reconfiguration approaches for treating the symptoms of geomorphic change under urbanization (e.g., rock protection) are outside the focus of this chapter, and for further information, the reader is referred to Vietz et al. (2016b).

Urbanization is not a binary process. Although it may sometimes appear like farmland one day and a suburb the next, it takes time for the urban landscape, and the surfaces that comprise it, to become established. There are nominally four phases of urban development, and each phase has a distinct influence on flow and sediment regimes as well as an independent, and compounding, impacts on the form and functioning of streams. These phases provide states of urbanization that must also be addressed differently if the goal of protecting or enhancing stream integrity is to be achieved. The states are (1) predevelopment (such as agriculture and other land uses prior to urbanization); (2) civil construction; (3) new developments (i.e., greenfield development); and (4) established developments (only minor infill occurs).

It is important to recognize that urbanization is not the only land use that must be considered in stream protection. Legacy impacts of predevelopment practices, such as land drainage, play a significant role before urban development begins (Vietz et al., 2016a). The civil construction phase is also highly influential, particularly with regard to sediment supply (as discussed in Chapter 10 on sediment and erosion control devices of this book). Even excess sediment supply to a stream and the deposition of sediments in the channel, however, can be readily mobilized by postconstruction stormwater runoff in a relatively short time frame (Russell et al., 2017). In this chapter, we focus on the latter two phases of urbanization (new and established developments) because they provide the greatest opportunities for water sensitive urban design (WSUD) to protect streams from excess stormwater runoff, manage ongoing sediment issues, and enhance stream integrity in the medium- to long term. We see WSUD as the approach and framework within which this can be achieved through the use of stormwater control measures (SCMs) such as tanks, rain gardens, wetlands, and biofiltration systems.

Although attempts to improve stormwater management through WSUD have been successful in reducing pollutant loads and providing some reductions in flow volume, current practice has commonly failed to arrest the geomorphic degradation of streams. Current approaches typically fail to restore the more important aspects of the flow regime (both high and low flows). This is due, in part, to the fact that WSUD has rarely been applied at a catchment scale sufficient to mitigate the magnitude and frequency of disturbances resulting from increased stormwater runoff from connected impervious areas.

## 12.2 CHANNEL FORM AND FUNCTION IN URBAN CATCHMENTS

### 12.2.1 Stream channels are a product of catchment inputs

The physical characteristics of stream channels are a function of the flow and sediment regime delivered to the channel (Gilvear, 1999; Knighton, 1998). Their morphology varies with time and space because of the changes in environmental

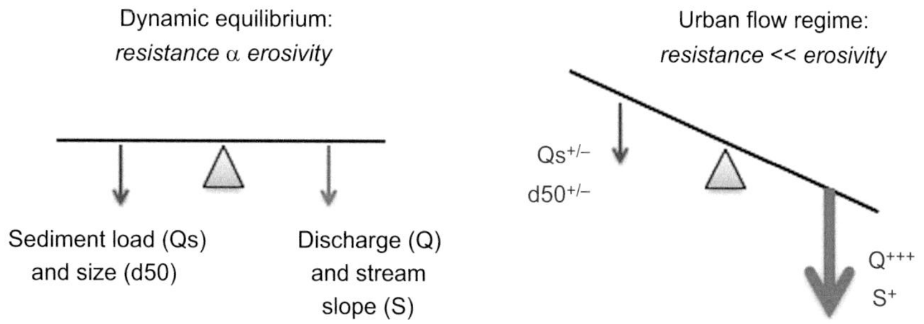

**FIGURE 12.1** Stream dynamic equilibrium and change under catchment urbanization based on an adaptation of Lane's (1955) balance considering the primary controls of erosion resistance (sediment load [Qs] and size [d50, the median sediment size of particles]) are proportional ($\alpha$) to the main drivers of erosion (discharge [Q] and stream slope [S]).

controls. Not only do channel features respond to channel inputs of flow and sediment and translate flow into erosive or sediment transport components, but also the channel itself is a deformable boundary. The altered boundary has a secondary influence on sediment and velocity, providing feedback that can accelerate change. For example, channel incision due to increased streamflow can increase both the channel capacity and the erosive energy that is contained within the channel, thereby contributing further to incision as larger runoff events are contained within the channel.

Urbanization of a catchment, and the resulting stormwater runoff, is arguably the most important driver of altered channel form and function. Urbanization of a channel increases flow and erosive forces by up to 10 times greater than for the predisturbance catchment (Burns et al., 2012). Sediment supply changes are more variable, with both increases and decreases observed across both space and time. Nevertheless, the dominance of altered hydrology, in conjunction with altered sediment supply, can combine to produce a double-edged sword exerting change on channel morphology and function (Fig. 12.1).

## 12.2.2 Impacts of urbanization on geomorphic functioning

Geomorphic instability has been documented in urban streams across multiple hydroclimatic settings (Fig. 12.2; Booth, 1990; Bledsoe and Watson, 2001; Leopold et al., 2005). Increased sediment transport capacity resulting from conventional stormwater management approaches (resulting in increased runoff volume and flow rate) has been well documented (Bledsoe, 2002; Grove and Ladson, 2006; Pomeroy et al., 2008). Enlargement of urban channels is likely where an increase in hydraulic discharge accelerates erosion, and the decrease in available sediment in the channel is unable to "keep pace" with the erosion. Erosion dominates over deposition throughout an urban catchment in almost all scenarios. The exception is poorly managed construction sites liberating large volumes of sediment (see Chapter 10 on sediment and erosion control devices). Despite civil construction sediment loads, the effect is short-lived in the stream once the unconsolidated sediments are overwhelmed by the dramatic increase in sediment transport capacity (Russell et al., 2017).

The erosion of riverbanks commonly increases following urbanization and this can be considered an important source of sediment for urban rivers (Gurnell et al., 2007; Trimble, 1997; Wolman, 1967). Trimble (1997) estimated that channel erosion provides about two-thirds of the total sediment yield from an urban catchment. In the Issaquah Creek catchment, Washington, Nelson and Booth (2002) found urban development to almost double sediment production, even though relatively little sediment was liberated directly from the urban areas. This source of sediment can be quickly enhanced by either the channel finding a stable state or channel interventions such as bank reinforcement. Sediments find some transient storages but are ultimately moved through the system to bays and estuaries.

Channel incision is a well-known geomorphic response to increased flow, decreased sediment load, an oversteepened channel gradient, or decreased sediment size (Lane, 1955). The channel evolution model for suburban streams (Hawley et al., 2012) highlights that incising channels tend to degrade vertically, lowering the bed, prior to a phase of lateral degradation, eroding the channel banks and widening the channel (Fig. 12.3). This process can be cyclical. The typical sequence of evolution from a disturbed to a recovered state progresses numerically from Stage 1 to Stage 5; however, that sequence can be interrupted in suburban streams with more prolonged and often cyclical periods of instability. Where flow and sediment regimes are not conducive to recovery, the return of in-channel features such as benches and bars (leading to a compound channel form) is unlikely within decades.

FIGURE 12.2 (a) Channel incision caused by excess urban stormwater plus stream management activities for infrastructure protection, at Mullum Mullum Creek, Melbourne, Australia. (b) Symptoms of channel degradation include exposing a buckled sewer pipe that has been undermined by channel downcutting. (c) Previously buried infrastructure exposed through channel incision. (d) Secondary downcutting and widening following an initial cycle of degradation, widening, and aggradation in an urban stream in Cincinnati, Ohio. *(a) Photo G. Vietz. (b) Inset of sewer pipe in channel. (b) and (c) photo R. Hawley. (d) Photo K. Cooper.*

## 12.2.3 Impacts of urbanization on channel morphology

Channels in established urban catchments are typically deeper, wider, and simpler. They experience both cross-sectional and planform changes from both indirect and direct modifications (Fig. 12.4). Under Pre-European conditions with forested land cover, the supply of water (Q) and sediment (Qs) are in balance such that stream networks are in a state of dynamic equilibrium. Following European settlement, watersheds were typically deforested and valleys were manipulated for agrarian uses, resulting in increased stormwater runoff and corresponding increases in stream discharge (Q+), as well as large increases in sediment yields from the watershed (Qs++). Under present-day suburban land uses, stormwater runoff and stream discharge is further increased (Q++) and sediment yields are often in flux during land development phases and in response to stream bank erosion (Qs+/−).

Streams in urban catchments have reduced physical habitat including less mobile bed substrate, fewer bars and riffles, less wood, and disengaged floodplains (Chin, 2006; Hawley and Bledsoe, 2013; Vietz et al., 2014b). Channel responses to urbanization are not universal in magnitude or trajectory (Chin, 2006; Fitzpatrick and Peppler, 2010). Some urban channels incise and have initially smaller width-to-depth ratios (W/D) (Park, 1977; Simon and Rinaldi, 2006; Vietz et al., 2014b), whereas others aggrade and develop larger W/D ratios (Chin and Gregory, 2001; Hawley et al., 2012). Although channel enlargement is common (Fig. 12.3), there is variation in enlargement ratios (posturban development vs. preurban development cross-sectional area). For example, enlargement ratios increase by 2–8 times in humid temperate urban streams (Hammer, 1972; Chin, 2006) and up to 14 in semiarid systems such as southern California (USA) (Hawley and Bledsoe, 2013).

Urbanization commonly reduces the thickness of sediments on the bed as transport dominates sediment supply (Vietz et al., 2014b). Changes in sediment size are common, but findings vary on the direction of change (Chin, 2006). Predominantly, streams draining urban catchments experience bed material coarsening (e.g., Finkenbine et al., 2000;

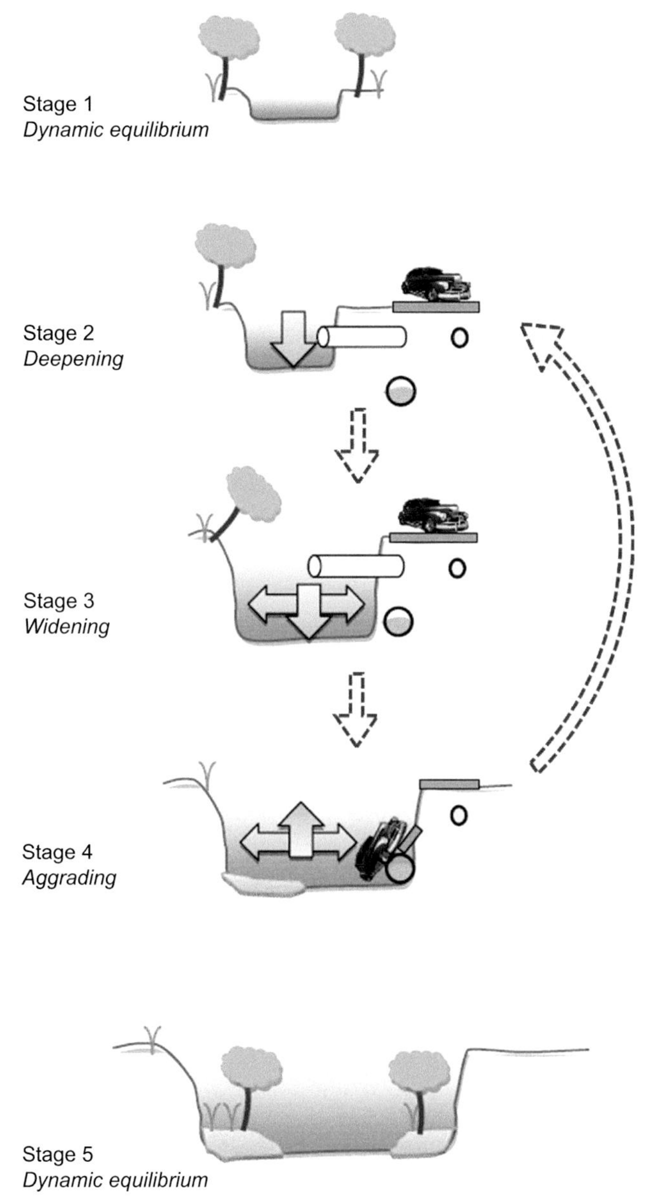


**FIGURE 12.3** Five-stage channel evolution model in suburban streams adapted from Schumm et al. (1984) and Hawley et al. (2012). The *red dashed arrows* with yellow fill underscore the perpetuating cycle of deepening, widening, and aggrading, which suburban streams commonly experience.

Pizzuto et al., 2000). For example, Robinson (1976) found that the 84th percentile particle (d84) in urban streams was approximately four times larger than in rural streams. Hawley et al. (2013a) measured the sediment coarsening in 12 streams with an average annual coarsening of ~2% per year in d50 (median sediment size of streambed particles) for every percentage increase in a catchment's total impervious area (TIA) up to a TIA of ~15%. For sites with TIA >20%, the relationship for bed coarsening was less distinct, suggesting complex and cyclical changes in streams draining more established suburbs, i.e., alternating from incision to aggradation (Fig. 12.3).

## 12.2.4 Variability in stream response

Susceptibility of streams to urbanization varies. Characteristics such as slope, substrate, riparian vegetation cover, and presence of bedrock or man-made hydraulic controls influence the robustness of channel morphology (Hawley and Vietz, 2016; Utz et al., 2016). For example, streams in western Washington's granular hillslope deposits exhibit considerably

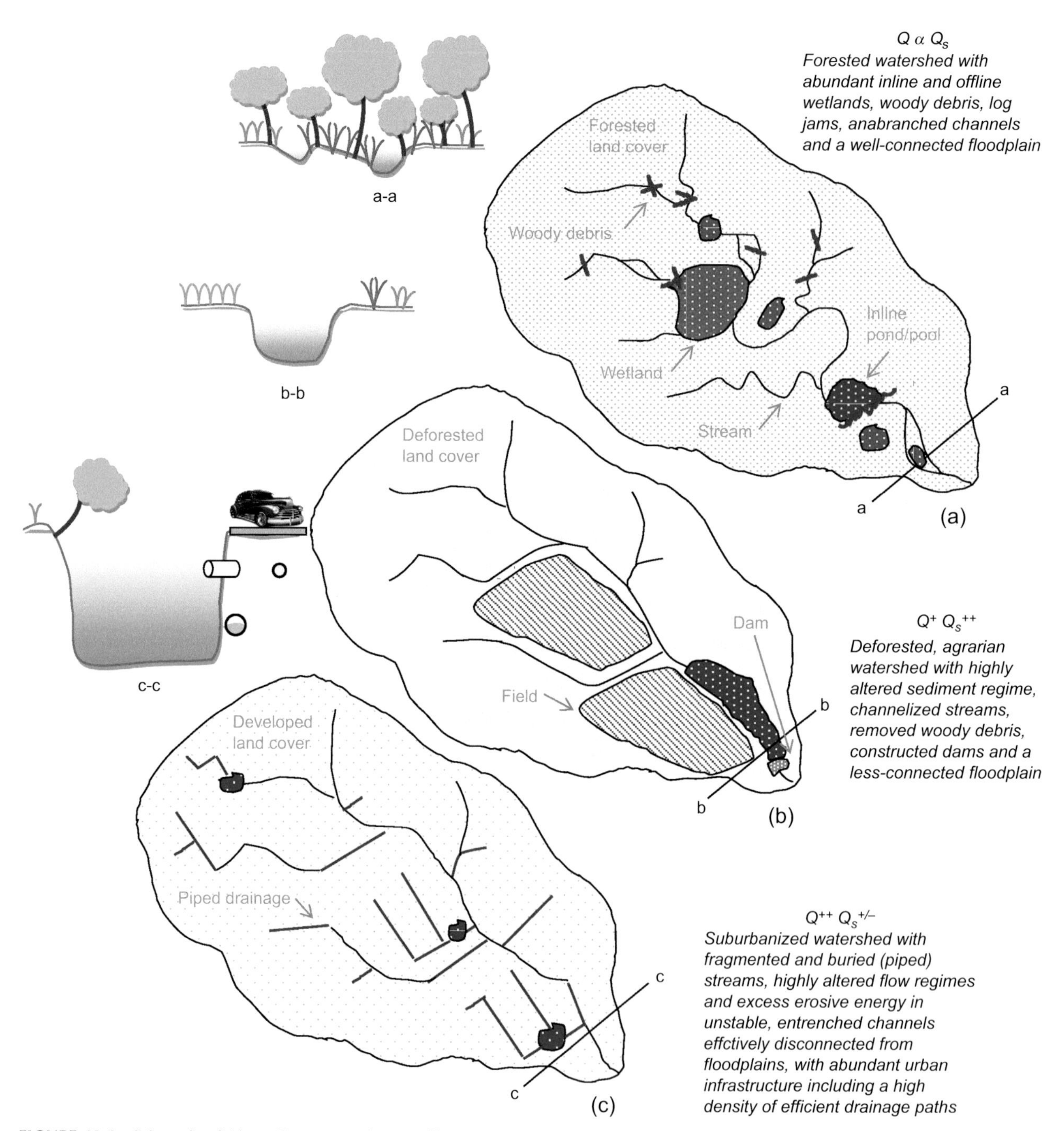


**FIGURE 12.4** Schematic of (a) pre-European settlement, (b) post-European settlement, and (c) present-day suburban hydrologic networks and corresponding cross sections in a hypothetical alluvial valley in a humid temperate catchment, with dominant responses and characteristics. $Q$, flow and $Qs$, sediment load. *Adapted from Hawley, R.J., Bledsoe, B.P., Stein, E.D., Haines, B.E., 2012. Channel evolution model of semiarid stream response to urban-induced hydromodification. Journal of the American Water Resources Association 48 (4), 722–744.*

greater change compared with those in cohesive silt–clay deposits (Booth and Henshaw, 2001). Steep channel slopes are not a direct determinant of the susceptibility of urban channels to incision, but Booth and Henshaw (2001) suggested that steeper gradients may increase the magnitude of change as channels with slopes > 4% exhibited the largest changes in bed deepening (>0.3 m/year). They found that susceptibility to incision is particularly dependent on substrate geology. The relative sensitivity of streams to urbanization propagates the uncertainty surrounding response, highlighting the importance of explicitly considering stream characteristics. Bedrock controls can reduce bed deepening and increase morphologic complexity (Vietz et al., 2014b).

## 12.2.5 Implications of channel form and function changes

The physical setting of a stream channel strongly influences how the flow regime translates into hydraulic conditions experienced by the riverine biota (Poff et al., 2010). The translation of the flow regime into small-scale hydraulic conditions has important implications for ecological processes (Newall and Walsh, 2005). Statzner and Higler (1985) state that "*physical characteristics of flow (*'stream hydraulics'*) are the most important environmental factor governing the zonation of stream benthos on a world-wide scale*" (p. 127).

The abundance and diversity of aquatic biota is closely associated with geomorphic aspects of channel morphology and functioning (Bledsoe, 2002). Local characteristics such as channel geometry, floodplain height, the presence of bars and benches, and streambed composition are all determinants of the impact of events, such as whether a given flow will create a bed-moving disturbance or an overbank flow (Poff and Zimmerman, 2010; Vietz et al., 2014b). For example, larger variations in bed elevation will provide fish and invertebrates with longer pool persistence during drying periods and refuge during higher "disturbance" flows. Bed sediments affect substrate diversity and hyporheic exchange (the movement of flow through and within the substrate sediments), which plays an important role in the chemical and biological functions of streams (Ryan and Boufadel, 2007). The pulses of sediment associated with bank failures can impact aquatic organisms by smothering streambed habitat and creating excessively turbid water. Streambed disturbance (i.e., mobilization of coarse-grained, habitat-forming particles on the streambed) has been documented as the dominant driver of biotic variability through time (Hawley et al., 2016, Fig. 12.5). Events that exceed the critical discharge for streambed disturbance ($Q_{critical}$) cause a direct disturbance on the benthic macroinvertebrates that inhabit those particles such as the caddisflies pictured in Fig. 12.5.

Changes to stream channels draining urban catchments also have implications for social and economic values. Better managing stormwater flows to stream channels can reduce flooding, loss of land, damage to infrastructure, safety risks, and the maintenance costs associated with incising channels (e.g., Fig. 12.2). Linking changes from urban streams to social values can provide significantly stronger cases for measures such as WSUD (Smith et al., 2016).

# 12.3 STORMWATER AS AN EFFECTIVE GEOMORPHIC AGENT

## 12.3.1 Urban stream disturbance regime

Urban stormwater runoff leads to a superabundance of flow and frequent disturbance. Flow alterations due to stormwater increase the quantity and rates of runoff and these can amplify the effectiveness of flows exceeding the threshold for

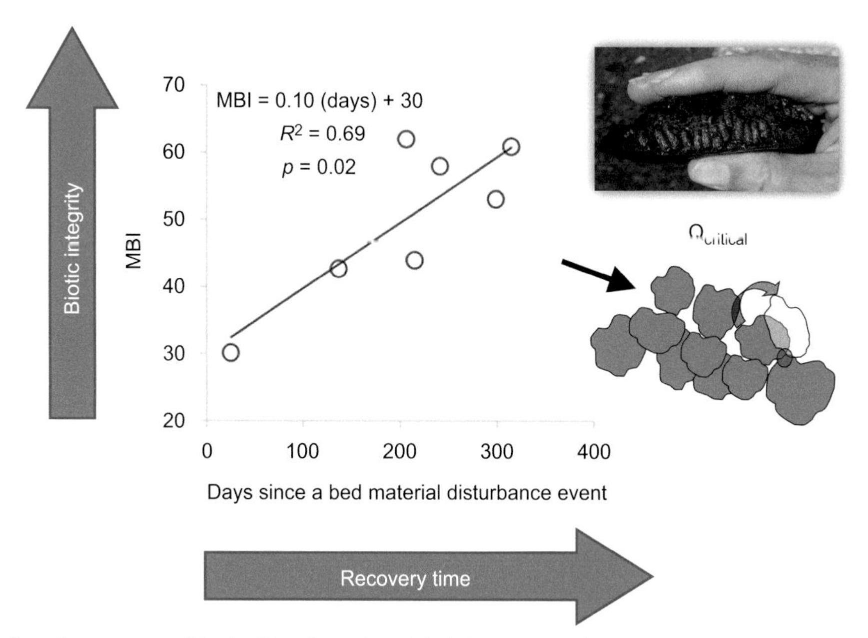


**FIGURE 12.5** Biotic integrity, as measured by the Macroinvertebrate Biotic Index (MBI), is greatest with the least bed disturbance. As time since a disturbance event increases, biotic integrity tends to increase proportionally. *Adapted from Hawley, R.J., Wooten, M.S., MacMannis, K.R., Fet, E.V., 2016. When do macroinvertebrate communities of reference streams resemble urban streams? The biological relevance of Qcritical. Freshwater Science 35 (3), 778–794, Freshwater Science.*

streambed erosion ($Q_{critical}$). These geomorphic changes can occur at very low levels of urbanization, with as little as 3%–4% of the impervious areas connected to the stream by stormwater pipes (Vietz et al., 2014b). This response highlights the immense challenge faced by waterway managers working in the urban context and how ambitious the goals need to be in developing WUSD options to address the problem. Climate change is likely to exacerbate degradation of urban stream morphology, within increased extremes of wet and dry. Even low flows can exacerbate erosion with greater drying and desiccation likely preparing banks for subsequent erosion, particularly where clay-rich banks are present.

### 12.3.2 How conventional stormwater drainage systems impact streams

There is now a well-recognized and explicit link between impervious surfaces connected directly to streams through stormwater pipes and the physical and ecological degradation of streams (Burns et al., 2012; Hawley et al., 2013b; Vietz et al., 2014b). Efforts to reduce the hydrologic impact of urbanization are most commonly reliant on detention basins and flow control ponds connected to stormwater pipes (Booth and Jackson, 1997). The conventional "peak standard" or "peak shaving" approach primarily focuses on flood control management whereby peak flows of pre- and posturban developments are matched. This has been found to be considerably less effective in reducing elevated sediment transport rates because flows may still exceed the threshold for sediment movement over longer durations (Bledsoe, 2002; Hawley et al., 2017; Fig. 12.6). While reducing the peak discharge rate, conventional detention basins may actually result in an *increase* in the duration of elevated flows, including increased periods of flows that exceed the $Q_{critical}$ for streambed erosion (Hawley et al., 2017). Booth and Henshaw (2001) investigated tributaries downstream of an intensive urban development in Washington State and the tributary with the detention basin exhibited the most drastic change in channel morphology.

The greater the duration of time flow is above $Q_{critical}$ (such as with Peak Control Detention), the more likely the bed sediments will be removed and incision will occur. Recognizing the practical limitations of perfectly matching the predeveloped hydrograph with SCMs, erosion control detention attempts to optimize the designs of SCMs (e.g., via more restrictive outlets) such that the cumulative erosion that would have occurred under predeveloped conditions would not be exceeded by postdeveloped conditions (i.e., see Hawley and Vietz (2016) for more detail).

## 12.4 SEDIMENT SUPPLY CONSIDERATIONS

Alterations to hydrology of streams in urban catchments and of the catchment landscape are accompanied by altered sediment dynamics (Wohl et al., 2015; Wolman, 1967). Sediment dynamics, which incorporate the spatial and temporal presence and variation in constituents of the bed and bank materials, is an ecologically important aspect of stream health. It is well established that reduced sediment supply to a stream increases the energy available to degrade the channel for a given flow, or alternatively stated, stream energy increases as sediment supply decreases (Schumm, 1977). In urban

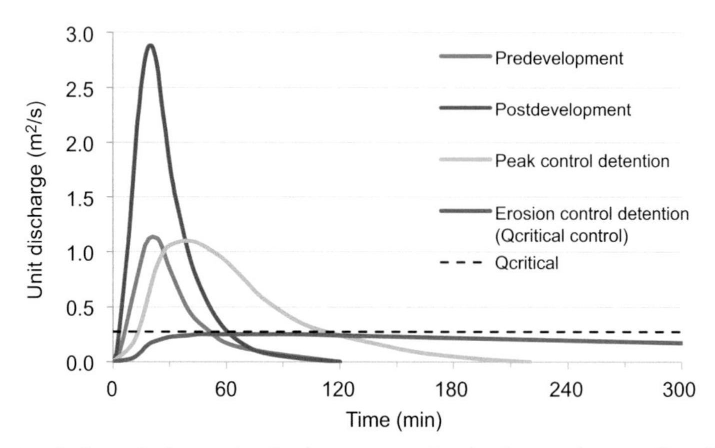


**FIGURE 12.6** Idealized 2-year storm hydrographs for preurban development, posturban development (no control), peak flow control via detention basin (conventional flood control management), and erosion control detention ($Q_{critical}$ control). *Adapted from Bledsoe, B.P., 2002. Stream erosion potential and stormwater management strategies. Journal of Water Resources Planning and Management 128 (6), 451–455.*

streams, bedload sediments are often depleted (Vietz et al., 2014b). Despite the importance of bedload sediment regimes on channel form, function, and condition and the importance of sediment considerations in WSUD, the role it plays in urban stream degradation has largely been ignored until recently.

The prevailing model of sediment dynamics related to land use of Wolman (1967) was revised 50 years later. Based on a global review of urban sediment studies, Russell et al. (2017) confirmed that sediment yields from urban construction can be an order of magnitude above yields from an agricultural catchment and almost two orders of magnitude greater than a forested catchment. The difference, however, was that the studies revealed that established urban catchments retained a relatively high sediment yield, greater than background levels preurbanization (Fig. 12.7). Although sediment loads are lower than during the construction phase, sediments are readily available through infill development and imported landscaping products, as well as the breakdown of surfaces (e.g., weathering of pavement materials). These recent findings are contrary to many considerations as it has been suggested that a reduction in sediment supply in urban environments is a result of erosion-resistant sealing of catchment surfaces (Gurnell et al., 2007).

Despite the legacy sediment loads in the channel following civil construction, and sediment loads still entering streams from established urban catchments, the greatly increased sediment transport capacity from an urban catchment is capable of overwhelming deposited sediments in the channel (Russell et al., 2017). This means that mobile sediments within the channel are reduced in load and depth to bedrock or in situ materials (Vietz et al., 2014b). Channel erosion may also provide sediments to the stream, particularly coarse-grained sediments, but this is relatively short-lived as the stream finds a new stable state. Minimizing frequent bed-sediment disturbance in streams and enabling a return of former channel morphology are dependent on flow regime management and the effectiveness of WSUD practices and SCMs in capturing and reducing stormwater runoff (Fig. 12.8).

The findings of Russell et al. (2017) have significant implications for the design and maintenance of SCMs and the protection of streams in urban catchments. Firstly, sediment supply from urban catchments may remain high and sediments, particularly fine-grained, may reduce the effectiveness of SCMs and increase maintenance. Secondly, the capture of sediments in SCMs (which can be both intended and unintended in detention ponds or biofilters, etc.) may exacerbate stream erosion if there are inadequate sediment load inputs to the channel (specifically coarse-grained sediments). Finally, SCMs can be damaged by sediments liberated during the construction phase if practices are not adequate to reduce sediment runoff.

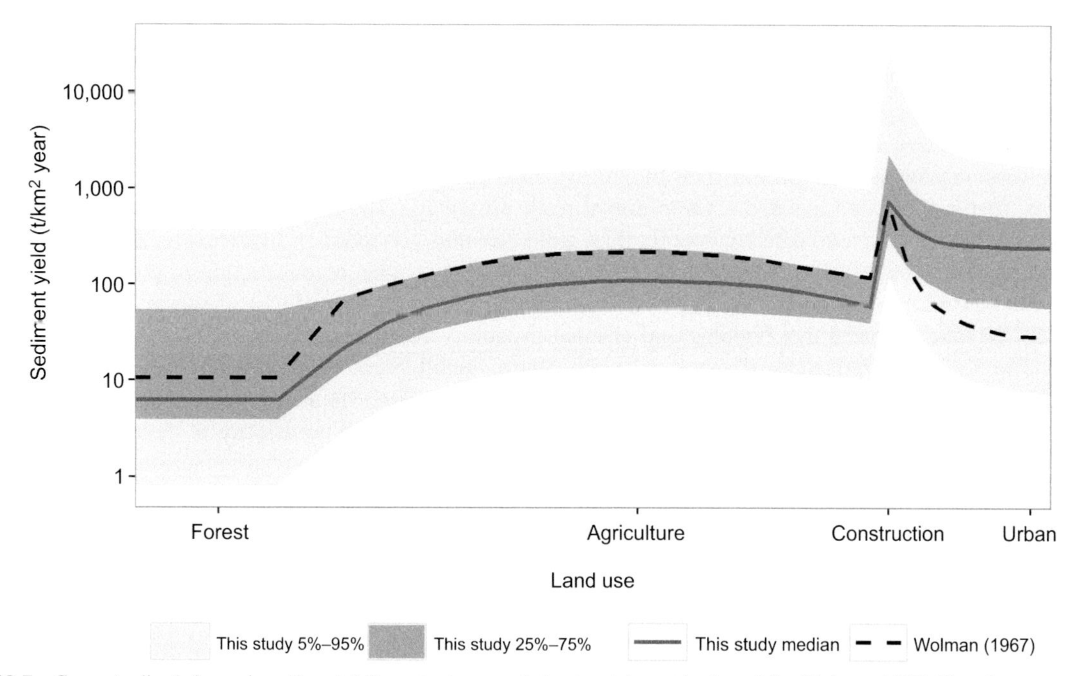


**FIGURE 12.7** Conceptualized change in sediment delivery to streams relative to catchment land use (after Wolman, 1967). Note the greater uncertainty regarding the urban phase. *From Russell, K., Vietz, G.J., Fletcher, T.D., 2017. Global sediment yields from urban and urbanizing watersheds. Earth-Science Reviews 168, 73–80.*

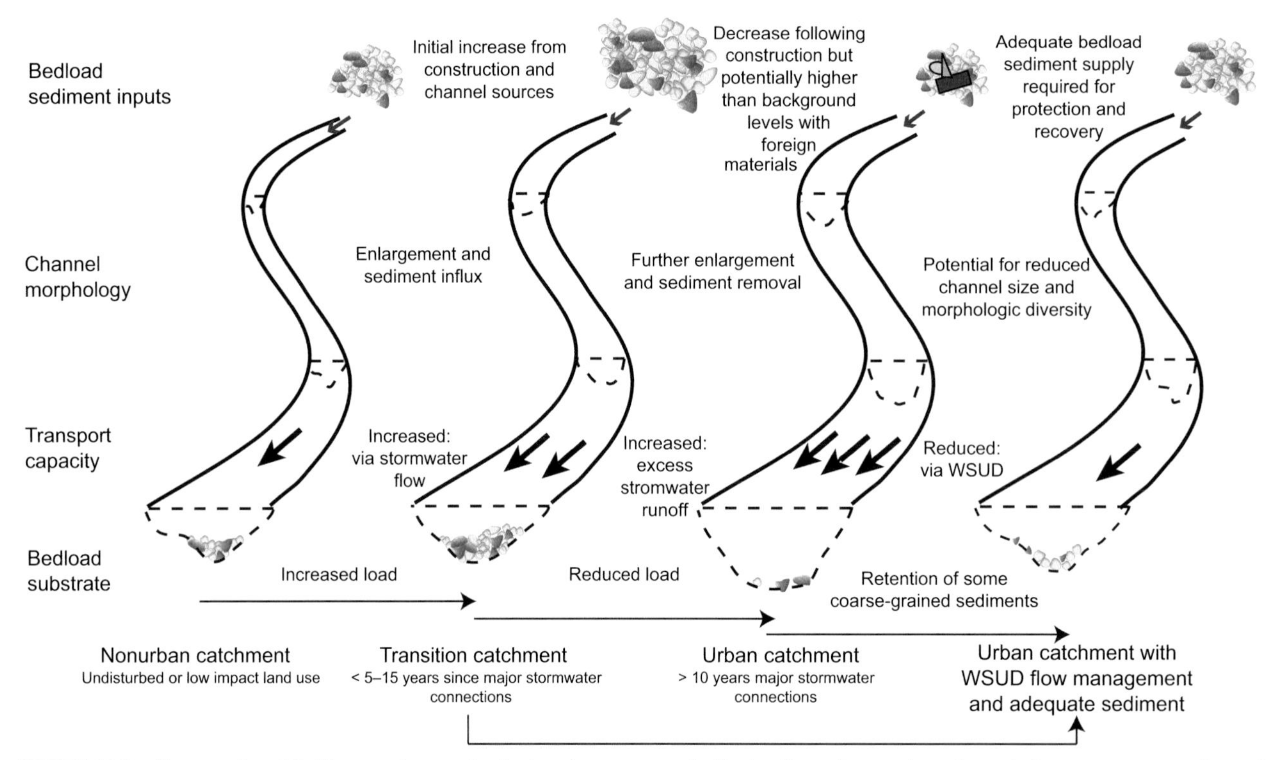


**FIGURE 12.8** Conceptual model of how catchment urbanization changes stream bedload sediment inputs, channel morphology, transport capacity, and bedload substrate. An urban catchment with WSUD provides greater opportunities for protection and recovery when sediment inputs are also considered.

## 12.5 RETHINKING STORMWATER MANAGEMENT TO BETTER MANAGE STREAM MORPHOLOGY

### 12.5.1 Improving your business case for WSUD by incorporating stream health

Rapid urban sprawl in major cities and centers throughout Australia with conventional stormwater drainage has negative impacts and costs for downstream waterways, which are rarely incorporated into business cases. On the flip side, alternative stormwater management in greenfield developments, such as stormwater harvesting and biofiltration, can generate local benefits, cost savings, and avoided costs to downstream waterways. These downstream waterway benefits include reduced fluvial flooding, increased amenity opportunities along streams, protection of biodiversity, and healthier receiving estuaries and bays as shown in Fig. 12.9. For example, the costs for channel reconstruction to convey increased urban stormwater can be AUD$2.5 M to AUD$6 M per kilometer and result in the loss of tens of species (Vietz et al., 2014a). Hawley et al. (2013b) estimated that flooding and channel instability could account for US$1.1 B in damages to state-funded roads in the urbanized areas across the United States, highlighting the potentially negative implications of conventional stormwater management. In Australia, Water by Design (2010) estimated that traditional stormwater management imposed rehabilitation and maintenance costs of AUD$8000–$60,000 per hectare of development.

While the perceived high cost of decentralized SCMs has hampered their widespread implementation, accounting for stream restoration costs may suggest otherwise. A case study by Hawley et al. (2013b) found that SCMs that address the cause of stream degradation are cost competitive compared with the costs of channel reconfiguration to combat the symptoms (Table 12.1). By coupling the multiple local scale benefits (within the development) with the avoided costs to the waterway downstream of the development, the business case for alternative stormwater management in suburbs is compelling.

### 12.5.2 WSUD systems for protection and rehabilitation of channel morphology

Conventionally designed, large SCMs have been found to exacerbate erosion by increasing the duration of time flow above $Q_{critical}$ (the discharge at which the bed is mobilized), leading to increased cumulative sediment transport capacities in

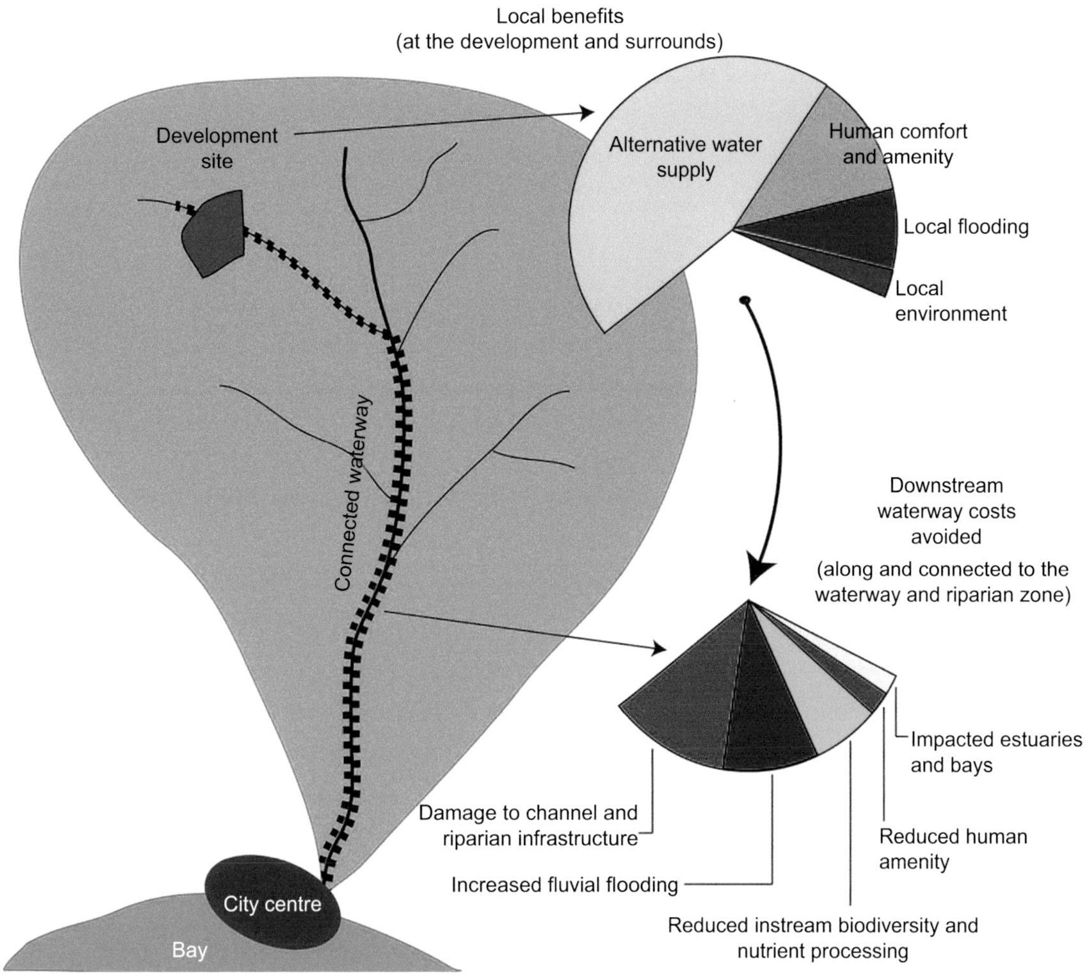

FIGURE 12.9 Benefits (and avoided costs) of alternatives to conventional urban drainage design at two scales: the local development and the downstream waterway. The relative size of each segment is a conceptual estimate of their relative magnitude. *From Vietz, G.J., Rutherfurd, I.D., Walsh, C.J., En Chee, Y., Hatt, B.E., 2014a. The unaccounted costs of conventional urban development: protecting stream systems in an age of urban sprawl. In: Vietz, G.J., Rutherfurd, I.D., Hughes, R.M. (Eds.), Australian Stream Management Conference, Catchments to Coast June 30 to July 2, 2014, Townsville, Queensland, pp. 518–524.*

streams (Hawley et al., 2013b). The focus of many managers and researchers is therefore on smaller, distributed SCMs (Fletcher et al., 2014; Vietz et al., 2016a).

SCMs to protect stream geomorphology display wide-ranging results and rely on adjusting outflows to reduce excess stormwater runoff below the critical discharge for bedload movement (Elliot et al., 2010). For example, Hogan et al. (2014) found small, decentralized SCMs to be effective in reducing stream flows in a range of urban developments (impervious surface cover 35%–40%). They suggested that SCMs contrasted with conventional drainage by maintaining stormwater flows at near predisturbance volumes. Conventional solutions lead to as much as a ninefold increase in total annual discharge. To achieve this would require significant loss of water volumes through infiltration and evapotranspiration. They recommended that management of runoff in small catchments such as these (~100–350 ha) could result in a significant reduction in stream channel erosion. Comparing larger structural approaches with distributed SCMs, Tillinghast et al. (2012) found that a combination of smaller SCMs (water tanks, biofilters) with small wetlands with multilevel outlets provided the greatest reduction in flow time above erosion flow rate thresholds. In the best case, stormwater runoff volumes were reduced by 41%, reducing erosion potential and increasing stream geomorphic stability. Rainwater tanks and biofilters are most effective when they provide for both infiltration and evapotranspiration or provide a slow release function (e.g., slow release to biofilters, Burns et al., 2012). This raises the question of the effectiveness of current SCM designs that are not commonly focused on the protection of stream geomorphology. Hydrologic metrics focused on geomorphic goals are important design input.

**TABLE 12.1 Comparison of Costs for Stream Channel Restoration (Direct Stream Channel Interventions) Relative to Catchment-Scale Stormwater Control Measures (SCMs, Decentralized Interventions) Highlighting the Cost Effectiveness of the Latter**

| | Typical Cost Ranges per Length of Stream ($US/m) | Sources |
|---|---|---|
| **Channel Restoration**[1] | | |
| Habitat rehabilitation in rural and suburban streams | $200–$1200 | Professional projects |
| Habitat rehabilitation and infrastructure protection | $1500–$5000 | Prof. projects; Water by Design (2010) and Sammonds and Vietz (2015) |
| Stream daylighting in suburban setting | $2000 | Prof. projects |
| Stream daylighting in urban corridor | $65,000 | Prof. projects; Hawley and Korth (2014) |
| **Catchment-Scale SCMs**[2] | | |
| Retrofits of conventional detention basins | $10–$70 | Prof. projects; Hawley et al. (2017) |
| New surface storage | $100–$500 | Prof. projects; Hawley et al. (2012) |
| New underground storage | $2000 | Hawley et al. (2013b) |
| New distributed green infrastructure | $5000 | King County (2013) |

[1]*Channel restoration costs from professional projects at 23 US sites across a gradient of geomorphic settings, as well as ranges provided by Australian case studies.*
[2]*Catchment-scale SCM costs from seven case studies, each with several competing scenarios. Solutions were optimized for flow regime restoration to reduce in-stream erosion and restore a more natural bed-sediment disturbance regime (e.g., see Hawley et al., 2012 for methods). Unit costs per effected drainage area were converted to equivalent stream length protected via a regionally appropriate drainage density of ~1.6 km/km$^2$ after Hawley et al. (2013a). For example, one catchment-scale SCM project that evaluated a range of detention basin retrofit and culvert restriction scenarios ranged in costs from ~$150,000 to $300,000 in a 3.3 km$^2$ watershed. Multiplying 3.3 km$^2$ by the regional average for stream drainage density of ~1.6 km/km$^2$ suggests that the SCMs would protect ~7.6 km of streams within the catchment. Dividing the total cost by 7600 m of stream results in the approximate unit cost range of ~$20–$40 per meter of stream protected. For comparison, a planning-level cost estimate to restore just ~1.4 km of the lowest section of the main channel in the watershed was ~$2.4 M (~$1700/km) and that would not have mitigated the remaining ~6 km of streams in the watershed.*

Recommendations made by Vietz et al. (2016a) to reduce excess stormwater runoff effects on stream channel morphology include identifying target flows to be prevented; implementing stormwater harvesting, such as tanks, where captured water can be reused (rather than just attenuating overflow); providing stormwater detention basins, where necessary, with restricting, multilevel offtakes to reduce flows above erosional thresholds; and developing combined stormwater and stream management strategies. As an example, Hawley et al. (2017) showed how this could be done through a simple retrofit of an existing detention basin outlet, reducing the cumulative erosive power (as represented by sediment transport capacity) by 40% compared with the preretrofit condition. The retrofit also reduced disturbance frequency, converting the 3-month, 6-month, and 1-year recurrence interval storms that caused runoff events that exceeded $Q_{critical}$, to events that no longer caused streambed mobilization. Additional ecosystem benefits included reduced flashiness and prolonged baseflows, all while providing a passive bypass of the restricted outlet for extreme events to facilitate a similar level of flood control performance as compared with the preretrofit condition.

Demonstrating that flow regime management can protect channel morphology in urban catchments requires results from applied field trials and is reliant on further case studies. After 5 years of postimplementation monitoring in San Diego County, it appears that geomorphically driven stormwater management targets based on the relative susceptibility of the stream setting to hydromodification (Bledsoe et al., 2012) are protecting receiving streams from excess erosion as compared with reference sites (e.g., ESA, 2016). Improving knowledge on addressing geomorphic change through flow regime management requires an effort comparable with that underway on water quality and ecological response projects such as Shepherd Creek, Ohio, USA (Roy et al., 2014), and Little Stringybark Creek, Melbourne, Australia (Walsh et al., 2015).

While a lot of attention is paid to the approaches for measuring ecological and geomorphic change in streams, the measurement of the "level" of urbanization can be just as important. Walsh et al. (2016) argue that much of the uncertainty in stream response to urbanization is the result of catchment variables used to define urbanization, providing a framework for consideration of the levels of impervious area and the role of the stormwater drainage network in delivering stormwater to streams.

Goals for returning hydrology to near-natural flow regimes have been considered "unreasonable" (Tillinghast et al., 2012), but this depends on water demands. Burns et al. (2012) argue that higher demand for harvested stormwater may make the aim of near-natural flow regime and stream channel protection more feasible. In existing catchments with dense urban developments lacking either conventional detention ponds or decentralized SCMs, experience suggests that creativity and pragmatism will be required to fit in geomorphically effective stormwater controls wherever space and funds allow (see Hawley et al., 2012). Regardless of the potential to reduce runoff volumes, the design of SCMs needs to be targeted toward specific geomorphic flow metrics (e.g., flow rates responsible for bed-sediment mobilization) to achieve goals for stream protection or restoration (Fletcher et al., 2014).

### 12.5.3 Addressing stream disturbance and informing WSUD

The determination of flow thresholds for disturbance and sediment movement is a long-running challenge. As geomorphic metrics aim to characterize stream erosion, they are most commonly focused on the higher flows that partially or completely fill the channel. Priority is commonly given to the "dominant discharge" or the "effective discharge" as they are the flows of sufficient magnitude and frequency that do the work to shape channels. For example, the 2-year recurrence flow ($Q_2$) has been used as an important threshold for channel change (McCuen and Moglen, 1988). Pickup and Warner (1976) found the most relevant discharge to be the 1.5-year Annual Recurrence Interval (ARI) or the 1.15−1.4-year ARI for bedload sediment. Booth et al. (2004) suggested that the half-year flood occurs often enough to transport streambed sediment in most alluvial channels. Wolman and Miller (1960) suggested that the flow threshold for the dominant discharge can be as little as the mean daily flow rate.

This variation points to the importance of considering sediment size and type in calculations and is a point that has long been raised (Sidle, 1988). Commonly, flow thresholds are based on hydrologic metrics that are not necessarily relevant to the sediment in the channel, even though the variability in thresholds for movement of different sediment sizes varies greatly. Some streams are more sensitive to flows of a given magnitude than other streams (Bledsoe et al., 2012). For example, a streambed of fine-grained sand will be mobilized by a much smaller flow than a bed containing cobbles. And a riverbank of cohesive silt and clay may require a larger flow again. Incorporating stream sensitivity into the hydrologic metrics may reduce the risks of a generic threshold being imposed on more sensitive streams and may improve the applicability of a metric across a broad range of stream types.

The critical mechanism for the initiation of the common channel evolution sequence in suburban streams (Fig. 12.3) is the transport of the particles that lie on the streambed. Without moving the sands, gravels, cobbles, and/or boulders that comprise the bed, streams cannot undergo the initiation of incision that often results in prolonged sequences of channel instability. Streambed mobilization also disturbs the benthic organisms that dwell on these substrates and the associated food webs that depend on such insects (e.g., insectivorous fishes, Knight et al., 2008).

The value of such a mechanistic understanding of these impacts is that the so-called "critical flow" ($Q_{critical}$) for streambed mobilization is something that can be determined for any alluvial stream channel. This can then be used to inform the management of stormwater runoff across its catchment. If simplicity is required and only one metric can be used to inform stormwater controls in a development, then $Q_{critical}$ is the most likely to prevent channel degradation and ecological disturbance.

Using standard methods of river mechanics and data from 195 sites across three distinct regions in California and Kentucky (United States) and Victoria (Australia), Hawley and Vietz (2016) determined the streamflow required to mobilize the bed sediment, and, as expected, larger particles generally required larger flows (see Fig. 12.10). To make this approach more generally applicable, discharge for each stream was standardized by the 2-year flow ($Q_2$) to account for unequal catchment sizes and climatic regimes. For example, $Q_{critical}$ developed for a boulder bed stream might be ~8 times larger than its $Q_2$ value and ~2.5 times larger than $Q_2$ in a cobble stream. This implies that streambed mobilization in boulder- and cobble-dominated streams is likely to be relatively infrequent (e.g., say once per decade) and suggests that these systems are probably more resistant to channel instabilities that those streams that are common in urban catchments.

By contrast, gravel- and sand-dominated streambeds have much less inherent resistance with mean $Q_{critical}$ estimates of 0.15 and 0.001 times $Q_2$, respectively, implying a much greater sensitivity to catchment urbanization. For example, the untreated urban runoff regime could double or even triple the particle mobilization frequency of gravel streams, which would otherwise occur in the order of a few times per year. Beyond creating channel instabilities and habitat impacts, the increased disturbance frequency could cause a shift in the types of organisms that inhabit those channels from a diverse mix of both long-lived and fast-growing taxa to a system largely lacking in long-lived organisms (e.g., Hawley et al., 2016).

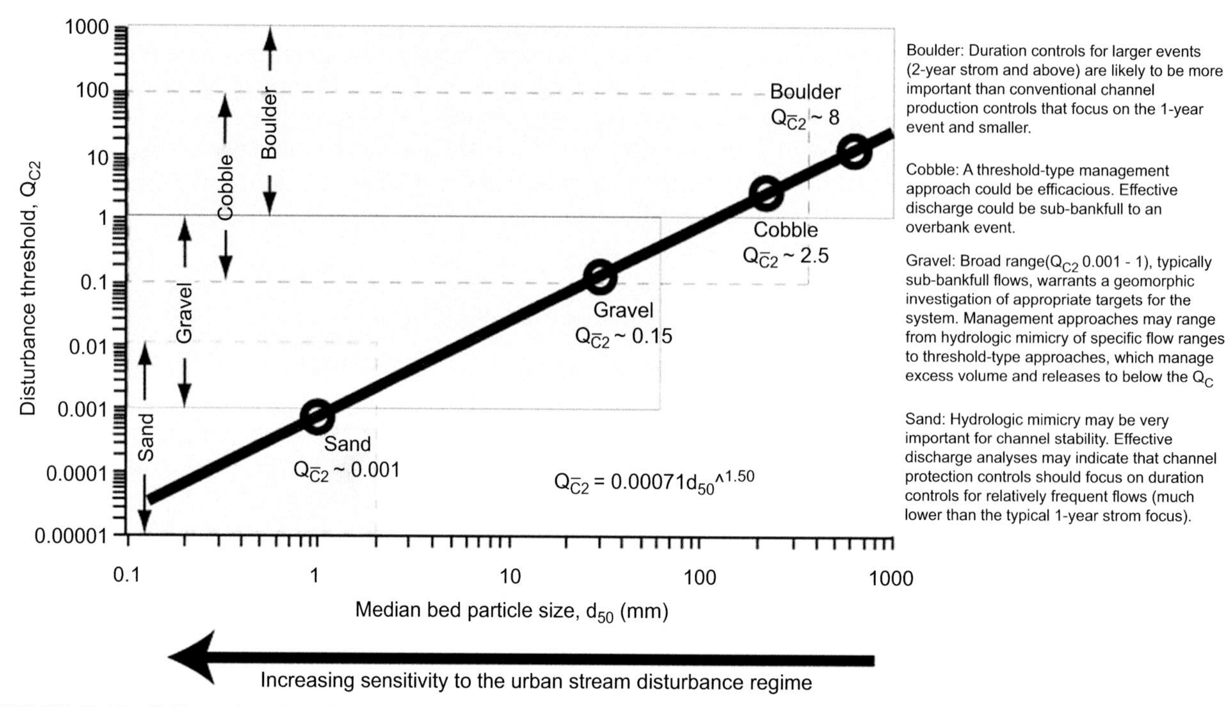


**FIGURE 12.10** Guidance for addressing the urban disturbance regime and bed-sediment mobilization: $Q_{c2}$ is the critical discharge ($Q_{critical}$) as a proportion of the 2-year ARI ($Q_2$). Mean threshold values for each size class ($Q_{\bar{c}2}$) are depicted for the nominal median particle for each size class. Given the broad range of $Q_{c2}$ values (note the range of each of the shaded rectangles is ~2–3 orders of magnitude), $Q_{\bar{c}2}$ should only be used in the absence of more detailed geomorphic assessments, which are strongly recommended. *From Hawley, R.J., Vietz, G.J., 2016. Addressing the urban stream disturbance regime. Freshwater Science 35 (1), 278–292.*

Channel instability is typically one of the defining traits of urban streams, but it is an impairment that can be prevented through smarter stormwater management. Fig. 12.10 demonstrates a simplification of the relationship between bed particle size and the magnitude of flows that create disturbance. This figure, however, is a simplification of the variables that drive bed disturbance. As suggested by Hawley and Vietz (2016), the physically based trends require investigation and a stormwater management approach tailored to the stream they are intended to protect, so a one-size-fits-all policy should be used with caution. Relatively simple field data programs combined with standard modeling approaches can facilitate the development of a "critical flow" for a stream or region that can be used to optimize SCMs to better protect receiving streams from excess channel erosion.

## 12.6 MANAGING SEDIMENT FOR STREAM PROTECTION AND RECOVERY

As suggested in Fig. 12.1, equilibrium streams are facilitated by a balance of erosion and resistance. One of the most important forms of resistance in many systems is the supply of sediment from upstream—the transport of sediment consumes erosive energy that could otherwise be expressed on the channel boundary. *Reduction or elimination* of the bedload sediment supply can thereby induce channel erosion and streambed coarsening, one of the commonly observed phenomena downstream of dams (e.g., Kondolf, 1997). As previously discussed, excess channel erosion impacts physical habitat and can disturb aquatic organisms. At the same time, *increased* suspended sediment loads (e.g., smothering of substrates, Wood and Armitage, 1997) can also cause habitat degradation in streams draining urban catchments. As a result, conflicts may exist between the need to restore or maintain coarse-grained sediment loads (i.e., bedload sediments), while at the same time reducing the concentrations of fine-grained sediment loads (suspended sediments) often associated with stormwater contaminants (e.g., nutrients and heavy metals). The latter have historically been the focus of WSUD treatment devices (CSIRO, 1999; Doyle and Douglas, 2012; Hatt et al., 2004).

Differentiation of coarse- and fine-grained sediments is complex but important in WSUD. A key management challenge in the future will be to continue to mitigate urban hydrologic disturbance and minimize stormwater pollution, while allowing coarse-grained sediment supply to streams to be maintained at rates consistent with natural, predeveloped levels (Fletcher et al., 2014). This requires specific identification and addressing of the stormwater flows that are likely to be mobilizing the coarse sediments (Hawley and Vietz, 2016), as well as strategic placement and design of SCMs. Otherwise, SCMs such as basins, gross pollutant traps, and wetlands, which do not consider stream geomorphology, may have the unintended consequence of exacerbating channel instability by trapping coarse-grained sediment fractions in some applications.

Replenishment of coarse-grained sediment downstream of dams has been trialed in the United States (Ligon et al., 1995) and Japan (Ock et al., 2013) with some success (Merz et al., 2005; Zeug et al., 2014). Ock et al. (2013) emphasized the importance of tailoring the replenishment method to the postdisturbance flow regime, which will be particularly important under a highly modified posturban flow regime.

Considerations are being made for redirection of coarse-grained sediments to streams in urban catchments, which could reduce maintenance costs and improve stream condition (Houshmand et al., 2014). A recent pilot study by Houshmand et al. (2014) tested coarse-grained sediment that had been trapped in a gross pollutant trap. In finding contaminant levels tested to be below guidelines, the sediments were transferred to a nearby stream that required replenishment. Reducing the sediment maintenance in SCMs, while maintaining appropriate sediment supply (in terms of both load and caliber) to enhance the condition of receiving streams, has both ecological and economic benefits (Houshmand et al., 2014). It is a major design challenge for the future of stormwater management (Russell et al., 2017).

Opportunities to maintain coarse-grained sediment supplies are greatest within new or peri-urban developments. Wherever possible, headwater sources of coarse-grained sediment could be protected, for example, by maintaining headwater stream connections and establishing corridors (Booth et al., 2010). In brownfield sites, urban renewal of land or drainage systems may enable re-engagement of headwater sources to supply coarse-grained sediment (e.g., through reinstatement of headwater channels and the removal of barriers such as weirs, etc.). The difficulty of returning appropriate sediment regimes in brownfield contexts reinforces the critical need to protect headwater streams during the development phase (Bernhardt and Palmer, 2007; Walsh et al., 2005). As is the case with the provision of sediment to assist stream recovery below rural dams, the potential for coarse-grained sediment provision in urban streams may be a novel management consideration in the protection and recovery of urban stream channels.

## 12.7 PERCEPTIONS AND PRACTICALITIES OF STREAMS IN URBAN CATCHMENTS

Development of rivers and streams in cities historically moves from a water supply focus to a drainage and flood control imperative (Brown et al., 2009). While attitudes are rapidly changing, the primary community concerns for urban streams are civil aspects (e.g., flood mitigation) and recreational and aesthetic values (Findlay and Taylor, 2006). Despite natural rates of erosion being an ecologically beneficial process, prevention of essentially all erosion is often the primary maintenance concern (Florsheim et al., 2008). If the goal is to provide a natural aesthetic, and particularly with some natural functioning in an urban stream, then the urban geomorphologist's role is a particularly challenging endeavor to demonstrate the benefits to engineers and the community of a "naturally" functioning stream system with appropriate rates of erosion. Erosion is a natural process, yet it is excess levels of erosion that are the concern.

Geomorphic dynamism could probably coexist with flooding and maintenance concerns, but only if appropriately understood and implemented within certain bounds. The success of appropriate rehabilitation will be difficult unless public expectations are tailored to an appropriate understanding of stream dynamics (Rhoads et al., 2008 p. 227). The steps to stakeholder confidence include highlighting the values of geomorphically functioning streams and improved understanding of channel response in urban environments so that outcomes and associated risks can be practically assessed.

There are a number of case studies in the United States (e.g., San Diego, California, and Northern Kentucky) where communities have taken steps to implement geomorphically relevant stormwater policies. These include catchment-scale WSUD projects that attempt to restore a more natural streambed disturbance and channel erosion regime through tailored SCMs. The communities that have shown arguably some of the most striking progress is the incorporation of the critical discharge ($Q_{critical}$) for streambed erosion. This serves as an accessible design target for engineers and managers as well as a tangible mechanism for ecologists, property owners, and the public.

In Australia, using bed disturbance metrics to drive stormwater controls is gaining traction. Duncan et al. (2014) used $Q_{critical}$ as the primary geomorphic metric driving stormwater controls required for stream protection from catchment urbanization. They measured bed-sediment sizes and used one-dimensional hydraulic models to quantify the discharge threshold with the aim of maintaining the duration of time above this threshold at preurbanization levels. The study highlighted the necessity to reduce excess stormwater runoff volumes (by up to 80%) rather than just modifying patterns of flow to achieve geomorphic and ecological protection metrics. The approach of defining stormwater controls on bed disturbance is being used increasingly throughout Melbourne, Victoria, and is gaining traction in other states of Australia, though implementation of these recommendations for stream protection is lagging. Nevertheless, the incorporation of the stormwater management strategy within the waterway management strategy is a significant step toward integrated management of streams through WSUD approaches.

## 12.8 CONCLUSION

Stream morphology is primarily a product of the hydrologic and sediment inputs. Catchment urbanization typically has a detrimental impact on the properties of a stream including shape, size, and sediment processes such as erosion and deposition, which tends to coincide with degraded ecological functioning. By incorporating geomorphic considerations into WSUD, SCMs can be optimized to mitigate the conventional impacts of catchment urbanization and better preserve stream morphology and ecological functions.

Using channel disturbance as a metric to drive stormwater controls for urban planners and developers provides a tangible WSUD approach. The mechanics of sediment mobilization is well understood and bed disturbance is explicitly linked to both ecological and physical degradation of streams. Sediment mobilization thresholds can be readily linked to components of the flow regime, which, in turn, inform the design of WSUD and the details of SCMs.

WSUD approaches must consider sediment supply both in the catchment and in the stream channel. Sediment supply in the catchment reduces following the initial spike associated with civil construction and channel degradation. An established catchment appears to maintain sediment supply, in part, from exogenous sediments such as landscape products and weathered pavements entering the stream. Changes in catchment sediments must be considered when designed SCMs, particularly where sediments may clog infiltration systems reducing effectiveness and increasing maintenance. In the stream, bedload sediments play an important role for stream condition, with sediment supply reducing the potential for erosion and enabling the persistence of bed substrates and depositional features such as bars and benches. Despite changes in sediment supply to the stream channel, the availability of mobile sediments (such as bedload) is dominated by the high sediment transport capacity generated by excess stormwater runoff. This leads to deepening stream channels with poor morphologic diversity and a lack of substrate sediments for biota.

For the protection of channel morphology, WSUD needs to focus on reducing volume and the duration of time above a stream-specific threshold. If WSUD is largely tailored to relatively small storms (<25 mm rain event), it would most likely need to be complemented by larger controls (e.g., extended detention basins, biofiltration basins, etc.) to protect channels from excess erosion (and meet standard flood control criteria). The more extensive the effort on catchment wide WSUD, the less the need for conventional flood control facilities (e.g., detention basins) to meet channel protection design criteria. This can be best achieved through distributed SCMs rather than centralized systems. A WSUD approach for protecting channel morphology could include house rainwater tanks with slow releases (to ensure some air space at the beginning of the big storms) and large rain gardens, intercepting a significant proportion of household impervious area, and streets with adequate rain gardens, wetlands, and infiltration basins, etc. With these intercepting impervious runoff, the geomorphic design criteria could feasibly be met. The benefits of these WSUD strategies need to be appropriate to community and stakeholder goals and expectations, such as neighborhood aesthetics, etc.

We must recognize that channel morphology and dimensions of streams draining urban catchments are unlikely to retained as similar to those with natural or even agricultural catchments pre-European forms or even postagricultural form. However, by focusing on the critical flow characteristics of preurbanization conditions, we may be able to restore some of the features and processes and the myriad array of social, economic, and ecological values that we expect from healthy functioning streams. There is clearly an important role for WSUD in the protection and management of stream morphology in urban catchments.

## ACKNOWLEDGMENTS

We would like to thank the Editors Ashok Sharma, Ted Gardner, and Don Begbie for their valuable comments on this chapter. Vietz is provided academic support by the Melbourne Waterway Research-Practice Partnership, a partnership between Melbourne Water and Melbourne University's Waterway Ecosystem Research Group. Hawley's time is provided by Sustainable Streams.

## REFERENCES

Bernhardt, E.S., Palmer, M.A., 2007. Restoring streams in an urbanizing world. Freshwater Biology 52, 738–751.

Bledsoe, B.P., 2002. Stream erosion potential and stormwater management strategies. Journal of Water Resources Planning and Management 128 (6), 451–455.

Bledsoe, B.P., Stein, E.D., Hawley, R.J., Booth, D., 2012. Framework and tool for rapid assessment of stream susceptibility to hydromodification. Journal of the American Water Resources Association 48 (4), 788–808.

Bledsoe, B.P., Watson, C.C., 2001. Effects of urbanization on channel instability. Journal of the American Water Resources Association 37 (2), 255–270.

Booth, D.B., 1990. Stream-channel incision following drainage-basin urbanization. Water Resources Bulletin 26 (3), 407–417.

Booth, D.B., Dusterhoff, S.R., Stein, E.D., Bledsoe, B.P., 2010. Hydromodification screening tools: GIS-based catchment analyses of potential changes in runoff and sediment discharge. In: Technical Report 605, March 2010 (Southern California Coastal Research Project).

Booth, D.B., Henshaw, P.C., 2001. Rates of channel erosion in small urban streams. In: Wigmosta, M., Burges, S. (Eds.), Land Use and Watersheds: Human Influence on Hydrology and Geomorphology in Urban and Forest Areas, pp. 17–38. AGU Monograph Series, Water Science and Application, Washington, DC.

Booth, D.B., Jackson, C.R., 1997. Urbanization of aquatic systems: degradation thresholds, stormwater detection, and the limits of mitigation. Journal of the American Water Resources Association 33 (5), 1077–1090.

Booth, D.B., Karr, J.R., Shauman, S., Konrad, C.P., Morley, S.A., Larson, M.G., Burges, S.J., 2004. Reviving urban streams: land use, hydrology, biology, and human behaviour. Journal of the American Water Resources Association 40 (5), 1351–1364.

Brown, R.R., Keath, N., Wong, T.H.F., 2009. Urban water management in cities: historical, current and future regimes. Water Science and Technology 59 (5), 847–855.

Burns, M.J., Fletcher, T.D., Walsh, C.J., Ladson, A.R., Hatt, B.E., 2012. Hydrologic shortcomings of conventional urban stormwater management and opportunities for reform. Landscape and Urban Planning 105 (3), 230–240.

Chin, A., 2006. Urban transformation of river landscapes in a global context. Geomorphology 79 (3), 460–487.

Chin, A., Gregory, K.J., 2001. Urbanization and adjustment of ephemeral stream channels. Annals Association of American Geographers 91, 595–608.

CSIRO, 1999. Uban Stormwater: Best Practice Environmental Management Guidelines, Stormwater Committee, Victoria. CSIRO Publishing, Collingwood, Australia.

Doyle, M.W., Douglas, S.F., 2012. Compensatory mitigation for streams under the Clean Water Act: reassessing science and redirecting policy. Journal of the American Water Resources Association 48 (3), 494–509.

Duncan, H.P., Fletcher, T.D., Vietz, G.J., Urrutiaguer, M., 2014. The feasibility of maintaining ecologically and geomorphically important elements of the natural flow regime in the context of a superabundance of flow. In: Waterway Ecosystem Research Group (Ed.), Stage 1 – Kororoit Creek Study, Melbourne Waterway Research-practice Partnership Technical Report 14.15. The University of Melbourne, Melbourne, Australia.

Elliot, A.H., Spigel, R.H., Jowett, I.G., Shankar, S.U., Ibbitt, R.P., 2010. Model application to assess effects of urbanisation and distributed flow controls on erosion potential and baseflow hydraulic habitat. Urban Water Journal 7 (2), 91–107.

ESA, November 9, 2016. Effectiveness assessment of the San Diego Hydromodification Management Plan. Prepared for the San Diego Regional Water Quality Control Board 41.

Findlay, S.J., Taylor, M.P., 2006. Why rehabilitate urban river systems? Area 38 (3), 312–325.

Finkenbine, J.K., Atwater, J.W., Mavinic, D.S., 2000. Stream health after urbanization. Journal of the American Water Resources Association 36 (5), 1149–1160.

Fitzpatrick, F.A., Peppler, M.C., 2010. Relation of urbanization to stream habitat and geomorphic characteristics in nine metropolitan areas of the United States. Scientific Investi- gations Report 2010-5056. US Geological Survey, Reston, Virginia.

Fletcher, T.D., Vietz, G.J., Walsh, C.J., 2014. Protection of stream ecosystems from urban stormwater runoff; the multiple benefits of an ecohydrological approach. Progress in Physical Geography 38 (5), 543–555.

Florsheim, J.L., Mount, J.F., Chin, A., 2008. Bank erosion as a desirable attribute of rivers. BioScience 58 (6), 519–529.

Gilvear, D.J., 1999. Fluvial geomorphology and river engineering: future roles utilizing a fluvial hydrosystems framework. Geomorphology 31, 229–245.

Grove, J., Ladson, A., 2006. Attacking urban areas with tanks: predicting the ecological and geomorphological recovery potential of urban streams. In: 30th Hydrology and Water Resources Symposium: Past, Present and Future, Institute of Engineers Australia, Launceston, Tasmania.

Gurnell, A.M., Lee, A., Souch, C., 2007. Urban rivers: hydrology, geomorphology, ecology and opportunities for change. Geography Compass 1 (5), 1118–1137.

Hammer, T.R., 1972. Stream channel enlargement due to urbanization. Water Resources Research 8 (6), 1530–1540.

Hatt, B.E., Fletcher, T.D., Walsh, C.J., Taylor, S.L., 2004. The influence of urban density and drainage infrastructure on the concentrations and loads of pollutants in small streams. Environmental Management 34 (1), 112–124.

Hawley, R.J., Bledsoe, B.P., 2013. Channel enlargement in semiarid suburbanizing watersheds: a southern California case study. Journal of Hydrology 496, 17–30.

Hawley, R.J., Bledsoe, B.P., Stein, E.D., Haines, B.E., 2012. Channel evolution model of semiarid stream response to urban-induced hydromodification. Journal of the American Water Resources Association 48 (4), 722–744.

Hawley, R.J., Goodrich, J.A., Korth, N.L., Rust, C.J., Fet, E.V., Frye, C., MacMannis, K.R., Wooten, M.S., Jacobs, M., Singha, R., 2017. Detention outlet retrofit device improves the functionality of existing detention basins by reducing erosive flows in receiving channels. Journal of the American Water Resources Association. https://doi.org/10.1111/1752–1688.12548.

Hawley, R.J., Korth, N.L., September 27–October 1, 2014. 'Stream Daylighting' as a strategy for CSO mitigation: saving $, reducing overflows, and restoring habitat by bringing streams back to the surface, Pages 3154–3163. In: Causey, P., Reeves, S. (Eds.), WEFTEC. Water Environment Federation, New Orleans, LA.

Hawley, R.J., MacMannis, K.R., Wooten, M.S., 2013a. Bed coarsening, riffle shortening, and channel enlargement in urbanizing watersheds, northern Kentucky, USA. Geomorphology 201, 111–126.

Hawley, R.J., MacMannis, K.R., Wooten, M.S., May 19–23, 2013b. How poor stormwater practices are shortening the life of our nation's infrastructure - recalibrating stormwater management for stream channel stability and infrastructure sustainability. World environmental and water resources congress, ASCE, Cincinnati, Ohio, pp. 193–207.

Hawley, R.J., Vietz, G.J., 2016. Addressing the urban stream disturbance regime. Freshwater Science 35 (1), 278–292.

Hawley, R.J., Wooten, M.S., MacMannis, K.R., Fet, E.V., 2016. When do macroinvertebrate communities of reference streams resemble urban streams? The biological relevance of Qcritical. Freshwater Science 35 (3), 778–794.

Hogan, D.M., Jarnagin, S.T., Loperfido, J.V., VanNess, K., 2014. Mitigating the effects of landscape development on streams in urbanizing watersheds. Journal of the American Water Resources Association 50 (1), 163–178.

Houshmand, A., Vietz, G.J., Hatt, B.E., 2014. Improving urban stream condition by redirecting sediments: a review of associated contaminants. In: Vietz, G.J., Rutherfurd, I.D., Hughes, R.M. (Eds.), Australian Stream Management Conference, Catchments to Coast June 30 to July 2, 2014, Townsville, Queensland, pp. 549–557.

King County, 2013. Development of a Stormwater Retrofit Plan for Water Resources Inventory Area 9: SUSTAIN Model Pilot Study. King County Department of Natural Resources and Parks, Seattle, Washington.

Knight, R.R., Gregory, M.B., Wales, A.K., 2008. Ecohydrology 1 (4), 394–407.

Knighton, D., 1998. Fluvial Forms and Processes - A New Perspective. John Wiley & Sons, New York, p. 383.

Kondolf, G.M., 1997. Hungry water: effects of dams and gravel mining on river channels. Environmental Management 21 (4), 533–551.

Lane, E.W., 1955. The importance of fluvial morphology in hydraulic engineering. Proceedings of the American Society of Civil Engineering 81 (paper 745), 1–17.

Leopold, L.B., Huppman, R., Miller, A., 2005. Geomorphic effects of urbanization in forty-one years of observation. Proceedings of the American Philosophical Society 349–371.

Ligon, F.K., Dietrich, W.E., Trush, W.J., 1995. Downstream ecological effects of dams. BioScience 183–192.

McCuen, R., Moglen, G., 1988. Multicriterion stormwater management methods. Journal of Water Resources Planning and Management 114, 414–431.

Merz, J.E., Chan, O., Leigh, K., 2005. Effects of gravel augmentation on macroinvertebrate assemblages in a regulated California river. River Research and Applications 21 (1), 61–74.

Mumford, L., 1961. The City in History. Its Origins, its Transformations and its Prospects, Harcourt. Brace & World, Inc., New York, p. 657.

Nelson, E.J., Booth, D.B., 2002. Sediment sources in an urbanizing, mixed land-use watershed. Journal of Hydrology 264 (1), 51–68.

Newall, P., Walsh, C.J., 2005. Response of epilithic diatom assemblages to urbanization influences. Hydrobiologia 532, 53–67.

Ock, G., Sumi, T., Takemon, Y., 2013. Sediment replenishment to downstream reaches below dams: implementation perspectives. Hydrological Research Letters 7 (3), 54–59.

Park, C.C., 1977. Man-induced changes in stream channel morphology. In: Gregory, K.J. (Ed.), River channel changes. John Wiley and Sons, Chichester, UK, pp. 121–144.

Pickup, G., Warner, R.F., 1976. Effects of hydrologic regime on magnitude and frequency of dominant discharge. Journal of Hydrology 29 (1–2), 51–75.

Pizzuto, J.E., Hession, W.C., McBride, M., 2000. Comparing gravel-bed rivers in paired urban and rural catchments of southeastern Pennsylvania. Geology 28 (1), 79–82.

Poff, L.N., Zimmerman, J.K.H., 2010. Ecological responses to altered flow regimes: a literature review to inform the science and management of environmental flows. Freshwater Biology 55 (1), 194–205.

Poff, N.L., Richter, B.D., Arthington, A.H., Bunn, S.E., Naiman, R.J., Kendy, E., Acreman, M., Apse, C., Bledsloe, B.P., Freeman, M.C., Henriksen, J., Jacobson, R.B., Kennen, J.G., Merritt, D.M., O'Keefe, J.H., Olden, J.D., Rodgers, K., Tharme, R.E., Warner, A., 2010. The ecological limits of hydrologic alteration (ELOHA): a new framework for developing regional environmental flow standards. Freshwater Biology 55, 147–170.

Pomeroy, C.A., Postel, N.A., O'Neill, P.E., Roesner, L.A., 2008. Development of storm-water management design criteria to maintain geomorphic stability in Kansas City Metropolitan Area streams. Journal of Irrigation and Drainage Engineering 134 (5), 562–566.

Rhoads, B., Garcia, M., Rodriguez, J., Bombardelli, F., Abad, J.D., Daniels, M., 2008. Methods for evaluating the geomorphological performance of naturalized rivers: examples from the Chicago metropolitan area. In: Darby, E., Sear, D. (Eds.), River Restoration: Managing the Uncertainty in Restoring Physical Habitat. John Wiley & Sons Ltd, Chichester, UK, p. 328.

Robinson, A.M., 1976. The effects of urbanization on stream channel morphology. In: Proceedings National Symposium on Urban Hydrology, Hydraulics, and Sediment Control, Conference, University of Kentucky, Lexington, pp. 115–127.

Roy, A.H., Rhea, L.K., Mayer, A.L., Shuster, W.D., Beaulieu, J.J., Hopton, M.E., Morrison, M.A., St. Amand, A., 2014. How much is enough? Minimal responses of water quality and stream biota to partial retrofit stormwater management in a suburban neighbourhood. PLoS One 9 (1), e85011.

Russell, K., Vietz, G.J., Fletcher, T.D., 2017. Global sediment yields from urban and urbanizing watersheds. Earth-Science Reviews 168, 73–80.

Ryan, R.J., Boufadel, M.C., 2007. Lateral and longitudinal variation of hyporheic exchange in a piedmont stream pool. Environmental Science and Technology 41 (12), 4221–4226.

Sammonds, M.J., Vietz, G.J., 2015. Setting stream naturalisation goals to achieve ecosystem improvement in urbanising greenfield catchments. Area 47 (4), 386–395.

Schumm, S.A., 1977. The Fluvial System. John Wiley & Sons, New York.

Sidle, R.C., 1988. Bed load transport regime of a small forest stream. Water Resources Research 24 (2), 207.

Simon, A., Rinaldi, M., 2006. Disturbance, stream incision, and channel evolution: The roles of excess transport capacity and boundary materials in controlling channel response. Geomorphology 79, 361–383.

Smith, R.H., Hawley, R.J., Neale, M.W., Vietz, G.J., Diaz-Pascacio, E., Hermann, J., Lovell, A.C., Prescott, C., Rios-Touma, B., Smith, B., 2016. Urban stream renovation: incorporating societal objectives to achieve ecological improvements. Freshwater Science 35 (1), 364–379.

Statzner, B., Higler, B., 1985. Stream hydraulics as a major determinant of benthic invertebrate zonation patterns. Freshwater Biology 16, 127–139.

Tillinghast, E.D., Hunt, W.F., Jennings, G.D., D'Arconte, P., 2012. Increasing stream geomorphic stability using storm water control measures in a densely urbanized watershed. Journal of Hydrologic Engineering 17 (12), 1381–1388.

Trimble, S.W., 1997. Contribution of stream channel erosion to sediment yield from an urbanizing watershed. Science 278 (5342), 1442–1444.

Utz, R.M., Hopkins, K., Beesley, L., Booth, D.B., Hawley, R.J., Baker, M.E., Freeman, M.C., Jones, K.L., 2016. Ecological resistance in urban streams: the role of natural and legacy attributes. Freshwater Science 35, 380–397.

Vietz, G.J., Rutherfurd, I.D., Fletcher, T.D., Walsh, C.J., 2016a. Thinking outside the channel: Challenges and opportunities for stream morphology protection and restoration in urbanizing catchments. Landscape and Urban Planning 145, 34–44.

Vietz, G.J., Rutherfurd, I.D., Walsh, C.J., En Chee, Y., Hatt, B.E., 2014a. The unaccounted costs of conventional urban development: protecting stream systems in an age of urban sprawl. In: Vietz, G.J., Rutherfurd, I.D., Hughes, R.M. (Eds.), Australian Stream Management Conference, Catchments to Coast June 30 to July 2, 2014, Townsville, Queensland, pp. 518–524.

Vietz, G.J., Sammonds, M.J., Walsh, C.J., Fletcher, T.D., Rutherfurd, I.D., Stewardson, M.J., 2014b. Ecologically relevant geomorphic attributes of streams are impaired by even low levels of watershed effective imperviousness. Geomorphology 206, 67–78.

Vietz, G.J., Walsh, C.J., Fletcher, T.D., 2016b. Urban hydrogeomorphology and the urban stream syndrome: treating the symptoms and causes of geomorphic change. Progress in Physical Geography 40 (3), 480–492.

Walsh, C.J., Fletcher, T.D., Bos, D.G., Imberger, S.J., 2015. Restoring a stream through retention of urban stormwater runoff: a catchment-scale experiment in a social-ecological system. Freshwater Science 34 (3), 1161–1168.

Walsh, C.J., Fletcher, T.D., Vietz, G.J., 2016. Variability in stream ecosystem response to urbanization: unravelling the influences of physiography and urban land and water management. Progress in Physical Geography 40 (5), 714–731.

Walsh, C.J., Roy, A.H., Feminella, J.W., Cottingham, P.D., Groffman, P.M., Morgan, R.P., 2005. The urban stream syndrome: current knowledge and the search for a cure. Journal of the North American Benthological Society 24 (3), 706–723.

Water by Design, 2010. A Business Case for Best Practice Urban Stormwater Management (Version 1.1). South East Queensland Healthy Waterways Partnership, Brisbane, Queensland.

Wohl, E., Bledsoe, B.P., Jacobsen, R.B., Poff, N.L., Rathburn, R.E., Wilcox, A.C., 2015. The natural sediment regime in rivers: broadening the foundation for ecosystem management. BioScience 65 (4), 358–371.

Wolman, M.G., 1967. A cycle of sedimentation and erosion in urban river channels. Geografiska Annaler. Series A, Physical Geography 49 (2/4), 385–395.

Wolman, M.G., Miller, J.P., 1960. Magnitude and frequency of forces in geomorphic processes. The Journal of Geology 68, 54–74.

Wood, P.J., Armitage, P.D., 1997. Biological effects of fine sediment in the lotic environment. Environmental Management 21 (2), 203–217.

Zeug, S., Sellheim, K., Watry, C., Rook, B., Hannon, J., Zimmerman, J., Cox, D., Merz, J., 2014. Gravel augmentation increases spawning utilization by anadromous salmonids: a case study from California, USA. River Research and Applications 30 (6), 707–718.

Chapter 13

# Urban Lakes as a WSUD System

Christopher Walker[1,2] and Terry Lucke[1]

[1]*Stormwater Research Group, University of the Sunshine Coast, Sippy Downs, QLD, Australia;*

[2]*Covey Associates Pty Ltd, Sunshine Coast, QLD, Australia*

## Chapter Outline



**ABSTRACT**

Constructed waterbodies serve a variety of purposes. Urban lakes are a contemporary constructed waterbody, which often serve to increase amenity and property values, fulfill a flood mitigation purpose, and may provide a source of land fill for development purposes. Although popular, urban lakes have degraded in many areas as a result of: (1) poor quality of inflow; (2) significant water level variation; (3) persistent stratification; (4) long stormwater residence times; and (5) high loading of organic carbon. The failure of many urban lakes to remain as a resilient and healthy ecosystem stems from poor design and a lack of pretreatment to runoff. Recent design guidelines have better informed designers as to the key design elements for urban lakes; however, management of these waterbodies remains reactive, rather than proactive. Utilizing the Ecosystem Health Paradigm as a template for urban lake management may allow for a more holistic assessment of urban lake health and better inform management requirements. This chapter looks at the history of urban lakes and how such systems are integrated and managed in the urban context.

**Keywords:** Ecosystem health; Eutrophication; Stormwater treatment; Urban lakes; Urban runoff.

## 13.1 INTRODUCTION

Constructed waterbodies are artificial water features that have played a significant role in history, serving to fill a range of functions (Mays et al., 2007), including the following:

- Potable water storage;
- Aquaculture;
- Detention; and
- Esthetic value.

Simply put, these waterbodies are a means to retain or detain a desired volume of water to satisfy a need or needs. Constructed waterbodies have been utilized by civilizations as far back as 3000 BC (Yang et al., 2005).

Approaches to Water Sensitive Urban Design. https://doi.org/10.1016/B978-0-12-812843-5.00013-7

Constructed waterbodies have also been built purely for ornamental and recreational purposes. In 18th and 19th century Europe, numerous shallow lakes were excavated for ornamental purposes, for aesthetics, or to attract waterfowl for game hunting (Sayer et al., 2008). With the industrial revolution and the creation of a wealthy mercantile class affluent landowners often built ornamental lakes to buffer their estates from the encroachment of urbanization (Binnie, 2004).

Modern constructed waterbodies built within development areas are commonly referred to as "urban lakes." Urban lakes are defined as a permanent body of open water of significant size (>0.1 ha) (Gibson et al., 2000). The size, depth, and retention time (i.e., the time it would take for runoff to displace the stored volume of water) of an urban lake vary according to the design requirements. The key roles of urban lakes are linked to aesthetics (Huser et al., 2016) and stormwater management (Hagare et al., 2015).

The desire for waterside living and the need for stormwater management have seen urban lakes become increasingly incorporated into urban developments. Urban lakes help to mitigate the density of urban developments by creating passive open space areas. They are generally designed to maximize the open water area, as this often results in increased property values (Walker, 2012; Walker et al., 2013; Sikorska et al., 2017).

In Australia during the 1980s and 1990s, urban lakes were often incorporated into developments to treat stormwater, with the assumption that the lakes would provide treatment functions typically associated with sediment ponds and constructed wetlands. This was a common assumption in South East Queensland (SEQ), Australia. This approach led to significant water quality issues, as the urban lakes were often not capable of processing nutrient loads typical of urban ecosystems. Urban lakes are now generally implemented as a "receiving environment" with stormwater treatment occuring upstream of the waterbody.

This chapter will explain how urban lakes fit into the water sensitive urban design (WSUD) philosophy and the design principles and requirements that are critical for lake management and maintainance of ecological health in such waterbodies. The term health, which has often been used ambiguously, will be assessed under the "Ecosystem Health" Paradigm (EHP). Ecosystem health expands aquatic health beyond physicochemical parameters, and incorporates biological and social indicators to determine overall health of an ecosystem (Walker, 2012).

## 13.2 URBAN LAKES AND URBAN DEVELOPMENT

Urban lakes provide a "pseudo-natural" aspect to urban developments, fulfilling a desire for a connection or proximity to nature and water. Furthermore, properties within close proximity to urban lakes often have higher values (Walker, 2012; Walker et al., 2013; Sikorska et al., 2017). Natural waterfront property is generally out of reach for most property investors (Cook et al., 2016) and urban lakes can provide a satisfactory alternative to natural waterfront property in many cases. Urban lakes have been established in the past to appeal to a broader range of investors.

The appeal of an urban lake often has a positive effect on the entire development and not just on the adjacent properties. Studies conducted in North America (Michael et al., 1996; Gibbs et al., 2002) found a strong correlation between property prices and lake clarity. This indicated that aesthetics were a significant driver of property value. In another study, Sikorska et al. (2017) found that lake accessibility played a more important role than water quality in property values in Poland. Although it may be arguable that water quality does not influence property value, it is clear that proximity and accessibility to open water does influence property value. If stakeholders can easily access and utilize the waterbody passively or actively, it will inherently increase values.

From a practical perspective, urban lakes can also provide a significant source of land fill for developers, which is extremely valuable for developments that need to be raised above a specific flood level. Urban lakes are typically constructed close to the lowest elevation point of a development for managing the increased flows (Schueler and Simpson, 2001; Walker, 2012) from the increase in impervious areas (CSIRO, 2006). Alternatively, a series of lakes can be constructed throughout a development to manage flows from smaller subcatchments. Either approach generally requires the waterbodies to be excavated, with the material often utilized for fill to ensure flood protection in other parts of the development. Removing or reducing the need to import fill into a development can be a significant cost benefit. Furthermore, in SEQ, achieving adequate flood protection needs to be unambiguously demonstrated prior to receiving a development approval.

Apart from economic drivers, urban lakes can provide intangible benefits such as fostering community well-being or social capital (Walker et al., 2012). Stakeholders have historically ranked proximity to natural settings, particularly water, as very desirable and associate them with a higher quality of life (Horwitz et al., 2001; Ogunseitan, 2005; Walker et al., 2013). Quantifying links between health, well-being, and urban lakes is difficult, but there is sufficient evidence that well-designed and established green spaces (e.g., parks, urban lakes, playing fields) have a positive impact on community health and well-being (Lee and Maheswaran, 2010).

In the early adoption of WSUD in Australia, urban lakes were utilized as a means of stormwater treatment. However, uncontrolled stormwater runoff has caused widespread adverse impacts on urban lake health (e.g., eutrophication). As problematic urban lakes become more common and research (Leinster, 2006; Walker et al., 2013; Walker, 2012) identified that urban runoff was the primary cause of lake degradation, urban lakes were no longer considered to be a viable means of treating urban stormwater runoff. This aspect will be discussed further within this chapter.

## 13.3 URBAN LAKES AND STORMWATER QUANTITY

Urban lakes/dams have historically played a significant role in regional flood management. For example, urban lakes can be used as stormwater retention/detention basins to attenuate stormwater runoff rates from urban developments. Stormwater basins collect stormwater and slowly release it at a controlled rate (attenuation) such that downstream areas are not flooded or eroded. This type of urban lake is generally constructed to reduce the cost of downstream channel upgrades; to reduce downstream flooding impacts; or to meet regulatory requirements limiting urban catchment outflow peaks (i.e., non worsening). Stormwater basins are usually designed to mitigate floods up to the 1% annual exceedance probability (AEP) flood level (i.e., a 1% AEP flood). These basins also provide community benefit such as recreation areas, scenic pond, and pollution control.

There are two general types of stormwater basins: detention basins and retention basins. Detention basins are usually dry when not being used for flood storage, whereas retention basins usually have a permanent pool of water, even when not acting as flood storage.

Detention basins are landscaped areas formed by the construction of simple dam walls or by excavation below ground level. They can also be designed to use any existing natural swales or depressions. These areas primarily serve to capture and temporarily store stormwater runoff from impervious surfaces, such as rooftops, roads, and pavements, to prevent excessive runoff and channel erosion in receiving environments. Detention basins are also effective at removing particulate-based contaminants via sedimentation but less effective for treatment of soluble pollutants where biological uptake of nutrients is required. Pollutant removal through sedimentation relies on strong affinity of metals, nutrients, and hydrocarbon contaminants to adsorb onto particulate matter such as clay and silt. Pollutant removal efficiency increases with increasing hydraulic residence-time (Persson and Wittgren, 2003; Hossain et al., 2005).

Retention basin–type urban lakes also treat stormwater runoff. They have a permanent ponding level of water in the lake at all times and as such, were often used in favor of dry detention basins, as the value of lakes to development/communities is often considered to be higher.

The water level in a stormwater pond is established by a low flow outlet. Generally, the outlet is incorporated as part of a metal or concrete structure called a riser. The outlet of a detention basin is usually level with the bottom of the basin so that all of the water eventually drains out and it remains empty between storms, whereas retention basins have an outlet or a weir at a higher point so that it retains a permanent pool of water.

One of the most important elements of maintaining basins is making sure the outlet is not blocked or clogged. Other maintenance activities include repairing erosion within the basin, removing sediment, and managing the vegetation. Repairing erosion early can save significant costs, especially the resulting sedimentation that will need to be removed from the basin.

The consequences of a storm exceeding a detention basin's or retention basin's operating design storm should be considered in the design and operation of a stormwater basin. Although such an occurrence may be rare, it is still possible, and the consequences of failure can be catastrophic because of the proximity to urban areas.

The design approach should be selected based on the target pollutants and site and economic constraints. Forebays (or sediment traps) at the inflow points to the detention area can be used to capture coarse sediment, litter, and debris, which will simplify and reduce the frequency of maintenance. Forebays can either be sized to hold the expected sediment volume between maintenance cycles or be designed to have sufficient capacity to detain or infiltrate (where possible) frequently occurring storm events (typically $<$ 1-year Average Return Interval) without discharge to the main detention area. Detention basins must also be designed to minimize the risk of mosquito breeding, which can be achieved by ensuring the basins are free draining.

## 13.4 FACTORS IMPACTING URBAN LAKE HEALTH

Many urban lakes, along with natural waterbodies, are considered degraded, with eutrophication (i.e., excessive concentrations of nutrients causing accelerated plant/algae growth) being a significant global problem (Wagner and Erickson, 2017). This widespread degradation of urban lakes has been broadly researched and is consistently linked to the impacts of urban runoff (Mitsch and Gooselink, 2000; Bayley and Newton, 2007; Walker et al., 2013). More specifically, three key factors associated with urban lake degradation are:

- Catchment runoff attributes (e.g., increased impervious area, loss of infiltration capacity, increased directly connected drainage infrastructure);
- Poor quality of inflow (e.g., high nutrient concentrations, high organic carbon load, high heavy metal concentrations); and
- Poor design attributes (e.g., lake depth, wind alignment, lack of riparian environment).

These factors have caused many urban lakes to shift from a stable state, dominated by submerged macrophytes and low turbidity, to a deteriorated, eutrophic state, explained by the "alternate stable state" theory (Scheffer and Carpenter, 2003). This alternate state is dominated by floating macrophytes, phytoplankton and/or cyanobacteria, potential odor issues, and turbid water (Scheffer and Carpenter, 2003). The mechanisms of recovery from the alternate stable state are difficult because of hysteresis effects (i.e., a transition which is not easily reversed), which presents difficulties in transitioning from the degraded state back to the stable ecosystem state (Beisner et al., 2003).

## 13.4.1 Key causes in urban lake degradation

Research on constructed urban lakes has identified five key causes of urban lake degradation (Taylor et al., 2007; Walker, 2012). These causes are as follows:

**Poor quality of inflow**—Untreated urban runoff is typically high in pollutant loads and is often of poor quality. Sediment and nitrogen and phosphorus species are largely the issue in Australia; however, heavy metals also pose an issue in many countries. The causes of poor runoff quality in urban catchments are varied but are predominantly associated with fertilizer use on lawns and gardens, higher sediment loads associated with erosion, and a lack of infiltration. Impervious surfaces can warm stormwater runoff as they absorb and retain heat at a higher capacity than natural ground cover. Therefore, stormwater runoff is often at a higher temperature than its receiving environment. This influences several biochemical processes (e.g., increases biological oxygen demand) and creates a more hospitable environment for microbiological contaminants (Morrison et al., 2005).

In urban developments, which contain commercial and light industrial uses (e.g., grocery stores and vehicle service centers), nutrients, sediment, heavy metals (e.g., due to construction materials, wearing of vehicle parts, and breakdown of paints), and other anthropogenic pollutants are often higher than in residential urban developments. Commercial and light industrial areas (e.g., car service centers, supermarkets) are increasingly incorporated into residential developments and serve to increase the quantity and variety of pollutants (Eves, 2007; Wagner and Erickson, 2017). In SEQ, Australia, where rainfall is infrequent and storm events often intense, this can cause significant impacts to waterways from catchment areas where pollutants have accumulated during dry periods (CSIRO, 2006).

It is well understood that directly connected stormwater drainage infrastructure intensifies the impacts of urban development. Hard stand areas directly connected to stormwater drainage infrastructure (e.g., concrete pipes) reduce the potential for infiltration through the soil and increase the volume of untreated stormwater discharged directly to a receiving environment (Walsh et al., 2007; Wong et al., 2006). Without a buffer or pretreatment systems that provide infiltration or stormwater treatment, receiving environments such as urban lakes are degraded over time.

**Water level variation**—Riparian vegetation is impacted by both prolonged inundation and reduction in water levels. Both events can lead to a loss of riparian vegetation, which, when it decays create conditions hospitable to nuisances such as flies and mosquitoes (Taylor et al., 2007). The riparian zone also influences ecological processes by providing critical habitat, regulating stream temperature, and serving as a source of leaf and woody litter, essential for many terrestrial and aquatic invertebrates (Ewel et al., 2001; Bond et al., 2008). The loss of the riparian system essentially removes critical habitats and buffers for urban runoff, thus reducing the resilience of the urban lake to eutrophication.

**Persistent stratification**—Stratification is the formation of two distinct layers in a lake, the epilimnion and hypolimnion (Fig. 13.1). The different layers form as a result of differing water density, associated with a temperature or salinity gradient (Walker, 2012). Although a common occurrence in natural lakes, stratification can cause issues in urban lakes if the design depth is greater than 3 m, particularly in subtropical and tropical climates. The hypolimnion layer (cooler and denser) in degraded urban lakes is typically of poor quality, as it often acts as a sink for organic matter and sediment causing low levels of dissolved oxygen (DO). More specifically the layers are defined as follows:

- *Epilimnion*: The surface layer of the lake. This layer may be up to 3 m deep, subject to lake clarity. The water temperature in this layer is determined by sunlight and wind mixing.

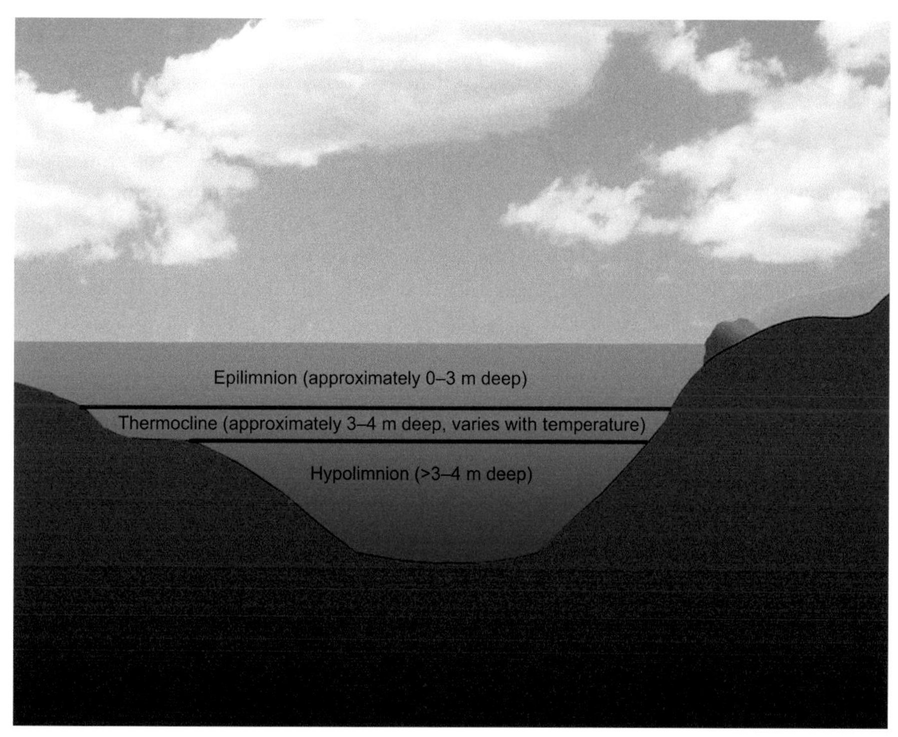


**FIGURE 13.1** Schematic of typical stratification layers in a lake.

- *Thermocline:* This layer is defined by an abrupt transition to cooler temperatures. The thickness of this layer is determined by temperature and seasonal variation.
- *Hypolimnion*: The bottom layer of the lake. This layer is generally >3 m deep. In degraded systems, this layer is often hypoxic because of ongoing decay of organic matter. The low levels of DO are associated with the decay of organic matter by aerobic (oxygen consuming) bacteria, and if water clarity is low, it can also result in reduced photosynthetic activity in submerged macrophytes. In low DO conditions, nutrients and heavy metals are typically desorbed from sediments and become bioavailable in the soluble form (e.g., NOx, $PO_4$) (Walker, 2012). In excessive concentrations, soluble nutrients can lead to, or exacerbate, lake eutrophication.

Stratification can be disrupted either by a severe wind event or by seasonal changes (e.g., air temperature cools the epilimnion below that of the hypolimnion). In such events, the water profile inverts and nutrient-rich waters within the hypolimnion rise to the surface of the waterbody and exacerbate the eutrophic state. Heavy metal concentrations may also impact the health of higher level species (e.g., macroinvertebrates, crustaceans, and fish) (Scheffer et al., 2001; Melbourne Water, 2005; Taylor and Breen, 2006; Taylor et al., 2007), which may lead to widespread mortality within an aquatic ecosystem.

**Stormwater residence time**—Stormwater residence time in urban lakes relates to the flushing frequency of the lake. Urban lakes have an optimal stormwater residence times that ensure that stormwater treatment is most effective. Urban lakes in cooler climates have an optimal stormwater residence time of up to 50 days, whereas lakes with a higher summer time water column temperature require a much shorter stormwater residence time (15−20 days) to limit the risk of pollution buildup and algal blooms, as higher algal growth rates are associated with higher water temperatures (Water by Design, 2012). Algal blooms are significant threats to both environmental and public health, as some species of algae and cyanobacteria are potentially toxic to mammals, notably, the *Anabaena* spp. (NHMRC, 2008).

An urban lake with limited flushing capacity is likely to experience a number of potential water quality issues. The length of the residence time is directly linked to the likelihood of an algal bloom, as algal cells that are not flushed from a waterbody can can multiply exponentially if nutrient concentrations are not limited (Burge and Breen, 2006; Taylor and Breen, 2006; Water by Design, 2012). Climatic conditions influence the rate of growth in algal species with warmer water, increasing metabolic activity (i.e., uptake of bioavailable nutrients by algal species). Prolonged stormwater residence time in urban lakes is usually a consequence of poor design. Poor designs often fail to take the catchment area and quantity of runoff from frequent events into account when sizing the lake system, e.g., small contributing catchment and large lake volume.

**High organic carbon loading**—In natural lake systems, organic matter is critical to microbial processes as it provides a food source for heterotrophic organisms (Ewel et al., 2001; Bond et al., 2008). Aquatic ecosystems not influenced by urban development are able to process organic carbon loads and do not generally create nutrient loads that are beyond the capacity of the lake to process and remain healthy. In urban lakes, organic carbon loads, particularly those sourced from floating macrophytes, add to nutrient loads that are already potentially high because of the impacts of urban runoff.

Microbial decomposition of floating macrophytes depletes DO levels within the hypolimnion (Fig. 13.1), which in turn creates higher concentrations of soluble nutrients and heavy metals (due to solubilization). Over time, a high mass of decomposing material may provide a continuous supply of nutrients to an urban lake, resulting in persistent eutrophication or algal blooms, regardless of the water quality entering the system (Taylor et al., 2007; Walker, 2012).

### 13.4.2 Consequences of urban lake degradation

Constructed urban lakes, react to increased nutrient loading and availability similar to that in natural lakes (Carpenter et al., 1998a, 1999). Scheffer et al. (2001) argued that natural lakes may degrade as a result of internal processes and external factors, but this may take centuries under natural conditions. In an urban lake environment, the shift to the alternate stable (degraded) state may take as little as 2 years (Bayley and Newton, 2007; Taylor et al., 2007; Walker, 2012).

Urban lakes typically begin in the stable state, which is relatively undisturbed (Scheffer et al., 2001; Walker, 2012). Over time, urban lakes with suboptimal designs will likely shift to an alternate stable state once some threshold point is reached. This alternate state is dominated by floating macrophyte growth or cyanobacteria and/or phytoplankton blooms and turbid water, which is consistent with a eutrophic to hypereutrophic system (Scheffer, 2004; Taylor et al., 2007; Walker, 2012).

Urban lakes follow the five-stage pattern of eutrophication described in Fig. 13.2. The alternate stable state may represent a progression to what Taylor and Breen (2006) described as a "catastrophic ecosystem change." At the final stage (Stage V), eutrophication is evident and the lake is degraded to the point where it may be no longer be capable of supporting higher trophic level species. Algal and cyanobacteria blooms become more frequent, as the loss of submerged macrophytes effectively removes competition for nutrient resources (Burge and Breen, 2006; Sirivedhin and Gray, 2006; Taylor et al., 2007).

Once a lake shifts to the alternate stable state, it can be extremely difficult to shift back to the original state. The resistance to recovery is described as the hysteresis effect (Beisner et al., 2003). A contributing factor to the hysteresis effect is that urban lakes act as a "sink" for pollutants, particularly for nutrients. Phosphorus, a limiting nutrient for algae and cyanobacteria, can buildup in such high concentrations within lake sediment that internal recycling can occur over decades. This can result in a long recovery time, regardless of the management strategy implemented (Coveney et al., 2005). Research has found that lakes in the alternate stable state may be resistant to recovery strategies, as many remediation efforts have had limited effectiveness (Carpenter and Cottingham, 1997; Carpenter et al., 1998a, 1999; Coveney et al., 2005; Wagner and Erickson, 2017).

As the mechanisms of urban lake degradation are now well understood and underpinned by the alternative stable state theory, a closer examination of the historical and contemporary role of urban lakes and stormwater treatment is warranted. This will be addressed in the next section.

## 13.5 URBAN LAKES AND STORMWATER TREATMENT

### 13.5.1 Historical view of urban lakes and stormwater treatment

In the early adoption of WSUD principles in Australia, urban lakes were utilized as stormwater treatment systems. The view was that urban lakes could provide treatment similar to that of sediment ponds (where sediment would settle out and be retained) or constructed wetlands (where submerged macrophytes process the nutrients from urban runoff).

Subsequently, many existing urban lakes were installed as the only means of urban runoff treatment, following the the now widely discredited mantra that "dilution is the solution to pollution" (Varis and Somlyody, 1997). Utilizing an urban lake allowed for more open water area to be realized, and offline treatment systems (e.g., bioretention basins and constructed wetlands) were not considered necessary. Greater areas of open water were considered to be a significant incentive to investors, given the higher aesthetic and economic values associated with greater areas of easily accessible open water (Walker, 2012; Walker et al., 2013; Sikorska et al., 2017).

An urban lake may initially be able to process pollutants associated with urban runoff, as described in Fig. 13.2, but this ability tends to be very brief. The often rapid nature of the shift to the alternative stable state demonstrates that the

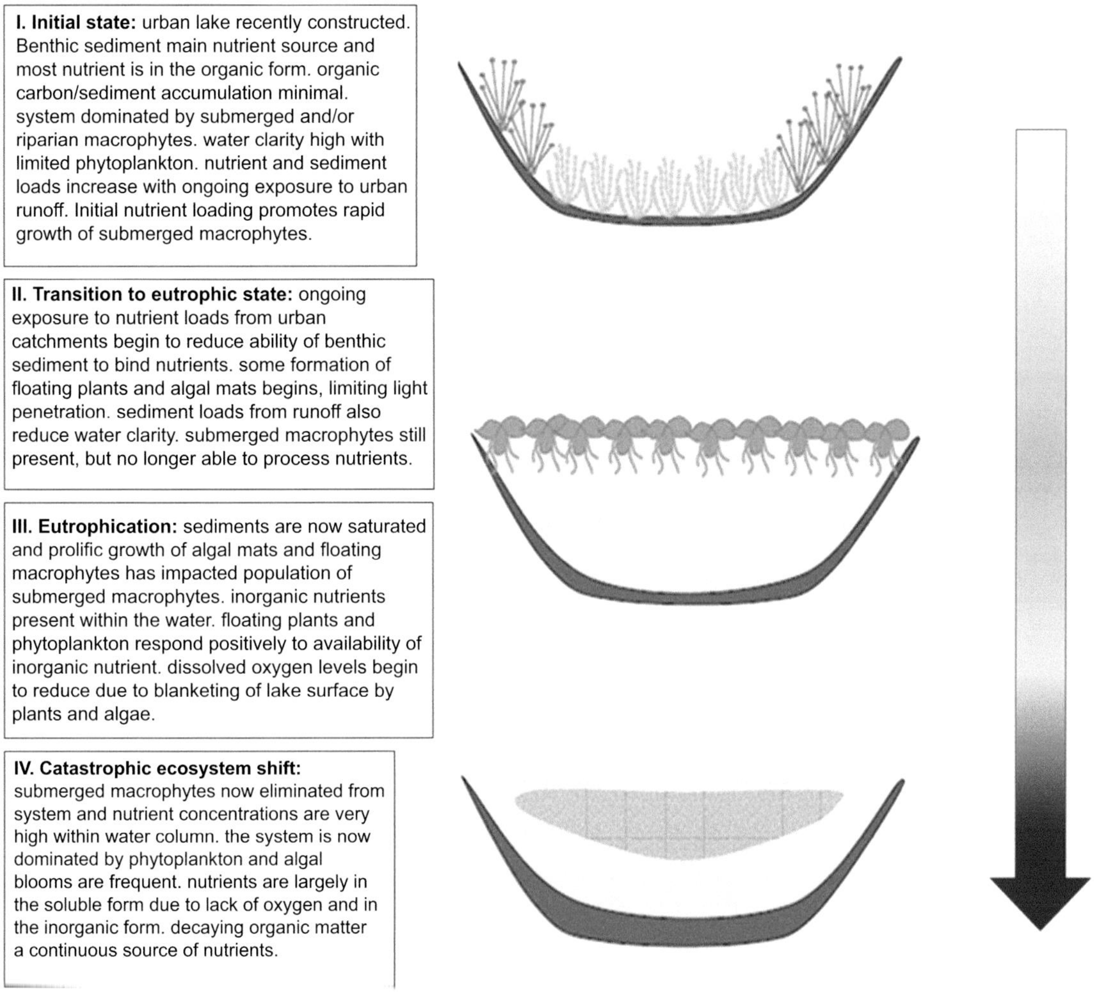

**I. Initial state:** urban lake recently constructed. Benthic sediment main nutrient source and most nutrient is in the organic form. organic carbon/sediment accumulation minimal. system dominated by submerged and/or riparian macrophytes. water clarity high with limited phytoplankton. nutrient and sediment loads increase with ongoing exposure to urban runoff. Initial nutrient loading promotes rapid growth of submerged macrophytes.

**II. Transition to eutrophic state:** ongoing exposure to nutrient loads from urban catchments begin to reduce ability of benthic sediment to bind nutrients. some formation of floating plants and algal mats begins, limiting light penetration. sediment loads from runoff also reduce water clarity. submerged macrophytes still present, but no longer able to process nutrients.

**III. Eutrophication:** sediments are now saturated and prolific growth of algal mats and floating macrophytes has impacted population of submerged macrophytes. inorganic nutrients present within the water. floating plants and phytoplankton respond positively to availability of inorganic nutrient. dissolved oxygen levels begin to reduce due to blanketing of lake surface by plants and algae.

**IV. Catastrophic ecosystem shift:** submerged macrophytes now eliminated from system and nutrient concentrations are very high within water column. the system is now dominated by phytoplankton and algal blooms are frequent. nutrients are largely in the soluble form due to lack of oxygen and in the inorganic form. decaying organic matter a continuous source of nutrients.

**FIGURE 13.2** The stages of eutrophication in a degrading urban lake. *Modified from Taylor, S., Breen, P., 2006. Chancellor Park Lakes – Validating Conceptual Models of Lake Behaviour: Monitoring Program Data Interpretation & Lake Management Recommendations, Ecological Engineering Pty. Ltd., pp. 1–117.*

stormwater treatment capacity in urban lakes is limited, transient, or absent entirely, with long recovery times between inputs (Leinster, 2006; Bayley and Newton, 2007, Walker, 2012). Shifts from the stable to alternate state have been frequently observed in urban lakes in Australia (Walker, 2012; Hagare et al., 2015).

In as little as 2 years, an urban lake may be unable to process nutrient loads from urban runoff. One reason is that these waterbodies do not have suitable surface areas for biofilm growth beyond that of the riparian vegetation (e.g., plant stalks on the lake edges). Biofilm coverage is an essential requirement for the sequestration of nutrients from stormwater (Borne et al., 2013; Winston et al., 2013). Biofilm helps process nitrogen in the water through nitrification, ultimately allowing uptake by plant species of nitrate. More importantly excess nitrate is converted into nitrogen gas via denitrification, as plant uptake may be limited. Refer to Fig. 13.3 for detail. Phosphorus can also be retained through binding processes that occur within the biofilm (e.g., adsorption) and uptake of orthophosphate is achieved by vascular macrophyte species. However phosphorus is largely managed via immobilization (i.e., bound to sediment).

As the key factors that cause a shift into the degraded (alternative stable) state were identified, the design and management of urban lakes evolved to better cater for the impacts of urbanization. For example, the Queensland state government, along with a number of Australian councils, has developed policies that designate urban lakes as "receiving environments" (DILGP, 2017). As a designated receiving environment, stormwater runoff must be appropriately treated prior to its discharge into an urban lake. In essence, it encourages an urban lake to be considered a natural aquatic ecosystem, and this affords it a similar level of protection. Such policy documents are now linked to urban lake design

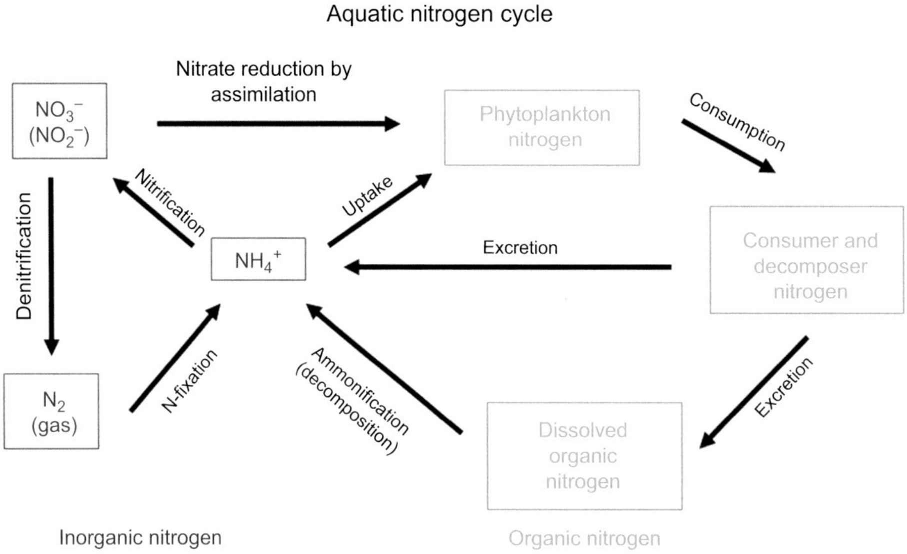

Aquatic nitrogen cycle

Nitrate reduction by assimilation

$NO_3^-$ ($NO_2^-$)

Phytoplankton nitrogen

Consumption

Denitrification

Nitrification

Uptake

$NH_4^+$

Excretion

Consumer and decomposer nitrogen

$N_2$ (gas)

N-fixation

Ammonification (decomposition)

Dissolved organic nitrogen

Excretion

Inorganic nitrogen

Organic nitrogen

FIGURE 13.3 Schematic of sources, sinks, transformations and flows in the aquatic nitrogen cycle. *Modified from http://www.esf.edu/efb/schulz/Limnology/Nitrogen.html.*

guidelines and management strategies. However, a remaining critical purpose of urban lakes is to provide flood attenuation. The contemporary design and management principles of urban lakes are discussed further in this chapter.

## 13.6 EVOLUTION OF URBAN DESIGN PRINCIPLES, MANAGEMENT, AND ROLE IN WSUD

Initial design strategies for urban lakes sought to maximize property values by creating greater areas of open water and reducing subdivision construction costs by providing a ready source of fill. Early designs focused exclusively on the economic benefits to the development and ignored the ecological role of urban lakes. This approach did not align with the principles of WSUD. As a result, it can be argued that many existing lake designs are in direct conflict with the core principles of WSUD, which seek to minimize or prevent environmental degradation, while simultaneously offering a benefit to the community in terms of esthetic and/or recreational value (BMT WBM, 2009).

### 13.6.1 Contemporary design principles

As the quality of many early urban lakes started declining, it became apparent that proactive, long-term strategies were required. As a result, key design principles for urban lakes were established to ensure that contemporary urban lakes in Australia are designed appropriately. Three well-known Australian guidelines include the following:

- Melbourne Water Constructed Lakes (2005) (https://www.melbournewater.com.au/Planning-and-building/Applications/Documents/Shallow-lake-systems-design-guidelines.pdf)
- Mackay Constructed Lakes guidelines (2008) (http://www.mackay.qld.gov.au/__data/assets/pdf_file/0008/14786/15.15_Constructed_Lakes_V2.pdf)
- Townsville Constructed Lakes (DesignFlow and RPS, 2010).

Although all three guidelines stress the importance of design, the Townsville guidelines (DesignFlow and RPS, 2010) also provide recommended steps for regulatory authorities to use when assessing urban lake design applications. The steps provide a framework to identify designs that avoid the issues associated with urban lake degradation. A summary of the key steps from the Townsville guidelines is provided in Table 13.1.

As seen in Table 13.1, urban lake design within Australia now considers a wide range of variables to ensure the risks associated with the five key causes of urban lake degradation are avoided. Waterbodies are now designed to cater to the increased understanding of urban lake ecosystems, and this must be demonstrated as part of a development approval process. An issue that still needs to be addressed, however, is the short-term and long-term management of existing urban lake systems.

**TABLE 13.1** Summary of Urban Lake Design Steps

| Design Steps | Detail | Purpose |
|---|---|---|
| Step 1 | Design intent | To establish purpose of waterbody, i.e., amenity, fill, flood mitigation, and to identify if concept lake design takes critical design principles (e.g., depth, inflows) into account. |
| Step 2 | Lake type | Identify if lake will be saline, brackish, or freshwater. Each lake type requires different management, with brackish lakes to be avoided due to high risk of cyanobacterial blooms. |
| Step 3 | Soil and groundwater assessment | Seeks to avoid issues associated with soils (acid sulfate, well-drained, or dispersive soils) and groundwater (poor quality groundwater, high in nutrient). |
| Step 4 | Lake inflows and outflows | Assesses quality of inflows and possible risks associated with prolonged stormwater residence time by assessing catchment area to lake volume ratio. |
| Step 5 | Define preliminary lake layout | Takes Steps 1–4 into consideration, as well as surrounding environments, infrastructure, etc., to establish a preliminary design. |
| Step 6 | Lake bathymetry | Identifies the depth and bed profile of the proposed urban lake to determine stratification/mixing issues. |
| Step 7 | Water balance model | Identifies potential for water level variation within lakes and if the waterbody will require frequent top up from stored supply. |
| Step 8 | Algal management | Takes the above into account to determine waterbody residence time and thus determines risk of algal blooms. This step also defines recirculation requirements and defines critical design requirements of recirculation systems (e.g., recirculating lake water through constructed wetlands to remove nutrients and algal cells). |
| Step 9 | Define final lake layout | Utilize the outcomes associated with Steps 1–8 to develop a final layout which takes into consideration the key design principles. |
| Step 10 | Design of inlets/outlets and other hydraulic structures | Ensures that the relevant structures are design appropriately for the catchment's hydrology and lake functions (e.g., mixing) and that maintenance issues are minimized (e.g., blockage of outlets), and other issues, such as climate change impacts, are considered. |
| Step 11 | Design of lake edge/batters | Ensures that appropriate edge profiles are established based on the lake type. For example, densely vegetated batters are encouraged for freshwater lakes. |
| Step 12 | Vegetated lake edge treatments | Defines the structure and diversity of the lake edge treatments and the establishment methodology. |
| Step 13 | Maintaining flood capacity | Ensure that specialist hydraulic advice in relation to the specific urban lake catchment is considered in the design. |
| Step 14 | Lake maintenance access | Ensures that suitable accesses into and around the lake is provided to allow for maintenance (e.g., floating macrophyte harvesting). |
| Step 15 | Vector control | Ensures that the design avoids habitat associated with mosquito breeding and considers approaches to manage risk. |
| Step 16 | Bird control | Establishes that design has considered risks associated with bird nesting and bird overpopulations. |
| Step 17 | Public safety | Lake design is established to minimize casual access or other associated risks to community. |
| Step 18 | Life cycle costing | Provides the asset manager with long-term maintenance costs associated with the urban lake. |

Adapted from DesignFlow and RPS, 2010. Townsville Constructed Lakes Design Guideline. Prepared for Townsville City Council.

### 13.6.2 Urban lake management

Although the reasons for urban lake degradation are well understood, the approach to managing urban lakes has failed to adequately characterize them for what they are: dynamic, integrated ecosystems. Urban lake monitoring programs are typically limited to measuring physicochemical indicators in a static state, which presents a monodisciplinary, incomplete measurement of lake health. This approach makes frequent use of short-term solutions (e.g., macrophyte harvesting) and is largely reactive. Management essentially requires a system to fail in the eyes of stakeholders before any intervention occurs. This approach does not consider the links between biotic and abiotic factors. The approach also ignores the local community as a component of the urban lake ecosystem, notwithstanding the inevitable interaction between the community and lake.

Reactive management intervention may even exacerbate problems in urban lakes. Many recovery strategies have actually been shown to reinforce the alternative stable state. For example, dredging lake sediment is one option, but it is an expensive exercise that causes a major disruption to both the lake environment and the urban community, and it potentially also impacts downstream ecosystems. Marotta et al. (2009) argued that while dredging removes the nutrient-laden sediment from a system, it also removes the aquatic macrophytes that take up nutrients and compete with algae and floating macrophytes. Furthermore, any uncollected dead floating macrophyte becomes a significant source of organic carbon on decay. The removal of these aquatic macrophytes, coupled with continuous nutrient laden input, can actually intensify eutrophication rather than limit it (Marotta et al., 2009).

Many current urban lake recovery strategies often involve retrofitting upstream catchment areas with stormwater quality improvement devices (SQIDs) to pretreat urban runoff. Some examples of typical SQIDs are shown in Fig. 13.4. SQIDs serve a range of uses, from treating primary pollutants (e.g., gross pollutant traps removing anthropogenic litter) to tertiary treatment (e.g., constructed wetland and bioretention basins remove nutrient from runoff).

Although commendable, these screening measures only serve to reduce future nutrient loads, and donot address nutrient cycling within the lake itself. Unfortunately, effective lake treatment strategies are limited.

FIGURE 13.4 Typical stormwater quality improvement devices. From top left, clockwise: dry detention/sediment basin, gross pollutant trap, constructed wetland, and bioretention basin.

Although valuable, physicochemical indicators serve to measure only a component of lake health, not the entirety. Management strategies for urban lakes derived from physicochemical monitoring alone and will often result in repeated interventions (e.g., macrophyte harvesting). Basing management strategies on such limited physicochemical observations rather than patterns of behavior limits the effectiveness of remediation strategies and may fail to identify the key issues driving the broader ecosystem health of the urban lake.

As urban lakes are often key features of residential developments, such an approach should be inclusive of the surrounding community in the assessment and indeed may help establish more adaptable and long-term management strategies (Foden et al., 2008; Walker, 2012). Although current guidelines assist in ensuring that important processes are considered in the urban lake design process, adaptive and ecosystem-based management principles for urban lake health are somewhat lacking.

## 13.7 URBAN LAKES AND ECOSYSTEM HEALTH

With the rapid increase in new urban developments, such as that occurring in SEQ, it is unlikely that the popularity of urban lakes will decline. The impacts of urban runoff on urban lakes have been evident for many years and will persist without holistic monitoring to ascertain the linkages (e.g., causal relationships) between biogeochemical indicators. Lake management is often reactive to one specific problem (e.g., prolific floating macrophyte growth) and common solutions often do not address the underlying causes of the issues. Many problems in urban lakes are not linked to a single factor. but rather are the result of complex interactions in the aquatic ecosystem.

Lake health is not dependent on a single parameter or indicator set, so effective monitoring and management of such systems may be more complex. There are a broad range of indicators that can be used to assess, and hence manage, the overall state of urban lake systems. Physicochemical indicators such as pH, DO, turbidity, and nutrients, are valuable, but provide an incomplete picture. Despite this, such indicators are the conventional norms for assessing urban lake health. Studies of aquatic ecosystems routinely conceptualize one or two indicators of ecosystem health, such as chemical or physical parameters, or assess a system from an economic or engineering perspective, rather than taking an integrated view (Rapport, 1998a; Walker, 2012; Wiegand et al., 2013). While the trophic status of an aquatic environment can be assessed using traditional physicochemical indicators, other indicators, such as macroinvertebrate diversity, constitute a significant component of biodiversity and are critical of a well-functioning aquatic ecosystem (Chessman et al., 2002).

Furthermore, research on urban lakes almost universally fails to link community well-being to urban lake ecosystem health (Walker et al., 2012; Wiegand et al., 2013). As urban lakes are often key features of residential developments, a more holistic approach should also be inclusive of the surrounding community and the links between quality of life and urban lake ecosystem health. When discussing ecosystem heath, it can be useful to compare it with the complexities of human health. Using physicochemical indicators as the sole measure of health is analogous to a medical doctor providing a diagnosis based on a patient's blood pressure. A conclusion is based on an incomplete assessment. This is typical to urban lake management, particularly in terms of community linkages. A simple approach to quantifying the link between a community and urban lake quality is to conduct a survey to investigate the values a community places on such a system (Walker et al., 2013; Hagare et al., 2015).

There is a substantial amount of research on integrating biological, chemical, physical, and social attributes (Rapport, 1998a,b,c; Howard and Rapport, 2004; Xu et al., 2004, 2005; Foden et al., 2008). However, little of this research is linked to urban lake ecosystems (Walker, 2012; Walker et al., 2013; Wiegand et al., 2013) and none is particularly recent. An approach that integrates the various disciplines associated with aquatic ecosystem health in urban lake research may provide novel and ultimately more effective solutions for the management of urban lakes. Broad assessments may also be costly, if not carefully planned. Water quality analysis alone may result in significant costs. Furthermore, it may be difficult to source the broad expertise to guide more holistic assessments.

### 13.7.1 The Ecosystem Health Paradigm

The idea of "health" has been extended to describe the state of many different ecosystems. However, the traditional methods attached to the concept remain the same. That is, health is measured through the degree of divergence from the "ideal" state (Pantus and Dennison, 2005). The EHP is an adaptable means of assessing system health and is defined on the case-by-case basis (Howard and Rapport, 2004). This increases the flexibility of the EHP approach, making it optimal for a variety of scenarios and not limited to any one particular area (Howard and Rapport, 2004); however, the approach is generally site specific, meaning results are not often translatable across systems. This approach has been used to assess the health of ecosystems such as rivers, lakes, and forests, typically linking human pressures on the ecosystem in the assessment, as well as the human uses and amenities derived from the system (Howard and Rapport, 2004; Xu et al., 2004). Prior to research by Walker (2012), the EHP had yet to be applied to subtropical urban lakes, despite the obvious similarity in fundamental ecosystem functions.

It has been well accepted (Mageau et al., 1995) that the EHP is represented by three key measures; vigor, resilience, and organization. These measures are briefly discussed below in the context of urban lakes.

**Vigor** often refers to the activity, metabolism, or production in an ecosystem. In the case of urban lakes, both high and low levels of vigor may be viewed as unhealthy states (Rapport, 1998b; Karr and Chu, 1999). Urban lakes that are eutrophic will be overly vigorous when systems that are oligotrophic (i.e., have very low productivity, such as glacial lakes) may be unable to sustain primary producers and benthic invertebrates that support diverse higher level species (Nürnberg and Shaw, 2006).

**Resilience** is an ecosystem's ability to cope with stressors and its ability to recover from disturbance. Despite disturbance, a resilient urban lake will display good biodiversity that is critical to many ecological functions. A healthy urban lake ecosystem has the means to buffer against stressors events (e.g., major storm events), minimize the disturbance to populations, and demonstrate ongoing resistance to shifting to the alternative (degraded) stable state.

**Organization** in an urban lake is defined by its complexity and the level of interactions (i.e., symbiosis, mutualism, and competition). Diversity and species richness are often indicators of good health. Indicator species, such as aquatic macroinvertebrates, are often used to develop sensitivity indices, such as the Stream Invertebrate Grade Number Average Level (SIGNAL2) or the Swan Wetlands Aquatic Macroinvertebrate Pollution Score (SWAMPS). SIGNAL2 (stream) and SWAMPS (wetlands) are often used to link faunal sensitivity to pollution and organization (diversity and abundance). However, they are rarely applied to constructed waterbodies (Walker, 2012).

Vigor, resilience, and organization all influence each other. These links characterize the complex interactions that occur in urban lakes. Given the setting of most urban lakes, in that they are close to residential communities, it is important to quantify the community as a part of the urban lake ecosystem and define the links between the two.

A broader, multifocal approach to managing and monitoring urban lakes presents a greater suite of information for managers to access, enabling a decision-making process that is not dominated by any one discipline. It presents a holistic assessment that may assist in establishing proactive and more adaptive management strategies. A suggested approach to monitoring urban lakes is provided below; however, it should be noted that the approach is broad and requires the collection of substantial data. This approach should be adopted when there are critical drivers to management:

- *Static and Dynamic State Water Quality Monitoring*—Regular physicochemical water quality monitoring is critical, as it allows for pattern identification and presents an inference of vigor or productivity. It is important to monitor a lake in both static and dynamic states (i.e., during storm events) to assess pollutant concentrations within and entering an urban lake so that impacts can be better assessed and management options can be better informed. Such data serve to infer the state of urban lakes, or ecosystem vigor, to an extent (i.e., whether or not the lake is in the alternate stable state). Such monitoring can also help to address resilience to some degree, as lakes are measured before, during, and after disturbances, including seasonal turnover and storm events. Storm event monitoring data are scarce, largely because of cost, but it is important to enable these events that represent the dynamic environmental state to be characterized, particularly for degraded systems requiring remediation. Systems with appropriate design are unlikely to require such monitoring, as upstream measures are in place to treat runoff. As part of the 2010 review of the Ecosystem Health Monitoring Program (EHMP) in SEQ, the need to include event monitoring as part of the program was identified (BMT WBM, 2010).
- *Floral/Faunal Indicators*—Floral and faunal surveys are useful in providing both a quantitative and a qualitative means of assessing organization. It enables managers to evaluate the impacts of water quality on organization and also to identify potential risks to lake health, such as the presence of weeds or other noxious species.
- *Screening Level Risk Assessment*—Identifying actual and perceived risks to the community living in association with an urban lake is arguably a critical component to applying the EHP, as well as also one that has had limited application to date. While an urban lake may not have a specific recreational use, it cannot be assumed that it will not be used for this purpose. Therefore, it is valuable to assess these waterbodies against hazards to public health using guidelines such as those provided by the National Health and Medical Research Council (NHMRC, 2008). The need to include human health monitoring as part of the SEQ EHMP has been identified (BMT WBM, 2010) and it bears merit to apply such assessments to urban lakes, subject to the proposed use (e.g., primary vs. passive recreation) given the proximity to, and use by, residential communities. This approach serves to identify and initially characterize the links between community and urban lakes, although the link is only characterized in a limited, one-way impact driven sense (i.e., risks lakes pose to the community).

- *Community Integration*—This is a fairly novel concept where urban lake monitoring is concerned. Community consultation and surveys are generally conducted in response to an existing issue within an urban lake, rather than as part of developing proactive, adaptable management strategies. This limits and inhibits community involvement, which can subsequently limit community responsibility and custodianship of the surrounding environment. Without including the community as key components of an assessment, managers cannot identify the attitudes and values the community has regarding urban lakes and cannot identify behaviors that may significantly impact lake health. This approach is important to fully characterize the link between communities and urban lakes and identify not only the impacts lakes may have on a community (and the reverse) but also the mutual benefits that can be realized. Two Australian studies (Walker et al., 2012; Hagare et al., 2015) surveyed the local community. Walker et al. (2012) found that the community placed a high value on an adjacent urban lake system and aesthetics were not the sole driver in determining value. Water quality and the creation of a sense of community were also two critical values that were important to residents. Hagare et al. (2015) found that not only did residents feel that the lakes added value to properties by increasing the aesthetic appeal, but also added value to the community by providing usable open space areas. Understanding and examining community values can allow for targeted and proactive maintenance strategies.

An integrated, interdisciplinary approach may serve to better assess urban lakes as an ecosystem and better infer the state of health than the current approaches to urban lake monitoring and management. This is particularly the case for degraded systems that required remediation. Although indicator assessments may not be as in-depth as specific measures of ecological processes, the scope is broader and indicators are linked, which serves to better characterize an urban lake as an ecosystem, enables a more robust measurement of health, and allows for a more informed approach to management and remediation.

Foden et al. (2008) provide a valuable critique on the assessment of natural aquatic ecosystems. In particular, they highlight the general lack of consistency and failure to conduct holistic, well-integrated assessments of such environments. However, there has been a substantial improvement in the monitoring and management of natural aquatic ecosystems in the last few years, a prime example being the ongoing evolution of the EHMP in SEQ. Fig. 13.5, adapted from Healthy Land & Water, characterizes the various integrated approaches taken in assessing aquatic environments to determine ecosystem health.

The success of the EHMP demonstrates the value of a holistic assessment and the excellent outcomes achieved through such assessments. Given the potential social and economic values urban lakes can represent, it is paramount that urban lake managers use the EHMP approach as a viable, tailored approach to better manage urban lake ecosystems.

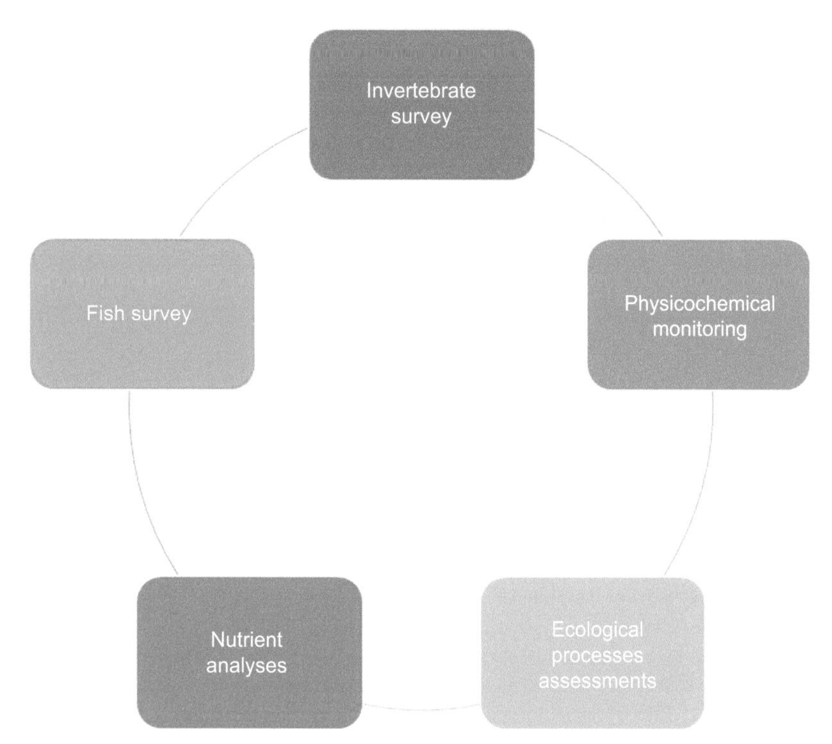


**FIGURE 13.5** Ecosystem Health Monitoring Program.

## 13.8 CONCLUSIONS

This chapter has discussed how urban lakes are integrated within the WSUD philosophy, and what design principles and requirements are critical for urban lake management and maintaining health in these waterbodies. Urban lakes are often included in developments for aesthetic benefits and stormwater detention purposes. They can also provide a significant source of land fill for developments where building pads need to be raised above a specified flood level. This chapter showed that the inclusion of an urban lake can have a positive effect on the appeal on an entire urban development, and not just on the immediately adjacent properties. Apart from economic drivers, urban lakes can provide intangible benefits such as fostering community well-being.

In the early adoption of WSUD in Australia, urban lakes were installed as a means of stormwater treatment. However, it was soon realized that untreated stormwater runoff has serious impacts on urban lake health (e.g., eutrophication), which often led to a shift to the alternative stable state. Consequently, in Queensland, urban lakes are no longer considered to be a viable means of treating urban stormwater runoff. Under the Queensland State Planning Policy (DILGP, 2017), urban lakes must only receive stormwater runoff that has been treated to the required standard, which varies with location within the state.

We identified three key factors that are associated with urban lake degradation:

- Hydrology of the urbanized catchment;
- Poor quality of inflow; and
- Poor lake design attributes.

Once an urban lake has degraded (to the alternative stable state), it can be extremely difficult to shift back to the original healthy state. As the key factors that caused lake degradation were progressively identified, the design and management of urban lakes evolved to better cater for the impacts of urbanization. Urban lakes in Australia are now often classified as "receiving environments," which means that stormwater runoff must be pretreated prior to its discharge into the lake.

Urban lakes are also designed to provide flood protection in that they collect stormwater, temporarily store it, and then slowly release it at a controlled rate such that downstream areas are not flooded or eroded.

As the quality of many early urban lakes started declining, it became apparent that proactive, long-term strategies were required to improve urban lake design and management. As a result, key design principles for urban lakes were established, and a number of urban lake design guidelines were developed. Urban lake design within Australia now considers a wide range of variables to ensure the risks associated with the five key causes of urban lake degradation (Section 13.4.1) are avoided.

While the reasons for urban lake degradation are now well understood, the approach used to manage urban lakes in the past has often failed to adequately acknowledge them for what they are—dynamic, integrated ecosystems. Urban lakes require an interdisciplinary approach that looks beyond the physicochemical quality and seeks to identify the links and relationships, inclusive of biogeochemical and socioeconomic processes within the ecosystem itself.

With the rapid increase in new urban developments in Australia cities, it is unlikely that the popularity of urban lakes will decline. Urban lakes are important assets that can play a key role in the social capital of urban developments. This chapter has shown that with careful design, management, maintenance, and understanding of lake dynamics, urban lakes can provide many long-term benefits to urban communities. Their widespread implementation should be encouraged in appropriate landscapes.

## REFERENCES

Bayley, M.L., Newton, D., 2007. Water quality and maintenance costs of constructed waterbodies in urban areas of South East Queensland. In: Proceeding from 5th International Water Sensitive Urban Design Conference- 'Rainwater and Urban Design', Sydney 2007.

Beisner, B.E., Haydon, D.T., Cuddington, K., 2003. Alternative stable states in ecology. Frontiers in Ecology and the Environment 1, 376–382.

Binnie, C.J.A., June 2004. The benefits of dams to society. In: Hewlitt, H. (Ed.), British Dam Society 13th Biennial Conference. British Dam Society, pp. 22–26.

BMT WBM, 2009. Evaluating Options for Water Sensitive Urban Design – A National Guide: Prepared by the Joint Steering Committee for Water Sensitive Cities: In Delivering Clause 92(ii) of the National Water Initiative. Joint Steering Committee for Water Sensitive Cities (JSCWSC), Canberra. http://155.187.2.69/water/publications/urban/water-sensitive-design-national-guide.html.

BMT WBM, 2010. Prepared for the South East Queensland Healthy Waterways Partnership, July, 2010. EHMP Review Report.

Bond, N.R., Lake, P.S., Arthington, A.H., 2008. The impacts of drought on freshwater ecosystems: an Australian perspective. Hydrobiologia 600, 3–16.

Borne, K.E., Fassman, E.A., Tanner, C.C., 2013. Floating treatment wetland retrofit to improve stormwater pond performance for suspended solids, copper and zinc. Journal of Ecological Engineering 54, 173–182.

Burge, K., Breen, P., 2006. Detention time design criteria to reduce the risk of excessive algal growth in constructed water bodies. In: From the 4th International Conference on Water Sensitive Urban Design Melbourne, pp. 1–8.

Carpenter, S.R., Cottingham, K.L., 1997. Resilience and restoration of lakes. Conservation Ecology 1, 1–13.

Carpenter, S.R., Bolgrien, D., Lathrop, R.C., Stow, C.A., Reed, T., Wilson, M.A., 1998a. Ecological and economical analysis of lake eutrophication by nonpoint pollution. Australian Journal of Ecology 23, 68–79.

Carpenter, S.R., Ludwig, D., Brock, W.A., 1999. Management of eutrophication for lakes subject to potentially irreversible change. Ecological Applications 9, 751–771.

Chessman, B.C., Trayler, K.M., Davis, J.A., 2002. Family- and species-level indices for macroinvertebrates of wetlands on the swan coastal plain, Western Australia. Marine and Freshwater Research 53, 919–930.

Cook, N., Davison, A., Crabtree, L. (Eds.), 2016. Housing and Home Unbound: Intersections in Economics, Environment and Politics in Australia. Routlidge, Abingdon, Oxon, New York, NY.

Coveney, M.F., Lowe, E.F., Battoe, L.E., Marzolf, E.R., Conrow, R., 2005. Response of a eutrophic, shallow subtropical lake to reduced nutrient loading. Freshwater Biology 50, 1718–1730.

CSIRO, 2006. Urban Stormwater: Best Practice Environmental Management Guidelines. CSIRO Publishing.

DesignFlow and RPS, 2010. Townsville Constructed Lake Design Guideline. Prepared for Townsville City Council.

Eves, C., 2007. Planned residential community developments: do they add value? Property Management 25, 164–179.

Ewel, K.C., Cressa, C., Kneib, R.T., Lake, P.S., Levin, L.A., Palmer, M.A., Snelgrove, P., Wall, D.H., 2001. Managing critical transition zones. Ecosystems 4, 452–460.

Foden, J., Rogers, S.I., Jones, A.P., 2008. A critical review of approaches to aquatic environmental assessment. Marine Pollution Bulletin 56, 1825–1833.

Gibbs, J.P., Halstead, J.M., Boyle, K.J., Huang, J.C., 2002. An hedonic analysis of the effects of lake water clarity on New Hampshire lakefront properties. Agricultural and Resource Economics Review 31, 39–46.

Gibson, G., Carlson, R., Simpson, J., Smeltzer, E., Gerritson, J., Chapra, S., Heiskary, S., Jones, J., Kennedy, R., 2000. Nutrient Criteria Technical Guidance Manual: Lakes and Reservoirs, first ed. United States Environmental Protection Agency.

Hagare, D., Maheshwari, B., Natarajan, S., Kaur, M., Daniels, J., 2015. Using lakes in urban landscapes for stormwater management: a study of water quality of Wattle Grove Lake in western Sydney and an overview of recreational and other value benefits to the local community. Journal of the Australian Water Association 42, 77–84.

Horwitz, P., Lindsay, M., O'Conner, M., 2001. Biodiversity, endemism, sense of place, and public health: inter-relationships for Australian inland aquatic systems. Ecosystem Health 7, 253–264.

Hossain, M.A., Alam, M., Yonge, D.R., Dutta, P., 2005. Efficiency and flow regime of a highway stormwater detention pond in Washington, USA. Water, Air, and Soil Pollution 164, 79–89.

Howard, J., Rapport, D., 2004. Ecosystem health in professional education: the path ahead. EcoHealth 1, 3–7.

Huser, B.J., Futter, M., Lee, J.T., Perniel, M., 2016. In-lake measures for phosphorus control: the most feasible and cost-effective solution for long-term management of water quality in urban lakes. Water Research 15, 142–152.

Karr, J.R., Chu, E.W., 1999. Restoring Life in Running Waters: Better Biological Monitoring. Island Press, Washington, D.C.

Lee, A.C.K., Maheswaran, R., 2010. The health benefits of urban green spaces: a review of the evidence. Journal of Public Health 33, 212–222.

Leinster, S., 2006. Delivering the final product - establishing vegetated water sensitive urban design systems. In: From the 4th International Conference on Water Sensitive Urban Design Melbourne, Australia, pp. 1–8.

Mackay City Council, 2008. Engineering Design Guidelines: Constructed Lakes. Planning Scheme Policy No. 15.15, 2007, pp. 1–46. https://www.melbournewater.com.au/Planning-and-building/Applications/Documents/Shallow-lake-systems-design-guidelines.pdf.

Mageau, M.T., Constanza, R., Ulanowicz, R.E., 1995. The development and initial testing of a quantitative assessment of ecosystem health. Ecosystem Health 1, 201–203.

Marotta, H., Bento, L., Esteves, F.A., Enrich-Prast, A., 2009. Whole ecosystem evidence of eutrophication enhancement by wetland dredging in a shallow tropical lake. Estuaries and Coasts 32, 654–660.

Mays, L.W., Koutsoyiannis, D., Angelakis, A.N., 2007. A brief history of urban water supply in antiquity. Water Science and Technology: Water Supply 7, 1–12.

Melbourne Water, 2005. Constructed Shallow Lake Systems: Design Guidelines for Developers, November, 2005 Version 2, pp. 1–20. http://www.mackay.qld.gov.au/__data/assets/pdf_file/0008/14786/15.15_Constructed_Lakes_V2.pdf.

Michael, H.J., Boyle, K.J., Bouchard, R., 1996. Water quality affects property prices: a case study of selected Maine lakes. Maine Agricultural and Forest Experiment Station 398, 1–18.

Mitsch, G.A., Gooselink, J.E., 2000. Wetlands, third ed. John Wiley and Sons, New York.

Morrison, C., Scott, S., Smith, J., 2005. Digital detection of wastewater in stormwater. Environmental Health 5, 88–94.

National Health and Medical Research Council, 2008. Guidelines for Managing Risks in Recreational Water, June 2005, pp. 1–219.

Nürnberg, G.K., Shaw, M., 2006. Productivity of clear and humic lakes: nutrients, phytoplankton, bacteria. Hydrobiologia 382, 97–112.

Ogunseitan, O.A., 2005. Topophilia and the quality of life. Environmental Health Perspectives 113, 143–148.

Pantus, F.J., Dennison, W.C., 2005. Quantifying and evaluating ecosystem health: a case study from Moreton Bay, Australia. Environmental Management 36, 757–771.

Persson, J., Wittgren, H.B., 2003. How hydrological and hydraulic conditions affect performance of treatment wetlands. Ecological Engineering 21 (4–5), 259–269.

Rapport, D.J., 1998a. Need for a new paradigm. In: Rapport, D.J., Costanza, R., Epstein, P., Gaudet, C., Levins, R. (Eds.), Ecosystem Health. Blackwell Science, Oxford, UK.
Rapport, D.J., 1998b. Dimensions of ecosystem health. In: Rapport, D.J., Costanza, R., Epstein, P., Gaudet, C., Levins, R. (Eds.), Ecosystem Health. Blackwell Science, Oxford, UK.
Rapport, D.J., 1998c. Defining ecosystem health. In: Rapport, D.J., Costanza, R., Epstein, P., Gaudet, C., Levins, R. (Eds.), Ecosystem Health. Blackwell Science, Oxford, UK.
Sayer, C.D., Davidson, T.A., Kelly, A., 2008. Ornamental lakes- an overlooked conservation resource? Aquatic Conservation: Marine and Freshwater Ecosystems 18, 1046–1051.
Scheffer, M., Carpenter, S.R., Foley, J.A., Folke, C., Walker, B., 2001. Catastrophic shifts in ecosystems. Nature 413, 591–596.
Scheffer, M., Carpenter, S.R., 2003. Catastrophic regime shifts in ecosystems: linking theory to observation. Trends in Ecology and Evolution 18, 648–656.
Scheffer, M., 2004. Ecology of shallow lakes. In: Usher, M.B. (Ed.), Population and Community Biology Series. Kluwer Academic Publishers, The Netherlands.
Sikorska, D., Sikorski, P., Hopkins, R.J., 2017. High biodiversity of green infrastructure does not contribute to recreational ecosystems services. Sustainability 9, 1–13.
Schueler, T., Simpson, J., 2001. Urban Lake management: Ellicott city. Watershed Protection Techniques 3, 747–750.
Sirivedhin, T., Gray, K.A., 2006. Factors affecting denitrification rates in experimental wetlands: field and laboratory studies. Ecological Engineering 26, 167–181.
Taylor, S., Breen, P., 2006. Chancellor Park Lakes – Validating Conceptual Models of Lake Behaviour: Monitoring Program Data Interpretation & Lake Management Recommendations. Ecological Engineering Pty. Ltd., pp. 1–117
Taylor, S., Cullen, E., Leinster, S., Eadie, M., 2007. Design and management responses to address common constructed urban lake sustainability issues. In: Presentation from the SIA Queensland State Conference "Mimicking Nature: Evolution or Revolution in Stormwater Management", Sunshine Coast, QLD, pp. 1–15.
The State of Queensland, 2017. Department of Infrastructure, Local Government and Planning. State Planning Policy, Brisbane, Queensland.
Varis, O., Somlyody, L., 1997. Global urbanisation and urban water: can sustainability be afforded? Water Science and Technology 35, 21–32.
Wagner, T., Erickson, L.E., 2017. Sustainable management of eutrophic lakes and reservoirs. Journal of Environmental Protection 8, 436–463.
Walker, C., 2012. An Ecosystem Health Assessment of Constructed Urban Lakes. PhD Thesis, University of the Sunshine Coast [online]. http://research.usc.edu.au/vital/access/manager/Repository/usc:7938.
Walker, C., Lampard, J.L., Roiko, A., Tindale, N., Wiegand, A., Duncan, P., 2013. Community well-being as a critical component of urban lake ecosystem health. Urban Ecosystems 16, 313–326.
Walsh, C.J., Waller, K.A., Gehling, J., MacNally, R., 2007. Riverine invertebrate assemblages are degraded more by catchment urbanization than riparian deforestation. Freshwater Biology 52, 574–587.
Water by Design, February 2012. Urban Lakes Discussion Paper: Managing the Risk of Cyanobacterial Blooms. Healthy Waterways Limited.
Wiegand, A.N., Walker, C., Duncan, P., Roiko, A., Tindale, N., 2013. A systematic approach for modelling quantitative lake ecosystem data to facilitate proactive urban lake management. Environmental Systems Research 2, 2–12.
Winston, R.J., Hunt, W.F., Kennedy, S.G., Merriman, L.S., Chandler, J., Brown, D., 2013. Evaluation of floating treatment wetlands as retrofits to existing stormwater retention ponds. Ecological Engineering 54, 254–265.
Wong, T.H.F., Fletcher, T.D., Duncan, H.P., Jenkins, G.A., 2006. Modelling Urban.
Xu, F.L., Lam, K.C., Zhao, Z.Y., Zhan, W., Chen, Y.D., Tao, S., 2004. Marine coastal ecosystem health assessment: a case study of the Tolo Harbour, Hong Kong, China. Ecological Modelling 173, 355–370.
Xu, F.L., Zhao, Z.Y., Zhan, W., Zhao, S.S., Dawson, R.W., Tao, S., 2005. An ecosystem health index methodology (EHIM) for lake ecosystem health assessment. Ecological Modelling 188, 327–339.
Yang, S.L., Zhang, J., Zhu, J., Smith, J.P., Dai, S.B., Gao, A., Li, P., 2005. Impacts of dams on Yangtze River sediment supply to the sea and delta intertidal wetland response. Journal of Geophysical Research 110, 1–12.

## FURTHER READING

Birch, S., McCaskie, J., 1999. Shallow urban lakes: a challenge for lake management. Hydrobiologia 395/396, 365–377.
Bolund, P., Hunhammar, S., 1999. Ecosystem services in urban areas. Ecological Economics 29, 293–301.
Carpenter, S.R., Caraco, N.F., Correll, D.L., Howarth, R.W., Sharpley, A.N., Smith, V.H., 1998b. Nonpoint pollution of surface waters with phosphorus and nitrogen. Ecological Applications 8, 559–568.
Findlay, C.S., Bourdages, J., 2000. Response time of wetland biodiversity to road construction on adjacent lands. Conservation Biology 14, 86–94.
Helfield, J.M., Diamond, M.L., 1997. Use of constructed wetlands for urban stream restoration: a critical analysis. Environmental Management 21, 329–341.
Hem, B., 2007. 'Entrepreneurs fill a niche by keeping ponds in subdivisions sparking'. The Houston Chronicle. From: http://www.redorbit.com/news/science/975505/entrepreneurs_fill_a_niche_by_keeping_ponds_in_subdivisions_sparking/index.html.
Kenna, T.E., 2006. Consciously constructing exclusivity in the suburbs? Unpacking a master planned estate in Western Sydney. Geographical Research 45, 300–313.

Lee, S.Y., Dunn, R.J.K., Young, R.A., Connolly, R.M., Dale, P.E.R., Dehayr, R., Lemckert, C.J., Mckinnon, S., Powell, B., Teasdale, P.R., Welsh, D.T., 2006. Impact of urbanisation on coastal wetland structure and function. Austral Ecology 31, 149–163.

Naselli-Flores, L., 2003. Man-made lakes in Mediterranean semi-arid climate: the strange case of Dr deep lake and Mr shallow lake. Hydrobiologia 506–509, 13–21.

Wong, T.H.F., Breen, P.F., Somes, N.L.G., 1999a. Ponds vs wetlands - performance considerations in stormwater quality management. In: From 1st South Pacific Conference on Comprehensive Stormwater and Aquatic Ecosystem Management Auckland, New Zealand, pp. 223–231.

Wong, T.H.F., Breen, P.F., Somes, N.L.G., Lloyd, S.D., 1999b. Managing urban stormwater using constructed wetlands. In: CRC for Catchment Hydrology Industry Report Series.

Yang, X., Wu, X., Hao, H., He, Z., 2008. Mechanisms and assessment of water eutrophication. Journal of Zhejiang University 9, 197–209.

Chapter 14

# Economics of Water Sensitive Urban Design

**Kym Whiteoak**
*RMCG, Melbourne, Australia*

## Chapter Outline



**ABSTRACT**

Economic assessment of water sensitive urban design (WSUD) investments is challenging, predominantly for reasons of data shortages and the broad range of nonfinancial benefit streams provided by WSUD that are difficult to rigorously quantify. However, being the language of decision-making in public policy, it is critical to communicate the merits of WSUD investments within an appropriate economic framework.

Such a framework is cost—benefit analysis (CBA) using a total economic value (TEV) framework. CBA is a decision-making framework that transparently orders information around costs and benefits of investments, providing a decision-maker with a decision rule on proceeding with an investment or otherwise. A TEV framework is used to identify and categorize all benefits accruing from an investment, including environmental and social benefits that may be difficult to quantify.

The remaining challenge to rigorous economic WSUD then becomes data availability of environmental and social benefits produced by the investments. A range of benefits produced by different types of WSUD investments can be quantified in dollar terms, including esthetic and environmental benefits.

**Keywords:** Choice modeling; Cost—benefit analysis; Cost-effectiveness analysis; Hedonic pricing; Multicriteria analysis; Nonuse value; Total economic value; Use value.

**Approaches to Water Sensitive Urban Design. https://doi.org/10.1016/B978-0-12-812843-5.00014-9**

## 14.1 INTRODUCTION

An interest in the economics of water sensitive urban design (WSUD) can be driven by a desire from practitioners to understand how their investments might be justified to decision-makers in convincing and realistic terms. Economics is the language of decision-making for public policy and public funding. As such, the ability to articulate the merits of WSUD investments in economic terms can be very useful, not only for justifying investments but also for meaningfully informing decision-making in integrated water management (IWM).

WSUD investments by councils, residential and commercial developers, and ultimately householders cost real dollars to construct and maintain. However, the benefits that these investments produce are often difficult to identify in financial terms. Rather, the benefits of WSUD are more readily measured in biophysical terms, such as stormwater pollutant abatement (kilograms of nitrogen, phosphorus, total suspended solids), or by the impact of these interventions on downstream waterway health. WSUD investments may also contribute to more vaguely defined but nonetheless important outcomes, such as "livability" or neighborhood esthetics. Moreover, while costs may be incurred by a single entity, benefits may accrue to a range of different parties (for example, rain gardens may contribute to urban cooling across a local area), providing another challenge in justifying investments.

There are real challenges in appropriately translating the benefits of WSUD investments into economic terms. However, applying a clear and transparent economic framework can be very useful in understanding the relative merits of a WSUD project, compared with its alternative. In addition, many economic methods exist, which have been used to measure a broad range of benefits that WSUD investments produce.

As usual, availability and quality of data remain significant challenges for good economic analysis in this subject matter. For example, high-quality data may be lacking in relation to the maintenance costs and long-term effectiveness of investments over time.

This chapter proposes an economic framework for considering WSUD investments, explores the tools for assessing different economic benefits produced by WSUD investments, and presents some of the evidence of these benefits. Lastly, key issues affecting good economics in this space are explored.

## 14.2 THE OVERARCHING FRAMEWORK FOR WSUD ECONOMICS—COST–BENEFIT ANALYSIS USING A TOTAL ECONOMIC VALUE FRAMEWORK

The guiding principles underpinning WSUD economic analysis are found in cost–benefit analysis (CBA). CBA is a way of ordering information such that we can answer the question: "do the benefits of the investment exceed its costs?"

An assessment of a project's benefits and costs requires consideration of a number of factors:

- **Timeframe**: costs and benefits of WSUD investments take place over the lifetime of the asset. As such, the timeframe of the analysis should be long enough to appropriately consider the lifetime of benefits and costs of the investment.
- **Discount rate**: costs and benefits produced today have different values to those in the future. Hence, we discount future costs and benefits to today's dollars, using an appropriate discount rate that appropriately reflects the time value of money. This allows us to compare options that incur costs and produce benefits at different times. The choice of discount rate (a percentage) will depend on the entity paying or benefitting but usually reflects their borrowing costs. For water businesses, this is sometimes defined by economic regulators as the weighted average cost of capital (WACC)[1].
- **Perspective**: from whose perspective is the assessment being undertaken—the household, a water business, local government, or all of these (a "whole of society" analysis)? This will influence the selection of costs and benefits for inclusion. For public policy decisions, a "whole of society" perspective is typically chosen (we elaborate in Section 14.2.4).
- **Scope**: defining the geographic boundaries of the analysis is important, and care must be made to consider impacts outside of the direct geographic region of the analysis (such as downstream waterway impacts). For public policy decisions, a "global" scope is typically chosen in that a full assessment of all plausible costs and benefits, to all affected parties, is considered.

In addition to these factors, perhaps the most important consideration for defining and estimating costs and benefits of a WSUD project is "compared to what?" Projects do not occur in a policy vacuum, so defining the scenario we are comparing a project against is critical. This is called defining the "base case."

---

1. For example, at the time of writing (June 2017), the WACC for water businesses in the state of Victoria, Australia, is 4.5% real.

### 14.2.1 The influence of context on economic framework—defining the base case

The base case is the scenario we describe to reflect what would have happened if the WSUD investment did not proceed. This may seem obvious at first blush but can be surprisingly easy to misdefine, particularly when considering the range of regulatory requirements that govern investment landscapes.

Defining the base case appropriately can have a significant impact on the results produced in an analysis. Very often, WSUD investments are implemented to meet regulatory requirements for stormwater management or water quality management. This regulatory context is extremely important in defining the base case. Importantly, the base case is rarely a "do nothing" scenario. The regulatory context often dictates that some form of intervention for stormwater management or water quality improvement is required.

As a hypothetical example, let us assume we are exploring the costs and benefits of rainwater tanks on every dwelling in a new suburban development. There are two potential base case options to compare our project against:

1. **Do nothing**: without the rainwater tanks, there would have been no other alternative WSUD investment. Water supply would have come from the potable network, and waterways would have received increased flow and pollutant loads (from stormwater runoff), affecting waterway health. There may be additional nuisance flooding.
2. **Alternative investment**: the (hypothetical) regulatory requirement for water quality improvement requires that pollutant loads be reduced from new suburban developments. If the rainwater tank investment does not proceed, the least cost means of meeting this requirement (by say bioretention basins) will need to be expanded, but no change in waterway health will be recorded.

There may also be regulations requiring an alternative water supply being provided for new suburban developments. If so, the base case would reflect this reality in the absence of the project.

These two alternative base case options reflect different contexts and will produce different costs and benefits. Depending on the situation, one will be more accurate than the other, but there will be situations in which either would be more appropriate.

### 14.2.2 Consideration of alternative frameworks—cost-effectiveness, cost–benefit, and multicriteria analysis

While a CBA framework is regarded as the preferred method for economic analysis, this preference is based on an assumption that appropriate and relevant data on costs and benefits are available. There may be scenarios in which other assessment tools are preferred.

**Cost-effectiveness analysis** (CEA) may be preferred if the outcome is fixed, and we are simply choosing the least cost means of achieving that outcome. For example, if we are exploring alternative interventions that could be used to meet stormwater quality regulations, the outcomes (i.e., pollution load abatement) may be fixed, and we are simply exploring the total costs of alternatives to achieve those outcomes.

The main drawback of CEA is when the alternative options also achieve positive outcomes, in addition to the primary one being explored. Taking the rainwater tank versus bioretention example from above, both may equally achieve a water quality target, but rainwater tanks also contribute water supply to households. In this case, cost savings to households can be subtracted from the total project cost. However, this is an imperfect solution for a number of reasons. It may misrepresent the actual costs of the project by subtracting other cost savings (especially if they attribute to different parties). Also, if an additional benefit can be identified elsewhere, subtracting this from project costs again may misrepresent actual project costs.

**Multicriteria analysis** (MCA) is often suggested as an alternative to CBA or CEA, particularly where significant benefits are difficult, or expensive, to quantify in dollar terms. In an MCA process, benefits that can be easily monetized are quantified, and a scoring system is developed for both this and the other outcomes that are considered important but difficult to monetize, for example, sustainability, livability, or waterway health outcomes. These outcomes are then weighted, and the option with the highest score is deemed the preferred option.

There are a number of challenges with MCA as a tool, including the difficulty of converting disparate outcomes into logically consistent scores. There is also the risk of double-counting outcomes. For example, does contribution to waterway health also contribute to sustainability performance? In addition, if an attempt is made to monetize some of the sustainability benefits of options, how is the weighting of the "sustainability" outcome adjusted in the MCA?

In addition, it must be recognized that the weighting criteria can be misused to produce an outcome preferred by the practitioner. Transparency on all weightings and the explanations for these decisions can assist in mitigating this risk.

Despite its limitations, MCA can be a useful approach in certain circumstances, such as when filtering a long list of WSUD options to produce a short list for more detailed analysis.

### 14.2.3 The preferred framework: cost–benefit analysis

CBA is generally preferred by economists for its logical consistency and single measurement metric (dollars). Of course, not every benefit can be perfectly measured in all cases, but if transparently undertaken, a clear understanding of a quantitative assessment can be made, with an accompanying description of the qualitative merits of the different options.

The main shortcoming of CBA as a tool is the challenge of appropriately identifying, quantifying, and monetizing the full range of benefits produced by WSUD investments. Although the costs of an investment are relatively straightforward to measure, the benefits can be far more difficult to estimate in dollar terms. Where the benefits can be monetized, it can be prohibitively expensive to fund the analysis required to produce these quantified values.

In subsequent sections, we explore the range of benefits from WSUD investments that can be quantified and monetized; describe the methods used for estimating them; and provide examples of attempts to quantify them.

### 14.2.4 Total economic value

In the sections following, we discuss the quantification of a range of different benefits that may accrue to different WSUD projects. However, a way of arranging these benefits in terms of economic values is required to structure an economic approach to valuing WSUD benefits. A total economic value (TEV) framework is a useful way to order this information.

TEV is a concept used in CBA that is often used in the context of environmental economics. It distinguishes between use values that often have a direct financial value, which are more easily quantifiable (such as water resource value or impacts on property prices), or an indirect use value (such as reduced urban heat island [UHI] effect), as well as nonuse values that reflect community preferences for outcomes that may not be directly experienced by community members (such as protection of endangered species and healthy aquatic environments for subsequent generations).

Use of a TEV framework allows us to order a full range of benefits produced by a WSUD project and ensures that all relevant values are considered. Fig. 14.1 demonstrates the TEV framework in the context of the WSUD.

## 14.3 ESTIMATING WSUD INVESTMENT PROJECT COSTS

Estimating project costs is relatively straightforward. The key challenge is to ensure that the full set of costs is included over the lifetime of analysis. Costs typically used include the following:

- **Capital costs**: CAPEX scaled to meet project objectives (such as pollution load abatement targets); these are typically estimated by project engineers;

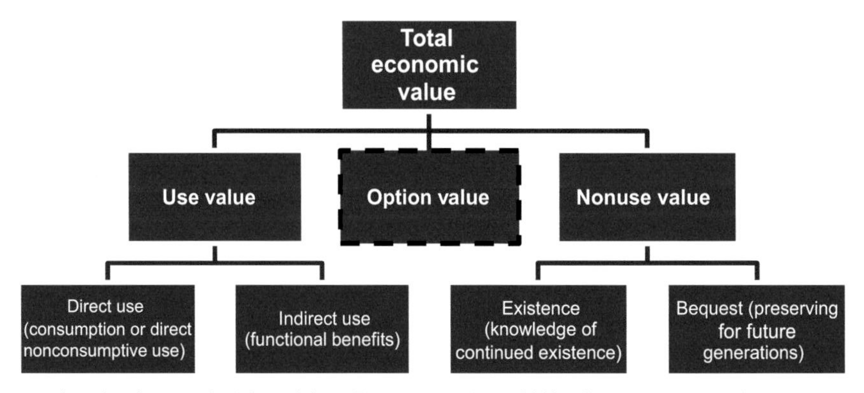


**FIGURE 14.1** Total economic value framework. *Adapted from Freeman III., A.M., 2003. The Measurement of Environmental and Resource Values, Theory and Methods, RFF Press Book, Published by Resources for the Future, Washington, DC.*

- **Maintenance costs**: OPEX typically estimated by project engineers; these should appropriately reflect the ongoing costs that will be incurred to meet the performance outcomes that the WSUD assets are expected to meet (this is discussed further below); and
- **Renewal costs**: WSUD assets have an estimated lifetime and need to be renewed at the end of this lifetime. The analysis period is often, but not always, chosen to reflect this asset lifetime.

Where analysis period does not match the asset lifetime (for example, if an analysis period is chosen at 20 years and the asset lifetime is 15 years), asset renewal is included and a residual value of the asset can be estimated at the end of the analysis period. This allows a like-for-like comparison of options.

## 14.3.1 Quantifying the benefits of WSUD investments

Defining, estimating, and appropriately attributing benefits produced by WSUD investments is more challenging than quantifying costs. The benefits may occur in different ways and accrue to different beneficiaries and may be expressed in social and environmental terms, rather than financial.

Nevertheless, there are ways of estimating many of the benefits that WSUD investments provide. Including them within economic analyses arguably makes for a more compelling case than simply listing a range of "qualitative" benefits, outside of the core quantitative assessment.

In this section, we discuss a range of benefits and examples of their estimation in different contexts.

### *14.3.1.1 Water resource value*

WSUD investments that provide a water supply (such as rainwater tanks or stormwater harvesting/reuse projects), or which reduce potable water use (such as urban street tree pits passively irrigated with road runoff), also produce a water resource value.

Estimating water resource value of a project is one of the more readily measurable benefits, as it is typically measured by water authorities and priced for users. As such, reducing or substituting a water use can be measured from the perspective of the water user through the price they would have paid for the water. It may also be measured from a whole of society perspective by the estimated cost of water supply over the long term (the *Long Run Marginal Cost* of water supply).

The benefit to a user may differ from the impact on the actual cost of water supply because of the way water charges are constructed and because charges may not accurately reflect the cost of water supply. For example, a retail water charge typically has a fixed and variable component—if a rainwater tank results in less water use from the potable system, the user only benefits from a reduction in the variable component, while their fixed charge remains unchanged. At the time of writing, a residential water bill in Melbourne, Australia, includes a potable fixed charge of $177.87 per year and a variable consumption charge of $2.64 per kL (step 1).[2] Substituting potable use with rainwater not only reduces the variable charge to the user but also defers system augmentation, which is not factored into the variable charge, and is therefore not captured by the user.

### *14.3.1.2 Flood protection*

It has long been argued that the potential exists for WSUD investments to produce meaningful flood mitigation benefits by retarding stormwater flows and thus reducing flood height and/or velocity. This would produce a measurable benefit in one of two ways:

- **Avoided cost**: if the WSUD infrastructure could reduce flooding with sufficient confidence, there may be a range of flood management costs that will no longer be needed to meet designated flood management standards.
- **Additional benefit**: conversely, if flood mitigation infrastructure is held constant and WSUD infrastructure is added in addition to this, the WSUD investment may produce additional flood mitigation benefits (reduced flood damage costs or the frequency of nuisance floods).

In practice, the author is not aware of any responsible authority that has reduced flood mitigation expenditure because of WSUD investments. This reflects a lack of confidence in the reliability or longevity of WSUD investments to produce the expected flood mitigation outcome over time.

---

2. Source: https://www.yvw.com.au/help-advice/help-my-account/understand-my-bill/fees-and-charges.

Flood mitigation infrastructure is typically long lasting (40 years or more), whereas WSUD infrastructure may have shorter lifetimes or rely on maintenance regimes that lack certainty. For example, domestic rainwater tanks may last 20–30 years but require ongoing maintenance and have pumps that often last less than 10 years. Furthermore, tanks are usually privately owned, and maintenance and management may be highly variable (Sharma et al., 2015).

In this context, it is understandable that flood management decision-makers have chosen not to reduce flood management infrastructure as a result of WSUD investments.

In contrast, it is likely that WSUD investments may make small but meaningful contributions to nuisance flood mitigation, the benefits of which can be calculated using MUSIC modeling and flood damages curves. This might be one of the many benefits the WSUD investments produce—but it is unlikely that small reductions in estimated flood damages alone would justify the WSUD expenditure, and the issue of reliability remains.

However, recent technological advancements and associated analysis have revisited this issue. For example, the capacity to monitor and control water levels in residential rainwater tanks has increased their potential to be used for flood mitigation outcomes for several reasons[3]:

- The ability to monitor tanks in real time allows for early detection of tank failure, and response measures can be enacted for maintenance; and
- The ability to control the tanks remotely can allow for tanks to be emptied prior to a predicted large rainfall event, potentially mitigating flood damage.

As such, recent studies have been revisiting the potential for WSUD investments to achieve significant flood mitigation benefits. For example,

- Melbourne Water and the Victorian Department of Environment, Land, Water and Planning have undertaken preliminary modeling of the flood mitigation potential of rainwater tanks on all dwellings in different catchments. This has explored different tank size, different catchment size, varying topography, and different levels of imperviousness. The data presented in Fig. 14.2 show the average modeled reduction in flood extent averaged across 20 catchments due to adding 2.5 and 5 kL tanks on every home (T1 and T2, respectively) across three scenarios: "Ex" (existing development, current climate), F1 (future development, current climate), and F2 (existing development, future climate).

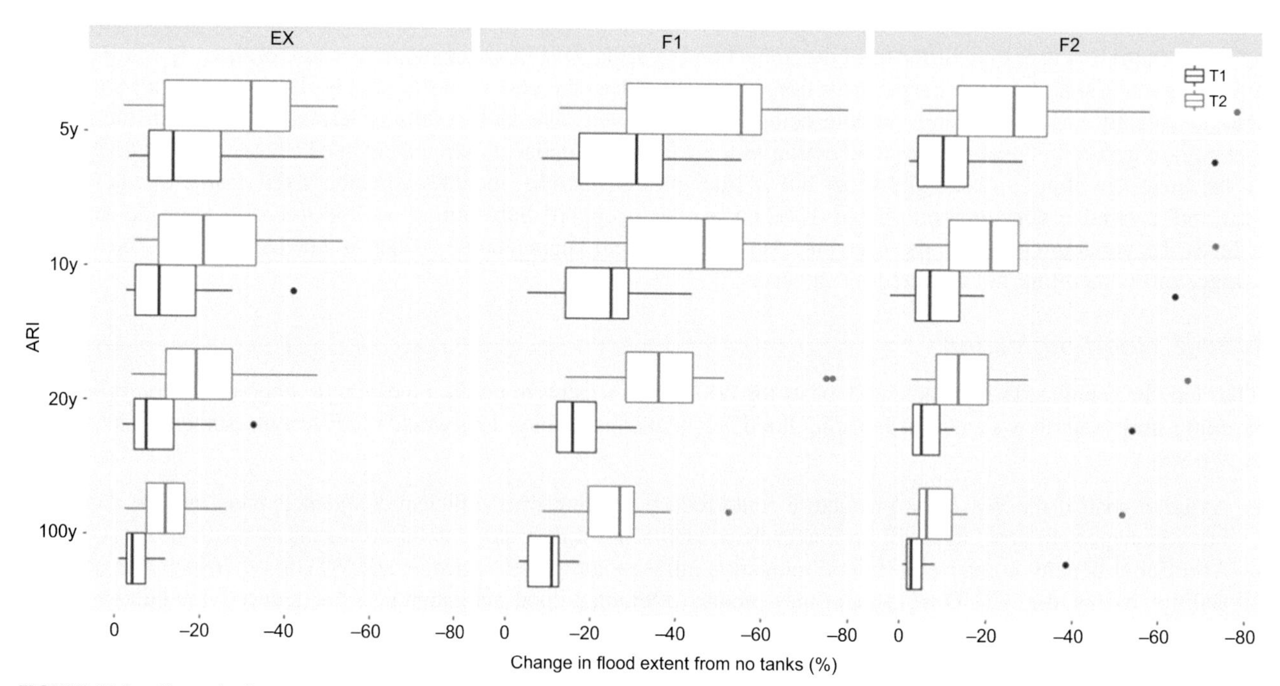


**FIGURE 14.2** Change in flood extent averaged across catchments, for each average recurrence interval (ARI). *From Moroka, December 2017. A Review of the Effectiveness of Distributed Storage to Decrease Flooding. Report prepared for Melbourne Water.*

3. http://www.awa.asn.au/AWA_MBRR/Publications/Latest_News/Water_sensitive_urban_development_launched.aspx

- The City of Melbourne has explored the potential for green infrastructure, distributed storages, and council drainage works to mitigate flood risk in CBD catchments, which have limited opportunities for stormwater capture and high costs of engineering works (Melbourne Water [unpublished]).
- Ballarat City Council has undertaken analysis that shows that rainwater tanks on residential and industrial allotments are modeled to produce up to a 20% reduction in peak flow for a 1 in 5-year flow event (E2Designlab and Water Technology, 2016).

While this work does not seek to measure the flood mitigation economic benefit of these modeled investments, it does provide the kind of physical evidence base upon which an economic analysis could be undertaken.

### *14.3.1.3 Waterway health improvement*

Many WSUD investments remove stormwater pollution from waterways, which can enhance both the esthetics of the waterway and their ecosystem health. Stormwater pollution from urban runoff contains a range of pollutants that are harmful to waterways. These can be treated a number of ways, including wetlands, bioretention, swales, and rainwater tanks.

With perfect information, the precise relationship between the WSUD investments and their impact on waterway health will be well understood, and the challenge will be appropriately estimating the value of the waterway health improvement.

In practice, understanding the precise impact of a WSUD investment and the health of downstream waterways is typically uncertain in a dynamic environment with many pressures affecting waterway health. Moreover, estimating the economic value of improvements in waterway health is extremely challenging.

However, a number of (imperfect) approaches have been used to estimate this value which we explore below.

### *14.3.1.4 Reduced pollution load: Melbourne Water's nitrogen abatement value*

Analysis in Melbourne found that nitrogen entering waterways and ultimately discharging to Port Phillip Bay was affecting the ecological health of the Bay, and efforts to reduce nitrogen volumes in stormwater and wastewater were implemented (Longmore, 2008).

Stormwater pollution load abatement standards for new urban developments were introduced, and for those developers who could not meet these standards on-site, a nitrogen abatement offset scheme was developed. The offset price is set at the estimated cost of nitrogen abatement incurred by Melbourne Water in constructing large-scale wetlands, and the money paid by developers is used to fund the construction of the wetlands.

Fig. 14.3 charts the increase in the nitrogen offset price over time, reflecting exhaustion of the lowest cost opportunities for wetland abatement and their replacement with higher cost options. The Melbourne Water offset scheme currently values nitrogen abatement at a capitalized value of $7226 per kg of nitrogen.[4]

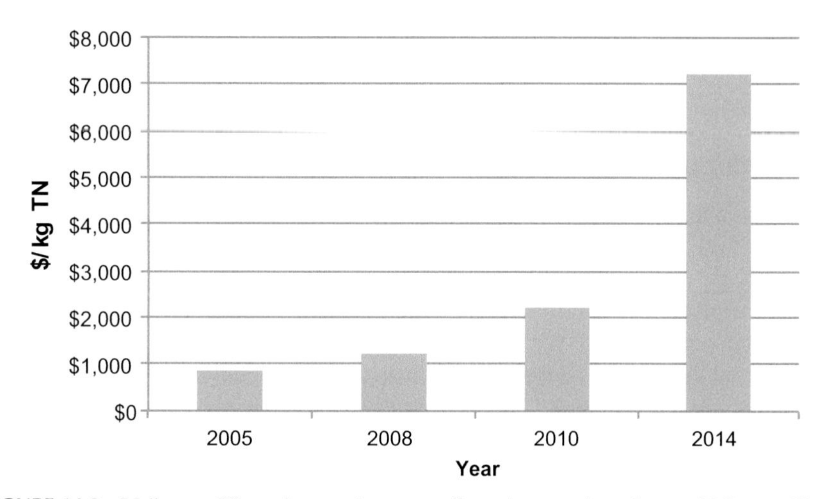


**FIGURE 14.3** Melbourne Water nitrogen abatement offset price over time. *Source: Melbourne Water.*

4. Source: https://www.melbournewater.com.au/Planning-and-building/schemes/offset/Pages/What-are-stormwater-quality-offsets.aspx.

While the unit of measurement is nitrogen, it is a proxy for a number of pollutants that are also removed from stormwater in wetlands and does not just apply to nitrogen abatement. This price can be seen as a market price for pollution load abatement, but for broad application has a number of limitations:

- The price reflects the cost of nitrogen abatement from wetlands in Melbourne. This may differ in other locations because of differences in topography, soil type, land value, and climate.
- By focusing only on wetlands to create offset prices, a range of alternative, potentially more cost-effective means of waterway pollution abatement have been ignored; this possibly would have the effect of inflating the price of nitrogen abatement.
- There is only a loose relationship between nitrogen abatement and actual waterway health outcomes. For example, a small WSUD investment may produce no meaningful impact on waterway health. Conversely, pollution abatement may produce a significant health improvement in a highly-valued waterway.

As such, while offsets are a useful contribution to measuring waterway health improvement value, it is worth exploring how much communities value changes to waterway health itself.

#### *14.3.1.5 Community willingness to pay for waterway health improvement and other water quality indicators*

Studies have been undertaken to explore how much communities are willing to pay for improved waterway health, including form and function. These studies use survey-based methods such as choice modeling and contingent valuation to elicit the value that respondents place on improvements to environmental outcomes.

The process involves describing different scenarios for improved waterway health and allocating different costs associated with achieving them. Respondents are then invited to choose their preferred option with the associated price. By varying the attributes that are being achieved in different scenarios (such as length of waterway restored) and the associated price, practitioners can calculate the price respondents are willing to pay for waterway restoration, provided a sufficiently large sample size has been used.

By way of example, Professors John Rolfe and Roy Brouwer undertook a metaanalysis of 19 studies estimating health improvement in Australian waterways from 2000 to 2010 (Rolfe and Brouwer, 2013). This produced an average willingness to pay (WTP) of $1.03 per km of waterway improved.

Recreational users may also be willing to pay for improvements to water quality indicators; for example, water clarity. Marsh and Baskaran (2009) used the same method to quantify recreational users' WTP for increased water clarity in the Karapiro catchment, New Zealand. They found that the mean annual WTP per household for water clarity from the current clarity (around 1 m) to see up to 1.5 m ($4.17), 2.0 ($21.03) and 4.0 m underwater ($65.82) (Marsh and Baskaran, 2009).

Use of this method in economic assessments for WSUD investments requires an understanding of the relationship between the WSUD investment itself and changes to waterway health. The nature and scale of these waterway health changes must also match the changes estimated in the source WTP studies. This is obviously a shortcoming of this estimation method.

#### *14.3.1.6 Community willingness to pay for stormwater reuse*

There is evidence that communities would be willing to pay more in water bills for increased wastewater recycling. However, no direct study of the community WTP for stormwater recycling, compared with traditional supplies, has been undertaken.

However, a recent study explored the community WTP for increased water recycling by surveying a statistically valid sample size of Sydney's water customers (MJA, 2014). The study explored their WTP for higher water bills in return for more recycled water to be used by others (industrial, municipal, residential, environmental flows).

The study considered not only the use of the recycled water but also the reduced discharge of treated wastewater to waterways and receiving waters. The results found strong and consistent WTP for more water recycling, estimated at between $450 and $1220 per mL.

While this is a rigorous and defensible economic nonuse value, transferring this value to WSUD investments requires some care. The original study explored recycled water only, although there is no reason to expect that the community would prefer wastewater recycling before stormwater reuse. The original study was undertaken in Sydney, and other geographic/social contexts may produce different results.

**TABLE 14.1 Community Willingness to Pay for Recycled Water Use in Sydney ($/kL)**

| End Uses for Recycled Water | Value ($/kL) |
|---|---|
| Residential (nonpotable) | $0.45–$1.22 |
| Environmental flows | $0.96–$1.35 |
| Council open space | $1.49–$1.51 |
| Business and industry | $2.06–$3.80 |

From Marsden Jacob Associates, March 2014. The Economic Viability of Recycled Water Projects - Technical Report 2 Community Values for Recycled Water in Sydney, Australian Water Recycling Centre of Excellence.

However, this study has the advantage of estimating a volumetric value for recycled water use, which can be applied to any project that has a reuse component. This makes it very easily transferred to other studies.

Table 14.1 summarizes the values for increased recycled water use in Sydney, which can be extrapolated (with care) to studies estimating stormwater reuse values.

## 14.3.2 Waterway restoration value—property price increase

There is a growing body of evidence that investments in the esthetics of urban waterways produce "use values" to local property owners, associated with the esthetic improvement produced by the investment.

A recent Australian case study tracked property prices within 200 m of a restoration project on Banister Creek in Western Australia, dating from the preintervention stage, during the intervention, and then in the years after the investment. The study used hedonic pricing, an economic tool that uses large data sets to isolate the impact of different factors on property prices. Full details can be found in Polyakov et al. (2016).

The results, graphed in Fig. 14.4, show that property prices declined temporarily during the construction phase of the investment, returned to parity in the 3 years afterward, and from 5 years thereafter increased in value by 4.7% per year, after controlling for other factors (such as property size and attributes).

## 14.3.3 Value of increased tree canopy cover—property price increase

As with waterway restoration projects, there is a large body of research linking the size, quality, and health of street trees and property price increases. For example, the size of the street tree canopy in public open spaces (street verges) is positively correlated with increased prices of adjacent private properties. Pandit et al. (2014) found that increasing street

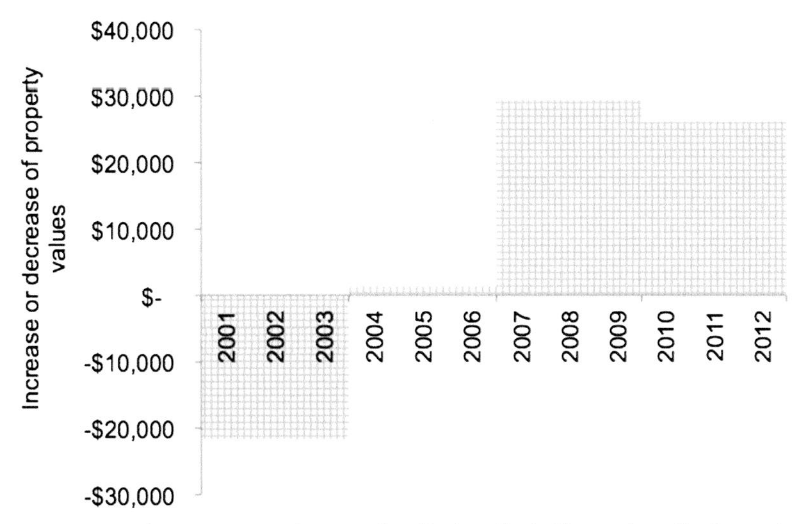


**FIGURE 14.4** Impact of waterway restoration on property prices over time, Banister Creek, Western Australia. *Source: https://watersensitivecities.org.au/wp-content/uploads/2016/07/IndustryNote_A1.2_livingstreams.pdf.*

tree canopy from a starting point of 20%–30% coverage produced a property price increase of around 1.8% of the median property price.

Plant et al. (2016) found that a 1% increase in street tree canopy cover within 100 m of a house increased its value by US$307–404. Interestingly, the same study found that, in most cases, an increase in tree canopy cover on the property itself produced a negative impact on property prices, perhaps reflecting maintenance costs and impacts of trees on property infrastructure.

Donovan and Butry (2010) studied value of trees in Portland, Oregon, finding that street trees add $8870 to sale price. The same authors explored rental impacts of street trees and found that an additional tree on a house's lot increased monthly rent by $5.62, and a tree in the public right of way increased rent by $21 (Donovan and Butry, 2011).

The impact of street tree size and quality on property prices is a reflection of esthetic impacts of WSUD investments and direct nonconsumptive use in the TEV framework described above.

### 14.3.4 Environmental flows

WSUD projects can contribute to environmental flows in stressed waterways, either directly by supplying water that can be used to augment waterway flows consistently with natural flow regimes or by substituting current water use that can allow additional releases from storages for environmental purposes.

Assessing the value of a contribution to environmental flows is a reflection of the value the community places on the environmental improvement produced by the environmental flow.

Studies have been undertaken to estimate the value of additional environmental flows using the choice modeling method, a survey-based tool in which respondents are repeatedly provided with different ecological outcomes with associated prices and asked to choose their preferred option, thus producing estimated value for different ecological outcomes.

Cooper et al. (2017) explored the WTP by Melbourne residents for additional environmental flows in Melbourne waterways, finding that the average household was willing to pay $1.13 for the ecological outcomes produced by a 1 GL increase in environmental flows—a total value of AU$1.84 m for the broader Melbourne population.

Sale et al. (2009) used the contingent valuation (survey) method to assess the amount that recreational users would be willing to pay to increase freshwater inflows into two South African estuaries, the Kowie and the Kromme. The study surveyed 150 respondents at each estuary site during December 2002 to January 2003, concluding that the value of freshwater inflows into the Kowie and the Kromme estuaries were around R0.072/m$^3$ and R0.013/m$^3$, respectively.

## 14.4 SHORTCOMINGS AND CHALLENGES

The key challenges with applying an economic framework to WSUD assessment are listed as follows:

- Lack of a rigorous evidence base for the full range of quantified benefits produced by the many different types of WSUD investments; and
- Inadequate understanding and/or incorporation of the likelihood of success of WSUD investments into economic analysis.

It is fair to say that not all benefits of WSUD outcomes can be perfectly accommodated within economic analysis. For example, the economic benefit of reducing the UHI effect is yet to be adequately assessed, but work is currently being undertaken to fill this data gap. Ongoing work by the Cooperative Research Centre for Water Sensitive Cities in Australia should provide some evidence on this issue.[5]

Perhaps the most significant remaining gap in research relates to the physical and mental health benefits of WSUD through urban greening and associated well-being. Much analysis has been undertaken to explore the relationship between green spaces and physical and mental health:

- Reduced morbidity and improved outcomes for physical health related to urban greening have been reported in studies by Mitchell and Popham (2007), Sugiyama et al. (2008), and Hunter et al. (2015).
- Improvements to mental well-being because of urban greening have been identified by Bratman et al. (2012), Hartig et al. (2014), and Cohen-Cline et al. (2015).

---

5. https://watersensitivecities.org.aarch/our-research-focus-2016-2021/integrated-research/irp2-wp6/

However, very little rigorous work has been undertaken to estimate the economic benefits of this relationship.

Filling this research gap would be highly challenging as it would require an evidence base that goes beyond correlation (people who use green spaces report higher levels of physical and mental health) to causation (an investment in WSUD would produce a quantifiable increase in physical and mental health of nearby residents).

If evidence of this causality could be gathered, along with data reflecting the scale of impact on physical and mental health produced by WSUD investments, the value of the investment could then perhaps be estimated based on a range of health-related economic indicators (such as health care and lost productivity costs, and quality-adjusted life years). Undertaking this work would involve time-series data and a large evidence base to ensure the validity of the causality.

Despite this, as this section demonstrates that there exists economic evidence for a broad range of benefits that can be used in economic analysis. These extend from water resource value to waterway health, to community preferences, and measured esthetic benefits that WSUD investments provide.

One key issue detracting from the credibility of evidence-based economic analysis of WSUD investments is lack of information on the ongoing performance of WSUD investments in the field. Anecdotal discussions undertaken with practitioners for this chapter revealed a general view within the industry that the actual performance of WSUD investments in the field is often unknown, due to a lack of monitoring by asset managers.

Practitioners generally accept that failure rates (defined as that proportion of assets which performs below that modeled by the practitioner) of WSUD investments made for water quality improvement are significant. However, defensible estimates of performance suffer from a lack of data. In broad terms, failure of a WSUD asset can occur from the following:

1. Poor design and implementation of the asset; or
2. Insufficient or inadequate maintenance and monitoring of the asset once it is implemented (this issue is dealt with in more detail in Chapter 22).

Consideration of this failure rate within the economics is important because ignoring this aspect assumes an outcome that does not reflect reality. Economic analysis that assumes perfect performance when, say, 30% of assets fail within 5 years will not only produce inaccurate results but also reduce the credibility of the many economic benefits of WSUD.

### 14.4.1 Benefit transfer

Another key issue in WSUD economics is drawing on economic values from existing studies that have quantified the same values that exist in a WSUD investment. It is not always possible, for reasons of time and budget, to develop primary estimates for the economic values of WSUD investments. This process, called "benefit transfer," allows us to fill gaps in economic analysis using preexisting studies.

Benefit transfer is a method by which environmental or other values developed in one or more source studies can be adapted and applied to a new site. There are different methods for undertaking benefit transfer, but where sufficient source studies exist, the development of transferrable value functions is preferred in the literature (for example, see Windle and Rolfe, 2014). This approach is generally preferred to a unit transfer approach because it allows the ability to make adjustments for other influences including site and population differences.

Using a benefit transfer function is considered to produce lower errors for transfers between less similar sites, allowing less focus on finding source studies that are strikingly similar to the transfer site.

Where multiple source studies exist, a metaanalysis can be used to consolidate a benefit transfer function, adding to the accuracy of transfer estimates by pooling data from a number of studies to allow systematic analysis (Rolf and Brouwer, 2010). Three key advantages of adopting a metaanalysis are identified (Rosenberger and Loomis, 2000):

1. More studies can be incorporated into the analysis;
2. Methodological differences are relatively easy to control; and
3. Subsequent value functions are easily adjusted for potential target sites.

Where a unit transfer approach is used, it is important to consider contextual issues that may affect the validity of transferring values from one context to another:

- Geographic context: Is the source study from the same country? If not, are they of similar socioeconomic status? Adjust for these (exchange rates, socioeconomic status such as property values).

- Timing: When was the source study undertaken? If more than a decade ago, the underpinning economic and biophysical economic conditions may have changed significantly. Be sure to adjust for inflation.
- Average and marginal values: Ensure that the source study is measuring a marginal change in value (e.g., the value of increasing tree canopy) if the current site is exploring a similar marginal change. Some studies seek to measure the total value of an environmental asset (e.g., the impact of a lake on property prices) rather than this marginal change.
- Other characteristics: Are biophysical conditions similar between source and current site? Are significant policy contexts different? Have other changes occurred in one context but not the other?

Benefit transfer can not only be highly useful but also easily misused by the inexperienced.

## 14.5 SUBURB-SCALE CASE STUDY

The following section is a case study of a regional approach to stormwater management within a new suburban growth area in Melbourne, Australia, to illustrate the application of WSUD economic analysis in practice. The case study is a real area; however, for confidentiality reasons the names have been withheld.

### 14.5.1 Township context

The case study scope includes the proposed development across three Precinct Structure Plans (PSPs)—separate subareas within the development zone—in a new suburban development adjacent to the existing township. High population growth is expected for the area—it is expected that there will be approximately 21,000 additional homes in the PSPs over the next 35 years, along with the supporting community and business infrastructure.

There are two creeks that abut the PSPs. Both have high environmental values and will be variously affected by increased volumes of treated recycled water and excess stormwater generated from the impervious surfaces created in the development through increased flow and pollutant load. Under the business-as-usual scenario, the environmental values of the waterways are expected to decline as the development proceeds, despite compliance with current regulations for stormwater quality standards.

One of the creeks has intermittent flow and has had relatively little impact from development. Analysis shows there is potential to rehabilitate the creek to a near-natural condition if expected increased stormwater and wastewater flows from development are significantly reduced and controlled. A reduction of 90% compared with forecast flows has been shown to replicate its ephemeral predevelopment flow conditions.

The second creek has experienced extensive modification because of upstream reservoirs, surrounding urban development, and the discharge of treated wastewater releases from the nearby wastewater treatment plant. However, the creek still retains good waterway values through the provision of habitat and amenity. The additional stormwater and recycled water releases to the second creek are likely to adversely impact these values. However, targeted releases of this additional water for environmental flows at certain times of the year could increase the ecological health of the waterway by providing a more variable and natural flow pattern.

IWM concepts including stormwater management and recycled wastewater management were explored to achieve multiple benefits: drinking water demand substitution, infrastructure cost savings to water and sewage services, environmental benefits associated with waterways and receiving waters (Port Phillip Bay), and reuse benefits (including to agricultural production).

Through an iterative process that refined the IWM options, seven options were developed for the township that included rainwater tanks, large-scale agricultural reuse, regional stormwater harvesting, recycled water reticulation via dual pipe schemes, and environmental flows.

The base case with which to compare these options was conventional supply of water and wastewater services, with treatment of stormwater for pollutant loads as per regulatory requirements. This would be undertaken with constructed wetlands, which would discharge to waterways.

Drawing in part on detailed economic analysis, a preferred option was produced for the township with the following key features:

- Regional stormwater harvesting to provide a new water supply.
- Recycled wastewater supplied for environmental flows in winter.
- Recycled wastewater supplied for agricultural irrigation in summer.

The high-level proposal for the regional stormwater harvesting scheme incorporates wetland treatment (as required under the best practice environment management for stormwater) prior to its storage in a redundant raw drinking water reservoir. The proposal for sewage includes wastewater treatment (as per the base case) followed by treatment to a *fit-for-purpose* water quality. The water will then either be used as environmental flows (to supplement winter flows) or used for agricultural irrigation.

The capture and reuse of stormwater and reuse of recycled water produced the following benefits:

1. Drinking water substitution; and
2. Significant waterway health improvement compared with no intervention beyond regulatory compliance for stormwater and wastewater treatment. The value of waterway health is not quantified in dollar terms—only the avoided expenditure by Melbourne Water and the nitrogen abatement values were estimated.

## 14.5.2 Analysis assumptions and inputs

Through an iterative process, the seven options were explored in the analysis, reflecting different scales and combinations of stormwater reuse, recycled wastewater, and rainwater tanks—all with different benefits and costs. These were compared with the base case of "business-as-usual" water management without IWM intervention.

An analysis period of 35 years was chosen, and all benefits and costs discounted to present day with a discount rate reflecting the water industry WACC of 4.5% (real).

A summary of economic assumptions used in the analysis is shown in Table 14.2. Details and explanations of these data points are provided below. Data points were compiled from the relevant water industry entities and from publically available literature.

## 14.5.3 Quantifying benefits for the case study

As part of the project economic analysis, a full range of economic benefits associated with the preferred option was explored, and those able to be defensibly quantified were included in the quantitative analysis. These were then allocated to the relevant entities, based on discussion with those entities (Table 14.3).

From this attribution process, an assessment of total benefit for each relevant entity was produced. The environmental benefit of waterway health was the key unquantified benefit identified. Cost transfers were accounted for reflecting potable water prices, bulk water prices paid and received by households, the water retailer, and the bulk water provider. After consideration of these, a total benefit to each entity is presented in Table 14.4.

## 14.5.4 Results

Four options were shortlisted for detailed analysis. In addition to the preferred option, the other three were:

- Combined recycled wastewater to residential and regional stormwater harvesting for a new water supply;
- Recycled wastewater to agriculture and stormwater reuse to residential; and
- Recycled wastewater to residential and stormwater discharge to creek.

A summary of the results is presented in Fig. 14.5.

The preferred option in this project was constructing a new water supply (stormwater treated to potable standards and distributed in the potable water mains) and recycled water used to irrigate urban and farm lands and supplement winter flows in the waterways. No option gave a positive economic return. However, the preferred option was the closest to zero, with an NPV (net present value) of -$7.7 million (a net loss) and a benefit–cost ratio (BCR) of 0.88 (first two bars in Fig. 14.5).

It was accepted by stakeholders that the "gap" between benefits and costs of $7.7 million over 35 years could reasonably be equated to the unquantified benefit of improved waterway health.

The final two bars in Fig. 14.5 show the benefits and costs for recycled water to dual pipe and stormwater to creek (a typical development option for Melbourne). The NPV of this option is -$53 million with a BCR of 0.49. Compared with the preferred option (the first two bars of Fig. 14.5), it is much less favorable.

The project has been approved by the relevant water authorities. It involved the input from the water retailer, the bulk water supplier, the waterway manager, local and state government, and consultants.

**TABLE 14.2 Key Economic Assumptions**

| Variable | Value | Beneficiary |
|---|---|---|
| Analysis period | 35 years | |
| Discount rate (real) | 4.5% | |
| Avoided cost of headwork augmentation (bulk supplier) | $493 per mL | Water retailer |
| Avoided cost of transfer augmentation (bulk supplier) | $220 per mL | Bulk water provider |
| Avoided cost of desalination water purchase | $430 per mL | Water retailer |
| Nitrogen abatement—stormwater | $7236 per kg (capitalized) | Bulk water provider |
| Nitrogen abatement—recycled water | $56.66 per kg | Bulk water provider |
| Avoided building code costs | Estimated based on forecast reduction in rainwater tanks | Households |
| Avoided best practice environmental management (regulatory) costs | Actual (estimated) change in developer service scheme costs | Developers |
| Avoided cost of waterway restoration | $7040 per km of waterway (for a 60% reduction in inflow)<br>$3575 per km of waterway (for a 25% reduction in inflow) | Bulk water provider |
| Avoided/deferred business-as-usual investments | Actual (estimated) cost for avoided water headworks and transfer and for sewage transfer and tailworks | Water retailer |
| Community willingness to pay for reuse | $225 per mL (residential)<br>$745 per mL (public open space)<br>$480 per mL (environmental)<br>$112.50 per mL (agriculture) | Whole of society |
| Agricultural benefits | $143 per mL (change in gross margin attributable to irrigated lucerne) | Whole of society |
| Short run pumping costs | $90 per mL | Water retailer |

**TABLE 14.3 Allocation of Quantified Benefits to Entities**

| Benefit | Value ($M PV) | Entity Attributed | Rationale |
|---|---|---|---|
| Nitrogen abatement value | $13.9 | Bulk water provider | Responsibility for nitrogen offset scheme |
| Long run marginal cost headworks | $4.9 | Water retailer | Reduced demand from the network |
| Long run marginal cost desalination | $4.2 | Water retailer | Reduced demand from the network |
| Long run marginal cost transfer | $2.2 | Water retailer | Reduced demand from the network |
| Western water variable transfer savings | $0.9 | Water retailer | Avoided cost |
| Avoided STP upgrade cost | $9.6 | Water retailer | Avoided cost |
| Avoided potable distribution | $2.7 | Water retailer | Avoided cost |
| Avoided building codes costs | $13.1 | New households | Avoided cost |
| Avoided cost of waterway restoration | $1.3 | Bulk water provider | Avoided cost |
| Community willingness to pay (WTP) for residential and commercial reuse of stormwater | $1.7 | Whole of society | Reflects broader societal value for reuse |
| Community WTP for environmental flows with reuse for stormwater | $3.6 | Bulk water provider | Reflects broader societal value for reuse |
| Increased agricultural value | $0.8 | Agricultural users | Reflects value of irrigation to agriculture |
| Total benefits | $58.9 | | |

**TABLE 14.4** Total Benefit for Each Entity Over a 35-Year Period

| Entity | Benefit (PV) |
|---|---|
| Water retailer | $27.1 m |
| Bulk water provider | $18.7 m |
| Developers | – |
| Whole of society | $1.7 m |
| New households | $10.8 m |
| Councils | – |
| Agriculture users | $0.5 m |
| Total benefits | $58.9 m |

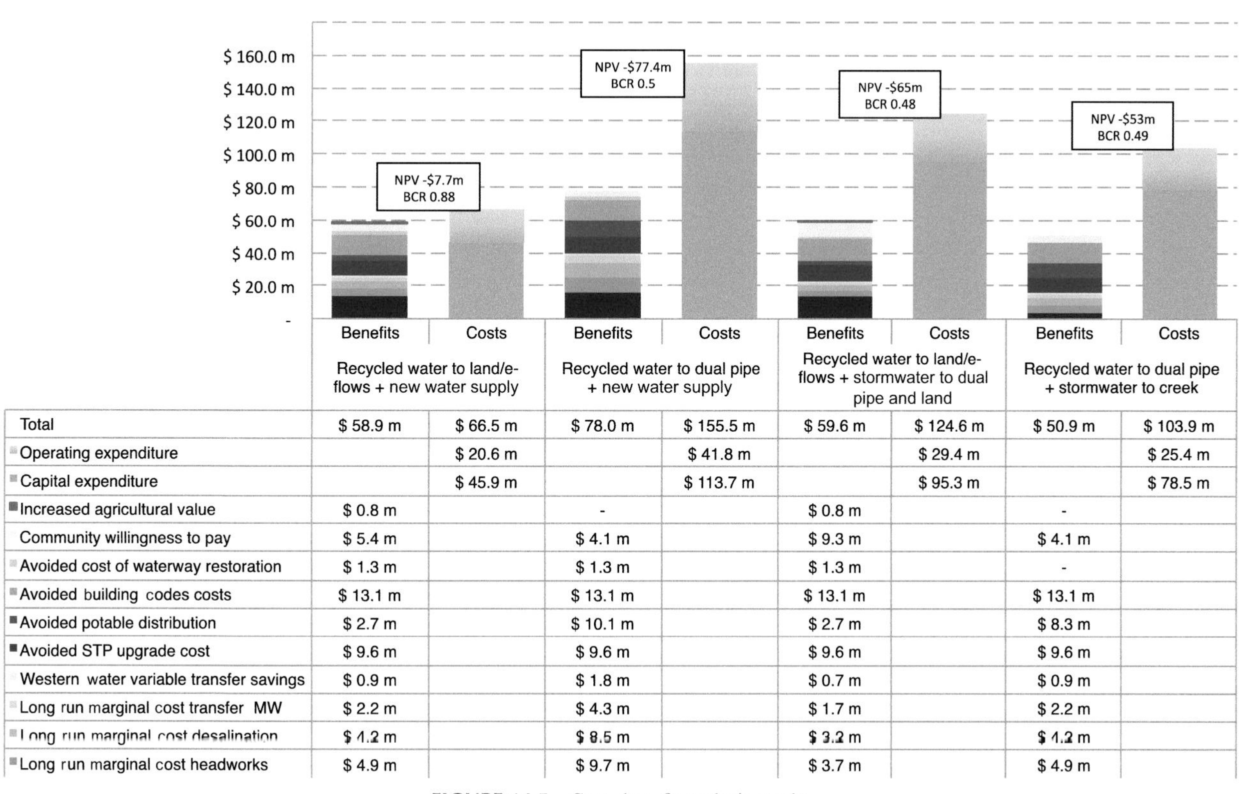


| | Recycled water to land/e-flows + new water supply: Benefits | Recycled water to land/e-flows + new water supply: Costs | Recycled water to dual pipe + new water supply: Benefits | Recycled water to dual pipe + new water supply: Costs | Recycled water to land/e-flows + stormwater to dual pipe and land: Benefits | Recycled water to land/e-flows + stormwater to dual pipe and land: Costs | Recycled water to dual pipe + stormwater to creek: Benefits | Recycled water to dual pipe + stormwater to creek: Costs |
|---|---|---|---|---|---|---|---|---|
| Total | $ 58.9 m | $ 66.5 m | $ 78.0 m | $ 155.5 m | $ 59.6 m | $ 124.6 m | $ 50.9 m | $ 103.9 m |
| Operating expenditure | | $ 20.6 m | | $ 41.8 m | | $ 29.4 m | | $ 25.4 m |
| Capital expenditure | | $ 45.9 m | | $ 113.7 m | | $ 95.3 m | | $ 78.5 m |
| Increased agricultural value | $ 0.8 m | | - | | $ 0.8 m | | - | |
| Community willingness to pay | $ 5.4 m | | $ 4.1 m | | $ 9.3 m | | $ 4.1 m | |
| Avoided cost of waterway restoration | $ 1.3 m | | $ 1.3 m | | $ 1.3 m | | - | |
| Avoided building codes costs | $ 13.1 m | | $ 13.1 m | | $ 13.1 m | | $ 13.1 m | |
| Avoided potable distribution | $ 2.7 m | | $ 10.1 m | | $ 2.7 m | | $ 8.3 m | |
| Avoided STP upgrade cost | $ 9.6 m | | $ 9.6 m | | $ 9.6 m | | $ 9.6 m | |
| Western water variable transfer savings | $ 0.9 m | | $ 1.8 m | | $ 0.7 m | | $ 0.9 m | |
| Long run marginal cost transfer MW | $ 2.2 m | | $ 4.3 m | | $ 1.7 m | | $ 2.2 m | |
| Long run marginal cost desalination | $ 1.2 m | | $ 8.5 m | | $ 3.2 m | | $ 1.2 m | |
| Long run marginal cost headworks | $ 4.9 m | | $ 9.7 m | | $ 3.7 m | | $ 4.9 m | |

**FIGURE 14.5** Cost–benefit analysis results.

## 14.6 CONCLUSION

Rigorous economic analysis of WSUD investments involves a transparent, consistent approach to assessing the merits of the investments against their costs.

CBA is generally preferred by economists for its logical consistency, transparency, and single measurement metric (dollars). Estimating project costs is relatively straightforward and included capital, maintenance, and renewal costs. The key challenge is to ensure that the full set of costs is included over the lifetime of analysis.

Defining, estimating, and appropriately attributing benefits is more challenging. The benefits may occur in different ways and accrue to different beneficiaries and may be expressed in social and environmental terms, rather than financial. Nevertheless, there are ways of estimating many of the benefits that WSUD investments provide. This chapter presented a TEV framework as a way to order this information in terms of economic values to structure an economic approach to valuing WSUD benefits.

The range of benefits from WSUD investments include water resource value; flood protection; waterway health improvement; reduced nutrient and pollution loads; augmentation of environmental flows; how much communities are willing to pay for improved waterway health and stormwater reuse; and the increase in property prices due, for example, to enhanced tree canopy cover and waterway restoration.

Expansion of the range of benefits rigorously quantified, along with incorporating better data on the expected performance of the WSUD assets, will further assist economic analysis of WSUD investments over time.

## REFERENCES

Bratman, G.N., Hamilton, J.P., Daily, G.C., 2012. The impacts of nature experience on human cognitive function and mental health. Annals of the New York Academy of Sciences 1249 (1), 118–136.

Cohen-Cline, H., Turkheimer, E., Duncan, G.E., 2015. Access to green space, physical activity and mental health: a twin study. Journal of Epidemiology and Community Health 69 (6), 523–529.

Cooper, B., Crase, L., Burton, M., June 2017. The Value of Melbourne's Environmental Water Entitlements: A Report on Preliminary Estimates Using Choice Modelling Prepared for Melbourne Water.

Donovan, G.H., Butry, D.T., 2010. Trees in the city: valuing street trees in Portland, Oregon. Landscape and Urban Planning 94, 77–83.

Donovan, G.H., Butry, D.T., 2011. The effect of urban trees on the rental price of single- family homes in Portland, Oregon. Urban Forestry and Urban Greening 10, 163–168.

E2Designlab and Water Technology, 2016. Ballarat WSUD for Frequent Flow Mitigation. Report prepared for Ballarat City Council (unpublished).

Freeman III, A.M., 2003. The Measurement of Environmental and Resource Values, Theory and Methods. RFF Press Book, Published by Resources for the Future, Washington, DC.

Hartig, T., Mitchell, R., de Vries, S., Frumkin, H., 2014. Nature and health. Annual Review of Public Health 35, 207–228.

Hunter, R.F., et al., 2015. The impact of interventions to promote physical activity in urban green space: a systematic review and recommendations for future research. Social Science and Medicine 124, 246–256.

Longmore, A.R., 2008. Port Phillip Bay Environmental Management Plan: Monitoring the State of Bay Nitrogen Cycling (2005-2006). Primary Industries Research Victoria, Marine and Freshwater Systems, Dept. of Primary Industries.

Marsden Jacob Associates, March 2014. The Economic Viability of Recycled Water Projects - Technical Report 2 Community Values for Recycled Water in Sydney. Australian Water Recycling Centre of Excellence.

Marsh, D., Baskaran, R., 2009. Valuation of water quality improvements in the Karapiro catchment: a choice modelling approach. In: Australian Agricultural and Resource Economics Society 53rd Annual Conference. Cairns, Australia.

Mitchell, R., Popham, F., 2007. Greenspace, urbanity and health: relationships in England. Journal of Epidemiology and Community Health 61 (8), 681–683.

Moroka, December 2017. A Review of the Effectiveness of Distributed Storage to Decrease Flooding. Report prepared for Melbourne Water.

Pandit, R., Polyakov, M., Sadler, R., 2014. Valuing public and private urban tree canopy cover. The Australian Journal of Agricultural and Resource Economics 58 (3).

Plant, L., Rambaldi, A., Sype, N., 2016. Property Value Returns on Investment in Street Trees: A Business Case for Collaborative Investment in Brisbane, Australia. Discussion Paper no 563. School of Economics, The University of Queensland, St Lucia, QLD, 4072. http://www.uq.edu.au/economics/abstract/563.pdf.

Polyakov, M., Fogarty, J., Zhang, F., Pandit, R., Pannell, D., 2016. The value of restoring urban drains to living streams. Water Resources and Economics. https://doi.org/10.1016/j.wre.2016.03.002i.

Rolf, J., Brouwer, R., March 2010. Testing for value stability with a meta-analysis of choice experiments: River health in Australia. Environmental Economics Research Hub Research Reports.

Rolfe, J., Brouwer, R., 2013. Design effects in a meta-analysis of river health choice experiments in Australia. Journal of Choice Modelling 5 (2), 81–97. https://doi.org/10.1016/S1755-5345(13)70053-8.

Rosenberger, R.S., Loomis, J., 2000. Using meta-analysis for benefit transfer: in-sample convergent validity tests for an outdoor recreation database. Water Resources Research 36, 1097–1107.

Sale, M., Hosking, S., Du Preez, M., 2009. Application of the contingent valuation method to estimate a recreational value for the freshwater inflows into the Kowie and the Kromme Estuaries. Water SA 35.

Sharma, A.K., Begbie, D., Gardner, T., 2015. Rainwater Tank Systems for Urban Water Supply – Design, Yield, Energy, Health Risks, Economics and Community Perceptions. IWA Publishing.

Sugiyama, T., Leslie, E., Giles-Corti, B., Owen, N., 2008. Associations of neighbourhood greenness with physical and mental health: do walking, social coherence and local social interaction explain the relationships? Journal of Epidemiology and Community Health 62 (5).

Windle, J., Rolfe, J., 2014. Applying benefit transfer with limited data: unit value transfers in practice. In: Johnston, R., et al. (Eds.), Benefit Transfer of Environmental and Resource Values: A Handbook for Researchers and Practitioners. Springer, Dordrecht, The Netherlands.

Chapter 15

# Optimization of WSUD Systems: Selection, Sizing, and Layout

Graeme C. Dandy[1], Michael Di Matteo[1,2] and Holger R. Maier[1]

[1]*School of Civil, Environmental and Mining Engineering, University of Adelaide, Adelaide, SA, Australia;* [2]*Water Technology Pty. Ltd., Adelaide, SA, Australia*

## Chapter Outline



## ABSTRACT

The planning and design of WSUD schemes is complicated by the fact that they may have multiple: (1) purposes and objectives; (2) components and design decisions for each component, and (3) spatial scales, which need to be considered. This provides motivation for the use of optimization methods that are aimed at finding the combination of design variables that achieve the best outcomes in terms of the multiple objectives. This chapter discusses how optimization methods can be used in conjunction with a range of simulation models to plan and design WSUD schemes. It also shows how optimization has been used to identify system trade-offs between a range of economic, social, and environmental indicators. Two case studies are presented that consider the selection, sizing, and layout of WSUD components for the purposes of water quality improvement and stormwater harvesting. A third case study demonstrates the use of multiobjective optimization in conjunction with visual analytics to select a portfolio of WSUD projects so as to satisfy the requirements of multiple stakeholders. Future developments are discussed, including (1) collaborative modeling and decision-making; (2) identifying, exploring, analyzing, and selecting multiobjective WSUD solutions; and (3) how optimization can support robust and collaborative WSUD planning and design approaches.

**Keywords:** Genetic algorithm; Lifecycle cost; Multicriteria analysis; Multiobjective optimization; Multistakeholder optimization; *MUSIC*; Simulation models; Stormwater harvesting; Trade-off analysis; *WaterCress*.

**Approaches to Water Sensitive Urban Design. https://doi.org/10.1016/B978-0-12-812843-5.00015-0**

## 15.1 INTRODUCTION

The planning and design of WSUD schemes is complicated by the following factors: (1) they may have multiple purposes and objectives; (2) they have many components and many design decisions for each component; and (3) multiple spatial scales may need to be considered.

The multiple purposes include:

- Flow regime improvement: water quality improvement, water quantity reduction, flood mitigation, and groundwater restoration;
- Water harvesting; and
- Increased urban amenity, vegetation, and recreation.

The components of a WSUD system include elements for collection, treatment, storage, distribution, and disposal of the stormwater. The many design decisions include the size, type, number and location of components, and how they are connected. These need to satisfy a number of practical constraints including cost, available land, and planning and other legal restrictions. Multiple spatial scales are also possible, including individual allotments, clusters of allotments, and precinct scale.

This combination of many possible components and a large number of options and spatial scales provides motivation for the use of optimization methods that are aimed at finding the combination of design variables that achieve the best outcome in terms of the multiple objectives identified for the scheme. This is further complicated by the need to consider the effects of climate change, as well as changes in land use and other activities in urban areas.

This chapter will discuss these issues in relation to the application of optimization techniques to assist in the planning, design, and project selection of WSUD schemes.

## 15.2 USE OF SIMULATION AND OPTIMIZATION MODELS FOR WSUD DESIGN

### 15.2.1 Traditional and optimization approaches to WSUD design

The traditional approach for the design of WSUD systems involves an experienced professional who will undertake the following steps:

1. Identify the purposes, objectives, and evaluation criteria of the scheme (e.g., cost, percentage removal of selected pollutants, volume retained on-site, harvested volume)
2. Select the scale at which the scheme will operate
3. Make a preliminary choice of the components to be used and the layout of these components
4. Undertake calculations (possibly using a spreadsheet) to obtain a preliminary estimate of the sizes and capacities of each component
5. Set up a detailed simulation of the preliminary design using a simulation package (or spreadsheet) and input data for selected design periods (including rainfall, evaporation, pollutant wash off, removal and retention rates, and selected operating rules for the system)
6. Run the simulation model to determine whether the preliminary design achieves satisfactory levels of the evaluation criteria
7. If the performance of the design is unsatisfactory, modify the design and return to Step 5; if satisfactory proceed to Step 8.
8. Document the chosen design and present it to the client

This process is shown schematically in Fig. 15.1.

There are two principal ways of incorporating optimization into this procedure. In the first one, the simulation model in Fig. 15.1 is replaced with an optimization model that has the explicit aim of identifying that combination of design variables that achieves the highest level of a stated objective or objectives subject to meeting all constraints on the system. This approach utilizes more traditional optimization techniques such as linear programming (LP) or gradient search methods.

The alternative approach follows a similar procedure to the traditional design approach (shown in Fig. 15.1) but with an optimization algorithm replacing the trial-and-error component of the procedure. This approach utilizes guided search techniques such as evolutionary algorithms (EAs) and is shown schematically in Fig. 15.2.

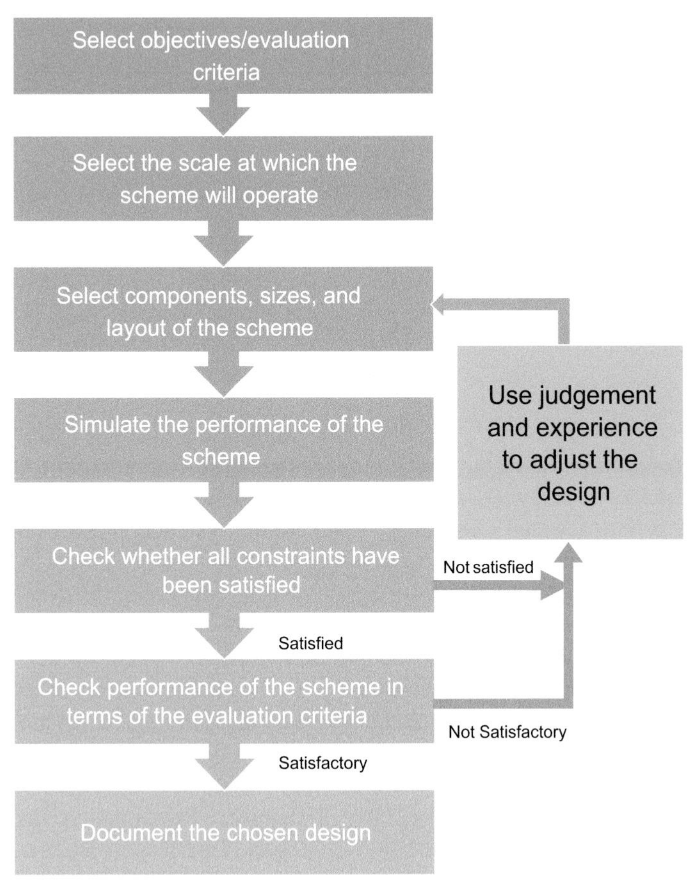


**FIGURE 15.1** Schematic of traditional design procedure for WSUD schemes.

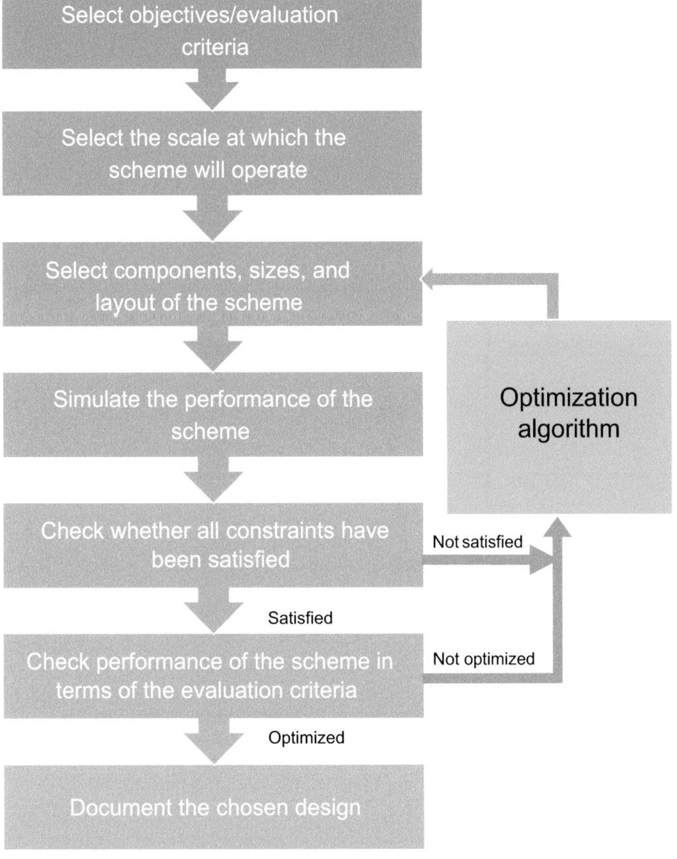


**FIGURE 15.2** Schematic of the optimization approach to WSUD design.

In either case, clear statements (usually in the form of equations) are required for the following: (1) the objectives and measures of effectiveness of the scheme; (2) the decision variables; and (3) constraints that limit choices in the design process.

The general multiobjective optimization problem formulation (Dandy et al., 2017) is given below:

$$\text{Max (or min.) } Z_k = f_k\,(x_1,\, x_2, \ldots, x_n)\ (k = 1, \ldots p) \tag{15.1}$$

subject to:

$$h_i(x_1, \ldots, x_n)\left\{\begin{array}{c} \le \\ = \\ \text{or} \ge \end{array}\right\} b_i(i = 1, \ldots, m) \tag{15.2}$$

where $x_j$ ($j = 1\ldots n$) are the decision variables, $Z_k$ ($k = 1\ldots p$) are the objective functions, $f_k$ (...) and $h_i$ (...) are any linear or nonlinear functions of the decision variables, and $b_i$ are known values.

Eq. (15.1) are the objective functions and Conditions (15.2) are the system constraints that can be of the less than or equal to, equal to, or greater than or equal to type.

In single objective optimization, $p = 1$ and there is only one objective function (for example, to minimize cost). Then, in principle, it is possible to identify a single solution that optimizes the objective function (e.g., the minimum cost design).

In multiobjective optimization, $p > 1$ and there is no single "best" solution. The purpose of multiobjective optimization is to identify the Pareto front (illustrated in Fig. 15.3). This is a line or surface that represents the most efficient trade-off between the various objectives. The choice between solutions on the Pareto front involves the input of values on the part of stakeholders as there is no technical reason to choose between solutions on the front. Fig. 15.3 shows the Pareto front for two objectives: in this case, cost and percentage removal of a pollutant (e.g., total suspended solids [*TSS*]). Cost is to be minimized and the percentage of pollutant removed is to be maximized. Points on the Pareto front represent the most efficient solutions in terms of these two objectives. Solutions not on the Pareto front are considered to be "dominated" by (or inferior to) the Pareto optimal solutions, as there is at least one solution on the Pareto front that performs at least as well as a dominated solution at a lower cost. Hence there is no reason to choose a solution that is not on the Pareto front (provided all relevant objectives have been considered). A choice between points on the Pareto front can be made by: (a) constraining one of the objectives (for example choosing the minimum cost design that achieves a specified level of pollutant removal); (2) placing a relative weight on the two objectives; or (3) asking the decision maker to choose from among a subset of these solutions directly.

A special case of the optimization model is where there is a single objective function that is a linear function of the decision variables (i.e., Eq. (15.1) is linear with $p = 1$) and a set of linear equalities or inequalities as constraints (i.e., all of the $f_k$ $(x_1, x_2, \ldots, x_n)$ and $h_i(x_j, \ldots, x_n)$ functions are linear in the decision variables). This problem can be solved using LP. There are a number of computer packages available for solving large LP problems with many thousands of variables and constraints. These include LINDO API (www.lindo.com) and GAMS (www.gams.com). LP works with a single solution at

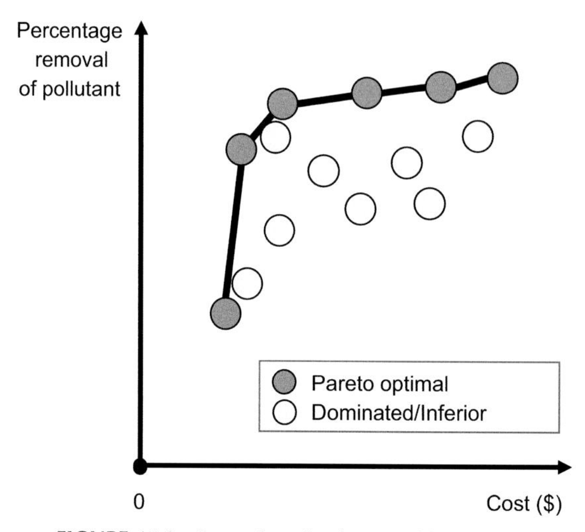


FIGURE 15.3 Pareto front for the two-objective case.

a time and progressively improves this until it finds the optimum solution. LP is guaranteed to find the optimal solution for a problem with a single linear objective and linear constraints.

The traditional optimization techniques for nonlinear problems are based on some type of gradient search, such as the "method of steepest ascent" (Taha, 2017). These methods work well in cases where the objective function is smooth and continuous and has a single optimum. Unfortunately, most practical engineering problems do not satisfy these requirements as there are usually multiple optima and/or discontinuities in the objective function.

Evolutionary algorithms (EAs) constitute another class of optimization techniques that can solve optimization problems with multiple optima and can handle any type of nonlinearity or discontinuity in the objective function, as well as logical decision variables. EAs are informed search techniques that use problem-specific knowledge to efficiently find better solutions to a problem (Russell and Norvig, 2003). In this sense, they may be contrasted with uninformed search procedures (such as trial-and-error) that are very inefficient in most cases.

EAs include genetic algorithms (Goldberg, 1989), scatter search (Glover et al., 2000), particle swarm optimization (Kennedy and Eberhart, 1995), differential evolution (Storn and Price, 1997), and ant colony optimization (Dorigo et al., 1996). These are also called metaheuristic techniques.

EAs work with a population of solutions and progressively improve these until near-optimal solutions are found. Some stochastic elements are included in the search process and this (together with the fact that a population of solutions is considered at any time) ensures a good coverage of the search space. There is no guarantee of finding the true optimum solution for a single objective case; however, these techniques work well with simulation models where each solution is simulated and then evaluated in terms of whether or not it satisfies the constraints and the level of the objective function obtained.

EAs have several advantages over traditional optimization techniques (Dandy et al., 2017; Liner and Maier, 2015) as they can:

1. Handle problems with discontinuous and nonlinear objective functions
2. "Bolt on" to existing simulation models; if a problem can be simulated, it can be optimized with an EA
3. Identity diverse solutions similar in objective performance but different in solution space, providing decision makers with a variety of solutions
4. Enable consideration of factors other than those mathematically formulated in the optimization problem
5. Enable "shopping" through the optimal and near-optimal solutions that represent the best trade-offs between objectives
6. Handle "discrete" decision variables where there are a fixed number of options for a decision
7. Have a significant advantage for multiobjective problems in that, as they work with a population of solutions, they can be configured to develop the full set of Pareto optimal solutions in a single run.

Examples of multiobjective EAs include NSGA II (Deb et al., 2002) and BORG (Hadka and Reed, 2012).

## 15.2.2 Simulation models for WSUD design

Traditional planning and design of WSUD schemes makes use of simulation models that usually model the hydraulic and hydrologic influence of WSUD options including infiltration, detention, retention, rainwater harvesting, and reuse. A number of these also model specific water quality parameters in the system (typically *TSS*, total nitrogen [TN], and/or total phosphorus [TP]). Table 15.1 shows a capability categorization of a number of stormwater system simulation tools used for WSUD conceptual design and is adapted from review studies in Lerer et al. (2015), Myers et al. (2013), and Bach et al. (2014). As can be seen, the capabilities of the tools are categorized based on: (1) the types of best management practices (BMPs) modeled; and (2) the questions addressed (adapted from Lerer et al., 2015), including:

- How much? (i.e., tools that provide quantitative answers, for example, to help size BMPs);
- Where? (i.e., tools that provide spatial answers, for example, using geographic information system data to help select sites within a catchment to place BMPs); and
- Which? (i.e., tools that include assessment methods to help select among BMP options).

Two of these models are used in the three case studies presented later in this chapter, namely WaterCress and MUSIC. WaterCress (Water-Community Resource Evaluation and Simulation System) is a water balance model that enables simulation of a water supply system. Each component of the system has an associated database, which contains all variables (e.g., demand, rainfall, and evaporation) necessary to enable quantities of water to be estimated and tracked through a specified water supply system (Clark et al., 2002). WaterCress is capable of integrating stormwater, wastewater, and

**TABLE 15.1** Capability Assessment of Stormwater Simulation Models Used for WSUD Conceptual Design

| Simulation Model | | SWMM | WaterCress | MUSIC | Urban-Developer | SUSTAIN | UrbanBeats |
|---|---|---|---|---|---|---|---|
| **Reference** | | **Rossman (2004)** | **Clark et al. (2002)** | **eWater (2011a)** | **eWater (2011b)** | **Lee et al. (2012)** | **Bach (2012)** |
| Types of BMPs | *Wetlands* | Yes | Yes | Yes | Yes | Yes | Yes |
| | *Bioretention* | Yes | No | Yes | Yes | Yes | Yes |
| | *Biofiltration* | No | No | Yes | Yes | No | Yes |
| | *Swales* | Yes | No | Yes | Yes | Yes | Yes |
| | *Porous pavement* | Yes | No | Yes | Yes | Yes | Yes |
| | *Sedimentation* | Yes | Yes | Yes | Yes | Yes | Yes |
| | *Green roof* | Yes | No | Yes | Yes | Yes | Yes |
| | *Infiltration* | Yes | Yes | Yes | Yes | Yes | Yes |
| | *Rainwater tanks* | Yes | Yes | Yes | Yes | Yes | Yes |
| | *Land use changes* | No | No | No | No | No | Yes |
| How much? | Hydraulic | Yes | No | No | Yes | Yes | No |
| | Hydrologic | Yes | Yes | Yes | Yes | Yes | Yes |
| | Water Quality Parameters | User Supplied | Salinity | TN, TP, *TSS* | TN, TP, *TSS* | *TSS* | TN, TP, *TSS* |
| | Stormwater harvesting | No | Yes | Yes | Yes | No | Yes |
| | Operation and control | Yes | Yes | No | Yes | No | No |
| | Costs | No | No | Yes | No | Yes | Yes |
| Where? | BMP siting (e.g., GIS) | No | No | No | No | Yes | Yes |
| Which? | Multiobjective assessment | No | No | No | No | Optimization | No |

*GIS*, geographic information system; *TN*, total nitrogen; *TP*, total phosphorous; *TSS*, total suspended solids. The capability assessment of UrbanBeats is based on its use with MUSIC (Model for Urban Stormwater Improvement Conceptualization).
Adapted from Bach P.M., Rauch W., Mikkelsen P.S., McCarthy D.T. and Deletic A., A critical review of integrated urban water modelling–urban drainage and beyond, Environmental Modelling and Software **54**, 2014, 88–107; Lerer S., Arnbjerg-Nielsen K. and Mikkelsen P., A mapping of tools for informing water sensitive urban design planning decisions—questions, aspects and context sensitivity, Water **7** (3), 2015, 993; Myers B., Pezzaniti D. and Gonzalez D., Hydrological Modelling of the Parafield and Cobbler Creek Catchment for Hazard Analysis Planning, Goyder Institute for Water Research Technical Report Series No. 13/3. 2013, South Australia, Adelaide.

mains water distribution, it can withhold water in a wetland until a target water quality is achieved and has been used for water systems optimization (Beh et al., 2014; Marchi et al., 2016).

MUSIC by eWater (2011a) is an integrated stormwater model that evaluates rainfall/runoff and pollutant generation and transport, pollutant removal performance of BMPs, and water balance analysis (Bach et al., 2014). MUSIC is used as a stormwater management conceptual design tool. It has been used in watershed-scale stormwater flow regime and water quality analysis (Burns et al., 2012) linked with a genetic algorithm for multiobjective WSUD systems optimization (Di Matteo et al., 2017a), and used with the UrbanBeats framework for WSUD planning considering land use changes (Bach et al., 2013). MUSIC algorithms simulate runoff based on models developed by Chiew and McMahon (1999) and urban pollutant load relationships based on analysis by Duncan (1999).

### 15.2.3 Use of optimization in WSUD design

A number of studies have been carried out that use optimization for WSUD design. The decisions considered include the selection, sizing, and location of WSUD components. Some of these studies are summarized in Table 15.2.

The studies have been classified according to the following criteria:

1. whether they are single objective or multiobjective;
2. the type of objective(s) considered (e.g., cost, water quality, reliability);
3. the purposes for which the stormwater scheme is designed, including reduction in flooding, flow retention, water quality improvement, and stormwater harvesting;
4. whether they are aimed at selection, sizing, or determining the location of WSUD devices;
5. the stormwater devices considered;
6. the optimization techniques used; and
7. the simulation model(s) used.

Each of these factors will be discussed in turn in this section.

The single objective studies are generally aimed at minimizing cost, subject to meeting constraints on the minimum percentage removal of specific pollutants and/or the minimum retention of a specified volume of stormwater on-site. In the case of Marchi et al. (2016), the benefit of harvested stormwater is included in the objective function as a reduction in cost of the alternative mains water. In addition, as net present value (*NPV*) is calculated over a number of climate change scenarios, two formal objectives, corresponding to the average and range of *NPV* over these scenarios, are used and the problem is solved using a multiobjective algorithm.

The multiobjective studies develop a Pareto optimal front, consisting of a number of solutions that represent the best trade-offs between cost and the other objectives, including water quality improvement, the volume of stormwater retained on-site, or the volume of stormwater harvested.

The purposes for which the stormwater systems are designed usually include water quality improvement, with flow retention being important in most of the studies. Only a few studies (Marchi et al., 2016; Di Matteo et al., 2017a,b; Di Matteo et al., 2018) consider stormwater harvesting as the primary purpose. Only one study (Zare et al., 2012) includes reduction in flood peak specifically. In this study, the expected reduction in flood damages is included as a benefit in the objective function.

The sizes of WSUD facilities are considered as the decision variables in most of the studies, with selection and location being considered in a number of the studies. The stormwater devices considered are diverse, with buffer strips, detention storages, sedimentation basins, and biofilters being common. Reichold et al. (2010) is rather different in that the decision variables considered are the percentages of impervious area in all subcatchments. These values are selected so as to minimize the change in the natural flow regime as characterized by 33 hydrological parameters.

A wide range of optimization techniques is used across the various studies. The majority use some form of EA (NSGA II, genetic algorithm, differential evolution, PACOA), which means the optimization technique interfaces with a simulation model (as shown in Fig. 15.2). Only one study (Loaiciga et al., 2015) uses linear programming and integer linear programming (in which some of the decision variables can only take integer values). In this case, the model is formulated to specifically include equations representing simulation of the stormwater devices, and no separate simulation model is used (as shown in Fig. 15.2). In the study by Jayasooriya et al. (2016), no formal optimization is used, as the search space is sufficiently small to enable trial-and-error to be used. The simulation models used include SWMM (Rossman, 2004), MUSIC (eWater, 2011a), WaterCress (Clark et al., 2002), HSPF (Bicknell et al., 2001), and AnnAGNPS (AGNPS, 2001).

In the next section, three of these studies (Marchi et al., 2016; Di Matteo et al., 2017a,b) are presented in detail.

**TABLE 15.2** Analysis of Optimization Studies for WSUD Component Sizing, Selection, and Layout

| Authors | Single or Multiobjective | Objectives | Purposes | Selection, Sizing, or Location of BMPs | Devices Considered | Optimization Technique Used | Simulation Model Used |
|---|---|---|---|---|---|---|---|
| Jayasooriya et al. (2016) | Single | Cost | Water quality | Sizing | Sediment basins and bioretention | Trial and error | MUSIC |
| Loaiciga et al. (2015) | Single | Cost | Retention, water quality | Sizing | Detention storages | Linear programming | Incorporated in LP |
| Loaiciga et al. (2015) | Single | Cost | Retention, water quality | Selection | Catchment basins | Integer linear programming | Incorporated in ILP |
| Lee et al. (2012) | Single | Cost | Retention, water quality | Selection and location | Pervious buffers, bio-retention and porous pavers | Scatter search, NSGA II | SWMM and purpose written model |
| Reichold et al. (2010) | Single | Flow regime change | Flow regime | Location | Impervious area | Genetic algorithm plus local search | SWMM, Ecological impact model |
| Zhen et al. (2004) | Single | Cost | Water quality | Sizing and location | Stormwater ponds | Scatter Search | AnnAGNPS |
| Di Matteo et al. (2017b) | Multi, multiple problem formulations | Cost, water quality (TN), harvesting capacity, amenity | Water quality, stormwater har-vesting, amenity | Project selection | Biofilters, wetlands, swales | Pareto Ant Colony Optimization Algorithm (PACOA) | MUSIC |
| Di Matteo et al. (2017a) | Multi | Cost, *TSS*, and volumetric reliability | Water quality, stormwater harvesting | Selection, sizing, and location | Biofilters, wetlands, sediment basins, ponds | NSGA II | MUSIC |
| Marchi et al. (2016) | Multi | *NPV* (Average and Range) | Stormwater harvesting | Sizing | Sedimentation basins and wetlands | NSGA II | WaterCress |
| Chichakly et al. (2013) | Multi | Cost, water quality (*TSS*) | Water quality | Sizing and location (within subbasins) | Bioretention, detention pond | Differential Evolution | Hydrological Simulation Program—Fortran (HSPF), Sensitivity Analysis |
| Zare et al. (2012) | Multi | Cost, water quality (*TSS* and BOD) and total SW runoff | Flooding, reten-tion, water quality | Sizing and location (within subbasins) | Bioretention, porous pavers, barrels (rain tanks), land use | NSGA II | SWMM |
| Lee et al. (2012) | Multi | Cost, flow volume, and pollutant load | Retention, water quality | Selection and location | Pervious buffers, bio-retention, and porous pavers | Scatter search, NSGA II | SWMM and purpose written model |
| Lee et al. (2005) | Multi | Cost, volume retained and pollutant retained | Retention, water quality | Selection and sizing | Detention ponds and on-site, wet weather controls | Premium Solver in Excel (Evolutionary solver, NLP) | Spreadsheet based on STORM |

## 15.3 CASE STUDIES

### 15.3.1 Optimum sizing of the components of a stormwater harvesting scheme

#### *15.3.1.1 Background*

This case study uses optimization to determine the size of the various components of the Parafield stormwater harvesting and aquifer storage scheme. The scheme is located in the Salisbury area, in the north of Adelaide, South Australia (Fig. 15.4). The Parafield scheme harvests an average of around 1100 ML/year of stormwater runoff from a catchment of 1580 ha of residential and industrial land and has an annual average stormwater yield distributed to users of approximately 870 ML/year. The scheme is an aquifer storage and recovery scheme in which stormwater is first stored in an in-stream basin (47 ML capacity), where coarse sediments are settled out. From here, water is pumped to a holding basin (48 ML capacity), where clay particles can settle. The next step is reached by a gravity pipeline and consists of a cleansing reedbed (referred to as a wetland in this study) that removes clay colloids and nutrients. From the wetland, water is injected into the aquifer and stored until it is needed. Currently, water is used by industrial and residential consumers for nonpotable purposes. Details of this study are given in Marchi et al. (2016).

#### *15.3.1.2 Formulation of optimization problem*

The aim of the study is to determine the dimensions of the in-stream basin, holding basin and wetland so as to maximize net present value (NPV). NPV consists of two components, benefits and costs. The costs considered include: the capital costs of the in-stream basin, holding basin, wetland, pumps, and distribution system; the maintenance costs for the in-stream basin, holding basin, wetlands, and pumps; and the operational costs of the pumps. The benefits considered include those associated with reduced usage of existing potable supplies, reduced salinity, and environmental benefits due to reduced usage of water from the River Murray, which is one of the existing sources of potable supply. A design life of 25 years and a discount rate of 6% were used throughout the study.

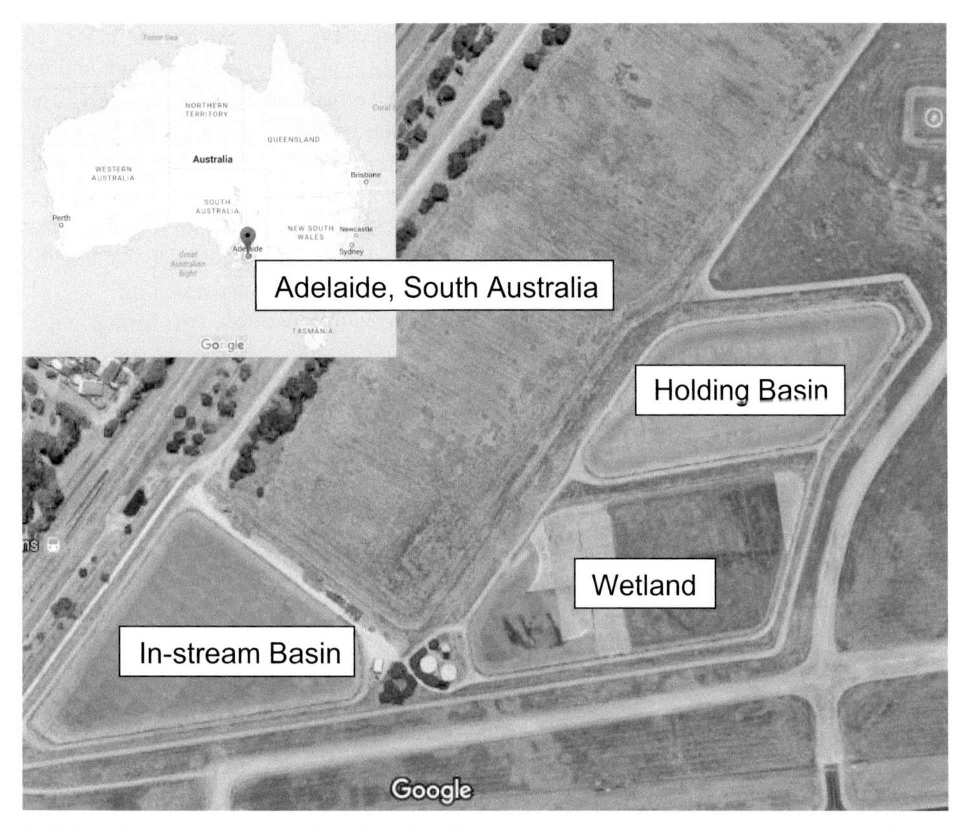


FIGURE 15.4 Location and configuration of the Parafield stormwater harvesting scheme. *Map data: Google.*

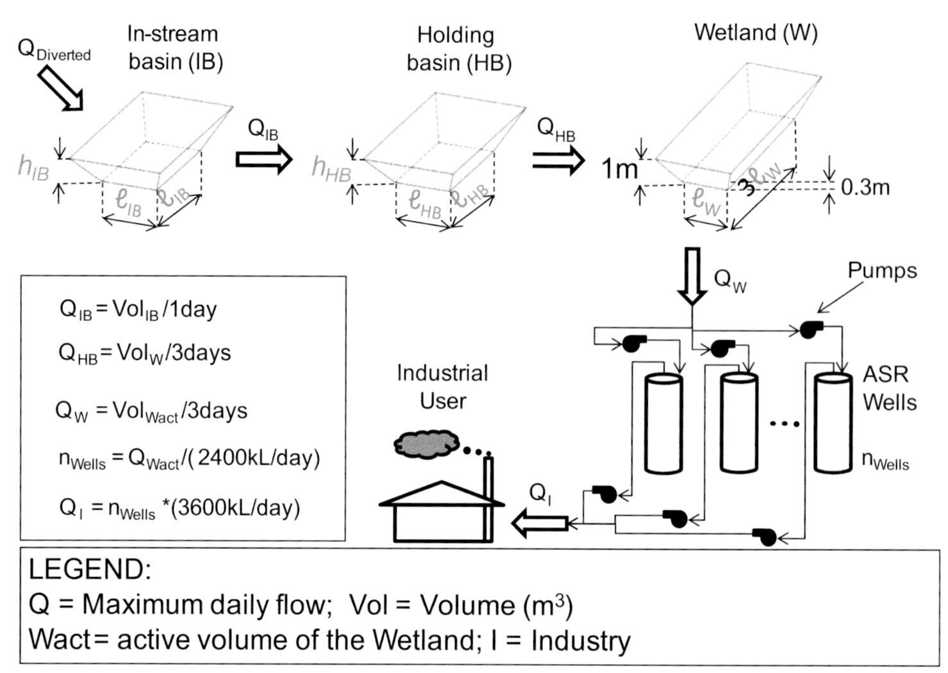


FIGURE 15.5 Summary of the decision variables (in *grey italics*) and their relationships. *I*, industry; *Q*, maximum daily flow; *Vol*, volume (m$^3$), *Wact*, active volume of the wetland. *Marchi, A., Dandy, G.C., Maier, H.R., 2016. Integrated approach for optimizing the design of aquifer storage and recovery stormwater harvesting schemes accounting for externalities and climate change. Journal of Water Resources Planning and Management 142 (4), with permission from ASCE.*

A summary of the decision variables and their relationships is given in Fig. 15.5. As shown in Fig. 15.5, there are five decision variables: the depths ($h_{IB}$, $h_{HB}$) and lengths of the sides ($\ell_{IB}$, $\ell_{HB}$) of the square in-stream and holding basins and the width of the rectangular wetland ($\ell_W$). As also shown in Fig. 15.5, the other design variables, such as pump sizes and the number of wells, are determined by the maximum daily flows from the in-stream basin to the holding basin ($Q_{IB}$), from the holding basin to the wetland ($Q_{HB}$), from the wetland to the aquifer ($Q_W$), and from the aquifer to the industrial user ($Q_I$). The values that the decision variables can take are as follows: $h_{IB}$, $h_{HB}$ can vary from 0.5 to 4 m in increments of 0.5 m; $\ell_{IB}$, $\ell_{HB}$ can vary from 20 to 200 m in increments of 10 m; and $\ell_W$ can vary from 20 to 150 m in increments of 10 m. The basins were assumed to be square, whereas the wetland was assumed to have a rectangular shape with a length to width ratio of 3:1. All side slopes were fixed at 1 in 4 and the maximum depth of all basins was 4 m.

To account for uncertainty in future rainfall due to climate change and variability, two formal objectives were used, namely the average and range of *NPV* values over a number of climate change scenarios as follows:

$$\text{Max } AvgNPV_j = \frac{\sum_{i=1}^{n} NPV_{j,i}}{n} \tag{15.3}$$

where $AvgNPV_j$ is the average NPV of design configuration $j$, $NPV_{j,i}$ is the NPV of design configuration $j$ for climate change scenario $i$, and $n$ is the number of climate change scenarios considered.

$$\text{Min } RngNPV_j = NPV_{\text{Max}} - NPV_{\text{Min}} \tag{15.4}$$

where $RngNPV_j$ is the range of the *NPV* of design configuration $j$ and $NPV_{\text{Max}}$ and $NPV_{\text{Min}}$ are the maximum and minimum values of *NPV*, respectively, for the design configuration considering different climate change scenarios.

The climate change scenarios were generated using a combination of scenarios from the Special Report on Emissions Scenarios (IPCC, 2007) and global circulation models (GCMs). Emissions Scenarios A2 and B1 were considered. Emissions scenario A2 correspond to a more pessimistic outlook (higher emissions), whereas B1 correspond to a more optimistic outlook (lower emissions). In addition, outputs from two GCMs were used. These were Mk3.5, developed by CSIRO (Australia) (Gordon et al., 2010), and ECHAM5, developed by the Max Planck Institute for Meteorology (Germany) (Roeckner et al., 2003). Mk3.5 was developed in Australia and usually forecasts a larger rainfall reduction resulting from climate change and ECHAM5 usually has better performances than other GCMs, according to the M Skill score (CSIRO, 2007). This resulted in a total of four climate change scenarios (i.e., Mk3.5-A2, Mk3.5-B1, ECHAM5-A2, and ECHAM5-B1), which were applied at planning horizons of 2030 and 2050.

### *15.3.1.3 Optimization process*

An outline of the optimization process is given in Fig. 15.6. As can be seen, values of the decision variables are generated using the optimization algorithm, resulting in a particular stormwater harvesting system design, the NPV, and hydrological performance of which are then calculated over the four climate change scenarios over the desired planning horizon. Hydrological performance is assessed with a simulation model of the system developed using WaterCress (Clark et al., 2002). The Non-Dominated Sorting Genetic Algorithm II (NSGA-II) (Deb et al., 2002) was selected as the multiobjective optimization engine, as it, or algorithms based on it, has been frequently applied in water supply and distribution problems and has been shown to give good results across a wide range of problems. As the optimization process progresses, the NSGA-II algorithm uses its search mechanisms (e.g., selection, crossover, mutation) to evolve better stormwater harvesting system designs based on the performance in terms of the objective functions (i.e., average and range of *NPV* over the four climate change scenarios), of previously selected designs. Optimization runs were completed for 2013 (reference scenario, without climate change), as well as for 2030 and 2050.

### *15.3.1.4 Results and discussion*

The outputs from the optimization included stormwater harvesting designs that maximized *NPV* for 2013 and provided optimal trade-offs between the average and range of *NPV*s over the four climate change scenarios considered for 2030 and 2050.

Table 15.3 shows decision variable values for selected optimal solutions for the three planning horizons evaluated. Operational costs and benefits become more important for designs with small basin sizes where capital and maintenance costs are lower. As these costs increase with the volume of stormwater available, this leads to a larger *NPV* range under different climate scenarios (e.g., Solutions 10, 11 and 19). Solutions 20, 34, and 37 reduce the range of *NPV* over climate

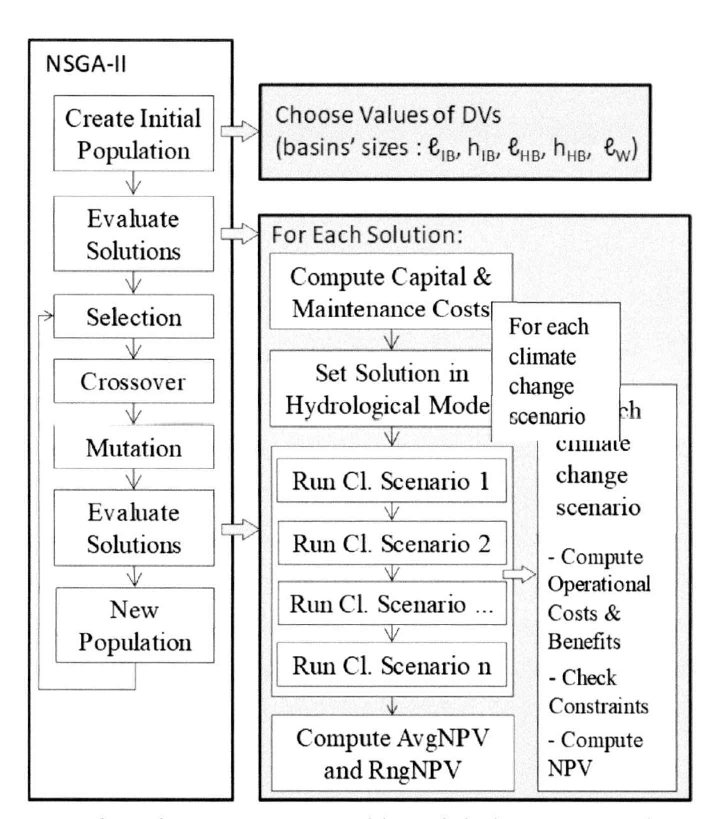


**FIGURE 15.6** Details of the stormwater harvesting system component sizing optimization process. *Marchi, A., Dandy, G.C., Maier, H.R., 2016. Integrated approach for optimizing the design of aquifer storage and recovery stormwater harvesting schemes accounting for externalities and climate change. Journal of Water Resources Planning and Management 142 (4), with permission from ASCE, with permission from ASCE.*

**TABLE 15.3 Details of Selected Solutions for the 2030, 2050, and 2013 Planning Horizons**

| | Sol. No. | *Avg NPV* ($m) | *Rng NPV* ($m) | $\ell_{IB}$ (m) | $h_{IB}$ (m) | $Vol_{IB}$ ($m^3$) | $\ell_{HB}$ (m) | $h_{HB}$ (m) | $Vol_{HB}$ ($m^3$) | $\ell_W$ (m) | $Vol_W$ ($m^3$) | No. Wells |
|---|---|---|---|---|---|---|---|---|---|---|---|---|
| 2030 Planning horizon | 1 | 3.64 | 1.22 | 80 | 3.0 | 25,536 | 80 | 4.0 | 37,205 | 70 | 15,841 | 4 |
| | 10 | 3.14 | 1.15 | 150 | 2.0 | 49,971 | 40 | 2.0 | 4651 | 70 | 15,841 | 4 |
| | 11 | 2.79 | 1.13 | 90 | 4.0 | 45,285 | 40 | 2.0 | 4651 | 70 | 15,841 | 4 |
| | 19 | 1.88 | 0.95 | 150 | 2.0 | 49,971 | 60 | 1.0 | 4101 | 70 | 15,841 | 4 |
| | 20 | 1.60 | 0.62 | 100 | 4.0 | 54,165 | 150 | 4.0 | 110,565 | 50 | 8321 | 2 |
| | 34 | 1.15 | 0.51 | 150 | 4.0 | 110,565 | 80 | 4.0 | 37,205 | 50 | 8321 | 2 |
| | 35 | 0.22 | 0.51 | 120 | 1.5 | 23,832 | 80 | 4.0 | 37,205 | 50 | 8321 | 2 |
| | 37 | 0.04 | 0.46 | 150 | 4.0 | 110,565 | 70 | 0.5 | 2593 | 50 | 8321 | 2 |
| 2050 Planning horizon | 38 | 3.51 | 2.65 | 100 | 2.5 | 30,333 | 130 | 2.0 | 38,131 | 70 | 15,841 | 4 |
| | 52 | 3.02 | 2.36 | 100 | 4.0 | 54,165 | 50 | 1.5 | 4722 | 70 | 15,841 | 4 |
| | 53 | 2.64 | 2.33 | 180 | 1.5 | 51,912 | 50 | 1.5 | 4722 | 70 | 15,841 | 4 |
| | 68 | 1.48 | 1.99 | 120 | 3.0 | 52,416 | 30 | 2.5 | 4083 | 70 | 15,841 | 4 |
| | 69 | 1.27 | 1.29 | 100 | 4.0 | 54,165 | 140 | 4.0 | 97,685 | 50 | 8321 | 2 |
| | 91 | 0.05 | 1.05 | 70 | 3.5 | 24,925 | 80 | 4.0 | 37,205 | 50 | 8321 | 2 |
| 2013 | – | 7.43 | 0.0 | 100 | 4.0 | 54,165 | 120 | 3.0 | 52,416 | 90 | 25,761 | 6 |

*h*, depth; *HB*, holding basin; *IB*, in-stream basin; *ℓ*, length; *Vol*, volume; *W*, wetland.
Marchi, A., Dandy, G.C., Maier, H.R., 2016. Integrated approach for optimizing the design of aquifer storage and recovery stormwater harvesting schemes accounting for externalities and climate change. Journal of Water Resources Planning and Management 142 (4), with permission from ASCE, With Permission From ASCE.

scenarios by increasing capital costs associated with storages and by reducing the size of the wetland, to have a more constant injection into the aquifer. The larger storage volumes were favored to harvest a sufficiently large volume to make the system viable.

Other notable results include:

- Savings in potable supply was by far the largest contributor to the benefits, being an order of magnitude greater than the benefits associated with lower salinity levels and environmental benefits in the River Murray. Capital cost was by far the largest contributor to costs (of the order of 4 times greater than the present value of maintenance and operational costs).
- Climate change had a significant impact on the best *NPV* that could be achieved. For 2013, the base case, the *NPV* of the optimal design was \$7.43 million, whereas the best average *NPV* for 2030 was \$3.64 million and for 2050 \$3.51 million. This was mainly due to reduced supply as a result of projected climate change, thereby reducing savings in potable supply significantly, which could not be offset by reduced capital, maintenance, and operational costs.
- There were significant differences in the optimal sizes of the basins and wetland that maximized *NPV* (or average *NPV*) for 2013, 2030, and 2050. For 2030 and 2050, the optimal sizes were generally smaller, corresponding to reduced rainfall and hence runoff. However, as mentioned above, the associated reductions in capital, maintenance, and operating costs were insufficient to counterbalance the lost benefits due to reduced water production.

## 15.3.2 Optimum selection and layout of best management practice WSUD components

### *15.3.2.1 Background*

This case study uses optimization to determine the size, type, and layout of the various distributed BMPs for a stormwater harvesting scheme, using a proposed greenfield residential development to be located in the north of Adelaide, South Australia (Fig. 15.7). The development is proposed to have 3342 medium density residential allotments with a catchment of 245 ha. Open space irrigation demand of between 61 and 122 ML/year is anticipated from the harvesting scheme, with the potential to supply an additional 40 ML/year to irrigate the grounds of a nearby school. There are four subcatchments

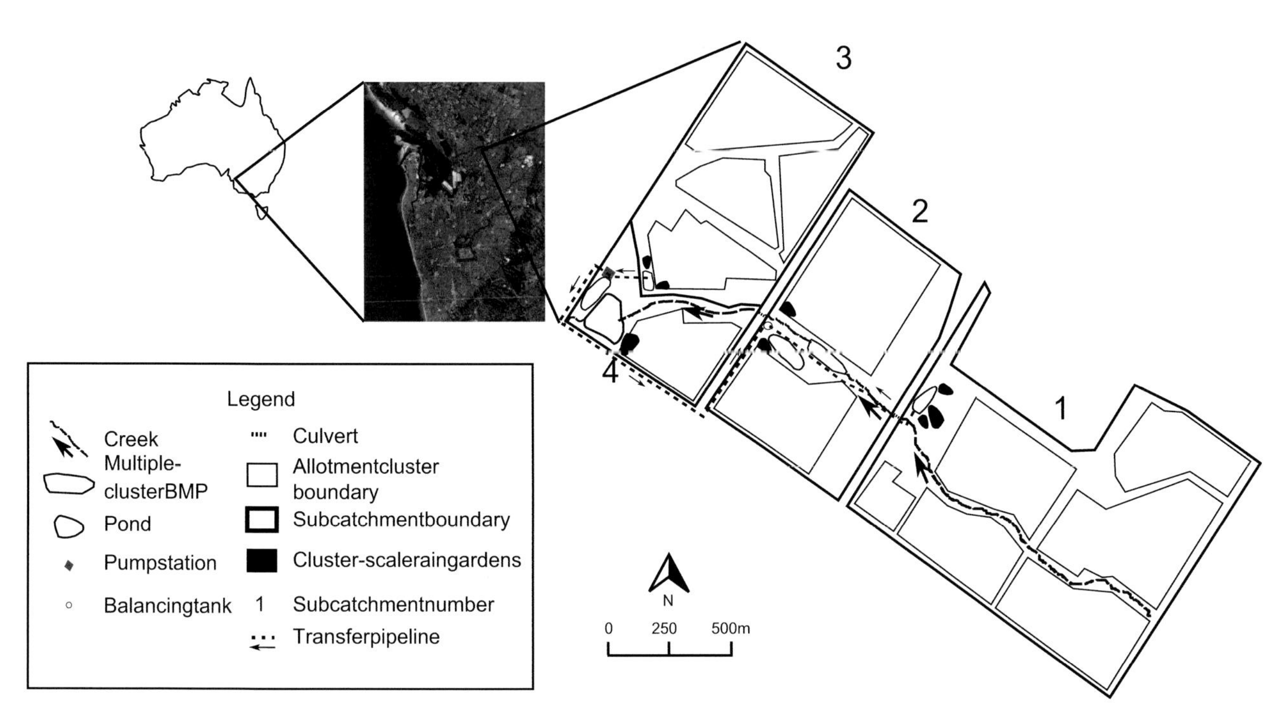


**FIGURE 15.7** Location of potential stormwater harvesting best management practices (BMPs) and transfer infrastructure for proposed housing development in Adelaide, South Australia. *Di Matteo, M., Dandy, G.C., Maier, H.R., 2017a. Multiobjective optimization of distributed stormwater harvesting systems. Journal of Water Resources Planning and Management. https://doi.org/10.1061/%28ASCE%29WR.1943-5452.0000756, with permission from ASCE.*

identified within the development. Subcatchments 1 and 2 (both with 50% impervious area) are delineated by major roads, and subcatchments 3 and 4 (with 60% and 55% impervious areas, respectively) are delineated by roads and a creek that conveys runoff downstream from subcatchments 1 and 2. Two large retarding basins on the creek, within subcatchments 2 and 4, can potentially have offline biofilters or wetlands to capture and treat runoff for harvesting, or otherwise act as sedimentation basins. Additionally, subcatchments 1, 2, and 3 have sites where runoff from clusters of allotments can be harvested using smaller biofilters. All harvested water must be stored in an open-water pond, at four possible sites, located adjacent to any biofilters/wetlands. Harvested water is transferred via gravity or pumped pipeline to a balancing tank located in subcatchment 2. The installment of BMP WSUD components in upstream subcatchments affects the runoff available to harvest in downstream subcatchments and runoff treatment performance of downstream BMPs. Details of the case study are available in Di Matteo et al. (2017a).

#### 15.3.2.2 Formulation of optimization problem

The objective of the study is to determine the size, type, and layout of BMP WSUD components within the harvesting system so as to maximize water quality improvement and harvested water volume, while minimizing system lifecycle costs. The costs considered include the capital costs of the stormwater BMPs, the pumps and the distribution system, the maintenance costs for the BMPs and the pumps, and the operational costs of the pumps. The water quality improvement is measured by *TSS* reduction, and harvested water is measured in average annual supply reliability.

To determine harvesting system costs, water quality improvement, and harvested water volume, three formal objectives are used, including lifecycle cost, volumetric reliability, and *TSS* reduction fraction, as follows:

$$\text{Min } LCC = LCC_{\text{harvest}} + LCC_{\text{transfer}} \tag{15.5}$$

where $LCC$ is the total lifecycle cost of a system, $LCC_{\text{harvest}}$ and $LCC_{\text{transfer}}$ are the capital and maintenance costs of the stormwater BMPs and transfer infrastructure, respectively.

$$\text{Max } R_V = \frac{\text{Volume of demand met}}{\text{Total demand volume}} \tag{15.6}$$

where $R_V$ is the volumetric reliability (i.e., the fraction of irrigation demand met per year, averaged over the simulation period).

$$\text{Max } TSSLoadRedn = 1 - \frac{TSS \text{ mass reaching catchment outlet}}{TSS \text{ mass generated by development}} \tag{15.7}$$

where *TSSLoadRedn* is the fraction of mass of *TSS* generated by the new development that is retained in BMPs within the development site per year, averaged over the simulation period.

The potential location of BMPs and ponds within a WSUD layout, the design parameters and their values to include as decision variables, are defined by the modeller. In this case study, there are 11 decision variables: three for the surface areas of small biofilters in subcatchments 1, 2, and 3; four for the type (biofilter, wetland, or sediment basin) and surface area (SA) of the BMPs located within the two retarding basins; and four for the surface areas of storage ponds in each catchment. The SA options were 0% (no BMP or pond), 33.3%, 66.6%, or 100% of the maximum area of a BMP or pond at a location. Maximum surface areas for various BMPs at locations in the catchment range from 0.35 to 2.2 ha, being constrained by the smaller of the amount of space available or 1.5% of the contributing impervious area for biofilters. The layout of BMPs determines the transfer infrastructure required.

The constraints include that the systems are required to meet water quality indicator targets including *TSS*, TN, and TP reductions of 80%, 45% and 45%, respectively. The water quality performance of a candidate WSUD layout is evaluated, after simulation in a MUSIC model, using Eq. (15.7) for *TSS* (as for the objective function), and using identical equations but with TN and TP, respectively, substituted in place of *TSS*. In addition, ponds are required to be adjacent to BMPs.

Three demand scenarios include: low irrigation demand (61.0 ML/year); high irrigation demand (for high amenity open space 122.0 ML/year); and high irrigation demand plus 40.0 ML/year export to a neighboring school for non-potable use (162.0 ML/year). Demand is requested from each storage pond within a system. Demand is distributed over ponds in the WSUD layout based on a simple algorithm, where, the irrigation demand within a subcatchment is supplied by a pond within that catchment, or else by the next downstream pond or upstream pond if no downstream ponds exist.

### *15.3.2.3 Optimization process*

An outline of the optimization process is given in Fig. 15.8. As can be seen, values of the decision variables are generated using the optimization algorithm, resulting in a particular stormwater harvesting system design. The lifecycle cost, volumetric reliability, and water quality improvement of this design are then calculated for a single demand scenario. A cost lookup table is used to evaluate Eq. (15.5). Hydrological and water quality performance are assessed with a simulation model of the system developed in MUSIC (version 6.1; eWater, 2009). The results of a MUSIC simulation of the WSUD layout are used to evaluate Eqs. (15.6) and (15.7), and the water quality performance constraints. As in the previous case study in this chapter, NSGA-II (Deb et al., 2002) was selected as the multiobjective optimization engine. MUSIC simulations used 10 years of hourly rainfall data. To save computational run time, candidate solutions were preemptively checked by the algorithm to ensure ponds were adjacent to BMPs in the layout, and if they were not, the system was not simulated in MUSIC. Optimization runs were carried out for the 61, 122, and 162 ML/year demand scenarios.

During the optimization process, the type of BMPs and size of BMPs and ponds (the decision variable values) are selected by the optimization algorithm (NSGA-II in Fig. 15.8) for a candidate WSUD layout. This process is repeated to generate a population of candidate WSUD layouts (solutions) by the algorithm. These solutions are then evaluated using the process in the right hand side of Fig. 15.8. By default, a junction is placed in the MUSIC model at locations where no BMP or pond (0% surface area) is selected. Once the initial population is evaluated, EA operators (selection, crossover, and mutation) are applied to generate a new population of WSUD layouts, which are subsequently evaluated. This step is repeated until the convergence criteria (e.g., maximum number of iterations) are achieved. The result of the optimization process is a set of Pareto optimal WSUD layouts, which need to be further analyzed.

### *15.3.2.4 Results and discussion*

The outputs from the optimization included stormwater harvesting system layouts that provided optimal trade-offs between lifecycle cost, *TSS* reduction, and volumetric reliability of supply for each demand scenario. Note that TN and TP reduction were considered as performance constraints rather than additional objectives. The optimization algorithm identified 77 Pareto optimal layouts over the three scenarios.

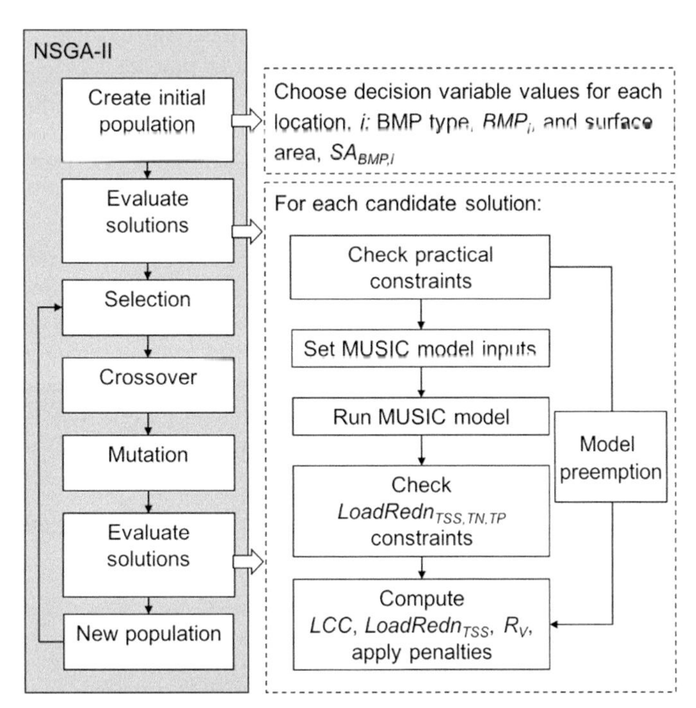


**FIGURE 15.8** Details of stormwater harvesting system layout case study optimization process. *Di Matteo, M., Dandy, G.C., Maier, H.R, 2017a. Multiobjective optimization of distributed stormwater harvesting systems. Journal of Water Resources Planning and Management. https://doi.org/10.1061/%28ASCE%29WR.1943-5452.0000756, with permission from ASCE.*

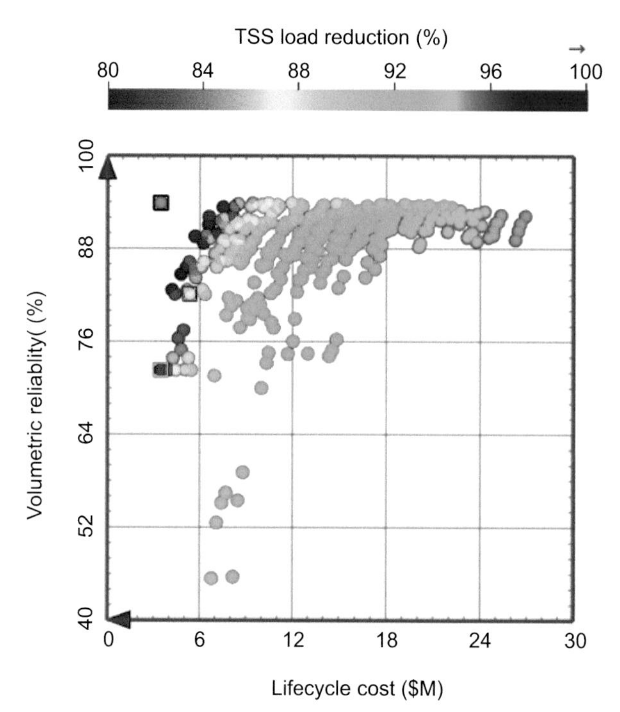


**FIGURE 15.9** Multiobjective performance trade-offs of Pareto optimal SWH systems. TSS, total suspended solids

Visualization of the trade-offs for the 61 ML/year demand scenario is shown in Fig. 15.9. As can be seen, if decision makers with a particular budget accept a slightly lower harvested water supply reliability, solutions with significantly higher *TSS* reduction are available as alternatives to solutions optimal with respect to cost and reliability only.

The optimal layouts generated by the optimization algorithm that distribute BMPs throughout the catchment perform better in terms of *TSS* reduction and volumetric reliability when compared with layouts with BMPs located only at the catchment outlet (subcatchment 4), which is a typical layout adopted in practice. As such, optimization results could support negotiation between decision makers for a distributed stormwater harvesting approach.

Optimization results can be used to explore the impact of the design decisions. Table 15.4 shows the values of the BMP type and BMP and pond SA decision variables of selected Pareto optimal WSUD layouts.

Distributed biofilters were available at cluster scale locations. The numbers (e.g., BMP 2, Pond 4) refer to the location of the subcatchment of the BMP. Based on MUSIC modeling, biofilters (B) were preferred over wetlands (W) in locations with high in-flows (i.e., "central" and catchment "outlet" locations). This indicates that, for the case study, systems with treatment occurring at locations with higher hydraulic loading rates performed better against multiple objectives, up to a point where the inflows exceeded the capacity of the treatment systems. Biofilters may have been preferred as they were assumed to have lower cost per unit pollutant reduction and volume yielded compared with wetlands. To supply the higher demand of Scenario 3, of the smaller biofilters, those strategically placed in a subcatchment with the highest impervious fraction (subcatchment 3), which otherwise directly contributes to the catchment outlet BMP (sub-catchment 4), provides the best return on investment for improvement in reliability and *TSS* reduction.

## 15.3.3 WSUD project portfolio optimization considering multiple stakeholders' objectives using visual analytics

### *15.3.3.1 Background*

This case study uses optimization and visual analytics to select portfolios of stormwater management projects that distribute costs and benefits equitably among multiple stakeholders within a catchment management region in a major Australian city (Fig. 15.10). A catchment management authority (CMA) identified sites for potential stormwater BMPs within a regional catchment to reduce the nutrient load from urban stormwater runoff flowing into a prominent marine body. The catchment covers an area comprising highly urbanized and peri-urban regions managed by three local councils. Because the potential sites for BMPs were within public open spaces managed by the councils, stormwater harvesting for

**TABLE 15.4 Decision Variable Values of Selected Solutions**

| Demand Scenario | Solution# | Biofilter Cluster 1 (ha) | Pond Cluster 1 (ha) | Biofilter Cluster 2 (ha) | *BMP* 2 (Central) | *SA* 2 (Central) (ha) | Pond 2 (Central) (ha) | Biofilter Cluster 3 (ha) | Pond Cluster 3 (ha) | *BMP* 4 (Outlet) | *SA* 4 (Outlet) (ha) | Pond 4 (Outlet) (ha) |
|---|---|---|---|---|---|---|---|---|---|---|---|---|
| 1 | 1 | – | – | – | B | 0.730 | 0.333 | – | – | B | 0.243 | – |
| | 3 | – | – | – | B | 0.486 | 0.666 | – | – | B | 0.243 | – |
| | 4 | – | – | – | B | 0.730 | 0.333 | – | – | B | 0.243 | 0.333 |
| | 7 | – | – | – | B | 0.730 | 0.666 | – | – | B | 0.243 | 0.333 |
| | 8 | – | – | – | B | 0.486 | 0.666 | – | – | B | 0.486 | 0.333 |
| | 16 | – | – | – | B | 0.730 | 1.00 | – | – | B | 0.730 | 1.00 |
| 2 | 17 | – | – | – | B | 0.243 | – | –– | – | B | 0.486 | 0.333 |
| | 19 | – | – | – | B | 0.243 | – | – | - | B | 0.486 | 0.666 |
| | 20 | – | – | – | B | 0.730 | 0.666 | – | – | B | 0.243 | 0.333 |
| | 27 | – | – | – | B | 0.730 | 1.00 | – | – | B | 0.486 | 1.00 |
| | 28 | – | – | 0.183 | B | 0.486 | 1.00 | – | – | B | 0.486 | 1.00 |
| | 29 | 0.306 | 0.5 | – | B | 0.243 | 1.00 | – | – | B | 0.243 | 1.00 |
| | 40 | 0.613 | 0.5 | 0.183 | B | 0.730 | 1.00 | – | – | B | 0.486 | 1.00 |
| | 41 | 0.613 | 0.5 | 0.183 | W | 1.00 | 1.00 | – | – | B | 0.486 | 1.00 |
| | 42 | 0.613 | 0.5 | 0.366 | W | 1.00 | 1.00 | – | – | B | 0.486 | 1.00 |
| 3 | 43 | – | – | – | B | 0.243 | – | – | – | B | 0.486 | 0.333 |
| | 44 | – | – | – | B | 0.730 | 0.333 | – | – | B | 0.243 | 0.333 |
| | 45 | – | – | – | B | 0.486 | 0.666 | – | – | B | 0.243 | 0.333 |
| | 47 | – | – | – | B | 0.486 | 1.00 | – | – | B | 0.243 | 0.333 |
| | 58 | – | – | – | B | 0.730 | 1.00 | 0.167 | 0.233 | B | 0.243 | 1.00 |
| | 63 | – | – | – | B | 0.730 | 1.00 | 0.167 | 0.350 | B | 0.730 | 1.00 |
| | 64 | 0.306 | 0.5 | 0.183 | B | 0.730 | 1.00 | – | – | B | 0.243 | 0.666 |
| | 77 | 0.613 | 0.5 | 0.183 | B | 0.730 | 1.00 | 0.333 | 0.350 | B | 0.243 | 1.00 |

Adapted from Di Matteo, M., Dandy, G.C., Maier, H.R, 2017a. Multiobjective optimization of distributed stormwater harvesting systems. Journal of Water Resources Planning and Management. https://doi.org/10.1061/%28ASCE%29WR.1943-5452.0000756. with permission from ASCE.

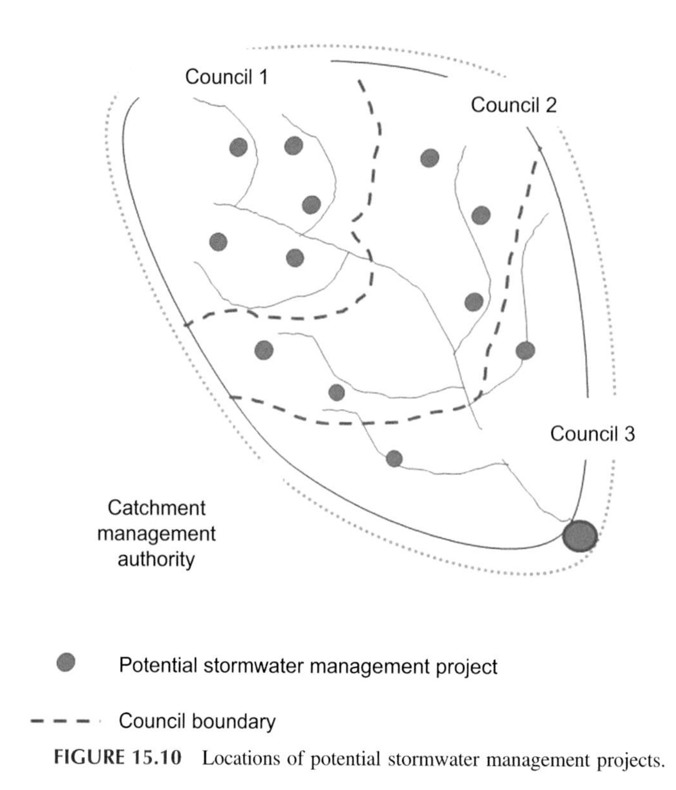


FIGURE 15.10 Locations of potential stormwater management projects.

irrigation of open spaces, increasing vegetation and public amenity value were considered to be important additional benefits. Seventy potential biofiltration, wetlands and swale projects were considered. These were distributed in open spaces throughout the three local council areas. It was desirable to identify a portfolio of 20 projects or fewer to take to conceptual design for possible funding. Details of the case study are available in Di Matteo et al. (2017b).

### 15.3.3.2 Multistakeholder optimization problem formulation

The objective of the study is to identify the portfolios of WSUD projects that are Pareto optimal (i.e., represent best trade-offs between costs and benefits) with respect to all stakeholders' individual values. This requires multiple optimization problem formulations to be solved. Each problem formulation consists of the projects, objectives, and constraints relevant to one stakeholder. Solving these optimization problems individually, with a suitable algorithm, produces multiple Pareto optimal sets of solutions.

Once the multiple optimization problems are solved, the multiple Pareto optimal sets are aggregated into one set. Then, the optimal portfolios that are suboptimal (or "dominated") by the Pareto set of one or more stakeholders are removed. This leaves only the jointly Pareto optimal portfolios for further consideration. Considering only the jointly optimal solutions assists with reaching consensus on a final equitable portfolio. This is because stakeholders do not have to accept a solution that is dominated in their preferred value space, nor do they have to explore and analyze results of a single problem formulation with aggregated or agreed on objectives that do not necessarily reflect their values.

A summary of the decision variables, the objectives and constraints of the four problem formulations (one for each stakeholder) are shown in Table 15.5. Binary (0−1) decision options for a number of decision variables, equal to the number of projects within the region governed by a stakeholder, represent the decisions on whether or not to fund each project. The 16 objectives in Table 15.5 are sorted in order of relative importance to individual stakeholders. The costs of projects considered include the capital expenses to be funded by the CMA and operating expenses to be funded by the local councils for projects within their region. Each stakeholder values water quality improvement (measured by TN reduction), stormwater harvesting capacity (measured in average annual volume harvested), and amenity (measured as a "green" score) of projects located within their region. As such, a portfolio containing projects only in one council region might provide benefits to that council and the CMA, but it would not be valuable to the other councils. The only constraints considered were the maximum number of projects allowed in a portfolio.

**TABLE 15.5 Decision Variable Values of Selected Solutions**

| Problem Formulation | Decision Variables | Objectives | Constraints |
|---|---|---|---|
| CMA | All projects | $CAPEX_{CMA}$<br>$TNRed_{CMA}$<br>$SWH_{CMA}$<br>$GREEN_{CMA}$ | ≤20 projects |
| Council 1 | Council 1 projects | $OPEX_1$<br>$SWH_1$<br>$GREEN_1$<br>$TNRed_1$ | ≤7 projects |
| Council 2 | Council 2 projects | $OPEX_2$<br>$GREEN_2$<br>$SWH_2$<br>$TNRed_2$ | ≤7 projects |
| Council 3 | Council 3 projects | $OPEX_3$<br>$SWH_3$<br>$GREEN_3$<br>$TNRed_3$ | ≤7 projects |

*CAPEX*, capital expenditure; *GREEN*, amenity "green" score; *OPEX*, operating expenditure; *SWH*, average annual volume of stormwater harvested; *TNRed*, total nitrogen reduction.
Adapted from Di Matteo, M., Dandy, G.C., Maier, H.R., 2017b. A multi-stakeholder portfolio optimization framework applied to stormwater best management practice (BMP) selection. Environmental Modelling and Software 97, 16–31, with permission from Elsevier.

### *15.3.3.3 Optimization process*

An outline of the optimization process used to identify the Pareto optimal solutions for each stakeholder is given in Fig. 15.11. In the first step, potential stormwater BMPs projects are identified in consultation with stakeholders. In the second step, a multicriteria analysis is undertaken, scoring each project against a number of criteria for each stakeholder. In the third step, an optimization algorithm is used to identify the Pareto optimal portfolios of projects, which represent the best trade-offs between the objectives. In the fourth step, the Pareto optimal set of solutions is visualized in a visual analytics package, and a stakeholder can "shop" through solutions to identify portfolios of interest.

The multistakeholder optimization framework used to identify the Pareto optimal solutions that distributed costs and benefits equitably between stakeholders is given in Fig. 15.12. In the first step, stakeholders select a portfolio that represents the "Best Alternative to a Negotiated Agreement" (BATNA) where they must individually fund all lifecycle costs of projects. In the second step, the approach in Fig. 15.11, described above, is used to identify the Pareto optimal portfolios for each stakeholder. In step three, analysts remove any portfolios that are dominated in one or more stakeholder objective space, leaving only portfolios that are Pareto optimal for all stakeholders, referred to as the "Joint-Pareto optimal set," or "Golden set."

In step four, the Golden set of Pareto optimal solutions is visualized in a multistakeholder trade-off space using a visual analytics package (Keim et al., 2008). Visual analytics software packages allow analysts and WSUD decision makers to engage in a process that involves information gathering, data preprocessing, knowledge representation, interaction, and decision-making based on model results. The BATNA for each stakeholder is plotted in their respective trade-off space and is used as a benchmark against which optimal solutions in the Golden set, where stakeholders share the capital and operational expenses, are compared (Di Matteo et al., 2018).

A variant of the Pareto Ant Colony Optimization Algorithm (Doerner et al., 2004) was used in this study. This algorithm identified 2537 portfolios containing optimal subportfolios that were solutions of the four individual problem formulations. The Golden set solutions represented 125 of the optimal portfolios that consisted of an optimal subportfolio from every stakeholder. The DiscoveryDV visual analytics package was used to explore and analyze and select solutions form the Golden set.

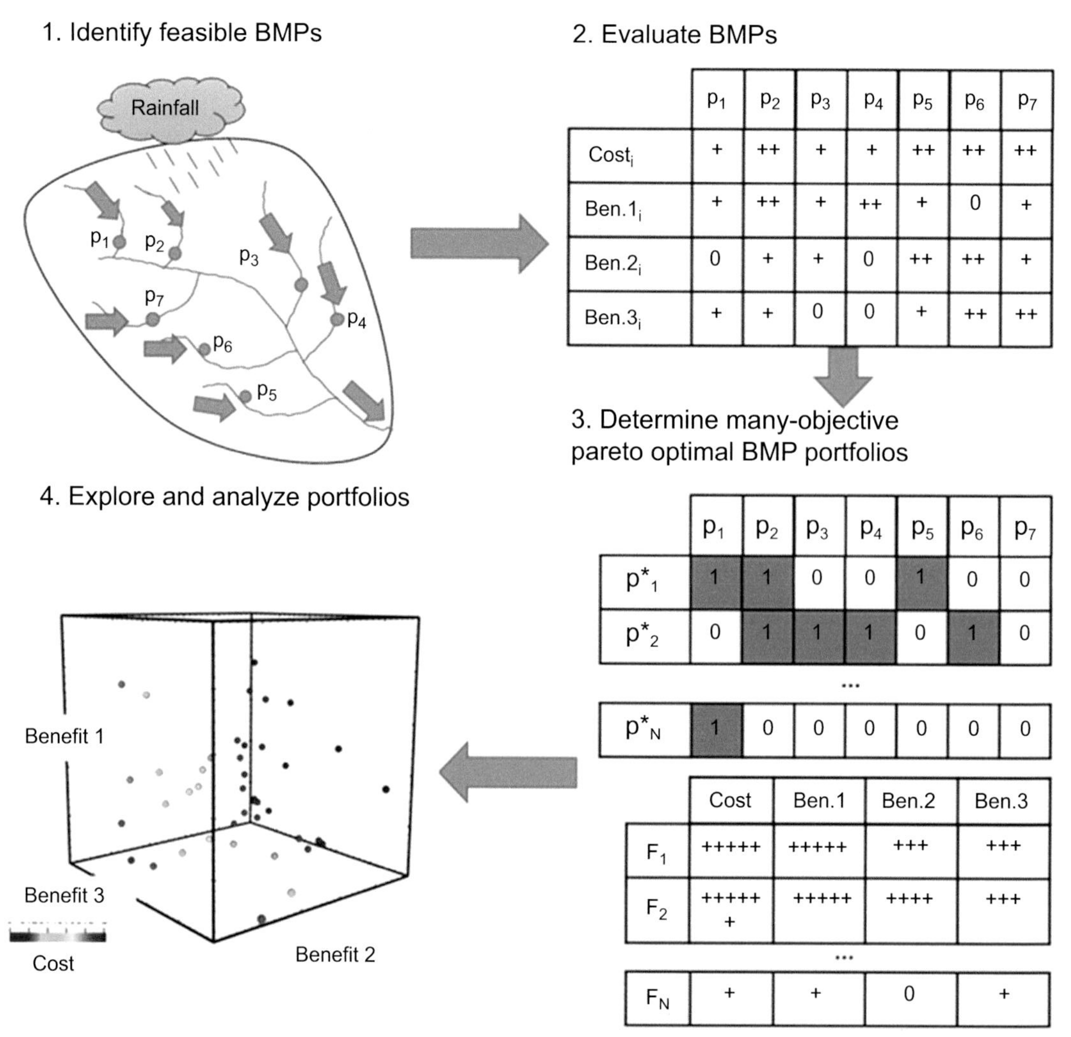


| | p1 | p2 | p3 | p4 | p5 | p6 | p7 |
|---|---|---|---|---|---|---|---|
| Cost$_i$ | + | ++ | + | + | ++ | ++ | ++ |
| Ben.1$_i$ | + | ++ | + | ++ | + | 0 | + |
| Ben.2$_i$ | 0 | + | + | 0 | ++ | ++ | + |
| Ben.3$_i$ | + | + | 0 | 0 | + | ++ | ++ |

| | p1 | p2 | p3 | p4 | p5 | p6 | p7 |
|---|---|---|---|---|---|---|---|
| p*$_1$ | 1 | 1 | 0 | 0 | 1 | 0 | 0 |
| p*$_2$ | 0 | 1 | 1 | 1 | 0 | 1 | 0 |
| ... | | | | | | | |
| p*$_N$ | 1 | 0 | 0 | 0 | 0 | 0 | 0 |

| | Cost | Ben.1 | Ben.2 | Ben.3 |
|---|---|---|---|---|
| F$_1$ | +++++ | +++++ | +++ | +++ |
| F$_2$ | ++++++ | +++++ | ++++ | +++ |
| ... | | | | |
| F$_N$ | + | + | 0 | + |

**FIGURE 15.11** Conceptual outline of a multiobjective optimization approach for stormwater management best management practice (BMP) selection. p is a project, p* is a Pareto optimal portfolio of projects, and F is a set of objective function values of a Pareto optimal solution.

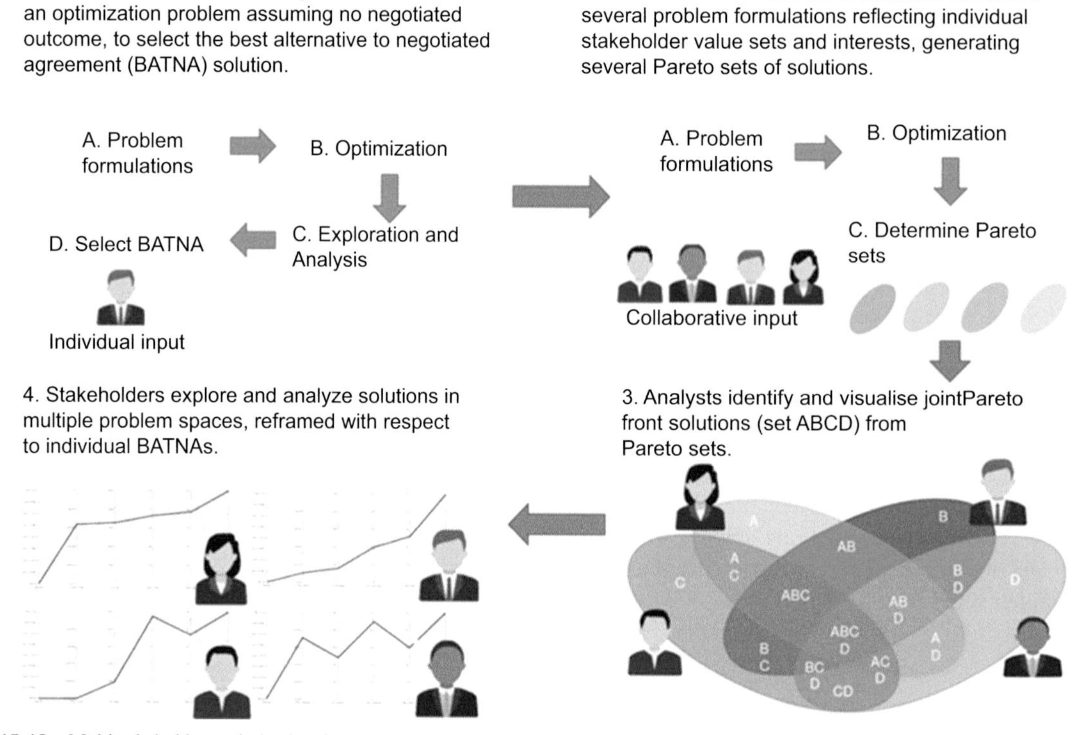


**FIGURE 15.12** Multistakeholder optimization framework incorporating multiple problem formulations to encourage a negotiated outcome. *From Di Matteo, M., Dandy, G.C., Maier, H.R., 2017b. A multi-stakeholder portfolio optimization framework applied to stormwater best management practice (BMP) selection, Environmental Modelling and Software 97, 16–31, with permission from Elsevier.*

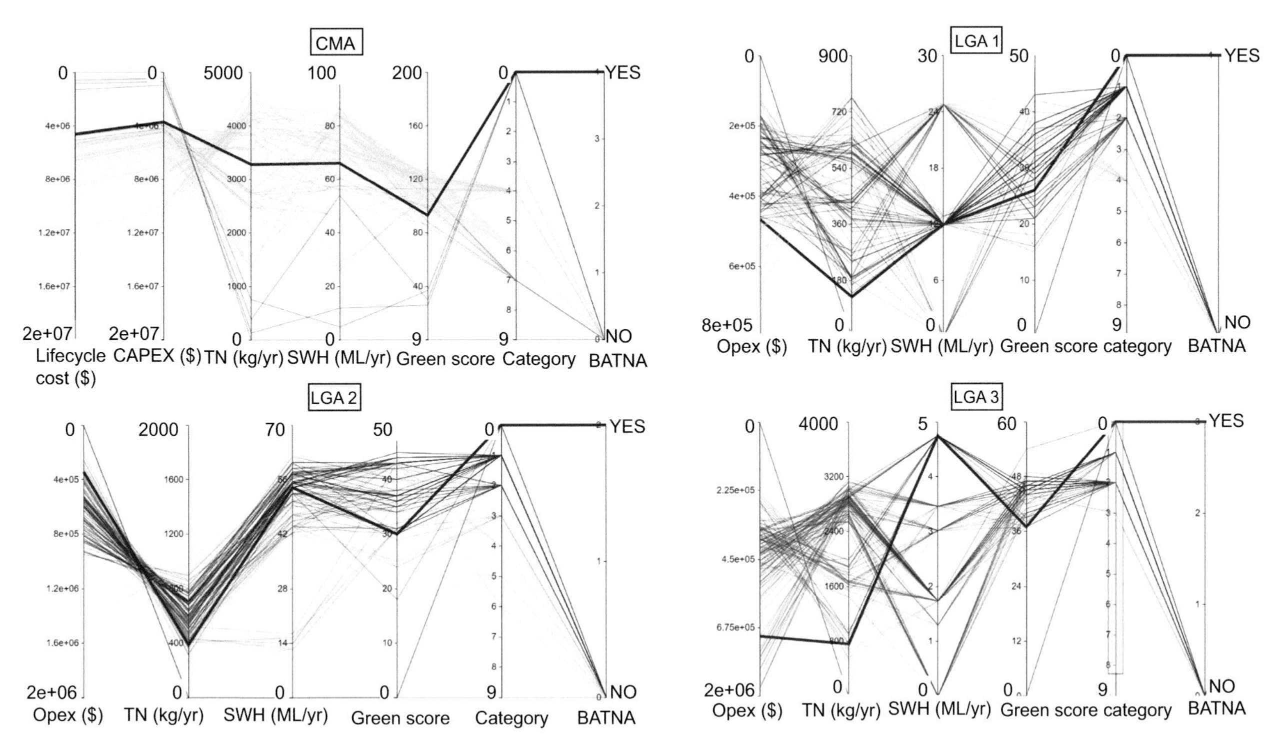


**FIGURE 15.13** Many-objective, multistakeholder trade-off space plot for the catchment management authority (CMA) and three local government authorities (LGAs). *From From Di Matteo, M., Dandy, G.C., Maier, H.R., 2017b. A multi-stakeholder portfolio optimization framework applied to stormwater best management practice (BMP) selection, Environmental Modelling and Software 97, 16–31, with permission from Elsevier.*

### *15.3.3.4 Results and discussion*

The Golden set (joint-Pareto optimal) portfolios were presented in multistakeholder trade-off space, as shown in Fig. 15.13, using the visual analytics software package. Each parallel coordinate plot shows the relative performance of the Golden set portfolios against an individual stakeholder's objectives (listed in Table 15.5). In this figure, values that are toward the top of each scale are better than values that are lower on corresponding scale. The BATNA portfolio for each stakeholder is shown as a dark bold line segment. The Golden set portfolios were categorized based on their performance against the BATNA for each stakeholder. Dark blue thin line segments (Category 1) represent portfolios that improve on all objectives at lower cost than a stakeholder's BATNA, whereas yellow to red (Category 2–8) indicate some compromise in benefits or cost compared to the BATNA. As can be seen, the local councils benefitted from cost sharing with the CMA to fund projects, as they have many dark blue (Category 1) portfolios. However, the CMA would need to compromise from their BATNA in order to reach a collaborative solution.

By searching through the Golden set portfolios, using the visual analytics package, portfolios of projects that may be preferred by each stakeholder were selected by analysts (Table 15.6). The selected portfolios were analyzed based on their performance in each stakeholder's value set of objectives, to identify viable portfolios that can be adopted or used as a benchmark for further negotiation between stakeholders. The three portfolios, "CMA low $," "CMA High Benefit," and "LGA 2 Low $," were identified as being viable for further consideration.

## 15.4 FUTURE DEVELOPMENTS

Optimization approaches generate a large number of best-performing WSUD solutions that provide decision makers with valuable insights into the problem at hand. This may assist with collaborative decision-making. For example, optimization results for sizing and selection of a distributed treatment train can provide data on which locations in a catchment and what types of BMPs provide the best returns on investment. This information can assist in negotiating for a distributed management approach (Di Matteo et al., 2017a). Where multiple stakeholders are involved in the WSUD planning process, optimization approaches can involve them in the problem formulation, selection of simulation model, and solution analysis to encourage buy-in into the final results (Wu et al., 2016; Di Matteo et al., 2017b). Design collaboration through model

**TABLE 15.6** Individual Stakeholder Selected Solutions

| | Selected Solution | | | | | | | |
|---|---|---|---|---|---|---|---|---|
| Solution Characteristic | CMA Low $ | CMA High Ben. | LGA 1 Low $ | LGA 1 High Ben. | LGA 2 Low $ | LGA 2 High Ben. | LGA 3 Low $ | LGA 3 High Ben. |
| **CMA Data** | | | | | | | | |
| $CAPEX_{CMA}$ | 3650 | 5710 | 4130 | 7450 | 5880 | 10,200 | 2970 | 6510 |
| $TN_{CMA}$ | 3802 | 3581 | 1357 | 3967 | 1753 | 4295 | 1345 | 4150 |
| $SWH_{CMA}$ | 51.17 | 83.22 | 69.1 | 76.93 | 84.45 | 78.16 | 29.12 | 32.12 |
| $GRN_{CMA}$ | 86 | 118 | 109 | 114 | 111 | 113 | 90 | 115 |
| BATNA Category | 3 | 4 | 6 | 4 | 6 | 4 | 5 | 6 |
| **LGA 1 Data** | | | | | | | | |
| $OPEX_1$ | 118 | 405 | 193 | 517 | 357 | 565 | 177 | 406 |
| $TN_1$ | 509.79 | 190.68 | 187.81 | 500.1 | 162.41 | 550.44 | 166.15 | 765.57 |
| $SWH_1$ | - | 24.78 | 11.95 | 24.78 | 24.78 | 24.78 | 11.95 | 11.95 |
| $GRN_1$ | 17 | 29 | 33 | 43 | 21 | 28 | 26 | 35 |
| $CAPEX_1$ | 639 | 1530 | 799 | 2120 | 1270 | 2350 | 641 | 1900 |
| BATNA Category | 3 | 1 | 1 | 1 | 2 | 4 | 1 | 1 |
| **LGA 2 Data** | | | | | | | | |
| $OPEX_2$ | 358 | 589 | 626 | 735 | 703 | 1160 | 248 | 529 |
| $TN_2$ | 384.58 | 605.27 | 685.65 | 706.57 | 681.71 | 793.51 | 270.37 | 376.57 |
| $SWH_2$ | 46.42 | 55.42 | 55.42 | 50.42 | 54.92 | 49.92 | 12.42 | 15.42 |
| $GRN_2$ | 25 | 43 | 44 | 31 | 45 | 39 | 22 | 34 |
| $CAPEX_2$ | 1210 | 2230 | 2430 | 3100 | 2850 | 5360 | 1030 | 2430 |
| BATNA category | 5 | 1 | 1 | 2 | 1 | 2 | 5 | 3 |

| **LGA 3 data** | | | | | | | | |
|---|---|---|---|---|---|---|---|---|
| $OPEX_3$ | 351 | 372 | 166 | 428 | 351 | 482 | 261 | 432 |
| $TN_3$ | 2907.4 | 2784.8 | 483.88 | 2760.5 | 908.93 | 2951.1 | 908.33 | 3007.9 |
| $SWH_3$ | 4.75 | 3.02 | 1.73 | 1.73 | 4.75 | 3.46 | 4.75 | 4.75 |
| $GRN_3$ | 44 | 46 | 32 | 40 | 45 | 46 | 42 | 46 |
| $CAPEX_3$ | 1800 | 1960 | 901 | 2230 | 1760 | 2520 | 1300 | 2180 |
| BATNA category | 1 | 2 | 5 | 2 | 1 | 2 | 1 | 1 |
| **Miscellaneous** | | | | | | | | |
| Directly connected impervious area (ha) | 680 | 649 | 243 | 711 | 316 | 769 | 241 | 737 |
| No. projects | 14 | 19 | 18 | 18 | 18 | 18 | 15 | 19 |
| LifeCycle cost | 4480 | 7080 | 5120 | 9130 | 7290 | 12,400 | 3660 | 7880 |

*CAPEX*, present value of capital expenses (2016$ '000); *CMA*, catchment management authority; *GRN*, green score (no units); *Hi. Ben.*, high benefits; *LGA*, local government authority; *LifeCycle Cost*, present value of lifecycle cost (2016$ '000).*Low $*, low costs; *OPEX*, present value of operating expenses (2016$ '000); *SWH*, average annual volume of stormwater harvested (ML/year); *TN*, total nitrogen reduction (kg/year).

Adapted from Di Matteo, M., Dandy, G.C., Maier, H.R., 2017b. A multi-stakeholder portfolio optimization framework applied to stormwater best management practice (BMP) selection. Environmental Modelling and Software 97, 16–31, with permission from Elsevier.

sharing, interactive problem formulation, and solution analysis can help to overcome decision-making biases and assist with arriving at consensus solutions. In addition, where traditional and trusted multicriteria analysis methods are used for BMP project selection, but which consider only a small number of project combination alternatives, portfolio optimization approaches can be applied to enhance these analyses to determine the globally optimal combinations of projects, ensuring the full cost and benefit trade-offs are available to assist the decision maker (Di Matteo et al., 2018).

WSUD systems typically perform multiple functions and can provide multiple benefits. Adding objectives to the optimization formulations can provide decision makers with even more insight into trade-offs between these multiple benefits, to identify optimal systems that balance stakeholder value preferences for the benefits achieved and costs required. However, the number of solutions that represent the Pareto front increases exponentially with the number of objectives, making solutions representing optimal trade-offs more difficult to identify, explore, and analyze. Therefore, metaheuristics that have been demonstrated to work on problems with large numbers of objectives can be used to identify Pareto optimal solutions (e.g., BORG; Hadka and Reed, 2012).

Future uncertainty has been considered in some WSUD optimization studies, such as the case study presented in Section 15.3.1 of this chapter (Marchi et al., 2016). However, there has been a lack of formal consideration of deep uncertainty and the formal assessment of the robustness of different WSUD designs in the face of a number of plausible future conditions. Such future conditions can be represented by scenarios associated with changes in climate, population, the economy, and/or technology using various metrics (Maier et al., 2016). The robustness of WSUD systems can be calculated postoptimization, as has been done in a number of integrated water resources studies (e.g., Paton et al., 2014; Beh et al., 2015). Alternatively, system robustness could be incorporated as a formal optimization objective, although this would most likely require the computational efficiency of the simulation model to be increased, using approaches such as metamodeling (Beh et al., 2017).

## 15.5 SUMMARY AND CONCLUSIONS

This chapter presents a general outline and case study applications of optimization approaches for the planning and design of WSUD schemes. It also contains a summary of selected WSUD scheme simulation models and optimization studies of WSUD schemes.

Optimization is suitable for the planning and design of complex WSUD schemes because these schemes may have multiple: (1) purposes and objectives; (2) components and design decisions for each component; and (3) spatial scales, which need to be considered. These complexities can be handled using an appropriate optimization approach, which is used to maximize or minimize mathematical formulations of one or more objectives (e.g., lifecycle cost, reliability, or volumetric retention) by evaluating and identifying the optimal combinations of decision variables (e.g., WSUD component sizes, types, and layout) that meet a set of constraints (e.g., water quality improvement targets).

Optimization approaches can include: (1) traditional manual trial-and-error using a simulation model; (2) optimization models that incorporate WSUD system simulation; and (3) metaheuristic search optimization algorithms that can be linked to user-supplied simulation models. Importantly, where the optimization method can be automated, large numbers of potential solutions can be evaluated, enabling the trade-offs between a number of objectives to be explored. This is not the case with traditional multicriteria evaluation methods, typically used in practice, where a smaller number of design alternatives is generated.

Three case studies were considered, using genetic algorithm variates and an ant colony optimization variant to optimize components of stormwater harvesting systems and to assist in the selection of WSUD projects to be funded. Through optimization, solutions that represented desirable trade-offs were selected for further consideration. For example, the benefits of stormwater harvesting systems could be quantified under multiple climate scenarios, or diminishing returns between cost and water quality improvement and stormwater harvesting reliability could be shown. This allows individual solutions with a desirable balance in performance metrics to be selected. In addition, relationships between design decisions and system performance were analyzed to encourage decision maker learning about the system. For example, decisions on the best dimensions of a wetland or basin, and locations of BMPs with in a catchment, could be supported by optimization results. Where multiple stakeholder interests are involved, the joint-Pareto optimal or "Golden set" of solutions can be identified by solving several problem formulations, each representing the interests of an individual stakeholder. The Pareto optimal solutions can then be explored, analyzed, and selected with the aid of a visual analytics package.

Future applications for optimization of WSUD schemes include (1) opportunities for model-based collaboration and decision-making to take place, (2) the identification, exploration, and analysis of trade-offs between large numbers of objectives and potential WSUD solutions using advanced algorithms and visual analytics techniques, and (3) comprehensive robustness testing of proposed designs under multiple future scenarios.

## REFERENCES

AGNPS, 2001. Agricultural Nonpoint Source Model. US Department of Agriculture. https://www.nrcs.usda.gov/wps/portal/nrcs/detailfull/null/?cid=stelprdb1042468.

Bach, P.M., McCarthy, D.T., Urich, C., Sitzenfrei, R., Kleidorfer, M., Rauch, W., Deletic, A., 2013. A planning algorithm for quantifying decentralised water management opportunities in urban environments. Water Science and Technology 68 (8), 1857–1865.

Bach, P.M., Rauch, W., Mikkelsen, P.S., McCarthy, D.T., Deletic, A., 2014. A critical review of integrated urban water modelling–urban drainage and beyond. Environmental Modelling and Software 54, 88–107.

Bach, P.M., 2012. UrbanBEATS e an Exploratory Model for Strategic Planning of Urban Water Infrastructure (Online). Melbourne. Available: www.urbanbeatsmodel.com.

Beh, E.H.Y., Dandy, G.C., Maier, H.R., Paton, F.L., 2014. Optimal sequencing of water supply options at the regional scale incorporating alternative water supply sources and multiple objectives. Environmental Modelling and Software 53, 137–153.

Beh, E.H.Y., Maier, H.R., Dandy, G.C., 2015. Adaptive, multi-objective optimal sequencing approach for urban water supply augmentation under deep uncertainty. Water Resources Research 51 (3), 1529–1551. https://doi.org/10.1002/2014WR016254.

Beh, E.H.Y., Zheng, F., Dandy, G.C., Maier, H.R., Kapelan, Z., 2017. Robust optimization of water infrastructure planning under deep uncertainty using metamodels. Environmental Modelling and Software 93, 92–105. https://doi.org/10.1016/j.envsoft.2017.03.013.

Bicknell, B.R., Imhoff, J.C., Kittle Jr., J.L., Jobes, T.H., Donigian Jr., A.S., 2001. Hydrological Simulation Program Fortran. HSPF Version 12 User's Manual. AQUA TERRA Consultants, Mountain View, California.

Burns, M.J., Fletcher, T.D., Walsh, C.J., Ladson, A.R., Hatt, B.E., 2012. Hydrologic shortcomings of conventional urban stormwater management and opportunities for reform. Landscape and Urban Planning 105 (3), 230–240.

Chichakly, K.J., Bowden, W.B., Eppstein, M.J., 2013. Minimization of cost, sediment load, and sensitivity to climate change in a watershed management application. Environmental Modelling and Software 50, 158–168.

Chiew, F.H.S., McMahon, T.A., 1999. Modelling runoff and diffuse pollution loads in urban areas. Water Science and Technology 39 (12), 241–248.

Clark, R.D.S., Pezzaniti, D., Cresswell, D., 2002. WaterCress—community resource evaluation and simulation system—a tool for innovative urban water system planning and design. In: Hydrology and Water Resources Symposium. Inst. of Eng, Melbourne, Australia.

CSIRO, 2007. Climate Change in Australia. Technical Report 2007. http://www.climatechangeinaustralia.gov.au/technical_report.php.

Dandy, G.C., Daniell, T.M., Foley, B., Warner, R.F., 2017. Planning and Design of Engineering Systems. In: CRC Press, Third Edition. Taylor & Francis Group, Boca Raton, Florida, ISBN 978-1-1380-3189-0. 448 p.

Deb, K., Pratap, A., Agarwal, S., Meyarivan, T., 2002. A fast and elitist multiobjective genetic algorithm: NSGA-II. IEEE Transactions on Evolutionary Computation 6 (2), 182–197.

Di Matteo, M., Dandy, G.C., Maier, H.R., 2017a. Multiobjective optimization of distributed stormwater harvesting systems. Journal of Water Resources Planning and Management. https://doi.org/10.1061/%28ASCE%29WR.1943-5452.0000756.

Di Matteo, M., Dandy, G.C., Maier, H.R., 2017b. A multi-stakeholder portfolio optimization framework applied to stormwater best management practice (BMP) selection. Environmental Modelling and Software 97, 16–31.

Di Matteo, M., Maier, H.R., Dandy, G.C., 2018. Many-Objective Portfolio Optimization Approach for Stormwater Management Project Selection Encouraging Decision Maker Buy-In, Environmental Modelling & Software. Submitted for publication.

Doerner, K., Gutjahr, W.J., Hartl, R.F., Strauss, C., Stummer, C., 2004. Pareto ant colony optimization: a metaheuristic approach to multiobjective portfolio selection. Annals of Operations Research 131 (1), 79–99.

Dorigo, M., Maniezzo, V., Colorni, A., 1996. Ant system: optimization by a colony of cooperating agents. IEEE Transactions on Systems, Man, and Cybernetics - Part B: Cybernetics 26 (1), 29–41.

Duncan, H.P., 1999. Urban Stormwater Quality: A Statistical Overview. Cooperative Research Centre for Catchment Hydrology, Melbourne, Australia.

eWater, 2009. MUSIC – Model for Urban Stormwater Improvement Conceptualization, User Guide Version 4. eWater Cooperative Research Centre, Canberra, Australia.

eWater, 2011a. MUSIC by eWater, User Manual. eWater, Melbourne, Australia.

eWater, 2011b. Urban Developer User Guide. eWater Cooperative Research Centre, Canberra.

Glover, F., Laguna, M., Marti, R., 2000. Fundamentals of scatter search and path relinking. Control and Cybernetics 29 (3), 653e684.

Goldberg, D.E., 1989. Genetic Algorithms in Search, Optimization, and Machine Learning. Addison-Wesley, Reading, Mass.

Gordon, H., O'Farrell, S., Collier, M., Dix, M., Rotstayn, L., Kowalczyk, E., Hirst, T., Watterson, I., 2010. The CSIRO Mk3.5 Climate Model. CAWCR Technical Report No. 021. The Centre for Australian Weather and Climate Research.

Hadka, D., Reed, P., 2012. Borg: an auto-adaptive many-objective evolutionary computing framework. Evolutionary Computation 21 (2), 231–259.

Intergovernmental Panel on Climate Change, 2007. Climate Change 2007: Synthesis report. Contribution of Working Groups I, II and III to the Fourth Assessment Report of the Intergovernmental Panel on Climate Change. IPCC, Geneva, Switzerland, p. 104.

Jayasooriya, V.M., Ng, A.W.M., Muthukumaran, S., Perera, B.J.C., September 2016. Optimal Sizing of Green Infrastructure Treatment Trains for Stormwater Management. water resources management.

Keim, D., Andrienko, G., Fekete, J.-D., Görg, C., Kohlhammer, J., Melançon, G., 2008. Visual analytics: definition, process, and challenges. In: Kerren, A., Stasko, J.T., Fekete, J.-D., North, C. (Eds.), Information Visualization: Human-Centered Issues and Perspectives. Springer Berlin Heidelberg, Heidelberg, pp. 154–175.

Kennedy, J., Eberhart, R., 1995. Particle swarm optimisation. In: Proc., IEEE Intl. Conf. on Neural Networks. IEEE Service Center, Piscataway, NJ, pp. 1942–1948.

Lee, J.G., Heaney, J.P., Lai, F-h, July 1, 2005. Optimization of integrated urban wet-weather control strategies. Journal of Water Resources Planning and Management 131 (4).

Lee, J.G., Selvakumar, A., Alvi, K., Riverson, J., Zhen, J.X., Shoemaker, L., Lai, F., 2012. A watershed-scale design optimization model for stormwater best management practices. Environmental Modelling and Software 37, 6–18.

Lerer, S., Arnbjerg-Nielsen, K., Mikkelsen, P., 2015. A mapping of tools for informing water sensitive urban design planning decisions—questions, aspects and context sensitivity. Water 7 (3), 993.

Liner, B., Maier, H.R., 2015. H2Optimization: An Introduction to Optimization and Operations Research for Infrastructure Professionals, Aldera. ISBN-10: 0996580409, ISBN-13: 978-0996580403. 94 p.

Loáiciga, H.A., Sadeghi, K.M., Shivers, S., Kharaghani, S., 2015. Stormwater control measures: optimization methods for sizing and selection. Journal of Water Resources Planning and Management. ISSN 0733–9496/04015006(10).

Maier, H.R., Guillaume, J.H.A., van Delden, H., Riddell, G.A., Haasnoot, M., Kwakkel, J.H., 2016. An uncertain future, deep uncertainty, scenarios, robustness and adaptation: how do they fit together? Environmental Modelling and Software 81, 154–164. https://doi.org/10.1016/j.envsoft.2016.03.014.

Marchi, A., Dandy, G.C., Maier, H.R., 2016. Integrated approach for optimizing the design of aquifer storage and recovery stormwater harvesting schemes accounting for externalities and climate change. Journal of Water Resources Planning and Management 142 (4).

Myers, B., Pezzaniti, D., Gonzalez, D., 2013. Hydrological Modelling of the Parafield and Cobbler Creek Catchment for Hazard Analysis Planning. Goyder Institute for Water Research Technical Report Series No. 13/3. South Australia, Adelaide.

Paton, F.L., Maier, H.R., Dandy, G.C., 2014. Including adaptation and mitigation responses to climate change in a multi-objective evolutionary algorithm framework for urban water supply systems incorporating GHG emissions. Water Resources Research 50 (8), 6285–6304. https://doi.org/10.1002/2013WR015195.

Reichold, L., Zechman, E.M., Brill, E.D., Holmes, H., May 1, 2010. Simulation-optimization framework to support sustainable watershed development by mimicking the predevelopment flow regime. Journal of Water Resources Planning and Management 136 (3). ISSN 0733–9496/2010/3-366–375.

Roeckner, E., Baum, G., Bonaventura, L., Brokopf, R., Esch, M., Giorgetta, M., Hagemann, S., Kirchner, I., Kornblueh, L., Manzini, E., Rhodin, A., Schlese, U., Schulzweida, U., Tompkins, A., 2003. The Atmospheric General Circulation Model ECHAM5. Max-Plank Institute for Meteorology.

Rossman, L.A., 2004. StormWater Management Model User's Manual Version 5.0. In: Laboratory, N.R.R. (Ed.). US Environmental Protection Agency, Cincinnati, Ohio.

Russell, S., Norvig, P., 2003. Artificial Intelligence. A Modern Approach, second ed. Prentice Education Inc., New Jersey.

Storn, R., Price, K., 1997. Differential evolution a simple and efficient heuristic for global optimization over continuous spaces. The Journal of Global Optimization 11 (4), 341–359.

Taha, H.A., 2017. Operations Research. In: An Introduction, 10th ed. Pearson Education Inc., Upper Saddle River, New Jersey.

Wu, W., Maier, H.R., Dandy, G.C., Leonard, R., Bellette, K., Cuddy, S., Maheepala, S., 2016. Including stakeholder input in formulating and solving real-world optimization problems: generic framework and case study. Environmental Modelling and Software 79, 197–213.

Zare, S.O., Saghafian, B., Shamsai, A., 2012. Multi-objective optimization for combined quality-quantity urban runoff control. Hydrology and Earth System Sciences 16 (12), 4531.

Zhen, X.Y.J., Yu, S.L., Lin, J.Y., July 2004. Optimal location and sizing of stormwater basins at watershed scale. Journal of Water Resources Planning and Management 130 (4), 339–347.

Chapter 16

# Infrastructure and Urban Planning Context for Achieving the Visions of Integrated Urban Water Management and Water Sensitive Urban Design: The Case of Melbourne

**Casey Furlong[1], Meredith Dobbie[2], Peter Morison[3], Jago Dodson[1] and Micah Pendergast[4]**

*[1]Centre for Urban Research, RMIT University, Melbourne, VIC, Australia; [2]School of Geography and Environmental Science, Monash University, Melbourne, VIC, Australia; [3]School of Ecosystem and Forest Sciences, University of Melbourne, Melbourne, VIC, Australia; [4]City of Port Phillip, Melbourne, VIC, Australia*

## Chapter Outline



**ABSTRACT**

Challenges associated with managing urban water, particularly in rapidly growing cities, have given rise to the visions of Integrated Urban Water Management (IUWM) and Water Sensitive Urban Design (WSUD). Our aim in this chapter is to understand the major barriers to achieving these visions. We begin with a literature review to clarify these visions, summarize previously identified barriers, and then explore these barriers through discussion of Melbourne, Australia, as a case study. This chapter draws on wide-ranging consultation with water industry experts in Melbourne to uncover the practical infrastructure and urban planning processes that are rarely covered in academic papers. We find that achieving the visions of IUWM and WSUD, as with all grand visions, cannot be considered as pass or fail, but rather as a journey of continual improvement. Progress is made when policy development, strategy and planning, and implementation of WSUD approaches, in both new and existing suburbs, are effectively coordinated. The primary barriers to achieving these visions in Melbourne are found to be (1) policy ambiguity around environmental, livability, and water security targets;

Approaches to Water Sensitive Urban Design. https://doi.org/10.1016/B978-0-12-812843-5.00016-2

(2) lack of policy guidance around appropriate and justifiable planning processes, in particular financial evaluations; (3) limitations on what developers can be required to do in new developments without more ambitious policy and legislation; and (4) limited resources available for projects in existing suburbs and the difficulty directing resources to the most beneficial uses.

**Keywords:** Barriers; Decision-making; Integrated urban water management; Low impact development; Urban planning; Water infrastructure planning; Water sensitive urban design

## 16.1 INTRODUCTION

Urban water management has traditionally involved the provision of water supply, sewerage, and drainage services through a network of buried pipes (Marlow et al., 2013). Across the developed world, the planning and management of these water infrastructure systems is conducted by state government departments and Water Service Providers (WSPs) such as government and privately owned water utilities and local government (municipalities) (Baietti et al., 2006; Marques and De Witte, 2011). Water infrastructure systems have generally been planned and managed reactively to maintain minimum service standards in existing suburbs and to service greenfield (previously undeveloped areas) development as cities expand geographically (Anderson and Iyaduri, 2003; Wilson et al., 2013; Mukheibir et al., 2014). Historically, WSPs have focused on "the protection of human health, ensuring reliable water supply and minimising flooding; often with minimal consideration of the environmental and ecological impacts" (Sharma et al., 2010, p. 1).

A series of challenges has led to the continual evolution of the water management field. The root cause of all of these challenges has been expanding human populations (Alcamo et al., 2007; Vörösmarty et al., 2000). Human population increases have resulted in urbanization, pollution, ecological degradation, water scarcity, livability, and climate change issues (Vörösmarty et al., 2010). Over time, the mandates of WSPs have changed to require them to address each new challenge that is faced by human populations (Brown et al., 2009).

The environmental role of WSPs became more widely recognized across the globe after the 1960 and 1970s, as part of a societal shift toward understanding that human activity was degrading the Earth's ecosystems (Vugteveen and Lenders, 2009; Wong, 2006a,b). The role of WSPs in this early environmental shift largely related to the construction and upgrading of sewage treatment plants to ensure that sewage was treated to a level that minimizes ecological damage to waterways and oceans (Brown et al., 2009). There has been gradual improvement in the level of environmental protection provided by WSPs ever since (Fam et al., 2014). In recent years, for some countries, this environmental shift has focused on stormwater quality improvements through stormwater management devices such as wetlands, rain gardens (biofiltration), rainwater tanks, and swales (Roy et al., 2008; Wong, 2006a,b). Some experts are now advocating for a focus on stormwater flow (quantity, frequency, and intensity) improvements to provide additional protection for urban waterways (Duncan et al., 2014).

In the 1990 and early 2000s, water security became a major focus of WSPs in many areas around the world. As water demands continued to increase, opportunities for additional traditional water sources, such as dams and river diversions, became increasingly limited (Vörösmarty et al., 2010). In many cases, sustained droughts further constrained traditional water supplies. This led to a focus on water efficiency, wastewater reuse, and desalination (Billi et al., 2007; Dolnicar et al., 2011; Grant et al., 2012). From a financial perspective, in many cases, potable reuse of wastewater is the most preferable water security option, although in some jurisdictions, such as Australia, community and political opposition constrain uptake (Australian Academy of Technological Sciences and Engineering, 2013; Dolnicar et al., 2011; Lazarova et al., 2001). Many scholars believe that, due to energy usage and the brine waste stream produced, desalination is less "sustainable" or "environmental" than other water security options (Bell, 2015). Therefore, these scholars argue for "fit-for-purpose" (nonpotable) reuse of both treated wastewater and treated stormwater (Ferguson et al., 2013; Turner et al., 2016), particularly when they are cheaper than other options when assessed on a total community cost basis (Marsden Jacob Associates, 2013). Direct potable reuse of treated urban stormwater is a relatively new concept that is currently being investigated around Australia (McArdle et al., 2011). For example, one of these projects has been constructed in Melbourne (Kalkallo Stormwater Harvesting Project) but is not yet operational. In such cases, extended monitoring periods are required to demonstrate to health regulators that produced water is safe for addition to potable networks (Furlong et al., 2016a).

Recently, there has been a global policy focus on the concept of "livability" (the attributes that make a location pleasant and healthy to live in) (Holmes, 2013), and this focus has become a priority for all levels of Australian government and WSPs (Hodge et al., 2014). The essential services that WSPs provide, such as water supply, sewerage, and drainage, are undoubtedly essential for livability, and there is a reasonable degree of clarity around what WSPs are expected to deliver (WSAA, 2014). However, there are also "nonessential" services that relate to livability, including urban greening, thermal comfort, community connection, local identity, natural environments/biodiversity, urban form/amenity, and leisure/recreation. There are multiple ways in which WSPs can, and do, contribute to these services (Holmes, 2013; WSAA, 2014). To the extent that WSPs are expected to contribute to these nonessential livability services, their mandate is extended beyond traditional water management functions (e.g., pipes, dams, and treatment plants) into the realm of urban planning.

The concept of livability has close links to environmental protection and water security. Firstly, environmentally degraded areas are not pleasant or healthy to live in. Secondly, the supply of alternative water sources (which mitigate water restrictions in times of drought) can have a major impact on livability through helping to maintain urban greening. For the purposes of structuring the later sections of this chapter (Sections 16.3.5 and 16.3.6), we define three interrelated categories of WSP objectives that have emerged in recent history: *environmental protection* (in regard to stormwater management to protect receiving waters); *livability* (in regard to amenity, greening, recreation, and health); and *water security* (in regard to provision of alternative water supplies).

The increasing focus on environmental protection, water security, and livability has required a shift in WSP policy, planning, and operations. An increasing focus on environmental and ecological elements requires WSPs to become transdisciplinary and develop environmental stewardship and biodiversity strategies and projects. Simultaneous consideration of water supply, sewerage, drainage, ecology, and livability services, as well as multiple "fit-for-purpose" water sources and uses, requires more complex and robust decision-making processes that consider trade-offs between social, environmental, and economic objectives. A wider role in urban planning outcomes requires collaboration between WSPs and a larger range of stakeholders, as well as more extensive understanding of community preferences.

For all of these reasons, practitioners and scholars have been promoting the need for a new "integrated" and "sustainable" water management paradigm (Bell, 2012; Marlow et al., 2013). Such a paradigm is broadly defined to include many different objectives and methods (Furlong et al., 2016b). It has been referred to by over 26 different terms, such as integrated water management, total water management, and integrated water cycle management (Furlong et al., 2015), and has associated concepts that vary between countries due to different challenges and personal experiences of proponents (Downs et al., 1991; Hering et al., 2015). Similarly, new paradigms for urban drainage are referred to by many different names such as sustainable urban drainage systems and best management practices (Fletcher et al., 2015).

This chapter focuses on two of these terms and their associated visions and practices: Integrated Urban Water Management (IUWM) and Water Sensitive Urban Design (WSUD). In much of the literature, it is common for scholars to make reference to these two terms in parallel as two synergistic ideologies but without clearly defining how they relate to one another (Ferguson et al., 2013; Sharma et al., 2012).

### 16.1.1 Focus of this chapter and sources of information

This chapter focuses on understanding the barriers that constrain the implementation of IUWM and WSUD.

Substantial research has been undertaken in relation to these barriers (Sharma et al., 2012; Mukheibir et al., 2014; Dobbie and Brown, 2013), which will be described in Section 16.2.4 of this chapter. However, such research does not include a description of the practical steps that water and urban planning practitioners actually go through during efforts to implement IUWM and WSUD. Therefore, we argue that a more complete picture of the barriers to achieving the IUWM/WSUD vision can be painted through a detailed explanation of how IUWM and WSUD are actually engaged with by practitioners, in this case by exploring the implementation of IUWM and WSUD in Melbourne as a case study.

Melbourne has a population of 4.6M and an area of 9990 $km^2$. Major challenges for Melbourne include drought and rapid urbanization. Population is expected to increase to 8M by 2050 (www.planmelbourne.vic.gov.au). In addition, Melbourne is located on Port Phillip Bay (Fig. 16.1) with its constrained inflow and outflow that makes both wastewater and stormwater treatment a priority to avoid an unacceptable buildup of pollution, especially nitrogen (Harris et al., 1996). For these reasons, Melbourne has been a focal point in the development of both the IUWM and WSUD ideologies as ways to reducing reliance on dams for water supply, minimizing the environmental damage from city expansion, and reducing nutrient loads into urban waterways and the bay (Brown and Clarke, 2007; Ferguson et al., 2013).

The development of this chapter has been somewhat guided by an overarching question of "what has gotten in the way of achieving the visions of IUWM and WSUD?" To answer this question, the authors have identified three research objectives:

1. Clarify the **visions** of WSUD and IUWM and determine to what degree they are similar (covered in Sections 16.2.1–16.2.3);

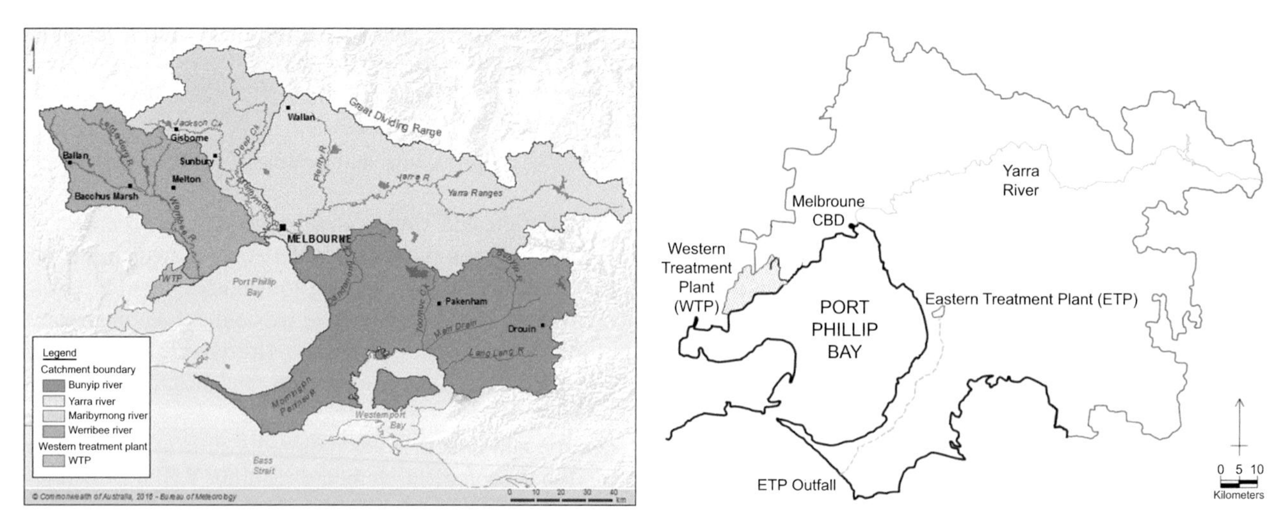


FIGURE 16.1 Maps showing Melbourne's major rivers (left) and sewage treatment plants discharging into Port Phillip Bay (right). *(Left) from www.bom.gov.au. (Right), adapted from Barker, F., Faggian, R., Hamilton, A.J., 2011. A history of wastewater irrigation in Melbourne, Australia. Journal of Water Sustainability 1(2), 31–50.*

2. Review the current **body of knowledge** on barriers to achieving the WSUD and IUWM visions (covered in Section 16.2.4); and
3. Explore **barriers** through discussion of the current IUWM/WSUD situation and practices in Melbourne (covered in Section 16.3).

To achieve these objectives, we have used a variety of information sources. A broad-ranging online literature search and review have been conducted into the meanings of, and barriers to, IUWM and WSUD. The bulk of the Melbourne case study narratives is based on findings from a wide-ranging industry consultation process relating to IUWM and WSUD implementation. This consultation process was largely conducted as part of PhD research (Casey Furlong, RMIT) into IUWM in Australia and resulted in consultation with 43 staff from 25 organizations working in the areas of IUWM and WSUD. This prior knowledge has been supplemented by additional consultation with five experts from Melbourne's water industry. Relevant qualitative data from these sources have been used in Section 16.3 to describe how IUWM/WSUD implementation processes work and the practical barriers to the achievement of the IUWM/WSUD visions.

## 16.2 LITERATURE REVIEW ON CONCEPTS AND BARRIERS

### 16.2.1 Understanding the concept of Integrated Urban Water Management

IUWM as a concept is in many respects complex, convoluted, and vague, both in terms of its meaning and its origins. It is generally referred to in literature and practice as a recent development, although many of IUWM's core principles can be traced back into ancient history. According to some researchers, there is evidence of integrated approaches to water management as far back as Spain in the Middle Ages, as well as Ancient Greece and Rome (Rahaman and Varis, 2005; Vivas et al., 2009).

A number of high-profile organizations, in particular the Global Water Partnership and the World Bank, have recently adopted the terminology of IUWM under the following definition:

> *Integrated urban water management (IUWM) offers a set of principles that underpin better coordinated, responsive, and sustainable resource management practice. It is an approach that integrates water sources, water use sectors, water services, and water management scales. It (1) recognises alternative water sources, (2) differentiates the qualities and potential uses of water sources, (3) views water storage, distribution, treatment, recycling, and disposal as part of the same resource management cycle, (4) seeks to protect, conserve and exploit water at its source, (5) accounts for non-urban users that are dependent on the same water source, (6) aligns formal institutions (organisations, legislation, and policies) and informal practices (norms and conventions) that govern water in and for cities, (7) recognises the relationships among water resources, land use, and energy, (8) simultaneously pursues economic efficiency, social equity, and environmental sustainability, and (9) encourages participation by all stakeholders.*
>
> Global Water Partnership (2012, p. 12)

This definition, and the way the term is used by the World Bank (Closas et al., 2012), gives IUWM such a broad meaning that it is almost synonymous with simply saying "good" water management. Essentially, IUWM means doing everything right. There is a high degree of consistency between such a view of IUWM and, for example, the OECD's 12 principles for good water governance, which include integrity, transparency, trade-offs between users, clear roles, scales of planning, policy coherence, etc (Akhmouch and Correia, 2016).

In the Australian context, the term IUWM is associated with specific planning methods and specific types of infrastructure projects. The planning approach that is associated with the term IUWM is the creation of IUWM strategies/plans that investigate future infrastructure options for specific geographical areas (often large residential or mixed use development sites), incorporating decision-making for water supply, sewerage, and stormwater into a single infrastructure planning process. In the vast majority of cases, these IUWM strategies result in the recommendations for some form of wastewater or stormwater reuse within the study area (Furlong et al., 2017b).

These water reuse projects are often referred to, either formally or informally, as IUWM projects. IUWM as a term is therefore more widely used and more specifically defined within the Australian water sector than it is globally. Some Australian experts argue that it specifically relates to integrated infrastructure strategies and water security through reuse. Others argue that it is more about livability or environmental considerations, such as improving amenity and recreational use of waterway corridors, contributing to urban greening, protecting waterways through wetlands and rain gardens, and supplying additional water specifically for watering of green spaces during drought (Furlong et al., 2016b).

## 16.2.2 Understanding the concept of Water Sensitive Urban Design

In comparison with IUWM, the meaning of WSUD is more well-defined and its origins easier to investigate largely because it began in a single country (Australia), rather than emerging as the result of a transnational evolution of complementary practices. WSUD first arose as a concept in the late 1980s in Western Australia (WA) in response to concerns about the negative impact of conventional urban development on the hydrological cycle of the Swan Coastal Plain, in the area around the city of Perth (Hedgecock and Mouritz, 1993). In particular, proponents wanted to raise the profile of water in the planning process; promote water recycling; and mitigate the impact of urban stormwater on urban waterways by providing at-source detention and retention of stormwater using landscape features (Hedgecock and Mouritz, 1993, p. 116).

To realize this vision, a water-sensitive design approach would include discontinuous areas of residential development with smaller lot sizes that were better integrated with more naturalistic public open space; narrower roads; smaller-scale, localized infrastructure; long-term management of residential development forming part of the approval process, enforced through controls on the land title; and increased consideration to retrofitting water-sensitive design principles (Hedgecock and Mouritz, 1993). WA produced its first WSUD guidelines in 1994 (Roy et al., 2008).

The other Australian states followed suit over the next 15 years, but their vision was not the "radical approach" of WA (Wong, 2006b, p. 214). Their focus shifted to the stormwater management aspect of the planning and design framework in response to a growing awareness globally of the impacts of urban stormwater quality on the ecological health of urban waterways (Wong, 2006a,b; Walsh et al., 2005). Thus, the vision of WSUD during the late 1990s and into the early 2000s was to protect natural systems within urban developments, integrate stormwater treatment into the landscape, protect water quality of waterways in urban developments, reduce runoff and peak flows, and add value while minimizing development costs (Stormwater Committee, 1999). To achieve this, WSUD integrated "best management practices" (e.g., rainwater tanks, roof gardens, swales and buffer strips, gross pollutant traps, sedimentation basins, constructed wetlands, porous paving, bioretention, and infiltration devices) to retain, treat, and/or store stormwater in the landscape, thereby creating multifunctional landscapes that offered visual and recreational amenity, protected the water quality of local waterways, reduced runoff and peak flows to these waterways, and minimized impervious areas and development costs of drainage infrastructure (Stormwater Committee, 1999; Lloyd, 2001; WBM, 2009).

More recently, the vision of WSUD has extended to include aquifer storage and recovery, graywater reuse, dual reticulation of treated wastewater, sewer mining, xeriscaping (water-efficient landscaping), water conservation, and urban heat island mitigation (WBM, 2009; Roy et al., 2008; Wong et al., 2012). Thus, WSUD now aspires to provide multiple water sources at allotment, neighborhood, and regional scales, using a mix of strategies, predominantly visible structures in the landscape, often in stormwater treatment trains. This has been informed by research that indicates that distributed stormwater treatments, rather than only end-of-catchment treatments, provide additional environmental and social benefits. The present vision reflects more closely the vision of those water practitioners in WA almost 30 years ago. However, what is still missing, but was present in the original vision, is the ambitious goal of seeking to alter urban planning guidelines (lot sizes, road widths, etc.) in accordance with water servicing considerations.

A commonly accepted definition for WSUD is as follows:

> *In Australia, Water Sensitive Urban Design (WSUD) has evolved from its early association with stormwater management to provide a broader framework for sustainable urban water management. It is a framework that provides a common and unified method for integrating the interactions between the urban built form (including urban landscapes) and the urban water cycle. It is increasingly practiced in new urban greenfield development areas and urban renewal developments linked to a broader Ecological Sustainable Development agenda. Key guiding principles of WSUD include: (1) reducing potable water demand through water efficient appliances and seeking alternative sources of water such as rainwater and (treated) wastewater reuse, guided by the principle of 'fit-for-purpose' matching of water quality and end uses, (2) minimising wastewater generation and treatment of wastewater to a standard suitable for effluent re-use opportunities and/or release to receiving waters, (3) treating urban stormwater to meet water quality objectives for reuse and/or discharge to surface waters [to support stream ecological health], and (4) using stormwater in the urban landscape to maximise the visual and recreational amenity of developments.*
>
> Wong (2006a,b, p. 2)

## 16.2.3 A combined IUWM/WSUD vision

In practice, the terms are generally used in very different circumstances (Fig. 16.2). The operational use of the term IUWM in Australia is largely related to the creation of "IUWM strategies" and wastewater and stormwater reuse projects. The operational use of the term WSUD in Australia is almost exclusively associated with the design and implementation of specific stormwater management technologies (rain gardens, wetlands, etc.), which sometimes include a component of stormwater harvesting and are normally intended to also improve urban design and amenity.

Based on these definitions, the concepts of IUWM and WSUD have many similarities, including environmental protection and deployment of decentralized water reuse infrastructure. The difference between the definitions of the two terms is that IUWM has a broader scope, which includes metropolitan-scale issues of water security, maintenance of environmental flows, energy, social equity, etc. Typically, WSUD is considered to be specifically related to (1) interaction between IUWM concepts and urban developments, making reference to visual and recreational amenity, and (2) stormwater management devices, such as rain gardens and wetlands, which are targeted directly at environmental protection of urban waterways and waterbodies. IUWM can be considered to be the urban water–centric variation of "Integrated Resource Management" (Biswas, 2004), whereas WSUD can be considered as the urban water–centric variation of "Low Impact Development" (Roy et al., 2008). Consideration of WSUD at a metropolitan or regional scale would in many respects meet the definition of IUWM, and consideration of IUWM at the local scale would in many respects meet the definition of WSUD.

Therefore, we argue that the visions of IUWM and WSUD in Melbourne have no substantive contradictions and can be seen as a singular IUWM/WSUD vision that combines all of the points listed in Sections 16.2.1 and 16.2.2.

## 16.2.4 Barriers to achieving the IUWM/WSUD vision

Now that the IUWM/WSUD vision has been clarified, it is possible to move on to the major focus of this chapter, which relates to the barriers to achieving this vision. Results from the literature review on implementation barriers are included in Table 16.1. The literature identifies at least 17 different barriers. It can be seen that the various implementation barriers

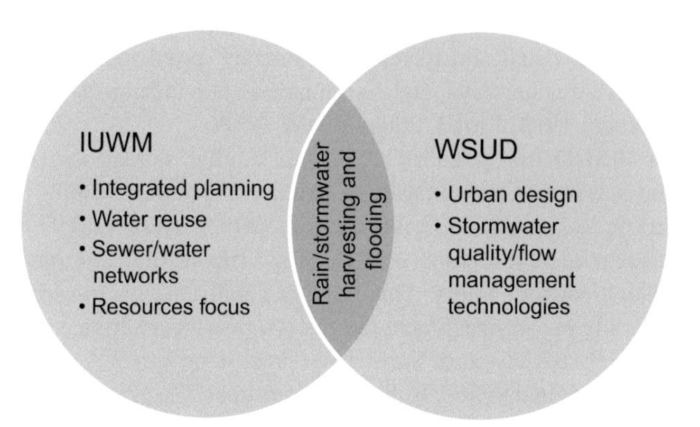


FIGURE 16.2 Schematic of how the terms Integrated Urban Water Management (IUWM) and Water Sensitive Urban Design (WSUD) are typically used in practice.

**TABLE 16.1** Barriers to Integrated Urban Water Management (IUWM) and Water Sensitive Urban Design (WSUD) Implementation Identified by Previous Research

| Barrier | Relating to Policy, Strategy, or Implementation? | Applies to IUWM or WSUD? | Supporting Research |
|---|---|---|---|
| Lack of long-term policy vision/strategies | Policy development | Both | Brown and Farrelly (2009), Morison and Brown (2011), Lee and Yigitcanlar (2010), and Sharma et al. (2016) |
| Funding limitations | Policy development | Both | Brown and Farrelly (2009), Sharma et al. (2016), and Roy et al. (2008) |
| Lack of regulatory incentives | Policy development | Both | Brown and Clarke (2007), Brown and Farrelly (2009), Lloyd (2002), and Sharma et al. (2012) |
| Lack of political will | Policy development | Both | Brown and Clarke (2007), Brown et al. (2009), Brown et al. (2011), and Morison and Brown (2011) |
| Lack of communication between practitioners | Strategy and planning | Both | Brown and Farrelly (2009) and Williams et al. (2010) |
| Financial assessment methodologies that ignore externalities (particularly indirect benefits) | Strategy and planning | Both | Furlong et al. (2017a), Lloyd (2002), Marlow et al. (2013), and Sharma et al. (2012) |
| Lack of communication between practitioners and regulators | Strategy and planning | Both | Brown and Farrelly (2009), Furlong et al. (2016a), and Williams et al. (2010) |
| Limited community engagement and empowerment | Strategy and planning | Both | Brown and Farrelly (2009), Lloyd (2002), Marlow et al. (2013), Sharma et al. (2012), and Sharma et al. (2016) |
| Fragmented roles and responsibilities | Strategy and planning | Both | Brown and Farrelly (2009), Lloyd (2002), Marlow et al. (2013), and Roy et al. (2008) |
| Limited past experience | Implementation | Both | Marlow et al. (2013) and Sharma et al. (2012) |
| Lack of evidence of capital, operations, and maintenance costs | Implementation | Both | Brown et al. (2009), Sharma et al. (2012), Sharma et al. (2016), and Roy et al. (2008) |
| Limited evidence of long-term performance efficacy | Implementation | Both | Brown et al. (2009), Lloyd (2002), Marlow et al. (2013), Sharma et al. (2012), and Roy et al. (2008) |
| Risk aversion by practitioners, developers, regulators, and community | All | Both | Brown et al. (2009), Brown et al. (2011), Dobbie and Brown (2012, 2013, 2014), and Sharma et al. (2012) |
| Lack of industry capability and capacity | All | Both | Brown, (2005), Brown and Farrelly (2009), Lloyd, (2002), Sharma et al. (2016), and Roy et al. (2008) |
| Being locked-in to a particular path due to previous decisions/investments | All | Both | Brown and Farrelly (2009) and Marlow et al. (2013) |
| Unsuitable governance protocols | All | Both | Brown et al. (2011), Furlong et al. (2016d), and van de Meene et al. (2011) |
| Organizational resistance | All | Both | Brown and Farrelly (2009), Morison and Brown (2011), Marlow et al. (2013), and Roy et al. (2008) |

identified in the literature apply to both IUWM and WSUD, providing additional evidence that it is justifiable to consider barriers to both simultaneously. The barriers to the implementation of IUWM and WSUD apply to different organizations within the water industry and across the policy, strategy, and implementation, a distinction which is further defined in Section 16.3.2.

Barriers that are more associated with water policy constrain IUWM and WSUD uptake upstream in the planning system. When policy is formulated, if no political interest and long-term vision is articulated, no money is set-aside nor financial mechanisms put in place, then there is a clear constraint on the ability of practitioners to later develop strategies and implement them (Morison and Brown, 2011).

Barriers that are associated with strategy development impact on organization's abilities to determine where and how to implement the policies that do exist. During strategy development, barriers to IUWM/WSUD uptake include organizations being unclear on what their role is, which is further compounded if there is a lack of communication with other organizations, regulators, and the community, and not having financial evaluation methods that consider indirect benefits (Furlong et al., 2017b).

Barriers that directly affect implementation include lack of knowledge and capacity, which impact on the planning of actual projects. Organizations lack sufficient knowledge to understand how systems will operate in terms of long-term performance and long-term cost. A lack of funding clearly impacts on an organizations' ability to implement projects; however, it is through upstream policy and strategy processes that systems for financing must be developed (Furlong et al., 2017a).

There are also a number of barriers that relate simultaneously to policy, strategy, and implementation. Previous infrastructure decisions can result in being locked into particular future actions. Unsuitable governance approaches, essentially determining who should do what, are important elements in all phases of planning. All the listed barriers in combination result in organizational resistance across all phases (Brown and Farrelly, 2009).

Risk aversion is a topic often singled out by scholars as a key issue (Dobbie and Brown, 2012). Some perceive that it is particularly developers who are risk averse because it is often the developers who have to provide the bulk of the funding for IUWM/WSUD projects. Other research has found that technical staff within government organizations are generally more risk adverse than their counterparts in the development industry (Dobbie and Brown, 2014). This is a very complex issue which is open to interpretation and analysis from a number of different perspectives. One of the factors at play may be that public sector actors are more required to consider social, environmental, and economic risks and uncertainties, as well as maintenance and performance, across a longer time period in comparison with the shorter-term focus of private developers.

## 16.3 EXPLORING BARRIERS USING MELBOURNE AS A CASE STUDY

### 16.3.1 Infrastructure and urban planning context for IUWM and WSUD in Melbourne

Efforts to achieve IUWM/WSUD exist within a complex "Water Infrastructure Planning" and "Urban Planning" context. Decision-making and planning processes used within the water sector and the urban planning sector affect and govern changes to urban outcomes, including both urban form and infrastructure (Furlong et al., 2016c). Identifying the specific planning processes through which each of these sectors has an impact on urban outcomes is a complex task. However, there are some clear examples for each that can serve as the starting point for further exploration and discussion.

In Melbourne, the major institutions involved in urban planning are the State Government Department of Environment, Land, Water and Planning (DELWP) (www.delwp.vic.gov.au), the Victorian Planning Authority (VPA) (www.vpa.vic.gov.au), and the 31+ municipal Local Government Areas (local councils). These institutions have a role in governing changes to Melbourne's urban form including metropolitan planning, planning schemes, local development controls (e.g., overlays), Precinct Structure Plans, and issue-specific plans, such as urban forest and public open space strategies. The role of DELWP is to set overall state planning policy for Melbourne and Victoria. The VPA is responsible for precinct structure planning in greenfield and designated major "infill" redevelopment sites. Local municipalities are responsible for selecting the areas to which various state-specified zones will apply and for managing development approval processes. Local government planning functions are required to implement Victorian Government policies by informing, guiding, and mandating public and private actions relating to urban development of the city.

The major institutions operating within Melbourne's water sector are DELWP, Melbourne Water (www.melbournewater.com.au), the three water retailers: City West Water (www.citywestwater.com.au), Yarra Valley Water (www.yvw.com.au), and South East Water (southeastwater.com.au), and the 31 + municipalities (Fig. 16.3). As with urban planning, DELWP also develops water policy. Melbourne Water is the bulk water and sewerage service provider, as well as large-scale drainage, catchment, and waterways manager. The water retailers provide the retail water and sewerage interface with residential and commercial customers. Municipalities have a role in local-scale drainage and some stormwater harvesting

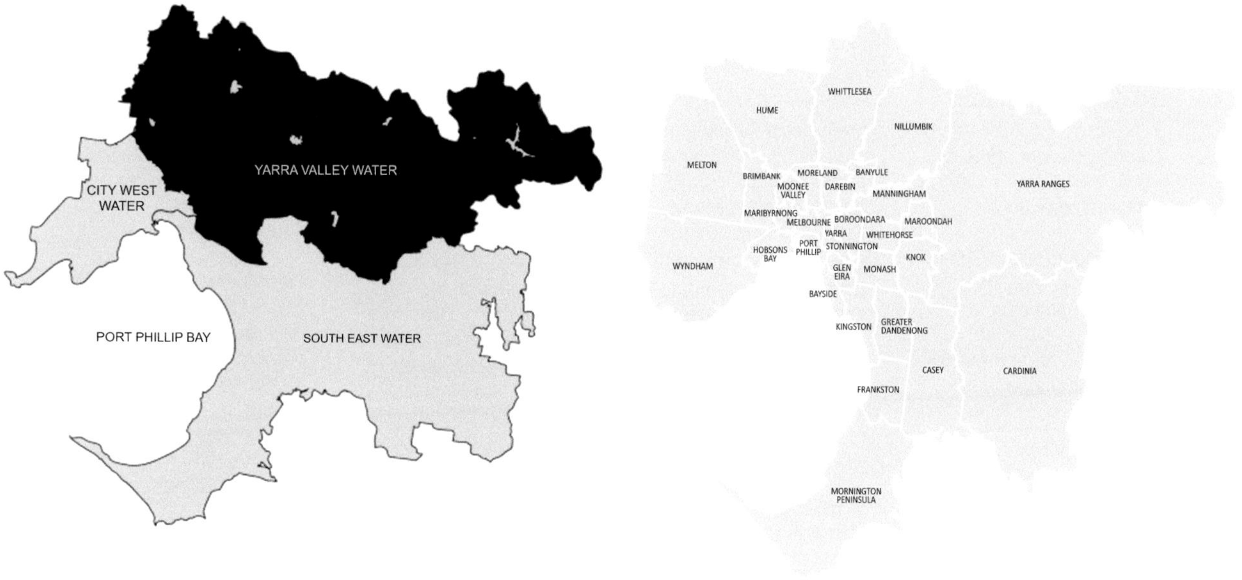

FIGURE 16.3 Borders of Melbourne's three "subregional" water retailers—City West Water, Yarra Valley Water, and South East Water (left) versus the 31 local councils/municipalities (right).

projects. In addition, the Essential Services Commission (www.esc.vic.gov.au), Department of Health and Human Services (www.dhhs.vic.gov.au), Environmental Protection Authority (www.epa.vic.gov.au), and Department of Treasury and Finance (www.dtf.vic.gov.au) each have regulatory roles. All the preceding institutions collectively manage the water supply, sewerage, and drainage systems for the city. This is done through development of city-wide and location-specific modeling and strategies to maintain current levels of service and determine future augmentations, which are used to inform ongoing construction and asset management efforts.

## 16.3.2 Conceptual framework used for Melbourne case study

IUWM/WSUD implementation requires the mandate of WSPs to expand beyond the traditional areas of water supply, sewerage, and drainage infrastructure systems. This expanded mandate plays out across a number of categories or what could be considered planning phases. A conceptual framework can be used to characterize such actions by distinguishing them as "constitutive" (overarching policies), "directive" (strategy and planning), or "operational" (action and infrastructure in new and existing areas) (Hill and Hupe, 2006). This characterization system, as it applies to IUWM/WSUD implementation, is outlined in Fig. 16.4.

Each of these IUWM/WSUD implementation categories involves WSPs collaborating with different stakeholders and through different processes and mechanisms. Policy development is often conducted primarily through collaboration between water utilities and DELWP. Strategies and plans have different focuses and require input from different stakeholders. Influencing greenfield development is largely done through collaboration between Melbourne Water, VPA, municipalities, and developers. Retrofitting in existing suburbs is generally conducted primarily through discussions between water utilities, municipalities, and customers and then financial approvals through the Essential Services Commission and Department of Treasury and Finance.

We will now discuss and explore these barriers through examination of the Melbourne case. For each of these categories, we explain how the IUWM/WSUD vision is actually implemented; how this implementation fits within the wider context of urban planning practice; and the practical barriers that are experienced by practitioners.

## 16.3.3 Policy development

Overarching policies and legislation that govern the implementation of IUWM and WSUD in Melbourne are generally created by the Victorian Government department concerned with water and the environment (currently DELWP), in consultation with Melbourne Water, the water retailers, and local government. Policy documents give the direction to be followed by strategy and implementation processes. Major policy documents from the state government since 2000 are listed in Table 16.2.

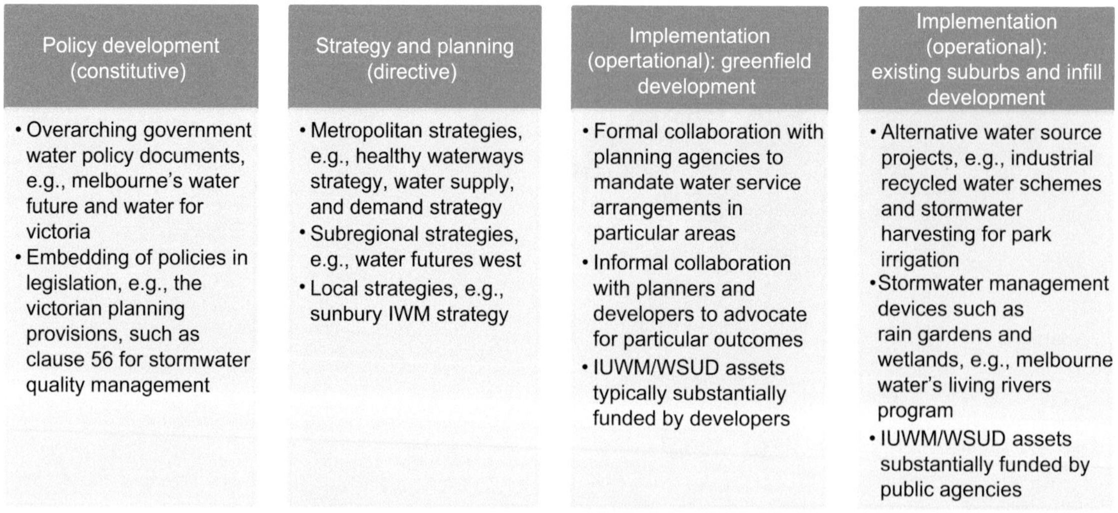

**Policy development (constitutive)**

- Overarching government water policy documents, e.g., melbourne's water future and water for victoria
- Embedding of policies in legislation, e.g., the victorian planning provisions, such as clause 56 for stormwater quality management

**Strategy and planning (directive)**

- Metropolitan strategies, e.g., healthy waterways strategy, water supply, and demand strategy
- Subregional strategies, e.g., water futures west
- Local strategies, e.g., sunbury IWM strategy

**Implementation (opertational): greenfield development**

- Formal collaboration with planning agencies to mandate water service arrangements in particular areas
- Informal collaboration with planners and developers to advocate for particular outcomes
- IUWM/WSUD assets typically substantially funded by developers

**Implementation (operational): existing suburbs and infill development**

- Alternative water source projects, e.g., industrial recycled water schemes and stormwater harvesting for park irrigation
- Stormwater management devices such as rain gardens and wetlands, e.g., melbourne water's living rivers program
- IUWM/WSUD assets substantially funded by public agencies

FIGURE 16.4 Major categories of Integrated Urban Water Management (IUWM)/Water Sensitive Urban Design (WSUD) actions undertaken by Water Service Providers.

**TABLE 16.2 Major Policy Documents Guiding Water Management in Melbourne 2000–16 (Links Provided Where Possible)**

| Year | Government Policy Document |
|---|---|
| 2002 | 21st Century Melbourne: a WaterSmart City, Strategy Directions Report |
| 2004 | Our Water Our Future: Securing our Water Future Together |
| 2005 | Sustainable Water Strategy: Central Region (www.water.vic.gov.au/planning-and-entitlements/sustainable-water-strategies/central-region-sustainable-water-strategy) |
| 2006 | Water Supply-Demand Strategy for Melbourne 2006–55 (southeastwater.com.au/SiteCollectionDocuments/AboutUs/Water_Supply-Demand_Strategy.pdf) |
| 2007 | Our Water Our Future: The Next Stage of the Government's Water Plan |
| 2013 | Melbourne's Water Future |
| 2016 | Water for Victoria (www.water.vic.gov.au/water-for-victoria) |

There has been substantial policy progress toward adoption of the IUWM/WSUD vision. Since 2002, government policy has generally shifted toward the adoption of IUWM/WSUD principles. Notably, the 2004 policy document paved the way toward establishing a policy that 5% of all water bills be set aside as "Environmental Contributions." Some of this money has been used to fund IUWM/WSUD projects as explained in Section 16.3.6.1. The latest two policy documents, Melbourne's Water Future and Water for Victoria, place a significant emphasis on IUWM and WSUD concepts, spanning wastewater and stormwater reuse, environmental protection of waterways, integrated planning methodologies, and the recreational, health, and amenity value of green space and community facilities. It is evident that the Victorian Government policy in the water portfolio is supportive of the IUWM/WSUD agenda.

Government policy is enacted by state and local government legislation, which affects the implementation of the IUWM/WSUD vision. Policy and legislation in the broader urban planning sphere have partially adopted the language and concepts of IUWM and WSUD. For example, the Victorian Planning Provisions (planningschemes.dpcd.vic.gov.au/schemes/vpps) were modified in 2006 to include within Clause 56 (1) a requirement for residential subdivisions to treat stormwater to minimum quality standards (80% reduction in suspended solids, 45% reduction in phosphorus, 45% reduction in nitrogen, etc.) and (2) an option for the water sector to mandate third-pipe installation in new developments, allowing for the supply of nonpotable recycled water (Fig. 16.5). This state legislation applies to greenfield development, as explained in Section 16.3.5.1, but does not cover small-scale development and redevelopment within existing suburbs, so a number of individual councils have created local policies to address this gap, as explained in Section 16.3.6.1.

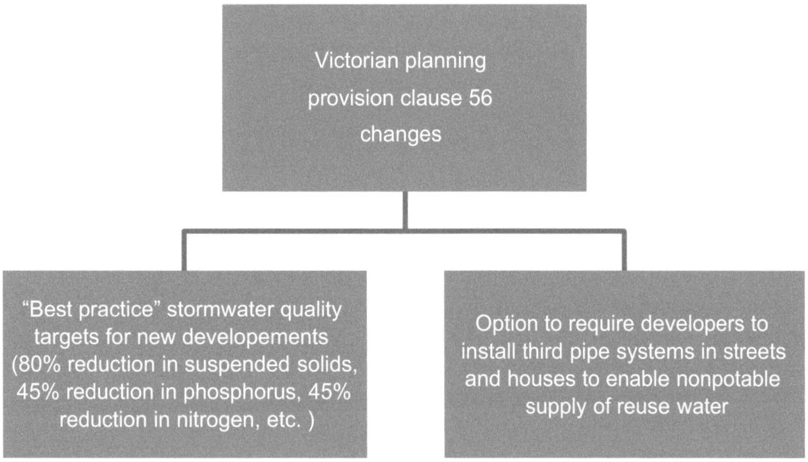

Victorian planning provision clause 56 changes

"Best practice" stormwater quality targets for new developements (80% reduction in suspended solids, 45% reduction in phosphorus, 45% reduction in nitrogen, etc. )

Option to require developers to install third pipe systems in streets and houses to enable nonpotable supply of reuse water

**FIGURE 16.5** Victorian Planning Provision 2006 changes that support Integrated Urban Water Management/Water Sensitive Urban Design implementation in greenfield developments.

More information is available online (www.melbournewater.com.au/Planning-and-building/Stormwater-management/WSUD_treatments/Pages/Stormwater-quality-objectives.aspx). The inclusion of stormwater quality and IUWM requirements within the Victorian Planning Provisions has had a significant impact on the adoption of WSUD technologies, as will be argued throughout Section 16.3.5.

Other aspects of Victoria's Planning Provisions require the water sector to be a party to the development of certain types of urban plans (e.g., Employment Cluster and Precinct Structure Plans) for priority development areas. This allows the water sector an avenue for pursuing IUWM/WSUD outcomes in greenfield and infill areas. Municipalities have also been given policy direction to produce stormwater management plans and later "Integrated Water Management" plans. The production of these plans has facilitated meaningful dialog between the water authorities and municipalities.

Although there has been significant progress in policy and legislation toward adopting the IUWM/WSUD vision, there is significant room for improvement. We offer three examples of barriers at the policy level.

Recently, Victoria's Essential Services Commission, which regulates water and electricity pricing, has determined that the three water retailers (see Fig. 16.3) should pay a flat fee to Melbourne Water rather than a variable fee based on quantity of water used. Such an action has removed a financial incentive for water retailers to construct IUWM/WSUD projects that produce alternative water supplies. According to one consulted expert, this creates a barrier to the achievement of the IUWM/WSUD vision.

A second example of a policy barrier is the lack of clear guidance as to the circumstances that should require stormwater quality (and quantity) management above the "best practice management guidelines" as currently specified in the Victorian Planning Provisions (see Fig. 16.5). Experts believe there are some high-value environmental areas of Melbourne that warrant additional stormwater management but do not have regulatory power to place additional requirements on developers. Existing stormwater quality targets (Fig. 16.5) are currently being reviewed by DELWP in a process that has now been ongoing for multiple years. Stormwater management needs to benefit the whole of community and ensure the protection of waterways now and in the future. Amending legislation related to stormwater management will need to find a way forward, in collaboration with the development industry, in an attempt to achieve the best possible overall outcomes for the community and environment. DEWLP's leadership, engaging with all institutions involved in stormwater management, is essential to a workable outcome in this space and bringing about greater industry confidence.

The third barrier at the policy and legislation level is the lack of clarity around financial evaluation requirements for IUWM/WSUD projects. For example, as many IUWM and WSUD projects do not pay for themselves, organizations and governments often have to subsidize projects, which are justified on the basis of indirect benefits. For example, if a recycled water scheme costs $10M and will recuperate $5M from customer charges over its lifetime, then it requires a $5M subsidy from various sources (e.g., government, other stakeholders, or charging to the wider customer base). However, there is currently a lack of clear policy in place to govern under what circumstances, and from whom, these subsidies should be provided. Particularly, what is required is a way to quantify social and environmental benefits in dollar terms to justify subsidies and cost-sharing arrangements. This issue is widely acknowledged within the water sector and has been discussed in numerous academic and industry documents (Furlong et al., 2017a; Marsden Jacob Associates, 2013). Many attempts have already been made to address this issue, and another major investigation has now commenced under the

Cooperative Research Centre for Water Sensitive Cities tranche two program of works (www.watersensitivecities.org.au). Consulted experts generally expressed a view that a consistent financial evaluation process for IUWM and WSUD projects would be preferable, but there was some disagreement around whether it will ever be possible to have all stakeholders agree on such an economic evaluation process.

### 16.3.4 Strategy and planning

Legislation, such as Clause 56 for stormwater quality treatment in greenfield developments (Fig. 16.5), is able to achieve real-world impact passively through regulatory approval and referral processes. However, many IUWM/WSUD policies, particularly in existing suburbs, cannot be implemented via passive and reactive processes, but rather require proactive action by water utilities, municipalities, and other stakeholders to design and fund projects. Such action requires complex planning and decision-making systems at multiple geographical scales. Determining which decisions should be considered at which scales is a crucial component for achieving the IUWM/WSUD vision. This section will explore barriers to the achievement of the IUWM/WSUD vision in strategy and planning at the metropolitan scale (whole of city) and also the subregional/local scale (see Fig. 16.3 for representation of subregional and local scales).

Strategy and planning processes relating substantially to IUWM and WSUD that occur at the **scale of Metropolitan Melbourne** include (1) water supply and demand strategies, (2) the Healthy Waterways Strategy, and (3) the Port Phillip and Western Port Bay Environmental Management Plan. These high-level strategic processes give guidance (and often targets) to planning processes at subregional and local scales.

According to the consulted experts, barriers to effective integration of IUWM/WSUD, which are common to all of these metropolitan-scale planning processes, include integration between different plans, the cultural and political struggle around prioritization of issues, and allocation of sufficient resources to ensure the meeting of environmental and social objectives in a city of increasing size. In some particular strategies, there have also been some technical challenges relating to conceptualizing the role of decentralized solutions in addressing whole of system goals (Furlong et al., 2016a) and also the linking of ecological objectives to infrastructure requirements (Furlong et al., 2017b). To address these issues, consulted experts highlighted the importance of sufficient community involvement; supportive policy; industry collaboration; and developing shared multidisciplinary language.

Most IUWM/WSUD-related planning processes that consider actual infrastructure outcomes are conducted at the **subregional or the local scales** (see Fig. 16.3). Such planning processes are generally referred to as "Integrated Water (Cycle) Management Strategies/Plans" and shall be referred to here as IUWM strategies. Melbourne Water recently commissioned a review of nine of these IUWM strategies (report not publicly available but results described in Furlong et al., (2017b)) and found a number of barriers to effective planning. These included inconsistent and insufficiently justified environmental targets, financial evaluation processes, and consideration of energy use, all of which relate to a lack of policy guidance as discussed in the previous section. The most consequential finding (according to the authors of the review) was that the IUWM strategies did not undertake scenario analysis and therefore failed to consider how water infrastructure options could be affected by uncertainties about population growth, water use, and climate projections (Furlong et al., 2017b). Therefore, according to the review, barriers to effective implementation of IUWM/WSUD at the subregional and local strategy and planning scale are generally related to (1) determining appropriate targets or levels of service and (2) planning process and methodologies that require more policy guidance.

### 16.3.5 Implementation: greenfield developments

This section relates to the implementation of IUWM and WSUD in greenfield areas of Melbourne. As it is such a broad and complex issue, the narratives provided are more easily understood when they are structured into subcategories. As outlined in the Introduction, aspects of IUWM/WSUD implementation can be divided into stormwater management (environmental protection) outcomes, livability (urban planning) outcomes, and alternative water sources (water security) outcomes.

#### *16.3.5.1 Stormwater management in greenfield developments*

In regard to stormwater management, greenfield developments are generally informed by Development Services Schemes (sometimes referred to as drainage schemes, for more information visit www.melbournewater.com.au/Planning-and-building/schemes/map/Pages/Drainage-schemes.aspx). These drainage schemes are typically produced well in advance of development and then refined as part of detailed design to determine the amount of stormwater quality treatment and flood management infrastructure required to meet the stormwater targets in Fig. 16.5.

A number of consulted experts expressed confidence that, in the vast majority of cases, such a process is effective at reaching the "best practice" stormwater quality treatment (Fig. 16.5) and flood retention targets required in the legislation, according to computer modeling. However, there are some uncertainties around the performance of constructed stormwater devices over extended periods, particularly in relation to construction and maintenance concerns (Duncan et al., 2014). Approximately 90% of greenfield developments are guided by a drainage scheme, and therefore Melbourne Water is confident that either stormwater targets have been met or developers have paid the stormwater quality offset fee to fund stormwater treatment assets elsewhere. In the remaining cases (~10%), there is some level of ambiguity because of the absence of a drainage scheme, meaning that municipalities are the responsible authority.

Acknowledging that some believe that the existing stormwater quality targets (Fig. 16.5) are insufficient to achieve desired environmental outcomes (a policy issue discussed earlier), a number of experts have sought to set higher environmental targets, which has led to inconsistencies between the processes used in IUWM strategies (a strategy issue), although this is not necessarily a problem. Research conducted for this chapter did not uncover any examples of developments that have implemented stormwater management targets in excess of the "best practice" guidelines (Fig. 16.5), although some may exist, but there are a number of planned greenfield development projects for which the water sector is actively considering and advocating for such action.

Consulted experts also argue, from an environmental perspective (e.g., protecting rivers and creeks), that distributed small-scale stormwater treatments (i.e., rain gardens and swales) are often more effective than end-of-catchment treatments (see www.melbournewater.com.au/planning-and-building/stormwater-management/wsud_treatments/pages/outlet-and-distributed-approaches.aspx). However, some municipalities and developers prefer end-of-catchment treatments (i.e., wetlands) because they are perceived as cheaper to operate, more easily marketed to consumers and community, and often become the responsibility of Melbourne Water rather than the municipalities. This rule that causes stormwater assets under a certain size to become a municipality asset and over a certain size to become a Melbourne Water asset (typically differentiated at a catchment threshold of 60 ha) is currently under review by relevant stakeholders. Issues around long-term performance of stormwater assets, as well as around ownership of stormwater assets of different sizes, are both relevant to ongoing issues of maintenance.

In summary, the major barriers to the implementation of stormwater management within greenfield developments can be considered to be (1) policy ambiguity around whether stormwater management targets should be made more stringent; (2) determining the correct mix of distributed and end-of-catchment treatments; and (3) ongoing questions relating to who is responsible for maintenance.

### 16.3.5.2 Livability in greenfield developments

Achievement of livability within greenfield areas is predominantly within the broader realm of the urban planning field. Street layouts, lot sizes, number and size of open spaces, and many more aspects that affect livability are guided by urban planning legislation. They are then implemented by developers with oversight from, and in consultation with, the VPA and local councils. There is, however, substantial overlap between the planning of public open space and the planning of stormwater and flooding management infrastructure because such infrastructure can often double as public open space to maximize recreational value. Consulted experts noted two significant barriers to the effective execution of this process.

Firstly, the legislation that guides the urban planning layout of public open space does not require coordination with the planning of stormwater/flooding infrastructure. Greenfield developments are typically required by the VPA to have ~6% active (sporting) open space, ~4% passive (walking) open space, and 5%–10% of the development's area for flood and stormwater assets. Small "passive" open space parks are generally laid out in a regularly spaced grid (approximately 400m by 400m), which is not influenced by topography or water management considerations. The larger parks with sporting and community facilities are generally intentionally placed as features on top of hills, whereas retarding basins and wetlands need to be placed at the lowest point in the development catchment. The problem with these practices is that (1) they do not produce green space corridors that have biodiversity and human health benefits through connectivity and (2) they do not facilitate stormwater harvesting projects for open space irrigation because the stormwater is collected far from the sporting and community facilities.

Secondly, there is a high degree of uncertainty in the planning and management of waterway corridors. There are multiple competing stakeholder interests involved, as well as some ambiguity around the legislation that guides this process. Two key examples of such ambiguity are whether extension of waterway corridor widths beyond minimum requirements counts toward a development's "passive" (no sporting facilities) open space quota; and which entity is responsible for landscaping and maintenance of this area, including bicycle and walking tracks.

Major barriers to the achievement of livability in greenfield developments, therefore, relate to the need for more effective coordination and partnerships between the VPA, Melbourne Water, municipalities, and developers to achieve the best possible urban planning outcomes. Achievement of highly livable developments (e.g., through the creation of well-planned green space and waterway corridors) if effectively implemented, with costs shared fairly, should be a positive outcome for all parties.

#### *16.3.5.3 Alternative water supply in greenfield developments*

Alternative water sources for greenfield areas include recycled wastewater (for nondrinking uses), rainwater harvesting from roofs, and stormwater harvesting (rainwater after it touches the ground).

Implementation of third-pipe recycled water supply for greenfield areas is now common practice in Melbourne. This is undertaken by Melbourne's water retailers (see Fig. 16.3 left) with the key factor being proximity to wastewater treatment (see Fig. 16.1 for the largest two wastewater treatment plants). The inclusion of "Integrated Water Management" provisions within the Victorian Planning Provisions (see Fig. 16.5) allows the water retailers to require developers to fund and construct third-pipe reticulation networks if they are deemed beneficial. Consulted experts noted that, in many cases, because of regulatory limits on the discharge of treated wastewater from small-scale sewage treatment plants, third-pipe supply of treated wastewater to greenfield developments (by the water retailers) is often either cost-neutral or actually saves money in comparison with other discharge options. Third-pipe systems are now being installed in many new developments around Melbourne's fringes. One growing barrier to this practice is the trend toward new houses having smaller gardens (smaller lots and/or bigger house footprints), which require less nonpotable water, although this is not preventing the ongoing implementation of such schemes. At present, there are no wastewater-to-potable schemes and none currently under consideration. Experts consider this to be due predominantly to political concerns around public opinion. In areas that do not have third-pipe recycled water networks, state building code legislation typically requires that new homes install either a rainwater tank or a solar hot water system (see www.vba.vic.gov.au/consumer-resources/other/standard-pages/rainwater-tanks).

Implementation of stormwater harvesting schemes, predominantly for on-site collection, treatment, and irrigation of parks, is also now a common practice in Melbourne. In some cases, these schemes completely, or almost, pay for themselves over extended periods, through reducing the need to purchase potable water (Furlong et al., 2017a). Research has found that in 2012, there were 108 schemes already existing (Ferguson et al., 2013). Therefore, the main barrier to the implementation of such schemes is the proximity of stormwater collection to potential stormwater uses, e.g., sporting and community facilities.

As mentioned in the introduction, one example exists (Kalkallo Stormwater Harvesting Project) of a stormwater-to-potable reuse scheme that has been constructed but is not being operated yet because of delays with the surrounding development, and once it is operational, it will require significant testing to provide evidence that the water is safe to drink (Furlong et al., 2016a). Currently, there are multiple other stormwater-to-potable schemes under consideration, but these experience the barriers of community perceptions around drinking such water, as well as health regulations. From a financial perspective, stormwater-to-potable schemes become cost-competitive to other water supply and drainage options if stormwater quality and quantity legislation is substantially strengthened (e.g., to require 90% of additional stormwater produced from developments to be retained). This is because such a stringent requirement is difficult and expensive to meet without stormwater-to-potable reuse (Furlong et al., 2017b).

Therefore, there are no substantive barriers to the implementation of nonpotable third-pipe wastewater or on-site stormwater reuse schemes in greenfield developments, as many already exist or are in development. There are, however, significant barriers to the implementation of potable wastewater/stormwater reuse schemes including the following:

1. The barrier to the implementation of stormwater-to-potable projects is that they only become cost-competitive (in comparison to other water supply and drainage infrastructure options) if mandatory stormwater retention targets are significantly increased; and
2. The barrier to the implementation of both potable wastewater and stormwater reuse schemes is community perceptions around drinking the treated water, and thus also political perceptions.

### 16.3.6 Implementation: existing suburbs

The following section relates to the implementation of IUWM/WSUD in existing suburbs and utilizes the same structure as used in Section 16.3.5.

### *16.3.6.1 Stormwater management in existing suburbs*

In the vast majority of cases, greenfield developments are being constructed to meet the "best practice" stormwater management guidelines (Fig. 16.5), but these guidelines do not apply to most small-scale development and redevelopment within existing suburbs. This is a major problem because as the densification of existing suburbs increases, so does the area of impervious surfaces, and therefore the quantity of stormwater and flood risk. Apart from this, stormwater management infrastructure in existing suburbs is more difficult to implement for a number of other reasons. There is an existing stormwater treatment deficit because suburbs have been developed long before stormwater management devices have become popular. In addition, given that streetscapes and parks are already developed, retrofitting of stormwater management is even more expensive.

Of Melbourne's 31 municipalities, 10 have now developed local policies that require new and renovating buildings to install stormwater management to treat on-lot stormwater to "best practice" standards, typically by a rainwater tank and/or rain garden. Three more municipalities have legislation under development. Melbourne Water receives no drainage contribution from infill development works but does, in some cases, receive some funding from developers choosing to pay the stormwater offset fees, rather than construct stormwater management devices on-site.

Due to limited ability to compel nearby private developers to provide funding, most stormwater management works on public land need to be funded by Melbourne Water's and/or municipalities' wider customer bases. Municipalities in Melbourne began developing Stormwater Management Plans in 2001. As mentioned earlier in the policy section, the Victorian Government in 2004 developed a policy that 5% of urban water bills be set aside as "Environmental Contributions." This money allowed Melbourne Water to set up a "Living Rivers" program in 2006 (www. melbournewater.com.au/livingrivers), which has since provided funding for municipality-owned stormwater management projects on public land in existing suburbs. Melbourne Water's Living Rivers program also funded municipalities to create "Integrated Water Management plans" (these plans were previously mentioned in the policy section), which largely focused on stormwater management. Creation of such plans facilitated easier communication between Melbourne Water and municipalities around stormwater management, partially because these plans generally set aside some municipality funding for planning and implementation.

The Living Rivers program has funded approximately 700 stormwater management assets, across approximately 200 site locations. For its first few years, these stormwater management projects were generally 100% funded by Melbourne Water, and the funds were allocated to municipalities in a somewhat opportunistic manner. For example, Melbourne Water would periodically inquire with municipalities about who was planning on doing road renewal works and whether they would be interested in installing rain gardens in their streets. In general, much of the focus has been on building capacity within local government, rather than focusing specifically on the areas of Melbourne with highest environmental value. More recently, the Living Rivers funding program has gradually shifted toward stricter requirements for municipalities to supply 50% of the funding required for proposed stormwater management projects, as well as more sophisticated multicriteria assessments to determine which local government stormwater project proposals are most worthy of funding.

A 2015 review (not publicly available) of the program found that (1) the Living Rivers funding program had a robust and transparent rationalization through improved municipality capacity to implement best practice stormwater management and (2) this investment lead to improved long-term waterway and bay condition outcomes, is cost-effective and efficient, and will result in improved value for Melbourne Water customers. Another recent audit (not publicly available) of approximately 90 stormwater assets funded by the Living Rivers program found that 75% are currently in good or moderate condition. Consulted experts noted that, in the past, many municipalities had a negative or cautious view of installing stormwater management devices because of maintenance concerns. However, most municipalities in Melbourne are beginning to view the technologies more favorably and starting to move to a "master planning" approach to better spatially plan stormwater management technologies within their jurisdictions.

There are a number of substantial "infill development" sites, where previous industrial or low-density suburbs are being extensively leveled and reconstructed as new developments. A key example of this is Fishermans Bend in Melbourne's inner city, Australia's largest urban renewal project (www.fishermansbend.vic.gov.au). In such cases, Melbourne Water, as the drainage authority, has a significant degree of influence over stormwater and flooding infrastructure but a limited ability to require developers to pay for such works. According to consulted experts, this is because of (1) the absence of a drainage scheme and (2) a lack of clarity around appropriate stormwater and flooding "levels of service."

In summary, we have found that the major barriers to the implementation of stormwater management technologies in existing suburbs are the following: (1) there is limited Melbourne Water and municipality funding, which is not sufficient to achieve best practice stormwater management targets in existing areas of Melbourne; (2) it is extremely challenging to direct this available funding to the most beneficial locations; (3) over half of Melbourne's municipalities have not developed regulations for small-scale property renovations and redevelopments to include on-lot stormwater management; and (4) there is a lack of clear policy guidance around required stormwater/flood works, and funding sources, to service large infill developments such as Fishermans bend.

### 16.3.6.2 Livability in existing suburbs

As in the case of greenfield development, achievement of livability objectives within existing suburbs predominantly falls within the realm of urban planning under the responsibility of municipalities but with some examples of Melbourne Water and the water retailers having an influence. In relation to existing suburbs, the main examples of water-sector influence on livability outcomes, other than stormwater management devices such as wetlands, relate to transforming concrete pipes and channels into natural-looking waterways (examples include Upper Stony Creek Transformation Project and Greening the Pipeline project) and landscaping the riparian edges of waterways (https://www.melbournewater.com.au/planning-and-building/waterway-management/pages/vegetation-management.aspx).

There are also some examples of the water sector attempting to have an influence on "urban greening," i.e., the attempt to increase the proportion of vegetation (plants, grass, and trees) within existing suburbs. Melbourne Water has been in discussion with DELWP and the Essential Services Commission to explore whether they have a role in funding urban greening for livability and urban cooling benefits. One of Melbourne's water retailers, City West Water, has also for a number of years been convening a group that includes all of Melbourne's western municipalities, and many other stakeholders, named "Greening the West" (www.greeningthewest.org.au). This group advocates for and supports greening, cooling, and livability outcomes.

These livability-targeted actions in existing suburbs require funding and have limited or no opportunity to transfer costs to private developers. Therefore, barriers to the achievement of livability outcomes within existing areas relate to (1) limited resources constraining waterway landscaping and naturalizing works and (2) ambiguity around which aspects of livability the water sector should be involved in and the resources, both human and financial, that they should be allocating.

### 16.3.6.3 Alternative water supply in existing suburbs

In comparison with greenfield developments, the provision of recycled wastewater and stormwater to properties in existing suburbs is far more challenging. This is because of the major cost involved in retrofitting an additional reticulation network, the alterations to internal domestic plumbing, and the absence of developers to whom the costs can be transferred. In some cases, water utilities have retrofitted these reticulation systems into industrial areas (e.g., Altona Recycled Water Project Stage 1), which is sometimes more favorable than residential because of higher and more consistent water usage and a smaller number of customer connections.

For infill developments, in a number of cases, Melbourne's water retailers are attempting to implement third-pipe reticulation networks in a similar way to greenfield developments. However, this is typically more expensive than in greenfield areas because of the absence of local wastewater treatment plants to supply the recycled water within Melbourne's inner city. A number of these schemes, such as wastewater reuse through sewer mining in Fishermans Bend, may be implemented. One example (Coburg Stormwater Harvesting Project) proposed to use stormwater harvesting for nonpotable third-pipe reuse; however, this project was canceled because of cost increases from original estimates (Furlong et al., 2016a).

For many existing residential areas, the promotion of rainwater tank installation is one of the only mechanisms by which the water sector is able to locally supplement residential water supply. Rainwater tank installation rates received a major boost after Melbourne's recent droughts, jumping from 11.6% of households in 2007 to 28.2% in 2010. This percentage has since climbed only slowly, reaching 31% in 2013 (Australian Bureau of Statistics, 2013). According to the consulted experts, some municipalities have attempted educational programs to promote rainwater tanks, but with mixed success. One community engagement campaign by Melbourne Water, the University of Melbourne, Knox City Council, and South East Water was successful in increasing uptake of rainwater tanks, although this was through substantial financial subsidies (more information can be found at southeastwater.com.au/SiteCollectionDocuments/CurrentProjects/Projects/Dobsons_Creek_Fact_Sheet.pdf).

As previously explained, the construction of stormwater harvesting schemes for irrigation of public open space is becoming increasingly common. For such schemes, being located within an existing suburb is not a barrier because, due to being located within open space, there are no roads and sidewalks to remove and reinstall. Therefore, there is minimal cost difference compared with their construction in greenfield developments. Thus, such schemes are sometimes able to come close to achieving full cost recovery over an extended period. The way this works is through municipalities constructing the schemes, with or without external funding assistance, and paying them off through reducing their potable water bills (Furlong et al., 2017a). As already stated, there were 108 stormwater harvesting schemes in Melbourne in 2012 (Ferguson et al., 2013). Therefore, in appropriate locations, these schemes are experiencing minimal barriers to implementation and are being constructed at a steady pace across Melbourne's parks, where topography and proximity to stormwater pipes permit. Melbourne Water's Living Rivers program has had a significant impact on the uptake of these schemes through funding assistance, capacity building, and ongoing support to local councils.

Major barriers to the implementation of alternative water supply in existing suburbs, therefore, relate to (1) high cost of retrofitting third-pipe reticulation and (2) inability to significantly influence the continuing uptake of rainwater tanks. There are no significant barriers to the implementation of stormwater to open space irrigation schemes, judging from the fact that their implementation is becoming increasingly common.

## 16.4 DISCUSSION

In this chapter we have sought to explore the relationship between IUWM and WSUD, review the current body of knowledge on barriers to IUWM/WSUD implementation, and then explore barriers through discussion of current practices within Melbourne.

Understanding the conceptual and practical relationship between IUWM and WSUD is a challenging task because of the nature of these concepts. IUWM is not a clearly defined term, and its meaning and scope vary widely in the minds of Australian experts. In addition, from reviewing available literature, IUWM appears to have a different meaning within Australia, where it is generally related to integrated planning of water services and water reuse. In contrast, internationally its meaning is often so broad as to mean nothing more specific than "good" water management. WSUD has a more clearly defined meaning in the sense that it has a strong association with designing urban developments with a focus on stormwater management, water reuse, and livability improvements. However, on a larger (e.g., city) scale, the meaning of WSUD is less distinct, with great similarity to how IUWM is used within Australia. After reviewing the two terms, we have formed the view that, although they are associated with different practices, there are no inherent contradictions between the two in terms of their core principles and vision.

Throughout the production of this chapter, we have observed there is ambiguity around where to draw the line between a purely livability-focused endeavor, such as landscaping of open space, and something that could be considered also an "IUWM" or "WSUD" endeavor, such as landscaping of a waterway. As this is predominantly a semantic rather than practical question, we do not intend to present a view on where such a line may be drawn, but it is potentially an issue that may need to be explored further in the future, in relation to what the water sector should, and should not, be involved in.

Previous research has effectively identified most of the major barriers to IUWM/WSUD implementation but has not reported actual issues in adequate detail. It is important in any analysis of the barriers to IUWM/WSUD implementation to understand not only what barriers exist but also where, when, and how these barriers are having an impact to achieve a clearer view of potential solutions. This demonstrates the importance of case studies to relate the barriers to specific contexts and practices. Without context, it is not possible to truly understand all the physical, financial, and planning process barriers that exist.

From our case study of practices within Melbourne, the most important barriers to the achievement of the IUWM/WSUD vision across the categories of policy, strategy, and implementation are shown in Table 16.3, which can be summarized into four points. Firstly, it is an extremely difficult task for policy developers, predominantly within DELWP, to determine appropriate environmental, livability, and water security targets that apply across all development types over the entire metropolitan region. Secondly, it is extremely difficult for all stakeholders to create, agree on, and consistently implement planning and decision-making processes, particularly those related to financial evaluation of options. Thirdly, without more ambitious targets or more widely and consistently adopted planning and financial evaluation processes, it is not possible to achieve outcomes in greenfield developments over and above that of the existing "best practice" standards (see Fig. 16.5), which some experts believe are inadequate. Finally, in existing suburbs, there are significant implementation constraints given: the limited opportunities to transfer costs onto private developers; the limited public resources available for projects; and the difficulty of directing these resources to most beneficial uses, due to competing stakeholder interests.

As highlighted by previous research, psychological barriers, such as the attitudes of customers, practitioners, policy makers, and regulators, are important. In particular, the present research highlights the importance of attitudes and beliefs of policy makers. Development of policy and legislation, if done effectively, can have the largest actual impact on outcomes because policy and legislation can impact across a large scale. For example, the development of the stormwater and Integrated Water Management provisions within Clause 56 of the Victorian Planning Provisions (Fig. 16.5), and to a

**TABLE 16.3 Summary of Major Barriers to Integrated Urban Water Management (IUWM) and Water Sensitive Urban Design (WSUD) Implementation Across Different Action Categories in Melbourne**

| Planning Focus | Major Barriers Identified |
|---|---|
| **Policy** | |
| Policy development | • Lack of financial incentives for water retailers to substitute potable water supplies with alternative water sources<br>• Inherent difficulty involved in decisions around whether to increase stormwater quality and quantity management targets<br>• Lack of guidance around holistic financial evaluation processes for IUWM/WSUD projects |
| **Strategy and Planning** | |
| Metropolitan scale | • Coordination difficulty in integrating various plans with each other<br>• Lack of funding required to achieve objectives in a city of increasing size<br>• Technical difficulty incorporating decentralized resources into overall supply and demand strategies<br>• Technical difficulty in linking ecological objectives to infrastructure objectives |
| Subregional and local scales | • Determining appropriate targets and levels of service<br>• Lack of guidance around planning process and methodologies |
| **Implementation: Greenfield Developments** | |
| Stormwater management | • Lack of regulatory mechanisms to increase stormwater management requirements above current best practice standards in high-priority areas<br>• Differing stakeholder drivers, benefits, and risks around distributed versus end-of-catchment treatments |
| Livability | • Inherent difficulty around collaboration between all stakeholders in regard to the planning of open space and waterway corridors |
| Alternative water supplies | • Stormwater-to-potable schemes are not cost-competitive in the absence of more stringent regulatory targets for stormwater<br>• Community and political perceptions are a major barrier to potable wastewater and stormwater reuse schemes (also applies to existing suburbs) |
| **Implementation: Existing Suburbs** | |
| Stormwater management | • Limited funding available because of inability to transfer costs to developers<br>• Difficulty in directing the available funding to most beneficial locations<br>• More than half of Melbourne's municipalities have not developed requirements for small-scale (re)development to include stormwater management<br>• Lack of policy guidance around required stormwater/flood works, and funding sources, to service large infill developments such as Fishermans Bend |
| Livability | • Limited funding available for modification of existing parks and waterway corridors<br>• Ambiguity around what the water sector should be involved in planning and/or funding |
| Alternative water supplies | • Expensiveness of retrofitting third-pipe reticulation<br>• Difficulty in continuing to influence the uptake of rainwater tanks |

lesser extent, the stormwater management Local Policy clauses developed by 10 of Melbourne's municipalities, could be considered to be the most significant actions taken to date to increase the implementation of IUWM/WSUD within Melbourne. Such action was only possible because state and local government policy makers supported the IUWM/WSUD agenda.

It logically follows that the most crucial factor determining whether IUWM/WSUD is implemented further within Melbourne and is whether policy makers increase legislative requirements for stormwater management, livability, and water reuse. However, it is important to recognize that the duty of policy makers is to attempt to weigh up the community benefits against the costs, which includes both the affordability of new housing and also water bills. If targets are to be increased in the future, as some experts argue they should, it will be the product of extensive good faith negotiation and mutual education between all relevant parties.

Therefore, we find that, as explained in the model given in Hill and Hupe (2006), success in the implementation of IUWM and WSUD as public policy requires a concerted alignment between the three worlds of policy, strategy, and implementation. Any conflict or discrepancy between them yields some form of implementation failure. In this case study of Melbourne, we see discrepancies between policy intents at the constitutional level, which create a lack of guidance at the strategy (directive) and implementation (operational) levels. Conversely, we see that strategy impediments (e.g., difficulty coordinating and linking multiple strategies) and implementation impediments (e.g., lack of knowledge and capability) are limiting the realization of clearly defined policies (e.g., protecting the environment through improved stormwater management).

## 16.5 CONCLUSION

Recalling the overarching question that has guided the development of this chapter ("what has gotten in the way of achieving the visions of IUWM and WSUD?"), we argue that the success of any vision of IUWM/WSUD exists on a continuum (rather than a binary of success or failure). The position of any city on such a continuum is also inherently subjective, in regard to a thinker's own beliefs and priorities. For example, an environmentalist may suggest that IUWM/WSUD is achieved when all stormwater and wastewater are treated and reused. An urban planner may suggest the vision is achieved when all neighborhoods have extensive green open space and all waterway corridors are pristine and allow for recreation. However, a social justice-oriented practitioner may argue that IUWM/WSUD can only make an overall positive contribution to society when affordability of water bills is maintained at the same time.

The case of Melbourne elaborates and illustrates barriers to IUWM/WSUD implementation that are already known. This chapter highlights the importance of context as different barriers will be significant in different cases. Understanding of barriers in general should be extended by more specific case studies. This can then lead to recommendations to guide a specific city toward implementing IUWM/WSUD. It is most unlikely that all of the findings from the Melbourne case would apply to another city.

However, we argue that the overarching proposition and conceptual framework (Section 16.3.2) outlined in this chapter are likely to apply in all circumstances, regardless of context. If a similar study was to be conducted on another city, it may be valuable to begin with the following three questions, shown in Fig. 16.6.

When seeking to achieve the IUWM/WSUD vision, there is no simple solution. It is enabled when all levels of policy, strategy, and implementation are working effectively in synchronicity. Policy development should ideally determine what the overall community wants to occur. Strategy and planning should ideally determine how, where, and when it is possible to achieve overarching objectives set out in policy. Greenfield development and retrofitting of existing suburbs should ideally follow through on strategic recommendations. The absence of clear policy guidance on any issue typically results in ineffective and/or inconsistent strategies and thus an implementation failure.

The achievement of the IUWM/WSUD vision should not be purely about achieving a complete set of physical indicators (e.g., percentage of stormwater treated, green space, water reused, etc.) but rather about the concerted achievement

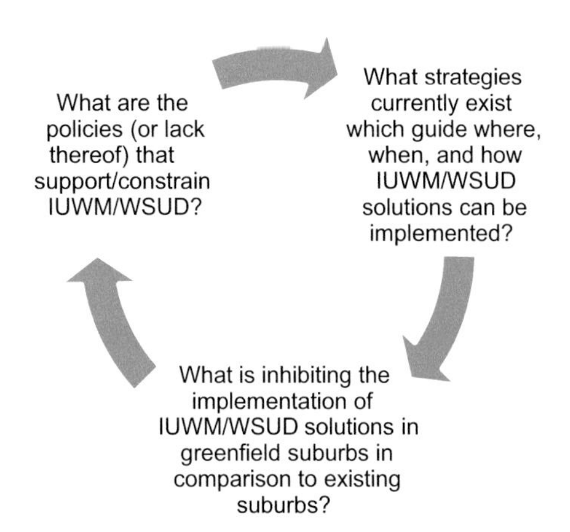


**FIGURE 16.6** Proposed preliminary questions for considering barriers to Integrated Urban Water Management (IUWM)/Water Sensitive Urban Design (WSUD) implementation in cities.

of effective policy development, strategy, and implementation processes that address the trade-offs between social, environmental, and economic benefits and achieve the overall most beneficial results for the wider community. The critical meaning of "integration" within IUWM and WSUD is as much related to this higher-level institutional coherence, as it is to the coordination between the physical water systems.

## ACKNOWLEDGMENTS

The authors wish to acknowledge the funding contributions of RMIT University and Melbourne Water Corporation, as well as the many experts who made time available for meetings over the past several years. Without these contributions, this chapter would certainly not have been possible.

## REFERENCES

Akhmouch, A., Correia, F.N., 2016. The 12 OECD principles on water governance – when science meets policy. Utilities Policy 43 (Part A), 14–20.

Alcamo, J., Flörke, M., Märker, M., 2007. Future long-term changes in global water resources driven by socio-economic and climatic changes. Hydrological Sciences Journal 52 (2), 247–275.

Anderson, J., Iyaduri, R., 2003. Integrated urban water planning: big picture planning is good for the wallet and the environment. Water Science and Technology 47 (7–8), 19–23.

Australian Academy of Technological Sciences and Engineering, 2013. Drinking Water through Recycling. Australian Water Recycling Centre of Excellence, Melbourne.

Australian Bureau of Statistics, 2013. Australian Bureau of Statistics. Available at: http://www.abs.gov.au/ausstats/abs@.nsf/Lookup/4602.0.55.003main+features4Mar%202013 [Online].

Baietti, A., Kingdom, W., Ginneken, M.V., 2006. Characteristics of Well-performing Public Water Utilities, s.l. World Bank: Water Supply and Sanitation Working Notes.

Barker, F., Faggian, R., Hamilton, A.J., 2011. A history of wastewater irrigation in Melbourne, Australia. Journal of Water Sustainability 1 (2), 31–50.

Bell, S., 2012. Urban water systems in transition. Emergence: Complexity and Organisation 14 (1), 44–57.

Bell, S., 2015. Renegotiating urban water. Progress in Planning 96, 1–28.

Billi, A., Canitano, G., Quarto, A., 2007. The Economics of water efficiency: a review of theories, measurement issues and integrated models. Water Use Efficiency and Water Productivity 57, 231.

Biswas, A.K., 2004. Integrated water resources management: a reassessment. Water International 29 (2), 248–256.

Brown, R., 2005, September 26. Impediments to integrated urban stormwater management: the need for institutional reform. Retrieved January 3, 2017, from Environmental Management.

Brown, R., Ashley, R., Farrelly, M., 2011. Political and professional agency entrapment: An. Water Resources Management 25, 4037–4050.

Brown, R., Clarke, J., 2007. Transition to Water Sensitive Urban Design: The Story of Melbourne, Australia. Facility for Advancing Water Biofiltration, Clayton.

Brown, R., Farrelly, M.A., 2009. Delivering sustainable urban water management: a review of the hurdles we face. Water Science and Technology 59 (5), 839–846.

Brown, R.R., Keath, N., Wong, T.H.F., 2009. Urban water management in cities: historical, current and future regimes. Water Science and Technology 59 (5), 847–855.

Closas, A., Schuring, M., Rodriguez, D., 2012. Integrated Urban Water Management - Lessons and Recommendations from Regional Experiences in Latin America, Central Asia, and Africa. The World Bank, Washington.

Dobbie, M., Brown, R., 2012. Risk perceptions and receptivity of Australian urban water practitioners to stormwater harvesting and treatment systems. Water Science and Technology: Water Supply 12 (6), 888–894.

Dobbie, M., Brown, R., 2013. Integrated urban water management in the water-sensitive city: systems, silos and practitioners' risk perceptions. Water 40 (4), 83–91.

Dobbie, M., Brown, R., 2014. Transition to a water-cycle city: sociodemographic influences on Australian urban water practitioners' risk perceptions towards alternative water systems. Urban Water Journal 11 (6), 444–460.

Dolnicar, S., Hurlimann, A., Grün, B., 2011. What affects public acceptance of recycled and desalinated water? Water Research 45 (2), 933–943.

Downs, P.W., Gregory, K.J., Brookes, A., 1991. How integrated is river basin management? Environmental Management 15 (3), 299–309.

Duncan, H.P., Fletcher, T.D., Vietz, G., Urrutiaguer, M., 2014. The Feasibility of Maintaining Ecologically and Geomorphically Important Elements of the Natural Flow Regime in the Context of a Superabundance of Flow: Stage 1 – Kororoit Creek Study. Technical Report 14.5. Melbourne Waterway Research-Practice Partnership, Melbourne.

Fam, D., Mitchell, C., Abeysuriya, K., Lopes, A.M., 2014. Emergence of decentralised water and sanitation systems in Melbourne, Australia. International Journal of Water 8 (2), 149–165.

Ferguson, B.C., Brown, R.R., Frantzeskaki, N., 2013. The enabling institutional context for integrated water management: lessons from Melbourne. Water Research 47, 7300–7314.

Fletcher, T.D., Shuster, W., Hunt, W.F., Ashley, R., Butler, D., Arthur, S., Trowsdale, S., Barraud, S., Semadeni-Davies, A., Bertrand-Krajewski, J.L., Mikkelsen, P.S., 2015. SUDS, LID, BMPs, WSUD and more—The evolution and application of terminology surrounding urban drainage. Urban Water Journal 12 (7), 525—542.

Furlong, C., Guthrie, L., De Silva, S., Considine, R., 2015. Analysing the terminology of integration in the water management field. Water Policy 17, 46—60.

Furlong, C., De Silva, S., Guthrie, L., 2016a. Planning scales and approval processes for IUWM projects. Water Policy 18 (3), 783—802.

Furlong, C., Guthrie, L., De Silva, S., 2016b. Understanding integrated urban water management as an ideology, method and objective. Water eJournal. https://doi.org/10.21139/wej.2016.016.

Furlong, C., De Silva, S., Guthrie, L., Considine, R., 2016c. Developing a water infrastructure planning framework for the complex modern planning environment. Utilities Policy 38, 1—10.

Furlong, C., Gan, K., De Silva, S., 2016d. Governance of integrated urban water management in Melbourne, Australia. Utilities Policy 43, 48—58.

Furlong, C., De Silva, S., Gan, K., Guthrie, L., Considine, R., 2017a. Risk management, financial evaluation and funding for wastewater and stormwater reuse projects. Journal of Environmental Management 191, 83—95.

Furlong, C., Brotchie, R., Considine, R., Finlayson, G., Guthrie, L., 2017b. Key concepts for integrated urban water management infrastructure planning: lessons from Melbourne. Utilities Policy 45, 84—96.

Global Water Partnership, 2012. Integrated Urban Water Management. GWP Tec Background Papers, Stockholm.

Grant, S.B., Saphores, J.D., Feldman, D.L., Hamilton, A.J., Fletcher, T.D., Cook, P.L., Stewardson, M., Sanders, B.F., Levin, L.A., Ambrose, R.F., Deletic, A., 2012. Taking the "waste" out of "wastewater" for human water security and ecosystem sustainability. Science 337 (6095), 681—686.

Harris, G., Batley, G., Fox, D., Hall, D., Jernakoff, P., Molloy, R., Murray, A., Newell, B., Parslow, J., Skyring, G., Walker, S., 1996. Port Phillip Bay Environmental Study Final Report. CSIRO, Canberra.

Hedgecock, D., Mouritz, M., 1993. Water sensitive residential design. Australian Planner 114—118.

Hering, J.G., Sedlak, D.L., Tortajada, C., Biswas, A.K., Niwagaba, C., Breu, T., 2015. Local perspectives on water. Science 349 (6247), 479—480.

Hill, M., Hupe, P., 2006. Analysing policy processes as multiple governance: accountability in social policy. Policy and Politics 34 (3), 557—573.

Hodge, K., Rodrigues, E., Blaise, M., Anstey, J., 2014. The Role of the Urban Water Sector in Contributing to Liveability. OzWater, Brisbane.

Holmes, M., 2013. Melbourne Water's Contribution to Liveability, s.l. Melbourne Water and the State Government of Victoria.

Lazarova, V., Levine, B., Sack, J., Cirelli, G., Jeffrey, P., Muntau, H., Salgot, M., Brissaud, F., 2001. Role of water reuse for enhancing integrated water management in Europe and Mediterranean countries. Water Science and Technology 43 (10), 25—33.

Lee, T., Yigitcanlar, S., 2010. Sustainable urban stormwatermanagement: water sensitive urban design perceptions, drivers and barriers. In: Rethinking Sustainable Development: Urban Management, Engineering, and Design. IGI Global, Hershey(PA), pp. 26—37.

Lloyd, S., 2001. Water Sensitive Urban Design in the Australian Context. Technical Report 01/7. Cooperative Research Centre for Catchment Hydrology, Melbourne.

Lloyd, S.W., 2002. Water sensitive urban design - a stormwater management perspective. CRC for Catchment Hydrology, Clayton.

Marlow, D.R., Moglia, M., Cook, S., Beale, D.J., 2013. Towards sustainable urban water management: a critical reassessment. Water Research 47, 7150—7161.

Marques, R.C., De Witte, K., 2011. Is big better? On scale and scope economies in the Portuguese water sector. Economic Modelling 28 (3), 1009—1016.

Marsden Jacob Associates, 2013. Economic Viability of Recycled Water Schemes. Australian Water Recycling Centre of Excellence, Brisbane.

McArdle, P., Gleeson, J., Hammond, T., Heslop, E., Holden, R., Kuczera, G., 2011. Centralised urban stormwater harvesting for potable reuse. Water Science and Technology 63 (1), 16—24.

Morison, P., Brown, R., 2011. Understanding the nature of publics and local policy commitment to Water Sensitive Urban Design. Landscape and Urban Planning 83—92.

Mukheibir, P., Howe, C., Gallet, D., 2014. What's getting in the way of a 'one water' approach to water services planning and management? Water: Journal of the Australian Water Association 41 (3), 67—73.

Rahaman, M.M., Varis, O., 2005. Integrated water resources management: evolution, prospects and future challenges. Sustainability: Science, Practice, and Policy 1 (1), 15—21.

Roy, A.H., Wenger, S.J., Fletcher, T.D., Walsh, C.J., Ladson, A.R., Shuster, W.D., Thurston, H.W., Brown, R.R., 2008. Impediments and solutions to sustainable, watershed-scale urban stormwater management: lessons from Australia and the United States. Environmental Management 42, 344—359.

Sharma, A., Burn, S., Gardner, T., Gregory, A., 2010. Role of decentralised systems in the transition of urban water systems. Water Science and Technology: Water Supply 10 (4), 577—583.

Sharma, A.K., Cook, S., Tjandraatmadja, G., Gregory, A., 2012. Impediments and constraints in the uptake of water sensitive urban design measures in greenfield and infill developments. Water Science and Technology 65 (2), 340—352.

Sharma, A.K., Pezzaniti, D., Myers, B., Cook, S., Tjandraatmadja, G., Chacko, P., Chavoshi, S., Kemp, D., Leonard, R., Koth, B., Walton, A., 2016. Water sensitive urban design: an investigation of current systems, implementation drivers, community perceptions and potential to supplement urban water services. Water 8, 272—286.

Stormwater Committee, 1999. Water sensitive urban design. In: Urban Stormwater: Best Practice Environmental Management Guidelines. CSIRO, Collingwood, pp. 47—62.

Turner, A., Mukheibir, P., Mitchell, C., Chong, J., Retamal, M., Murta, J., Carrard, N., Delaney, C., 2016. Recycled water—lessons from Australia on dealing with risk and uncertainty. Water Practice and Technology 11 (1), 127—138.

van de Meene, S., Brown, R., Farrelly, M., 2011. Towards understanding governance for sustainable urban water management. Global Environmental Change 21 (3), 1117–1127.

Vivas, G., Gómez-Landesa, E., Mateos, L., Giráldez, J.V., 2009. Integrated water management in an ancestral water scheme in a mountainous area of southern Spain. s.l. In: World Environmental and Water Resources Congress 2009: Great Rivers.

Vörösmarty, C.J., Green, P., Salisbury, J., Lammers, R.B., 2000. Global water resources: vulnerability from climate change and population growth. Science 289 (5477), 284–288.

Vörösmarty, C.J., McIntyre, P.B., Gessner, M.O., Dudgeon, D., Prusevich, A., Green, P., Glidden, S., Bunn, S.E., Sullivan, C.A., Reidy Liermann, C.E., Davies, P.M., 2010. Global threats to human water security and river biodiversity. Nature 467 (7315), 555–561.

Vugteveen, P., Lenders, H.J.R., 2009. The duality of integrated water management: science, policy or both? Journal of Integrative Environmental Sciences 6 (1), 51–67.

Walsh, C.J., Fletcher, T.D., Ladson, A.R., 2005. Stream restoration in urban catchments through redesigning stormwater systems: looking to the catchment to save the stream. Journal of the North American Benthological Society 24 (3), 690–705.

WBM, B., 2009. Evaluating Options for Water Sensitive Urban Design-A National Guide. Joint Steering Committee for Water Sensitive Cities, Canberra.

Williams, N., Rayner, J., Raynor, K., 2010. Green roofs for a wide brown land: Opportunities and barriers for rooftop greening in Australia. Urban Forestry & Urban Greening 9 (3), 245–251.

Wilson, G., Edwards, P., McGrath, J., Baumann, J., April 2013. Integrated Water Management planning in Melbourne's north. Water 118–122.

Wong, T., Allen, R., Beringer, J., Brown, R.R., Deletic, A., Fletcher, T., Gangadharan, L., Gernjak, W., Jacob, C., O'Loan, T., Reeder, M., Tapper, N., Walsh, C., 2012. Blueprint 2012-stormwater Managment in a Water Sensitive City. CRC for Water Sensitive Cities, Clayton.

Wong, T., 2006a. An overview of water sensitive urban design practices in Australia. Water Practice and Technology 1 (1).

Wong, T., 2006b. Water sensitive urban design-the journey so far. Australian Journal of Water Resources 213–222.

WSAA, 2014. The Role of the Urban Water Industry in Contributing to Liveability, s.l. WSAA.

Chapter 17

# Integrating WSUD and Mainstream Spatial Planning Approaches: Lessons From South Africa

**Elizelle Juanee Cilliers and Hildegard Edith Rohr**

*Urban and Regional Planning, Research Unit for Environmental Sciences and Management, North-West University, Potchefstroom, South Africa*

## Chapter Outline



**ABSTRACT**

Within the notion of broader sustainability and resilience thinking, the concepts of green infrastructure planning and water sensitive urban design (WSUD) are gaining importance. This chapter argues that these concepts should form part of mainstream spatial planning approaches to help ensure its success and legal acknowledgment. As such, this paper considers green infrastructure and WSUD and focuses on the current reality in South Africa in terms of need of WSUD, its application, and implementation and discusses the various unique planning considerations. Based on the current reality, the chapter describes lessons from South Africa in terms of (1) the need for context-driven design guidelines, (2) the importance of social benefits related to WSUD in South Africa, (3) the need of the emerging middle-class South African to be familiar with WSUD approaches, and (4) integrating WSUD as part of mainstream spatial planning.

**Keywords:** Design; Green infrastructure; Planning; South Africa; Spatial; Water sensitive.

## 17.1 THE INTERFACE BETWEEN GREEN INFRASTRUCTURE AND WSUD

> *Among the many things I learnt as a president, was the centrality of water in the social, political and economic affairs of the country, the continent and the world.*
>
> Nelson Mandela (2002)

The concept of green infrastructure has emerged internationally as a way of understanding how green assets and ecological systems function as part of the fabric of urban infrastructure that supports and sustains society and builds resilience (Harrison et al., 2014, p. 67) and secure the provisioning of ecosystem services in human-dominated

Approaches to Water Sensitive Urban Design. https://doi.org/10.1016/B978-0-12-812843-5.00017-4

city landscapes. According to Ahern (2011, p. 159), green infrastructure is "spatially and functionally integrated systems in support of sustainability." Articulating the value of green infrastructure in terms of social, environmental, and economic considerations is an emerging research theme (Dobbs et al., 2011; James et al., 2009; Soares et al., 2011), with much of this research arguing that green infrastructure is fundamental to the concept of sustainable cities (Ahern et al., 2014, p. 255). However, valuing green infrastructure is complex, as it cannot always be related to a quantifiable economic value (Rics, 2004). Unlike the market for most tangible goods, the market does not yield an observable unit price for environmental quality (Cilliers et al., 2015, p. 353). "The fact that green spaces are not articulated in direct monetary terms is one of the most important reasons for their vulnerability to urban pressures" (More et al., 1988, p. 141; Luttik, 2000) and often the reason why green infrastructure and related approaches are neglected in the planning and decision-making process (Bertaud, 2010). As such, the concept of green infrastructure became dominant in recent planning research to help put it on par with other infrastructure (transport, communication, water supply, and wastewater systems) (Pauleit et al., 2011) and to illustrate the added community value of including green infrastructure as part of the broader planning process.

Within the multidisciplinary and multibeneficial role of natural assets lies the interface between green infrastructure and water sensitive urban design (WSUD). Similar to the discussion on public and intellectual discourse on stormwater and wastewater management (Seung-Hyun, 2015) is the argument that green infrastructure could provide a variety of solutions (in contrast to the monofunction of gray infrastructure), such as afforestation and forest conservation, reconnecting rivers to floodplains, wetlands restoration, and conservation, water harvesting options, green space planning, permeable pavements, riparian buffers, establishing flood bypasses, and green roofs initiatives (Green Cape, 2016, p. 31). Similar to green infrastructure planning, WSUD approaches aim to consider the environment in conjunction with infrastructure planning, design and management at the earliest possible stage of the decision-making process. WSUD is a multidisciplinary approach to urban water management and aspires to manage the urban water cycle more sustainably, focusing on the interaction between the urban built form and water resources management (Wong, 2006), thereby addressing water security concerns (Carden et al., 2013). By considering all major aspects of the water cycle and their interaction with urban design, WSUD aims to be the medium through which sustainable urban water management can be achieved. It has also been argued that WSUD can satisfy water-related needs at the lowest cost to society, while minimizing environmental and social impacts.

## 17.2 THE STATE OF WSUD IN SOUTH AFRICA

South Africa is a rapidly urbanizing country facing complex water management challenges, including significant resource shortages, environmental issues, and fragmented institutional structures (Carden et al., 2013). At 490 mm per annum, South Africa's rainfall is half the global average (WWF South Africa, 2013). According to the Department of Water and Sanitation, South Africa experienced its worst drought in 1983 with the national average dam level at 34% (Writer, 2015). In December 2016, the dam level increased to 54%. To compound matters, the country's water distribution is split between east and west, with 43% of South Africa's total rainfall occurring on only 13% of the land area (Writer, 2015) resulting in 8% of the land area producing 50% of the surface water, as illustrated in Fig. 17.1. Moreover, 21% of South Africa receives less than 200 mm rain a year (WWF South Africa, 2013).

Wetlands comprise only 2.4% of South Africa's land surface area, and yet the 2011 National Biodiversity Assessment revealed that 48% of the countries wetlands types are critically endangered (SANBI, 2016). According to Kleinhans and Louw (2007), land use practices and activities are known to influence the drivers of ecosystem condition (hydrology, geomorphology, water quality), which in turn determine habitat attributes and their biological responses to stress. Sustainable water supply has and always will in future be a major concern in South Africa because of the countries low average rainfall, limited underground aquifers, critical state of wetlands, high reliance on water transfers from neighboring countries (Water Resources Group, 2009), and most (98%) (Turton, 2008) of its surface water resources are already fully allocated (Fisher-Jeffes et al., 2012).

To make matters worse, rapid population growth and the lack of strict spatial planning and land use management plans and policies (in favor of protecting valuable ecological infrastructure) have resulted in many local municipalities being characterized by low-density urban areas and unplanned sprawling rural settlements. These development characteristics encourage car use, exacerbate social segregation, increase greenhouse gas emissions, and lead to the loss of natural resources (SACN, 2016). As a result, the typical South African city is resource intensive and suffers from inefficiencies across sectors (Turok and Borel-Saladin, 2014), placing tremendous stain on the country's water and wastewater distribution systems integrity. The National Research Council (2006, pp. 7–10) defines system integrity as consisting of physical, hydrological, and water quality integrity. Like many other developing countries "South Africa has made great strides in addressing the inequalities of the past in provision of water, but unfortunately the focus on expanding services

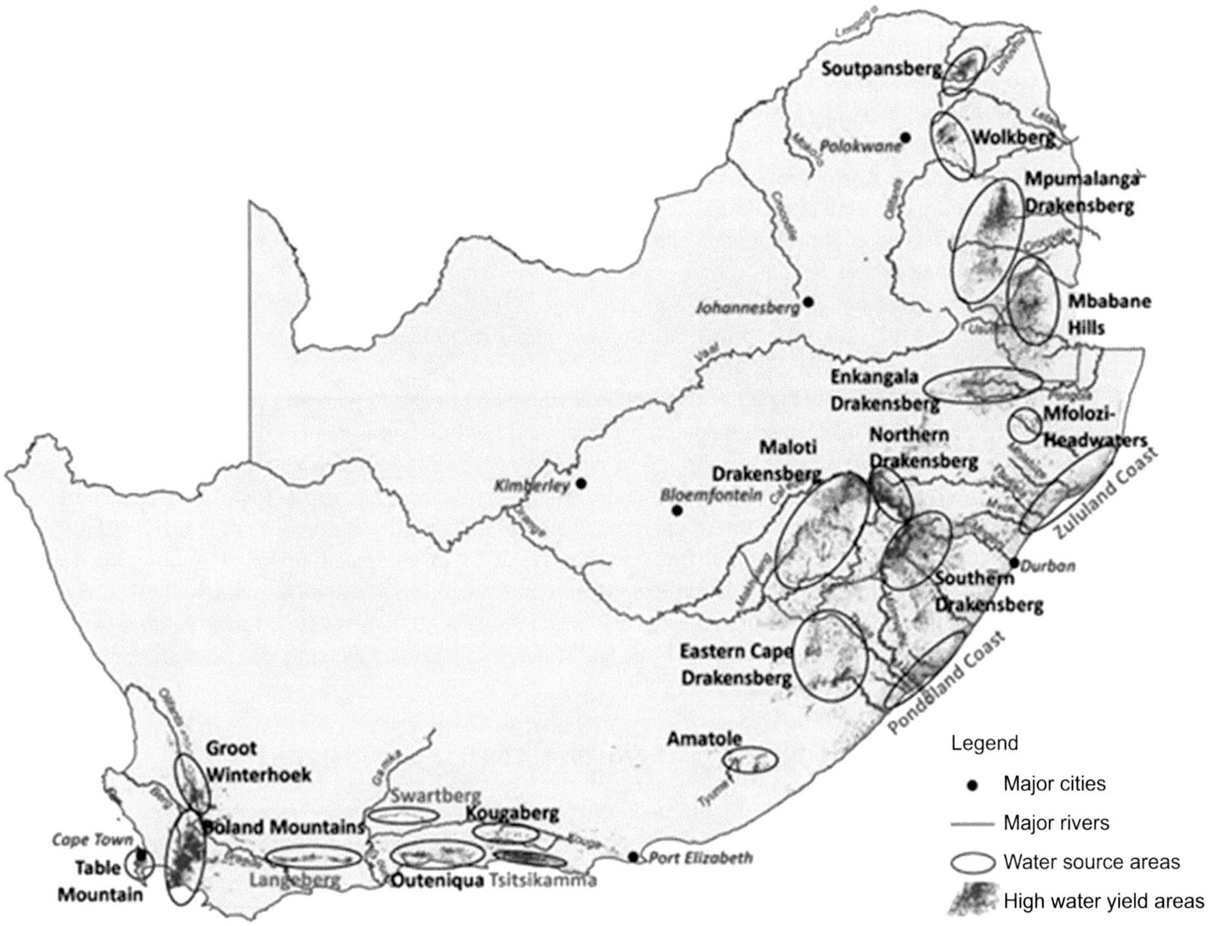

**FIGURE 17.1** South Africa's water source areas. *From WWF South Africa, 2013. An Introduction to South Africa's Water Source Areas. 2013 Report. Available at: http://awsassets.wwf.org.za/downloads/wwf_sa_watersource_area10_lo.pdf.*

provision has often been at the expense of adequate operation and maintenance of existing infrastructure" (Van Zyl, 2014, p. 1). The lack of asset management has resulted in a loss of physical integrity of the distribution systems, reflected in the 37% of urban water which is lost (as nonrevenue water) once it enters the engineered distribution (WWF South Africa, 2013).

However, the need and focus on extending provision of water service to unserviced communities has resulted in the perception that this is a less important component of the distribution system (Van Zyl, 2014, p. 35). As such, water is recognized as a strategic resource under stress (Turton, 2008) because in the short term, water pollution poses the greatest risk to societies (Ashton, 2000; Turton, 2008).

Current water use statistics illustrate that 59% of water in South Africa is used by agricultural activities, 25% by urban activities, 5.7% by mining and industrial activities, and 4.3% by rural activities (Writer, 2015). Although the urban water use is fairly low in comparison with the other sectors, South Africans have a general disregard for the value of water, which has resulted in very poor usage habits. In several municipalities, the average daily consumption ranges between 400 and just below 900 litres/person/day (L/p/d) (DWS, 2012), which is extremely high and irregular for a water-scarce country. Without intervention, the current consumer trends will increase the demand for water significantly over the next 30 years. Based on rising population, economic growth projections and current efficiency levels, demand for water in South Africa will be 17% greater than the available supply by 2030, representing a 2.7–3.8 billion $m^3$ water deficit.

In the long term, economic development will be limited by water availability and quality (Fisher-Jeffes et al., 2012). Various possibilities exist to address these coming issues including some technical solutions, enhancing efficient governance, improved stakeholder collaboration, and sustainability objectives that engender partnership and climate consciousness. WSUD approaches, linked and integrated into mainstream urban planning approaches, might hold the key to solving some of these problems.

WSUD is a relatively new concept in South Africa, although a fair amount of research has been conducted in recent years, leading to the 2014 framework published by the South African Water Research Commission (WRC) entitled "Water Sensitive Urban Design for South Africa: Framework and Guidelines" (Armitage et al., 2014). The framework introduced the philosophy of WSUD in South Africa and defined water sensitivity in the local context as "... the management of the country's urban water resources through the integration of the various disciplines of engineering, social and environmental sciences, whilst acknowledging that: Republic of South Africa (RSA) is water scarce; access to adequate potable water is a basic human right; the management of water should be based on a participatory approach; water should be recognized as an economic good; and water is a finite and vulnerable resource, essential to sustaining all life and supporting development and the environment at large" (Armitage et al., 2014, p. ii). Although an all-encompassing definition, it does not identify any concrete objectives relating to the planning or implementation of WSUD. The same is true for the broader application of green infrastructure and its implementation within the South African environment (Cilliers and Cilliers, 2016).

In this sense, the approach to planning is part of the problem, where an integrated planning context is envisioned but not enforced. For example, the 2001 White Paper on spatial planning and land use management (Ministry of Agriculture and Land Affairs, 2001) refers to the provision of life support systems that "require interferences with the landscape where the natural resources, like bio-mass, energy resources, minerals, water, land-space, are to be found." Green infrastructure planning and WSUD are encapsulated under a broad umbrella as part of a national solution but not realized in either policy formulation ("water" only being referred to three times within the 2013 promulgated Spatial Planning and Land Use Management Act [SPLUMA] of South Africa) or implementation and execution at the local municipal level, these being primarily concerned with land use management and related issues. Pretorius (2012) concluded that "Spatial Development Frameworks (as a component of strategic spatial planning) and water resource management and planning on a local level are not effectively integrated."

## 17.3 THE APPLICATION OF WSUD IN A SOUTH AFRICAN CONTEXT

Given the widespread poverty and inequality in the country, South Africa faces various challenges in the delivery of services to the disadvantaged. Given the major disparities between rich and poor, a legacy of the apartheid era where policy enforced separate development for different ethnic groups (Fisher-Jeffes et al., 2012), a major focus on rural planning and design, is required today. Hence, WSUD is referred to as water sensitive design (WSD) in South Africa to allow for a broader focus on the rural environment (Carden et al., 2016). However, for the purposes of this paper, the term WSUD will be used. Although service delivery in a "green" or water-sensitive manner adds another layer of administrative complexity, green infrastructure, and WSUD might actually deliver more equitable services and water supply to the communities in question (Fisher-Jeffes et al., 2012). Nonetheless, when considering WSUD in the South African context, various challenges dictate its success including the following:

1. **The lack of accountability regarding water and water usage**: The role and status of water is not appreciated by all people and many residents do not recognize South Africa to be a water-scarce country, leading to a culture of dependency (Rodda et al., 2016).
2. **Inadequate funding and limited capacity for water management**: The provision of stormwater management in South Africa is largely funded from property rates, implying that stormwater departments have to compete with many other departments (housing, transport, etc., with often more pressing needs) when advocating for funding. Consequently most stormwater departments in South Africa are chronically underfunded, with some estimated to be receiving as little as 10% of what is actually required for maintenance (Carden et al., 2013). Furthermore, water tariffs are not cost reflective. According to the nonfinancial census of 2013 (Statistics South Africa, 2013), 5.27 million households received free basic water and 3.10 million households received free basic sanitation. Many municipalities in South Africa run their water/sanitation service at a loss. Currently, no municipality in South Africa charges for stormwater services (Fisher-Jeffes et al., 2012), placing more pressure on other sources of income to finance these services.
3. **Separated water management structures**: A case study of four major metropolitan municipalities (City of Cape Town, City of Johannesburg, eThekwini, and City of Tshwane) in South Africa identified that management of stormwater was separated from that of water and sanitation (Fisher-Jeffes et al., 2012). Generally speaking, all towns in South Africa are designed with separate stormwater infrastructure focused on getting resources in and out of a city as quickly as possible. This perspective of dealing with water services prioritizes quantity (flow) management with little or no emphasis on the preservation of the environment (Armitage et al., 2014). This too has had a major impact on water quality, as nonpoint sources of pollution contaminate stormwater. Unlike combined stormwater and wastewater

infrastructure, which treats water before discharge, separate systems discharge contaminated stormwater back to receiving waterbodies with little or no treatment at all. South Africa has no national or provincial standards for pollutant removal from stormwater (Armitage et al., 2013, p. 14). As water and sanitation departments generate more income than stormwater departments, the former has effectively more power in the city decision-making process (Fisher-Jeffes et al., 2012), thereby inhibiting integrative water cycle ambitions.

4. **Inadequate integration between different spheres concerned with water management**: Spatial Development Frameworks (SDFs) are often not being formulated because of a lack of transdisciplinary planning approaches. Delays in finalizing planning and regulatory instruments with legal force further inhibit the potential for cross-departmental coordination and integration (Fisher-Jeffes et al., 2012). The typical fragmented silo management approach leads to unsynchronized planning approaches where planning at a city-wide level (such as strategic or spatial planning) is often not coordinated with infrastructure planning, which is being carried out at a line function level (as with water service departments).
5. **The threat of wastewater treatment**: Wastewater treatment works (WWTW) in South Africa are generally in a poor condition, with many being hydraulically overloaded (Mema, 2010). According to the 2011 Green Drop assessment (Department of Water Affairs, 2011), almost 40% of all WWTW were in a critical state. Inadequate wastewater treatment poses a major threat to the water sector. Due to the scarcity of water, pollutants need to be treated to ever higher standards before discharge to waterbodies that become downstream supply sources.
6. **Service backlogs**: The 2011 South African Census (Statistics South Africa, 2012) revealed that 46% of households in South Africa have access to piped water and about 85% have access to water that is of a Reconstruction and Development Programme (RDP)—acceptable level. National standards are prescribed under section 9(1) of the Water Services Act, 1997 (Act No. 108 of 1997). A basic (or RDP) household water supply is defined by the Strategic Framework for Water services as either 25 L per person per day or 6000 L per household per month supplied to the following criteria: (1) minimum flow rate of not less than 10 L/min; (2) within 200 m of a household, however, according to the DWS (2015), Cabinet approved that a basic water supply be amended from within 200 m to within the yard, although this has yet to be promulgated; (3) interruption of less than 48 h at any one time and a cumulative interruption time of less than 15 days/year; and (4) at a potable standard (SANS 241). Similar trends were visible for the sanitation supply backlog, where in 2011, 31.3% of households had below the national standard for basic sanitation (Statistics South Africa, 2012). Basic level of domestic sanitation implied: (1) a ventilated improved pit latrine, which is a dry toilet facility; (2) the preferred temporary sanitation solution is a chemical toilet; and (3) bucket toilet is unacceptable. Political pressure to provide full waterborne sanitation as basic level of sanitation is further impacting on the cost and timely delivery of basic services linked to overall municipal viability.
7. **Unique social dimension**: Water provision and management is considered more of a social concern than an environmental or economic concern in South Africa. This is because of the government's promise to provide free basic services for the poor (Fisher-Jeffes et al., 2012), conflating water provision with broader social equity challenges. Additionally, the lack of technical capacity and low staff skills further compounds the matter. In 2013, Water Services Authorities in South Africa were assessed, and only 3.3% were rated as low vulnerability. The areas of highest vulnerability (46% of utilities) reflected a lack of technical staff capacity (numbers), lack of operation and maintenance of assets (hence infrastructure fails and service delivery suffers), and low levels of staff skill, wastewater and environmental safety, and revenue collection (Department of Water Affairs, 2013). As such, many municipalities are unable to run a successful and sustainable water service business because of lack of capacity and skills.
8. **Misperceptions regarding WSUD**: Given the widespread poverty and inequity in the country, politicians are more likely to question: (1) how reasonable it is to place so much emphasis on the environment when so many people are living in inhumane conditions (despite ecosystem services being promoted in terms of its social value); (2) the riskiness of new untested measures such as WSUD in comparison with tried and tested infrastructure; and (3) the longer timeframes associated to WSUD compared with the urgent need to provide basic services to the people (Fisher-Jeffes et al., 2012).
9. **Misplaced planning focus**: Infrastructure-focused targets have neglected longer-term sustainability requirements, and inappropriate, unsustainable higher levels of service are often provided for short-term political gain.

Another challenge relates to the regulatory (policy and legislative) framework applicable to WSUD and planning in South Africa. Various overlapping policies and legislation guide the sections of the planning and management of water and related services in South Africa. This poses a major challenge in itself when dealing with implementation and coordination of such services. Table 17.1 captures some of the applicable frameworks, policies, and legislation along with the implementation challenges.

**TABLE 17.1 Regulatory Challenges Related to Water Sensitive Urban Design (WSUD) in South Africa**

| Guiding Policy and Legislation | Objective of Policy and Legislation | Implementation Challenge |
|---|---|---|
| The Constitution of the Republic of South Africa (Schedule 4, Part B) | Determines that the provision of stormwater services in urban areas is the responsibility of the local municipality | Many municipalities treat stormwater as "hazardous water" that needs to be disposed of as rapidly as possible to prevent damage to road structures. However, this paradigm fails to recognize a broad range of regulations and has resulted in the fragmented "silo management" of the urban water cycle. |
| National Environmental Management Act (NEMA) | Guarantees citizens the right to an environment that is not harmful to their health or well-being | NEMA places a responsibility on developers to prevent practices that have harmful effects on the environment. However, it seems that governments at local level have limited jurisdiction or authority on environmental affairs, often referring them to provincial governments. |
| National Water Act (NWA), (Part 4, Section 19.1) | Places the responsibility of controlling water pollution on the land owner | The regulatory framework is at odds with the current approach to stormwater management, which actively conveys often polluted stormwater to the nearest watercourse, posing a potential public health risk and undermining citizens' rights to a healthy environment. |
| The National Water Services Act, No. 8 of 1997 (NWSA) | Water service delivery is the responsibility of local government in their role as Water Services Authorities, who are required to complete a Water Services Development Plan (WSDP) every 5 years, encapsulating responsibilities and tasks required in service delivery | The Act neither spells out local government's role in water resource protection nor its responsibilities for integrated water resource management (Haigh et al., 2010). Stormwater is considered a part of the provision of roads and infrastructure. Hence it is seldom comprehensively dealt with in municipal WSDPs. |
| National Development Plan (NDP) | Sets a broad strategy and goals for the development of a desirable future for South Africa | Water is one component addressed in the plan, but the document does not, nor is meant to, deal with the details of managing water. It does, however, include the provision of affordable, sufficient, and safe water to meet the needs of the population, while ensuring limited negative environmental impacts. |
| Second National Water Resource Strategy (NWRS2) | Addresses various national goals and provides strategies to manage water resources at catchment scale | It neither deals with nor sets a vision for the management of water within an urban setting. This lack of a vision makes WSUD hard to achieve. |
| White Paper on Climate Change | Encourages development and use of WSUD to capture water in the urban landscape and to minimize pollution, erosion, and disturbance | It notes that "urban infrastructure planning must account for water supply constraints and impacts of extreme weather related events"; however, an integrated planning approach is still lacking. |

## 17.4 SPATIAL PLANNING CONTEXT IN SOUTH AFRICA

Planning as a profession has evolved from a designing art to a management and social science (Zhang, 2006, p. 12). The focus of spatial planning is to enable and translate future visions based on evidence, local distinctiveness, and community-derived objectives into policies, priorities, and programs (The Royal Town Planning Institute, 2007). It aims to achieve land and resource allocations and their redistribution among various interest groups to achieve a balance between the segments of society and society as a whole, and between current and future costs and benefits (Zhang, 2006, p. 25). Twenty years ago, Campbell (1996) predicted that planners will face tough decisions in coming decades about "where they stand on protecting the green city, promoting the economically growing city, and advocating social justice." Today, this is still true as conflicts between these goals and visions are not superficial, personal, or merely conceptual or temporary, caused by the untimely confluence of environmental awareness and economic recession. Rather, they are related to "the historic core of planning and are a leitmotif in the contemporary battles in both our cities and rural areas" (Campbell, 1996). The current notion of sustainable development is central to the core objectives of spatial planning, encapsulated in the recent UN Development Agenda "Transforming our World – the 2030 Agenda for Sustainable Development," along with the identified Sustainable Development Goals (United Nations, 2015). Planning is now more than ever required to guide sustainable future initiatives, create feasible spatial realities, and manage land use and space change effectively.

This is also true for South Africa where the apartheid legacy left trajectories of unbalanced development, inequity, and unsustainable space. Informal and traditional land use development processes are still poorly integrated into formal systems of spatial planning and land use management (SPLUMA, 2016), resulting in harmful and in many cases irreversible degradation of land and water resources across the country. South Africa, as in many other countries, "calls for cross-sectoral coordination and integrated planning approaches echoed across different fields of planning" (Tornberg, 2011).

South Africa's planning history has seen a considerable array of legislation. Much of the legislation responsible for managing land uses predates 1994 (the apartheid era when urban development was strictly separated according to different ethnic groups). The post-apartheid era saw a great focus on integrated development with the Local Government Transition Act Second Amendment (97 of 1996) requiring Integrated Development Plans (IDPs) as a legal requirement. The White Paper on Local Government (1998) provided elements and methodology to realize such requirements (Pretorius, 2012). However, some of the post-1994 legislation was recently found unconstitutional by the Constitutional Court (such as sections of the Development Facilitation Act [DFA], 1995). As such, 2013 saw the promulgation of a new SPLUMA. It was designed to assist with effective and efficient planning and land use management (SACN, 2015, p. 4), focusing on local land use planning procedures. The Act also aimed to specify the relationship between spatial planning and land use management systems and other categories of planning, such as planning for **water and environmental resources**.

Acknowledgment of other related services versus the actual integration of such services as part of mainstream planning is a very different reality. Until recently, the vision of an integrated, holistic planning process was not realized. In fact green infrastructure and related WSUD approaches were often neglected or sacrificed (Cilliers and Cilliers, 2015). This may be the result of broad regional-scale planning that had yet to be translated to the level of local government tasked with implementation of the process (Cilliers and Cilliers, 2016). Nevertheless, SPLUMA now enables integrative, trans-disciplinary planning approaches as it requires all municipalities to produce Spatial Development Frameworks (SDFs), which is instrumental in embedding integrated green infrastructure across all three tiers of government. Although SPLUMA does not specifically focus on water planning and management as part of spatial planning, it does require integrative, collaborative planning, and WSUD could well be realized within this new spatial planning legislative framework when integrated into new SDFs.

## 17.5 SOUTH AFRICAN APPROACHES TO WSUD

WSUD (or WSD in South Africa) is a system-based approach that promotes the development of blue-green infrastructure, including sustainable drainage systems (SuDS), alternative water resources, and Water Conservation and Water Demand Management techniques (Carden et al., 2016). The segregated spatial reality in South Africa makes approaches to WSUD more complex, as different approaches are needed in formal and informal settlements. Informal settlements are characterized by high population densities and limited infrastructure, thus resulting in very different needs to that of formal urban areas, where water technologies have been implemented for some time. Such complexity is captured in the framework for water-sensitive settlements (Armitage et al., 2014) and expressed as "two histories, one future."

According to the research of Fisher-Jeffes et al. (2012), the WSUD approach requires a combination of tools, trials, and tactics to achieve long-term success. The South African approaches to WSUD are still fairly focused on tools, with only limited references to trials and tactics. The WRC of South Africa is the driver of the tools and recently commissioned a project to develop a framework and set of national guidelines for implementing WSUD in South Africa, more specifically to enhance water resource protection, water conservation, and water/stormwater reuse in urban areas (Fisher-Jeffes et al., 2012). Other resource materials and tools developed over the years include the South African Guidelines for SuDS, the Guidelines for Human Settlement Planning and Design, the Complete Streets Design Guideline Manual of the City of Johannesburg, the Standard Detail for Roads and Stormwater of the City of Tshwane, and the Urban Design policy of the City of Cape Town (Water Sensitive Urban Design South Africa, 2016). The City of Cape Town was instrumental in the passing of the first municipal bylaw in the country that required the treatment of stormwater prior to its discharge to the receiving waters (Fisher-Jeffes et al., 2012). Various academic papers and student research have been conducted in recent years, further expanding the literature base and resource materials for WSUD in South Africa. For examples, see Fisher-Jeffes et al. (2013), Carden and Armitage (2013), Ward et al. (2012), Vice and Armitage (2011), Fisher-Jeffes et al. (2012), Rohr (2012), and Lottering et al. (2015).

Some local case studies and practices were also initiated in recent years in South Africa, such as the following:

- The 10.5 ha Green Point Urban Park in Cape Town (Fig. 17.2) opened in 2011 as a "people's park" that provides recreational, educational, and ecological facilities (City of Cape Town, 2017). Irrigation needs of the park are supplied from artesian wells in the suburb. It is estimated that this system saves 580 ML of potable water annually. The artesian water (under pressure) is also used to drive a hydroelectric water wheel to generate electricity for use in the park and to showcase renewable energy, part of the environmental education and demonstration center. Unique gardens and wetland areas showcase biodiversity and endangered vegetation types. The Green Point Urban Park is an example of the implementation of socially specific, people-oriented WSUD strategies, with the ultimate aim of increasing both amenity and biodiversity value (Carden et al., 2013).
- The first major permeable paving scheme of the Western Cape was implemented in 2010 at the City of Cape Town's Grand Parade (open plaza) (Fig. 17.3). Only 15% of the total area of the site was required to be laid with permeable pavers to handle the stormwater runoff for the entire parade surface. While the scheme has improved flooding conditions by attenuating sheet flows, no provision was made to capture and use the stormwater runoff that flows to a low-lying portion of the site, where it infiltrates into the permeable surfaces. It is subsequently collected and discharged by a perforated drainage pipe into the stormwater disposal system (Carden et al., 2013). This case is an example of an integrated approach toward planning and engineering systems and the added value in terms of WSUD.
- Water recycling initiatives in Bloemfontein (South Africa) at the Qala Phelang Tala Canaan Project (Fig. 17.4) where innovative solutions focused on domestic food production in combination with water recycling (Cilliers and Cilliers, 2016). The project presents a range of examples such as rainwater harvesting, maize plantations, and a hanging pumpkin patch to shade the patio, thus focusing on the household scale, and ways to adopt greener approaches (Cleangreenfs, 2014). This project serves as an example of how WSUD initiatives can be introduced to and maintained by local communities, especially with reference to rural environments.

**FIGURE 17.2** Green Point Urban Park (City of Cape Town, 2017). *From https://en.wikipedia.org/wiki/Green_Point_Common#/media/File:Green_Point_%26_Mouille_Point_From_Signal_Hill.jpg.*

**FIGURE 17.3** City of Cape Town's Grand Parade (Urban Water Management, 2017). *From https://en.wikipedia.org/wiki/Grand_Parade_(Cape_Town)#/media/File:Cape_Town_City_hall_and_Grand_Parade_panorama.JPG.*

**FIGURE 17.4** Qala Phelang Tala Canaan Project (Startlivinggreen, 2017).

FIGURE 17.5 Tshwane (City of Tshwane, 2012).

- Development and conservation of greenbelts and natural assets in Tshwane (100 km North of Johannesburg) (Fig. 17.5), especially within residential settlements, was initiated by Tshwane's Parks, Horticulture, and Cemetery branch. The initiative generally focuses on previously disadvantaged areas, including Atteridgeville, Soshanguve, and Ga-Rankuwa, and includes the development of nurseries for seedling production, forested nature areas, conservation areas and bird sanctuaries, and the rehabilitation of wetlands and bushveld. Many of these projects are focused on school greening with the goals of environmental education and creating environments within schools that are conducive for learners, staff, and residents around the area (City of Tshwane, 2012). This case stresses the need and importance to sensitize local authorities and communities in terms of the value of natural assets and the long-term benefits it holds for both cities and their inhabitants.
- South Africa has also seen various green roof initiatives being implemented, such as the Green Roof Pilot Project of eThekwini Municipality in Durban (South Africa), part of the Municipal Climate Protection Programme (Fig. 17.6). It was initiated in 2004 and focused on impacts of climate change, such as higher temperatures, and increases in the frequency and severity of floods and droughts that are expected as the result of climate change (Greenroofs, 2008). Two adjoining flat topped roofs at the Engineering Services building have also been planted with 12 different varieties of vegetation in small, tailor-made planting trays to monitor the growth patterns of the various vegetation types throughout the year (Armitage et al., 2014, p. 10). Green roof case studies are also found in Cape Town with a garden on the roof of the Dorp Street offices of the Western Cape Department of Environmental Affairs and Development Planning in the CBD of Cape Town (Armitage et al., 2014, pp. 4−11).
- Some other local initiatives include the Johannesburg Eco City initiative, where ecological principles were adopted in urban design and renewal processes (UNEP, 2007); the integration of environmental issues in the newest IDP for Durban; and the development proposal of Verkykerskop (Fig. 17.7), a small-scale agricultural town in the Free State Province that successfully integrates sustainable agricultural and environmental management into the larger fabric of the town (DEA, 2012). The Verkykerskop Project serves as an example of how integrated planning approaches, considering new urbanism, green urbanism, and green infrastructure planning could result in more sustainable and efficient living environments.

Although limited, and in most instances not fully integrated as part of the spatial planning process, these case studies do offer some insights and lessons for further endeavors when considering WSUD in a local context. Understanding the rural context is especially important for planners and authorities, as most of South Africa's "remaining" freshwater ecosystem priority areas (wetlands, estuaries, free-flowing rivers, and areas with high groundwater recharge potential) are located

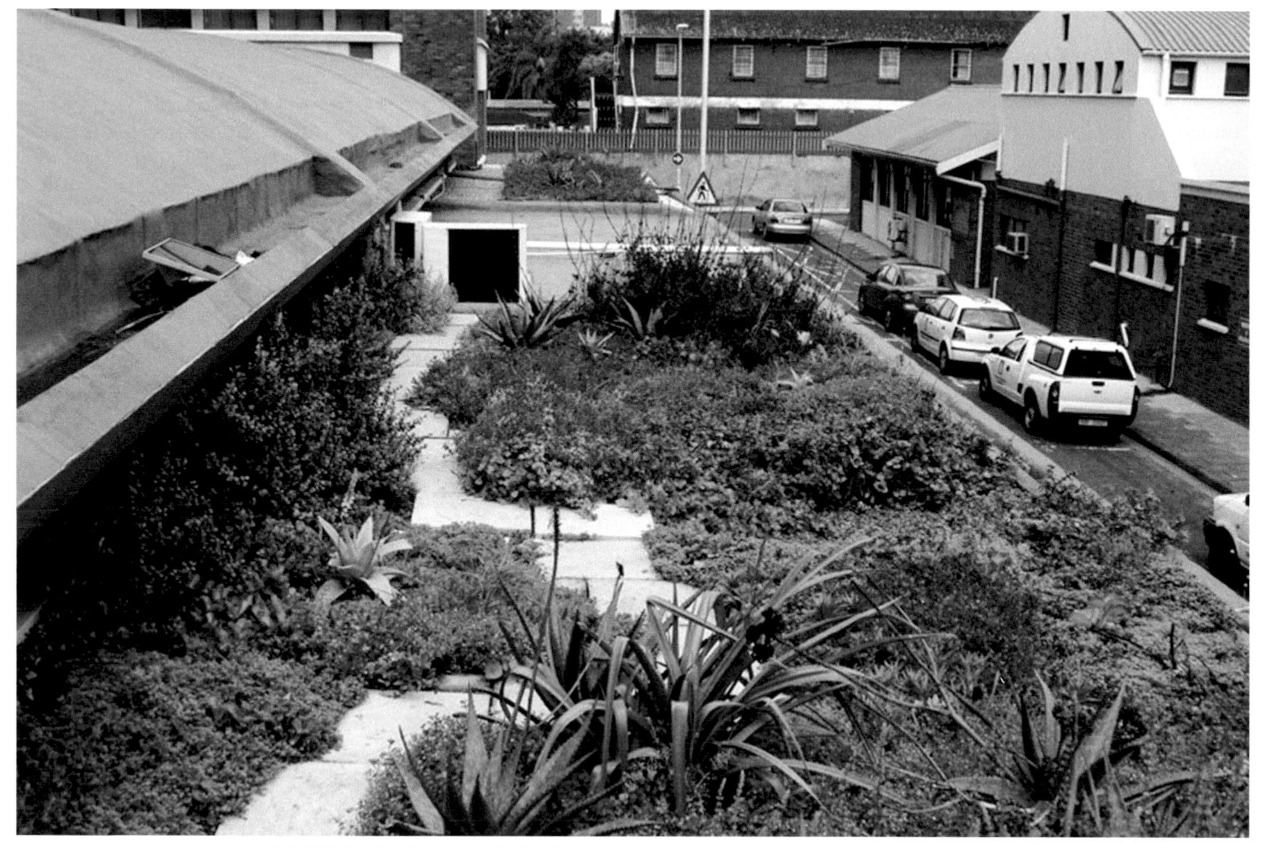

FIGURE 17.6 Green Roof Pilot Project of eThekwini Municipality (Greenroofs, 2008).

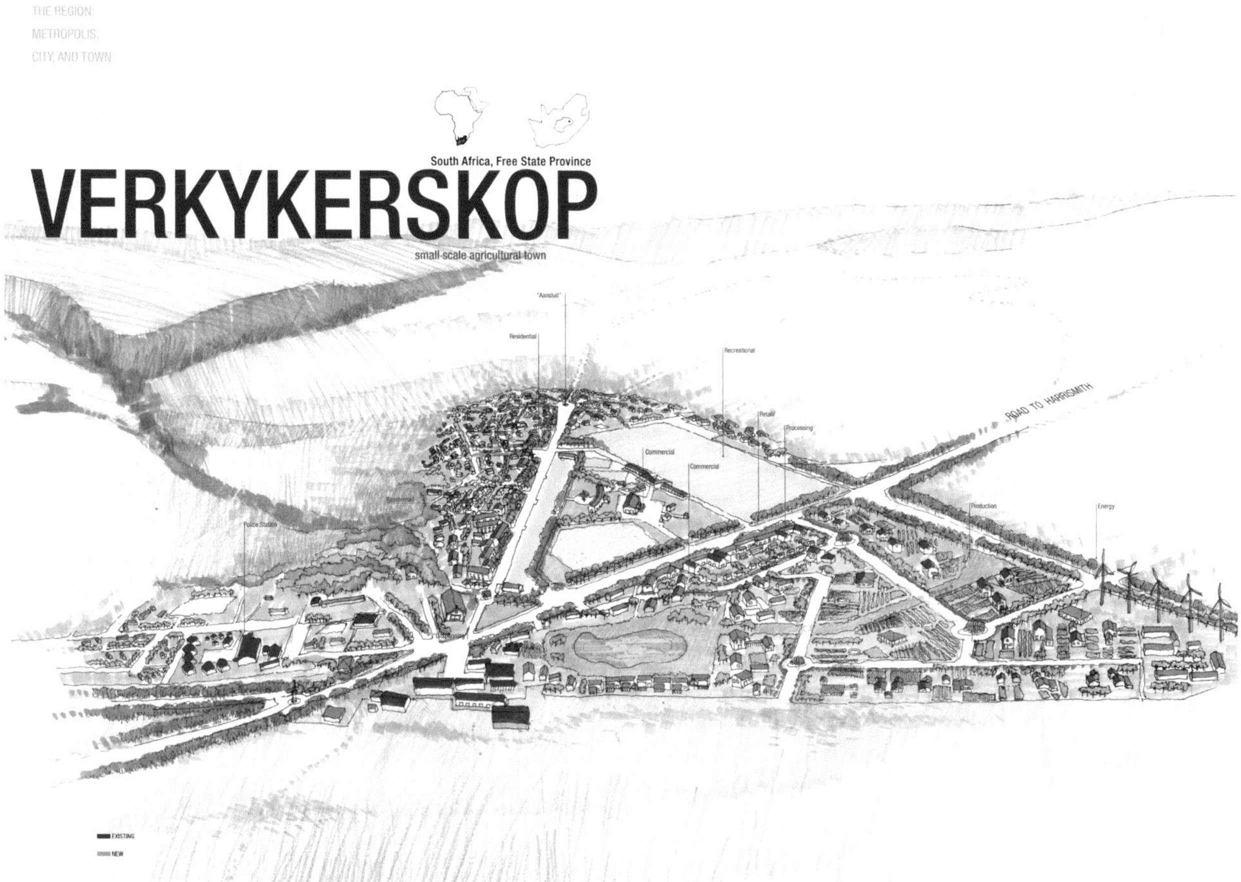
THE REGION:
METROPOLIS,
CITY, AND TOWN
South Africa, Free State Province
VERKYKERSKOP
small scale agricultural town
"Aansluit"
Residential
Recreational
Retail
Processing
Commercial
Commercial
ROAD TO HARRISMITH
Production
Energy
Police Station
EXISTING
NEW

FIGURE 17.7 Development proposal of Verkykerskop (Urban Green File, 2012, p. 19).

outside of urban footprints. Unfortunately, many municipal spatial planning tools, such as the Spatial Development Framework, lack detail planning because of the already unstructured and unmanaged characteristics of most rural settlements. This too has led to unplanned and unmanaged uptake of valuable lands, especially those valued for their ecosystem services. The incremental introduction of spatial planning and land use management in rural area, as mandated by SPLUMA, create an opportunity to implement WSUD initiative in rural areas. Currently the WRC is funding a research project titled Securing Water Sustainability through "Innovative Spatial Planning and Land Use Management Tools – Case study of two local municipalities", which specifically focuses of community-orientated and cost-effective ways to introduce WSUD in rural settlements.

Based on the local case studies, it is evident that planners and authorities should rethink services provision (levels of services) in (especially) rural areas, as the dispersed and sprawling characteristics of rural settlements make the provision, operations, and maintenance of water and wastewater distribution systems extremely expensive. As such, green infrastructure or "off-grid" services should be considered.

## 17.6 LESSONS FROM SOUTH AFRICA

Lessons from South Africa are based on green infrastructure and specific WSUD cases, including but not limited to the issues discussed in the following sections.

### 17.6.1 There is a need for unique, context-driven design guidelines and considerations

The received planning "wisdom" in South Africa has always been northern in origin (Watson, 2004, p. 252) as South Africa was annexed by the British and officially became their colony in 1815. While not criticizing the validity of the South African planning literature, there is a need to understand the assumptions on which these theories are based (Watson, 2004, p. 253) so that they can be applied in the local context. In fact, some of these assumptions do not hold in all contexts, and planners are often urged to be innovative when dealing with planning challenges. The local enthusiasm for creating "African places" is gaining importance, but then so too should the tools, trials, and tactics be developed in support of context-driven planning. Much of the planning theories deal with environmental processes at a regional scale, which are often not translatable in a practical way to the local governments tasked with their implementation (Cilliers and Cilliers, 2016).

### 17.6.2 Socioeconomic growth is subject to WSUD approaches

Water demand is driven by growth of the high-consuming lower middle class and above segments of society. Collectively, these segments have increased from 61% of the population (55 million) in 2000 to 69% in 2009 and have resulted in a rising per capita consumption because of wider use of showers, toilets, and landscaping of residential areas. Household water use in 2030 is projected at 3.6 billion $m^3$, with the wealthiest quintile of the population accounting for half of this use. South Africa's socioeconomic growth will be restricted if water security, resource quality, and associated water management issues are not resolved in a timely manner. WSUD should no longer be regarded as a luxury, but rather should be prioritized as a necessity in interests of the country as a whole.

### 17.6.3 The emerging African middle class are reliant on WSUD approaches

The (re)development and upgrading of informal settlement areas in a water-sensitive manner pose several challenges, such as limited budgets, increasing population, and a National Housing Policy advocating for only basic water supply and sanitation services for these areas. WSUD should no longer be the domain of the upper socioeconomic class as it is equally important to the poor communities in need of quantity and quality water. WSUD not only entails far more than retrofit of urban systems to be more water sensitive but also includes a social dimension to environmentally educate communities. As such, informal settlement development should attempt to "leapfrog" the stages through which the formal settlement areas have developed, thereby avoiding the need to retrofit these areas at some time in the future. Using water-sensitive technologies should also result in a range of secondary benefits for these communities, helping to address some of the misperceptions of authorities regarding the social advantages of WSUD. WSUD approaches should form part of national priorities, recognizing that advocating WSUD principles in policies will be confronted by challenges of density, scale of demand, and political sensitivities concerning the perceived quality of the engineering options it represents. The focus of providing WSUD in South Africa should be framed as a social component and justified in terms of equity and provision of services to all people (Fisher-Jeffes et al., 2012).

### 17.6.4 WSUD should be integrated as part of mainstream spatial planning

Research of Carden et al. (2013) refers to the transdisciplinary nature of WSUD and that "the adoption of WSUD should extend further than the purely engineering approach, such as the construction of specific features such as wetlands or retention basins." However, WSUD should be extended even further and included as part of mainstream spatial planning. This relates to different planning scales, ranging from community-level planning (with localized social benefits related to the WSUD approaches) to city-level planning (with economic and environmental benefits translated into cost savings and revenue). "Urban environments have little national legal protection, leaving the responsibility, 'moral' obligation and initiative to municipalities to ensure that their urban environments are sustainably managed, and included in planning strategies" (Du Toit and Cilliers, 2015). Green infrastructure should be considered at every scale of the planning process, from the strategic framework right down to neighborhoods, streets, and individual houses (Scottish Government, 2011, p. 2). If formalized as part of mainstream spatial planning, WSUD will probably gain the legal support it needs to be fully integrated into the urban development cycle. And as mentioned before, incorporating green infrastructure and WSUD approaches into spatial planning approaches requires multidisciplinary collaboration (Cilliers and Cilliers, 2016).

From a planning perspective, the integration of WSUD and mainstream spatial planning are dependent on the following:

- Acknowledgment of the different planning scales (site, neighborhood, city, and regional scale);
- Inclusion as part of formal policies and legislation guiding spatial planning and future developments;
- Multi- and transdisciplinary collaboration between different stakeholders to inform the planning, design, and implementation process;
- Utilizing planning tools to determine water demand and supply for different timeframes; and
- Raising awareness of the mutual benefits and spin-offs of including WSUD as part of mainstream spatial planning to change perceptions and build a case in favor of integrated planning approaches.

## 17.7 CONCLUSIONS

There is no single resource as essential to sustaining life as water (UNDP, 2005). Green infrastructure planning, along with WSUD approaches, is gaining importance within the milieu of broader sustainability and resilience thinking. This is further accentuated in South Africa where water demand is increasing, and water supply is likely to decrease over the coming years. Recently, some WSUD initiatives have been initiated in South Africa, but an integrated planning approach is still lacking.

South Africa needs to develop its own unique WSUD approach to fit the local context and challenges. Such an approach should include the following:

1. Enhance water accountability among citizens;
2. Address the inadequate funding required to support water management activities;
3. Initiatives to integrate water management structures;
4. Initiatives to integrate the different disciplines and spheres of responsibility charged with the planning and management of water;
5. Address the current service backlogs;
6. Address the environmental and health issues related to inadequate wastewater treatment;
7. Justifying and selling WSUD primarily as a social tool in the South African environment;
8. Addressing the misperceptions regarding WSUD and, indeed, the overall green infrastructure planning approaches in South Africa; and
9. Refocus the short-term planning vision in favor of long-term, sustainable practices.

With the promulgation of the SPLUMA of 2013, spatial planning in South Africa is now positioned to facilitate transdisciplinary planning approaches, an essential requirement for successful WSUD approaches. This entails the development of tools, trials, and tactics (Fisher-Jeffes et al., 2012) to fit the local context. Various tools have been developed in recent years, and there is some reference to trials in urban and rural areas where WSUD approaches have been successfully implemented. From a spatial planning perspective, more tactics should be developed to prove the benefit of WSUD in terms of environmental, social and economic terms. This could involve inter alia, the valuation of green assets, measured in terms of direct and indirect benefits to communities and cities.

Currently the tools, trials, and tactics of WSUD are approached in an ad hoc manner in South Africa. They need to be integrated and planned on a national level to fully realize the objectives and possibilities of WSUD.

Clearly, spatial planning structures could be utilized to coordinate such initiatives and align objectives of the various disciplines and planning departments.

Finally, based on the unique social complexities characterizing the South African nation, a unique context-driven WSUD approach is needed. Such an approach would be strongly linked to the socioeconomic growth possibilities and the emerging African middle class who are in immediate need to understand and master the WSUD approaches and water-sensitive technologies, which should mitigate the effects of water scarcity and reverse water pollution behaviors. In this sense, WSUD can be instrumental in developing social and intergenerational equity, as it has the potential to "connect" spatially divided communities and settlements through linking of open spaces and promoting these spaces to showcase blue-green infrastructure and creating "liveable" spaces, as in the case study of rural Vaalharts area.

Land use that was once considered a local issue is now recognized as a force of global importance (Sustainable Cities Institute, 2012). The effective integration of water resource management and planning in strategic spatial planning is key to sustainable, equitable, and viable communities (Pretorius, 2012). It is essential to invest in green infrastructure and adopt dynamic, integrated, and forward-thinking solutions. WSUD provides such an innovative solution to the land-water challenge and should be integrated as part of mainstream spatial planning to capitalize on outcomes from the full range of social, environmental, economic and legal benefits. Such an approach and regional strategy should facilitate cross-sectoral and transdisciplinary integration, highlighting the links between policy making and service delivery at national, regional, and local scales.

## ACKNOWLEDGMENTS

The financial assistance of the National Research Foundation (NRF) and Water Research Commission (WRC) toward the research outcomes described in this chapter is acknowledged. Opinions expressed and conclusions arrived at are those of the authors and are not necessarily to be attributed to the NRF.

## REFERENCES

Ahern, J., 2011. From fail-safe to safe-to-fail: sustainability and resilience in the new urban world. Landscape and Urban Planning 100 (4), 341–343.

Ahern, J., Cilliers, S., Niemela, J., 2014. The concept of ecosystem services in adaptive urban planning and design: a framework for supporting innovation. Landscape and Urban Planning 125 (2014), 254–259.

Armitage, N., Fisher-Jeffes, L., Carden, K., Winter, K., Naidoo, V., Spiegel, A., Mauck, B., Coulsen, D., 2014. Water Sensitive Urban Design (WSUD) for South Africa: Framework and Guidelines. Report to the Water Research Commission by Urban Water Management Research Unit Departments of Civil Engineering. Environmental & Geographical Science, Social Anthropology, and Political Studies, University of Cape Town. WRC Report No. TT 588/14.

Armitage, N., Vice, M., Fisher-Jeffes, L., Winter, K., Spiegel, A., Dunstan, J., 2013. Alternative Technologies for Stormwater Management. The South African Guidelines for Sustainable Drainage Systems. Report to the Water Research Commission by University of Cape Town. WRC Report No. TT 588/13.

Ashton, P., 2000. Integrated Catchment Management: Balancing Resource Utilization and Conservation. AWIRU.

Bertaud, A., 2010. The Study of Urban Spatial Structures. http://alain-bertaud.com.

Campbell, S., 1996. Green cities, growing cities, just cities? Urban planning and the contradictions of sustainable development. Journal of the American Planning Association 62 (3), 296–312.

Carden, K., Armitage, N.P., 2013. Assessing urban water sustainability in South Africa – not just performance measurement. Water SA 39 (3), 345–350.

Carden, K., Fisher-Jeffes, L.N., Coulson, D., Armitage, N.P., August 2013. Integrating design, planning and management - water-sensitive urban settlements. Water and Wastewater, IMIESA Conference proceedings.

Carden, K., Ellis, D., Armitage, N.P., 2016. Water Sensitive Cities in South Africa – Developing a Community of Practice. Urban Water Management Research Unit, University of Cape Town, South Africa.

Cilliers, E.J., Cilliers, S.S., 2015. From green to gold: a South African example of valuing urban green spaces in some residential areas in Potchefstroom. Town Planning Review 67, 1–12.

Cilliers, E.J., Cilliers, S.S., May 2016. Planning for Green Infrastructure: Options for South African Cities. South African Cities Network.

Cilliers, E.J., Timmermans, W., Van Den Goorbergh, F., Slijkhuis, J.S.A., 2015. Green place-making in practice: from temporary spaces to permanent places. Journal of Urban Design 20 (3), 349–366.

City of Cape Town, 2017. Green Point Park. Available at: http://www.capetown.gov.za/capetownstadium/green-point-park.

Cleangreenfs, 2014. UFS Students to Think Green in Urban Planning. Available at: http://cleangreenfs.co.za/wp/?p=5.

City of Tshwane (CoT), 2012. Spatial Development Framework. Available at: http://www.tshwane.gov.za/AboutTshwane/CityManagement/CityDepartments/City%20Planning,%20Development%20and%20Regional%20Services/MSDF%202012/Chapter%204.pdf.

DEA (Department of Environmental Affairs), 2012. About the Green Economy. Available at: https://www.environment.gov.za/projectsprogrammes/greeneconomy/about.

Dobbs, C., Escobedo, F.J., Zipperer, W.C., 2011. A framework for developing urban forest ecosystem services and goods indicators. Landscape and Urban Planning 99.

Du Toit, M., Cilliers, S., 2015. Urban ecology. In: du Plessis, A. (Ed.), Environmental Law and Local Government in South Africa. Juta, Cape Town, pp. 753–780.

DWS Department of Water Affairs, 2011. Green Drop Report 2011. Available at: https://www.dwa.gov.za/Documents/GD/GDIntro.pdf.

DWS (Department of Water and Sanitation), 2012. Blue Drop Report 2012. http://www.dwa.gov.za/Documents/BD2012/Limpopo.pdf.

DWS Department of Water Affairs, May 2013. Strategic Overview of the Water Sector in South Africa. Department of Water Affairs (DWA).

DWS Department of Water Affairs, 2015. Strategic Overview of the Water Sector in South Africa. Department of Water and Sanitation (DWS).

Fisher-Jeffes, L.N., Carden, K., Armitage, N.P., Spiegel, A., Winter, K., Ashley, R., 2012. Challenges facing implementation of water sensitive urban design in South Africa. In: Proceedings of the 7th International Conference on Water Sensitive Urban Design, Melbourne, Australia.

Fisher-Jeffes, L., Carden, K., Armitage, N.P., 2013. The future of urban water management in South Africa – achieving the goal of Water Sensitive Cities. In: Southern African Region YWP Conference, 16–18 July 2013. Stellenbosch, South Africa.

Green Cape, 2016. Water Market Intelligence Report. Western Cape Department of Agriculture. Available at: http://greencape.co.za/assets/GreenCape-Water-Economy-MIR-2016.pdf.

Greenroofs, 2008. Greenroofs.com: The Greenroof and Greenwall Project Database: EThekwini Municipality Green Roof Pilot Project. Available at: http://www.greenroofs.com/projects/pview.php?id=1002.

Haigh, E.H., Fox, H.E., Davies-Coleman, H.D., 2010. Framework for local government to implement integrated water resource management linked to water service delivery. Water SA 36 (4).

Harrison, P., Bobbins, K., Culwick, C., Humby, T., La Mantia, C., Todes, A., Weakley, D., 2014. Resilience Thinking for Municipalities. University of the Witwatersrand, Gauteng City-Region Observatory.

James, P., Tzoulas, K., Adams, M.D., Annett, P., Barber, A., Box, J., 2009. Towards an integrated understanding of green space in the European built environment. Urban Forestry and Urban Greening 8, 65–75.

Kleinhans, C., Louw, M., 2007. River EcoClassification: Manual for EcoStatus Determination (Version 2). Module A: EcoClassification and EcoStatus Determination. WRC Report No. TT 329/08. Water Research Commission, Pretoria, South Africa.

Lottering, N., Du Plessis, D., Donaldson, R., 2015. Coping with drought: the experience of water sensitive urban design (WSUD) in the George Municipality. Water SA 41 (1), 01–08.

Luttik, J., 2000. The value of trees, water and open space as reflected by house prices in The Netherlands. Landscape and Urban Planning 48 (3), 161–167.

Mema, V., 2010. Impact of Poorly Maintained Waste Water and Sewage Treatment Plants: Lessons from South Africa. Council for Scientific and Industrial Research, Pretoria. Available at: http://www.ewisa.co.za/literature/files/335_269%20Mema.pdf.

Ministry of Agriculture and Land Affairs, July 2001. White Paper on Spatial Planning and Land Use Management. Wise Land Use.

More, A.A., Stevens, T., Allen, P.G., 1988. Valuation of urban parks. Landscape and Urban Planning 15, 139–152.

National Research Council, 2006-Committee on Public Water Supply Distribution, National Reseach Council, 2006. Drinking Water Distribution System: Assessing and Reducing Risks. National Academic Press, Washington, DC.

United Nations, 2015. Resolution Adopted by the General Assembly on 25 September 2015. Available at: http://www.un.org/ga/search/view_doc.asp?symbol=A/RES/70/1&Lang=E.

Pauleit, S., Liu, L., Ahern, J., Kazmierczak, A., 2011. Multifunctional green infrastructure to promote ecological services in the city. In: Niemela, J. (Ed.), Urban Ecology: Patterns, Processes and Applications. Oxford University Press, New York, pp. 272–285.

Pretorius, H., 2012. A Practical Assessment of Spatial Development Frameworks in Terms of Water Resources for Development. Dissertation submitted for the degree Magister Artium et Scientiae in Urban and Regional Planning and the North-West University.

Rics, 2004. On the Value of Sustainability: Meeting of the Minds. Asset Strategies.

Rodda, N., Stenström, T.A., Schmidt, S., Dent, M., Bux, F., Hanke, N., Buckley, C.A., Fennemore, C., 2016. Water security in South Africa: perceptions on public expectations and municipal obligations, governance and water re-use. Water SA 42 (3), 456–465.

Rohr, H.E., 2012. Water Sensitive Planning: An Integrated Approach towards Sustainable Urban Water System Planning in South Africa (Dissertation, North-West University).

SACN (South African Cities Network), 2015. SPLUMA as a Tool for Spatial Transformation. SACN Programme: Inclusive Cities. Available at: http://www.sacities.net/wp-content/uploads/2015/SPLUMA-as-a-tool-for-spatial-transformation_final.pdf.

SACN (South African Cities Network), 2016. State of South African Cities Report 2016, Johannesburg: SACN: s.n. 978-0-620-71463-1.

SANBI, 2016. A Framework for Investing in Ecological Infrastructure in South Africa. South African National Biodiversity Institute, Pretoria, pp. 1–22.

The Scottish Government, 2011. Green Infrastructure Design and Placemaking, p. 28.

Seung-Hyun, K., 2015. Green Infrastructure as Water Sensitive Urban Design Strategy for Sustainable Stormwater Management. National Institute of Environmental Research, Korea.

Soares, A.L., Rego, F.C., McPherson, E.G., Simpson, J.R., Peper, P.J., Xiao, Q., 2011. Benefits and costs of street trees in Lisbon. Portugal. Urban Forestry and Urban Greening 69–78.

SPLUMA, 2016. Spatial Planning and Land Use Management Act. Government Notice 559 in Government Gazette 36730 Dated 5 August 2013. Commencement Date: 1 July 2015 [Proc. No. 26, Gazette No. 38828 dated 27 May 2015].

Startlivinggreen, 2017. Qala Phelang Start Living Green. Available at: http://startlivinggreen.co.za/wp/volunteers4change/.

Statistics South Africa, 2012. Census 2011. Stats SA, Pretoria.

Statistics South Africa, 2013. Non-financial Census of Municipalities. Stats SA, Pretoria.
Sustainable Cities Institute, 2012. Land Use. The National League of Cities. Available at: http://www.sustainablecitiesinstitute.org/Documents/SCI/Topic_Overviews/Land%20Use%20-%20Full%20OverviewNew_NLC.pdf.
The Royal Town Planning Institute, 2007. Shaping and Delivering Tomorrow's Places: Effective Practice in Spatial Planning. Available at: http://discovery.ucl.ac.uk/10828/2/10828_Executive_Summary.pdf.
Tornberg, P., 2011. Making Sense of Integrated Planning: Challenges to Urban and Transport Planning Processes in Sweden.
Turok, I., Borel-Saladin, J., 2014. Is urbanisation in South Africa on a sustainable trajectory? Development Southern Africa 31 (5), 675–691.
Turton, A., November 18, 2008. Three Strategic Water Quality Challenges that Decision-Makers Need to Know about and How the CSIR Should Respond. Centre for Scientific and Industrial Research (CSIR).
UNDP (United Nations Development Programme), 2005. Status of Integrated Water Resources Management (IWRM) Plans in the Arab Region.
UNEP, 2007. Eco-cities. Available at: http://www.unep.org/training/programmes/Instructor%20Version/Part_2/Activities/Lifestyle_Patterns/Consumption/Strategies/Eco_City.pdf.
Urban Green File, 2012. Paradise found. Urban Green File Magazine 16 (11), 19–25.
Urban Water Management, 2017. Water Sensitive Design. Available at: http://www.uwm.uct.ac.za/uwm/wc/grand-parade.
Van Zyl, J., 2014. Introduction to Operation and Maintenance of Water Distrubution Systems, 1st ed. Republic of South Africa: Water Research Commision.
Vice, M.A.P., Armitage, N.P., 2011. A 'systems thinking' assessment of the management of a constructed wetland A case study in the City of Cape Town, South Africa. In: Proceedings of the 12th International Conference on Urban Drainage. Porto Alegre, Brazil.
Ward, S., Lundy, L., Shaffer, P., Wong, T., Ashley, R., Arthur, S., Armitage, N., Walker, L., Brown, R., Deletic, A., Butler, D., 2012. Water sensitive urban design in the city of the future. In: Proceedings of the 7th International Conference on Water Sensitive Urban Design, Paper 315, 8 pp., Melbourne, Australia, 21-23 February 2012.
Water Resources Group, 2009. Charting Our Water Future: Economic Frameworks to Inform Decision-Making. The Barilla Group, the Coca-cola Company, the International Finance Corporation, McKinsey & Company, Nestlé S.A., New Holland Agriculture SABMiller Plc, Standard Chartered Bank, and Syngenta AG.
Water Sensitive Urban Design South Africa, 2016. Water Sensitive Urban Design South Africa. Urban Water Management. Available at: http://wsud.co.za/.
Watson, V., 2004. Teaching planning in a context of diversity. Planning Theory and Practice 5 (2), 252–253.
Wong, T., 2006. Water sensitive urban design - the journey thus far. Austrailian Journal of Water Resources 3 (10), 213–222.
Writer, S., 2015. Who Is Using All the Water in South Africa? Business Tech. Available at: https://businesstech.co.za/news/general/104441/who-is-using-all-the-water-in-south-africa/.
WWF South Africa, 2013. An Introduction to South Africa's Water Source Areas. 2013 Report. Available at: http://awsassets.wwf.org.za/downloads/wwf_sa_watersource_area10_lo.pdf.
Zhang, T., 2006. Planning theory as an institutional innovation: diverse approaches and nonlinear trajectory of the evolution of planning theory. City Planning Review 30 (8), 9–18.

## FURTHER READING

Moyogo, 2017. Tshwane. Flickr.com/photos/moyogo/2006852021/.
WESSA (Wildlife and Environment Society of South Africa, 2012. South Africa's Water Resources: WESSA Position Statement. Available at: http://wessa.org.za/uploads/images/position-statements/Water%20Resources%20-%20WESSA%20Position%20Statement%20-%20Approved%202013%20.pdf.

Chapter 18

# The Role of WSUD in Contributing to Sustainable Urban Settings

**Beau B. Beza[1], Joshua Zeunert[2] and Frank Hanson[3]**

[1]*School of Architecture and Built Environment, Deakin University, Geelong, VIC, Australia;* [2]*UNSW, Sydney, NSW, Australia;* [3]*Victorian Planning Authority, Melbourne, VIC, Australia*

## Chapter Outline



**ABSTRACT**

Water sensitive urban design (WSUD) is a concept widely accepted and partially acted on throughout Australia's federal and state governments. The concept, however, is currently applied at the local municipal level by linking with existing corridor and precinct structure plans, urban planning/legislative frameworks (e.g., Clause 56.07—04, Victoria), and service provisions (e.g., drainage guidelines). However, to realize WSUD outcomes at the applied municipal level, coordination and cooperation of developer, client, consultants, the local government and respective water authority is essential. Unfortunately, much of the planning and design-related WSUD material is focused on stormwater management. In terms of WSUD, *management* at the municipal level is a key term to reflect upon, as the handling and control of water is essential so that its design and use minimizes risk to built features and people. The other aspect to management, *directing*, has been largely relegated in WSUD to a service role, where harvested stormwater is used for maintenance-related issues such as greening roads and street verges, open space areas, and a city's landscape designs and features. To move away from this predominantly service role and into a more progressive application of WSUD, the human/water interaction may be better achieved through the merging of contemporary insights of sustainability with those of WSUD when realizing built outcomes.
To act more sustainably, applications of WSUD in suburban and city settings will need to shift toward a wider *dynamic* use where, for example, harvested stormwater is used to produce environmental outputs that contribute to and enhance urban sustainability, rather than simply maintain the built setting. Relatively new concepts such as healthy and livable suburbs potentially help provide contemporary links with WSUD, which can be used to promote sustainability and provide economic and/or social benefits to communities. This chapter examines the role WSUD has in contributing to urban sustainability through a discussion of its current role and its potential for enhancing sustainability.

**Keywords:** LID; SuDS; Urban planning; Water-sensitive city; Water sensitive urban design.

## 18.1 THE CURRENT ROLE OF WSUD IN URBAN DEVELOPMENT

The use of water as a resource in urban development and its contribution to a sustainable urban setting can be loosely divided into a number of design intervention categories. However, it may be best to first consider these "categories" more

**Approaches to Water Sensitive Urban Design. https://doi.org/10.1016/B978-0-12-812843-5.00018-6**

in terms of a "spectrum of use." Each category may then be drawn upon, in a respective situation, to achieve ideally, water sustainability, or preferably, positive environmental outcomes. Referring to the latter point, minimal water is directed away from natural systems to accommodate people in the urban setting, and more water is allowed to enter into the environment through natural means. In terms of relating this "spectrum" thinking to built outcomes that fulfill the principles of water sensitive urban design (WSUD) described in Chapter 1, one needs to understand the scales at which urban development operates. In the Australian state of Victoria, urban development and its visioning occurs at three different scales: A **growth area scale** that revolves around the planning of urban spaces that meet the needs of 100,000−400,000 people; **precinct scale** that envisions services and facilities for 7000−30,000 people; and the **subdivision/permit scale** that lays out the homes, stores/shops, open spaces, roadways, etc., for populations of 500−5000 people (PSPG, 2009).

At the growth area scale, the application of WSUD revolves around larger state or national policy concepts of urban sustainability, such as "city as catchment" and "urban water cycle management." This type of conceptual thinking considers the "[...] coordinated management of all components of the water cycle including water consumption, rainwater, stormwater, wastewater and groundwater, to secure a range of benefits for the wider catchment" (TW-CAAC, 2014, p. 2) and adjacent land areas. In this sense, water catchments (also called watersheds in North America), aquatic flows, and their effect on the environment are considered part of the growth area planning process.

At the precinct scale, the "pen to paper" ideas on the layout of urban settings begin to take shape and start to influence the flow of water, catchment water quality and its harvest/replenishment potential, and the degree to which urban development and WSUD impact the environment. At this spatial scale, state-based legislated planning and environmental and vision documents take effect (Choi, 2016) and largely determine the extent to which WSUD measures are acted upon.

The "subdivision/permit" scale begins to realize the actual urban impacts on the environment. Importantly, this scale is where the application of WSUD measures conceived at the policy level occurs. In this sense, WSUD measures can be *performance* based, but they are usually left to the authority and/or developer to decide on the means to achieve this end. An alternative way to achieve positive environmental outcomes at this scale is to *prescribe* the WSUD means to achieve a desired outcome (e.g., Melbourne Planning Scheme, 2017a).

To draw attention to the various WSUD measures that can be applied at the precinct and subdivision levels, this discussion returns to the "spectrum of use" idea presented above. The spectrum alludes to a palette of WSUD measures one may draw from to achieve positive environmental outcomes. The spectrum includes biophysical elements such as swales, wetlands, rainwater/stormwater/detention tanks and basins, and/or performance/prescription-based measures. To simplify this discussion, we use the terms "soft" design measures and "hard" design approaches that can be employed to achieve positive outcomes. Table 18.1 lists the range of soft and hard WSUD measures one can use to create positive environmental outcomes.

The soft design approach revolves around more natural design and constructed measures and elements to achieve positive outcomes. For example, if the intention is to move water slowly through a setting and to use it in the filtering and/or recharging of subsurface systems, then a grass swale and bioretention system may be sufficient to achieve these desired effects (see Wong, 2006). Soft elements may also be recommended for use through performance-based design guidelines or stormwater management documents. Hard approaches to environmental enhancement may involve the installation of drains, pipes, and storage tanks; for example, to effectively capture rainwater and store it for later use. Hard approaches can be prescribed through legislation or identified in a contract or development agreement. However, irrespective of which WSUD approach(s) is applied at a given spatial scale, the intent in using hard and soft means "[...] is for the built environment to 'integrate' within the natural fabric" of the setting (EOFWSUD, 2009, pp. 2−5).

## 18.1.1 (Urban) sustainability

Unfortunately, the term and the conceptual understanding of "sustainability" has been distorted, diluted, corrupted, misconstrued, and claimed without substantiation (Cronon, 1995; Nordhaus and Shellenberger, 2004; Freyfogle, 2006; Jensen and McBay, 2009). It has now come to reflect, in decision-making arenas, an economic underpinning (Van der Ryn and Cowan, 1996); that is, economic interests and agendas mainly influence sustainability outcomes. This argument is well articulated by the chairman of "Natural Capital Committee" in the United Kingdom, who states "The environment is part of the economy and needs to be properly integrated into it so that growth opportunities will not be missed" (Natural Capital Committee, 2014, p. 4). This perception of the environment and how it "fits" into an economic agenda reflects a market-led economy (Harvey, 2005; Hackworth, 2007; Springer et al., 2016; Allmendinger, 2016) and within this market-led focus, the life-sustaining elements of water, air, and soil are usually treated as externalities.

A market-led focus is pronounced in the Australian urban setting, where sustainability takes on an abstract form through a general disconnect between natural and urban environments. Gleeson and Beza (2014) argue that, in Australia,

**TABLE 18.1** Soft and Hard Water Sensitive Urban Design (WSUD) Measures

| | WSUD Measure | |
|---|---|---|
| | Mainly Outdoor Examples and Uses | |
| Item | Soft Measure | Hard Measure |
| (Bio)retention ponds | ✓ | |
| Check dams (to control stormwater runoff) | ✓ | ✓ |
| Designation of development/nondevelopment areas and protection of vegetation, green corridors, and watercourses | ✓ | ✓ |
| Designation of hard surfaces including roads, roof tops/building designs, drainage systems as water collection, and/or distribution devices | | ✓ |
| Educational programs | ✓ | |
| Filtration systems: soft, for example, being passive plants, soil mediums in bio-filtration/rain gardens and hard being UV or other mechanical filtration techniques | ✓ | ✓ |
| Grey/reticulated water for facilities use and/or for watering vegetation | | ✓ |
| Infiltration systems (e.g., wetlands, porous pavement, rain gardens/biofiltration) | ✓ | ✓ |
| Legislation/regulation/policy | ✓ | ✓ |
| Rain/grey/storm/black water storage tanks/devices | | ✓ |
| Retarding basins | ✓ | ✓ |
| Stream and riparian rehabilitation | ✓ | ✓ |
| Streetscapes | ✓ | ✓ |
| Swale(s) and vegetative buffer strips | ✓ | |
| Constructed wetlands | ✓ | |

Adapted from EOFWSUD, July 2009. Evaluating Options for Water Sensitive Urban Design - A National Guide. Prepared by the Joint Steering Committee for Water Sensitive Cities. The Australian Government, Department of the Environment and Energy. Available from: https://www.environment.gov.au/system/files/resources/1873905a-f5b7-4e3c-8f45-0259a32a94b1/files/wsud-guidelines.pdf; Morison, P.J., Brown, R.R., 2011. Understanding the nature of publics and local policy commitment to water sensitive urban design. Landscape and Urban Planning 99, 83–92; Wong, T., Brown, R., Deletic, A., November 2008. Water Management in a Water Sensitive City. Water (Australian Water Association), pp. 52–62.

this market-led focus is used as an arbiter of public opinion regarding the creation, use, and management of spaces within the urban environment. Public opinion, however, can be "advanced" in a number of ways and be (re)directed to focus on developing positive environmental outcomes. For example, WSUD is a relatively new water-focused sustainability measure in Australia and across the globe, which revolves around water being used in a responsible and forward-thinking manner.

## 18.1.2 Use of WSUD in new urban settings

In the urban setting, Dolman et al. (2013) suggest that the use of WSUD measures in Australia have evolved from climate change impacts, growth in the urban footprint, increasing resource use, increases in flood risk, and a growing scarcity of fresh water in cities, exacerbated by cyclical droughts. In terms of WSUD measures, the thinking underpinning this sustainability concept revolves around three different, but interrelated, engineered water systems:

*i) the potable water supply consisting of a piped system delivering water treated to drinking water standard [...], ii) the sewage system consisting of a piped system collecting and transporting wastewater to treatment plants; and iii) the stormwater drainage system consisting of various elements, from natural waterways through to constructed channels and underground piped systems mainly transporting stormwater (Wong, 2006, p. 213).*

Gradually, however, thinking by the water cycle decision makers is transitioning from "*the water supply city*" through to "*the water-sensitive city*" (WSC) (Brown et al., 2009). The WSC is positioned at a larger conceptual scale, above the three urban scales previously mentioned (i.e., growth area, precinct, and subdivision). Brown et al. (2009) present a typology of six conceptual cities that help to identify and describe the hydrosocial policy drivers used to influence positive water behavior and outcomes (i.e., urban water management) through a range of educational and/or built environment means (see also Wong, 2006; Choi, 2016). An illustration of six conceptual cities by Brown et al. (2009) is provided in Chapter 1; these are summarized below as the following:

- Water supply city
- Sewered city
- Drained city
- Water city
- Water cycle city
- Water-sensitive city.

What Brown et al. (2009) begin to reveal in the latter stages of their city water typology is a link between water and urban livability/sustainability. The term "link," however, is probably inadequate to fully capture society's growing concern for water as a life-sustaining element. Dolman et al. (2013), for example, explain that in the Netherlands, water management, land use, and urban development have been "mutually dependent" (p. 86). These authors also point out that this mutual dependence is reinforced by the Dutch government's policy to ensure its citizens are aware of the inherent dangers of living with water and that is must be accommodated "[...] by building adaptively with water in mind and by investing in green infrastructure" (p. 90). Hence, WSUD is a key element in both "green" and "blue" infrastructure (see Dreiseitl, 2015). In terms of the latter, we frame blue infrastructure as the incorporation and use of the various WSUD measures (see Table 18.1) one may apply in the urban setting to promote sustainability.

This mutual dependency of living with water demonstrates parallels with Wong's (2006) description of ecologically sustainable development (ESD), which "[...] go[es] beyond the protection of the environment from the impacts of pollution, to protecting and conserving natural resources" (p. 213). With a majority of the earth's population now living in cities, WSUD needs to play a key role in "[...] the integration of urban planning with the management, protection and conservation of the urban water cycle, that ensures urban water management is sensitive to hydrological and ecological process" (NWI, 2004, p. 30).

An issue faced by designers trying to incorporate WSUD measures is that with infill developments, current water infrastructure systems may be at capacity or soon reach capacity (Sharma et al., 2016). Reaching system capacity in infill developments results from increased water consumption and waste generation associated with the increased residential population. Sharma et al. (2016) discuss the three options urban designers and decision makers are faced with: first, to accept the current urban situation and "do nothing" to "[...] overcome the problem" (p. 10); second, enhance system capacity through the "construction of new drainage systems in addition, to, or replacing, the existing pipe and channel network" (p. 10); and lastly, "employ WSUD measures [...] [with] progressive on-site flow management measures for new re-development sites [...]" (p. 10).

In terms of these three options, the "do nothing" approach is not a viable option. Furthermore, small incremental WSUD measures may not be sufficient to turn the tide. What needs to happen is a leapfrog approach to water cycle management, as argued by Wong (2016) where he states "[t]he essence of a leapfrogging approach is about integration, so that from the outset, responses to water challenges are multi, rather than mono, functional" (p. 6). Applying this concept will help "[...] cities to avoid the environmental, social and economic vulnerabilities that come from managing the water cycle in a segmented way" (p. 5). Both developed and developing urban areas can benefit from this type of thinking. Applying the concept of leapfrogging to WSUD in the "developed" setting may be applied at the growth area, precinct, and subdivision scales. However, the WSC may be best supported by applying WSUD measures at the subdivision scale, as the aggregate of small-scale WSUD interventions may result in collective outcomes.

## 18.1.3 Living with and applying WSUD

For the WSC to be effective at a growth area scale, Wong et al. (2008) present "tentative" principles to support the development of this concept. An important dimension is a city's resilience to water system disturbances "[...] such as floods, droughts and waterway health degradation [...]" (p. 54). Equally important (and missing from principles by Wong et al. (2008)) is the public's ability to live and work with the measures called for in a WSC. In this sense, a WSC needs to be supported by the public's "buy in" to these measures.

In Melbourne, Australia, for example, a large-scale urban visioning document, Plan Melbourne Refresh (2015) (now superseded by Plan Melbourne 2017-2050 (2017)), presented a number of strategic environmental principles to "[...]

facilitate sustainability actions to address pressures such as population growth, climate change, and changing community expectations" (p. 69). Key elements in these principles are water optimization, its retention in the landscape, stormwater and wastewater treatment, and use of WSUD measures that "[...] support greener suburbs and cleaner waterways" (p. 70). This document drew attention to waterways and "use[s] the planning system to better recognise and support healthy catchments, waterways and bays" (p. 70). It also called for the implementation of both hard and soft WSUD measures to be embedded in the state's urban planning process. However, despite a focus on water in the *Plan Melbourne Refresh* document, the delivery of positive environmental gains in urban sustainability largely revolved around the concept of the 20-minute suburb, which is still a feature of *Plan Melbourne 2017—2050*. This aims to improve community health and well-being by advocating that frequently used destinations are within a 20-minute walking distance of a dwelling. Although this is an admirable goal, it fails to capture the importance of water's integration into new (greenfield) and existing (brownfield) sites.

## 18.2 WSUD AND URBAN PLANNING

The issue arising with the "*Plan Melbourne*" example is that despite calls for WSUD to be integrated and utilized within planning frameworks, it is not happening. Wong (2006) argues that a major reason for the failure to fully integrate WSUD into planning is because of organizational fragmentation in the decision-making roles, which consequently "[...] impedes integrated approaches to urban water cycle management" (p. 219).

To overcome this lack of integration, in 2004, the Australian National Water Initiative (NWI) was agreed to by the country's state and territories. The agreement itself sets forth an agenda to establish a framework that leads these states and territories to develop and implement measures that use water responsibly. However, Choi (2016) observed that despite the states and territories adhering to the NWI agreement, the WSUD-related measures remain discretionary. Notwithstanding Choi's (2016) "discretionary" comment, many states and territories do have WSUD measures woven into their respective planning frameworks (e.g., Melbourne Planning Scheme, 2017a,b). A key point of Choi's (2016) reflection is that a hierarchy exists within the planning regulatory system, and in terms of the application of WSUD, policy is regarded as a "soft law" and lacks proper penalties if a local authority chooses to ignore WSUD measures.

The implication of this "soft law" consideration is that new governance models may need to be put in place (Wong et al., 2008) to fully implement WSUD measures. For example, Legacy (2014) in her discussion of deliberative planning efforts in Vancouver, Canada, outlines how the government and community groups worked together for 7 years to produce a vision for the city that obtained long-term support and implementation for over 14 years. In effect, the interests of market-led decision-making were trumped "[...] by [implementing] a 'bottom-up' metropolitan governance framework [...]" that allowed for a single planning board to be created so that "[...] local governments [could] coordinate metropolitan-wide services and achieve metropolitan-wide policy goals" (Legacy, 2014, p. 79). The term "local" supports the Cooperative Research Centre for Water Sensitive Cities (CRCWSC, 2014) and Williams' (2016) suggestion that the municipal level is the most appropriate place to legislate for implementation of WSUD measures.

The NWI (2004) also calls for the implementation of statutory state- and territory-based water plans that address "[...] surface water and ground water management [...]" (p. 7). These water plans call for the linking of municipal land planning with WSUD (e.g., South Australian Water Plan, 2010) or through alignment of land use planning the water cycle and water-sensitive planning to achieve improved sustainability outcomes (e.g., Water for Victoria, 2016).

What one may conclude from the different foci of the two example water plans from Victoria and South Australia is that water and its "use" in the built environment is applied differently throughout Australia, and indeed throughout the world. This is evident in a suite of terminology, including WSUD, sustainable urban drainage systems (SuDS), low impact development (LID), integrated urban water management, water cycle management, and WSC, whereby each revolve around water and its appropriate use in an urban setting (see Table 18.2). Each water-based term, however, articulates a conceptual framework and application that is sufficiently different from each other that it creates fragmentation at a global scale. The result of this separation means that global awareness, benchmarking, knowledge sharing, and discourse are also fragmented, reducing the effectiveness and missing synergistic opportunities for improved environmental water cycle management (Zeunert, 2017, pp. 100—117).

In the United States, for example, LID is an intended feature of new urban developments (and redevelopment) that focus on implementing natural and/or constructed means to encourage water to percolate into the ground (EPA, 2012). By encouraging infiltration, flooding, combined sewer overflow, and/or hydrologic pollution is expected to be reduced. In the United Kingdom, SuDS is the term used to signify a sustainable use of water and revolves around building open air drainage systems in new and/or existing developments to filter out pollutants by slowly moving water through a site to a

**TABLE 18.2 Terms Used to Describe the Use and Focus of Water in New and Existing Urban Development**

**Applications and foci of using water sustainably in Australia, New Zealand, UK, USA, and Canada (this table is not exhaustive).**

| Term | Country | Focus (Sustainable Use of Water in the Urban Setting Achieved Through the Use of the following application) |
|---|---|---|
| Water sensitive urban design (WSUD) | Australia (the Middle East and the Netherlands) | "The integration of urban planning with the management, protection, and conservation of the urban water cycle, that ensures urban water management is sensitive to natural hydrological and ecological processes" (NWI, 2004, p. 30). |
| Low impact urban development and design (LIUDD) | New Zealand | Applying the concept of ecological carrying capacity to reduce impacts on the environment and its processes when making decisions on realizing new and/or existing urban developments over different spatial scales and temporal dimensions. |
| Sustainable urban drainage systems (SuDS) | United Kingdom | Realizing open air drainage systems in new and/or existing developments that slowly move water through a site to a collection point. At this collection point, water is encouraged to percolate into the ground. |
| Low impact development (LID) | USA and Canada | Implementing natural and/or constructed means that encourage water to percolate into the ground. |

Adapted from British Geological Survey, 2017. Sustainable Drainage Systems (SuDS). British Geological Survey, Natural Environment Research Council. Available from: http://www.bgs.ac.uk/suds/; EPA, March 2012. Benefits of Low Impact Development: How LID Can Protect Your Community's Resources. United States Environmental Protection Agency; Office of Wetlands, Oceans, and Watersheds, Washington, DC; NWI, 2004. Intergovernmental Agreement on a National Water Initiative. Australian Government, Productivity Commission. Available from: http://www.pc.gov.au/inquiries/current/water-reform/national-water-initiative-agreement-2004.pdf; Van Roon, M., van Roon, H., 2009. Low Impact Urban Design and Development: The Big Picture. An Introduction to the LIUDD Principles and Methods Framework (Landcare Research Science Series No. 37). The University of Auckland: National Institute of Creative Arts and Industries. Manaaki Whenua Press, Lincoln; Zeunert, J., 2017. Landscape Architecture and Environmental Sustainability: Creating Positive Change through Design, Bloomsbury, London.

collection point. The water is then designed to percolate into the ground at the collection point. A good description of SuDS is contained in the British Geological Survey (2017) report available at http://www.bgs.ac.uk/suds/. Table 18.2 summarizes these WSUD terms, and their focus as used around the world. A more complete description of the various WSUD terminologies is provided in Chapter 1.

An interesting dimension to these varying water cycle applications in the urban setting is whether or not water is considered as a core or as a peripheral element in urban development. For example, Dolman et al. (2013) refer to "water" as being a central design consideration, and feature, when realizing urban settings in Australia and the Netherlands. However, the use of water as a sustainability measure, in say the application of SuDS, is as a peripheral rather than a core feature in the development of settings. That is, water is viewed as part of the urban drainage and management system.

Regardless of the relative position of water in the Australian national or local agenda, water and its use in the built environment is moving toward "[...] 'integrated urban water cycle planning and management' (IUWCM) and urban design'" (Wong et al., 2008, p. 52). Within IUWCM are principles of WSUD, described in Chapter 1, but these are now expanded to "[...] minimising the import of potable water, minimising the export of wastewater, the improvement of stormwater quality, and the management of wastewater and stormwater [...]" (Wong et al., 2008, p. 52). All of these have been promoted by the setting of targets in the NWI to achieve the creation of water-sensitive Australian cities (see NWI, 2004, p. 20). Unfortunately, it is important to note that the National Water Commission (NWC), established to assist with the implementation of the NWI was abolished in 2014. The responsibilities of the NWC "[...] have now been transferred to other agencies" (Australian Government, 2017, npn), which may contribute further to a fragmentation of decision-making regarding water suggested by Wong (2006) earlier in this chapter.

The WSC is therefore the ultimate goal of legislators and WSUD is the objective to help achieve this desired result. Wong et al. (2008) described nine tentative principles one may use in the realization of the WSC. At the conceptual level, three "pillars" help shape the core concept (see Wong et al., 2008, p. 56):

1. City as catchment: Water for human and environmental consumption is achieved from a variety of centralized and decentralized infrastructure means.
2. Cities providing ecosystem services: Realized features, such as (re)constructed wetlands, are strategically located throughout the city to aid in both the built and natural environment's enhancement.

3. Cities comprising water-sensitive communities: Community through to individual behavior revolves around using water in an environmentally responsible and considered way.

   We propose a fourth pillar:

4. Cities as productive water systems: The city as catchment (i.e. pillar 1) is used to strategically influence WSUD measures to provide water supply systems for 'productive' urban and peri-urban irrigation/water intensive activities (e.g. urban and peri-urban agriculture and industrial/commercial activities (see Zeunert, 2017: 229).

WSUD is argued to have evolved from stormwater management to now include principles of ESD and is applied at a range of urban-related scales (see Wong, 2006). For example, considerations of water sustainability are now embedded in the state of Victoria's urban planning visioning documents (e.g., Plan Melbourne 2017-2050 (2017) at the growth area and precinct scales, which in turn are implemented through local planning legislation (e.g., Melbourne Planning Scheme (2017b) at the site scale). Unfortunately, despite these gains in water being explicitly considered in many planning visioning documents and legislation, Choi (2016) reports that "[…] all [Australian] jurisdictions lack targeted WSUD policy for urban infill and lot scale developments" (p. 5).

Overall, our discussion of WSUD thus far suggests that it represents the building blocks for creating a WSC (Dolman et al., 2013). The main weakness is that despite local municipal legislation being a well-targeted place for the implementation of WSUD measures, "[t]here is also no policy framework to support catchment scale planning and to avoid downstream LGAs [Local Government Areas] bearing the burden of upstream development activities" (Choi, 2016, p. 7). For example, in the State of Victoria, there are 31 municipal councils in metropolitan Melbourne and 48 rural councils in the state. Each largely makes development decisions based on their own respective planning schemes that are influenced by State-level planning provisions. To help with this "building block" analogy, examples of realized urban developments using WSUD measures applied in the State of Victoria are presented in the next section.

## 18.3 DRIVERS AND IMPEDIMENTS TO THE APPLICATION OF WSUD

There are a range of drivers and impediments that affect the adoption of WSUD measures in urban development. These include (1) a fragmented planning system that neither coherently supports nor enforces the use of WSUD measures and (2) a negative perception by developers and/or local governments of WSUD's management and maintenance costs. For an excellent detailed account of drivers and impediments to applying WSUD, see Sharma et al. (2016) and Dolman et al. (2013). An interesting dimension to the uptake of WSUD measures by councils and/or the public is the proximity to a water body (such as a lake or waterway). Morison and Brown (2011), for example, report positive relationships between a community's and local government's understanding of issues of water quality and their spatial proximity to a water body (see also Brody et al., 2004). They also noted that support for applying WSUD measures is affected by the socioeconomics of the area, exposure to community assets (such as wetlands), the presence of environmental/community groups, and messaging (e.g., a municipality focused on either roads, people, or the environment) (Morison and Brown, 2011). The following case studies discuss drivers and impediments for enacting WSUD measures in the State of Victoria, Australia, and its capital city, Melbourne.

### 18.3.1 WSUD examples in Victoria

Applying WSUD and environmentally sympathetic water systems in planned greenfield and growth areas of metropolitan Melbourne is increasingly becoming "business as usual." Initially treated as "one-off" opportunities to solve local drainage problems, deployment of increasingly sophisticated WSUD practices and techniques has seen their adoption grow. In this sense, WSUD has grown from a "nice to have add on" that was used as a marketing edge by innovative developers to a practice now used in most major urban developments (e.g., the current Aurora estate to accomodate 25,000 residents, http://www.places.vic.gov.au/land-and-housing/aurora). WSUD measures are now embedded in the current Victorian Planning Authority's (VPA) "Precinct Structure Planning Guidelines for Victoria - Draft for Consultation" (March 2017) (see PSPG, 2017).

Despite the specification of WSUD measures in VPA guidelines, there have been some "teething" issues in their implementation by the property development industry and local municipal authorities, for both large- and small-scale urban development projects. However, in the experience of the authors, as landscape architects and urban designers, the initial concerns regarding the implementation of WSUD have generally been addressed, with the delivery of some high-quality urban landscapes (e.g., University of Sydney rain garden project, Sydney, Australia (see Zeunert, 2017, pp. 109, 100−117)) and public realm settings (e.g., Salisbury Wetlands, Adelaide, Australia (see Zeunert, 2017, pp. 60−61)).

Some of the pioneering WSUD developments in housing estates in metropolitan Melbourne were implemented by the Victorian Government land development agency, The Urban and Regional Land Authority (URLC), which became VicUrban (in 2003), Places Victoria (in 2011), and is now known as Development Victoria. Among its early WSUD-oriented estates was Lynbrook (http://www.places.vic.gov.au/land-and-housing/lynbrook) in the south eastern suburbs of greater Melbourne (City of Casey local authority) and Aurora in the northern suburbs of Melbourne in the City of Whittlesea local authority.

## 18.3.2 Lynbrook estate, new urban setting

Lynbrook is a suburban fringe area approximately 43 km south-east of Melbourne and has a residential population of around 9000 (City of Casey, 2017). The Lynbrook development started to take shape in the mid-1990s and was promoted by the state government as "[t]he first residential community in Melbourne to fully integrate water sensitivity into its design [...], [which] incorporate[d] an innovative system that filters stormwater at street level [by using, for example, planted swales], then sends it along meandering watercourses for more filtration before it flows into Lynbrook Lake" (Lynbrook, 2017, npn). Fig. 18.1, shows an example of a median strip swale used in the development.

The realization of this water-sensitive form of estate design radical for the mid-1990s required a cooperative process between the URLC (now Development Victoria), Melbourne Water, and The City of Casey (the local municipal authority). An extensive communications process was also implemented to inform the builders, contractors, and future residents about the benefits of this new way of treating urban stormwater. Some early apprehension, however, came from the City of Casey where it was concerned with the maintenance costs of this new form of public realm, which included swales wetlands with biofiltration elements. Contractors building Lynbrook also had to be educated about not stockpiling materials and parking vehicles in the swales and the Lynbrook's residents needed to learn to "appreciate" or become accustomed to a new WSUD-based development aesthetic (Bruce, 2000).

## 18.3.3 Aurora estate, new urban setting

As Lynbrook achieved recognition for its WSUD approach in the early 2000s, planning for another estate in Victoria named Aurora began in 2001. Aurora is a 700 ha development approximately 25 km north of Melbourne, with 7500 dwellings planned to accommodate 25,000 people within a 634 ha area (Aurora Community Association, 2017). Aurora was ambitiously argued to be one of the most innovative residential communities in Australia employing, at the time, best practice WSUD. The development was also Victoria's first 6-star energy-rated community. Places Victoria (see Aurora, 2017, npn) describe the constructed houses as having the following features:

- Solar orientation of dwellings for maximum comfort and energy efficiency;
- Solar hot water;

FIGURE 18.1 Planted central median swale used in the Lynbrook development.

FIGURE 18.2 Rain garden with boardwalk used in the Aurora estate.

- 4- and 5-star heating, cooling, and appliances;
- AAA-efficient water fixtures;
- Reticulated recycled water for laundry use, toilet flushing, and garden watering, resulting in 45% less potable water use compared with the equivalent average home.

In addition to these precinct scale approaches to WSUD, Aurora implemented planted and grass swales along with filtration, infiltration, and rain garden systems as an integral part of the estate design (Fig. 18.2).

Many of these features are now considered as standard in *Plan Melbourne 2017–2050,* but as mentioned previously, they are yet to be fully incorporated into legislation throughout Victoria. The Aurora estate was also "[...] instrumental in the development of an industry-wide Sustainable Community Rating tool (Green Star – Communities), which was developed in partnership with the Urban Development Institute of Australia, the Property Council of Australia and the Municipal Association of Victoria" (Aurora, 2017, npn).

Despite the generally positive learnings listed on Places Victoria Aurora website (http://www.places.vic.gov.au/land-and-housing/aurora), the estate also had many challenges that affected the realization of WSUD measures on ground. For example, the development of Aurora on a gently undulating basalt plain with heavy clay soils, stoney rises, major stands of remnant river red gum trees (*Eucalyptus camaldulensis*), and a fragmented land ownership pattern presented significant issues for the development of a comprehensive WSUD program. These issues were initially difficult for the marketers to address in this master-planned community.

The site's main watercourse, a very subtle waterway which, prior to development, was an ephemeral creek, is known as Edgars Creek. The increasing impervious area associated with urban development increased stormwater runoff and a more reliable water flow has resulted in the creek becoming an important landscape element with a Growling Grass Frog (*Litoria raniformis*) habitat area included in the open space network. A network of swales, rain gardens, and wetlands filter stormwater runoff before it passes into the revitalized and naturalized Edgars Creek. The maintenance of these systems—as with Lynbrook—has required a continual education process for all concerned. Unfortunately, despite this education process, people still "mistake" the swales as parking areas (Fig. 18.3). The City of Whittlesea (the local municipal authority) has had to adapt its parks, gardens, and verge maintenance practices to include swales planted with dense native vegetation.

Aurora was also designed as a higher residential density (up to 30 allotments/ha in key locations) to support a high-frequency train service (yet to be implemented). Hence, there was a deliberate decision to reduce both lot sizes (many lots are about 300 $m^2$ in size) and street cross sections compared with current practices at the time. This "squeezing" of space meant a conflict between the provision of swales, rain gardens, and other WSUD techniques, the space required for on street parking, and the placement of rubbish and recycling bins.

FIGURE 18.3 The commonplace practice of mistaking a swale for a car parking area.

### 18.3.4 WSUD and the Victorian Planning Authority (from concept to application)

These local examples of applying WSUD measures have become exemplars for the broader development industry in Victoria, which has taken the key learnings, (referred to above), and begun to apply them across a broad spectrum of development projects. However, to be truly effective, a catchment-wide approach including government "buy in" with strong support from the private sector is required. To assist with this approach, the key agency responsible for supporting the planning and application of a state-wide agenda to WSUD is the VPA (VPA, 2017; https://vpa.vic.gov.au/).

A key role of the VPA is the production of Precinct Structure Plans (PSPs), which are used "[...] to set the blueprint for development and investment that will occur [in growth areas] over many years" (PSPG, n.d., p. 2). Complementary to PSPs are "PSP Guidelines" that are intended to be "tailor made" and support local site characteristics, including natural and built environment elements that incorporate WSUD measures in greenfield sites. Integral to these considerations in preparing guidelines is the preparation of a comprehensive analysis of the planned area, including an assessment of its natural systems and drainage requirements. In practice, this involves a concerted effort in bringing together the often-competing interests of developer, engineering, water management, open space and natural systems, and cultural heritage proponents to create a more integrated approach to water management. This effort has the potential to create new and/or enhance existing linear open spaces that, in the Aurora example, can impart a strong local sense of place, while performing a vital urban drainage, wetland, and habitat function.

The PSP Guidelines include provisions for integrating drainage, biodiversity, heritage, and conservation as part of an open space network (PSPG, 2017). Supporting this provision is an "Integrated Water Management Plan that identifies catchments, natural and constructed waterways, major pipelines, key overland flow paths, retarding basins and stormwater quality treatment assets (e.g., wetlands, sedimentation basins) as well as any water sensitive design and stormwater harvesting features" (PSPG, 2017, p. 45). Key considerations in developing an integrated water management plan include (1) providing reliable water services for the environment; (2) responding to site-specific needs and qualities of a precinct; (3) ensuring that "waterways and water infrastructure [...] form key structural elements of precinct design" (p. 48); and (4) "water and other related infrastructure" (p. 48), which are provided for in new developments. Another important consideration of these WSUD elements is their reinforcement in the PSP guideline that they must "[...] be integrated with the broader precinct design to support the amenity and character of town centres, community hubs and the broader open space network" (p. 46). Hence, what these considerations and elements identify is that WSUD measures in the state's new urban developments are now "business as usual."

## 18.4 LIMITING FACTORS FOR WSUD, IUWCM, AND WSCs

Following the discussion in the previous sections, we identify several overarching barriers, which, if addressed, present the opportunity to increase the reach and effectiveness of WSUD, along with IUWCM and WSCs. The following discussion elaborates on a simple hierarchy of development sites where WSUD measures may be applied in existing urban settings; addressing market-led decisions on how WSUD is implemented; state and local government-level WSUD legislation; calling for the national standards on WSUD; and going beyond the top-down approach in applying WSUD.

### 18.4.1 Beyond new developments

Universal application of WSUD is constrained given the legislative and market focus that is predominantly on new development projects, such as Lynbrook and Aurora. While this approach is important in motivating WSUD in new developments, it greatly limits its reach by largely ignoring existing settlements and their expansive land areas and water systems. As urban population growth in new developments is widespread in Australia, these clearly present easier opportunities to implement WSUD than existing contexts. However, the universal realization of WSUD requires some form of incentives across a range of urban contexts.

One way of addressing this shortcoming is through a strategic review of existing city-wide and municipal-level urban settings to identify key sites suitable for WSUD interventions. Using this data, cities and/or municipalities could develop a hierarchy of sites whereby, through redevelopment initiatives (e.g., tax concession), WSUD measures could be applied to create positive environmental outcomes, such as "[...] improve[ing] the health of waterways and catchments" (Water for Victoria, 2016, p. 12). Australia's predominantly low-density suburban typology (which in Victoria is currently at 16 dwellings per ha) and relative abundance of open and green space present greater ease of implementation than countries having longer urban histories and greater densities of urban settlement. In this sense, many municipalities in Australia possess underutilized public green spaces with irrigation-intensive lawn areas suitable for retrofitting size-appropriate hard or soft WSUD measures such as biofiltration/rain gardens/wetlands that could be filled by diverted stormwater runoff from ubiquitous impervious areas of roads, roofs, and pavements.

### 18.4.2 Beyond market-led systems

One of the "biggest" barriers to overcome in the implementation of WSUD is the "leave it to the market" paradigm currently embedded in politics, contemporary governance, and built environment practice (Harvey, 2005; Hackworth, 2007; Leitner et al., 2007; Springer et al., 2016; Allmendinger, 2016). In this sense, sustainability within a market-led approach to development is incompatible (see Klein, 2014). For example, the Aurora development illustrated how market/sustainability conflict arose when plot density increased and WSUD measures were then compromised. Positive environmental outcomes in a market-led system are actually often "less bad" than those of "business as usual" (McDonough and Braungart, 2002), despite not achieving water or environmental net balance or regenerative outcomes (Lyle, 1994; McDonough and Braungart, 2013; Zeunert, 2017). Hence, for WSUD to increase its effectiveness, like many other sustainability pursuits, it must find ways of maneuvering or bypassing market-led decision-making systems.

Furthermore, while guidelines and discretionary measures (such as the NWI) are useful starting points, there has been about seven decades of global environmental discourse advocating stronger measures be implemented to effect positive environmental outcomes. Many of these have been previously mentioned (i.e., WSUD, LID, SuDS, LIUDD) and each revolves around water and its consumption in the urban setting. However, without mandatory legislative requirements at the local government level, as called for by the CRCWSC (2014) and Williams (2016), WSUD measures will remain voluntary and may be compromised in practice, in favor of more immediately favorable outcomes.

### 18.4.3 Beyond State Legislative Requirements

State- and municipal-level implementation of WSUD can create complex, diffuse, confusing, and arguably ineffective means to achieve its realization, which is highlighted in Wong's (2006) and Zeunert's (2017) discussion of a fragmentation of knowledge and decision-making when attempting to implement WSUD measures. When legislative measures are applied at state and municipal levels, complexity and variation can be significant. However, to effectively and efficiently apply WSUD, the status quo of calling for measures to be applied at the municipal level may not be all that is required—lateral thinking may also be needed. For example, to increase the penetration of WSUD one may ask to what extent can WSUD and/or improved water performance requirements be achieved at a federal level. One example that could potentially cut across the numerous and complex state and local governance systems is to embed WSUD into the *Standards Australia* system, which is an Australian national set of "[...] documents setting out specifications, procedures and guidelines" that "[...] are based on industrial, scientific and consumer experience and are regularly reviewed to ensure they keep pace with new technologies" (Standards Australia, 2017, npn). In this sense, a national set of WSUD standards applied across Australia may encourage rapid uptake and uniformly promote the creation of positive environmental outcomes in the urban setting.

While increased (and uniform) legislative measures and standards are crucial for WSUD, award-winning and outstanding examples of WSUD (see Zeunert, 2017, pp. 100–117; Zeunert, 2016) do not result from legislative measures alone. Such exemplary WSUD projects, such as the Salisbury Wetlands (see Zeunert, 2017, pp. 60–61, 229) and "Waterproofing the South," both in Adelaide, Australia, have not relied on guidelines, legislative mechanisms, or market-led structures for their realization. Rather, such projects have gone far beyond the "business-as-usual" approach to applying WSUD to produce outcomes that substantially contribute to urban water sustainability in practice, not just in rhetoric.

## 18.5 CONCLUSION

Urban development is now considered an element of sustainability and part of a global push to improve environmental outcomes. An important mechanism to promote sustainability is to implement WSUD measures in the development of new and existing urban centers. This chapter has examined the philosophical underpinning and the planning and legislative instruments that encourage WSUD in urban developments. However, for WSUD to be effective, it must be integrated into developments and move beyond its roots in stormwater management to now contribute to the protection and enhancement of the natural environment.

Application of WSUD at three scales of urban development is likely to assist its greater integration: (1) the growth area scale; (2) the precinct area scale; and (3) subdivision/permit area scale. At the growth area scale, WSUD may first be conceived in national- and/or state-level policy, where high-level "visions" of sustainability and ways forward are presented (e.g., Australia's NWI). Sustainability and the direct application of WSUD may then be directly acted upon at the precinct area scale where the "pen to paper" ideas of laying out an urban setting begin to take shape. Then at the subdivision/permit area scale, the realization of WSUD and its cumulative contribution to urban and environmental sustainability can be assessed. At this small scale, visions and direct application of WSUD may then be revisited based on data obtained from examination at the subdivision/permit level.

This holistic "urban-scale" approach to WSUD begins to relate to IUWCM, which is a contemporary view of water management that encompasses a complete water cycle. In Melbourne, two case studies by the State's land development authority present several realized WSUD measures, with many of these measures becoming commonplace in subsequent developments throughout Melbourne's metropolitan region. However, a notable application and integration of WSUD comes from an overseas example where the Dutch have successfully conveyed a message of mutual dependence between water management, land use, and urban development to its people. To learn from this national-level example of IUWCM, Australia could also implement a national WSUD "standard" for its states and territories to apply. At this national level, WSUD measures could be developed at a scale above the previously mentioned growth, precinct, and subdivision areas. And with this potential action, a national-level agenda for WSUD could help curtail or decrease the organizational fragmentation of decision-making relating to water cycle management.

Despite this national standard suggestion, WSUD measures are easier to implement in greenfield developments. However, even in greenfield sites, the application of hard and soft WSUD measures is hindered by a fragmentation of government decision-making roles and their lack of consistency in calling for use of WSUD. Hence, the drivers or impediments for WSUD include the visioning of urban and environmental sustainability, developing consistent planning instruments and legislation that call for and enforce its application, and an educated public and development market that values WSUD as a key element to new and existing urban settings.

## REFERENCES

Allmendinger, P., 2016. Neoliberal Spatial Governance. Routledge, New York.

Aurora, 2017. Aurora. Places Victoria. Available from: http://www.places.vic.gov.au/land-and-housing/aurora.

Aurora Community Association, 2017. About Aurora. Aurora Community Association. Available from: http://www.aurora.asn.au/about-aurora/.

Australian Government, 2017. National Water Initiative. The Australian Government, Department of Agriculture and Water Resources. Available from: http://www.agriculture.gov.au/water/policy/nwi.
British Geological Survey, 2017. Sustainable Drainage Systems (SuDS). British Geological Survey, Natural Environment Research Council. Available from: http://www.bgs.ac.uk/suds/.
Brody, S.D., Highfield, W., Alston, L., 2004. Does location matter? Measuring environmental perceptions of creeks in two San Antonio watersheds. Environment and Behavior 36 (2), 229–250.
Brown, R.R., Keath, N., Wong, T.H.F., 2009. Urban water management in cities: historical, current and future regimes. Water Science and Technology 59 (5), 847–855.
Bruce, D., 2000. Water, Water Everywhere. Monash University, Monash Magazine. Available from: http://www.monash.edu.au/pubs/monmag/issue6-2000/pg4.html.
Choi, L., 2016. Towards a water sensitive policy framework for Australia's cities. In: Paper Published in the 2016 International Low Development Conference, June 26-29, 2016, Beijing, China. https://10times.com/ilidc-beijing.
City of Casey, 2017. Community Profile, Lynbrook. The City of Casey. Available from: http://profile.id.com.au/casey/about?WebID=260.
CRCWSC, 2014. Statutory Planning for Water Sensitive Urban Design. The CRC for Water Sensitive Cities. Available from: https://watersensitivecities.org.au/wp-content/uploads/2016/05/FS_B5-1_StatutoryPlanningWSUD.pdf.
Cronon, W., 1995. Uncommon ground: toward reinventing nature. In: Cronon, W. (Ed.), Uncommon Ground: Rethinking the Human Place in Nature. W.W. Norton and Co., New York, pp. 69–90.
Dolman, N., Savage, A., Ogunyoye, F., 2013. Water-sensitive urban design: learning from experience. Proceedings of the Institute of Civil Engineers 166 (ME2), 86–97.
Dreiseitl, H., 2015. Blue-green social place-making: infrastructures for sustainable cities. Journal of Urban Regeneration and Renewal 8 (2), 161–170.
EOFWSUD, 2009. Evaluating Options for Water Sensitive Urban Design - A National Guide. Prepared by the Joint Steering Committee for Water Sensitive Cities. The Australian Government, Department of the Environment and Energy. Available from: https://www.environment.gov.au/system/files/resources/1873905a-f5b7-4e3c-8f45-0259a32a94b1/files/wsud-guidelines.pdf.
EPA, March 2012. Benefits of Low Impact Development: How LID Can Protect Your Community's Resources. United States Environmental Protection Agency; Office of Wetlands, Oceans, and Watersheds, Washington, DC.
Freyfogle, E., 2006. Why Conservation Is Failing and How it Can Regain Ground. Yale University Press, New Haven.
Gleeson, B., Beza, B.B., 2014. The public city: a new urban imagery. In: Gleeson, B., Beza, B.B. (Eds.), The Public City: Essays in Honour of Paul Mees. Melbourne University Press, Melbourne, pp. 1–12.
Hackworth, J., 2007. The Neoliberal City: Governance, Ideology, and Development in American Urbanism. Cornell University Press, Ithaca.
Harvey, D., 2005. A Brief History of Neoliberalism. Oxford University Press, Oxford.
Jensen, D., McBay, A., 2009. What We Leave Behind. Seven Stories Press, New York.
Klein, N., 2014. This Changes Everything: Capitalism vs. The Climate. Allen Lane, London.
Legacy, C., 2014. Public plan-making: a deliberative approach. In: Gleeson, B., Beza, B.B. (Eds.), The Public City: Essays in Honour of Paul Mees. Melbourne University Press, Melbourne, pp. 74–88.
Leitner, H., Peck, J., Sheppard, E., 2007. Contesting Neoliberalism: Urban Frontiers. The Guilford Press, New York.
Lyle, J., 1994. Regenerative Design for Sustainable Development. John Wiley, NY.
Lynbrook, 2017. Lynbrook. Places Victoria. Available from: http://www.places.vic.gov.au/land-and-housing/lynbrook.
McDonough, W., Braungart, M., 2002. Cradle to Cradle. North Point Press, NY.
McDonough, W., Braungart, M., 2013. The Upcycle: Beyond Sustainability - Designing for Abundance. North Point Press, New York.
Melbourne Planning Scheme, 2017a. Integrated Water Management, Clause 56.07. The Planning Schemes Online. Available from: http://planning-schemes.delwp.vic.gov.au/schemes/vpps/56_07.pdf.
Melbourne Planning Scheme, 2017b. Stormwater Management (Water Sensitive Urban Design), Clause 22.23. The Planning Schemes Online. Available from: http://planning-schemes.delwp.vic.gov.au/schemes/melbourne/ordinance/22_lpp23_melb.pdf
Morison, P.J., Brown, R.R., 2011. Understanding the nature of publics and local policy commitment to water sensitive urban design. Landscape and Urban Planning 99, 83–92.
Natural Capital Committee, March 2014. The State of Natural Capital: Restoring Our Natural Assets (Second Report to the Economic Affairs Committee). Natural Capital Committee, London. http://socialsciences.exeter.ac.uk/media/universityofexeter/collegeofsocialsciencesandinternationalstudies/leep/documents/2014_ncc-state-natural-capital-second-report.pdf.
Nordhaus, T., Shellenberger, M., 2004. The Death of Environmentalism. The Breakthrough Institute. Available from: http://www.thebreakthrough.org/images/Death_of_Environmentalism.pdf.
NWI, 2004. Intergovernmental Agreement on a National Water Initiative. Australian Government, Productivity Comission. Available from: http://www.pc.gov.au/inquiries/current/water-reform/national-water-initiative-agreement-2004.pdf.
Plan Melbourne 2017-2050, 2017. Plan Melbourne 2017-2050: Metropolitan Planning Strategy. State of Victoria Department of Environment, Land, Water and Planning, Melbourne. http://www.planmelbourne.vic.gov.au/.
Plan Melbourne Refresh, October 2015. Plan Melbourne Refresh: Discussion Paper. State Government of Victoria, Melbourne. http://www.planmelbourne.vic.gov.au/__data/assets/pdf_file/0006/377313/Plan-Melbourne-Refresh-Discussion-Paper_WEB_FA-R2.pdf.
PSPG, n.d. Precinct Structure Planning Guidelines: Overview of Growth Area Planning (Part One). The Growth Areas Authority. Available from: https://vpa-web.s3.amazonaws.com/wp-content/uploads/2016/06/PSP_Guidelines_PART_ONE.pdf.pdf.

PSPG, 2009. Precinct Structure Planning Guidelines, One: Overview of Growth Area Planning. State Government of Victoria, Growth Areas Authority, Melbourne. https://vpa-web.s3.amazonaws.com/wp-content/uploads/2016/06/PSP_Guidelines_PART_ONE.pdf.pdf.

PSPG, March 2017. Precinct Structure Planning Guidelines for Victoria (Consultation Draft - *to Be Published 2017*). Victorian Planning Authority, Melbourne.

Sharma, A.K., Pezzaniti, D., Myers, B., Cook, S., Tjandraatmadja, G., Chacko, P., Chavoshi, S., Kemp, D., Leonard, R., Koth, B., Walton, A., 2016. Water sensitive urban design: an investigation of current systems, implementation drivers, community perceptions and potential to supplement urban water services. Water 8 (7) (Special issue: Urban Drainage and Urban Stormwater Management) Open Access.

Springer, S., Birch, K., MacLeavy, J., 2016. The Handbook of Neoliberalism. Routledge, New York.

South Australian Water Plan, June 2010. Water for Good: A Plan to Ensure Our Water Future to 2050. Office for Water Security, Adelaide, ISBN 978-1-921528-34-7. http://www.environment.sa.gov.au/files/bc6d668d-11bc-40fa-8c5c-a1d400fe4b90/water-for-good-summary-plan.pdf.

Standards Australia, 2017. Standards Australia: What Is a Standard? Standards Australia. Available from: http://www.standards.org.au/StandardsDevelopment/What_is_a_Standard/Pages/default.aspx.

TW-CAAC, 2014. Total Watermark – City as a Catchment (Update 2014). City of Melbourne, Melbourne. https://www.melbourne.vic.gov.au/SiteCollectionDocuments/total-watermark-update-2014.pdf.

Van der Ryn, S., Cowan, S., 1996. Ecological Design. Island Press, Washington, DC.

Van Roon, M., van Roon, H., 2009. Low Impact Urban Design and Development: The Big Picture. An Introduction to the LIUDD Principles and Methods Framework (Landcare Research Science Series No. 37). The University of Auckland: National Institute of Creative Arts and Industries. Manaaki Whenua Press, Lincoln.

VPA, 2017. Victorian Planning Authority: about. The Victorian Planning Authority. Available from: https://vpa.vic.gov.au/about/.

Water for Victoria, 2016. Water for Victoria: Water Plan. The State of Victoria Department of Environment, Land, Water and Planning, Melbourne, ISBN 978-1-76047-348-8. https://www.water.vic.gov.au/__data/assets/pdf_file/0030/58827/Water-Plan-strategy2.pdf.

Williams, D., 2016. Room for improvement: influence of statutory land use planning on the adoption of water sensitive urban design practices in Australia. In: Stormwater Conference. CRC for Water Sensitive Cities Web Site. Available from: https://watersensitivecities.org.au/content/pc787-swconf16-dwilliams/.

Wong, T., 2006. Water sensitive urban design - the journey thus far. Australian Journal of Water Resources 10 (3), 213–222.

Wong, T., June 2016. Human Settlements: A Framing Paper for the High-Level Panel on Water. Australian Water Partnership/eWater Ltd., Canberra.

Wong, T., Brown, R., Deletic, A., November 2008. Water Management in a Water Sensitive City. Water (Australian Water Association), pp. 52–62.

Zeunert, J., 2016. Adelaide botanic garden wetland. World Landscape Architect (26), 4–7.

Zeunert, J., 2017. Landscape Architecture and Environmental Sustainability: Creating Positive Change through Design. Bloomsbury, London.

Chapter 19

# WSUD and Urban Heat Island Effect Mitigation

**Elmira Jamei[1] and Nigel Tapper[2]**

*[1]Course Chair of Building Design, College of Engineering and Science, Victoria University, Melbourne, VIC, Australia; [2]Urban Climate Research Group Leader, School of Geography and Environmental Science, Monash University, Melbourne, VIC, Australia*

## Chapter Outline



**ABSTRACT**

Urbanization exacerbates the urban climate and contributes to the development of the urban heat island (UHI) effect, adversely impacting public health and thermal comfort. This chapter provides evidence for the effectiveness of green infrastructure (GI) and water sensitive urban design (WSUD) strategies as mechanisms to address some of the challenges faced by contemporary urban environments, including extreme heat events, drought, UHI, and thermal discomfort. Through several examples in various countries and climates, the application of GI in different forms (parks, street trees, green roofs, and green walls) and WSUD approaches were investigated and ranked as the most effective strategies in mitigating the increased urban air temperature. This chapter highlights the necessity of integrating climate knowledge into planning practices and raises awareness among urban planners of the importance of critically assessing planning policies before developing a new neighborhood.

**Keywords:** Green infrastructure; Urban climate; Urban heat island effect; Water sensitive urban design (WSUD)

## 19.1 INTRODUCTION—URBANIZATION AND THE URBAN HEAT ISLAND EFFECT

Urban areas cover only 3% of the Earth's surface, but they produce approximately 80% of the gross world product, consume approximately 78% of the world's energy, and are home to over half of the world's population. By 2050, this urban population is expected to increase to 6.3 billion people (United Nations, 2014). Given this rapid urbanization, the quality of urban life and public health is affected by anthropogenic heat flux, heat-absorbing construction materials, reduced vegetation coverage, and the exacerbated microclimate of the cities. Urban development also changes the hydrology of landscapes, reduces infiltration of water into soils, increases surface runoff, and reduces the evaporation fluxes from cities. This can often lead to increased urban air temperature and thermally uncomfortable outdoor spaces.

As a result of urban transformation and the consensus on global warming, urban temperatures are expected to increase, and heat waves will become more frequent, more intense, and longer lasting. According to the Australian Bureau of

**Approaches to Water Sensitive Urban Design. https://doi.org/10.1016/B978-0-12-812843-5.00019-8**

Meteorology (2013a), the heat waves that occurred in Australia in 2013 and 2014, with 3 days of temperatures above 37°C, will become more common in summer.

A city has a reciprocal relationship with its prevailing local climatic condition. The extent of urban climate modification caused by urbanization depends on the regional climate, the urban design parameters, the geographical characteristics of the city, and the magnitude of man-made alterations.

The complex morphology of urban areas alters the energy balance in cities. Spatial scale plays an important role in understanding the effect of urban transformation on local climate. Oke (in Kim and Penn, 2004) classified three spatial scales of interest for urban climate studies.

1. Microscale, which refers to the surface energy balance of individual elements, such as plants, buildings, gardens, and streets with land areas of up to 100 $m^2$;
2. Local or neighborhood scale, which refers to the areas such as private gardens, streetscapes, and local public parks, which cover an area of 20 $m^2$ to 10,000 $m^2$; and
3. Mesoscale, which refers to the city and regional climatic processes that occur across horizontal areas, such as citywide networks of parks, street trees, reserves, green wedges, and gardens that commonly exceed 10,000 $m^2$.

The vertical layer of urban climate is also classified into two layers: the urban canopy layer (UCL) and the urban boundary layer (UBL).

The UCL covers the areas between the ground and the rooftops of buildings (i.e., buildings, gardens, streets, squares, and parks). The energy exchange in this layer depends on the characteristics of the site and varies across short distances. The majority of urban transformations and human activities occur in the UCL. Accordingly, the energy exchange in this layer has been the topic of research for urban planners and urban designers for many years.

The UBL is a mesoscale phenomenon, the characteristics of which are affected by the nature of the urban surface. The UBL is influenced by regional climatic processes, land topography, and precipitation. Fig. 19.1 shows a schematic of the extent of each layer in urban areas and urban climate studies.

At the mesoscale, the city modifies regional climate by altering the cloud cover, precipitation, solar radiation, air temperature, and wind speed. Consequently, the climate in urban areas is different from the climate in suburban and rural areas. At the micro- and local scales, the city alters the urban climate through the geometry of streets, green infrastructures (GIs), and building materials.

Understanding the process of energy exchange and energy balance in urban areas is essential to explore the modification of urban climate caused by urbanization. The energy balance in urban environments is influenced by the energy gains and losses and the level of the energy stored in urban elements, such as buildings, cars, and pavements (Fig. 19.2).

$$\text{Energy gains} = \text{energy losses} + \text{energy storage}.$$

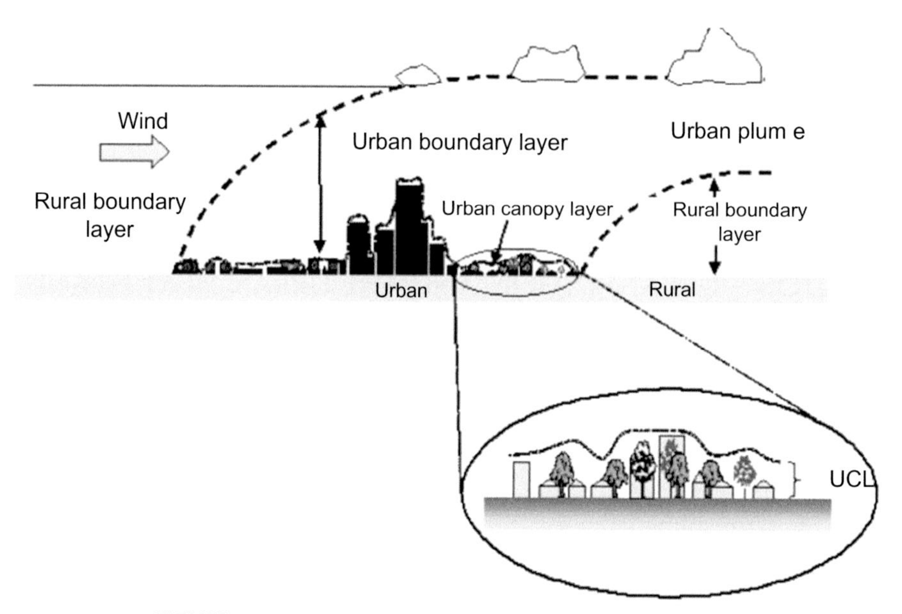


**FIGURE 19.1** Components of the urban atmosphere (Voogt, 2012).

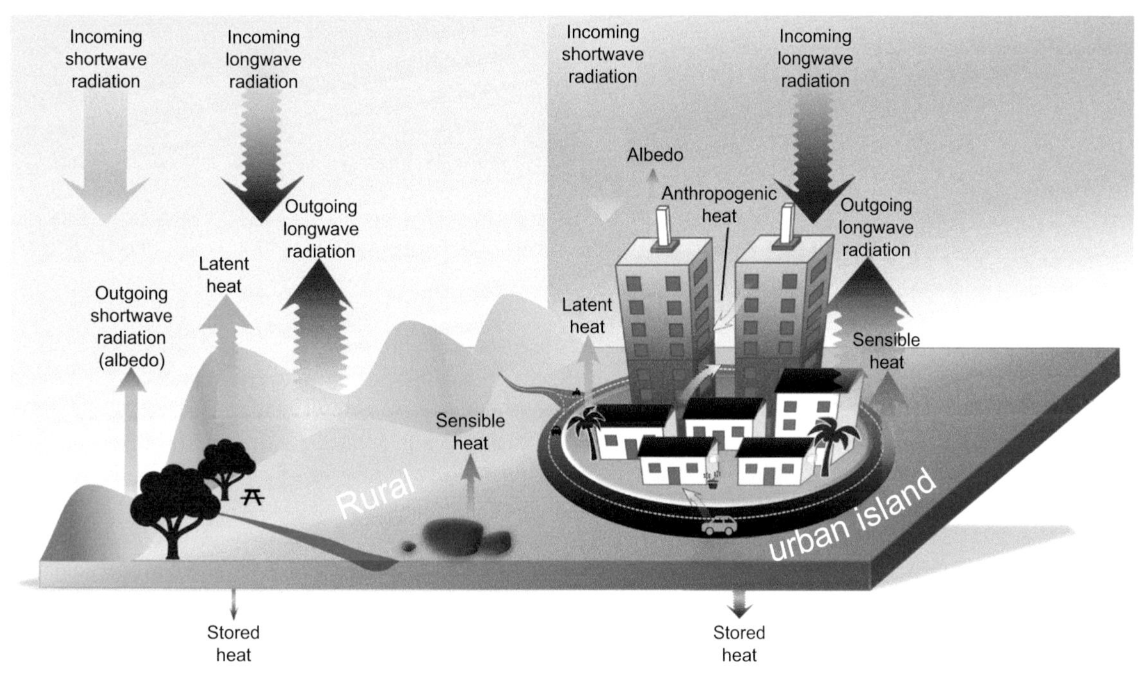


**FIGURE 19.2** Schematic depiction of energy flux in urban area. *Graphic by Alison Vieritz, adapted from Oke, T.R., 1988. The urban energy balance. Progress in Physical Geography 12, 471–508.*

Energy gains include the sum of the net radiative flux (Qr) emitted by opaque elements and the level of anthropogenic heat (QT) created by human activities, transportation, power generations, and other heat sources.

Energy losses occur by the sensible heat ($Q_E$) caused by heat convection between opaque surfaces and the air. The level of evapotranspiration (ET) from urban surfaces also influences energy losses via the latent heat flux ($Q_L$). The energy balance of the surface air is calculated as

$$Q_r + Q_T = Q_E + Q_L + Q_S + Q_A \tag{19.1}$$

where, $Q_r$, net radiative flux; $Q_T$, anthropogenic heat; $Q_E$, sensible heat; $Q_L$, latent heat; $Q_S$, stored energy; and $Q_A$, net energy transferred to or from the system through lateral advection of sensible of heat.

One of the consequences of high-energy gain and low-energy loss is the urban heat island (UHI) phenomenon. UHI refers to a city-scale phenomenon of the temperature (T) difference between urban and surrounding rural areas. It is often quantified as the difference in air temperature between high-density urban sites (e.g., the central business district or CBD) and rural locations (i.e., UHI intensity = T urban−T rural).

The intensity of the heat island is at a maximum overnight when the absorbed heat is gradually released to the atmosphere. It also varies spatially and temporally depending on the local microclimates, geography, and urban development. These urban−rural temperature differences are strongest under clear and calm conditions (Tapper, 2014).

The adverse effect of the UHI phenomenon has been widely documented in the literature. The UHI effect contributes to storms/precipitation events (Bornstein and Lin, 2000), increases energy consumption in cites (Santamouris et al., 2015), and leads to a high rate of heat-related mortality (Hondula et al., 2014). UHI indirectly increases energy consumption for cooling (Preetha and Thomas, 2011), reduces the air quality in cities (Stone, 2012), and threatens the ecosystem by warmer water flowing from the cities (Judd, 2011).

A study conducted by an international team of economists quantified the combined impact of the UHI effect on urban economies. Based on the analysis of 1692 cities, the total economic costs of climate change for cities this century could be 2.6 times higher when heat island effects are considered than when they are eliminated (Estrada et al., 2017). Luke Howard was the first researcher who documented UHI and found that the temperature in London was 3.7°C higher than the temperature in the suburbs (Howard, 1818). Since then, this phenomenon has been reported in many other urban areas worldwide, such as the Greater Athens area (Assimakopoulos et al., 2007), Nicosia (Theophilou and Serghides, 2015), Malacca (Jamei et al., 2015), and Melbourne (Jamei and Rajagopalan, 2017; Coutts et al., 2007). Researchers have examined the characteristics and magnitude of the UHI effect and found that its intensity can reach as high as 10 K (10°C)[1], depending on the urban properties and local climatic conditions (Rosenzweig et al., 2011; Mihalakakou et al., 1998).

Although the UHI phenomenon is explicitly linked to the physical characteristics of a city, such as land cover and morphology, its magnitude and variability are a function of nonlinear interactions among multiple factors. This could be a main reason for different UHI intensities in different cities located in various countries with diverse geographical conditions.

Many factors contribute to the development of the UHI (Oke et al., 1991), such as decreased long-wave radiation loss, increased thermal storage in the building fabric, released anthropogenic heat, decreased effective albedo of the system caused by multiple reflections, and reduced evaporating surfaces in urban areas. Albedo is defined as a measure of the reflectance or optical brightness of a surface (Taha, 1997).

The increase of the UHI intensity triggers heat-related diseases such as heat stroke and cardiorespiratory stress, cardiovascular stress, thermal exhaustion, and premature deaths, weakness, consciousness disturbances, cramps, and fainting. It can even exacerbate preexisting chronic diseases such as diabetes, respiratory failure, as well as cardiovascular, cerebrovascular, neurological, and renal diseases to the point of causing death. Monash University produced an UHI effect vulnerability map for Melbourne, Australia, and indicated that vulnerability to heat in urban areas largely depends on the urban form, the sociodemographic characteristics of the resident population, and the capacity to adapt to the heat, which is largely place-specific (Loughnan et al., 2012b, 2014). The research also showed that elderly people with low socioeconomic status and residents in high-density older housing stocks with limited surrounding vegetation are particularly at risk due to the UHI effect.

Health impacts from extreme heat events have been recorded in Chicago, USA (31% mortality increase) (Whitman et al., 1997), Paris, France (130% mortality increase) (Dhainaut et al., 2003), Moscow, Russia (60% mortality increase) (Revich, 2011), and Melbourne, Australia (62% mortality increase) (Carnie, 2009). The cause of heat-related deaths is not from the maximum daytime air temperature, but rather the high minimum nocturnal temperatures for consecutive days, which leads to increased physical stress on human bodies and increased mortality rate. In the future, it is anticipated that many cities will encounter catastrophic heat events as the frequency, intensity, and duration of heat events are predicted to increase along with climate change (Alexander and Arblaster, 2009).

Identification of the factors that contribute to the development of the UHI effect and the methods to address, mitigate, and adapt to this urban climate have received considerable attention from different perspectives (e.g., climatology, environmental sciences, material design, building design, energy fuels, medicine, urban planning, and urban design).

Thermal comfort is defined as "the condition of the mind that expresses satisfaction with the thermal environment" (Ashrae, 1997). It has been also described as "the absence of thermal discomfort, in which individuals feel neither too warm nor too cold" (McIntyre, 1980). One of the methods to create thermally comfortable and sustainable urban environments is promoting climate sensitive and water sensitive urban designs (WSUD).

Outdoor thermal comfort is determined by meteorological factors (i.e., air temperature, humidity, radiation, and air movement), personal factors (i.e., insulation and clothing value), and rate of metabolism, which in turn is affected by body shape, age, and gender. These factors affect thermal comfort by altering the heat exchange process of the human body through convection, conduction, and radiation. Therefore, to assess human thermal comfort, heat budget models that consider all heat exchange components must be developed.

Understanding what humans perceive as thermally comfortable outdoor settings should be considered in designing urban spaces. Recent research in Australia showed that a comfortable air temperature in Melbourne is 21.5°C, but 25.7°C in Adelaide (Loughnan et al., 2012a). However, human thermal comfort is better calculated by thermal indices rather than air temperature, as the indices include all the parameters that have an impact on the thermal comfort—air temperature, wind speed, relative humidity, mean radiant temperature, clothing values, activity type, and physiology (age, sex, weight, and height). Mean radiant temperature represents the uniform surface temperature of a fictional enclosure (Matzarakis and Mayer, 2000). Mean radiant temperature plays a key role in the thermal comfort level of pedestrians in an urban setting. Urban dwellers may experience uncomfortable thermal conditions under the direct sun and feel more thermally comfortable under the shade of buildings and trees. The difference between the sunny and shaded space is the level of mean radiant temperature that pedestrians receive from the environment. This means that the air temperature does not vary from the sunny spot to the shaded spot. However, less mean radiant temperature is received by pedestrians.

A number of thermal comfort indices have been developed over the last 30 years based on the human heat budget model (Höppe, 2002). These indices enable researchers to evaluate the thermal consequences of urbanization. Thermal

1. The temperature difference in Kelvin is the same as the temperature change in Celsius ($\Delta K = \Delta C$).

indices, such as the physiological equivalent temperature (PET), were developed and successfully tested for outdoor settings. Therefore, PET can be used to analyze the thermal impacts of mitigation strategies and provide insights to improve urban design guidelines. PET is defined “as the air temperature at which, in a typical indoor setting (without wind and solar radiation), the heat budget of the human body is balanced with the same core and skin temperature as under the complex outdoor conditions to be assessed” (Höppe, 1999).

Greater insights into what limits human thermal comfort and exacerbates the responses to the UHI are needed to inform the design of resilient spaces. Reducing the heat exposure through implementing WSUD and urban greening provides healthier outdoor environments for human activities. They are therefore identified as effective strategies in mitigating the heat island effect. In fact, these strategies restore a more natural hydrology in urban areas, increase the overall evaporation rates in cities, and contribute to lower air temperatures at the local scale. WSUD and urban greening can also be utilized to improve the pedestrian thermal comfort level by providing a higher level of shading and increasing evaporative cooling in outdoor spaces.

The following sections present case studies from the literature on some of the most effective strategies to mitigate the increased urban air temperature and increase pedestrian thermal comfort through different GIs and WSUD approaches.

## 19.2 HEAT MITIGATION STRATEGIES (GREEN INFRASTRUCTURE AND WSUD)

Infrastructure is a fundamental physical and organizational structure built to serve the society. Gray infrastructure (man-made infrastructure, buildings, pavements) provides a certain level of protection from various types of risks, but it also adversely affects the city environment in many ways. GI, on the other hand, provides resilience to climate extremes and supports urban livability. GI is where nature plays a role in the augmentation of urban services and mitigates climate extremes, such as flood, drought, and heat waves. GI can be incorporated in urban areas on a local to large scale, for example, city parks, street trees, green walls, and green roofs.

GI cool cities through three processes: shading, evapotranspiration (ET), and the alteration of wind pattern (Oke et al., 1989). Shading cools the atmosphere by intercepting solar radiation and thereby limiting the increase in air or surface temperature. ET refers to the transpiration of water from plants and evaporation from soils and waterbodies (Kotzen, 2003). The absorbed solar energy is converted into the latent heat of evaporation; thus, the plant canopy and the temperature of its surroundings are cooled (Taha et al., 1988). However, this is not the case for the impervious, often dark-colored urban surfaces, which absorb solar radiation, increase temperature, and reradiate heat in the long-wave band much like a domestic electric heater (Taha et al., 1988). Trees also change the wind behavior, and their capacity to change the wind velocity depends on the tree type (Bonan, 1997). For example, a deciduous tree can reduce wind speeds by 30%–40% (Oke et al., 1991). Therefore, deciduous trees contribute to improved pedestrian thermal comfort, as they reduce the wind speed caused by interacting high-rise buildings (urban canyons) in city centers.

Urbanization modifies the urban water balance and therefore the urban surface energy balance. Conventional stormwater drainage systems are designed to transport stormwater rapidly away from the urban areas. One of the methods in reducing heat retention in cities is to retain more of the water within cities by decreasing runoff, such as implementing WSUD that utilizes green or blue (free water) spaces.

The main concept behind WSUD is to minimize the reliance on centralized water supply systems and/or provide the opportunity to retain the water in cities, particularly under drought and extreme heat conditions. WSUD achieves this through techniques such as retaining, treating, reusing, and diverting stormwater, using technologies such as biofiltration systems, wetlands, rainwater tanks, rain gardens, and stormwater harvesting systems (Wong and Brown, 2009).

The other objective of WSUD is maximizing the use of artificial waterbodies (e.g., man-made wetlands and lakes). Natural waterbodies provide a downwind cooling benefit, as the air is cooled by evaporation over the waterbody and moves laterally—a sort of negative advection. For example, Saaroni and Ziv (2003) showed air temperature reductions up to 40 m downwind of a pond in Israel. The maximum effect was observed at midday, when the temperature drop reached 1.6°C (Saaroni and Ziv, 2003).

Another key WSUD practice is capturing stormwater from impervious surfaces such as roads and roofs and using the water for irrigating the landscapes in urban areas. Irrigation directly affects the thermal comfort by transpiration and increases the heat storage in the subsoil. The maximum cooling impact of irrigation is observed during heat waves. The maintenance of GI in cities is directly related to the level of reliable water for irrigation. Irrigation is an excellent method to distribute water into the urban landscapes and increase the ET rate locally.

Although GI of all types effectively mitigates the heat island effect at a range of scales, the critical role of integrated urban water management (IUWM) and irrigation on the cooling of urban landscapes has been often underestimated. Widespread irrigation can be applied anytime and therefore is more effective. Redirecting the runoff from impervious to pervious surfaces also increases the ET rate, but this only occurs after rain. Biofiltration systems, rainwater tanks, and stormwater harvesting systems can be used to capture, store, and treat stormwater and graywater.

The impact of irrigated GI on cooling was evaluated during a heat wave in Phoenix, Arizona. The maximum daytime air temperature was reduced by up to 1.0°C as a result of well-irrigated vegetation (Grossman-Clarke et al., 2010). The nighttime cooling rate was also increased with irrigation. However, a further increase in the irrigation level was not effective in providing additional cooling. This result indicated that adding water in well-watered neighborhoods is not an appropriate strategy to achieve the best thermal condition. A similar finding of the nonlinear relationship between ET and cooling was reported in Melbourne, Australia (Demuzere et al., 2014). In contrast, studies conducted in Portland, Oregon, and Canberra, Australia, showed that high proportions of irrigated landscaping reduced the maximum afternoon temperature by up to 4°C compared with that in a paired urban landscape with no vegetation (House-Peters and Chang, 2011; Mitchell et al., 2008).

Greening the cities is one of the key measures in microclimate cooling, but thinking how we can keep a city green is also critical. For Melbourne, Australia, this endeavor will be carried out through a major shift in the use of water. For example, during the last drought in South Eastern Australia (2006–10), parks and gardens dried out, stimulating planning for the capture and recycling of stormwater. While 500 billion litres of water fall on the city every year, only 1% is being captured (Agarwal, 2000). The inner-city neighborhoods with street gardens adopted a recycled water strategy to irrigate the vegetation, and the temperature difference between the gray and GIs was reported as 10°C (Coutts et al., 2013b).

## 19.3 COOLING EFFECT OF WATERBODIES

Urban waterbodies or "urban blue space" refers to the substantial bodies of surface water located in urban areas. They encompass rivers, ponds, canals, and sustainable drainage systems. Urban designers place a high priority on designing public open spaces with waterbodies, as they play an important role in creating thermal comfort for urban dwellers (Tominaga et al., 2015).

Published studies (Tominaga et al., 2015; Steeneveld et al., 2014) highlight the role of evaporative cooling where absorbed solar energy is transformed into latent heat via the production of water vapor. The thermal properties of water (specific heat capacity and enthalpy of vaporization) give it a high thermal inertia, moderating the air temperature during the day and acting as a thermal buffer.

Few studies have been conducted on the cooling effect of waterbodies compared with the cooling effect of GI. Most of the waterbody studies explored the large cooling impact on the temperature of the water surface, rather than the air temperature. However, an analysis of 27 studies demonstrated that waterbodies can provide, on average, a cooling impact of 2.5 K (2.5°C) on their surrounding environment (Völker et al., 2013). Large waterbodies provide the maximum cooling effect near their boundaries and in downwind areas (Theeuwes et al., 2013).

The magnitude and distribution of the cooling effect caused by waterbodies are influenced by the size, spread, and distance of the influenced area from the waterbodies. A single large waterbody creates a larger cooling impact compared with several smaller, regularly shaped waterbodies with the same total volume of water (Steeneveld et al., 2014).

The geometry of the waterbodies is also another important factor. A study in China (Zhu et al., 2011) showed that square or round geometries provide greater cooling impact compared with that from irregular-shaped waterbodies. The importance of the width of a waterbody in creating cooling was demonstrated in a study conducted in China, which measured the temperature and humidity levels next to a river. The study concluded that a river width of 40 m produced significant and stable effects of reduced temperature and increased humidity in the surrounding area (Zhu et al., 2011).

Deeper waterbodies display an additional impact on the thermal condition of outdoor spaces compared with shallow waterbodies. However, during heat waves and periods of high UHI intensities, even shallow waterbodies can generate a substantial cooling impact Therefore, the creation of shallow ponds as part of a sustainable drainage system is strongly recommended to urban developers (Abis and Mara, 2006). While deep waterbodies demonstrated the largest cooling impact, shallow waterbodies utilize the thermal capacity of the entire water column for thermal exchange. Therefore, in the context of compact urban development, distributing shallow waterbodies, ponds, swales, and water gardens (which utilize all their available thermal capacities) across the urban areas is suggested. However, shallow waterbodies require the constant input of water to maintain a source of free water. They also do not create nighttime heating or increase the level of humidity as much as that from large waterbodies (Newman and Herbert, 2009).

## 19.4 COOLING EFFECT OF URBAN GREENING

### 19.4.1 Trees

Trees serve numerous and diverse purposes. They have great social impacts, such as improved public health, increased community interaction (Van Dillen et al., 2012), and increase property values (Pandit et al., 2012). Trees encourage contact among community residents, promote physical activity, decrease stress, and stimulate social cohesion. Research has shown that lower levels of crime and enhanced public safety are other outcomes of a green community. For example, urban areas with a substantial number of trees have reduced crime levels by approximately 50% in the United States (Kuo and Sullivan, 2001).

Street trees improve air quality, reduce emissions, airborne pollutants and noise, decrease stormwater runoff, provide shading, and mitigate the intensity of the UHI effect. Large trees reduce air pollution between 60 and 70 times more than smaller trees (McPherson et al., 1994). In Melbourne, Australia, planting 100,000 trees is claimed to have reduced carbon emissions by 1M tonnes (Moore, 2009).

Urbanization means increased impervious surfaces and substantial soil compactness, which in turn results in less water infiltration into the soil and increased stormwater runoff and peak flow rates. Trees play an important role in reducing stormwater runoff. The leaves and branches of the trees intercept, absorb, and store water before evaporating from the surface of the trees (Armson et al., 2013) and enhance infiltration into the soil. Mature deciduous trees, such as sweet gum (*Liquidambar styraciflua*), intercept between 2 and 3 kL of water per year (depending on the rainfall level), whereas evergreen trees, such as pines, intercept up to 15 kL of water per year (Seitz and Escobedo, 2011).

Trees also contribute significant economic benefits for communities and local governments. A 10% increase in the tree canopy coverage can reduce the overall heating and cooling energy use of dwellings by 5%−10% (US$50 to $90). For example, street trees in Sacramento, California, contributed to a significant reduction in the summertime electricity use by 185 kWh (or 5.2%) per household (Donovan and Butry, 2009). In Auburn, Alabama, a 10% increase in the shade coverage reduced electricity consumption by 1.29 kWh/day (Pandit and Laband, 2010).

Treescaping also contributes to increased business income by creating a sense of identity in neighborhoods, increased thermal comfort levels, and a more favorable environment for shopping (Wolf, 2005). In Alabama, 75% of residents indicated that the presence of street trees influenced their choice of a new residence (Zhang et al., 2007).

Tree species with large crowns, short trunks, and dense canopies contribute to the optimal thermal benefits for the community, as they provide the maximum level of shading. Eight percent of the cooling effect of a tree is due to its shading effects (Shashua-Bar et al., 2009), which can reduce daytime air temperature in cities by 5−20°C (Killicoat et al., 2002).

Mean radiant temperature, one of the key parameters that influence outdoor thermal temperature, is directly reduced by tree shading. During a typical central European summer day, the shading effect of trees can reduce the mean radiant temperature by up to 30°C (Mayer et al., 2008).

Different tree species provide different amounts of radiation interception because of their varied structural morphology (Dixon and Mote, 2003). Trees remove a large amount of incoming short-wave radiation by reflection and transmission through the leaves. Leaves reflect 10% of visible and 50% of solar infrared radiation and transmit 10% of visible and 30% of solar infrared radiation (Brown and Gillespie, 1995) (Fig. 19.3).

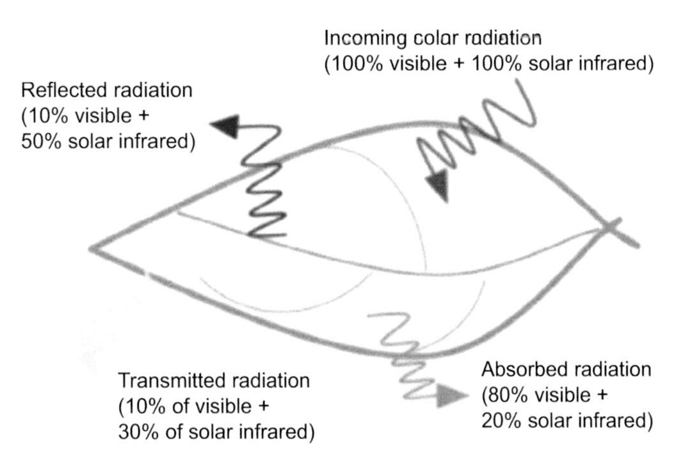


**FIGURE 19.3** Partitioning of solar radiation by a tree leaf. Solar radiation is absorbed (80% visible, 20% solar infrared) (green), reflected (10% visible, 50% solar infrared) (blue), and transmitted (10% visible, 30% solar infrared) (yellow) by plant leaves (Brown and Gillespie, 1995).

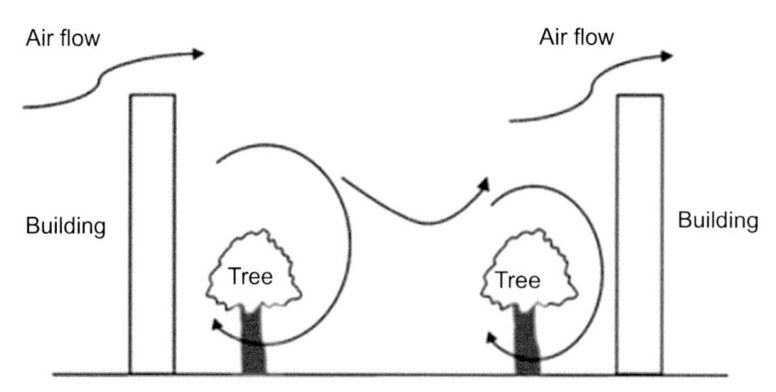


FIGURE 19.4 Wind flow in a street canyon with trees (Kong et al., 2017).

Transpiration occurs when trees release water vapor to the air from leaf stomata during photosynthesis. Transpiration causes latent heat loss by the conversion of liquid water to vapor, thereby resulting in the cooling of the leaf and the surrounding environment.

Transpiration has been considered as one of the major mechanisms to dissipate energy load on leaves and is one of the most important regulating ecosystem services. Pataki et al. (2011) found that tree transpiration differs greatly among species for urban forests in the Los Angeles metropolitan area because of the different morphologies of tree species.

Street trees decrease the turbulence intensity and alter the average wind speed (Fig. 19.4), thereby affecting human comfort, particularly in the areas with high-rise buildings. The presence of four sidewalk street trees reduced the wind speed under the canopy by up to 50% (Park et al., 2012), whereas trees with dense canopies reduced the mean wind speed by up to 90% compared with open areas (Heisler et al., 1994).

Low soil water availability and high levels of heat stress caused by the UHI effect compromise the cooling performance of a tree and limit its effectiveness in mitigating UHI. High-resolution thermal mapping and airborne thermal remote sensing were conducted in two Melbourne, Australia, municipalities in the summer of 2011–12 under hot, clear, and calm conditions to provide further information about the optimal landscape design and to identify which individual tree canopies required further irrigation (Coutts et al., 2013b). The findings of the study showed that irrigating the trees has a significant role in land surface temperature reduction. This reinforces the importance of irrigation to maximize the cooling efficiency of existing street trees.

One of the WSUD strategies that provides landscape irrigation is tree pits. Tree pits allow stormwater to flow into the root zone of the trees and become stored in the growing media. A study in Melbourne, Australia, monitored the cooling performance of street trees supported by tree pits (Coutts et al., 2013a). Microscale observations of the air temperature, humidity, soil moisture, and leaf-scale physiology data were conducted to measure the performance of the trees (Fig. 19.5).

This WSUD approach resulted in an increase in the water availability for the roots of the trees and enhanced the health of the tree. The soil moisture increased rapidly, becoming saturated after rain, but the sandy soil of the tree pits meant it

FIGURE 19.5 Tree monitoring (left) and leaf-scale observations and soil moisture monitoring (right), Melbourne (Coutts et al., 2013a).

also dried out rapidly. This addresses the stormwater management objectives, but it does not necessarily enhance tree health. The researchers concluded that for optimal growth and cooling efficiency of street trees, the soil moisture level should be maintained between field capacity and wilting point by irrigation. However, irrigation scenario modeling (Järvi et al., 2011) showed that regular irrigation limits the capacity of tree pits to decrease runoff, as the soil is always moist and therefore limits infiltration. A balance is therefore needed between the urban hydrology objectives and microclimate cooling benefits. Effective mitigation of heat island impacts through enhanced ET from the trees requires a well-maintained soil moisture level. If irrigating urban trees, it's important to bear in mind that the aim is to increase the soil moisture level without unduly compromising the benefits from reducing runoff.

## 19.4.2 Park cool islands

Parks create the largest cooling effect in urban areas. Their cooling efficiency depends on the size of the park, plant species, the level of sky obstruction (sky view factor [SVF]), the irrigation level, and the geographical features of the reference point.

Oke was among the first researchers to present evidence of park cool islands (PCI) in northern American cities (Oke, 1982). His study showed that urban parks were generally 1−3°C cooler than their surrounding urban areas. A study on 61 city parks in Taiwan concluded that urban parks, on average, are 0.81 K (0.81°C) cooler than their surrounding built-up areas at noon during summer (Chang et al., 2007). An urban park in Melbourne, Australia, was 2.5°C cooler than the Melbourne CBD during nighttime conditions (Stewart and Oke, 2012). Infrared photographs along a green pedestrian canyon in (hot/humid) Singapore showed that large urban parks appeared as "cold spots" and efficiently mitigated the heat island effect (Forsyth et al., 2005). This was supported by a study on 10 parks in Singapore, which confirmed the air temperature in parks could be 8−12°C cooler than that in the surrounding built-up areas (Hondula et al., 2014). In hot and dry climates, parks play an even more vital role in improving thermal comfort. Field measurement in 21 parks in hot, dry Addis Ababa, Ethiopia, showed significant improvement in pedestrian thermal comfort and reduction in the air temperature (up to 7°C) as a result of appropriate selection of plant types (Whitman et al., 1997).

The geographic and built form characteristics of the park's surrounding area also play a significant role in the park cooling effect. The features include the compactness, the magnitude of anthropogenic heat, and the construction material of the buildings. A study in Athens, Greece (hot, dry Mediterranean climate), showed that a park's cooling effect was insignificant because of the heavy traffic in the surrounding area (Zoulia et al., 2009). In Sweden, the park's cooling effect was also decreased but because of the wind blockage by the surrounding high-rise buildings (Upmanis and Chen, 1999).

Generally, PCI diminishes with increasing distance from the edge of the park (Yan et al., 2014). Some researchers found that the cooling effect extends for the same distance as the width of the park (Spronken-Smith, 1994; Watkins et al., 2002). Field measurements in hot, humid Singapore found that the average air temperature in the areas closer to the park was 1.3 K (1.3°C) lower than surrounding areas and that the cooling effect decreases with increasing distance from the park (Yu and Hien, 2006).

Although previous researches showed that PCI is more significant as the park area increases, field measurements also verified the cooling effect of parks with small footprints (Watkins et al., 2002). The magnitude of PCI also depends on the interval between the green pockets and it may be insignificant for small, isolated urban parks. But, even small green areas with sufficient intervals performed well in cooling the cities (Wong and Nichol, 2013).

The cooling effect of urban parks also depends strongly on the level of irrigation. The impact of irrigation on PCI was investigated in an urban park located in Footscray, Melbourne (Coutts et al., 2013c). The site was irrigated prior to a warm sunny period. Surface temperature, soil moisture, and ET rates were measured using sets of instruments and chambers. A section of a park was kept dry as a comparison reference site. Two irrigation events were monitored and a significant reduction in the surface temperature (up to 10°C) was observed as a result of evaporative cooling and the changes in soil heat storage. The level of cooling after the second irrigation increased three times compared with the cooling effect over the dry, nonirrigated area (Fig. 19.6). However, the PCI effect was more significant during the nighttime, as the vegetated areas cooled immediately after the sunset compared with the surrounding impervious areas (Fig. 19.7). In contrast, the park acted as a heat island during the day because of the dry, hard surface, even after irrigation. It was only after the second irrigation that the park began to act as a significant PCI.

## 19.4.3 Green roofs

A green roof is defined as "a living layered vegetated roofing system that consists of a waterproofing membrane, insulation, growing medium, and a layer of vegetation itself" (Carnie, 2009).

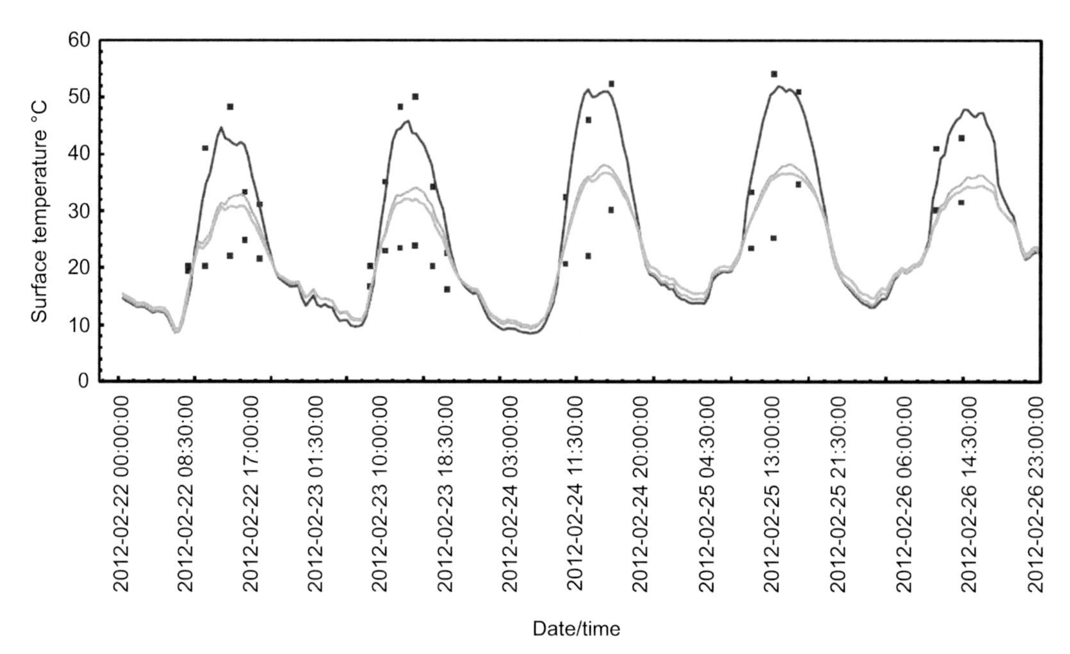


FIGURE 19.6 Surface temperature of the irrigated (*blue and green lines*) and dry zones (*red line*) of the park following the second irrigation on February 22, 2012, using automatic (*solid lines*) and manual instrumentation (*dots*) (Coutts et al., 2013c).

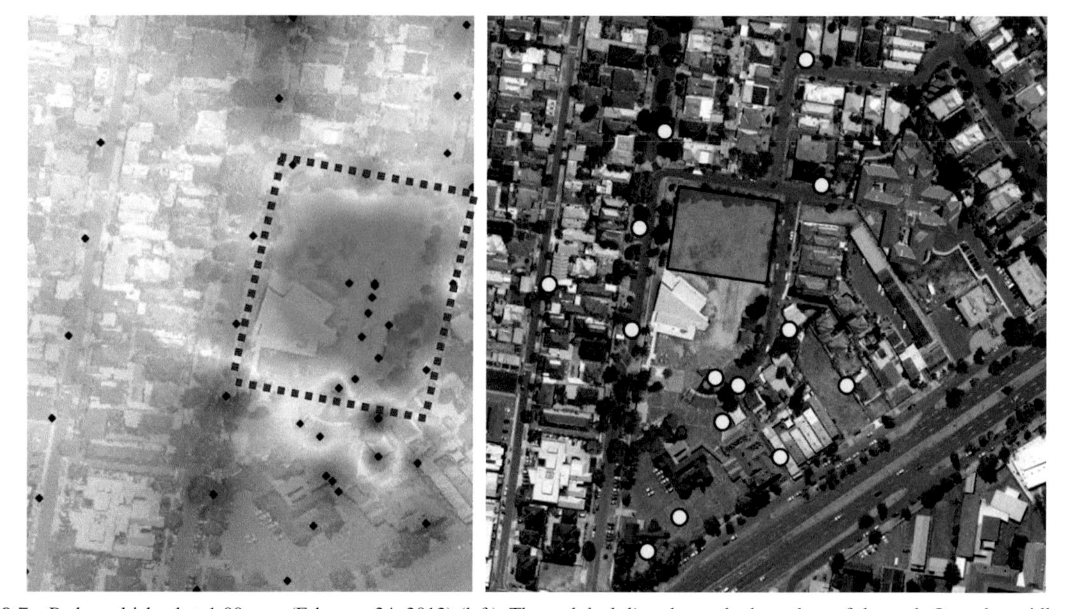

FIGURE 19.7 Park cool island at 1:00 a.m. (February 24, 2012) (left). The *red dash line* shows the boundary of the park. It cools rapidly after sunset compared with the surrounding built-up urban landscape. The location of the oval at Footscray Park, Melbourne, and the location of the instruments that monitored the surface temperature are shown in an aerial photo (right) (Coutts et al., 2013c).

Green roofs are divided into two categories: intensive and extensive types (Alexander and Arblaster, 2009). Intensive green roofs are characterized by thick-growing media (>200 mm), with soil weights exceeding 300 kg/m$^2$. The deep soil in intensive green roofs allows the growth of a wide range of plants, thereby providing great planting choices.

Extensive green roofs are the basic form of green roof and are limited to small plants. The thickness of the growing medium is generally <150 mm, with soil weights between 60 and 150 kg/m$^2$. Less effort is needed to construct, irrigate, and maintain this type of green roof. Given their light weight and low level of maintenance, extensive green roofs are popular in cities. Fig. 19.8 demonstrates the difference in the thickness and properties of various layers in intensive and extensive green roofs.

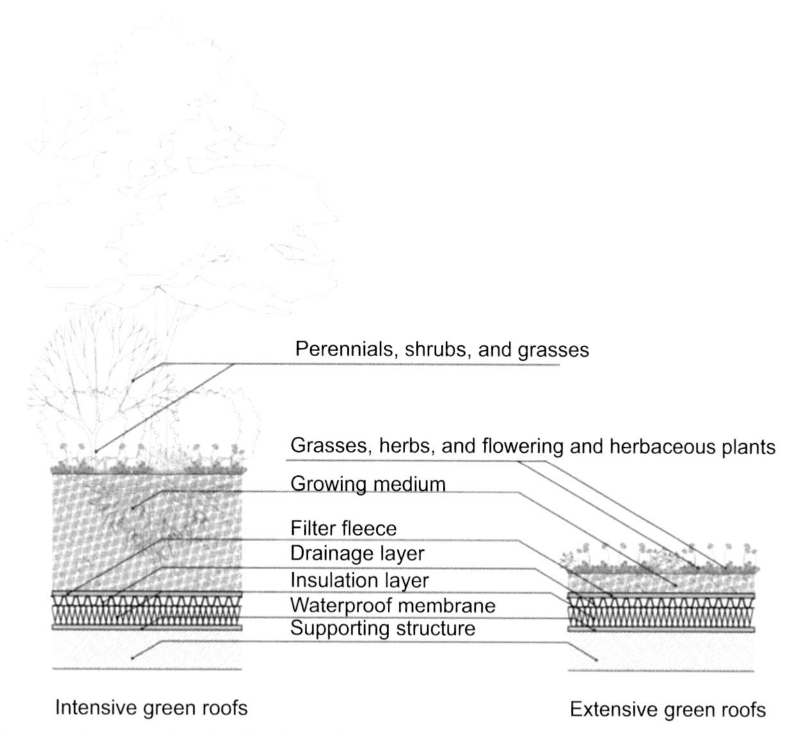


**FIGURE 19.8** The layered construction details of intensive versus extensive green roofs (Brudermann and Sangkakool, 2017).

In an urban environment where the available ground space is limited, roofs offer a substantial surface for the implementation of UHI mitigation strategies. German cities initiated green roof policies as early as the 1970s. In Munich, for example, all suitable flat roofs with a surface area of over 100 $m^2$ must follow green roof policies. In Copenhagen, Denmark, installing green roofs for all newly constructed roofs with a pitch less than 30 degrees is compulsory.

Green roofs provide several environmental benefits including stormwater flow reduction, air quality improvements, building energy saving, and mitigation of the UHI effects (Doug et al., 2005; Vijayaraghavan, 2016). The thermal properties of green roofs decrease the temperature of the roof proper by 10−15°C, thereby reducing building heating and cooling needs. The impact of green roofs on energy consumption has been quantified in offices located in four different climate zones (Sailor et al., 2012). Increasing the leaf area index, shading, and ET of the vegetation played a more important role in energy savings than the change in growing media depth. Similar findings were reported from a study in Italy that monitored the performance of green roofs in well-insulated buildings (D'Orazio et al., 2012). In New York, green roofs reduced daily average temperatures by 0.3°C and afternoon temperatures by 0.6°C at pedestrian level. In Tokyo, green roofs had negligible impact on the pedestrian-level air temperature because of the height of the buildings. But in Hong Kong, green roofs were found to be effective in cooling the air temperature at pedestrian level by 0.4−0.7°C.

ET from green roofs is cited as a key mechanism for cooling in mitigating the UHI effect. In fact, green roofs support high ET rates if they are irrigated well. Some types of green roofs require supplementary irrigation during warm and sunny days. However, there needs to be a balance between rainfall retention for stormwater management objectives and urban climate objectives. For instance, extensive roofs that are usually designed with thin substrates and have drought-tolerant *Sedum* species may improve stormwater management, but they do little to improve the urban microclimate. Therefore, the ideal green roof is the one which is irrigated from a sustainable water source and has a diverse range of plant species.

The ET and surface energy balance of a green roof were quantified though an experimental study to understand the role of evaporative cooling on the microclimate at roof level (Coutts et al., 2013d). A clear perspex chamber was installed on the roof, and an infrared gas analyzer measured the change in concentration of the water vapor and then calculated the ET flux (Fig. 19.9). The measurements were conducted under different climate scenarios, and the soil moisture and relevant meteorological data were monitored to understand the parameters influencing ET. The atmospheric heating from the green roof was compared with the conventional roof to identify the effectiveness in cooling the microclimate at roof level.

An example of ET response from the study is shown in Fig. 19.10. On December 22, 2011, the maximum temperature was 29°C, and the ET rate was low, with the average value of 100 $W/m^2$, but with individual values as high as 153 $W/m^2$.

**FIGURE 19.9** Perspex flux chamber with infrared gas analyzer (IRGA) used to measure the evapotranspiration rate (left). The vegetation type (*Sedum rubrotinctum*) used to conduct the experiment on the green roof (right) (Coutts et al., 2013d).

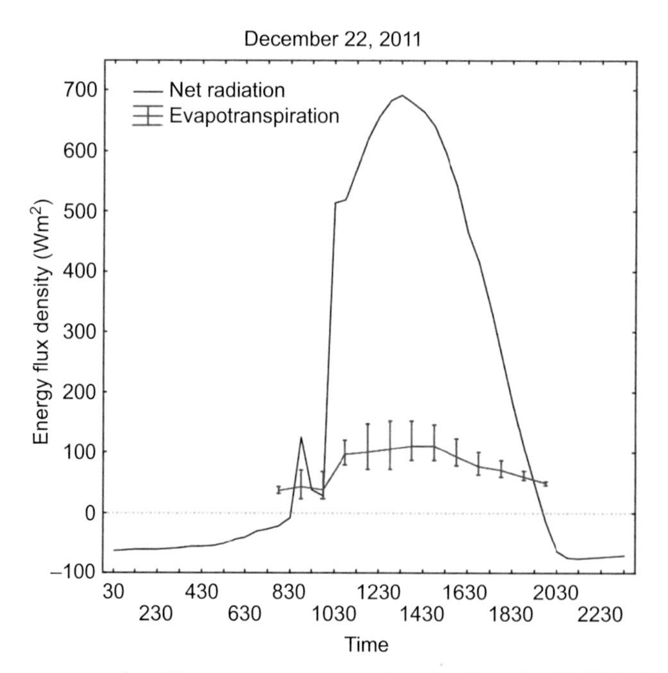


**FIGURE 19.10** Net radiation and evapotranspiration of the vegetated experimental roof on December 22, 2011. *Bars indicate* the range of observations (Coutts et al., 2013d).

The study concluded that only a small proportion of available energy at the surface of the vegetated experimental roof (up to 700 W/m$^2$) was used in ET.

The findings of this study also indicate that although green roofs may be one of the best measures to mitigate UHI, ET of green roofs is low unless the soil moisture is high and the vegetation is transpiring actively. The study also suggests the need to take special consideration in selecting the soil type for green roofs, as the thin substrates will not store large amounts of water, and the vegetation may conserve water by limiting transpiration (a drought avoidance strategy). Consequently, evaporative cooling effects can be small.

## 19.4.4 Green walls

Green walls are defined as climbing plants grown either directly against the walls or on support structures integrated to external building walls. Similar to other types of GI, green walls are promoted because of their benefits in reducing the building energy consumption and adaptation to a warming climate.

The plants on the facades reduce the temperature of the building wall surfaces by absorbing incoming solar radiation and providing thermally resistive air gaps between foliage and the external building wall (Akbari and Taha, 1992). Green walls cause significant reduction in cooling demands, particularly in hot climates. In cold climates, green facades act as an exterior insulation and decrease the magnitude of heat loss through the building envelope, with a significant reduction in building energy use. Green walls also bring some additional benefits, such as improving air quality, reducing energy consumption, and mitigating greenhouse gas emissions.

The efficiency of green walls in cooling largely depends on climatic conditions, the type of the green wall, and vegetation species. In one experimental study, the temperature regulation for three scenarios was explored: direct facade greening system, indirect facade greening system, and living wall system (Fig. 19.11).

For all the greening systems analyzed in this study, no air temperature differences were recorded at more than 10 cm from the front of the facades. However, the vertical greening systems reduced the surface temperatures behind the green layer more than that of the bare wall scenario. Depending on the planter boxes, the direct greening system and the living wall system were the most effective in reducing the wind speed and hence affecting the thermal resistance of the building envelope. The living wall system had the most impact on the thermal resistance because of the extra air cavity that was created.

To achieve the optimal cooling impact (from the surface to the air temperature in front of the wall), it is important to identify the best orientation and direction for the planting of the greenery on the walls. A study in China examined cooling and shading effects of direct green wall systems, and it was shown that planting the west-facing walls reduce the air temperature by 2.0°C (Di and Wang, 1999).

The cooling effect of the green walls also changes with the season. In the hot months of July and August in the northern hemisphere, the green ivy reduced the energy absorbed by the wall and indoor temperature accordingly. In June, when the outdoor temperature is lower, the thermal load of the building is low and the green wall acts as a barrier and thereby the indoor temperature of the building is higher than a building with no ivy green wall. The cooling effect of the same green wall may result in different outcomes in the southern hemisphere.

FIGURE 19.11 Examples of direct façade greening (a), indirect façade greening (b), and living wall (c), with corresponding temperature profiles (Perini et al., 2011).

The cooling effect of the green wall also depends on the characteristic of its surrounding area. In a study conducted in Singapore (Wong et al., 2010), the average surface temperature of green walls was 4.4°C cooler than bare walls. The study also showed up to a 3.3°C reduction in the ambient air temperature in front of the walls (measured 0.15 m away). The cooling impact disappeared within a distance of 0.6 m.

Green walls are one of the WSUD strategies and an effective method in mitigating the UHI effect. All types of green walls reduce the surface temperature and provide a greater benefit for pedestrian thermal comfort and pedestrian-level air temperature compared with green roofs, which mainly contribute to a reduction in radiative loading and cooler air temperature at roof level. The effectiveness of green walls and green roofs in reducing the surface and air temperature depends on the level of the irrigation. The green wall will be actively transpiring (if well irrigated) compared with a bare wall, whose energy balance will either be partitioned into absorbed by the building materials or atmospheric heating. The irrigation supply for green walls can be sourced from bioretention systems or rainwater diverted through a vertical green wall system.

Table 19.1 lists other studies carried out on the efficacy of urban greening in mitigating UHI and improving thermal comfort worldwide over the past years. The majority of the research was carried out in regions likely to exhibit high summer UHI and occurred mainly in subtropical and tropical regions, such as East Asia, North America, and the European part of the Mediterranean Sea. The Middle East and North Africa are less explored in the literature but are no less exposed to the UHI effect.

Several studies explored increasing the number of trees (shading effects) and parks; some studies installed grass on horizontal surfaces and highlighted the benefits of green roofs and walls to mitigate the UHI effect. Other authors emphasized the critical role of irrigation, waterbodies, and evaporative cooling in mitigating the UHI effect. Moreover, many researchers analyzed the impact of combined methods in reducing UHI.

The vegetation studied in Table 19.1 are in the forms of street trees, green roofs, and green walls. Through the studied examples in different countries characterized with different climates, vegetation showed appreciable reduction of air temperatures within urban areas. Street trees and urban parks contributed to lower air temperature at pedestrian level, whereas green walls and green roofs reduced the air temperature at roof level.

With respect to pedestrian thermal comfort, mean radiant temperature is the main parameter that determines the outdoor comfort level. To achieve a better pedestrian thermal comfort, GI should be combined with low-albedo materials on the ground surface to avoid the reradiation of solar radiation to pedestrians.

## 19.5 URBAN GREENING CASE STUDY

Australia is one of the mostly urbanized nations in the world with >80% of its population living in urban areas (ABS, 2013). Melbourne, Victoria, is the second largest city in Australia. Three-quarters of all Victorians reside in Melbourne, which is home to 4.25 million people and covers an area of 10,000 km$^2$ with 31 local government districts. Its population is expected to reach 6.5 million by 2050. Over the past 2 decades, Melbourne has undergone a transformation via city center regeneration and outer suburban development. Moreover, the city is currently experiencing a rapid growth in its central business area.

In Melbourne, the complex relationship between heat, human health, and urban areas has received particular attention following the incidence of several heat waves, with air temperature above 43°C for three consecutive days in 2009, 2013, and 2014. Melbourne currently experiences temperatures >35°C for 9 days per annum, but climate modeling predicts these temperatures will occur for 11 days per year by 2030, increasing to 20 days per year by 2070 (Cleugh et al., 2011).

To accommodate Melbourne's ongoing population growth, an additional 600,000 new dwellings must be constructed, of which 316,000 dwellings will be built close to the CBD. To plan the future growth of the city, "Plan Melbourne" was released in May 2014 by the Victorian government. This plan outlines the vision for the future growth of Melbourne by 2050 (Victorian Government, 2014) and specifically lists the actions that must be adopted through short-, medium-, and long-term planning strategies.

Structure plans provide guidance to communities, planners, business, governments, and developers about the appropriate directions and opportunities for future changes. The structural plans for Melbourne aim to change City North from a low-rise to a medium-rise neighborhood. Some of the strategies presented for the future of City North include a new street hierarchy, increasing the building heights, increasing the tree canopy coverage, and installing green roofs (Victorian Government, 2012). Figs. 19.12 and 19.13 illustrate the potential built form and urban forest strategies in City North by 2050.

To investigate the impact of future structural plans on UHI and pedestrian thermal comfort, a three-dimensional microclimatic modeling system, ENVI-met, was used. ENVI-met is one of the most sophisticated microclimate models

**TABLE 19.1 Studies Conducted on the Effectiveness of Green Infrastructure in Mitigating Urban Heat Island (UHI) and Improving Thermal Comfort**

| City | Climate | Findings |
|---|---|---|
| Tel-Aviv, Israel | Mediterranean | Shading from the increased tree canopy coverage contributed to 80% of the overall cooling effect (Shashua-Bar and Hoffman, 2000) |
| Bangalore, India Tropical | Tropical | 5.6°C temperature difference between the sections of the streets with trees and without trees (Vailshery et al., 2013) |
| Hong Kong | Humid-subtropical | 15% increase in tree coverage resulted in 5.6°C temperature reduction (Ng et al.,2012) |
| Saga, Japan | Humid-subtropical | 20% increase in the number of the trees resulted in 2.27°C reduction in maximum temperature (Srivanit and Hokao, 2013) |
| Singapore | Hot-humid | From 1.5 to 2.8°C lower air temperature under tree canopies (Nichol, 1996) |
| Kumamoto, Japan | Humid-subtropical | 3.8°C lower air temperature under tree canopies (Saito et al., 1990) |
| Serdang, Malaysia | Hot-humid | Trees with high values of leaf area index contribute to lower air temperatures (Shahidan et al., 2010) |
| Ghardai, Algeria | Subtropical-desert | Shading trees significantly improve the pedestrian thermal comfort (Ali-Toudert and Mayer, 2007) |
| Beijing, China | Humid-continental | Tree canopy coverage and shading level significantly influence the thermal comfort (Yan et al., 2012) |
| Sao Paulo, Brazil | Humid-subtropical | Thermal comfort index (Temperature of Equivalent Perception, TEP) was improved by 10°C (Johansson et al., 2013) |
| Mendoza, Argentina | Mid latitude desert | Green infrastructure along the street significantly improved the thermal condition (Correa et al., 2012) |
| Shanghai | Humid-subtropical | Dense tree or grass on the pavement contributed to a significant reduction (Yang et al., 2011) |
| Sao Paulo, Brazil | Humid-subtropical | 12°C reduction in thermal comfort index (physiological equivalent temperature, PET) level (Spangenberg et al., 2008) |
| London, UK | Temperature-oceanic | 1°C reduction in the air temperature as a result of installing green roofs (Virk et al., 2015) |
| Toronto, Canada | Semi continental | 0.4°C reduction in the air temperature as a result of installing green roofs (Berardi, 2016) |
| Madrid, Spain | Mediterranean | Only a moderate effect of green roofs on the surrounding microclimate, but a large contribution when combined with vegetation at pedestrian level (Alcazar et al., 2016) |
| Tehran, Iran | Dry-summer subtropical | Average air temperature above the green roof was 3.06–3.7°C cooler than that of the reference roof (Moghbel and Erfanian Salim, 2017) |
| Adelaide, Australia | Mediterranean | Significant cooling effects in summer as a result of green roofs, green walls, street trees and other water sensitive urban design strategies (Razzaghmanesh et al., 2016) |
| Padua, Italy | Humid sub-tropical | The "Green ground" scenario allows up to 1.4 and 3°C decrease in air temperature during the night and day, respectively (Noro and Lazzarin, 2015) |
| Paris, France | Oceanic climate | 0.79°C reduction in the air temperature as a result of green pavement (Hendel et al., 2016) |
| Nottingham, UK | Oceanic climate | Green wall enabled 6.1°C temperature reduction in sunny days compared with bare wall (Cuce, 2017) |
| Cosenza, Italy | Oceanic climate | Vegetated roof was able to halve summer daily temperature excursions (Bevilacqua et al., 2017) |
| Kuala Lumpur, Malaysia | Tropical | Urban greening resulted in 4°C reduction in the air temperature (Aflaki et al., 2017) |
| Phoenix, United States | Warm humid continental | Urban trees reduced the air temperature by about 1–5°C (Upreti et al., 2017) |
| Montreal, Canada | Continental | Tree cover reduced the air temperature at the tree level by 4 and 2°C at 60 m from the ground (Wang and Akbari, 2016) |

*Continued*

**TABLE 19.1 Studies Conducted on the Effectiveness of Green Infrastructure in Mitigating Urban Heat Island (UHI) and Improving Thermal Comfort—cont'd**

| City | Climate | Findings |
|---|---|---|
| Freiburg, Germany | Mediterranean | Trees on grasslands lead to 2.7 K (2.7°C) reduction in the air temperature (Lee et al., 2016) |
| Hong Kong | Humid-subtropical | Roadside trees reduced the thermal comfort index (PET) to 29 4°C in urban areas (Tan et al., 2017) |
| Hong Kong | Humid-subtropical | Trees with a large crown, short trunk, and dense canopy are the most efficient in mitigating the UHI effect (Kong et al., 2017) |
| Taipei, Taiwan | Humid-subtropical | Urban parks were 0.81 K (0.81°C) cooler than their surrounding built-up areas (Bowler et al., 2010) |
| Athens, Greece | Subtropical —Mediterranean | The average nighttime and daytime park cool island varied between 0.7 K (0.7°C) and 2.6 K (2.6°C), respectively (Skoulika et al., 2014) |
| Melbourne, Australia | Oceanic | 2.5°C temperature difference between an urban park and its surrounding areas (Torok et al., 2001) |
| Singapore | Hot-humid | Large urban parks significantly mitigated the UHI effect (Forsyth et al., 2005) |
| Taipei, Taiwan | Humid-subtropical | The park cooling effect largely depends on the evapotranspiration rate of the trees inside the park (Chang and Li, 2014) |
| Addis Ababa, Ethiopia | Subtropical-highland | The park cooling effect is a variable of the park area (Feyisa et al., 2014) |
| Florida, United States | Humid-subtropical | Lower air temperature under the shade of the trees compared to the surrounding areas (Sonne and Vieira, 2000) |
| Netherlands, Europe | Temperate | Improvement in the thermal comfort index (PET) by 5 K (5°C) as a result of grassland (Klemm et al., 2015) |
| Sacramento and Vancouver | Mediterranean-oceanic | Frequent irrigation produced from 1 to 2°C and 5 to 7°C cooling effect in Vancouver and Sacramento, respectively (Spronken-Smith, 1998) |
| Athens, Greece | Subtropical Mediterranean | Park cooling effect was reduced because of congested areas and traffic in the surrounding area of a large urban park (Zoulia et al., 2009) |
| Gothenburg, Sweden | Oceanic | High-rise buildings in surrounding area of urban parks reduce the cooling impact of parks (Upmanis and Chen, 1999) |
| Sacramento and Vancouver | Mediterranean-oceanic | The park cooling impact extended equal to the width of the park (Spronken-Smith, 1994) |
| Athens, Greece | Subtropical—Mediterranean | The greatest zone of influence for park cooling effect extended downwind from the park (Dimoudi and Nikolopoulou, 2003) |
| Singapore | Hot-humid | Areas close to parks were on average 1.3 K (1.3°C) cooler than the surrounding areas (Yu and Hien, 2006) |
| Mexico | Humid-subtropical | The park cooling impact with the area of 500 ha extended equal to the width of the park (2 km) (Jauregui, 1990) |
| Japan | Humid-subtropical | The park cooling impact with the area of 100 and 400 $m^2$ extends to 300 and 400 m, respectively (Honjo and Takakura, 1990) |
| Tokyo, Japan | Humid-subtropical | The park cooling impact with an area of 0.6 $km^2$ was found to be 1.58°C, extending for 1 km (Ca et al., 1998) |

and includes all the energy and radiative processes that occur in the urban environment (Bruse, 2008). ENVI-met simulates the interaction among surfaces, plants, and air in an urban environment with a typical resolution of 0.5–10 m in space and 10 s in time.

The ENVI-met model was validated against field measurements and adjusted to generate simulated outputs close to the measured data (Jamei and Rajagopalan, 2017). Fig. 19.14 shows modeling inputs of the different aspects of the future structural plans for City North.

**FIGURE 19.12** Potential built form of City North by 2050, showing the transformation from a low rise (gray color) to medium rise (white color) (Victorian Government, 2012).

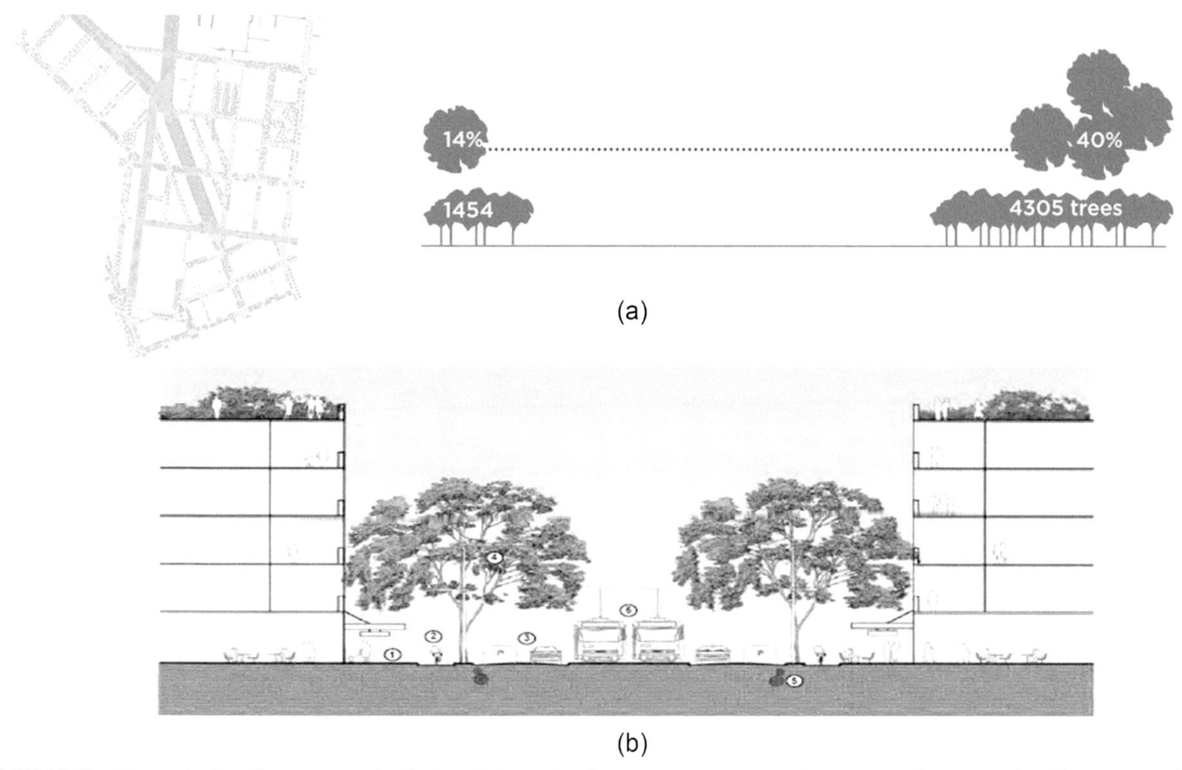


**FIGURE 19.13** Proposed urban forest strategy in City North: increasing the tree canopy coverage from 14% to 40% (a) and installing green roofs (b) (Victorian Government, 2012).

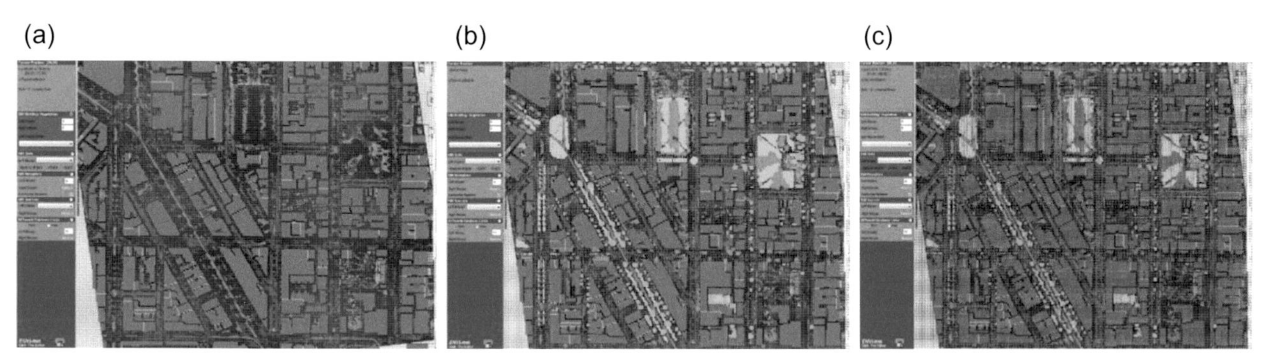


**FIGURE 19.14** Buildings (a), vegetation (b), and soil (c) inputs used in the ENVI-met model to assess climate impacts of the City North structural plans (Jamei and Rajagopalan, 2017).

Two other scenarios were also explored to further mitigate the UHI and improve thermal comfort: (1) integrating existing open public spaces with parks; and (2) increasing tree canopy coverage by a further 10%–50% (Fig. 19.15).

The simulations predicted the cooling effect of trees varied from 0.1 to 1.6°C during the day, provided they are adequately irrigated. It was also found that increasing the tree canopy coverage decreases the received mean radiant temperature by 2°C. This reduction in the mean radiant temperature led to an improvement in the pedestrian thermal comfort.

Green roofs created cooling at roof level but had negligible cooling effects at pedestrian level. Pedestrian thermal comfort, as measured by the PET, slightly improved as a result of the future structural plans (Fig. 19.16). The simulation for the proposed two greening scenarios also showed further improvements in PET level.

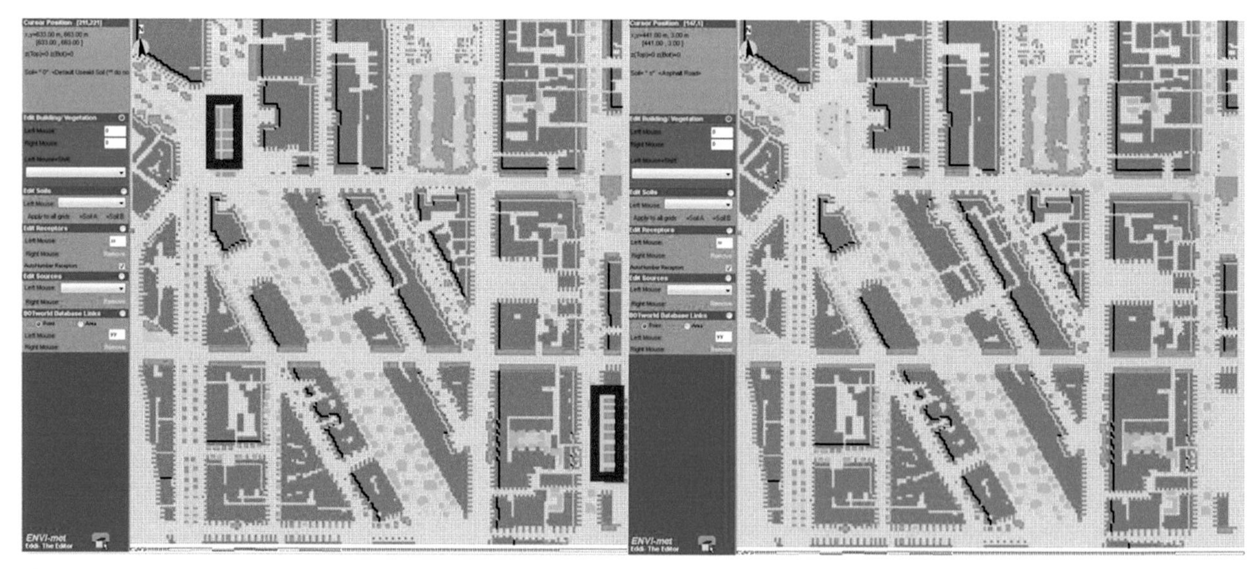

**FIGURE 19.15** Simulation of the proposed scenarios: (1) integrating the existing open public spaces with parks (left) and (2) increasing the tree canopy coverage to 50% (right). *Red boxes* show the location of the proposed open green public spaces. Green infrastructure (green), built-up areas (Gray) (Jamei and Rajagopalan, 2017).

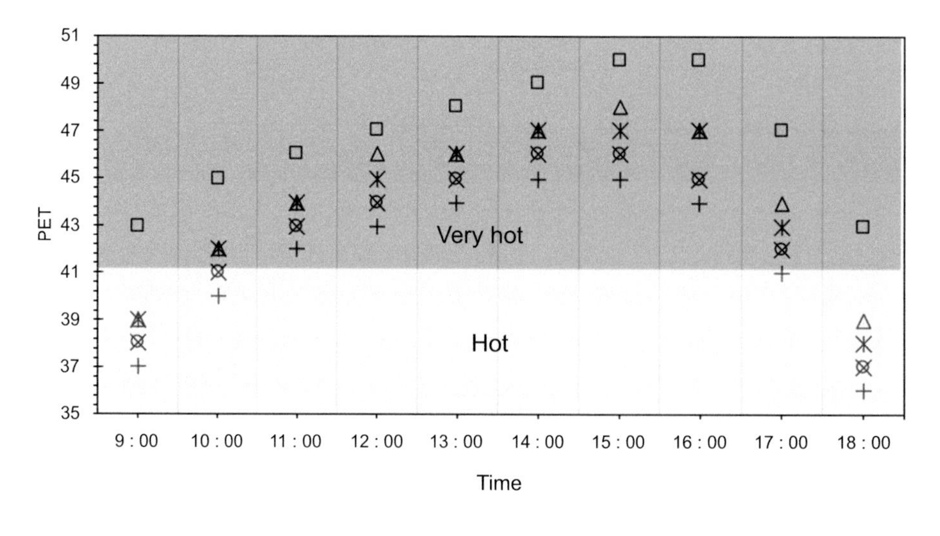


**FIGURE 19.16** Psychological equivalent temperature (°C) predicted on a hot summer's day for City North, Melbourne, under existing condition, future developments, and proposed greening scenarios (Jamei and Rajagopalan, 2017).

Fig. 19.16 shows the thermal condition under the existing condition and future development scenarios. In the existing condition, the thermal comfort (PET) is within the "very hot" range during the whole day. Future developments were simulated, and the outcomes show an improvement under different stages of the development. However, even after the full implementation of future developments, the thermal condition was found in the "very hot" range at certain hours of the day. Therefore, the proposed greening scenarios were modeled to examine if the PET level can be further reduced. The greening scenarios were (1) increasing the tree canopy coverage from 40% to 50% and (2) integrating public realms with small urban parks.

Implementing the greening scenarios resulted in a 5.1°C improvement in the PET level compared with the one under the existing condition on an extremely hot summer day. Fig. 19.16 summarizes the variation of PET at the different stages of future development in City North. The study proposes that future works are needed to determine if the vegetation type, spatial configuration, and percentage of GI can further improve the pedestrian thermal comfort during the problematic times of the day (from 11:00 a.m. to 5:00 p.m.) when the thermal comfort is in the "very hot" range.

## 19.6 WSUD CASE STUDY

An example of the WSUD approach to cool the urban environment and decrease heat discomfort is Mawson Lakes, Adelaide, South Australia. Mawson Lakes is a mixed urban, residential, and commercial development, in the City of Salisbury, housing a resident population of nearly 10,000 people (Tjandraatmadja et al., 2014).

Mawson Lakes is located 12 km north of Adelaide's CBD (Fig. 19.17). The suburban neighborhood is characterized by wetlands, artificial lakes, and irrigated public green spaces. One of the unique characteristics of Mawson Lakes compared with its surrounding neighborhoods is the high level of available water for irrigating the vegetation. There is a dual water supply system in this suburb which supplies potable and recycled water through separate reticulation systems. Stormwater runoff from the impervious surfaces (e.g., roads) is used to top up the waterbodies in winter, but recycled water from a large sewage treatment plant (Bolivar) and captured stormwater from nearby Parafield are needed to maintain levels in the hot, dry summers of this Mediterranean climate (Broadbent et al., 2017b). The irrigated green spaces and all the waterbodies create an oasis effect, whereby the warm air passes over the "wet surface" and becomes cooler and more humid. This exiting airflow then cools the surrounding areas covered by impervious surfaces.

Fig. 19.17 shows the details of the WSUD features installed in Mawson Lakes. Broadbent et al. (2017b) undertook comprehensive field measurements to understand the influence of water availability and urban design parameters on human thermal comfort and mitigating the UHI effect. Fixed station field measurements, bicycle transects, and high-resolution thermal images were used to capture the intra-air temperature difference at various locations. Aerial imagery and Light Detection and Ranging (LiDAR) data were collected to characterize the different types of urban landscape. Land cover classification, normalized difference vegetation index (NDVI), surface radiative temperature, and sky view factor (SVF) were generated to assist in understanding the key drivers of urban microclimate across Mawson Lakes.

The NDVI is a measure of live green vegetation (Gao, 1996), and the SVF denotes the ratio between the radiation received by a planar surface and that from the entire hemispheric radiating environment and is calculated as the fraction of sky visible from the ground up. SVF is a dimensionless value that ranges from 0 to 1.

Fig. 19.18 shows that there is a significant spatial and temporal intrasuburban variability of air temperature in Mawson Lakes. The range of spatial variability was around 2.0°C for day and night averages. A daytime cooling effect associated with lakes and wetlands, especially north/north east of the Spacial Data Modeling Language (SDML) sites (sites 4, 5, 6, and 8), is clearly visible in Fig. 19.18(a). Urban sites were generally warmer, but there was spatial variability in the air temperature across urban locations, which indicates that wind speed has a more important role in daytime air temperature variability. Natural sites that were not located near waterbodies (sites 23, 28, and 29) showed higher air temperatures, which indicate that unirrigated grass is not an effective land cover for reducing daytime air temperature.

At night, the cooler sites were located away from built-up areas, including the two sites in the Salisbury wetland (sites 24 and 27), the sites in the grassed areas east of the suburb (sites 23, 28, and 29), and the sites located along the river channel (sites 26 and 30).

The study also concluded that lake and wetland are, on average, cooler than the residential areas and open green space by 1°C during the daytime. Open green space was cooler than the other sites by 0.6°C during the nighttime. The most important factors in daytime cooling were wind speed, tree cover fraction, SVF, and soil water availability. The influencing factors for nighttime cooling were the fraction of pervious ground cover, tree cover fraction, and wind speed.

In a related study of Mawson Lakes, the role of IUWM in passive evaporative cooling was assessed (Broadbent et al., 2017a). The research investigated the cooling potential of several purpose-managed irrigation scenarios based on recycled water. Heat wave conditions were simulated using the SURFEX (land-surface modeling scheme) (Masson et al., 2013),

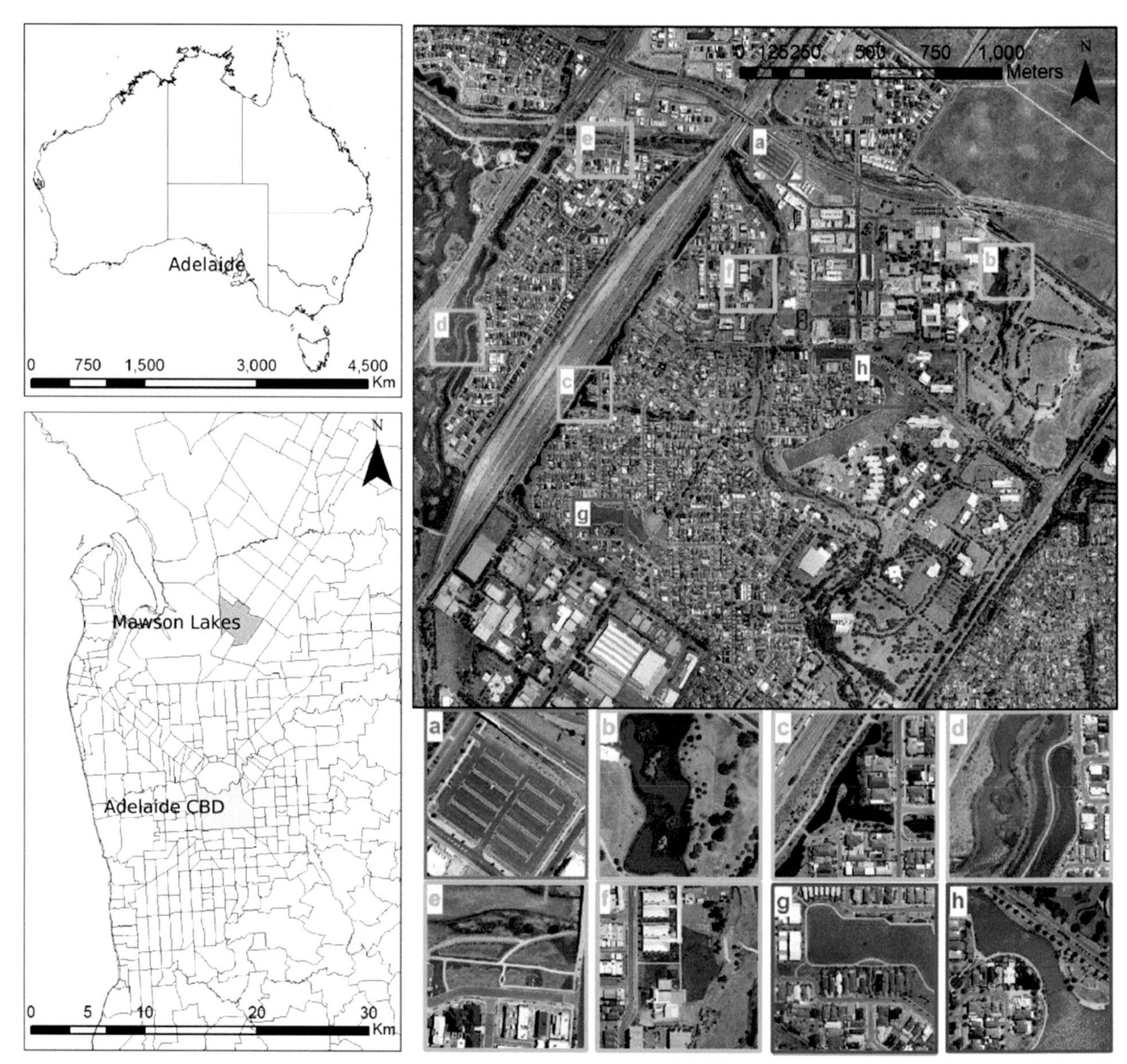


FIGURE 19.17 The location of Adelaide and Mawson Lakes in South Australia with examples of the water sensitive urban design features applied: (a) pervious swales, (b and c) watercourses, (d) wetlands, (e and f) irrigated green space, and (g and h) artificial lakes (Broadbent et al., 2017b).

and the effects of a range of irrigation scenarios with different irrigation rates and timing were evaluated. The range of water usage was incrementally increased in each scenario.

The hot and dry summer, which triggers the heat wave events, is the main reason to initiate such heat mitigation strategies, but because of the limited water supply, particularly in summer (when the average rainfall is only 12.4 mm), the implementation of these strategies is a challenging task. To address this issue, the local government in Adelaide proposed utilizing the integrated water management systems approach for efficiently mitigating the heat in the future developments.

The simulation outputs indicate that the diurnal average temperature reduced by 2.3°C with the magnitude of cooling proportional to the pervious (irrigated) fraction. However, because of the limitation in the simulation approach and the lack of atmospheric mixing that was captured, the simulation outputs are only valid for an isolated patch of irrigation. Therefore, the real cooling effect may be lower if a coupled atmospheric model was used. The study also showed that there is a nonlinear relationship between the irrigation and cooling effect (20 L/m$^2$ d). The study also indicated that although irrigation increased the humidity rate, it improved the pedestrian thermal comfort during the heat wave conditions. The outcomes of this study can inform the practitioners on using the water to achieve an optimal cooling during heat wave conditions.

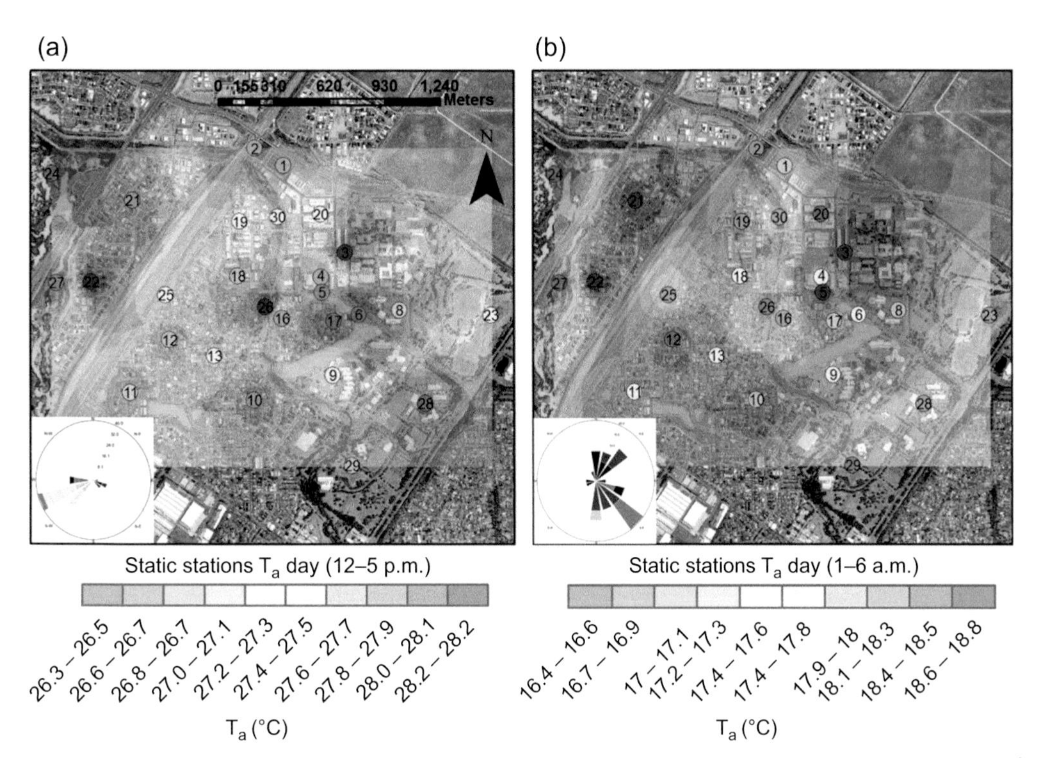


FIGURE 19.18 The average air temperature for fixed stations during the day (12–5 p.m.) with the average wind speed 2.5 $ms^{-1}$ (a) and night (1–6 a.m.) with the average wind speed 5.3 $ms^{-1}$ (b) (Broadbent et al., 2017b).

## 19.7 CONCLUSION

The UHI observed in cities impacts on public health via heat stress and heat-related illnesses. In Australia, where extreme weather conditions and prolonged drought are not uncommon, urban developments occur with low water availability together with projected future climate change. This will almost certainly have adverse implications for the health and well-being of urban dwellers.

Given the rapid growth of Australian cities, alternative growth management policies can have a significant beneficial effect on the incidence and severity of the UHI effect. These policies should inform urban planners as to how best to mitigate the UHI effects. With the emergence of smart cities and smart growth concepts, a major question that remains is "what type of smart growth policy can effectively contribute to UHI mitigation?"

However, urban planners need to make a trade-off between fewer suburbs with a high UHI impact (which accompanies infill development) and more suburbs with a moderate UHI effect (which accompanies urban sprawl and higher infrastructure and transport costs). The integration of GI with WSUD and policies on a mandatory minimum fraction of green space (trees, parks, green walls, green roofs, etc.) could be the key answer to this question. We argue that new design concepts for urban infill developments should reintroduce the green WSUD approaches and biodiversity to the urban-built environment.

This chapter highlights the necessity of integrating climate knowledge into planning practices and raises awareness among urban planners and landscape architects of the importance of critically assessing planning policies before developing a new neighborhood. Future policies have little impact on the form and structure of the already-built neighborhood or a city. Therefore, planning policies should control the urban development and initiate urban greening strategies to mitigate the UHI effect.

This chapter also argues that modification of the water balance in cities is one of the main causes of high urban air temperatures and poor pedestrian thermal comfort. Increasing vegetation coverage is the key approach for mitigating UHI impacts. As the magnitude and intensity of the UHI effect varies across different climates, a specific urban greening strategy (e.g., parks, street trees, green roofs, and green walls) may be required for each particular climate. The cooling benefit of GI is also associated with positive economic impact, environmental and air quality benefits, city character, heritage preservation, and the promotion of preventive public health measures, such as active transportation.

This chapter provides evidence for the effectiveness of GI and WSUD strategies as mechanisms to address some of the challenges faced by contemporary urban environments, including extreme heat events, drought, UHI, and thermal discomfort. In Australian cities with rapid urbanization growth, aging population, and anticipated climate change, the implementation of these strategies is becoming even more critical. However, it has to be noted that the evaporative cooling effect may not be strong under humid climatic condition. In humid climates, trees with a high level of tree canopy coverage, which can promote shading, will be more beneficial in reducing the air temperature at pedestrian level and decreasing the surface radiative temperatures.

The combination of WSUD and GI methods will assist in maximizing the cooling benefits of the existing vegetation. For instance, stormwater harvesting can support the fit-for-purpose water for irrigating the landscape and contribute to healthy vegetation that can provide shading, evaporative cooling, and improved thermal comfort and microclimate. WSUD strategies are most efficient cooling methods under hot and dry climatic conditions.

Inter- and intraurban climates differ from one city to another. For instance, the predominant climatic condition in a city may be temperate climate, whereas another city may be subtropical or tropical climate. Some parts of the city may experience cooler air temperature than other parts. Therefore, customized policies and guidelines are needed in urban planning and design, which are tailored for the particular context of each region. To implement fit-for-place design solutions, the local context has to be taken into account because each city offers a unique set of characteristics. Modified design of WSUD, particularly for Australian cities where water shortage has always been a major challenge, is required in addition to the measured and defined benchmarks against which UHI mitigation is assessed.

This chapter provided a comprehensive review on the cooling benefits of diverse green and WSUD infrastructures and presents some recent case studies in Australian cities that utilized these methods to mitigate the UHI effect and improve the local climate. However, only limited research was conducted on how vegetation should be incorporated in each particular city. Furthermore, the majority of these studies have been conducted through modeling approaches. Therefore, future research should focus on conducting experimental and field measurement-based studies to achieve the optimal cooling through exploring the role of vegetation type, required green ratio per built-up ratio, and the best spatial distribution for different GI. Vegetation type is of particular interest, as there is a direct link between the vegetation type and the water required for irrigation. Further research is also needed on the optimal design of individual WSUD features, efficient irrigation systems with highest level of evaporative cooling, and seasonal effects of each WSUD strategy.

## REFERENCES


(ABS), A. B. O. S, 2013. Department of Transport, Planning and Local Infrastructure Preliminary Projections 2014.

Abis, K.L., Mara, D., 2006. Temperature measurement and stratification in facultative waste stabilisation ponds in the UK climate. Environmental Monitoring and Assessment 114, 35–47.

Aflaki, A., Mirnezhad, M., Ghaffarianhoseini, A., Ghaffarianhoseini, A., Omrany, H., Wang, Z.-H., Akbari, H., 2017. Urban heat island mitigation strategies: a state-of-the-art review on Kuala Lumpur, Singapore and Hong Kong. Cities 62, 131–145.

Agarwal, A., 2000. Drought? Try capturing the rain. Center for Science and Environment 1–16.

Akbari, H., Taha, H., 1992. The impact of trees and white surfaces on residential heating and cooling energy use in four Canadian cities. Energy 17, 141–149.

Alcazar, S.S., Olivieri, F., Neila, J., 2016. Green roofs: experimental and analytical study of its potential for urban microclimate regulation in Mediterranean–continental climates. Urban Climate 17, 304–317.

Alexander, L.V., Arblaster, J.M., 2009. Assessing trends in observed and modelled climate extremes over Australia in relation to future projections. International Journal of Climatology 29, 417–435.

Ali-Toudert, F., Mayer, H., 2007. Effects of asymmetry, galleries, overhanging façades and vegetation on thermal comfort in urban street canyons. Solar Energy 81, 742–754.

Armson, D., Stringer, P., Ennos, A., 2013. The effect of street trees and amenity grass on urban surface water runoff in Manchester, UK. Urban Forestry and Urban Greening 12, 282–286.

Ashrae, A.H.-F., 1997. American Society of Heating, Refrigerating and Air-Conditioning Engineers. Inc. Atlanta.

Assimakopoulos, M., Mihalakakou, G., Flocas, H., 2007. Simulating the thermal behaviour of a building during summer period in the urban environment. Renewable Energy 32, 1805–1816.

Australian Bureau of Meteorology, 2013a. Climate data online. http://www.bom.gov.au/climate/data/.

Australian Bureau of Statistics, 2013b. Regional Population Growth Australia.

Berardi, U., 2016. The outdoor microclimate benefits and energy saving resulting from green roofs retrofits. Energy and Buildings 121, 217–229.

Bevilacqua, P., Mazzeo, D., Bruno, R., Arcuri, N., 2017. Surface temperature analysis of an extensive green roof for the mitigation of urban heat island in southern Mediterranean climate. Energy and Buildings 150, 318–327.

Bonan, G.B., 1997. Effects of land use on the climate of the United States. Climatic Change 37, 449–486.

Bornstein, R., Lin, Q., 2000. Urban heat islands and summertime convective thunderstorms in Atlanta: three case studies. Atmospheric Environment 34, 507–516.

Bowler, D.E., Buyung-Ali, L., Knight, T.M., Pullin, A.S., 2010. Urban greening to cool towns and cities: a systematic review of the empirical evidence. Landscape and Urban Planning 97, 147–155.

Broadbent, A.M., Coutts, A.M., Tapper, N.J., Demuzere, M., 2018. The cooling effect of irrigation on urban microclimate during heatwave conditions. Urban Climate 23, 309–329.

Broadbent, A.M., Coutts, A.M., Tapper, N.J., Demuzere, M., Beringer, J., 2017b. The microscale cooling effects of water sensitive urban design and irrigation in a suburban environment. Theoretical and Applied Climatology 1–23.

Brown, R.D., Gillespie, T.J., 1995. Microclimatic Landscape Design: Creating Thermal Comfort and Energy Efficiency. Wiley.

Brudermann, T., Sangkakool, T., 2017. Green roofs in temperate climate cities in Europe – an analysis of key decision factors. Urban Forestry and Urban Greening 21, 224–234.

Bruse, M., 2008. ENVI-met V3. 1, a Microscale Urban Climate Model.

Ca, V.T., Asaeda, T., Abu, E.M., 1998. Reductions in air conditioning energy caused by a nearby park. Energy and Buildings 29, 83–92.

Carnie, J., 2009. January 2009 Heatwave in Victoria: An Assessment of Health Impacts. Victorian Government Department of Human Services, Melbourne.

Chang, C.-R., Li, M.-H., Chang, S.-D., 2007. A preliminary study on the local cool-island intensity of Taipei city parks. Landscape and Urban Planning 80, 386–395.

Chang, C.-R., Li, M.-H., 2014. Effects of urban parks on the local urban thermal environment. Urban Forestry and Urban Greening 13, 672–681.

Cleugh, H., Cleugh, H., Smith, M.S., Battaglia, M., Graham, P., 2011. Climate Change: Science and Solutions for Australia. CSIRO.

Correa, E., Ruiz, M.A., Canton, A., Lesino, G., 2012. Thermal comfort in forested urban canyons of low building density. An assessment for the city of Mendoza, Argentina. Building and Environment 58, 219–230.

Coutts, A.M., Beringer, J., Tapper, N.J., 2007. Impact of increasing urban density on local climate: spatial and temporal variations in the surface energy balance in Melbourne, Australia. Journal of Applied Meteorology and Climatology 46, 477–493.

Coutts, A., Beringer, N., Jason, T., Daly, E., White, E., Broadbent, A., Pettigrew, J., Harris, R., Gebert, L., Nice, K., Hamel, P., Fletcher, T., Kalla, M., 2013a. Determine the microclimate influence of harvesting solutions and Water Sensitive Urban Design at the micro-scale, Green cities and microclimate. CRC for Water Sensitive Cities 14–15.

Coutts, A., Beringer, N., Jason, T., Daly, E., White, E., Broadbent, A., Pettigrew, J., Harris, R., Gebert, L., Nice, K., Hamel, P., Fletcher, T., Kalla, M., 2013b. Determine the microclimate influence of harvesting solutions and water sensitive urban design at the micro-scale, Green cities and microclimate. CRC for Water Sensitive Cities 6–7.

Coutts, A., Beringer, N., Jason, T., Daly, E., White, E., Broadbent, A., Pettigrew, J., Harris, R., Gebert, L., Nice, K., Hamel, P., Fletcher, T., Kalla, M., 2013c. Determine the microclimate influence of harvesting solutions and water sensitive urban design at the micro-scale, green cities and microclimate project 5: footscray primary school irrigation study. CRC for Water Sensitive Cities 12–13.

Coutts, A., Beringer, N., Jason, T., Daly, E., White, E., Broadbent, A., Pettigrew, J., Harris, R., Gebert, L., Nice, K., Hamel, P., Fletcher, T., Kalla, M., 2013d. Determine the microclimate influence of harvesting solutions and water sensitive urban design at the micro-scale, green cities and microclimate project 7: assessment of evapotranspiration from rain gardens in little stringybark creek. CRC for Water Sensitive Cities 16–17.

Cuce, E., 2017. Thermal regulation impact of green walls: an experimental and numerical investigation. Applied Energy 194, 247–254.

D'Orazio, M., Di Perna, C., Di Giuseppe, E., 2012. Green roof yearly performance: a case study in a highly insulated building under temperate climate. Energy and Buildings 55, 439–451.

Demuzere, M., Coutts, A.M., Göhler, M., Broadbent, A.M., Wouters, H., Van Lipzig, N., Gebert, L., 2014. The implementation of biofiltration systems, rainwater tanks and urban irrigation in a single-layer urban canopy model. Urban Climate 10, 148–170.

Dhainaut, J.-F., Claessens, Y.-E., Ginsburg, C., Riou, B., 2003. Unprecedented heat-related deaths during the 2003 heat wave in Paris: consequences on emergency departments. Critical Care 8, 1.

Di, H., Wang, D., 1999. Cooling effect of ivy on a wall. Experimental Heat Transfer 12, 235–245.

Dimoudi, A., Nikolopoulou, M., 2003. Vegetation in the urban environment: microclimatic analysis and benefits. Energy and Buildings 35, 69–76.

Dixon, P.G., Mote, T.L., 2003. Patterns and causes of Atlanta's urban heat island–initiated precipitation. Journal of Applied Meteorology 42, 1273–1284.

Donovan, G.H., Butry, D.T., 2009. The value of shade: estimating the effect of urban trees on summertime electricity use. Energy and Buildings 41, 662–668.

Doug, B., Hitesh, D., James, L., Paul, M., 2005. Report on the Environmental Benefits and Costs of Green Roof Technology for the City of Toronto.

Estrada, F., Botzen, W.W., Tol, R.S., 2017. A global economic assessment of city policies to reduce climate change impacts. Nature Climate Change 7 (6), 403.

Feyisa, G.L., Dons, K., Meilby, H., 2014. Efficiency of parks in mitigating urban heat island effect: an example from Addis Ababa. Landscape and Urban Planning 123, 87–95.

Forsyth, A., Musacchio, L., Fitzgerald, F., 2005. Designing Small Parks: A Manual Addressing Social and Ecological Concerns. J. Wiley.

Gao, B.-C., 1996. NDWI — a normalized difference water index for remote sensing of vegetation liquid water from space. Remote Sensing of Environment 58, 257–266.

Grossman-Clarke, S., Zehnder, J.A., Loridan, T., Grimmond, C.S.B., 2010. Contribution of land use changes to near-surface air temperatures during recent summer extreme heat events in the Phoenix metropolitan area. Journal of Applied Meteorology and Climatology 49, 1649–1664.

Heisler, G.M., Grimmond, S., Grant, R.H., Souch, C., 1994. Investigation of the influence of Chicago's urban forests on wind and air temperature within residential neighbourhoods. In: Chicago's Urban Forest Ecosystem: Results of the Chicago Urban Forest Climate Project, vol. 19.

Hendel, M., Gutierrez, P., Colombert, M., Diab, Y., Royon, L., 2016. Measuring the effects of urban heat island mitigation techniques in the field: application to the case of pavement-watering in Paris. Urban Climate 16, 43–58.

Hondula, D.M., Georgescu, M., Balling, R.C., 2014. Challenges associated with projecting urbanization-induced heat-related mortality. The Science of the Total Environment 490, 538–544.

Honjo, T., Takakura, T., 1990. Simulation of thermal effects of urban green areas on their surrounding areas. Energy and Buildings 15, 443–446.

Höppe, P., 1999. The physiological equivalent temperature – a universal index for the biometeorological assessment of the thermal environment. International Journal of Biometeorology 43, 71–75.

Höppe, P., 2002. Different aspects of assessing indoor and outdoor thermal comfort. Energy and Buildings 34, 661–665.

House-Peters, L.A., Chang, H., 2011. Modeling the impact of land use and climate change on neighborhood-scale evaporation and nighttime cooling: a surface energy balance approach. Landscape and Urban Planning 103, 139–155.

Howard, L., 1818. The Climate of London: Deduced from Meteorological Observations, Made at Different Places in the Neighbourhood of the Metropolis. W. Phillips, sold also by J. and A. Arch.

Jamei, E., Jamei, Y., Rajagopalan, P., Ossen, D.R., Roushenas, S., 2015. Effect of built-up ratio on the variation of air temperature in a heritage city. Sustainable Cities and Society 14, 280–292.

Jamei, E., Rajagopalan, P., 2017. Urban development and pedestrian thermal comfort in Melbourne. Solar Energy 144, 681–698.

Järvi, L., Grimmond, C., Christen, A., 2011. The surface urban energy and water balance scheme (SUEWS): evaluation in Los Angeles and Vancouver. Journal of Hydrology 411, 219–237.

Jauregui, E., 1990. Influence of a large urban park on temperature and convective precipitation in a tropical city. Energy and Buildings 15, 457–463.

Johansson, E., Spangenberg, J., Gouvêa, M.L., Freitas, E.D., 2013. Scale-integrated atmospheric simulations to assess thermal comfort in different urban tissues in the warm humid summer of São Paulo, Brazil. Urban Climate 6, 24–43.

Judd, M.D., 2011. Experience with UHF partial discharge detection and location in power transformers. In: Electrical Insulation Conference (EIC), 2011. IEEE, pp. 201–205.

Killicoat, P., Puzio, E., Stringer, R., 2002. The economic value of trees in urban areas: estimating the benefits of Adelaide's street trees. In: Proceedings of 3rd National Street Tree Symposium. Adeliade University Australia, pp. 90–102.

Kim, Y.O., Penn, A., 2004. Linking the spatial syntax of cognitive maps to the spatial syntax of the environment. Environment and Behavior 36, 483–504.

Klemm, W., Heusinkveld, B.G., Lenzholzer, S., Jacobs, M.H., Van Hove, B., 2015. Psychological and physical impact of urban green spaces on outdoor thermal comfort during summertime in The Netherlands. Building and Environment 83, 120–128.

Kong, L., Lau, K.K.-L., Yuan, C., Chen, Y., Xu, Y., Ren, C., Ng, E., 2017. Regulation of outdoor thermal comfort by trees in Hong Kong. Sustainable Cities and Society 31, 12–25.

Kotzen, B., 2003. An investigation of shade under six different tree species of the Negev desert towards their potential use for enhancing micro-climatic conditions in landscape architectural development. Journal of Arid Environments 55, 231–274.

Kuo, F.E., Sullivan, W.C., 2001. Environment and crime in the inner city: does vegetation reduce crime? Environment and Behavior 33, 343–367.

Lee, H., Mayer, H., Chen, L., 2016. Contribution of trees and grasslands to the mitigation of human heat stress in a residential district of Freiburg, Southwest Germany. Landscape and Urban Planning 148, 37–50.

Loughnan, M., Coutts, A., Tapper, N., Beringer, J., 2012a. Identifying summer temperature ranges for human thermal comfort in two Australian cities. In: WSUD 2012: Water Sensitive Urban Design; Building the Water Sensiitve Community; 7th International Conference on Water Sensitive Urban Design. Engineers Australia, p. 525.

Loughnan, M., Nicholls, N., Tapper, N.J., 2012b. Mapping heat health risks in urban areas. International Journal of Population Research 2012.

Loughnan, M., Tapper, N., Phan, T., 2014. Identifying vulnerable populations in subtropical Brisbane, Australia: a guide for heatwave preparedness and health promotion. ISRN Epidemiology 2014.

Masson, V., Le Moigne, P., Martin, E., Faroux, S., Alias, A., Alkama, R., Belamari, S., Barbu, A., Boone, A., Bouyssel, F., 2013. The SURFEXv7. 2 land and ocean surface platform for coupled or offline simulation of earth surface variables and fluxes. Geoscientific Model Development 6, 929–960.

Matzarakis, A., Mayer, H., 2000. Atmospheric conditions and human thermal comfort in urban areas. In: Proc. 11th Sem. Environmental Protection. Environment and Health, Thessaloniki, pp. 155–165.

Mayer, H., Holst, J., Dostal, P., Imbery, F., Schindler, D., 2008. Human thermal comfort in summer within an urban street canyon in Central Europe. Meteorologische Zeitschrift 17, 241–250.

McIntyre, D., 1980. Indoor Climate. Applied Science Publishers, London.

McPherson, G.E., Nowak, D.J., Rowntree, R.A., 1994. Chicago's Urban Forest Ecosystem: Results of the Chicago Urban Forest Climate Project.

Mihalakakou, G., Santamouris, M., Asimakopoulos, D., 1998. Modeling ambient air temperature time series using neural networks. Journal of Geophysical Research: Atmospheres 103, 19509–19517.

Mitchell, V., Cleugh, H., Grimmond, C., Xu, J., 2008. Linking urban water balance and energy balance models to analyse urban design options. Hydrological Processes 22, 2891–2900.

Moghbel, M., Erfanian Salim, R., 2017. Environmental benefits of green roofs on microclimate of Tehran with specific focus on air temperature, humidity and $CO_2$ content. Urban Climate 20, 46–58.

Moore, G., 2009. Urban trees: worth more than they cost. In: Proc. 10th National Street Tree Symp. Univ. Adelaide/Waite Arboretum, Adelaide, pp. 7–14.
Newman, L., Herbert, Y., 2009. The use of deep water cooling systems: two Canadian examples. Renewable Energy 34, 727–730.
Nichol, J.E., 1996. High-resolution surface temperature patterns related to urban morphology in a tropical city: a satellite-based study. Journal of Applied Meteorology 35, 135–146.
Ng, E., Chen, L., Wang, Y., Yuan, C., 2012. A study on the cooling effects of greening in a high-density city: an experience from Hong Kong. Building and Environment 47, 256–271.
Noro, M., Lazzarin, R., 2015. Urban heat island in Padua, Italy: simulation analysis and mitigation strategies. Urban Climate 14, 187–196.
Oke, T., Johnson, G., Steyn, D., Watson, I., 1991. Simulation of surface urban heat islands under 'ideal' conditions at night Part 2: diagnosis of causation. Boundary-Layer Meteorology 56, 339–358.
Oke, T.R., 1982. The energetic basis of the urban heat island. Quarterly Journal of the Royal Meteorological Society 108, 1–24.
Oke, T.R., 1988. The urban energy balance. Progress in Physical Geography 12, 471–508.
Oke, T.R., Crowther, J., Mcnaughton, K., Monteith, J., Gardiner, B., 1989. The micrometeorology of the urban forest [and discussion]. Philosophical Transactions of the Royal Society of London B Biological Sciences 324, 335–349.
Pandit, R., Laband, D.N., 2010. Energy savings from tree shade. Ecological Economics 69, 1324–1329.
Pandit, R., Polyakov, M., Sadler, R., February 2012. The Importance of tree cover and neighbourhood parks in determining urban property values. In: 2012 Conference (56th), pp. 7–10.
Park, M., Hagishima, A., Tanimoto, J., Narita, K.-I., 2012. Effect of urban vegetation on outdoor thermal environment: field measurement at a scale model site. Building and Environment 56, 38–46.
Pataki, D.E., Mccarthy, H.R., Litvak, E., Pincetl, S., 2011. Transpiration of urban forests in the Los Angeles metropolitan area. Ecological Applications 21, 661–677.
Perini, K., Ottelé, M., Fraaij, A.L.A., Haas, E.M., Raiteri, R., 2011. Vertical greening systems and the effect on air flow and temperature on the building envelope. Building and Environment 46, 2287–2294.
Preetha, P., Thomas, M.J., 2011. Partial discharge resistant characteristics of epoxy nanocomposites. IEEE Transactions on Dielectrics and Electrical Insulation 18.
Razzaghmanesh, M., Beecham, S., Salemi, T., 2016. The role of green roofs in mitigating Urban Heat Island effects in the metropolitan area of Adelaide, South Australia. Urban Forestry and Urban Greening 15, 89–102.
Revich, B., 2011. Heat-wave, air quality and mortality in European Russia in summer 2010: preliminary assessment. Ekologiya Cheloveka/Human Ecology 3–9.
Rosenzweig, C., Solecki, W.D., Hammer, S.A., Mehrotra, S., 2011. Climate Change and Cities: First Assessment Report of the Urban Climate Change Research Network. Cambridge University Press.
Saaroni, H., Ziv, B., 2003. The impact of a small lake on heat stress in a Mediterranean urban park: the case of Tel Aviv, Israel. International Journal of Biometeorology 47, 156–165.
Sailor, D.J., Elley, T.B., Gibson, M., 2012. Exploring the building energy impacts of green roof design decisions–a modeling study of buildings in four distinct climates. Journal of Building Physics 35, 372–391.
Saito, I., Ishihara, O., Katayama, T., 1990. Study of the effect of green areas on the thermal environment in an urban area. Energy and Buildings 15, 493–498.
Santamouris, M., Cartalis, C., Synnefa, A., Kolokotsa, D., 2015. On the impact of urban heat island and global warming on the power demand and electricity consumption of buildings — a review. Energy and Buildings 98, 119–124.
Seitz, J., Escobedo, F., 2011. Urban forests in Florida: trees control stormwater runoff and improve water quality. City 393.
Shahidan, M.F., Shariff, M.K.M., Jones, P., Salleh, E., Abdullah, A.M., 2010. A comparison of *Mesua ferrea* L. and *Hura crepitans* L. for shade creation and radiation modification in improving thermal comfort. Landscape and Urban Planning 97, 168–181.
Shashua-Bar, L., Hoffman, M.E., 2000. Vegetation as a climatic component in the design of an urban street. Energy and Buildings 31, 221–235.
Shashua-Bar, L., Pearlmutter, D., Erell, E., 2009. The cooling efficiency of urban landscape strategies in a hot dry climate. Landscape and Urban Planning 92, 179–186.
Skoulika, F., Santamouris, M., Kolokotsa, D., Boemi, N., 2014. On the thermal characteristics and the mitigation potential of a medium size urban park in Athens, Greece. Landscape and Urban Planning 123, 73–86.
Sonne, J.K., Vieira, R.K., 2000. Cool neighborhoods: the measurement of small scale heat islands. In Proceedings of 650–655.
Spangenberg, J., Shinzato, P., Johansson, E., Duarte, D., 2008. Simulation of the influence of vegetation on microclimate and thermal comfort in the city of São Paulo. Revista da Sociedade Brasileira de Arborização Urbana 3, 1–19.
Spronken-Smith, R.A., 1994. Energetics and Cooling in Urban Parks.
Spronken-Smith, R.A., 1998. The thermal regime of urban parks in two cities with different summer climates. International Journal of Remote Sensing 19, 2085–2104.
Srivanit, M., Hokao, K., 2013. Evaluating the cooling effects of greening for improving the outdoor thermal environment at an institutional campus in the summer. Building and Environment 66, 158–172.
Steeneveld, G., Koopmans, S., Heusinkveld, B., Theeuwes, N., 2014. Refreshing the role of open water surfaces on mitigating the maximum urban heat island effect. Landscape and Urban Planning 121, 92–96.
Stewart, I.D., Oke, T.R., 2012. Local climate zones for urban temperature studies. Bulletin of the American Meteorological Society 93, 1879–1900.

Stone, G., 2012. A perspective on online partial discharge monitoring for assessment of the condition of rotating machine stator winding insulation. IEEE Electrical Insulation Magazine 28.

Taha, H., 1997. Urban climates and heat islands: albedo, evapotranspiration, and anthropogenic heat. Energy and Buildings 25, 99–103.

Taha, H., Akbari, H., Rosenfeld, A., Huang, J., 1988. Residential cooling loads and the urban heat island—the effects of albedo. Building and Environment 23, 271–283.

Tan, Z., Lau, K.K.-L., Ng, E., 2017. Planning strategies for roadside tree planting and outdoor comfort enhancement in subtropical high-density urban areas. Building and Environment 120, 93–109.

Tapper, N.E.A., 2014. Urban Populations' Vulnerability to Climate Extremes, Mitigating Urban Heat through Technology and Routledge.

Theeuwes, N., Solcerová, A., Steeneveld, G., 2013. Modeling the influence of open water surfaces on the summertime temperature and thermal comfort in the city. Journal of Geophysical Research: Atmospheres 118, 8881–8896.

Theophilou, M., Serghides, D., 2015. Estimating the characteristics of the Urban Heat Island effect in Nicosia, Cyprus, using multiyear urban and rural climatic data and analysis. Energy and Buildings 108, 137–144.

Tjandraatmadja, G., Cook, S., Chacko, P., Myers, B., Sharma, A., Pezzaniti, D., 2014. Water Sensitive Urban Design Impediments and Potential: Contributions to the SA Urban Water Blueprint-Post-implementation Assessment and Impediments to WSUD. Goyder Institute for Water Research, Adelaide.

Tominaga, Y., Sato, Y., Sadohara, S., 2015. CFD simulations of the effect of evaporative cooling from water bodies in a micro-scale urban environment: validation and application studies. Sustainable Cities and Society 19, 259–270.

Torok, S.J., Morris, C.J., Skinner, C., Plummer, N., 2001. Urban heat island features of southeast Australian towns. Australian Meteorological Magazine 50, 1–13.

United Nations, 2014. Department of Economic and Social Affairs, Population Division, Urban Agglomerations.

Upmanis, H., Chen, D., 1999. Influence of geographical factors and meteorological variables on nocturnal urban-park temperature differences-a case study of summer 1995 in Göteborg, Sweden. Climate Research 13, 125–139.

Upreti, R., Wang, Z.-H., Yang, J., 2017. Radiative shading effect of urban trees on cooling the regional built environment. Urban Forestry and Urban Greening 26, 18–24.

Vailshery, L.S., Jaganmohan, M., Nagendra, H., 2013. Effect of street trees on microclimate and air pollution in a tropical city. Urban Forestry and Urban Greening 12, 408–415.

Van Dillen, S.M., De Vries, S., Groenewegen, P.P., Spreeuwenberg, P., 2012. Greenspace in urban neighbourhoods and residents' health: adding quality to quantity. Journal of Epidemiology and Community Health 66, e8.

Victorian Government, 2012. City North Structure Plan, Planning for Future Growth.

Victorian Government, 2014. Plan Melbourne Metropolitan Planning Strategy.

Vijayaraghavan, K., 2016. Green roofs: a critical review on the role of components, benefits, limitations and trends. Renewable and Sustainable Energy Reviews 57, 740–752.

Völker, S., Baumeister, H., CLAßEN, T., Hornberg, C., Kistemann, T., 2013. Evidence for the temperature-mitigating capacity of urban blue space — a health geographic perspective. Erdkunde 355–371.

Voogt, J.A., 2012. Urban Heat Islands: Hotter Cities. 2004.

Virk, G., Jansz, A., Mavrogianni, A., Mylona, A., Stocker, J., Davies, M., 2015. Microclimatic effects of green and cool roofs in London and their impacts on energy use for a typical office building. Energy and Buildings 88, 214–228.

Wang, Y., Akbari, H., 2016. The effects of street tree planting on Urban Heat Island mitigation in Montreal. Sustainable Cities and Society 27, 122–128.

Watkins, R., Palmer, J., Kolokotroni, M., Littlefair, P., 2002. The London Heat Island: results from summertime monitoring. Building Services Engineering Research and Technology 23, 97–106.

Whitman, S., Good, G., Donoghue, E.R., Benbow, N., Shou, W., Mou, S., 1997. Mortality in Chicago attributed to the July 1995 heat wave. American Journal of Public Health 87, 1515–1518.

Wolf, K.L., 2005. Trees in the small city retail business district: comparing resident and visitor perceptions. Journal of Forestry 103, 390–395.

Wong, T., Brown, R.R., 2009. The water sensitive city: principles for practice. Water Science and Technology 60, 673–682.

Wong, N.H., Tan, A.Y.K., Chen, Y., Sekar, K., Tan, P.Y., Chan, D., Chiang, K., Wong, N.C., 2010. Thermal evaluation of vertical greenery systems for building walls. Building and Environment 45, 663–672.

Wong, M.S., Nichol, J.E., 2013. Spatial variability of frontal area index and its relationship with urban heat island intensity. International Journal of Remote Sensing 34, 885–896.

Yan, H., Wang, X., Hao, P., Dong, L., 2012. Study on the microclimatic characteristics and human comfort of park plant communities in summer. Procedia Environmental Sciences 13, 755–765.

Yan, H., Fan, S., Guo, C., Wu, F., Zhang, N., Dong, L., 2014. Assessing the effects of landscape design parameters on intra-urban air temperature variability: the case of Beijing, China. Building and Environment 76, 44–53.

Yang, F., Lau, S.S.Y., Qian, F., 2011. Thermal comfort effects of urban design strategies in high-rise urban environments in a sub-tropical climate. Architectural Science Review 54, 285–304.

Yu, C., Hien, W.N., 2006. Thermal benefits of city parks. Energy and Buildings 38, 105–120.

Zhang, Y., Hussain, A., Deng, J., Letson, N., 2007. Public attitudes toward urban trees and supporting urban tree programs. Environment and Behavior 39, 797–814.

Zhu, C., Li, S., Ji, P., Ren, B., Li, X., 2011. Effects of the different width of urban green belts on the temperature and humidity. Acta Ecologica Sinica 31, 383–394.

Zoulia, I., Santamouris, M., Dimoudi, A., 2009. Monitoring the effect of urban green areas on the heat island in Athens. Environmental Monitoring and Assessment 156, 275–292.

## FURTHER READING

Alberti, L.B., 1966. Ten Books on Architecture.

Armson, D., Stringer, P., Ennos, A., 2012. The effect of tree shade and grass on surface and globe temperatures in an urban area. Urban Forestry and Urban Greening 11, 245–255.

Bartfelder, F., Köhler, M., 1987. Experimentelle Untersuchungen zur Funktion von Fassadenbegrünungen. Berlin-Verlag.

Bosselmann, P., 1984. Sun, Wind, and Comfort, a Study of Open Spaces and Sidewalks in Downtown San Francisco.

Bosselmann, P., Arens, E., Dunker, K., Wright, R., 1995. Urban form and climate: case study, Toronto. Journal of the American Planning Association 61, 226–239.

Coutts, A.M., Tapper, N.J., Beringer, J., Loughnan, M., Demuzere, M., 2013e. Watering our cities. Progress in Physical Geography 37, 2–28.

Chen, H., Ooka, R., Huang, H., Tsuchiya, T., 2009. Study on mitigation measures for outdoor thermal environment on present urban blocks in Tokyo using coupled simulation. Building and Environment 44, 2290–2299.

Gagge, P., Fobelets, A.P., Berglund, L., 1986. ASHRAE Transactions 92, 5–1.

Ghaffarianhoseini, A., Berardi, U., Ghaffarianhoseini, A., 2015. Thermal performance characteristics of unshaded courtyards in hot and humid climates. Building and Environment 87, 154–168.

Grimmond, C., Oke, T.R., 2002. Turbulent heat fluxes in urban areas: observations and a local-scale urban meteorological parameterization scheme (LUMPS). Journal of Applied Meteorology 41, 792–810.

Heydecker, W.D., Ernest, G., 1929. Neighborhood and Community Planning.

Jamei, E., 2016. Impact of Urban Development on the Microclimate and Pedestrian Thermal Comfort in Melbourne. Doctor of Philosophy Deakin University.

Jamei, E., Rajagopalan, P., Seyedmahmoudian, M., Jamei, Y., 2016. Review on the impact of urban geometry and pedestrian level greening on outdoor thermal comfort. Renewable and Sustainable Energy Reviews 54, 1002–1017.

Knowles, R., 1981. Sun Rhythm Form.

Liao, J., Wang, T., Jiang, Z., Zhuang, B., Xie, M., Yin, C., Wang, X., Zhu, J., Fu, Y., Zhang, Y., 2015. WRF/Chem modeling of the impacts of urban expansion on regional climate and air pollutants in Yangtze River Delta, China. Atmospheric Environment 106, 204–214.

Mazumder, A., Taylor, W.D., 1994. Thermal structure of lakes varying in size and water clarity. Limnology and Oceanography 39, 968–976.

Pal, S., Xueref-Remy, I., Ammoura, L., Chazette, P., Gibert, F., Royer, P., Dieudonné, E., Dupont, J.-C., Haeffelin, M., Lac, C., 2012. Spatio-temporal variability of the atmospheric boundary layer depth over the Paris agglomeration: an assessment of the impact of the urban heat island intensity. Atmospheric Environment 63, 261–275.

Palladio, A., 1570. Four Books on Architecture.

Reps, J., 1969. Town Planning in Frontier America.

Rykwert, J., 1976. The Idea of a Town.

Sailor, D.J., Elley, T.B., Gibson, M., 2011. Exploring the building energy impacts of green roof design decisions – a modeling study of buildings in four distinct climates. Journal of Building Physics 35, 372–391.

Santamouris, M., 2014a. Cooling the cities—a review of reflective and green roof mitigation technologies to fight heat island and improve comfort in urban environments. Solar Energy 103, 682–703.

Santamouris, M., 2014b. On the energy impact of urban heat island and global warming on buildings. Energy and Buildings 82, 100–113.

Smith, K.R., Roebber, P.J., 2011. Green roof mitigation potential for a proxy future climate scenario in Chicago, Illinois. Journal of Applied Meteorology and Climatology 50, 507–522.

Tan, J., Zheng, Y., Tang, X., Guo, C., Li, L., Song, G., Zhen, X., Yuan, D., Kalkstein, A.J., Li, F., 2010. The urban heat island and its impact on heat waves and human health in Shanghai. International Journal of Biometeorology 54, 75–84.

Wang, Y., Akbari, H., 2015. Development and application of 'thermal radiative power' for urban environmental evaluation. Sustainable Cities and Society 14, 316–322.

Chapter 20

# The Role of Green Roofs and Living Walls as WSUD Approaches in a Dry Climate

**Simon Beecham[1], Mostafa Razzaghmanesh[2], Rosmina Bustami[1] and James Ward[1]**

[1]*Natural and Built Environments Research Centre, University of South Australia, Adelaide, SA, Australia;* [2]*ORISE Fellow at US Environmental Protection Agency, Edison, NJ, United States*

## Chapter Outline



**ABSTRACT**

The addition of green infrastructure, including green roofs and living walls, into buildings is part of a new approach to urban design aimed at resolving current problems associated with built environments. Green roofs and living walls are becoming an important component of water sensitive urban design systems, and their use around the world has increased in recent years. Green roofs can cover the impermeable roof areas that densely populate our urban areas, and through doing so, can provide many environmental, economic, and social benefits. In addition to roofs, there are a number of bare walls that have the potential to be transformed into vegetated, living walls. Living walls can potentially improve air quality, reduce pollution levels, reduce temperatures inside and outside of buildings, reduce building energy usage, and improve human health. Despite such benefits, both green roofs and living walls are relatively new technologies, and there are several research gaps and practical barriers to overcome before these systems can be applied more widely. Furthermore, specific design criteria need to be developed for a range of climatic conditions to develop resilient green infrastructure. Consequently, several field experiments comprising both intensive and extensive green roof test beds, as well as living walls, have been

**Approaches to Water Sensitive Urban Design. https://doi.org/10.1016/B978-0-12-812843-5.00020-4**

recently established. In these recent research studies, stormwater quality and quantity, hydrological behavior, plant performance, and thermal benefit have been investigated. The findings of these studies can be used to identify the key elements of resilient green roof and living wall systems.

**Keywords:** Green Infrastructure; Green Roofs; Green Roof Hydrology; Living Wall; Low Impact Development; Stormwater Quality; Water Sensitive Urban Design; Thermal Performance.

This chapter will describe several experimental and modeling studies that have been conducted on both living walls and green roofs. In terms of field experiments, the quality and quantity of runoff from both intensive and extensive green roofs are discussed, as is the use of fertilizers in green roof and living wall systems. In terms of modeling applications, the effects of different scenarios of adding green roofs to a typical urban environment are explained using industry available tools such as the commonly used ENVI-met software (Huttner and Bruse, 2009). Finally, in terms of green roof and living wall design, methods of optimizing plant performance are described including plant selection, media type and depth, irrigation management, and other important design factors. In addition, opportunities to recycle and reuse outflow water from green roofs and living walls for end uses such as irrigation and toilet flushing are explored using current design standards.

This chapter should assist urban planners and designers in developing resilient green infrastructure models for cities with dry climates around the world.

## 20.1 WATER SENSITIVE URBAN DESIGN

Over recent decades, the hydrologic cycle of water has changed significantly because of continuous changes in Australian green spaces from forest or similar vegetation to urban environments (ANZECC, 2000). Australia is one of the most urbanized countries in the world with 85% of its inhabitants living in towns or cities (Skinner, 2006). The growth rate of urbanization has led to changes of green spaces with large impervious areas such as roofs, car parks, roads, highways, and paving. This in turn has led to changes in the urban hydrologic cycle. In an investigation by Razzaghmanesh et al. (2012), various studies from Europe, North America, Asia, Australia, and New Zealand were compared to understand how green roofs could be adapted to meet water sensitive urban design (WSUD) objectives in Australia. It was found that green roofs are used as an important WSUD infrastructure around the world, but that this technology is very much in its infancy in Australia. Furthermore, specific design criteria needs to be developed for the wide range of climate conditions found in Australia. WSUD, however, is not a single technology but rather it is a systems approach (Beecham, 2003). Australian cities have developed guidelines on developing water-sensitive cities. For example, the Adelaide 30-year Plan (South Australia Government, 2010) articulates a vision for the Adelaide community beyond 2037. The purpose of the plan is to promote Adelaide as a city that is recognized worldwide as livable, competitive, and resilient to climate change. Generally, WSUD can be used as a strategy for incorporation across a wide range of urban development scales, including residential homes, roads, vehicle parking areas, subdivisions, multistorey units, commercial and industrial areas, and public land. Green roofs, living walls, permeable pavements, and wetlands are some of the commonly used WSUD technologies that aim to improve water quality reduce flood risk, and enhance biodiversity in urban areas (Fig. 20.1). There is also a concept

**FIGURE 20.1** One Central Park in Sydney Australia with its 1200 $m^2$ of living wall.

known as *blue roofs* which is an unvegetated system designed to retain water above an impermeable membrane, either temporarily or permanently. Blue roofs can provide a number of benefits including storage of rainfall to reduce runoff impacts, and storage for reuse purposes such as toilet flushing and irrigation. Because of the risk and consequences of leakage into buildings, blue roofs are seldom used in practice, and are considered beyond the scope of this chapter.

## 20.2 GREEN ROOF AND LIVING WALL CONCEPTS

### 20.2.1 Extensive green roofs

Extensive green roofs are those in which the depth of the growing media is generally <150 mm, although Table 20.1 provides different definitions of green roof depth from various authors (Berndtsson et al., 2010; Fassman and Simcock, 2012). Extensive green roofs are lightweight structures with drought-tolerant, self-seeding vegetated cover. The vegetation has to cope with little or no irrigation during the roof's operational life. Generally, they are constructed on roofs with slopes up to 33%. Because of their relatively low weight, it is often possible to retrofit them on existing structures without installing extrastructural support. Fig. 20.3(a) shows a schematic of a typical extensive green roof.

### 20.2.2 Intensive green roofs

Intensive green roofs are those where the growing media depth is generally >150 mm and are often covered with shrubs, grassed areas, and occasional trees (FLL, 2002). They are usually installed on flat roofs (Fig. 20.2). Regular maintenance, such as watering, weeding, and fertilizing, is needed to keep intensive green roofs alive (Berndtsson et al., 2008). They can also be used as amenity areas. Fig. 20.3(b) shows a schematic of a typical intensive green roof.

### 20.2.3 Categorization of green walls

Green wall is a general term for a vertical wall covered in vegetation. Other terms for plants grown on vertical landscaping include vertical greenery system, vertical garden, green vertical system, vertical green, bioshader, and vertical landscaping. It is commonly agreed that these can be divided into two categories according to the installation and growing methods for these systems. These categories are green façades and living walls (Dunnett and Kingsbury, 2008; Köhler, 2008; Manso and Castro-Gomes, 2015; Safikhani et al., 2014a) (see Fig. 20.4).

In direct green façades, the vegetation is rooted in the ground and grows vertically on the wall. For indirect green façades, climbing plants grow vertically on trellises, cables, or mesh support systems without attaching to the surface of the building (Fig. 20.5). A living wall on the other hand is a building envelope system where plants are planted and irrigated on a structure attached to the wall without relying on a ground level rooting media (Fig. 20.6). Popular systems include living walls with modular boxes, felt pockets, planter boxes, and hydroponic systems. Green façades have received more attention in Europe (Dunnett and Kingsbury, 2008; Pérez et al., 2014), whereas living walls are generally more popular in Asia (Pérez et al., 2014).

**TABLE 20.1 Green Roof Nomenclature Based on Media Thickness, as Described by Various Authors**

| Intensive (mm) | Extensive (mm) | References |
|---|---|---|
| >150 | ≤150 | Fassman-Beck et al. (2013) and Fassman and Simcock (2012) |
| 150–1200 | 50–150 | Kosareo and Ries (2007) |
| >500 | – | Köhler et al. (2002) |
| 150–350 | 30–140 | Mentens et al. (2006) |
| >100 | <100 | Hien et al. (2007) |
| >300 | – | Berndtsson et al. (2006) |
| >100 | 20–100 | Graham and Kim (2005) |

Modified from Berndtsson, J.C., Bengtsson, L., Jinno, K., 2010. Runoff water quality from intensive and extensive vegetated roofs. Ecological Engineering, 35, 369–380.

FIGURE 20.2 Intensive green roof at the Fairmont hotel in Singapore.

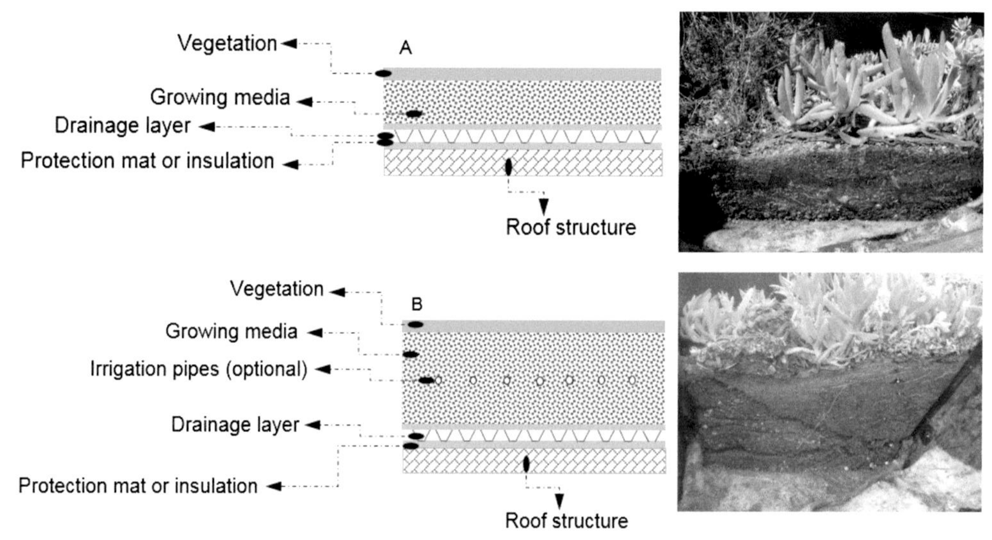


FIGURE 20.3 Schematics and photos of (a) an extensive (depth $\leq$ 150 mm) green roof profile and (b) an intensive (depth > 150 mm) green roof profile (Razzaghmanesh, 2015).

Apart from being aesthetically pleasing, green walls provide environmental, social, and economic benefits which are attributed to their design, plant choice, density of vegetation, and location. Examples of these environmental, social, and economic benefits are listed in Table 20.2.

Investigations into green façades have taken place since the 19th century, but there has been an increasing number of guidelines and other publications since the 1980s (Köhler, 2008). It has only been since the early 2000s that studies have

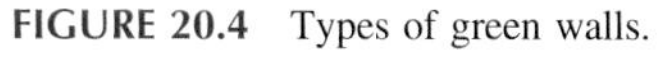


**FIGURE 20.4** Types of green walls.

**FIGURE 20.5** Green façade at the Hotel Boss in Singapore.

begun into living walls. There are a number of factors to consider in selecting a suitable green wall. Initial installation and maintenance costs for green façades are lower than for living walls, making them a more economical choice. However, living walls offer a wider selection of plants and are often considered more aesthetically pleasing than green façades. Living walls also have the capacity to provide instant cooling upon their installation, whereas green façades may take several years to colonise the entire wall (Dunnett and Kingsbury, 2008; Ottelé et al., 2011; Perini and Rosasco, 2013).

Recent studies have confirmed that both green façades and living walls can mitigate the urban heat island (UHI) effect by creating a microclimate. One of the key benefits of green walls over other shading devices is the capability of plants to repartition solar radiation and sensible heat into latent heat during the transpiration process (Scarpa et al., 2014). In addition, green walls act as a layer of insulation, thereby reducing building heating demand in cooler climates.

**FIGURE 20.6** Living wall at the Marina Barrage in Singapore.

**TABLE 20.2 Environmental, Economic, and Social Benefits of Green Walls**

| Category | Benefits |
|---|---|
| Environmental | Temperature reduction from shading<br>Improved air quality<br>Carbon dioxide sequestration from photosynthesis<br>Sound attenuation<br>Improved flora biodiversity in urban areas |
| Economic | Reduced energy demand and increased energy efficiency of buildings<br>Suitable for retrofitting projects |
| Social | Increased human health and well-being<br>Improved building esthetic |

Hui, S.C.M., Zhao, Z., 2013. Thermal regulation performance of green living walls in buildings. In: Joint Symposium 2013: Innovation and Technology for Built Environment, Hong Kong, pp. 1–10; Safikhani, T., Abdullah, A.M., Ossen, D.R., Baharvand, M., 2014a. A review of energy characteristic of vertical greenery systems. Renewable and Sustainable Energy Reviews 40, 450–462; Sheweka, S., Magdy, A.N., 2011. Living walls as an approach for a healthy urban environment. Energy Procedia 6, 592–599.

In various green façade studies, the maximum surface temperature differences between a green façade and a bare control wall were 15.2°C in Lleida, Spain (Pérez et al., 2011), 11.3°C in Nagoya, Japan (Koyama et al., 2013), 12.6°C in Chicago, USA (Susorova et al., 2014), and 9.9°C in Bangkok, Thailand (Sunakorn and Yimprayoon, 2011). Similar positive temperature differences have also been measured for living walls compared with bare control walls. For example, in the Mediterranean climate of Adelaide, South Australia (Fig. 20.7), the maximum temperature difference recorded between a living wall and a control wall was 14.9°C in the summer of 2015 (Razzaghmanesh and Razzaghmanesh, 2017). In experiments conducted in Italy a maximum temperature difference of 20°C was recorded between a living wall and its bare control wall (Mazzali et al., 2013).

Both experimental and modeling studies have shown that living walls provide more cooling benefits compared with green façades, including providing instant cooling to the building surface following its installation (Jaafar et al., 2013; Ottelé et al., 2011; Perini and Rosasco, 2013; Safikhani et al., 2014b; Wong et al., 2010). In humid, tropical Singapore, living walls were up to 7°C cooler than green façades (regardless of the percentage of green cover) (Wong et al., 2010), whereas in Malaysia living walls were 0.5–1.5°C cooler than a green façade (Safikhani et al., 2014b).

A simulation exercise for Mediterranean climates showed that green façades/living walls could save up to 43% of the cooling energy costs. It was shown also that a living wall with planter boxes could reduce heating energy demand by up to 6% compared with only a 1% saving for green façades (Ottelé et al., 2011).

Mazzali et al. (2013) demonstrated that a living wall's overall outgoing heat flux was measured at $-87\ W/m^2$ compared with the incoming heat flux of $30\ W/m^2$ on its bare wall. This measured difference in net energy flux from the wall is presumably due to the significant shielding effect of the green cladding that reduces the amount of incoming energy from

**FIGURE 20.7** Experimental living wall at the University of South Australia.

**FIGURE 20.8** Modular box-type living wall in Xi'an, China.

the sun. It is also due to other factors, including the type of vegetation, the latent heat of evaporation, and the reflection coefficient for solar radiation. In a Mediterranean climate, their experiment recorded a 12–20°C surface temperature reduction on sunny days and a 1–2°C temperature reduction on cloudy days. This demonstrated yet again the potential of green walls to mitigate the UHI effect, at least locally.

The installation and maintenance costs of green wall systems can be expensive, as confirmed by cost-benefit analyses undertaken by several researchers. Installing and maintaining living walls costs approximately US\$150/$m^2$ compared with US\$45/$m^2$ (in today's costs) for green facades (Perini and Rosasco, 2013). However, some unquantifiable benefits such as increased flora biodiversity, aesthetic values, and UHI mitigation, were not included in the analysis.

A lifecycle assessment by Ottelé et al. (2011) for different green wall types indicated that direct green façades are the most economically sustainable, followed by modular pot living wall systems (Fig. 20.8).

## 20.3 GREEN ROOF ELEMENTS

The outer layer of a green roof system consists of vegetation. In theory, almost any plant species could be used for green roof applications, providing they are suited to the climatic region, grown in a suitable substrate, and are given adequate irrigation (Oberndorfer et al., 2007; Rowe et al., 2012). Wind stress resulting from eddy formation around tall buildings may need to be accounted for in the selection of plants. Visibility and accessibility are other selection criteria. Although *Sedum* remains the most commonly used genus for green roofs in cold climates, the range of green roof vegetation is wide. Researchers have tested many herbaceous and woody taxa under various rooftop conditions since the 1980s (Durhman et al., 2004; Monterusso et al., 2005).

Heinze (1985) compared combinations of various *Sedum* species, grasses, and herbaceous perennials, planted in two substrate depths in simulated roof platforms. Slow-growing sedum performed well in thin substrates. Grass and herbs had better performance in deeper substrates. It should be noted that *Sedum* is classified as a weed in many states of Australia.

## 20.3.1 Growing media

Growing media is the supporting layer for vegetation in a green roof system. The media should be lightweight, well drained, have good moisture storage capacity, and should be able to resist biological breakdown over time (Getter and Rowe, 2006). An optimum growing media is: 80%–90% (by volume), of lightweight inorganic aggregate (LWA), and 10%–20% organic matter (OM). LWA provides a porous media for water and gas exchange, whereas the OM provides nutrient supply and retention, as well as promoting a rootzone ecology essential for plant growth (Friedrich, 2005; Fassman and Simcock, 2012).

A commonly used standard for growing media properties is the FLL (Forschungsgesellschaft Landschaftsentwicklung Landschaftsbau) guideline (FLL, 2002). Other standards for various countries are listed in Table 20.3. Note that the inorganic matrix can include scoria, ash, pumice, sand, coir, pine bark, chemically inert porous foams, and recycled materials such as crushed bricks and roof tiles.

## 20.3.2 Root barrier

Root barriers are often used in green roofs to protect the roof's waterproofing membrane (Fig. 20.9) from plant root growth. The most common type of root barrier is a thin polyethylene sheet, laid over the waterproofing membrane. This may not be required if the waterproofing membrane is certified as root-resistant. The root barrier must also be resistant to the humic acids produced when plants decompose. Separation sheets are sometimes installed between the waterproofing layer and root barrier to provide additional protection and to separate materials that are not compatible (Green Roof Australia, 2010; Carpenter, 2014).

## 20.3.3 Drainage layer

The primary role of a green roof drainage layer is to remove the excess water from rainfall as quickly as possible, and to refill external storages for future irrigation use. Basically, green roof drainage systems can be divided into two classes: aggregate drains and geocomposite drains. These may be combined or used separately in conjunction with the drain outlets (Wingfield, 2005). Aggregate drainage layers less than 100 mm in depth should be freely drained. With deeper layers, drainage restriction (by textural layering) can be used to increase the water-holding capacity of the overlying media. A number of granular materials considered suitable include gravel, lava and pumice, expanded clay and slate, and recycled materials such as crushed roofing tiles or bricks.

Geocomposite drains are any drains composed of two or more materials, one of which is a geosynthetic (Carpenter, 2014; Wingfield, 2005). Geocomposite drains may also include heavy duty high-density polyethylene with excellent load-bearing capacity to retain and drain water (Fig. 20.10). Depending on the product chosen, the drainage layer can often take the weight of a pedestrian or even vehicular traffic, with a design life of 50 years.

## 20.3.4 Insulation layer or protection mat

Protection mats are often used to protect the building's waterproof membrane from damage during installation of the green roof. The most common materials used are water-permeable, hard-wearing, and dense synthetic fibers, polyester, and polypropylene. Protection matting is installed directly on top of the waterproofing layer for root-resistant membranes or on top of the root barrier layer, providing further protection against root penetration, as well as doubling as a separation sheet (Green Roof Australia, 2010).

**TABLE 20.3 Tests for Developing Local Recycled Growing Media in Different Climates**

| Researchers | Country | Purpose | Materials | Tests or Targets | Standards |
|---|---|---|---|---|---|
| Molineux et al. (2009)[a] | United Kingdom | Characterizing alternative recycled waste materials for use as green roof growing media in the United Kingdom | Crushed red brick, clay and sewage sludge, paper ash, carbonated limestone | • pH<br>• Particle size distribution<br>• Loose bulk density<br>• Particle density<br>• Leachate analyses | **1.** Particle size distribution: BS EN 933-2:1996<br>**2.** Loose bulk density and void spaces: BS EN 1097-3:1998<br>**3.** Particle density and water-holding capacity: (BS EN 13055-1:2002)<br>**4.** Leachate analysis: BS EN 12457-3:2002 |
| Fassman and Simcock (2008, 2012) | New Zealand | Development and implementation of locally sourced extensive green roof substrate in New Zealand<br>Moisture measurements as performance criteria for extensive green roof substrates | Blend of 70% 4–10 mm pumice, 10% 1–3 mm zeolite, 15% pine bark fines plus mushroom farm waste, and 5% peat and installed at a depth of 70 mm<br>70% v/v 4–10 mm pumice, 10% v/v 1–3 mm zeolite, and 20% organic matter at a 100 mm depth are recommended to maintain plants without irrigation (excluding drought conditions) and minimize weeds while preventing runoff from storms with less than 25 mm of rainfall | • Retention of a design storm<br>• Wet system weight<br>• Dry bulk density<br>• Minimum target permeability<br>• Minimum plant cover | **1.** FLL (2002, 2008): German guidelines for green roofs<br>**2.** AS/NZS 1170 |
| Rayner (2010) | Australia | Choosing substrates for Australian green roofs | Gravels, sands, topsoil, scoria (various grades), crushed clay brick, bottom ash (enviroagg) products, pumice, perlite, recycled plastics (chips and beads), light expanded clay granules, foam flakes (urea formaldehyde) and vermiculite | • Porosity (air-filled porosity)<br>• Permeability/hydraulic conductivity<br>• Water-holding capacity<br>• Particle size distribution | **1.** Australian Standards: AS 3743 Potting Mixes<br>**2.** FLL Guidelines (2002, 2008): German guidelines for green roofs |

[a]*The addition of organics also significantly reduced the pH of the recycled aggregates making growing conditions for plants more favorable in these substrates.*

**FIGURE 20.9** Installation of a roof barrier sheet in an experimental green roof in Australia. The drainage layer is the black open matrix.

FIGURE 20.10 Example of a synthetic green roof drainage material showing the raised 3D structure creating drainage channels. The material is high-density polyethylene.

### 20.3.5 Roof structure

It is generally possible to install a green roof on any roof type irrespective of the material and slope, subject to safety and load-bearing considerations. However, green roofs are generally installed on concrete roof decks because of structural integrity, ease of design, durability, and amenity considerations (Carpenter, 2014). As described by Munby (2005), it is often quite difficult to obtain as-constructed structural drawings for existing buildings, especially for those over 10 years old. A structural survey is therefore recommended in the majority of cases to determine a building's roof load-bearing capacity before designing the retrofit of a green roof (Castleton et al., 2010).

## 20.4 LIVING WALL ELEMENTS

There are several factors to be considered in determining the choice of green wall, especially in dry climates. This includes their design and performance, as well as installation and maintenance costs. The choice of green wall type (green façades or living walls) will very much affect installation and maintenance costs. Perini and Rosasco (2013) took into consideration the yearly maintenance costs of pruning and irrigation. They found that annual pruning of green façades will generally begin 4 years after installation. However, the environmental and social benefits of living walls were found to start immediately on installation. Selection of plant species and growing media are among the core considerations in designing green wall structures. These have to be carefully evaluated to suit the location and environment of the green wall, thus maximizing the benefits delivered by the system.

Installation and maintenance of green façades are generally less complicated than living walls; therefore, the following sections will briefly discuss green façades but will focus more heavily on living wall elements.

### 20.4.1 Vegetation

Vegetation plays the critical role in any green wall system. While nonvegetative building cladding only contributes to shading, the thickness of vegetation combined with transpiration processes contribute to both shading and temperature regulation of a building and its microclimate. Plant selection depends largely on the building orientation and climatic conditions including local weather. Evergreen and native plants are often preferred for minimum maintenance and to prolong the lifetime of the system. The canopy cover, canopy thickness, and plant morphology are among the factors contributing to the cooling benefits of green walls (Cameron et al., 2014; Manso and Castro-Gomes, 2015; Stav and Lawson, 2012). Other associated variables that contribute to thermal cooling benefits include vegetation height, leaf reflectivity, and leaf emissivity (Stav and Lawson, 2012).

Flexibility in plant selection allows the creation of attractive patterns, colors, texture, and thickness (Manso and Castro-Gomes, 2015). Such systems are generally more popular with designers and building residents. Living walls offer a wide selection of growing methods and plants, unlike green façades, which are limited to climbing plants. Perennial and native plants usually minimize maintenance and irrigation needs for living wall installations (Perini and Rosasco, 2013; Mårtensson et al., 2014). A living wall system with low water usage is often preferable in terms of its lifecycle performance (Natarajan et al., 2014).

Drought-tolerant succulent carpets have also been used in living wall applications following their successful use in green roofs (Manso and Castro-Gomes, 2015). However, as living walls are vertically orientated they are not well suited for succulent plants, which often require near horizontal growing surfaces. A "decision tree" developed by Perini et al. (2013) recommended evergreen shrubs for living walls, while evergreen climbing species were preferred for green façades.

### 20.4.2 Growing substrate

Research into the selection of growing media in living walls is limited. Selection of a substrate media is important because it influences plant root growth and the subsequent growth of the living wall plants (Jørgensen et al., 2014; Weinmaster, 2009). Popular substrates for living walls include rock wool, coir, peat, and potting soil (Weinmaster, 2009), as well as hydroponic media, which supply nutrients through the irrigation water. Lightweight materials are generally preferred because of the increasing weight of growing vegetation.

## 20.5 GREEN ROOF HYDROLOGY

An important strategy in sustainable urban drainage systems and WSUD is "at-source" runoff control (Berndtsson et al., 2010; Roehr and Fassman-Beck, 2015). Green roofs are viewed as a best management practice to attenuate peak runoff flows in urban areas (Palla et al., 2010). Therefore, one of the more important issues in green roof studies is their hydrology. The following section describes recommendations on green roof hydrology based on previous studies.

### 20.5.1 Rainfall and runoff relationship in green roofs

Most researchers have used the water mass balance equation to study the hydrologic behavior of green roofs (Mentens et al., 2003). A steady state form is given by (Vilareal and Bengtsson, 2005)

$$P + I - E - Q - D + \Delta S = 0 \tag{20.1}$$

where, P, precipitation, E, evapotranspiration, Q, runoff, D, deep percolation, ΔS, water storage in the system, and I, irrigation (summer irrigation if need be).

By neglecting D, as deep percolation rarely occurs in green roof systems, the equation becomes

$$Q = P + I - E - \Delta S \tag{20.2}$$

In most water balance studies of green roofs, the main objective is to estimate the various components of this equation.

### 20.5.2 Runoff comparison between green and conventional roofs

Green roofs provide a substantial opportunity to reduce both runoff volume and peak discharge from roofs (Fassman-Beck et al., 2013). The most significant differences between the outflow hydrographs from conventional roofs and green roofs are the peak flow and the time of runoff movement on the surface (Berndtsson et al., 2010). Fig. 20.11 shows the rainfall and conventional roof and green roof runoff as measured by Razzaghmanesh and Beecham (2014) from an experiment in Adelaide, Australia.

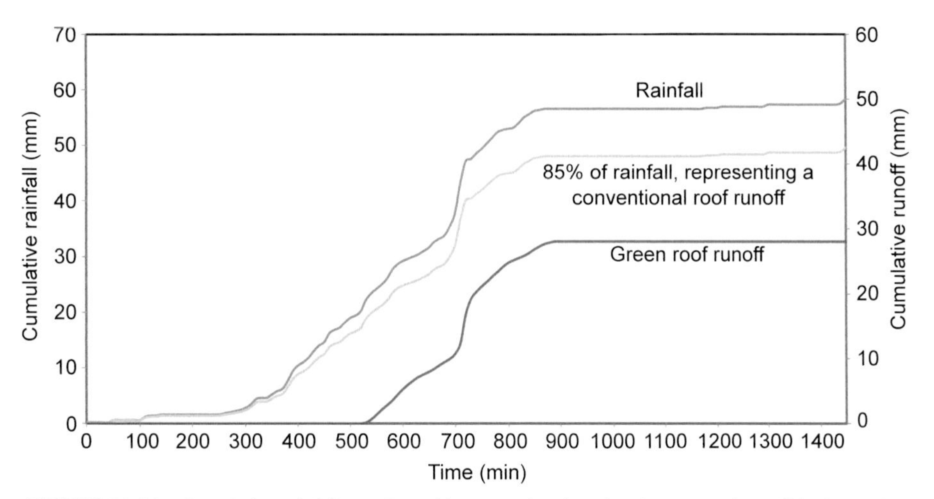


**FIGURE 20.11** Cumulative rainfall together with conventional roof and green roof runoff hydrographs.

### 20.5.3 Retention capacity of green roofs

Hydrological studies of green roofs, especially studies focusing on water retention in green roofs, began in Germany several decades ago (Mentens et al., 2006). There has since been rapid growth of the green roof industries in both Germany and North America. In a study of two green roofs undertaken in Portland, Oregon, USA, by Hutchinson et al. (2003) the precipitation retention was calculated as 100% in the ( non snow) warm seasons and 69% on average.

Voyde et al. (2010) studied the hydrology of a green roof in Auckland (NZ) and found, a 66% retention of precipitation over a one year period. They concluded that, regardless of rainfall properties, green roofs can significantly reduce runoff and in particular the maximum runoff rate. For some rainfall events green roofs could retain 82% of average rainfall and reduce peak flows by up to 93%.

Four extensive green roofs and three conventional (control) roofs were investigated by Fassman-Beck et al. (2013) in Auckland, New Zealand, over a 2-year period. Runoff reductions of over 50% were measured from green roofs with substrate depths of 50–150 mm and >80% plant coverage. Runoff rarely occurred from storms with <25 mm of rainfall. Peak discharge rates were 60%–90% less than those from conventional roofs, and did not vary seasonally.

### 20.5.4 Rainfall events

Rainfall is one of the most important factors in the water mass balance equation. In this section, rainfall characteristics such as design rainfall intensity, duration, and frequency are discussed. As Mentens et al. (2006) discussed, according to the German guideline (FLL, 2002), a design rainstorm is defined as an event having rainfall of 300 L/s/ ha during 15 min, equivalent to 27 mm in 15 min. Furthermore, peak runoff during design rainstorm events is defined as the amount of runoff during the last 5 min of rainfall (FLL, 2002).

Carter and Rasmussen (2006) found an inverse relationship between the depth of rainfall and the percentage of rain retained. For small storms (<25 mm), 88% was retained, for medium storms (25–75 mm), 54% was retained, whereas for large storms (>75 mm), 48% was retained. The moisture conditions of the roof materials before the storms were not given. Similarly, Simmons et al. (2008) found that all small rain events (<10 mm) were completely retained by the green roofs. However retention also depends on rainfall intensity.

Villarreal-Gonzalez and Bengtsson (2005) found that water retention by green roofs depended to a large extent on rainfall intensity. The lower the intensity, the larger the retention. For a rainfall intensity of 24 mm/h and roof slopes of 2 and 14 degrees, retention was 60% and 40% of the simulated precipitation, respectively. As rainfall intensity increased to 80 mm/h, the retention reduced to 21% and 10%, respectively. Considering the high permeability of the green roof media, this response was unexpected. However, Bengtsson (2005) also found that the water storage capacity of a green roof was related to the rainfall intensity and that the vertical percolation process through the growth media dominated the rainfall–runoff relationship. In a 24-month study by Razzaghmanesh and Beecham (2014) in Adelaide, Australia, the experimental intensive and extensive green roofs were able to retain 100% of all the rainfall from 1-year Average Recurrence Interval (ARI) events with a duration of less than 7 h (Fig. 20.12).

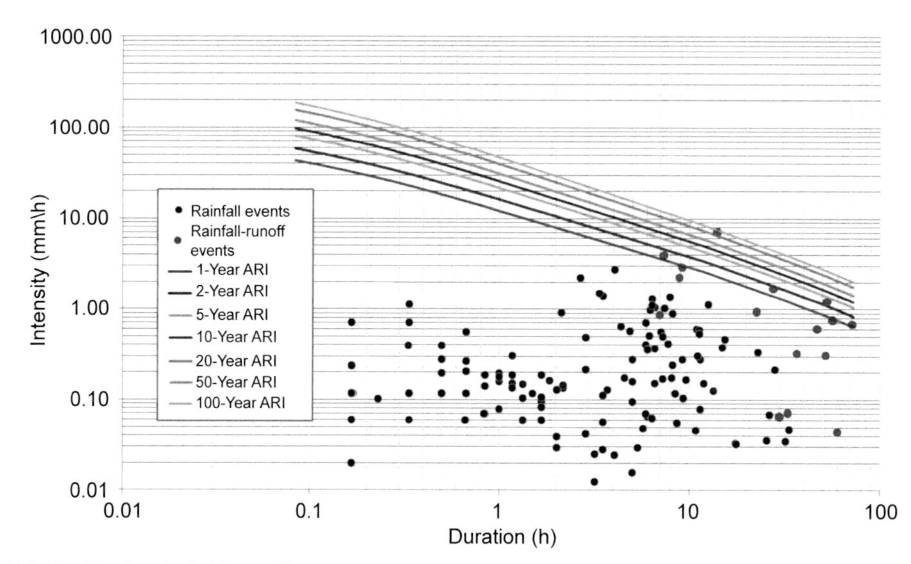


**FIGURE 20.12** Monitored rainfall-runoff events compared with Adelaide's Intensity-Duration-Frequency curve (IDF curves).

## 20.5.5 Effect of design parameters on green roof hydrological performance

In this section, the influence of parameters such as slope, substrate material and depth, and vegetation cover on green roof hydrology is discussed, based on published studies from various countries with different climates.

### *20.5.5.1 Slope*

The published literature is inconclusive regarding the effects of slope on green roof hydrology. Some researchers argue that an increase in slope can cause an increase in green roof runoff. Others posit there is no relationship between slope and water retention in green roofs (Berndtsson et al., 2006; Mentens et al., 2006). On the other hand, others believe higher roof slopes reduce outflow and improve the water retention properties of green roof systems (Köhler et al., 2002). Generally, studies on the effect of slope on green roofs have been combined with examining the effect of other factors. VanWoert et al. (2005) tested the effects of two slopes, 2% and 6.5%, and growing media with 2.5, 4.0, and 6.0 cm depths for a range of rainfall events. The test beds with 2% slope and 4 cm media depth had the greatest mean retention at 87%. They concluded that outflow runoff is decreased with less slope and deeper growing media. Similar results were reported by Getter et al. (2007) who constructed green roof beds with four slopes (2%, 7%, 15%, and 25%) and found the maximum retention value (86%) was in the 2% slope roof. Moreover, time for runoff initiation increased and peak discharges reduced for all slopes.

### *20.5.5.2 Vegetation*

The amount of transpiration from green roofs depends on the local climate and the type of vegetation. Some studies have been undertaken to examine the plant species and/or their combinations to improve the performance of green roofs. Dunnett et al. (2008) examined the influence of vegetation composition on runoff in two green roof experiments. In the first experiment, an outdoor lysimeter was used to investigate the quantity of runoff and evapotranspiration from trays containing 100 mm growing media, and a combination of grasses and forbs with bare substrate. In their second experiment, conducted in a laboratory, simulated rainfall was used to understand how much water was retained by different vegetation types. In both cases, a combination of vegetation worked best in terms of water retention. In another study, Schroll et al. (2010) monitored runoff from conventional roofs with impervious surfaces, roofs with a half-impervious surface area, and a vegetated roof. During winter rainfall events, vegetation had no influence on roof water retention. In contrast, in summer periods, vegetated roofs retained more water than the other two roof types.

### *20.5.5.3 Substrate*

Substrates or growing media are another important factor in the design of green roofs in that they can improve the water retention performance of these systems.

Beck et al. (2011) evaluated changes in water discharge quality and quantity from an extensive green roof after adding biochar, which is a carbon-based soil amendment promoted for its ability to retain nutrients in soils. Green roof trays with and without biochar were planted with *Sedum* or ryegrass and subjected to two sequential 74 mm/h rainfall events using a rainfall simulator. The 7% biochar treatment showed increased water retention, reduced peak runoff rates, and a significant decreases in the discharge of nitrogen, phosphorus, and dissolved organic carbon.

## 20.5.6 Hydrological modeling of green roofs

Carter and Jackson (2007) observed that there were few studies that have examined the impact of green roof applications on the hydrology of an urban catchment. Consequently, they used local green roof rainwater retention data to model the hydrologic effects of green roofs in an urban catchment at a variety of spatial scales. Spatial analysis identified areas of the catchment where green roofs would significantly reduce the fraction of total impervious area. Subsequent hydrological modeling demonstrated that appropriately located green roof implementation could significantly reduce peak runoff rates, particularly for small storm events.

Palla et al. (2012) developed a conceptual linear reservoir using a simple mechanistic (HYDRUS-1D) model to simulate the hydrologic behavior of green roof systems in an urban environment at the University of Genova, Italy. This model was calibrated and validated using data collected from an experimental green roof site. The hydrographs from the hydrologic model reproduced, with acceptable accuracy, the experimental measurements, as confirmed by the Nash–Sutcliffe efficiency index being generally greater than 0.60. Indeed, they concluded that the layered roof structure and each green roof component could be designed to improve the hydrologic process. The mechanistic model, based on a single porosity approach, was able to adequately describe the variably saturated flow within the thin stratigraphy.

## 20.6 STORMWATER QUALITY

The quality of discharges from green roofs is important as it is often reused or discharged to receiving waters. Previous studies have shown that stormwater quality can be a major challenge for green roof designers. This section summarizes the results of previous studies that have investigated green roof water quality.

### 20.6.1 Green roof water quality

A green roof can act as a sink for the contaminants that occur in rainfall such as nitrogen and phosphorus species. An investigation of water quality in the outflows from green roofs was started in Germany some time ago and continued by other researchers. NO, $NO_2$, $NO_3$, $NH_3$, TKN, TN, and $NH_4$ are various forms of nitrogen which are often investigated in green roof systems. For green roofs, the $PO_4$ forms of total phosphorus (Tot-P) are often studied because $PO_4$ is the most bioavailable form. Anions, cations, and OM are examples of other commonly studied constituents, as shown in Table 20.4. The growing media of green roofs, applied fertilizers, applied water for summer irrigation, and air pollutants are the major sources of green roof pollutants.

### 20.6.2 Factors affecting water quality

Berndtsson et al. (2010) concluded that the factors potentially influencing green roof runoff quality can be summarized as follows:

- Type of material used (composition of soil, material of drainage and/or underlying hard roof material, pipe material);
- Soil thickness;

**TABLE 20.4** Summary of the Water Quality Constituents Measured in Green Roof Outflows Across the World

| Region | Country | Researcher | Nutrients | Cations | Anions | Heavy Metals | Organic Matter |
|---|---|---|---|---|---|---|---|
| Europe | Sweden | Berndtsson et al. (2006, 2008), Berndtsson et al. (2010) | $NO_3$, $NH_4$, $PO_4$, Tot-N, Tot-P | Cr, Fe, K, Mn | | Cd, Cu, Pb, Zn | DOC |
| | Germany | Steusloff (1998) | | | | Cd, Cu, Pb, Zn | |
| | | Köhler et al. (2002) | $NO_3$, $PO_4$ | | | Pb, Cd | |
| | Estonia | Teemusk and Mander (2007) | Tot-P, Tot-N, $PO_4$, $NO_3$, $NH_4$ | Ca, Mg | $SO_4$ | | BOD, COD |
| North America | USA | Monterusso et al. (2004) | $NO_3$, Tot-P | | | | |
| | | Hathaway et al. (2008) | TKN, $NO_3$, $NO_2$, $NH_3$, Tot-N, Tot-P, and OP | | | | |
| | | Bliss et al. (2009) | P, Tot-N | | $SO_4$ | Pb, Zn, Cd | COD |
| | | Alsup et al. (2010) | | Cu, Fe, Mn | | Cr, Ni, Pb, Zn, Cd | |
| | | Carpenter and Kaluvakolanu (2011) | $NO_2$, P, TSS | | | | |
| | | Gregoire and Clausen (2011) | TKN, $NO_3 + NO_2$, $NH_3$, Tot-P, and $PO_4$ | | | Cd, Cr, Pb, Zn | |
| | Canada | Van Seters et al. (2009) | Tot-P, PAH | Ca, Mg | | | |
| Australia | Australia | Razzaghmanesh et al. (2014a), Beecham and Razzaghmanesh (2015) | $NO_3$, $NO_2$, $NH_3$, $PO_4$ | B, Ca, K, Mg, Fe | Na, Cl | Cd, Cu, Pb, Mn, Mo and Zn | |

- Type of drainage;
- Maintenance/chemicals used (including fertilizers);
- Type of vegetation and seasonal growth pattern;
- Dynamics of precipitation;
- Wind direction;
- Local pollution sources; and
- Physicochemical properties of pollutants.

## 20.7 GREEN ROOF VEGETATION GROWTH FACTORS

### 20.7.1 Vegetation diversity

Green roofs often provide a harsh and stressful growing environment which only a limited range of plant species are able to tolerate. However, ecological theory suggests that highly diverse or species-rich vegetation might be more resistant and resilient to severe environmental stress (Nagase and Dunnett, 2010). To assess this hypothesis, a complex experiment was set up to investigate the influence of vegetation diversity on plant survival following an imposed drought. Twelve species were selected from the three major taxonomic plant groups commonly used for green roofs (forbs, *Sedums*, and grasses) and planted in combinations of increasing diversity and complexity, overlaid with three watering regimes. Greater survivability and higher visual rating occurred with a diverse plant mix under dry conditions (irrigated every 3 weeks). Drought tolerance in *Sedums* was superior to that of forbs and grasses, which in turn were little different to each other. It was recommended that irrigation may be unnecessary if *Sedum* species alone are used for an extensive green roof, as they can survive more than 3 weeks without watering, in UK growing conditions. However, if forbs or grasses are used, frequent irrigation is required to maintain good visual quality.

Oberndorfer et al. (2007) explained that climatic condition is also one of the major factors influencing selection of green roof plants. The nature of rainfall and extreme temperatures may restrict the use of certain species or it may necessitate the use of irrigation. Native plants are generally considered ideal choices for landscapes because of their adaptations to local climates. However, many native plants appear to be unsuitable for conventional extensive green roof systems because of the harsh environmental conditions, and typically shallow substrate depths. Local policies for biodiversity and nature conservation may affect the establishment of locally distinctive and representative plant communities. Monterusso et al. (2005) found only 4 of the original 18 native prairie perennial species remained growing in the 10 cm substrate after 3 years. In comparison, all nine nonnative species of *Sedum* thrived. In addition, Dunnett et al. (2008) reported that the most commonly used green roof vegetation is a mix of predominantly exotic (nonnative) *Sedum* (stonecrop) species that have very low maintenance requirements. The species used were generally low growing carpeting plants that were drought-resistant and capable of growing on thin substrate depths. They concluded that green roof vegetation with greater structural and species diversity may provide different benefits than *Sedum*-dominated roofs.

### 20.7.2 Substrates

Oberndorfer et al. (2007) argued that the depth of the substrate has a significant effect on vegetation diversity and the range of suitable species. Substrate depths between 20 and 50 mm have rapid rates of desiccation and high diurnal temperature variations. However, they can still support simple *Sedum*—moss communities. Increasing substrate depths to 70—150 mm supported a more diverse mixture of grasses, geophytes, alpines, and drought-tolerant herbaceous perennials, but these were also more vulnerable to weed invasion.

In another study by Rowe et al. (2012), the effect of green roof media depth on *Crassulacean* plant succession was investigated over 7 years in south central Michigan, USA. 25 succulents (various species of *Graptopetalum*, *Phedimus*, *Rhodiola*, and *Sedum*) were grown in three media depths (25, 50, and 75 mm). Deeper media depths generally produced greater survival rates and better coverage (Bates et al., 2013).

### 20.7.3 Suitable plants for Australian green roofs

Williams et al. (2010) suggested that a major barrier to the widespread uptake of green roofs in Australia, and regions with similar dry climates, is the lack of plants that can both survive and be aesthetically attractive under local climatic conditions. To survive in temperate northern hemisphere climates, green roof plants need to be adapted to heat, cold, sun, wind, and water deficit, and be tolerant to some root inundation (Snodgrass and Snodgrass, 2006). Similar criteria will apply to roofs in southern hemisphere cities, although the plants will generally not be required to tolerate freezing conditions. However, plants will often be required to have much greater tolerance to longer, and more extreme, periods of water-deficit stress. While *Sedums* are widely used in northern hemisphere green roofs, there are concerns about their suitability for southern hemisphere countries. Some *Sedum* species can switch their photosynthetic pathway from C3 to a C4 crassulacean acid metabolism (CAM) during periods of water stress (Borland and Griffiths, 1990; Castillo, 1996). This allows the plants to open their stomata at night and store carbon dioxide for subsequent photosynthesis during the day. This reduces water loss and increases water use efficiency. However, *Sedum* species have a relatively weak CAM potential (Castillo, 1996; Razzaghmanesh et al., 2014b,c) and may need relatively low temperatures, either during the day or at night, to perform $CO_2$ exchange when water stressed (Kluge, 1977).

In recommending suitable plants for south eastern Australian climates, Farrell et al. (2012) evaluated the effects of severe drought (113 days without water) on the growth, water use, and survival of five succulent species (*Sedum pachyphyllum*, *Sedum clavatum*, *Sedum spurium*, *Disphyma crassifolium*, and *Carpobrotus modestus*) planted in three different green roof growing media with different water-holding capacities. Water use determined survival under severe drought conditions, with the higher water use species (*D. crassifolium* and *C. modestus*) dying at least 15 days earlier than *Sedum* species, which are conservative water users. Farrell et al. (2012) found that to maximize survival, green roofs in continuous or seasonally hot and dry climates should be planted with species that have high leaf succulence and low water use, in substrates with high water-holding capacity.

In the tropical climate of Queensland, Australia, Perkins and Joyce (2010) found that the native plants *Myoporum parvifolium* (creeping myoporum) and *Eremophila debilis* (winter apple) and the exotic plant, *Sedum sexangulare* (tasteless stonecrop), displayed the greatest survival rates and coverage on an extensive green roof.

## 20.8 THERMAL PERFORMANCE OF GREEN ROOFS

In urban environments, vegetation has largely been replaced by dark and impervious surfaces such as asphalt, road highways, and roofs (Oberndorfer et al., 2007). This brings more serious environmental problems such as flooding and UHI effects (Carter, 2011). The UHI effect is expressed as higher urban temperatures due to the repartitioning of solar radiation into sensible heat rather than the latent heat emboded in evapotranspiration (Bacci and Maugeri, 1992). Consequently, towns and cities are significantly warmer than their surrounding suburban and rural areas, particularly at night. The UHI effect depends principally on the modification of energy balance, as well as the effect of urban canyons (Landsberg, 1981), the thermal properties of building materials (Montávez et al., 2000), the substitution of green areas with impervious surfaces that limit evapotranspiration (Takebayashi and Moriyama, 2007; Imhoff et al., 2010), and decreases in the urban albedo (Akbari and Konopacki, 2005).

Many studies have established a correlation between an increase in green areas and a reduction in local temperatures (Takebayashi and Moriyama, 2007), suggesting the augmentation of urban vegetation as a possible mitigation strategy for the UHI effect. As densely urbanized areas have few residual spaces that can be converted into green areas, one solution is to convert conventional dark, flat roofs into vegetated areas. Approximately 20%−40% of the impervious area in an urban environment is occupied by roofs (Akbari et al., 2003; Carter and Jackson, 2007; Kingsbury and Dunnett, 2008).

Increasing the fraction of vegetated surfaces in urban areas is associated with increasing albedo (Solecki et al., 2005), which is the reflection of incoming short-wave radiation away from a surface. A regional simulation model using a uniformly distributed green roof coverage of 50% showed temperature reductions as great as 2°C in some areas of Toronto, Canada (Bass et al., 2002).

### 20.8.1 Urban heat island effects

UHI effects have a significant impact on building energy consumption and outdoor air quality (Mirzae and Haghighat, 2010). There are several approaches to study the UHI effect, including multiscale models, empirical observations, and simulation techniques. Because of the complexity of UHI effects, multiscale modeling is not generally feasible. Instead, observations or theoretical approaches have most often been employed. However, the causes of UHI effects differ in different climates, and even with different city features. Therefore, general conclusions cannot be made based on limited empirical monitoring data. With recent progress in computational tools, simulation methods have often been used to study UHI effects.

## 20.8.2 Available numerical models

There are two common numerical models used for urban microclimate studies and particularly UHI investigations namely: RayMan (Matzarakis et al., 2007) and ENVI-met (Huttner and Bruse, 2009). ENVI-met is a three-dimensional non-hydrostatic model for the simulation of surface—plant—air interactions. It is often used to simulate urban environments and to assess the effects of green infrastructure on urban and built environments. It is designed for microscale applications with a typical horizontal resolution of 0.5—10 m and a typical temporal resolution of 24—48 h, with a time step of 1—5 s. This resolution enables the analysis of small-scale interactions between individual buildings, surfaces, and plants. The RayMan model estimates the radiation fluxes and the effects of clouds and solid obstacles on short-wave radiation fluxes. The model, which takes complex structures into account, is suitable for utilization and planning purposes at both the local and regional scales.

## 20.8.3 Green roof mitigation of urban heat island effects

Susca et al. (2011) evaluated the thermal effects of vegetation at both the urban and building scales. The UHI effect was monitored in four areas of New York City, USA, and an average temperature difference of 2°C was found between the most vegetated and the least vegetated areas. Green roofs showed a potential for decreasing the use of energy for cooling and heating and, as a consequence, reducing peak energy demands.

In another study by Alexandri and Jones (2008), a two-dimensional microscale model was used to study the thermal effects of covering the building envelope with vegetation for various climates and urban canyon geometries. The effect of temperature decreases on outdoor thermal comfort, and energy savings were investigated and it was found that plants on the building envelope can be used to ameliorate the UHI effect. From this quantitative research, it was shown that there is potential for lowering urban temperatures when the building envelope is covered with vegetation. It was also shown that air temperature decreases at roof level can be up to 26.0°C with an average daytime decrease of 12.8°C. At ground level, decreases in temperature reached up to 11.3°C maximum with an average daytime decrease of 9.1°C. Overall the hotter and drier a climate is, the greater the effect of vegetation on urban temperatures. However, it has been pointed out that humid climates can also benefit from green surfaces, especially when both walls and roofs are covered with vegetation. For example, Alexandri and Jones (2008) recorded an 8.4°C maximum temperature decrease for humid Hong Kong. Temperature decreases due to vegetation are primarily affected by the amount and geometry of the vegetation itself rather than by the building orientation in street canyons, in hot periods. If applied to the whole city scale, green roofs can mitigate increased urban temperatures and, especially for hot climates, bring temperatures down to more comfortable levels. They can also reduce energy cooling costs for buildings by 30%—100%.

Skelhorn et al. (2014) tested seven green space scenarios that might be applied at a block or neighborhood level and investigated the resulting microclimate changes that can be achieved through such applications for Manchester, a city in temperate North West England. The research employed ENVI-met to compare the changes in air and surface temperatures on a warm summer's day in July 2010. The modeling demonstrated that, even in in temperate cities, a 5% increase in mature deciduous trees can reduce mean hourly surface temperatures by 1°C over the course of a summer's day.

Perini and Magliocco (2014) investigated the effects of several variables that contribute to the UHI effect and outdoor thermal comfort in dense urban environments. The study was conducted using the three-dimensional ENVI-met model. The effects of building density (% of built area) and canyon effect (building height) on potential mean radiant temperature and Predicted Mean Vote (PMV) distribution were quantified. PMV is a method of describing thermal comfort. The influence of several types of green areas (vegetation on the ground and on roofs) on temperature mitigation and on comfort improvements was investigated for different atmospheric conditions and latitudes in a Mediterranean climate. It was found that vegetation on the ground and on roofs mitigated summer temperatures, decreased the indoor cooling load demand, and improved outdoor comfort. The results of this study also showed that vegetation is more effective in environments with higher temperatures and lower relative humidity.

Coutts et al. (2012) explained how the combination of excessive heating driven by urban development, low water availability, and future climate change impacts could threaten human health and amenity for urban dwellers. They reviewed the literature to demonstrate the potential of WSUD to help improve outdoor human thermal comfort in urban areas and support climate sensitive urban design (CSUD) objectives within the Australian context. They further argued that WSUD provides a mechanism for retaining water in the urban landscape through stormwater harvesting and reuse while also reducing urban temperatures through enhanced evapotranspiration and surface cooling. It was also shown that WSUD features are broadly capable of lowering temperatures and improving human thermal comfort and, when integrated with vegetation, have the potential to meet CSUD objectives. However, the degree of benefit (the intensity of cooling and improvements to human thermal comfort) depends on a multitude of factors including local environmental conditions, the design and placement of the systems, and the nature of the surrounding urban landscape.

The ability of two types of extensive and intensive green roofs to reduce the surrounding microclimate temperature was monitored by Razzaghmanesh et al. (2016). The results showed that green roofs have significant cooling effects in summer time and could behave as an insulation layer to keep buildings warmer in the winter. Furthermore, different scenarios of adding green roofs to the Adelaide urban environment were investigated using the ENVI-met model. The scenario modeling of adding green roofs in a typical urban area supported the hypothesis that this can lead to reductions in energy consumption in the Adelaide urban environment. In addition, an increased use of other WSUD technologies, such as green walls and street trees, together with the adoption of high-albedo materials is recommended for achieving the optimum efficiency in terms of reducing urban temperatures and mitigating UHI effects.

## 20.9 CONCLUSION

Green roofs and living walls are increasingly important components of WSUD systems and have become widely used around the world in recent years. Green roofs can cover current impermeable roofs that densely populate our urban areas and by so doing, can provide many environmental, economic, and social benefits. Despite such benefits, green roofs are relatively new in regions with a dry climate, and there are several research gaps and practical barriers to overcome before these systems can be applied more widely. Furthermore, specific design criteria need to be developed for both green roofs and living walls for a range of weather conditions to develop climate-resilient systems. Improving water quality is one of the objectives of WSUD. However, some WSUD components, including green roofs and living walls, might indeed act as pollutant sources particularly during the early years of plant establishment. Nitrogen, phosphorus, potassium, chloride, and heavy metals have all been detected in both green roof and living wall outflow samples collected in several studies. These are believed to be largely sourced from applied fertilizers, the growing media components, or irrigation water application. Even so, pollutant concentrations in the outflow from green roofs and living walls are normally within the recommended ranges for nonpotable reuse such as toilet flushing and urban irrigation, but are seldom within the guideline ranges for potable consumption. Some of the environmental and social benefits of green roofs and living walls derive from the selection of aesthetically pleasing plants.

In the design of green roof systems in a dry climate, the size of the plants, the combinations of green roof layers, and the time of planting should all be taken into consideration. Stormwater quantity control is another main objective of WSUD. Regardless of the configuration, both extensive and intensive roof profiles can retain significant volumes of stormwater runoff. The peak attenuation of the recorded events is usually in the range of 15%–100%. This indicates that green roofs can, if designed appropriately, serve as effective source control structures. Intensive green roofs have generally displayed more capacity for retaining water, which is important in dry climates where the retained water may reduce the need for supplementary irrigation.

Both green roofs and living walls are able to mitigate the UHI phenomenon. In various macroscale studies, the addition of green roof areas reduced electricity consumption. This is a very important strategy with respect to developing low carbon, resilient, and livable cities. A longitudinal study over a 5–10-year period would be ideal for examining the changing performance over time of green roofs and living walls. In addition to better understand plant growth mechanisms, further investigations using monoculture plantings might provide important information to understand individual plant water requirements, water use efficiencies, evapotranspiration rates, and cooling potential.

## REFERENCES

Akbari, H., Konopacki, S.J., 2005. Calculating energy-saving potentials of heat-island reduction strategies. Energy Policy 33, 721–756.

Akbari, H., Shea Rose, L., Taha, H., 2003. Analyzing the land cover of an urban environment using high-resolution orthophotos. Landscape and Urban Planning 63, 1–14.

Alexandri, E., Jones, P., 2008. Temperature decreases in an urban canyon due to green walls and green roofs in diverse climates. Building and Environment 43, 480–593.

Alsup, S., Ebbs, S., Retzlaff, W., 2010. The exchangeability and leachability of metals from select green roof growth substrates. Urban Ecosystems 13, 91–111.

Australian and New Zealand environment and conservation council (ANZECC), 2000. Australian Guidelines for Urban Stormwater Management.

Bacci, P., Maugeri, M., 1992. The urban heat island of Milan. Il Nuovo Cimento - B 15C, 417–424.

Bass, B., Krayenhoff, E.S., Martilli, A., Stull, R.B., Auld, H., 2002. The impact of green roofs on Toronto's urban heat island. In: Proceedings of the First North American Green Roof Conference: Greening Rooftops for Sustainable Communities. 20–30 May, Chicago. IL, pp. 292–304.

Bates, A.J., Sadler, J.P., Mackay, R., 2013. Vegetation development over four years on two green roofs in the UK. Urban Forestry and Urban Greening 12, 98–108.

Beck, D., Johnson, G., Spolek, G., 2011. Amending greenroof soil with biochar to affect runoff water quantity and quality. Environmental Pollution 1–8.

Beecham, S., 2003. Water sensitive urban design - a technological assessment, waterfall. Journal of the Stormwater Industry Association 17, 5–13.

Beecham, S., Razzaghmanesh, M., 2015. Water quality and quantity investigation of green roofs in a dry climate. Water Research 70, 370–384.

Bengtsson, L., 2005. Peak flows from thin sedum-moss roof. Nordic Hydrology 36, 269–280.

Berndtsson, J., Emilsson, T., Bengtsson, L., 2006. The influence of vegetated roofs on runoff water quality. The Science of the Total Environment 355, 48–63.

Berndtsson, J.C., Bengtsson, L., Jinno, K., 2008. First flush effect from vegetated roofs during simulated rain events. Hydrology Research 39, 171–179.

Berndtsson, J.C., Bengtsson, L., Jinno, K., 2010. Runoff water quality from intensive and extensive vegetated roofs. Ecological Engineering 35, 369–380.

Bliss, D.J., Neufeld, R.D., Ries, R.J., 2009. Storm water runoff mitigation using a green roof. Environmental Engineering Science 26, 407–417.

Block, A.H., Livesley, S.J., Williams, N.S.G., 2012. Responding to the Urban Heat Island, a Review of the Potential of Green Infrastructure. Victorian Centre for Climate Change Adaptation Research, Melbourne, Victoria.

Borland, A.M., Griffiths, H., 1990. The regulation of CAM and respiratory recycling by water-supply and light regime in the C3 intermediate *Sedum telephium*. Functional Ecology 4, 33–39.

Cameron, R.W.F., Taylor, J.E., Emmett, M.R., 2014. What's "cool" in the world of green façades? How plant choice influences the cooling properties of green walls. Building and Environment 73, 198–207.

Carpenter, S., 2014. Growing Green Guide, a Guide to Green Roofs, Walls and Facades in Melbourne and Victoria, Australia.

Carpenter, D.D., Kaluvakolanu, P., 2011. Effect of roof surface type on stormwater run-off from full-scale roofs in a temperate climate. Journal of Irrigation and Drainage Engineering 137, 161–169.

Carter, J.G., 2011. Climate change adaptation in European cities. Current Opinion in Environmental Sustainability 3, 193–198.

Carter, T., Jackson, R., 2007. Vegetated roofs for storm water management at multiple spatial scales. Landscape and Urban Planning 80, 84–94.

Carter, T.L., Rasmussen, T.C., 2006. Hydrologic behaviour of vegetated roofs. The American Water Resources Association 42, 1261–1274.

Castillo, F.J., 1996. Antioxidative protection in the inducible CAM plant Sedum album L following the imposition of severe water stress and recovery. Oecologia 107, 469–477.

Castleton, H.F., Stovin, V., Beck, S.B.M., Davison, J.B., 2010. Green roofs; building energy savings and the potential for retrofit. Energy and Buildings 4, 1582–1591.

Coutts, A.M., Tapper, N.J., Beringer, J., Loughnan, M., Demuzere, M., 2012. Watering our cities: the capacity for water sensitive urban design to support urban cooling and improve human thermal comfort in the Australian context. Progress in Physical Geography 1, 2–28.

Dunnett, N., Kingsbury, N., 2008. Planting Green Roofs and Living Walls. Timber Press Inc., Portland, Oregon.

Dunnett, N., Nagase, A., Booth, R., Grime, P., 2008. Influence of vegetation composition on runoff in two simulated green roof experiments. Erban Ecosystem 11, 385–398.

Durhman, A., VanWoert, N., Rowe, D., Rugh, C., May, E., 2004. Evaluation of Crassulaceae species on extensive green roofs. In: 2nd North American Green Roof Conference, Toronto, Canada.

Farrell, C., Mitchell, R.E., Szota, C., Rayner, J.P., Williams, N.S., 2012. Green roofs for hot and dry climates: interacting effects of plant water use, succulence and substrate. Ecological Engineering 49, 270–276.

Fassman, E., Simcock, R., 2008. Development and Implementation of Locally-Sourced Extensive Green Roof Substrate in New Zealand. World Green Roof Congress, London, UK.

Fassman, E., Simcock, R., 2012. Moisture measurements as performance criteria for extensive living roof substrates. Journal of Environmental Engineering 138, 841–851.

Fassman-Beck, E., Voyde, E., Simcock, R., Hong, Y.S., 2013. 4 Living roofs in 3 locations: does configuration affect runoff mitigation? Journal of Hydrology 490, 11–20.

Feng, H., Hewage, K., 2014. Energy saving performance of green vegetation on LEED certified buildings. Energy and Buildings 75, 281–289.

FLL, 2002. Guideline for the planning, execution and upkeep of green-roof sites. In: Colmanstr. 32, 53115 Bonn, Germany, Forschungsgesellschaft Landschaftsentwicklung Landschaftsbau E.V. 94.

FLL, 2008. Guidelines for the planning, construction and maintenance of green roofing. In: Forschungsgesellschaft Landschaftsentwicklung Landschaftsbau E.V. Colmanstr. 32, 53115 Bonn, Germany.

Friedrich, C.R., 2005. Principles for selecting the proper components for a green roof growing media. Greening Rooftops for Sustainable Communities, Washington D.C., Green Roofs for Healthy Cities.

Getter, K.L., Rowe, D.B., 2006. The role of green roofs in sustainable development. Horticultural Science 41, 1276–1286.

Getter, K.L., Rowe, D.B., Andresen, J.A., 2007. Quantifying the effect of slope on extensive green roof stormwater retention. Ecological Engineering 31, 225–231.

SA Government, 2010. The 30 Year Plan for Greater Adelaide. Department of Planning and Local Government, South Australian Planning Strategy.

Graham, P., Kim, M., 2005. Evaluating the stormwater management benefits of green roofs through water balance modeling. In: The First North American Green Roof Infrastructure Conference Awards and Trade Show, Chicago, Illinois.

Green Roof Australia, 2010. GR 101, Green Roof Design and Installation. Green Canopy Design, Sydney, Australia.

Gregoire, B., Clausen, J., 2011. Effect of a modular extensive green roof on stormwater runoff and water quality. Ecological Engineering 37, 963–969.

Hathaway, A.M., Hunt, W.F., Jennings, G.D., 2008. A field study of green roof hydrologic and water quality performance. American Society of Agricultural and Biological Engineers 51 (1), 37–44.

Heinze, W., 1985. Results of an experiment on extensive growth of vegetation on roofs. Rasen Grünflachen Begrünungen 16 (3), 80–88.

Hilten, R.N., Lawrence, T.M., Tollner, E.W., 2008. Modelling stormwater runoff from green roofs with HYDRUS-1D. Journal of Hydrology 358, 288–293.

Hui, S.C.M., Zhao, Z., 2013. Thermal regulation performance of green living walls in buildings. In: Joint Symposium 2013: Innovation and Technology for Built Environment, Hong Kong, pp. 1–10.

Hutchinson, D., Abrams, P., Retzlaff, R., Liptan, T., 2003. Stormwater Monitoring Two Ecoroofs in Portland, Oregon, USA, City of Portland. Bureau of Environmental Services. http://www.portlandonline.com/shared/cfm/image.cfm?id=63098.

Huttner, S., Bruse, M., 2009. Numerical modelling of the urban climate – a preview on ENVI-met 4.0. In: The Seventh International Conference on Urban Climate. Yokohama, Japan.

Imhoff, M.L., Zhang, P., Wolfe, R.E., Bounoua, L., 2010. Remote sensing of the urban heat island effect across biomes in the continental USA. Remote Sensing of Environment 114, 504–513.

Jaafar, B., Said, I., Reba, M.N.M., Rasidi, M.H., 2013. Impact of vertical greenery system on internal building corridors in the tropic. Procedia - Social and Behavioral Sciences 105, 558–568.

Jørgensen, L., Dresbøll, D.B., Thorup-Kristensen, K., 2014. Root growth of perennials in vertical growing media for use in green walls. Scientia Horticulture 166, 31–41.

Kingsbury, N.L., Dunnett, N., 2008. Planting Green Roofs and Living Walls. Rev. and Updated Ed., second ed. Timber Press, Portland, Oregon.

Kluge, M., 1977. Is sedum a CAM plant. Oecologia 29, 77–83.

Köhler, M., 2008. Green facades — a view back and some visions. Urban Ecosystems 11 (4), 423–436.

Köhler, M., Schmidt, M., Grimme, F.W., Laar, M., 2002. Green roofs in temperate climates and in the hot-humid tropics – far beyond the aesthetics. Environmental Management and Health 13, 382–391.

Kosareo, L., Ries, R., 2007. Comparative environmental life cycle assessment of green roofs. Built Environment 42, 2606–2613.

Koyama, T., Yoshinaga, M., Hayashi, H., Maeda, K.I., Yamauchi, A., 2013. Identification of key plant traits contributing to the cooling effects of green façades using freestanding walls. Building and Environment 66, 96–103.

Landsberg, H.E., 1981. The Urban Climate. New York Academic Press.

Manso, M., Castro-Gomes, J., 2015. Green wall systems: a review of their characteristics. Renewable and Sustainable Energy Reviews 41, 863–871.

Mårtensson, L.-M., Wuolo, A., Fransson, A.-M., Emilsson, T., 2014. Plant performance in living wall systems in the Scandinavian climate. Ecological Engineering 71, 610–614.

Matzarakis, A., Rutz, F., Mayer, H., 2007. Modelling radiation fluxes in simple and complex environments, application of the RayMan model. International Journal of Biometeorological 51, 323–334.

Mazzali, U., Peron, F., Romagnoni, P., Pulselli, R.M., Bastianoni, S., 2013. Experimental investigation on the energy performance of living walls in a temperate climate. Building and Environment 64, 57–66.

Mentens, J., Raes, D., Hermy, M., 2003. Effect of orientation on the water balance of green roofs. Greening Rooftops for Sustainable Communities Chicago 363–371.

Mentens, J., Raes, D., Hermy, M., 2006. Green roofs as a tool for solving the rainwater runoff problem in the urbanized 21st century? Landscape and Urban Planning 77, 217–226.

Mirzae, P.A., Haghighat, F., 2010. Approaches to study urban heat island abilities and limitations. Building and Environment 45, 2192–2201.

Molineux, C.J., Fentiman, C.H., Gange, A.C., 2009. Characterising alternative recycled waste materials for use as green roof Growing media in the UK. Ecological Engineering 35, 1507–1513.

Montávez, J.P., Rodríguez, A., Jiménez, J.I., 2000. A study of the urban heat island of granada. International Journal of Climatology 20, 899–911.

Monterusso, M.A., Rowe, D.B., Rugh, C.L., Russell, D.K., 2004. Runoff water quantity and quality from green roof systems. Acta Horticulture 639, 369–376.

Monterusso, M.A., Rowe, D.B., Rugh, C.L., 2005. Establishment and persistence of Sedum sand native taxa for green roof applications. HortScience 40, 391–396.

Munby, B., 2005. Feasibility study for the retrofitting of green roofs. In: Civil and Structural Engineering. University of Sheffield, Sheffield, UK.

Nagase, A., Dunnett, N., 2010. Drought tolerance in different vegetation types for extensive green roofs: effects of watering and diversity. Landscape and Urban Planning 97, 318–327.

Natarajan, M., Rahimi, M., Sen, S., Mackenzie, N., Imanbayev, Y., 2014. Living wall systems: evaluating life-cycle energy, water and carbon impacts. Urban Ecosystems 18 (1), 1–11.

Hien, W.N., Puay Yok, T., Yu, C., 2007. Study of thermal performance of extensive rooftop greenery systems in the tropical climate. Building and Environment 42 (1), 25–54.

Oberndorfer, E., Lundholm, J., Bass, B., Coffman, R., Doshi, H., Dunnett, N., Gaffin, S., Kohler, M., Liu, K.K.Y., Row, B., 2007. Green roofs as urban ecosystems: ecological structures, functions, and services. BioScience 57, 823–833.

Ottelé, M., Perini, K., Fraaij, A.L.A., Haas, E.M., Raiteri, R., 2011. Comparative life cycle analysis for green façades and living wall systems. Energy and Buildings 43 (12), 3419–3429.

Palla, A., Gnecco, I., Lanza, L.G., 2009. Unsaturated 2D modelling of subsurface water flow in coarse-grained porous matrix of a green roof. Journal of Hydrology 379, 193–204.

Palla, A., Gnecco, I., Lanza, L., 2010. Hydrologic restoration in the urban environment using green roofs. Water 2, 140–154.

Palla, A., Gnecco, I., Lanza, L.G., 2012. Compared performance of a conceptual and a mechanistic hydrologic model of a green roof. Hydrological Processes 26, 73–84.

Pérez, G., Rincón, L., Vila, A., González, J.M., Cabeza, L.F., 2011. Green vertical systems for buildings as passive systems for energy savings. Applied Energy 88 (12), 4854–4859.

Pérez, G., Coma, J., Martorell, I., Cabeza, L.F., 2014. Vertical Greenery Systems (VGS) for energy saving in buildings: a review. Renewable and Sustainable Energy Reviews 39, 139–165.

Perini, K., Magliocco, A., 2014. Effects of vegetation, urban density, building height, and atmospheric conditions on local temperatures and thermal comfort. Urban Forestry and Urban Greening 13 (3), 495–506.

Perini, K., Rosasco, P., 2013. Cost-benefit analysis for green facades and living wall systems. Building and Environment 70, 110–121.

Perini, K., Ottelé, M., Haas, E.M. l, Raiteri, R., 2013. Vertical greening systems, a process tree for green façades and living walls. Urban Ecosystems 16 (2), 265–277.

Perkins, M., Joyce, D., 2010. Australian plant for extensive green roofs. In: Green Roofs Conference 2010.

Rayner, J., 2010. Choosing substrates for Australian green roofs. In: Green Roof Conference Adelaide.

Razzaghmanesh, M., 2015. Developing Resilient Green Roofs for Adelaide. University of South Australia (Ph.D. Dissertation), 314 p.

Razzaghmanesh, M., Beecham, S., 2014. The hydrological behaviour of extensive and intensive green roofs in a dry climate. The Science of the Total Environment 499, 284–296.

Razzaghmanesh, M., Razzaghmanesh, M., 2017. Thermal performance investigation of a living wall in a dry climate of Australia. Building and Environment 112, 45–62.

Razzaghmanesh, M., Beecham, S., Kazemi, F., 2012. The role of green roofs in water sensitive urban design in South Australia. In: 7th International Conference on Water Sensitive Urban Design, Melbourne, Australia.

Razzaghmanesh, M., Beecham, S., Kazemi, F., 2014a. Impact of green roofs on stormwater quality in a South Australian urban environment. The Science of the Total Environment 470, 651–659.

Razzaghmanesh, M., Beecham, S., Kazemi, F., 2014b. The growth and survival of plants in urban green roofs in a dry climate. The Science of the Total Environment 476, 288–297.

Razzaghmanesh, M.S., Beecham, Brien, 2014c. Developing resilient green roofs in a dry climate. The Science of the Total Environment 490, 579–589.

Razzaghmanesh, M., Beecham, S., Salemi, T., 2016. The role of green roofs in mitigating urban heat island effects in the metropolitan area of Adelaide, South Australia. Urban Forestry and Urban Greening 15, 89–102.

Roehr, D., Fassman-Beck, E., 2015. Living Roofs in Integrated Urban Water Systems. Routledge, London, UK.

Rowe, D.B., Getter, K.L., Durhman, A.K., 2012. Effect of green roof media depth on Crassulacean plant succession over seven years. Landscape and Urban Planning 104, 310–319.

Safikhani, T., Abdullah, A.M., Ossen, D.R., Baharvand, M., 2014a. A review of energy characteristic of vertical greenery systems. Renewable and Sustainable Energy Reviews 40, 450–462.

Safikhani, T., Abdullah, A.M., Ossen, D.R., Baharvand, M., 2014b. Thermal impacts of vertical greenery systems. In: Environmental and Climate Technologies, 14, vol. 1. Riga Technical University, pp. 5–11.

Scarpa, M., Mazzali, U., Peron, F., 2014. Modeling the energy performance of living walls: validation against field measurements in temperate climate. Energy and Buildings 79, 155–163.

Schroll, E., Lambrinos, J., Righetti, T., Sandrock, D., 2010. The role of vegetation in regulating stormwater runoff from green roofs in a winter rainfall climate. Ecological Engineering.

Sheweka, S., Magdy, A.N., 2011. Living walls as an approach for a healthy urban environment. Energy Procedia 6, 592–599.

Simmons, M.T., Gardiner, B., Windhager, S., Tinsley, J., 2008. Green roofs are not created equal: the hydrologic and thermal performance of six different extensive green roofs and reflective and non-reflective roofs in a sub-tropical climate. Urban Ecosystem 11, 339–348.

Skelhorn, C., Lindley, S., Levermore, G., 2014. The impact of vegetation types on air and surface temperatures in a temperate city: a fine scale assessment in Manchester, UK. Landscape and Urban Planning 121, 129–140.

Skinner, C.J., 2006. Urban density, meteorology and rooftops. Urban Policy and Research 24 (3), 355–367.

Snodgrass, E.C., Snodgrass, L.L., 2006. Green roof Plants: A Resource and Planting Guide. Timber Press, Portland, Oregon, USA.

Solecki, W.D., Rosenzweig, C., Parshall, L., Pope, G., Clark, M., Cox, J., Wiencke, M., 2005. Mitigation of the heat island effect in urban New Jersey. Environmental Hazards Human and Policy Dimensions 6, 39–49.

Stav, Y., Lawson, G., 2012. Vertical vegetation design decisions and their impact on energy consumption in subtropical cities. In: Pacetti, M., Passerini, G., Brebbia, C.A., Latini, G. (Eds.), The Sustainable City VII: Urban Regeneration and Sustainability. WIT Press, Ancona, Italy, pp. 489–500.

Steusloff, S., 1998. Input and Output of Airborne Aggressive Substances on Green Roofs in Karlsruhe. Urban Ecology, Springer-Verlag, Berlin, Germany.

Sunakorn, P., Yimprayoon, C., 2011. Thermal performance of a biofacade with natural ventilation in the tropical climate. Procedia Engineering 21, 34–41.

Susca, T., Gaffin, S.R., Dell'Osso, G.R., 2011. Positive effects of vegetation: urban heat island and green roofs. Environmental Pollution 159, 2119–2126.

Susorova, I., Azimi, P., Stephens, B., 2014. The effects of climbing vegetation on the local microclimate, thermal performance, and air infiltration of four building facade orientations. Building and Environment 76, 113–124.

Takebayashi, H., Moriyama, M., 2007. Surface heat budget on green roof and high reflection roof for mitigation of urban heat island. Building and Environment 42, 2971–2979.

Teemusk, A., Mander, U., 2007. Rainwater runoff quantity and quality performance from a greenroof: the effects of short-term events. Ecological Engineering 30, 271–277.

Van Seters, T., Rocha, L., Smith, D., MacMillan, G., 2009. Evaluation of green roofs for runoff retention, runoff quality, and leachability. Water Quality Research Journal of Canada 44, 33–47.

VanWoert, N., Rowe, B., Andresen, J., Rugh, C., Fernandez, T., Xiao, L., 2005. Green roof stormwater retention: effects of roof surface, slope, and media depth. Journal of Environmental Quality 34, 1036–1044.

Vilareal, Bengtsson, 2005. Response of a sedum green-roof to individual rain events. Ecological Engineering 25, 1–7.

Villarreal-Gonzalez, E., Bengtsson, L., 2005. Response of a sedum green-roof to individual rain events. Ecological Engineering 25, 1–7.

Voyde, E., Fassman, E., Simcock, R., 2010. Hydrology of an extensive living roof under sub-tropical climate conditions in Auckland, New Zealand. Journal of Hydrology 394, 384–395.

Weinmaster, M., 2009. Are green walls as "Green" as they look? An introduction to the various technologies and ecological benefits of green walls. Journal of Green Building 4 (4), 3–18.

Williams, N.S., Rayner, J.P., Raynor, K.J., 2010. Green Roofs for a wide brown land: opportunities and barriers for roof top greening in Australia. Urban Forestry and Urban Greening 9, 245–251.

Wingfield, A., 2005. The Filter, Drain, and Water Holding Components of Green Roof Design. http://www.greenroofs.com/archives/gf_mar05.htm.

Wong, N.H., Kwang Tan, A.Y., Chen, Y., Sekar, K., Tan, P.Y., Chan, D., Chiang, K., Wong, N.C., 2010. Thermal evaluation of vertical greenery systems for building walls. Building and Environment 45 (3), 663–672.

Chapter 21

# Greening and Cooling the City Using Novel Urban Water Systems: A European Perspective

Martina Winker[1], Simon Gehrmann[2], Engelbert Schramm[1], Martin Zimmermann[1] and Annette Rudolph-Cleff[2]

[1]*ISOE – Institute for Social-Ecological Research, Frankfurt/Main, Germany;* [2]*Technical University of Darmstadt, Dept. for Urban design and Development, Faculty of Architecture, Darmstadt, Germany*

## Chapter Outline



## ABSTRACT

In today's cities water appears as drinking water, wastewater, rainwater, and runoff, as well as natural and artificial waterbodies. These water streams play a key role in the urban metabolism. The management of the water streams is challenging, especially in dense urban areas and in the context of climate change. Moreover, additional requirements have evolved including adapting to climate change, improving the quality of urban life, creating urban cooling and green areas in the cities, increasing resource protection, and flood protection and prevention.

To tackle these challenges, current water infrastructure needs to strongly adapt or even transform its essential character. It has to become more flexible regarding its response time to adaptation and provide services more targeted toward the specific local needs. Here, recent innovations in water infrastructures, also called novel urban water systems, come into the picture. They provide possibilities able to react both faster and more specifically, and to build strong bridges to other technical infrastructure and urban planning.

Water sensitive urban design (WSUD) focuses on the management of all water streams within the city. Although the focus of this approach is mostly identified in stormwater management, WSUD also includes the sustainable management of domestic wastewater. When it comes to the projects built under WSUD design principles, stormwater is usually considered, whereas communal/domestic wastewater is often not taken into account. This chapter argues that there are specific cases in novel urban water systems where an active integration of wastewater into the WSUD concept should be considered, as it provides clear advantages and benefits for both. Moreover, it provides further details on the integration of novel urban water systems into a WSUD approach and shows examples where such integration is already practiced.

**Approaches to Water Sensitive Urban Design. https://doi.org/10.1016/B978-0-12-812843-5.00021-6**

**Keywords:** Climate adaptation; Green infrastructure; Graywater; Novel urban water systems; Reuse; Service water; Stormwater; Wastewater

## 21.1 EXISTING CHALLENGES

In today's cities water appears as drinking water, wastewater, rainwater, and runoff, as well as natural and artificial waterbodies. These water streams play a key role in the urban metabolism. The management of the water streams is challenging, especially in dense urban areas and in the context of climate change. Without sufficient water supplies, urban living would be limited or completely impossible. This means urban development and water are closely linked. At the same time, water requirements and the handling of water in cities and municipalities have a major influence on the natural water cycle: they are changing the cycle and the involved ecosystems. From history, we know that this sensible balance needs close attention. The task of water infrastructure is to manage the balance to provide sufficient access to the resource, to organize the distribution in the city, as well as between city and nature, and to guarantee the discharge and treatment of used water and stormwater back to nature. This chapter provide an overview on the water discussions going on in Europe with a special focus on Germany. At certain points, discussions and progress from other continents are also included.

Local sources, such as rainwater, which mostly means the runoff from roofs, were always considered as an important resource for ancient cities, until a high anthropogenic pressure led to changes in the system because of the need for flood protection and hygiene improvements. A brief look at European history shows that forerunners of the modern water supply systems could be found 3500 years ago, e.g., at the island of Crete in Greece, or in ancient Rome (Hodge, 2002, p. 347). Those systems were based on the construction of small creeks, aqueducts, or canals to convey water to the urban areas. Centuries later, the urban water supply was usually dependent on local water sources as well to fulfill the growing demands (Geels, 2005, p. 370).

With the increase in urban density with industrialization, hygiene aspects became more and more important, as outbreaks of diseases such as cholera and typhus fever occurred (Gleick, 2003, p. 276; Domènech, 2011, p. 295). One of the biggest concerns was the contamination of the local water sources with human and animal faeces. To improve this situation, a piped water supply and a sewer network were developed (see Dingle, 2008, p. 9; Domènech, 2011, p. 295). The cities were equipped with big central sewer networks and water pipes to bring the different waters from one point to another, usually to/from the peri-urban areas, where the infrastructure treatment facilities were located. Most of the early European sewer systems conveyed the stormwater together with the wastewater in a combined sewer system (in contrast to separate systems) to beyond the cities' boundaries. The possibility to transport freshwater water to the city, and to discharge wastewater beyond it, made the urban areas more and more independent from their existing local water sources.

In the present day, the framing conditions of this sensitive balance are undergoing a challenging shift. The influence of climate change, socioeconomic change, and demographic change is becoming more important. Moreover, additional requirements have evolved including adapting to climate change, improving the quality of urban life, creating urban cooling and green areas in the cities, increasing resource protection, and flood protection and prevention.

To tackle these challenges, current water infrastructure needs to strongly adapt or even transform its essential character. It has to become more flexible regarding response time to adaptation, and provide services more targeted toward the specific local needs. Here, recent innovations in water infrastructures, also called novel urban water systems (Schramm et al., 2017), come into the picture. They provide possibilities which are able to react both faster, and more specifically, and to build strong bridges to other technical infrastructure and urban planning (Libbe et al., 2017, Fig. 21.1).

However, as investigations in Germany have shown, such bridging between various sectors is not easy to achieve (Schramm et al., 2017). Experts mostly picture the role of technical infrastructure planning (mainly occurring subsurface) as a subsidiary service of urban planning (above ground), which is realized under traditional framework conditions. In the case of water infrastructure, this planning task is mostly focused on the implementation of the discharge of stormwater and wastewater into the existing sewer system. In such a situation, a forward-looking integrated planning approach involving novel urban water systems is not feasible as urban planning has already progressed too far.

Cities are highly dependent on the technical infrastructure for runoff management and flood protection. Most of the urban areas are characterized by sealed, highly impervious surfaces, which are supposed to manage the runoff. Basically, this means fast conveyance to the combined sewer system or the stormwater system (where this is separated from the sewer system). In recent years, climate change and the expected attendant intensification of the hydrological cycle, which means an increase of large and intense rain events and a corresponding increase in urban runoff, has resulted in additional pressure

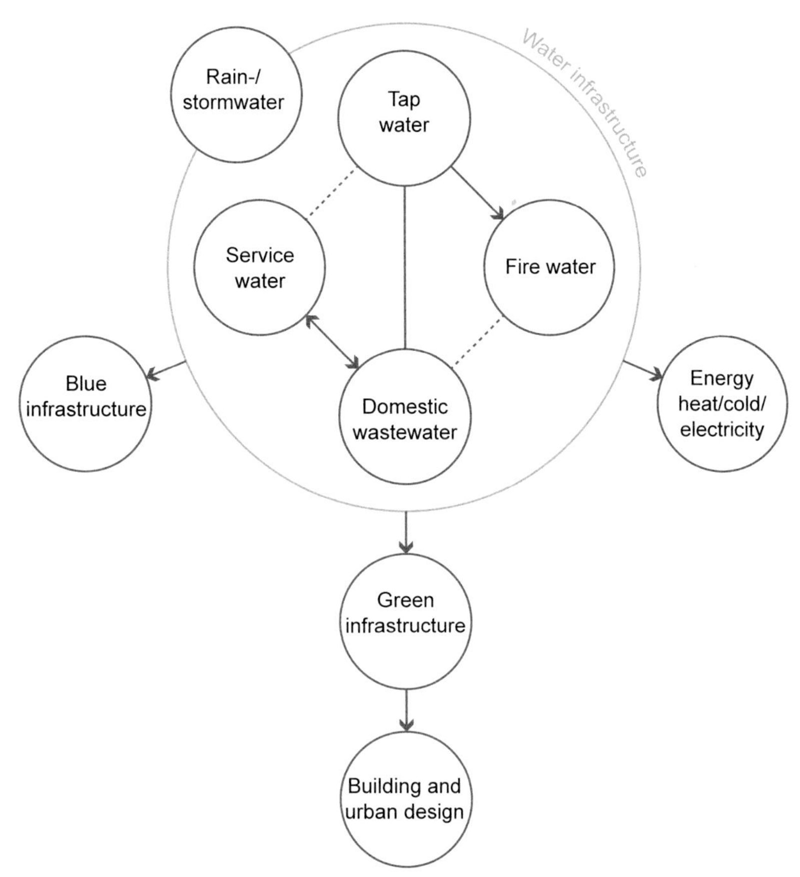


**FIGURE 21.1** Connections between water infrastructure and other infrastructures as seen from the water perspective. *From Winker, M., Trapp, J.H., Libbe, J., Schramm, E. (Eds.), 2017. Wasserinfrastruktur: Den Wandel gestalten. Technische Varianten, räumliche Potenziale, institutionelle Spielräume. Edition Difu – Stadt Forschung Praxis, 16. Difu, Berlin.*

on the water infrastructure. Hence, these rain events may well cause the infrastructure to fail, resulting in sewage overflows, water pollution, and localized flooding.

Furthermore, a large fraction of sealed areas leads to heat islands, as they prevent the city from "breathing." Dark sealed areas, such as asphalt and concrete materials, absorb the solar radiation and store it as "sensible heat," which results in significant air temperature difference, compared with that in unsealed areas. The impervious coverage can be as high as 90% in dense urban areas (Schueler, 2000). The temperature of urban surfaces can differ by up to 20°C between vegetated and sealed areas (Chan et al., 2017). Consequences of higher temperatures, in combination with other factors, may lead to greater evaporation and hence more annual rainfall. Facts supporting this theory include urban solitary trees transpire more than trees in rural areas or in a forest (Kjelgren and Montague, 1998); an increased convection because of the heat and the higher amount of aerosols in urban areas (Jin et al., 2005; Shepherd, 2005) and particulate matter; condensation nuclei for cloud formation; and finally, the higher number of precipitation anomalies observed for coastal cities like Houston (Shepherd and Burian, 2003). This phenomenon could even increase the pressure on the existing technical infrastructure. Chinese research additionally indicates that heat islands might contribute up to 30% to the climate warming in China (Zhao, 2011; Ren et al., 2008). In Europe, the number of the so-called "tropical" nights (nights where the mean temperature >20°C) increases significantly (Fallmann et al., 2017). In contrast to this, green, open, vegetated spaces store water in the soil, resulting in groundwater recharge, and supporting evaporation and transpiration, which in turn decreases the local temperature by converting sensible heat into latent heat. Therefore, when referring to green areas in this chapter, the green vegetation is always understood to be interacting with its soil.

To face these challenges, it is important to shift the existing planning governance toward an innovative, overarching sector approach, which involve approaches that are not "business as usual." New approaches or methods of planning, and cooperation between the different stakeholders are required very early on in the process to take account of the contribution of water infrastructure to improving climate protection, or to reducing impacts of climate change.

In addition to rainwater, domestic wastewater streams can be seen as new water sources to tackle these effects. Wastewater is already generated in the urban areas and needs treatment prior to discharge. Moreover, by using treated wastewater for watering the green spaces in the city, nutrients are also delivered to the plants. Irrigation with treated wastewater helps to establish and conserve green spots to cool and green cities in the times of their most urgent need: in summertime when rainfall is largely absent and hot periods are more frequent.

Water sensitive urban design (WSUD) focuses on the management of all water streams within the city. The approach can be seen as going from the existing system to a "green" paradigm. It is mainly used to describe design approaches where urban stormwater management aims to create a nature-oriented water cycle, while contributing to the amenity of the city. Although the focus of this approach is mostly identified in stormwater management, WSUD also includes the sustainable management of domestic wastewater. When it comes to the projects that were built under WSUD design principles, stormwater is usually considered, whereas communal/domestic wastewater is often not taken into account (Barton and Argue, 2007). This chapter argues that there are specific cases in novel urban water systems where an active integration of wastewater into the WSUD concept should be considered, as it provides clear advantages and benefits for both.

## 21.2 PROSPECTS FOR URBAN WATER MANAGEMENT DUE TO NOVEL WATER INFRASTRUCTURE CONCEPTS

Novel urban water systems provide a paradigm shift for urban water management. They comprise new technologies such as heat recovery from graywater or nutrient recovery from urine, as well as new management approaches such as source separation and water reuse (Otterpohl et al., 1999; Larsen et al., 2009; Kluge and Schramm, 2016). Source separation means that different wastewater streams are collected, discharged, and treated separately according to their specific needs. Source-separated wastewater streams and their respective products for further use are as follows:

- Management of rainwater and stormwater → service water (nonpotable water) for household purposes, park and garden maintenance, and landscape management;
- Management of graywater (water originating from shower/bath sinks) → service water (purposes as above) and generation of heat/cold on a building scale;
- Management of blackwater (toilet water) → electric power/gas supply, generation of heat in a community scale, and nutrient recovery. Blackwater can also be split further into faecal matter and urine, which, when diluted with water, is called brownwater and yellowwater respectively.

This means that service water can evolve from different raw water sources, which are mostly rainwater, stormwater, and graywater. Rainwater (roof runoff) can often be used directly as it is not contaminated with most urban pollutants, whereas graywater needs treatment according to its proposed end use (DWA, 2017).

The following text points out when the source of nonpotable service water is of specific relevance. Wastewater of industrial origin is not addressed in this chapter

### 21.2.1 New handling of stormwater

As soon as rainwater touches the urban surface, it becomes stormwater. The movement of stormwater from one point to another can be described as runoff, which needs appropriate management to avoid any kind of damage. While many countries, for example, Australia, install rainwater tanks to capture the runoff from their roofs for potable and nonpotable uses, the majority of the European cities are not constructed for using rainwater, nor are considering it as a resource.[1]

Engineers have traditionally focused runoff management firstly on flood protection, and then on the sustainable management during the last few centuries. This resulted in the construction of large centralized infrastructure, which either carries stormwater together with domestic and industrial wastewater in a combined sewer system or in separate sewers. Both approaches can be described as "out of sight, out of mind," whose focus was to transport stormwater out of the city as fast as possible.

After the industrialization era, most of the urban water supply in European cities was sourced from groundwater, which was reticulated through large-scale, centralized distribution grids. This changed the perception of rainwater from an important water source to a problem for daily life, as the connection between rainwater and tapwater was no longer clearly visible. Today, our cities are characterized by dense and impermeable areas, which are equipped with technical infrastructure underneath the visible surface, which allows the use and discharge of water wherever and whenever needed.

---

1. http://www.kompetenz-wasser.de/fileadmin/user_upload/pdf/veranstaltungen/Wasserwerkstatt/WW37_Regenwassermanagement.pdf.

In the case of a combined sewer network, the stormwater runoff undergoes the same treatment as the domestic wastewater at the treatment plant before it is discharged to the local waterbodies. As large amounts of runoff dilute the wastewater, this can significantly influence the effectiveness of the biological treatment processes (Wilderer, 2007; Otterpohl, 2001). A separate stormwater system usually does not contain a biological treatment step, although the urban runoff might be heavily polluted by oil, petrol, heavy metals, nutrients, sediment, and other pollutants (USEPA, 2005; Hvitved-Jacobsen et al., 2010). Those systems are sometimes equipped with detention or retention and sedimentation basins, which are designed to filter large pollutants and reduce the peak runoff before the stormwater is discharged to the receiving waters. Nevertheless, in both cases, the centralized infrastructure (separate or combined) carries the water away from the urban area.

Several authors acknowledge the advantages of centralized water infrastructure for reliable water supply, flood control, food production, and hydro-energy use (Gleick, 2003). However, in recent years, especially in the context of climate change, water shortages and more frequent flooding have increased concerns about the environmental impacts, and social costs of large-scale centralized projects. These concerns have entered the water policy debate (see McCully, 1996; Sauri and Del Moral, 2001).

The centralized infrastructure approach is becoming highly vulnerable in some areas because of the age of the sewers and the fact that the centralized infrastructure prevents the sustainable recharge of local aquifers. The “out of sight, out of mind strategy” may work well in case of engineered flood protection, but has a wide range of negative side effects for the urban areas, such as its impacts on the local water system. This becomes even more evident with emerging climate change.

The impact of the climate change can be described as an intensification of the hydrological cycle for most parts of the world (for example, see Huntington, 2006). Recent calculations of the Intergovernmental Panel on Climate Change forecast a relatively constant amount of annual rainfall for Germany during the next decades, but anticipate large changes in the monthly distribution. It is expected that the rainfall intensity will increase during the winter period and decrease significantly during summer. Extended rainfall in winter will lead to increased runoff, as well as increased nutrient leaching from soils, leading to water quality problems in the watershed. In summertime, less rainfall may well stress European agriculture, the natural vegetation, and the recharge of the aquifers (information from various unpublished source). In that context, the German Ministry of Education and Research indicates that the amount of drinking water required will increase along with irrigation water for agriculture and cooling water for industry (BMBF, 2014). These changes, will drive a re-design of our cities’ water infrastructure, which has to be adjusted to the new situation.

In recent years, engineers along with architects, urban planners, and other disciplines have started to question and rethink the existing centralized approach (see Langergraber and Muellegger, 2005; Domènech, 2011). Stormwater came into focus as it can provide a lot of benefits to the urban space when managed properly. Design approaches, which include sustainable runoff management, and the local use of rainwater can be considered within the context of “WSUD.” While the term “WSUD” is most common in Australia, the same principles can be found in the Guidelines of Low Impact Design, which is common in the United States, Sustainable Urban Design, which is more common in Europe, and other design principles all over the world (Hoyer et al., 2011).

According to the common principles of these various design approaches, separate systems for stormwater collection, become more important. Urban stormwater runoff should be infiltrated locally. Visible elements such as rain gardens or designed retention areas will increase and improve the green perception of the urban space. Moreover, a more engineered-driven approach can incorporate stormwater as an important resource into the European water supply, and its use for energy and cooling issues (Fig. 21.2).

This new system for the management of rainwater and stormwater will result in the following major challenges that have to be addressed by a design change:

1. The European urban sewer system usually collects wastewater streams together with stormwater, which increases the water masses. This not only affects the treatment effectiveness, but it is also is necessary to control the hydraulic load entering the sewage treatment facility to avoid damage to the treatment train. Therefore, the combined sewer infrastructure is equipped with overflow structures, which spill diluted sewage into waterbodies when the conveyance capacity of the sewer is reached. This results in pollution of the waterbodies, algae blooms, and may pose health risks (Gervin and Brix, 2001; Gasperi et al., 2008).
2. The existing sewers have been designed to transport domestic wastewater and a limited amount of stormwater runoff based on statistical analysis of past rainfall events. In many areas, the sewers are 50 years old or even older. This will require large future investments to upgrade or renew the existing infrastructure system.
3. One of the most visible consequences of climate change in Germany is an increase in heavy rain events while the total amount of annual rainfall remains more or less constant. This intensifies the above-mentioned problems dramatically.

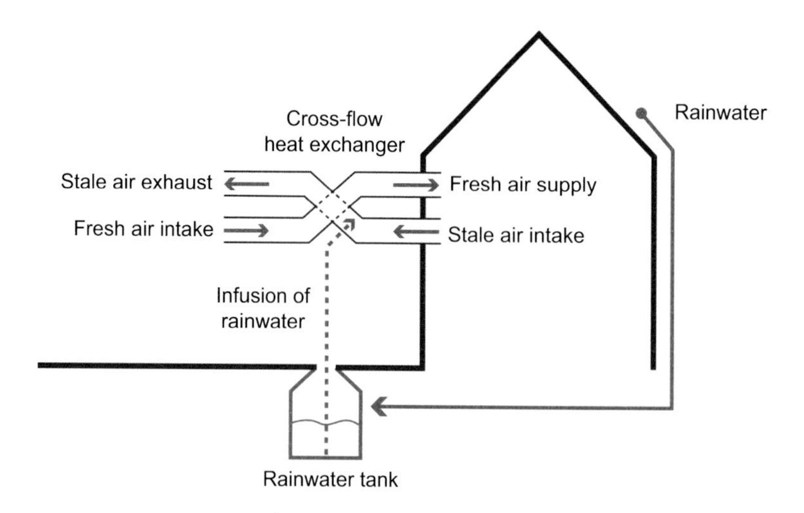

FIGURE 21.2 Adiabatic cooling with rainwater in the high school Gymnasium Frankfurt Riedberg, Frankfurt am Main. Top: Schematic showing how the adiabatic cooling is applied. Bottom: Picture of the high school. *From (Top) Gehrmann referred to Mall GmbH; (Bottom) Winker, 2018.*

In recent years, the designed capacity of the urban sewer systems for stormwater management was reached several times in a number of German cities. The vulnerability of these systems becomes clear when the city gets flooded as a result of the overloaded sewer system. This happened in Frankfurt/Main in 2016 after 3 days of rainfall where private houses, train stations, tunnels, and streets were flooded (Fig. 21.3).

4. Even those cities equipped with separate stormwater systems usually discharge their stormwater into the rivers, which are usually well removed from the place where the stormwater is generated. Hence, the runoff is carried away many kilometers and is not recharging the local aquifers. Falling groundwater levels, which contributes to the heat island effects (see Fig. 21.4), are the direct result, which significantly influence the microclimate of urban areas. Those "hot spots" can be seen on satellite images of dense urban areas. This also leads to the reduced night cooling of those places, and the number of hot nights per year increase (Fallmann et al., 2017; Trusilova et al., 2008; Beniston et al., 2007; Beniston, 2009).

One way to improve the microclimate within the city is to use the stormwater locally, i.e., infiltration, retention, irrigation. By improving green spaces, increasing the amount of free water surfaces, and unsealing urban surfaces, allows the city to begin "breathing" again. Green spaces, when designed properly, allow local infiltration of the stormwater, which in turn can improve the vegetation. Forests, parks, lakes, rivers, and features such as water retention playgrounds, fountains, or rainwater gardens not only influence the cities microclimate, but also attract the citizens and change their perception from gray to green. (e.g., Jim and Chen, 2006; netWORKS 4, 2017).

**FIGURE 21.3** Flooding of streets from overflowing combined sewers in Frankfurt/Main, Germany, after heavy rainfallin summer. Left: Gehrmann, 2016; right: Feuerwehr Frankfurt A.M., 2016. *From Left: Gehrmann, 2016; right: Feuerwehr Frankfurt A.M., 2016.*

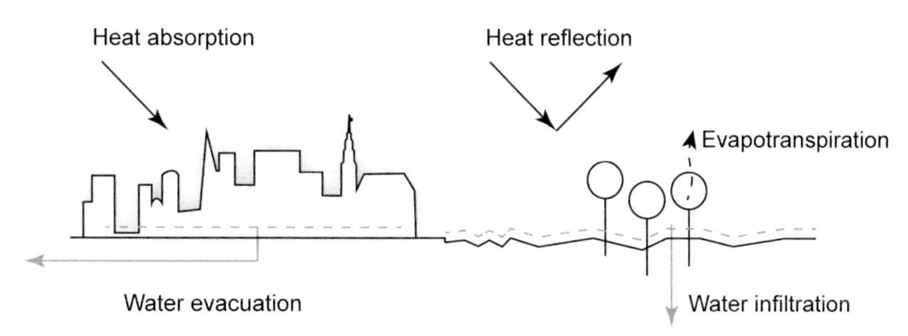


**FIGURE 21.4** Illustration of the heat island effect due to falling regional water tables. Capillary rise of water to source cooling evaporation at night is reduced. *Source: Gehrmann.*

From architectural and technical points of view, it is not a big challenge to maximise stormwater for those measures. Retention areas within communities offer possibilities to become playgrounds while cooling down the neighborhood at the same time. When they are designed to hold the water for several days, this effect might create a significant influence on the local climate and improve the livability of urban areas.

Although this can be realized easy within new designed projects, it is often difficult to implement these kinds of ecosystem services in dense urban areas such as Frankfurt. Retrofitting of those areas not only requires technical solutions, but also has to consider implementation costs, local policy, different stakeholders, and the environmental impacts, which makes it much more complicated than for new projects.

A specific German opportunity for new urban development in recent years is the disengagement of French and American military troops. Large military areas and/or brownfield sites are being considered for redevelopment for civil use. In this context numerous projects are taking place, which are intended to create a livelihood for thousands of citizens, everywhere in Germany. Those areas are usually polluted as a result of military action over decades. These sites were often build near the city boundaries, in some cases located close to groundwater protection zones, making the local infiltration of runoff impossible. Any use of stormwater not conveyed to the sewer (or a technical treatment plant) must be considered extremely carefully to protect the ground water sources. Hence these projects need special attention regarding their redevelopment (e.g., Mannheim Benjamin Franklin Village, Fliegerhorst Oldenburg) and the integration of stormwater to establish ecosystem services.

Pollution of the runoff also happens—but usually less intense—because of runoff from streets or garbage dumping sites in urban areas. Therefore, awareness of the different land uses generating runoff within the stormwater catchment is necessary to avoid harm to people, especially children. The most polluted part of the urban runoff, the first flush, usually contains most of the pollutants. This water should be diverted from the mainstream and can be stored in specifically designed wetlands or storage tanks before they receive appropriate treatment.

Diversion of the first flush in dense urban areas which are highly polluted (e.g., industrial zones, industry, train stations, etc.), might not be sufficient. In such cases, all the runoff should be considered as wastewater, and treated to the appropriate standard for the intended use.

However, local infiltration of detained/retained stormwater runoff—where possible—reduces the amount of stormwater that enters the combined/separated sewer network, and therefore reduces the pressure on the infrastructure, refills the local aquifers, and improves the microclimate.

## 21.2.2 Separation and diversification of urban wastewater streams

Novel urban water infrastructure includes new combinations of technical modules for the management of rain/stormwater, graywater, and blackwater, as well as heat and energy generation, partially from source-separated wastewater streams (see Schramm et al., 2017). Additionally, they generate new possibilities for greening and cooling the city. Moreover, novel urban innovations also allow for new urban decentralized water supply and wastewater discharge (Paris and Huber, 2016; see Fig. 21.5):

1. **Integration of a new system in existing infrastructure**. In this case, only parts of the wastewater (e.g., the graywater) are treated and reused locally, with the balance, including blackwater, discharged into the central network.
2. **Island system, islandisation**. In this case, the wastewater is treated directly at its place of production. The generated service water and irrigation water, as well as the recovered energy is used locally. Residues are treated locally or transported to a central treatment plant. The concept is a fully self-sufficient system.

These new options (Fig. 21.5) can be implemented on a range of scales, such as house, block, or quarter level. A typical quarter can range from 2000 to 5000 inhabitants in Germany, with up to 15,000 to 50,000 inhabitants in China. No general definition exists applicable all over the world. The authors define the house and block level as decentralized and the quarter

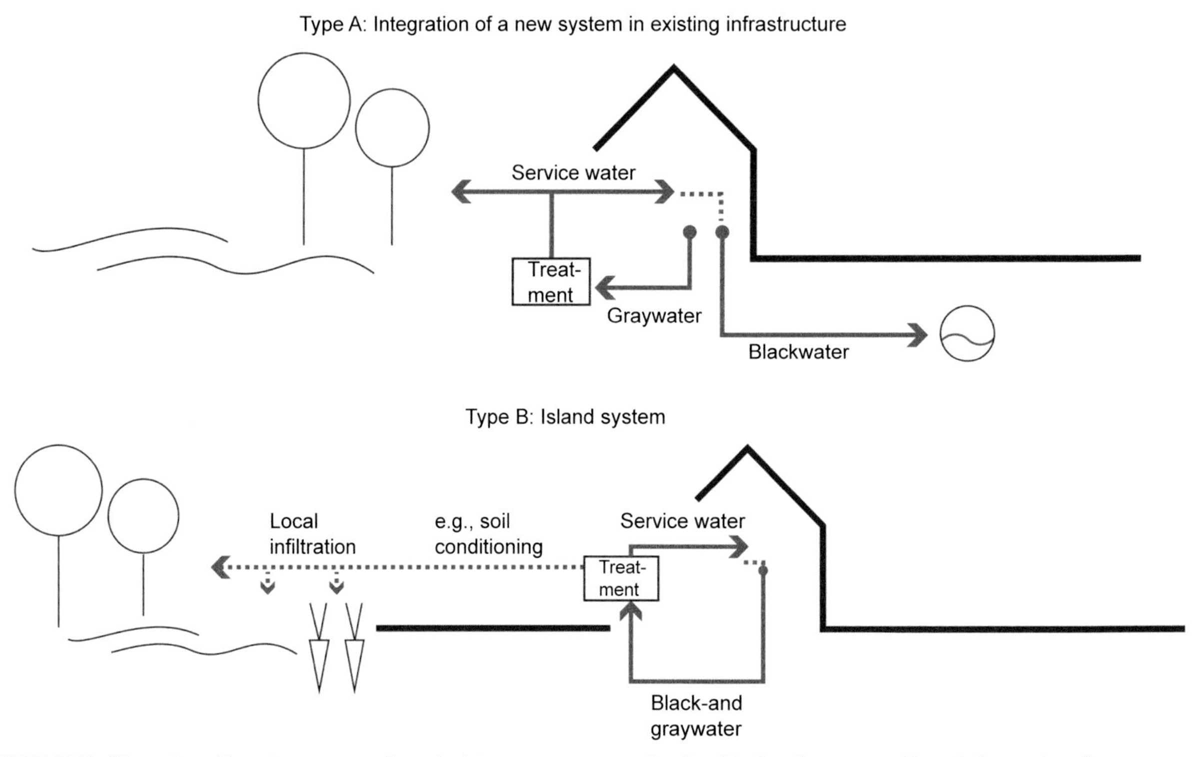


FIGURE 21.5 Illustration of how the two types of novel urban water systems can be placed in the urban context. Type A: Integration of a new system in existing infrastructure. Only the graywater is reused. Type B: Island system with a closed water cycle. *From Gehrmann, modified of Paris, S., Huber, H., 2016. Wasser wiederverwenden. In: Kluge, T., Schramm, E. (Eds.), Wasser 2050. Mehr Nachhaltigkeit durch Systemlösungen. oekom Verlag, Munich, pp. 211–224; Modified of Merten, N., Hackner, T., Meuler, S., 2011. Wastewater Management in Cities of the Future – One Size Does Not Fit All. Gwf International S1.*

level as semicentralized scale (Schramm et al., 2017). To provide a better understanding what this means in practice, some specific designs of novel urban water systems are presented in the following text. To provide a holistic view on the water streams in the innovative approaches, the handling of rainwater and stormwater (where relevant) is also described.

#### *21.2.2.1 Example 1: ConvGrey system graywater separation for energy recovery and water reuse*

This example fits the Type A in Fig. 21.5. Only slightly polluted graywater streams, which include wastewater from showers, basins, and washing machines, are collected separately from the blackwater and kitchen wastewater. The blackwater and kitchen wastewater are discharged to the conventional sewer, whilst waste heat from the graywater is recovered, and the graywater itself is treated for reuse (Fig. 21.6). The management of the graywater stream can be established on a house or block level, as well as at a central location for a whole quarter. The recovered energy is used for heating water (for showers, etc.), while the treated graywater can be used for indoor purposes such as toilet flushing, or for outdoor purposes such as garden and other green space irrigation or car washing.

The rainwater is collected and used locally for green roofs, etc., as well as infiltration measures. Runoff from streets is separated and treated in the local area. This water stream could be used for maintaining creek flows, which was isolated from its natural catchment.

The advantage of ConvGrey is that the direct interaction with the inhabitants, and behavior changes required of users are minor. Moreover, such concepts are already being discussed by investors and implemented at house/block level (Hefter et al., 2015). The required technical units are available on the market. The legal and institutional framing is relatively secure, and it is easy to receive a full overview as existing implementation on house/block level show (Kunkel et al., 2017; Kerber et al., 2016; Löw, 2011; Nolde, n.d).

Expanding this concept from the house/block scale to a whole quarter could happen relatively easily as a number of successful implementation examples already exist at the block level, e.g., in Frankfurt/Main (Winker et al., 2017) and Berlin (Hefter et al., 2015; Löw, 2011). These concepts are also discussed in other research projects.[2]

#### *21.2.2.2 Example 2: Hamburg Water Cycle*

The concept of the Hamburg Water Cycle (HWC; Fig. 21.7) intends to achieve a self-sufficient system (Type B in Fig. 21.5). Blackwater and graywater are collected and discharged separately at the quarter level. A vacuum system is used to transport blackwater and hence only requires small volumes of water for transport of the waste. Graywater is treated and used as service water (ie toilet flushing) or discharged into the environment. Energy is recovered from the graywater.

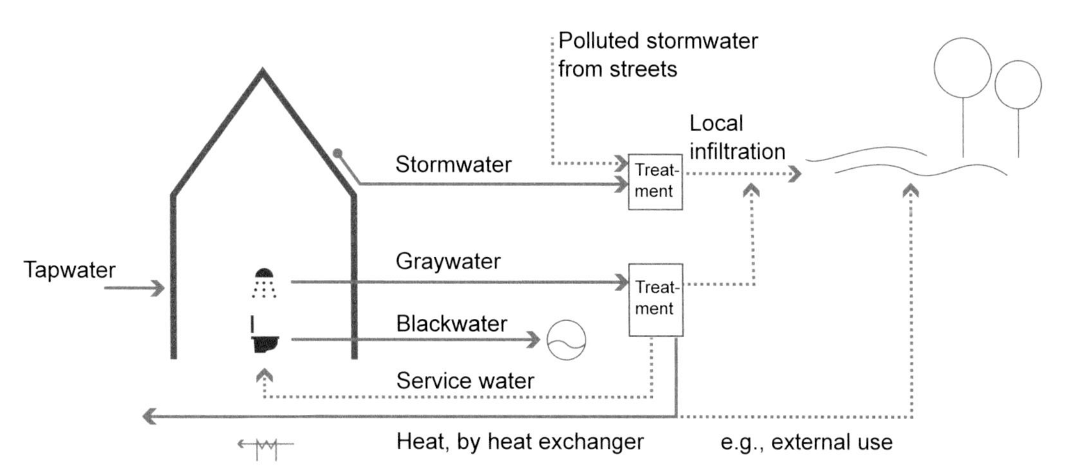


**FIGURE 21.6** Schematic of ConvGrey as designed jointly by researchers, planners, and environmental engineers for a quarter of the City of Frankfurt/Main, Germany. *From Gehrmann, referred to Davoudi, A., Milosevic, D., Scheidegger, R., Schramm, E., Winker, M., 2016. Stoffstromanalyse zu verschiedenen Wasserinfrastruktursystemen in Frankfurter und Hamburger Quartieren. netWORKS-Papers, 30. Deutsches Institut für Urbanistik, Difu, Berlin.*

2. http://www.twistplusplus.de/twist-de/index.php.

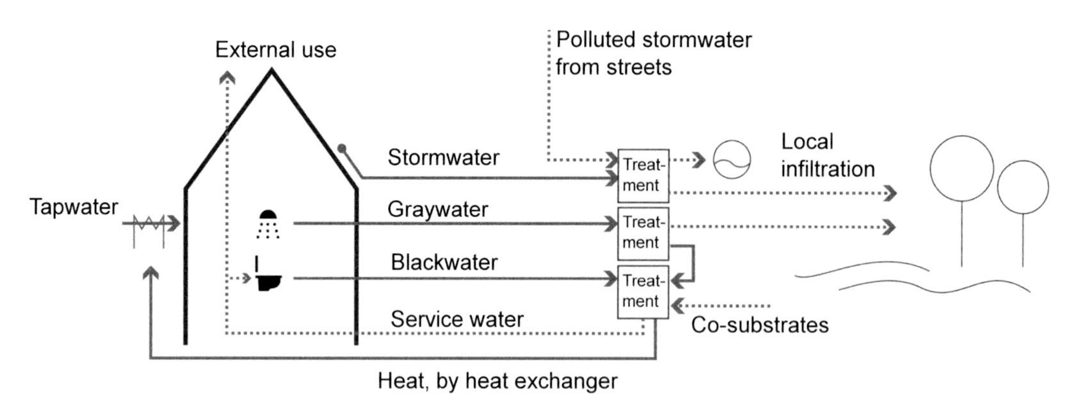


**FIGURE 21.7** Schematic of the Hamburg Water Cycle concept at urban quarter scale. *From Gehrmann, referred to Davoudi, A., Milosevic, D., Scheidegger, R., Schramm, E., Winker, M., 2016. Stoffstromanalyse zu verschiedenen Wasserinfrastruktursystemen in Frankfurter und Hamburger Quartieren. netWORKS-Papers, 30. Deutsches Institut für Urbanistik. Difu, Berlin.*

Blackwater is treated together with organic co-substrates (e.g., lawn cuttings and other organic waste) in an anaerobic process to generate biogas for energy production. The residues can be composted and used in agriculture. The liquid component of the treated blackwater goes into the graywater treatment (see Giese and Londong, 2016; Davoudi et al., 2016). This concept involves a complete disconnection of the rainwater and stormwater from the wastewater sewer and effectively closes the water cycle at a local level.

The water cycle also considers aspects of landscaping and urban planning: first, the stormwater runoff is led through open raceways, ditches, and cascades before it reaches a large stormwater retention basin designed to mimic a natural lake. This artificial pond strongly enhances the appearance and image of the quarter. The pond can provide cooling effects during daytime in summer. However, it is unclear if the lake provides warming effects during nighttime. At same time, it contributes to flood prevention in that the retention basin is sized to include storage capacity from heavy rainfall events.

The HWC concept has been implemented in Jenfelder Au, a new quarter in Hamburg, Germany, with more than 600 flats, and documented by an accompanying research project (Project KREIS, funded by the German Federal Ministry for Research and Education: see Giese and Londong, 2016 for details). The technical solution promises less treatment effort than the conventional system, a better use of reusables from the highly concentrated blackwater, better removal of pharmaceutical residues and pathogens, and hydraulic relief for the sewer system. The biogas production improves the independence of the quarter from energy imports and reduces its $CO_2$ footprint.

It was planned to use the treated graywater as environmental flows in a neighboring creek to restore habitat and fishway, but the water authority was apprehensive about water quality problems. Therefore, the full HWC concept has not been implemented. Instead, the collected graywater is transported in the sewers to the central water treatment plant.

### *21.2.2.3 Example 3: Semizentral concept with Resource Recovery Center*

The Semizentral concept also follows Type B water system and aims for self-sufficiency. The Semizentral pilot project was developed as a sustainable infrastructure solution for urban China. The implementation of this project in Qingdao is based on the separation of domestic graywater and blackwater on a community scale. Treatment occurs in one unit for the whole quarter, called a Resource Recovery Center (RRC). The unit comprises three technical modules: a graywater module, a blackwater module, and an energy module (Fig. 21.8), with the core wastewater treatment based on membranes. The treated graywater is reused as service water for domestic purposes (as toilet flushing in the hotels); the treated blackwater is reused for irrigation of urban green space and public cleaning (e.g., washing down streets). The residuals from the graywater and blackwater treatment are mixed with co-substrates/organic waste and fed into the energy module for biogas production. The residues (biosolids) from the anaerobic treatment are used as a soil conditioner after dewatering.

Stormwater is collected separately in Qingdao and discharged to the local lakes, creeks, and rivers. The use of rainwater, especially the catchment of stormwater, depends strongly on the local settings and climate conditions.

The Semizentral concept was developed to provide flexibility in infrastructure planning and modular construction for new urban areas in the fast growing cities of China. The ideal scale is around 50,000 inhabitants. The system is particularly attractive for areas facing water shortages, or dependency on high-cost drinking water produced from the desalination of seawater.

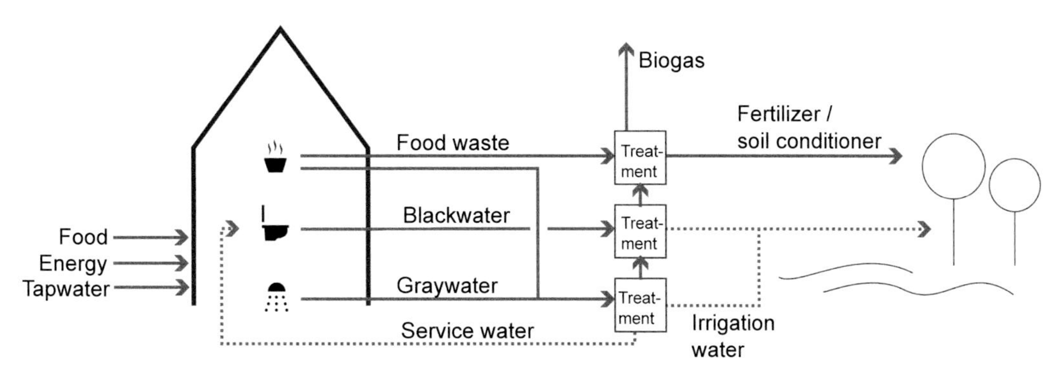


**FIGURE 21.8** Schematic of the Semizentral concept with Resource Recovery Center in China. *From Gehrmann, referred to Zimmermann, M., Winker, M., Schramm, E., 2018. Vulnerability analysis of critical infrastructures in the case of a semi-centralized water reuse system in Qingdao, China. International Journal of Critical Infrastructure Protection (submitted).*

The implementation phase is finalized and the system is fully operating. The RRC serves the World Horticulture Exhibition (WHE) village of the Qingdao, China, and was built in 2014. Domestic wastewater from 12,000 inhabitants comprising two residential areas, two hotels, and the WHE office area is collected and treated at the RRC (Tolksdorf et al., 2016; Zimmermann et al., 2018). Service water is only provided to one of the hotels for toilet flushing. Nevertheless, in Qingdao, the demand for the product water is very strong, especially in summer when the demand for irrigation water is substantially higher than the amounts produced.

## 21.2.3 Diversification of water types for urban water supply

In Central Europe, two types of water with different quality requirements are usually discussed: drinking water originating mainly from natural water sources such as groundwater or surface water; and service/irrigation water, which is usually sourced from rainwater/stormwater. In some areas of extreme water shortage, as it is the case in Israel (Gross et al., 2015), Spain (Domènech, 2011), or Australia (Barker et al., 2011), treated wastewater is often used for irrigation purposes.

In the long run, novel urban water systems not only allow a diversification of the wastewater system but also influence the provisioning system (Domènech, 2011). New sources for service and irrigation water are obtained. Treated graywater becomes a major driver. For example, in Germany, about 50% of the discharged wastewater is graywater in origin, which amounts to about 60 L/p/d (liters/person/day) (DWA, 2008). Usually the local graywater treatment systems (e.g., membranes, sequencing batch reactors, constructed wetlands) produce a water quality that fulfills the standards of the European Union bathing water directive (2006/7/EC). For details, see also Nolde (2005) and DWA (2017).

Novel urban water systems also support the reuse of water for cooling and greening of the city. The major advantages are as follows:

- Wastewater is not discharged to a remote central treatment location. By treating all or some wastewater streams locally or semicentrally in the city, the product water can readily be used as a new urban water, given it is much closer to the places of demand.
- The concept of source separation makes it easier to specifically design the water system according to the needs of the quarter for irrigation and cooling water, aesthetic purposes, and ecosystem services such as water features and green areas.
- Novel urban water systems provide new water sources by recycling the water already in use. This may relieve stressed traditional potable water sources from additional demands.
- Wastewater is available 24/7 with a consistent quality and quantity. This consistency will improve, if the treatment is on a larger scale, e.g., quarter.
- Wastewater can provide nutrients for plant fertilizing purposes, provided the treatment system is customized, e.g., treating graywater in a membrane unit, but not installing a phosphorus precipitation device. Wastewater (except rainwater) has to be treated in all cases. The level of treatment depends on the end uses; e.g., wastewater used for supplying urban waterbodies needs a much higher level of treatment than waste water used for irrigation of public green space or golf courses.
- Rainwater, stormwater, and service water can create new visible design elements such as rain gardens or designed retention areas, which can improve the green aesthetics of the urban space.

- Rainwater and service water are important resources for future water supply, and can be used for heating and cooling at the same time.
- Wastewater contains carbon or/and heat. Therefore, it can be used for heating (via digester gas or heat exchanger). All wastewater treatment systems are mainly below the ground level and, as such, act as heat or cold buffers. Last but not least, rainwater and service water can be used for cooling of buildings.

## 21.2.4 Interfaces and consequences for WSUD due to novel water infrastructures

A variety of components support the greening and cooling in a city. Nevertheless, how much those elements are self-sustaining depends very much on the characteristics of the specific location, e.g., during long periods without rainfalls. But even in areas with regular rainfall, supply might be difficult, as evaporation can be higher than rainfall. As an example, four cities are discussed where the authors collected experimental data on novel urban water systems showing four cases with very different framing conditions (Table 21.1).

When examining the water balance, it becomes clear that the water naturally available is not sufficient to support the green city areas nor use it directly for cooling. It is in these situations where service water and stormwater can play a large part (see Table 21.2).

As discussed previously it takes certain techniques to treat raw water according to its level of pollution before it can be used for green infrastructure (e.g., parks, green spaces, roof top greening) and blue infrastructures (e.g., waterbodies such as urban lakes). Table 21.3 provides an overview of which measures need to be taken to achieve water quality fit for purpose. It should be noted that certain measures not only provide the required treatment but also are the green infrastructure themselves, such as constructed wetlands or retention soil filters, as well as intermediate blue infrastructures such as stormwater reservoirs. This means they provide not only the respective "technical service" but also the services and characteristics of green (and blue) infrastructures related to greening and cooling.

By treating wastewater within the city and reusing it— whole reuse cascades are envisaged in advanced technical scenarios (Schramm et al., 2017)—the pressure on the natural water resources, as well as on water supply systems, can be reduced. This applies particularly during hot days of the year when the water demand is naturally high: larger demands of domestic water supply (e.g., for additional showering) coincide with the seasonal demands for irrigation water in a period where rainfall is low and groundwater levels are at their natural lower limit. Hence, application of wastewater can help to reduce the demand on the natural water resources. A first reuse option is to treat the domestic wastewater for reuse for toilet flushing, or as wash water for washing machines. However, this chapter has focused on reuse for the greening and cooling of cities.

An obvious application when it comes to using wastewater for greening and cooling the city is public open space irrigation. A number of (older) pilot projects for novel urban water systems using this end use have already been implemented. Examples include Block 6 in Berlin (Nolde, 2005); Lübeck-Flintenbreite in Germany (OtterWasser, 2009);

**TABLE 21.1 Overview of the Different Characteristics and Water Patterns Existing in Four Cities Chosen for Demonstration Purposes**

| City | Frankfurt/Main, Germany | Hamburg, Germany | Qingdao, China | Outapi, Namibia |
|---|---|---|---|---|
| Location within the country | In the center of Germany | Northern Germany, close to the North Sea | Eastern China on the coast | Northern Namibia, close to the border to Angola |
| Water usage (L/p/d) | 143 (Hessisches Statistisches Landesamt, 2012) | 111 (Hamburg Wasser, 2011; oral information) | 110 (City of Qingdao, 2013) | 26 (Informal settlements) 91 (Formal settlements) (Woltersdorf et al., 2017) |
| Annual rainfall (mm/year) | 642 (City of Frankfurt am Main, n.d.) | 750 (BSU, 2006) | 582 (City of Qingdao, 2013) | 472 (Sturm et al., 2009) |
| Annual evaporation potential (mm/year) | 532 (HydroTeamRC, 2017) | 653 (BfG, n.d.) | 1153 (HydroTeamRC, 2017) | 4296 (HydroTeamRC, 2017) |
| Water balance | Slightly positive: 50–100 mm (BfG, n.d.) | Positive: 200–300 mm (BfG, n.d.) | Negative | Negative |

**TABLE 21.2 Overview of the Components That Are Able to Green and Cool the City**

| Component | Application of service water | Water sources | | | | Heat recovery |
|---|---|---|---|---|---|---|
| | | Arid climate | | Humid climate | | |
| | | SW | RSW | SW | RSW | |
| Green roof | | | ○ | | ● | |
| Green facades | ✔ | ◐ | | ◑ | | |
| Living walls | ✔ | ◐ | | ◑ | | |
| Gardens, allotment gardens | ✔ | ◐ | | ◑ | | |
| Parks (any size) | ✔ | ◐ | | ◑ | | |
| Natural wastewater/rainwater treatment and infiltration units (for details see Table 27.3) | ✔ | ● | ○ | ● | ● | |
| Technical cooling | ✔ | ◐ | | ◑ | | ✔ |
| Heat storage | | ● | ○ | ● | ○ | ✔ |
| Storage of cold | | ● | ○ | ● | ● | (✔) recovery of cold |
| Lakes, ponds | | ◐ | | ◑ | | |
| Feeding of running waters | ✔ | ◐ | | ● | | |

*RWS*, rainwater and stormwater; *SW*, service water/recycled water; *black*, primary application; *white*, secondary additional application; *semicircles* were chosen when two options are possible (to complete the circle).
For Each Component, Information is Provided if Service Water Can Be Applied and Which Water Resources are Available Depending on the Specific Climate.

**TABLE 21.3 Components Supporting Green Infrastructure (Indirect Components)**

| Component | Resource | Purpose/Technical Service | Blue or Green Infrastructure |
|---|---|---|---|
| Source separation | Domestic wastewater | cf. Table 21.1; to separate the wastewater streams to receive graywater, which can be upgraded to service water | |
| Adequate engineered treatment of the wastewater streams | Domestic wastewater | Technical treatment units to, e.g., upgrade graywater to service water. Technical options are numerous | |
| Wastewater treatment close to nature | Domestic wastewater streams | Irrigation, domestic use, cooling, cleaning of streets, etc. | Green appearance, sometimes also water surfaces |
| Stormwater reservoir | RSW | Irrigation | Blue/green appearance possible, depending on design and water flow |
| Infiltration such as retention soil filter, etc. | RSW, SW | Improvement of green infrastructure | Green/blue appearance possible, depending on chosen design |
| Groundwater recharge | SW, RSW | Also an indirect component, as a sustaining/higher groundwater level supports vegetation, and by this, results in an improvement to green infrastructure | Green appearance possible |
| Feeding of watercourses | SW, RSW, other | Improvement of blue infrastructure; adiabatic cooling | Blue appearance |

*RSW*, rainwater and stormwater; *SW*, service water/recycled water.

FIGURE 21.9 Top: Graywater treatment with integrated courtyard design in Klosterenga, Oslo, Norway. Bottom left: Constructed wetland for treatment of graywater in the ecovillage of Flintenbreite, Lübeck, Germany. Bottom right: Ferrocement tank for rainwater harvesting in Epyeshona, Namibia. *From (Top) Ulrich, 2008; (Bottom left) Wendland, 2010; (Bottom right) CuveWaters.*

Klosterenga in Oslo, Norway (Jenssen, 2005, see also Fig. 21.9); and Semizentral/RCC in Qingdao, China (Tolksdorf et al., 2016, see also Fig. 21.10). Nevertheless, they were not established with urban green infrastructure as their major focus. Until recently, the link to urban green has been more or less a "collateral benefit" and largely dependent on the design focus of the planners, the spaces available, the technology selected, costs, and the evaluation of the potential health hazard resulting from contact by the community with the treated wastewater.

However, a change of perspective is happening. A number of research projects are currently investigating how service water can be applied via locally implemented wastewater treatment to support urban green projects (e.g., netWORKS 4, Kuras, Interes-I, Roof Water-Farm, Green4Cities[3]). The project Semizentral is looking proactively into this within an additional research activity. From the beginning of the project, it was planned that the treated blackwater would be used for irrigation purposes. What was originally conceived as a demonstration-scale perspective has evolved into a major driving force over the last two summers. In addition, Roof Water-Farm picks up this link by upgrading the existing system in Block 6 (Berlin) to include graywater and rainwater treatment for toilet flushing and irrigation of gardens, with a greenhouse and aquaponic system for rooftops (see footnote #3 for details, and Million et al., 2014).

By maintaining the existing green infrastructure and enhancing the establishment of additional infrastructure, direct cooling effects can be achieved. First of all, there is the shading effect of trees and bushes. Second, green spaces do not heat up and radiate heat as much as sealed areas such as roads, rooftops, and facades. Third, the water evaporation by plants

3. netWORKS 4: www.networks-group.de; KURAS: http://www.kuras-projekt.de/; Interes-I (internal information, preliminary project phase); Roof Water-Farm: http://www.roofwaterfarm.com/; Green4Cities: http://www.green4cities.com/.

**FIGURE 21.10** Top left: Recreation area in the top of the Resource Recovery Center (RRC) in Qingdao, China. The service water originates from graywater. Top right: Illegal garden next to the RRC. The garden is watered with service water. Bottom left: Water trucks being filled with service water in Qingdao. Bottom right: Use of treated wastewater at an agricultural irrigation site in Outapi, Namibia. *Source for all pictures from RRC/Qingdao: Winker, 2017; (Bottom right) CuveWaters.*

results in an active cooling effect. Hence, by implementing green elements, a direct effect on the microclimate can be achieved as green facades and green roofs result in a measurable temperature reduction (Thiele, 2015). Moreover, they provide a contribution toward aesthetics and quality of life in urban areas in the seasons of the year when natural green canopies can turn brown because of insufficient water supply and heat stress. The amenity value are extended in neighborhoods where such areas are preferred points for residents to spend free time and get together, especially during long and hot summer days (Mell et al., 2013; Wang et al., 2014).

Nutrients are also often provided with the irrigation water. This has to be considered when plants are selected and can also be used in a proactive way for food production, e.g., Roof Water-Farm and other applications where wastewater is used for agricultural irrigation (cf. Million et al., 2014). Here, a careful design is required between wastewater treatment, service water provision, and food production. Examples include avoiding production of crops eaten raw such as salads/carrots, and by planting crops where the harvested produce does not get in direct contact with the water, e.g., tomatoes, peppers, or egg plants (cf. UBA, 2017).

Last but not least, the wastewater treatment unit, or parts thereof, can function as a green or blue area itself. This can be the case when natural treatment elements, or elements close to natural systems, are implemented (see Table 21.3). A typical example is a constructed wetland planted with reeds, or a planted soil filter. New innovative elements can be also be considered, such as algae production on house walls[4] or intensive green walls that function as wastewater treatment and evaporation units at the same time.

Service water can also allow replenishment of waterbodies that have had their long-term sustainability adversely impacted, e.g., by urban expansion where small creeks become disconnected from their catchment and no longer receive

4. http://www.biq-wilhelmsburg.de/.

sufficient water. This means they need to be fed with, for example, pumped groundwater (as occurs at a location in Frankfurt/Main), or they become dry during certain times. By supplying them with service water, they can reach a better state of ecological health. Threshold values need to be defined for the service water discharge, taking into account the relevant chemical parameters and temperature in the seasonal cycle. In addition to the ecological benefit, costs of the original water supply can be saved.

Further options are available to support the cooling of the city with service water and relieve the pressure on natural resources. A major element indirectly linked to WSUD is the cooling of buildings, referred to here as "technical cooling." A shift in the highest energy demands has occurred from winter to summer seasons in Germany (AGFW, 2012). One reason is the intensive running of air conditioning and ventilation (Koch et al., 2017). Therefore, the construction of cooling networks (Kältenetze) with cogeneration of heat, cold, and electricity via environmental cooling (e.g., groundwater) comes into consideration. For example, the first modules of a storage tank for collected rainwater, which can be used as a cooling unit on a building, are now available on the market (see Fig. 21.2).

Moreover, the experiences of using service water supplied from graywater and blackwater in Qingdao (China) show the water can be used for cleaning and rinsing of streets—which also delivers a minor cooling effect in summer—and cleaning of waste bins. In addition, flushing of blocked wastewater channels, and other cleaning activities in cities, is now possible, where potable water was once required.

Further connections in a wider sense are also possible, e.g., for urban agriculture and gardening. These activities provide green space in the city. Private initiatives in commercial gardens are expanding (in addition to the traditional allotment gardens and urban large field agriculture). In addition, hydroponic and aquaponic system start-ups have occurred (Roof Water-Farm, TomatenFisch,[5] HypoWave[6]). These developments cause a certain level of greening in existing neighborhoods and changes them from gray to green infrastructure. Here, sufficient water supply and fertile soil/production sites are a major issue. By developing these approaches, not only are the greening and cooling benefits emphasized but also food production becomes closer to the consumer. People living in the city then develop a much more direct and emotional relationship to food production and agriculture.

Despite the potential benefits described above, certain restrictions have to be taken into account. Service water can differ in its source and required levels of treatment. Service water from wastewater usually contains higher nutrient loads than service water from rainwater and stormwater. This can be of advantage when it comes to irrigation (depending on the plants and their specific nutrient requirements). But it can also cause problems; e.g., algae grow when the water body is stagnant and the temperature increases. In addition, salt concentrations in the service water from wastewater can require specific treatment; e.g., its use for technical cooling requires precipitation of salts to avoid equipment failures. Rainwater and stormwater usually do not pose problems regarding nutrient loads or salts as, at least from a European perspective, the use of fertilizers for lawns is less common compared with other parts in the world. Nevertheless, stormwater from urban areas can be polluted with sediments, rubber particles, oil, petrol, sediment, heavy metals, and other material (for details, see Chapter 3). In that case, it is necessary to apply adequate treatment before its use as service water is possible.

Another aspect—which was indirectly discussed above—is the availability of the water sources. While the quantity of rainwater and stormwater available depends very much on the seasons and can vary greatly—for example, heavy rainfalls in summer—wastewater is continuously available. However, to use it effectively, certain preconditions for its capture (separate pipes) within the building have to be in place. Hence, although it is theoretically available everywhere in the city at any time, its practical availability needs to be assessed.

When it comes to the practical considerations, especially wastewater as a water source, the specific property rights are important to consider. In most cases, the wastewater is produced in private buildings or buildings owned by private entities. When the wastewater reaches a sewer, it is already mixed and becomes the responsibility of a public entity, such as a water utility. Hence, to apply such systems, both property rights and the coordination between private and public entities have to be considered.

Last but not least, systematically applying such transformations can cause unexpected consequences. For example, a more rational local use of rain/stormwater and wastewater within the city will result in their greening and cooling. On the other hand, limiting runoff and discharge of water from the city to the local waterbodies (excluding the planned augmentation of waterbodies with service water) might result in the smaller watercourses becoming dry in summer, as they are often fed to a large extent with the treated effluent from wastewater treatment plants and stormwater runoff. Hence, it is important to undertake a proper analysis of the respective urban water balance before specific measures are chosen, and established on a large scale using public funding.

---

5. http://www.tomatenfisch.igb-berlin.de/.
6. http://www.hypowave.de/: or http://www.isoe.de/en/projects/current-projects/wasserinfrastruktur-und-risikoanalysen/hypowave/.

# 21.3 CONSEQUENCES AND REQUIRED CHANGES IN THE CONTEXT OF URBAN DESIGN

## 21.3.1 Shift in infrastructural design: from gray to green

The increasing importance of nature-based solutions in flood protection and/or wastewater treatment (for details, see Tables 21.2 and 21.3) show the paradigm shift from gray to green infrastructure. Ecosystem services such as the cooling effects from vegetation reduces the dependence on engineered technology, and complements urban infrastructure. Nature-based solutions can do even more, as they offer the opportunity to gain an attractive public space and urban landscape by resource-sensitive design, which not only increases resilience, but also increases the quality of life for people. Good project examples start precisely at the point where technology is integrated into a holistic planning approach, centering on environmental design and public space, and ultimately on human needs.

## 21.3.2 Shift in planning: cross-sectoral perspective

The perspective of the specialist engineer is not enough in resource-sensitive design. There is a need for integrated solutions for a sustainable urban development due to the various disciplines and topics involved (for details, see Fig. 21.1). Modern approaches consider rainwater and stormwater as an integrated element of planning, independent of the scale. In recent years, planning consultants, such as Ramboll Studio Dreiseitl (a leading WSUD landscape designer in Europe, the USA, and Asia), integrated different disciplines, such as urban planners, engineers, and hydrologists, into their projects to achieve a design which answers not only architecture and landscape design but also flood protection, biodiversity, and livelihood. This approach shows the future direction of urban planning (for instance, Potsdamer Platz, Berlin; Bishan-Ang Mo Kio Park in Singapore). Projects that are planned from the beginning under consultation of different professions can help reestablish the perception of rainwater and wastewater back toward an important element of life. Moreover, these modern approaches include the consideration of ecosystem services and nature-based solutions. They have to take into account climate change and promote climate adaptation, community-based, and participatory planning.

### *21.3.2.1 Ecosystem services and nature-based solutions*

Gray, blue, and green infrastructures are currently the central elements of sustainable planning, in addition to long-term land management. Gray infrastructure encompasses the technical infrastructure below ground such as pipes and pumps. Both need to be reconciled in location-specific solutions that address predicted risks. At the same time, these solutions are designed to open up new ways to create livable urban landscapes. Especially the interaction between water management and urban design in WSUD concepts demonstrates how it is possible to use decentralized networks as part of the design of livable and sustainable cities.

The Copenhagen Strategic Flood Masterplan of 2013 demonstrated that water management can be highly attractive for a city and the design of its streets. This project can be seen as a new generation of "blue-green infrastructures" combining mobility, recreation, health, biodiversity, and economic viability.[7]

The Netherlands has a long experience with flood protection and is a pioneer in innovative projects. According to the "Waterplan 2" Masterplan, which was developed by the Municipality of Rotterdam in 2007, water retention and storage areas should be integrated into dense urban spaces as "Water Squares." These "water squares" usually provide a public open space, but can become large volume water collection basins in heavy rainfall events by collecting and detaining the runoff from its surroundings. One of the these projects, "The Water Square Benthemplein" in Rotterdam (see Fig. 21.11), designed by "De Urbanisten" and "Studio Marco Vermeulen" and finished in 2013, demonstrates how those kind of projects work: This low-lying basin relative to its surrounding environment is used as a football field, and its perimeter is equipped with large steps for seating. During dry weather conditions, this structure is used as a high-quality public space. During heavy rain events, this low-lying plaza becomes flooded with stormwater and acts as a detention basin with a capacity of 1700 $m^3$. By releasing the runoff slowly to the drainage system, the peak flows are reduced and the pressure on the existing infrastructure and waterbodies reduced significantly.[8]

7. http://www.landezine.com/index.php/2015/05/copenhagen-strategic-flood-masterplan.
8. http://www.urbanisten.nl/wp/?portfolio=waterplein-benthemplein.

FIGURE 21.11 The Water Square Benthemplein in Rotterdam, Holland is located in a low-lying area which gets flooded during heavy rain. It detains stormwater from the surrounding areas and releases the runoff slowly to the drainage system. *From Bolik, 2017.*

### *21.3.2.2 Climate change adaptation*

The issue of climate-sensitive planning to ensure adaptation and mitigation goals are reached is gaining importance at the level of urban planning. The aim is to develop actions plans for particularly affected areas (i.e., hot spots) by means of an overall urban view of the urban heat islands, while taking into account socioeconomic aspects and other factors that influence the vulnerability of different urban areas. This includes the typologies of settlement and urban open space, as well as demography, which can also change in parallel with climate change, and can either exacerbate or mitigate its effects (Karmann-Woessner, 2015). Without a major change in the urban planning of European cities, it is expected that the urban heat island effects will become worse in the coming years, especially in the context of climate change. Research by the city planning office of Karlsruhe (Germany) shows that today 14% of the urban area is affected by hot spots. It is expected that this fraction will increase up to 32% in 30 years, and up to 95% in 100 years (see Fig. 21.12). Greening and cooling the city in the WSUD context will play a major role in the adaptation of German cities to climate change.

With adaptation to climate change, all levels of government are concerned, from the United Nations and the European Union to the municipalities. In Germany, legislation has been enacted to ensure climate protection, and adaptation to climate change is incorporated into the *Federal Building Code* (*German*: 'Baugesetzbuch, BauGB'). One way to meet this requirement is to draw upon the urban planning framework as an informal planning tool for cities and municipalities. Parallel to federal initiatives, the states are developing their own adaptation or research activities: e.g., "climate change and

FIGURE 21.12 Hot spot affected areas in Karlsruhe Germany. 2010 (left), 2046 (middle), and 2090 (right). *From Gehrmann modified of Stadtplanungsamt Karlsruhe, Städtebaulicher Rahmenplan Klimaanpassung (2015).*[9]

9. More details regarding Karlsruhe are available: https://www.karlsruhe.de/b3/bauen/projekte/klimaanpassung.

consequences for the water cycle" (KLIWA),[10] "climate change – consequences, risks, and adaption" (KLARA),[11] and climate change and model-based adaption in Baden-Württemberg (KLIMOPASS),[12] to name a few since 2010. The position paper "Adaptation to Climate Change" by the Association of German Cities and Towns (DST, 2012) marked the beginning of strategic planning in climate change adaptation. The subject of climate adaptation is a cross-sectional task at all spatial levels. It is important to combine long-term strategic concepts on the formal and informal level with location-specific measures. Outside the municipal planning law, a compelling communication strategy is important for the implementation phase, because the majority of the individual measurements require the commitment of private owners.

Informed discussion in cities and municipalities is particularly important for the definition of planning goals, and for the sustained success of investments and project measures, as it is often necessary to negotiate "goal conflicts" in the planning process. One example is the balance between urban densification and climate protection. Internal redevelopment is an important development goal in cities to avoid further urban sprawl and to promote dense, compact urban structures. However, densification is in potential conflict with the goal of climate adaptation through the integration of urban open space and green structures in cities.

### *21.3.2.3 Climate sensitive planning*

Climatic problems, and therefore control requirements, arise in particular in the existing building stock. For this reason, the instruments in urban development in existing districts—with all competing demands—continue to gain importance in urban regeneration and urban planning. The German federal building code offers a "simplified procedure" to receive building permission (Article 3, BauGB), which is frequently used for urban development in existing districts. The benefit for this simplified procedure is that in accordance with the law (e.g., developing plans), and no environmental assessment or environmental report needs to be drawn up. However, for the future development, and to ensure that the German urban space can successfully tackle the challenges of climate change, this Article should be modified.

For the improvement of urban microclimate, the reduction of heat islands, and the retention of water, the following measures are useful at the different planning scales:

- For city areas, the maintenance and creation of cold-air areas and cold-air snowboards, as well as large open spaces and forest areas, is important. For heat-exposed areas, connected parks and green areas, the preservation of rivers and open waterbodies, as well as the creation of artificial water surfaces, and large-scale retention/detention areas are central design elements.
- For urban districts, decommissioning and unsealing impervious areas are considered to be consistent measures to avoid hot spots. Shading of parking lots and streets, public places, and buildings is highly recommended. Pocket parks and the greening of inner courtyards also create an attractive residential environment. The design of water surfaces in public spaces, e.g., water playgrounds, fountains, and medium-scale retention/detention basins, offers additional cooling effects. It is important that irrigation of the surrounding green areas should be linked in to the ecological design of all water surfaces.
- For the buildings, the greening of facades and rooftops and improved protection against summer heat load (e.g., shading structures) are regarded as central elements of heat reduction. Increasing the surface reflection (albedo) through light surfaces can also be used.

The selection of measures should be based on the criteria of effectiveness and local feasibility within the instruments of city planning. The measures and their effects can be validated by literature or quantified by modeling. The transferability of local measures, and thus their relevance, depends to a great extent on the type of urban space considered. The research project UrbanReNet (for details, see Fig. 21.13), was funded by the German Ministry of Economic Affairs and Energy in the years 2012–15. It analyzed the potentials of interconnecting different urban typologies in the context of their energy needs by considering active and passive measures (Hegger et al., 2012; Hegger and Dettmar et al., 2014).

The aim of climate adaptation is particularly challenging in cultural heritage buildings and requires respectful design of urban space and protective measures. Unsealing pavements, shading by large trees, and integrated water surfaces in the public space are proven measures in cooling the city. However, these approaches are in conflict with the requirements of monument protection, and the historical purpose of classical and baroque squares. One possible solution is a diffuse source of water mist from water bodies as shown in Place de la Bourse in Bordeaux, France (Fig. 21.14).

10. orig. title: "KLIWA - Klimaveränderung und Konsequenzen für die Wasserwirtschaft." www.kliwa.de.
11. orig. title: "KLARA - Klimawandel – Auswirkungen, Risiken, Anpassung." http://www4.lubw.baden-wuerttemberg.de/servlet/is/244207/.
12. orig. title: "KLIMOPASS - Klimawandel und modellhafte Anpassung in Baden-Württemberg." http://www4.lubw.baden-wuerttemberg.de/servlet/is/244199/.

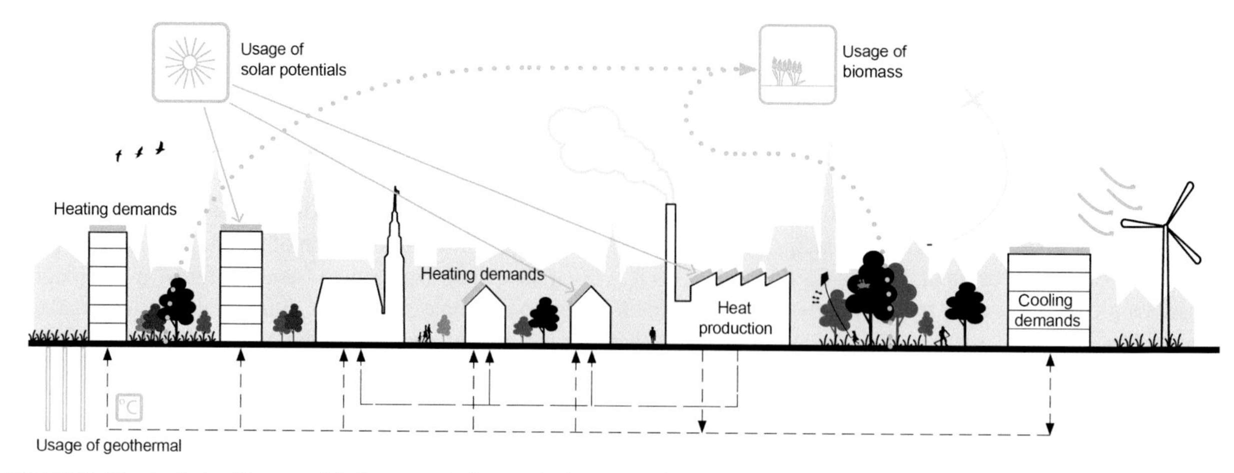


**FIGURE 21.13** Analysis of the potential of connected urban typologies to provide complementary energy needs. *From Hegger, M., Dettmar, J., Martin, A., Boczek, B., Drebes, C., Greiner, M., Hesse, U., Kern, T., Mahlke, D., Meinberg, T., Al Najjar, A., Schoch, C., Schulze, J., Sieber, S., Stute, V., Sylla, O., Wurzbacher, S., Zelmer, A., 2012. Eneff:Stadt | Forschungsprojekt "UrbanReNet" Vernetzte regenerative Energiekonzepte im Siedlungs- und Landschaftsraum - Schlussbericht (Final report), Darmstadt, Germany.*

**FIGURE 21.14** Place de la Bourse in Bordeaux, France. *Source: Padberg.*

### 21.3.2.4 The role of landscape architecture

To face the future challenges, given the expected intensification of heavy rainfall events in winter and less rain in summer (BMBF, 2014), it is necessary to develop a holistic framework that involves all the affected disciplines. For decades, architects in Europe have optimized the buildings from an energy perspective and, in recent years, added the carbon dioxide equivalent to their designs to assess the rating of their green buildings. The building itself has already reached a point of optimization with little opportunity for significant gains. Engineers working on sustainable infrastructure combine with landscape architects designing green spaces, which includes green farming, urban parks, and highly livable spaces, to produce a healthy environment. In practice, projects where all the planning disciplines are working together right from the beginning are still rare. From an architectural point of view, it is already possible to create zero energy buildings which are essentially independent from the urban environment. Rainwater as well as wastewater, are indifferent to buildings' boundaries. It might impact the building's roof or facade, which might be designed "green," and then continues its way through the roof drainage system into the drainage infrastructure.

Urban planners, architects, engineers, landscape architects, and hydrologists should work together to develop holistic design approaches. Landscape architecture offers a powerful instrument to ensure flood protection beyond the gray infrastructure located underneath the urban areas. Green infrastructure in the context of WSUD is an element that can be easily integrated into buildings. Architects, and engineers, and landscape architects can utilize synergistic effects to create an optimized community.

Green spaces allow the stormwater to infiltrate to the soil. Vegetated areas can improve the local climate. Buildings surrounded by green areas needs less cooling energy. This is of particular interest in the southern parts of Europe with hot dry summers. By considering green spaces as an important design element, landscape design becomes a major focus of interest. And to keep green areas green during long and hot summer months, collected rainwater and treated wastewater streams are obvious irrigation sources.

When addressing flood protection issues, designing green spaces can easily help to improve both water conveyance and local infiltration. Stormwater swales, for example, can be designed as broad hollows or channels, which can be used for recreational activities, or playgrounds, during dry weather conditions. These types of swales can easily be designed to handle high stormwater flows during the rainy season. This reduces significantly the need for gray infrastructure and the pressure on the centralized treatment facilities, as they become hydraulically more independent from stormwater. Nearly all projects that are in the planning phase can consider local infiltration of the stormwater by using rain gardens, detention/retention ponds, or existing local waterbodies (for details, see Tables 21.2 and 21.3). Engineers then can help to make the link to service water supplies.

## 21.4 CONCLUSION

Integrated planning concepts emphasize the opportunities that arise in the design of livable urban environments. The great opportunity for our cities and their quality of life lies in WSUD and its combination with engineering services and ecosystem services: this conclusion is strongly supported from the pilot projects we have reviewed.

However, the revision of our concepts for urban space and landscape needs further conceptual work. Climate change in particular is bringing new challenges to the European cities. Innovative concepts, which include stormwater management as shown for the cities of Rotterdam, Karlsruhe, and Frankfurt in the chapter, can are now being implemented in most new European urban developments. By implementing novel urban water systems linked to WSUD, the toolkit for WSUD can be extended. This applies to arid areas where rainwater is only available in certain seasons, and which would require huge storage tanks to provide sufficient water for those dry periods. The portfolio of urban water resources is broadened, and applies to both semiarid and semihumid climates. The components and measures we have described allow increasing the amount of available urban water.

It is necessary to distinguish local water sources (produced and consumed onsite or at the quarter level) and consider their suitability regarding water quality and quantity over time. For example, it is possible to mine sewers to produce service water for local nonpotable end uses. But under a WSUD perspective, other uses are added. Those include irrigation of parks, lawns or urban gardens, infiltration into the local aquifer, implementation of green roofs, living walls, or algae growth facades. This provides green and cool urban areas directly or indirectly without relying on additional water supply from the surrounding areas. In cases where nature-based solutions such as constructed wetlands or retention soil filters are possible, the wastewater treatment unit itself can function directly as the green area or cooling unit.

The authors are confident that when following this water-sensitive path, many more options and measures will be identified. However, it is important to identify synergistic effects across technical discipline areas to motivate the respective players to undertake action when designing/redesigning urban spaces.

## REFERENCES

AGFW (Ed.), 2012. Schnittstelle Stadtentwicklung und technische Infrastrukturplanung. Ein Leitfaden von der Praxis für die Praxis. Der Energieeffizienzverband für Wärme, Kälte und KWK e.V. (AGFW), Frankfurt am Main, Germany.

Barker, F., Faggian, R., Hamilton, A.J., 2011. A history of wastewater irrigation in Melbourne, Australia. Journal of Water Sustainability 1 (2), 31–50.

Barton, A., Argue, J., 2007. A review of the application of water sensitive urban design (WSUD) to residential development in Australia. Australasian Journal of Water Resources 11 (1).

BauGB: Article 3. Baugesetzbuch (BauGB), Bunderepublik Deutschland October 2017. https://www.gesetze-im-internet.de/bbaug/__5.html.

Beniston, M., 2009. Trends in joint quantiles of temperature and precipitation in Europe since 1901 and projected for 2100. Geophysical Research Letters 36 (7).

Beniston, M., Stephenson, D., Christensen, O., Ferro, C., Frei, C., Goyette, S., Halsnaes, K., Holt, T., Jylhoenen, K., Koffi, B., Palutikof, J., Schoell, R., Semmler, T., Woth, K., 2007. Future extreme events in European climate: an exploration of regional climate model projections. Climate Change 81 (1), 71–95.

BfG, n.d. Hydrologischer Atlas Deutschland. German Federal Institute of Hydrology (BfG). http://geoportal.bafg.de/mapapps/resources/apps/HAD/index.html?lang=de.

BMBF, 2014. Strategien für die Wasserwirtschaft im Zeichen des klimatischen und demographischen Wandels 2014. German Federal Ministry for Education and Research (BMBF). http://www.oowv.de/fileadmin/user_upload/oowv/content_pdf/nawak/BMBF-Projektblatt-INIS-NAWAK.pdf.

BSU, 2006. Dezentrale naturnahe Regenwasserbewirtschaftung. Agency for Urban Development and Environment, Freie und Hansestadt Hamburg (BSU), Hamburg.

Chan, S.Y., Chau, C.K., Philipp, C., Chan, E.H.W., Yung, H.K., 2017. On the study of shading effect of different paving materials inside a park. In: World Sustainable Built Environment Conference (WSBE) 2017 Hong Kong, Hong Kong, 5-7 June 2017, pp. 1354–1361.

City of Frankfurt am Main, n.d. Statistisches Porträt Frankfurt am Main 2012. http://www.frankfurt.de/sixcms/detail.php?id=2811&_ffmpar%5b_id_inhalt%5d=7526.

City of Qingdao, 2013. Statistical Yearbook of Qingdao for 2012. China.

Davoudi, A., Milosevic, D., Scheidegger, R., Schramm, E., Winker, M., 2016. Stoffstromanalyse zu verschiedenen Wasserinfrastruktursystemen in Frankfurter und Hamburger Quartieren. netWORKS-Papers, 30. Deutsches Institut für Urbanistik Difu, Berlin.

Dingle, T., 2008. The life and times of the Chadwickian solution. In: Troy, P. (Ed.), Troubled Waters: Confronting the Water Crisis in Australia's Cities. ANU E Press, Canberra, ACT Australia, pp. 7–18.

Domènech, L., 2011. Rethinking water management: from centralised to decentralised water supply and sanitation models. Documents D'Analisi Geografica 57.2, 293–310.

DST, 2012. Positionspapier Anpassung an den Klimawandel, Empfehlungen und Maßnahmen der Städte. Association of German Cities and Towns (DST), Berlin. http://www.staedtetag.de/imperia/md/content/dst/positionspapier_klimawandel_juni_2012.pdf.

DWA, 2008. Neuartige Sanitärsysteme (NASS), DWA-Themenband. German Association for Water, Wastewater and Waste e.V. (DWA), Hennef, Germany.

DWA, 2017. Merkblatt DWA-M 277. Hinweise zur Auslegung von Anlagen zur Behandlung und Nutzung von Grauwasser und Grauwasserteilströmen. German Association for Water, Wastewater and Waste e.V. (DWA), Hennef, Germany.

Fallmann, J., Wagner, S., Emeis, S., 2017. High resolution climate projections to assess the future vulnerability of European urban areas to climatological extreme events. Theoretical and Applied Climatology 127 (3–4), 667–683.

Gasperi, J., Garnaud, S., Rocher, V., Moilleron, R., 2008. Priority pollutants in wastewater and combined sewer overflow. The Science of the Total Environment 407 (1), 263–272.

Geels, F.W., 2005. Technological Transitions and System Innovations: A Co-evolutionary and Socio-Technical Analysis. Edward Elgar Publishing, Cheltenham.

Gervin, L., Brix, H., 2001. Removal of nutrients from combined sewer overflows and lake water in a vertical-flow constructed wetland system. Water Science and Technology 44 (11–12), 171–176.

Giese, T., Londong, J. (Eds.), 2016. Kopplung von regenerativer Energiegewinnung mit innovativer Stadtentwässerung - Synthesebericht zum Verbundforschungsvorhaben KREIS. Schriftenreihe des b.is, Vol. 30, Berlin.

Gleick, P.H., 2003. Water use. Annual Review of Environment and Resources 28 (1), 275–314.

OtterWasser GmbH, 2009. Ecological Housing Estate, Flintenbreite, Lübeck, Germany – Draft. Case Study of Sustainable Sanitation Projects. Sustainable Sanitation Alliance (SuSanA). http://www.susana.org/en/resources/case-studies/details/59.

Gross, A., Maimon, A., Alfiya, Y., Friedler, E., 2015. Greywater Reuse. Boca Raton, London, New York.

Hefter, T., Birzle-Harder, B., Deffner, J., 2015. Akzeptanz von Grauwasserbehandlung und Wärmerückgewinnung im Wohnungsbau. Ergebnisse einer qualitativen Bewohnerbefragung. netWORKS-Paper, 27. Deutsches Institut für Urbanistik Difu, Berlin.

Hegger, M., Dettmar, J. (Eds.), 2014. Energetische Stadtraumtypen Strukturelle und energetische Kennwerte von Stadträumen. Fraunhofer IRB Verlag.

Hegger, M., Dettmar, J., Martin, A., Boczek, B., Drebes, C., Greiner, M., Hesse, U., Kern, T., Mahlke, D., Meinberg, T., Al Najjar, A., Schoch, C., Schulze, J., Sieber, S., Stute, V., Sylla, O., Wurzbacher, S., Zelmer, A., 2012. Eneff:Stadt | Forschungsprojekt "UrbanReNet" Vernetzte regenerative Energiekonzepte im Siedlungs- und Landschaftsraum - Schlussbericht (Final report), Darmstadt, Germany.

Hessisches Statistisches Landesamt, 2012. Statistische Berichte. https://www.destatis.de/GPStatistik/servlets/MCRFileNodeServlet/HEHeft_derivate_00001260/QI1-3j10.pdf;jsessionid=B9D93EAE80AE296617E79DB4FFDBB162.

Hodge, T.A., 2002. Roman Aqueducts and Water Supply. Gerald Duckworth & Co. Ltd., London.

Hoyer, J., Dickhaut, W., Kronawitter, L., Weber, B., 2011. Water Sensitive Urban Design – Principles and Inspiration for Sustainable Stormwater Management in the City of the Future. Jovial Verlag, Hamburg.

Huntington, T., 2006. Evidence for intensification of the global water cycle: review and synthesis. Journal of Hydrology 319, 83–95.

Hvitved-Jacobsen, T., Vollertsen, J., Nielsen, A.H., 2010. Urban and highway stormwater pollution. CRC Press Taylor and Francis Group, Boca Raton.

HydroTeamRC, 2017. Evapotranspiration Web Viewer. Esri, USGS, NOAA, University of Montana, USA. http://www.arcgis.com/apps/OnePane/main/index.html?appid=b1a0c03f04994a36b93271b0c39e6c0f.

Jenssen, P.D., 2005. Decentralised urban greywater treatment at Klosterenga Oslo. In: von Bohemen, H. (Ed.), Ecological Engineering - Bridging between Ecology and Civil Engineering. Æneas Technical Publishers, The Netherlands, pp. 84–86. http://www.susana.org/en/resources/library/details/248.

Jim, C.Y., Chen, W.Y., 2006. Perception and attitude of residents toward urban green spaces in Guangzhou (China). Environmental Management 38 (3), 338–349.

Jin, M., Shepherd, J.M., King, M., 2005. Urban aerosols and their variations with clouds and rainfall: a case study for New York and Houston. Journal of Geophysical Research: Atmospheres 110 (D10), 1–12.

Karmann-Woessner, A., 2015. Ziele und Grenzen der Klimaanpassung in der kommunalen Praxis am Beispiel Karlsruhe. In: General Meeting oft he German Academy for Urban Planning (DASL), 20 November 2015, Rheinfelden.

Kerber, H., Schramm, E., Winker, M., 2016. Transformationsrisiken bearbeiten: Umsetzung differenzierter Wasserinfrastruktursysteme durch Kooperation. netWORKS-Papers, 28. Deutsches Institut für Urbanistik Difu, Berlin.

Kjelgren, R., Montague, T., 1998. Urban tree transpiration over turf and asphalt surfaces. Atmospheric Environment 32 (1), 35–41.

Kluge, T., Schramm, E., 2016. Wasser 2050. Mehr Nachhaltigkeit durch. Systemlösungen oekom Verlag, Munich.

Koch, M., Hesse, T., Kenkmann, T., Bürger, V., Haller, M., Heinemann, C., Vogel, M., Bauknecht, D., Flachsbarth, F., Winger, C., Wimmer, D., Rausch, L., Hermann, H., Stieß, I., Birzler-Harder, B., Kunkis, M., Tambke, J., 2017. Einbindung des Wärme- und Kältesektors in das Strommarktmodell PowerFlex zur Analyse sektorübergreifender Effekte auf Klimaschutzziele und EE-Integration. Final report of FKZ 0325708, funded by BMWi.

Kunkel, S., Utesch, B., Winker, M., Felmeden, J., 2017. Wärmerückgewinnung und Betriebswassernutzung - Umsetzung einer Systemalternative in Frankfurt a. M. In: Winker, M., Trapp, J., Libbe, J., Schramm, E.E. (Eds.), Wasserinfrastruktur: Den Wandel gestalten. Technische Varianten, räumliche Potenziale, institutionelle Spielräume. Edition Difu - Stadt Forschung Praxis, 16. Difu, Berlin, pp. 99–115.

Langergraber, G., Muellegger, E., 2005. Ecological sanitation – a way to solve global sanitation problems? Environment International 31 (3), 433–444.

Larsen, T.A., Alder, A.C., Eggen, R.I.L., Maurer, M., Lienert, J., 2009. Source separation: will we see a paradigm shift in wastewater handling? Environmental Science and Technology 43 (16), 6121–6125.

Libbe, J., Schramm, E., Winker, M., Deffner, J., 2017. Integrierte infrastrukturplanung. In: Winker, M., Trapp, J.H., Libbe, J., Schramm, E. (Eds.), Wasserinfrastruktur: Den Wandel gestalten. Technische Varianten, räumliche Potenziale, institutionelle Spielräume. Edition Difu – Stadt Forschung Praxis 16. Difu, Berlin, pp. 81–90.

Löw, K., 2011. An Innovative Greywater Treatment System for Urban Areas – International Transferability of a German Approach, Installed in GIZ's Headquarters in Eschborn (Master thesis). HFWU – Hochschule für Wirtschaft und Umwelt Nürtingen-Geislingen, Germany.

McCully, P., 1996. Silenced Rivers: The Ecology and Politics of Large Dams. Zed Books, London.

Mell, I.C., Henneberry, J., Hehl-Lange, S., Keskin, B., 2013. Promoting urban greening: valuing the development of green infrastructure investments in the urban core of Manchester, UK. Urban Forestry and Urban Greening 12 (3), 296–306.

Merten, N., Hackner, T., Meuler, S., 2011. Wastewater Management in Cities of the Future – One Size Does Not Fit All. Gwf International S1.

Million, A., Bürgow, G., Steglich, A., Raber, W., 2014. Roof water-farm. Participatory and multifunctional infrastructures for urban neighborhoods. In: Roggema, R., Keffe, G. (Eds.), Proceedings 6th AESOP Sustainable Food Planning Conference. VHL, Velp, pp. 659–678.

netWORKS 4, 2017. Resilient networks: Beiträge von städtischen Versorgungssystemen zur Klimagerechtigkeit. https://networks-group.de/de/networks-4/das-projekt.html.

Nolde, E., 2005. Greywater recycling systems in Germany – results, experiences and guidelines. Water Science and Technology 51 (10), 203–210.

Nolde, E., no date. Die klima-positive Recyclinganlage. http://nolde-partner.de/system/files/final_web.pdf.

Otterpohl, R., 2001. Stand der Technik und Entwicklungen für den urbanen Bereich. In: Wilderer, P.A., Paris, S., Wiesner, J. (Eds.), DESAR Kleine Kläranlagen und Wasserwiederverwendung. Technische Universität München, Berichte aus Wassergüte- und Abfallwirtschaft, München, pp. 23–41.

Otterpohl, R., Albold, A., Oldenburg, M., 1999. Source control in urban sanitation and waste management: ten systems with reuse of resources. Water Science and Technology 39 (5), 153–160.

Paris, S., Huber, H., 2016. Wasser wiederverwenden. In: Kluge, T., Schramm, E. (Eds.), Wasser 2050. Mehr Nachhaltigkeit durch Systemlösungen. oekom Verlag, Munich, pp. 211–224.

Ren, G., Chu, Y., Zhou, J., Zhang, A., Guo, J., Liu, X., 2008. Urbanization effects on observed surface air temperature in North China. Journal of Climate 21, 1333–1348.

Sauri, D., Del Moral, L., 2001. Recent developments in Spanish water policy. Alternatives and conflicts at the end of the hydraulic age. Geoforum 32 (3), 351–362.

Schramm, E., Kerber, H., Trapp, J.H., Zimmermann, M., Winker, M., 2017. Novel urban water systems in Germany: governance structures to encourage transformation. Urban Water Journal. https://www.tandfonline.com/doi/full/10.1080/1573062X.2017.1293694.

Schueler, T.R., 2000. The Importance of Imperviousness. Reprinted in the Practice of Watershed Protection. Center for Watershed Protection, Ellicott City, MD.

Shepherd, J.M., 2005. A review of current investigations uf urban-induced rainfall and recommendations for the future. Earth Interactions 9 (12), 1–27.

Shepherd, J.M., Burian, S.J., 2003. Detection of urban-induced rainfall anomalies in a major coastal city. Earth Interactions 7 (4), 1–17.

Sturm, M., Zimmermann, M., Schütz, K., Urban, W., Hartung, H., 2009. Rainwater harvesting as an alternative water resource in rural sites in central Northern Namibia. Physics and Chemistry of the Earth 34, 776–785.

Thiele, M., 2015. Klimaschutzpotenzialanalyse von Dach-, Fassaden- und Straßenbaumbegrünung (Master thesis). Hochschule für Nachhaltige Entwicklung Eberswalde, Germany.

Tolksdorf, J., Lu, D., Cornel, P., 2016. First implementation of a SEMIZENTRAL resource recovery center. Journal of Water Reuse and Desalination 6 (4), 466–475.

Trusilova, K., Jung, M., Churkina, G., Karstens, U., Heimann, M., Claussen, M., 2008. Urbanization impacts on the climate in Europe: numericalexperiments by the PSUG NCAR mesoscale model (MM5). Journal of Applied Meteorology and Climatology 47 (5), 1442–1455.

UBA, 2017. Recommendations for deriving EU minimum quality requirements for water reuse. In: Scientific Opinion Paper. German Environmental Agency (UBA), Dessau.

USEPA, 2005. National Management Measures to Control Nonpoint Source Pollution from Urban Areas. US Environmental Protection Agency (USEPA).

Wang, Y., Bakker, F., De Groot, R., Wörtche, H., 2014. Effect of ecosystem services provided by urban green infrastructure on indoor environment: a literature review. Building and Environment 77, 88–100.

Wilderer, P., 2007. Zentrale vs. dezentrale Abwasserbehandlung – für und wider. 82. Darmstädter Seminar. Abwassertechnik, Darmstadt.

Winker, M., Trapp, J.H., Libbe, J., Schramm, E. (Eds.), 2017. Wasserinfrastruktur: Den Wandel gestalten. Technische Varianten, räumliche Potenziale, institutionelle Spielräume. Edition Difu – Stadt Forschung Praxis, 16. Difu, Berlin.

Woltersdorf, L., Zimmermann, M., Deffner, J., Gerlach, M., Liehr, S., 2017. Benefits of an Integrated Water and Nutrient Reuse System for Urban Areas in Semi-Arid Developing Countries. Resources, Conservation and Recycling. https://doi.org/10.1016/j.resconrec.2016.11.019.

Zhao, Z.-C., 2011. Impacts of Urbanization on Climate Change, 10,000 Scientific Difficult Problems: Earth Science (In Chinese), 10,000 Scientific Difficult Problems Earth Science Committee. Science Press, pp. 843–846.

Zimmermann, M., Winker, M., Schramm, E., 2018. Vulnerability analysis of critical infrastructures in the case of a semi-centralized water reuse system in Qingdao, China. International Journal of Critical Infrastructure Protection (submitted).

Chapter 22

# WSUD Asset Management Operation and Maintenance

Jack Mullaly

*Ideanthro, PO Box 386, Sherwood, QLD, Australia*

## Chapter Outline



**ABSTRACT**

As the stormwater components of water sensitive urban design have gained traction in Australia, large numbers of stormwater control measures (SCMs) have been constructed. Examples of SCMs include trash racks, bioretention systems, and swales. SCMs must be appropriately managed if they are to function as designed. As a reasonably new asset class, management of SCMs around Australia is far from perfect. However, many local governments, and some private sector asset owners, have made considerable progress in developing processes and practices to manage and maintain their SCMs. They have done this by: (1) locating existing assets; (2) determining the condition of existing assets; (3) determining which department should be responsible for maintenance; (4) building a business case; (5) determining if on-ground work should be undertaken by contractors or in-house crews; (6) briefing upward to management; and (7) continually improving. Despite efforts to appropriately manage SCMs, large challenges exist. These will need to be addressed and SCM asset management will need to become mainstream if SCMs are to be appropriately managed. This chapter describes these challenges, operation and maintenance requirements, and a process to develop a mechanism for ongoing operation and maintenance of SCMs.

**Keywords:** Asset management; Asset register; Bioretention system; Cost efficiency; Local government; Operation and maintenance; Stormwater control measure; Water sensitive urban design.

## 22.1 INTRODUCTION

Since its beginnings in the 1990s, water sensitive urban design (WSUD) has grown from an initial concept to a practice that is applied broadly across Australia, and around the world. Although only a small proportion of the urban form has adopted WSUD measures to date, its implementation nonetheless covers a large geographic area. As WSUD has been applied more widely, it has given rise to an entirely new type of infrastructure asset, the WSUD asset. Like all other forms of infrastructure, these WSUD assets must be appropriately managed throughout their life cycle if they are to deliver the outcomes for which they were installed.

The growth of WSUD has not occurred across all the water cycle evenly nor in urban design and landscape architecture. To date, WSUD has gained most traction in the stormwater sphere (Fletcher et al., 2015). As a result, it is within this niche that most WSUD assets have been constructed. Stormwater assets constructed in accordance with WSUD principles are known collectively as stormwater control measures (SCMs), although a wide array of other terms (e.g., stormwater quality improvement devices and best management practices) are also used (Fletcher et al., 2015). However, SCMs are not simply another type of drainage asset. Many types of SCM incorporate a landscaped or ecological element, not seen in traditional drainage assets. These types of asset require markedly different operation, maintenance, and renewal approaches if they are to function as intended.

In this chapter, we will explain what it takes to appropriately manage SCMs. Our exploration of asset management in this chapter will

- describe the types of SCMs;
- present typical operation and maintenance requirements for SCMs;
- address what is currently known about the useful operational life of SCMs;
- outline the challenges associated with implementing appropriate asset management for SCMs;
- present a protocol for establishing an initial, bare-bones SCM asset management system in an organization that has not previously managed this type of asset; and
- examine emerging trends in SCM asset management.

### 22.1.1 The components of good asset management

Asset management, as defined by the Asset Management Council of Engineers Australia (2014), is *the lifecycle management of physical assets to achieve the stated outputs of the enterprise.* For SCMs, this roughly equates to managing these assets across their life cycle to maximize the stormwater management outcomes for which they are constructed, as well as secondary outcomes such as amenity, urban cooling, and water supply; all the while minimizing costs. Allbee and Rose (n.d.) list the key elements of good asset management:

- developing and maintaining an asset register;
- understanding the condition of the asset base and how this affects the functions and outcomes delivered;
- determining failure modes;
- determining residual life of existing assets;
- determining life cycle and replacement costs;
- setting target levels of service; and
- optimizing investment and determining a funding strategy.

## 22.2 TYPES OF STORMWATER CONTROL MEASURE

After more than 20 years of WSUD in Australia, a wide range of SCMs now exist. In broad terms, SCMs can be classified into two categories: structural and vegetated. Structural SCMs include trash racks, proprietary gross pollutant traps, gully baskets, and other similar systems that use mechanical means to manage stormwater, and do not contain a vegetated component, nor make use of soil. Vegetated SCMs include bioretention systems, swales, wetlands, and other such systems that typically include vegetation. As a minimum, they rely on ecological processes to facilitate stormwater treatment.

Some of the more common structural SCMs are:

- **Trash racks**—Metal racks containing evenly spaced bars placed in drains and creeks, or at pipe outlets to capture litter and debris (Fig. 22.1).
- **Gully baskets**—Fabric or metal mesh baskets placed in gully pits in roads and drains below other paved areas to capture litter and debris.

FIGURE 22.1 Trash rack in an urban drain shortly after a major storm.

FIGURE 22.2 Coarse sediment forebay (concrete, center) adjacent to a maintenance access ramp (concrete, left) and bioretention basin (behind trees, right).

- **Coarse sediment forebays**—Aboveground areas that slow stormwater flows, allowing coarse sediment (>1 mm diameter) to settle out (Fig. 22.2).
- **Underground sediment traps**—Underground concrete sumps that capture and slow water, allowing coarse sediment to settle out.
- **Proprietary gross pollutant traps**—Proprietary devices, typically located underground that use screens to capture organic and man-made litter and debris (Fig. 22.3).
- **Proprietary cartridge and media filters**—Proprietary devices, typically located underground, which use specially formulated media to remove nutrients and metals from stormwater.
- **Rainwater tanks**—Above- or below-ground tanks collecting water from roofs for reuse (Fig. 22.4).

Some of the more common vegetated SCMs are:

- **Bioretention systems**—Shallow vegetated areas that capture stormwater and filter it vertically through a sandy loam media, prior to it infiltrating into the surrounding soil, or being retained and routed to the drainage network (Fig. 22.5).
- **Swales**—Shallow grass or vegetated channels that convey and infiltrate water and facilitate coarser sediment to settle out of stormwater (Fig. 22.6).

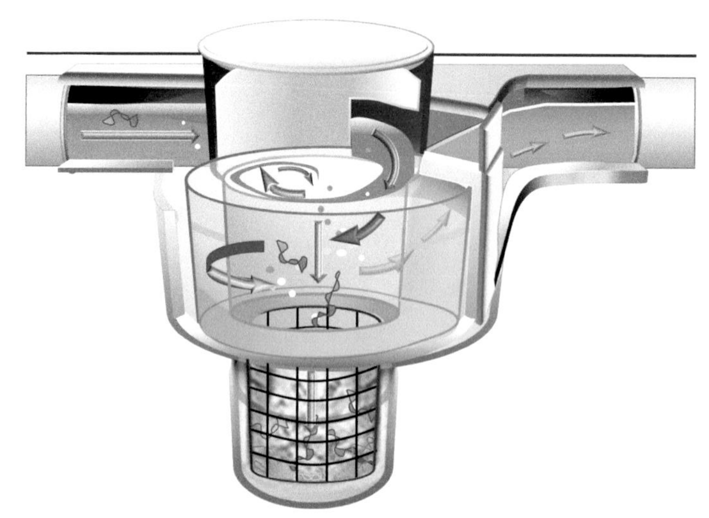

FIGURE 22.3 Functional schematic of a gross pollutant trap.

FIGURE 22.4 A rainwater tank on a suburban house block. *Sharma, Begbie and Gardner 2015.*

**FIGURE 22.5** A bioretention basin collecting and treating runoff from the adjacent car park.

**FIGURE 22.6** Vegetated swale in a residential street. It conveys and infiltrates stormwater and captures sediment.

FIGURE 22.7 A constructed stormwater treatment wetland with dense emergent vegetation, overlooked by a viewing platform.

FIGURE 22.8 A sediment basin located adjacent to a riparian zone and surrounded by recently planted vegetation.

- **Constructed wetlands**—Shallow water bodies filled with dense emergent vegetation that remove sediments, nutrients, and metals from stormwater using settlement, filtration, adsorption, plant uptake, and biogeochemical reactions processes (Fig. 22.7).
- **Sediment basins**—Open water bodies, surrounded by vegetation, which are designed to capture coarse sediments (by settlement) but not fine sediments ($<1$ mm diameter) (Fig. 22.8).

FIGURE 22.9 Passively irrigated garden bed in a car park.

FIGURE 22.10 Turf buffer strip adjacent to a footpath (sidewalk) in a residential area.

- **Passively irrigated gardens**—Gardens adjacent to impervious surfaces that are designed to collect and use the runoff from these surfaces (Fig. 22.9).
- **Buffer strips**—Vegetated areas or strips adjacent to impervious surfaces that intercept stormwater prior to it entering the drainage network (Fig. 22.10).

Less common SCMs, or those that were once common, but which are now used less frequently, include:

- **Sand filters**—Similar to bioretention systems. The filter media have a coarser (sandier) texture than that in bioretention systems (sandy loam), and the surface is unvegetated (Fig. 22.11).
- **Infiltration trenches**—Sand or gravel trenches that capture stormwater and encourage it to infiltrate into the surrounding soil.

## 22.3 OPERATIONAL AND MAINTENANCE REQUIREMENTS FOR STORMWATER CONTROL MEASURES

Operational and maintenance requirements for SCMs vary with asset type. However, there are strong commonalities between the various structural SCMs, as well as between the various vegetated SCM types.

FIGURE 22.11 Sand filter in a residential subdivision.

In general, operation and maintenance of structural SCMs focus on cleaning of accumulated pollutants, and repair of damaged components. If unmanaged, they will often quickly cease to function effectively. However, once operations and maintenance resumes, the function of the asset is often restored at little additional cost. Structural SCMs typically require:

- regular inspections;
- cleaning to remove accumulated pollutants; and
- repair of structural defects, such as corrosion of metal components.

The frequency of operational and maintenance activities for structural SCMs is typically dictated by factors such as:

- pollutant loads generated by the catchment;
- size of the asset and its storage capacity relative to the catchment size;
- visibility of the asset; and
- potential for stored pollutants to degrade, releasing soluble pollutants into passing stormwater.

Operational and maintenance requirements of vegetated SCMs differ from that of structural SCMs, in that it typically involves far less cleaning of accumulated pollutants. Rather, it focuses on vegetation management. Of course, litter and sediments still need removal, scours must be managed, and structures unblocked, but these activities are secondary to the vegetation management. If weeds are allowed to proliferate, they can often outcompete the desirable, efficacious vegetation (see Fig. 22.12). When this occurs, the cost of repairing the system typically exceeds the cost of the maintenance that would have prevented the weed invasion in the first place. In this respect, there is a strong business case for regular, proactive management of vegetated SCMs.

The risk of weeds proliferating varies between the different types of vegetated SCMs. Those containing only turf are usually at less risk of damage from invading weeds as the system can easily be mowed. Systems with a canopy are also resilient, as shade from the canopy plus leaf litter inhibits weed germination and growth (see Fig. 22.13). Systems with only understory plantings are most at risk of weeds proliferating as they receive copious sunlight to allow vigorous growth, and weeds, once established, are expensive to eradicate.

The frequency of maintenance of structural SCMs is typically dictated by factors such as:

- the rate at which weeds invade the system, and their ability to outcompete the desirable vegetation;
- the rate at which litter and debris accumulate, structures block, and scours form; and
- visibility of the asset.

**FIGURE 22.12** A bioretention basin before (a) and after becoming infested with weeds (b).

**FIGURE 22.13** Inside a bioretention basin with a tree canopy and vigorous understory inhibiting weed growth.

As a reasonably new asset class, it is likely that best practice operation and maintenance of SCMs will continue to evolve over time to develop the most efficient and effective methods. Reliability centered maintenance (RCM) may provide an appropriate framework through which this can happen. RCM determines what functions must be preserved through maintenance and which are less critical. Maintenance is conducted for those functions that must be persevered, whereas those that are less critical are allowed to run to failure.

Many jurisdictions and regional bodies have developed detailed guidelines for the maintenance and operation of SCMs. Some examples include:

- Maintaining Vegetated Stormwater Assets (Water by Design, 2012a)
- Rectifying Vegetated Stormwater Assets (Water by Design, 2012b)
- WSUD Maintenance Guidelines—A Guide for Asset Managers (Melbourne Water, 2013b)
- WSUD Maintenance Guidelines—Inspection and Maintenance Activities (Melbourne Water, 2013c)

- WSUD Audit Guidelines (Browne et al., 2017)
- Chapter 31 of The SuDS Manual (CIRIA, 2015).

Collectively, these guidelines cover maintenance activities, rehabilitation of failed assets, recommended inspection frequencies, condition assessments and auditing, and record keeping.

Both Melbourne Water and Water by Design also provide guidance on the cost of maintaining and operating selected SCMs:

- Water Sensitive Urban Design Life Cycle Costing Data (Melbourne Water, 2013a)
- Guide to the Cost of Maintaining Bioretention Systems (Water by Design, 2015).

Proprietary device manufacturers can also provide information on the recommended maintenance of their products.

## 22.4 LIFESPAN OF STORMWATER CONTROL MEASURES

All infrastructure, no matter how well designed or constructed, eventually reaches the end of its useful life. SCMs are no different. As a relatively young class of asset, our knowledge of the useful life of SCMs is still developing. This is particularly true for vegetated SCMs.

In the case of structural SCMs, end-of-life typically occurs when the materials or components from which they are constructed degrade and fail to such a degree that routine repairs are no longer cost-effective. The time at which this occurs is influenced by the materials used and the environment in which the asset is located. For example, an asset containing metal components located in a predominantly dry environment is likely to have a very different life span to one located in a wetter or saline environment.

The end-of-life for vegetated SCMs is dictated by: (1) the degradation of the materials from which the asset's structural components were built, for example degradation of concrete outlet structures; (2) major reduction in the system's ability to capture and treat pollutants; and (3) major changes in the appearance of the system. For example, in the case of a vegetated swale receiving only surface runoff from an adjacent car park, the time to end-of-life will be influenced by (1) the degradation rate of the material used to construct the overflow outlet; (2) the rate at which sediment accumulates at the interface of the car park surface and the swale; and (3) the changes in aesthetics as plants grow, die, and are replaced by invasive species. In the case of sediment accumulation, sediment may accumulate at the edge of the system over time to such an extent that no water enters the swale, and hence no stormwater management function occurs (see Fig. 22.14). This is not an insurmountable problem as the sediment, which was prevented from entering the waterway, can be removed, restoring the swale's function.

In general, though, many types of vegetated SCMs have existed for insufficient time to observe well-designed, constructed, and managed assets reaching the end of their useful life. Until that occurs, asset management professionals must

FIGURE 22.14 An uneven road surface (left) and sediment accumulation (center) prevents water from entering this treed bioretention swale.

rely on estimates of asset life developed from first principles. For example, the "bioretention end-of-life" case study is a cautionary tale regarding interpreting estimates of asset life span.

**Bioretention end-of-life case study**

It is common to hear estimates of bioretention life span of less than 10 years. It is claimed that the filter media must be replaced if the system is to continue to function. This is expensive because replacing bioretention filter media requires a complete system rebuild. It may also not be accurate.

Early bioretention research identified a theoretical limit to the amount of certain pollutants that bioretention systems could treat, in particular, dissolved phosphorus and metals. This limit was postulated because the removal mechanism is adsorption/immobilisation by fine particles and organic matter in the filter media. As there is a finite quantity of adsorption sites per unit of filter media (Water by Design, 2014), it was concluded that there was also a finite limit to the amount of these pollutants that a bioretention system could immobilize. Accelerated laboratory scale testing supported this hypothesis.

Glaister et al. (2014), for example, used accelerated laboratory scale testing to show that phosphorus breakthrough occurred after 6 months to 2.5 years, whereas Hatt et al. (2011) estimated zinc breakthrough to occur after 12–15 years (Deletic et al., 2014). However, such breakthrough has not been observed in bioretention systems installed in real-world conditions. Glaister et al. (2013) studied six field-scale bioretention systems and found no sign of phosphorus breakthrough after 12 years. Hatt et al. (2011) estimated that the time to heavy metal breakthrough varied more than 10-fold, depending on the depth of the filter media, the size of the bioretention system relative to its contributing catchment, and the nature of the catchment land use (which affected stormwater composition).

This case study highlights the uncertainty in estimating the life span of vegetated SCMs using empirical mass balance estimates. At present, although their usable life span can be estimated, it is not known for certain.

## 22.5 ASSET MANAGEMENT FOR STORMWATER CONTROL MEASURES—THE CHALLENGES

Understanding how and when to operate and maintain a small number of SCMs is one challenge. Appropriately estimating their useful life span, and planning for this time, is another. An even greater challenge comes from trying to manage a large and growing asset base, with numerous SCMs of diverse types, and in varying conditions of repair. Achieving this to a satisfactory standard requires a detailed suite of processes and practices to be established.

Local governments are the largest single owner of SCMs. As such, establishing processes and practices within local government has been the focus of much of the initial effort in developing maintenance processes. A significant amount of progress has been made. However, local governments do not own all the SCMs. In fact, local governments may own less than half of all SCMs (Mullaly and Mackenzie, 2011). The remainder are owned by private companies, body corporates, individuals, other government agencies, and water utilities. Although ownership by other government agencies and utilities presents many of the same challenges as ownership by local governments, the distributed ownership of SCMs in private business and individual ownership present a very different challenge.

There are four main items to be addressed in establishing processes and practices to manage SCMs:

- **The quantity and type of assets**—An understanding of the number and type of SCMs is fundamental to appropriately managing them. Without this understanding, assets will be overlooked and left unmanaged, their condition will deteriorate, and budgets will not be developed.
- **The condition of assets**—A sound understanding of asset condition is the primary driver of the type and frequency of on-ground works. Without this understanding, on-ground works are likely to be ineffective, and asset condition and function are likely to degrade.
- **Responsibilities and skill sets**—SCMs, and especially vegetated SCMs, are a new type of infrastructure. In large organizations such as local governments, a decision must be made as to which department is to be responsible for managing them. Failure to assign this responsibility to the department with the correct skill set results in substandard management outcomes, and higher costs.
- **A maintenance budget**—An appropriate budget for managing SCMs facilitates cost-effective on-ground works. Without such a budget, works are typically restricted to reactive fix-ups in response to asset failures and public complaints. Without a budget, regular maintenance cannot be completed, and asset condition and function inevitably degrades.

## 22.6 THE PATH TO INITIAL SUCCESS

Although asset management of SCMs around Australia is far from perfect, many local governments, and some private sector asset owners, have made significant progress in developing processes and practices to manage and maintain their SCMs. All had to take an important first step toward managing their SCMs. This first step is not easy. Consider the following scenario:

> *You work for a local government. Your local government has required new urban developments to install SCMs for the past five years, and now has a moderately sized, but growing base of SCM assets to maintain. You are tasked with establishing a management and maintenance program. To do this, you need to know what assets you have and the condition they are in. You also need to know how much it costs to maintain them, but without prior experience, you do not have a good estimate of the costs. Without a defensible cost estimate, you cannot make a budget bid in the yearly council budget cycle. Without a budget, you cannot collect any meaningful data on how much maintenance costs. You are stuck in a self-reinforcing, vicious cycle. How do you establish a maintenance program?*

Breaking out of this cycle is the key. In this section, we will describe a process that owners of SCMs, in particular local governments, can break out of this cycle and establish an inaugural SCM maintenance and management program. This process can be used in any part of the world.

The process is built based on the author's observations as to how selected local governments throughout Australia have undertaken this process. The process followed by each local government was slightly different and did not follow a "cookie cutter" approach. Instead, these local governments followed their instincts and overcame challenges as they occurred. It is only in hindsight that the commonalities in their paths have been distilled into a process to guide others.

Any local government (or utility) wishing to establish its first SCM management and maintenance program should use the following process as a guide, while responding to its own local situation and specific conditions. The process comprises seven steps:

1. Locating existing assets;
2. Determining the condition of existing assets;
3. Determining which department should be responsible for maintenance;
4. Building a business case;
5. Determining if on-ground work should be undertaken by contractors or in-house crews;
6. Briefing upward to management; and
7. Implementing and continually improving.

### 22.6.1 Locating existing assets

The first step in developing an SCM maintenance and management program is, as far as practicable, to identify, locate, and collect data on all the SCMs owned by the local government. The process can be applied with a few changes to the private sector or to other public agencies. Consider at the beginning whether to collect information only on assets owned by the local government, or also those in private ownership. Identifying both upfront will save significant duplication of effort in the future.

A list of asset types and the type of data to be collected on each must be collated. The complexity of this information will vary. At the more complex level, an asset classification manual or asset register can be created, which describes all the asset types of interest, how they perform, how to identify them, and what information to collect, in detail. Creating such a manual takes time. To make rapid progress, a simple list of asset types and data to be collected will suffice. Once data collection begins, it may be necessary to revisit the list, or asset classification manual, to adjust based on the characteristics of the identified assets.

Once the asset types and data requirements are known, a method to identify the assets is required. The first step in this process is to identify how and why assets have been created in the past. For example, assets may have been constructed by the developer in accordance with planning regulations, or they may be built by the local authority itself, or by private residents. Long-term staff are an excellent source of this information. They will usually be able to provide valuable information about how and why assets were constructed.

Once the pathways through which assets are delivered are understood, a process can be developed to identify all the SCMs. For example, in a local government where land developers have constructed assets, it may be necessary to inspect development approvals and development plans. Other potential sources of information include Council's own programs of works and historical aerial photographs.

These sources of information should be thoroughly searched to identify the assets. This is often labor-intensive and moderately time-consuming. As the SCMs are located, they should be given an asset identification number and the information stored in a spatial database.

## 22.6.2 Determining the condition of existing assets

Once the number and type of SCMs in the local government area are known, the condition of the assets needs be determined. In broad terms, this is a continuation of basic data collection, but with one key difference. A large amount of basic data can be collected by desktop techniques. Condition inspection requires asset inspection in the field.

Condition inspection is important for three reasons. First, the condition of SCMs is a very good indicator of function. For example, a trash rack with broken bars would be considered to be in poor condition and would trap less litter than one in good condition. A bioretention system heavily infested with weeds would be classed as moderate to poor condition and deliver very little aesthetic benefit (see Fig. 22.12), although it may still deliver acceptable treatment performance.

The second reason that condition inspection is important is that, as discussed previously, on-ground costs for vegetated SCMs in particular varies greatly with asset condition.

Third, the act of field inspection provides the opportunity to verify data collected in the desktop survey. It is not unusual to adjust the asset database after field inspections.

For these three reasons, determining the condition of the asset base is critical for the later step of building a business case to support a bid for maintenance budgets.

Before inspecting the SCMs, a condition rating system should be developed for each type of SCM present in the local government area. When combined with a scoring system to determine overall condition, a quick and simple picture of the overall condition of the asset class can be gleaned.

The scoring system must be designed in such a way that it groups assets with similar defects (and hence similar operational, maintenance, and rehabilitation requirements) under the same score. This will assist later development of the business case. Scoring systems that arbitrarily assign weightings, and/or group systems with no physical similarity in condition, provide a false sense of data accuracy and will not support future business case development, nor an efficient spend of limited resources.

It is beyond the scope of this chapter to describe in detail any one specific scoring system. However, developing a robust scoring system will be facilitated by:

- Referring to guidelines with inspection forms such as *Maintaining Vegetated Stormwater Assets* (Water by Design, 2012a).
- Referring to asset management resources such as the International Infrastructure Management Manual (Institute of Public Works Engineering Australasia, 2015). These manuals were developed for structural infrastructure assets such as road pavement and stormwater pipes. Although their frameworks will be useful, their detail will not be sufficiently nuanced, without adjustment, for managing vegetated SCMs.
- Speaking with other local governments.

The data collected from condition inspections should be recorded in a spatial database.

Once the type, quantity, and condition of all SCMs is known, developing a maintenance and management program is required. This involves four tasks:

- Determining which (local government) department should be responsible for maintenance;
- Building a business case;
- Determining whether the on-ground work should be done by contractors or in-house crews; and
- Briefing upward to management.

We discuss these tasks in a linear fashion, but they are far more likely to be undertaken nearly simultaneously, and in an iterative fashion.

## 22.6.3 Determining which department should be responsible for maintenance

The next task is to determine which department within the local government is best suited to managing and maintaining the SCMs. It is likely that some types of SCMs (e.g., structural SCMs) are more suited to one department (e.g., roads and pipes), whereas other types (vegetated SCMs) are more suited to another (e.g., parks and gardens).

Most local governments that establish a maintenance and management program have owned SCMs for many years. Without a dedicated maintenance budget, it is unlikely proactive maintenance and management will have been undertaken. However, it is likely that they will have completed some on-ground works, usually in response to public complaints as asset condition deteriorated. If an effective maintenance and management program is to be established, these existing processes and responsibilities must be temporarily placed to one side to answer the question:

*What skillsets are needed, and which department/s are most suited to undertake this work based on the maintenance activities required?*

Answering this question may initially be challenging. Some successful local governments have used field visits to several moderate to good condition SCMs to assist. Representatives from departments that could feasibly be appropriate for running the management and maintenance program inspected the systems and then answered the question above. Doing so helped to separate discussions of long-term, proactive maintenance, from the reality of unbudgeted reactive works in response to current complaints.

In most local governments, there will be three departments (or at least three skill sets) that may be appropriate for managing the program. These departments are:

- Stormwater drainage—typically a structural skill set developed to clean stormwater drains and maintain conveyance capacity;
- Parks—typically a horticulture skill set focused on maintaining ornamental species in parks and gardens; and
- Natural areas—typically an ecological and bushland management skill set applied in reserves and riparian corridors.

Answering the above question will give a good indication of which departments needs to be involved, but it does not answer which department should be responsible. To do this, local governments sometimes consider:

*Of the maintenance activities identified, which are most frequent, and which occur only sporadically?*

Doing so directs focus toward the activities (and hence the department) that need to be involved most frequently.

### 22.6.4 Building a business case

For vegetated SCMs, it is far cheaper to regularly and proactively maintain them than to rehabilitate them after they have degraded. This fact can be used to develop a business case to support a "whole-of-council" budget bid for SCM maintenance.

One valuable tool in constructing the business case is to develop a simple Excel spreadsheet model running a net present value analysis (say over 20 years). When combined with the GIS database (previously developed) and asset condition information, this model can be used to explore various maintenance and rehabilitation scenarios. More importantly, it will almost certainly clearly demonstrate that regular, proactive maintenance is much cheaper, and more effective, than periodic rehabilitation.

The next step is to determine the magnitude of the budget request. When most local governments develop their first SCM budget (or expand an existing maintenance budget to include vegetated SCMs), a large proportion of the asset base may be in poor to very poor condition. This will require substantial rehabilitation before maintenance can commence. This will almost certainly be a costly and time-consuming task.

To constrain the first-year budget request to a "reasonable" quantum, many local governments only request that amount of money that is required to maintain that portion of the asset base that is currently in good condition, plus the money required to rehabilitate some fraction of the assets that are in poor condition. In the second year, slightly more money for maintenance is requested (to cover good condition assets plus those recently rehabilitated) plus an amount to rehabilitate another fraction of the remaining poor assets.

In this manner, a budget program to maintain and rehabilitate the entire asset base can be developed over a period of several years.

### 22.6.5 Determining whether the on-ground work should be done by contractors or in-house staff

There is no single clear answer to the question of using contractors or council staff to undertake maintenance and/or rehabilitation tasks, but it must be answered before the budget bid is made. The decision will influence which department is best suited to run the program.

Engaging suitably skilled contractors avoids the need to train new staff, as occurs with a decision to create a new in-house crew. On the other hand, in-house crews allow council to build internal expertise and retain greater control over how works are completed. However, if existing crews are used, resources can become overstretched (between SCMs and other asset types), resulting in a poor on-ground outcome. Creating a new crew requires hiring new staff and undertaking training, which takes time to implement, and increases the fixed overheads costs for council.

### 22.6.6 Briefing up to management

After compiling a business case and internal delivery arrangements, departmental management must be briefed to submit the budget bid. The exact way this occurs is highly dependent on the individual local government. As such, it is not possible to specify a process to follow. Nevertheless, it is informative to consider the approach taken by one Council.

Mountain View Regional Council (a fictitious name) identified four departments with an interest in the maintenance and management of SCMs: Stormwater Drainage, Parks and Gardens, Natural Areas, and Environmental Policy. Through field inspections and in office discussions, officers of these four departments agreed that the Natural Areas department was best suited to run the program. A simple business case was developed, and officers from each of the four departments simultaneously briefed their own departmental managers. They recommended the Natural Areas department run the program and explained the benefits of proactive maintenance as detailed in the business case. In addition, each officer informed their manager that officers in the other three departments were making identical recommendations. They recommended that the managers from the four departments meet to formalize the decisions made to date. From there, the Environmental Policy department made a budget bid to the full Council and secured funding for an initial maintenance program.

### 22.6.7 Implementing and continually improving

If the budget bid is successful, it is very important to deliver on expectations in the first year, to help ensure funding continues in later years. The knowledge and data collected from the inaugural maintenance and rehabilitation activities can be used to inform the creation of a robust asset management approach for SCMs.

## 22.7 THE FUTURE OF STORMWATER CONTROL MEASURE ASSET MANAGEMENT

Asset management of SCMs will, in the future, likely center around two key issues. The first is addressing the challenge of how to cost-effectively and affordably implement WSUD, in sufficient quantities, to achieve the desired waterway health outcomes. This will require consideration of:

- The total amount of SCMs required in a city, and the investment required to maintain them; and
- The broader issues of aging asset bases, and how to fund their maintenance and replacement.

For WSUD, this points to one hard reality. New SCM technologies that cost substantially less to construct and maintain are required.

Irrespective of success in the first, will be the need to mainstream and make permanent appropriate asset management of SCMs.

### 22.7.1 The total amount of SCMs required in a city, and the investment required to maintain them

Using a city the size of Brisbane, Australia, as an example (1.1 M people, with an urban footprint of 380 km$^2$ (Brisbane City Council, 2017)), meeting stormwater quality targets specified in the State Planning Policy (State of Queensland, 2017) will require between 11 and 17 million square meters of bioretention area. Using contemporary bioretention technologies and planting pallets, this translates into a maintenance bill of between AUD $20M and AUD $85M per annum (or $17 to $78 per resident per annum). This is a substantial investment and many times greater than is currently invested in SCM maintenance by local governments the size of Brisbane. Of course, a true WSUD approach would not rely solely on just one type of SCM, but the example nonetheless gives an indication of the scale and cost of SCMs required to achieve a water sensitive city.

### 22.7.2 Low maintenance cost stormwater control measures

Considerable progress has been made in recent years in developing low maintenance cost SCMs, particularly the common bioretention system.

Initial cost saving initiatives focused on providing maintenance access and dedicated maintenance areas. For example, coarse sediment forebays (see Fig. 22.2) were installed on bioretention systems in the expectation that capturing and storing such sediment separately would make it easier, and cheaper, to remove the sediment. However, this approach failed to reduce maintenance costs because a visual eyesore was quickly created, and sediment cleanout frequencies increased. Even cleaning sediment from a dedicated, concrete forebay is an expensive task.

Other efforts to develop low-maintenance SCMs have centered around the concept of Zero Additional Maintenance WSUD (ZAM WSUD). In ZAM WSUD, the system installed does not increase maintenance over and above that already occurring. For example, Manningham Council, in Melbourne, Australia, developed a series of low maintenance bioretention systems for use in residential road reserves (see Figs. 22.15 and 22.16). The Council recognized that standard road reserves receive regular mowing (by residents) and street sweeping (by Council). By installing divots in the kerb and channel (to collect sediment), a grated inlet (to collect litter), a coarse sand surface layer (to prevent clogging), and turf and

FIGURE 22.15 A Zero Additional Maintenance WSUD bioretention system in Mannnigham, Melbourne. Note the screened stormwater inlet grates, the turf planted verge that is mown by local residents, and the sediment collection divots full of leaves.

FIGURE 22.16 Sediment collected in divots awaiting removal by a mechanical street sweeper, in Maningham, Melbourne.

FIGURE 22.17 The Hoyland St bioretention system (Brisbane, Australia) showing a dense canopy of trees and shrubs, which allow the system to act as a self-sustaining, natural bushland ecosystem.

the absence of exposed structures (to allow mowing), Manningham's system requires no additional maintenance to that required for a standard residential verge.

Another promising approach to reducing maintenance costs revolves around mimicking natural processes, such as those that suppress weed growth.

The Hoyland St bioretention system (Fig. 22.17) in Brisbane is the second oldest bioretention system in Australia. Constructed in 2001, it received maintenance for only the first 4 years of its life. Thereafter, the only maintenance was clearing of debris from the outlet structure and mowing of the grass perimeter. The system has declined less in 12 years without maintenance, than a standard bioretention design would decline in 1 year without maintenance. This resilience is possible because the dense tree/shrub canopy inhibits weed growth, leaf litter replenishes organic matter stores, and the deep root system supplies stored water for the plants, carbon for the denitrification process, and maintains the permeability of the filter media. In short, it functions as a small ecosystem of bushland. It takes only a small amount of regular work to supplement these natural processes and maintain the condition of the bioretention system.

It is likely that to reduce long-term maintenance requirements and costs, future SCMs will be designed with ZAM WSUD principles in mind, and make use of ecological processes to build resilient ecosystems. Local governments will be required to progressively refine planning requirements to ensure that the appropriate design of low maintenance assets are installed in the appropriate locations. An example of such a planning document in Australia is Whitsunday Regional Council's Stormwater Quality Guideline (2016). The guideline outlines a variety of bioretention "typologies" as well as where they are to be used to reduce maintenance costs and integrate with the surrounding landscape. Typologies include:

- Streetscape—low profile—grass and trees typology bioretention systems (see Fig. 22.18) for use in residential streets with street trees;
- Streetscape—high profile—garden typology bioretention system (see Fig. 22.19) for use in civic spaces with high amenity value.

## 22.7.3 Mainstreaming stormwater control measure asset management

The majority of SCM asset management effort to date has focused managing this class of asset appropriately, for the first time. The process identified earlier (Section 22.6), described how local governments have created an initial asset register, assessed asset condition, and developed a case for their first budget. To manage SCMs in the long term, however, we must mainstream and make permanent SCM asset management.

Some of the major barriers to mainstreaming WSUD, and hence SCM asset management, are institutional (Brown and Farrelly, 2009; Wong and Brown, 2009). Werbeloff et al. (2013) describe how a practice is considered institutionalized when (1) it is widely used within a particular sector or social system and (2) has a foundation enabling it to persist. Specifically, institutionalization requires cultural, structural, and practice change (Werbeloff et al., 2013). This is exactly what must occur if SCMs are to be managed appropriately in the long term.

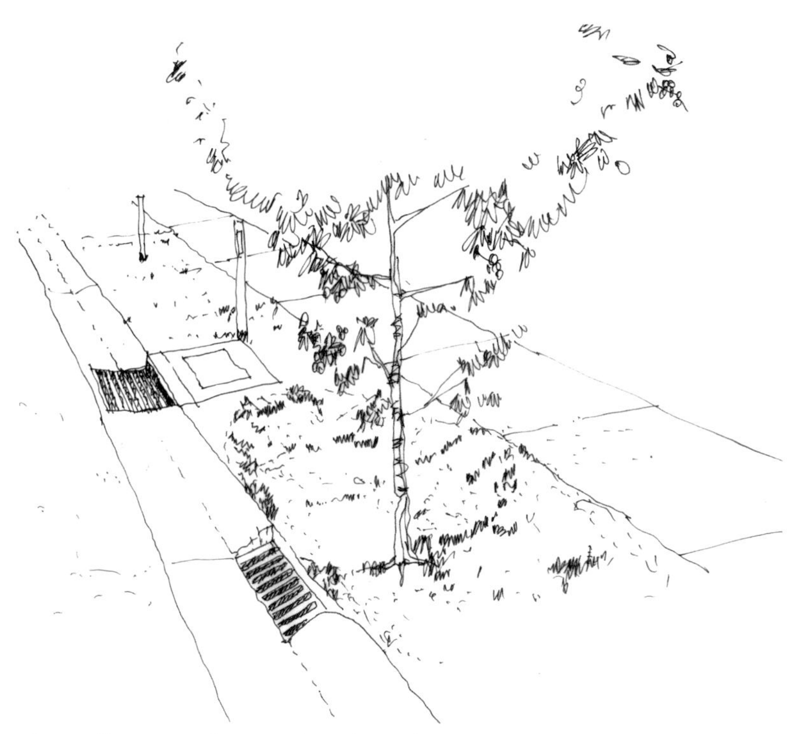

**FIGURE 22.18** Streetscape—low profile—grass and trees typology bioretention system to be used in residential streets. *Image – Whitsunday Regional Council.*

**FIGURE 22.19** A bioretention basin on the Gold Coast similar to the Streetscape—high profile—garden typology bioretention system to be used in civic spaces with high amenity value.

Examples what form this **may** take include:

- **Cultural change**—recognition of SCMs as a vital asset class on a par with more traditional infrastructure such as roads, water supply, sewerage, etc.
- **Structural change**—development of dedicated SCM maintenance teams and reporting frameworks
- **Practice changes**—implementation of new and more effective maintenance practices.

## 22.8 CONCLUSION

As the stormwater components of WSUD have gained traction in Australia, SCMs have grown rapidly in number. Like all forms of infrastructure, SCMs must be appropriately managed if they are to function as designed.

SCMs are a very broad asset class, including such devices as trash racks, coarse sediment forebays, bioretention systems, constructed wetlands, swales, and more. However, in general terms, SCMs can be considered as being either structural (systems that use mechanical means to manage stormwater and do not contain a vegetated component) or vegetated (systems that rely on ecological processes to facilitate stormwater treatment). Maintenance activities and frequencies vary significantly between structural and vegetated type SCMs, but less so within each designation.

As a reasonably new asset class, management of SCMs around Australia is far from perfect. However, many local governments, and some private sector asset owners, have made significant progress in developing processes and practices to manage and maintain their SCMs. Those that have made progress have typically addressed the following important elements of SCM asset management:

1. Locating existing assets;
2. Determining the condition of existing assets;
3. Determining which department should be responsible for maintenance;
4. Building a business case;
5. Determining if on-ground work should be undertaken by contractors or in-house crews;
6. Briefing upward to management; and
7. Implementing and continually improving.

Despite efforts to appropriately manage SCMs, large challenges exist. These will need to be addressed. Chief among these is the sheer number of SCMs (and thus the necessary maintenance expense) required to achieve water quality outcomes, let alone broader water sensitive outcomes. Future SCM asset management will need to focus on developing technologies and maintenance approaches that deliver water sensitive outcomes for substantially less cost than current approaches. Early examples of this include Manningham City Councils ZAM WSUD systems and bioretention systems that mimic natural bushland such as Brisbane City Council's Hoyland St bioretention system. Ultimately, SCM asset management will need to become mainstream in the manner that asset management for other infrastructure classes currently is.

## REFERENCES


Allbee, S., Rose, D., (n.d.). The Fundamentals of Asset Management, Powerpoint Presentation, GHD Inc.

Asset Management Council, 2014. Framework for Asset Management, Engineers Australia.

Brisbane City Council, 2017. Brisbane Community Profiles. Available online: https://www.brisbane.qld.gov.au/about-council/governance-strategy/business-brisbane/business-opportunities/brisbane-community-profiles.

Brown, R., Farrelly, M., 2009. Challenges ahead: social and institutional factors influencing sustainable urban stormwater management in Australia. Water, Science and Technology 59 (4), 653–660.

Browne, D., Godfrey, M., Markwell, K., Boer, S., 2017. WSUD Audit Guidelines. Stormwater Victoria. http://musicauditor.com.au/HostedFiles/WSUD%20Audit%20Guidelines%20Draft.pdf.

CIRIA, 2015. The SuDS Manual. https://www.ciria.org/Resources/Free_publications/SuDS_manual_C753.aspx.

Deletic, et al., 2014. Biofilters and Wetlands for Stormwater Treatment and Harvesting. Cooperative Research Centre for Water Sensitive Cities, Monash University.

Fletcher, T.D., Shuster, W., Hunt, W.F., Ashley, R., Butler, D., Arthur, S., Trowsdale, S., Barraud, S., Semadeni-Davies, A., Bertrand-Krajewski, J., Mikkelsen, P.S., Rivard, G., Uhl, M., Dagenais, D., Viklander, M., 2015. SUDS, LID, BMPs, WSUD and more – the evolution and application of terminology surrounding urban drainage. Urban Water Journal 12 (7), 525–542.

Glaister, B.J., Cook, P.L.M., Fletcher, T.D., Hatt, B.E., 2013. Long-term phosphorus accumulation in stormwater biofiltration systems at the field scale. In: 8th International Conference on Water Sensitive Urban Design. Gold Coast, Australia.

Glaister, B.J., Fletcher, T.D., Cook, P.L.M., Hatt, B.E., 2014. Co-optimisation of phosphorus and nitrogen removal in stormwater biofilters: the role of filter media, vegetation and saturated zone. Water Science and Technology 69, 1961–1969.

Hatt, B.E., Steinel, A., Deletic, A., Fletcher, T.D., 2011. Retention of heavy metals by stormwater filtration systems: breakthrough analysis. Water Science and Technology 64, 1913–1919.

Institute of Public Works Engineering Australasia, 2015. International Infrastructure Management Manual.

Melbourne Water, 2013a. Water Sensitive Urban Design Life Cycle Costing Data. Melbourne Water. https://www.melbournewater.com.au/sites/default/files/Life%20Cycle%20Costing%20-%20WSUD.pdf.

Melbourne Water, 2013b. WSUD Maintenance Guidelines – A Guide for Asset Managers. Melbourne Water. https://www.melbournewater.com.au/sites/default/files/WSUD-Maintenance-manager-guidelines.pdf.

Melbourne Water, 2013c. WSUD Maintenance Guidelines – Inspection and Maintenance Activities. Melbourne Water. https://www.melbournewater.com.au/sites/default/files/WSUD-Maintenance-Inspection-and-maintenance-activity-guidelines.pdf.

Mullaly, J., Mackenzie, D., 2011. Creating a WSUD future: managing logan city Council's water sensitive urban design assets, 2011. In: Stormwater Queensland Conference.

Sharma, A., Begbie, D., Gardner, T., 2015. Rainwater Tank Systems for Urban Water Supply: Design, Yield, Energy, Health Risks. Economics and Social Perceptions. IWA Publishing, London, UK.

State of Queensland, Department of Infrastructure, Local Government and Planning, 2017. State Planning Policy.

Water by Design, 2012a. Maintaining Vegetated Stormwater Assets. Healthy Waterways Ltd. http://hlw.org.au/u/lib/mob/20141014104448_50d476147252a7c87/2012_maintainingvegetatedassets-5mb.pdf.

Water by Design, 2012b. Rectifying Vegetated Stormwater Assets (Draft). Healthy Waterways Ltd. http://hlw.org.au/u/lib/mob/20141014104209_53b4d8e4f8a537087/2012_rectifyingguideline-44mb.pdf.

Water by Design, 2014. Bioretention Technical Design Guidelines. Healthy Waterways. http://hlw.org.au/u/lib/mob/20150715140823_de4e60ebc5526e263/wbd_2014_bioretentiontdg_mq_online.pdf.

Water by Design, 2015. Guide to the Cost of Maintaining Bioretention Systems. Healthy Waterways Ltd. http://hlw.org.au/u/lib/mob/20150213075847_0731a67b612832545/wbd_2015_guide-cost-maintain-bioretention_mq_online.pdf.

Werbeloff, L., Brown, R., Brodnik, C., 2013. Mainstreaming WSUD.

Whitsunday Regional Council, 2016. Stormwater Quality Guideline. North Queensland, Australia. https://www.whitsunday.qld.gov.au/DocumentCenter/View/3020.

Wong, T., Brown, R., 2009. The water sensitive city: principles for practice. Water Science and Technology 60 (3), 673–682.

Chapter 23

# Capacity Building for WSUD Implementation

**Rob Catchlove**[1], **Susan van de Meene**[2] **and Sam Phillips**[3]

[1]*Wave Consulting, Southbank, VIC, Australia;* [2]*School of Social Sciences, Monash University, Clayton, VIC, Australia;* [3]*Department of Environment, Water and Natural Resources, for the Adelaide and Mt Lofty Ranges NRM Board, SA, Australia*

## Chapter Outline



**ABSTRACT**

This chapter discusses the need, theory, and practice of building capacity across all industries associated with water sensitive urban design (WSUD). Capacity building programs were developed on the premise that, in designing, building, and maintaining WSUD assets, specific entities are necessary to focus on building the personal, intraorganizational and interorganizational capacity of the industry. These entities enable a faster adoption process, improve the quality of the assets constructed, and increase the life of asset. Successful capacity building results in the efficient delivery of assets, an improved return on the investment, and reduces the risk of asset failure. The chapter starts with an overview of why capacity building is necessary, and what it hopes to achieve. The chapter discusses the latest theories that underpin capacity building, and uses a case study of South Australia to follow the process of developing a business case, and implementing a capacity-building program. The chapter presents capacity building to an end, healthy cities and healthy waterways, bays, and marine environments. Finally, the chapter discusses some potential future issues that are expected to influence capacity building in the future.

**Keywords:** Capacity building; Culture; Implementation; Leadership; Skills; South Australia; Water sensitive urban design.

## 23.1 INTRODUCTION

Capacity building for water sensitive urban design (WSUD) has developed into an effective strategy for improving the design and implementation of WSUD technologies and assets, and helping industry recognize the critical role of the whole

**Approaches to Water Sensitive Urban Design. https://doi.org/10.1016/B978-0-12-812843-5.00023-X**

WSUD process in city shaping (Chesterfield, C., 2014, pers comms). Since the early 2000s, targeted capacity building programs (CBPs) have been established in major cities across Australia and have improved the profile of WSUD across the water sector (e.g., Keath and White, 2006a). Our understanding of capacity building for WSUD across individual, organizational, and institutional areas has deepened through academic research and evaluation of real-world CBPs. The complexities of WSUD capacity building in practice are being further investigated and can provide feedback into program design and implementation (Morison, 2010, 2011).

This chapter describes the topic of WSUD capacity building in depth, exploring what capacity building is, why it is needed, and what it aims to achieve. The methods for capacity building and its benefits are also described. The chapter then turns its focus to capacity building in practice, outlining the existing CBPs in Australia and the United Kingdom and providing a detailed case study of the South Australian CBP. The South Australian case study explores how to establish a stand-alone WSUD CBP, build a business case, its governance, and evaluation. The staged process undertaken in South Australia is documented. Finally, the future opportunities and challenges for WSUD capacity building are discussed.

## 23.2 WHAT IS CAPACITY BUILDING?

Capacity building for WSUD is a process used to improve the ability of urban water practitioners to plan, design, implement, and maintain WSUD assets, within the context of smarter and more efficient city building and the associated institutions and disciplines that play a role in city building. It is often considered to be teaching individuals new knowledge, for example about new technologies and tools. However, as theory and practical experience has demonstrated, capacity building needs to help practitioners apply their knowledge to their own situations (Farahbakhsh and Lewis, 2007 cited in Farahbakhsh et al., 2009). Therefore, capacity building needs to look beyond the individual and encompass intra-organizational, interorganizational, and broader administrative areas (Brown et al., 2006; Heslop and Hunter, 2007; Morison, 2010; van de Meene et al., 2010; Morison and Brown, 2011).

Capacity building is not a new concept. It has been used extensively in the international development, health, public administration, and community sectors (Grindle and Hilderbrand, 1995; Bolger, 2000; Crisp et al., 2000; Craig, 2007), and more recently, it has been used in the natural resources (Robins, 2008) and water management sectors (e.g., Alaerts et al., 1999; Brown et al., 2006; Bos and Brown, 2014). The general approach is to focus on individuals, organizations, and the supporting administrative framework. Robins (2007) identifies four types of capital to be built through capacity building strategies: human (knowledge, skills, experience); social (social norms and structural networks); institutional (governance arrangements); and economic (infrastructure and financial resources). We note that human capital exists at the individual level; social and economic capitals exist at the organizational and interorganizational levels, whereas institutional capital occurs in the administrative areas. It can be argued therefore that building capacity should be undertaken with different foci, and a wide variety of types of capital should be developed.

### 23.2.1 Why is capacity building needed?

Over the past three decades or so, urban water practice has changed significantly (Wong, 2006). For example, new stormwater treatment technologies (see, e.g., Bratieres et al., 2008), assessment and modeling tools (Bach et al., 2013; de Haan et al., 2013), and technical guidelines have been developed (e.g., Stormwater Committee, 1999; Department of Planning and Local Government, 2009), and this knowledge continues to expand. These technical developments are underpinned by social values. The urban water system was initially developed to protect public health and reduce flood risk by providing water supply, sewerage, and drainage services. Today the water system also needs to deliver on a wide variety of social values including protecting environmental health, providing social amenity, mitigating carbon emissions, and reducing the urban heat island effect, which is likely to be exacerbated by climate change (Brown et al., 2009; Sharma et al., 2016).

These changes in urban water practice have been reflected in the professionals who now work on urban water projects. Where urban water services were once the domain of engineers, there are now town planners, asset managers, natural resources managers, environmental scientists, ecologists, economists, and community engagement professional who work together with engineers (Brown, 2005; Brown et al., 2009; van de Meene et al., 2011). To work effectively, these professionals need to work across their disciplines, overcoming differences of professional language and work methods to deliver high-quality WSUD projects (McIntosh and Taylor, 2013).

Concurrently urban water practice has increased in complexity, which together with uncertainty from climate change and a rapid pace of change is likely to continue into the future (Edelenbos and Teisman, 2013). Therefore, urban water professionals need to continually learn and develop new strategies for implementing WSUD. Capacity building is such a strategy that has been advocated by many in the academic field (e.g., Brown et al., 2006; Keath and White, 2006a; Farahbakhsh et al., 2009; van de Meene et al., 2010; Morison, 2011; Sharma et al., 2016) and CBPs have been implemented in Australia since the early 2000s to help advance WSUD practice.

Furthermore, capacity building has also been advocated as a strategy to overcome the entrenched barriers inhibiting mainstream implementation of sustainable urban water management strategies as identified by Brown and Farrelly (2009a,b), Leidl et al. (2010) and Sharma et al. (2016). These barriers include a lack of knowledge and understanding of new practices; undeveloped regulatory frameworks; poor organizational commitment; lack of a clear technical and economic justification for implementation; lack of community and political will and organizational commitment; lack of system monitoring and evaluation; inadequate financial and human resources; poor communication and lack of experience working across organizational departments; unclear and fragmented roles and responsibilities; and technocratic path dependency (Brown and Farrelly, 2009a,b; Leidl et al., 2010; Sharma et al., 2016). Without adequate capacity in these areas, achieving change and widespread implementation of WSUD will be difficult and slow (Bettini et al., 2015b).

Lastly, WSUD assets that require maintenance, as well as the number of jurisdictions that require WSUD in the urban development process, have increased across all of Australia.

### 23.2.2 Capacity building objectives

Capacity building is needed to address the increased complexity of urban water management and, the numerous, systemic barriers outlined above. This section discusses the capacity attributes that should be developed: the objectives of CBPs for individuals and organizations, and the policy and regulatory settings.

The basic knowledge of WSUD is a key requirement for individuals (Urrutiaguer et al., 2010; Sharma et al., 2016). This may involve knowledge about designing WSUD systems, how the system fits together, and the different technical components such as filter media and plant selection for the local context. More broadly, individuals will need diverse knowledge and skills to understand and operate in the professional WSUD workplace, as well as a multidisciplinary outlook that values different disciplines and their contributions (Urrutiaguer et al., 2010; van de Meene et al., 2011). It has been argued that a systems-perspective of the water sector, as well as an understanding of how the water sector contributes to societal well-being, is needed to effectively implement WSUD (van de Meene et al., 2010). Other personal attributes include resilience, which help individuals cope when they are continually challenged to learn and adopt new practices (van de Meene and Brown, 2009). Resilience is influenced by an individual's ability to learn, plan, and develop from previous experiences, and it affects an individual's response to change (Marshall and Marshall, 2007). A desire to contribute to a greater social good (van de Meene et al., 2010) and being committed to creating change (Cettner et al., 2014) are important motivating factors, which contribute to individuals being able to overcome the hurdles of implementing WSUD. Taking responsibility for their work, being open to new approaches, and being willing to take risks are other important attributes (van de Meene et al., 2011). Effectively, these attributes reinforce those attributes of successful watershed leaders identified by Wolfson et al. (2015, p. 86). That is, leadership, communication, collaboration, policy, and planning skills are required, together with an ability to integrate and solve problems. These attributes are common to many multidisciplinary or "wicked" public policy problems (Weber and Khademian, 2008).

Organizational capacity, and its relationship with designing, constructing, and maintaining WSUD, has been the focus of a number of studies (e.g., Morison, 2010; Bos and Brown, 2014). Organizational capacity for WSUD has been defined as the "ability to anticipate and influence change, make informed and intelligent policy decisions, attract, absorb, and manage resources, and evaluate current activities to guide future action" (Morison, 2010, p. 65). Extending this definition and examining transitional change explicitly, Bos and Brown (2014, p. 189) focus on organizational capacity that enables "implementation of innovative ideologies and practice, without detracting from routine practices." Innovative strategies are more readily implemented when the organization is flexible and able to respond and adapt to changes (van de Meene et al., 2016). This may involve a flexible interpretation of the guiding policy and regulatory frameworks to facilitate innovative approaches for WSUD (Bettini et al., 2015a). Implementing WSUD is likely to be enhanced when the executive supports it (Urrutiaguer et al., 2010) and encourages coordination and collaboration across organizational departments (van de Meene and Brown, 2009; van de Meene et al., 2010; Floyd et al., 2014; Bettini et al., 2015a; Sharma et al., 2016). Stakeholders who are willing to engage with others are more likely to effectively implement WSUD, and building trust is an important component of any effective relationship (van de Meene et al., 2011; Dobbie et al., 2016). Implicit in the organizational capacity definition is the need for them to continually learn. This occurs in organizations where learning is valued by leadership, and therefore adequate resources (staff and time) are provided to engage in learning and reflection (van de Meene and Brown, 2009; van de Meene et al., 2010).

A supportive administrative and regulatory framework is important for implementing WSUD (Farahbakhsh et al., 2009). Such a framework is likely to have a clear vision (Farrelly et al., 2012; Cettner et al., 2014; Floyd et al., 2014), clear and coordinated administrative arrangements, and use a variety of policy instruments (van de Meene et al., 2010, 2011). Indeed, having clear roles and responsibilities is an ideal attribute frequently identified in the literature (van de Meene and Brown, 2009; van de Meene et al., 2010, 2011). The administrative framework should specify the arrangements for ownership, authority, legitimacy, and accountability (Bettini and Head, 2013). Effective policy instruments are likely to differ in different settings and together will form a progressive framework (Farahbakhsh et al., 2009). For example, financial incentives will likely stimulate innovation by leading organizations, whereas regulations will be needed to bring lagging organizations up to a basic standard (van de Meene et al., 2011). Other tools that can be used to develop a supportive policy framework include establishing performance targets with regular and effective monitoring and evaluation (Farrelly et al., 2012). The processes used to engage with and involve stakeholders in decision-making are important for implementing WSUD (Sharma et al., 2016). Processes that involve negotiated collective decision-making are more likely to develop the capacity for the water sector to act in times of change (Bettini et al., 2015a). Finally, being aware of the political context surrounding WSUD policy development and implementation is an important capacity attribute (Honadle, 1981). Such awareness can facilitate changes in the policy and regulatory framework to facilitate WSUD implementation.

Lastly, it must be stressed that a CBP is simply a means to an end. The end goal is the higher-level objectives of WSUD.

## 23.2.3 Tools and techniques

WSUD capacity building stretches across all areas of the technical and institutional dimensions of the industry and the discipline. Fig. 23.1 illustrates how these strategies to build capacity can in some way be linked to the delivery of assets on the ground.

Firstly, the location of the asset itself, broadly, requires capacity building at many levels. There needs to be a strategy to guide the process and define the objectives and targets. There also needs to be a cross-department team to consider where and how the assets should be delivered, and what local opportunities exist to retrofit or build new assets in the public realm.

WSUD often fundamentally changes a local urban landscape, where once were underground drains, now we have a three-dimensional green asset. This physical change requires more effort and strategy in community engagement. From a WSUD capacity perspective, it means that the open space planners and landscape architects and community engagement officers are now engaged in an area that was traditionally dealt with by the engineering departments. The community

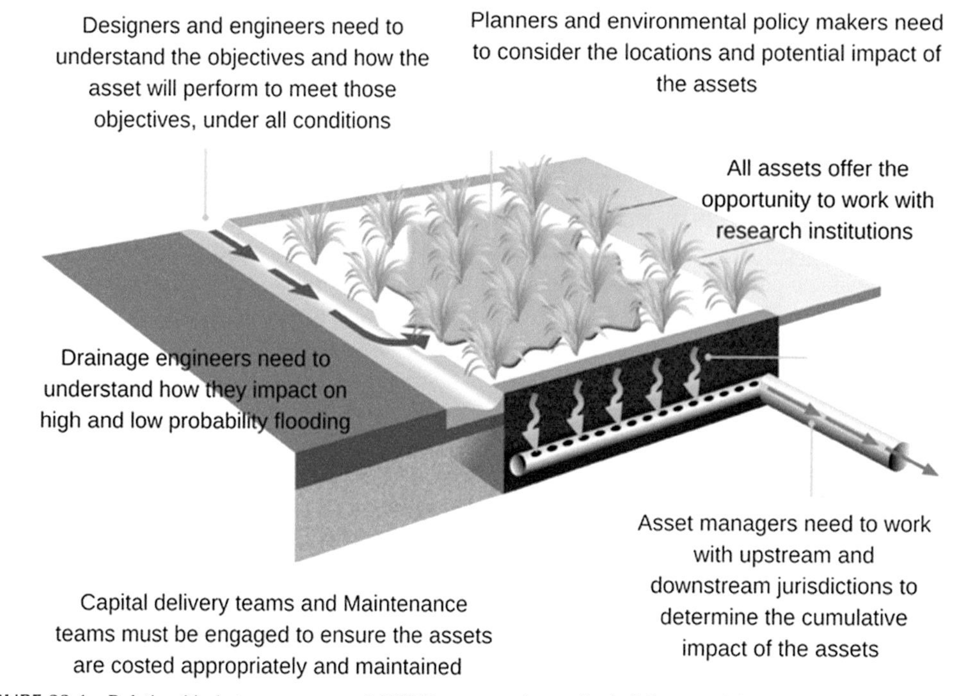


FIGURE 23.1 Relationship between on ground WSUD assets and capacity building. *Modified figure from City of Kingston.*

acceptance and community engagement associated with WSUD is often located in the public realm, where there are a lot of other competing demands.

## 23.2.4 The benefits of WSUD capacity building

The impacts of capacity building for WSUD are multiple and varied. Identifying and measuring the benefits of WSUD capacity building is undertaken through program evaluation. Program evaluation is a key component of effective implementation; it can clarify how successful capacity building strategies were implemented (Sobeck and Agius, 2007). For the existing stand-alone CBPs, evaluation of events or the entire program is typically undertaken by seeking feedback from participants (see for e.g., Dahlenburg and Lees, 2004; Keath and White, 2006b). Alternatively, longitudinal evaluations are undertaken across a broader range of program or organizational capacity attributes (Dahlenburg and Lees, 2004; Eggleton et al., 2012; Bos and Brown, 2014). Annual reporting of the CBPs to their steering committee or executive management is also important (Alluvium and Kate Black Consulting, 2012).

Postevent feedback collection data have invariably provided positive feedback (Dahlenburg and Lees, 2004; Keath and White, 2006a). Participants stated that they value the experiential learning undertaken through site visits (Dahlenburg and Lees, 2004) and, in the case of the UK CBP "Construction Industry Research and Information Association" (CIRIA—http://www.ciria.org/), a survey identified that 98% of participants say that "susdrain" (the United Kingdom term for WSUD) is a credible and neutral resource (CIRIA, 2016).

One such comment from participants was "After attending the seminars and workshops, council staff stated that they felt more confident in bringing about change in their organizations and presenting WSUD ideas to their colleagues and managers" (Dahlenburg and Lees, 2004, p. 606).

Broader program evaluation also reveals the development of the CBPs and their potential impact on industry. The industry perception of Clearwater (the Victorian-based WSUD CBP established in 2001), for example, shifted from an unknown industry actor to being "seen as a relevant and central authority on best practice urban water management." In other words, a driver for advancing the implementation of WSUD (Keath and White, 2006a, p. 238).

Self-assessment of the impacts of CBPs more clearly reveal the impact of capacity-building strategies across a variety of indicators. The CBPs with longitudinal evaluations include the WSUD in Sydney Program (Dahlenburg and Lees, 2004); the Cooks River Sustainability Initiative (CRSI) in Sydney (Bos and Brown, 2014); and the Living Rivers Program in Melbourne (Eggleton et al., 2012). A key common benefit identified was improved coordination across councils, as the programs allowed both time and focus for the coordination. This often resulted in cross-departmental working groups (Eggleton et al., 2012), improved communication among staff, and knowing whom to consult, and when to do so (Bos and Brown, 2014). The increases in senior management support and commitment also varied, with the lower-WSUD performing councils developing greater improvements in senior management support (Bos and Brown, 2014). Organizational commitment improvements were observed, for example, through councils' WSUD staff who worked alone now being included in the planning phases (Dahlenburg and Lees, 2004).

Another key benefit of the CBPs was the improvement in staff knowledge and skills related to WSUD. This benefit was observed across the three CBPs evaluated (Dahlenburg and Lees, 2004; Eggleton et al., 2012; Bos and Brown, 2014). Other relevant skills such as community engagement were also developed (Bos and Brown, 2014). Despite this, resources continue to be a limiting factor in advancing WSUD implementation in local government (Eggleton et al., 2012). The impact of CBPs on resource access was varied, with the CRSI program having a positive impact on resource access, both through external and council funding to WSUD (Bos and Brown, 2014).

This discussion has highlighted the importance of continual and longitudinal evaluation of capacity building events and programs to identify the impact and benefits of capacity building. Assessing benefits of these programs provides opportunities for participants to provide feedback as well reviewing the CBP more systematically to identify the benefits of the program, and to ensure that future planning is relevant for the industry needs (Eggleton et al., 2012; Bos and Brown, 2014). The main measured benefits of the CBPs are improvements in intraorganizational coordination, improved commitment from senior managers and elected officials, and improvements in WSUD knowledge.

## 23.2.5 How do you build capacity?

There are many ways to build capacity. Urrutiaguer et al. (2010, p. 2335) argue that there "is no single approach to building institutional capacity." Capacity-building strategies need to reflect their implementation context, and be flexible, iterative, and participatory (Leidl et al., 2010). Allowing time for reflection strategies and their subsequent reinvention is important (Leidl et al., 2010).

The most familiar capacity building strategy is training and education. This can be used to introduce individuals to new knowledge and tools. They can be delivered in person or online (Wolfson et al., 2015). A training program may be delivered in a traditional classroom setting or through site visits and may contribute to continuing professional development. Site visits offer participants the opportunity to see the real-world context of WSUD and ask questions of the site host, who is often an industry professional closely involved in the project. This may provide unique insights into the project development and implementation and offer opportunities for participants to ask questions and receive immediate responses (Dahlenburg and Lees, 2004). Experiential learning through study tours (Catchlove et al., 2012) and cooperative design competitions (Dahlenburg, 2006) can also provide real-world opportunities for learning in context. Together, training, site visits, and other experiential learning settings are proposed to improve trust and confidence in the technical feasibility of WSUD technologies (Brown and Farrelly, 2009a). A side benefit of face-to-face training is the informal social networking that occurs. Informal networks provide an important part of overall system adaptive capacity (Larson et al., 2013) and they can facilitate learning, reflection, and sharing of information and ideas (Schiffer and Hauck, 2010; Larson et al., 2013).

Providing information and tools such as guidelines also facilitate capacity building. The more detailed and comprehensive websites are often associated with specific WSUD CBPs (e.g., CIRIA in the United Kingdom and Clearwater in Melbourne). These websites aim to be "go to" places for the industry to obtain up to date information and news, case studies, publications, and other relevant resources. Additionally, relevant guidelines can be provided online, with practice notes to facilitate their use (e.g., Water by Design in Queensland—http://hlw.org.au/initiatives/waterbydesign).

The specific CBPs referred to above (i.e., CIRIA, Clearwater, Water by Design) are examples of targeted CBPs that focus on WSUD and other related issues. For example, CIRIA also has a large construction focus and Water by Design has explicit priority areas related to waterway remediation, and erosion and sediment control. These are issues that overlap with WSUD. A specific WSUD-focused CBP has the opportunity to provide strategies to enhance awareness, understanding, and confidence around WSUD (Brown and Farrelly, 2009a) in a coordinated way that is needed to make an impact (Urrutiaguer et al., 2010). An important aspect of stand-alone CBPs is the supportive legislation (Alluvium and Kate Black Consulting, 2012). Having supportive legislation lends support to the CBPs' existence and strengthens the argument for obtaining funding from relevant government and other agencies.

A capacity building strategy with a different structure was implemented by Melbourne Water, Australia, through its Yarra River Action Plan (Morison, 2010; Urrutiaguer et al., 2010). The Yarra River Action Plan was a state government policy aimed at reducing the pollutants entering the Yarra River from stormwater under the control of local government. Rather than adopting a traditional grant-funding model, Melbourne Water adopted a partnership approach (Edwards et al., 2007). This partnership approach involved Melbourne Water working closely with the local governments to identify stormwater quality improvement projects. A multi criteria assessment tool (Urrutiaguer et al., 2010) was developed to assess each project proposed by local governments, and a capacity needs analysis (Bolton et al., 2007) was used to assess the existing capacity of local governments.

Local government representatives considered that the capacity assessment process was useful for them to identify new opportunities for improving stormwater quality. It was also considered to be a mechanism (through peer-to-peer local government competition) to improve organizational commitment to stormwater quality improvement (Bolton et al., 2007). Consultants were involved to undertake the design of the projects, as well as helping build local government capacity through workshops (Edwards et al., 2007). Furthermore, a WSUD officer was part-funded by Melbourne Water to support councils during the implementation of the stormwater quality improvement projects (Edwards et al., 2007; Morison, 2010). The dedicated WSUD officer together with key WSUD officers in local government organizations facilitated the stormwater quality improvement projects, which were often considered innovative and novel by the local government (Edwards et al., 2007). As a CBP, the Yarra River Action Plan was generally considered to be effective. An evaluation of the impact of this program identified that capacity across 83% of local councils had been improved (Eggleton et al., 2012). However, its effectiveness at stimulating local implementation of WSUD was considered by local government stormwater officers to be questionable (Morison, 2010). It should be noted that evaluating the impacts of capacity building and other similar programs (e.g., research programs) can be difficult as the impacts are varied, often intangible and/or hidden, and depend on the starting capacity of the participating individuals and organizations (Bos and Brown, 2014; Bos and Farrelly, 2015).

Other partnership approaches to WSUD capacity building have been implemented in New Zealand and the Cooks River in Sydney. The partnership approach to WSUD capacity building involved researchers, consultants, and local government organizations (Heslop and Hunter, 2007). The program was focused on the low impact urban design and development (LIUDD) concept (similar to WSUD) and had an explicit aim of facilitating the uptake and implementation of LIUDD approaches and policies in New Zealand (Heslop and Hunter, 2007). Researchers provided the case study councils with detailed understanding of the legislative framework and investigated the mechanisms and strategies local government were using to implement LIUDD/WSUD (Heslop and Hunter, 2007). Researchers identified that policy tools outside of the legislative framework were used to facilitate the implementation of LIUDD/WSUD and specific capacity building strategies were used to build capacity across individual, organizational, and interorganizational capacity spheres (Heslop and Hunter, 2007). These insights were used to develop a capacity building assessment framework for New Zealand (Heslop, 2010).

The CRSI was a governance experiment with a central objective of capacity building through partnerships (Bos and Brown, 2014). CRSI was funded by the New South Wales government with the aim of improving the health of the Cooks River, and also building capacity within and among councils (Bos et al., 2013b). A partnership between a university and eight local governments was formed to trial a multidisciplinary, participatory approach to develop local, adaptive water management plans for six subcatchments within the Cooks River catchment (Bos et al., 2013a; Bos et al., 2013b, p. 1710). The program and supporting organizational structure were intentionally established so that participants (municipal staff and other stakeholders) could explore their different perspectives, roles, and interdependencies and how these could complement each other (Bos et al., 2013a). The program and organization purposefully stimulated interaction among different stakeholders at local and catchment scales, and a project manager and four officers provided continual support (Bos et al., 2013a). Additionally, a set of cross-municipal committees at different hierarchical levels and departments were formed to facilitate knowledge sharing, perspectives, and insights from new experiences (Bos et al., 2013a,b). A detailed organizational capacity assessment was undertaken with six municipalities that revealed varied organizational development outcomes; four municipalities improved their capacity whereas two remained stable (Bos and Brown, 2014). The main areas of improvement included raising the profile of WSUD across organizational levels, developing professionals' willingness and confidence to implement WSUD, improving intraorganizational cooperation, and a shift in WSUD values and beliefs (Bos and Brown, 2014).

## 23.2.6 Who needs capacity building?

Fig. 23.1 illustrates how the different stakeholders interact around a WSUD feature and their different capacity interests. The key stakeholders of local government, drainage authority, state government, consultants, and developers each have different responsibilities and roles to play when developing and implementing WSUD projects. The capacity attributes needed by these stakeholders to effectively implement WSUD projects have been identified above in Section 23.2.2.

## 23.2.7 Assessing the needs of industry

Just as capacity building needs to target different levels, so too do capacity assessments. Organizational capacity assessment has been the most widely implemented method (see, e.g., Bolton et al., 2007; Morison, 2010; Bos and Brown, 2014). Bettini et al. (2015b) developed a tool to map the institutional context across an urban water sector, which is then used to identify areas where capacity building interventions can be implemented. The tool comprises three phases: (1) context gathering, (2) identifying the institutional setting, and (3) characterizing the institutional setting. This processes focused on accessing the tacit knowledge held by professionals through a focus of institutions as "rules in use" (as per Ostrom, 2005), identifying actors who were maintaining, creating, or disrupting these institutions, and the resulting system dynamics (Bettini et al., 2015b). In the field of adaptive capacity, Gupta et al. (2010) developed an institutional capacity assessment wheel, which identifies six categories of adaptive capacity attributes: variety, learning capacity, room for autonomous change, leadership, resources, and fair governance. The assessment tool can be used to "assess and inform social actors about how their institutions influence different aspects of adaptive capacity and where there may be room for discussion and reform" (Gupta et al., 2010, p. 464).

The **organizational capacity** assessment tools have drawn on organizational change and policy implementation research. The tools have focused on local government organizations as these are the key delivery mechanisms for WSUD in Australia. The main attributes assessed include local government commitment to WSUD, organizational resources, and knowledge; and how organizations implement WSUD, that is, the different sections of council that are involved in WSUD implementation, strategy, and culture (see, e.g., Bolton et al., 2007; Morison, 2010; Bos and Brown, 2014).

Understanding **institutional capacity** is undertaken in a similar way, although the variety of organizations assessed is larger, and the issues considered are more systemic. The attributes assessed in mapping institutional context include: professional practices, identity, and mission (the purpose water management in society, and the policy agenda articulated by government); beliefs and cognitive frames (ways that individuals filter feedback, opportunities, emerging issues, and pressures from the broader contextual environment); governance setting (regulations, incentives, etc.); public discourse; inter- and intraorganizational relations; space for innovation and learning indicated by a culture supportive of experimentation; and strategic managerial support for innovation (Bettini et al., 2015b, p. 70).

Implementing the organizational capacity assessment tools are typically undertaken through surveys and interviews or workshops. The participants of the surveys are generally local government officers involved in WSUD implementation (e.g., engineers, planners, environment officers). Capturing the perspective of executives and elected officials is also considered important to evaluate the organization's commitment to WSUD (Morison et al., 2010). Workshops and face-to-face interviews provide opportunities to validate the survey data collected (Morison et al., 2010) as well as explore the key issues raised in more detail. Another advantage of workshops is that they offer a forum for the local government officials to discuss the key issues, challenges, and solutions in an open and constructive way, which in turn can raise awareness across the organization and possibly stimulate organizational change to improve processes (Bolton et al., 2007).

Institutional capacity mapping also involves interviews and workshops, where participants initially describe the institutional setting surrounding the institutional response to water shortages in the case study cities of Perth and Adelaide (Bettini et al., 2015b). From the interviews and workshops, the institutional context was mapped, including the dynamics that facilitated and constrained change (Bettini et al., 2015b). Similar methods were advocated when assessing adaptive capacity (Gupta et al., 2010).

## 23.2.8 Where is WSUD capacity building happening?

Targeted WSUD CBPs are common across major cities in Australia. WSUD capacity building is also active in the United Kingdom, New Zealand, United States, and Europe. Additionally, capacity building strategies have also been implemented through targeted policies and funding programs such as the Yarra River Action Plan in Melbourne (Edwards et al., 2007; Morison, 2010; Urrutiaguer et al., 2010) and the New Zealand LIUDD program (Heslop and Hunter, 2007). The stand-alone CBPs with an independent source of funds are considered to be a primary means for building water industry capacity for WSUD (Brown and Farrelly, 2009a). The features of the main programs are presented in Table 23.1, which focuses on Australia and includes CIRIA in the UK, which is a long-standing CBP and since 2012 has had a sustainable urban drainage focused section, susdrain.org. CIRIA is membership funded and is an independent organization focusing on improving construction practices, including sustainable drainage (CIRIA, 2017). Initiated in 1960, it has assisted sustainable urban drainage professionals for more than 20 years, and more recently focused on sustainable drainage and established the susdrain.org website and community in 2012 (CIRIA, 2016). CIRIA has established networks that help facilitate information sharing and idea exchange among professionals and, also organizes events that further facilitate relationship building and networking across the industry (CIRIA, 2017). Additionally, CIRIA undertakes research and thus positions itself as an industry knowledge leader and also conducts professional training (CIRIA, 2017). The susdrain website hosts numerous technical resources (guidelines, case studies, fact sheets, research results, etc.), and as a subset of CIRIA has positioned itself as a knowledge leader and broker (Susdrain, 2017a). There are five key public (Environment Agency) and private partners and over 25 supporting organizations (both public and private) (Susdrain, 2017b) and project support comes directly from CIRIA (Susdrain, 2017a). Between 2012 and 2016, the susdrain website had 79,000 visitors and distributed its newsletter to over 11,000 people, and 930 people attended a susdrain event in person (CIRIA, 2016; CIRIA, 2017). As the specialist WSUD subgroup of CIRIA, susdrain focuses on promoting WSUD to local authorities and lead flood agencies (CIRIA, 2016).

The CBPs have different governance structures and funding models, with some being externally funded (e.g., Clearwater) and others relying on user pays fees for the services they provide (e.g., Water by Design, though it also receives state government funding) or membership (e.g., CIRIA). It is also apparent from Table 23.1 that over time the programs can change in terms of funding. It appears that a secure source of funding contributes to program longevity and stability (Dahlenburg and Lees, 2004). The CIRIA program is a key example of this.

## 23.2.9 2000–15—Lessons learned

Capacity building for WSUD has advanced substantially since 2000. Initially, the academic research grappled with identifying the barriers to WSUD implementation, and how these could be addressed through CBPs (Keath and White, 2006a). Through the establishment of specific, targeted CBPs such as Clearwater in Melbourne, Water by Design in South East Queensland, and the WSUD in Sydney Program, the relevance of capacity building initiatives was demonstrated (Context, 2004, cited in Keath and White, 2006a).

A key landmark in WSUD capacity building was the publication of a chapter on institutional capacity in Australian Runoff Quality (Brown et al., 2006). This chapter brought together the concept of institutional capacity and clearly explained its components in the form of four nested spheres (individuals, intraorganizational, interorganizational, and administrative and regulatory framework) (Fig. 23.2). Its publication in a traditionally technical manual for engineering

TABLE 23.1 Key Features of WSUD Capacity Building Programs in Australia, the United Kingdom, and the United States

| Name | Scope/ Objective | Started | Governance | Host Organization | Funding | Strategies | Guiding Legislation |
|---|---|---|---|---|---|---|---|
| Clearwater, Melbourne | • Initially focused on building stormwater management capacity of local government and industry professionals.<br>• Extended to WSUD/water sensitive cities. | 2002 | • Steering Committee with representatives from Melbourne Water; Water Sensitive Cities CRC.<br>• Program delivered by **four** full time staff. | Melbourne Water | Recurrent funding from Melbourne Water. Three Victorian Government agencies provide additional project specific funding. | • Website http://www.clearwater.asn.au.<br>• Web based tools and resources, e.g., Case Studies, Fact Sheets, Presentations, Video Gallery and Image Gallery, Papers, and Reports.<br>• Issue based events program—Hot Topics.<br>• Technical Site Visits Program.<br>• Technical Training Program.<br>• Guidelines e.g., Melbourne Water MUSIC Guidelines; Precinct Structure Plan Guidelines; Developing a Strategic Approach to WSUD Implementation. | Victoria Planning Provision. Clause 56.07–4 Urban Runoff Management Objectives and Standard (C25). Including water quality and runoff quantity. |
| New Waterways, Perth | • Enable excellence in integrated water cycle management and build capacity of government and industry practitioners. | 2006 | • Partnership delivered via an MOU among the following organizations who are represented on a Board of Management: Dept of Planning; Dept of Water; WALGA*, UDIA (WA); Swan River Trust.<br>• Program delivered by one full time Program Manager. | Western Australian Department of Water | Not currently available | • Website http://www.newwaterways.org.au.<br>• A series of tools and resources such as picture library, weblinks, etc.<br>• A series of lecture based seminars covering the following topics: Retrofitting for WSUD. Introduction to stormwater practices. Better Urban Water Management—Everything you need to know. | • State Planning Policy 2.9.<br>• Planning Bulletin 61 Urban Stormwater Management. |

Continued

**TABLE 23.1** Key Features of WSUD Capacity Building Programs in Australia, the United Kingdom, and the United States—cont'd

| Name | Scope/ Objective | Started | Governance | Host Organization | Funding | Strategies | Guiding Legislation |
|---|---|---|---|---|---|---|---|
| Water by Design, Brisbane | • Support the uptake of WSUD in South East Queensland by Government and Industry. | 2005 | • Steering committee with representatives from State Govt; Local Govt (LGA and three councils), Industry (SIA, AILA, and UDIA). Project Reference Groups established as required.<br>• A Scientific Expert Panel provides scientific advice, as required.<br>• Program delivered by **three** permanent staff. | Healthy Waterways Partnership Renamed Healthy Land and Water ... from 2017 | Initial funding AUD$3M over 3 years. Now reduced. Currently jointly funded by Local Governments of South East Queensland, User-pay contributions for various products and services. | • Website http://waterbydesign.com.au/ a series of web-based tools and resources. e.g., Case Studies, Fact Sheets, etc.<br>• Online Community of Practice—technical advice and support.<br>• Guidelines, e.g., Business Case Costs and Benefits; Concept Design Guideline; Stormwater Harvesting Guidelines.<br>• Issue-based events and projects to address knowledge gaps, e.g., Meeting the Proposed Stormwater Management Objectives in Queensland: A Business Case. | • Mandated WSUD Design targets: Water Quality; Waterway erosion; Reduced frequency flow disturbance. |
| WSUD Program in Sydney | • Encourage the uptake of WSUD within the Greater Sydney area at the local level. | 2002 Currently under review and redevelopment | • Project Reference Groups are established, as required. e.g., Water Sensitive Cities CRC Consortium.<br>• A Scientific Expert Panel is available via Sydney Metropolitan CMA's Board.<br>• Program delivered by **one** full time staff. | Hosted initially at the Sydney Metropolitan CMA 2006. Currently hosted by Dept Local and Land Services | $100,000 per annum since its inception. Funding via National Grant Funding. Income derived from training provides seed funding for issue based projects. | • Website http://www.wsud.org.<br>• Houses a series of tools and resources. E.g. Case studies, picture library, fact sheets, etc.<br>• Technical Training Program using existing modules: Introduction to WSUD. Climate Change and Sydney's Stormwater Infrastructure. Licensing and regionalization of Clearwater and Water by Design technical training modules.<br>• Technical Site Visit Tour Program.<br>• Issue-based projects and events program. | • Environmental Planning and Assessment Act, 1979—State Environmental Planning.<br>• Local Government Act, 1993s.<br>• Implemented via Development Control Plans governed by Local Environment Plans. |

| | | | | | | | |
|---|---|---|---|---|---|---|---|
| CIRIA, UK (Construction Industry Research & Information Association) & susdrain.org | • An impartial, independent, and not-for-profit body, providing business improvement services and research; facilitates relationships and collaborative activities among organizations with common interests. | CIRIA established 1960; susdrain.org established in c. 2012 | • Executive Board, responsible for the overall business. Comprises senior construction professionals.<br>• Council meets biennially to set strategic direction.<br>• Expert panels for key issues. | Stand-alone organization | Membership fees.<br>Users pay for training, publications;<br>Hiring of meeting rooms. | • Website: www.ciria.org & www.susdrain.org providing case studies.<br>• Training provided face-to-face & online.<br>• Bookshop.<br>• Briefings published monthly on topical issues. | Information not available |
| Washington Stormwater Center | • Helps NPDES permittees and stormwater managers to navigate stormwater management challenges through providing training, research, information, technical assistance. | 2010 | • Curriculum Steering Committee.<br>• Advisory committee (until June 2011).<br>• Management team of Director, Deputy Director.<br>• Technical, research, program staff (total 12 although not all full time). | Joint arrangement between City of Puyallup, Washington State University and University of Washington | Initial funding provided by Washington Department of Ecology, transitioning to user pays for training.<br>Likely in-kind support from partner organizations. | • Website http://www.wastormwatercenter.org/, a series of web-based tools and resources. e.g., Case Studies, research results, technical guidance, etc.<br>• Provides basic facilitation for local government officers in relation to NPDES permits.<br>• Training in low impact development moving to online and limited face-to-face delivery.<br>• Gateway to information on emerging best management practice technologies.<br>• Resources for catchment planning, public outreach and involvement, NPDES permitting process and requirements, Stormwater Channel (a YouTube video channel), research outcomes, examples of policy instruments, management plans, etc., lunchtime webinars. | • Washington State legislation House Bill 2222, since codified in RCW 90.48.545. |

*AILA*, Australian Institute of Landscape Architects; *CMA*, Catchment Management Authority; *LGA*, Local Government Association; *NPDES*, National Pollutant Discharge Elimination Scheme; *SIA*, Stormwater Industry Association; *UDIA*, Urban Development Institute of Australia; *WALGA*, Western Australian Local Government Association. Reproduced from Alluvium and Kate Black Consulting, 2012; CIRIA, 2016, 2017; Susdrain, 2017a; Washington Stormwater Center, 2017.

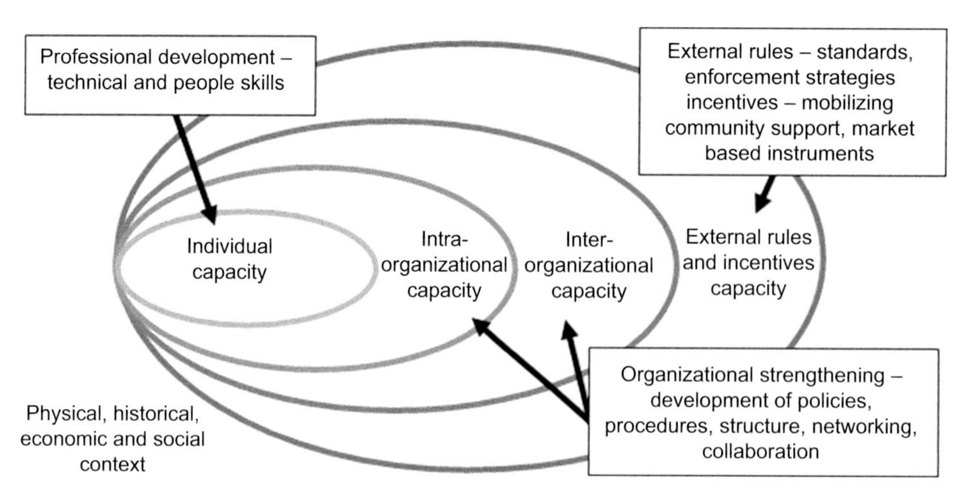


FIGURE 23.2 Framework for individual, intraorganizational, interorganizational, and administrative and regulatory capacites. *Reproduced from Brown, R.R., Mouritz, M., Taylor, A., 2006. Institutional capacity, in Australian runoff quality: a guide to water sensitive urban design. In: Wong, T.H.F. (Ed.), Engineers Australia, Barton, Australian Capital Territory, pp. 5-1–5-21.*

design raised awareness of the socioinstitutional setting for WSUD application. Additionally, Brown et al. (2006) identified capacity building interventions across each of these four spheres. This model of institutional capacity has influenced other CBPs internationally (e.g., Heslop and Hunter, 2007).

As further research has been conducted, our understanding of the factors influencing organizational, and specifically local government, capacity has improved. Local government capacity and commitment were identified as key attributes influencing the potential for WSUD to be implemented through an intergovernmental partnership–funding arrangement (Morison and Brown, 2010). More recently, a specific program evaluation was undertaken in Sydney for the CRSI (Bos and Brown, 2014). Both of these evaluations provide clear evidence of the benefits of CBPs (Eggleton et al., 2012; Bos and Brown, 2014). Although this academic knowledge has been developing, industry has been designing and implementing purpose-specific WSUD CBPs. Some of these have now entered their second decade and continue to lead the industry into better implementation of WSUD, such as Water by Design (QLD) and Clearwater (Vic). The number of WSUD CBPs is also growing (see Table 23.1). Beyond the scope of organizational capacity, institutional capacity is complex and multifaceted.

From this summary of the progress in WSUD capacity building, we posit that our understanding of water sector capacity has improved, and many capacity building interventions have been developed, most notably, those as stand-alone programs. Additionally, capacity assessment tools have been developed and implemented, which will further contribute to stand-alone CBPs and intergovernmental program implementation.

## 23.3 CASE STUDY—SOUTH AUSTRALIA

The state of South Australia is used here as a case study to illustrate a real-world example of how the industry has approached the scoping, design, delivery, and evaluation of a WSUD CBP. It is a useful case study as there are acute water, urban growth and city design issues, and the CBP, **Water Sensitive SA**, is the most recent in Australia to commence.

A steering committee oversaw the development of the business case, and a new steering committee was established to oversee the implementation of the program. The Adelaide and Mount Lofty Ranges Natural Resources Management Board was critical to their success.

### 23.3.1 The business case

WSUD CBPs require a clear business case to secure funding and resources.

The business case for WSUD capacity building (as opposed to the case for WSUD itself) addressed the following drivers:

- Improving the quality of the design, construction, and maintenance of assets;
- Protecting the investment in WSUD assets through improved monitoring, maintenance, and evaluation to assist future designs;

- Meeting practitioners' need for professional development and knowledge on new technologies;
- Supporting and assisting the delivery of government policy; and
- Transferring research to practice.

The business case can be a short or long exercise and may be done with or without stakeholder consultation. We recommend the following approach in developing a business case:

- Complete a report documenting the business case;
- Collect and collate good quality data throughout the process to assist in building an evidence-based business case;
- Engage with practitioners to ensure that the needs of the industry are clearly understood, and a program is targeted to the needs of the industry; and
- The champions, or coordinators, of a campaign to establish a CBP must also engage at a strategic level with the key potential funding bodies.

South Australian water and NRM agencies completed a business case for WSUD capacity building in 2012, and we suggest it is an excellent exemplar of a comprehensive business case process (see watersensitivesa.com).

To establish a CBP, one of the leading WSUD advocates in South Australia, the *Adelaide and Mt Lofty Ranges NRM Board*, determined that it would be beneficial to clearly articulate a business case. The case could be used to not only identify the appropriate structure and work plan for a CBP, but also to assess the needs of the industry, and become a tool to secure funding to implement the program.

In May 2012, the Adelaide and Mt Lofty Ranges NRM Board engaged consultants (Alluvium Consulting and Kate Black Consulting) to develop the business case for the implementation of a CBP for water sensitive urban design for South Australia. The project was overseen by a steering committee that represented a range of stakeholders across the South Australian WSUD industry.

The business case objectives included an assessment of the need for a WSUD CBP in SA; a structure that addressed the assessed needs; an implementation plan that set out the "who, what, when, where, why and how" of the WSUD CBP; and the benefits to SA of implementing the CBP.

This business case assessed the needs for a WSUD CBP for SA according to the following three themes:

- **Policy implementation**: The capacity of the industry to implement the State Government's Water for Good (2009) plan and The 30-Year Plan for Greater Adelaide (2010)—http://livingadelaide.sa.gov.au/
- **Managing existing infrastructure**: The expertise of industry to manage construction and maintenance of urban water infrastructure, and transitioning the whole city stormwater infrastructure to a new standard over the life of the assets
- **Meeting industry needs**: Meeting the needs of the existing WSUD industry and facilitating a more efficient and increased uptake of WSUD.

The business case was informed by three major inputs: a literature review; a review of interstate WSUD CBPs; and industry opinion obtained through workshops, tours, and online surveys. Between June and September 2012, eight research activities, both qualitative and quantitative in nature, were used to collect data from stakeholders. These engagement activities included an online practitioner survey, workshops, site visits, one-on-one interviews, an online discussion forum, and sector-specific forums and meetings. The "Water Sensitive SA" brand, including a dedicated website (www.watersensitivesa.com), was created to provide a platform for this engagement.

The engagement process resulted in feedback from approximately 400 people, associated with 16 industry associations, 100 community organizations, and 22 local councils. Gaps in industry capacity (skills, knowledge, and governance) identified from this engagement were:

- Institutional capacity. The ability of various agencies and sectors to work together, collaborate on projects, and discuss and debate the technical, political, and socioeconomic issues.
- Lack of project experience. Compared to other urban centers around Australia, there were relatively few examples that could be used for training and other educational uses.
- Engineering guidelines. A lack of awareness and application of existing WSUD technical guidelines, and a desire to localize several of the interstate guidelines to South Australia's biophysical context.
- Coordinated approach to training. Although several organizations offered technical WSUD training, there was no central place for practitioners to mix with other disciplines and discuss issues.
- Advocacy. Limited advocacy at an industry-wide level to create momentum toward achieving more water-sensitive outcomes at the local level.

- Policy. The absence of a state policy to guide and/or mandate WSUD for all types of development was considered to inhibit the uptake and implementation of WSUD.
- Life-cycle costing. A lack of understanding of both the capital (CAPEX) and operational costs (OPEX) associated with WSUD was identified as a key barrier to its adoption.
- Monitoring and evaluation. Further monitoring and evaluation was required to establish the benefits of WSUD.

The Business Case documented the potential benefits of a CBP (which is different from the overall benefits of WSUD). These benefits were both of a direct and indirect character, but the majority were not readily quantifiable in a financial sense.

The first benefit was improved design and construction of assets. Based on a 1%, 5%, or 10% improvements in efficiency, the potential savings in WSUD costs for SA was estimated at $120,000–$1,200,000 per annum through the improved design, construction, operation, and maintenance that a CBP would enable.

The second benefit was the support of practitioners. Supporting champions within a region or within individual organizations delivering WSUD, is a key strategy to growing the WSUD industry across South Australia. Supporting passionate individuals is an important strategy in creating momentum, as it empowers and reenergises practitioners and shows institutional support for the work they are doing.

There are also benefits that a CBP can deliver by avoiding potential risks that are expected to occur without a program in place. The two most immediate risks from NOT implementing a CBP are the lack of industry support for WSUD policy at a State level and a possible industry backlash. The ability to attract professionals to South Australia and to secure further grants (Australian Government and others) will also be at risk without a CBP.

An additional cost of not implementing a CBP is the financial liability that poor WSUD design, construction, operation, and maintenance poses to local councils. If a system needs resetting or modifying due to the lack of knowledge and skills of the designers, or inadequate consultation with stakeholders, the cost of WSUD will increase. This is an avoidable cost and a compelling reason to support a CBP.

The business case was finished by documenting a recommended structure, host organization, and program plan over the first 3 years. It was informed by administration systems from similar programs in other jurisdictions.

## 23.3.2 Matching needs with capacity building program

A comprehensive survey was conducted to gather in depth knowledge about WSUD information needs of practitioner. A critical success factor of a WSUD CBP is the ability to address the specific needs of the industry.

Any survey of practitioners transverses several disciplines (engineering, planning, landscape architecture, maintenance, community engagement), is generally difficult, and requires considerable effort to reach a representative sample of an entire population or industry group.

The survey conducted for the South Australian business case was promoted through a variety of industry groups, government agencies, professional bodies, social media, and personal networks. They included:

- Australian Institute of Architects
- Australian Institute of Landscape Architects
- Australian Institute of Project Management
- Australian Water Association
- Civil Contractors Federation
- Consult Australia
- Engineers Australia
- The Goyder Institute for Water Research
- Housing Industry Association
- Hydrological Society of SA
- Institute of Public Works Engineering Australia
- Local Government Association of SA
- Planning Institute of Australia
- South Australian Local Government Supervisory Officers' Association
- Stormwater Industry Association SA
- Urban Development Institute of Australia.

The survey resulted in 346 responses, which was considered by the steering committee (made up of representatives from Adelaide and Mount Lofty Ranges NRM Board, Department of Environment, Water and Natural Resources, South Australian Murray-Darling Basin NRM Board, Stormwater Industry Association, Institute of Public Works Engineering Australia, Local Government Association of South Australia, and Environment Protection Authority South Australia) to be very successful.

## 23.3.3 Governance

The governance structures for a WSUD CBP were based on three key principals:

- Not for profit;
- Commercial or limited liability company; and
- Hosted within a government agency.

Each of these governance structures comes with different management structures, as well as financial reporting obligations. The key funders will most likely determine the structure of the entity.

Some pros and cons for the governance structures are shown in Table 23.2, with a focus on how they apply to WSUD CBP.

Following the completion of the business case, the Adelaide and Mt Lofty Ranges NRM Board initiated a "Bridging Program Manager" project to obtain additional funding and resolve finer details of the program delivery model.

The Bridging Program Manager project identified four delivery model options:

1. Program staff employed directly by the Department of Environment, Water and Natural Resources on behalf of the Adelaide and Mt Lofty Ranges NRM Board.
2. Program staff employed directly by another WSUD stakeholder organization.
3. Program legally constituted as an independent not-for-profit entity.
4. Program delivered by a service provider directly contracted to the Adelaide and Mt Lofty Ranges NRM Board.

The fourth option was determined to be the most practicable option at the time (2014), and a service provider was contracted to run *Water Sensitive SA* following a public tender process.

Program governance and overview was established in the form of a volunteer steering committee, who were appointed through an open tender EOI process, representing the interests of South Australian WSUD stakeholders.

Water Sensitive SA was established as an independent entity, with "arm's length" separation from any one funder. This has given it the freedom to establish its own credibility, independent of the Adelaide and Mt Lofty Ranges NRM Board (or any other funder), and to pursue advocacy positions without any direct influence from government entities.

**TABLE 23.2** Pros and Cons of Different Governance Structures

| Structure | Pros | Cons |
|---|---|---|
| Not for profit | • Increases chance of relationships and partnerships.<br>• Can attract philanthropic funds.<br>• Ensures program manager is cost conscious. | • Unable to commercialize any new courses, materials, or ideas. |
| Commercial or limited liability company | • Ensure clear focus on financial sustainability.<br>• Enables a stand-alone brand to be established. | • Legal difficulties to partner with some industry groups and government agencies.<br>• May restrict delivery of some important resources or information because of high cost/low return on investment. |
| In house within a government agency | • Start-up costs provided by host agency.<br>• Usually provides for secure funding. | • Must report financial and business performance using reporting protocols of the host agency.<br>• Can be seen as an extension of the government agency, rather than as an independent entity.<br>• Difficult to advocate for policy changes that contradict position of host agency or state government. |

### 23.3.4 Funding

There are three options for funding a WSUD CBP:

- Sole funder;
- Multiple funding partners; and
- Customer or retail model.

Other options include crowd funding, philanthropic funding, and levy options. A levy has been used in the Upper Parramatta River Trust—http://pandora.nla.gov.au/tep/40487 that delivered capital works, as well as capacity building across the catchment.

**Sole funder**. The sole funder model is conceptually simple in that only one organization, with sufficient funds, is involved in funding the CBP, thereby reducing the administration and reporting requirements associated with multiple funders. The main risk is that a change in support or strategic direction by the funder may result in closure of the whole CBP.

**Multiple funders**. A CBP with multiple funders is more resilient to organizational change and funding cycles, but it requires more work in convincing funders to agree to a contribution, service the different administration requirements of multiple partners, build trust, and report back to them.

**Customer funding**. Various CPB around Australia rely on a cost recovery model for individual training sessions. However, none have yet demonstrated a financial model whereby customers provide sufficient revenue to support a whole program. For example, a staff of 1.5 people running a CBP would need of the order of AUD$150,000 per year to cover costs. If on average participants paid $150 for a training session, then a program would require 1000 paying customers per year. The program would need to deliver a training course each week, with 20 fully paid tickets, to reach that revenue goal. After allowing for public holidays, etc., 25 or 30 fully paid tickets per event would be required. This is a lot to organize, and a lot to expect in terms of patronage from full paying customers.

The process of securing funding should not be underestimated. It requires constant effort, a good understanding of the industry and technical issues associated with WSUD, and a good understanding of the strategic and political environment in which funds are collected and spent.

There is research (Dahlenburg and Lees, 2004) to suggest that grant funded projects or a program based on one-off grants is not sustainable, and once funding ceases, the benefits are quickly lost to the funders, participants, and co-ordinators that are within the CBP.

### 23.3.5 Business planning/strategic direction

A critical component of a WSUD CBP is the development, and implementation of a business or strategic plan. The steering committee and program manager are usually responsible for developing the strategic direction, and following its agreement, the program manager will draft a 1–3-year business plan. Funding is set at this point, so that strategy and the business plan are bounded by the available funds.

WSUD CBPs often grapple with the degree to which their strategic direction encompasses advocacy compared to increasing the capacity of practitioners and organizations. The risk of taking on an advocacy role is that the program will be seen to be aligned with some parts of the industry, and not a neutral program and a trusted resource for all the industry. This risk can be managed.

For Water Sensitive SA, a 3-year business plan was created and signed off by the steering committee. The plan has been the guiding document for the delivery of the program.

The business plan includes:

- Vision
- Program goals
- Priority projects
- Key performance indicators (KPIs)
- Implementation targets
- Financial plan
- Risks
- Communication strategy
- Action plan.

## 23.3.6 Resourcing

Resourcing is a critical component and closely tied to funding. Typically, with capacity building and labor-intensive programs, 80%–90% of funding is required for staff, with the remaining funds available for communications and events.

WSUD CBPs must be conscious of the skill set and personality of the program manager and staff. Experience from Australia has shown that good program managers have reasonable knowledge of WSUD but are not technical experts. Rather, they must have skills across policy, communications, advocacy, event management, and civil engineering to be a credible and knowledgeable program manager.

Taylor et al. (2008) documented the drivers of environmental champions, which reinforces and supports the fact that drivers of change (staff within a WSUD CBP are very much drivers of environmental change) need a range of skills that are useful in running a successful program. These include the ability to be a good networker, sell a vision, and get others to buy into that vision, manage and lead people, engage with people across disciplines, identify the windows of opportunity, trust others, have an authentic leadership style, and have a mentor.

Activities can increase when grant funding, i.e., funding that is different to and above the usual operating budget. This extra funding enables the WSUD CBP to grow in staff numbers to deliver specific programs, which are policy or research related. WSUD CBP managers must be cognizant of the opportunities to secure these grants as they arise.

## 23.3.7 Communications

Communications are critical for WSUD CBPs, as much of the strategic plan relates to engaging practitioners, industry groups, and disseminating research. Communications are fundamental to all aspects of a CBP, including the marketing of training events, content management on a website, and coordinating advocacy.

The Water Sensitive SA program developed a brand and website very early in its journey, which then enabled it to launch a series of events, engage the wider industry under one banner, and start to clearly establish a program that is known to support practitioners and raise awareness of WSUD in South Australia.

There is a need to continuously share and engage with the community in a timely manner. Practitioners are subject to information from many internal and external sources, which increases the difficultly for a WSUD CBP to cut through the noise and get information through to practitioners. It is even more difficult to reach those that are not interested or aware of WSUD.

Water Sensitive SA has used a range of communication channels to reach practitioners and share the work and outcomes of the program. These include:

- A dedicated website;
- Blog on website;
- A regular newsletter;
- Printed banners at industry events;
- Partnerships with other industry associations;
- Twitter; and
- Face-to-face meetings.

Example communications from Water Sensitive SA are shown in Fig. 23.3.

**FIGURE 23.3** Examples of Water Sensitive SA's methods to share WSUD capacity building events and activities.

### 23.3.8 The first three years

Water Sensitive SA commenced in 2014 and used the business case to guide its strategic direction and work plan. The plan was broadly to deliver an information portal, localize of interstate WSUD guidelines, deliver technical training programs, and facilitate a Community of Practice. The plan was to engage practitioners at all levels of learning—self-directed, directed, mentoring and peer-to-peer, and rely on a well-recognized local leader to run the program.

The Adelaide and Mt Lofty Ranges NRM Board provided a significant administrative and in-kind support, as well as leadership to ensure that the program started. The Board acted as the hosting organization, and a steering committee was established reflecting representatives from industry, government, academia, and experts in water, green infrastructure, and science.

The Water Sensitive SA brand, created in 2012 during the business case period, was retained, and a logo created. A program manager and communications expert were engaged to run the program.

In the first 3 years, the focus was on establishing brand recognition and becoming the primary source of technical information on WSUD; and delivering training on key topics identified by the industry and practitioners, and advocating for WSUD across government and industry.

### 23.3.9 What has been learned

In early 2017, an evaluation of the Water Sensitive SA program was undertaken. The key elements considered in the review included:

- Overall impact of the WSUD CBP program;
- Perception of stakeholder satisfaction;
- Engagement with practitioners;
- Engagement with leaders;
- Development and maintenance of partnerships;
- Delivery of business plans; and
- Securing funding.

The evaluation report was based on two data sets: perceptions of funding partners collected through 16 one-on-one interviews with individual funding partners, and perceptions of practitioners gathered through an online survey. Overall the funders and practitioners were very supportive and positive about the work to date and would like to see the program continue.

One of the tasks for Water Sensitive SA was to advocate for changes to the state planning scheme, so that WSUD is mandated in new developments. The feedback from funders and practitioners was that it has not met expectations in achieving change in this area. Another change objective of Water Sensitive SA was achieving "culture change," which is a long-term goal. Feedback from interviews reports a perception that little progress has been achieved to date.

The key results from the online survey are:

- 84% of practitioners are satisfied or very satisfied with the program to date; and
- 96% of practitioners believe that the program should continue.

The funding partners were reasonably consistent in their perceptions of the program, and these are represented graphically in Fig. 23.4.

## 23.4 FUTURE OF WSUD CAPACITY BUILDING

There are many factors that in the future may influence the tools, structure, and effectiveness of WSUD CBPs.

### 23.4.1 Extreme events and environmental shocks

The core outcome of WSUD is to significantly improve the environmental values of the creeks, waterways, lakes, bays, and gulfs, where stormwater flows into after rainfall events. There is a high probability that in the future there will be more extreme events across the world that will highlight the relationship between cities, rainfall, and receiving waters. Extreme rainfall event will impact on the visual acuity of bays, beaches, and rivers, highlighting the value and need for WSUD. An example of an extreme rainfall event and its visual impact occurred during the 2011 floods that followed the Millennium drought in Australia.

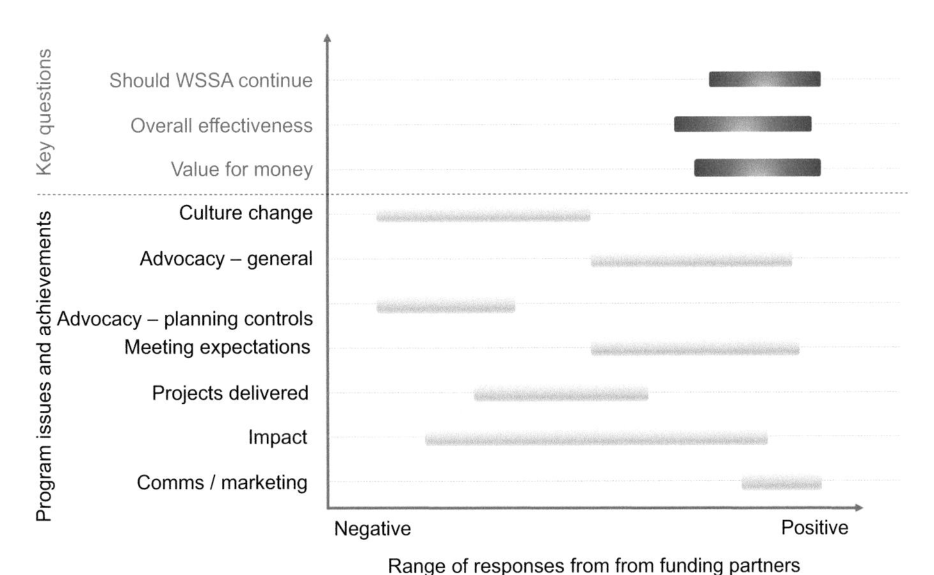


**FIGURE 23.4** Perceptions of WSUD capacity building based on qualitative survey of stakeholders: Water Sensitive SA (WSSA).

Algal blooms, black water events, odors, dead or sick marine life, reports of toxic marine life, flooding, sick swimmers, and many more major events, are all possible, in all major capital cities, and will create an immediate driver and support for WSUD and the associated capacity building needs. Chapters 6–12 cover various aspects of WSUD and associated ecological health benefits and should be viewed for more detail on these issues.

Most of the urban water supply in Australian cities comes from surface impoundments and we argue that the storage levels in these dams can be key triggers in driving future change in WSUD CBPs. When their storage levels are consistently low, as they were during the almost decade long Millennium drought of the 2000s, a vigorous public conversation about the value of water emerges, with a focus on how to capture local rainwater and stormwater (Van den Honert and McAneney, 2011; Rowley, 2016). Future droughts, and the associated change in the cities' water storages, are predicted to be significant driver of support for WSUD and capacity building.

### 23.4.2 Climate change

Climate change is forecast to fundamentally change the hydrological regime in cities around the world (Bureau of Meteorology, 2016; IPCC, 2014). Climate change is an important issue that links to WSUD and capacity building, as it results in more extreme events and highlights the need for more resilient infrastructure in cities.

The update to Australian Rainfall and Runoff (ARR, 2016) has explicitly stated that climate change can, and should, be considered for stormwater assets that are designed to perform over the long term.

More extreme weather and catastrophic events that result in damage to property, public infrastructure, and loss of life would create a large impetus for changes to the way cities and water assets are built, and would most likely result in increased adoption and delivery of WSUD across urban areas.

### 23.4.3 WSUD changes

The future design and construction of WSUD may change over time as new research is published and new technology identified, which may decrease the complexity of the systems and probably decrease the per unit cost of the technology. These changes will change the way WSUD capacity building is structured and delivered, creating a need for new training course.

The development of modular systems that reduce the cost and reduce the impact of construction is one such potential change. This would then shift the needs of WSUD capacity building to assist practitioners identify how to understand the benefits and select good quality products, consider the suitability of the products to the problems in their region, and understand how to effectively monitor and maintain those products.

Along with design and product changes is the potential for computing power/artificial intelligence to reduce the lead time between identifying the appropriate locations of WSUD assets, analyze a catchment to size the appropriate asset, installing the infrastructure, and then using a variety of remote sensing and telemetry systems to monitor the WSUD assets. This would result in a change in the type of topics and knowledge required by practitioners.

One such example is the use of aerial imagery over a period of time, at a specific location. Google Earth and Nearmap are two such services that allow practitioners to review the change in the catchment land use, and change in the planning and construction of a local site (such as a road reserve), with relative ease. WSUD CBPs could help practitioners understand the data, the process, and the results and application to designing and constructing WSUD assets.

### 23.4.4 Advancements in training

The methods of capacity building, and approach to the education sector more broadly, has changed dramatically over the past decade. Universities now routinely use online forums, webinars, stream lectures live, and allow for online submissions for assignments. We argue that graduates of today will be very comfortable in accessing information using these techniques in their professional careers.

Hence the tools and techniques that WSUD CBPs use today will need to adopt these new technologies to deliver training and distribute knowledge bases. For example, Water Sensitive SA has introduced two interactive features online: Online forums and an Interactive Map. At this stage, both products are attracting only low engagement (i.e., number of people posting questions and discussing issues online), but we believe they will become part of the standard tool kit that all WSUD CBPs rely on as practitioners become more comfortable with these formats.

### 23.4.5 Economic evaluation

Changes to the methodologies accounting for the externalities of WSUD, by explicitly internalizing costs and benefits, will have another significant change to WSUD CBPs (see Chapter 14 for further discussion on the economics of WSUD approaches).

This may occur due to two potential changes. Firstly, WSUD CBPs may need to communicate more of the theory and practice of economics, as it applies to WSUD, to help practitioners increase the suite of justifications to adopt WSUD in their jurisdiction. Secondly, as the price signals of mains water (as measured in $/kL) and asset construction (as measured in $\$/m^2$) change, the uptake and adoption of WSUD may rapidly change.

A parallel example to this is the way price signals, as observed through the cost of photovoltaic panels and the associated feed-in-tariffs for the panels, have changed the adoption of that technology across Australia. The first feed-in-tariffs were set at various levels in different states of Australia and were typically about 60 cents for each kilowatt-hour that an individual household exported to the grid. This created an incentive to spend capital on this infrastructure, as not only will the photovoltaic panels replace the need to import electricity to run appliances in a home, but also when there was excess power it could become a revenue source, usually at a rate higher than was being paid for imported electricity. As more people adopted photovoltaic panels, and the cost to the energy retailers, networks, and government increased, state governments ended these high feed-in-tariffs and replaced them with reduced tariffs.

### 23.4.6 Technology and the "Internet of Things"

Technology and the "Internet of things" is a disrupter of many industries and sectors and has the potential to be a disrupter of the WSUD industry as well. The collection of data, analysis of large datasets, and the potential to present the data in real time to responsible authorities is a potential disrupter.

This means that WSUD practitioners may have more data available, more quickly, and be able to make decisions in a more timely manner. For example, it is quite possible that real-time urban microclimate data may be available across a whole city, and provided to practitioners (and the public) in real time. The implications for a WSUD CBP are that practitioners will need to learn more about where these data come from, how to access it, how to incorporate it into the design process, and how to use the data for proactive maintenance.

It is also possible that there may be real-time monitors of flows and pollutants in thousands of stormwater drains across a city. It would allow authorities to identify hotspots, identify poor performing assets, and calibrate the effectiveness of WSUD designs.

## 23.5 CONCLUSIONS

WSUD capacity building is a critical component of improving the quality, quantity, and value of water resources and amenity in urban landscapes. The first WSUD CBPs were developed in Australia in 2005, and since then a program has emerged to support practitioners in each capital city. The knowledge, systems, and impact of the capacity building industry are still evolving, due to its relatively short history and limited personnel involved to date.

The conclusions from almost two decades of experience in this sector, and in particular the most recent experience in South Australia, are as follows:

- The theory of WSUD capacity building is clearly outlined in Brown et al. (2006), and this work underpins the delivery of capacity building across all CBPs in Australia.
- A clear business case is required to convince government agencies, local government, nongovernment, noncorporate entities such as industry associations, and corporations to create new programs.
- Regular evaluation of the program and practitioner needs is critical to ensuring the program is focused on the most pressing issues.
- There is a lack of evidence linking CBPs to environmental change, which is consistent with many other capacity building programs in other sectors (e.g., sustainable transport, waste management, and emissions reductions). However, evidence is emerging from CBP evaluations that demonstrate potential multiple benefits across elements of organizational capacity (see for example Bos and Brown, 2014).
- There is variability between WSUD CBPs across Australia, in terms of their scope and level of resourcing.
- CBPs typically have between one and four staff members, with the optimal number being determined by local program considerations.
- Governance structures require a diverse representation and clear KPIs to guide the programs.
- Funding is an ongoing problem in each state, and the funding mechanisms often go through different phases according to the state of the industry and the policy drivers at the time.
- Although new methods of capacity building are emerging, the key techniques of networking, site visits, and web portals to host locally specific information, guidelines, and case studies are all still very important in effectively delivering a program.
- CBPs will most likely evolve in the future, as new drivers such as climate change, technology, economic evaluation, and advances in training techniques impact on the structure and delivery of the programs.

There is no silver bullet or easy fix to improving the capacity of the industry and designing, building, and maintaining WSUD assets. The best advice is to be aware of upcoming windows of opportunity, where there is strategic alignment between the industry, the policy environment, and available funds. This will assist in increasing the adoption of WSUD and in conjunction with other key drivers of change in urban enviornments may enable a faster adoption and higher profile to a local WSUD CBP.

## ACKNOWLEDGMENTS

The authors acknowledge several individuals, groups, and government agencies that helped in discussing WSUD capacity building and the way ahead: *Mellissa Bradley (Water Sensitive SA), Kathryn Bothe (Water Sensitive SA), Water Sensitive SA Steering Committee, Andrew O'Neil (Healthy Waterways—Water by Design), Carol Jadraque (Clearwater) Jacquie White (Clearwater), Leonie Duncan (Alluvium Consulting Australia), Paul Shaffer (CIRIA), Kate Black (Kate Black Consulting) and Megan Farrelly (Monash University).*

## REFERENCES

Alaerts, G.J., Hartvelt, F.J.A., Patorni, F.M., 1999. Water Sector Capacity Building: Concepts and Instruments: Proceedings of the Second UNDP Symposium on Water Sector Capacity Building, Delft 1996. A. A. Balkema Publishers, Rotterdam, The Netherlands.

Alluvium, Kate Black Consulting, 2012. A Business Case for a Water Sensitive Urban Design Capacity-building Program for South Australia. Alluvium, Melbourne, Australia.

Bach, P.M., McCarthy, D.T., Urich, C., Sitzenfrei, R., Kleidorfer, M., Rauch, W., Deletic, A., 2013. A planning algorithm for quantifying decentralised water management opportunities in urban environments. Water Science and Technology 68 (8), 1857–1865.

Bettini, Y., Brown, R.R., De Haan, F.J., 2015a. Exploring institutional adaptive capacity in practice: examining water governance adaptation in Australia. Ecology and Society 20 (1), 47.

Bettini, Y., Brown, R.R., De Haan, F.J., Farrelly, M.A., 2015b. Understanding institutional capacity for urban water transitions. Technological Forecasting and Social Change 94, 65–79.

Bolton, A., Edwards, P., Lloyd, S., Lamshed, S., 2007. Needs analysis: an assessment tool to strengthen local government delivery of water sensitive urban design. In: Proceedings of 13th International Rainwater Catchment Systems Conference and 5th International Water Sensitive Urban Design Conference, Sydney, Australia, 21–23 August 2007.

Bos, J.J., Brown, R.R., 2014. Assessing organisational capacity for transition policy programs. Technological Forecasting and Social Change 86, 188–206.

Bos, J.J., Farrelly, M.A., 2015. Industry Impact of the CaWSC Research Program: Insights from Five Case Studies. Cooperative Research Centre for Water Sensitive Cities, Melbourne, Australia.

Bos, J.J., Brown, R.R., Farrelly, M.A., 2013a. A design framework for creating social learning situations. Global Environmental Change 23, 398–412.

Bos, J.J., Brown, R.R., Farrelly, M.A., de Haan, F.J., 2013b. Enabling sustainable urban water management through governance experimentation. Water Science and Technology 67 (8), 1708–1717.

Bratieres, K., Fletcher, T.D., Deletic, A., Zinger, Y., 2008. Nutrient and sediment removal by stormwater biofilters: a large-scale design optimisation study. Water Research 42, 3930–3940.

Brown, R.R., 2005. Impediments to integrated urban stormwater management: the need for institutional reform. Environmental Management 36 (3), 455–468.

Brown, R.R., Farrelly, M.A., 2009a. Challenges ahead: social and institutional factors influencing sustainable urban stormwater management in Australia. Water Science and Technology 59 (4), 653–660.

Brown, R.R., Farrelly, M.A., 2009b. Delivering sustainable urban water management: a review of the hurdles we face. Water Science and Technology 59 (5), 839–846.

Brown, R.R., Mouritz, M., Taylor, A., 2006. Institutional Capacity, in Australian Runoff Quality: a guide to water sensitive urban design. In: Wong, T.H.F. (Ed.), Engineers Australia, Barton, Australian Capital Territory, pp. 5-1–5-21.

Brown, R.R., Keath, N., Wong, T.H.F., 2009. Urban water management in cities: historical, current and future regimes. Water Science and Technology 59 (5), 847–855.

Bureau of Meteorology and CSIRO, 2016. State of the Climate 2016.

Catchlove, R., van de Meene, S., Duncan, L., Phillips, E., 2012. Insights for future capacity building: reflections on a study tour for water sensitive cities and the value of experiental learning. Water: Journal of the Australian Water Association 39 (1), 42–47.

Cettner, A., Ashley, R., Hedstrom, A., Viklander, M., 2014. Assessing receptivity for change in urban stormwater management and contexts for action. Journal of Environmental Management 146, 29–41.

CIRIA, 2016. Proposal 3044 September 2016-2018. Construction Industry Research and Information Association, London. Available at: http://www.susdrain.org/files/resources/project_info/ciria_proposal_3044_susdrain_2016_2018_160404_.pdf.

CIRIA, 2017. About CIRIA. Construction Industry Research and Information Association, London. Available at: http://www.ciria.org/CIRIA/About/About_CIRIA/About/About_CIRIA.aspx.

Craig, G., 2007. Community capacity-building: something old, something new...? Critical Social Policy 27 (3), 335–359.

Crisp, B.R., Swerissen, H., Duckett, S.J., 2000. Four approaches to capacity building in health: consequences for measurement and accountability. Health Promotion International 15 (2), 99–107.

Dahlenburg, J., 2006. The sustainable water challenge: building capacity, sharing innovations and support networks. In: Deletic, A., Fletcher, T. (Eds.), 7th International Conference on Urban Drainage Modelling and 4th International Conference on Water Sensitive Urban Design, vol. 2, pp. 147–154. Melbourne, 5–7 April 2006.

Dahlenburg, J., Lees, S., 2004. Building the capacity in water sensitive urban design amongst local councils in Sydney. In: International Conference on Water Sensitive Urban Design: Cities as Catchments, Adelaide, 21-25 November 2004, pp. 604–613.

de Haan, F.J., Ferguson, B.C., Deletic, A., Brown, R.R., 2013. A socio-technical model to explore urban water systems scenarios. Water Science and Technology 68 (3), 714–721.

Dobbie, M.F., Brown, R.R., Farrelly, M.A., 2016. Risk governance in the water sensitive city: practitioner perspectives on ownership, management and trust. Environmental Science and Policy 55, 218–227.

Edelenbos, J., Teisman, G., 2013. Water governance capacity: the art of dealing with a multiplicity of levels, sectors and domains. International Journal of Water Governance 1, 89–108.

Edwards, P., Lamshed, S., Francey, M., 2007. Organisational change: transdisciplinarity in urban stormwater quality management programs, keynote address. In: Rainwater 2007, 5th International Conference on Water Sensitive Urban Design. Engineers Australia, Sydney, Australia.

Eggleton, S., Catchlove, R., Morison, P., 2012. Assessing and responding to local government's capacity to deliver sustainable stormwater management, keynote address. In: Rainwater and Urban Design 2007. Engineers Australia, Barton, A.C.T.

Farrelly, M.A., Rijke, J., Brown, R.R., 2012. Exploring operational attributes of governance for change, Keynote address. In: 7th International WSUD Conference, Melbourne, Australia, February 21-23, 2012.

Floyd, J., Iaquinto, B.L., Ison, R., Collins, K., 2014. Managing complexity in Australian urban water governance: transitioning Sydney to a water sensitive city. Futures 61, 1–12.

Grindle, M.S., Hilderbrand, M.E., 1995. Building sustainable capacity in the public sector: what can be done. Public Administration and Development 15, 441–463.

Gupta, J., Termeer, C., Klostermann, J., Meijerink, S., van den Brink, M., Jong, P., Nooteboom, S., Bergsma, E., 2010. The Adaptive Capacity Wheel: a method to assess the inherent characteristics of institutions to enable the adaptive capacity of society. Global Environmental Change 13, 459–471.

Heslop, V.R., 2010. Sustaining Capacity: Building Institutional Capacity for Sustainable Development, Thesis, Planning. University of Auckland, Auckland.

Heslop, V., Hunter, P., 2007. The capacity to change: partnering with local government to strengthen policy, Keynote address. In: Rainwater and Urban Design 2007. Engineers Australia, Barton, A.C.T.

Honadle, B.W., 1981. A capacity-building framework: a search for concept and purpose. Public Administration Review 41 (5), 575–580.

IPCC, 2014. Summary for policymakers. In: Field, C.B., Barros, V.R., Dokken, D.J., Mach, K.J., Mastrandrea, M.D., Bilir, T.E., Chatterjee, M., Ebi, K.L., Estrada, Y.O., Genova, R.C., Girma, B., Kissel, E.S., Levy, A.N., MacCracken, S., Mastrandrea, P.R., White, L.L. (Eds.), Climate Change 2014: Impacts,Adaptation, and Vulnerability. Part A: Global and Sectoral Aspects. Contribution of Working Group II to the Fifth Assessment Report of the Intergovernmental Panel on Climate Change. Cambridge University Press, Cambridge, United Kingdom and New York, NY, USA, pp. 1–32.

Keath, N., White, J., 2006a. Building the capacity of local government and industry professionals in sustainable urban water management. Australian Journal of Water Resources 10 (3), 233–238.

Keath, N., White, J., 5-7 April 2006b. Building the capacity of local government and industry professionals in sustainable urban water management. In: Deletic, A., Fletcher, T. (Eds.), 7th International Conference on Urban Drainage Modelling and 4th International Conference on Water Sensitive Urban Design, vol. 2, pp. 171–179. Melbourne.

Larson, S., Alexander, K.S., Djalante, R., Kirono, D.G.C., 2013. The added value of understanding informal social networks in an adaptive capacity assessment: explorations of an urban water management system in Indonesia. Water Resources Management 27, 4425–4441.

Leidl, C., Farahbakhsh, K., FitzGibbon, J., 2010. Identifying barriers to widespread implementation of rainwater harvesting for urban household use in Ontario. Canadian Water Resources Journal 35 (1), 93–104.

Marshall, N.A., Marshall, P.A., 2007. Conceptualizing and operationalizing social resilience within commercial fisheries in northern Australia. Ecology and Society 12 (1), 1. Available at: http://www.ecologyandsociety.org/vol12/iss1/art1/.

McIntosh, B.S., Taylor, A., 2013. Developing T-shaped water professionals: building capacity in collaboration, learning, and leadership to drive innovation. Journal of Contemporary Water Research and Education 150 (1), 6–17. Available at: https://doi.org/10.1111/j.1936-704X.2013.03143.x>.

Morison, P.J., 2010. Management of Urban Stormwater: Advancing Program Design and Evaluation. thesis. School of Geography and Environmental Science, Monash University, Melbourne.

Morison, P., 2011. Re-thinking institutional capacity-building: lessons from Australia. In: Proceedings of the 12th International Conference on Urban Drainage. International Water Association, Porto Alegre, Brazil.

Morison, P.J., Brown, R.R., 2010. Avoiding the presumptive policy errors of intergovernmental environmental planning programmes: a case analysis of urban stormwater management planning. Journal of Environmental Planning and Management 53 (2), 197–217.

Morison, P.J., Brown, R.R., 2011. Understanding the nature of publics and local policy commitment to Water Sensitive Urban Design. Landscape and Urban Planning 99, 83–92.

Morison, P.J., Brown, R.R., Cocklin, C., 2010. Transitioning to a waterways city: municipal context, capacity and commitment. Water Science and Technology 62 (1), 162–171.

Ostrom, E., 2005. Understanding Institutional Diversity. Princeton University Press, Princeton, New Jersey, USA.

Robins, L., 2007. Capacity-building for natural resource management: lessons from the health sector. EcoHealth 4, 247–263.

Robins, L., 2008. Making capacity building meaningful: a framework for strategic action. Environmental Management 42, 833–846.

Rowley, S., November 1, 2016. Australias lesson for a thirty California. New York Times. Available online at: https://www.nytimes.com/2016/11/01/opinion/australias-lesson-for-a-thirsty-california.html?mcubz=0&_r=0.

Schiffer, E., Hauck, J., 2010. Net-map: collecting social network data and facilitating network learning through participatory influence network mapping. Field Methods 22 (3), 231–249. Available at: http://journals.sagepub.com/doi/abs/10.1177/1525822X10374798.

Sharma, A., Pezzaniti, D., Myers, B., Cook, S., Tjandraatmadja, G., Chacko, P., Chavoshi, S., Kemp, D., Leonard, R., Koth, B., Walton, A., 2016. Water sensitive urban design: an investigation of current systems, implementation drivers, community perceptions and potential to supplement urban water services. Water 8, 272–286.

Sobeck, J., Agius, E., 2007. Organizational capacity building: addressing a research and practice gap. Evaluation and Program Planning 30, 237–246.

Stormwater Committee, 1999. Urban Stormwater: Best Practice Environmental Management Guidelines. Stormwater Committee, Prepared for the Stormwater Committee with assistance from Environment Protection Authority, Melbourne Water Corporation, Department of Natural Resources and Environment and Municipal Association of Victoria, CSIRO Publishing, Collingwood, Australia.

Susdrain, 2017a. About susdrain. Construction Industry Research and Information Association, London. Available at: http://www.susdrain.org/about.html.

Susdrain, 2017b. SuDS Directory: Supporters. Construction Industry Research and Information Association, London. Available at: http://www.susdrain.org/suds-directory/supporters/index.html.

Urrutiaguer, M., Lloyd, S., Lamshed, S., 2010. Determining water sensitive urban design project benefits using a multi-criteria assessment tool. Water Science and Technology 61 (9), 2333–2341.

van de Meene, S.J., Brown, R.R., 2009. Delving into the 'institutional black box': revealing the attributes of future sustainable urban water management regimes. Journal of the American Water Resources Association 45 (6), 1448–1464.

van de Meene, S.J., Brown, R.R., Farrelly, M.A., 2010. Capacity attributes of future urban water management regimes: projections from Australian sustainability practitioners. Water Science and Technology 61 (9), 2241–2250.

van de Meene, S.J., Brown, R.R., Farrelly, M.A., 2011. Towards understanding governance for sustainable urban water management: a practice-oriented perspective. Global Environmental Change: Human and Policy Dimensions 21 (3), 1117–1127.

van de Meene, S.J., Head, B., Bettini, Y., 2016. Toward Effective Change in Urban Water Policy: The Role of Collaborative Governance and Cross-scale Integration. Cooperative Research Centre for Water Sensitive Cities, Melbourne, Australia.

Van den Honert, R.C., McAneney, J., 2011. The 2011 Brisbane floods: causes, impacts and implications. Water 3 (4), 1149–1173.

Washington Stormwater Center, 2017. About Us. Washington Stormwater Center, Puyallup, Washington. Available at: http://www.wastormwatercenter.org/about-us.

Weber, E.P., Khademian, A.M., 2008. Wicked problems, knowledge challenges, and collaborative capacity builders in network settings. Public Administration Review 68 (2), 334–349.

Wolfson, L., Lewandowski, A., Bonnell, J., Frankenberger, J., Sleeper, F., Latimore, J., 2015. Developing capacity for local watershed management: essential leadership skills and training approaches. Journal of Contemporary Water Research and Education 156, 86–97.

Wong, T.H.F., 2006. Water sensitive urban design - the story thus far. Australian Journal of Water Resources 10 (3), 213–221.

## FURTHER READING

Canadian International Development Agency, 2000. Capacity Development: Why, What and How. Canadian International Development Agency, Gatineau, Quebec.

Cooperative Research Centre for Water Sensitive Cities, 2013. Specifying the Urban Water Governance Challenge (Project A3.1 Milestone Report). Cooperative Research Centre for Water Sensitive Cities, Melbourne, Australia. Available at: http://watersensitivecities.org.au/wp-content/uploads/2015/01/A3-1_Specifying-the-urban-water-governance-challenge.pdf.

Farahbakhsh, K., Despins, C., Leidl, C., 2009. Developing capacity for large-scale rainwater harvesting in Canada. Water Quality Research Journal of Canada 44 (1), 92–102.

Government of South Australia, 2009. Water Sensitive Urban Design Technical Manual for the Greater Adelaide Region. Government of South Australia, Adelaide, Australia.

Chapter 24

# Community Perceptions of the Implementation and Adoption of WSUD Approaches for Stormwater Management

**Rosemary Leonard[1], Sayed Iftekhar[2], Melissa Green[3] and Andrea Walton[3,4]**

*[1]School of Social Sciences and Psychology, University of Western Sydney, Penrith, NSW, Australia; [2]Centre for Environmental Economics & Policy (CEEP), University of Western Australia; [3]Josh Byrne & Associates, Fremantle, WA, Australia; [4]CSIRO Land and Water, Brisbane, QLD, Australia*

## Chapter Outline



**ABSTRACT**

In this chapter we will explore five dimensions of people's attitudes to and engagement with WSUD systems. The dimensions are visibility, recreation and other amenity, economic considerations for residents, place attachment, social capital, and community engagement. The chapter will cover community perception based on sociotechnical assessment of selected developments designed using WSUD philosophies. The analysis using social capital and place attachment provides a conceptual framework for understanding support for WSUD systems and reveals the potential for a cycle of prosocial and proenvironmental behavior. Not only are social capital and place attachment created from the benefits of WSUD, but also increased social capital and place attachment supports the ongoing management of WSUD. Interventions that increase awareness of WSUD's benefits, or initiate and support development of community groups, strengthen social capital within a community and support WSUD over the long term.

**Keywords:** Amenity; Community engagement; Economics; Place attachment; Recreation; Social capital; WSUD.

## 24.1 INTRODUCTION

As Radcliffe describes in Chapter 1, WSUD addresses all stages of the urban water cycle including water supply, wastewater, stormwater, groundwater, urban design, and environmental protection (Joint Steering Committee for Water

**Approaches to Water Sensitive Urban Design. https://doi.org/10.1016/B978-0-12-812843-5.00024-1**

Sensitive Cities, 2009); however, in this chapter the particular focus is stormwater and community perceptions of the implementation of WSUD approaches to stormwater management. To set the context, we need to recognize that most people in Australia live in urban areas where there is no pressing reason for the general public to have any understanding of their water system in general, or their stormwater system in particular. Water appears from taps or falls as rain and disappears down drains rarely causing any concern or inconvenience to the public unless there is a flood or local ponding causing nuisance to local residents.

The invisibility of our water, sewerage, and drainage systems is no accident. It is an achievement of over a century of water culture dominated by what Sofoulis (2005) calls "Big Water Dreaming" whereby large scale water projects such as the Snowy Mountains Scheme and Warragamba Dam for Sydney have been lauded a triumph of man harnessing nature. These projects have given us the ability to create a myth of infinite water without effort or responsibility by the water user. Large-scale utilities are usually given a monopoly on bringing water to our gate and taking sewage away in a largely invisible underground process. The monopoly granted to water utilities is usually justified in terms of the huge expense of maintaining and growing the massive infrastructure required for "big water" where there are severe consequences of a major failure. For example, the extreme water shortage during the Millennium Drought in Australia led to severe water restrictions, urgent investment in desalination and water recycling, and the search for alternative water supplies including rainwater tanks and stormwater harvesting.

A positive side of this monopoly is that engineers and scientists within the utilities have taken their role as guardians of the water supply and public health very seriously, and alarm is raised when scientific staffing is reduced (Robertson, 2016). If the public are largely unaware of their mains water system, they are even less likely to be aware of stormwater systems, especially those with a WSUD character. It is thought that the public understanding of WSUD systems may be limited or inaccurate (Roy et al., 2008), whereas Morison and Brown (2011) suggest that WSUD jargon may prevent necessary connections with the public. Even in iconic WSUD developments such as Mawson Lakes and New Haven Village in Adelaide, South Australia, residents often have a low level of awareness of the innovative water systems (Tjandraatmadja et al., 2008; Leonard et al., 2014b; Sharma et al., 2012).

Because WSUD systems are often not managed by the water utilities, they threaten their monopoly and also, perhaps legitimately, the technical experts become concerned about water infrastructure that they cannot monitor. For example, it is not surprising that when the Commonwealth Scientific and Industrial Research Organization (CSIRO), Australia asked experts from utilities in SE Queensland about their priorities for research on rainwater tanks, the experts focused on tank maintenance and the public health problems such as Dengue fever from mosquitoes, or poor water quality that might arise from inadequate maintenance. They also raised the issue of the regulatory control that might be imposed to manage the perceived risk (Leviston et al., 2013). This concern with maintenance contrasts with other concerns that might increase the use of tanks, such as potable water substitution, helping manage climate variability, bush fire control, concerns about the chlorine and fluoride in mains water, and water supply for gardens and well-being. Although the concern about health risk in that region needs to be addressed, the emphasis on regulatory control reveals a predisposition of technocrats to reassert control over the water supply. There is, however, a general lack of information on the performance of WSUD projects over their life cycle. This includes a lack of data on maintenance and operating costs (Urrutiaguer et al., 2010) and a lack of knowledge about construction and maintenance practices (Sharma et al., 2012).

Community engagement is another area of contrast between advocates of centralized water systems "big water" and WSUD. Within "big water," the ideal is for the system to work seamlessly and invisibly without the public's involvement. Leviston et al. (2013) found that technical experts saw themselves as separate from, and higher in status than, the general public. "The dominant depiction made by technical experts of community members suggests that community risk perception is inaccurately caricatured as relatively nonaccepting, emotion-focused and driven, focused on health concerns, and with a lack of trust and confidence in scientific, policy, and management processes" (Leviston et al., 2013, p. 1162). Although the results actually revealed that technical experts and lay people had very similar concerns, the technical experts perceived a sharp division between their rarefied knowledge and that of the wisdom of the community (Leviston et al., 2013). In contrast, and perhaps conflict, there is growing evidence for the importance of public engagement and acceptance of WSUD projects. For example, Tjandraatmadja et al. (2008) showed that poor stakeholder engagement and management of community expectations were barriers to the successful implementation of WSUD developments. Issues included a lack of familiarity with WSUD development systems and the requirement for a greater involvement of community in the management of decentralized systems. Sharma et al.'s (2012) review of the literature found that the majority of barriers to WSUD developments are social and institutional rather than technical. The technical knowledge exists, but there is a lack of practical experience within institutions to aid uptake. From their review of nine WSUD sites, Sharma et al. (2012) concluded that community engagement and consultation is essential for successful implementation.

Reflecting on the importance of community engagement, the term "hydrosocial contract" has been coined to describe the pervading values and often implicit agreements, between communities, governments, and businesses as to how urban water should be managed (Turton and Meissner, 2000). Brown, Keath, and Wong (2008) propose that the hydrosocial contract in the Water Sensitive City is adaptive and underpinned by flexible institutional regimes and coexisting and diverse infrastructure. To establish the Water Sensitive City, Brown et al. (2008) suggest that there will need to be a sociotechnical overhaul of conventional approaches to water management, which will increase the importance of community engagement. However, water authorities, local governments, government departments, and private industries have evolved to deliver conventional water services. Hence integrated water management concepts such as WSUD require a reevaluation of roles and responsibilities, which can be challenging to the existing management paradigm (Mitchell, 2006). Traditional authoritarian processes can hinder innovation and change, and entrench existing historical administrative, political, and economic values (Brown and Clark, 2007).

Despite the public's low levels of awareness of WSUD systems, research suggests that people generally have positive attitudes to these innovations. For attitudinal research to be valid, explanation of the systems is required. For example, through diagrams of WSUD systems in surveys (Leonard et al., 2015a) or by face-to-face explanations of WSUD systems in focus groups or interview formats (Leonard et al., 2015b). Findings clearly support high levels of acceptance of WSUD schemes to reuse stormwater for nonpotable purposes, whether surveying residents of areas with WSUD installations or the general population (Leonard et al., 2015a,b; Mankad and Walton, 2015). Further, participants who had received information from experts in a focus group preferred to use the treated stormwater in the mains system, i.e., for potable purposes (Mankad et al., 2015). Their reasons were twofold: they were assured that the stormwater could be properly treated to potable standards, and they viewed the mains water system as cheaper and fairer than a third pipe distribution system that would only be available in new developments. These results suggest that public participation and education could significantly change support for WSUD schemes, even those involving potable water (Leonard and Alexander, 2012).

In this introduction we have highlighted a disjuncture between the traditional culture and operation of water systems that are large, essentially invisible, and controlled by experts and WSUD systems which are small, local and require community engagement. Despite the public's general lack of awareness of WSUD systems, they have a largely positive attitude to these innovations. Fortunately, there are a range of intrinsic benefits associated with many WSUD projects that make them attractive to the community such as improved aesthetics; increased livability and improved health of communities; adding a distinct visual character and identity to the area; sustainability; expression of heritage values; and more opportunities for the community to engage in public life (CRC for Water Sensitive Cities, 2011). However, even while enjoying the benefits of a master-planned WSUD community, people are unlikely to appreciate the system and all its benefits unless it is explained to them.

In this chapter we will explore five dimensions of people's attitudes to and engagement with WSUD systems. These dimensions are visibility; recreation and other amenity; economic considerations for residents; place attachment; and social capital for community engagement. To illustrate the issues, the analysis will draw on examples from Adelaide, South Australia (Christie Walk, Adelaide CBD; Lochiel Park, Campbelltown; Mawson Lakes Salisbury; Springbank Waters, Salisbury; Harbrow Grove; and Mile End), which were part of a more comprehensive study of WSUD in the Adelaide area (Sharma et al., 2016). They can be contrasted with four sites in northern Europe (Hammarby Sjöstad, Stockholm, Sweden; EVA Lanxmeer, Culemborg, Holland; Rieselfeld Freiburg, Germany and Vauban, Freiburg, Germany) (Table 24.1). These places are of particular interest because of the contrast in the rainfall. Adelaide has a Mediterranean climate with hot dry summers, often struggling for an adequate water supply, whereas northern Europe needs to deals with excess water and ancient water and drainage systems in the old cities.

## 24.2 VISIBILITY OF WSUD SYSTEMS

There is no doubt that the greatest drawcard of WSUD management of stormwater for the public in dry climates is the abundance of green space and waterways that are often created by the availability of alternative water supplies. Water is an essential component of green space. When it can be seen, water is pleasing to the human eye; people often prefer landscapes that contain bodies of water, or even just water features (Lothian et al., 2010). Just as importantly, even when water cannot be seen, it is enjoyed. Water is what keeps vegetation alive, and water is what enables vegetation to cool the environment through transpiration (Ely and Pitman, 2014). Some researchers believe that the human preference for lush, green landscapes is the result of evolution, and stems from the fact that these environments appear fertile and potentially survival enhancing to people (Orians and Heerwagen, 1992; Lothian et al., 2010). They vary from a dramatically formal style such as the open lakes of Mawson Lakes to a more natural style with wetlands at Lochiel Park, both in South Australia. Larger scale green spaces increase biodiversity so residents and visitors can enjoy the birdlife and sounds of

**TABLE 24.1 Case Studies of WSUD Development**

| Site | Description | Features | Sources |
|---|---|---|---|
| Christie Walk Adelaide CBD | Medium density<br>54 residents<br>27 dwellings | **WSUD:** Shared underground tank that collects stormwater and rainwater, site gradients restrict runoff, and roof garden filters water. Water is used for toilets and gardens with vegetables.<br>Other features: Very strong community with numerous groups—governance, maintenance, community garden, shared laundry, meeting/activity room, plans for further innovation, welcoming committee, inclusion of all ages and renters, community education program for schools, etc. | Leonard et al. (2014b)<br>Sharma et al. (2016)<br>McClean and Onyx (2009)<br>Oliphant (2004) |
| Lochiel Park Campbelltown Adelaide SA | 15 ha<br>109 dwellings | **WSUD:** Stormwater filtered bioretention gardens and swales, wetlands, catchment runoff collected; MAR, purple pipe system to public spaces and private homes, community gardens. But there have been many problems. The residents' group is very active in trying to get problems with the site fixed—engaging with utility local and State govt. The many problems have brought the residents together.<br>Other features: Home orientation, solar, rainwater tanks, bike track to city, native plants, home efficiency monitoring. Active community groups. | Leonard et al. (2014b)<br>Sharma et al. (2016)<br>Edwards and Pocock (2011) |
| Mawson Lakes Salisbury Adelaide SA | 600 ha<br>4000 homes, 10,000 residents, commercial Technology Park, schools, Uni | **WSUD:** Focus is on WSUD: Uses MAR plus adds treated blackwater to greywater system; vegetated channels to reduce flood risk; wetlands; and purple pipes (dual reticulation) to all homes and public spaces. Open lakes and fountains are not a good model because of high evaporation.<br>Other features: Limited other sustainability features, e.g., passive solar, solar hot water. Limited community engagement, but there is an environmental group and community newsletter. | Leonard et al. (2014a,b)<br>Sharma et al. (2016)<br>Sharma et al. (2012)<br>Myers et al. (2013)<br>http://www.lendlease.com/australia/projects/mawson-lakes.aspx<br>Radcliffe et al. (2017) |
| Springbank Waters Salisbury Adelaide SA | | **WSUD:** MAR, large wetlands with birdlife that dry out in summer, vegetated channels to reduce flood risk; ponds with fountains, purple pipe to public parks.<br>Other features: strong contrast with surrounding industrial area; low income area; general knowledge and appreciation of WSUD but no community group for maintenance. | Leonard et al. (2014b)<br>Sharma et al. (2016)<br>Myers et al. (2013)<br>Radcliffe et al. (2017) |
| Harbrow Grove Adelaide | Development of narrow degraded piece of land used by local youth as a bike track | **WSUD:** Community park with large underground tank; stormwater from the road is captured, filtered through swales, and reused for park irrigation. Problem with leakage from the pond, so it is often dry.<br>Other features: Community unaware of the WSUD features, but the park is attractive and well used. | Leonard et al. (2014b)<br>Sharma et al. (2016)<br>Myers et al. (2013) |
| Mile End Adelaide | Retrofit to historic suburb of 4000 people mainly residential but some commercial | **WSUD:** Bioretention gardens on the roadside filter water from road and retain water underground to support green infrastructure.<br>Other features: Public are appreciative that flooding has stopped; some are annoyed at the traffic restrictions; almost all are unaware of the WSUD elements. | Leonard et al. (2014b)<br>Sharma et al. (2016)<br>Myers et al. (2013) |
| Hammarby Sjöstad Stockholm Sweden | Rehabilitated industrial site<br>Medium–high density<br>20,400 residents in 9000 flats plus 11,000 business spaces | **WSUD:** Highly integrated and sophisticated water, energy and waste system; stormwater is treated separately from sewerage and used for public water features and parks, then drains into the harbor; relieves strain on sewerage system. | Jernberg et al. (2015)<br>Leonard (2016) site visit |

*Continued*

**TABLE 24.1 Case Studies of WSUD Development—cont'd**

| Site | Description | Features | Sources |
|---|---|---|---|
| | | Other features: The engineering is impressive but social dimension is undeveloped. The Glashaus provides community education, but there is no community meeting place or residents association. | |
| EVA Lanxmeer Culemborg Holland | 24 ha<br>250 dwellings 40,000 $m^2$ commercial space, urban farm<br>Medium & low density & commercial | Excess water and flooding of the sewerage system was the main challenge; stormwater is separated from sewerage and deep ponds were excavated; ponds now used for recreation; groundwater that is used for drinking is protected by keeping an orchard over the main stormwater collection area.<br>Other features: Lanxmeer coherent and integrated Ecoframework: energy, water, landscape, mobility, supply chain management, and communication and education. Foundation group negotiated and found solutions to regional government's objections, then developed it. A new residents group now runs the heating system and maintains public areas. | van Timmeren Delft Uni (2007)<br>www.evacentrum.com/info.html<br>Leonard (2016) site visit |
| Rieselfeld Freiburg Germany | 70 ha<br>Mixed use (1000 jobs), 4200 homes for about 10,000 people<br>Medium density (4 story flats and villas) | First place in Germany to dispose of stormwater on-site (now it is a city requirement)—vegetated channels, community gardens, extensive wetlands with educational walks.<br>Other features: Claims to be the first low-energy and eco-district of this size in the world. Social workers first on-site. Designed to avoid social problems of nearby high rise area. Youth multimedia center supported by volunteer cafe, sports center, community gardens, only fruit trees planted in streets, nine types of recycling, residents helped to form groups. 5 km/h speed limit on side streets and kids have priority. No through streets, only a tram goes down the middle. | www.energy-cities.eu/IMG/pdf/0902_19_Rieselfeld_engl.pdf<br>Leonard (2016) Rieselfeld site visit |
| Vauban Freiburg Germany | 38 ha<br>2000 dwellings 5000 residents<br>High and medium density | **WSUD:** Sewerage and stormwater are kept separate: graywater is cleaned in biofilm plants and returned to the water cycle. Sewage is transported through vacuum pipes into a biogas plant and the biogas generated is used for cooking.<br>Other features: High insulation standards, high solar power, waste used for biogas, good energy management, car restrictions, strong citizen voice in design with "construction communities"; artists community; Forum Vauban lobbied hard for sustainable development agenda and the community center. | Review by Sustainability Victoria<br>www.vauban.de/info/abstract.html<br>www.forum-vauban.de/tasks.shtml<br>Leonard (2016) site visit |

frogs at night. The beauty of such places counteracts stereotypes of suburban development as monotonous, bland, and tasteless (Forsyth and Crewe, 2009). Even in small-scale developments such as Christie Walk, Mile End and Harbrow Grove, South Australia, with purpose-built green space for water filtration, people appreciate the increased greenery. Stormwater systems not only provide greener public spaces but also supply water to residents (via third pipes) to keep their gardens green, so whole suburbs become green oases.

The visual appeal of WSUD developments is further highlighted by the contrast between those spaces and the surrounding areas especially during periods of seasonal low rainfall or protracted drought. In the southwest of Australia the climate has become hotter and drier over the past 30 years. Those in Adelaide (South Australia) have been made acutely aware of their growing city and inadequate rainfall with water being pumped from the lower reaches of the River Murray for over 60 years at considerable cost to the taxpayers. The quality and quantity of this water has been questionable, and it is vulnerable to pollution and overuse along the whole Murray-Darling river system. It is therefore perhaps unsurprising that Adelaide has been a front-runner in the development of WSUD stormwater management and the contrast between areas of Salisbury Local Government Area with multiple WSUD systems and other regions is dramatic, especially in the dry summer months. Perth, Western Australia, has a similar climate, but has been able to delay action on securing the city's water supply because of large natural shallow aquifers. Recent action has focused mainly on the construction of desalination plants. Currently, however, a managed aquifer recharge system using recycled wastewater is being implemented (www.watercorporation.com.au/water-supply/ongoing-works/groundwater-replenishment-scheme), but because it will not be distributed through the mains water system, it does not require active engagement of residents. In the eastern states of Australia, rainfall is more reliable, although the 10-year "millennium drought" increased awareness of the water supply that had been taken for granted, and thus increased interest in WSUD schemes.

As a corollary to people's pleasure in the green space and waterways, residents could be particularly dissatisfied when problems decreased the appearance of an area. For example, the water feature at Harbrow Grove initially could not retain water while the lake at Springbank Waters dried out in summer. In Mile End, public gardens were bare for a long period during the construction, and there was community concern that the plants would die in summer because of lack of water, a concern that was due in part to lack of understanding of the water retention aspects of the bioretention installations. Mawson Lakes residents objected to muddy water in the lakes that did not have vegetation to filter the water. Even when the systems are functioning well, the public spaces need regular maintenance to maintain their appearance. Littering, graffiti on signs, and weed infestations all detract from the visual amenity of an area, and there were complaints in Mawson Lakes and Springbank Waters (Leonard et al., 2014b).

There can, also, be tensions between the esthetic appeal and WSUD operations. An important example is Mawson Lakes where the large open lake, free of vegetation, creates a dramatic entrance to the lot development, and is the site of the more expensive housing (Fig. 24.1). However, because of the hot dry climate in summer, there are high rates of evaporation that could be seen as a waste of a precious resource. Fountains are another feature much appreciated by residents and visitors. Although they oxygenate the water keeping the lake visually healthy, they further facilitate evaporation. The alternative, vegetated wetlands, not only decrease evaporation but also filter the water. However, complaints about aesthetics arise when the wetlands were drained for maintenance. The maintenance is necessary to remove sediments that reduce the wetlands effectiveness in flood mitigation, and to remove pollutants arising from their role in water filtration. But it does lead to unpleasant sights and odors.

Less visible aspects of WSUD systems were less likely to be appreciated. WSUD's ability to improve quality of water runoff, however, was less well understood and underappreciated. A few respondents from Springbank Waters discussed the importance of clean water discharge in maintaining the seagrass ecosystem in Barker Inlet, with consequent benefits for fish stock (Leonard et al., 2014b). At the Mile End development, only 1 of the 32 residents surveyed talked about the filtering function of the rain gardens in delivering cleaner water to the river. In terms of understanding ecological processes, residents near WSUD wetlands are generally aware of the cleansing function of reed beds, but the design intent of landscaped swales was not clear to Harbrow Grove residents and the benefits of the rain gardens at Lochiel Park were largely overlooked by the residents. Overall, awareness of offsite effects of water quality was generally not given significant attention by participants, perhaps reflecting Adelaide public's inadequate understanding of environmental linkages. Even the environmentally conscious population at Christie Walk was unaware of the environmental benefits of keeping rainwater and stormwater on-site. However, there was strong support for these measures when the function of the WSUD feature was explained by the researchers, resulting in a new level of community appreciation of the benefits that they provide (Leonard et al., 2014b).

**FIGURE 24.1** View of the main lake at Mawson Lakes.

### 24.2.1 Communication to increase awareness

Although there may be many limitations to public understanding of WSUD, there has been a significant increase in the role of communities for defining the WSUD problem and participating in developing the WSUD strategies. Wong (2007) gives examples of the incorporation of public art and the implementation of community participatory action models. Demonstration sites that effectively use the media have been noted by Roy et al. (2008) as ways to increase public awareness and reduce skepticism or resistance to WSUD. Even simple techniques such as putting blue dye into recycled graywater can increase residents' awareness that the blue water is for nonpotable use (Sant Cugat regional council in Spain: Water Sensitive Cities, 2009).

Two examples of effective public communication come from Perth, Western Australia, and Singapore. Perth has been using water from shallow aquifers as an important water source but, because of overuse, water levels are declining with adverse consequences on groundwater dependent ecosystems and variations in the water table have led to seawater intrusion into the aquifer (Government of Western Australia, 2009). The water supplier, the Water Corporation of Western Australia, has piloted a Managed Aquifer Recharge scheme to replenish the aquifer with treated wastewater and stormwater. The treatment process is very thorough, involving microfiltration, reverse osmosis, ultraviolet, and other treatments. Community engagement has been a key part of the pilot project and an interactive website is a major component of the community engagement process. The key features of the website are transparency and two-way communication. All deviations from ideal operating levels are reported online and available to the public, as is the action taken by the Water Corporation to address the problem. The public can make comments or ask questions about the process. There is also a walk-through demonstration plant for the community. The pilot has been successful and avoided the community backlash which has occurred in other places, so the State government has proceeded with the full program (Government of Western Australia, 2016).

The success of the NEWater recycling facility in Singapore has been credited to the Public Utility Board (PUB) for creating a focus on water literacy amongst citizens. Despite processing wastewater for potable uses, the NEWater scheme seems to have wide public acceptance (Water Sensitive Cities, 2009). Numerous techniques were used in the successful awareness raising campaign of Singapore's Water Reclamation Scheme, such as documentaries, media releases, internet websites, school visits, and a very sophisticated visitor's center (Po et al., 2003). PUB states that part of the reason the Singaporean public has been accepting the NEWater scheme is because the water authority decided to change the existing communication paradigm to one where creative communication methods, such as street style magazines and creative ideas for people to engage with water recreationally, were used to encourage Singaporeans to take ownership over their water supply (P.U.B, 2008). Visitors to NEWater in Singapore are invited to watch videos of Californian residents speaking about the benefits of the California Water 21 reuse scheme (Khan and Gerrard, 2006). Elements of the paradigm shift in PUB's water communication included a Water Wally mascot, a magazine style annual report, a street style magazine revolving around water aimed at youth, use of glamorous celebrities in promotion, attractive packaging for NEWater bottles, and an open, honest, and timely approach in all media releases (P.U.B, 2008).

## 24.3 RECREATION AND OTHER AMENITY

### 24.3.1 Recreation and health benefits

The high social and recreational value attributed to green space and waterways in developments is an important role for WSUD stormwater systems. From the study of six WSUD installations in Adelaide (Table 24.1), recreational amenity was widely enjoyed at the various sites. Participants felt that the open green spaces were an asset to the community as they provided areas for walking and cycling and created a country atmosphere in the city. Smaller areas provided opportunities for community gardens to grow vegetables, picnic and barbeque areas, and children's playgrounds. WSUD stormwater systems also supported more formal recreation by keeping sports fields green. There was even a model yacht club at Mawson Lakes. Even when small WSUD systems provide street trees rather than large green spaces, such as at Mile End, there are very obvious benefits in cooling the bitumen roads, footpaths, and the surrounding homes, which encouraged people to walk, run, and cycle in their neighborhood (AECOM, 2017).

Different types of parks tend to have different types of benefits (Schebella et al., 2014). One type of WSUD green space that is particularly useful for stimulating exercise as well as making the city more accessible are linear parks with walkways and cycleways, which connect parts of a city that are difficult to navigate by car. Schebella et al. (2014) found that linear parks (e.g., Torrens River Linear Park) facilitated significantly more physical activity than traditional community parks (e.g., Thorndon Park). Nonetheless, community parks were found to facilitate significantly more nonphysical benefits, including mental health, environmental, and social benefits. Schebella et al. (2014) also found that irrigated parks were

FIGURE 24.2 Map showing all green space within Campbelltown coded into six park classifications (Schebella et al., 2014). *Aqua*, Sports Park; *Dark green*, Natural Resource Park; *Light green*, School; *Pink*, Linear Park; *Purple*, Neighborhood Park; *Red*, Community Park.

associated with a range of physical activities, such as sports and low-intensity activities (e.g., slow-paced walking). The linear park they studied was not irrigated, but clearly, linear parks are more of a technical challenge for WSUD installation because of the distances to be covered (Fig. 24.2).

The psychological and physical benefits of WSUD features have the potential to increase well-being and quality of life among its residents (Schebella et al., 2014). Maller et al. (2008) conducted a metaanalysis that showed improvements in mental health, longevity, and life satisfaction, and enhanced social integration, derived from time in outdoor green spaces. Chen et al. (2014) found that urban green spaces reduced mortality from heat stress. They estimated that by doubling the tree leaf canopy in Sydney, there would be up to 28% fewer heat-related deaths during heat waves. Street trees also play a part in reducing cardio-metabolic health problems (Taylor et al., 2015) and contribute positively to our mental well-being (Kardan et al., 2015).

The green spaces also provided an avenue for developing social capital through opportunities for people to interact in the attractive public spaces. People who take regular walks, often with a dog, had ample opportunity to talk to their fellow residents. Taking children to use the playgrounds is also a good opportunity for contact. The opportunities for participation and connection are mainly through sharing recreational spaces, but also through public events. For example, Mawson Lakes has unique activities such as their annual carp fishing competition to remove carp from the lake. It is not easy to discover the extent or strength of networks promoted by these spaces and events, and many connections are likely to be private friendships.

## 24.3.2 Food production

The ample water available from WSUD water supply systems enables community gardens to provide vegetables and fruit, and such gardens are a feature of many sustainable developments. The food gardens ranged in size from Christie Walk's small patch to Aurora's four large areas (Moreland Energy Foundation et al., 2011) and EVA-Lanxmeer's urban ecological farm that sells produce in its village shop and supports an apple picking festival in autumn. In addition to community gardens, all new street plantings in Rieselfeld are fruit-bearing trees. The gardens are places for community education about organic processes and food production. They are also places of social connection as people work side by side and share the produce. In particular, they are places where people can connect across cultural and language and socioeconomic barriers. For example, people of non–English speaking background renting at Lochiel Park often did not join the residents association, but they did work in the community garden and could share vegetables and herbs with their multicultural neighbors. As people working in food gardens meet and interact on a regular basis, food production provides a stronger basis for the development of social capital than unstructured green space (http://www.abc.net.au/gardening/stories/s845472.htm).

## 24.3.3 Benefits of third pipe systems to the home

In some WSUD developments, treated stormwater is piped to individual homes for nonpotable uses such as toilet flushing and gardening. People appreciate having a supply that is not subject to water restrictions and usually charged at a lower tariff than mains water irrespective of the production cost. In Mawson Lakes people liked being able to maintain their lawns and gardens green, and being able to wash their cars during water restrictions imposed by the extended drought. Within the home, toilet flushing and laundry are significant contributors to water consumption, which can be reduced through stormwater reuse. Studies in Adelaide show support for expanded use of recycled water and rainwater (e.g., for toilet, laundry, and hot water service), or retrofitting older suburbs (e.g., Springbank Waters) for household applications. The purple pipe infrastructure, carrying recycled water, was widely known and perceived as providing security against water shortages (Leonard et al., 2014b). The Adelaide case studies document the high regard with which participants viewed an alternative water source that is independent of the centralized mains water grid. All participants with access to alternative water, such as recycled water, felt positively about it because of the increased supply and the improved water security this offered. Participants acknowledged that although the cost of alternative water was high, the security of a guaranteed water supply, regardless of fluctuating rainfall, was highly valued by most people. Similarly, Hurlimann and McKay (2006) found support for recycled water, but their requirements for the water such as color, odor, and price depended on how it was to be used, and they recommend consultation with the community to ensure recycled water is fit for purpose.

Although stormwater use outside the home is seen as relatively low risk, use inside the home can raise concerns. Indeed, the Joint Steering Committee for Water Sensitive Cities (2009) found that stormwater reuse is typically seen as moderate to high risk depending on degree of management measures and adherence to Australian Guidelines for Water Recycling (www.environment.gov.au/system/files/resources/9e4c2a10-fcee-48ab-a655-c4c045a615d0/files/water-recycling-guidelines-augmentation-drinking-22.pdf). Differences in perceptions of external and internal uses of stormwater have been noted in a number of studies. Discoloration can be a problem when stormwater has been stored in an aquifer. Dzidic and Green (2012) investigated discoloration and turbidity for different end uses (toilet, laundry, garden) and reported that, overall, participants were accepting of some level of esthetic degradation for all nonpotable uses. Participants were accepting of using nonpotable groundwater for the watering of public open space and household gardens, regardless of the esthetic attribute. However, acceptability decreased as use became more personal, e.g., clothes washing was less acceptable than use in the toilet, which aligns with other research demonstrating that acceptability decreases as uses become more "personal" (Mankad and Tapsuwan, 2011).

Further, Roy et al. (2008) argue that there are multiple layers of risk and risk aversion that may cause resistance to WSUD by both practitioners and the general public. These can include risks at the institutional level, such as loss of revenue, loss of functionality, risk of failure, and risk of disease, or at the community level where it may be perceived as unattractive or ineffective (Roy et al., 2008). However, it is the actual health risks and perceptions that are likely to be the strongest barriers to WSUD acceptance and that is related to the degree of personal contact with the water (Mankad and Tapsuwan, 2011). Once purple pipe water enters the home, there is an increased risk of problems arising from faulty installation of domestic plumbing. For example, in several cases in Mawson Lakes the purple pipes were cross-connected with the potable supply, so nonpotable water was coming out of the kitchen taps. This was of particular concern because the stormwater had been mixed with treated sewerage (Leonard et al., 2014b). Although the problem was identified and rectified, such serious errors can undermine public confidence in purple pipe systems. Thus residents need to have a high degree of trust in the installation and maintenance of WSUD systems, and any faults that occur can aggravate their concerns.

Attitudinal research on acceptance of alternative water sources (Dolnicar et al., 2010; Hurlimann and Dolnicar, 2010) has consistently shown that provided personal contact with recycled water stays low or limited, acceptance for use will be high. Residents in the Adelaide case studies did not seem overly knowledgeable about the precise nature of the water source, but were happy to use it, and did not cite any risks or concerns that would prevent them using it, other than the cost. For the majority of cases, personal contact with recycled water remained low, as the water was specifically designated for gardens and toilets in the medium to large housing developments and neighborhood parks. Therefore, use of the alternative water seemed highly acceptable to residents. In Christie Walk, the smallest of the WSUD developments, recycled water was more used more widely including laundry as well as gardens and toilets; however, acceptance for use remained high within this group. This is arguably because of residents' high level of knowledge of the WSUD features and their "hands-on" involvement with the day-to-day running of the WSUD systems and related features (Leonard et al., 2014b).

## 24.3.4 Flood mitigation

Reducing the risk of flooding is another important role of WSUD stormwater systems (Department of Planning and Local Government, 2010; Joint Steering Committee for Water Sensitive Cities, 2009). In the European context where there is ample rain and often ancient water systems, stormwater disposal into the sewerage system is common. However, as cities grow the volume of wastewater and stormwater is often too great for the old combined sewer systems, resulting in sewer overflows causing flooding from an unpleasant mix of sewage and stormwater. New WSUD systems such as in Reiselfeld in Germany, the first in the region, and EVA Lanxmeer in Holland, allowed for the reinfiltration of stormwater on-site thereby greatly reducing the risk of flooding. Indeed in EVA Lanxmeer, the regional authorities would not approve development until the stormwater problem had been solved. Even though Australia is a much drier continent, local stormwater management for flood prevention has been one of the most frequent applications of WSUD (Sharma et al., 2013). From the residents' point of view, it is a benefit most clearly recognized by those who have experienced previous flooding events. Generally people in cities take for granted that they will not be inconvenienced by heavy rains. The newly constructed neighborhood-scale WSUD system in Mile End was perceived to be effective in flood mitigation. Interviewees observed after heavy rains that "it (rain gardens) seems to be working," except where organic litter blocked the drains. Similarly, a significant rainfall event also occurred in Harbrow Grove postinstallation, with no reports of flood damage. Mawson Lakes and Springbank Waters, both locations where swampy ground had been reengineered as housing sites, have functioned well without significant flood events over the long term, which was cited as an achievement by long-term residents familiar with the historic patterns of serious flooding (Leonard et al., 2014b).

## 24.3.5 Poor amenity

WSUD installations are still relatively new and malfunctions are not uncommon (Sharma et al., 2012). In all the Adelaide sites we surveyed, there were problems to be overcome (Leonard et al., 2014b). These engineering problems become social problems because stakeholder perceptions of a WSUD feature achieving its stated function are linked to acceptance. The more effective the functioning of the feature, for example, providing greenscape, water capture, treatment and storage; or distribution of recycled water, the more appreciative residents and workers were of the feature (Leonard et al., 2014b). Conversely, when ongoing operations of the WSUD feature were suboptimal, stakeholder acceptance was lower. Effective ongoing functioning seemed to be dependent on correct initial design and installation, as well as adequate ongoing maintenance. Poor design and installation caused significant cost issues, frustration, and ill feeling toward the developer or final caretaker (the local council). The cost to residents of rectifying the problems depends on the type of governance and the size of the development. In developments where a relatively small number of residents share the title over common areas, residents will have greater expense than more populous areas where the council has responsibility for common areas. In situations where problems were ongoing and cause for community concern and frustration, many of these problems were evident from the beginning. Resulting tension can jeopardize not only community acceptance but also council enthusiasm for future expansion of WSUD features to other sites (Leonard et al., 2014b).

Good ongoing management and maintenance also underpin perceived effectiveness. In larger developments, where alternative water systems were on a wider scale and servicing many homes directly, upkeep was considered very important. There was discontent among participants when WSUD installations around the home, and in communal areas, were not maintained adequately or in a timely manner. In particular, participants from the large- and medium-sized developments were unhappy when management of the housing development shifted from the developer (initial caretaker) to the local council (final caretaker). During this transition period, some participants felt that aspects of the maintenance and upkeep for the development went unattended because the new caretaker had not been adequately informed of certain issues, arrangements and priorities. Participants recommended that for future large—medium-scaled WSUD developments, the developers and subsequent caretakers must create a comprehensive "handover" plan to prevent overlooking important maintenance issues. It was also important to residents that once the developer has vacated the site, the new caretaker set up a visible presence in the area so that residents had somewhere to report future problems as they occurred (Leonard et al., 2014b). These concerns suggest that Councils need to take care at the approval stage that they will have both the human and financial resources to maintain the site after the developer has left.

## 24.4 ECONOMIC CONSIDERATIONS FOR RESIDENTS AND COMMUNITIES

As can be seen from the previous sections there is a wide variety of visual, recreational, health, biodiversity, and other benefits that accrue from WSUD installations. Given the multiple benefits it is difficult to assess the total value that people might place on WSUD installations. It is particularly difficult because most of the WSUD installations for stormwater (such as rain gardens and bioretention swales, green roofs, living streams, constructed wetlands) are multifunctional. One way of assessing the value of WSUD to local residents and communities more broadly is to apply a dollar value that can provide a financial argument for WSUD installations to residents, developers, and governments, all of whom can benefit from different aspects of WSUD. Economic considerations in WSUD are presented in more detail in Chapter 14 of this book; however, a snapshot of the economics of diverse types of benefits associated with WSUD services is provided in this section.

Although the economic value of some WSUD systems could be estimated based on existing market prices (e.g., the economic value of water savings from stormwater management), for many of these services it is very difficult to use existing market information to calculate economic value as either there is no existing market or the market is not well developed. In these cases, economists often use other methods to calculate economic value.

The three most common methods to calculate the nonmarket values of WSUD services are: revealed preference (such as hedonic pricing and travel cost methods); stated preference (such as choice experiment and contingent valuation); and benefit transfer. In a revealed preference approach, observed behavior is used to calculate people's willingness to pay (WTP). For example, in a hedonic pricing method, prices for heterogeneous goods (such as houses) are analyzed to calculate the environmental or amenity benefits of a project or asset. Travel cost method uses the costs of travel and time to calculate the recreation benefits of a project. In a stated preference approach such as a choice experiment, community surveys are used where respondents are presented with a series of options with different amount of costs. From the choices made by individual respondents, their WTP for different services are calculated. Although the revealed preference approach is more reliable as it is based on real purchasing decisions, the stated preference approach is suitable to estimate values for services (current or hypothetical) with no market information. However, both of these methods could be quite expensive and time-consuming. A third approach is to use benefit transfer, where WTP estimates from existing studies are used to calculate the values for a service in a new area. Benefit transfer is useful when there is already a good reliable set of estimates (Iftekhar et al., 2017b). There is a growing body of literature estimating the nonmarket values of WSUD using these methods in Australia and elsewhere. A brief summary of some of their findings is presented below.

Rain gardens are a common element of WSUD in Australia. For example, Melbourne Water has registered more than 10,000 rain garden projects by 2015 in Melbourne. Other States also have favorable policies to install rain gardens. However, there are only a few studies looking at the nonmarket values of rain gardens. Bowman et al. (2012) found that residents in Ames, Iowa were willing to pay to have rain gardens close to their houses. Recently, Iftekhar et al. (2017a) presented preliminary results from their choice experiment analysis of nonmarket values of rain gardens in Sydney and Melbourne. They have reported that people have positive WTP for more rain gardens at road intersections and for rain gardens bordered with hedge trees, but have less preference for trees on rain gardens. In Sydney, the establishment of rain gardens has net positive benefits only if the nonmarket values of pollution reduction benefits are factored in.

Some studies have looked at the nonmarket values of water quality improvement and improving water supply through stormwater management. For example, Brent et al. (2017) reported results from a choice experiment conducted in Sydney and Melbourne. They estimated people's WTP for different services or benefits from local stormwater management. They found that people were willing to pay to avoid water restrictions (median WTP AUD$155 in Melbourne and AUD$242 in Sydney); for improvements in local stream health (AUD$234 in Melbourne and AUD$229 in Sydney); and for decreased peak urban temperatures (AUD$45 in Melbourne and AUD$54 in Sydney). There was no significant difference between the two cities in WTPs except for water restrictions.

Rainwater tanks are a good example of how evaluation on just one dimension can be misleading. Although tanks are usually seen solely as a water source, if most homes in an area had substantial tanks then there is a positive benefit for reducing stormwater runoff. Tanks are significant because they are a commonly used rain harvesting technology in Australia (Iftekhar et al., 2016; Sharma et al., 2015). According to a 2013 estimate, 31% and 47% of the detached houses in Melbourne and Brisbane, respectively, had rainwater tanks installed (Australian Bureau of Statistics, 2013). There have been a number of studies conducted in different cities in Australia (Khastagir and Jayasuriya, 2011; Tam et al., 2010; Christian Amos et al., 2016) estimating the net benefits of rainwater tanks as a water source. It has been found that there is substantial variation in the value of rainwater tanks in that harvesting rainwater is beneficial in Gold Coast, Brisbane, and Sydney, but not in Melbourne, Adelaide, Perth, and Canberra (Tam et al., 2010). Similarly, the Queensland Competition Authority (2012) found that maintaining the compulsory installation of rainwater tanks actually has negative net benefit.

However, they often focused on economic value of water saved and did not consider the other types of benefits to the residents (Castonguay et al., 2016). For example, in a recent study, focusing on nonmarket values of rainwater tanks in Perth, Australia, Zhang et al. (2015) found that there is a significant positive effect of rainwater tanks on house prices, and if this value were included, the net benefit of rainwater tanks would be substantially higher.

There are numerous studies exploring the economic benefits of green infrastructures (see Symons et al., 2015). Green infrastructures provide a number of important ecosystem services. Using hedonic pricing analysis approach, Rossetti (2013) estimated the nonmarket values of urban green space across mainland Australia. They estimated the relationship between house prices and Enhanced Vegetation Index (EVI) at the postcode level and found that a one standard deviation increase in the EVI leads to an increase in housing prices of A$32,000 or A$58,000 (depending on the statistical method used). In another study in Perth, Pandit et al. (2013) found that the impact of trees on house price depended on the location. For example, a broad-leaved tree on the street verge has positive impact on house price (the median property price increase by about AUD$17,000), but broad-leaved trees located on the property or on neighboring properties do not have any significant impact on house price. In Blacktown, Sydney, AECOM (2017) found that a 10% increase in the size of the canopy across the suburb could increase the property prices by 7.7% (AUD$55,000 for an average house). In Germany, Kolbe and Wüstemann (2015) showed that increasing the area that is parkland by 1% point within a 500-m buffer around accommodation would result in a rise in apartment prices of 0.1%.

It is important to remember that efficient decision-making would require consideration of both benefits and costs of implementing water sensitive urban designs and options. A few studies have included nonmarket benefits in the cost-benefit analysis. For example, Water by Design (2010) noted that several nonmarket benefits could be included, namely: the value of pollution removal, avoided costs associated with downstream waterway rehabilitation and maintenance, potential increase in property values due to amenity benefits, and avoided development costs. They argued that if these benefits are considered then WSUD are likely to be cost-effective across a range of residential development scenarios in Queensland. Polyakov et al. (2017) assessed the costs and amenity benefits of an urban drainage restoration project in Perth, Western Australia and showed that total nonmarket values of amenity benefits could easily cover the cost of the restoration project. In another study, Carter and Keeler (2008) estimated benefits (both social and private) of green roofs in Georgia, USA. Green roofs perform a number of environmental functions such as: absorption of rainfall, reduction of roof temperatures, improvement in ambient air quality, and provision of urban habitat. They estimated the social benefits (using a 4% discount rate over 40 years) of installing green roofs (flat roofs constitute 176,234 $m^2$ or 7.4% of impervious surface in the watershed) was over US$24 million, which was 12.14% higher than the nonmarket benefits of traditional roofing. A further example is from Vandermeulen et al. (2011) who estimated the net benefit of installing a green cycle belt in Bruges, Belgium. They observed that the benefit–cost ratio for the project was 1.22. However, this estimate was sensitive to various factors including the number of users of the greenbelt. In a recent study, Mekala et al. (2015) developed a business case for Stony Creek Rehabilitation Project in the City of Brimbank, Victoria, Australia. They estimated that the potential public benefits of avoided health costs was about AUD$75,000 per annum and potential private benefits of AUD$3.9 million capitalized in the house price.

This section provides a snapshot of the economic estimates of diverse types of benefits associated with water sensitive urban designs. We recognize that in many cases, the benefits are highly contextualized and depend on a range of factors such as the estimation methods; temporal and spatial scale scope of the project; comparative benchmark; and distribution of benefits and costs across various groups of stakeholders (CRC WSC, 2014). Nevertheless these examples clearly show that WSUD installations can provide positive economic benefits, especially when the whole spectrum of positive effects is taken into account. Thus it is essential to incorporate this information into the decision-making process for urban development.

## 24.5 PLACE ATTACHMENT

### 24.5.1 Introducing place attachment

Attachment to place refers to a dependence on or an emotional bond with biophysical aspects of the landscape. These emotional bonds can influence residents' ability and willingness to address local problems, participate in their communities, and work to improve and protect them (Manzo and Perkins, 2006). Devine, Wright, and Howes (2010) argue that forming attachment to a place is as much a social process as a psychological process. In a context of change, individuals adopt specific beliefs and attitudes, contingent on levels of trust, which position them among influential groups or institutions. Given that many WSUD systems service defined locations such as a neighborhood, place attachment could be an important driver for the support for WSUD systems, especially where the WSUD system improves the ambiance of the area through greenery or water features. However, there are likely to be variations in the degree of attachment ranging from a superficial enjoyment of the pleasant ambiance, to a strong personal identification. Such variations in attachment are likely to influence people's engagement and general support for WSUD schemes. Generally, people associate positively with WSUD features such as improved aesthetics, greener surroundings, and improved local habitat (Tjandraatmadja et al., 2008). The amenity offered by the WSUD features was often reported as a significant reason why respondents chose to live there (Leonard et al., 2014b).

## 24.5.2 Diversity in place attachment

There is potentially a wide range in the levels of attachment and the extent to which the place attachment is associated with a WSUD installation. The Adelaide case studies (see Table 24.1) illustrate some of the diversity of positions. Christie Walk, that encompassed community-placed WSUD features, had the strongest positive place attachment. There was a clear place attachment and ownership in the way participants discussed their housing community. The site's eco-credentials were a major incentive for most of the residents and they were very proud of their water conservation strategies, and this contributed to a wider sense of pride in their eco-complex, including their areas of greenery, solar energy supply, and the unique construction of their houses, which were made from natural and renewable materials (Leonard et al., 2014b). At Mile End, in contrast, residents' place attachment was associated with the cosmopolitan main street, cafe culture, and heritage housing. In Mile End, many residents were unaware of the WSUD features of the roadside gardens and felt themselves ill-informed and separated from the decision-making process in relation to the bioretention gardens. This sentiment meant that many residents were unhappy with the changes to their streetscape and perceived the newly installed bioretention gardens as an obstruction to street parking space, rather than as a novel and beneficial stormwater harvesting and filtering system (Leonard et al., 2014b). These concerns might have been overcome with more interactive communication between the local Council and the residents.

In Harbrow Grove, another area with retrofitted WSUD installation, the transformation from a waste ground to a highly attractive and well-used park greatly increased a positive place attachment based on the indirect benefits of WSUD amenity (Fig. 24.3). However, residents were unaware of any WSUD water management benefits. Residents were happy with the green space that WSUD created and spoke about the space being popular for families living in the area, as it also provided

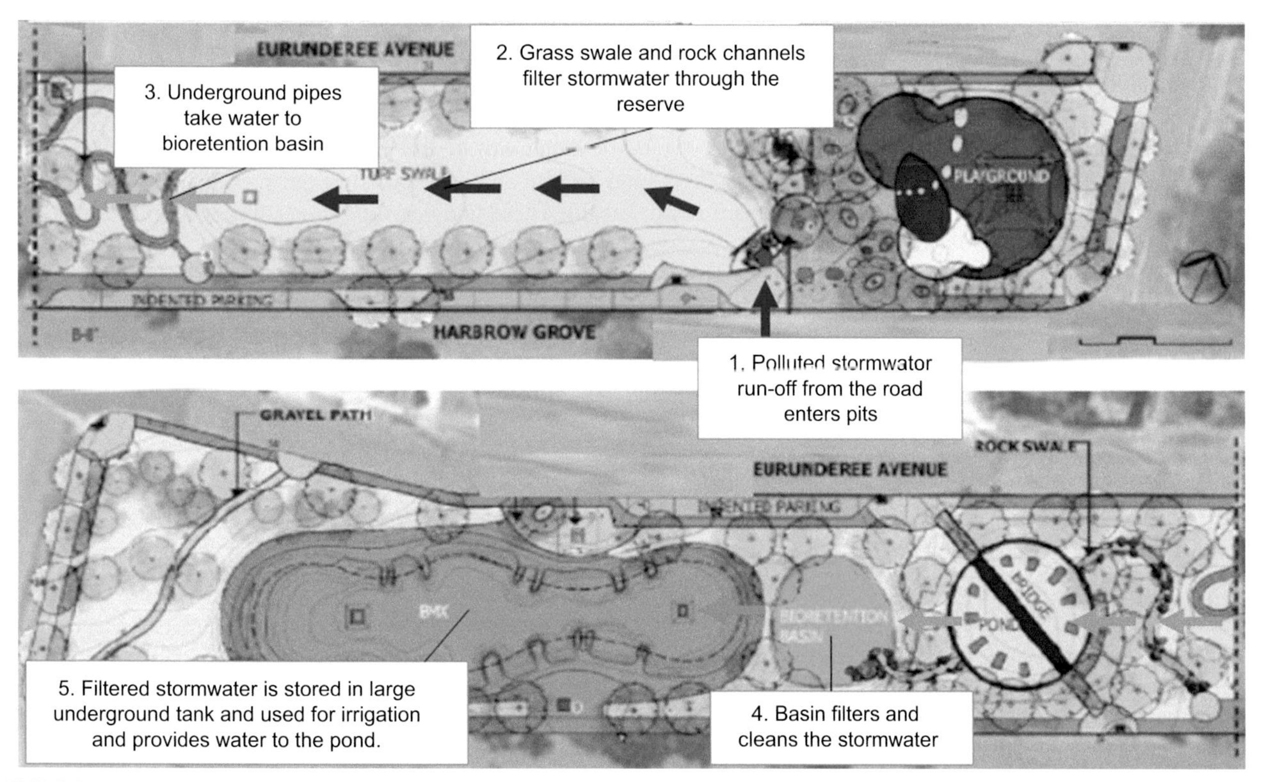


FIGURE 24.3 Diagram of Harbrow Grove Reserve WSUD features. From https://www.marion.sa.gov.au/page.aspx?u=701&c=4050#stormwater_management. Used with permission from City of Marion.

FIGURE 24.4 A garden within a U-shaped block of flats, Reiselfeld, Freiburg Germany.

a social activity hub. However, there was no knowledge of the unique and indeed quite advanced technological features of their WSUD system. In contrast, most people from Springbank Waters had a high level of awareness of their site's WSUD features contributing to water savings. Springbank Waters, a low socioeconomic area located near an industrial zone, had large wetlands, which acted to distinguish the estate from its industrial surrounds. There was a strong place attachment related to the watercourses and birdlife. Residents also were aware, and proud of, the water saving and filtration features of their WSUD system.

Mawson Lakes is the largest and most impressive of Adelaide's WSUD sites situated 12 km from the CBD. Residents regarded Mawson Lakes as a community with very good amenities, on-site recreational activities, attractive landscape, and well served by transport infrastructure. Because of the wide range of WSUD features, the area could appeal to diverse groups by providing areas that were formal and informal places for both active and passive recreation by adults and children. The gardens, parks, lakes, and creeks were key features that were admired and enjoyed by the local community and visitors. Residents enjoyed the status associated with living in such an exclusive complex that was well known around the Adelaide area and beyond. The residents of Lochiel Park were drawn there because of the environmental and sustainability aspects of the development and planned to stay in the long term. A strong sense of place was embodied in the WSUD features. Residents enjoyed the open spaces, trees, and birdlife and were proud of the sustainability concept that underpinned the Lochiel Park development (Leonard et al., 2014b). These findings were in keeping with other studies of Lochiel Park (Edwards and Pocock, 2011).

In three of the European case studies (see Table 24.1), efforts were made to build in place attachment by engaging future residents in the design of the development. In Vauban and EVA Lanxmeer, future residents could join a community group to work through the design, and in Reiselfeld, locals were consulted from the earliest stages. Further, the designs gave residents control over common spaces. Although these spaces were open to the public, the residents in the surrounding apartments of townhouses, which overlooked the space were involved in their design and maintenance (Fig. 24.4).

## 24.5.3 Shared value of water saving

An important aspect of place attachment can be the feeling that one is living near like-minded people. In the Adelaide case studies, participants expressed shared values around water savings. Residents at the case study sites were troubled by unnecessary or excessive water usage and the wastage of a scarce resource. Within the general context of responsible stewardship of water, stories emerged about watering plants with stormwater overflow because "we are crazy about using every drop" (Christie Walk) and the residents of Lochiel Park reported on their struggle to have the treated stormwater system made functional. At both localities, the water utility (SA Water) was criticized for sending mixed messages to the wider community. Initially teaching South Australians how not to waste water during the drought, and then backing off the educational campaign when rains increased. Even where people were unaware of their WSUD systems, they had feelings of regret at seeing water going out to the Gulf of St Vincent during heavy rainfall events (Mile End) and concern when broken water mains were watering the street rather than vegetation (Harbrow Grove). In contrast, Mawson Lakes residents appreciated that they did not have to engage in water saving because of their extensive WSUD system. Study participants were not always been able to articulate how water saving approaches such as aquifer storage and recovery works; specify the exact sources of their lake water; and detail the underground technology of WSUD systems, nor analyze the trade-offs of various water policy alternatives, but their in-principle support for water saving and reuse was unwavering during the research project (Leonard et al., 2014a,b).

### 24.5.4 A sense of pride

A sense of pride in one's place of residence can be an important element of place attachment. Pride shows how people value the particular place to which they belong. Educational tours were one way in which the developments demonstrated a strong sense of pride, as locals were eager to showcase the innovations to visitors. Vauban, Reiselfeld, EVA Lanxmeer, and Christie Walk all had educational tours for schoolchildren, the general public, and interstate and overseas visitors. Local government officials and water industry professionals also often visited these complexes. For Vauban and Reiselfeld, their municipality, Freiberg, has embraced ecotourism and multiple innovations and tours are available in the area (www.upperrhinevalley.com).

At Mawson Lakes, there was an overall sense of pride within the community for their recycled water status, especially when, showcasing "world firsts" when overseas visitors came to inspect their cutting edge urban water system. Although WSUD may not have been the dominant factor in deciding to live at Mawson Lakes, residents embraced the WSUD features of the complex and spoke favorably of living with the design with its benefits of amenities and aesthetics. At Springbank Waters, there was a sense of pride regarding the water features and a high regard for the Salisbury Council for their management of the area. At the local school they said, "the children always want to go that way and make sure the fountains are on." Even in Lochiel Park where many of the WSUD systems were not working properly, residents felt a sense of pride at being part of experiments in innovation and were open to sharing their knowledge with others. At Mile End, people's sense of pride focused on the historical and cultural features of the area. In contrast, the WSUD installations were not salient. For Harbrow Grove residents, there was no particular sense of pride until their innovative system was explained to them. They then realized they had something quite special and were more tolerant of its empty pond (Table 24.1).

The case of Harbrow Grove illustrates the importance of knowledge of the WSUD systems to increase pride and attachment and clearly underlines the need for community education. Further, our research also highlighted that those residents who were highly involved in the day-to-day management of the water sensitive features, or those who felt well informed and consulted were most likely to express their sense of pride in the green credentials of their housing development. It was as though participants felt vicarious praise for their eco-living through other members of society, and were proud to be representatives of living with the benefits of WSUD.

## 24.6 SOCIAL CAPITAL AND COMMUNITY ENGAGEMENT

### 24.6.1 Understanding community through social capital

Community is a word often used colloquially to refer to people who live in the same place, but it is better used for genuine groups where people interrelate. Not that a community needs to be a single cohesive group where everyone needs to know everyone else, but there needs to be something that exists at the collective level. This might be a residents association, or a number of community groups, or a physical community center or a town center where people see each other regularly and have opportunities to interact through multiple overlapping personal networks. According to social identity theory, personal connections are not necessary (Tajfel and Turner, 2004). It is sufficient to strongly identify with a place, which could be prompted by nonrelational features such as its environment, past associations, or the contrast with other places. Another common assumption is that community forms spontaneously, but again this needs to be questioned, especially in new estates where everyone is a newcomer. The features of a community we have described probably needs some support from thoughtful urban design and active community development.

WSUD initiatives can be recognized as "common pool resources" in the narrow sense that a neighborhood may share the maintenance and benefits of an installation. Or in the broader sense that the water supply is a finite resource that needs to be shared across the city. In the broadest sense the environment is a *common pool resource* for everyone on the planet.

Ostrom (1990) argued that management of common pool resources is widespread and only becomes a problem when "participants may simply have no capacity to communicate with each other, no way to develop trust, or no sense that they must share a common future" (1990, p. 21). Because Ostrom formed her argument, Putnam (2000) has popularized the concept of social capital and the literature has grown exponentially (Halpern, 2005). The definitions of social capital are now many and varied, but the common thread is social networks and the resources that can accrue from them (Rostila, 2010). The concepts of trust and a sense of shared resources are also common themes (Leonard and Onyx, 2004; Putnam, 2000). Thus the issue of the management of common pool resources can be understood as being problematic in the absence of social capital. In the worst cases it leads to the overuse of a resource and the "tragedy of the commons" (Hardin, 1968).

There have been a number of studies that have identified a positive relationship between social capital and environmental action, primarily in rural communities of developing countries (Pretty and Ward, 2001; Anderson et al., 2002; Adger, 2003). In Australia, Onyx et al. (2004) found a strong relationship between social capital and concern for the environment in the remote mining town of Broken Hill. Morrison et al. (2011) reported that social capital was a better predictor of landholders' participation in environmental programs, than psychological theories that focused on attitudes. Some of the most powerful evidence, however, comes from Portney and Berry (2010) who examined 27 USA cities using the Social Capital Benchmark Survey. They found that those cities most committed to pursuing sustainability policies tended to be more participatory places with respect to signing petitions, participating in demonstrations, belonging to local reform groups, and joining neighborhood associations.

## 24.6.2 Communication about WSUD

### 24.6.2.1 Orientation for new residents

As described in Section 24.3, there are some risks that arise from WSUD systems and some form of orientation for the residents is important. A lack of information can lead to risks for both the WSUD system and for the new residents. For example, in Springbank Waters a hole in a recycled water purple pipe was spreading water around, but the local residents new to the area did not know it was not a safe place for children to play. Another renter in Mawson Lakes was not told that the purple pipes coming into his home were not for potable water. At Lochiel Park, residents did not look after the plantings in the biofiltration beds. However, there are also some good examples of orientations. In Caroline Springs a suburb 25 km west of the Melbourne CBD, Victoria, the developer, Delfin, have dedicated community workers who present "Welcome Home" workshops, welcome packs, and BBQs (Johnson, 2012). At Lochiel Park, the developer also had social events for purchasers and potential purchasers. These informal meetings often resulted in a residents group being formed. At The Ecovillage at Currumbin, Queensland (www.theecovillage.com.au), the preparation included Resident Training Courses on conflict resolution and other social and management skills, so that residents should be able to work together more harmoniously. However, such developer-organized orientations usually cease when all the lots are sold. Often, information is also lost when homes change hands. In Australia, renters, in particular, can miss out on such an orientation. Not having attended the social functions for purchasers at Lochiel Park, renters felt excluded from the residents group even though this was not the intention of the owners. Most proactive was the residents group at Christie Walk where all new residents, owners, and renters are given an induction program about the site, and its maintenance, and improvement groups. Almost all residents choose to join a group due to their shared sustainability values and to take part in the social life of the site. In the European case studies, there were housing co-ops that provided a valuable contact point for both renters and home buyers.

### 24.6.2.2 Internal communication

A sense of community can be developed through a variety of tools for internal communication. Although regular face-to-face meetings are feasible for small developments and are the main communication method in Christie Walk, larger developments rely more on newsletters, newspapers, email lists, and websites. A combination of methods is not unusual. In Mawson Lakes, the newsletter is the initiative of just one resident, and it is a quixotic combination of important environmental information, community announcements, and anecdotes about her dog (https://issuu.com/mawsonlakesliving). The Caroline Springs intranet supports 61 cultural groups by advertising their news and events. It has a section where residents can report local damage or malfunction (e.g., streetlights not working), and report antisocial behavior and dangerous driving to the police. By way of bridging social capital, it includes people from surrounding suburbs, who like to use services and attend events within the development.

### *24.6.2.3 Community centers*

Community centres have huge potential for community development due mainly to the practical support they can provide for diverse activities and organizations, as well as a source of information about local events and services. Further, they have a symbolic role as a visible reminder that this is a community and that people "do things together." The city of Freiberg in Germany is a place that has changed its mind about community centres. In the 1990s, the Vauban community committee had to wage a long campaign to retain one old building for use as a community center. And they were required to raise the funds for the renovation or do it themselves. Ten years later, in the new sustainable suburb of Rieselfeld, the Glashaus provides an eye-catching town center and supports multiple activities. Such centres can provide information, training courses, and advice services to residents on sustainability (e.g., Vauban, Rieselfeld). For small developments such as Christie Walk, a single meeting room and kitchen can serve as a functional community center. In contrast, Hammarby in Sweden is disappointing because of the lost opportunity. The social dimension was not a priority during the planning stages of this large suburb, and there is no community center. Although the sustainability information center (a small Glashaus) is a feature of the suburb, it is too small for meetings or other community development activities (http://www.stockholmvattenochavfall.se/om-oss/vara-kontor-och-anlaggningar/glashusett).

### *24.6.2.4 Engagement beyond the development*

Master planned estates can be gated communities that enjoy keeping a social distance from the surrounding suburbs. Although this may create a sense of identity and safety to those within the estate, it can create long-term divisions that could be problematic in the future. We argue that it is more beneficial to create a sense of identity within the estate and to create connections with the surrounding suburbs. These two things are not in opposition to each other, as previous research has found that bonding and bridging social capital are more likely to be positively related to each other than negatively related (e.g., Leonard and Bellamy, 2010). Publicly accessible green spaces from WSUD attract people from surrounding areas, but it is unclear how much contact they have with local residents. Examples of more proactive connection beyond the development include festivals such as the Mawson Lakes carp fishing, the EVA Lanxmeer apple picking, and the Caroline Springs community intranet that specifically includes people from surrounding suburbs so they can join in community events (Johnson, 2012). For the purposes of promoting positive attitudes to environmental sustainability, it seems particularly important that outsiders have the opportunity to visit sustainable developments. In a number of developments such visits are actively promoted (e.g., EVA-Lanxmeer, Christie Walk, Ecovillage at Currumbin, Vauban, and Rieselfeld). However, only in Christie Walk, Ecovillage at Currumbin, and EVA Lanxmeer are site visits provided by a community group. Tours at the other sites are provided by local government. Indeed, green tourism is now a mainstay of the City of Freiberg (www.upperrhnevalley.com).

The presence of social capital and community groups influence the level of community understanding of WSUD features. For example, in Christie Walk and Lochiel Park, which had multiple organizations with groups that distributed information and were involved in education activities, the residents were generally well informed. Mawson Lakes was the largest site and respondents had widespread levels of knowledge notwithstanding the community magazine and the environment group's providing avenues for community education. In contrast, Springbank Waters, Mile End, and Harbrow Grove had low levels of knowledge about their WSUD sites and lacked formal organizations to keep residents informed.

## 24.6.3 Creating social diversity

New estates with WSUD systems can be more expensive to purchase because of the cost of the infrastructure and the public landscape ambience created. Many developments are conscious of the need to avoid forming white, middleclass ghettos where only the relatively affluent can enjoy the benefits of a sustainable development. Such an outcome would defeat a common aim of WSUD developments, which is to act as models to be replicated in other places. Developments can counter this trend by either formally setting aside specific homes for public housing (e.g., Lochiel Park and EVA-Lanxmeer) or informally encouraging diversity by having large differences in plot sizes or home sizes (e.g., Mawson lakes and EVA-Lanxmeer). One innovative approach from the not-for-profit housing cooperative in EVA Lanxmeer was to sell low-cost housing with the proviso that, if the home were sold, then half the profit would be returned to the cooperative.

Of course, creating a diverse community is not only about the cost of housing. As Leonard et al. (2014b) found in Lochiel Park, having public housing tenants and homeowners living in proximity did not create interaction between the two groups. Only after a specific invitation by the homeowners did some tenants feel included. Activities that openly appreciate cultural diversity and make people feel welcome are needed to promote integration. Community gardens are a powerfull site of connection across cultures. Public spaces that are accessible and "all ability" playgrounds encourage participation by people with disabilities or frailty. In Rieselfeld, they specifically wanted to avoid the problems associated with marginalized young people that had occurred in the nearby suburb. After ongoing consultation, a community center with the interests of young people and a sports center were built, and children's games have priority on the side roads.

### 24.6.4 The crucial role of organized community groups

In three sites in Adelaide, Australia, (Christie Walk, Lochiel Park, and Mawson Lakes), and three European sites (Vauban, EVA Lanxmeer, and Rieselfeld), social capital was clearly present in the formal organizations that had emerged to support education about, and maintenance of, the WSUD system. There are two different models for such groups. In places where there is a body corporate or community title, the group can consist of owners regardless of whether they reside at the site. Alternatively, in all sites regardless of the title both owners and renters who reside there can form residents groups. In some of these sites, group activity extended beyond the WSUD features and encompassed care for the broader natural environment. For example, in Christie Walk resident groups were very active on a broad range of proenvironmental activities in addition to activities associated with the ongoing management of their water features. These included monitoring their water consumption and holding regular working bees for maintenance of their WSUD installations. In Lochiel Park, the groups that had formed were focused on lifestyle and recreation activities, as well as their WSUD features. For example, exercise groups and dog-walking groups used the green space provided by the WSUD. Both new residents and long-term residents of Lochiel Park described a strong sense of community, which they depicted as having active community groups, caring neighbors, and extensive opportunities for social engagement and gatherings. The *Friends of Lochiel Park* group, with approximately 20 active members, was the main representative community group with a mandate to tend gardens in common areas, manage weeds, mulch gardens with support from Council, and educate residents. This group had also added lobbying to their role and were actively pursuing the necessary corrections to the recycled water system, and a reduction in the price of purple pipe water for residents. Through their shared norms and active groups, both Christie Walk and Lochiel Park could mobilize social capital to maintain and improve their sites and liaise and lobby external bodies such as councils, water utilities, and government departments when necessary.

In Mawson Lakes, the local council is responsible for maintaining the common space including the WSUD systems; however, there is also a local environmental group of 25 active members that voluntarily helped to keep the lake and wetlands clean of rubbish, and maintain the community gardens with support from Council. Also the community magazine educated residents about water use, appropriate plants for the purple pipe water, and appropriate behavior around the lakes and generally promoted an environmental ethos in the community. Given the high number of residents within the development (Table 24.1), and the varying reasons why people chose to live at Mawson Lakes, the social connections among residents were not as strong as in Lochiel Park or Christie Walk. However, residents did give the impression that they had a shared interest in seeing Mawson Lakes properly maintained and prospering, and they displayed some level of shared capital and social cohesion (Leonard et al., 2014b).

In contrast, community groups functioning to support the WSUD was dormant in the other three Adelaide sites. In Springbank Waters, the local school with its active community liaison officer was the hub of social capital. In the early stages of the Springbank Waters development, the developer and the City of Salisbury had actively involved the school community in a variety of activities including planning proposals for the various water features; planting vegetation; water testing in local waterways; and naming of local landmarks. This involvement contributed to the school feeling a sense of ownership of the wetlands. However, this involvement did not endure after the development was complete, and there was no replacement residents' group. In Mile End and Harbrow Grove, local groups did exist, for example, churches and Neighborhood Watch (https://www.nhwa.com.au), but they were not associated with the WSUD systems. In both communities, the WSUD features had been retrofitted into a well-established community, and the local residents had no active involvement in the care or maintenance of their WSUD features (Leonard et al., 2014b).

#### *24.6.4.1 Community stakeholder group*

Well before civil works commence, stakeholder groups can be formed to inform the progress of the development. Stakeholder groups can include those who have an interest in the site, those with expertise in sustainable development such as academics, and those with practical knowledge such as conservation groups. In EVA-Lanxmeer, the stakeholder group was essential for gaining approval for the development. They succeeded (where the local government had failed) in reassuring the regional authorities that the problem of extra stormwater runoff from increased impervious areas could be dealt with and not overload the combined sewerage system. In Vauban, constant vigilance by the stakeholder group was needed to ensure sustainable design practices. In Aurora, Victoria, community stakeholder group and local council provided input into the identification of community facilities and services that needed to be provided in the development (Moreland Energy Foundation et al., 2011).

Community groups also provide a possible point of consultation for councils and other authorities (Carlson et al., 2014). However, in most cases they were not well utilized. Residents in the six Adelaide case studies consistently reported being dissatisfied with the lack of consultation leading to a subsequent mismatch of expectations and information on WSUD features. Residents often felt a lack of control in decision-making related to the WSUD features. Their involvement with the everyday functioning of the WSUD was limited to those located on their individual properties. The smaller street-scale initiatives in Mile End were received negatively after weak public input, as the subsequent reduction in on-street parking came as a surprise to residents. In contrast, only residents of Christie Walk did not mention lack of consultation as a barrier to acceptance, because of their integrated involvement in the activation and maintenance of the WSUD systems. While facing difficulties with accessing information, they were able to overcome this problem.

In most WSUD sites, there were initial installation problems that reduced amenity and had the potential to reduce community acceptance of WSUD. In Christie Walk, the residents quickly found and repaired faults in the initial installation. They also lobbied and worked with the city council and water utility to try, unsuccessfully, to increase their WSUD credentials by treating their wastewater to a standard that allowed its use to irrigate a local park. Lochiel Park has faced the most "teething" problems especially with their purple pipe system, which were exacerbated by disputes between the developer (a State government entity), council, and the water utility about responsibility for the repairs. Residents had been actively lobbying for 2 years, a struggle that increased their community cohesion. On the positive side, they have reduced weeds, improved the gardens, and are now making wall mosaics to beautify their site. As part of their activism, they were successful in reducing the price of purple pipe water that had become higher than the price of mains water for those with low water use. In Harbrow Grove where there is no residents' group, people are critical of the empty pond and the reduced amenity, but they are unaware that council is working to rectify the problem nor do they understand why it is taking so long. Similarly, in Mile End where residents are most critical of the WSUD installations, the long period in which the gardens have lain bare has not been explained. Even so, there was potential to form a residents group as some residents were already monitoring and reporting any problems. For example, deciduous autumn leaves blocked the stormwater drains and individuals reported the problem to the council although they could have cleared the drains themselves and avoided the minor flooding that occurred. The absence of a resident's group lead to a failure of imagination, and the self-confidence to initiate community action.

### *24.6.4.2 Community governance*

Once the WSUD development has been built, arrangements need to be made to maintain the site, especially to maintain its environmental credentials. A community governance body can develop from the stakeholder group (e.g., Christie Walk) or be a new entity, (e.g., EVA-Lanxmeer, Vauban). In Australia, the most common form of the community governance is a body corporate that is required under the strata title act. Such groups not only maintain the infrastructure but also can add to the quality of life of residents and increase the green credentials of the development. They can raise money for community projects and negotiate with local authorities and companies for better services or resources. In a comprehensive study of the role of bodies corporate with strata and community title in Queensland, NSW, and Victoria, Warnken et al. (2009) found that such bodies could play a valuable role in supporting the use of decentralized water systems, which allowed urban densification without increasing the strain on centralized urban water and sewerage systems. They made a number of recommendations including appropriate insurance, training of strata managers, and metering of individual dwellings. There would need to be legislative changes to license small wastewater plants and an appropriate inspection regime. For larger developments, they recommended legislation changes so that small water utilities, rather than bodies corporate, could manage these sites.

The handover from the developer to the local authority can be facilitated by the local community governance organization. After all, it is the local community that understands the local issues and has a vested interest in ensuring that the site is well maintained. The residents group at Lochiel Park successfully lobbied local council to resist the handover from the developer until a range of problems had been addressed. When such a group does not form spontaneously, it is worth engaging a community development worker to assist in the process of recruiting committee members and ensuring they are knowledgeable about community management and the requirements of the site.

## 24.7 CONCLUSION

In keeping with previous research, there was a positive community attitude to WSUD and high acceptance of treated reclaimed water for nonpotable uses (Leonard et al., 2015a; Hurlimann and Dolnicar, 2010; Mankad and Tapsuwan, 2011) including an appreciation of improved aesthetics, greener surroundings, and improved local habitat (Centre for Water Sensitive Cities, 2011). Maintenance problems can detract from the efficacy of WSUD sites and potentially reduce public support for WSUD installations. However, place attachment and high social capital can ameliorate the problems in the WSUD systems or at least the negative attitudes they can trigger potential for a virtuous cycle of support for WSUD.

Social capital and place attachment provide a conceptual framework for understanding support for WSUD systems. They also show the potential for a virtuous cycle of prosocial and proenvironmental behavior. A virtuous cycle is a self-reinforcing loop, where each event in the cycle provides a positive feedback or beneficial effect for the next. In the context of neighborhood-scale WSUD, the cycle operates as follows:

1. Visual, recreational, and other amenities can provide opportunities for social interaction and the development of social capital.
2. Amenity can also promote increased place attachment as people appreciate the appearance and their interactions with the place. But conversely, WSUD system problems can decrease place attachment.
3. There is a strong interconnection between place attachment and social capital. Social interactions evoke meaning to the space as a "place" embellished with emotional connections, strengthening residents' attachment to the community and physical setting.
4. Both the place attachment and social capital foster greater appreciation of WSUD. This is enhanced by better knowledge and understanding of WSUD, and the formation of community groups that take some responsibility for maintaining the quality of the space. They address any system problems either directly or indirectly through lobbying the relevant authorities.
5. Ultimately the care and stewardship devoted to the WSUD site maintains and improves its amenity, and the social activities around the site increase social capital.

### 24.7.1 Points of intervention: enhancers and detractors of the virtuous cycle

A WSUD installation will not necessarily lead to a virtuous cycle of support, but our analysis suggest points at which advocates of WSUD, such as developers, local councils, and state government might intervene to encourage its development. One point of intervention is the appreciation of the WSUD feature. Although WSUD features that improved aesthetics, greenscape, recreational amenity, and increased resident control over their own water supply have instant appeal, the less visible outcomes, such as improving the quality of runoff or flood mitigation, are often overlooked by the community. When WSUD features are retrofitted, the local community is less likely to care about the presence of WSUD features because they chose to live in the area for other reasons. In the retrofitting case, the water authorities or council will have to be proactive in creating involvement within the community, so that the community perceives the added benefit of the features, tolerates the construction inconvenience, and ultimately accepts the features as an improvement to their area.

The need for education on WSUD systems and water use was a common point of agreement for further action across the sites. Lack of understanding about WSUD system operations; lack of community information about WSUD features and how to use water sustainability; and lack of industry and government knowledge about WSUD technology emerged as three separate, but related, dimensions that participants identified as requiring intervention. Participants were aware of the deficits in their knowledge and called for more community education programs. Christie Walk is a place where formal resident outreach played a role in wider community education about innovative and best water practices. They were also looking to the government for more support for this education to occur at the wider level. In Christie Walk and Lochiel Park, residents expressed a belief that they could serve as exemplars to build community awareness and expertise about sustainable water management in wide Adelaide if information outlets and information transfer mechanisms were strengthened.

A second point of intervention is the establishment of groups that have some sense of responsibility and appreciation of the WSUD feature. Developers often promote a sense of community by inviting existing local residents and home purchasers to local events. From such events, a residents' group might emerge spontaneously, but the developers could take this a step further and support the formation of such a group. They could point out the value for a formal channel of communication between residents, council, and the developers, particularly to address ongoing maintenance at the time of the handover from the developer to the council. Councils could require such initiatives as part of the approval process. In places where WSUD installations are being retrofitted, councils could consider mobilizing the existing social capital of the community by contacting existing groups. Or if there are no appropriate groups, they invite residents to form one to participate, for example, in monitoring the system.

A third point of intervention is supporting the groups that engage with the ongoing monitoring and maintenance of the site by providing training, materials, and ongoing information about the site, and being responsive to problems and initiatives as they arise. In particular, participants recommended that for future large−medium-scaled WSUD developments, the developers and subsequent caretakers (usually local councils) should create a comprehensive "handover" plan to prevent overlooking important maintenance issues. It was also important to residents that once the developer had vacated from the site, the new caretaker sets up a visible presence in the area so that residents have somewhere to report future problems, as and when they were identified.

The role of social capital and place attachment are closely linked to community acceptance and can be used to overcome barriers, enhance benefits, and ultimately to foster acceptance. Not only are social capital and place attachment created from the benefits of WSUD, especially the indirect benefits, but also increased social capital and place attachment supports the ongoing management of WSUD. Interventions that initiate and support development of community groups strengthen social capital within a community and support WSUD over the long term. The case studies highlighted that the presence of organized community groups that effectively mediated the relationship between the WSUD installations and the community, with their influence being stronger in smaller developments through shared involvement in running and maintaining the WSUD installations.

## REFERENCES


Adger, W.N., 2003. Social capital, collective action, and adaptation to climate change. Economic Geography 79, 387−404.

Aecom, 2017. Green Infrastructure: A Vital Step to Brilliant Australian Cities. Brilliant Australian Cities.

Anderson, C.L., Locker, L., Nugent, R., 2002. Microcredit, social capital and common pool resources. World Development 30, 95−105.

Australian Bureau Of Statistics, 2013. Rainwater Tanks (Online). Available: http://www.abs.gov.au/ausstats/abs@.nsf/Lookup/4602.0.55.003main+features4Mar%202013.

Bowman, T., Tyndall, J.C., Thompson, J., Kliebenstein, J., Colletti, J.P., 2012. Multiple approaches to valuation of conservation design and low-impact development features in residential subdivisions. Journal of Environmental Management 104, 101−113.

Brent, D.A., Gangadharan, L., Lassiter, A., Leroux, A., Raschky, P.A., 2017. Valuing environmental services provided by local stormwater management. Water Resources Research 4907−4921.

Brown, R.R., Clarke, J.M., 2007. Transition to water sensitive urban design: The story of Melbourne, Australia. Facility for Advancing Water Biofiltration, Monash University, Melbourne.

Brown, R., Keath, N., Wong, T., 2008. Transitioning to Water Sensitive Cities: Historical, Current and Future Transition States. 11th International Conference on Urban Draingage, Edinburgh, Scotland, UK.

Water By Design, 2010. A Business Case for Best Practice Urban Stormwater Management (Version 1.1). South East Queensland Healthy Waterways Partnership, Brisbane.

Carlson, C., Asce, M., Barreteau, O., Kirshen, P., Foltz, K., 2014. Storm water management as a public good provision problem: survey to understand perspectives of low-impact development for urban storm water management practices under climate change. Journal of Water Resource Planning and Management.

Carter, T., Keeler, A., 2008. Life-cycle cost−benefit analysis of extensive vegetated roof systems. Journal of Environmental Management 87, 350−363.

Castonguay, A.C., Urich, C., Iftekhar, M.S., Deletic, A., 2016. Modelling urban transition: a case of rainwater harvesting. In: 8th International Congress on Environmental Modelling and Software, Toulouse, France.

Chen, D., Wang, X., Thatcher, M., Barnett, G., Kachenko, A., Prince, R., 2014. Urban vegetation for reducing heat related mortality. Environmental Pollution 192, 275−284.

Christian Amos, C., Rahman, A., Mwangi Gathenya, J., 2016. Economic analysis and eeasibility of rainwater harvesting systems in urban and peri-urban environments: a review of the global situation with a special focus on Australia and Kenya. Water 8.

CRC For Water Sensitive Cities, 2011. Project 8: Demonstration and Integration through Urban Design: Literature and Practice Review. CRC for Water Sensitive Cities, Melbourne.

CRC WSC, 2014. Strategies for Preparing Robust Business Cases. CRC for Water Sensitive Cities, Melbourne.

Department Of Planning And Local Government, 2010. Water Sensitive Urban Design Technical Manual for the Greater Adelaide Region. Government of South Australia, Adelaide.

Devine Wright, P., Howes, Y., 2010. Disruption to place attachment and the protection of restorative environments: A wind energy study. Journal of Environmental Psychology 30, 271−280.

Dolnicar, S., Hurlimann, A., Nghiem, L.D., 2010. The effect of information on public acceptance−the case of water from alternative sources. Journal of Environmental Management 91, 1288−1293.

Dzidic, P., Green, M., 2012. Outdoing the Joneses: understanding community acceptance of an alternative water supply scheme and sustainable urban design. Landscape and Urban Planning 105, 266−273.

Edwards, J., Pocock, B., 2011. Comfort, Convenience and Cost: The Calculus of Sustainable Living at Lochiel Park. Adelaide.

Ely, M., Pitman, S., 2014. Green Infrastructure: Life supports for human habitats. Government of South Australia Department of Environment, Water and Natural Resources, Adelaide.

Forsyth, A., Crewe, K., 2009. New Visions for Suburbia: Reassessing Aesthetics and Place-making in Modernism, Imageability and New Urbanism. Journal of Urban Design 14, 415–438.

Government of Western Australia, 2009. Gnangara Sustainability Strategy Situation Statement. Government of Western Australia, Perth.

Government of Western Australia, 2016. Western Australia's water supply and demand outlook to 2050. Government of Western Australia, Perth.

Halpern, D., 2005. Social Capital. Polity Press, Cambridge.

Hardin, G., 1968. Tragedy of the commons. Science 162, 1243.

Hurlimann, A., Dolnicar, S., 2010. When public opposition defeats alternative water projects - the case of Toowoomba Australia. Water Research 44, 287–297.

Hurlimann, A.C., Mckay, J.M., 2006. What attributes of recycled water make it fit for residential purposes? The Mawson Lakes experience. Desalination 187, 167–177.

Iftekhar, M., Urich, C., Schilizzi, S., Deletic, A., 2016. Effectiveness of incentives to promote adoption of water sensitive urban design: a case study on rain water harvesting tanks. In: 8th International Congress on Environmental Modelling and Software (IEMS). Toulouse, France.

Iftekhar, M.S., Fogarty, J., Polyakov, M., Zhang, F., Burton, M., Pannell, D., 2017a. Non-market values of water sensitive urban design: a case study on rain gardens. In: Australian Agricultural and Resource Economics Society (AARES) National Conference. Brisbane, Australia.

Iftekhar, M.S., Polyakov, M., Ansell, D., Gibson, F., Kay, G., 2017b. How economics can further the success of ecological restoration. Conservation Biology 31, 261–268.

Jernberg, J., Hedenskog, S., Huang, C.C., 2015. Hammarby Sjöstad, an Urban Development Case Study of Hammarby Sjöstad in Sweden, Stockholm. China Development Bank Capital.

Johnson, L.C., 2012. Creative suburbs? How women, design and technology renew Australian suburbs. International Journal of Cultural Studies 15, 217–229.

Joint Steering Committee For Water Sensitive Cities, 2009. Evaluating Options for Water Sensitive Urban Design: A National Guide. Developed in Accordance with the National Water Initiative Clause 92 (ii).

Kardan, O., Gozdyra, P., Misic, B., Moola, F., Palmer, L.J., Paus, T., Berman, M.G., 2015. Neighborhood greenspace and health in a large urban center. Scientific Reports 5, 11610.

Khan, S.J., Gerrard, L.E., 2006. Stakeholder communications for successful water reuse operations. Desalination 187, 191–202.

Khastagir, A., Jayasuriya, N., 2011. Investment evaluation of rainwater tanks. Water Resources Management 25, 3769–3784.

Kolbe, J., Wüstemann, H., 2015. Estimating the Value of Urban Green Space.

Leonard, R. 2016, unpublished. Site visits and key informant interviews at Rieselfeld, Freiburg, Germany; Vauban, Freiburg, Germany; EVA Lanxmeer, Culemborg, The Netherlands; Hammarby Sjöstad Stockholm, Sweden.

Leonard, R., Alexander, K., 2012. Assessment of alternative water options in Adelaide: the MARSUO and optimal water resource mix projects. In: In Begbie, D.K., Kenway, S.J., Biermann, S.M., Wakem, S.L. (Eds.), Science Forum and Stakeholder Engagement: Building Linkages, Collaboration and Science Quality. Urban Water Security Research Alliance. CSIRO, Brisbane, QLD.

Leonard, R., Bellamy, J., 2010. The Relationship between Bonding and Bridging Social Capital among Christian Denominations across Australia. Nonprofit Management and Leadership 20, 445–460.

Leonard, R., Onyx, J., 2004. Social Capital & Community Building: Spinning Straw into Gold. Janus Publishing, London.

Leonard, R., Walton, A., Koth, B., Green, M., Spinks, A., Myers, B., Malkin, S., Mankad, A., Chacko, P., Sharma, A., Pezzaniti, D., 2014b. Community Acceptance of Water Sensitive Urban Design: Six Case Studies. Goyder Institute for Water Research, Adelaide.

Leonard, R., Mankad, A., Alexander, K., 2015a. Predicting support and likelihood of protest in relation to the use of treated stormwater with managed aquifer recharge for potable and non-potable purposes. Journal of Cleaner Production 29, 248–256.

Leonard, R., Walton, A., Farbotko, C., 2015b. Using the Concept of Common Pool Resources to Understand Community Perceptions of Diverse Water Sources in Adelaide. Water Resources Management, South Australia.

Leviston, Z., Browne, A.L., Greenhill, M., 2013. Domain-based perceptions of risk: a case study of lay and technical community attitudes toward managed aquifer recharge. Journal of Applied Social Psychology 43, 1159–1176.

Lothian, A., Marks, J., Szili, G., Rofe, M.W., 2010. Perceptions of Water in the Urban Landscape (Doctoral dissertation). Wakefield Press.

Maller, C., Townsend, M., St Leger, L., Henderson-Wilson, C., Pryor, A., Prosser, L., Moore, M., 2008. Healthy Parks, Healthy People: The Health Benefits of Contact with Nature in a Park Context. Deakin University and Parks Victoria, Melbourne.

Mankad, A., Tapsuwan, S., 2011. Review of socio-economic drivers of community acceptance and adoption of decentralised water systems. Journal of Environmental Management 92, 380–391.

Mankad, A., Walton, A., 2015. Accepting managed aquifer recharge of urban storm water reuse: the role of policy-related factors. Water Resources Research 51, 9696–9707.

Mankad, A., Walton, A., Alexander, K., 2015. Key dimensions of public acceptance for managed aquifer recharge of urban stormwater. Journal of Cleaner Production 89, 214–223.

Manzo, L.C., Perkins, D.D., 2006. Finding common ground: the importance of place attachment to community participation and planning. Journal of Planning Literature 20, 335–350.

Mcclean, S., Onyx, J., 2009. Institutions and Social Change: implementing co-operative housing and environmentally sustainable development at Christie Walk. Cosmopolitan Civil Societies: An Interdisciplinary Journal 1.

Mekala, G.D., Jones, R.N., Macdonald, D.H., 2015. Valuing the benefits of creek rehabilitation: building a business case for public investments in urban green infrastructure. Environmental Management 55, 1354–1365.
Mitchell, V.G., 2006. Applying integrated urban water management concepts: a review of Australian experience. Environ Manage 37, 589–605.
Moreland Energy Foundation, Net Balance, Green Spark Consulting, 2011. Business Models for Enabling Sustainable Precincts Melbourne: Sustainability Victoria.
Morison, P.J., Brown, R.R., 2011. Understanding the nature of publics and local policy commitment to Water Sensitive Urban Design. Landscape and Urban Planning 99, 83–92.
Morrison, M., Oczkowski, E., Greig, J., 2011. The primacy of human capital and social capital in influencing landholders' participation in programmes designed to improve environmental outcomes*. Australian Journal of Agricultural and Resource Economics 55, 560–578.
Myers, B., Chacko, P., Tjandraatmadja, G., Cook, S., Umapathi, S., Pezzaniti, D., Sharma, A., 2013. The Status of Water Sensitive Urban Design in South Australia. Goyder Institute for Water Research, Adelaide. Technical report Jan 2013.
Oliphant, M., 2004. Inner City Residential Energy Performance. UEA report for SENRAC, Adelaide.
Onyx, J., Osburn, L., Bullen, P., 2004. Response to the environment: social capital and sustainability. Australasian Journal of Environmental Management 11, 212–219.
Orians, G.H., Heerwagen, J.H., 1992. Evolved responses to landscapes. In: Barkow, J.H., Cosmides, L., Tooby, J. (Eds.), The adapted mind: Evolutionary psychology and the generation of culture. Oxford University Press, New York.
Ostrom, E., 1990. Governing the Commons: The Evolution of Institutions for Collective Action. Cambridge University Press, New York.
Pandit, R., Polyakov, M., Tapsuwan, S., Moran, T., 2013. The effect of street trees on property value in Perth, Western Australia. Landscape and Urban Planning 110, 134–142.
Po, M., Kaercher, J.D., Nancarrow, B.E., 2003. Literature Review of Factors Influencing Public Perceptions of Water Reuse. CSIRO Land and Water Technical Report, Canberra.
Polyakov, M., Fogarty, J., Zhang, F., Pandit, R., Pannell, D.J., 2017. The value of restoring urban drains to living streams. Water Resources and Economics 17, 42–55.
Portney, K.E., Berry, J.M., 2010. Participation and the pursuit of sustainability in U.S. Cities. Urban Affairs Review 46, 119–139.
Pretty, J., Ward, H., 2001. Social capital and the environment. World Development 29, 209–227.
P.U.B. 2008. PURE. P.U.B. annual report 2007/08 Singapore: P.U.B., Singapore's national water agency.
Putnam, R., 2000. Bowling Alone: The Collapse and Revival of American Community. Simon & Schuster, New York.
Queensland Competition Authority, 2012. Assessment of Proposed Repeal of Water Saving Regulations. Queensland: Queensland Competition Authority. Office of Best Practice Regulation.
Radcliffe, J.C., Page, D., Naumann, B., Dillon, P., 2017. Fifty years of water sensitive urban design, Salisbury, South Australia. Frontiers of Environmental Science and Engineering 11.
Robertson, J., 2016. 'Worst thing… in decades': Fears for Sydney's drinking water. Sydney Morning Herald.
Rossetti, J.O.E., 2013. Valuation of Australia's Green Infrastructure: Hedonic Pricing Model Using the Enhanced Vegetation Index (Honours Honours Thesis). Monash University.
Rostila, M., 2010. The facets of social capital. Journal for the Theory of Social Behaviour 41, 308–326.
Roy, A.H., Wenger, S.J., Fletcher, T.D., Walsh, C.J., Ladson, A.R., Shuster, W.D., Thurston, H.W., Brown, R.R., 2008. Impediments and solutions to sustainable, watershed-scale urban stormwater management: lessons from Australia and the United States. Environmental Management 42, 344–359.
Schebella, M., Weber, D., Brown, G., Hatton Macdonald, D., 2014. The Importance of Irrigated Urban Green Space: Health and Recreational Benefits Perspectives. Goyder Institute for Water Research Technical Report Series Adelaide, South Australia.
Sharma, A., Cook, S., Tjandraatmadja, G., Gregory, A., 2012. Impediments and constraints in the uptake of water sensitive urban design measures in greenfield and infill developments. Water Science and Technology 65, 340–352.
Sharma, A.K., Tjandraatmadja, G., Cook, S., Gardner, T., 2013. Decentralised systems – definition and drivers in the current context. Water Science and Technology 67, 2091.
Sharma, A.K., Begbie, D., Gardner, T., 2015. Rainwater Tank Systems for Urban Water Supply. IWA Publishing.
Sharma, A., Pezzaniti, D., Myers, B., Cook, S., Tjandraatmadja, G., Chacko, P., Chavoshi, S., Kemp, D., Leonard, R., Koth, B., Walton, A., 2016. Water sensitive urban design: an investigation of current systems, implementation drivers, community perceptions and potential to supplement urban water services. Water 8.
Sofoulis, Z., 2005. Big water, everyday water: a sociotechnical perspective. Continuum: Journal of Media and Cultural Studies 19, 445–463.
Symons, J., Jones, R., Young, C., Rasmussen, B., 2015. Assessing the Economic Value of Green Infrastructure: Literature Review.
Tajfel, H., Turner, J.C., 2004. The Social Identity Theory of Intergroup Behavior. Political psychology: Key readings. Psychology Press, New York, NY, US.
Tam, V.W., Tam, L., Zeng, S., 2010. Cost effectiveness and tradeoff on the use of rainwater tank: an empirical study in Australian residential decision-making. Resources, Conservation and Recycling 54, 178–186.
Taylor, M.S., Wheeler, B.W., White, M.P., Economou, T., Osborne, N.J., 2015. Research note: urban street tree density and antidepressant prescription rates—a cross-sectional study in London, UK. Landscape and Urban Planning 136, 174–179.
Tjandraatmadja, G., Cook, S., Sharma, A., Diaper, C., Grant, A., Toifl, M., Barron, O., Burn, S., Gregory, A., 2008. ICON Water Sensitive Developments.
Turton, A.R., Meissner, R., 2000. The Hydro-Social Contract and its Manifestation in Society: A South African Case Study. African Water Issues Research Unit Occasional Paper. African Water Issues Research Unit.

Urrutiaguer, M., Lloyd, S., Lamshed, S., 2010. Determining water sensitive urban design project benefits using a multi-criteria assessment tool. Water Science and Technology: A Journal of the International Association on Water Pollution Research 61, 2333–2341.

Van Timmeren, A., Kaptein, M., Sidler, D., 2007. Sustainable urban decentralisation: EVA Lanxmeer, Culemborg, The Netherlands. ENHR: Sustainable Urban Areas, 2007 Rotterdam. ENHR.

Vandermeulen, V., Verspecht, A., Vermeire, B., Van Huylenbroeck, G., Gellynck, X., 2011. The use of economic valuation to create public support for green infrastructure investments in urban areas. Landscape and Urban Planning 103, 198–206.

Warnken, J., Johnston, N., Guilding, C., 2009. Exploring the Regulatory Framework and Governance of Decentralised Water Management Systems: A Strata and Community Title Perspective. National Water Commission, Canberra.

Wong, T., 2007. Water Sensitive Urban design - The story thus far. BEDP Environment Design Guide. Royal Australilan Institute of Architects.

Zhang, F., Polyakov, M., Fogarty, J., Pannell, D.J., 2015. The capitalized value of rainwater tanks in the property market of Perth, Australia. Journal of Hydrology 522, 317–325.

Chapter 25

# Post Implementation Assessment of WSUD Approaches: Kansas City Case. WSUD Systems Scale Monitoring and Watershed Level Model Validation

Deborah J. O'Bannon[1] and Yanan Ma[2]

[1]*University of Missouri-Kansas City, Kansas City, MO, United States;* [2]*AECOM-Water, Kansas City, MO, United States*

## Chapter Outline



**ABSTRACT**

A rich, longitudinal data set was collected on six stormwater control measures (SCMs) in Kansas City, Missouri, USA. Detailed inflows, pool level, and outflows were measured over 57 rain events in a 3-year period. The hydrology data, together with the drainage characteristics, were included in a multiple linear regression (MLR) model after pretreatment with the Principal Component Analysis, which created scaled, normalized compound variables as inputs to the MLR. The models were calibrated and verified with field data, and then extrapolated to the entire 40 ha catchment containing 135 SCMs. This expanded model was compared to independent field data, and the stipulated stormwater removal target of 1135 $m^3$ from a 36 mm storm. The model results indicated that the current SCM design was very conservative and had exceeded the performance expectations.

**Keywords:** Catchment; Infiltration; Statistics; Stormwater; Stormwater control measures; Urban.

## 25.1 INTRODUCTION

An urban area of Kansas City, Missouri was selected as a study area to see if an intensive implementation of decentralized "green" Stormwater Control Measures (SCMs) could perform similarly to traditional storage (grey) infrastructure in reducing stormwater flows into a combined sewer collection system (Simon, 2016; Pitt et al., 2010, Chapter 7). The



**Approaches to Water Sensitive Urban Design. https://doi.org/10.1016/B978-0-12-812843-5.00025-3**

original infrastructure remediation plan for the study area called for two large underground stormwater storage tanks (13,000 $m^3$ total capacity). The cost of the alternative, decentralized SCM construction was comparable to the proposed grey infrastructure. The well-defined boundaries of the SCMs, which were defined by concrete curbing and channeling, facilitated installation of flow measuring instrumentation of inlets and outlets. The study area was a 40 ha urban area in which 135 SCMs were installed in 2012. This demonstration area (see Fig. 25.1) drains to the combined sewer system. The SCM solutions were designed to capture rainfall events of up to 36 mm of rainfall per 24 h. The SCMs in the study area can be broadly categorized into two groups—unconnected SCMs (rain gardens) and connected SCMs, which include drains to the combined sewer collection system.

Of the 40, 20 ha drain into the SCMs, which comprise 83 unconnected SCMs, 5 connected SCMs, 47 bioretention SCMs, 1 bioswale, and almost 500 $m^2$ of porous sidewalk. All SCMs have an engineered soil layer to improve infiltration. The unconnected SCMs infiltrate or overflow (although no overflows were observed in this study) their inflows. The connected SCMs (Smart Drain technology) largely infiltrate the captured runoff: the remainder drains to the combined sewer system. The bioretention SCMs have both infiltration and storage capabilities. This storage functions as detention, where the excess water from the SCM is stored below grade and slowly overflows through a small orifice to the collection system for treatment. (Kansas City Water Services Department, 2013; Dods, 2016).

## 25.2 LITERATURE REVIEW

### 25.2.1 Stormwater control measures monitoring, field studies, and models

SCM stormwater capture and infiltration performance has been reported by a number of investigators (Selbig and Balster, 2010; Welker et al., 2013; Jones and Wadzuk, 2013; Wadzuk et al., 2017; Aguilar and Dymond, 2016; Hunt et al., 2006, 2008, 2012; Wilson et al., 2015; Dietz and Clausen, 2005; Carmen et al., 2016; Potter, 2007; Line et al., 2012; Guo and

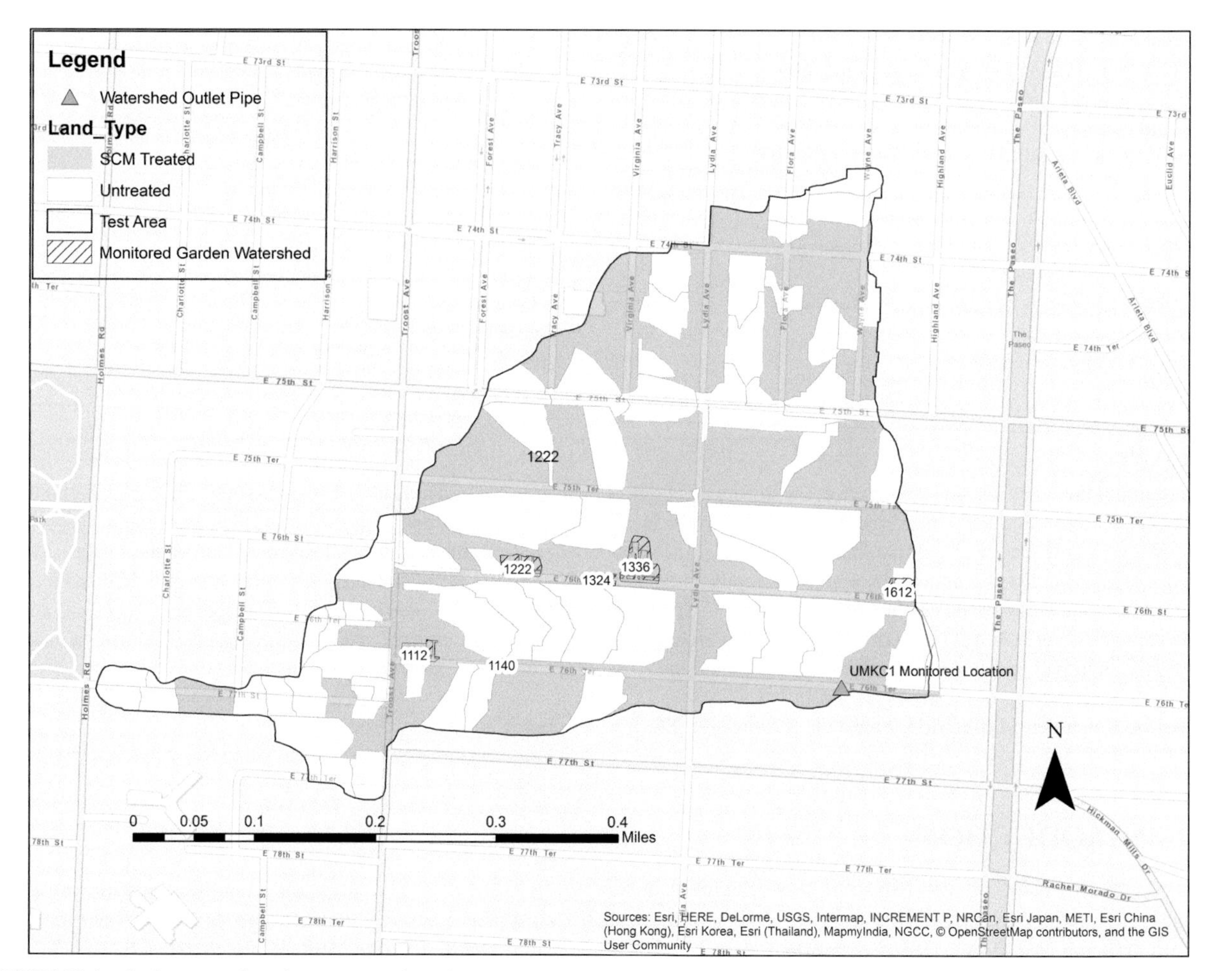


FIGURE 25.1 Study area and catchment areas of monitored (stormwater control measures) SCMs (including the catchment contributions of all constructed SCMs) and catchment outflow pipe measurement. The monitored SCMs are indicated by numbered text boxes and show the catchment of each monitored SCM.

FIGURE 25.2 Profiles of stormwater control measures (SCMs) (a) unconnected SCM, (b) connected SCM, (c) bioretention SCM with storage. *Graphics by Alison Vieritz.*

Luu, 2015; Zhang and Guo, 2013; Qin et al., 2013; Hekl and Dymond, 2016; Hood et al., 2007; Ma et al., 2018; Mangangka et al., 2015). SCMs are gaining acceptance as a functional alternative to grey (pipe) stormwater improvements. SCMs are effective at intercepting storm runoff and attenuating the peak flows to receiving streams or combined sewer systems.

Stormwater modeling has evaluated the effect of SCMs in urban settings, using well-accepted models, such as SWMM (Damodaram and Zechman, 2013; Mancipe-Munoz et al., 2014, EPA, 2016) and SUSTAIN (EPA, 2014), which was used in a paired catchment study (Bedan and Clausen, 2009).

Measured and monitored data can help define or modify low impact development (LID) guidelines. However, a sufficiently robust data analysis approach framework is lacking (Ahiablame et al., 2012a,b). The authors recommended that future LID research should develop an easy-to-use decision tool that can incorporate LID units.

## 25.3 FIELD METHODS

The demonstration area established in Kansas City was chosen to assess if SCMs could provide sufficient stormwater capture and absorption capacity to reduce stormwater inflows into a combined sewer collection system. There were concerns that the compacted clayey soil in the urban area would respond poorly to green engineering SCM intervention. Six SCMs were chosen for intense data collection from 2011 to 2015. They had some minor design differences, which helped the City design SCMs for other green solution deployments. Inflows and outflows to each SCM, as well as SCM pool elevations, were continuously monitored using a combination of weirs, flumes, pressure transducers, and data loggers to provide the data on their hydraulic efficacy. Fig. 25.1 shows the locations of the six SCMs and their contributing catchment areas in the 40 ha catchment. Each SCM was labeled by the closest house number (e.g., 1112, 1140, 1222, 1324, 1325, and 1336).

The hydraulic performance of the 40 ha area was earlier analyzed using the SWMM model, which identified a need for 1135 $m^3$ of stormwater detention to prevent combined sewer overflows. The city decided to install SCMs in lieu of the stormwater detention tanks (13,000 m^3) and sought to quantify their efficacy for reducing stormwater flow to the combined sewers. Cross section flow schematics of the three major SCM types are shown in Fig. 25.2.

### 25.3.1 Monitored stormwater control measuress

The monitored SCMs included four unconnected SCMs (traditional rain gardens) and two connected SCMs using the Smart Drain technology. Example photographs and plan drawings of monitored SCMs, including the location of monitoring equipment are shown in Figs. 25.3–25.5 below.

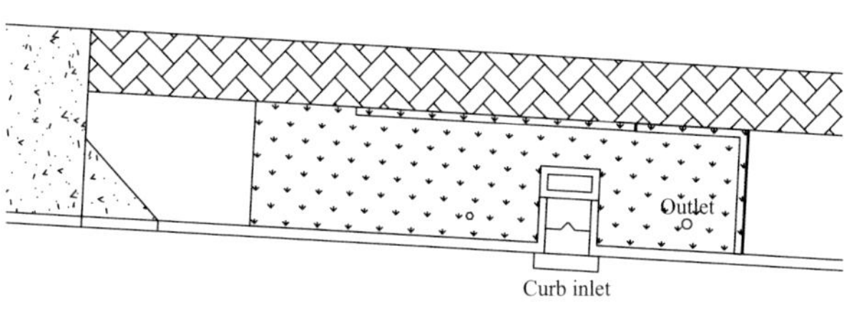


FIGURE 25.3 Photograph and plan drawing (SCM length = 13.6 m, SCM width = 2.5 m) of connected SCM 1140. SCM, stormwater control measure.

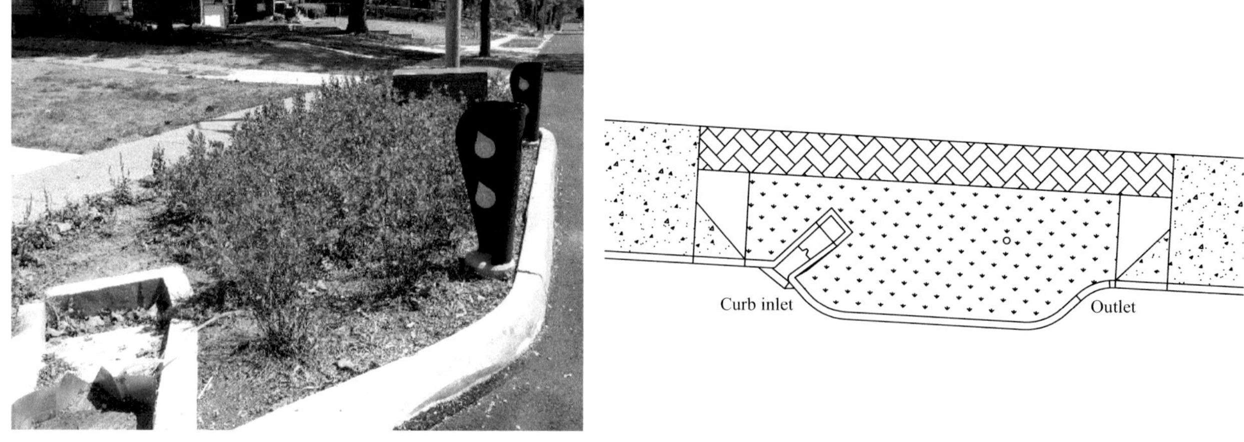


FIGURE 25.4 Photograph and plan drawing (SCM length = 13.8 m, SCM width = 2.4 m) of unconnected SCM 1324. Safety bollards were installed on this SCM type. SCM, stormwater control measure.

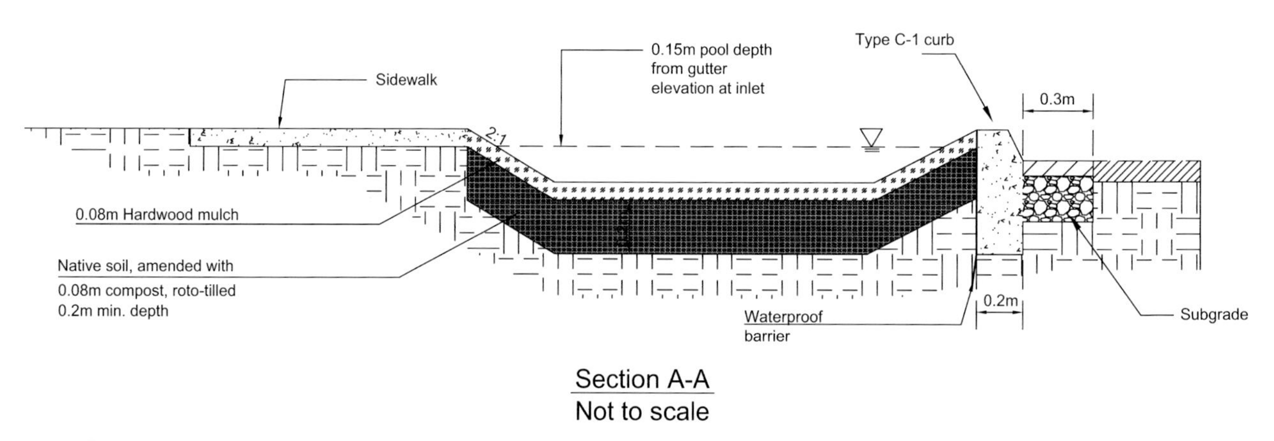


FIGURE 25.5 Typical cross-section of an unconnected stormwater control measures, showing relative pool depth and soil amendments.

## 25.3.2 Rain data Collection

The monitoring commenced in March 2011 and ended in December 2015: a total of 54 months. A tipping bucket rain gauge with data logger was installed near the study area. The average distance from the rain gauge to monitored sites was 600 m. A total of 57 rain events were used in this analysis. Discrete rainfall events were defined as having an intervening dry period of >10 h.

## 25.3.3 SCM inflow measurements

The SCMs in the demonstration area have well-defined boundaries formed by concrete curbing. Stormwater inflow to the SCMs was from the road pavement. Each monitored SCM had a 0.5H flume (Fig. 25.6(a)) installed to measure inflow, with the water level measured using pressure transducers (±3 mm resolution). The flumes emptied freely into the inlet structure, as shown in Fig. 25.6(b). The sediment basin downstream of the flume shown in Fig. 25.6(a) was emptied frequently. Water levels were converted into discharge (L/s) using a standard flume rating curve (Gwinn and Parsons, 1976).

## 25.3.4 Stormwater control measures outflow measurements

Outflow monitoring was undertaken using 22 degrees V-notch overflow weirs (Fig. 25.7(a) below) installed in four of the six SCMs. They were instrumented with pressure transducer and data logger. These four SCMs were unconnected to the sewer, did not have underdrains, and yet did not experience any overflow during the 54-month monitoring period. That is, all the stormwater inflow to these SCMs was absorbed.

FIGURE 25.6 0.5H flumeinlet: street side (a) and outlet (b).

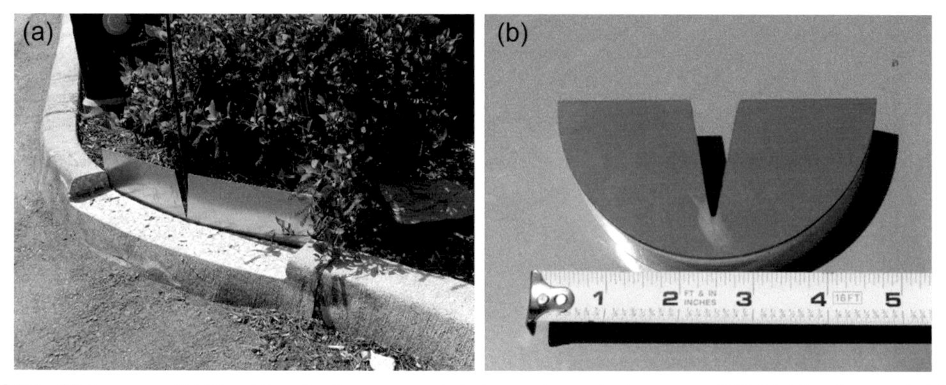


FIGURE 25.7 22 degrees outlet V-notch weirs for unconnected stormwater control measures (SCMs) (a) and connected SCMs (b).

Two of the six monitored SCMs were constructed with capillary flow–based Smart Drain underdrains (SmartDrain.com). 22 degrees V-notch weirs were installed in the outlet pipes of the drainage system. Fig. 25.7(b) shows the V-notch weir prior to its installation into a 150-mm underdrain, located downstream of the Smart Drain. A weir rating curve (USBR, 1997) allowed conversion of weir water levels into discharge (L/s).

A typical event hydrograph is show in Fig. 25.8. The graph shows the rain data for the event, the depth measurements at the inlet flume, and the pool depth inside the SCM as a function of time. The pool depth quickly returns to zero because of infiltration into the media. High absorption rates were observed in the SCMs for most of the rain events.

The absorption volume for the two connected SCMs with Smart Drain underdrains (sites 1222 and 1140: see Fig. 25.1) was calculated by subtracting outflow to sewer from the inflows. Fig. 25.9 shows data from a connected SCM for May 27, 2013 storm (same date as Fig. 25.8) where the rainfall, inlet flume water depth, and outlet weir water depth were plotted as a function of time. There were only 4 L of outflow for this 60-mm rain event. Figs. 25.8 and 25.9 are just two of the 255 hydrographs measured in this study.

The measured volume of storm runoff captured by the four unconnected SCMs equals the inflow volume, as no overflows were measured from these SCMs (sites 1324, 1325, 1336, and 1112) in the study period. However, for the connected SCMs, (sites 1140 and 1222), the water volume captured equals the absorption volume. The captured volume in both cases is absorbed and removed from the catchment's runoff.

## 25.3.5 Catchment outflow pipe measurement

The outflow from the entire 40 ha test catchment was monitored within the combined sewer pipe (as indicated in Fig. 25.1), which receives sanitary and stormwater flows. The meter used in this monitoring was the ISCO 2150 Area Velocity Module (http://www.teledyneisco.com/en-us/waterandwastewater/Pages/2150-Area-Velocity-Module.aspx). It uses continuous wave Doppler technology to measure mean flow velocity. The data from the meter included both velocity and flow area to allow calculation of discharge ($m^3/s$).

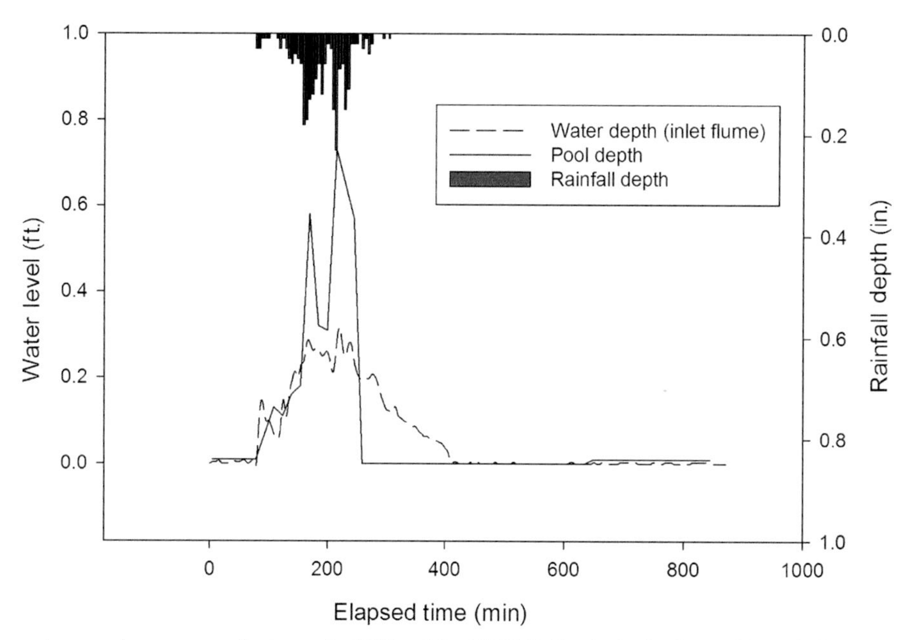


FIGURE 25.8 Stormwater control measure monitoring at site 1325 on May 27, 2013. Total rain depth was 60 mm(2.41 in.), and the inlet volume was 20.7 $m^3$.

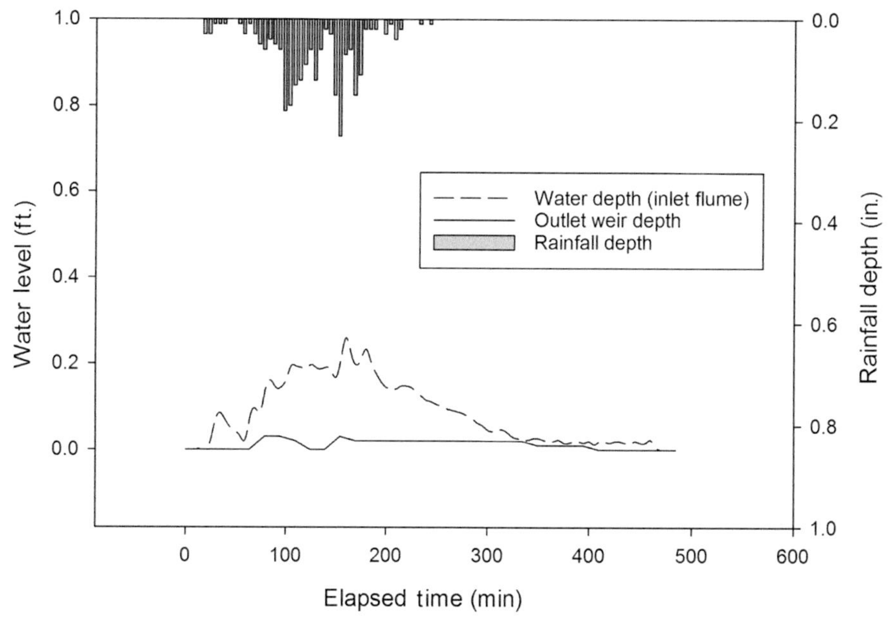


FIGURE 25.9 Stormwater control measure monitoring at site 1222 on May 27, 2013. Total rain depth was 60 mm and the inlet volume was 11.25 $m^3$ and the outlet volume was 0.0038 $m^3$.

## 25.4 MODEL FRAMEWORK

A statistical model was constructed (in two parts) and tested to analyze the extensive hydrological data from the six SCMs for 57 rainfall events (Fig. 25.10). The empirical model (Ma, 2013) complements a deterministic model developed for the same catchment and data, which is described in Chapter 7 (Talebi and Pitt, 2018). Statistical methodologies such as Principal Component Analysis (PCA) transformations are being adopted for data analysis by civil engineers, where highly dependent variables can be converted to new independent variables while retaining the field data's variances. The statistical model was subjected to a classic verification (using reserved data not used in model development) process to demonstrate

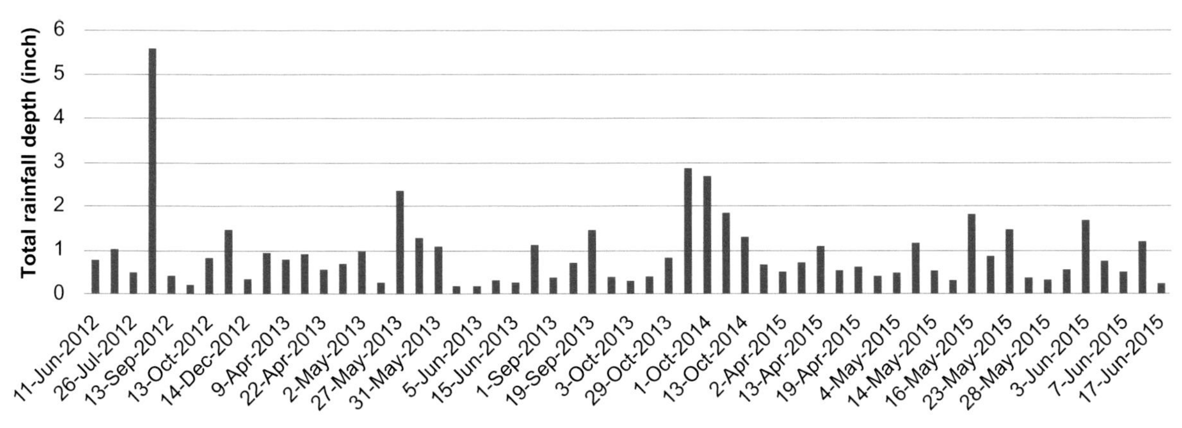


**FIGURE 25.10** Rainfall events measured over the 36 month monitoring period from June 2012–June 2015. The average rainfall depth was 24 mm with an average duration of 540 min. The average antecedent dry period was 7 days, and peak rainfall intensity was 32 mm/h.

**TABLE 25.1 Data Sets for the Unconnected SCM Model I (Four SCMs Not Connected to Urban Drainage System)**

| Site# | Total Number of Monitored Events | Data Loss From Equipment Failure or Inlet Flooding | Events Reserved for Validation of Model I | Events Used in Development of Model I |
|---|---|---|---|---|
| 1324 | 51 | 3 | 6 | 42 |
| 1325 | 49 | 6 | 7 | 36 |
| 1336 | 24 | 5 | 2 | 17 |
| 1112 | 29 | 0 | 8 | 21 |
| Total | 153 | 14 | 23 | 116 |

**TABLE 25.2 Data Sets for the Connected SCM Model II (Two SCMs Connected to Urban Drainage System via SmartDrain)**

| Site# | Total Number of Monitored Events | Data Loss From Equipment Failure or Inlet Flooding | Events Reserved for Validation of Model II | Events Used in Development of Model II |
|---|---|---|---|---|
| 1222 | 51 | 6 | 8 | 37 |
| 1140 | 51 | 11 | 3 | 37 |
| Total | 102 | 17 | 11 | 74 |

the model's usefulness. The statistical approach took advantage of the rich, measured rainfall and inflow data to the SCMs, which yields a range of catchment runoff values because of the stochastic nature of rainfall data. The range of runoff is a valuable planning tool because it captures the variance of the data used to build the model.

The objective of the model was to quantify the hydrological impact of the SCMs on the catchment outflow. The field data had various time and length scales, different magnitudes, and some variables that were related to each other, which caused covariance issues. The use of PCA (Everitt and Hothorn, 2011) to reduce, combine, and reorganize these heterogeneous data into statistically consistent master variables permitted the use of traditional multiple linear regression (MLR) to validly analyze the master variables that contain the field data. The data were divided into those used in model development (calibration) and a smaller data set reserved for model verification. The exact data distributions are presented in Tables 25.1 and 25.2.

Two separate models were constructed: one for the four monitored, unconnected SCMs (Model I) and the other for the two monitored, connected SCMs (Model II). The variables used in these models were: catchment area (Drainage), street slope (Street), catchment slope (Slope), impervious area percentage (Imp), total rainfall depth (RainTotal), rainfall duration (Storm), antecedent dry days (Dry), and rainfall peak intensity (Peak_I). The objective of the models was to predict the volume of stormwater captured by the SCMs. Therefore, the output of Model I was the total captured volume (inflow) for each unconnected SCM for each rain event, whereas the output of Model II was the captured volume (absorption) for each connected SCM at each rain event. There were 57 rain events in this data set.

Model I was built on data from four sites. Model II has the same analysis variables as Model I; however, only data from only two sites were available.

### 25.4.1 Collinear data treatment

The catchment slope had a very strong linear correlation with impervious area percentage, and a strong linear correlation with catchment area in Model I. Street slope and catchment slope had a very strong linear correlation. Total rainfall depth had a strong linear relationship with duration. The peak intensity had a moderate relationship with total rainfall depth. An r value between $|0.5|$ and $|0.7|$ is in a collinearity (multicollinearity) threshold range (Dormann et al., 2013). Collinearity existed in some of both Model I and Model II predictor variables. High collinearity between variables occurs when those data share substantial amounts of information, e.g., RainTotal and Peak_I. Linear regression requires independent data: the existence of collinearity indicates that the data are not independent and direct linear regression is invalid. PCA is a common way to remove the collinearity in variables.

In Model II, because there were only two sites' data, there were only two different values in catchment area, street slope, catchment slope, and impervious area percentage. Those four variables naturally show very strong linear relationships. Variables in the Model II set (connected SCMs) were particularly problematic because there were only two SCMs, so that data were easily linked to the locations and could not be classified as independent. The rainfall total depth also showed a strong correlation with peak intensity. PCA was used to create independent variables for both Model I and Model II.

### 25.4.2 Statistical model overview

A basic assumption of MLR is the independence of variables. The PCA is applied to the original dataset to transform the correlated variables into nonlinear-correlated principal scores. The new principal scores formed the new variables for MLR (Shi et al., 2011; Guillén-Casla et al., 2011; Eldaw et al., 2003; Salas et al., 2011; Nasir et al., 2011).

The hybrid model from PCA and MLR was used for the forecasting of streamflows based on hydrologic predictors, atmospheric predictors, and oceanic predictors. Potential predictors of the flow data were selected based on the correlation coefficient, r. A 95% confidence level was used as the decision criterion. The PCA and MLR approach was applied on the model development for an individual site with a good performance (Salas et al., 2011). The PCA and MLR hybrid model was compared with general MLR in a long-range forecasting study on the Nile River flow using climate data by Eldaw et al. (2003). This study demonstrated that the hybrid model from PCA and MLR approaches had a higher accuracy of forecasting streamflow for the Nile River. The correlation relationship between predictors may impact on the accuracy of the model. PCA can determine orthogonal variables from the original variables to remove the correlation. An uncertainty model was used for the Ballona Creek catchment in California, which was calibrated and verified, where most of the field data were contained within the 95% confidence intervals (Muleta et al., 2013).

Numerical rain garden hydrological performance studies have evaluated SCM efficiency. SCM performance was modeled with uncertainty analysis by Park et al. (2011). Water quality parameters were modeled at the outlet of SCMs, and three independent data sets were used to check model predictions. The authors concluded that the Latin Hypercube Sampling methods were the most efficient for characterizing the uncertainty of SCM performance.

### 25.4.3 Principal component analysis

Collinearity in the Model I and Model II data sets required that PCA be applied before MLR. Each model dataset was randomly split as a training data set and a test data set. The training data set contained 80% of the original data points, whereas the test data set used 20% of the original data (see Tables 25.1 and 25.2). An independent test data set can validate the regression model and help to resolve any bias introduced in the training/test data split (Bravo and Irizarry, 2010; Spade, 2016).

The inflow data were normalized by a square root transformation to satisfy the normality needs of the regression.

### 25.4.4 Principal component analysis and multiple linear regression hybrid model

The principal component (PC) coefficients for Model I are presented in Table 25.3, which show which field data are included in each PC.

The PCA is a data transformation process. It creates a new independent, uncorrelated variable data set. The new PC data sets retains the original data variances. For example, PC6-I contains the data from RainTotal, Storm, Dry, and Peak_I to predict SCM inflow. Six PCs explained 100% of the original data variances; therefore PC7-I and PC8-I are unnecessary. These six PC scores were the input variables for the Model I MLR to predict SCM captured inflow for the unconnected SCMs.

The MLR coefficients for each PC value for Model I are listed in Table 25.4, where the residual standard error was 19.1 with 82 degrees of freedom, the multiple $r^2$ was 0.67, the adjusted $r^2$ was 0.64, the F statistic was 26.9 with 7 and 82 degrees of freedom, and its probability of exceedance was $2.2 \times 10^{-16}$.

Model II followed a similar process to Model I: the only difference was the response term is absorption volume rather than the inflow to each site. Data from connected SCM #1140 and #1222 were used for model development. The PC coefficients are presented in Table 25.5.

For Model II, five PCs can explain 100% of the original data variances; therefore, PC6-II, PC7-II, and PC8-II were not needed. Again, the next step was MLR. Five PC scores were the inputs to Model II to predict captured volume for the connected SCMs and are shown in Table 25.6. Here, the residual standard error was 12.96 with 49 degrees of freedom, the multiple $r^2$ was 0.606, the adjusted $r^2$ was 0.56, the F statistic was 15.7 with 5 and 49 degrees of freedom, and its probability of exceedance was $2.46 \times 10^{-9}$.

**TABLE 25.3 Principal Component Coefficient Matrix for Model I**

| | PC1-I | PC2-I | PC3-I | PC4-I | PC5-I | PC6-I | PC7-I | PC8-I |
|---|---|---|---|---|---|---|---|---|
| Drainage | −0.514 | 0.155 | 0.383 | | −0.166 | | −0.386 | 0.616 |
| Street | −0.352 | −0.592 | −0.156 | 0.118 | | | 0.622 | 0.312 |
| Slope | −0.497 | −0.421 | | 0.135 | −0.103 | | −0.396 | −0.623 |
| Imp | 0.318 | −0.472 | −0.462 | | 0.141 | | −0.555 | 0.366 |
| RainTotal | 0.289 | −0.355 | 0.559 | | | −0.69 | | |
| Storm | 0.139 | −0.269 | 0.338 | −0.68 | −0.258 | 0.515 | | |
| Dry | −0.324 | | | −0.46 | 0.819 | −0.108 | | |
| Peak_I | 0.235 | −0.16 | 0.434 | 0.531 | 0.451 | 0.496 | | |

**TABLE 25.4 Multiple Linear Regression Coefficients for Model I**

| | Estimate | Standard Error | t Value | Probability (>\|t\|) |
|---|---|---|---|---|
| (Intercept) | 51.5 | 2.05 | 25.2 | $2.00 \times 10^{-16}$ |
| PC1-I | 7.69 | 1.28 | 6.03 | $4.86 \times 10^{-8}$ |
| PC2-I | −13.9 | 1.57 | −8.86 | $1.66 \times 10^{-13}$ |
| PC3-I | 6.26 | 1.67 | 3.75 | $3.3 \times 10^{-4}$ |
| PC4-I | −6.06 | 1.90 | −3.19 | $2.02 \times 10^{-3}$ |
| PC5-I | −6.42 | 2.49 | −2.58 | $1.18 \times 10^{-2}$ |
| PC6-I | −23.5 | 5.08 | −4.64 | $1.36 \times 10^{-5}$ |

**TABLE 25.5 Principal Component Coefficient Matrix for Model II**

| | PC1-II | PC2-II | PC3-II | PC4-II | PC5-II | PC6-II | PC7-II | PC8-II |
|---|---|---|---|---|---|---|---|---|
| Drainage | −0.491 | | | | | | −0.452 | 0.739 |
| Street | −0.491 | | | | | 0.674 | 0.544 | |
| Slope | −0.491 | | | | | | −0.544 | −0.671 |
| Imp | 0.491 | | | | | 0.736 | −0.452 | |
| Rain | | −0.705 | | | 0.703 | | | |
| Storm | | −0.531 | 0.493 | 0.437 | −0.53 | | | |
| Dry | | | 0.697 | −0.71 | | | | |
| Peak_I | 0.133 | −0.456 | −0.498 | −0.55 | −0.473 | | | |

**TABLE 25.6 Multiple Linear Regression Coefficients for Model II**

| | Estimate | Std. Error | t Value | Probability (>\|t\|) |
|---|---|---|---|---|
| (Intercept) | 34.3 | 1.72 | 20.0 | $2.00 \times 10^{-16}$ |
| PC1-II | 2.95 | 0.914 | 3.24 | $2.13 \times 10^{-3}$ |
| PC2-II | −8.86 | 1.29 | −6.86 | $9.08 \times 10^{-9}$ |
| PC3-II | −2.93 | 1.54 | −1.91 | $6.22 \times 10^{-2}$ |
| PC4-II | 3.53 | 2.14 | 1.65 | $1.05 \times 10^{-1}$ |
| PC5-II | 7.98 | 5.18 | 1.54 | 0.130 |

## 25.4.5 Stormwater control measures model equations

The regression equation that predicts the absorbed stormwater capture (SCM inflow) by the unconnected SCMs (Model I) is (as per Table 25.4):

Square root of inflow volume = 51.55 + 7.69(PC1-I) − 13.9(PC2-I) + 6.26(PC3-I) − 6.06(PC4-I) − 6.42(PC5-I) − 23.5(PC6-I).

The regression equation for predicting the absorbed stormwater capture (SCM inflow—outflow) by the connected SCMs (Model II) is (as per Table 25.6):

Square root of absorption volume = 34.34 + 2.96(PC1-II) − 8.86(PC2-II) − 2.94(PC3-II).

The PC values are recombinations of the field data. The PCA process retains the nature of the data and creates new, independent variables, which were then used in MLR.

## 25.4.6 Model validation

Internal validations were performed separately for both models. The test data sets of SCM−rain event pairs were randomly selected out for the development and validation processes (see also Tables 25.1 and 25.2). Then the same PCA coefficient matrices (see Tables 25.3 and 25.5) were applied, respectively, to the data pair set aside for the validation. The new PC values were then calculated and input into the MLR equations to predict captured stormwater runoff for the validation rainfall data.

The results for Model I showed that 92% of the test data points were within the 95% prediction interval. For Model II, 93% test data points were within the 95% prediction interval. The internal test data set validation indicated that the two models have good predictive power.

## 25.4.7 Catchment prediction model

The two tested MLR models developed from the Kansas City site study data, based on six individual SCMs, were then expanded and applied to the entire 40-ha catchment containing 135 SCMs. The modeling goal was to predict the reduction in stormwater runoff from the SCM construction. The two models, one for unconnected SCMs and one for connected SCMs, were used, and the model output from the catchment's collection system was compared with the flows measured in the catchment outlet pipe (see Fig. 25.1).

Eight rainfall events with catchment outflow data were reserved for testing performance of the models that were expanded to include 135 SCMs. These eight events were not used in the development of the Model I or II regression equations.

Although the total drainage area of the monitored catchment is 40 ha, only 20 ha drains through SCMs. Table 25.7 summarizes the details of the three types of SCMs (see also Fig. 25.2). The average SCM catchment was 0.15 ha, the average impervious area was 48%, the average catchment slope was 6.6%, and the average street slope was 7.5%.

Model I was used for prediction and validation for bioretention cells with subsurface storage. These bioretention units have their underdrain pipe connected to the combined sewer system; hence the storage functions as a detention reservoir, which releases water to the sewer pipe very slowly. Although these bioretention units are technically connected to the collection system, they release water so slowly that their outflows do not contribute to the storm hydrograph, as observed in the catchment pipe measurements. Fig. 25.11 shows the installation of a below-grade storage pipe (see also Fig. 25.2(c)).

The four variables included in Model I and II were available for all 135 SCMs in the watershed: catchment slope, street slope, impervious area percentage, and drainage area. The data for the other 128 SCMs and the eight storms were coded into the PC values.

**TABLE 25.7** Three Categories of Stormwater Control Measures (SCMs) in the Monitored Catchment, and the Area Treated by Each Type of Device

| Design Plan Component | Number of Storm Water Control Units in Demonstration Area | Drainage Area for Each Unit (ha) | Total Area Treated by These Devices (ha) |
|---|---|---|---|
| Unconnected SCM | 83 | 0.13 | 10.9 |
| Bioretention with underdrain | 47 | 0.18 | 8.5 |
| Connected SCM | 5 | 0.10 | 0.51 |
| Total number of SCM units | 135 | | 20.0 |

**FIGURE 25.11** Photo of bioretention cell with below-grade storage *Photo by David Dods.*

For the unconnected SCM group, Model I was used to predict catchment performance. For unconnected SCM sites, the assumption (based on field data) was that all the inflow to those SCMs was absorbed and did not contribute to the sewer flow during the storm hydrograph.

A similar process to the model validation was followed to apply the MLR model to the entire catchment. The new PC values, updated for the rain data and additional SCM site data, were then calculated and inputed into the MLR equations to predict stormwater runoff captured by the SCMs for the entire catchment.

Model II was used for prediction and validation of the connected SCMs (Smart Drain). There is no underground storage, and any infiltration into the SCM Smart Drain goes directly to the sewer system. Model II predicts the absorbed volume captured by the connected SCMs (SCM stormwater inflow − SCM outflow to sewer).

### 25.4.8 Monitored site validation

Eight rain events were available for the catchment-wide mass balance validation. One of the monitored **unconnected** SCM locations (site 1325) is presented in Fig. 25.12, and shows the model fit for specific SCM performance during those eight rain events.

Monitored values that fall within the 95% prediction interval boundary from the model indicate a good model fit. Overall, there was an 86% model fit to the field data. A model fit graph for **connected** SCM 1222 is presented in Fig. 25.13.

The prediction values are very conservative compared to the monitored data. The monitored data are closer to the 95% prediction interval upper level boundary.

### 25.4.9 Catchment level model prediction

For the entire catchment, the water mass balance should follow the equation:

$$\text{Total rainfall volume } - V_{\text{untreated}} - V_{\text{SCM captured}} = V_{\text{pipe}}$$

Where $V_{\text{untreated}}$ is the infiltration volume from the untreated catchment (areas not draining to SCMs), $V_{\text{SCM captured}}$ is the volume of water captured and absorbed by all types of SCMs, and $V_{\text{pipe}}$ is the volume of water from a rainfall event measured in the outflow sewer pipe from the entire catchment.

The average rainfall runoff percentage was 42% before any of the SCMs were constructed. Six storms were available in the preconstruction period with flow monitoring in the pipe. Therefore, 100% − 42% is 58%, which is the fraction of rain

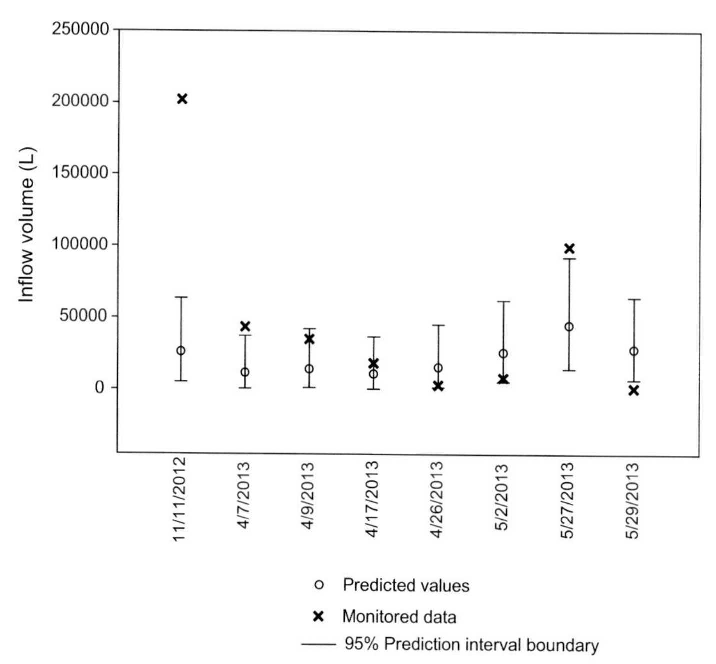


FIGURE 25.12 Predicted and monitored data for Unconnected SCM site 1325.

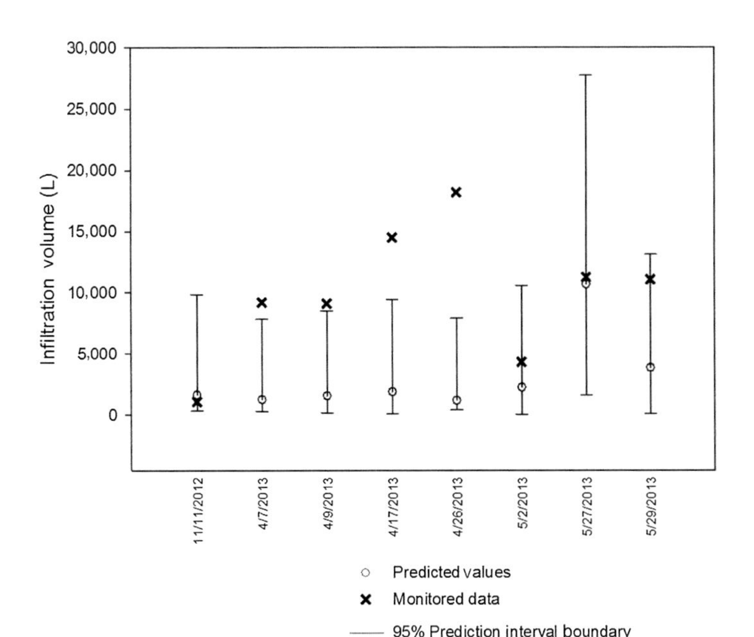


**FIGURE 25.13** Predicted and monitored data for Connected SCM site 1222.

**TABLE 25.8 Entire Catchment Mass Balance for Monitored Events**

| | Rain (mm) | SCM Capture (%) | Natural Absorption (%) | Sewer Flow (%) | Mass Sum (%) |
|---|---|---|---|---|---|
| November 11, 2012 | 37 | 12 ± 25 | 30 ± 1 | 11 | 53 |
| April 7, 2013 | 25 | 5 ± 20 | 30 ± 1 | 7 | 42 |
| April 9, 2013 | 20 | 8 ± 26 | 30 ± 1 | 15 | 53 |
| April 17, 2013 | 23 | 5 ± 21 | 30 ± 1 | 9 | 44 |
| April 26, 2013 | 18 | 10 ± 30 | 30 ± 1 | 9 | 49 |
| May 2, 2013 | 26 | 15 ± 30 | 30 ± 1 | 7 | 52 |
| May 27, 2013 | 60 | 18 ± 30 | 30 ± 1 | 6 | 54 |
| May 29, 2013 | 33 | 15 ± 24 | 30 ± 1 | 17 | 62 |

infiltrated in the preconstruction period. After construction, 51% of the catchment area does not drain to SCMs. The natural absorption by the untreated 20-ha catchment was 30% of the rainfall (calculated as: 58% × 51%). This 30% natural absorption value was used for the mass balance analysis of each runoff event in post-SCM implementation.

Table 25.8 summarizes the mass balance components for each event. For the 11 November 2012 event, 12% (±25%) of the rainfall was capture by the SCMs, 30% of the rainfall was absorbed by non-SCM draining areas, and 11% of the rainfall was measured in the sewer pipe at the outlet of the catchment. The volume of rain accounted for by the three categories is 53%. The overall performance of the post−SCM construction catchment is the substantial reduction in measured outflow in the pipe from the entire catchment, which was 42% (preconstruction) to 11% on November 11, 2012. Table 25.8 and Fig. 25.14 show that the data-driven, statistical model is able to explain 51% (±25%) of the total water mass balance after expanding the model to the entire catchment. The relaxed model fit suggests that the model can be applied to catchments beyond the Kansas City locale.

## 25.5 RESULTS OF CATCHMENT MODEL PREDICTION

The model prediction for the entire catchment underpredicts the mass balance. The average prediction values can only explain half of the water mass balance for the entire catchment (see Table 25.8; Fig. 25.14). One of the possible reasons for

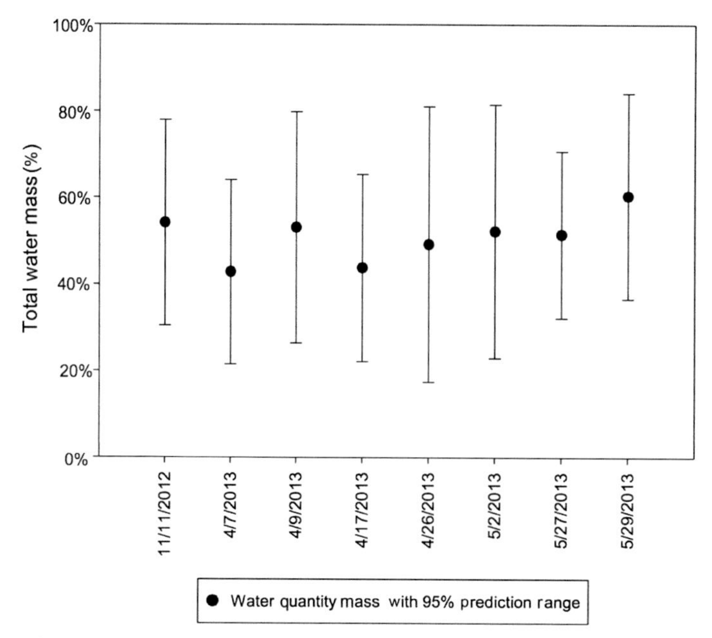


FIGURE 25.14 Mass balance with 95% prediction interval for monitored events.

this lower prediction is that the monitored sites are not sufficiently typical to represent the behavior of the other SCMs, particularly for those monitored sites that have below average inflow capture compared to the others (see Table 25.9). Three of the monitored SCMs have much smaller contributing drainage areas than the average for the whole catchment. Another possible explanation is that there was a roof downpipe disconnection campaign in the early 2010s. Any houses that redirected their roof runoff from the collection system to their lawns would have increased the untreated area absorption.

The six monitored SCM sites have relatively smaller catchment areas compared to the average SCM catchment area from the entire SCM deployment area. They had a slightly higher impervious percentage than the average value, but a lower catchment slope and a higher street slope than the entire catchment. Therefore, it is very possible that those six monitored sites are not sufficiently typical to represent the 135 units based on the four SCM features.

The model prediction, because it is conservative, can still be used to assess the performance of the SCMs and engineering design runoff reduction goals. The monitored SCMs performed very well in the study. That is, the unconnected SCMs absorbed all the water that flowed into them. Even the connected SCMs absorb most of their inflows. Observations following significant rain events (including events >130 mm) since the monitoring study ended have not shown any outflows from the SCM devices. Overall, the SCMs have confidently met the engineering design goal of capturing 1135 $m^3$ from a 36-mm rainfall event.

**TABLE 25.9 Monitored Sites Compared to Average Catchment SCM**

| Site | SCM Drainage Area (ha) | Impervious Percentage (%) | Watershed Slope (%) | Street Slope (%) |
|---|---|---|---|---|
| SCM Watershed Average | 0.15 | 48 | 6.5 | 2.8 |
| 1324 | 0.03 | 66 | 3.6 | 1.7 |
| 1325 | 0.03 | 67 | 3.7 | 1.9 |
| 1336 | 0.21 | 40 | 5.5 | 1.9 |
| 1612 | 0.08 | 33 | 3.6 | 4.4 |
| 1112 | 0.11 | 70 | 7.9 | 8.0 |
| 1140 | 0.01 | 86 | 2.1 | 1.5 |
| 1222 | 0.13 | 29 | 6.6 | 7.5 |

**TABLE 25.10** SCM Watershed-Wide Capture Analysis

| Date | Rain Total (mm) | Mean SCM Capture of Rainfall (%) | SCM Capture of Rainfall ($m^3$) |
|---|---|---|---|
| Nov 11, 2012[a] | 37 | 0.12 | 1800 |
| April 7, 2013 | 25 | 0.05 | 500 |
| April 9, 2013 | 20 | 0.08 | 660 |
| April 17, 2013 | 23 | 0.05 | 460 |
| April 26, 2013 | 18 | 0.1 | 730 |
| May 2, 2013 | 26 | 0.15 | 1570 |
| May 27, 2013[a] | 60 | 0.18 | 4370 |
| May 29, 2013 | 33 | 0.15 | 2000 |

[a]*Rainfall events exceeding the 36 mm design storm.*

Two of the storm events reported in Table 25.10 exceed the 36-mm design storm. The model predicted that the SCMs captured more than the required 1135 $m^3$, thereby keeping that volume from entering the collection system. In fact, in this case study, SCM performance does not drop off after 36 mm of rain—the SCMs maintain their performance up to at least 60 mm of rain.

## 25.6 CONCLUSIONS

An advanced linear regression model was developed and verified with field data for a variety of SCM installations in an urban catchment in Kansas City, Missouri, USA. The MLR model, using PCA preprocessing, is based on catchment characteristics that are likely to be applicable to other urban areas (in the United States) with similar slopes and soil types. The PCA transformation was essential to transform stochastic data with covariance, which would otherwise be problematic for use in other regression models.

The model provides very conservative projections for SCM performance. The model is statistically valid, but generally underpredicts the catchment water balance. The conservative prediction indicates that it is a tool that can be used reliably to estimate stormwater removal/capture by SCMs in a catchment. The performance of SCMs over a 3-year period and over a range of rainfall events indicates that the deployment of SCM is effective in preventing combined sewer overflows. The SCMs continue to perform well (i.e., absorb all the inflow) up to rainfall totals of 60 mm (measured data) and even up to 150 mm (qualitative observation in 2017).

High capture effectiveness of the SCMs was not expected at the time of construction because of the compacted, clay soils, and SCM absorption behavior at other locations should be field-verified before applying the model. The modeling results, coupled with field observations, support the efficacy of green solutions to reduce urban runoff. Although the study was located in a combined sewer catchment, the model of SCM performance is not limited to combined sewer areas.

## ACKNOWLEDGMENTS

The work reported in this document was funded by the US Environmental Protection Agency The document has been subjected to the Agency's peer and administrative reviews and has been approved for publication. However, any opinions expressed in this report are those of the authors and do not necessarily reflect the views of the Agency. Mention of trade names does not constitute endorsement or recommendation for use.

## REFERENCES

Aguilar, M.F., Dymond, R.L., 2016. Evaluation of variability in response to the NPDES phase II stormwater Program in Virginia. Journal of Sustainable Water in the Built Environment 2 (1), 04015006.

Ahiablame, L.M., Engel, B.A., Chaubey, I., 2012a. Representation and evaluation of low impact development practices with L-THIA-LID: an example for site planning. Environment and Pollution 1 (2), 1–13.

Ahiablame, L., Engel, B., Chaubey, I., 2012b. Effectiveness of low impact development practices: literature review and suggestions for future research. Water, Air and Soil Pollution 223 (7), 4253–4273.

Bedan, E.S., Clausen, J.C., 2009. Stormwater runoff quality and quantity from traditional and low impact development watersheds. Journal of the American Water Works Association 45 (4), 998–1008.

Bravo, H.C., Irizarry, R.A., 2010. Lecture 5: Model Selection and Assessment. http://www.cbcb.umd.edu/~hcorrada/PracticalML/pdf/lectures/selection.pdf.

Carmen, N.B., Hunt, W.F., Anderson, A.R., 2016. Volume reduction provided by eight residential disconnected downspouts in durham, North Carolina. Journal of Environmental Engineering 142 (10), 05016002.

Damodaram, C., Zechman, E.M., 2013. Simulation-optimization approach to design low impact development for managing peak flow alterations in urbanizing watersheds. Journal of Water Resources Planning and Management 139 (3), 290–298.

Dietz, M.E., Clausen, J.C., 2005. A field evaluation of rain garden flow and pollutant treatment. Water, Air and Soil Pollution 167 (1), 123–138.

Dods, D., 2016. Personal Communication.

Dormann, C.F., Elith, J., Bacher, S., Buchmann, C., Carl, G., Carré, G., García Marquéz, J.R., Gruber, B., Lafourcade, B., Leitão, P.J., Münkemüller, T., McClean, C., Osborne, P.E., Reineking, B., Schröder, B., Skidmore, A.K., Zurell, D., Lautenbach, S., 2013. Collinearity: a review of methods to deal with it and a simulation study evaluating their performance. Ecography 36 (1), 27–46.

Eldaw, A.K., Salas, J.D., Garcia, L.A., 2003. Long-range forecasting of the Nile River flows using climatic forcing. Journal of Applied Meteorology 42 (7), 890–904.

EPA, 2014. System for Urban Stormwater Treatment and Analysis Integration (SUSTAIN). https://www.epa.gov/water-research/system-urban-stormwater-treatment-and-analysis-integration-sustain.

EPA, 2016. Storm Water Management Model (SWMM). https://www.epa.gov/water-research/storm-water-management-model-swmm.

Everitt, B., Hothorn, T., 2011. An Introduction to Applied Multivariate Analysis with R. London.

Guillén-Casla, V., Rosales Conrado, N., León-Gonzáles, M., Pérez-Arribas, L., Polo-Díez, L., 2011. Principal component analysis (PCA) and multiple linear regression (MLR) statistical tools to evaluate the effect of E-beam irradiation on ready-to-eat food. Journal of Food Composition and Analysis 24 (3), 456–464.

Guo, J.C.Y., Luu, T.M., 2015. Hydrologic model developed for stormwater infiltration practices. Journal of Hydrologic Engineering 20 (9), 06015001.

Gwinn, W.R., Parsons, D.A., 1976. Discharge equations for HS, H, and HL flumes. Journal of the Hydraulics Division 102 (1), 73–88.

Hekl, J.A., Dymond, R.L., 2016. Runoff impacts and LID techniques for mansionization-based stormwater effects in fairfax county, Virginia. Journal of Sustainable Water in the Built Environment 2 (4), 05016001.

Hood, M.J., Clausen, J.C., Warner, G.S., 2007. Comparison of stormwater lag times for low impact and traditional residential development. Journal of the American Water Works Association 43 (4), 1036–1046.

Hunt, W.F., Jarrett, A.R., Smith, J.T., Sharkey, L.J., 2006. Evaluating bioretention hydrology and nutrient removal at three field sites in North Carolina. Journal of Irrigation and Drainage Engineering 132 (6), 600–608.

Hunt, W.F., Smith, J.T., Jadlocki, S.J., Hathaway, J.M., Eubanks, P.R., 2008. Pollutant removal and peak flow mitigation by a bioretention cell in urban Charlotte, N.C. Journal of Environmental Engineering 134 (5), 403–408.

Hunt, W.F., Davis, A.P., Traver, R.G., 2012. Meeting hydrologic and water quality goals through targeted bioretention design. Journal of Environmental Engineering 138 (6), 698–707.

Jones, G.D., Wadzuk, B.M., 2013. Predicting performance for constructed storm-water wetlands. Journal of Hydraulic Engineering 139 (11), 1158–1164.

Kansas City Water Services Department (KCWSD), 2013. Kansas City's Overflow Control Program- Middle Blue River Basin Green Solutions Pilot Project Final Report.

Line, D.E., Brown, R.A., Hunt, W.F., Lord, W.G., 2012. Effectiveness of LID for commercial development in North Carolina. Journal of Environmental Engineering 138 (6), 680–688.

Ma, Y., 2013. Watershed-Level Analysis of Urban Rain Garden Performance. University of Missouri-Kansas City, Kansas City.

Ma, Y., Nall, J., O'Bannon, D., 2018. Assessment of orifice-controlled flow monitoring device for rain garden performance. Journal of Sustainable Water in the Built Environment 4 (2), 05018002.

Mancipe-Munoz, N.A., Buchberger, S.G., Suidan, M.T., Lu, T., 2014. Calibration of rainfall-runoff model in urban watersheds for stormwater management assessment. Journal of Water Resources Planning and Management 140 (6), 05014001.

Mangangka, I.R., Liu, A., Egodawatta, P., Goonetilleke, A., 2015. Performance characterisation of a storm water treatment bioretention basin. Journal of Environmental Management 150, 173–178.

Muleta, M.K., McMillan, J., Amenu, G.G., Burian, S.J., 2013. Bayesian approach for uncertainty analysis of an urban storm water model and its Application to a heavily urbanized watershed. Journal of Hydrologic Engineering 18 (10), 1360–1371.

Nasir, M.F.M., Samsudin, M.S., Mohamad, I., Awaluddin, M.R.A., Mansor, M.A., Juahir, H., Ramli, N., 2011. River water quality modeling using combined principal component analysis (PCA) and multiple linear regressions (MLR): a case study at Klang river, Malaysia. World Applied Sciences Journal 14 (14), 73–82.

Park, D., Loftis, J.C., Roesner, L.A., 2011. Performance modeling of storm water best management practices with uncertainty analysis. Journal of Hydrologic Engineering 16 (4), 332–344.

Pitt, R., Voorhees, J., Clark, S., 2010. "Integrating green infrastructure into a combined sewer service area model. In: World Environmental and Water Resources Congress, pp. 2950–2963.

Potter, K.W., 2007. Field Evaluation of Rain Gardens as a Method for Enhancing Groundwater Recharge. R/UW-BMP-002. University of Wisconsin Final Report.

Qin, H.-P., Li, Z.-X., Fu, G., 2013. The effects of low impact development on urban flooding under different rainfall characteristics. Journal of Environmental Management 129, 577–585.

Salas, J.D., Fu, C., Rajagopalan, B., 2011. Long-range forecasting of Colorado streamflows based on hydrologic, atmospheric, and oceanic data. Journal of Hydrologic Engineering 16 (6), 508–520.

Selbig, W.R., Balster, N., 2010. Evaluation of Turf-grass and Prairie-vegetated Rain Gardens in a Clay and Sand Soil, Madison, Wisconsin, Water Years 2004-08. USGS Scientific Investigations Report, pp. 2010–5077.

Shi, G.-L., Zeng, F.G., Feng, Y.-C., Want, Y.-Q., Liu, G.X., Zhu, T., 2011. Estimated contributions and uncertainties of PCA/MLR–CMB results: source apportionment for synthetic and ambient datasets. Atmospheric Environment 45 (17), 2811–2819.

Simon, M., 2016. EPA's Summary Report of the Collaborative Green Infrastructure Pilot Project for the Middle Blue River in Kansas City, MO. EPA/600/R-16/085. Ohio: Cincinatti.

Spade, D., January 22, 2016. Personal Communication.

Talebi, L., Pitt, R., 2018. Water sensitive urban design (WSUD) Approaches in sewer system overflow management. In: Water Sensitive Urban Design Netherlands: Amsterdam.

USBR, 1997. U.S. Department of the Interior, Bureau of Reclamation. Water Measurement Manual. Available from:, third ed. http://www.usbr.gov/tsc/techreferences/mands/wmm/index.htm.

USEPA (US Environmental Protection Agency), 2009. SUSTAIN. A Framework for Placement of Best Management Practices in Urban Watersheds to Protect Water Quality. Office of Research and Development National Risk Management Research Laboratory-Water Supply and Water Resources Division. EPA-600-R-09–095.

Wadzuk, B.M., Lewellyn, C., Lee, R., Traver, R.G., 2017. Green infrastructure recovery: analysis of the influence of back-to-back rainfall events. Journal of Sustainable Water in the Built Environment 3 (1), 04017001.

Welker, A.L., Mandarano, L., Greising, K., Mastrocola, K., 2013. Application of a monitoring plan for storm-water control measures in the Philadelphia region. Journal of Environmental Engineering 139 (8), 1108–1118.

Wilson, C.E., Hunt, W.F., Winston, R.J., Smith, P., 2015. Comparison of runoff quality and quantity from a commercial low-impact and conventional development in Raleigh, North Carolina. Journal of Environmental Engineering 141 (2), 05014005.

Zhang, S., Guo, Y., 2013. Explicit equation for estimating storm-water capture efficiency of rain gardens. Journal of Hydrologic Engineering 18 (12), 1739–1748.

Chapter 26

# WSUD Implementation in a Precinct Residential Development: Perth Case Study

Josh Byrne[1], Melissa Green[2] and Stewart Dallas[3]

[1]*School of Design and Built Environment, Curtin University, Bentley, WA, Australia;* [2]*Josh Byrne & Associates, Fremantle, WA, Australia;* [3]*School of Engineering and Information Technology, Murdoch University, Murdoch, WA, Australia*

## Chapter Outline



**ABSTRACT**

The delivery of Integrated Urban Water Management (IUWM) and Water Sensitive Urban Design (WSUD) is taking place at a range of development scales as part of sustainable urban development and water sensitive city aspirations. Understanding the delivery process, and learning from on-ground implementation of such initiatives is crucial to inform future developments, so they may build on previous successes. The precinct scale provides an opportunity to test a mixture of design and technology that is suitable to the local conditions and water requirements. Here we detail a case study in WSUD implementation, the WGV infill development site, located in the suburb of White Gum Valley within the City of Fremantle, near Perth, Western Australia. WGV presents an innovative approach to urban infill, with a range of sustainable water, energy, and urban greening strategies implemented in the 2.2 ha medium density site of mixed building typologies. The site is guided by the One Planet Living Framework, and key to meeting these criteria is the innovative approach to water management. Initiatives include water efficiency measures inside and outside of homes, rainwater harvesting, a community bore, stormwater management, and precinct greening. The technical and decision-making details of these initiatives are presented in this chapter.

**Keywords:** Alternative water; Community bore; Integrated Urban Water Management; Knowledge sharing; Precinct greening; Rainwater harvesting; Urban infill development; Water efficiency; Water savings; Water Sensitive Urban Design.

**Approaches to Water Sensitive Urban Design.** https://doi.org/10.1016/B978-0-12-812843-5.00026-5

## 26.1 INTRODUCTION: PRECINCT APPROACHES TO INTEGRATED URBAN WATER MANAGEMENT

Many urban areas of Australia strive to create a more productive, liveable, sustainable, and resilient city. As part of this, a model of integrated water management is preferred to protect against water stresses and accommodate population growth, economic uncertainty, and environmental balance. Increasingly, Australian urban developments are incorporating Integrated Urban Water Management (IUWM) with Water Sensitive Urban Design (WSUD) at a range of development scales such as greenfield, infill, and retrofit (Sharma et al., 2012a). Here, we define IUWM as the long-term holistic planning that integrates multiple water sources with various stakeholders and urban planning. WSUD is a subset of IUWM focusing on incorporating green infrastructure to improve liveability and environmental outcomes (Furlong et al., 2016). These approaches aim to replace use of drinking water for nonpotable consumption and reduce the strain on centralized sources with alternative sources such as rainwater harvesting, stormwater harvesting, groundwater extraction and treatment, greywater collection and treatment, and wastewater collection and treatment (Byrne, 2016; Sharma et al., 2012b). Such schemes have considerable benefits in promoting a more natural water cycle, local source diversification, and resource efficiency and providing decentralized solutions (Marlow et al., 2013). They also have the potential to increase biodiversity, ecological health, landscape esthetic, and amenity, which can add to the distinct character, identity, and sustainability of a place (Johnstone et al., 2012; Lehmann, 2010). Moreover they assist in managing public health, urban microclimates, and heat mitigation (Johnstone et al., 2012).

Despite the numerous benefits, changing from a traditional centralized water delivery system toward an integrated approach remains slow and requires institutional capacity to manage uncertainty and risk, financial considerations, flexibility with changing technology, and community acceptance (Marlow et al., 2013). The term "hydrosocial contract" has been used to describe "the pervading values and often implicit agreements between communities, governments, and business on how water should be managed" and is often shaped by dominant cultural perspectives and historically embedded water values (Wong and Brown, 2008, p. 3). Wong and Brown (2008) suggest that transforming toward a water-sensitive city needs to focus on "how" to ensure a connection between IUWM, urban design, and social and institutional systems. Therefore, good working relationships between stakeholders, well-communicated processes, and demonstration projects can all compliment sound technical knowledge and assist in ensuring that future implementation of IUWM is successful.

Perth stakeholders have expressed a desire to become a more water sensitive city, with traditional approaches to water management no longer sufficient to meet competing demands (Rogers et al., 2015). The southwest of Australia is experiencing a significant drying trend with reduced rainfall, particularly between May and July, and a reduction in the number and size of runoff events to fill surface reservoirs (BOM and CSIRO, 2016). Compared with other major cities in Australia, much of Perth's urban water supply comes from groundwater, seawater desalination, and some surface water. The uptake of desalination, currently supplying almost half of the urban water supply in Perth (BoM, 2017), has largely secured Perth's drinking water supply, but at a cost of being the most energy intensive scheme of all capital cities in Australia (Bureau of Meteorology, 2016). As Perth continues to grow in size and population, resilient water supply solutions are needed (Rogers et al., 2015). A heavy reliance on desalination will leave Perth vulnerable because of rising energy costs, rising infrastructure costs, a rising greenhouse gas footprint, and a cultural disconnection between the biophysical realities of the water cycle and our role in it.

Integrated urban water systems can provide long-term sustainable solutions at a variety of scales (i.e., lot, cluster, precinct, or district), depending on local sources and needs. Typically the general business case becomes more attractive when certain thresholds are met. However, each application is likely to have its own unique drivers that influence scale and technology choice (Diaper et al., 2007). Various fit-for-purpose applications and scales include:

- Rainwater harvesting for direct fit-for-purpose uses with localized storage at household and cluster scales (Ho et al., 2008; Hunt et al., 2011; Coombes et al., 2003; Gurung and Sharma, 2014; Umapathi et al., 2013; Sharma et al., 2015)
- Stormwater harvesting for aquifer recharge and groundwater recovery and reuse at household, cluster, and precinct scales (Hunt et al., 2005; Dillon et al., 2014; Page et al., 2015).
- Groundwater extraction and treatment to fit-for-purpose uses for various drinking and nondrinking purposes at household, cluster, and precinct scales (Lieb et al., 2006; Dhakal et al., 2015).
- Greywater collection and treatment to fit-for-purpose and reuse at household, cluster, and precinct scales (Nolde, 2000; Priest et al., 2004; Friedler and Hadari, 2006; Ghisi and de Oliveria, 2007; Evans et al., 2008; Evans et al., 2009; Mohamed et al., 2013).
- Wastewater collection and treatment for reuse at household, precinct, and district scales (Ho et al., 2001; Gardner, 2003; Radcliffe, 2006; Sharma et al., 2012b; Chong et al., 2013; Kunz et al., 2016).

In this chapter, we examine the application of both IUWM and WSUD initiatives at a precinct scale, focusing on the 2.2 ha WGV infill residential development in the suburb of White Gum Valley within the City of Fremantle, Perth, Western Australia[1]. WGV, led by the Western Australian land development agency LandCorp, presents a unique case study on how to implement alternative water management initiatives as part of the overall journey for greater Perth to become a water sensitive city. According to the Cooperative Research Centre for Water Sensitive Cities (2017), WSUD is commonly implemented at the precinct scale, as it enables application of an appropriate mixture of design and technologies to achieve a spatially relevant, climate-resilient, and resource-sensitive outcome.

## 26.2 PROJECT OVERVIEW

### 26.2.1 Biophysical context

Fundamental differences exist between Perth's Swan Coastal Plain and other cities on the east coast of Australia where many WSUD tools have been designed and implemented to suit the local conditions (Dunnicliff-Wells, 2014). Here, we provide an overview of the biophysical uniqueness of Perth and illustrate the importance of implementing site-specific WSUD and IUWM initiatives.

Perth is situated on the Swan Coastal Plain that stretches 650 km from north to south and is underlain by two principal groundwater systems—the Gnangara[2,3] and Jandakot groundwater systems (Davidson, 1995, refer to Fig. 27, p. 55). The former comprises four main aquifers to the north of the Swan River, whereas the latter groundwater system comprises three aquifers to the south. The upper, shallow, unconfined aquifers are referred to as the Gnangara and Jandakot mounds (respectively), and both are principally recharged by rainfall. In addition to supporting many wetlands and lakes, these two mounds provide the bulk of Perth's water for both potable supply and domestic irrigation. It is estimated that there are over 180,000 residential or "backyard" bores that draw from these two shallow aquifers, with a total annual extraction of approximately 31 GL/a (Gigalitres per annum) (DoW, 2014). Groundwater levels across both the Gnangara and Jandakot mounds have been in decline for the last 40 years due to a combination of declining rainfall, increasing groundwater abstraction, and expanding pine plantations (DWER, 2017).

The soils of the Swan Coastal Plain (Davidson, 1995, refer to Fig. 23, p. 46) are predominantly sandy and highly permeable, allowing stormwater to rapidly infiltrate and recharge the superficial aquifer (Dunnicliff-Wells, 2014). This natural characteristic has historically enabled stormwater runoff in the metropolitan area to be infiltrated close to, or on site, using a combination of soakwells and open drainage sumps. It is this characteristic that has required a different approach to stormwater management when compared with the hydrogeological conditions of the eastern Australian states, especially for managing nutrients and other stormwater pollutants. The recent development of the nutrient modeling tool UNDO (Urban Nutrient Decision Outcomes) by the State's water agency has been required to account for this unique hydrogeological feature (DoW, 2015).

Perth has seen a major decline in annual rainfall over the last two decades, whereas heatwave conditions are predicted to increase (BoM and CSIRO, 2016) and mains water demand over the next 40 years is projected to double (DoW, 2014), despite the current water sources being highly constrained (Water Corporation, 2015). With a population in excess of 1.9 million people (ABS, 2015) and growing, Perth has adopted a centralized approach to water service delivery (Bettini et al., 2015), like many modern cities around the world, and it presents an interesting example of the inherent vulnerabilities at this scale. The growing population, coupled with the long-term decline in rainfall, has seen a shift away from surface water reservoirs in Perth to an increased dependency on groundwater extraction (46% of supply, Water Corporation, 2017a), significant reliance on large-scale seawater desalination (47% of supply, Water Corporation, 2017a), and more recently, the increasing use of indirect potable reuse using recycled water via the Water Corporation's groundwater replenishment scheme with a target of 115 GL (or 20% of total supply) by 2060 (Water Corporation, 2017a).

Approximately 70% of all water supplied by the Perth Integrated Water Supply Scheme (IWSS) is consumed by residential customers (Water Corporation, 2010). A recent survey suggests that Perth residents use an average of 340 L per person per day, compared with 166 L in Melbourne, and 194 L in South East Queensland, making Perth a prolific water user (AWA, 2017). Around 40% of the water supply is used for domestic irrigation. This is despite watering restrictions (that have been in place since 2001, Water Corporation, 2009) limiting residential garden irrigation with

---

1. http://www.landcorp.com.au/Documents/Corporate/Innovation%20WGV/Innovation-WGV-Waterwise-Development-Exemplar.pdf.
2. http://www.water.wa.gov.au/water-topics/groundwater/understanding-groundwater/gnangara-groundwater-system.
3. Refer to map http://www.water.wa.gov.au/planning-for-the-future/allocation-plans/swanavon-region/gnangara-groundwater-areas-allocation-plan.

FIGURE 26.1 An aerial computer-generated image of the WGV development with the precinct boundary indicated by the *dashed yellow line*. *Source: LandCorp.*

mains water to 2 days per week, and bore water to 3 days per week, within set time windows. Of the 291 GL of potable water supplied through the Perth IWSS (Water Corporation, 2015), only around 50% is used for potable purposes (Water Corporation, 2010).

### 26.2.2 Development site

WGV is a 2.2 ha medium density, mixed residential infill development located in Fremantle, Perth's port city, some 14 km southwest of Perth central business district, with a build out target of around 100 dwellings. The structure plan area comprises Lot 2089 Stevens Street (2.1 ha subdivided into 28 residential lots and 11.7% public open space) and Lot 2065 Hope Street (0.16 ha City of Fremantle reserved for public open space for the purposes of drainage, otherwise known as a sump). Previously a school site, the development has been led by the Western Australian land development agency LandCorp[4]. WGV incorporates many leading urban design characteristics such as alternative water management (both supply and disposal), climate sensitive considerations, diverse building typologies, solar energy generation and storage, and creative urban greening strategies to demonstrate design excellence. An aerial, computer-generated image of the WGV development is shown in Fig. 26.1.

WGV is the first Western Australian residential project to achieve national recognition for One Planet Living (OPL), a holistic framework that incorporates 10 principles based on the metrics of ecological and carbon footprinting (Bioregional, 2017). Key to achieving this is the implementation of WSUD and water efficiency measures, which were introduced at the structure planning phase, and aim to achieve a 60%–70% reduction in scheme water consumption across the various dwelling typologies. These initiatives, to be discussed in this chapter, include integrated stormwater management, rainwater harvesting systems for internal use, a community bore, a waterwise irrigation system, water efficient fixtures/appliances, real-time monitoring, and low water use landscaping. Design guidelines and a resident information kit assist residents in meeting the OPL vision for the site.

### 26.2.3 Planning and urban design characteristics

Defined as a medium density development, the multi-typology site consists of detached houses, group dwellings and apartments. Key to the project are the underlying principles of sustainability and innovation, and the variety of housing

4. https://www.landcorp.com.au/Residential/White-Gum-Valley/.

options demonstrate a unique approach to urban infill. All lots have been sold and at completion, the final make up will include:

- Detached residential dwellings (under construction, some completed and occupied): 23 single residential lots ranging in size from approximately 250–350 $m^2$.
- Gen Y Demonstration House (construction completed and now occupied): Three single-bedroom apartments on a 250 $m^2$ block, incorporating sustainable design principles. The apartments also feature a novel strata-managed solar panel and battery system, as well as a shared rainwater system supplying toilets and cold water inlets to washing machines.
- Sustainable Housing for Artists and Creatives (SHAC) (construction completed and now occupied): A combined initiative between Access Housing and SHAC to deliver a variety of housing options for professional artists working in the Fremantle area.
- Evermore apartments (due for completion in mid 2018): 24 one-, two-, and three-bedroom apartments will incorporate a number of sustainability initiatives. In addition to following the One Planet Living framework, Evermore will be the first private apartment development in the state to utilize shared solar energy and battery storage.
- Baugruppen project (still in planning phase): Australia's first Baugruppen project will test the German model of affordable housing and cooperative living. Approximately 20 parties will develop their own innovative multiunit housing, with assistance from leading architects, to ensure design meets their long-term needs.
- Group housing site (still in planning phase), which will include up to six detached houses with survey strata to be compliant with the design guidelines.

WGV structure planning was finalized in 2013, with civil works taking place in 2014 and construction commencing in 2015, with total construction approximately 50% complete (as of mid 2018). The site is also guided by an overall vision that seeks to create an infill development that is site responsive, built on the local context, and leverages off the site's attributes to ensure future residents, and the surrounding community, all benefit (LandCorp, 2017).

## 26.3 INTEGRATED URBAN WATER MANAGEMENT INITIATIVES

The suite of initiatives ultimately selected for implementation at WGV was derived after consideration of a broad range of possible options at both lot and precinct scales. This selection process took place at the structure planning stage and was initiated by the project consultants. It took into account a range of factors such as cost, maintenance considerations, water savings, sustainability, and economies of scale as part of a high-level business case. Various alternative water sources were considered within the suite of IUWM initiatives and included rainwater, groundwater, greywater, and recycled wastewater. For example, various fit-for-purpose sources of water for irrigation were evaluated, and although private (on lot) greywater reuse has merit, it was ultimately excluded in favor of the community bore scheme that could satisfy both public and private irrigation demands in a robust and innovative manner. This process of conducting a feasibility assessment assisted in informing stakeholders and the client of an overall approach to reducing mains water consumption.

Once the initiatives had been selected, it was then a matter of determining how best to achieve their successful implementation. This was achieved via two principal strategies—design guidelines (DGs) and a Sustainability Package. The DGs consisted of measures for the detached residential (only) dwellings, which were either "mandatory" and termed "development controls," or "recommended" and termed "design guidance." These guidelines are intended to achieve climate responsive design, WSUD, environmental performance, and liveability, in a pragmatic manner. Importantly, the DGs attempt to capture the project's aspirations without becoming a commercial encumbrance to the developer. The "recommended" options are provided on and above the mandated measures in an effort to further support the uptake of these initiatives. Adherence to the DGs is achieved via the submission of building plans to the WGV Estate Architect for approval. All these initiatives are "open source"[5] with the concepts and technical information being made available to others for utilization.

The Sustainability Package provided by LandCorp was a financial incentive of AU$10,000 to upgrade or complement the minimum standards for three key initiatives as described in the DGs. The three initiatives consisted of a solar PV upgrade (to enable the homes to operate at Net Zero Energy), an advanced shade tree (to contribute precinct tree canopy coverage), and the supply and installation of a 3 kL rainwater tank with pump, pressure tank, controls and water meter, plumbed to toilets and cold water inlet to the washing machine. The rainwater component of the package built on the

5. https://www.landcorp.com.au/Documents/Projects/Metropolitan/White%20Gum%20Valley/WGV%20Design%20Guidelines%20February%202016.pdf.

mandatory requirement (design control) in the DGs to have rainwater-ready plumbing. Similarly, to ensure that the potential for greywater reuse was not extinguished as a future option, the recommendation that all buildings incorporate greywater-ready plumbing was included as design guidance within the DGs (the cost to retrofit this drainage is typically cost-prohibitive).

Eligibility for the package was based on all three items being included on the one property. To best ensure the correct installation of high-quality equipment, with a 12-month maintenance and trouble-shooting period included, a vetting of potential suppliers was undertaken and nominated suppliers were selected to supply and install the items.

Delivery of initiatives for the multi-residential lots in WGV also follows particular implementation pathways. The Gen Y Demonstration House is compliant with the DGs, as well as having its unique strata-billing system for water and energy consumption. The other multi-residential lots have been guided by the OPL framework, with an Action Plan developed to accompany EOIs detailing how the multi-residential lots will respond to the OPL criteria. LandCorp was responsible for delivering all the common infrastructure and public areas and ensuring each met the OPL framework.

## 26.3.1 Water efficiency

The efficient use of water is the primary foundation on which all other IUWM initiatives at WGV build. This ensures that the maximum potential from each individual water source is achieved, as well as minimizing the resulting volume of "wastewater" that has cost and greenhouse gas emission implications at a water utility level.

The principal water efficiency measures at WGV comprise both in-house and ex-house initiatives. In-house water efficiency measures are, as a minimum, those stipulated within the National Construction Code (NCC) and in some instances more efficient fixtures are specified. For example, showers heads may provide up to 9 L per min (the minimum three Stars WELS rating and NCC Vol. 3, Section B1.6), but a maximum of 7.5 L per min is prescribed as mandatory within the WGV DGs. The ex-house measures as outlined in Table 26.1 are based on industry best practice for landscape design and adopt techniques specifically tailored to the local climate and geology.

In addition to the above measures, smart metering is employed with real-time data logging for leak detection, as well as providing user feedback to support efficient water-use behavior.

### *26.3.1.1 Estimated water savings*

Estimated water savings through improved efficiency have been derived from a variety of sources with the Perth Residential Water Use Study (PRWUS) undertaken in 2008/09 by the Water Corporation (2010) providing the comparative "Perth average water use" benchmark, as well as the basis on which the various in-house and ex-house demands are proportioned.

**TABLE 26.1 Water Efficiency Measures Adopted Outside the House**

| Development Controls | Design Guidance |
|---|---|
| An automatic irrigation system including a rain sensor using a programmable controller must be connected to the meter provided by LandCorp. The water source will be supplied by a community bore that will only operate during set time periods each day. The bore will not operate during the Winter Sprinkler Ban or on days where sufficient rain has occurred. | Consider establishing irrigation for the first two summers and then for extended dry hot periods only. |
| Water efficient in-line drip irrigation must be installed for all garden beds. | Consider adopting hydrozoning principles, which involves grouping plants with similar water needs together in an effort to be more water efficient. |
| Private water bores are not permitted. | Consider incorporating irrigation control technologies such as evapotranspiration sensors or soil moisture sensors to ensure efficient watering of landscaping when the community bore system is active or operational. |
| Any outdoor swimming pool or spa must be supplied with an accredited cover that reduces water evaporation. | Consider grading to create micro swales and basins to help to recharge the soil moisture and reduce runoff from stormwater. |
| Indoor and outdoor taps must not to be connected to the community bore supply. | |
| Spray irrigation may be used on turf areas only. | |

"Embedded" water efficiency gains are expected to be achieved from the increased dwelling density across the development when compared to the Perth average, as this in turn leads to reduced landscape areas requiring irrigation. Planned dwelling density varies from R35 (medium density) to R80 (high density) with anticipated landscape areas being based on the required lot sizes, plot ratios, and open space requirements under the WA Residential Design Codes (Department of Planning WA, 2015) plus relevant local government planning policies and development DGs.

Water savings through efficiency measures are expected to range between 51% for single residential and 54% for multi-residential on a per person basis with the variation between typologies resulting from the difference in landscaping areas, the specific initiatives being deployed, and assumed occupancy. For example, the water savings for the efficiency measures for single residential dwellings (R35) have been estimated at 25 kL/person per year for embedded density savings, 17 kL/person per year for in-house water efficiency measures, 5 kL/person per year for ex-house water efficient landscapes, 4 kL/person per year for smart metering, and 3 kL/person per year through improved resident water-use behavior. Details on how these figures were established are presented below.

- Embedded density savings: Irrigation volumes were established for an average lot size of 288 $m^2$, with 92 $m^2$ of open space comprised 35 $m^2$ of paving, 30 $m^2$ of turf, and 27 $m^2$ of garden bed (in line with WAPC, local government, and DG requirements), assuming two 10-mm irrigation events per week from September through May as per the Water Corporation sprinkler roster (Water Corporation, 2018). This figure (on a kL/person per year basis) was compared with the Perth average ex-house volumes (Water Corporation, 2010), which can be assumed to be indicative of typical Perth low density housing (zoning often R20 with blocks averaging 450 $m^2$, Falconer et al., 2010). The difference between these volumes equates to 25 kL/person per year.
- **In-house water savings:** Water savings resulting from the inclusion of minimum specified water efficient fixtures for all new dwellings under the Building Code of Australia (BCA) WA addition 2.3.1 standards at the time (BCA, 2014) (now NCC) were estimated. The additional savings from efficiency gains achieved through the inclusion of enhanced water efficient fixtures and appliances over and above the BCA as required under the development DGs were then calculated. This results in a difference of approximately 17 kL/person/year.
- **Ex-house water savings:** Water savings achievable with a water-efficient landscape were estimated from the enhanced efficiency achievable with the measures described in Table 26.1. Data were derived from a range of peer reviewed publications, which indicated that savings in irrigation demand of approximately 30%−40% over conventional landscape treatment and irrigation was achievable. An average value of 5 kL/person per year was used.
- **Smart metering:** A 2-year trial of smart metering undertaken by the Water Corporation in Kalgoorlie indicated a 11% reduction in the overall residential water use (Beal and Flynn, 2014). A conservative value for the potential water saving of 4 kL/person per year has been used given all dwellings are new with modern low flow fixtures.
- **Behavior change:** A 3 kL/person per year savings was estimated as the result of planned resident education and support (behavior change) initiatives integrated with the OPL Action Plan.

In addition to water efficiency measures, further reduction to mains water consumption at WGV will be achieved via the use of fit-for-purposes nondrinking water sources outlined in the following sections.

## 26.3.2 Rainwater harvesting

Rainwater is a relatively high-quality, nonpotable alternative water source that can play a significant role in displacing mains water within the Perth metropolitan area. This runs contrary to the broader public opinion in Perth. For rainwater harvesting to be effective there are two principal elements that need to be in place: sufficient roof catchment area connected to an appropriately sized rainwater tank, and the tank must be plumbed into the house for internal nonpotable demands, e.g., toilet flushing, cold water for washing machine (Sharma et al., 2015). Determining the appropriate end use is particularly relevant in low rainfall or Mediterranean (winter-wet, summer-dry) climates such as Perth, where it stands to reason that a lack of rain, combined with high summer irrigation requirements, would suggest rainwater is not viable for irrigation purposes (Byrne, 2016). Thus, although rainfall is likely to be adequate to meet toilet demand during the limited seasonal rainy period, it is an impractical proposition for garden watering, given the inverse relationship between rainfall and landscape water demand.

Simple measures to ensure the best possible quality of water such as rain heads and first flush devices are important, as is a level of periodic maintenance for sustained performance. But these are of a secondary order of importance compared to the two key elements described previously.

**TABLE 26.2 Alternate Water Sources (Rainwater)**

| Development Controls | Design Guidance |
|---|---|
| All toilets and washing machine cold taps are to be installed with dual plumbing to allow for the future connection to an alternative water source (e.g., rainwater), without breaking the fabric of the building. | Install a 3 kL rainwater tank with pump, pressure tank, and mains water backup valve connection to the dual plumbing as an alternative water source. |
| Provide sufficient space for future installation of a rainwater tank (minimum capacity of 3 kL) close to a rainwater downpipe/s with a minimum roof catchment area of 70 m$^2$, an external power outlet, a garden tap or mains water takeoff point, and the dual plumbing pipe work. | |

The important role that rainwater could play in the WGV development was captured by embedding mandatory provisions within the DGs as outlined in Table 26.2, with the supply and installation of a 3 kL rainwater tank, along with associated equipment, made available to residents through the Sustainability Package.

### 26.3.2.1 Estimated water savings

The contribution of rainwater to mains water savings for the single residential dwellings has been estimated as 10 kL/person per year with modelling undertaken based on recent rainfall data, a 3 kL rainwater tank, a minimum roof catchment area of 70 m$^2$, and typical residential toilet and washing machine consumption using the methodology of Hunt et al. (2011).

### 26.3.2.2 Governance considerations and potential issues

It is important to note that it is not possible to ensure the performance of the rainwater systems over the longer term, for example because of a lack of maintenance (Moglia et al., 2014). However, the inclusion of water meters, as specified under the sustainability package program, helps to address this.

The footprint required for a rainwater tank on relatively small urban blocks can be an issue. Some owners have chosen to install belowground rainwater tanks (with the additional cost at their own expense) as a means to avoid this constraint. The sandy soils typical of the Swan Coastal Plain make excavation of belowground tanks relatively inexpensive.

## 26.3.3 Community bore

Groundwater can be considered a sustainable source of fit-for-purpose water provided it is of suitable quality, and its extraction is replenished. Sustainable extraction is estimated by a water balance that demonstrates that it is recharged by direct infiltration on-site within a typical year (Department of Water, 2009). Allowances need to be made for the rainwater that is taken out of the system flow, as well as losses to runoff and evaporation/evapotranspiration. This methodology is described further in Section 26.3.4.

The community bore is a precinct-scale, nonpotable water supply scheme for the irrigation of public and private green space and is key to the overall mains water savings for the development (Fig. 26.2). The community bore scheme supplies groundwater from the superficial aquifer via a third pipe system. It is centrally controlled, and DGs stipulate an irrigation controller and individual meters for optimal efficiency. The community bore leads to urban greening, which are beneficial for community well-being and ecosystem functioning.

The community bore ensures that no mains water has to be used outside the home. It operates alongside a best practice water efficient irrigation system that employs sub-mulch drip-line irrigation, and water saving technologies including rain sensors and hydrozoning principles. Similar irrigation requirements have been mandated in the design for private lots.

### 26.3.3.1 Estimated water savings

The mains water savings from the community bore for single residential dwellings has been estimated at 15 kL/person per year. This was based on the assumption that all ex-house irrigation demand could be satisfied by the community bore. The demand was derived by initially subtracting those savings likely to be achieved through new house design and enhanced ex-house water efficiency, from Perth's average residential garden and lawn watering figure of 41 kL/person per year (Water Corporation, 2010).

**FIGURE 26.2** Community bore pump set. *Source: Josh Byrne & Associates.*

#### *26.3.3.2 Governance considerations and potential issues*

All lots are required to connect to the community bore scheme. Cost recovery for its operation is obtained by the City of Fremantle using a Special Area Rate, which is added to the standard property rates. This method of nonvolumetric billing was the easiest for the City to manage, but it may lead to abuse and greater water consumption. Metering each lot connection and the inclusion of water efficiency measures in the DGs are intended to address this.

A summary of the various mains water savings initiatives at WGV compared to the Perth residential average is provided in Fig. 26.3. Monitoring of actual water use is underway and will provide accurate data on the actual savings achieved.

A summary of the various water saving initiatives for single residential dwellings at WGV, including approximate costs for supply and installation, and associated energy intensity of the alternative water sources is provided in Table 26.3.

A schematic of the overall water balance at WGV is presented in Fig. 26.4.

### 26.3.4 Stormwater management

Stormwater management at WGV takes advantage of the site's topography which varies by 8 m between the two main ridges and a central depression and the naturally permeable soils as described previously. Community consultation indicated this physical form was valued and the decision was made early not to fill and level the site.

Drainage from all lots is infiltrated through soakwells for up to the 20-year annual recurrence interval (ARI) event with stormwater generated in larger events infiltrated via underground infiltration/drainage cells as shown in Fig. 26.5. Stormwater generated within the road network up to the 5-year ARI event is conveyed through a piped network to one of 10 sets of underground drainage chambers located beneath the roads and road reserves as shown in Fig. 26.6. Stormwater from larger events (>5 year ARI) is conveyed via the road reserves to the infiltration chambers.

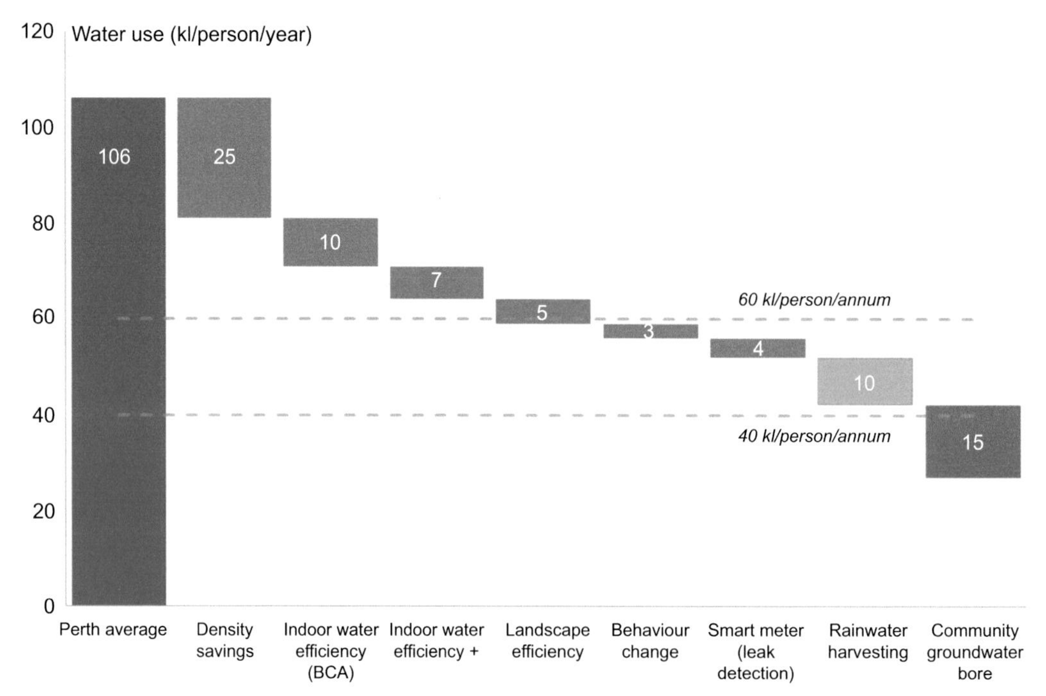


FIGURE 26.3 Projected mains water savings for single residential dwellings at WGV presented in a "decumulative" fashion. *Source: Josh Byrne & Associates.*

**TABLE 26.3 Summary of Water Saving Initiatives for Single Residential Dwellings at WGV**

| Initiative | Water Saving[a] (kL/person /year) | Installed Cost per Dwelling (AU$) | Ongoing Cost (AU$/dwelling/year) | Energy Intensity (kWh/kL) |
|---|---|---|---|---|
| Embedded density savings | 25–28 | 0 | 0 | – |
| In-house water efficiency | 17 | 550 | 0 | – |
| Dual plumbing | – | 1000 | 0 | – |
| Water efficient landscape | 4–5 | 500 | 0–50 | – |
| Smart metering | 4 | 800 | 78 | – |
| Behavior change | 3 | – | – | – |
| Plumbed rainwater | 0–10 | 5000 | 50–150 | 1.2–1.3 |
| Community bore | 10–15 | 3800 | 270 | 1.5–2.0 |

[a]*Range of water savings for the embedded, water efficient landscaping, rainwater and community bore reflect the modeling for the single and multi-res scenarios.*

In effect, the design of the stormwater system ensures 100% infiltration of all stormwater generated on-site in up to the 1 in 100 year storm event. To investigate the interaction of the site with the superficial aquifer, a site water balance was undertaken. Using recent annual rainfall statistics from the Bureau of Meteorology, the quantity of water to fall on the site in an average year was calculated to be 16,112 kL. Not all of this rainfall will make it to groundwater, as some will be lost to evaporation and some will be taken up by plant roots or held in the soil. As such, a recharge factor of 0.5 is applied, which is the estimated recharge factor applicable to residential sites in Perth (Department of Water, 2009). After an assumed volume of 840 kL was deducted for rainwater intercepted by rainwater tanks, a net recharge to the local aquifer of 2,216 kL/a is anticipated after the annual demand of 5,000 kL from the community bore is met. The water balance between the site and the local groundwater supply is shown in Fig. 26.4.

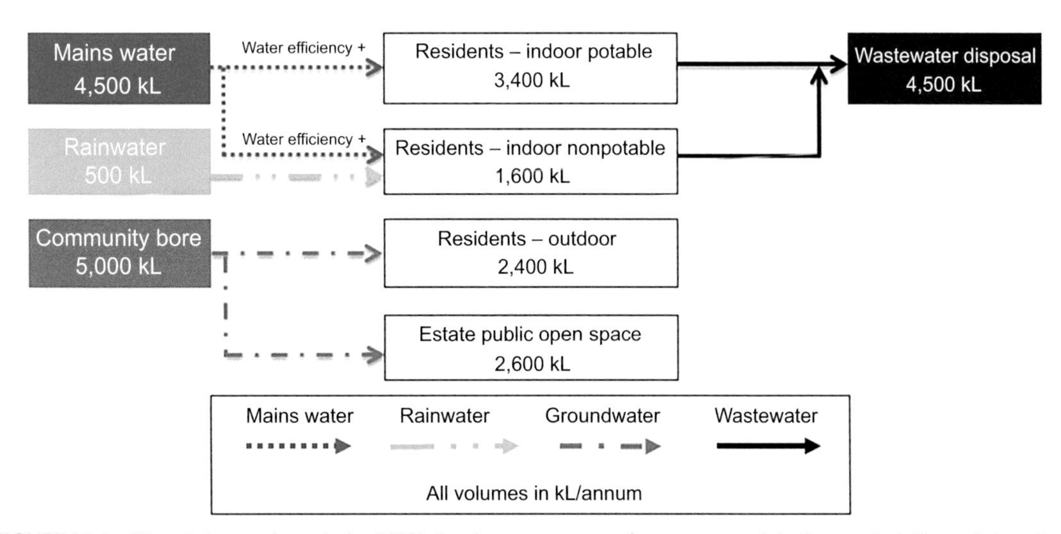


**FIGURE 26.4** Water balance schematic for WGV showing water streams from source to sink. *Source: Josh Byrne & Associates.*

**FIGURE 26.5** Underground drainage chambers for stormwater infiltration. *Source: TABEC Engineering Consultants.*

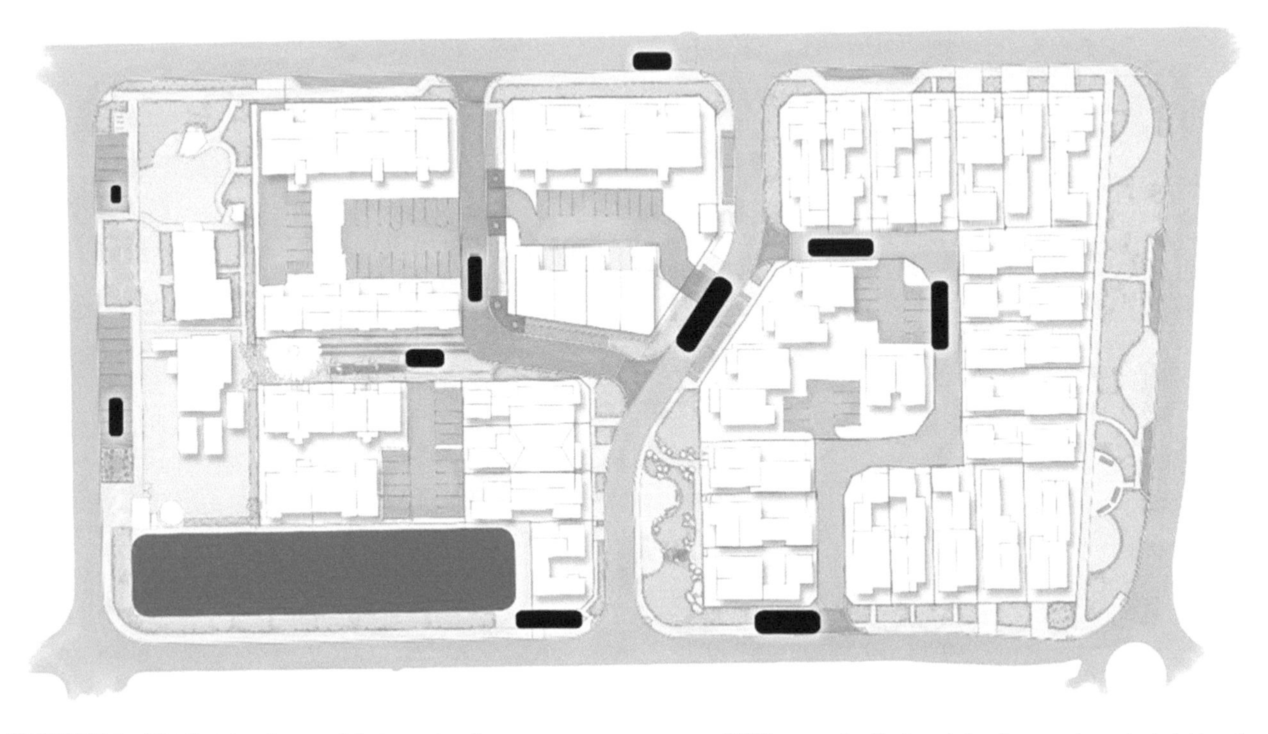

**FIGURE 26.6** Distributed underground drainage sites for stormwater management at WGV—note: the distributed chambers are shown in dark blue, the light blue area is the historical sump. *Source: Josh Byrne & Associates.*

Comprehensive infiltration of stormwater has been achieved through a combination of drainage cells, flush kerbs, ephemeral winter-wet depressions (common features of the Swan Coastal Plain), and the use of damp land native plants as a simple and effective WSUD solution to minimize stormwater runoff and improve localized infiltration. Micro swales and vegetated basins have also been incorporated into the design, and private owners are encouraged to take similar approaches.

Importantly, no stormwater from the development is discharged to the existing council-owned drainage sump located in the southwest corner of the site as shown in Fig. 26.6. Investigations identified that the sump was already at capacity and could not be utilized for additional stormwater runoff generated by the WGV development. Such drainage sumps are common across the Swan Coastal Plain and represent the historic, engineering response to managing stormwater resulting from urban development. Although LandCorp was under no obligation to remediate the sump, its size and location meant it would sterilize a substantial portion of highly visible land to both new and existing residents of the area. Initial community consultation suggested that the 1960s-style drainage sump was both an eyesore and a potential hazard to children. Through a unique collaboration between LandCorp and the City of Fremantle, the sump was reconstructed for stormwater management as well as enhanced environmental, educational, and amenity values.

An assessment of the total water cycle and stormwater behavior of the neighboring White Gum Valley catchment (12 ha), as shown in Fig. 26.7, was undertaken to size the sump and drainage cells appropriately.

Stormwater runoff generated from approximately 6 ha of roads and road reserves within the larger 12 ha catchment surrounding WGV, passes through gross pollutant traps before entering a sump chamber that was filled with drainage cells, before infiltrating into the aquifer. The sump is equipped with a water level sensor, and in conjunction with the community bore's groundwater level sensor, it provides real-time data on the efficiency of stormwater recharge and seasonal aquifer level fluctuations. Importantly, the stormwater from surrounding areas that infiltrates at the sump does not form part of the WGV net positive groundwater recharge balance, as it comes from outside the WGV precinct catchment.

**FIGURE 26.7** Extent of larger catchment draining to the historical sump. *Source: Hyd2o Hydrological Consultants.*

**FIGURE 26.8** Historical WGV Sump before (a) and after (b) conversion. *Source: Josh Byrne & Associates.*

Redevelopment of the sump has created a high profile demonstration of how these once poorly utilized assets can be remediated to become both a functional and useable space for residents, ensuring beneficial community outcomes from urban infill (Fig. 26.8).

## 26.3.5 Precinct greening

Sustainable stormwater management has been integrated with the streetscape and landscape design. This approach means that natural vegetation and habitat are doubly valued for their service to the function of the precinct, as well as to biodiversity.

Road verges were widened in key areas to retain existing trees, and the use of rear access lots enables verge flora to be maximized. The road design responds to retained trees and new plantings, which will combine with existing vegetation to recreate the canopy that was previously in place across the site. Fremantle City's canopy target of 20% (City of Fremantle, 2015; now the Greening Fremantle Strategy 2020; City of Fremantle, 2017) is being exceeded at WGV.

Retention and enhancement of tree canopy is vital to reducing the impacts of Urban Heat Island (UHI) effect, where an increase in urban density and canopy loss can lead to increased nighttime temperatures due to heat retention. Heat waves and extreme heat events can exacerbate the UHI and lead to it becoming a serious health hazard (WHO, 2016). The urban forest can have a direct effect on regulating local temperatures through evapotranspiration, shading, and cooling and providing comfort for pedestrians at street level (Brown et al., 2013; City of Melbourne, 2014). Shading provided by trees planted around perimeters and in median strips of parking lots and streets can be an effective way to cool a community (US EPA, 2008) and cooling effects may be felt up to 1 km away from a park boundary (WHO, 2016). Studies have demonstrated that increasing overall vegetation cover can effectively reduce land surface temperature and tree cover is effective for daytime surface temperature reduction (Coutts and Harris, 2013). Further, Coutts and Harris (2013) found a 10% increase in total vegetation cover resulted in a 1°C reduction in land surface temperature during the day in Melbourne. Chapter 19 of this book provides further details on the UHI effect and mitigation strategies.

The historic laneways of the broader White Gum Valley suburb and surrounding areas have been referenced at the WGV site through the inclusion of activated laneways, providing highly visible and useable spaces for future residents. These laneways were included in the initial subdivision design for WGV and have subsequently been included in the project DGs (Fig. 26.9).

### *26.3.5.1 Landscaping*

Community consultation in the planning of WGV identified the retention of trees as a key objective, and this became a significant part of the design process. The project's low-key road network and carefully aligned subsoil drainage system have been integral to the retention of 25% of the existing mature trees on the site (Fig. 26.10).

The landscape design at WGV aims to provide an environment that captures and enhances the suburb's sense of place, while strengthening biodiversity, local food production, and community cohesion. Tree, shrub, and lawn species have been carefully selected for the public domain to ensure they are suited to Perth's hot, dry summers (Fig. 26.11). Landscaping and

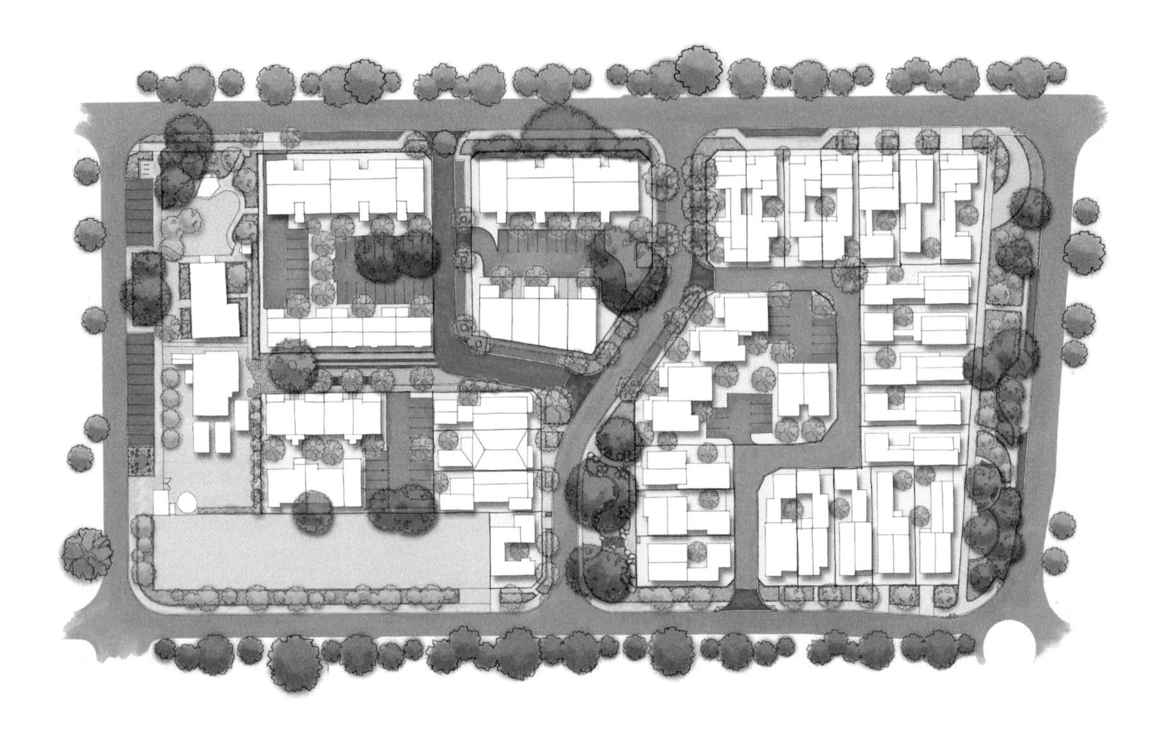

**FIGURE 26.9** WGV subdivision plan showing lot layout, roads, laneways, and verges. *Source: Josh Byrne & Associates.*

**FIGURE 26.10** Retention of Mature Trees in verges for shade and biodiversity value. *Source: Josh Byrne & Associates.*

FIGURE 26.11 Low water use landscaping in the public domain using local native species that are well adapted to the site conditions (source: Josh Byrne & Associates).

**TABLE 26.4 Water Efficiency via Landscaping**

| Development Controls | Design Guidance |
|---|---|
| Trees planted in front gardens of north facing properties must be deciduous to maintain solar passive design principles. | Consider native ground covers, shading and cooling potential of trees/shrubs, trees with high canopy cover, solar access, and waterwise native or endemic species via recommended plant list. |
| Turf/lawn is to be limited to a maximum of 50% of the landscaped area of the site (excluding building area). All turf/lawn is to be an approved waterwise variety. | |
| Artificial turf is not permitted in front gardens or verges. | |

street trees include a mixture of native species for year-round shade and deciduous trees for areas where summer shade and winter sun is required. Table 26.4 details the development controls and design guidance used to ensure water efficiency is achieved via landscaping.

### *26.3.5.2 Natural environment, vegetation, and native wildlife*

A number of initiatives were implemented to ensure that the natural environment would be protected and restored, as part of adhering to the OPL framework.

**Tree preservation:** The nature of the development meant that many existing trees would have to be removed. However, by clever design, 25% of the mature trees were retained. They will contribute to an urban green strip formed by additional native tree and groundcover plantings, seating areas, and informal footpaths.

**Revegetation:** To revegetate the site after development, a diverse range of native and exotic trees and shrubs will feature within the public open spaces and road reserves, to support the local ecology and increase local biodiversity. These species provide habitat and food for native animals, birds, and insects.

The tree canopy target for the site is 30% at 15 years post construction, with a tree canopy diameter of 6m. This target will match the percentage tree canopy coverage measured prior to the redevelopment in 2014.

**Native wildlife:** To help preserve/restore native wildlife species, an intensive fauna survey was undertaken prior to timber harvesting and site clearing. Habitat boxes will be included in mature trees within the development and adjacent road reserves. DGs encourage private owners to develop native verge gardens, frog-friendly gardens, and constructed fauna habitats.

## 26.4 INDUSTRY ENGAGEMENT AND KNOWLEDGE SHARING

The implementation of IUWM initiatives at WGV provides benefits not just for the local residents but also for the wider community, industry, and WSUD/Low Carbon Living research community. Specifically, this has been achieved via knowledge-sharing initiatives such as metering and data acquisition; industry tours; DGs for residents; and sump revitalization and urban greening initiatives that are accessible to all. These industry engagement and knowledge-sharing initiatives are detailed below.

### 26.4.1 Performance monitoring and data acquisition

All lots at WGV have been provided with dual metering—one for mains water and one for the community bore connection—to demonstrate mains water use (and savings) as well as provide prompt feedback of leaks or water supply issues. Further, rainwater meters have been provided for detached dwellings to obtain data on the contribution of rainwater in reducing mains water consumption in Perth. They also help ensure that the equipment is working, and actually delivering mains water reduction as planned. Collecting and disseminating data is a key component of the WGV project as it aims to demonstrate its water and energy saving initiatives via sharing evidence-based information with water and land development industries and the research community. This will be achieved through data logging over a 3-year research and performance monitoring phase that is now underway (and potentially extended). Meter readings are being hosted on a web interface and are accessible to residents, authorized Water Corporation and City of Fremantle staff for research, billing, and management purposes. The monitoring and research component, in effect a "living laboratory," allows innovations to be tested in real life settings and potentially influence long-term behavior change.

### 26.4.2 Community and industry engagement

The WGV site is one of LandCorp's "Innovation through Demonstration" initiatives[6]. Specifically, the learnings from the precinct's water and energy initiatives are openly shared with both the broader community, and industry stakeholders, in a bid to make sustainable urban design accessible to all. On-site tours are one of the main avenues for knowledge sharing and have been well received by a variety of attendees. The various WSUD and greening initiatives, such as the community bore, pocket park, and revitalized sump, provide improved amenity and recreation for the community.

### 26.4.3 Waterwise development exemplar

WGV achieved Water Corporation Waterwise Development status[7] in 2015 in recognition of its water-saving initiatives. Central to the WGV Waterwise Development Exemplar project is a 3-year program of research activities that will quantify the mains water savings (the target is 70%), monitor the performance of the various water-based initiatives, critically assess the performance of the stormwater drainage design requirements for the site via real-time data capture, and analyze data and report on the performance findings to a broad audience that includes industry, local government, and the community.

### 26.4.4 Resources

Knowledge-sharing resources have assisted others in learning about the various WSUD components and have also played a key role in ensuring the initiatives at WGV are captured, documented, and available to inform future developments. These include the community bore resource kit, supported by Water Corporation, and a WGV water initiatives episode featured in the "Density × Design"[8] video series on sustainable urban infill development.

## 26.5 CONCLUSION

The WGV residential development is demonstrating how a cross-disciplinary collaborative approach to urban infill can achieve a high level of water efficiency in a sustainable and innovative manner, as well as increasing liveability. This was achieved, in part, by adopting a suite of integrated WSUD initiatives with prior knowledge of their upfront and ongoing

6. https://www.landcorp.com.au/Our-Work/Innovation-Through-Demonstration/.
7. https://www.watercorporation.com.au/about-us/news/media-statements/media-release/wgv-to-be-was-most-water-efficient-development.
8. https://densitybydesign.com.au/wgv/.

cost, their biophysical suitability to the site's characteristics, sustainability, social acceptability, and ease of implementation. The provision of an overarching sustainability framework—One Planet Living—in conjunction with development guidelines and a Sustainability Package has ensured that water efficient systems and behavior have been embedded.

Specifically, stormwater is effectively managed on site via a combination of drainage cells, flush kerbs, winter-wet depressions, micro swales, and carefully selected vegetation to minimize stormwater runoff and improve localized infiltration. These WSUD initiatives mean that no stormwater from the site is discharged into the historical sump located adjacent to the precinct (instead, it is functioning for the broader suburb). An integrated approach to water management suggests that mains water consumption will be reduced by 60%−70%, with the use of plumbed rainwater systems for in-house nonpotable use and a community bore to provide groundwater for the irrigation of public and private green space. It is the combination of these initiatives and the underlying philosophy of the project that ensures that retained natural vegetation and additional precinct greening approaches not only effectively service the precinct but are also valued for biodiversity and their contribution to enhanced liveability.

Ongoing data logging and monitoring will allow the performance of the various WSUD initiatives to be assessed. The greater learnings of the site are therefore yet to come as the precinct becomes populated and fully operational. The collaborative nature of the project will enable this data to be shared with industry. WGV presents a unique case study, and it is anticipated that it will provide an example of how to implement alternative water management initiatives as part of Perth's journey to become a water-sensitive city.

## ACKNOWLEDGMENTS

The initiatives at WGV have been made possible through the financial and in-kind support of Water Corporation, Department of Water and Environmental Regulation (DWER), LandCorp, City of Fremantle, CRC for Water Sensitive Cities, Curtin University, CRC for Low Carbon Living, the Urban Development Institute of Australia (UDIA) and Josh Byrne & Associates. These organisations have demonstrated their commitment to making a difference in the way that water is managed in residential development in Western Australia.

## REFERENCES

Australian Water Association, 2017. Do Perth Residents Really Have the Highest Water Consumption Rate? http://www.awa.asn.au/AWA_MBRR/Publications/Latest_News/Do_Perth_residents_have_highest_water_consumption_rate.aspx.

Australian Bureau of Statistics (ABS), 2015. 2016 Census Quick Stats. Greater Perth. http://www.censusdata.abs.gov.au/census_services/getproduct/census/2016/quickstat/5GPER?opendocument.

BCA, 2014. Building Code of Australia Part 3.12 WA Additions.

Beal, C., Flynn, J., 2014. The 2014 Review of Smart Metering and Intelligent Water Networks in Australia and New Zealand. Griffith University. http://hdl.handle.net/10072/65782.

Bettini, Y., Brown, R.R., de Haan, F.J., 2015. Exploring institutional adaptive capacity in practice: examining water governance adaptation in Australia. Ecology and Society 20 (1), 47−66.

Bioregional, 2017. One Planet Living. http://www.bioregional.com/oneplanetliving/.

Brown, H., Katscherian, D., Carter, M., Spickett, J., 2013. Cool Communities: Urban Trees, Climate and Health. Curtin University. http://ehia.curtin.edu.au/local/docs/CoolCommunities.pdf.

Bureau of Meteorology (BoM), 2016. National Performance Report 2014−15: Urban Water Utilities, Part a. Bureau of Meteorology, Melbourne. http://www.bom.gov.au/water/npr/docs/2014-15/National-performance-report-2014-15_part-A_lowres.pdf.

Bureau of Meteorology (BoM), 2017. National Water Account 2016: Urban Regions Overview. http://www.bom.gov.au/water/nwa/2016/urban/index.shtml.

Bureau of Meteorology (BoM) and CSIRO, 2016. State of the Climate. http://www.bom.gov.au/state-of-the-climate/index.shtml.

Byrne, J.J., 2016. Mains Water Neutral Gardening: An Integrated Approach to Water Conservation in Sustainable Urban Gardens (Doctoral thesis). Murdoch University, Murdoch, Western Australia.

Chong, M.N., Ho, A., Gardner, T., Sharma, A., Hood, B., 2013. Assessing decentralised wastewater treatment technologies: correlating technology selection to system robustness, energy consumption and GHG emission. Journal of Water and Climate Change 4 (4), 338−347. https://doi.org/10.2166/wcc.2013.077.

City of Fremantle, 2015. Green Plan 2020. https://www.fremantle.wa.gov.au/sites/default/files/sharepointdocs/Green%20Plan%202020-C-000476.pdf.

City of Fremantle, 2017. Greening Fremantle: Strategy 2020. https://www.fremantle.wa.gov.au/sites/default/files/sharepointdocs/Greening%20Fremantle%20Strategy%202020-C-000625.pdf.

City of Melbourne, 2014. Urban Forest Strategy. Making a Great City Greener 2012-2032. City of Melbourne, Melbourne, Australia.

Coombes, P.J., Kuczera, G., Kalma, J.D., 2003. Economic, water quantity and quality impacts from the use of a rainwater tank in the inner city. Australian Journal of Water Resources 7 (2), 111−120.

CRCWSC, 2017. https://watersensitivecities.org.au/content/project-d5-1.

Coutts, A., Harris, R., 2013. Urban Heat Island Report: A Multi-scale Assessment of Urban Heating in Melbourne during an Extreme Heat Event: Policy Approaches for Adaptation. Victorian Centre for Climate Change Adaptation Research.

Davidson, W.A., 1995. Hydrogeology and Groundwater Resources of the Perth Region, Western Australia. In: Western Australia Geological Survey, Bulletin, vol. 142. Perth Western Australia.

Department of Water (DoW), 2009. Perth Regional Aquifer Modelling System (PRAMS) Model Development: Application of the Vertical Flux Model, Hydrological Record Series, Report No. HG27. Government of Western Australia, DoW, Perth. https://www.water.wa.gov.au/__data/assets/pdf_file/0004/4459/84333.pdf.

Department of Water (DoW), 2014. Environmental Management of Groundwater from the Gnangara and Jandakot Mounds Annual Compliance Report to the Office of the Environmental Protection Authority July 2012 to June 2013.

Department of Water (DoW), 2015. Urban Nutrient Decision Outcomes (UNDO) Tool. http://www.water.wa.gov.au/planning-for-the-future/water-and-land-use-planning/undo-tool.

Department of Planning (WA), 2015. State Planning Policy 3.1 Residential Design Codes. https://www.planning.wa.gov.au/Residential-design-codes.aspx.

Diaper, C., Tjandraatmadja, G., Kenway, S.J., 2007. Sustainable Subdivisions: Review of Technologies for Integrated Water Services. CRC for Construction Innovation.

Dillon, P., Page, D., Dandy, G., Leonard, R., Tjandraatmadja, G., Vanderzalm, J., Rouse, K., Barry, K., Gonzalez, D., Myers, B., 2014. Managed Aquifer Recharge and Urban Stormwater Use Options: Summary of Research Findings. Goyder Institute for Water Research Technical Report Series No. 14/13, Adelaide, South Australia. ISSN: 1839−2725. http://www.goyderinstitute.org/publications/technical-reports/.

Dhakal, R.S., Syme, G., Andre, E., Sabato, C., 2015. Sustainable water management for urban agriculture, gardens and public open space irrigation: a case study in Perth. Agricultural Sciences 6 (7), 676−685.

Dunnicliff-Wells, N., 2014. Understanding the Hydrology of the Swan Coastal Plain. CRC for Water Sensitive Cities. https://watersensitivecities.org.au/content/understanding-hydrology-swan-coastal-plain/.

Department of Water and Environmental Regulation (DWER), 2017. Gnangara Groundwater System. http://www.water.wa.gov.au/water-topics/groundwater/understanding-groundwater/gnangara-groundwater-system.

Evans, C., Radin-Mohamed, R., Anda, M., Dallas, S., Mathew, K., Jamieson, S., Milani, S., 2008. Greywater recycling in Western Australia: policy, practice and technologies. In: Proceedings of the Onsite and Decentralised Sewerage and Recycling Conference, 12−15 October 2008, VIC.

Evans, C.S., Anda, M., Dallas, S., 2009. An assessment of mains water savings achieved through greywater reuse at a household scale in Perth, Western Australia. In: Proceedings of the 6th International Water Sensitive Urban Design Conference and Hydropolis No. 3, Perth, pp. 464−471.

Falconer, R., Newman, P., Giles-Corti, B., 2010. Is practice aligned with the principles? Implementing new urbanism in Perth, Western Australia. Transport Policy 17 (5), 287−294.

Friedler, E., Hadari, M., 2006. Economic feasibility of on-site greywater reuse in multi-storey buildings. Desalination 190 (1), 221−234.

Furlong, C., Gan, K., De Silva, S., 2016. Governance of integrated urban water management in Melbourne, Australia. Utilities Policy 43, 48−58.

Gardner, E.A., 2003. Some examples of water recycling in Australian urban environments: a step towards environmental sustainability. Water Science and Technology: Water Supply 3 (4), 21−31.

Ghisi, E., de Oliveria, S.M., 2007. Potential for portable water savings by combining the use of rainwater and greywater in houses in southern Brazil. Building and Environment 42 (4), 1731−1742.

Gurung, T.R., Sharma, A., 2014. Communal rainwater tank systems design and economies of scale. Journal of Cleaner Production 67, 26−36. https://doi.org/10.1016/j.jclepro.2013.12.020.

Ho, G., Dallas, S., Anda, M., Mathew, K., 2001. On-site wastewater technologies in Australia. Water Science and Technology 44 (6), 81−88.

Ho, G., Anda, M., Hunt, J., 2008. Rainwater harvesting at urban land development scale: mimicking nature to achieve sustainability. In: Proceedings of IWA World Water Congress, 7−12 September 2008, Vienna.

Hunt, J., Anda, M., Mathew, K., Ho, G., Priest, G., 2005. Emerging approaches to integrated urban water management: cluster scale application. Water Science and Technology 51 (10), 21−27.

Hunt, J., Anda, M., Dallas, S., Ho, G., 2011. Rainwater harvesting systems for local government buildings, three case studies. In: Proceedings of International Conference on Integrated Water Management, 2−5 February 2011, Perth.

Johnstone, P., Adamowicz, R., de Haan, F.J., Ferguson, B., Wong, T., 2012. Liveability and the Water Sensitive City - Science-Policy Partnership for Water Sensitive Cities. Cooperative Research Centre for Water Sensitive Cities, Melbourne, Australia, ISBN 978-1-921912-17-7.

Kunz, N.C., Fischer, M., Ingold, K., Hering, J.G., 2016. Drivers for and against municipal wastewater recycling: a review. Water Science and Technology 73 (2), 251−259.

LandCorp, 2017. White Gum Valley. https://www.landcorp.com.au/Residential/White-Gum-Valley/Living-options/.

Lehmann, S., 2010. Green urbanism: formulating a series of holistic principles. S.A.P.I.EN.S (Online) 3 (2). Online since 12 October 2010. http://sapiens.revues.org/1057.

Lieb, D., Brennan, D., McFarlane, D., 2006. The Economic Value of Groundwater Used to Irrigate Lawns and Gardens in the Perth Metropolitan Area. Technical Report. CSIRO: Water for a Healthy Country National Research Flagship, Canberra. In: https://publications.csiro.au/rpr/download?pid=procite:32b8e0e6-a8c4-4762-9bc1-baf8045a43ae&dsid=DS1.

Marlow, D.R., Moglia, M., Cook, S., Beale, D.J., 2013. Towards sustainable urban water management: a critical reassessment. Water Research 47 (20), 7150−7161. https://doi.org/10.1016/j.watres.2013.07.046.

Moglia, M., Tjandraatmadja, G., Delbridge, N., Gulizia, E., Sharma, A.K., Butler, R., Gan, K., 2014. Survey of Savings and Conditions of Rainwater Tanks. Smart Water Fund and CSIRO, Melbourne, Australia.

Mohamed, R.M.S.R., Kassim, A.H.M., Anda, M., Dallas, S., 2013. A monitoring of environmental effects from household greywater reuse for garden irrigation. Environmental Monitoring and Assessment 185 (10), 8473–8488.

Nolde, E., 2000. Greywater reuse systems for toilet flushing in multi-storey buildings – over ten years' experience in Berlin. Urban Water 1 (4), 275–284.

Page, D., Gonzalez, D., Sidhu, J., Toze, S., Torkzaban, S., Dillon, P., 2015. Assessment of treatment options of recycling urban stormwater recycling via aquifers to produce drinking water quality. Urban Water Journal 13 (6), 657–662.

Priest, G., Anda, M., Mathew, K., Ho, G., 2004. Domestic greywater reuse as part of a total urban water management strategy. In: Proceedings of the International Sustainability of Water Resources Conference, 13–14 November 2004, Perth.

Radcliffe, J.C., 2006. Future directions for water recycling in Australia. Desalination 187 (1), 77–87.

Rogers, B.C., Hammer, K., Werbeloff, L., Chesterfield, C., 2015. Shaping Perth as a Water Sensitive City: Outcomes of a Participatory Process to Develop a Vision and Strategic Transition Framework. Cooperative Research Centre for Water Sensitive Cities, Melbourne, Australia.

Sharma, A., Cook, S., Tjandraatmadja, G., Gregory, A., 2012a. Impediments and constraints in the uptake of water sensitive urban design measures in greenfield and infill developments. Water Science and Technology 65, 340–352.

Sharma, A., Chong, M.N., Schouten, P., Cook, S., Ho, A., Gardner, T., Umapathi, S., Sullivan, T., Palmer, A., Carlin, G., 2012b. Decentralised Wastewater Treatment Systems: System Monitoring and Validation. Urban Water Security Research Alliance Technical Report No. 70. http://www.urbanwateralliance.org.au/publications/UWSRA-tr70.pdf.

Sharma, A.K., Begbie, D., Gardner, T., 2015. Rainwater Tank Systems for Urban Water Supply – Design, Yield, Energy, Health Risks, Economics and Community Perceptions. IWA Publishing. http://www.iwapublishing.com/books/9781780405353/rainwater-tank-systems-urban-water-supply.

Umapathi, S., Chong, M., Sharma, A.K., 2013. 2013. Evaluation of plumbed rainwater tanks in households for sustainable water resource management: a real-time monitoring study. Journal of Cleaner Production 42, 204–214. https://doi.org/10.1016/j.jclepro.2012.11.006.

U.S. Environmental Protection Agency, 2008. Reducing Urban Heat Islands: Compendium of Strategies. Draft. https://www.epa.gov/heat-islands/heat-island-compendium.

Water Corporation, 2009. Water Forever: Towards Climate Resilience, WC, Perth. https://www.watercorporation.com.au/-/media/files/about-us/planning-for-the-future/water-forever-50-year-plan.pdf.

Water Corporation, 2010. Perth Residential Water Use Study (PRWUS) 2008/2009: Final Report, WC, Perth.

Water Corporation, 2015. Annual Report, WC, Perth. https://www.watercorporation.com.au/-/media/files/about-us/our-performance/annual-report-2015/water-corporation-annual-report-2015.pdf.

Water Corporation, 2017a. Sources. https://www.watercorporation.com.au/water-supply/rainfall-and-dams/sources.

Water Corporation, 2018. https://www.watercorporation.com.au/save-water/watering-days.

Wong, T., Brown, R., 2008. Transitioning to water sensitive cities: ensuring resilience through a new hydro-social contract. In: 11th International Conference on Urban Drainage, Edinburgh, Scotland, UK, pp. 1–10.

World Health Organisation, 2016. Urban Green Spaces and Health. WHO Regional Office for Europe, Copenhagen. http://www.euro.who.int/__data/assets/pdf_file/0005/321971/Urban-green-spaces-and-health-review-evidence.pdf?ua=1.

## FURTHER READING

Water Corporation, 2014. H2Options Water Balance Tool. https://www.water.wa.gov.au/__data/assets/pdf_file/0016/5272/98576.pdf.

Water Corporation, 2017b. How Many Litres of Water Do People Use in Perth Each Year? https://www.watercorporation.com.au/home/faqs/saving-water/how-many-litres-of-water-do-people-in-perth-use-each-year.

Chapter 27

# WSUD "Best in Class"—Case Studies From Australia, New Zealand, United States, Europe, and Asia

**Stephen Cook**[1], **Marjorie van Roon**[2], **Lisa Ehrenfried**[3], **James LaGro, Jr.**[4] **and Qian Yu**[5]

[1]*CSIRO Land and Water, Clayton, VIC, Australia;* [2]*University of Auckland, Auckland, New Zealand;* [3]*Yarra Valley Water, Mitcham, VIC, Australia;* [4]*University of Wisconsin-Madison, Madison, WI, United States;* [5]*China Institute of Water Resources and Hydropower Research, Beijing, China*

## Chapter Outline



**Approaches to Water Sensitive Urban Design. https://doi.org/10.1016/B978-0-12-812843-5.00027-7**

**ABSTRACT**

Over the last three decades, there has been a growing recognition of the need to ensure that the design and management of the built environment is sensitive to minimizing disturbance to natural hydrology and the water quality of the receiving environments. The practice of water sensitive urban design (WSUD) has emerged along with a number of other approaches to mitigate risks such as flash flooding, pollution of fresh and marine receiving waters, and water scarcity. The drivers for adopting WSUD practices, and the approaches used, depend on the specific context. Practices have continued to evolve and be refined, which has in large part been due to lessons learnt from pioneering developments that trialled innovative WSUD approaches. The cases reviewed in this chapter discuss some examples of leading-edge WSUD approaches that have provided the practical demonstration of the challenges and benefits to implementing WSUD. They can be used not only to refine standards and guidelines but also to build confidence in the WSUD approaches. A synthesis of findings from the case studies revealed the importance of WSUD being integrated across different urban functions, stakeholders, and levels of government. It is also shown that the benefits of WSUD often extend beyond the primary objective of improved urban stormwater management, reflecting the multifunctional nature of many WSUD approaches. These co-benefits often help to build public understanding and engagement in the benefits of WSUD. The cases also identified the importance of using economic instruments that reflect the true cost of different stormwater management approaches, which in turn can create the financial incentives for the adoption of WSUD approaches.

**Keywords:** Integrated urban water management; Low impact urban design and development and design; Water sensitive urban design.

## 27.1 INTRODUCTION

The rapid growth of the global urban population over the last 50 years has altered landscapes through the construction of buildings, roads, and other impervious surfaces that prevent the infiltration of rainwater, resulting in increased flash flooding and polluted runoff draining directly to urban waterways and the oceans. The conventional approach to managing water in modern cities focuses on the provision of clean drinking water for all purposes, appropriate treatment and discharge of wastewater, and the rapid conveyance of floodwaters away from the urban area. These water management paradigms have helped to improve public health and enable the growth of cities. However, there are limitations to conventional approaches to managing stormwater. Stormwater management paradigms prevalent in the mid-to-late 20th century, for example, resulted in increased flash flooding and polluted runoff flows draining directly to waterways and the oceans (Debo and Reese, 2002). Some of the challenges emerging from current urban water management approaches include the following: inefficient use of resources, degraded ecosystems, water scarcity, and urban flooding.

Additionally, urban water management systems must consider future pressures, which include climate change, rising sea levels, and the continuing growth of global, urban population. For the first time in history more than 50% of the world's population lives in urban settlements, with the urban population projected to grow to 70% of the global population by 2050 (Vimal et al., 2015). A more integrated approach to managing urban water systems is not only feasible but also essential. Yet, the policies, institutions, and practices that affect urban water management are highly variable, worldwide, as demonstrated in scholarship on urban water management typologies (Brown et al., 2008). As cities transition from a focus on just the provision of safe and secure water services to creating water sensitive, resilient and equitable cities, multi-functional infrastructure and integrative urban design will play increasingly important roles in this paradigm change.

The term water sensitive urban design (WSUD) was first used in Australia during the early 1990s, as practitioners started to explore and formalize approaches for more integrated water management (Lloyd et al., 2002). WSUD takes a holistic perspective in managing cost-effective water services in a way that protects public health, while also mitigating the environmental impacts of urban development, and provides for improved community amenity. WSUD aims to minimize the impact of urbanization on the natural water cycle, and its principles can be applied from the scale of a single household up to a whole subdivision (Lloyd et al., 2002).

In Australia, the National Water Commission (2004, p. 30) defined WSUD as follows:

> *The integration of urban planning with the management, protection and conservation of the urban water cycle that ensures urban water management is sensitive to natural hydrological and ecological processes.*

Davies (1996) proposed that, fundamentally, WSUD strives to maintain the water balance and water quality of an urbanized environment in much the same state as prior to urbanization. Wong (2006) highlights the importance of bringing prominence to the management of water within the urban design process, and that WSUD integrates understanding of social and physical systems.

Wong (2006) further identified that the objectives of WSUD include the following:

- Reducing potable water demand through water-efficient appliances and reuse of rainwater and graywater;
- Minimizing wastewater generation and its treatment to a standard suitable for effluent reuse and/or release to receiving waters; and
- Preserving the hydrological regime of catchments.

Implementing WSUD not only varies with the specific constraints and opportunities of each project but can also include rainwater harvesting, rain gardens, bioretention systems, green roofs, pervious pavements, aquifer storage and recovery (ASR), wastewater recycling, gross pollutant traps (GPTs), swales, constructed wetlands, and demand management.

WSUD can also be viewed as a nested concept within ecologically sustainable development, but where the focus is on the interactions between the urban built form and the integrated urban water cycle (Wong, 2006). WSUD takes an integrated perspective where the implications of different technologies are assessed across the whole urban water cycle (Wong, 2006). For example, stormwater harvesting and reuse for irrigation could reduce environmental impacts of stormwater discharge to waterways, replenish groundwater, while also reducing demand for mains water. Fletcher et al. (2015) undertook a comprehensive review of terminology associated with urban stormwater management, which included WSUD, Best Management Practices (BMPs), Integrated Urban Water Management, Low Impact Urban Design and Development (LIUDD), Low Impact Development (LID), Green Infrastructure (GI), and Sustainable Drainage Systems (SUDS). Their review identified that while the concepts are all underpinned by the principles of reducing disturbance to natural hydrology and mitigating the water quality impacts of urbanization, there are subtle differences in the scope and focus of terms (Fletcher et al., 2015).

The objective of this chapter is to review a selection of leading examples of WSUD from around the world, which highlights the reasons for the adoption of WSUD approaches in different development contexts and countries. The review then explores how the WSUD approaches have been implemented including funding models, regulatory implications, management approaches, community engagement, and summary of outcomes. Sharma et al. (2016) highlighted that WSUD approaches are still relatively novel in many contexts when compared with traditional water management approaches. A significant impediment to the widespread and successful adoption of WSUD is the lack of quantitative understanding on how these approaches perform in different development contexts (Roy et al., 2008; Sharma et al., 2012). This lack of performance monitoring and assessment can impede the development of improved guidelines and governance frameworks to achieve sustainable urban water management objectives. The case studies presented in this chapter can help build the knowledge base for explaining the various successes in implementing WSUD in different contexts.

## 27.2 CASE STUDIES

The seven case studies presented in this chapter have been selected as representative examples of WSUD best practice in different countries and development contexts. They showcase a diversity of WSUD approaches that have been implemented for reasons ranging from reducing the demand for imported drinking water to mitigating the impact of urban runoff on an environmentally sensitive waterway. The authors recognize that the list in Table 27.1 is not an exhaustive list of best practice WSUD from around the world. Rather, the case studies selected represent a range of WSUD case studies, based on the authors' knowledge and experience, which can be used to illustrate various lessons to be learnt from previous WSUD implementation.

### 27.2.1 Case study 1: Long Bay, North Shore, Auckland, New Zealand

#### *27.2.1.1 Overview*

Long Bay is situated 30 kms north of the Auckland Central Business District on the northeast coast of New Zealand (Fig. 27.1). The Long Bay development (Fig. 27.2), scheduled to construct over 2500 houses and accommodate around 6000 people, implements many WSUD principles and methods. The primary focus has been on minimizing the impacts of urbanization on the receiving water and optimizing stream corridor linkage to existing downstream parks, while contributing to solving Auckland's housing shortage. The receiving waters include the Long Bay—Okura Marine Reserve and recreational areas that attract up to 25,000 visitors in a day. Most of the land within the development area is within the 2.5-km long Vaughans Stream catchment (360 ha) with a minor proportion in the lower catchment of the adjacent Awaruku Stream. This is a greenfield residential development with a minor commercial center. Housing densities range

**TABLE 27.1 Selected Water Sensitive Urban Design (WSUD) Case Studies From Around the World**

| Name | Location | Main Driver for WSUD | WSUD Approaches |
|---|---|---|---|
| Long Bay | Auckland, New Zealand | Maintain predevelopment hydrology and minimize impacts on local water quality. | Integrated, whole-of-catchment approach that is supported by the following source control measures: rainwater tanks, rain gardens, ponds, and artificial wetlands. |
| Regis Park | Auckland, New Zealand | Prevent further degradation of water quality in the Tamaki Estuary. | Catchment-based perspective to reduce environmental impacts through Low Impact Urban Design and Development approach, which included stormwater pits and rain gardens that feed a drip irrigation system, and on-site waste recycling, which is also used for irrigation. |
| Green Roofs | New York City, USA | Reduce combined sewer overflow. | Green roofs as part of overall approach to reduce runoff while also providing co-benefits such as increased thermal comfort. |
| Imperviousness fee | Germany | Provide equitable and transparent economic incentive to reduce runoff. | WSUD approaches to enable infiltration and reduce runoff connectivity with drainage system. Approaches implemented include rainwater tanks, green roofs, and pervious pavements. |
| Sponge City Construction | Wuhan, China | Address problems with urban flooding, water pollution, and water security. | Depends on city context, such as green roofs, wetlands for stormwater capture and reuse, and permeable pavements. |
| Lochiel Park | Adelaide, Australia | Provide example of best practice sustainable urban development. | Integrated approach that includes rainwater harvesting and reuse for hot water use, bio-retention, and stormwater harvesting and reuse. |
| Sydney Olympic Park | Sydney, Australia | To demonstrate green credentials and community support for hosting "green" Olympic Games. | Locally integrated approach to providing water services, which includes stormwater treatment and harvesting, wastewater recycling, and demand management. |

from large lot residential (>2500 m$^2$) in the catchment headwaters to low and medium-density urban in the mid-catchment and apartments in the village center located in the lower catchment. Stakeholders include the community-based Great Park Society, other Long Bay residents, Auckland Council (local government that provides plan and policy directives, grants consents, and owns and manages the adjoining Regional Park), Todd Property Group (developer), house construction companies, the Auckland public, the Department of Conservation (for management of the Marine Reserve), Long Bay College, and Long Bay Primary School.

### *27.2.1.2 Water sensitive urban design approach*

A holistic design approach has been applied at Long Bay using the catchment as the unit for spatial and infrastructural planning. This has allowed for the layout of the development in terms of urban earthworks requirements; land uses; housing density; impervious surface cover; intermittent stream protection versus piping; vegetation cover, protection, and restoration; public open space protection; and infrastructure (roads and drainage) to be planned to fulfill the objective of hydrological neutrality and minimal water quality effects on receiving waters. The urban form limits the disturbance and land use intensification of ridgelines and retains intermittent streams and terrestrial ecology, in recognition that these are the primary determinants of catchment-wide aquatic ecosystem functionality and health. This urban form is complementary to engineered stormwater management infrastructure that is required to be installed in neighborhoods within the catchment as development progresses. These engineered devices for at-source stormwater control include the following: rainwater tanks, rain gardens, ponds, and artificial wetlands. To achieve the above objectives, the development area has been divided into two parts (Heijs and Kettle Consulting Ltd, 2009), that is, upper Vaughans catchment that drains to Vaughans Stream and

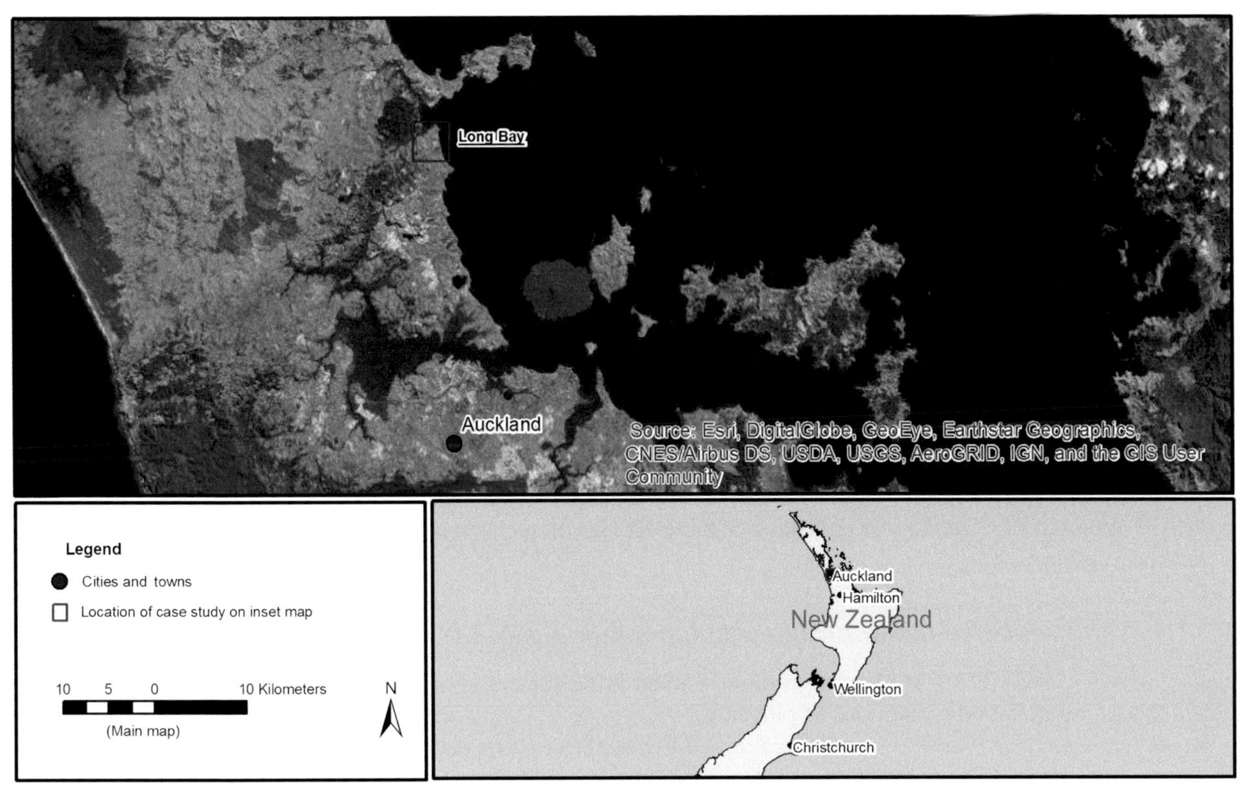


FIGURE 27.1 Case study location map—Long Bay, New Zealand.

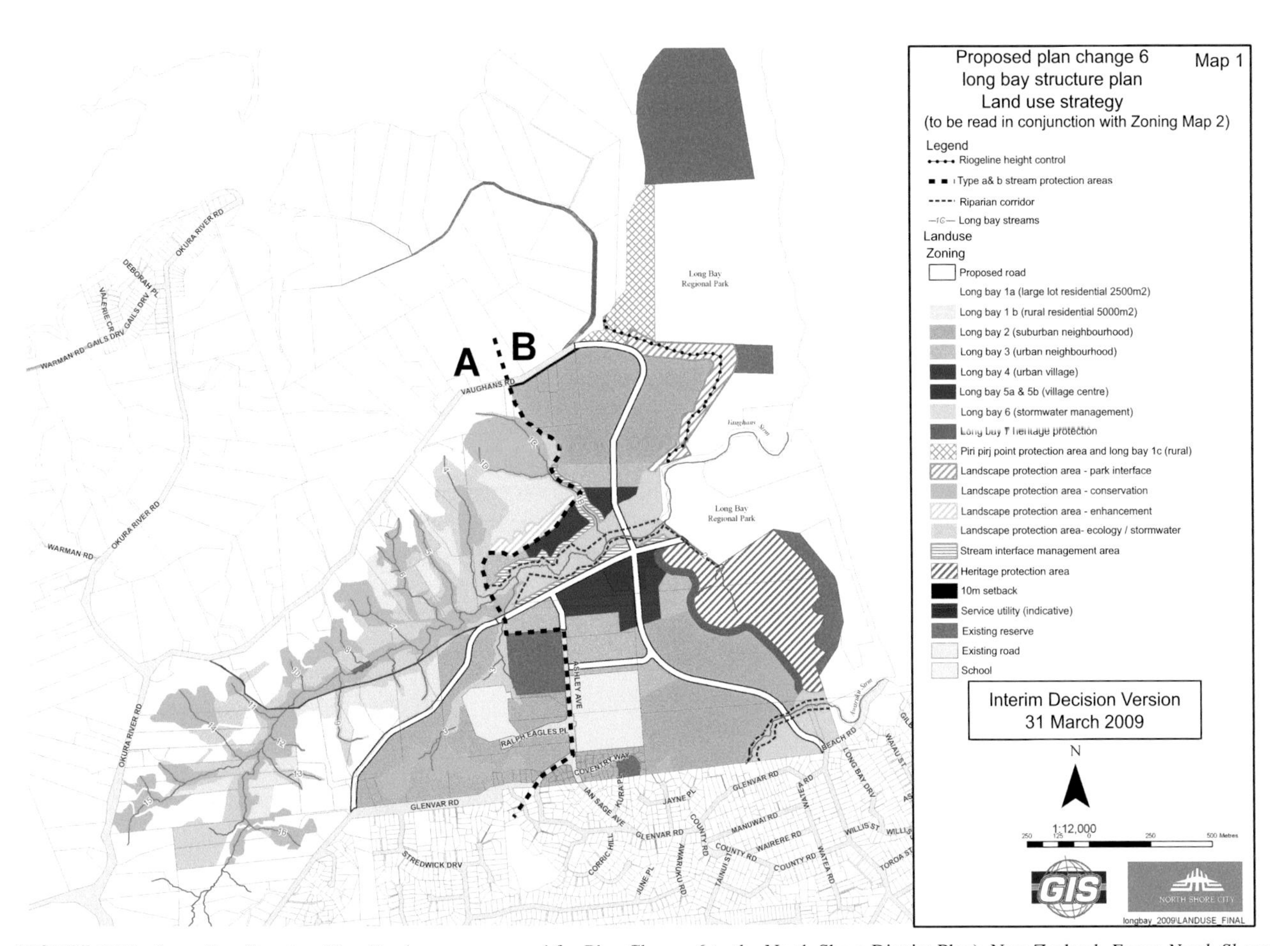


FIGURE 27.2 Long Bay Structure Plan (land use as proposed for Plan Change 6 to the North Shore District Plan), New Zealand. *From: North Shore City Council (2009).*

lower catchments of both Vaughans and Awaruku Streams that drain to the ocean. A higher level of protection has been afforded to the upper catchment. The urban form dictates a very low density of development in the catchment headwaters, with gradual intensification of development and density with progression toward the sea. In the more intensively developed lower catchment neighborhoods, there is a greater tolerance of intermittent stream destruction and more extensive earthworks. Development is concentrated in those parts of the catchment where adverse effects on the natural environment can be minimized.

### 27.2.1.3 Drivers for the adoption of WSUD approaches

The primary driver for the adoption of a WSUD approach at Long Bay was the protection and maintenance of the ecology, hydrological regime, and water quality of streams and the marine receiving waters. Councils have sought to protect Vaughans Stream, which flows through the 111 ha Long Bay Regional Park that is visited annually by more than 1.3 million people for outdoor recreation including swimming, canoeing, and picnics. The near-shore marine environment along the 1-km length of the Regional Park has the status of a "no take" Marine Reserve under the Marine Reserves Act 1971 (New Zealand Government, 1971). This status provides New Zealand's highest level of protection to the marine ecosystem but does not prevent recreational use provided no harm is done to marine life or habitat. The primary driver for the WSUD approach has been for the protection of these ecological, recreational, and esthetic values and assets treasured by Auckland residents.

### 27.2.1.4 Water sensitive urban design implementation process

*Funding arrangements:* The cost of subdivision and urban infrastructure is largely born by a developer contribution to council that is reflected in the sale price of properties.

*Regulatory context and management structure:* An Environment Court decision in 1996 allowed development of the catchment to commence. The design process was carried out under a "Structure Planning" process by the North Shore City Council, which, in 2010, merged with six other councils to form the Auckland Council (population 1.7 million). The Structure Plan was embedded in the North Shore District Plan (North Shore City Council, 2004) giving it Statutory status. Following the formation of Auckland Council, the North Shore District Plan section on Long Bay was incorporated firstly into the Auckland Council District Plan North Shore Section (Auckland Council, 2010a) and then into the Auckland Unitary Plan (Auckland Council, 2016) as part of the Long Bay Precinct.

*Stakeholder engagement and communication:* The original North Shore City Council carried out extensive public consultation, especially with local residents and park users, during development of the Structure Plan from 2000 onward. The Structure Planning process and subsequent changes to the North Shore District Plan provided intermittent public submission opportunities followed by public hearings. The Long Bay Great Park Society expressed strong opposition to any form of urban development at this site. Dissatisfied submitters had the opportunity to lodge appeals with the Environment Court. However, legal costs limited this to the developer and council who were in disagreement over subdivision layout.

*Ongoing arrangements for WSUD maintenance post development:* The ownership and maintenance of streets, drainage systems, and parks are taken over by council post development with ongoing maintenance costs covered by council income from local rates.

*Evaluating success:* As the development is not yet completed (2018), the receiving water outcomes have yet to be determined. The main impediment to implementation was the 15-year process and considerable expenditure by council to defend the Structure Plan through the public consultation and Environment Court processes. Nonetheless, Long Bay has strongly influenced the Auckland Council Water Sensitive Design manual (Auckland Council, 2015) and the Auckland Unitary Plan (Auckland Council, 2016).

### 27.2.1.5 Summary and key findings

Long Bay has integrated land use planning with integrated catchment management planning to help secure the ecological and valued societal resources, both aquatic and terrestrial. It has succeeded like no other New Zealand development in bringing about the integration of Water Sensitive Design and Urban Planning, while drawing on the interplay of diverse actors including the public, stakeholders, professionals, politicians, and the courts. This has been achieved at considerable cost including time spent in justifying an approach that will be the "new normal" in the future developments.

## 27.2.2 Case study 2: Regis Park, Auckland, New Zealand

### *27.2.2.1 Overview*

The Regis Park development provides an example of LIUDD principles which extend beyond WSUD (van Roon, 2011). The primary focus of the Regis Park, greenfield, low-density residential development has been on demonstrating an alternative approach to urban form, infrastructure, and ecosystem enhancement. The overarching subdivision management objective was improving rather than degrading the natural aquatic and terrestrial environments of this previously pastoral landscape. Regis Park occupies five minor subcatchments in the headwaters of the Otara Stream catchment (Fig. 27.3), which is home to the much larger Flat Bush greenfield development. Current stakeholders include the Regis Park Residents Society Incorporated (http://www.regispark.co.nz/pages/about-us.php), Auckland Council, and the residents. Additional stakeholders in the recent past included the now superseded Manukau City and Auckland Regional Councils and Regis Holdings Ltd who designed and constructed Regis Park.

### *27.2.2.2 Water sensitive urban design approach*

The Regis Park low-density development of 66 household allotments (Fig. 27.4) occupies approximately 33 ha which is distributed over five subcatchments. Small, low-flow headwater streams descend from common steep ridgelines and converge downstream of the site. Prior to development, stream gullies were colonized by willow and poplar trees with pasture grass understory. There were negligible wetland vegetation remnants. Surrounding hillsides were pasture. As recommended by the LIUDD concept, a holistic design approach has been applied at Regis Park based on the sub-catchments as the units for spatial and infrastructure planning. This resulted in a layout where roads, driveways, linear infrastructures, and houses dominated the ridgelines, whereas the wide stream valleys were undisturbed by urban earth-works and revegetated with native trees and wetland plants. Road and driveway widths are appropriate to (low) traffic volumes, thereby minimizing impervious surfaces. In 2004, earthworks on the site were carried out along ridgelines, followed by the planting of native tree and wetland seedlings throughout 60% of the site. This vegetation is now several meters tall. At-source stormwater is managed on-site through the following techniques. The steepness of the roads meant that many device types, such as swales, were unsuitable. Street stormwater drains to catchpits, and from there is diverted to

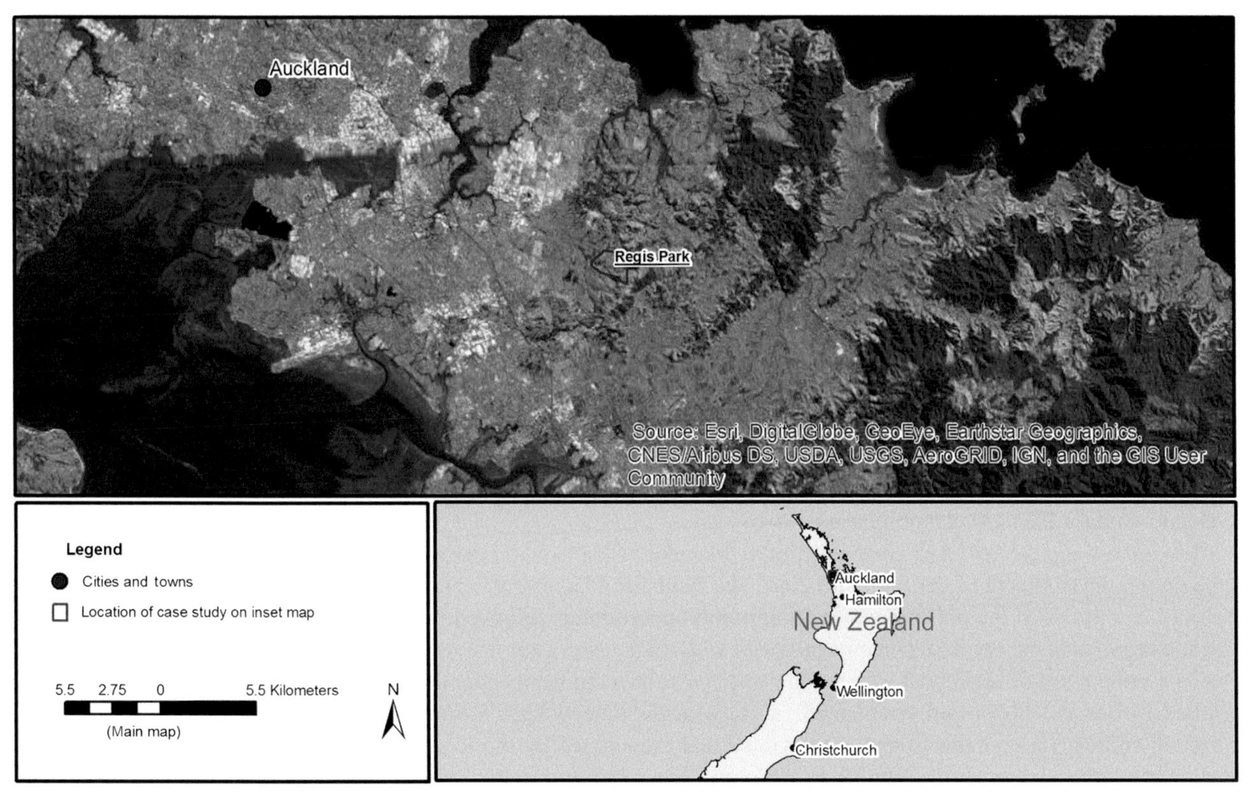


FIGURE 27.3 Case study location map—Regis Park, New Zealand.

FIGURE 27.4 Regis Park low-density development, New Zealand. *Source: Marjorie van Roon.*

a drip irrigation network across revegetated slopes. Stormwater from roofs and paved surfaces on the residential lots flows to a rain garden behind each house. Overflows from the 66 rain gardens are also directed to the drip irrigation network. Sewage is treated within the development using a recirculating packed bed reactor (Scott et al., 2003). The treated effluent from this also goes to drip irrigation. Thus, the urban form and infrastructure at Regis Park minimizes hydrological and water quality effects on receiving waters and promotes biodiversity. As with the Long Bay development, a higher level of protection has been afforded to the upper catchment of Flat Bush where Regis Park is located. Development is concentrated in those parts of the Flat Bush/Otara Stream catchment where the adverse effects on the natural environment can be minimized.

### 27.2.2.3 Drivers for the adoption of WSUD approaches

The primary driver for adoption of a WSUD approach at Regis Park, and indeed the whole of Flat Bush, was to prevent further degradation of water quality in the Tamaki Estuary, which has been subject to the cumulative effects of urbanization over the previous century. Successive local councils have sought to create sustainable and liveable neighborhoods while ensuring that waterways and riparian corridors are protected and enhanced. The unique catchment-based design at Regis Park came about from a partnership between a land developer and a landscape architect who together envisioned a catchment-based development, which promoted aquatic and terrestrial ecosystem enhancement. The Flat Bush Structure Plan (Auckland Council, 2010b) requirements for residential densities and riparian corridor revegetation provided some of the impetus for the Regis Park site design, but these rules have been interpreted very differently in adjacent developments.

### 27.2.2.4 Implementation process for Regis Park

*Funding arrangements:* The cost of subdivision and urban infrastructure was born by the developer and reflected in the sale price of properties with communal space. Roads only were transferred to council following development.

*Regulatory context and management structure:* Extension of the Auckland urban limits to allow urban development of the Otara Stream catchment at Flat Bush was followed by the preparation of both a Structure Plan and an Integrated Catchment Management Plan. Plan preparation was initially led by former Manukau City Council, which was merged into the new Auckland Council in 2010. The Structure Plan was embedded in the Manukau District Plan giving it statutory status. A Resource Consent Application (Scott et al., 2003) under the Resource Management Act 1991 (New Zealand Government, 1991) for Regis Park was granted in 2003. Regis Park is now part of the Flat Bush Precinct under the Auckland Unitary Plan (Auckland Council, 2016).

*Stakeholder engagement and communication:* Manukau City Council carried out public consultation with local residents during development of the Structure Plan. The Structure Planning process, and subsequent changes to the Manukau District Plan, provided intermittent public submission opportunities followed by hearings.

*Ongoing arrangements WSUD maintenance post development:* The ownership and maintenance of streets were taken over by Council post development with maintenance costs covered from council rates. The rain gardens at Regis Park were installed by the developer and maintained by individual homeowners. Stormwater and wastewater infrastructure and revegetation areas on communal land are operated and maintained by the Regis Park Residents Incorporated Society, a legal entity with some similarity to a Body Corporate (See: http://www.regispark.co.nz/pages/about-us.php). For 5 years post development, the developer employed a landscape contractor to care for communal revegetation areas.

*Evaluating success:* One indication of success for the Regis Park development is the health of stream ecosystems. At intervals between 2005 and 2017, stream biotic indices for benthic macroinvertebrate communities have been measured during early summer (Van Roon and Rigold, 2016; van Roon, 2017). Improvements in stream health coincide with the maturation of riparian revegetation zones, although forest weeds remain a problem.

#### 27.2.2.5 Summary and key findings

Regis Park demonstrates that the merging of urban design, urban planning, integrated catchment management, and catchment revegetation can bring about enhancement of aquatic and terrestrial ecology, thereby optimizing the sustainability, esthetics and liveability of low-density residential neighborhoods. This is brought about by a combination of urban layout (clustering of buildings to optimize creation of communal open space) and at-source stormwater management.

### 27.2.3 Case study 3: Green Roofs, New York City, United States

#### 27.2.3.1 Overview

In 2007, the then Mayor of New York City (NYC), Michael Bloomberg, announced a vision of making most of New York's rooftops green by the year 2050 as part of the PlaNYC Strategic Plan (Bloomberg and Holloway, 2010). This was part of an overarching Green Infrastructure Plan, which has the purpose of reducing combined sewer overflows, as well as providing other co-benefits such as amenity and mitigation of urban heat island effect. The discharge of untreated sewage (combined wastewater and stormwater) during high-rainfall events has contributed to poor water quality in the receiving waterways and harbor. The main 20-year objective of NYC Green Infrastructure Plan is to capture the first 25 mm of rainfall on 10% of the impervious areas in combined sewer catchments using detention or retention methods (Bloomberg and Holloway, 2010). This plan is supported by funding and tax incentives (described below), which have increased the adoption of green roofs in NYC. Fig. 27.5 depicts the location of green roofs in NYC, which have been registered under the Department of Environment's Green Infrastructure database.

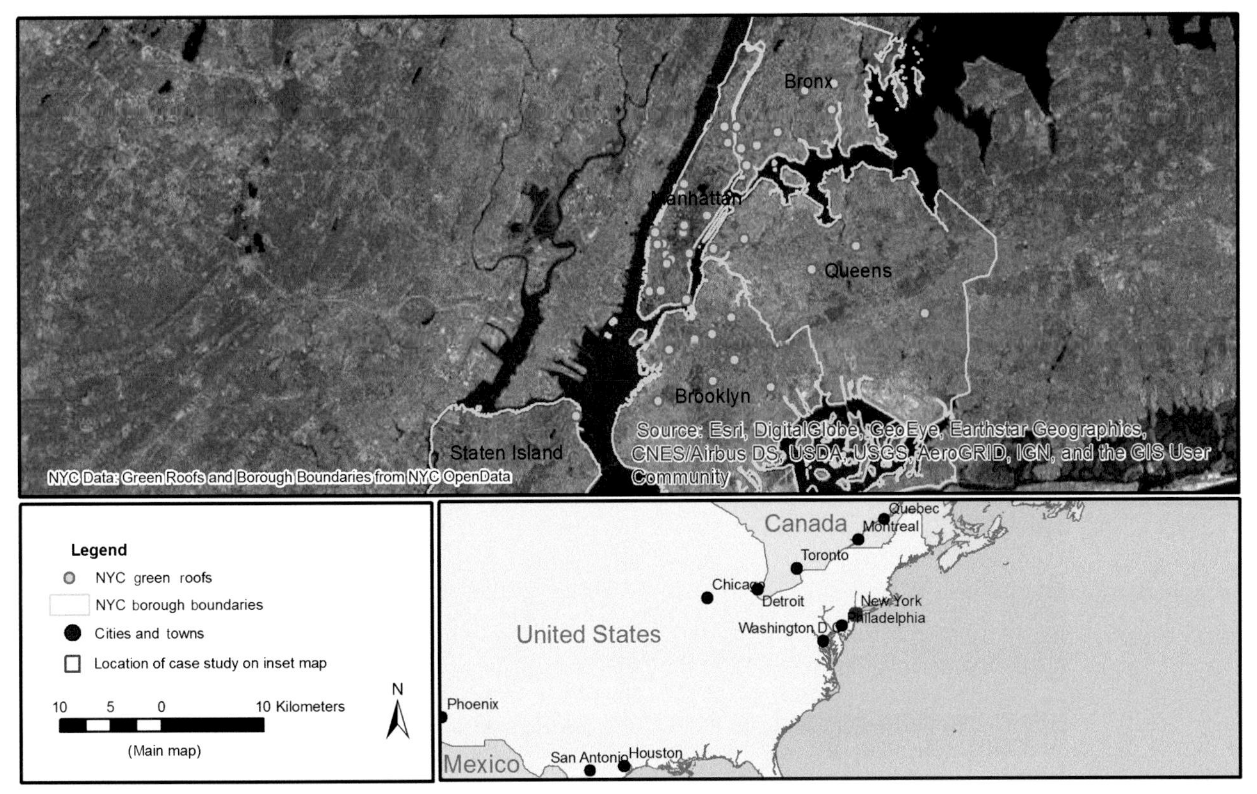


FIGURE 27.5 Location of Green Roofs, New York City, United States. *Source: NYC Green Infrastructure Initiative, NYC OpenData.*

### 27.2.3.2 Water sensitive urban design approach

Examples of green roofs in NYC include Sky Island, which is located on the 35-storey Visionaire apartment building in Battery Park City, NYC (Fig. 27.6). The apartment building was designed to achieve a platinum green building rating under the Leadership in Energy and Environmental Design (LEED), which is highest rating level under the scheme (https://new.usgbc.org/leed). The approach facilitated urban agriculture on a rooftop terrace and included sustainability outcomes such as increased stormwater retention and graywater recycling (Mark K. Morrison Landscape Architecture, 2017).

Green roofs have also been installed at Columbia University (Fig. 27.7) to provide environmental benefits, including mitigating the urban heat island effect in warmer months, improving building insulation to reduce energy for heating and

**FIGURE 27.6** Green roof at the Visionaire apartment building, New York City. *Mark K., 2017. Morrison Landscape Architecture, Green Roof Design, Undated, From: http://www.markkmorrison.com/greenroof/.*

**FIGURE 27.7** Columbia University, New York green roof research station. *From: Gaffin, S., Khanbilvardi, R., Rosenzweig, C., 2009. Development of a green roof environmental monitoring and meteorological network in New York city. Sensors 9 (4), 2647.*

cooling, and reducing stormwater runoff. Columbia University has installed more than 1500 $m^2$ of green roofs, which are estimated to reduce stormwater runoff by around 1.6 ML per year (Rugh, 2013).

### 27.2.3.3 Drivers for water sensitive urban design adoption

The key driver for the adoption of green infrastructure in NYC is the need to alleviate pressure on the combined sewer system (CSS). The combined sewers cannot cope with stormwater inflows during high-rainfall events, which results in sewer overflows, with untreated wastewater polluting the city's waterways (Rugh, 2013). At the city scale, green roofs are intended to reduce the peak flow during rainfall events by detaining and infiltrating precipitation. It is also thought that green roofs can provide filtration of runoff before it is discharged to waterways, improve local biodiversity, and also provide a more esthetic landscape and a recreational space for residents (NYC Mayor's Office of Sustainability, 2013).

There are also other, more local-scale drivers for the adoption of green roofs. At Colombia University the drivers for green roofs included the following (Rugh, 2013):

- *Improved building thermal performance*—Green roofs can improve building thermal performance, reducing the energy required to heat or cool buildings. Green roofs can reduce the energy required for cooling during summer by up to 40% (Spala et al., 2008).
- *Moderating urban heat island effect*—Green roofs can mitigate the heat effect of an urban environment; green roofs keep surface temperatures close to ambient air temperatures, compared with that of concrete and bituminized surfaces which absorb solar energy and reemit it as sensible heat (Gaffin et al., 2009). See Chapters 19–21 for more details on the Urban Heat Island Effect and mitigation strategies including green roofs and other green infrastructure.

### 27.2.3.4 Water sensitive urban design implementation process

*Funding arrangements:* The NYC government created incentives to encourage the addition of green roofs in the city. The Green Roof Tax Abatement program provides a 1-year tax relief of $4.50 per square foot ($48.45/$m^2$) of green roof installed (up to $100,000 or the building's tax liability, whichever is the lesser) (NYC Mayor's Office of Sustainability, 2013).

*Regulatory context and management structure:* Green roofs are encouraged in NYC through the PlaNYC strategic plan and one-off tax relief. This commitment to green buildings has resulted in Local Law 86 (2005) that requires buildings in NYC that receive City funding to achieve an LEED rating level of Certified or Silver. This rating typically includes a 30% reduction in potable water use compared with similar uncertified building types and a stormwater management plan that captures and treats 90% of runoff using BMPs. Green roofs are typically managed as part of the building's overall facilities management.

*Ongoing arrangements WSUD maintenance post development:* Green roofs require regular inspection to ensure that waterproof membranes are not leaking to avoid damaging the underlying building structure. The green roof vegetation will also require regular maintenance, which will include weed control and possibly irrigation during dry periods (Rugh, 2013).

*Evaluating success:* Colombia University has an extensive monitoring program for three green roof installations in NYC. The monitoring program aims to quantify the hydrological impact of adding green roofs to reduce runoff in a highly urbanized area. The results have shown that the three green roofs detained 36%, 47%, and 61% of total annual rainfall of 1300 mm (Carson et al., 2013). However, rainfall detention declined with increasing intensity of the rainfall events, whereas the hydrological effectiveness of the green roofs was strongly influenced by the depth of the pervious matrix supporting the vegetation.

### 27.2.3.5 Summary and key findings

Green roofs in NYC have been encouraged by the City government as an approach to reduce the water quality impacts from CSS overflows. The installation of green roofs, which is part of an overall green infrastructure strategy, has been supported by government building laws and tax relief that helps to offset the green roof installation costs. Monitoring has shown that green roofs can provide a significant reduction in roof runoff, but the effectiveness is influenced by the depth of the pervious storage matrix, and by antecedent soil moisture conditions. In a heavily urbanized area, like NYC, green roofs offer a range of other benefits to building owners and the community, which include urban heat mitigation, enhancing local biodiversity, neighborhood beautification, and enhanced opportunities for recreation and urban agriculture.

## 27.2.4 Case study 4: imperviousness fee in Germany

### 27.2.4.1 Overview

In many German cities, an imperviousness fee has been introduced on properties to reflect the volume of stormwater generated on the property that is discharged to the CSS or drainage system. The CSS, which is common throughout southern Germany, but less prevalent in north of the country, collects both runoff and wastewater in the same pipe, which is then transported to the wastewater treatment plant for treatment prior to discharge to a receiving waterbody. Prior to the introduction of the imperviousness fee, the wastewater fee was based on properties' water demand. However, a series of federal and state court rulings, starting in the 1970s, and gaining momentum in the early 2000s, deemed this practice unfair, as it did not reflect the actual amount of stormwater running off a property. It was decided there was the need for increased transparency and equitable rate structure for stormwater services (Buehler et al., 2011). The fee is calculated based on the impervious area that is connected and the likelihood of runoff entering the CSS or drainage discharge system. Hence, by minimizing connection to the discharge system, households can minimize the fee. This has encouraged the adoption of green infrastructure (Zhang et al., 2017). It is asserted that the introduction of Individual Parcel Assessments for estimating stormwater contributions will result in a user-pays approach that will drive more efficient stormwater practices at the property level (Keeley, 2007).

The introduction of the fee was part of a larger shift in Germany toward LID and green infrastructure. Since the introduction of the fee, a decrease in stormwater runoff has been observed in the cities reported in this case study. At least part of his decrease can be attributed to the imperviousness fee, together with other regulations such as mandatory green roofs and requirements for runoff infiltration systems in new buildings (Fehring, 2012; DDV, 2016).[1]

### 27.2.4.2 Water sensitive urban design approach

A range of WSUD approaches have been adopted by property owners in Germany to achieve source control of stormwater. Installation of rainwater harvesting systems by households has been popular. Herrmann and Schmida (2000) reported that there are more than 100 commercial manufactures of rainwater tanks in Germany, with the leading manufacturer of prefabricated concrete tanks installing more than 100,000 in Germany over a 10-year period. The majority of these tanks are used for nonpotable purposes (Zhang et al., 2017). The popularity of rainwater harvesting systems has also resulted in economic benefits with more than 4000 jobs created across Germany, as well as saving an estimated 75 GL of drinking water per year (FAZ, 2006).

Green infrastructure approaches have also been widely adopted in Germany, such as the replacement of paved car parks with pervious pavements. Germany is considered to be a world leader in the adoption of green roofs (Vijayaraghavan, 2016), with 10% of its buildings estimated to have installed green roofs (Saadatian et al., 2013). It is estimated that more than 10,000 ha of roofs will be "greened" by 2017 (Ansel, 2017). The direct infiltration of stormwater is now encouraged near the source, where permission is generally not required for the infiltration of mildly polluted stormwater (Nickel et al., 2014). However, where the stormwater is from large areas or from roads with high traffic flows, some municipalities require permits and pretreatment prior to discharge (Nickel et al., 2014; Niederschlagswasserfreistellungsverordnung, 2014). The imperviousness fee has encouraged the adoption of WSUD approaches as both public and private land owners seek to minimize their imperviousness fee, while also achieving flood mitigation and ecological benefits.

### 27.2.4.3 Drivers for the adoption of WSUD approaches

In Germany, a primary driver for the adoption of the imperviousness fee was to implement a more equitable and transparent approach to stormwater charges. Additional benefits are the reduction of sewer overflows during high-rainfall events and reduction of the costs of treating wastewater volume that can be attributed to runoff. CSSs were traditionally used, there has now been a shift toward separate stormwater and wastewater systems in new developments (Brombach and Dettmar 2016). Furthermore, with 72% of Germany's water supply being from groundwater and spring water, the replenishment of groundwater sources from stormwater infiltration is important (UBA, 1997; Umweltbundesamt, 1998).

These WSUD initiatives were further supported by European environmental policy, which set ambitious goals through its Water Framework Directive released in 2000. This directive promoted more integrated and user/polluter-pays approaches to all aspect of water management so that water service fees were reflective of the true cost of providing services (European Commission, 2016).

1. This review, in addition to literature, was informed by interviews conducted with staff at Stadtentwaesserung Dresden and Stadtwerke München.

While the initial driver for the introduction of the fee was of a legal nature, it became recognized there were a range of co-benefits that followed on from the financial incentives to reduce urban imperviousness and stormwater runoff. These co-benefits include heat island mitigation and reduced energy demand for building heating and cooling (Saadatian et al., 2013), groundwater replenishment (Aevermann and Schmude, 2015), reduced noise and air pollution (Vijayaraghavan, 2016), and flood mitigation and reduced potable water demand (Zhang et al., 2017). For example, in the City of Hamburg, the imperviousness fee was introduced in parallel with a major project addressing integrated stormwater management that aimed to restore the natural water cycle, reduce water pollution, and achieve flood mitigation (Bertram et al., 2017). The combined projects resulted in the increased prominence and perceived value of stormwater management in Hamburg, as well as synergies in data collection needed for both projects.

German property owners are increasingly aware and supportive of the benefits of reducing runoff through WSUD approaches. As stated above, rainwater tanks are installed in homes to reduce potable water demand while also providing environmental benefits (Herrmann and Schmida, 2000). The uptake of WSUD approaches such as green roofs is being encouraged by a range of grant programs and regulations in different cities. In Munich, for example, nearly 20% of all roofs are now green (2 $m^2$ per citizen or 300 ha in total), which can be attributed to a regulation, in place since the 1990s, that all new buildings with flat roofs over 100 $m^2$ must be green (Ansel and Appl, 2011). The initial adoption of green infrastructure was also encouraged with subsidies. For example, the Berlin Senate provided direct subsidies to increase green areas and reduce the coverage of impervious area in the dense inner city using technologies such as green roofs (Nickel et al., 2014). This program resulted in more than 65,000 $m^2$ of green roofs and approximately 740,000 $m^2$ of green courtyards and facades, mainly on private properties, at a cost of €16.5 million (Nickel et al., 2014). While the program has now been discontinued, there are continuing programs to encourage the adoption of green roofs, particularly on public buildings (Ansel, 2017).

### *27.2.4.4 Water sensitive urban design implementation process*

*Funding arrangements:* The mechanism for determining how the imperviousness fee is calculated and levied is set by the responsible municipal government and, therefore, varies between German cities. A review of stormwater charges for the most populous cities in Germany showed that on average €1.10 was charged for every $m^2$ connected to the municipal sewerage system, which ranged from €0.66 per $m^2$ in Stuttgart to €1.825 per $m^2$ in Berlin (Oelmann et al., 2014). There are two main approaches for estimating the fee. The first approach uses the assumed impervious area based on the building type and lot area. This approach is efficient to administer and does not require site-specific data. If the property owner disagrees on the impervious area assessment, it is their responsibility to provide the evidence to support a reduced fee. The other approach is based on measurement of the actual impervious area for each property and its connection to drainage system. This approach uses high-resolution aerial imagery together with other data such as building permits. This approach is more data intensive, but more accurately reflects the actual impervious area connected to the drainage system. It provides an economic incentive for adoption of WSUD to reduce impervious area and engenders greater community confidence that the fee has been accurately calculated (Fehring, 2012).

*Regulatory context and management structure:* Stormwater in Germany is managed at the local municipality level, thereby allowing implementation of the fee to be determined locally. Some smaller administrative bodies (e.g., councils of small townships) were concerned about the additional administrative and cost burdens in calculating an accurate imperviousness fee for each property, with some pushing back against the court decisions that mandated the implementation of the fee. However, they were eventually overruled by the State and Federal Governments. Nonetheless, it has been shown that calculating fees for stormwater discharge by measurement at the property level has the highest implementation and administration costs when compared with other forms of stormwater funding such as flat fees or government grants (Keeley, 2007).

*Stakeholder communication and engagement:* The successful introduction of the imperviousness fee relied on good communication with the community. This was achieved through a number of means including public information sessions, telephone helplines, brochures[2], and media campaigns (Thimet, 2014). Elected officials also requested information about the fee and its benefits. A survey in the State of Baden-Württemberg (in southern Germany) found that when elected council officials were first confronted with the task of implementing the fee, more than 80% indicated they needed more information (Fehring, 2012).

2. Examples for detailed brochures used to inform the community about the different options to minimize connected imperviousness (in German):Munich: https://www.muenchen.de/rathaus/dam/jcr:3bc23f13-0e76-43e4-8669-c568c1db3677/Brosch%C3%BCre_Regenwasser_versickern.pdf Dresden: https://www.stadtentwaesserung-dresden.de/fileadmin/shared/user_upload/pdf/broschueren/Ratgeber-Regenwasser.pdf.

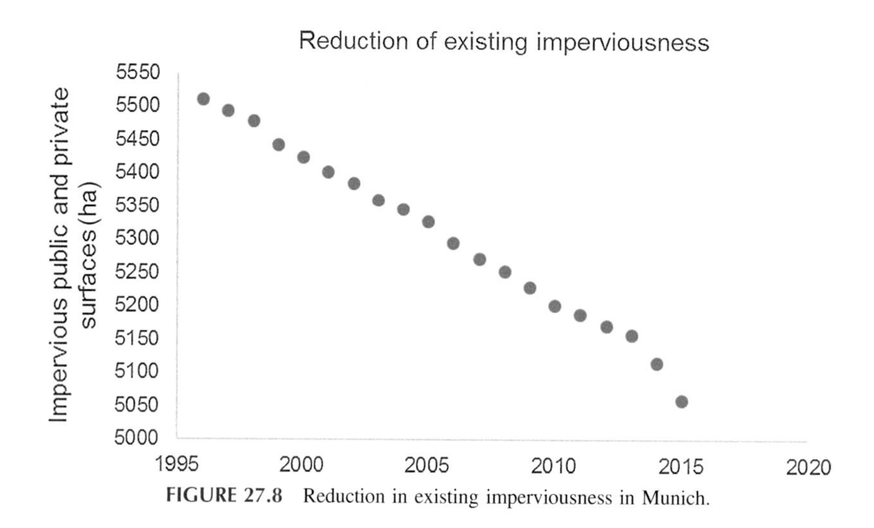

FIGURE 27.8 Reduction in existing imperviousness in Munich.

*Evaluating success:* Since the introduction of the fee, a reduction in imperviousness has been observed in many cities; for example, in Munich, where between 1997 and 2015, 4.5 $km^2$ of impervious surfaces have been disconnected from the drainage system (Fig. 27.8). This avoided approximately 3000 ML of runoff into the discharge system per year. The impact of the imperviousness fee in reducing runoff through WSUD has also been observed in a number of other cities, such as Berlin, Cottbus, Giessen, and in the State of Baden-Württemberg (Keeley, 2007).

#### *27.2.4.5 Summary and key findings*

The introduction of the imperviousness fee in Germany has been part of a broader shift to more integrated management of stormwater and the adoption of WSUD approaches, such as rainwater tanks and green roofs, has resulted in a reduction in the impervious area and runoff flowing to combined sewers. An additional benefit of the imperviousness fee is that the data gathered to calculate the fee are used in other important applications. For example, in Hamburg, new local stormwater management and flood protection measures are developed based on this information (Bertram et al., 2017). The fee has also resulted in greater awareness and involvement of the urban community in managing runoff as there is an economic incentive for property owners to reduce flows from their land.

### 27.2.5 Case study 5: Sponge City Construction, Wuhan, China

#### *27.2.5.1 Overview*

The Sponge City Construction (SCC) initiative was launched in China in 2015 (Li et al., 2017). The SCC initiative was designed to address the impacts of rapid urbanization in China, particularly flooding, degraded local water quality, and water shortages (Li et al., 2017; Wang et al., 2017). The principles of Sponge Cities incorporate elements of LID and Green Infrastructure from the United States, WSUD from Australia, and SUDS from the United Kingdom (Li et al., 2017). The SCC initiative selected 30 cities as pilot cities to demonstrate SCC approaches, which included Wuhan City. Chapter 1 provides a more detailed account of the Sponge City initiative in China.

Wuhan is the capital of the Hubei province in Central China, with a population of around 8.5 million. The city is located on a floodplain at the intersection of the Yangtze and Han rivers (Dai et al., 2017) (Fig. 27.9). Wuhan is renowned for its abundant water resources, which have allowed the development of a significant industrial sector. However, urbanization and industrial activities have also resulted in increased flooding and degraded water quality (Dai et al., 2017).

The hydrology problems faced in Wuhan are typical of the issues faced in many Chinese cities undergoing rapid urbanization. While economic progress in China has resulted in a rapid increase in the standard of living, it is now recognized that this has been accompanied by environmental degradation and increased public health risks (Zhang et al., 2010). The problems associated with urbanization and the increase in impervious areas include the following reduced groundwater recharge, overextraction of groundwater, waterway degradation, and urban flooding. These problems cannot be satisfactorily addressed by traditional stormwater management approaches (Workman, 2017). Indeed, these traditional approaches may exacerbate the problems. Instead, SCC encourages the widespread adoption of permeable surfaces and green infrastructure to help address problems by trying to recreate the natural hydrology in urban areas.

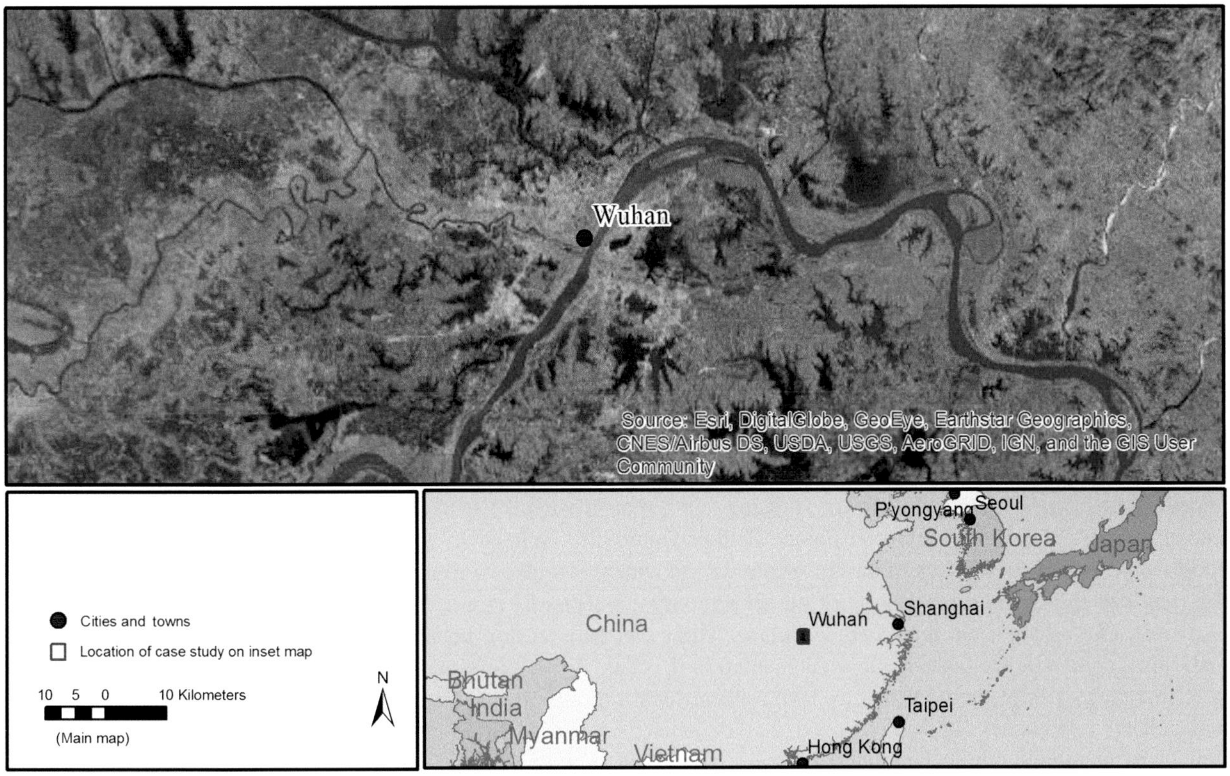

FIGURE 27.9 Case study location map—Wuhan, China.

### *27.2.5.2 Water sensitive urban design approach*

The approach taken to implement SCC depends on the context of each specific city and the water management problems they face. China's Ministry of Housing and Urban–Rural Construction issued draft technical guidelines for the implementation of SCC in 2014, which recommended potential approaches that can be applied depending on the local context (Jia et al., 2017). The targets for reductions in annual runoff and rainwater reuse are determined for each city depending on local environment and climate. WSUD approaches are targeted at frequent rainfall events, with smaller runoff volumes, to increase infiltration of precipitation and recharge of groundwater (Jia et al., 2017). Cities aim to reduce runoff through approaches such as green roofs, wetlands for stormwater capture and reuse, and permeable pavements that allow infiltration of runoff to groundwater (Dai et al., 2017). The objectives are to reduce flood risk, recycle runoff to improve water security, and recharge groundwater. However, Workman (2017) has highlighted it can be difficult to balance these competing priorities.

The first SCC example in Wuhan was built on the site of a reclaimed rubbish dump, which covered 3000 ha. The sponge city concepts implemented included pervious pavement, rainwater gardens, and rainwater harvesting for irrigation (Dai et al., 2017). Rainwater harvesting for irrigation is estimated to save more than US$200,000 per year, and 600 people who had previously been displaced by the rubbish dump have now moved back to the area surrounding the new park (Dai et al., 2017). More than 400 SCC projects have been designed and launched in Qingshan District, Wuhan City. The target of SCC projects in Wuhan is to better manage storm events with a return period of less than 50 years, to reduce annual runoff by 70% and improve local water quality. The Gangcheng School, located in Qingshan District, provides an example of SCC project in Wuhan City. This school, which is located in a low-lying area, has used pervious pavements and underground stormwater storage to reduce flooding.

### *27.2.5.3 Drivers for the adoption of WSUD approaches*

The primary drivers for adoption of the SCC were to address the worsening problems of urban flooding, water pollution, and water security. In Wuhan, climate change and further urbanization are projected to exacerbate existing flood risks (Dai et al., 2017). The seriousness of the flood risk was demonstrated in 2016, when more than 30 people were killed in an urban flood (Dai et al., 2017). Water pollution is also a critical issue for Wuhan as more than one-third of the rivers exceed prescribed water quality standards (Dai et al., 2017). Water pollution is caused by partially treated domestic and industrial wastewater being discharged to the river together with pollutants from contaminated road surfaces and agriculture.

#### 27.2.5.4 Implementation process for sponge cities

*Funding arrangements:* Initial government funding of between US$50−100 million was provided to the first 16 sponge cities, to "kick-start" the program of works (Workman, 2017). This was only 20% of the estimated budget, with the remainder coming from the city government using approaches such as private−public partnerships (PPP). There are questions around how the PPP funding model will work for funding green infrastructure when there is no obvious way to recoup costs (Workman, 2017). In Wuhan, a total of US$2.4 billion is needed to fund the sponge cities projects. It is expected that only US$230 million will come from central government, with the remainder made up of funds from PPPs and long-term loans, together with contributions from municipal and district governments (Dai et al., 2017).

*Regulatory context and management structure:* The SCC policy was launched by China's central government, with the scheme managed and guided by the following ministries: Ministry of Housing and Urban-Regional Development, Ministry for Water Resources, and Ministry of Finance (Li et al., 2017). While these ministries help to select cities and guide the process, the on-ground implementation is managed at the city level (Li et al., 2017). The implementation is supported by guidelines developed by the Ministry of Housing and Urban-Regional Development. However, local guidelines are also needed that reflect priorities and constraints of the specific city context. In many cases, it was difficult to achieve the ambitious targets given the lack of design standards and codes, as well as a shortage of appropriately trained professionals. This can lead to poor implementation of the sponge city concepts (Li et al., 2017).

*Stakeholder engagement and communication:* There can be a tension between the top−down nature of sponge cities goals and targets with the bottom−up implementation (Li et al., 2017; Workman, 2017). The ambitious national targets can be challenging to implement at the city level, where local ordinances and practices tend to favor traditional stormwater management approaches. It is recognized that there is the need to build stakeholder ownership and engagement in implementing sponge city to build trust and raise awareness of risks (Dai et al., 2017).

*Ongoing arrangements WSUD maintenance post development:* Compared with traditional stormwater management, the measures implemented as part of a sponge city will require more frequent and different types of maintenance practices (Li et al., 2017). The maintenance tasks become more complicated when assets are distributed over a large area, particularly if they are located on private land, as many of them are (Li et al., 2017).

*Evaluating success:* Wuhan is one of the first 16 pilot cities implementing sponge city concepts. Initial results are encouraging, reporting reduced runoff and provision of an alternative irrigation water source. However, as the program is still a "work in progress" it is too early to evaluate success in achieving the stated sponge city objectives, such as mitigating flood risk and reducing local water pollution.

#### 27.2.5.5 Summary and key findings

SCC is an ambitious program of work to address current challenges of stormwater management in China's cities. This grand vision from the central Chinese government has been implemented to address many of the critical environmental challenges that face China's cities after decades of unconstrained urban growth with the hope that SCC can help to restore ecosystems, improve local water quality, reduce flooding, and provide alternative water sources. However, there are a number of challenges in achieving this vision. Thus far, it has been difficult to attract private sector interest in augmenting government funding. Although some cities have attracted private funds for SCC projects, it has not met the shortfall, so there is the need to consider other innovative funding models for SCC projects. In addition, there has been conflict between the central government SCC policies and municipal stormwater policies. Another barrier is the need to build the technical and administrative capabilities to successfully implement policies and regulations associated with sponge cities.

### 27.2.6 Case study 6: Lochiel Park, Adelaide, Australia

#### 27.2.6.1 Overview

The Lochiel Park residential development is located approximately 10 km from the Adelaide CBD in South Australia (Fig. 27.10). The 15 ha development was designed to showcase sustainable living for medium-density urban development

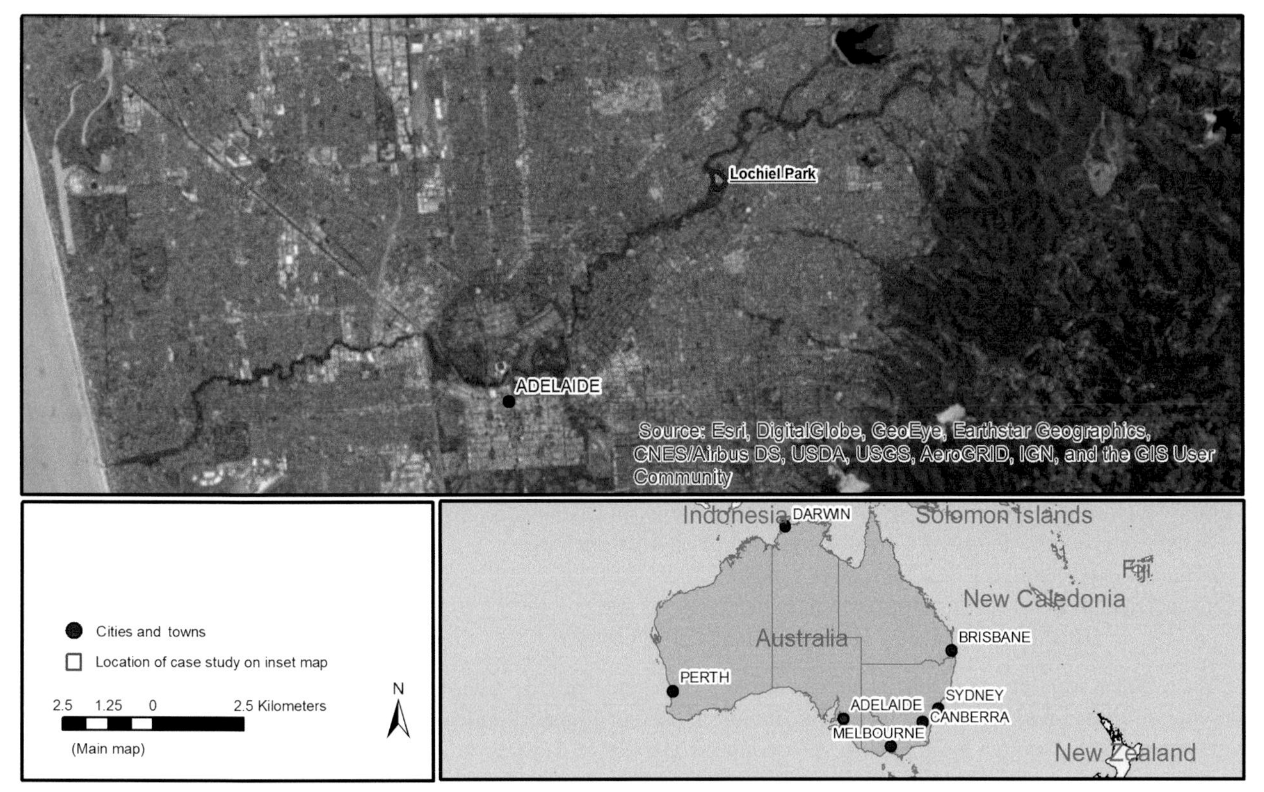


**FIGURE 27.10** Case study location map—Lochiel Park, South Australia, Australia.

(Edwards and Pocock, 2009) and provide a nationally significant example of Ecologically Sustainable Development to encourage its adoption as a mainstream practice (Blaess et al., 2007). Renewal SA (the state government land developer) was tasked with delivering the vision for Lochiel Park. The development with 106 medium-density dwellings is now largely completed.

### *27.2.6.2 Water sensitive urban design approach*

The WSUD approach was guided by the high-level sustainability objectives that were set for Lochiel Park. These objectives were formulated based on benchmarking against international and national leading sustainable developments (Berry et al., 2013). The final objectives were based on reducing resource consumption compared to the average Adelaide home (2004 levels). In particular,

- 66% reduction in energy used;
- 74% reduction in greenhouse gas emissions; and
- 78% reduction in potable water use.

To achieve the water savings target, a range of initiatives were implemented, including the following:

- Demand management;
- Rainwater tanks for hot water supply; and
- The use of stormwater harvesting and ASR for nonpotable uses.

Renewal SA developed a Development Masterplan and Urban Design Guidelines to guide the development of Lochiel Park and ensure that all services were aligned with sustainability principles, including WSUD. The Masterplan specified the delivery of community level infrastructure, whereas the design guidelines specified the sustainability principles and practices to be incorporated in individual homes (Berry et al., 2013).

The stormwater harvesting scheme collected water from the surrounding catchment, which was directed to a wetland for initial treatment, before being injected to an aquifer for storage and further treatment, and then extracted and distributed

FIGURE 27.11 Lochiel Park streetscape with bioretention pit and Southern Wetland. *Source: Stephen Cook, CSIRO.*

for nonpotable uses via a dual reticulation network. The Lochiel Park wetlands had the following design objectives (Ecological Engineering, 2006):

- water from the wetland is of a suitable quality for aquifer injection;
- pollutant export to the River Torrens is reduced;
- no increase in flood risk to properties in the catchment; and
- provide an attractive, community accessible, landscape feature.

The wetland was designed to reduce stormwater pollutant loads by filtering runoff through shallow vegetation and allowing sufficient residence time for sedimentation and other biogeochemical processes. Prior to entering the wetland, stormwater passed through a GPT using a Continuous Deflective Separation system (Cook et al., 2003). The GPT was designed to remove debris, large sediment (>5 mm), and oil and grease prior to runoff reaching the wetland (Ecological Engineering, 2006).

Bioretention pits and swales are used at Lochiel Park to treat stormwater before it reaches the wetland and provide a landscape feature. Fig. 27.11 shows a bioretention pit at Lochiel Park. 250 $m^2$ of bioretention systems were installed in the 4.3 ha of urban area at Lochiel Park.

### 27.2.6.3 Drivers for the adoption of WSUD approaches

The purpose behind Lochiel Park was to provide a leading example of sustainable urban development in South Australia. The development set ambitious objectives for sustainability performance relative to average Adelaide households and included the 78% reduction in mains water use. Garnaut (2008) highlighted that a development such as Lochiel Park provides a credible example to the development and construction industries on the practical ways to deliver more ecologically sustainable developments. In addition, Lochiel Park should provide an incubator for research that can assist in setting targets and developing guidelines for WSUD development, which will move urban development in South Australia well beyond the *business as usual* approach (Garnaut, 2008). To ensure the capture and dissemination of knowledge developed at Lochiel Park, such as rainwater harvesting for hot water supply, there are ongoing monitoring projects to quantify delivery of the WSUD and other sustainability features specified in the development objectives (see: http://www.unisa.edu.au/lochiel-park).

### 27.2.6.4 Implementation process for Lochiel Park

*Funding arrangements:* The capital costs of constructing the WSUD elements at Lochiel Park were provided by Renewal SA, with ongoing operation and maintenance to be covered by local government. Sharma et al. (2016) highlighted that budgets for ongoing operating and maintenance of WSUD assets are often not considered in the planning stage. This uncertainty in these costs often results in a reluctance by local government to assume responsibility for constructed WSUD features.

*Regulatory context and management structure:* Many of the WSUD features at Lochiel Park went well beyond the existing regulatory requirements for water sensitive development. For example, Berry et al. (2013) reported that the success of solar-heated rainwater for hot water demand gave the state government the policy confidence to establish sustainability requirements that went well beyond current industry norms and regulatory standards.

*Stakeholder engagement and communication:* Renewal SA invested considerable effort in facilitating the development of a cohesive community, which included urban design to encourage community interaction via medium-density dwellings and public open space, a community website, and community-based groups (Edwards and Pocock, 2009). The research found that the sustainability ethos helped develop a sense of community as Lochiel Park residents came to terms with the unconventional systems (Edwards and Pocock, 2009). A survey of Lochiel Park residents also showed that many were attracted by the sustainability features such as WSUD, along with its location and local amenities (Edwards and Pocock, 2009).

*Ongoing arrangements WSUD maintenance post development:* The ongoing maintenance for public open space has now been handed over from the developer (Renewal SA) to the Campbelltown City Council. The operation and maintenance of the stormwater harvesting and recycling scheme has been handed over to SA Water (the state's water service provider). As part of this role, SA Water conducts final inspections of all plumbing work to ensure there are no cross connections between the potable and nonpotable supply networks.

*Evaluating success:* Chao et al. (2015) undertook a comprehensive postoccupancy monitoring program to determine actual residential water and energy usage and found that Lochiel Park households used on average 36% less mains water than the average Adelaide household. This reduction did not reach the stated target of 78% reduction of average household water use as the stormwater harvesting and recycling scheme was not operational at the time of analysis.

The recycled water system was delayed by a number of years due to installation issues, as well as an extended period of testing to validate the quality of the water that would be delivered. Residents had to use drinking water for nonpotable purposes (toilet, laundry, and garden use), much to the frustration of some of the residents (Sharma et al., 2016).

#### 27.2.6.5 Summary and key findings

Lochiel Park was designed to push the boundaries on sustainable development and included a number of innovative WSUD approaches to achieve the targeted 78% reduction in drinking water use compared with the average Adelaide home. Delays in the recycled water scheme becoming operational meant that this target was initially not met. The delays, due to a combination of installation and water quality validation issues, highlighted the unexpected barriers that can be faced when moving beyond current, well-understood urban water supply practices. It is likely that subsequent developments will benefit substantially from the well-documented experiences of Lochiel Park.

### 27.2.7 Case study 7: Olympic Park, Sydney, Australia

#### 27.2.7.1 Overview

Sydney Olympic Park is located 16 km from the Sydney CBD in New South Wales, Australia. It is situated in the City of Parramatta, which is the heart of the rapidly developing western suburbs of Sydney (Fig. 27.12). The area was initially an industrial area that was planned for major urban renewal. However, the area was developed as a sporting precinct when Sydney hosted the 2000 Olympics Games. The Games were promoted as the "Green Games," with a focus on implementing best environmental practices in developing the Games venue (Kearins and Pavlovich, 2002). This included allocating nearly two-thirds of the 640-ha site to green space, which incorporated WSUD elements. Following the Olympic Games, the area became a focus for Sydney's sporting and cultural events, as well as allowing commercial and high-rise residential developments.

#### 27.2.7.2 Water sensitive urban design approach

The Sydney Olympic Park took a local, integrated approach to providing water services, which includes stormwater harvesting, treatment and reuse, wastewater recycling, and demand management (Sydney Olympic Park Authority, 2017). The integrated water management approach incorporated aspects such as landscape planting and building design. The following outlines the WSUD approach implemented at the Sydney Olympic Park development.

There are three wetlands that collect stormwater runoff, which is pretreated by GPTs. These ponds allow for sediments and other pollutant to settle out and are planted with aquatic plants to assist with the removal of pollutants and enhance the local ecology (Fig. 27.13). The wetlands provide habitat for the endangered Golden Bell frog species and a variety of waterbirds (Sydney Olympic Park Authority, 2017). Stormwater exiting the wetlands is stored in a disused brick quarry. From there, the stormwater can be sent to the treatment plant (detail below) prior to its nonpotable uses, such as irrigation, or discharged to the nearby Parramatta River (Sydney Olympic Park Authority, 2017).

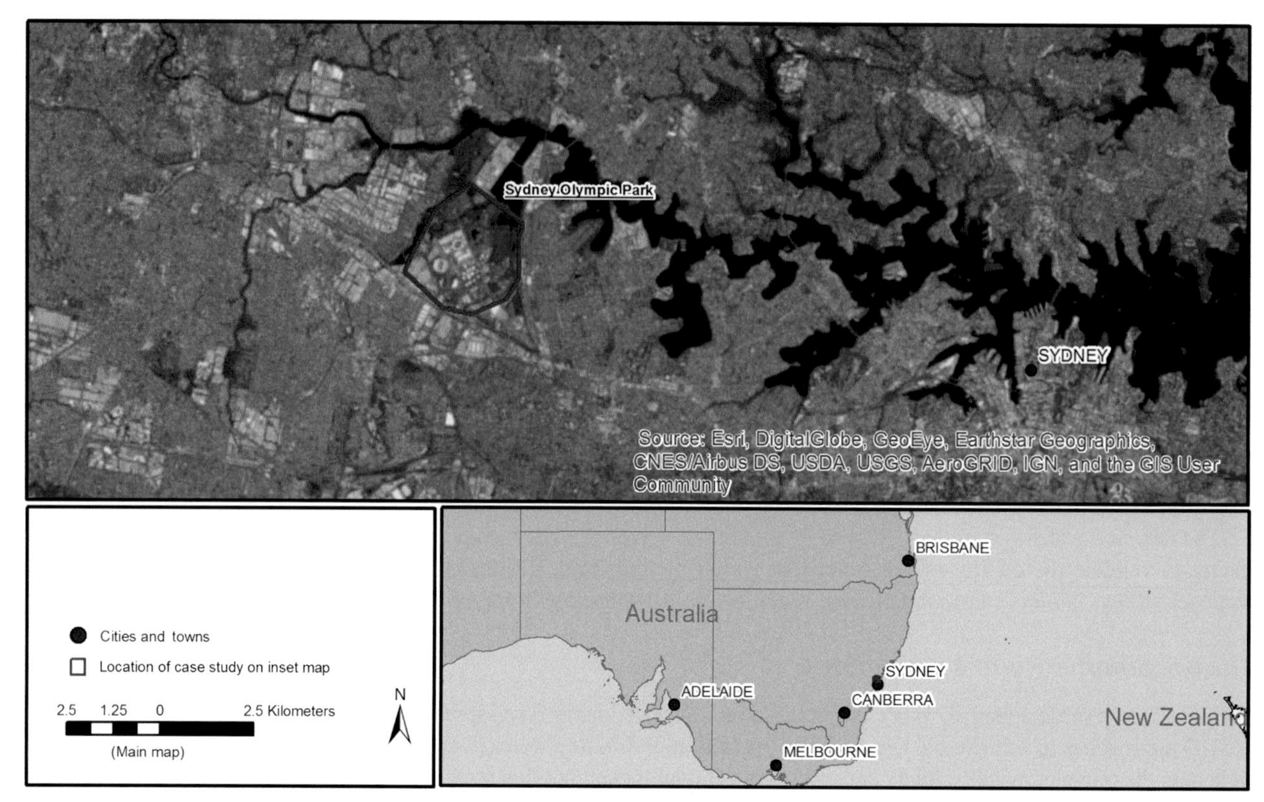


FIGURE 27.12 Case study location map—Sydney Olympic Park, New South Wales, Australia.

The stormwater management is part of the broader Water Reclamation and Management Scheme (WRAMS), which was one of Australia's first water recycling schemes (Sydney Olympic Park Authority, 2017). This scheme takes sewage from the sporting, commercial and high-rise residential developments and treats it using advanced biological treatment processes, microfiltration, and UV disinfection (Barton and Argue, 2007). The Homebush development has dual pipe

FIGURE 27.13 Wetlands at Sydney Olympic Park, New South Wales, Australia. *Source: https://www.flickr.com/photos/alexeyg/14976501462 CCA.*

reticulation, which allows for nonpotable end uses alongside the drinking water supply. The recycled water is used for toilet flushing and clothes washing, as well as external uses such as garden irrigation and car washing (Barton and Argue, 2007). The recycled water scheme can be augmented with stormwater, which is drawn from the brick pit storage and further treated with reverse osmosis before being added to the nonpotable supply system. The recycled water scheme services around 20,000 people and has achieved almost 100% treatment and reuse, reducing wastewater discharge by almost 3 ML per day (Barton and Argue, 2007; Sydney Olympic Park Authority, 2017).

The Sydney Olympic Park also implemented design and management practices to reduce water demand, which included drought-tolerant landscaping and efficient irrigation practices (Sydney Olympic Park Authority, 2017). Demand management, in combination with the WRAMS, has meant that less than 5% of the water used at the development is drinking water (Sydney Olympic Park Authority, 2017).

### 27.2.7.3 Drivers for the adoption of WSUD approaches

The key driver for the adoption of WSUD approaches at the Sydney Olympic Park was the need to demonstrate green credentials as part of the promise made to the Olympic Games Committee to deliver a "Green Games." This meant there was a focus on implementing best environmental practices (Kearins and Pavlovich, 2002). The Olympic movement started placing increasing emphasis on the environmental sustainability in the 1900s, which meant Sydney's bid to host the games needed to demonstrate how the games would be sustainable (Mitchell et al., 2008).

### 27.2.7.4 Implementation process for Sydney Olympic Park

*Funding arrangements:* The design and construction of the WRAMS cost around AU$16 million (in 2000), with an estimated AU$1.5 million operating and maintenance budget over the 25-year life cycle of the asset (UGL, 2014). The current cost of the recycled water is set at AU$0.15 less than drinking water supplied by Sydney Water, with an additional connection fee. However, it is recognized that this does not reflect the true cost of supplying the recycled water, with the need to review pricing structures of drinking and recycled water to ensure they reflect their true costs.

*Regulatory context and management structure:* A unique aspect of this development was that the Sydney Olympic Park Authority (SOPA) not only manages the whole precinct but also received regulatory approval to operate as a water business. Mitchell et al. (2008) highlighted that this approval to operate as a water authority overcame a number of major institutional barriers that were in place at the time, which included the need to ensure adequate monitoring and management of decentralized systems and integrating decentralized approaches with the centralized system to guarantee minimum levels of service.

*Stakeholder engagement and communication:* SOPA ensures that customers are aware of the contribution they are making to reducing drinking water demand and wastewater discharge to waterways and the ocean (Sydney Olympic Park Authority, 2017). Waitt (2003) highlighted that an important component of building public support for Sydney's Olympic Games bid was the leaving of a lasting public legacy. The site was a government-owned brownfield site, heavily contaminated from previous industrial uses. The remediation of this land including the water sensitive approach to public open space, in an area of Sydney that was historically socially disadvantaged, thereby provided a way to build community support for the Olympic Games bid.

*Ongoing arrangements WSUD maintenance post development:* SOPA owns the water recycling and stormwater systems, but there is a private sector contract to maintain and operate the systems for the first 25 years (Mitchell et al., 2008; UGL, 2014).

*Evaluating success:* The Sydney Olympic Park precinct has been successful in delivering a recycled water scheme where nearly 100% of the wastewater is treated and recycled in the local area. The WSUD approach taken at Sydney Olympic Park was part of the successful remediation of a site that was located on disused, heavily contaminated, industrial lands. The remediation not only created a valuable public space in a part of the city that had previously been neglected but also helped to restore biodiversity by creating habitats for a range of species including the iconic Golden Bell frog.

### 27.2.7.5 Summary and key findings

The driver for adopting WSUD approaches was to demonstrate sustainability credentials as part of Sydney's successful bid to host the 2000 Olympics. A unique enabling aspect for the Sydney Olympic Park was the creation of a single entity that was responsible for delivering the development, which in turn allowed it to become a licensed water service provider. This approach circumvented major regulatory impediments at the time and ensured an integrated approach between the planning and design of the built environment with water sensitive features.

## 27.3 DISCUSSION AND CONCLUSIONS

Water sensitive approaches to urban development emerged only about 25 years ago, when it was recognized that there was a need to better plan and manage the urban water cycle to avoid disturbance to natural hydrology, reduce pollution of waterways and oceans, and reduce the demand for mains drinking water. The practice of WSUD continues to evolve and be refined, which has in large part been due to lessons learnt from pioneering developments that trialled innovative WSUD approaches. The cases reviewed in this chapter are some of the leading-edge examples of WSUD and provide the practical demonstration of the challenges to and benefits of implementing WSUD. These lessons can be used not only to refine standards and guidelines but also to build momentum for the continued growth and evolution of WSUD.

In Long Bay (New Zealand), there was a long consultation/legal process to defend the structure plan and to ensure the techniques and protocols for managing the adverse impact of the urban development would protect the ecological, recreational, and esthetic values of the receiving environment. This process in Long Bay has strongly influenced the water sensitive design approaches in current Auckland Council design manuals and the 2016 Auckland Unitary Plan. The innovative nature of the Lochiel Park (Australia) approach meant that there were also significant delays in delivering the nonpotable water supply because of the need to validate the quality of the recycled water to ensure that it would reliably meet environmental and health regulations. This delay led to some community disappointment, which highlights the difficulty when WSUD expectations move beyond current, mainstream water supply practice and challenge existing regulatory frameworks.

The cases highlight that for WSUD implementation to be successful, there is the need for integration across different urban functions, stakeholders, and levels of government. In Regis Park (New Zealand) a holistic perspective was applied to protect and enhance the water quality of receiving waters. This approach brought together urban design and planning with integrated catchment management and vegetation planning to ensure that the urban layout and infrastructure delivered stormwater management objectives. In Germany, the introduction of the imperviousness fee at the municipal level reflected national and European support for introduction of a polluter pays principle and the shift to a more integrated approach to managing stormwater. In Wuhan (China), the agenda for the ambitious sponge city program was set by the central government. However, it is implemented at the city level. To help achieve the overarching national objective, local guidelines and supporting policies need to be developed that reflect the local contexts of each selected city. The top—down polices and strategic plans can encourage water sensitive developments, but location-specific WSUD approaches need to be determined from the bottom—up to ensure it is appropriate to local limitations and opportunities and is aligned with community values. The green roofs program in NYC (USA) was first declared by the Mayor, and the objectives were subsequently incorporated into the city's strategic plan, local building codes, and tax incentives, but the actual on-ground delivery is often determined by the individual property owner.

In addition to the primary objectives of the WSUD approach, it was frequently recognized that co-benefits followed on from taking a more water sensitive approach to urban design. For example, the green roof program in NYC was motivated by the need to reduce combined sewer overflows but also ended up providing a range of other benefits such as urban heat island mitigation and reduced energy for heating and cooling buildings. The benefits of WSUD often extend beyond objectives associated with improved urban stormwater management, which reflects the multifunctional nature of many WSUD approaches. These co-benefits, such as improved amenity and recreational spaces, can help build public understanding and engagement in the benefits of WSUD. In some WSUD approaches, such as a green roof, the stormwater functions might not be immediately apparent to the general public, but the benefits to improved urban esthetics and heat mitigation will be. In the Sydney Olympic Park (Australia) development, the WSUD approach implemented not only provided for reduced stormwater and wastewater discharge and less demand for mains water but also provided an area of high-value public open space in a historically socially disadvantaged, environmentally degraded part of the City.

Economic instruments that reflect the true cost of different stormwater management approaches can create a financial incentive for the adoption of WSUD approaches. In Germany, the use of the imperviousness area fee provided a fair and transparent way for property owners to contribute toward the costs of managing combined sewer flows. However, it also encouraged property owners to reduce their fees through the adoption of water sensitive approaches that reduce runoff, such as rainwater tanks and permeable pavements. In NYC, the adoption of green roofs has been encouraged through a tax relief that offset the cost of installing a green roof.

The growth of the global urban population, climate change impacts, and need to protect ecosystems that are already under pressure means that water sensitive approaches to urban development will continue to be an important component of ensuring future cities are sustainable and liveable. WSUD will also increase resilience to shocks, such increased severity and frequency of extreme weather events. The case studies documented in this chapter can help disseminate lessons learnt from WSUD implementation in different countries and can be adapted in other contexts to help in the refinement of WSUD principles and practices.

## REFERENCES

Aevermann, T., Schmude, J., 2015. Quantification and monetary valuation of urban ecosystem services in Munich, Germany. Zeitschrift für Wirtschaftsgeographie 59, 188.

Ansel, W., 2017. Director of Deutscher Dachgärtner Verband. L. Ehrenfried.

Ansel, W., Appl, R., 2011. An International Review of Current Practices and Future Trends: Green Roof Policies. From: http://www.igra-world.com/images/news_and_events/IGRA-Green-Roof-Policies.pdf.

Auckland Council, 2010a. Auckland Council District Plan Operative Manukau Section 2002. From: http://www.aucklandcouncil.govt.nz/EN/planspoliciesprojects/plansstrategies/DistrictRegionalPlans/manukaucitydistrictplan/Pages/districtplantexthome.aspx.

Auckland Council, 2010b. Auckland District Plan North Shore Section. 17B Long Bay Structure Plan. From: http://www.aucklandcity.govt.nz/council/documents/districtplannorthshore/text/section17b-longbaystructureplan.pdf.

Auckland Council, 2015. Water Sensitive Design for Stormwater. Guideline Document. Auckland, New Zealand, Auckland Council 2015/004.

Auckland Council, 2016. Auckland Unitary Plan Operative in Part: Chapter I North Precincts 1519 Long Bay Precinct. From: http://unitaryplan.aucklandcouncil.govt.nz/pages/plan/Book.aspx?exhibit=AucklandUnitaryPlan_Print.

Barton, A.B., Argue, J.R., 2007. A review of the application of water sensitive urban design (WSUD) to residential development in Australia. Australasian Journal of Water Resources 11 (1), 31–40.

Berry, S., Davidson, K., Saman, W., 2013. The impact of niche green developments in transforming the building sector: the case study of Lochiel Park. Energy Policy 62 (Suppl. C), 646–655.

Bertram, N.P., Waldhoff, A., Bischoff, G., Ziegler, J., Meinzinger, F., Skambraks, A.-K., 2017. Synergistic Benefits between Stormwater Management Measures and a New Pricing System for Stormwater in the City of Hamburg. Water Science and Technology.

Blaess, J., Rix, S., Bishop, A., Donaldson, P., 2007. Lochiel Park–a nation leading Green Village. In: Proceedings of the State of Australian Cities Conference. Australia, Adelaide.

Bloomberg, M.R., Holloway, C., 2010. NYC Green Infrastructure Plan: A Sustainable Strategy for Clean Waterways. The City of New York, New York City.

Brombach, H., Dettmar, J., 2016. In the mirror of the statistics: sewage and rainwater treatment in Germany. DWA German Association for Water, Wastewater and Waste 3, 176–186.

Brown, R., Keath, N., Wong, T., 2008. Transitioning to Water Sensitive Cities: Historical. Current and Future Transition States 11th International Conference on Urban Drainage, Edinburgh, Scotland, UK.

Buehler, R., Jungjohann, A., Keeley, M., Mehling, M., 2011. How Germany became Europe's green leader: a look at four decades of sustainable policymaking. The Solutions Journal 2 (5), 51–63.

Carson, T.B., Marasco, D.E., Culligan, P.J., McGillis, W.R., 2013. "Hydrological performance of extensive green roofs in New York City: observations and multi-year modeling of three full-scale systems. Environmental Research Letters 8 (2), 024036.

Chao, P.R., Umapathi, S., Saman, W., 2015. Water consumption characteristics at a sustainable residential development with rainwater-sourced hot water supply. Journal of Cleaner Production 109 (Suppl. C), 190–202.

Cook, T., Watkins, R., Watling, R., 2003. The effectiveness of continuous deflective separation (CDS) pollutant traps in reducing geochemical input into urban wetlands: a comparative study of to contrasting stormwater catchments, Perth WA. In: Advances in the Regolith, CRC, pp. 80–81.

Dai, L., van Rijswick, H.F.M.W., Driessen, P.P.J., Keessen, A.M., 2017. Governance of the sponge city programme in China with Wuhan as a case study. International Journal of Water Resources Development 1–19.

Davies, J., 1996. Water sensitive urban design progress in Perth. Hydrology and Water Resources Symposium 1996: Water and the Environment. Preprints of Papers, Institution of Engineers, Australia.

DDV, 2016. Kommunale Gründach-Initiativen Haben Konjunktur. GründachAktuell, Deutscher Dachgärtner Verband. From: http://www.dachgaertnerverband.de/infomaterial/images_dynamic/DDV-Magazin_Gruendach-Aktuell-2-2016_40.pdf.

Debo, T., Reese, A., 2002. Municipal Stormwater Management. CRC Press, Boca Raton.

Ecological Engineering, 2006. Lochiel Park- Functional Design of Southern Wetland System, Prepared for the Land Management Cooperation.

Edwards, J., Pocock, B., 2009. Comfort, Convenience and Cost: The Calculus of Sustainable Living at Lochiel Park. University of South Australia.

European Commission, 2016. Introduction to the New EU Water Framework Directive. From: http://ec.europa.eu/environment/water/water-framework/info/intro_en.htm.

FAZ, 2006. Deutschlands Kommunen spalten die Abwassergebuehr. From: http://www.fabry.eu/pdf/FAZ_Bericht_Regenwasser_07.01.06.pdf.

Fehring, B., 2012. Der Stand der Umstellung auf die gesplittete Abwassergebühr in Baden-Württemberg. University of Applied Science Ludwigsburg.

Fletcher, T.D., Shuster, W., Hunt, W.F., Ashley, R., Butler, D., Arthur, S., Trowsdale, S., Barraud, S., Semadeni-Davies, A., Bertrand-Krajewski, J.-L., Mikkelsen, P.S., Rivard, G., Uhl, M., Dagenais, D., Viklander, M., 2015. SUDS, LID, BMPs, WSUD and more – the evolution and application of terminology surrounding urban drainage. Urban Water Journal 12 (7), 525–542.

Gaffin, S., Khanbilvardi, R., Rosenzweig, C., 2009. Development of a green roof environmental monitoring and meteorological network in New York city. Sensors 9 (4), 2647.

Garnaut, R., 2008. The Garnaut Climate Change Review. Cambridge, Cambridge.

Heijs, J., Kettle Consulting Ltd, 2009. Low impact design in the Long Bay structure plan; what happened?. In: Proceedings of Water New Zealand Stormwater Conference. Water New Zealand, Auckland.

Herrmann, T., Schmida, U., 2000. Rainwater utilisation in Germany: efficiency, dimensioning, hydraulic and environmental aspects. Urban Water 1 (4), 307–316.

Jia, H., Wang, Z., Zhen, X., Clar, M., Yu, S.L., 2017. China's sponge city construction: a discussion on technical approaches. Frontiers of Environmental Science and Engineering 11 (4), 18.

Kearins, K., Pavlovich, K., 2002. The role of stakeholders in Sydney's green games. Corporate Social Responsibility and Environmental Management 9 (3), 157–169.

Keeley, M., 2007. Using individual parcel assessments to improve stormwater management. Journal of the American Planning Association 73 (2), 149–160.

Li, H., Ding, L., Ren, M., Li, C., Wang, H., 2017. Sponge city construction in China: a survey of the challenges and opportunities. Water 9 (9), 594.

Lloyd, S., Wong, T., Porter, B., 2002. The planning and construction of an urban stormwater management scheme. Water Science and Technology 45 (7), 1–10.

Mark, K., 2017. Morrison Landscape Architecture. Green Roof Design. Undated, From: http://www.markkmorrison.com/greenroof/.

Mitchell, C., Retamal, M., Fane, S., Willetts, J., Davis, C., 2008. Decentralised Water Systems – Creating Conducive Institutional Arrangements. Enviro '08 Melbounre.

National Water Commission, 2004. Intergovernmental Agreement on a National Water Initiative—between the Commonwealth of Australia and the State Governments of New South Wales, Victoria, Queensland, South Australia, the Australian Capital Territory and the Northern Territory, 2004.

New Zealand Government, 1971. Marine Reserves Act. From: www.legislation.govt.nz.

New Zealand Government, 1991. Resource Management Act. From: www.legislation.govt.nz.

Nickel, D., Schoenfelder, W., Medearis, D., Dolowitz, D.P., Keeley, M., Shuster, W., 2014. German experience in managing stormwater with green infrastructure. Journal of Environmental Planning and Management 57 (3), 403–423.

Niederschlagswasserfreistellungsverordnung, 2014. Bavarian Law on Stomwater Runoff Infiltration. From: http://www.gesetze-bayern.de/Content/Document/BayNWFreiV-1.

North Shore City Council, 2009. Long Bay Structure Plan. North Shore City Council, North Shore, New Zealand.

North Shore City Council, 2004. North Shore District Plan: Proposed Plan Change 6. North Shore City Council, Which since 2010 Has Been Part of Auckland Council. From: http://www.aucklandcouncil.govt.nz/EN/planspoliciesprojects/plansstrategies/DistrictRegionalPlans/northshorecitydistrictplan/plan-changes-home/Pages/home.aspx.

NYC Mayor's Office of Sustainability, 2013. Green Roof Tax Abatement. From: http://www.nyc.gov/html/gbee/html/incentives/roof.shtml.

Oelmann, M., Czichy, C., Waldhoff, A., Bischoff, G., Ziegler, J., 2014. Studie zum RISA Querschnittsthema Finanzierung, Teil I: Kostenprognose der RISA.

Roy, A.H., Wenger, S.J., Fletcher, T.D., Walsh, C.J., Ladson, A.R., Shuster, W.D., Thurston, H.W., Brown, R.R., 2008. Impediments and solutions to sustainable, watershed-scale urban stormwater management: lessons from Australia and the United States. Environmental Management 42 (2), 344–359.

Rugh, C., 2013. Columbia University Greens Up with Green Roofs. From: https://strategicstoryteller.com/wp-content/uploads/2013/03/XFA_Columbia_LW_Mar-April2013.pdf.

Saadatian, O., Sopian, K., Salleh, E., Lim, C.H., Riffat, S., Saadatian, E., Toudeshki, A., Sulaiman, M.Y., 2013. A review of energy aspects of green roofs. Renewable and Sustainable Energy Reviews 23 (Suppl. C), 155–168.

Scott, D.J., Kaye, B., Ltd, K.C., 2003. Overview Assessment of Environmental Effects - Resource Consent Application to Manukau City Council for Regis Park Development, vol. 1. Resource Consent Application to Manukau City Council for Regis Park Development.

Sharma, A., Pezzaniti, D., Myers, B., Cook, S., Tjandraatmadja, G., Chacko, P., Chavoshi, S., Kemp, D., Leonard, R., Koth, B., Walton, A., 2016. Water sensitive urban design: an investigation of current systems, implementation drivers, community perceptions and potential to supplement urban water services. Water 8 (7), 272.

Sharma, A.K., Cook, S., Tjandraatmadja, G., Gregory, A., 2012. Impediments and constraints in the uptake of water sensitive urban design measures in greenfield and infill developments. Water Science and Technology 65 (2), 340–352.

Spala, A., Bagiorgas, H.S., Assimakopoulos, M.N., Kalavrouziotis, J., Matthopoulos, D., Mihalakakou, G., 2008. On the green roof system. Selection, state of the art and energy potential investigation of a system installed in an office building in Athens, Greece. Renewable Energy 33 (1), 173–177.

Sydney Olympic Park Authority, 2017. Urban Water Reuse & Integrated Water Management. Undated, From: http://www.sopa.nsw.gov.au/__data/assets/pdf_file/0019/344620/urban_water_reuse_brochure_2006.pdf.

Thimet, J., 2014. Gesplittete Abwassergebühr oder Neues von der teuer erkauften Gerechtigkeit. From: https://www.bay-gemeindetag.de/kxw/common/file.aspx?data=yUb2ROUJJZnruUf+LzLu6yU42yoZ9uAmJ9/TlpZ6Gi5i3teMZJm9LLFz2LlzO9g1D0Yp+EcaEI9EZhJc+Yq3ZXGsmVL6BADhxsFjdxgqfbMWsn29D38grQ.

UBA, 1997. Daten zur Umwelt. E. S. Verlag. Berlin, Umweltbundesamt.

UGL, 2014. A World Class Integrated Approach to Water Conservation. From: https://uglcdn.ugllimited.com/Asset/cms/Case_Study/Water/Municipal-Wastewater/WRAMS_Water_Case_Study/016_WRAMS_Water_CaseStudy_V8_WEB.pdf.

Umweltbundesamt, 1998. Grundwasser in Deutschland. From: files/medien/publikation/long/3642.pdf.

van Roon, M., 2011. Low impact urban design and development: catchment-based structure planning to optimise ecological outcomes. Urban Water Journal 8 (5), 293–308.

van Roon, M., 2017. Justifying water sensitive development: science informing policy and practice. In: International Conference for Sustainable Development. Rome.

Van Roon, M.R., Rigold, T., 2016. Urban form and WSUD in Auckland residential catchments determine stream ecosystem condition. Urban Water: planning and Technologies for Sustainable Management. In: 9th International Conference of Novatech. Lyon, France.

Vijayaraghavan, K., 2016. Green roofs: a critical review on the role of components, benefits, limitations and trends. Renewable and Sustainable Energy Reviews 57 (Suppl. C), 740–752.

Vimal, M., Auroop, R.G., Bart, N., Dennis, P.L., 2015. Changes in observed climate extremes in global urban areas. Environmental Research Letters 10 (2), 024005.

Waitt, G., 2003. Social impacts of the Sydney Olympics. Annals of Tourism Research 30 (1), 194–215.

Wang, H., Cheng, X.T., Man, L., Li, N., Wang, J., Yu, Q., 2017. Challenges and future improvemnets to China's sponge city construction. In: Proceedings of International Low Impact Development Conference China ASCE, pp. 339–351.

Wong, T.H.F., 2006. Water sensitive urban design - the journey thus far. Australasian Journal of Water Resources 10 (3), 213–222.

Workman, J., 2017. The hard road to 'soft' cities. The Source (International Water Association) 7, 28–32.

Zhang, D., Gersberg, R.M., Ng, W.J., Tan, S.K., 2017. Conventional and decentralized urban stormwater management: a comparison through case studies of Singapore and Berlin, Germany. Urban Water Journal 14 (2), 113–124.

Zhang, J., Mauzerall, D.L., Zhu, T., Liang, S., Ezzati, M., Remais, J.V., 2010. Environmental health in China: progress towards clean air and safe water. The Lancet 375 (9720), 1110–1119.

# Index

*Note:* 'Page numbers followed by "f" indicate figures, "t" indicate tables.'

## D



## E

## F



## G

## H



## I

## Q



## R

## Y



## Z

Printed in the United States
By Bookmasters